ENERGY TECHNOLOGY VIII

"New Fuels Era"

Proceedings of the
Eighth Energy Technology Conference
March 9-11, 1981, Washington, D.C.

Edited by
Dr. Richard F. Hill

Government Institutes, Inc.
August 1981

PUBLISHER'S NOTE

The specialized nature and timeliness of this material demands prompt publication. Therefore, to expedite publication, the text has been reproduced directly from a reduction of the original typed copy. This format eliminates the considerable time and expense of detailed text editing and composition often required of publication processes, while conserving both paper and energy.

The publisher, in reproducing the papers as received from the authors, disclaims any and all responsibilities for the contents of individual papers. Publication of this book does not signify that the contents necessarily reflect the views and policies of Government Institutes, Inc.

Government Institutes, Inc. is indebted to many individuals and organizations for the preparation and publication of this most timely contribution to the field of energy technology. We are particularly grateful to Dr. Richard F. Hill, Dean of the College of Science and Engineering at the University of Bridgeport, who served admirably as Conference Chairman and Editor of these proceedings, Charlene Ikonomou of our publications staff for the collection and organization of the individual papers, and Loretta Pitts for her time and patience in assisting in the final production.

Most importantly, we would like to express deep appreciation to all the authors who contributed to these proceedings. Our appreciation for their efforts will be shared by the thousands who will read and utilize the information contained in these proceedings.

Thomas F. P. Sullivan
President
Government Institutes, Inc.

August 1981

966 Hungerford Drive, #24, Rockville, Maryland 20850.

Library of Congress Catalog Card No. 80-66431

International Standard Book Number: 0-86587-008-X

Printed and bound in the United States of America

PREFACE

The Eighth Energy Technology Conference occurred soon after a new Administration assumed responsibility for the federal government. The primary objective of the new Administration is to dramatically improve the U.S. domestic economy by decreasing the federal budget, reducing taxes, eliminating some regulations and stabilizing growth in money supply. Each of these four efforts will significantly affect the development and use of the various energy resources.

Since the Seventh Energy Technology Conference, Congress passed and President Carter signed the Energy Security Act which, along with nine other titles, established the U.S. Synthetic Fuels Corporation; President Reagan completed the decontrol of domestic oil prices; and the U.S. consumer decreased the use of energy, especially petroleum products. Prior to the Ninth Energy Technology Conference (which will be held February 16-18, 1982) the industrial sector is expected to assume a greater—and the government a lesser—role in the development and application of new energy technologies. Progress is being made.

The papers presented herein describe impressive technology developments since the rude awakening to our energy insecurity in 1973. They also define the solid technological foundation on which must be built a new, stable energy future.

These proceedings are organized into five parts:

Policy
Energy Analysis, Planning and Regulation
Using Energy Effectively
Technology for Fossil, Nuclear and Geothermal Resources
Technology for Renewable Resources

In addition, a Subject Index has been prepared to assist in locating areas of particular interest throughout the 144 papers.

Again this year, both the Conference and the Exposition were the largest to-date. We appreciate all comments and suggestions as to how future conferences and proceedings can be made more useful.

Richard F. Hill
Editor

TABLE OF CONTENTS

Table of Contents

Table of Contents

Table of Contents

Table of Contents

Table of Contents

Table of Contents

Table of Contents

Table of Contents

SUBJECT INDEX

AUTHOR INDEX

INVOCATION

Almighty God, Lord of the Universe, we pause as we are about to reflect on the great forces of the Cosmos. As we do so, we ask that we find in this moment of prayer the light to discern Your presence both in our midst and in the physical and spiritual forces of our universe.

In Your providence, You have blessed us in our times with the discovery of vast forces of energy and power hidden in the earth below and in the skies above. As we come to grasp the magnitude of these forces obscured for so long from our ancestors, let us appreciate these marvels—the products of Your handiwork and of the talents of men and women blessed by You with talent and dedication. Let us appreciate them with wonder and excitement but also with gratitude and humility, knowing that with the discovery of your great wonders of creation, there lies always the subtle danger of misuse and perversion.

Guard us then against the pride and hubris that have marred the efforts of others in their search for truth. Bless us rather with a sense of responsibility as we strive to harness the powers You have placed in the bowels of the earth and in the far reaches of the heavens.

May our efforts be rooted not in greed or in the desire to dominate but in dedication to the care and concern of our brothers and sisters of our present generation and of generations to come, ever sensitive to the moral and spiritual values that are at the heart of happiness in our human family. Let our pursuit of the far-reaching potential of the seas, the land, the wind, and the atmosphere be for the advancement of industry and art, the beauty of lands and waters, the promotion of physical prosperity and spiritual well being of all our brothers and sisters everywhere.

Thus, may our coming together this week at this Energy Technology Conference contribute to the enrichment and not the impoverishment of the gifts with which You have endowed us.

Monsignor Thomas M. Duffy
Archdiocese of Washington

TAPE RECORDINGS AVAILABLE

Although we have tried to make this volume of proceedings as complete a record of the 8th Energy Technology Conference as possible, it still does not convey the full educational impact of the dialogue between the speakers and their audiences. For this reason, all Conference sessions—including the workshops, panel discussions and question-and-answer periods—are captured on cassette tape recordings, available separately or as a set. For further information, contact:

Government Institutes, Inc.
966 Hungerford Drive, Suite 24
Rockville, MD 20850
(301) 251-9250

IS YOUR ET SERIES COMPLETE?

To enhance your knowledge of energy technology and to gain an historical perspective on the growth of the energy industries in the United States and abroad, your energy bookshelf should contain a complete set of Energy Technology Proceedings I through VIII. Available individually or as a specially discounted set, the Energy Technology Series represents the most complete information on all major energy sources--fossil, nuclear, hydro, renewables and others—compiled in the last eight years. Furthermore, the Proceedings reflect the slow but evident development of a national energy policy over these crucial years. With a list of contributing authors that reads like a Who's Who in the international energy community, the Energy Technology Conference Proceedings are sure to become an invaluable element of your energy reference library. Order the copies you've missed by using the form below or calling us at (301) 251-9250.

Complete and mail to: **Government Institutes, Inc., 966 Hungerford Drive, #24, Rockville, MD 20850.**

____	**ET8: New Fuels Era**	$48.00
____	**ET7: Expanding Supplies & Conservation**	45.00
____	**ET6: Achievements in Perspective**	38.00
____	**ET5: Challenges to Technology**	38.00
____	**ET4: Confronting Reality**	30.00
____	**ET3: Commercialization** (microfiche only)	15.00
____	**ET2: International**	25.00
____	**ET1: Energy R & D** (microfiche only)	15.00
____	**Energy Technology Series** (includes 1-7)	169.00

Add $2.00 on invoiced orders. GI pays shipping and handling on prepaid orders. Maryland residents add 5% sales tax. ________

TOTAL enclosed ________

Name ____

Company ____

Address ____

____

[ ] Check payable to Government Institutes.
[ ] Bill company, P.O. No. ______________
[ ] Credit Card:

[ ] American Express [ ] VISA [ ] Master Card

No. _ _ _ _ _ _ _ _ _ _ _ _ _ _ _ _

Signature ______________________________

Expiration Date ____________________

PART I

POLICY

The primary purposes of the Energy Technology Conference and these Proceedings are to describe and interrelate evolving technologies for extracting, converting, delivering and using the various energy resources. Always underlying—and often complicating—the interactive utilization of these technologies and resources is a complex array of laws and regulations that promote, control or restrict specific applications.

Within the law, economics and the perceived risks of the marketplace will guide the development and use of technology by the business sector. But predicting future laws, regulations and economics is necessary to anticipate the business risks.

These three papers present the views of prominent energy leaders on the policies that are likely or needed.

THE NEAR-TERM FUTURE FOR THE ENERGY RESOURCE INDUSTRIES

Armand Hammer
Chairman of the Board
Chief Executive Officer
Occidental Petroleum Corporation

It's a pleasure for me to be with you today at this important Energy Technology Conference and Exposition which has become an annual tradition here in Washington.

On my way into town from the airport, I noticed a billboard advertising the movie, "The Formula," in which one of the main characters is an oil tycoon played by Marlon Brando. Brando sports a shaved head, rimless specs and a hearing aid in this movie. According to the pre-release publicity, Brando said he decided on this approach after seeing a magazine photograph of a real oil magnate, Dr. Armand Hammer.

In reply, let me say that I have my own hair, as you can see, and I do not wear rimless specs or a hearing aid.

It also was reported that for playing in the movie Brando would be paid $250,000 a day for each day he worked. At those prices I would have been happy to play the role myself!

As you know, the theme I have been asked to discuss with you today is "The Near Term Future of the Energy Resource Industries." I believe that the future of these industries rests with our determination and resourcefulness to put ourselves on the road to energy independence.

Speaking of being resourceful, there's the story of the man who was hard of hearing. When hearing aids first came out, he consulted with a doctor who advised him to get one. Upon hearing that they cost $150, the man said he could not afford one. About a month later, the doctor saw the man and noticed he had a string hanging from a button behind his ear to the handkerchief pocket of his jacket. "What the heck is that?" the doctor asked. The man replied: "This is my new hearing aid and it only cost me a few cents." "How does it work?" the doctor asked. The man answered, "Well, when people see me they think I'm wearing a hearing aid and they yell at me loud enough so I can hear them!"

I have been involved in a number of different ventures, climaxed by my acquisition of Occidental Petroleum Corporation 24 years ago, the greatest adventure of them all.

In the 1920's, I controlled the first trading concessions between the United States and the Soviet Union. I returned to the United States in 1930 and during the next 26 years, I was an art dealer in New York; a whiskey distiller in Kentucky, building an unknown brand—J.W. Dant—into the number one Kentucky bourbon in the United States; and a Black Angus cattle breeder in New Jersey. In 1956, I moved to California with my wife Frances, ostensibly to retire.

Life is often filled with a number of accidental but opportune events. Something happens and we find ourselves following a path we never expected to pursue. An accountant friend of mine told me about the tax advantages of oil investments; that is, the write-offs that could be claimed by those in a high tax bracket when you hit a dry hole, which seemed to occur most of the time. He suggested that I look into Occidental Petroleum, a small, struggling oil company that appeared to be going nowhere but down.

Occidental at the time had three full-time employees, and the stock had fallen from $14 to $.18 a share on the Los Angeles Stock Exchange. We were offered as many shares as we wanted. There were 600,000 shares outstanding; after looking at their books, I estimated the net worth of the company at about $34,000. I said, "Your stock isn't worth 18 cents—you're on the verge of insolvency." They said they had acquired two promising leases and needed $50,000 to drill each well. My wife and I loaned them the money for a 50-percent interest, figuring that if they were dry holes we could qualify for a tax write-off. To our surprise, both wells came in and one opened a small oil field. I'll never forget the first time I went to the site; I brought along a Polaroid camera because I had never seen an oil well before. The stock jumped to a dollar a share. I began buying the stock on the open market until I became the largest shareholder. They made me president of the company and the rest is history.

In slightly more than two decades, Occidental has grown to become the 21st largest corporation on the Fortune 500 list and the 11th largest oil company, with 1980 sales totaling $12.5 billion and profits of $710 million after taxes. I was particularly heartened last year when Fortune Magazine announced its 25th anniversary of the Fortune 500 Directory; the magazine announced that "the champion climber (in the history of the Directory) has been Occidental Petroleum. In 1954, Oxy's sales were only $3,000."

Many people look upon OXY as a "maverick oil company" because we go our own way and often take calculated risks. Oil and gas exploration is our main business, and I believe we do it very well. We do no refining or retailing of petroleum products as such. We were a late entry into the North Sea, but now we have the major interest in two of its most productive fields, Piper and Claymore. In Libya, we moved in, when other American oil companies had given up, and made major oil discoveries. In the Peruvian jungle, out of 47 wells we found only 3 dry holes, a record hard to beat. Nineteen companies followed us into Peru, investing billions in exploration without success. This year we are investing $835 million in overall exploration and see great potential in some of the new leases we have acquired and have planned.

As you know, however, Occidental means more than just oil and gas. Our Hooker Chemical Division is a major U.S. producer of industrial and agricultural chemicals. Our Island Creek Coal Division is the fourth largest coal company in the United States. In addition, we are active in minerals mining, including uranium, agricultural products, energy research and shale oil, which I will say more about in a few minutes.

If you look closely at our makeup, you can discern our directions for the near-term future. Each of these seemingly disparate divisions actually serves to reinforce each other's interests as we look forward to the future energy demands of the nation and the world.

Years ago, King Faisal of Saudi Arabia said that oil was much too valuable to burn. He was right. If you ask the average American what the energy crisis means to him, he will probably indicate that his greatest concerns are energy for moving his automobile and energy for heating and electrifying his home. Even in the halls of Congress, it is the direct consumption of energy that gets most of the attention. I am reminded of the little boy who ran to his father and excitedly said: "Wow! You oughta see the great lawn mower our neighbors have. It doesn't need gas or anything. You just push it!"

Indeed, the most neglected area of our energy crisis is the use of oil and gas for petrochemical feedstocks. From oil and gas, human ingenuity has developed a great variety of goods and materials that we can no longer do without. And there are no readily available substitutes at affordable prices. In this regard, OPEC is sending us and the rest of the world a message, but we don't seem to be receiving it very well.

In energy, I believe our country faces the worst crisis since the Republic was formed. For the first time we find ourselves dependent on a group of small countries, some of which are unfriendly, for a commodity that we cannot do without.

We and our western allies are trapped in the center of a storm of controversy in the Middle East that involves oil, petrodollars, the spectre of Soviet influence and the rising time of religious fundamentalism. There seems to be little or no stability in the governments of most of the oil producing countries to guarantee western interests.

In the past seven years, the price of oil has increased more than tenfold, from three dollars a barrel to an official price of $34 a barrel and a spot market price of $40 a barrel. Just prior to the Iranian Revolution in late 1978, the Department of Energy forecast that the price of oil would reach $30 by the year 1990. The $30 price, in fact, was reached just one year later. There is no limit to where it will go so long as we have no alternative sources of energy supply. Last year we Americans spent more than $80 billion overseas to pay for imported oil, even though we used 20 percent less oil than we did the year before. I predict the price of oil could reach $100 a barrel in three to five years, unless something is done.

The soaring costs of oil are one of the major causes of double-digit inflation, high interest rates and unemployment in this country and throughout most of the world. The price of energy is the primary component in our world economic cycle. Other price increases stem from the price of oil. Foreign trade is affected.

More tragically, food production is reduced, particularly in the Third World countries of Africa and Asia, thus leading to widespread starvation. Some of the Third World nations now use so much of their budgets to pay for foreign oil that there is not enough left to pay for purchases of badly needed fertilizers, seed, and farm equipment needed to feed the populations. I saw mass starvation during my early days in Russia and it is a scene I hope never to see again.

Winston Churchill once said, "Things don't get better by being left alone. Unless they are adjusted, they explode with shattering detonation."

What are we going to do about it?

I believe there are four things we should do, starting right now, to help get us out of our predicament and put us back on the road to energy independence.

First, we should look to our neighbors to the South. We must continue to improve our relations with Mexico, and I was most pleased when President Reagan took that initiative. The oil fields in Mexico are huge. Occidental geologists were invited there several years ago by President Lopez Portillo. When they returned, they reported to me that Mexico has some of the largest oil deposits in the world — equivalent to the largest fields in the Middle East.

The time for us is propitious. We are in an excellent position to help Mexico — and they certainly can help us.

American businessmen must reassess their attitudes toward Mexico. Mexico needs the help now. According to the latest statistics, Mexico has 50 percent unemployment.

I propose that other Americans follow the lead of our subsidiary, Hooker Mexicana, which has been fully "Mexicanized"; that is, 51 percent of the shares is owned by the Mexican public.

All the Hooker Mexicana employees, save one, are Mexican citizens trained by our American staff. We are doing more business and our profits are higher with the 49 percent interest than they were when the subsidiary was fully owned by Occidental.

Although my proposal is not a "quid pro quo" for Mexican oil, I believe that in return for our helping them to build their industrial base, we could make a deal with the Mexicans to increase their oil production. The Mexicans are now producing about 2.3 million barrels of oil per day and they expect to go about 1.7 million barrels this year. I feel this production could be doubled in three to five years, and we could expect to get an additional 2 million barrels per day above the 700,000 barrels per day we are now receiving.

Second, another part of our direction for the future is far greater use of coal. In the short term, we need to substitute coal wherever we find oil being burned to heat a boiler. It is estimated that 50 percent of the oil burned by utilities could be saved by converting to coal. If we had 100 percent substitution, which I believe is feasible, we could save upwards of 1-1/2 million barrels of oil a day and more, if we extended that to industrial boilers. There are environmental problems, to be sure, but they can be met with scrubbers, bag houses, and even newer environmental control technologies. We should help the utilities by giving them accelerated depreciation, other tax advantages, and perhaps non-recourse loans. It is two-thirds cheaper to burn coal, on a dollars-per-BTU basis, than to burn oil. The thing to do is to give companies the incentives.

Within three years, there will not be a single boiler in Japan that will burn anything but coal. Japan has no energy natural resources to speak of. The country must import just about all of its oil, natural gas and coal. We have a huge supply of coal in this country — enough to last 100 years or more. If Japan can do it, with tight environmental restrictions, we ought to be able to do it too — and we should do it.

The most recent World Coal Study has forecast that the United States has the opportunity with its enormous coal reserves to become the Saudi Arabia of coal exporters. During 1980 we saw a dramatic increase in U.S. coal deliveries overseas. They totaled some 90 million tons — a one-third increase over 1979, as countries in western Europe and Japan convert their power generating utilities from oil and gas to coal at a much faster rate than here in the United States. The exports are estimated to be worth some $4 billion.

Despite the substantial increases in foreign demand for coal, the industry is threatened by inadequate port facilities, particularly on the East Coast. We all read in the newspaper about as many as 120 coal ships sitting at anchor outside Norfolk Harbor waiting to be loaded. The waiting time averages as much as 60 days at times and substantial demurrage charges are applied. Something must be done in a hurry to clear this bottleneck. The coal companies themselves are willing to make major investments; Occidental, as an example, is spending $50 million to expand its coal loading facilities at Baltimore Harbor. However, the coal companies still remain captives of the railroads which control operations at the port sites.

Third, we must start immediately to develop a sensible synthetics fuel industry in the United States. Despite large cuts in the Department of Energy, President Reagan's budget still emphasizes synfuel production. And we are blessed by major resources in shale oil and coal to do the job.

If we proceed promptly and vigorously to develop a synfuels industry and make much greater use of our coal reserves, OPEC would realize that in 10 years or so we could be virtually independent of their oil. I believe they would consider that it would be to their advantage to put a cap on oil prices. The OPEC nations know they need to preserve their markets for the oil they hope to sell in the long term.

Now, I have been a voice crying in the wilderness over the past 10 years promoting shale oil, but today I firmly believe its time has come and it is the most immediate answer to our energy dilemma.

There is the equivalent of 1.8 trillion barrels of oil locked up in the Rocky Mountains in Colorado, Utah and southwestern Wyoming. This is two-and-a-half times the reserve of oil in the free world. As you know, the oil can be released if the shale is heated to temperatures of 900 degrees Farenheit.

Our Occidental engineers developed a proprietary modified in situ process for recovering the oil underground in commercial quantities and in conformity with reasonable environmental considerations. With our process, we mine only one-fourth as much rock compared to surface methods for shale oil recovery and we retort that above ground. The remaining shale is rubblized and retorted in place. Thus, because we need to do less mining, we can increase capacity faster and we have only one fourth as much spent shale to dispose of after we have extracted the oil.

Another great advantage with shale oil is you know there are no dry holes!

One hundred and four patents have been issued for the process and many more are being processed. We have already invested $100 million of our own funds in the past eight years.

I am sure that most of you are familiar with our process, so let me bring you up to date on our progress. Our last retort—retort number six—was as high as a 30-story building and an acre in area. It produced over 50,000 barrels of oil, which was 40 percent of the oil in place. As any oil man knows, anytime you get 40 percent, you are well above average.

We have completed the excavation of retorts seven and eight. We rubblized seven about 3 weeks ago and are about to rubblize eight. We hope to fire them up sometime this coming winter.

In partnership with Tenneco, we are also proceeding with plans to construct a commercial plant capable of producing 94,000 barrels a day by 1992, with production beginning in the mid-80's. Tenneco agreed to pay us $110 million for a 50-percent partnership. In 1981 dollars, the oil could be sold in the $30 to $40 range with an adequate return, making it competitive with world prices for conventional petroleum. The recoverable reserves in our 5,000-acre C-b tract are estimated at 1.2 billion barrels for 30 years of operation. Since the total costs for such an ambitious project may total close to $3 billion 1980 dollars, the partners are seeking a guaranteed loan from the government to help finance part of our investment, since there can be unforeseen regulatory, legal and institutional problems. One key to this development, of course, is the government's willingness to provide loan guarantees, which generally do not require any flow of funds from the Treasury. Very few companies are willing to take a risk to finance such a project. Thus, without the government guarantees for at least part of the initial investment, it will be difficult to persuade the financial community to provide the necessary investment capital to finance the development of this industry.

The government controls 600 such 5,000-acre tracts which could be put up for bidding. Once the industry is on its way, however, I believe competition will take care of the cost and government incentives can be modified or withdrawn entirely.

In order to accelerate a shale oil program, I have offered our process royalty free to the government for defense purposes. We have offered to license our patents to anybody for a reasonable fee. Hopefully, such a move will prompt more energy companies to get into the shale oil business.

Fourth, I also propose that we move ahead on developing coal liquefaction. In our opinion, liquefaction is more expensive and further down the road than shale oil, but it is nevertheless a viable option. The Germans flew their Messerschmidts and dove their war machines with gasoline produced from liquefied coal during World War II. Today, gasoline and other petroleum products are produced on an increasing scale from coal in South Africa. South Africans today can pull up to the pump and fill their tanks with Sasol. The country plans to double and triple its production of Sasol in order to make itself independent of foreign imports.

When I am told all of this cannot be done so fast, and that there are risks, I remind the speaker of how Franklin Roosevelt and the American people rallied to the crisis in World War II when the Nazis were sinking our ships bringing in natural rubber. I was in the alcohol business then and I recall how we converted our plants almost overnight to develop a synthetic rubber industry that was so vital to the war effort. Who would have believed we could develop a synthetic rubber industry and the Manhattan Project in the space of just a few years? When President Roosevelt was warned by timid members of Congress that he was going too fast with a crash program that might make costly mistakes, he replied, "If we can't act now, we won't even have the opportunity to make mistakes."

I personally have been very impressed by President Ronald Reagan during his early days in office and I believe our energy future will be brighter under his leadership.

In my lifetime, I have been a man of many dreams, and have been fortunate that so many of them have come true. As we move ahead into the decade of the eighties, I plan to continue my efforts in the field of energy and food production for the nation and the world and for world peace which is so intimately connected with these vital issues. In this country, I think we all need to work harder to bring about the changes necessary in our energy economy so that we can begin to turn back the tide which has eroded our confidence in the future of our country. There are still many obstacles in our path. But I firmly believe we have the opportunity in our grasp to find a way around them, and by all of us working together, we can continue to make this country grow and remain the greatest nation on the face of the earth.

STATE OF ENERGY, 1981
ENERGY AND THE ECONOMY

J. Peter Grace*
President & Chief Executive Officer
W. R. Grace & Company

The subject of energy goes hand-in-hand with the economy, so this paper concerns itself with inflation, the eroding dollar, and world money markets.

This paper will present a series of charts to provide some ideas on the status of our nation's economy and what we ought to be doing to produce more energy here at home and reduce our foreign oil dependence.

Charts 1-3 state the problem we have in the U.S.—too little investment, productivity and growth, and too much inflation and government. This is not only bad for the economy but critical to our National Security and world leadership.

---

*This paper is derived from the remarks made by Mr. Grace at the Energy Technology Conference, State of Energy address delivered March 9, 1981. Mr. Grace's remarks were transcribed and edited by the staff of Government Institutes.

CHART 1

ECONOMIC PERFORMANCE IN EIGHT COUNTRIES, 1962 - 1980

	(1)	(2)	(3)	(4)
	1962 - 1980			
	Average Investment As % Of GNP	Government Spending As % Of GNP (a)	Real GNP (Avg. Ann. % Increase)	Productivity (Avg. Ann. % Increase)
Japan	32.5 %	8.7 %	7.9 %	7.8 %
Belgium	21.5	15.0	3.9	6.6
Netherlands	23.6	16.8	4.1	6.4
Italy	20.6	15.4	4.1	5.6
France	22.9	13.8	4.4	5.4
Germany	20.6	17.5	3.6	5.2
United Kingdom	18.4	18.7	2.3	2.7
United States	17.8	20.6	3.5	2.2

Japan, columns (1) and (2): 3.7X

United States, columns (1) and (2): 0.9X

Japan Versus U.S.

83%	Higher Investment Rate
58%	Lower Gov't Expenditures
126%	Higher GNP Growth
3.5X	Productivity Growth

(a) Federal, State and Local Current Spending Excluding Transfer Payments and Capital Spending.

Chart 1 shows that for 18 years (1962 to 1980) we have fallen behind in economic performance. We are at the bottom of the list in terms of average investment as a percentage of GNP; second to last in real growth in GNP; absolute last in productivity increase, but we top everyone in government spending as a percentage of GNP.

CHART 2

THE RETARDED U. S. ECONOMY

115 Million Japanese Buy More New Plant and Equipment than 220 Million Americans

so

Machinery in Japan is on Average 6-7 Years Newer

so

Toyota Makes 9 Engines Per Employee/Day Vs. Ford's 2

and

Automated Warehouses Allow Toyota to Carry 1-Hour In-Plant Inventory Vs. Ford's 3 Weeks.

Chart 2 shows that 115 million Japanese buy more new plants and more equipment than 220 million Americans. Machinery in Japan is on an average 6-7 years newer than ours. Toyota makes 9 engines per employee/day versus Ford's 2 engines per employee/day. Automated warehouses allow Toyota to carry one-hour in-plant inventory versus Ford's three-weeks' in-plant inventory.

CHART 3

THE DETERIORATING U.S. ECONOMY

	(1)	(2)	(3)	(4)
	7 Years Ending			% (Deterioration
	1966	1973	1980	1966-1980
(1) Federal Spending as % of GNP (Average %)	18.8 %	20.4 %	22.0 %	(17.0)%
(2) Unemployment Rate (Average %)	5.3	4.6	6.8	(28.3)
(3) Real GNP (Avg. Ann. % Change)	4.5	3.5	2.4	(46.7)
(4) Real Business Investment (Avg. Ann. % Change)	7.9	3.6	2.0	(74.7)
(5) Productivity (Avg. Ann. % Change)	3.4	2.1	0.5	(85.3)
(6) Inflation (Avg. Ann. % Change)	1.5	4.6	9.2	(513.3)
(7) Federal Deficit (Average, $ Billions)	$(1.3)	$(9.7)	$(41.8)	(3,115.4)

The statistics shown in Chart 3 indicate a significant increase in federal spending as a percentage of GNP over the time period 1959-1989. The unemployment rate, real growth in GNP, real business investment, productivity increase, inflation and the federal deficit all suffered during the same time span.

CHART 4

"FOR MORE THAN NINE YEARS, RESTRICTIVE PRICE CONTROLS HAVE HELD U. S. OIL PRODUCTION BELOW ITS POTENTIAL, ARTIFICIALLY BOOSTED ENERGY CONSUMPTION, AGGRAVATED OUR BALANCE OF PAYMENTS PROBLEMS AND STIFLED TECHNOLOGICAL BREAKTHROUGHS. PRICE CONTROLS HAVE ALSO MADE US MORE ENERGY-DEPENDENT ON THE OPEC NATIONS -- A DEVELOPMENT THAT HAS JEOPARDIZED OUR ECONOMIC SECURITY AND UNDERMINED PRICE STABILITY AT HOME."

President Reagan
January 28th, 1981

In the order eliminating price controls on U. S. Oil Production and Marketing.

The problems of the U.S. economy are often blamed on OPEC and energy prices. The next 15 charts (4-18) put this in perspective.

Inflation, now, is up 513.3% and the federal deficit, of course, is up over 3,000%. We often blame our troubles on OPEC and energy prices but President Reagan has reminded us that for more than 9 years, we have held U.S. oil production down with restrictive price controls. I think that this is a key issue. The quotation on Chart 4 shows that President Reagan recognizes this as a major issue.

CHART 5

U. S. ENERGY CONSUMPTION

	(1)	(2)	(3)	(4)	(5)	(6)	(7)	(8)
					Quads. Percentage Distribution By Source			
	Quad. Btu	Value Total ($ Bil.)	Value Per MMBtu (¢ or $)	Value As A % Of GNP	Oil	Gas	Coal	Hydro, Nuclear And Other
(1) 1950	33.6	$ 8.7	25.9 ¢	3.1 %	39.6 %	17.8 %	38.3 %	4.3 %
(2) 1960	44.1	13.7	31.1	2.7	45.2	28.1	22.9	3.8
(3) 1970	66.8	22.0	32.2	2.2	44.2	32.6 H	18.9	4.4
(4) 1975	70.7	75.7	$ 1.07	4.9	46.3	28.2	18.0	7.5
(5) 1978	78.2	111.8	1.43	5.2	48.6 H	25.6	17.6 L	8.2
(6) 1980	76.5	225.0	2.94	8.6	44.8	27.2	20.4	7.7
(7) 1985	80.9	479.0	5.92	11.1	40.2	25.3	22.7	11.7
(8) 1990	91.0	755.0	8.30	10.5	33.7	23.4	28.9	14.0

End of the Period of Illusion, but, 5X in 15 years.

Slow progress from 1980 and grossly inadequate re 1950

Chart 5, "U.S. Energy Consumption," documents some interesting facts. These figures are in quadrillion BTU's; note that in 1980, U.S. energy consumption was 76.5 quads—down from 78.2 in 1978. There had been real growth up through 1978; however, the monetary value between 1978 and 1980 went from $111 billion to $225 billion; from $1.43 per MMBtu to $2.94 per MMBtu—double! We expect it to go to $8.30 per MMBtu in 1990. From 2.2% of GNP in 1970, we predict it will be 11.1% in 1985. We think our energy situation will change by the year 1985—that will be the end of our period of illusion. We have been living under illusions that once again we will have cheap energy in this country. The value of energy as a percentage of GNP increased 5 times during the 15-year period, 1970 to 1985. In coal consumption (quadrillions as a percentage of distribution by source), it was 22.9% in 1960 and dropped to 18% in 1975. Now in 1980, it is up to 20.4 and we expect it to go to 28.9 in the 1980's. That is slow progress and is grossly inadequate versus the 38.3% in 1950.

CHART 6

ENERGY TRADE BALANCE

		(1)	(2)	(3)	(4)	(5)	(6)	(7)	(8)	(9)	
		Net Exports/(Imports)									
		Quadrillion Btu's				Billion $ Current				$ Per MMBtu	
		Oil	Gas	Coal	Total	Oil	Gas	Coal	Total		
(1)	1950	(1.2)	-	0.8	(0.4)	$(0.5)	-	$ 0.3	$(0.2)	50¢	
(2)	1960	(3.6)	(0.2)	1.0	(2.7)	(1.7)	-	0.4	(1.3)	48	
(3)	1970	(6.9)	(0.8)	2.0	(5.7)	(3.4)	$(0.2)	1.0	(2.6)	46	13X in a Decade
(4)	1975	(12.5)	(0.9)	1.7	(11.7)	(26.6)	(1.1)	3.2	(24.5)	$ 2.09	
(5)	1978	(17.1)	(0.9)	0.7	(17.3)	(41.8)	(2.0)	1.6	(42.2)	2.44	
(6)	1980	(13.5)	(1.0)	2.0	(12.5)	(75.4)	(4.0)	4.5	(74.9)	6.00	
(7)	1985	(13.6)	(2.6)	2.3	(13.9)	(130.0)	(30.5)	7.5	(153.0)	11.00	
(8)	1990	(12.8)	(3.5)	2.6	(13.7)	(197.0)	(67.5)	11.5	(253.0)	18.50	
	Increase/ (Decrease) 1980-1990										
(9)	Amount	0.7	(2.5)	0.6	(1.2)	(121.6)	(63.5)	7.0	178.1	12.50	
(10)	Percent	5.2%	(250.0)%	30.0%	(9.6)%	(161.3)%	(1,587.5)%	155.6%	237.8%	208.3%	

1	Quadrillion Btus
= 1	Million billion Btus
= 172.4	Million barrels of oil
= 1	Trillion cubic feet of gas
= 45.5	Million tons of coal

(253.0) = 186.2% Of 1980 Defense Budget

Chart 6 depicts our energy trade balance and, of course, this is terribly important. Note that in 1978, we imported 17.1 quadrillion BTU's of oil, then dropped to 13.5 BTU's in 1980. In 1990, we expect to import 12.8 quadrillion BTU's of oil, a reduction of 5.2%. However, the monetary value of our oil imports will increase from $75.4 billion in 1980 to $130 billion in 1985 and almost $200 billion in 1990. So, although we are dropping the quantity of our imports, their costs are increasing dramatically. At this rate, we will spend almost $200 billion in oil imports and another $67.5 billion in gas imports in 1990. Meanwhile, we will be exporting $11.5 billion worth of coal, about 2.6 quadrillion BTU's—a real trade balance problem.

The cost per million BTU's was $.46 in 1970 and $6.00 in 1980. We predict an eventual increase to $18.50 per million BTU's.

CHART 7

OIL, GAS, COAL PRICE COMPETITION
($ or ¢ Per MMBtu)

		(1)		(2)		(3)	(4)	(5)
		Oil (Low-Sulfur Residual)		Natural Gas		Coal (Low-Sulfur)(a)	Percent Coal Higher/(Lower) Than Oil	Percent Coal Higher/(Lower) Than Gas
1950		32¢		26¢		50¢	56.3%	92.3%
1960		35		36		52	48.6	44.4
1970		49		39		70	42.9	79.5
1975	9.9X	$ 2.07	7.1X	97	3.4X	$ 1.65	(20.3)	70.1
1978		2.27		$ 1.49		1.95	(14.1)	30.9
1980		4.83		2.75		2.40	(50.3)	(12.7)
1985		8.45		8.00		4.15	(50.9)	(48.1)
1990		13.75		16.50		5.80	(57.8)	(64.8)

(a) Includes higher capital and operating costs associated with burning coal.

Only misguided government policies and regulations could produce this result.

Chart 7 shows oil, gas and coal competition. In 1970, oil (low-sulfur residual) was $.49 per million BTU's and in 1980, it increased 9.9 times to $4.83. Natural gas increased from $.39 to $2.75 (7.1 times). Coal increased from $.70 per million BTU's to $2.40 (3.4 times). Therefore, coal was higher than oil by 56.3% in 1950; in 1980, it was 50.3% lower than oil and 12.7% lower than gas. We expect that differential for coal to continue and in 1990 to be 57.8% lower than oil and 64.8% lower than gas. This is due only to misguided government policies. We have 3.5 trillion tons of coal, and we don't need to increase our balance of payments to these huge, adverse figures when we have as much energy indigenous to our country as coal provides.

CHART 8

ENERGY PRICE UNCERTAINTIES AND OPPORTUNITIES

		(1)	(2)	(3)	(4)	(5)	(6)	(7)
					$ or ¢ Per MMBtu			
		Oil ($/Bbl)	Gas ($ or ¢/ Mcf)	Coal ($/Ton)	Oil	Gas	Coal	Oil As % Of Coal
(1)	1950	$ 2.51	6¢	$ 4.84	43¢	6¢	18¢	238.9%
(2)	1960	2.88	14	4.69	50	14	18	277.8
(3)	1970	3.15	17	6.26	54	17	26	207.7
(4)	1980	27.00	$ 1.70	23.00	$ 4.66	$ 1.70	$1.03	452.4
(5)	1985 Low	42.50	5.25	30.00	7.33	5.25	1.36	539.0
(6)	Medium	55.00	6.25	35.00	9.48	6.25	1.59	596.2
(7)	High	80.00	8.50	50.00	13.79	8.50	2.27	607.5
(8)	1990 Low	60.00	12.50	40.00	10.34	12.50	1.90	544.2
(9)	Medium	89.50	15.50	47.00	15.43	15.50	2.24	688.8
(10)	High	135.00	23.00	75.00	23.28	23.00	3.57	652.1
		125% Low to High		87.5% Low to High				

Chart 8 shows the energy price uncertainties for the future. In 1985, oil will vary in price per barrel from $42.50 low, $55.00 medium, and $80.00 high. The price differential depends on the possibility of a complete blow-up in the Middle East. In 1990, we are looking for a low of $60.00, a medium of $89.50, and a high of $135.00. There is a spread of 125% between the low and the high because of all the political uncertainties in the Middle East. We expect coal to increase from $23.00 a ton in 1980 to $35.00 in 1985 and $47.00 in 1990. That means that the cost of oil compared to coal in terms of million BTU's was 452.4% higher in 1980 and, on the mean, will increase to 596.2% in 1985 and 688.8% in 1990.

CHART 9

UTILITY OIL AND COAL COSTS

		(1)	(2)	(3)	(4)	(5)	(6)	(7)
		Price		Volume		Cost		Added Cost To Burn Oil
		Residual Oil ($/Bbl)	Coal ($/Ton)	Residual Oil (Mil. Bbls.)	Coal (Mil. Tons)	Residual Oil ($ Millions)	Coal ($ Millions)	($ Millions)
(1)	1950	$ 2.00	$ 6.38	75.4	91.9	$ 150.8	$ 586.3	$ 29.4
(2)	1960	2.17	6.26	85.3	176.6	185.1	1,105.5	24.5
(3)	1970	2.45	7.08	335.5	320.2	822.0	2,627.0	183.2
(4)	1975	12.69	17.63	506.1	406.0	6,422.4	7,157.8	3,877.2
(5)	1980	25.75	28.75	401.0	545.0	10,325.8	15,668.8	6,950.0

		Price		Volume		% Of Btu Volume	
		Residual Oil ($ or ¢ Per MMBtu)	Coal ($ or ¢ Per MMBtu)	Residual Oil (Quadrillion Btu)	Coal (Quadrillion Btu)	Residual Oil	Coal
(6)	1950	32¢	27¢	0.6	2.4	20.0%	80.0%
(7)	1960	34	30	0.6	4.2	12.5	87.5
(8)	1970	40	31	2.1	7.2	22.6	77.4
(9)	1975	$ 2.01	81	3.2	8.8	26.7	73.3
(10)	1980	4.15	$ 1.35	2.5	11.6	17.7	82.3

> This $7 billion bonus for burning coal increases to about $15 billion per year by 1990. Since we produce only 40% of our oil, this adds to our Balance of Trade problems, and the $15 billion is about 60% of our last 3 years average annual Trade Deficit.

Chart 9 shows that the cost of residual oil increased from $2.00 per barrel in 1950 to $2.17 in 1960 and $2.45 in 1970, an increase of only $.45 in 20 years. Then, the price jumped to $12.69 in 1975 and to $25.75 in 1980. Coal increased only from $6.38 per ton in 1950 to $7.08 in 1970. In 1975, the cost of coal was $17.63, and it increased to $28.75 in 1980. The added cost to burn oil in this country is about $7 billion more than coal, and it may increase to $15 billion in 1990. We produce about 40% of our oil, so all of this additional payment for oil adds to our balance of trade problems. The $15 billion in 1990 alone is 60% of our last three years' average annual trade deficit.

CHART 10

ECONOMICS OF OIL IMPORT DEPENDENCE

	(1)	(2)	(3)	(4)	(5)	(6)	(7)	(8)
	Price				Volume			
			Domestic Higher/ (Lower) Than Imported					
	Domestic	Imported	Amount	Percent	Domestic	Imported	Total Supply	% Imported Of Total
	-------($ or ¢ Per Barrel)-------				--(Million Barrels Per Day)--			
1950	$ 2.51	$ 2.51	-	-	5.9	0.6	6.5	9.2 %
1960	2.88	2.88	-	-	8.0	1.6	9.6	16.7
1970	3.18	2.96	22¢	7.4%	11.3	3.2	14.5	22.1
1975	7.67	12.47	$(4.80)	(38.5)	10.0	5.8	15.9	36.5
1978	10.02	14.32	(4.30)	(30.0)	10.3	8.0	18.3	43.7
1980	23.50	32.69	(9.19)	(28.1)	10.2	6.3	16.5	38.2
1985	54.75	55.50	(75)¢	(1.4)	9.4	5.9	15.4	38.3
1990	89.00	90.00	$(1.00)	(1.1)	8.9	5.5	14.5	37.9

Massive incentive to import oil and neglect domestic sources.

With price decontrol import dependence declines.

Chart 10 shows that the domestic price of oil was 38.5% lower than that of imported oil in 1975; in 1980, it was 28.1% lower. We are forecasting them to be on par in 1985 and 1990 due to the energy program of the new Administration. The volume of imported oil was only 9.2% in 1950, and it got up to 43.7% in 1978. It's now down to 38.2%, and we expect it to stay at about 38%, although we still have to contend with that huge $197 billion trade imbalance from importing oil.

CHART 11

U. S. LOW GASOLINE PRICES, HIGH OIL CONSUMPTION
(Current $)

		(1)	(2)	(3)
		Premium Gasoline		
		Price Per Gallon	% Of U. S. Price	Per Capita Oil Consumption
(1)	Italy	$ 3.91	285.4 %	10.4
(2)	France	3.32	242.3	13.3
(3)	Spain	2.89	210.9	7.7
(4)	Germany	2.75	200.7	14.1
(5)	United Kingdom	2.44	178.1	9.2
(6)	Japan	2.20	160.6	14.7
(7)	Average, 6 High Price Countries	$ 2.92	213.1 %	11.6
(8)	United States	1.37	100.0	28.9
(9)	Canada	1.04	75.9	26.3

6 Countries-113% Higher Prices. U. S.-149% Higher Consumption.

So That With 5.1% Of World Population, The U. S. Consumes 49.3% Of The World's Gasoline, 9.7X Its Share.

One of our problems is the high oil consumption here in the United States and the low gasoline prices compared to the rest of the world, according to Chart 11. Seven countries have higher prices for gasoline than the United States. The United States has 149% higher consumption on a per capita per person basis; so, with 5.1% of the world's population, the United States consumes 49.3% of the world's gasoline—9.7 times our share. This is really high, particularly when we are using a gradually disappearing resource. In 40 to 50 years, there will certainly be a world shortage of oil—we are using too much!

CHART 12

GASOLINE PRICES AND TAXES
Europe And The U. S.

		(1)	(2)	(3)	(4)	(5)
		Price		Tax		
					As Percent Of	
		At Pump	Pre-Tax	Amount	Pump Price	Pre-Tax Price
(1)	Italy	$ 3.91	$ 1.38	$ 2.53	64.7%	183.3%
(2)	Denmark	2.84	1.23	1.61	56.7	130.9
(3)	France	3.32	1.50	1.82	54.8	121.3
(4)	Seven European Countries	2.96	1.38	1.58	53.4	114.5
(5)	United Kingdom	2.44	1.17	1.27	52.0	108.5
(6)	Netherlands	2.56	1.26	1.30	50.8	103.2
(7)	Belgium	2.91	1.55	1.36	46.7	87.7
(8)	Germany	2.75	1.52	1.23	44.7	80.9
(9)	United States	1.37	1.22	0.15	10.9	12.3
(10)	Seven European Countries As Multiple Of United States	2.2X	1.1X	10.5X	4.9X	9.3X

Note on Chart 12 that seven European countries have a gasoline tax of $1.58 while we here in the United States have a tax of $.15. The gasoline tax of these seven countries thus is 10.5 times higher than that of the United States.

CHART 13

AVERAGE RETAIL PRICE OF
REGULAR GASOLINE IN THE U. S.
(¢ Or $ Per Gallon)

		(1)	(2)	(3)
		Price	Tax	Tax As % Of Price
(1)	1960	31.1 ¢	10.1 ¢	32.5 %
(2)	1965	31.2	10.5	33.7
(3)	1970	35.7	11.1	31.1
(4)	1975	57.2	11.8	20.6
(5)	1976	59.5	12.0	20.2
(6)	1977	63.1	12.4	19.7
(7)	1978	65.7	12.6	19.2
(8)	1979	87.7	13.5	15.4
(9)	1980	$ 1.22	14.4	11.8

In 1960 U. S. was taxing at 275% of the 1980 rate.

Chart 13 shows that we have gone down in our tax percentage of the price of gasoline. The tax as a percentage of price was 32.5% in 1960 and 33.7% in 1965; it now is only 11.8%. Therefore, in 1960, we were taxing at 275% of the 1980 rate.

CHART 14

CONSUMER COSTS OF ENERGY AND TAXES, 1970-1980
(Billions Of Current Dollars)

		(1)	(2)	(3)	(4)	(5)	(6)	(7)
		Energy			Federal Taxes			
		Gasoline	All Other	Total	Personal Income	Social Security	Excise, Estate And Gift	Total
(1)	1970	$ 22.4	$ 19.8	$ 42.2	$ 88.8	$ 49.3	$ 19.4	$ 157.5
(2)	1971	23.9	21.5	45.4	85.7	54.4	20.6	160.7
(3)	1972	25.4	23.8	49.2	102.7	62.7	21.0	186.4
(4)	1973	28.6	26.7	55.3	109.5	79.5	21.8	210.8
(5)	1974	36.6	31.7	68.3	126.4	89.8	21.3	237.5
(6)	1975	40.4	37.4	77.8	120.8	94.1	21.3	236.2
(7)	1976	44.0	42.7	86.7	141.5	106.5	22.6	270.6
(8)	1977	48.2	48.8	97.0	162.7	118.5	24.7	305.9
(9)	1978	52.7	54.2	106.9	189.4	137.2	23.8	350.4
(10)	1979	68.4	63.3	131.7	225.7	159.0	24.1	408.8
(11)	1980	89.1	66.9	156.0	251.0	172.2	35.7	458.9
	Amount Increase							
(12)	1970-1980	$ 66.7	$ 47.1	$ 113.8	$ 162.2	$ 122.9	$ 16.3	$ 301.4
(13)	1975-1980	48.7	29.5	78.2	130.2	78.1	14.4	222.7

Federal Taxes Up By A Multiple Of 2.8X Energy

Chart 14 shows the consumer costs of energy and taxes during 1970 through 1980. The current and former Administrations have both blamed inflation on the rising cost of energy, but it is worthwhile to note that between 1975 and 1980, the cost of federal taxes increased 2.8 times the cost of energy prices.

CHART 15

THE PRICE OF A BARREL OF IMPORTED OIL
IN MAJOR COUNTRIES
(1970 = 100.0)

	(1)	(2)	(3)	(4)
Country (Currency)	1970	1980	Total % Change	U. S. Price Change As Multiple Of Change In Other Countries
Switzerland (Swiss Franc)	100.0	429.1	329.1 %	3.1X
Germany (Mark)	100.0	550.5	450.5	2.2
Netherlands (Guilder)	100.0	607.0	507.0	2.0
Japan (Yen)	100.0	700.0	600.0	1.7
France (French Franc)	100.0	844.1	744.1	1.3
United States (Dollar)	100.0	1,104.4	1,004.4	1.0

The U. S. is paying 3-Fold the Swiss increase in the price of oil due to depreciation of the U. S. $, despite its 79% energy independence

Chart 15 shows a comparison of the price of a barrel of imported oil in six major countries. Due to fiscal policies and the weak dollar, the Swiss paid only 329.1% a barrel more for imported oil in 1980 than in 1970. The United States, however, paid an increase of 1,004.4%, so we are paying threefold the Swiss increase in the price of oil due to the depreciation of the U.S. dollar, despite the U.S.'s 79% energy independence. The U.S. price change as a multiple of the change in the other countries is 3.1 times of Switzerland; 2.2 times of Germany; 2.0 of the Netherlands; 1.7 times of Japan and 1.3 times of France.

CHART 16

OIL IMPORT DEPENDENCE
(Second Half 1979)

		(1) Value of Oil Imports as % of Value of Energy Consumption	(2) Value of Oil Imports as % of Value of Total Exports: Amount	(3) Value of Oil Imports as % of Value of Total Exports: Rank
(1)	Japan	74.0%	32.6%	2
(2)	Italy	67.0	18.5	3
(3)	France	59.0	16.0	4
(4)	Germany	53.0	11.0	5
(5)	U. S.	21.0	32.9	1

U. S. Least Dependent On Oil Imports
But
Least Able To Pay Through Exports

We have a low dependency on imported oil as a percentage of our total energy consumption (see Chart 16). We are only at 21.0% dependency compared to Japan's 74.0%. They are almost 4 times as dependent on imported oil, yet the value of their oil imports as a percentage of the value of total exports is the same as ours. We are not running our foreign trade situation very intelligently when Japan has 74.0% dependency on oil imports and has the same percentage of the value of their exports in oil imports.

CHART 17

COMPARISON OF TV COMMERCIALS AND DOMESTIC OIL PRICES
(Current $)

	(1)	(2)	(3)	(4)
			1979 Increase Over 1975	
	1975	1979	Percent	Each As Multiple Of Upper Tier Oil
	Cost of 30-Second Commercials			
(1) World Series	$ 52,200	$100,000	91.6%	9.4 X
(2) Super Bowl	110,000	200,000	81.8	8.4
(3) Academy Awards	80,000	145,000	81.3	8.4
(4) Wide World of Sports	15,200	26,000	71.1	7.3
(5) Average Prime Time	30,700	51,000	66.1	6.8
(6) Summer Olympics (a)	42,000	65,100	55.0	5.7
(7) Winter Olympics (a)	34,500	44,800	29.9	3.1
	Cost of a Barrel of Oil			
(8) Lower Tier Oil	$ 5.03	$ 6.00	19.3%	2.0 X
(9) Upper Tier Oil	12.03	13.20	9.7	1.0

(a) 1976-1980

For the 1981 Super Bowl, a 30-Second Commercial Cost $270,000 or 35.0% Above 1979.

Chart 17 shows the price of a 30-second commercial during the World Series, the Super Bowl and the Academy Awards. These costs increased much faster than the price of upper tier oil. Actually, the cost of a 30-second commercial during the Super Bowl this year was $270,000 or 35.0% above 1979. If energy prices increased like that, people would really scream. These commercial prices have gone up much faster, yet the media talks about the obscene profits of the energy companies.

CHART 18

COMPARISON OF MEDIA AND OIL COMPANY PROFITS
(Latest 2 Years)

	(1)	(2)	(3) % Return On Capital 1978-79
(1)	Dow Jones	Media Company	26.5 %
(2)	Washington Post	Media Company	24.8
(3)	Times Mirror	Media Company	18.8
(4)	CBS	Media Company	18.4
(5)	McGraw Hill	Media Company	18.3
(6)	ABC	Media Company	17.4
(7)	Gannett	Media Company	17.4
(8)	Exxon	Oil Company	14.5
(9)	Standard (California)	Oil Company	13.9
(10)	Standard (Indiana)	Oil Company	13.5
(11)	Mobil	Oil Company	13.3
(12)	Atlantic-Richfield	Oil Company	12.1
(13)	N. Y. Times	Media Company	11.7
(14)	Gulf	Oil Company	11.4
(15)	Shell	Oil Company	11.2
(16)	Texaco	Oil Company	10.4
	Average Returns		
(17)	8 Media Companies		19.2
(18)	8 Oil Companies		12.5

Chart 18 compares media and oil company profits and shows that media companies make much higher returns on capital. The media never talks about return on capital, only about amount of profit. If the capital happens to be 10 times in one company versus another, there is no reason to look just at their profits. The real issue is the rate of return, and the media companies make a much higher rate of return than do the energy companies.

RECOMMENDED ACTIONS

The following six major steps should be taken:

1. Balance the budget over the course of the business cycle. Halt net deficit spending over the cycle, the prime cause of a too rapidly expanding money supply in relation to real growth in goods and services—which ultimately leads to higher prices.

2. Cut the top personal income tax rate from 70 percent to 36 percent. This reduction in tax rates would cut government revenues by only 4.9 percent. In the process, economic activity would be greatly stimulated and the tax base expanded to quickly overcome revenue shortfalls.

3. Adjust all personal income and corporate profits for inflation before taxation. Only real gains would be taxed as opposed to nominal gains brought about strictly by inflation; government revenues would no longer automatically benefit from inflation.

4. Eliminate the capital gains tax and thus get rid of this disincentive to investment. Countries that have no capital gains tax such as Japan and Germany—where economic growth has been exceptionally strong—provide convincing evidence of the benefits that accrue from a zero capital gains tax.

5. Manage agricultural exports to strengthen the U.S. economy. The United States, together with Canada, Australia and Argentina, controls more than 80 percent of the world's food grain exports. These countries should work to sustain agricultural product prices in line with those of gold and oil in world markets.

6. Eliminate price controls and reduce regulation on all forms of energy.

One primary concern must be to create a favorable climate for increased investment, particularly in small companies—those employing fewer than 100 workers. These small companies account for about 82 percent of all new jobs. Inadequate savings and investment have been the root causes of lagging productivity, slow growth and persistent unemployment in this country.

Enacting legislation which would cut taxes through such means as individual rate reductions, indexing personal income to inflation, freezing Social Security rates and allowing realistic depreciation schedules on plant and equipment should be the priority of the 97th Congress.

The 1980 elections changed the composition of the House and Senate significantly, producing a climate more receptive to enacting measures designed to create jobs, reduce inflation, rationalize government spending and boost industrial production.

LUNCHEON SPEECH
8TH ENERGY TECHNOLOGY CONFERENCE

S. David Freeman
Chairman of the Board, Tennessee Valley Authority

This 8th Annual Energy Technology Conference focuses on the 1980's as a "new fuels era." I think that theme is highly appropriate, and I can't imagine a better topic for my remarks today than the question of how we can go about adjusting our national energy system to the new era that is now dawning.

First, we need to recognize that the 1980's will be an era of new fuels not by choice but by necessity. The low-cost oil and gas that made possible the explosive growth of the American economy in this century have been largely used up. There are, of course, more reserves to be found, but they are going to be very expensive to develop. Most experts doubt that we can do much better than maintain current levels of domestic petroleum production. Yet it is imperative that we find ways to reduce sharply our continuing national dependence on imported oil.

Seven years after the Arab oil embargo of 1973-74, America remains a Nation whose energy security is tied to a very fragile oil lifeline.

Despite some recent progress in reducing imports, we are still dependent on petroleum from abroad for 40 percent of our oil, a quantity accounting for nearly 20 percent of all our energy supplies.

The ill effects of this dependence have been stated over and over again. But its true dangers are not apparent to the consumer when there is plenty of gas at the pump. Like addicts, we grow numb to the ill effects of our addiction. Let me briefly spell out the problems this addiction presents.

Imported oil is sapping the life out of our economy. Payments to the OPEC cartel and other producing nations sent more than $60 billion overseas in 1980, more than $275 for every man, woman, and child in the U.S. Continuing price increases guarantee that figure will be higher in 1981. The rise in gasoline and heating oil prices in the last two years has already added at least two percentage points to the national inflation rate.

A new oil embargo would have far more serious effects than even the continuing run-up in world oil prices. A complete cutoff for six months would virtually shut down the U.S. economy, bringing to a standstill the world's greatest industrial complex. Our oil import dependence thus casts a long shadow of almost certain depression over the U.S. should access to foreign supplies be abruptly curtailed.

Even beyond such grave economic effects, imported oil poses a tremendous danger, because it calls into question our very national security. The U.S. dependence on Arab nations of the Middle East for more than a third of our total imports is already making our vital peace-making role between the Arabs and Jews increasingly difficult to carry out. The Middle East is volatile. There is an active war in progress in the region between two major oil producers, Iran and Iraq. The need to protect our oil lifeline greatly increases the possibility of U.S. military intervention and even world war.

There appears to be universal agreement that continued dependence on large quantities of imported oil represents a major continuing threat to this country's independence. It presents a clear and present danger to our survival that can be dealt with only by forceful action on the part of government and citizens alike.

I want to speak today of the role that electric power can play in bringing coal and uranium--our largest source of domestic fuels--into active use in transportation and industry to substitute for oil. I believe this approach requires much more public attention that it has received if it is to be implemented. Let me say first that using our petroleum more efficiently--energy conservation--should come ahead of fuel substitution. Conservation is not only the quickest and easiest way to reduce petroleum imports, it is also by far the cheapest, at least for the first several million barrels. And I'm sure I don't need to remind you that conservation energy is environmentally benign.

The evidence indicates that the American people are learning to take advantage of the savings conservation can afford. Gasoline lines and the doubling of gasoline prices over the last two years have led people to begin demanding fuel-efficient automobiles in huge numbers. Gas-guzzlers have become a glut on the market. And thanks in part to a fuel-efficiency law passed in 1975, Detroit is finally beginning to meet the demands of the marketplace for more fuel-efficient cars.

Conservation is also taking hold outside the transportation sector. Hundreds of thousands of homes heated by oil have been weatherized and insulated in recent years as people have come to realize the opportunities for reducing energy waste. At the same time, industry has turned more and more of its attention to plugging leaks and increasing the efficiency with which it uses energy. The net result of all of these conservation initiatives, together with an economic recession, has been an oil import reduction of more than 20 percent from the record level of 1978.

Yet despite this progress, imports could rise again as the U.S. economy begins to grow. And oil prices will also increase. By 1990, most experts believe the price of crude oil will be triple what it is today, $100 a barrel or more. If America is still trying to import more than 6 million barrels of oil a day in 1990, we could well be shipping more than $200 billion a year overseas.

This astronomical price simply reflects the growing scarcity of all high-grade fossil fuels. If we allow our dependence on large quantities of imported petroleum to continue, a steady drain on the U.S. economy will likewise continue, and so will the national security problem. At some point, we must pay the price of stronger conservation measures and wholesale substitution of other domestic energy sources if we are to maintain both our prosperity and our security.

Conservation is not the only form of unconventional energy that can play a role in re-establishing U.S. energy self-sufficiency. Increasingly, the wood in our forests can be converted to liquids that can help run our cars. And, to the extent petroleum is used simply as a source of heat, active solar hot water heating and passive solar space heating are particularly attractive substitutes. Decentralized energy can save money and imported oil today, not just after the turn of the century, as some would have it.

But in order to deal effectively with the imported oil threat, we will have to confront it at its source: the demand for liquid fuels for transportation and industrial uses. Conservation and the other forms of decentralized energy can only go so far in reducing the U.S. appetite for imported oil in these areas. Eventually, we must begin displacing imported oil with homegrown fuel of another kind. Fortunately, there exists a uniquely flexible form of energy that is ideally suited to this purpose. I am speaking, of course, of electricity.

In the middle years of this century, from the 1930's through the 1960's, electricity made possible a dramatic improvement in people's lives. The light bulb replaced the kerosene lantern, and pumps and motors eased the burden of manual labor on the farms and in the factories of America. TVA's example in slashing prices helped make electricity dirt cheap in virtually every village and hamlet.

It is true that people often used electricity wastefully during prior years, but that practice is now rapidly coming to an end. As we enter the fuel-efficient era of the 1980's, electricity can again solve problems that could make our daily lives miserable. Electricity can provide a reliable fuel for electric cars and mass transit to get people back and forth to work; for electric trains to move people and freight between cities; and to replace oil in power plants, factories, and commercial establishments.

This substitution of kilowatts for barrels of oil may not seem quite as dramatic as lighting up the countryside, but it may well be more important. What is at stake is the very survival of a civilization that prizes mobility and independence for the individual citizen.

Today, we know that electricity is a high-grade form of energy, and we can't afford to allow it to be wasted--or used in applications where its unique qualities of cleanliness, flexibility, and precision are not required. Where energy in less premium forms--solar, wood, garbage, and all the relatively diffuse renewable energy sources--can serve the purpose, these should be our first choice. The Tennessee Valley Authority is actively promoting all of these energy sources where they are applicable.

But it is not wasteful to substitute electricity for oil. In fact, it is vitally important that we actively promote the increased use of electricity in applications where it can substitute for scarce petroleum--and where maximum efficiency has already been achieved through strict conservation techniques. No other form of energy offers so much opportunity for innovation as a petroleum substitute, whether in transportation, residential heating, or a host of industrial applications. And no other type of energy is available for these purposes in so convenient and environmentally acceptable a form.

This indeed is the crux of the matter. We possess vast reserves of coal and uranium in this country, but we lack the means to take advantage of them quickly except in the form of electricity. There is no immediate energy shortage except in the area of liquid fuel. And electric power production represents the only near-term way of converting the coal and uranium we do have into a form that can take the place of imported oil.

Synthetic fuels from coal are still on the drawing boards. Such plants will not produce a sizeable quantity of fuel until well into the 1990's. For most of the 1980's and the 1990's, the only real replacements for oil are conservation--which we are already pushing hard--and electricity.

Even if synthetic fuels were available today, we would be well advised to choose electricity over synthetics as a substitute for oil wherever electricity is feasible. A number of detailed studies have concluded that electricity constitutes a more energy efficient use of our large coal reserves than synthetic fuel in most applications. This is particularly true of transportation and home heating, which together consume the equivalent of far more than our current level of oil imports.

One study found, for instance, that the overall energy efficiency of an electric car is about 15 percent on the road and 10 percent around town. This included all the energy lost in coal mining, power production, transmission and consumption. The comparable figures for an internal combustion vehicle powered by synthetic liquid fuels were 10 percent efficiency on the highway and 6 percent efficiency around town. In other words, the electrically powered system was found to be at least 50 percent more efficient in the use of basic coal than a synthetic gasoline powered car.

Electricity can also be used to heat homes efficiently if a heat pump is used. The heat pump is a particularly useful electrical device because it is actually a form of solar collector--it transfers heat from the surrounding outdoor air indoors. It already does so with an efficiency several times that of electrical resistance heating. And it offers exciting possibilities for further development. We are just discovering the wide range of applications for this device, from miniature room units to heat pump water heaters.

Increased efficiency of end-use may improve both the economics and the energy balance of electricity in other areas as well. These include precision industrial applications such as treatment, welding and separation of metals and the brave new world of lasers. Energy transitions have traditionally provoked technological innovation. If that continues to hold true, we could be in for a truly exciting period as more and more electricity is used very efficiently as part of the modernization of American industry.

At least four specific areas offer major opportunities to carry out this substitution of electricity for petroleum. The most significant oil savings involve transportation, which currently consumes the equivalent of all U.S. oil imports.

Electric cars offer perhaps the most exciting prospect for displacing oil. A variety of innovative battery and motor configurations for reliable cars with an 80-100 mile range are now being developed. Every time the price of oil jumps up, the economics of the electric car improves. At least one major manufacturer, the General Motors Co., is gearing up for mass production in the 1980's. In fact, General Motors expects to market its first commercial vehicle in the mid-1980's.

Each electric car uses about .5 kilowatthours per mile, Thus, 10 million electrics averaging 10,000 miles a year would require some 50 billion kilowatthours, the output of 8 large power plant units. On the TVA system, we estimate up to 400,000 electrics could be operated without building additional plants, since most of the charging would be done at night, during off-peak periods. Additional sales of 2 billion kilowatthours would result, creating a market for almost 1 million tons of coal a year to substitute for imported oil. .

Another extremely promising way of displacing imported oil is by electrifying major railroad trunk lines. This already makes economic sense in many parts of the country, given the rapidly increasing cost of diesel fuel. TVA is negotiating with the L&N and Southern lines to electrify the track between Atlanta and Cincinnati. This would require up to 200,000 kilowatts of capacity, by some estimates. And electrifying 26,000 miles of track nationally would require about 5,000,000 kilowatts of new capacity.

Direct industrial substitution of electricity for oil offers additional hope of permanently weaning this country from imported energy supplies. The urgent need to find alternative means of powering the Nation's industrial machine makes this sector a logical target for a concentrated conversion campaign. As oil and gas prices rise rapidly with continuing decontrol, electricity is likely to look more and more attractive for this purpose.

The French have forged way ahead of us in industrial applications of electricity as well as in encouraging residential use of heat pumps. They have managed to increase their industrial use of electricity each year over the last 10 years. Their goal is to supply nearly half of all energy used in industry from nuclear-powered generating plants by the year 1990, and they are progressing toward that goal. Virtually every kilowatthour they use is a kilowatthour that will not have to be burned in the form of imported oil in an industrial boiler.

Petroleum has been found so pervasively useful that it is employed, even after a decade of rapid price hikes, in virtually every application imaginable, from direct steam production to the petrochemical industry. Electricity could be an energy-efficient replacement for oil in microwave, infrared, induction and even resistance heating applications that have been little explored to date because they have not been considered cost-effective.

One of the most pressing needs is to displace imported oil in the production of electricity itself. Nearly 200,000 megawatts, or 36 percent, of total U.S. electricity production still comes from oil or gas. Two-thirds of this capacity is oil-fired. Fortunately, TVA uses only small amounts of these premium fuels in its power plants. But unless utilities can raise the capital to build coal and uranium-powered plants, their dependence on oil and gas will continue, at steadily increasing cost.

I lump oil and natural gas together deliberately because they are substitutable for a great many uses, and because the price of natural gas will equal or even exceed the price of oil within a decade. Homeowners now being encouraged to hook up to gas heat or even to switch from electric to natural gas heat will soon regret the move as natural and synthetic gas prices rise much more steeply than electricity. Homeowners may complain now about the cost of heating an electrically heated home, but people heating with gas or oil will be clamoring for electric heat pumps before this decade is over.

I, of course, realize that electricity is not the sole answer to our imported oil problem. Domestic production of petroleum can fill most of our needs for several decades. And we must continue to improve fuel efficiency. The 50-mile-per-gallon automobile can be a major part of America's future.

We can also displace substantial quantities of imported oil through synthetic fuels from renewable resources. The prospects of getting alcohol from wood are particularly promising. This could provide a major new industry for the Tennessee Valley and other heavily wooded areas once remaining technical problems are resolved. TVA is developing a program to perfect the technology for making liquid fuel from wood. In the meantime, I think this country should go slow in making alcohol from grain, so as not to disrupt food supplies. We can't afford to rob the breadbasket just to continue the joy ride.

Improved efficiency and alcohol from wood can make important contributions, and we can in the future obtain some synthetic gasoline from coal. But in my view, electricity offers the greatest promise of all as a replacement for oil imports. For transportation, it not only provides the only short-term means of substituting our domestic coal and uranium for oil, but also promises substantial financial savings. TVA projections indicate that the fuel cost of running an electric car will be about $450 annually in 1990, for instance, compared to more than $900 for an internal combustion vehicle. Similar savings will make electric trains and increased industrial use of electricity more and more economical.

The substitution of electricity for oil in applications where electricity is an efficient use of energy provides an expanded market for electricity that could begin to be sizeable by 1990. Yet we continue to hear charges in the Tennessee Valley and elsewhere that we are overbuilding generating capacity. In the context of an emerging national energy shortage, these are likely to seem in retrospect much like the charges of old that there would be no market for TVA's hydropower. TVA will have the electricity to make economic development possible and to help wean the region it serves away from its dangerous dependence on imported oil, oil that remains subject to arbitrary pricing and arbitrary cutoff.

Many other utilities are not likely to be as fortunate. Cancellations and long-term deferrals of new generating plants have become epidemic in the last couple of years.

Up to a point, this trend was efficient and cost-effective. TVA has itself deferred the construction of four large nuclear units. Deferrals represented an appropriate adjustment to new realities of lower growth in electrical demand and increased awareness of conservation opportunities.

But now the pendulum is swinging too far the other way. Cancellations are now beginning to occur simply because utilities can't raise the money, even for plants for which there will be an economical market. And utilities are not ordering the electric power plants the nation needs for the 1990's.

The central problem is that the price of electricity is being held at an unrealistically low level. Current prices will not provide adequate funds for the new construction that is needed to allow electricity to take its rightful place in the energy supply system.

The price of oil has already been decontrolled and the price of natural gas is being decontrolled. That is drawing huge amounts of capital into the production of those scarce and depleting resources. I think it is time we gave some serious thought to regulatory reforms that will provide similar financial backing for the production of electricity, which is the only form of energy in which we can put our vast domestic coal and uranium supplies to immediate use.

In summary, the additional electric power we need to replace oil and gas threatens to become increasingly scarce in the 1980's and 1990's. I have identified a number of opportunities for economically substituting electricity for oil and gas within a framework of strict energy efficiency. Together, in my view, they can create a market for every kilowatthour of electricity this Nation's electric utilities can generate as well as every kilowatthour they and their ratepayers can save. But we can't give OPEC an electric jolt if we don't build the electric power plants to make it possible.

# # #

PART II

ENERGY ANALYSIS, PLANNING AND REGULATION

The national and international energy systems are complex combinations of energy sources, technologies and users. Almost innumerable and continuously changing physical, economic, political and social forces modify and drive the systems.

This section examines various energy systems techniques, analyses, and results for the U.S. and the world.

RESULTS OF THE SEVEN-YEAR GLOBAL ENERGY STUDY

Wolf Häfele
Deputy Director
International Institute for Applied Systems Analysis (IIASA)
Laxenburg, Austria,
Program Leader for the Energy Systems Program

For the past seven years we at IIASA have been studying the energy problem. Quite early in the process, we learned that the "energy problem," though the subject of much discussion and debate, is indeed something that escapes easy definition. Is it a problem of prices, a problem of environment, resources, life-styles, technology or simply politics? To all these possibilities the answer is, undoubtedly, yes, all these aspects are part of the energy problem. And it is not too difficult to analyze any one aspect of the problem at a time; but when everything is taken together the problem becomes a difficult one, a complex one, a burning one. It is in fact, a typical systems problem, a question of synthesis.

It is thus essential to organize one's thinking early in the game, to explore the problem's many aspects, and to organize them so that they can be dealt with scientifically. By this I do not mean to suggest that the energy problem should be considered an exclusively "scientific" problem. It contains moral aspects and political aspects that do not lend themselves immediately to scientific treatment in the rigorous sense of the term. But it is precisely because of these aspects, not in spite of them, that one should identify what can, and what cannot, be dealt with on a scientific basis. Making that distinction is not a trivial task. The early stages of the Energy Systems Program devoted much exploratory and interactive thinking to this question, and it took us some time to organize our thoughts to the point where we were confident we could approach the problem constructively.

"The original version of this paper first appeared in IIASA REPORTS, Volume 3, Number 1, 1981, pp. 49-68, Copyright© 1981 International Institute for Applied Systems Analysis. The paper summarizes a study conducted by the Energy Systems Program Group of the International Institute for Applied Systems Analysis and reported in two books published by Ballinger Publishing Company, Cambridge, Massachusetts: Energy in a Finite World: Volume 1. Paths to a Sustainable Future: Volume 2. A Global Systems Analysis: Report by the Energy Systems Program Group of the International Institute for Applied Systems Analysis, Wolf Häfele, Program Leader."

Our decision was to concentrate on the natural-science aspects, on engineering and economic approaches, although we have always kept it clearly in mind that there is much, much more to the energy problem - the politics, the moral questions, etc. But our choice of perspective was based on the nature of the Institute. IIASA is an East-West institute, and it is expected to provide a service - a service to its National Member Organizations (NMOs) and ultimately to the whole world. In this case, the service takes the form of clarifying a factual basis upon which political issues may be settled. I say "may be settled," for of course no analysis can resolve all the difficulties with which we are faced. Still, to agree on some sort of factual basis would be extraordinary.

This concentration on the natural-science aspects of the energy problem means that the scenarios we developed - I shall elaborate on them later - necessarily anticipate a future of a relatively surprise-free nature, or more concretely, of modest economic growth. In particular, we have decoupled the energy problem from the monetary problem by carrying out the analysis in constant US dollars.

The factual basis, then, is the platform on which political issues, such as setting environmental standards or determining development policies for nuclear power, can be settled. Figure 1 illustrates this idea. The balance-of-payments issue is used to caricature

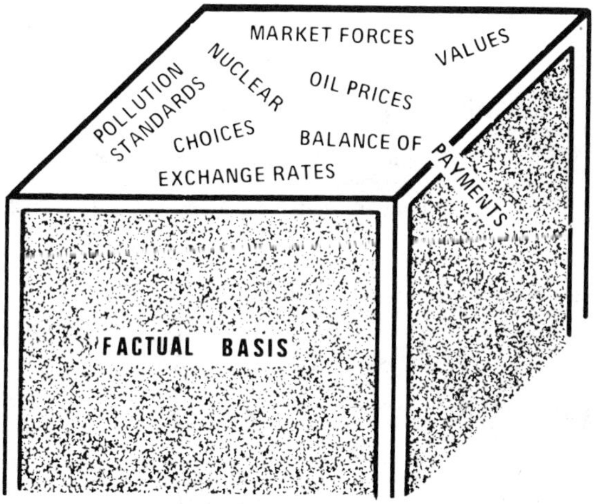

FIGURE 1 Providing a factual basis for energy policy

what happens when wishes become a driving force independent of the supporting facts - attempted resolutions unsupported by the facts cannot survive long. We set out to provide a service by identifying the factual basis necessary for informed energy decisions. That is, we began by asking what can be accomplished, thus providing the essential background for addressing ourselves later to politics and other problems.

CHOOSING A TIME FRAME

A first and fundamental question in our study is the question of a time frame of 50 years for several reasons. When we started our investigations, a number of studies were already under way, the Workshop on Alternative Energy Strategies (WAES 1977), for example, and the work of the World Energy Conference (WEC 1978). It is reasonable and appropriate for these studies to approach the year 2000 from the front, so to speak. However, we chose to consider the question the other way around. That is, after the year 2000 a major transition must take place, a transition not only in terms of energy supply patterns but also in terms of the infrastructures underlying both supply and demand patterns. This will require time, and we are therefore led directly to the question of what

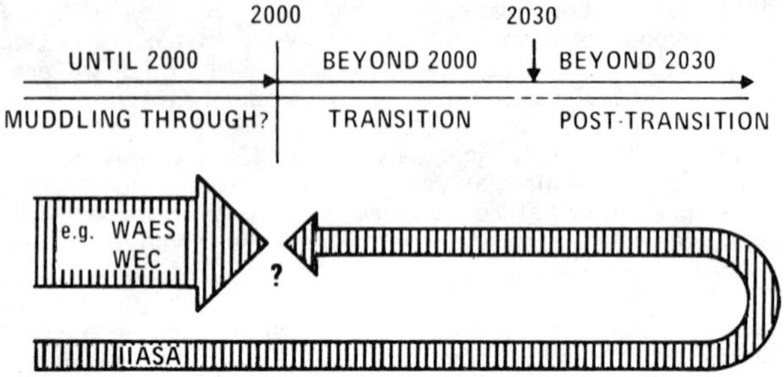

FIGURE 2 Time phases of energy and IIASA's approach to a global energy strategy. (WAES stands for the Workshop on Alternative Energy Strategies. WEC stands for the World Energy Conference.)

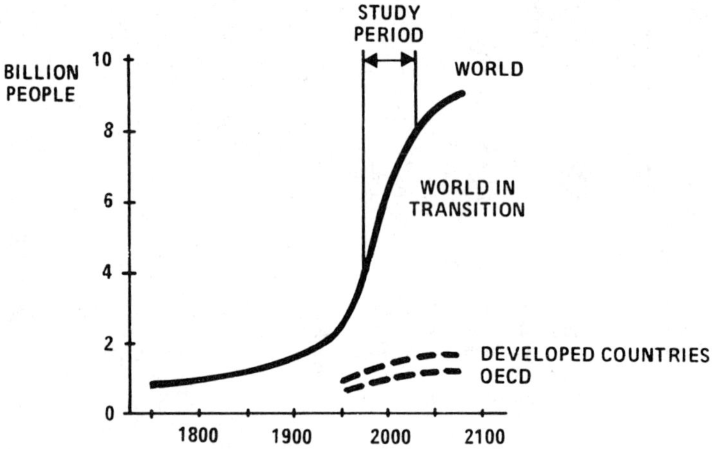

FIGURE 3 The past and future trends in world population. (OECD stands for the Organization for Economic Cooperation and Development.)

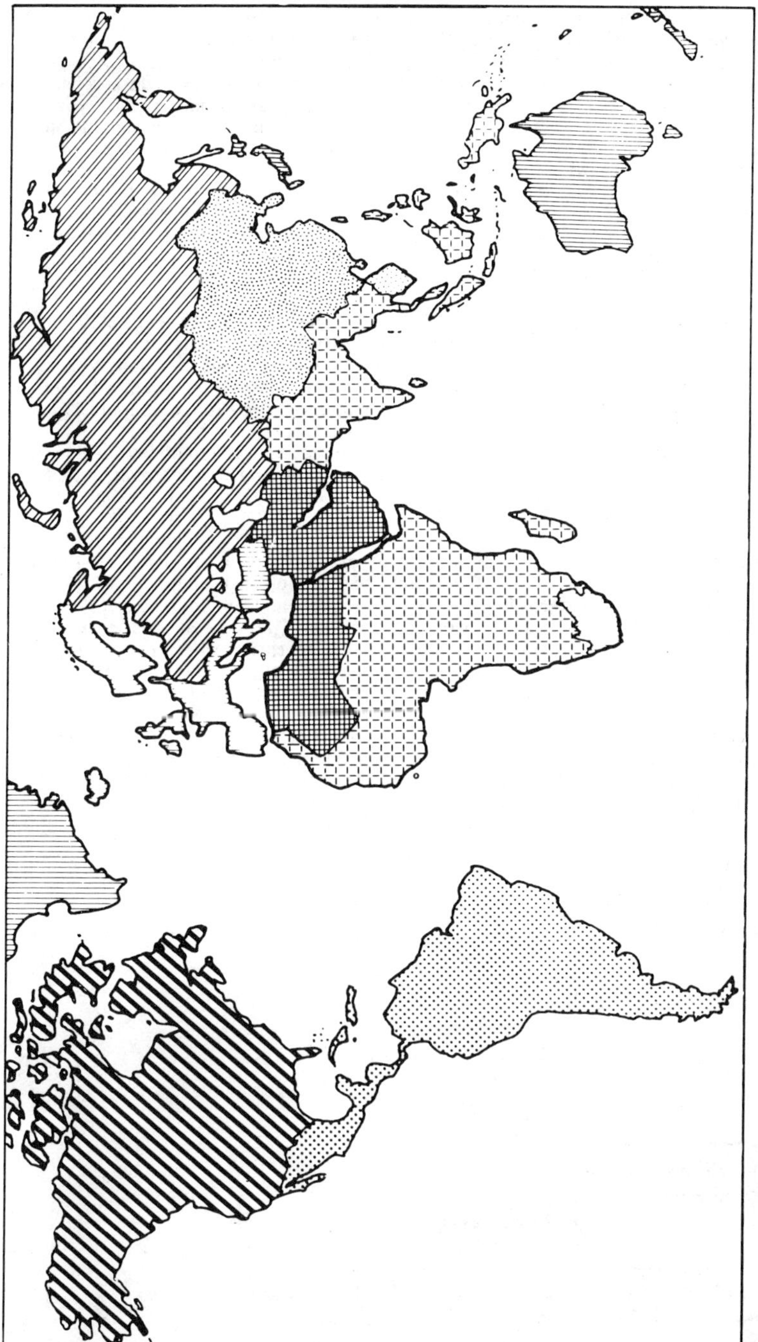

Region I (NA) North America

Region II (SU/EE) Soviet Union and Eastern Europe

Region III (WE/JANZ) Western Europe, Japan, Australia, New Zealand, S. Africa, and Israel

Region IV (LA) Latin America

Region V (Af/SEA) Africa (except Northern Africa and S. Africa), South and Southeast Asia

Region VI (ME/NAf) Middle East and Northern Africa

Region VII (C/CPA) China and Centrally Planned Asian Economies

FIGURE 4 The seven IIASA world regions

happens after the year 2000. More specifically, will we be sufficiently prepared to master the more immediate transition around the year 2000 so that we are ready for the substantial transition that will take place during the twenty-first century? Thus, our idea was to go far out into the future and to look back at the year 2000 through the year 2030. In a sense, the question mark shown in Figure 2 describes the spirit of our project appropriately. Now, as is shown in Figure 3, the year 2000 coincides with a very critical period in the evolution of the globe and its population. There has been some sort of an equilibrium in the past, that is, only modest population growth, with a billion or so people around the year 1800. But our study period from 1980 to 2030 coincides with what is anticipated will be the steepest increase ever in the global population, with the projection for the year 2030 being on the order of 8 billion people. World population is the force ultimately driving the energy problem, and the growth pattern in Figure 3, therefore, only reinforces the importance of extending our analysis as far as 2030.

As indicated in Figure 3, population growth will not be the same the world over. In the industrialized countries the population growth will be quite modest; thus, it is in the developing countries where most of the growth will occur. Because of this, we can expect shifts in political structures and accompanying shifts in the allocation of resources. It was therefore necessary to introduce some level of geographical disaggregation our analysis.

We felt it would be inappropriate to consider, for example, the 153 member nations of the United Nations; in the end, we grouped the countries of the world into seven world regions chosen on the basis of national energy resources and economic structures but not on the basis of geographic proximity.

The regions are shown in Figure 4 and can be characterized briefly as follows: Region I (NA), North America, has a developed market economy and is rich in resources. Region II (SU/EE), the Soviet Union and Eastern Europe, is developed and rich in resources but differs from Region I in that it has a planned economy. Region III (WE/JANZ) is essentially the member countries of the Organization for Economic Co-operation and Development (OECD) minus North America and has a developed market economy but not many resources. Region IV (LA), Latin America, is a developing region with market economies and many resources. Region V (Af/SEA), Africa (except Northern Africa and South Africa) and South and Southeast Asia, is also a developing region with mostly market economies but not many resources. The countries of the Middle East and Northern Africa, Region VI (ME/NAf), are a special case since they are rich in oil and gas resources. And, finally, Region VII (C/CPA), China and the centrally planned Asian economies, is a developing region with centrally planned economies, but it is not so rich in resources.

THE CONCEPTUAL APPROACH

I have already stressed the necessity of conceptualizing, and the diagram presented in Figure 5 represents our thinking, or conception, at the beginning of our study.

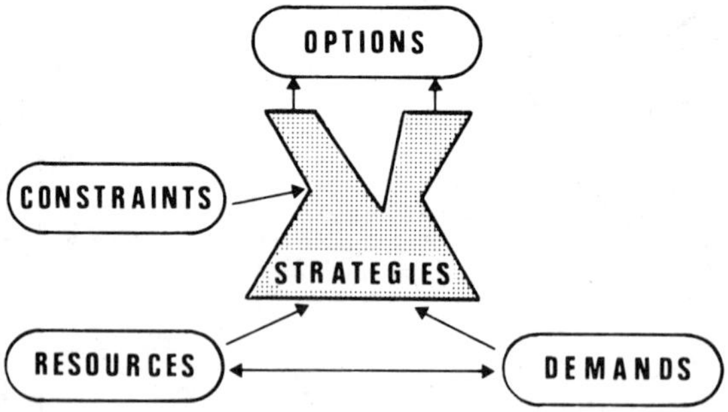

FIGURE 5 The approach to energy systems

The diagram is meant to be read from the bottom, which corresponds to the present, to the top, which corresponds to the period after 2030. It represents the fact that at last a few long-term options are available. Specifically, we can count on coal as a source of energy, or on solar power, or on nuclear power. The immediate question then becomes, what are the conditions and features for exercising these options? In particular, how will energy demand evolve over the next decades? Or, how will the resource picture develop? In this area there is a strong tendency to consider the resources from a traditional perspective, and it may well be more appropriate to change substantially our attitudes toward resources and the associated, relevant infrastructure.

Given the basic features of demand development and resource characteristics, the idea was then to define and compose strategies for realizing the possible long-term options. And here the point was not so much to optimize as it was to discover what is feasible. Feasibility turned out in our studies to be of much more immediate concern than optimality. Put another way, the constraints (mentioned at the left of Figure 5) played a very important role in our considerations. Principal constraints were the environmental and public-health risks associated with different energy supply technologies. These were studied in cooperation with the International Atomic Energy Agency (IAEA), and, I may say, it was a particularly enjoyable and productive exercise. We looked into the climate problems - the case of waste heat and carbon dioxide - and above all we looked into the problem of time; do we have enough time to make it?

Identifying energy strategies cannot be done qualitatively according to the slogan "try harder and you'll make it." One must identify more clearly what is meant by trying harder. A quantitative analysis is required, and it is important to be precise about what we mean by a quantitative analysis.

Three types of mathematical model are listed in Figure 6. The first model, or the first type of model, is built upon the physical and chemical laws of nature, which precisely

1. MODELS FORMULATING LAWS OF NATURE
   ——→ VERIFIABLE PHYSICS.
2. MODELS BUILT UPON TIME SERIES
   ——→ SHORT-TERM ECONOMIC ANALYSIS.
3. MODELS FOR A QUANTITATIVE CONCEPTUALIZATION OF COMPLEX SYSTEMS WITH EMPHASIS ON CONSISTENCY ——→ SCENARIOS TO PUT THINGS INTO PERSPECTIVE.

FIGURE 6 Three kinds of mathematical model.

describe a very large part of reality. In fact, it is tempting to say that they represent reality entirely, and whether we refer to them as reality or as only a model is merely a philosophical question. However, they make up only one type of mathematical model. The second category includes short-term econometric models that use intelligent processing of time-series data. These models also describe reality, perhaps not with the same precision as those in the first category and with a time horizion of only two to three or four years. The third type of mathematical model is the one identified with writing scenarios, and it is this approach that we adopted.

I stress at this point that, in writing scenarios, we are not trying to make predictions. In fact, by concentrating explicitly on defining a factual basis, our results indicate, in a sense, a potential and upper limit to what can be done without political constraints. But, in general, one should not view scenario writing as an attempt to predict the future. Rather it is a way of organizing one's thoughts and the information available, and its basis is a rigorous insistence on two things: internal consistency and global comprehensiveness. For example, in the past there have been many, many exercises at a national level, each of which has essentially concluded that any excess of energy demand over energy supply would be met by imports. Now, if every nation goes through a similar exercise, and practically every one does, the question arises of whether one given barrel of imported oil has been, at least analytically, appropriated by several different parties. That is, are the world's resources really as sufficient as they appear based on the collection of uncoordinated national studies? It is precisely such weaknesses that can be avoided with the help of carefully constructed scenarios.

TWO SCENARIOS

In constructing our scenarios, the principal tool used to ensure internal consistency on a global scale over 50 years was the set of quantitative computer models listed in Figure 7. At the top of the figure are shown the two key characteristics that essentially define any scenario - economic growth and population growth.

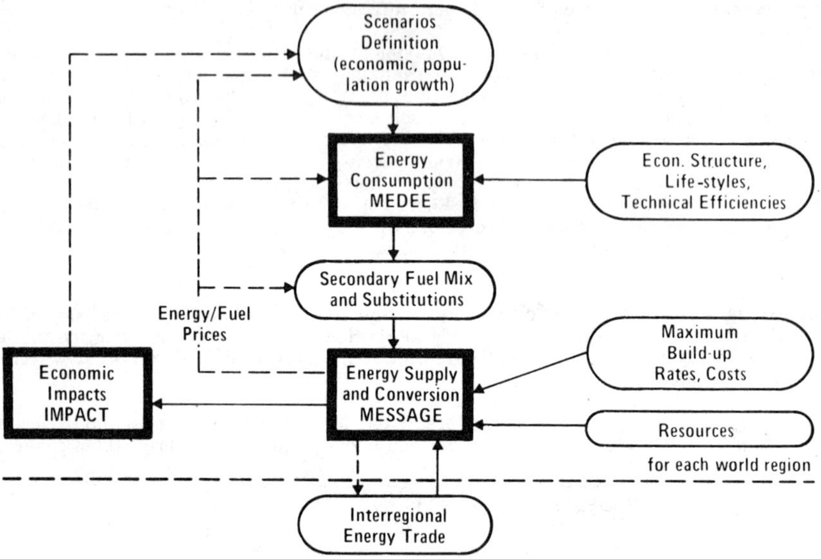

FIGURE 7 A simplified representation if IIASA's set of energy models. (MEDEE stands for Modele d'Evolution de la Demande d'Energie. MESSAGE stands for Model for Energy Supply Systems Alternatives and their General Environmental Impact. IMPACT is the acronym representing a model that calculates the required direct and indirect (energy-related) costs of new energy facilities and that thus provides the basic information for assessing whether or not an economy can afford a given energy scenario.)

Our assumptions about population growth are shown in aggregated form in Figure 3; in our scenarios we used the population figures forming this aggregate.

TABLE 1 Assumptions about the growth rate of gross domestic product (GDP) for the High and Low Scenarios from 1975 t0 2030 (percent per year).

IIASA	region	High Scenario	Low Scenario
I	(NA)	2.87	1.68
II	(SU/EE)	3.91	2.99
III	(WE/JANZ)	.93	1.88
IV	(LA)	4.37	3.48
V	(Af/SEA)	4.32	3.27
VI	(ME/Naf)	5.09	3.57
VII	(C/CPA)	3.77	2.64
World		3.44	2.37

However, in the case of economic growth rates, we distinguished two principal sets of assumptions, as shown in Table 1. These two sets of assumed economic growth rates in turn define two scenarios, which we labeled the High scenario (corresponding to the set of higher growth rates) and the Low scenario (corresponding to the lower growth rates). In both scenarios, however, the growth rates assumed for the developing countries outpace those assumed for the developed countries. It is important to note that the growth rates assumed for the developing regions do not match the aspirations that these countries express.

RESULTS ON ENERGY DEMAND

Having defined the boundaries of our scenarios - the population assumptions, the economic growth rates, and the seven world regions - we now turn to interior detail, the purpose of the MEDEE model, described briefly in Figure 8 and in more detail in Lapillone (1978). (Figure 7 showed its position in the overall model set.) Basically we had to go

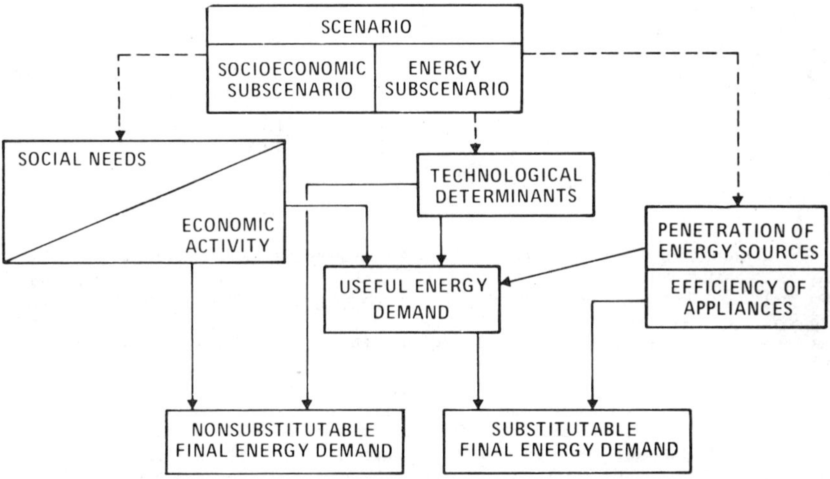

FIGURE 8 The MEDEE approach.
(MEDEE stands for Modele d'Evolution de la Demand d'Energie.)

through a detailed accounting of energy end uses in the various sectors, such as transportation, industry, or household. For instance, we had to identify what room temperatures will be appropriate in India by the year 2000. Of course, these are assumptions but in any exercise like this one, such assumptions must be made. It is crucial to make them explicit and to document them. We have done this at several levels; in a book for the general reader, in a comprehensive technical report, and, at a more detailed level, in a series of research reports and working papers.*

In developing the demand figures, we assumed substantial energy conservation. Thus a clear, qualitative result of the overall analysis is that, without energy conservation, it cannot be done. The issue is not whether to pursue energy conservation, but rather how much energy conservation must be realized.

I do not want to go into the details of the analysis here, but I do want to show you some central results.

---

*Energy Systems Program Group of the International Institute for Applied Systems Analysis, Wolf Häfele, Program Leader (1981) Energy in a Finite World: Volume 1. Paths to a Sustainable Future: Volume 2. A Global Systems Analysis. Cambridge, Massachusetts: Ballinger. Volume 1 is the book for the general reader, Volume 2 is the comprehensive technical report; the latter provides detailed listings of the supporting literature.

• Currently, in North America, Region I, the final energy per capita, as shown in Table 2, is on the order of 7.89 kilowatt-years per year (kW-yr/yr)(or, in brief, 7.89 kW/cap). For the Low scenario we arrived at a modest increase by the year 2030 to a figure around 8.37 kW/cap, which is not very high. In fact, it essentially implies zero energy growth over the next 50 years. In the High scenario the increase is a little higher.

TABLE 2 Final (commercial) energy consumption per capita for the High and Low scenarios from 1975 to 2030 (kilowatt-years per year per capita, abbreviated in the text as kW/cap).

IIASA region	Base year 1975	High scenario		Low scenario	
		2000	2030	2000	2030
I (NA)	7.89	9.25	11.63	7.95	8.37
II (SU/EE)	3.52	5.47	8.57	4.98	6.15
III (WE/JANZ)	2.84	4.46	5.70	3.52	3.90
IV (LA)	0.80	1.75	3.31	1.28	2.08
V (Af/SEA)	0.18	0.42	0.89	0.32	0.53
VI (ME/NAf)	0.80	2.34	4.64	1.76	2.46
VII (C/CPA)	0.43	0.93	1.87	0.64	0.93
World	1.46	1.96	2.86	1.58	1.83

• The numbers for Region II (SU/EE) are not arbitrary numbers. They were calculated in close cooperation with the Academy of Sciences of the USSR in Moscow, and we therefore consider them reliable and realistic: 8.57 kW/cap in 2030 in the High scenario and 6.15 kW/cap in the Low scenario.

• Of special note is the situation in Region V (Af/SEA), where the consumption level currently is only 0.18 kW/cap for commercial energy (see Table 2). However, there is necessarily consumption of noncommercial energy on top of this, some 0.3 kW/cap from dung, wood and other sources. But these noncommercial sources have supply problems of their own and, in the end, one must substitute commercial energy for the noncommercial sources. Therefore, parts of the 2000 and 2030 figures for Region V include something that is already there. In general, the Region V figures seem low; however, we were assisted in this analysis by representatives of the Indian Statistical Office, thus incorporating the wisdom and judgment of people from the region.

• In Region IV (LA), where current consumption is only 0.8 kW/cap, we see in Table 2 that the High-scenario 2030 consumption level is approximately 3.3 kW/cap, exceeding the average consumption in Region III (WE/JANZ) in 1975. This comparison reflects the high expectations and the high expected growth rates for Latin America; it provides a yardstick for considering the distribution of energy consumption over the regions.

TABLE 3 Primary (commercial) energy consumption by IIASA regions for the High and Low scenarios from 1975 to 2030 (terawatt-years per year).

IIASA region	Base year 1975	High scenario		Low scenario	
		2000	2030	2000	2030
I (NA)	2.65	3..89	6.02	3.31	4.37
II (SU/EE)	1.84	3.69	7.33	3.31	5.00
III (WE/JANZ)	2.26	4.29	7.14	3.39	4.54
IV (LA)	0.34	1.34	3.68	0.97	2.31
V (Af/SEA)	0.33	1.43	4.65	1.07	2.66
VI (ME/NAf)	0.13	0.77	2.38	0.56	1.23
VII (C/CPA)	0.46	1.44	4.45	0.98	2.29
Total[a]	8.21[b]	16.84	35.65	13.59	22.39

[a]Columns may not sum to totals because of rounding.
[b]Includes 0.21 TW-yr/yr of bunkers - fuel used in international shipments of fuel.

• Turning to aggregate primary energy consumption, we find in Table 3 a current rate of 8.2 terawatt-years per year (TW-yr/yr) for the world as a whole. (One TW-yr/yr is a large energy unit; it equals roughly a billion tons of coal per year or 14 million barrels of oil per day, which is greater than Saudi Arabia's current possible production rate.) Most of the 8.2 TW-yr/yr goes to Regions I, II and III, with only a minor fraction going to the four other regions.

• In the High scenario, where primary energy consumption in 2030 reaches 36 TW-yr/yr, the assumed higher growth rates are associated with the additional benefits of innovation and support for equalizing social differences. Thus, by 2030 Regions I (NA), II (SU/EE), and III (WE/JANZ) account for a much smaller share of the global primary energy consumption than they did in 1975. In the Low scenario, where primary energy consumption reaches 22 TW-yr/yr in 2030, the trend toward equalization across regions can also be seen, although it is less pronounced than in the High scenario. The two numbers 36 TW-yr/yr and 22 TW-yr/yr are not meant to represent extremes in either direction, but rather are assumed to cover a middle ground. Still, their magnitudes indicate that an increase in energy supply by a factor of around three or four will be required over the next 50 years.

ENERGY RESOURCES

It is on the basis of energy demand, then, that we have to comtemplate the supply problem. Do we have enough energy resources, particularly fossil resources? Typically, the answer is, "yes and no." Orginally, when I was more naive, I thought that looking into the resource problem would not be too difficult and that the numbers would be well established. Not so: it proved to be a most complex problem. And our fundamental concern was to look at the problem in terms of the right categories - not the traditional ones, but those of tomorrow. Table 4 illustrates my point. According to traditional wisdom, 1000 TW-yr is a very good indication of global fossil resources, and it is consistent with the 1091 TW-yr shown in the table as the global total of what may be referred to as conventional fossil resources. Furthermore, the 560 TW-yr of category 1 coal listed in the table equal some 600 billion tons of coal equivalent (tce), essentially the conventional component of coal resources as, for instance, reported at the Detroit World Energy Conference (WEC 1974). The same can be said for oil, with 264 TW-yr, and gas, with 267 TW-yr. But when one goes to higher-cost categories - and here I mean not only monetary costs, but also environmental-impact and social-difficulty costs - one gains additional resources amounting to a threefold increase: that is, not 1000, but 3000 TW-yr is the more appropriate figure. However, this does not mean that the additional resources have the same nature as the first 1000 TW-yr. The difficulties that accompany category 2 and 3 resources are significant, and I shall return to them later. And, of course, there is the key question: how do we use these 3000 TW-yr most intelligently, if at all?

TABLE 4 Global energy resources and their costs (terawatt-years).

Resource	Cost category 1	2	3	Total
Coal	560	1019	-	1579
Oil	264	200	373	837
Gas	267	141	130	538
Total	1091	1360	503	2954

NOTES: Cost categories represent estimates of costs either at or below the stated volume of recoverable resources (in constant 1975 US$). For coal - category 1: 25$, and category 2: 25 - 50$ per metric ton of coal equivalent. For oil and natural gas - category 1: 12$, category 2: 12 - 20$, and category 3: 20 - 25$ per barrel of oil equivalent.

We looked, not only into fossil resources, but also into alternatives for supplying energy, and Table 5 gives a brief summary of them. In the case of the renewables, it is important to realize that, while wood, for instance, may have an infinite potential, there is a finite limit constraining the possible annual production level: some 2.5 TW-yr/yr is a good figure. When all the renewables are added, within appropriate limits, they total about 6 TW-yr/yr, and certainly do not exceed 14 TW-yr/yr, a large - but not very large - number. Oil and gas production is limited to 8 to 12 TW-yr/yr with a question mark, and, to recognize even greater uncertainty, the production potential for coal is listed at 10 to 14 TW-yr/yr with two question marks.

The case of nuclear energy requires a more detailed discussion. If we continue to use only burners, the total resource that we can exploit is only 300 TW-yr - much smaller than the 3000 TW-yr I associated with fossil resources. This means that, by the year 2030, we could produce only 12 TW-yr/yr, and this for only a few years, after which there would be a rapid decline and the uranium resource would be exhausted. This description is, of course, a simplification, but it captures the essential points of a nuclear future based solely on burners.

TABLE 5 Alternatives to fossil resources for supplying energy: resources and production potentials.

Source	Production (TW-yr/yr)	Resource (TW-yr)
Renewables		
wood	2.5	∞
hydro	1-1.5	∞
total	6-14	∞
Oil and gas	8-12(?)	1 000
Coal	10-14(??)	2 000(?)
Nuclear		
burner	12 for 2020	300
breeders	⩽ 17 by 2030	300 000
fusion	2-3 by 2030	300 000
Solar		
soft	1-2	∞
hard	2-3 by 2030	∞

∞ indicates infinite supply.

However, if breeders are introduced, the most efficient possible energy strategy would lead to a nuclear-energy production level by 2030 of 17 TW-yr/yr, which could continue well beyond 2030 in view of the associated resource potential of 300 000 TW-yr. The potential for fusion is also 300 000 TW-yr, but here the build-up must be more limited and very slow. If fusion finally does come into use, its major contributions must come after the year 2030. It is difficult to conceive that by 2030 fusion could offer more than 2 to 3 TW-yr/yr of caloric power.

The situation for solar energy is also noteworthy. For the localized, decentralized use of solar power, which Table 5 labels "soft," it is difficult to conceive of more than 1 or 2 TW-yr/yr, although the resource is indeed essentially infinite. The other category of solar power that Table 5 shows is "hard" solar, a classification perhaps best typified by a large centralized facility located in the Sahara Desert. In the final analysis, production in this category could be very large. Still, the hard-solar option takes time, and it will be difficult to bring to reality. To expect more that 2 to 3 TW-yr/yr by the year 2030 would be unrealistic. Again it is time, and not resources, that is the principal constraint during at least the next 50 years.

COMBINING THE ENERGY RESOURCES

The next question, then, is: how do we combine all these various resource potentials? Table 6 shows the primary energy supply mixes associated with each of the two scenarios. Because there is so much discussion of optimality - of optimal supply mixes and optimal strategies - I want to stress again that we do not offer either of the scenarios as an optimal future; rather, both represent plausible and feasible futures. It is feasibilty, not optimality, that is the immediately pressing constraint.

TABLE 6 The primary (commercial) energy sources used in the High and Low scenarios from 1975 to 2030 (terawatt-years per year).

Primary source	Base year 1975	High scenario 2000	High scenario 2030	Low scenario 2000	Low scenario 2030
Oil	3.62	5.89	6.83	4.75	5.02
Gas	1.51	3.11	5.97	2.53	3.47
Coal	2.26	4.95	11.98	3.93	6.45
LWR	0.12	1.70	3.21	1.27	1.89
FBR	0.00	0.04	4.88	0.02	3.28
Hydro	0.50	0.83	1.46	0.83	1.46
Solar	0.00	0.10	0.49	0.09	0.30
Other	0.21	0.22	0.81	0.17	0.52
Total[a]	8.21	16.84	35.65	13.59	22.39

[a]Columns may not sum to totals because of rounding.

Significantly, as can be seen from Table 6, the production and consumption of oil in both scenarios go up, not down, compared to 1975. The relative share of oil declines, but the absolute numbers go up.

The trend is even more pronounced in the case of coal: from 2 TW-yr/yr in 1975 to 12 TW-yr/yr in 2030 in the High scenario and to 6.5 TW-yr/yr in 2030 in the Low scenario. So, Carroll Wilson is absolutely right in concentrating on studying coal use over the next 50 years (Wilson 1980). Many of you might tell me that it will be impossible to produce 12 billion tons of coal per year. My first answer is yes, it will be impossible. But, because we are globally comprehensive and consistent, we must then necessarily ask what primary energy source will assume the burden. Relieving the pressure on one resource can only increase the pressure on another. One may therefore argue that the High scenario is impossible, and that the Low scenario should be considered. But, in this case, it is important to understand the implications of lower energy-use levels in the developing parts of the world. If we insist on being globally comprehensive, there is no escape.

Let us look at the oil situation; after all, it is our most pressing problem. In the High scenario, as shown in Figure 9 for the world's market economies, known reserves of conventional oil will be exhausted by the year 2010. This is essentially the man in the street's perception of the energy situation, and he is absolutely correct. This means, first, that we must seek new reserves of conventional oil; Mexico, for instance, is a case in point. However, our High scenario already takes into account all the Mexicos still to come, and yet production never gets above 25 million barrels per day. Therefore, we must turn to the unconventional oil sources, such as the Athabasca tar sands in Canada or the Orinoco heavy crudes in Latin America. At the same time, we should recognize the completely new geopolitical patterns that will develop when the unconventionals begin to play such a major role. However, just changing to unconventional oil sources is not enough. Even after all this, it will still be necessary to import oil from Region VI - the member countries of the Organization of the Petroleum Exporting Countries (OPEC) and particularly the member countries of the Organization of Arab Petroleum Exporting Countries (OAPEC) - in order to meet the demand for liquids.

It is a tough picture that we draw, but one that we cannot ignore. It means, among other things, that oil trading will play a key role in the world energy situation. Today, as shown in Figure 10, we have two net oil-consuming regions, three supplier regions, and two regions that are exactly self sufficient. In order to have a feasible match between demand and supply in 2030 in our High scenario, we have assumed that Region I (NA), Region IV (LA), Region II (SU/EE), as well as Region VII (C/CPA), will be self sufficient. Only Region VI (ME/NAf) will be an exporter, and these exports will go to the imports of Region III (WE/JANZ), a developed region, and Region V (Af/SEA), a developing region, with an associated set of political implications somewhat different from those of today. While it is not our duty here to perform the political analysis, this example illustrates how we serve by identifying the factual basis necessarily underlying global politics in the decades to come.

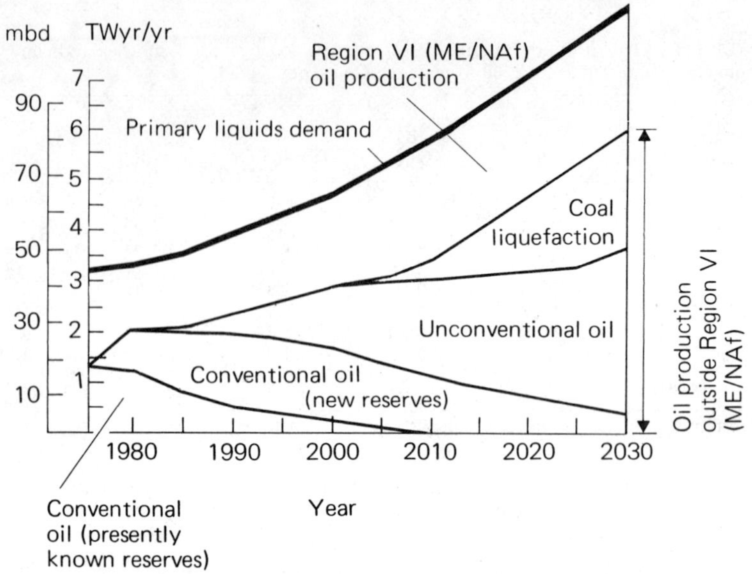

FIGURE 9 Oil production for the world's market economies in the High scenario from 1975 to 2030. (mbd is million barrels per day. TW-yr/yr is terawatt-years per year.)

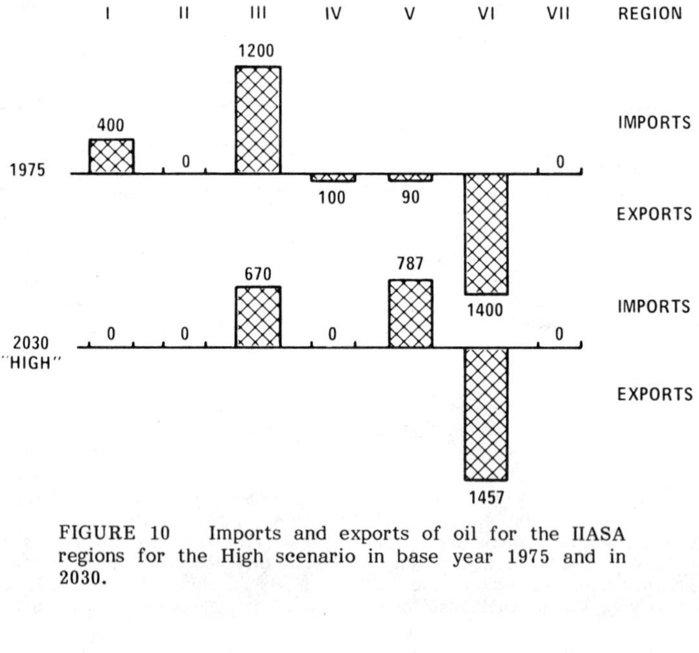

FIGURE 10 Imports and exports of oil for the IIASA regions for the High scenario in base year 1975 and in 2030.

Now the question is, "where are we by the year 2030?" Table 7 shows that "in between" is the most appropriate answer. In the High scenario 68 percent of categories 1 and 2 oil is consumed by 2030, although only 1 percent of category 3 oil is consumed. Some 370 TW-yr of category 3 oil would be left, but, at a consumption rate of 30 to 50 TW-yr/yr, it will not last forever. The same is essentially true for natural gas, although the situation is slightly more favorable. Moreover, for the near future our more detailed analysis indicate that the potential for natural gas is remarkable. Other analytical results point to the same conclusion. In the case of coal, only 61 percent of category 1 resources will have been consumed, leaving more than 1000 TW-yr of coal still in the ground.

TABLE 7 The cumulative uses by 2030 of fossil fuels in the High and Low scenarios.

Fossil fuel	Total resource available (TW-yr)	Total consumed as percentage of total available	
		High scenario	Low scenario
Oil			
categories 1 & 2	464	68	57
category 3	373	1	0
Natural gas			
categories 1 & 2	408	49	36
category 3	130	0	0
Coal			
category 1	560	61	40
category 2	1019	0	0

NOTE: For definition of cost categories see Table 4 in this paper.

Translating the quantitative findings into a qualitative conclusion, we say that in the next 50 years it will not be the resources that limit us. Instead, it will be the build-up rates associated with doing something new: time will be the constraint. In the second half of the next century the fossil resources will indeed come to an end, and we will have to take this into account when developing strategies for the next 50 years, because time will always be the limiting factor.

CONCLUSIONS

Before summarizing the principal insights that emerge from the analysis, I should note that the two scenarios I have just discussed were not the only ones we investigated; for example, we examined one assuming very low energy demand and a nuclear moratorium and one based on an all-out nuclear effort until the year 2030.

The question that arises after all this analysis is, how do we translate it? What does it boil down to? What do we have to say to someone who wants to understand the essential meaning of all this? We have identified qualitatively a number of important conclusions, the most important being that it can be done - the world's energy needs can be met. There are difficulties and expenses involved, but we cannot subscribe to the idea that the world is doomed to failure because nature has not given us the necessary endowments for 8 billion people, or even a larger number, such as 10 or 12 billion. It will be difficult to do it, and it will be at an expense, but it can be done.

With respect to demand, the important conclusions can be summarized as follows:

- Only radical changes in life-style and the structure of the economy can lead to a very low energy demand.
- The demand for liquid fuels is a principal driving force in the energy problem. We have an energy problem within the energy problem.
- The conservation measures implied by our Low scenario are strict, but probably more realistic than those of the very low-demand scenario.
- Our High scenario projects growth rates that may be considered moderately satisfactory but that will transfer the hardship to the supply side.
- Over the next 50 years, under any set of circumstances, economic growth rates will be limited.

The institutional implications deserving particular emphasis are the following:

• The hard/soft controversy is essentially a political issue and not a factual one.
• The realities of political, social and institutional problems will make the situation grimmer than has been described in our two scenarios.
• Society has not yet developed adequate mechanisms for dealing with the risks associated with energy-supply techniques.

With respect to supply, there are these conclusions:

• Fossil fuels will continue to be available but will become increasingly unconventional and expensive.
• Renewable energy sources can contribute in an important, albeit limited, way to meeting demand.
• The oil-exporting countries will continue to dominate the oil market.
• Accordingly, an international coal market must be developed.
• Coal liquefaction must be installed with a strategic outlook. It must serve as a bridge to the future.
• Energy investments will grow significantly but will not be a large portion of the gross domestic product (GDP) in the developed countries.

Beyond the 50 years of the detailed analysis, the message can be summarized in two conclusions:

• It is indeed possible to have a sustainable global energy system. It will take time, but nature has provided us with the possibility.
• In order to arrive at this future, the build-up and operation of a sustainable energy system must make prudent use of the carbon atom.

REFERENCES

Lapillone, B. 1978. MEDEE 2: A Model for Long-Term Energy Demand Evaluation. RR-78-17. Laxenburg, Austria: International Institute for Applied Systems Analysis.

Wilson, Carroll L. 1980. COAL - Bridge to the Future. Report of the World Coal Study. Cambridge, Massachusetts: Ballinger.

Workshop on Alternative Energy Strategies. 1977. Energy: Global Prospects 1985 - 2000. New York: McGraw Hill.

World Energy Conference. 1974. World Energy Conferences Survey of Energy Resources, 1974. New York.

World Energy Conference. 1978. World Energy Resources 1985 - 2000. Executive Summaries of Reports on Resources Conservation and Demand of the World Energy Conference, September 1977. Guildford, UK:IPC Science and Technology Press.

COMMERCIALIZATION - THE FACTS SHOW THAT IT TAKES A LONG TIME

B. J. THARPE AND J. C. GRAF
ADVANCED ENERGY PROGRAMS DEPARTMENT
GENERAL ELECTRIC COMPANY

With the advent of the oil embargo and the ensuing long lines at filling stations in 1973 and 1979 there has been increased interest in both government and private circles in the introduction of new alternative solutions to the energy problem. This process is referred to as commercialization and is one which has long been studied and evaluated by business and business school researchers alike. Perhaps its least remembered attribute is the frequent requirement for considerable patience and perseverance on the part of the management which is sponsoring the endeavor.

This paper first examines the commercialization process, reviews a number of examples both in and out of the energy field and then takes a close look at what might be done to stimulate innovation and to establish a climate which promotes the commercialization of new products. Establishment of such an environment must be a function of the Federal Government. Recommendations for improvement are proposed.

Commercialization of new products generally requires at least two major ingredients, first an innovator or entrepeneur who is motivated by the technological challenge and unafraid of failure and secondly establishment of a patient and supportive environment for the development. Often this takes the form of a separate organizational structure which has direct access and the support of senior management (often the chief executive officer). By so configuring the organization and by providing support at the top, the normal short term measurements can be relaxed and proper emphasis can be placed on the development. An added important ingredient is to provide for the freedom to advocate support for the project and to allow for the key members of the team to be selected based on their support and commitment rather than merely assigned.

Given all of the above conditions, the road to successful commercialization can still best be characterized as rough, long and rocky. Studies have shown that over 70 percent of all new product investments by U.S. Corporations go to efforts that are never commercialized (1). Another study from the same source polled 51 companies and revealed that 58 new product ideas were required to yield one successful product. Most of the failures occur fairly late in the development process and often are related to lack of an adequate perception of the market need as compared to failures in the research or technological part of the process. Fully three-quarters of the products which have been successfully commercialized have been in response to a well perceived and well defined market need (2).

To fully appreciate the extent and the complexity of the commercialization process it is instructive to investigate the elements of the process and secondly to examine the interactions as described by the development of the Technology Distribution System which may be used to visually describe the interactions. The phases of commercialization may be expressed as:

- Concept Exploration
- Screening
- Product Development
- Market Testing
- Commercial Introduction
- Product Acceptance

Concept exploration is conducted to identify all possible ideas, concepts or solutions. It may be initiated by as simple a process as brainstorming but should be supportable by basic research. The concepts should also be examined to make certain that commercial potential exists. In general, business evaluations are not an important part of this phase of the process. These are incorporated during Phase II, Screening. In this phase financial criteria are developed and examined. In particular, market demand is considered to establish that a sufficient need exists. Included in the evaluation process would be engineering experiments, market forecasts and user surveys. Phase III is Product Development which includes the development of prototypical units, the initiation of cost projections, layout of plants and definition of tooling. Resources required by the developing company will be identified at that time. A preliminary market plan will be a part of this phase of the process. This constitutes a business assessment which has sometimes been broken out as a separate phase in the process.

The fourth phase is Market Testing. This may often be initiated regionally to limit cost and exposure and also to provide an opportunity for fine tuning of the product and the promotional approach. The Commercial Introduction phase will usually include a complete marketing, advertising and sales promotion effort in addition to the classical production tasks plus quality assurance, distribution and service considerations. Finally, if the introduction is successful, the market is expanded and profitable sales are attained. Hopefully, these profits will be great enough to more than repay all "sunk" costs - in real terms.

Obviously the venture may be aborted in any of the steps and successful completion of each is required to move on to the next. It is apparent that the sooner the phased system discovers some non-recoverable problem the less expensive is the endeavor. Another not uncommon reason to abort is that a more promising approach has been found.

Let's next examine some specific examples of new product development and gather some insight into the historical times which have been required to commercialize a product. Mueller (3) cites three studies of the time lag from first working model to commercial success. In the first study of over 200 inventions the time was found to be from 33 to 38 years. In a second study of 75 innovations the average time lag was found to be 37 years. In a third study of 35 innovations the average interval was 33 years.

---

(1) Introduction to the Analysis of Commercialization Conditions, Booz Allan Applied Research Division, Booz, Allen and Hamilton Inc. Bethseda, MD, July 1976

(2) Why Innovations Fail

(3) Robert K. Mueller. The Innovation Ethic. American Management Association, Inc. 1971

A more definitive time table is provided by Dean (4). In his article which argues that no corporate management can wait for innovation to meet the normal short time operating constraints and goals of a business, a number of specific innovations are examined. These are given in Table I.

Table I. New Product Development Duration

	Start of Development	Year of 1st Realization	Duration Years
Heart Pacemaker	1928	1960	32
Hybrid Corn	1908	1933	25
Hybrid Small Grains	1937	1955	18
Green Revolution Wheat	1950	1966	16
Electro Photography	1937	1959	22
Input-Output Economic Analysis	1936	1964	28
Organo Phosphorous Insecticides	1934	1947	13
Oral Contraceptives	1951	1960	9
Magnetic Ferrites	1933	1955	22
Video Tape Recorder	1950	1956	6

In these examples it is clear that the average interval is less than that defined by Mueller, however, it is still approaching twenty years and it is not clear that the definition of "first realization" is equivalent to commercialization. Most would agree that commercialization implies a profitable position and share of the marketplace not usually thought of as synonymous with first realization.

An even more extensive evaluation of elasped time to new product development was gathered by Adler (5). He makes the point that many marketing organizations operate on a hypothesis that 2 to 3 years is an adequate new development time period. As a result underestimating the time for orderly market development and growth is one of the major causes of product failure. Adler's examples are presented in Table II.

In addition to consumer goods, Adler also took a look at a number of industrial goods. These are shown in Table II(c). I would question whether it would be proper to define either the Page Master or Dictet as new products as opposed to evolutionary products but the choice is obviously judgmental. Nevertheless, it is again obvious that the time period for development is much longer than the 2-3 year time period often considered adequate by the marketing team.

For further evaluation of the process of commercialization, let's examine a new energy product which has been relatively recently developed and with which General Electric has been deeply involved - the gas turbine. Analysis of corporate expenditures shows that twenty-five years of investment in research and development were required prior to achieving a cost competitive product. These investments were more responsible for achieving reductions in installed costs than were reductions through quantity production. Nevertheless, several hundreds of millions of dollars have been invested in plant and equipment since 1949.

---

(4) Robert C. Dean, Jr., The Temporal Mismatch-Innovation's Pace vs. Management's Time Horizon, Research Management, May 1974, pp 12-15

(5) Lee Adler, Time Lag in New Product Development, New Product Development, Edited by J. O. Eastlack, Jr., Jack Trukes and Partners, American Management Association, 1968, pp. 9-16

Table II(a). New Product Development Time - Consumer Packaged Goods

Product	Company	Date Development Started or Idea Born	Test or Initial Markets	Large-scale or National	Elapsed Time	Reference
Birdseye frozen foods	Birdseye Division, General Foods	1908	1923		15 years	Printers Ink, 5/29/64
Ban (roll-on deodorant)	Bristol-Myers	About 1948	1954	March, 1955	6 years	Printers' Ink, 6/5/59 Sponsor, 4/16/56
Calm powder deodorant in aerosol can	Alberto-Culver	1959		February, 1964	5 years	Printers' Ink, 1/24, 64
Chlorodent (tooth paste)	Lever Brothers	1930s	March, 1951	Early 1952	Between 11 and 21 years	Tide, 3/28/52
Citroid (cold compound)	Grove Laboratories	1954-55		1956	1 to 2 years	Advertising Agency, 10/26/56
Coldene (cold-remedy liquid)	Pharma-Craft	1954	1955	1956	1 Year	Printers' Ink, 2/7/58
Crest (flouride tooth paste)	Procter & Gamble	1945	January, 1955	January, 1956	10 years	Advertising Age, 8/1/60
Decaf (decaffinated instant coffee)	Nestle	1947	1953		10 years	Tide, 1/25/57 Nestle Company
Flav-R-Straws	Frontier Foods Corp. & others	1953	April, 1956	Early 1957	3 years	Food Business, 4/57
Gerber (strained baby foods)	Gerber	1927	1928		1 year	Business Decisions That Changed Our Lives, Sidney Furst, Milton Sherman (Random House, 1964). p. 167
Hills Brothers (instant coffee)	Hills Brothers	1934	1956		22 years	New York Times, 11/16/56
Johnson liquid shoe polish containers that are also applicators	S.C. Johnson	1957	February, 1960	Early 1961	3 years	Printers' Ink, 7/14/61
Lustre Creme (liquid shampoo)	Colgate-Palmolive	1950		June, 1958	8 years	Drug Trade News, 5/19/58
Marlboro (filter cigarettes)	Philip Morris	May, 1953	March, 1955		2 years	Advertising Age, 2/28/55
Maxim (concentrated instant coffee)	General Foods	1954	May, 1964		10 years	Printers' Ink, 5/1/64
Minute Maid (frozen orange juice)	Minute Maid	1944	1946		2 years	Sales Management, 4/1/49 Advertising Age, 3/14/49
Minute Rice	General Foods	1931		First Quarter, 1949	18 years	Food Field Reporter, 12/13/54 Advertising Agency, 11/49
Purina Dog Chow	Ralston-Purina	1951	February, 1955	April, 1957	4 years	Wall Street Journal, 1/2/58
Red Kettle (dry-soup mixes)	Campbell Soup	Before 1943		August, 1962	19 years plus	Food Field Reporter, 8/28/61 Advertising Age, 9/24/62
Stripe (tooth paste)	Lever Brothers	1952	1957	Early 1958	5-1/2 to 6 years	New York Times, 1/15/58

Table II(b). New Product Development Time - Other Consumer Goods

Product	Company	Date Development Started or Idea Born	Test or Initial Markets	Large-scale or National	Elapsed Time	Reference
Bendix (washer/dryer)	Bendix	Prior to World War II		March, 1953	12 years plus	Fortune, 3/53
Eversharp ("Fountain Ball" ball pen)	Eversharp	January, 1958	September 1958	January, 1959	8 months	Sales Management, 1/2/59
Fairchild (Mark IV 8 mm sound projector)	Fairchild Camera & Instrument	Late 1961	August, 1963		2 years	New Products, New Profits, American Management Association, 1964
Floron (plastic floor tile)	Pabco Products	1947-48		October, 1953	5 to 6 years	Sales Management, 1/1/55
GE (electric tooth brush)	General Electric	1958-59	October, 1961	April, 1962	3 to 4 years	Printers' Ink, 7/20/62
Polaroid Land Camera	Polaroid Corp.	1945-46	1947-48		2 years	Business Week, 9/3/49 and 1/19/63; Printers' Ink, 5/29/64; Standard & Poor's Corporate Records, 1965, p. 9745
Polaroid Color-pack Camera	Polaroid Corp.	1948	January, 1963	May, 1963	15 years plus	Business Week, 1/19/63
Scripto Tilt Tip (ball pen)	Scripto	1959	April, 1961	Mid-1961	2 years	Advertising Age, 1/30/61
Sinclair (Power X gasoline; Extra Duty Motor Oil)	Sinclair Oil	Late 1952		April, 1953	6 months	Printers' Ink, 6/18/54
Smith Corona (portable electric typewriter)	Smith Corona	1952	Early 1957		5 years	New York Times, 11/17/57
Sunbeam (electric tooth brush)	Sunbeam Corp.				5 years	Sales Management, 9/4/64
Talon (zippers)	Corporate Predecessor of Talon, Inc.	1883	1913	1918	30 years	Business Decisions That Changes Our Lives, Sidney Furst, Milton Sherman (Random House, 1964), p. 115
Television	(Many)	1884	1939	1946-47	55 years	Federal Communications Commission, "Broadcast Primer," Bulletin 2-B, 1961

Table II(c). New Product Development Time - Industrial Goods

Product	Company	Date Development Started or Idea Born	Test or Initial Markets	Large-scale or National	Elapsed Time	Reference
Dictet (portable recording machine	Dictaphone Corp	1954	Early		20 months	A Critical Look At The Purchasing Function, Robert F. Loglet, American Management Association Bulletin No. 13, 1961, pp. 113-21
Isothalic (chemical component to improve house paints)	Oronite Corp. (subsidiary of Standard Oil of California	1951	Late 1957-early 1958		6 to 7 years plus	Sales Management, 9/19/58
Krilium (soil conditioner)	Monsanto	1939	May, 1952		12-1/2 years	Fortune, 12/52
Page Master (selective pocket-paging system)	Stromberg Carlson	1955		March, 1957	2 years	American Management Association Bulletin No. 13, 1961, pp. 122-41
Penicillin	(Many)	1928	1943		15 years	Business Week, 3/3/45
Transistors	Bell Laboratories	1940	1955-56		15 to 16 years	Business Week, 3/26/60
Xerox (electrostatic copying machines)	Xerox Corp.	1935	1950		15 years	Forbes, 9/15/62; Fortune, 7/62; New York Times, 12/10/61; Standard & Poor's Corporate Records, 1965, p. 4472.

Initial installations of the gas turbine in the utility business, in pipeline pumping stations and in the petrochemical industry were followed by at least a fifteen year period of evaluation prior to commercial acceptance in the marketplace. One of the major advantages of gas turbines was the relatively short production cycle and therefore the ability to deliver capacity to the customer on a fast turnaround basis. It was this capability that was a significant factor in the high growth rate in the post 1965 Northeast blackout period. Prior to this domestic occurrence, major acceptance had been in the international market. Another unexpected event, the OPEC oil embargo has been a similar negative effect on the volume of sales of these units although it has been more dramatic in the United States than it has been off-shore.

The long growth pattern and R&D investment cycle is shown pictorially in Figure 1. Market and technology development items which have figured extensively in the historical development period are summarized in Table III.

Table III. Major Factors Influencing Gas Turbine Market Development

- Higher firing temperatures (1450°F to 1900°F)
- Increased turbine and compressor efficiencies
- Equipment packaged into self-contained units to minimize installation time
- The replacement of forged buckets by castings
- Heat regeneration and combined steam/gas plants
- Ability to operate on multiple fuels (distillate, crude, residual)

Although the initiation period stretches back to the early 1900's, a very similar growth pattern for steam-turbines also occurred. Sizable investments (greater than $100 Million) in research, development, test and manufacturing has affected a growth in efficiency of these systems from 4 to 40 percent. Some of the factors of the ongoing research and development efforts are summarized in Table IV.

Table IV. Major Factors Influencing Steam Turbine Market Development

- Higher Steam Pressures (100 to 5000 psia)
- Higher Temperatures (500° to 1050°F)
- Higher Speeds (1800 to 3600 rpm)
- Increased Ratings (0.5 to 1100 MW)
- Increased Bucket Length (17" to 52")
- Advanced Cooling Techniques
  - Air (prior to 1940)
  - Hydrogen (1940-77)
  - Oil (1955-56)
  - Cryogenics (198?)
  - Water (1956-77)
- Improved Materials (ferrite alloys, forgings, insulation, bearings)
- Improved Production Techniques

The growth pattern for steam turbines is shown in Figure 2. Note that steam turbines also appear to have been adversely affected by the 1973 OPEC oil embargo. As in initial commercialization, corporate commitment is necessary to weather the impact of such unpredictable events.

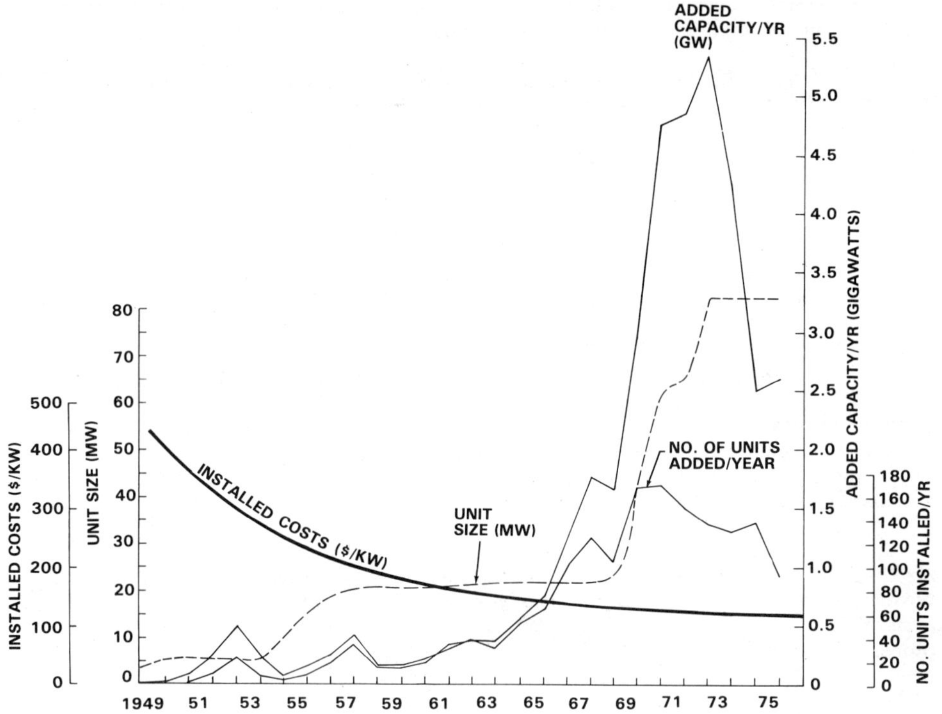

Figure 1. Gas Turbine Growth Patterns

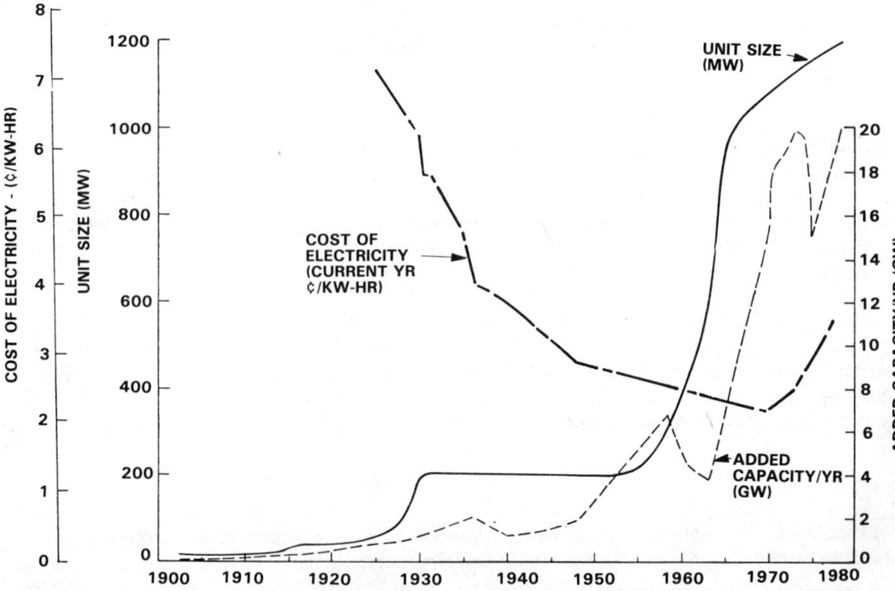

Figure 2. Steam Turbine Growth Patterns

In examining the commercialization of these two energy products it was found that some common elements emerge. These are summarized in Table V.

Table V. Common Elements of New Product Commercialization

- Early product introductions were not competitive
- Continuing research and development improved performance
- Many technological paths were unsuccessful
- Early market setbacks were common
- Prototypes, field trials and early product proof testing provided reliability and confidence
- Unpredictable events affected later market growth
- 20 to 25 years was typical for R&D thru initial commercialization for normal development and market forces

Having reviewed the many examples of the time required for commercialization and thereby establishing that the period is quite lengthy, let's next consider what can be done to stimulate innovation with particular emphasis on the Government's role.

In a recently completed study by the Research and Policy Committee of the Committee for Economic Development (6) specific recommendations were proposed for actions by the Government to stimulate innovation. The recommendations were prepared by a blue ribbon group of businessmen and heads of leading universities who are intimately familiar with the process of commercialization. There is little question but that one of the roles of Government is to maintain a climate that encourages innovation in all sectors of the economy (7). Highest on the priority list for stimulating technological progress and innovation would be tax changes which increase capital formation and thereby improve the climate for investment and innovation. A more rapid capital recovery allowance is the highest priority alternative tax measure. The reasoning is that allowing depreciation of capital plant and equipment over a shorter perior of time than that stipulated by current tax law would immediately stimulate the introduction of innovation embodied in the new plant and equipment. An alternative would be to index depreciation allowances to replacement costs which are rising due to inflation. However, rapid depreciation is considered to be less burdensome administratively.

Our views are slightly different than the CED report but in general parallel their latest suggestions quite closely. The second highest priority is a reduction in the corporate tax rate. Broad based incentives such as these have the highest probability of stimulating technological progress. The cost of commercialization is composed of the costs of concept exploration, screening, product development, market testing and commercial introduction. These lead to the potential for product acceptance if the other parameters are successfully completed.

There are a number of other general and specific tax changes which would be conducive to stimulating technological innovation. The general tax changes may be characterized as creating increased capital formation which could be used for innovative endeavors and would be a signal to business that government is serious in strengthening the private sector. The four other general incentives recommended are:

---

(6) Stimulating Technological Progress. A Statement by the Research and Policy Committee of the Committee for Economic Development, January 1980

(7) Industrial Innovation and Public Policy Action, Report of a Colloquium, National Academy of Engineering, National Academy Press, Washington, DC, 1980

- Regular rate reductions on income taxes
- Reduction in the tax on capital gains, and
- Reduction in maximum marginal tax rate
- Elimination of the double taxation of dividends

Some additional selective incentives which are recommended for consideration to stimulate innovative developments are:

- Flexible depreciation for R&D structures and equipment to assure up-to-date research facilities in industry. This flexible depreciation would allow depreciating the R&D assets fully in the first year of their life while maintaining the investment tax credit. One of the rationales for this recommendation is that it would bring U.S. tax policy in this area more in line with that of our major competitors in the industrialized world, i.e., Japan, West Germany, France, Canada and the United Kingdom.
- Faster depreciation for patents
- Flexible depreciation of purchased patents and other intangible properties
- Require only directly related R&D expenses to be deducted from foreign-source income

Given that the combined pressures of inflation and foreign competition have made tax reform the highest priority activity for Government action, there are other opportunities for Government to improve the climate for innovation. These include reducing regulatory barriers and reforming patent policy.

Although almost all agree that certain regulatory functions are necessary for the preservation of the health and safety of the country there are disincentives to innovation caused by regulations. These include a shifting of the allocation of resources, an increase in the waiting time for return on investment and an increase in uncertainty.

Obviously the first consideration with respect to regulation is to question whether any is required.

If it is concluded that regulations are required, then it is recommended that Congress and the regulatory agencies examine each proposed regulation for its effect on innovations and recommend only those whose social benefits outwiegh their full costs. Careful review of the regulation should be undertaken to make certain that they provide minimum distortion of the market system and are also <u>not</u> designed to provide a zero risk society.

An example where regulation is being phased out which appears to be quite successful is the airline industry where open market competitive forces are dictating economic decisions. It seems quite likely that other parts of the transportation industry might be similarly deregulated with a resulting positive effect on the business environment.

The recommendations for improvements to the patent procedure to stimulate innovation made by the Committee for Economic Development were quite extensive. The role of patents was considered to be very important in establishing a basis in rights to stimulate businessmen to sponsor research and to provide the incentive for investment in facilities to produce innovative products. Specific recommended changes in patent policy include:

- First-to-file Patent System. The U.S. and Canada are unique in industrialized countries in not following this system. Instead, the U.S. system provides for interference, a quasi-judicial procedure for establishing who was first by examining in detail what occurred prior to the filing dates. It is believed that the first-to-file approach would save much time while still providing a defense to the individual who can show that he was the first to invent and to reduce to practice the invention. As a reference, the European Economic Community has such a procedure.
- Re-examination of Patents. It is recommended that the Patent and Trademark Office be allowed to examine the references for a defendant which could result in a decision that a patent should not have been awarded without the expense of bringing a suit. It is encouraging to note that this concept was included in a Senate bill passed in the most recent Congress.
- Arbitration of Patent Disputes. Because of the extremely high cost of protracted patent litigation, it is recommended that voluntary arbitration of patent disputes be allowed including both considerations of validity and infringement.

Several other recommendations on patent policy were included in the CED report including:

- Establishment of a single patent appeals court (this was included in House bill H.R. 3806 passed on September 15, 1980).
- Provision for copyrighting of computer software.
- Allowance for enforcement of U.S. patents on processes when those processes are used abroad and the goods imported into the U.S.
- Extending the life of a patent for a time equal to the length of governmental regulatory delay.
- Provide title to inventions and patents made under government title to the contractor who made the invention.
- Provide a license to all domestic manufacturers on a royalty-free basis for all government-owned patents.

Although each of the above recommendations are not individually significant, they become important when all are considered for each contributes to the climate for stimulating innovation. The recommendations for action recently proposed by the President and the actions taken by Congress provide an encouraging sign that the need for stimulating technological innovation is perceived by the Government.

In summary, examination of numerous examples of the time period required for new product development substantiates that the commercialization process is longer than most would initially anticipate. As a result, considerable commitment and patience is required by management to achieve successful commercialization. Furthermore, the significant relationship between new product development, productivity and economic well-being of a country demands that considerable effort be expended to make certain that innovation and commercialization of new technologies be stimulated and that an environment be developed to foster the stimulation. Specific recommendations are made for government actions to provide that innovative environment. Foremost among these are tax changes to provide for a more rapid capital recovery allowance and for corporate tax rate reductions.

FINANCING THE SYNTHETIC FUELS INDUSTRY

J.G. Stabback and G.A. Cumming
Global Energy & Minerals Group
The Royal Bank of Canada

I am delighted to have been invited to Washington to speak on the important issue before us this morning, that of the synthetic fuels industry. My remarks will be confined to the financing aspects and, in this regard, I propose to discuss both the macroeconomic problem of capital availability and the microeconomic aspects of financing a specific project.

The world of the 1980's will present a vastly different financing environment compared with the 1970's. The two OPEC price eruptions have created enormous problems in oil-importing countries. For these countries, balance of payments deficits have become endemic and governments and industry are scrambling to develop indigeneous energy resources. Energy-related investment is escalating dramatically throughout the oil-importing world. Projects being pursued to lessen foreign energy dependence tend to be extremely capital-intensive; consequently, we are witnessing an unparalleled demand for capital by the energy industry around the world.

In Canada, for example, energy investment will be the dominant component of fixed capital formation for the remainder of this century. In current dollar terms, we estimate energy spending north of the border, in a country of only 23 million people, may approach 1,400 billion dollars in the next twenty years. This implies that energy investment will expand from 5% to about 9% of gross national product, thereby somewhat squeezing out other competing demands on the goods and services of the country. Clearly Canada will be looking to the world capital markets for a significant portion of this capital.

In the developing world, continuous massive deficits will be encountered and these countries will be calling frequently on world capital markets. No country in the world will require more capital than the United States. Prospective investments in the energy sector, and in the industrial section in order to modernize American plant and equipment, will be enormous. At the same time, this country will likely suffer current account problems as the dependence on foreign oil persists throughout the decade. Virtually all international oil price forecasts project the real price rising throughout the 1980's. Consequently, OPEC will maintain very high current account surpluses. The 1978 surplus was about 8 billion dollars; last year's figure was approximately 110 billion dollars. The scale of the recycling problem as OPEC surpluses accumulate is increasing steadily. Private lending institutions, which serve as intermediaries, are confronted with a growing recycling problem and the volatility in the Middle East contributes to an evermore tottering arrangement.

The solution, difficult though it be, is becoming clearer. Energy policy in this country, after such a long and acrimonious debate, distinctly appears to an onlooker to be pointed in the right direction. The move to decontrol oil prices will provide a considerable inducement for consumers to restrain their demands and for suppliers to aggressively explore and develop new sources of oil and other forms of energy. The synfuel programme is a bold and imaginative gesture which ultimately will contribute to solving America's energy demand - supply imbalance.

However, both the conventional and non-conventional segments of the energy sector will have to compete aggressively for limited capital resources. The United States traditionally has maintained a relatively low savings rate which will tend to exacerbate the capital availability problem in the 1980's. Fortunately, a growing consensus appears to be emerging that this country must remove the impediments to saving and investing, if the country is to restore its international competitiveness and energy balance. Such a move requires somewhat reduced personal and governmental spending in order to make room for investment. Enhancing the rate of saving and investment requires a long-term commitment, but, analogously, the benefits tend to be long lasting. The Japanese with their extraordinarily high rate of savings appear to have clearly understood this economic relationship.

There are several other somewhat related factors worth considering. Balance sheets of commercial banks are becoming increasingly mismatched as banks accept more short-term deposits to finance large-scale long-term projects such as synfuel plants. In a volatile interest rate environment, the mismatching of the life of assets and liabilities exposes banks to large roll-over risks. Thus, financial intermediaries will seek to lengthen the term of their deposits and shorten the term of assets, such as synfuel loans.

Secondly, banks have been increasing the weight of energy loans in their overall portfolios. For diversification reasons, limits will emerge on how far the banking sector will weigh its portfolio with synfuel or energy loans. For any single bank, the problem is far more acute, especially given the enormous scale of certain synfuel projects. For example, the Royal Bank of Canada, although the fourth largest bank in North America, could not prudently swallow the financing of a multi-billion dollar synfuel project, just as many project sponsors face with trepidation the prospect of proceeding alone with one of these mega-projects. The huge energy investment requirements of North America, in which synfuels figure prominently, will also place a strain on equity markets. Sponsors will have to rely on debt funding for the larger part of their financing needs. Nevertheless, many firms will require substantial new volumes of equity to support the huge debt. In recent years, equity markets have not always been a buoyant and robust source of capital. Unless synthetic fuel projects prove to be a profitable investment, equity markets will not yield the necessary volume of capital.

An interesting adjunct to this line of reasoning is that banks, in order to fund huge volumes of energy debt financing, will themselves be forced to go to the equity market to expand their capital base. In so doing, banks will be competing directly in the equity markets with the energy sponsors whom they are attempting to fund. The very large capital requirements for a synfuel project will result in the largest banks, those with sufficient capital bases, leading the way in synfuel financing.

Thus, in aggregate terms, the portfolios of banks may become top-heavy with certain forms of energy investments and technological risks may dictate the need for innovative forms of finance and some form of governmental guarantees. Capital will be available to credit-worthy borrowers but financial spreads will likely be higher in the future, reflecting the acute demand for capital.

I would like to focus attention more closely on financing specific projects. I propose to highlight a few Canadian projects which you may find quite interesting and instructive. My remarks will then conclude with several more general project financing considerations.

In the synthetic fuel sector, Canada is not starting from scratch. The Canadian tar sands located in Alberta are exceedingly large. An estimated one trillion barrels of oil lie embedded in sand, a deposit larger than total middle Eastern oil reserves. As with your oil shale, only a small portion of these deposits, 10 to 20% with present technology, can be recovered. They are also costly to develop and they entail lengthy development periods.

The first synfuel plant, the 45,000 barrel-per-day Suncor plant, has been operational for a surprising fourteen years. The mammoth syncrude plant, rated at over 125,000 barrels-per-day, has been on stream for three years. Already 9% of Canadian oil demand is met by domestic synthetic oil. Two additional Canadian projects are awaiting final cabinet approvals. These plants, each rated at 140,000 barrels-per-day, could bring total domestic synthetic supply up to 25% of national demand within six years.

The first tar sands plant, owned by the Sun Oil Company, turned dirt in 1965 and was completed two-and-a-half years later at the then unheard of cost of one-fifth of a billion dollars. From today's perspective, that project was a terrific gamble. The Canadian price of oil was less than $3 per barrel and no one had ever commercially extracted the heavy bitumen from the isolated Athabaska region—a region known principally for its mercilessly cold winters. Suncor, as the company is known today, financed the plant entirely by what we term "corporate finance." The corporate balance sheet was put on the line and considerable internal funds were committed. The vision and gamble succeeded. As international and domestic prices rose, the precarious investment was recovered. Valuable technology was proved and this concrete commercial demonstration acted as a spur to other developers. The only sad note, I regret to say, is that the domestic administered price granted to Suncor has recently been more than halved from $38 a barrel to $17.75 by the new Canadian National Energy Programme. This move has been justified on the grounds that the investment required for expansion has been recovered and receipt of the $38 price was unnecessary for continued profitability. The lesson in this case is that political risk remains a prominent consideration even in stable, democratic Western countries.

Indicative of things to come, by 1978, when Syncrude—the second tar sands plant—became operational:

1. Rated output was up almost threefold but costs were up about elevenfold in comparison with the earlier plant,

2. Project risks were spread among six participants (not one as in the Suncor case), and

3. The governments of Canada, Ontario and Alberta had to be brought in on the financing to ensure a go-ahead.

Technical problems continue to plague this operation although the bottlenecks are gradually — 2 1/2 years later — being eliminated. The lessons from this plant are indeed instructive and warrant consideration and attention by project sponsors and interested parties in this country. At its peak, 8,500 workers were involved, 50 million man-hours in total were expended, and macroeconomic pressure on the Alberta economy could be noticeably felt. A noteworthy feature of Syncrude is that the go-ahead was taken when world energy prices were only 45% of the present level and the Syncrude plant is now a very valuable asset providing 6% of Canada's oil requirement in a manner which is profitable to the owners.

Continuing our earlier progression, the proposed Alsands project rated at 140,000 barrels per day or 10% larger than Syncrude, is estimated to cost, when built, 4 1/2 to 5 times as much as Syncrude — in the 9-11 billion dollar vicinity.

The fourth project, the Cold Lake Plant sponsored by Esso Resources Canada, a Subsidiary of Exxon, is possibly the most exciting. Whereas the other synthetic projects in Canada are surface mining of tar sands, Cold Lake is an "in situ" process whereby the bulk of the resource, which is embedded too deeply to be recoverable by surface mining, can be captured by a huff-and-puff steam drive process. This constitutes an important technical breakthrough. Exact project costs depend upon the commencement date, but twelve billion dollars is a recent estimate. That equates to roughly eighty thousand dollars per barrel of installed capacity, about 180 times the capital costs of Middle Eastern oil. That is a frightening figure to a lender but is a telling statistic about the horrendous cost of energy security.

Taking this as the starting point, let us examine some of the risks and uncertainties inherent with synthetic fuel plant financing. One of the most critical variables affecting the economic viability of a project is the price of energy. High oil and gas prices are required to support the enormous capital costs of synfuel projects. In the 1980's, even following an allowance for conservation gains and inter-fuel substitution, the world's demand for crude oil appears quite likely to outstrip the available supply, thereby placing sustained upward pressure on the real price of oil. Further re-enforcing this momentum is the capacity of a few key countries, principally Saudi Arabia, Kuwait and Qatar, to restrict production so as to counter any demand constraint or supply success achieved in the West. The vice-like grip of OPEC appears quite solid and, interestingly enough, serves as a mild guarantee for the economic viability of synfuel plants by maintaining a high price for energy.

Project costs are another extremely critical variable, for example, the Cold Lake Project is escalating at over two billion dollars per year. Regulatory delay can be very large indeed. Overruns on multi-billion dollar projects are especially dangerous as they may run in the hundreds of millions of dollars or indeed exceed one billion dollars, posing a threat to even the largest sponsors. Overruns may be attributed to inflation, to more stringent socio-environmental requirements, to strikes, etc. Technological risk is also significant in part simply because there is not a single large-scale, commercial synfuel plant operating in this country. Technical failures and difficulties are certain in scaling-up to commercial size for first-of-a-kind synfuel plants. For example, in Canada, extremely cold weather presented innumerable difficulties, leading to substantial interruptions and, consequently, subsequent adjustments in plan design.

From a lender's perspective, measures which alleviate or reduce potential risk are particularly important. Thus a large equity investment, proven competent sponsors and contractors, and creditworthy established product buyers play a prominent role in attracting capital. Various forms of product purchase contracts become essential when the project scale is so large as to present excessive risk for the sponsor. In effect, such purchase guarantees shift, or distribute, risk from the lender and project sponsor to the buyer. In essence, the objective is to design a total financial package which reduces the sponsoring companies' risks while presenting the lenders with sufficient credit support.

As was mentioned earlier, political risk is an ever present concern. The Canadian Suncor Case, where the regulated selling price was cut dramatically overnight is an excellent, albeit discouraging example. The alterations in the Synthetic Fuel Corporation and synfuel policy in this country are also a telling example. However, we happen to believe that in the longer run the American synfuel industry will probably be strengthened by the Reagan policy direction. Specifically, compelling greater resort to the private market and not pushing as many, or as exotic, projects is likely to avoid white elephants, thereby engendering greater public and marked confidence in the synthetic fuels industry. What is also required for more forthcoming bank financing is a stable policy environment.

One positive feature which synfuel projects typically exhibit is that the risk concerning reserve size tends to be smaller than with conventional oil deposits. In part, this is due to the very large size of the deposit in the case of oil shale, tar sands or coal.

Overall risks are large in number and ever-present. They are also more significant than with conventional energy projects. These factors are extremely important to bankers. However, we at the Royal Bank of Canada believe that soundly conceived and well-managed projects do not present unacceptable levels of risks and this assessment forms the basis for why we are gearing up, in a major way, to finance synfuel development.

From a synfuel perspective, the 1980's promise to be an exciting and innovative decade. The necessity and challenge of synthetic fuels is becoming evermore apparent, and if envisaged global trends materialize, rewards will be present for the country, for energy developers and for consumers from a sound synthetic fuel development programme.

EFFECTS ON THE UNITED STATES ECONOMY OF SYNFUELS INDUSTRY DEVELOPMENT

Gary S. Stacey, Battelle, Columbus Division
Peter V. King, Westinghouse Electric Corp.
W. Halder Fisher, Battelle, Columbus Division (1)

The experience in the United States with the problems that stem from our high dependence on foreign sources for oil and oil products has led to more active consideration of methods by which we can become less dependent. The development of domestic alternatives such as conservation and substitution is one approach. Another is the development of new "synfuel" sources of supply such as shale oil and coal liquefaction which will entail considerable capital outlay and at the same time create additional output and employment. Developing industry valued at many billions of dollars obviously has the potential for creating a variety of effects at the national level. This work was directed at estimating such effects.

The changes examined arise from substituting domestic alternatives for imported crude oil. The general plan for substitution is based on proposals by President Carter. At the present time, the United States is heavily dependent upon foreign countries for imported oil. That dependence is shown by the fact that almost 6.7 million barrels per day (mbpd) or almost 40 percent of all oil requirements for 1980 were filled from foreign sources.

The Carter energy proposal recommended that 4.5 mbpd of the imported oil be replaced by 1990 through substitution, conservation, and the development of domestic alternative sources. The work summarized here was conducted in examination of the effects on the economy of reducing these imports and developing domestic capabilities to replace them. The analysis is based on using a model of

(1) Important contributions to this work were also made by Edward J. Honton, Ann R. Buhr, and F. Jere Bates, from Battelle, Columbus Division.

the United States economic system into which changes corresponding to the above-mentioned conservation, substitution, and new energy technologies are introduced. From this, changes in employment, prices, gross national product, and total output are estimated. Also, the balance of payments effects are addressed qualitatively.

APPROACH

The approach taken was to utilize Battelle's intput/output model of the United States economy in 1979 as a baseline. Simulations were made characterizing the changes that would be necessary to switch from imports to domestic synthetic fuel production. The simulations may be characterized broadly as capital and operating simulations.(2) In these simulations, conservation and substitution was introduced to account for up to 2.0 mbpd reduction in demand for imported crude. The remaining 2.5 mbpd would be filled by new technologies. Those considered were shale oil and coal liquefaction. Current estimates of the input requirements for producing 2.5 mbpd of crude equivalents from these sources in a 30-percent shale oil, 70-percent coal liquefaction mix were used to create two new separate economic sectors. As conservation, substitution, and domestic production take place, imports of crude oil are assumed to be reduced by the same amounts in the simulation.

The capital simulation considers the effects during a "peak" year of capital formation for these industries. Based on an estimated 10-year construction period the peak is assumed to entail the creation of one-fifth of all the capital required in a single year. At that same time, one-half of the necessary conservation and substitution will be complete. In order to avoid the complications that otherwise would be introduced by real-time demographic trends, unrelated technological changes, etc., the model treats 1979 as if it were the year of peak capital formation activity. Thus, the differences that emerge between "baseline" 1979 and "capital formation" 1979 result solely from the changes introduced by the simulation.

The operations simulation is based on full operation of the new technologies, after capital formation is completed and with all conservation and substitution goals achieved. Again, in order to avoid the complications introduced by real-time changes, the model treats 1979 as if it were the first year of full operation of the two synthetic fuels sectors. Thus any differences that emerge between "baseline" 1979 and "operations" 1979 result solely from the changes introduced by the simulation.

In addition to the 70/30 mix of synthetic fuels that was proposed in the President's program, a parallel set of simulations was run. In this second set of simulations the controlling synfuel mix was set at 50/50--that is, the formation and operation of equal-sized shale oil and coal liquefaction industries was assumed.

CAPITAL AND INTERMEDIATE COEFFICIENTS FOR SYNTHETIC FUEL INDUSTRIES

To examine the economic impacts of the new synthetic fuel industry, technical and capital coefficients were established. These coefficients indicate in a quantified manner the technical and capital natures of the new industries. There are many different production processes being investigated at this time. Each process involves a different technological framework and therefore different coefficients. Thus, a "typical" or average process must be assumed

---

(2) A definition of the simulations is presented in the Addendum.

for each industry. The assumptions and methods used in obtaining the coefficients follow.

Shale oil is assumed to be retorted entirely above ground. Half the shale to be retorted comes from underground shaft mines, the other half from surface strip mines. All shale oil plants will be located in the Colorado-Utah-Wyoming area near the minemouth. For an initial small industry of up to 1.25 million barrels per day, adequate water supplies can be obtained from area streams. In some cases water may have to be piped several miles to the plant, with a corresponding increase in the price of delivered water. A price for water, including an average amount of pumping, is included as a cost of production. Land restoration costs were also included as costs of production as part of an overall environmental cost of one dollar per ton of rock mined.

Initial detailed data for the shale oil process were obtained from Oil-Shale-1974: An Economic Evaluation Using 30-Gallon Shale and Producing 50,000 Barrels per Calendar Day of Shale Oil (Report No. 75-3, U.S. Department of Interior, Bureau of Mines, September 1974). The technological inputs detailed in this report were distributed over the Battelle input/output sectors. Prices were updated, sector-by-sector, to 1979. Further updating to incorporate more recent technological thinking was undertaken using the broader sectors and data from the Bechtel Study Report, Resource Requirements, Impacts, and Potential Constraints Associated with Various Energy Futures (DOE Contract No. EX-77-C-3794, August 1978). Additional detail was derived where necessary by studying both coal mining and oil refinery technology. Finally, the proposed coefficients were reviewed and judgmentally modified by knowledgeable Battelle staff to obtain final values. For shale oil, these coefficients suggest a capital-to-output ratio of 1.96. In other words, $1.96 of capital must be invested to produce a dollar's worth of annual output. The output price under normal profit conditions is estimated to be $26.60 per barrel (1979 price).

Coal liquefaction technology is currently being used in South Africa. However, several new and more efficient technologies have been investigated in recent years. Of the process types being studied, hydrogenation appears to be the most economical at this time. The best hydrogenation technology now available appears to be the Exxon Donor Solvent method. Accordingly, this process was chosen as the typical technology to be quantified in coefficient form.

It is assumed that the plant will be a minemouth operation. Coal will be mined by the coal-mining sector and delivered to the coal liquefaction plant. The resulting crude will be transported to existing refineries. Half the plants will be in the western U.S. and half in the eastern part of the country. Adequate water for processing the coal will be available locally to the liquefaction plants, provided that they are properly spaced within the coal fields. If the coal liquefaction industry begins processing many millions of barrels of crude oil per day, water scarcities will drive the input price of water above the level used in this study.

The basic sectoral data detail for the coefficients came from the process evaluation report Liquid Fuel from Coal Plant (Report No. 74-27, U.S. Department of Interior, Bureau of Mines). The prices in this report were updated to 1979. Additional more current information was obtained from the EPA's Technology Assessment Report for Industrial Boiler Applications: Synthetic Fuels (Preliminary Report, June 1979). The coefficients thus obtained were reviewed and, where necessary, judgmentally modified. The capital/output ratio for coal liquefaction is estimated to be $2.60 of capital invested to obtain a dollar's worth of annual output. This high capital intensity combined with a higher operations cost results in a 1979 price for coal liquids of $33.70 per barrel.

ADDITIONAL ASSUMPTIONS

(1) A barrel of crude oil produces a barrel of refined output.

(2) All synthetic fuel outputs are treated as crude oil equivalents.

(3) No employment is lost due to conservation and substitution even though total domestic outputs fall. Actually one could expect some temporary increase in employment to help enact the conservation changes.

RESULTS

THE 70/30 MIX

Some of the major results are summarized in Table 1. The employment effects of the full operation scenario exceed the effects during construction of the facilities. Obviously, employment effects are felt in the shale oil and liquefaction industries but all industries that support those industries must expand. In addition, the coal mining industry must expand considerably to provide the coal required by the liquefaction process. Equally important, however, it must expand to support the substitution of coal for imported oil in fueling electrical utility output, an important part of the conservation/substitution aspects of the simulations.

TABLE 1

SUMMARY OF SIMULATED EFFECTS OF SUBSTITUTING DOMESTIC CONSERVATION, SUBSTITUTION, AND PRODUCTION FOR IMPORTED CRUDE

		Replacement of Imported Oil with Domestic Alternatives*	
	Reliance on Imported Oil (Baseline)	Peak Capital Formation Year	Full Operation Year
Change in Employment	0.0	+677,000	+743,000
Change in GNP (Billions of 79 $)	0.0	+12.2	+29.6
Average Price Pass-through Impact**	1.0	N/A	+1.00022
Change in Total Domestic Output (Billions of 79 $)	0.0	+19.1	+14.8

* 70/30 mix of Coal Liquefaction/Shale Oil.

** Weighted impact on consumer prices of using higher priced synthetics.

The change in GNP (over baseline) is estimated to be +$12 billion during the year of peak capital formation and +$30 billion in the year of full operation. These changes in GNP arise from several major sources. First, as conservation and substitution take place, the new technological configuration that this embodies allows the fulfillment of final demand requirements with reduced resources, thereby directly reducing GNP. Since we assume no non-simulation changes in final demand, only the new investment in synthetic fuels constitutes an increase in GNP. It more than offsets the above-mentioned reduction. Finally, there is a further increase because of the reduced purchase of imported oil. Reduced oil imports more than offset the increased imports of other raw materials, etc., and substantially add to the growth of GNP.

The cost increases attributable to the replacement of imported oil by higher-cost synthetics are largely offset by the economies achieved through conservation. As a result, prices paid by consumers are expected to rise--from this cause only--by about one-fiftieth of one percent (0.02%).

Total domestic output differs from GNP because it is a measure of transactions including all interindustry transactions and involves multiple-counting. It is, however, a good measure of the total activity that an individual sector of the economy would experience, and therefore can be used to identify the sectors that would be stressed as a result of the creation of domestic synthetic fuel capacity. Table 2 shows the 10 sectors that will have to expand output most (in percentage terms) between the baseline and the peak capital formation situations. Mining Machinery (Sector 10.03) is the high growth sector both because of the assumed substitution of coal for oil in generating electricity and because of the requirements for machinery which this sector produces in the liquefaction and shale oil industries. The needs for Primary Metals (Sectors 7.01 and 7.03) arise from their use in a wide variety of capital-producing sectors.

TABLE 2

INDUSTRIES EXPERIENCING GREATEST INCREASE IN TOTAL DOMESTIC OUTPUT DURING SIMULATED CAPITAL FORMATION (70/30 MIX)

Sector	Name	Percent Increase In Output
10.03	Mining Machinery	21
8.05	Fabricated Structural Metal Products	18
12.05	New Construction--Non-Residential Building	12
9.02	General Industrial Machinery and Equipment	7
10.05	Material Handling Machinery Except Trucks	6
2.04	Coal Mining	6
7.03	Primary Aluminum	4
12.03	Industrial Controls, etc.	3
7.01	Primary Iron and Steel	3
8.04	Non-Electric Heating Equipment	2

Industries that would experience the greatest increase in total output during the full operation of the shale oil and coal liquefaction industries are shown in Table 3. There are, of course, the increases in output from the coal liquefaction (10.1 billion 79 $) and shale oil industries (4.3 billion 79 $) themselves. These are supported by large increases in Coal Mining (Sector 2.04) and Mining Machinery. Agriculture, Forestry and Fisheries (Sector 1.04) rise because of the need for and use of these products in land resotration that would be required by strip-mining coal and oil shale.

TABLE 3

INDUSTRIES EXPERIENCING GREATEST INCREASE IN TOTAL DOMESTIC OUTPUT DURING SIMULATED OPERATION (70/30 MIX)

Sector	Name	Percent Increase in Output
2.04	Coal Mining	26
10.03	Mining Machinery	17
8.05	Fabricated Structural Metal Products	5
1.04	Agriculture, Forestry and Fisheries	3
9.02	General Industrial Machinery and Equipment	3
8.04	Non-Electric Heating Equipment	2
9.01	Engines and Turbines	2
5.06	Miscellaneous Chemical Products	2
19.02	New Construction--Non-Residential Building	2
4.04	Wooden Containers	2

In addition to these increases, the petroleum refining and related products industry output would experience a decrease in output in both simulations brought about as a result of the conservation and substitution effort.(3) Such supporting industries as pieplines and metal barrels and drums would also show declines.

THE 50/50 MIX

In addition to the simulations that consider the 70/30 mix of coal liquefaction and shale oil production, a 50/50 situation also was analyzed. This is a less likely eventuality in the short-run because all the shale oil plants must be located in a small region which places areal strains on the local community, labor force, water supplies, and the environment. Nevertheless, the 50/50 simulations, in comparison with the 70/30 ones, show slightly lower employment, decreased GNP, smaller price effects, and reduced total domestic output. This pattern is due primarily to the fact that coal liquefaction, which has greater capital requirements, is reduced in the 50/50 mix.

INTERNATIONAL TRADE

U.S. exports to OPEC countries have dropped from a balance with imports in 1972 and 1973 to the current situation in which exports are about one-half the value of imports (for 1977 exports were $13.9 billion and petroleum and petroleum product imports were $31.2 billion). In the full operation simulation imports are reduced by 53 percent. Assuming a commensurate 50-percent reduction in exports to OPEC countries, this would constitute a reduction in GNP, partly offsetting the increase in GNP resulting from reductions in imports.

In 1977, the last year for which data are available, U.S.

---

(3) It should be noted that petroleum refining capacity has been assumed as transferable between various qualities of petroleum and synfuel inputs. This is not true in many instances, so that the assumptions may understate required capital formations in the refinery sector.

exports to OPEC countries were distributed in the following pattern:

Capital goods	46%
Industrial supplies	15%
Automotive products	10%
Consumer goods	6%
Other non-agricultural products	10%
Agricultural products	12%

If these proportions continue to hold, capital goods exports would be the most heavily affected by any decrease in exports to OPEC countries financed by our oil imports. In other words, while a reduction in oil imports would help our balance of payments, it might well be followed by a reduction in our exports. The net effect of all this is impossible to forecast. However, if we speculate that OPEC's need for capital and other goods from the U.S. is relatively inelastic with respect to our petroleum imports, the net effect could still be favorable for the U.S. balance of trade.(4)

CONCLUSION

The simulations showed many positive effects on the United States economy, with increases in employment, GNP and total output. However, some industries would require significant expansion, particularly Mining Machinery 21% and Fabricated Structural Metal Products 18%. These industries may find difficulties in meeting this demand if any additional pressures arise.

Rapid expansion in these industries and others causes inflationary pressures which were largely offset by the conservation assumption. This significant impact assumed for conservation will probably occur, largely through escalating energy prices.

Other areas of major concern that were not directly assessed are transportation, water resources, and the environment. For the initial small minemouth plants, adequate transportation and water would be available, however environmental issues remain uncertain.

ADDENDUM--DESCRIPTION OF SIMULATIONS

SIMULATION 1-A

This simulation assumed half the production of synthetic fuels will be coal liquids and the remaining half would be shale oil. Conservation of 0.625 million barrels of refinery products per day would occur and an additional 0.375 million barrels per day of oil would be replaced through the use of coal in the electric utilities. Overall, the reduction in crude oil imports amounted to 1.0 million barrels per day. The first part of the simulation modeled the effect of conservation over all sectors and substitution in the electric sector. The second part of the simulation modeled the increase in total domestic output that would occur as a result of the increase in new capital investment necessary to construct the new synfuel industries.

SIMULATION 1-B

This simulation had identical conservation and substitution

---

(4) In fact, if <u>all</u> exports to OPEC countries were cut off due to our reduction by 50 percent of imported oil, the net effect would still be slightly beneficial to the U.S. balance of payments.

effects as for Simulation 1-A. New capital investment under this scenario assumes that the mix of synfuel industries is 70 percent coal liquids and 30 percent shale oil.

SIMULATION 2-A

In this simulation full conservation and substitution of coal for oil by electric utilities would occur. This means that 1.25 millon barrels of oil per day would be conserved by all sectors proportionate to each sector's use of oil. The electric sector would substitute coal for an additional 0.75 million barrels of oil per day. The first part of the simulation modeled the effect on the baseline economy of enacting this level of conservation and substitution.

The second part of the simulation implemented the operation of the new synthetic fuel industries. Intermediate coefficients for the two new synfuel industries were placed in the input/ouput matrix. Half the production was assumed to be shale oil and the other half was coal liquids. The resulting changes in the economy were then observed by running the model.

The third part of the simulation entailed the determination of the effect of adding new replacement investment for the synthetic fuel industries into final demand. Each step of Simulation 2-A was accumulated onto the previous step so that this effect of replacement investment was calculated assuming full conservation and substitution along with the actual operation of the synfuel industries also taking place.

SIMULATION 2-B

The fourth simulation also had the same three steps as Simulation 2-A. Identical conservation and substitution effects were assumed to have taken place. The second part again involved operating the synfuel industries. In this simulation 70 percent of the production was coal liquids and the remainder was shale oil. The final third step added the effects of replacement capital necessary for this same mix of synfuels.

Evaluation and Comparison of Socioeconomic and Environmental Impacts of Commercial-Scale Synfuels Plants at Six Sites in Kentucky

Gaylord M. Northrop
Vice President
The Center for the
Environment & Man, Inc.
Hartford, Conn. 06120

Charles A. D'Ambra
Associate Scientist
The Center for the
Environment & Man, Inc.
Hartford, Conn. 06120

Harry Enoch
Asst.Dir., Tech. Assess.
Bur.of Energy Research
Ken. Dept. of Energy
Lexington, Ky. 40578

ABSTRACT

Under DOE Contract EY-77-C-02-4575, managed by Mr. Thomas Ruppel and Dr. Earl Evans of the Pittsburgh Energy Technology Center, CEM has applied its Site Evaluation for Energy Conversion Systems (SELECS) methodology to evaluate the socioeconomic and environmental impacts of commercial-scale coal liquefaction plants at six potential sites in Kentucky.* Five of the sites were in western Kentucky; one was in the northeast part of the state. The six sites to be evaluated were selected by the staff of the Bureau of Energy Research of the Kentucky Department of Energy. The sites are:

- Maceo, Daviess County
- Curdsville, Daviess County
- Caseyville, Union County
- Rockport, Muhlenberg County
- Calvert City, Marshall County
- Maysville, Mason County

At each site, SELECS was used to evaluate socioeconomic and environmental impacts of two commercial-scale coal liquefaction processes, each capable of processing about 26,000 tons of coal per stream day. The processes are SRC-I, a direct liquefaction route with solid boiler fuel as the principal product, and Coal-to-Methanol-to-Gasoline, an indirect liquefaction route, with transportation fuel as the primary product. To show comparable impacts of equal sized facilities, a two-gigawatt coal-fired steam-electric power plant and an 849-employee automobile parts manufacturing plant were also evaluated at each site. Impacts of each facility in its operating mode were evaluated at one or two nearby cities or towns, and four to six counties where significant impacts might be expected due to proximity of the plant and the likelihood that the nearby towns and surrounding counties would acquire many new residents and businesses.

At each of the six sites, CEM evaluated 11 Impact Categories:

- Population
- Community Services
- Bonding Capacity
- Housing
- Land Use
- Water Availability
- Air Quality
- Water Quality
- Endangered & Threatened Species
- Aesthetics & Important Sites
- Tax Rates

SELECS indicated that the Coal-to-Methanol-to-Gasoline process would have somewhat lower adverse impacts than SRC-I, and Calvert City and Curdsville were the sites where a plant would have the least adverse impact on nearby towns and the surrounding region. The Rockport and Caseyville sites appeared to have the most adverse impacts.

This paper presents the results of an application of the SELECS Level 1 methodology, which was computerized for this study. This study demonstrates the utility of the SELECS methodology as a "screening tool" for alternative sites for a specified process and/or alternative processes at a specified site.

1.0 The SELECS Methodology

The SELECS methodology for site evaluation for energy conversion systems was described at the Sixth Energy Technology Conference in a paper based on the methodology development and User's Guide. [Ref. 1,2] SELECS is a quantitative methodology in which relative changes in socioeconomic and environmental Impact Elements are determined, normalized, and then used to find a relative importance value from a subjectively-determined Relative Importance Curve. One or several weighted Relative Importance Curves may constitute an Impact Category, and where more than one are involved, they are summed and normalized.

2.0 Application of SELECS

To demonstrate the applicability of the SELECS methodology as a workable, useful "screening tool" for alternative energy conversion processes at one or more sites, the U.S. DOE Pittsburgh Energy Technology Center contracted with CEM to work in conjunction with the Bureau of Energy Research of the Kentucky Department of Energy (BER/KDOE) in evaluating the potential impact of two commercial-scale coal liquefaction processes at six sites in Kentucky, as shown in Figure 2-1. [Ref. 3]

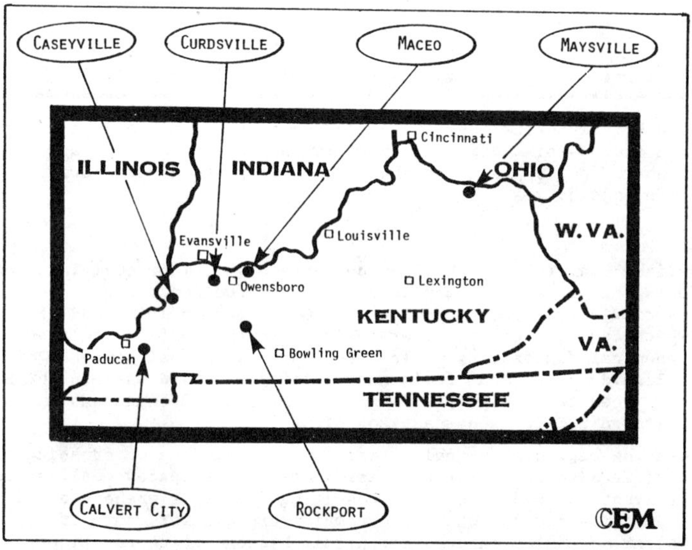

Figure 2-1. Typical locations of potential sites for coal liquefaction facilities.

The application of SELECS took place in 1980. It began with a field survey and information gathering by BER/KDOE and CEM staff, and two coordination and review meetings. Additional regional data were acquired by telephone and mail. Plant process information was provided to CEM by BER/KDOE for the two 26,000 tons per stream day coal liquefaction processes (SRC-I and Coal-to-Methanol-to-Gasoline). [Ref. 3,4,5,6,7] To provide a comparison of the socioeconomic and environmental impacts of implementing the coal liquefaction processes with facilities of equal coal consumption and number of employees, plant characteristics were also provided by BER/KDOE for a two-gigawatt coal-fired steam-electric power plant and an 849-employee auto parts manufacturing plant. Six sites were specified by BER/KDOE--five in western Kentucky and one in the northeast of the state. In that order, they are: Maceo, Curdsville, Caseyville, Calvert City, Rockport, and Maysville.

CEM prepared Process Data Sheets for the four processes, and Site Data Sheets for the seven towns and cities near the six sites, and the 25 counties constituting the regions that might be impacted: 20 counties in Kentucky, two each in Indiana and Ohio, and one in Illinois.

The SELECS Level 1 Methodology described in Reference 1 was computerized, revised, and extended to evaluate impacts in one or more towns and cities near the site, as well as the several counties that constitute the region of 30 to 40 miles radius around a site. The analysis at each of the six sites involved determining (1) socioeconomic impacts in six Impact Categories on the nearest city or town, as well as the total region, and (2) environmental impacts in five Impact Categories throughout the total region. These analyses were replicated for the four processes, thus constituting 24 individual site-and-process analyses. Detailed "SELECS Site Evaluation" reports for each of the six sites, along with an overall summary of the SELECS evaluations, are contained in Reference 3.

3.0 THE SRC-I AND COAL-TO-METHANOL-TO-GASOLINE PROCESSES

This section briefly describes some of the characteristics of the two coal liquefaction processes. Additional details for these processes and the two-gigawatt electric power plant and the automobile parts manufacturing plant are found in Reference 3.

3.1 SRC-I

Solvent Refined Coal (SRC) technology converts high sulfur coal into low sulfur, low ash fuels. It is a noncatalytic direct liquefaction route. There are two principal variations in SRC technology: SRC-I produces solid and liquid fuel products which can be used as boiler fuels or further refined to transportation fuels and chemical feedstocks, while SRC-II processes coal somewhat differently to produce liquid and gaseous fuels and feedstocks. Data used in the SELECS analyses have been taken from preliminary SRC-I plant designs and extrapolation of the results to SRC-II is not advisable.

SRC-I processing begins with coal pretreatment consisting of crushing and drying to meet particle size and moisture specifications. Prepared coal is mixed with a recycle solvent and hydrogen. The resulting slurry is preheated and delivered to the liquefaction reactor where the complex organic structure of coal is broken down to smaller molecules through exothermic hydrogenation reactions. Reactor temperature is typically between 435 and 470°C (815 to 878°F). Product slurry enters a gas separation unit which removes volatile materials. The larger organic molecules in the gaseous fraction condense upon cooling and are upgraded by hydrotreating to naphtha and fuel oil. The low molecular weight gaseous hydrocarbons are consumed in the SRC-I process for plant energy requirements.

The slurry leaving the gas separation unit contains liquid organics, mineral residue (ash) and unreacted coal. Solids and liquids are separated, then recovered solids are sent to a gasifier to generate hydrogen for the liquefaction reactions. The liquid stream enters a fractionation unit which has three product streams: the solvent fraction which is used to slurry the prepared coal feed, a wash solvent which is recycled to the solid/liquid separation unit, and the SRC-I product, which solidifies upon cooling. [Ref. 5]

A number of important auxiliary processes are associated with SRC production, including coal receiving and storage, water supply, wastewater treatment, hydrogen generation, steam/power generation (by combustion of by-product fuel gases), a cooling system, oxygen generation, acid gas removal and processing, solid waste disposal, by-product recovery, and product and by-product storage. The principal by-product is sulfur. The SELECS analyses in this paper are based on a plant with four trains similar to the single 6000 tpsd train SRC demonstration plant planned for Newman, Kentucky. [Ref. 6]

3.2 Coal-to-Methanol-to-Gasoline

Coal-to-methanol-to-gasoline is classified as an indirect liquefaction technology because the coal is first gasified and the gases are converted to liquid products. Indirect liquefaction processes are inherently less efficient that direct liquefaction processes; however, using Fischer-Tropsch synthesis, it has been practiced commercially since 1955 in South Africa. For the SELECS analyses herein, a six train facility designed for DOE by Badger Plants, Inc., has been scaled down to a two train facility converting 24,700 tpsd of coal to 55,830 barrels per stream day of gasoline.

Coal processing begins with washing and crushing to a specified size range. Prepared coal is fed to an oxygen-blown gasifier operated at 1540 to 1650°C (2800 to 3000°F) and 3.5 x 10^6Pa (500 psig). The synthesis gas produced is cleaned (acid gas removal) and shifted to the hydrogen-carbon ratio required for methanol synthesis. Methanol is formed by a catalytic reaction between carbon monoxide and hydrogen. [Ref. 7] A Mobil Oil Corporation process is used to convert methanol to a variety of organic compounds, including aromatics. The distribution of compounds and the physical properties of the mixture are similar to those of petroleum-derived gasolines. The Mobil process uses a shape-selective zeolite catalyst in the methanol to gasoline reactor. Auxiliary processes at the coal to gasoline facility would include coal receiving and storage, water supply, wastewater treatment, steam and power generation, a cooling system, oxygen generation, acid gas removal and processing, solid waste disposal, and product and by-product storage. Steam and power are produced in coal-fired boilers, and air emissions control systems would be installed.

Environmental and socioeconomic considerations in the siting of synfuel plants have been described in a paper presented at the Seventh National Conference on Energy and the Environment. [Ref. 8]

4.0 Comparison of the Characteristics of Two Selected Sites

To perform a SELECS analysis, the analyst prepares Site and Process Data Sheets. These data are then used in approximately 80 computational steps to evaluate about 30 Impact Elements. The process is described step-by-step in the User's Guide. [Ref. 2] Refinements to SELECS are found in Appendix B of Reference 3. As in any well-structured methodology for environmental impact analysis, SELECS serves as a checklist of items to be considered and as a well-organized structure for bookkeeping. Preparing site descriptions in a consistent format is part of the SELECS methodology. Two summarized examples are given below, one for the Curdsville site where adverse impacts are small, and another for the Caseyville site, where SELECS indicated there likely would be significant adverse impacts. The contrast in the socioeconomic "carrying capacity" of the two regions is self-evident from the descriptions, It is understandable that SELECS--used to screen potential sites--would place Curdsville and Caseyville at opposite ends of the spectrum of magnitude of adverse socioeconomic impacts. Complete descriptions of the characteristics of all six sites are given in the "SELECS Site Evaluation" reports in Reference 3.

4.1 Curdsville

Curdsville is a small community of about 50 homes, located in north central Kentucky, on the western border of Daviess County. The nearest significant municipality is Owensboro, 12 miles to the east. Owensboro's population in 1978 was 57,970. Population density in the region ranged from 174 residents/sq mi in all of Daviess County (population 80,300 in 1977) to 2700 residents/sq mi within Owensboro.

Daviess County occupies 462 square miles, including 21.4 square miles within the boundaries of Owensboro. Residential areas account for 2.1 percent of the county land area. Agricultural use comprises 77.5 percent and 56.8 percent is considered to be prime farmland. The total amount undeveloped (including agriculture) is 96 percent, which is similar to the surrounding counties.

Per capita income in Daviess County was $6926 in 1977. The largest employment sectors are manufacturing and wholesale and retail trade. Property taxes generated $1,417,895 in 1979 for Daviess County.

There are 191 police officers employed in all of Daviess County. The local fire departments have 119 paid employees. The 2 hospitals have 591 beds. There are presently 85 doctors and 40 dentists practicing in the county. The county school district has an enrollment of 10,891 students and employs 540 teachers. In 1970, there were 25,374 housing units in Daviess County. The average number of occupants per unit was 3.2 and 820 units were available for sale or rent.

Curdsville is located near the confluence of the Green River and Panther Creek. The area climate has wide variations in temperature and precipitation. The mean annual temperature is 56°F.

4.2 Caseyville

Caseyville, a very small community, is located in western Kentucky on the Ohio River, in Union County. The nearest significant municipality is Sturgis, 5 miles to the east. The population of Sturgis in 1976 was 2087, and has been declining. Population density in the region ranged from 53 residents/sq mi in all of Union County (population 18,000 in 1978) to 1459 residents/sq mi within Sturgis.

Union County occupies 340 square miles, including 1.43 square miles within the boundaries of Sturgis. Residential areas account for 1.86 percent of the county land area. Agricultural use comprises 93 percent and 65.8 percent is considered to be prime farmland. The total amount undeveloped (including agriculture) is 95.1 percent, which is similar to the surrounding counties.

Per capita income in Union County was $5826 in 1977. The largest employment sectors are service industries, manufacturing, wholesale and retail trade and mining. Property taxes generated only $38,247 in 1979 for Union County.

There are 18 police officers employed in all of Union County. The local fire departments have 9 paid employees. The one hospital has 54 beds. There are presently 5 doctors and 4 dentists practicing in the county. The county school district has an enrollment of 3269 students and employs 146 teachers. In 1970, there were 4909 housing units in Union County. The average number of occupants per unit was 3.7 and 191 units were available for sale or rent.

Caseyville is located in low hills near the Ohio River. The area climate has wide variations in temperature and precipitation. The mean annual temperature is 56°F.

5.0 Comparison of Socioeconomic and Environmental Impacts

The more than 30 Impact Element algorithms used in the evaluation were divided (unequally) among 11 Impact Categories:

- Population
- Community Services
- Bonding Capacity
- Housing
- Land Use
- Water Availability
- Air Quality
- Water Quality
- Endangered & Threatened Species
- Aesthetics & Important Sites
- Tax Rates

Weighted, summed, and normalized Impact Element values give Impact Category scores on a scale of 0 to 1. Using an arbitrary level of 0.2 and above as representing "significant" adverse impacts, Table 5-1 shows a matrix of the number of Impact Categories indicating significant adverse impact, as a function of process, site, and whether the impact is for the nearest town or city, or for the entire region.

TABLE 5-1

NUMBER OF SELECS IMPACT CATEGORIES WITH SIGNIFICANT ADVERSE IMPACT SCORES

Process	Impact Coverage	Site					
		Maceo	Curdsville	Casey-ville	Rockport	Calvert City	Maysville
SRC-I	Nearest Town/City	1	1	6	4	0	2
	Total Region	3	2	3	2	3	3
Coal-to Methanol to-Gasoline	Nearest Town/City	1	1	6	4	0	2
	Total Region	2	1	2	1	1	2
Coal-Fired Electric Power Plant	Nearest Town/City	1	1	4	4	0	1
	Total Region	3	2	3	2	2	3
Manurac-turing Plant	Nearest Town/City	1	0	6	3	0	1
	Total Region	2	0	2	1	1	0

Note: Impact Category score judged significant at 0.2 and above, on a scale of 0 = No Impact and 1.0 = Maximum Adverse Impact.

The table indicates that the SELECS results might be used to rank-order the six sites as follows:

Least Adverse Impacts
- Calvert City
- Curdsville

Mid-Range
- Maceo
- Maysville

Most Adverse Impacts
- Rockport
- Caseyville

Also, it appears that the SRC-I process, as characterized by the process data used, has somewhat more regional adverse impacts than the Coal-to-Methanol-to-Gasoline process. For comparative purposes,the two-megawatt power plant with its smaller work force and approximately equal coal requirements has almost equal numbers of Categories with adverse impacts as the two coal liquefaction processes. At Caseyville, the manufacturing plant, with employment comparable to the coal liquefaction plants, has numbers of Categories with adverse impacts essentially equal to that of the coal liquefaction plants. However, at the other five sites the impacts are usually less.

The above analysis is much too cursory to be used for decision making, of course. There are also the questions of determining for each site which Impact Elements exhibit significant adverse impacts, and what the actual levels of impacts are, both for Impact Elements and Categories. Since SELECS is most appropriately used for screening sites and/or processes, the step of analyzing how changes can be made in site or process characteristics to mitigate adverse impacts would also be considered. For example, the plant developers and state and federal agencies might cooperate with the nearby community agencies in planning and funding the orderly and timely growth of police, fire, medical, educational, etc. services. Arrangements might be made for property tax payments on a basis concurrent with construction. The plant construction schedule might be stretched to hold down the peak in the required construction force. Tradeoffs might be investigated to mitigate water requirements, or reduce air emissions or water effluents. In short, SELECS can and should be used at an early stage when it can serve to highlight potential significant adverse impacts, and then show the mitigating effects of changes in concept and/or design tradeoffs.

In the remainder of this section,graphical comparisons are made of magnitudes of adverse impacts for ten Impact Categories, by site and process. (The Water Quality Category is omitted because all processes assumed impoundment of processed effluent.) The discussion of impacts is extremely brief, because of length limitations on this paper.

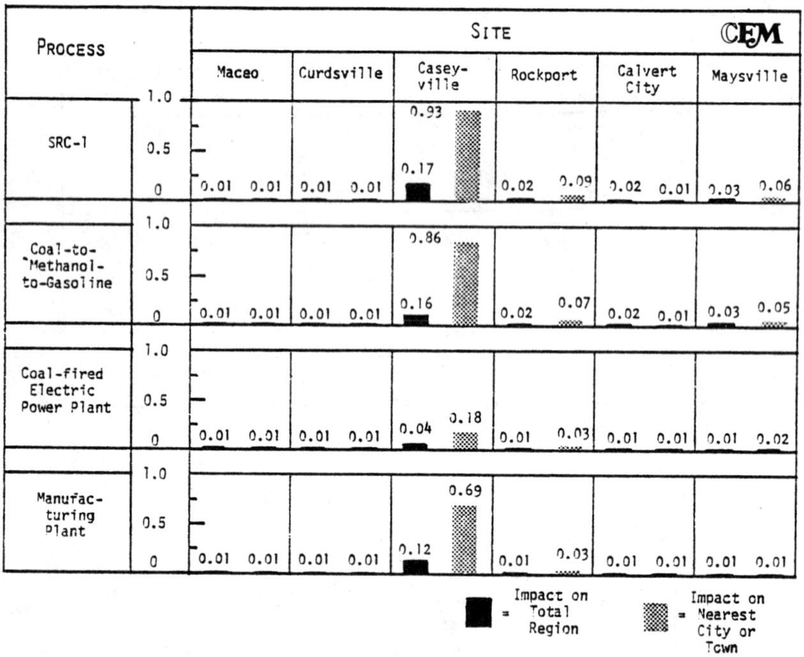

Figure 5-1. Comparison of SELECS impact scores for Category 1: Population.

For the SELECS analysis, the operational employment is: 900 for SRC-I; 798 for Coal-to-Gasoline; 470 for the electric power plant; and 849 for the manufacturing plant. Only at the Caseyville site would these employment levels cause significant socioeconomic adverse impacts.

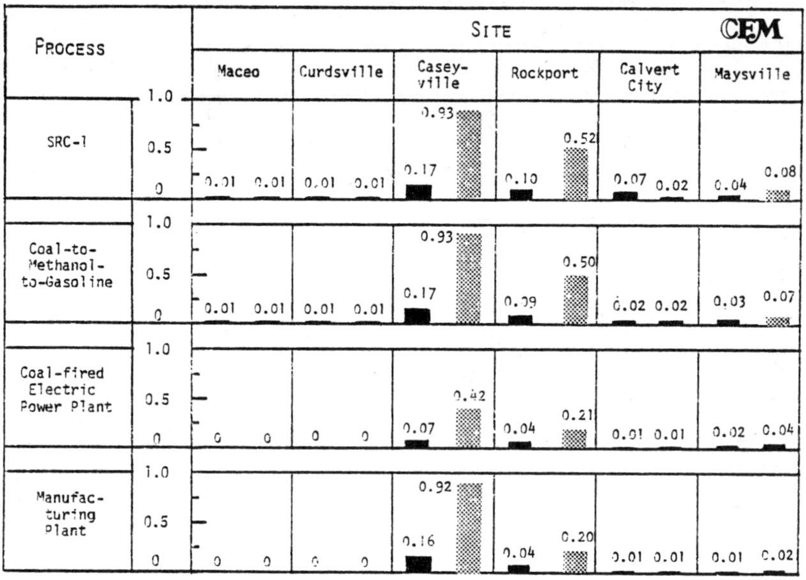

Figure 5-2. Comparison of SELECS impact scores for Category 2: Community Services.

■ = Impact on Total Region

▒ = Impact on Nearest City or Town

All sites except Caseyville and Rockport could accomodate the expansion of community services without significant adverse impacts.

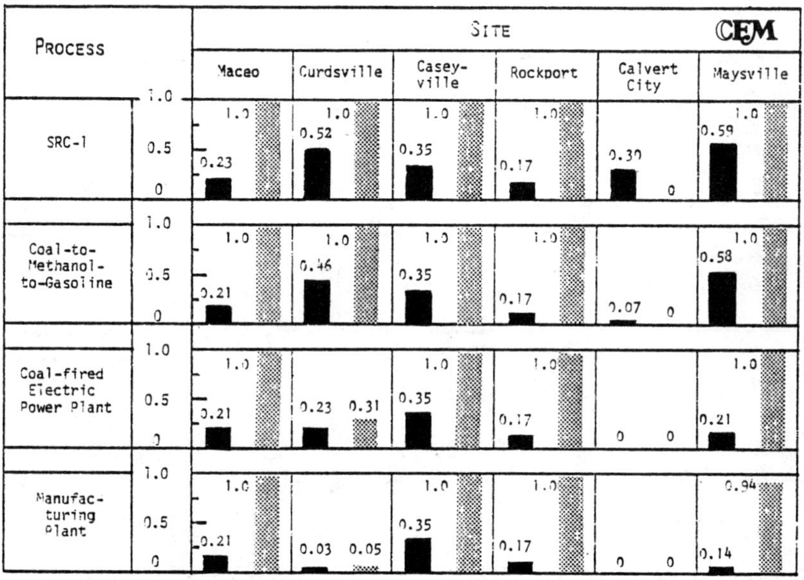

Figure 5-3. Comparison of SELECS impact scores for Category 3: Bonding Capacity.

■ = Impact on Total Region

▒ = Impact on Nearest City or Town

Bonding to pay for increased municipal or county capital expenditures will be a problem at all sites except Calvert City. Bonding capacity is set legally to be a percentage of taxable property. However, in Kentucky, a state bonding authority can issue industrial development bonds if local bonding capacity is exhausted.

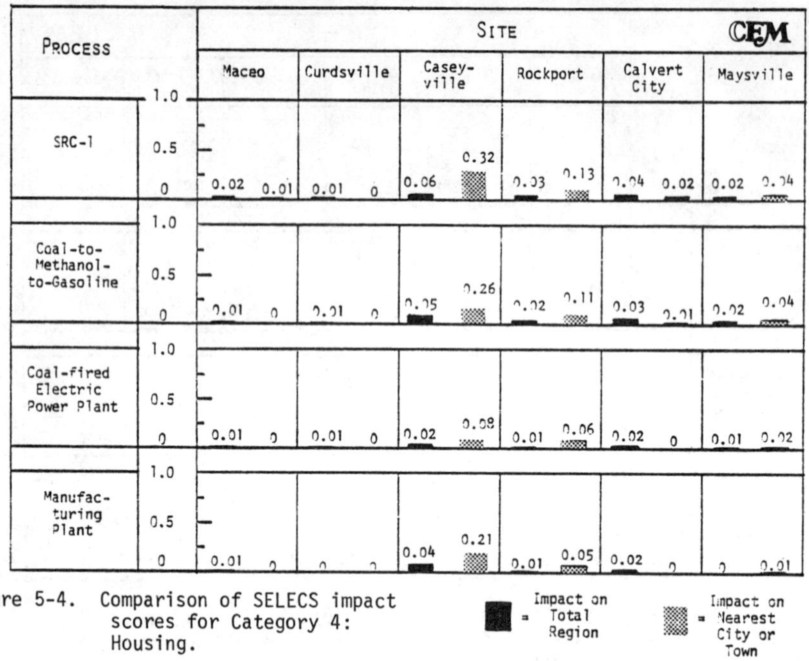

Figure 5-4. Comparison of SELECS impact scores for Category 4: Housing.

Sturgis, near the Caseyville site, would require about a 50% increase in housing units, if either coal liquefaction process were implemented. At Rockport, the increase would be about 25%. At the other sites the housing impact is relatively minor.

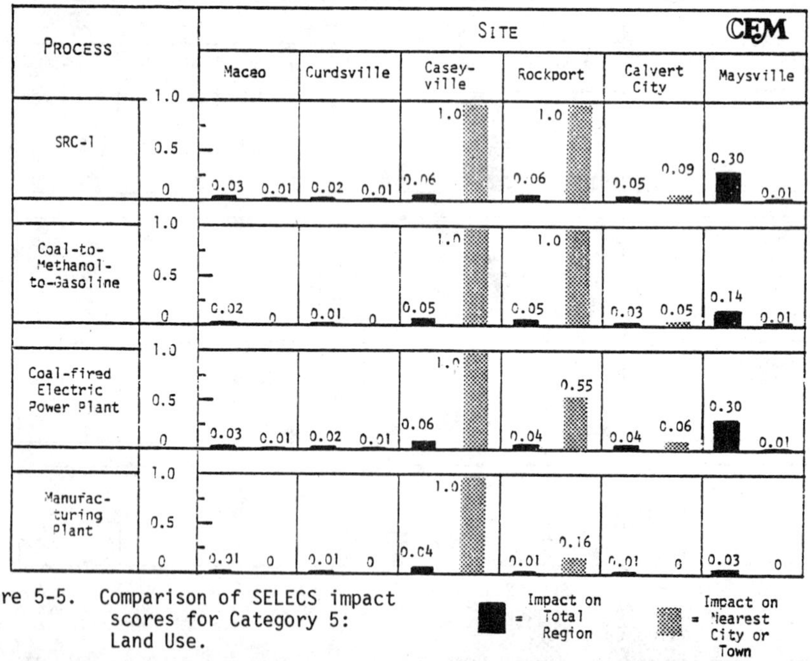

Figure 5-5. Comparison of SELECS impact scores for Category 5: Land Use.

Caseyville, Rockport and the Maysville region exhibit the most significant land use adverse impacts. Both Sturgis and Central City (serving the Caseyville and Rockport sites, respectively) would likely have to expand their present boundaries. At Maysville, a sizeable amount of prime farmland may be taken.

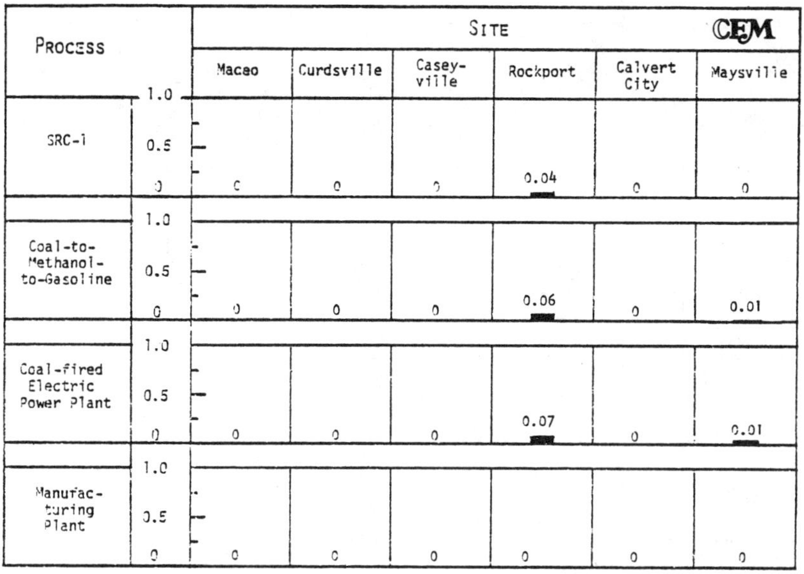

Figure 5-6. Comparison of SELECS impact scores for Category 6: Water Availability
Water availability is based on plant and residential requirements on the average and 7-day 10-year low flows of the nearby stream. Only at the Rockport site does there appear to be a problem of any significance.

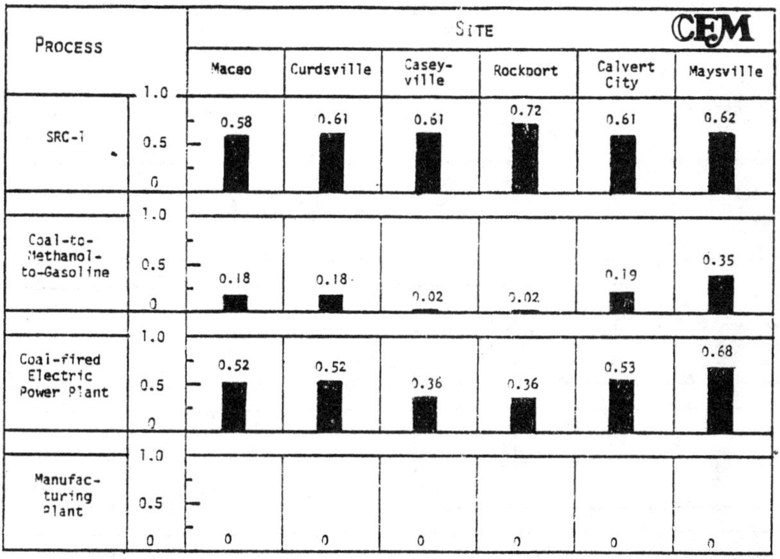

Figure 5-7. Comparison of SELECS impact scores for Category 7: Air Quality.
The coal burning facilities must meet Prevention of Significant Deterioration (PSD) standards, to limit degradation of air quality in areas where particulate or sulfur dioxide concentrations are below the National Ambient Air Quality Standards (NAAQS). Air quality monitoring indicates that the Maceo, Curdsville, Calvert City and Maysville are particulate non-attainment areas, and emission offsets may be required. Regional variations in Air Quality scores are due to applicability of the PSD standards. High scores "flag" the need to take special measures to achieve acceptable air quality, rather than a reason to reject the site.

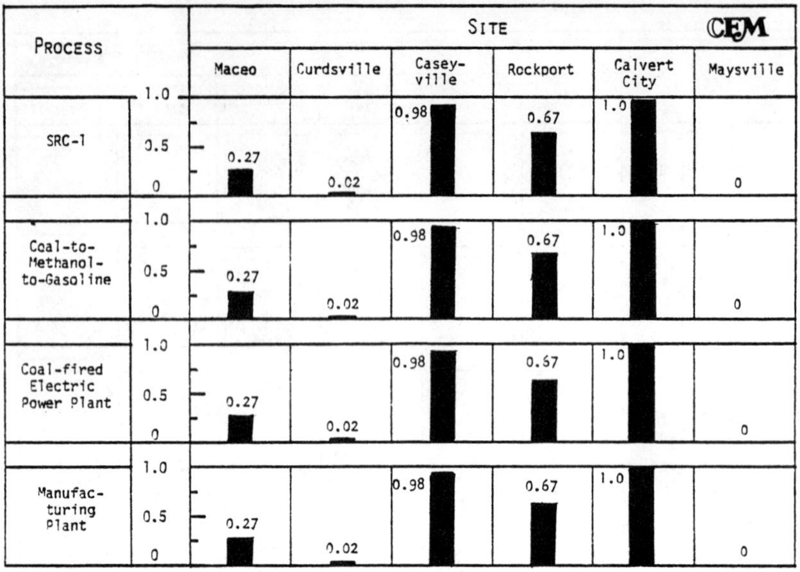

Figure 5-8. Comparison of SELECS impact scores for Category 9: Endangered and Threatened Species.

A special SELECS algorithm for Endangered and Threatened Species was designed to incorporate data provided by the Kentucky Nature Preserves Commission, which located 9 species within 6 miles of the Caseyville site; 14 species within 10 miles of the Rockport site (none closer than 6.7 miles); and 10 species within 9 miles of Calvert City. No species on the federal endangered species list have been reported within 10 miles of any site.

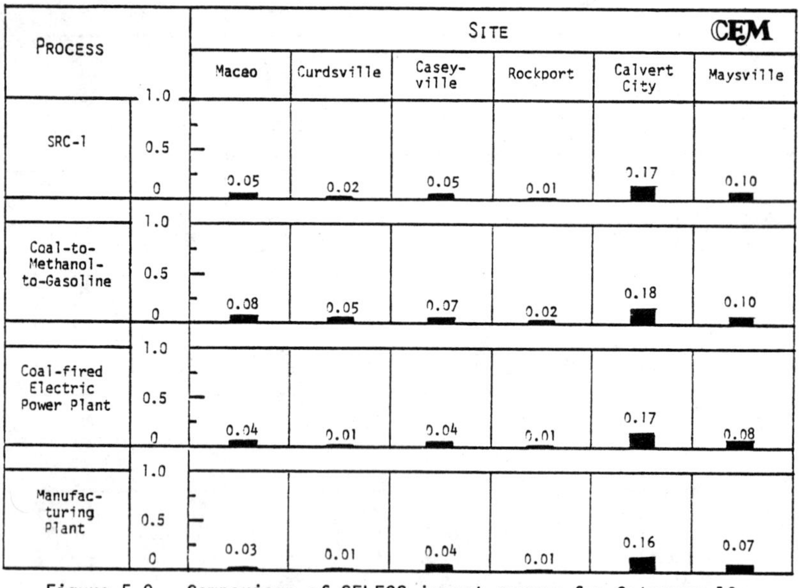

Figure 5-9. Comparison of SELECS impact scores for Category 10: Aesthetics and Important Sites.

This Impact Category considers the visual impact and proximity of the plant to sites of historical, cultural, natural beauty, or other local importance. Calvert City had the highest score because it is visually available to a large number of viewers. Building color, location, and vegetative screening can ameliorate this impact.

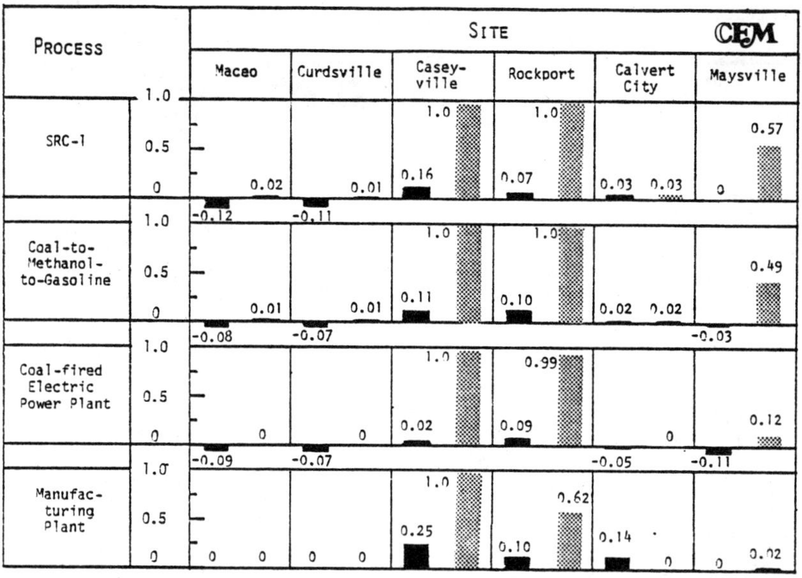

Figure 5-10. Comparison of SELECS impact scores for Category 11: Tax Rates.

■ = Impact on Total Region ▒ = Impact on Nearest City or Town

The change in average tax rates for municipalities and counties can be adverse or beneficial (indicated by negative impact score in the figure above). The county where the plant is located may derive a substantial tax base increase. Outlying counties and municipalities may not fare as well, unless special provisions are made to share tax revenues. Special federal and/or state programs to assist communities may also be available to further ameliorate adverse impacts on tax rates.

6.0 CONCLUSIONS

The SELECS methodology for determining socioeconomic and environmental impacts of synfuels plants and other facilities at potential sites has been clearly demonstrated to be comprehensive, orderly, easy to understand, straightforward to apply, and adaptable to the special considerations of the Kentucky locale.

The SELECS process has been shown to be capable of distinguishing the relative differences in impacts among the six sites and the two synfuel processes and two other facilities specified by the Bureau of Energy Research of the Kentucky Department of Energy. It has provided a comprehensive basis for narrowing down potential sites in an objective, open, and selective fashion. The potential adverse impacts of alternative synfuel processes, and other facilities of comparable dimensions, have been well delineated.

The SELECS methodology is intended as a screening tool for choosing the preferred site for a specified process, or to select from among alternative processes or process tradeoffs at a specific site. SELECS does not replace a detailed site-and-process-specific environmental impact statement (EIS) or assessment (EIA). Rather, it provides a guide for acquiring important socioeconomic and environmental data and information; a checklist of important issues to be considered; and a set of preliminary assessments and procedures which can be used as a guide during the analysis and writing of a full EIS.

SELECS provides a methodological framework within which further expansion, improvements, and refinements can take place to achieve greater comprehensiveness. For example, expansion of SELECS to encompass the construction phase of activity is expected to begin shortly. SELECS is intended for application by local, regional, state, and federal agencies, as well as industry, environmental groups, and educational institutions.

7.0 ACKNOWLEDGMENTS

More than three dozen people at CEM, BER/KDOE, PETC/U.S. DOE, Area Development Districts, Chambers of Commerce, and municipal, county, and state agencies generously offered their time and cooperation in providing the data, information, and ideas that contributed to the success of this study. The authors express their thanks and appreciation to all.

8.0 REFERENCES

1. Northrop, G.M., R.L. Scott, and C.A. D'Ambra. *SELECS: A Workable Methodology for Evaluating Socioeconomic and Environmental Impacts of Energy Conversion Facilities at Potential Sites*, Proceedings of the Sixth Energy Technology Conference, February 26-28, 1979, Washington, D.C.

2. Northrop, G.M., R.L. Scott, and C.A. D'Ambra. CEM Report 4231-625 (DOE Report No. COO-4571-1), *Environmentally-Acceptable Fossil Energy Site Evaluation and Selection: Methodology and User's Guide, Volume 1*, The Center for the Environment and Man, Inc., Hartford, Connecticut and Cameron Engineers, Inc., Denver, Colorado.

3. Northrop, G.M. and C.A. D'Ambra. CEM Report 4231-704, *Application of the SELECS Methodology to Evaluate Socioeconomic and Environmental Impacts of Commercial-Scale Coal Liquefaction Plants at Six Potential Sites in Kentucky*, The Center for the Environment and Man, Inc., Hartford, Connecticut, November 1980.

4. U.S. Department of Energy. *Draft Environmental Impact Statement: SRC-I Solvent Refined Coal Demonstration Plant* (review copy), August 1980.

5. Shields, J.J., H.J. Hopkins, E.E. Weiss and C. Thompson. *Environmental Assessment Report: Solvent Refined Coal (SRC) Systems*, EPA-600/7-79-146, Hittman Associates, June 1979.

6. Southern Company Services, Inc. *SRC-I Coal Refinery, Task 7-Environmental Analysis*, July 1979.

7. Badger Plants, Inc. *Conceptual Design of a Coal-to-Methanol-to-Gasoline Commercial Plant, Volume III, Environmental*, Second Interim Final Report, FE-2416-43 (Vol. 3), March 1979.

8. D'Ambra, C.A., and G.M. Northrop. *Selecting Sites for Synfuels Plants*, Proceedings of the Seventh National Conference on Energy and the Environment, December 1-3, 1980, Phoenix, Arizona.

---

*This paper is based on work performed under Contract DOE-EY-77-C-02-4575 for the U.S. Department of Energy. Neither the U.S. Department of Energy or the Kentucky Department of Energy, nor any of their employees, contractors, subcontractors, or their employees makes any warranty, expressed or implied, or assumes any legal liability or responsibility for the accuracy, completeness or usefulness of any information, apparatus, product or process disclosed, or represents that its use would not infringe privately owned rights. Processes and locations used in applications of the SELECS methodology contained herein are for demonstration purposes only and do not represent an endorsement.

MARKET OUTLOOK FOR LOW-AND MEDIUM-BTU COAL GASIFICATION

Michael T. Eckhart
General Electric Co.

Robert L. Church
Booz, Allen & Hamilton, Inc.

Low- and Medium-Btu gases (LBG and MBG) can be produced from coal with a wide variety of available and emerging technologies. There are three major potential markets for LBG and MBG, including industrial fuel, chemical feedstocks and electric utility applications such as repowering existing oil-fired units and new Integrated Gasifier Combined Cycle (IGCC) plants.

Coal gasification is not new. The technology was developed initially in the 1800's, and it reached peak utilization about 1920 with over 11,000 installations in operation. The ultimate market collapse occured in the late 1940's when cheap, clean natural gas swept the country.

The eventual return to coal as a mainstay of U.S. energy supply systems seems inevitable when a comparison between domestic reserves and consumption is made. Currently 74% of our consumption of energy--gaseous and liquid fuels--is supplied by just 7% of our domestic energy reserve base. Synthetic fuels provide a technological bridge between coal--constrained by end use and transportation factors--and the fuels demanded by the market. Some claim the transition to synthetic fuels is already underway. The current situation is characterized by 49 active LBG and MBG projects in the United States today:

- Twelve gasifiers operating or in the startup phase
- Thirteen DOE "PIN" studies
- Fourteen DOE synfuel studies under the two alternate fuels awards

- Seven Commercial projects
- Two DOE demonstration projects
- The TVA project in Alabama.

Low- and Medium-Btu gasification has been favored by DOE as an attractive means of reducing petroleum dependence. Thirty-five percent of the funding under the two DOE Alternate Fuels Award went to Low- and Medium-Btu gasification resulting in 47 percent of the prospective fuel production.

The prospects for Low- and Medium-Btu gasification seem always 5 to 10 years away. Over the past 20 years, synthetic fuel cost estimates have increased as fast as conventional fuels prices, seeming always to stay 50 percent higher than the competing fuels. In addition, a host of new technologies such as direct combustion of powdered coal, fluidized bed combustion, enhanced oil and gas recovery and many others are competing for many of the same technological niches. The final answer is not in yet, but there are some indications of what the answer is likely to be.

Current Activity

While most of the operational experience with Low- and Medium-Btu coal gasification is with industrial single user plants, there has been a shift in interest. Projects currently under consideration and driven by private sector money are in three markets:

- Multiple user Medium-Btu plants
- Chemical feedstock applications
- Utility Integrated Gasifier Combined Cycle (IGCC) plants.

A large portion of the government involvement, directed toward reduced dependency on foreign oil, is in repowering of oil fired electrical generation facilities and toward single user industrial facilities.

The present immaturity of the market is best demonstrated by the existence of over forty technologies for gasification of coal. A technological shakeout has only just begun, eleven technologies are currently being considered in the studies underway and of those eleven there are three attracting the most attention:

- Texaco
- Koppers-Totzek
- Lurgi.

The most important new information available is the preliminary results of DOE's "PIN". Nine of the thirteen studies have been completed and the preliminary results are enlightening. The nine studies analyzed site specific economics for eleven different installations including:

- Four utility repowering situations
- One IGCC plant
- Three Medium-Btu industrial fuel gas plants
- Two Low-Btu industrial fuel gas plants
- One chemical feed stock application.

One of the benefits of so many similar studies adjusted to a common dollar basis and reflecting similar if not identical plant scope, is the ability to test some of the common "rules-of-thumb" use by designers. Of considerable interest to most potential purchasers of coal gasificaiton technology are the prospects for economies-of-scale in plant size. Currently many designers are using an equation to predict these economies that was derived from experience of the petrochemical industry. Capital costs are predicted to follow the relationship:

$$PC_2 = PC_1 \frac{C_2}{C_1}^{.70} \qquad (1)$$

PC=Plant Cost
C =Plant Capacity.

If this relationship holds for coal gasification plants then plant costs versus plant capacity when plotted on a log-log scale should form a straight line with a slope of .70. Exhibit 1 shows the results obtained for the studies completed. The lines do appear straight, but the best estimate of the slope is .85 not .70. This indicates much less significant scale economies than many designers have been assuming in their plant design calculations.

The plausible explanation for less dramatic scale economies relates to the modular nature of proposed coal gasification plants. Most manufacturers offer only a limited number, many only a single, size gasifier. To increase the plant size, several gasifiers and gas cleanup trains are used, resulting in limited economies of scale for such components as air separation plants (for MBG) and coal storage and handling equipment.

There is considerable variation in the projection for capital costs of Low- and Medium-Btu gasification plants, as shown in Exhibit 2. Plant cost is expressed as a capacity cost (millions of dollars per billion Btu of daily capacity) and plant size is presented on a log scale. All costs are constant 1980 dollars. The data was normalized to reflect cold clean gas at the battery limits of the gasification plant. All extra systems such as gas transmission piping, combined cycle gas turbines or extra equipment such as methanol plants was removed from the capital costs. Many of these adjustments were made on the basis of engineering judgment and several feasibility studies did not provide sufficient information to allow good approximation of the cost of cold clean gas. The Mobay Chemical Corp. study and the Northern Resources study both contain capital equipment not directly related to the production of cold clean gas.

- The Mobay study contains facilities to separate Hydrogen and Carbon Monoxide from the stream of fuel gas for use as chemical feedstock.

- The Northern Resources study contains additional gas cleanup equipment to clean Hydrogen Sulfide from a stream of refinery gas that is subsequently blended with cold clean coal gas.

Even with these caveats there remains a large range in the estimates compared to the expected value shown as a solid line for Low-Btu plants and Medium-Btu plants without oxygen plants and the dashed line representing Medium-Btu plants with an

EXHIBIT 1
Plant Capital Costs

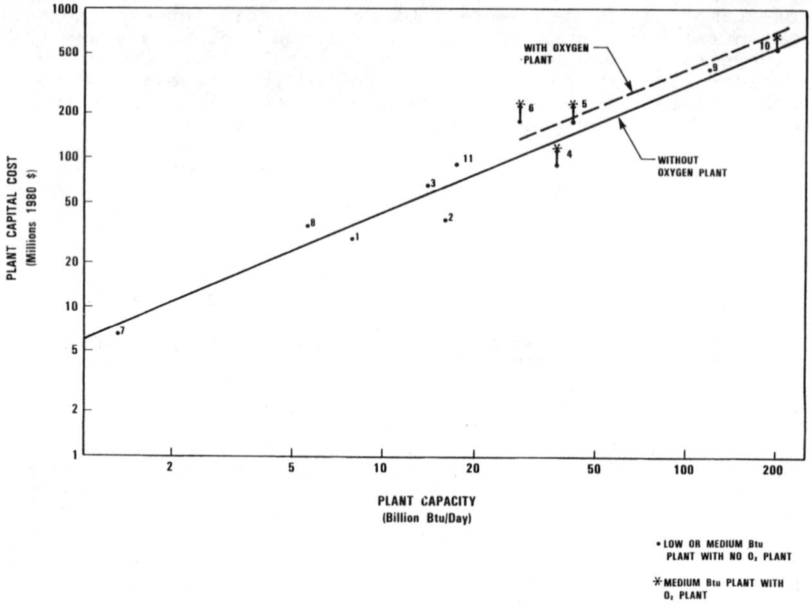

EXHIBIT 2
Plant Capacity Cost

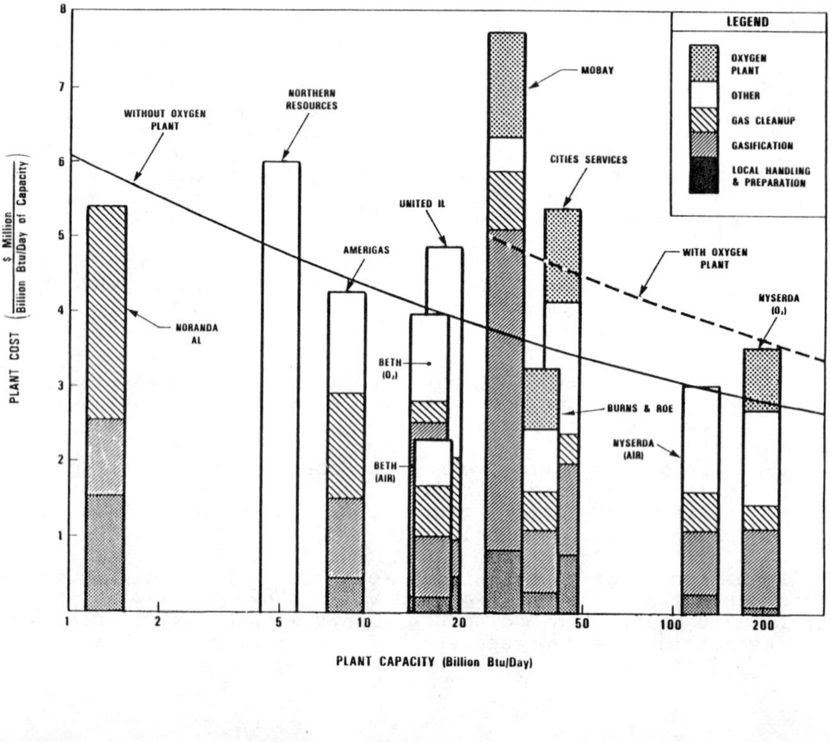

oxygen plant. These two curves were derived from the scale economies projections shown in Exhibit 1 and were converted from total plant cost to capacity cost.

Even more striking than the variation in total plant costs is the variation in major system costs. Some of the variation such as differences in coal handling and preparation can be explained by site or study specific differences:

- The Consolidated Edison study sponsored by the New York State Energy Research and Development Authority was able to use existing coal handling equipment
- The Noranda Aluminum Study allocated all cost to coal handling, gasification and cleanup with no costs in a miscellaneous category.

But many costs, such as gasifier systems, seem to reflect basic differences in the capital cost of the equipment.

These differences are made easier to assimilate by presenting the major systems as a percent of total plant cost, as in Exhibit 3. The major systems are shown here together with a band showing one standard deviation from the mean and the range encountered in the study. The oxygen plant is broken out separately because it is not included at all in Low-Btu applications and it is an option in Medium-Btu plants to purchase an oxygen plant or to buy oxygen "across-the-fence" on long term contract. The extreme variation in gasifier cost is apparent in this exhibit, ranging from 10 to 68 percent of total system cost. Despite all of the research and development that has gone into coal gasificaton systems, the gasifier remains the cost item with most variability.

Required gas selling price, shown in Exhibit 4, normalized to a consistent financial base, ranged from a low of $4.20/MM Btu for the Bethleham Steel Low-Btu application to a high of $8.60/MM Btu for the Mobay Chemical application that included chemical feed stock separation. A key to viable plant economics will be efficient utilization of by products, primarily waste heat. The available credits can offset as much as 25% of the required selling price, as shown in the exhibit.

A significant result is that capital costs are dominated by operating and coal cost, which together make up three quarters of the gas cost. The significance is that there is little opportunity to improve gasifier system economics through equipment cost reductions. In a typical case, a 20 percent reduction in the cost of the gasifier components would translate into a 1.5 percent reduction in accounted gas costs.

External Factors

The external factors that will drive the market for coal gasification are the ones that affect the price of competing fuel, residual fuel oil and natural gas. Current estimates indicate a 4 to 8 percent annual real increase in oil prices, dependent primarily on the political stability of the Middle East. Coupled with increases in world oil prices are the institutional factors--natural gas price de-regulation and fuel use regulations--and the availability of new natural gas.

All of these uncertainties combine to create a market place with high risk. The issue is not what is most likely to happen to future energy prices, but the viability of investing up to one billion dollars in a situation controlled by political, not cost, considerations.

EXHIBIT 3

MAJOR SYSTEM COST AS A PERCENT OF TOTAL INSTALLED COST

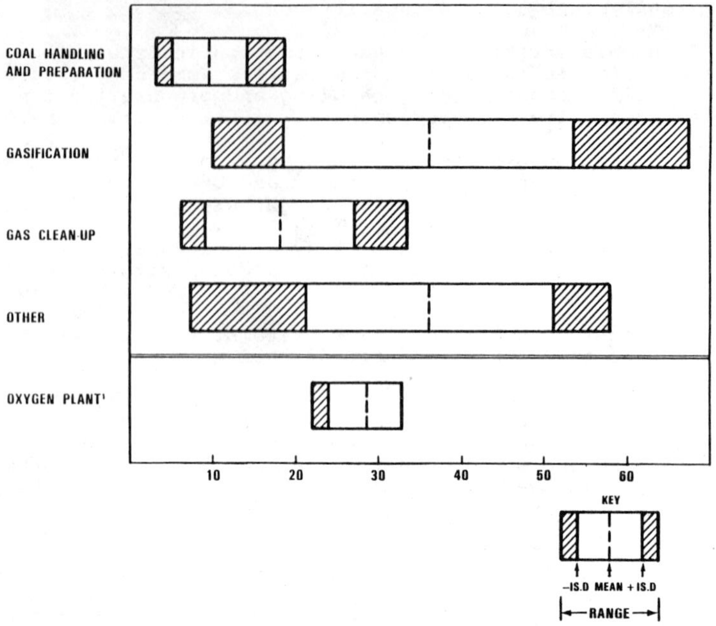

EXHIBIT 4

PRODUCT COST ($/MMBtu)

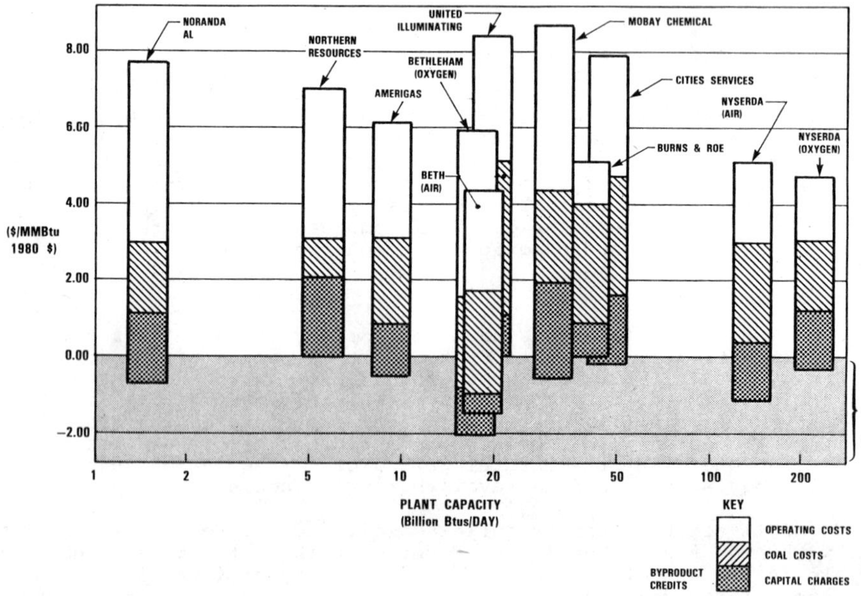

Considering the political, economic and institutional uncertainty, we believe the market drivers are the basics. The realistic markets will be the ones least affected by the political uncertainties:

- The chemical industry will be driven by the cost of feedstocks, estimated to increase by a factor of 20 to 40 between 1970 and 1990, in an environment where small differences in feedstock costs can mean the difference between profit and loss.
- For the electric utility business, the IGCC makes sense as a technology that can serve as a baseload coal unit in areas of degraded air quality.
- Also in the electric utility market are the potential repowering applications, displacing fuel oil or natural gas with coal gasification. This market is only attractive if legislative mandate forces utilities to convert.
- By increasing the capital available for projects, the SFC and DOE programs provide an opportunity for businesses that otherwise wouldn't be viable, the third party ventures.

This leaves two industries that currently are interested in coal gasification, but are not believed to be major long term participants in the Low- and Medium-Btu market:

- The gas industry will cannibalize existing products by developing Low- and Medium-Btu gasification, but is already the viable market for High-Btu gasification projects.
- Manufacturers will continue to view coal gasification as a discretionary investment competing for funds on a return on investment basis.

Summary

It is clear that a reemergence of LBG and MBG is underway, currently led by government funds. While the technological shakeout has not yet occured, private sector interest is focused on only a few of the possible technologies. The economics still appear marginal at best, and indicate that coal prices and operating costs will be key factors in early plants. The strongest long term markets for Low- and Medium-Btu coal gasification are anticipated to be chemical feedstocks and electric utility IGCC's.

APPENDIX A
FINANCIAL ASSUMPTIONS

Plant Capacity Factor	.80
Debt Fraction	40%
Construction time	3 years
Plant Life	25 years
Fuel Price Escalation Rate	11%
Coal Cost Escallation Rate	11%
Operating Cost Escallation Rate	10%
Inflation Rate	10%
Return on Equity	25%
Dept Rate	20%
Investment Tax Credit	20%

Results: Fuel prices in 1980$ assuming 1980 plant startup.

AN EVALUATION OF REGIONAL HAZE VISIBILITY IMPACTS

David Nochumson
Los Alamos National Laboratory
Los Alamos, New Mexico

ABSTRACT

Regional haze visibility impacts related to future energy development are evaluated in this paper. The visibility impacts are estimated for Air Quality Control Regions located in the continental US and are summarized by Federal region. National Energy Plan II high and low oil price scenarios are evaluated in this study to the year 2000. Visibility impacts are estimated for the scenario years of 1990 and 2000 and are compared to the visibility in the baseline year of 1975. The indices of atmospheric visual air quality that are used in this study include the total light extinction coefficient, the median visual range, and the 90th percentile visual range (poor visibility). Light-scattering by particulate sulfates, by fine particulate matter, by particulate nitrates, and by air molecules is accounted for in the calculation of the total light-extinction coefficient. The effect of relative humidity on the enhancement of light-scattering is accounted for in the calculations. The regional visibility impacts are displayed on national maps.

INTRODUCTION

Atmospheric visual air quality is an aesthetic air-quality value. The appearance of scenic vistas is affected by the quality of the atmosphere between it and the observer. Atmospheric particles and gases that are generated by man's activities can absorb and scatter light and cause a degradation of the color, texture, line, and form of a scene. The color and texture are the first to be degraded, followed by line and form. A haze caused by light-scattering by atmospheric aerosols can make a scene appear washed out. A colored plume across one's field of view can diminish one's appreciation of a scenic vista. Meteorological conditions of fog, precipitation, and high relative humidity also can impair visibility.

The US Congress, through the Clean Air Act Amendments of 1977, has declared that existing impairments to visual air quality in mandatory Class I Federal areas must be remedied and that no future impairment in these areas is to result from man-made air pollution.

The US Environmental Protection Agency (EPA) recently has published its rules to implement the Congressionally mandated visibility goal. The rules, published in the Federal Register on December 2, 1980, involve a phased approach (1). In Phase I, the control of impairment that can be traced to a single major source or a small group of sources is required. Regional hazes and urban plume problems are to be addressed in future phases. The states, the Federal land managers, and the EPA are to have important roles in implementing the rules. The states with mandatory Class I Federal areas are to have key responsibility for developing the program and making any substantive decisions. The rules will require 36 states to revise their State Implementation Plans (SIP) within 9 months of the promulgation of the regulations. The revisions will include Best Available Retrofit Technology (BART) requirements; the development of a long-term strategy (10-15 years) for making reasonable progress toward remedying existing impairment and preventing future impairment; and measures that will supplement the states' new source review program. Although BART requirements are part of the visibility rules, preliminary analyses have indicated that no existing sources will need to install additional controls under these rules (1). The states may consider energy and economic impacts when evaluating sources that have visibility impacts on integral vistas of mandatory Class I Federal areas. Integral vistas are defined in the EPA's visibility rules to mean "a view perceived from within mandatory Class I Federal area of a specific landmark or panorama located outside the boundary of the mandatory Class I Federal area."

In this study, the contribution by anthropogenic sources to regional haze was estimated. Regional haze can be caused by air contaminants transported in an air mass over distances of hundreds to thousands of kilometers. It can have a large number of emission sources contributing to it, including those from urban areas. Generally, it is vertically homogeneous to the top of the mixing layer and horizontally homogeneous over distances of hundreds of kilometers. Regional haze, however, cannot be readily traced back to its sources of origin.

This discussion includes an evaluation of potential regional visibility impacts related to the National Energy Plan (NEP) II high oil-price scenario. The NEP II low oil-price scenario was also evaluated, but there were only small differences in regional visibility impacts between the low and high scenarios. The NEP II high scenario assumes a 50% growth in primary energy supply between 1977 and 2000. The projected primary energy supply for the year 2000 for the high scenario is 117 quadrillon Btus: 34.5% from domestic oil and gas, 32.8% from coal, 14.3% from nuclear power, 8.4% from hydroelectric, solar, and geothermal power, and 10.1% from imported oil and gas. It should be emphasized that this projection is not a prediction of future energy supply and demand but is one scenario of many possible future outcomes. The estimates of potential future regional visibility impacts are not predictions of future effects but are projections based on the assumptions of the scenario, regulatory requirements, and the study methodology. The evaluation is most useful for indicating problem areas that could result from meeting future energy needs.

Anthropogenic regional visibility impacts are estimated by Air Quality Control Region (AQCR) and are summarized by Federal Region. The light-extinction coefficient, the median visual range, and the 90th percentile visual range* are used as indices of regional visual air quality. Light-scattering by anthropogenically related particulate sulfates, fine particulate matter, particulate nitrates, and

---

*The upper percentile visual range corresponds with poor visibility.

air molecules are accounted for in the calculation of the total light-extinction coefficient. The contribution by particulate nitrates was based on regional light-extinction and emission budgets. A more detailed analysis that included long-range transport calculations conducted by Brookhaven and Pacific Northwest Laboratories was used in estimating the contribution by sulfates and primary fine particulates. The effect of relative humidity on the enhancement of light-scattering also is accounted for in the calculation. The potential impacts from the major air pollution emission source categories were estimated in this study.

Sulfur oxide (SO_x) emissions from copper smelters were assumed to be effectively controlled (90% control) by 1990. Because copper smelters are a major source of SO_x emissions in the West, the extent they are controlled should have a substantive effect on regional visibility in the West. New and existing sources are assumed to meet existing EPA regulatory requirements. The methodology used in this study and the results are discussed in the following sections.

This work is part of a larger study [the Department of Energy's Regional Issue Identification and Assessment (RIIA) program] that has evaluated potential environmental, socioeconomic, and institutional impacts caused by the NEP II scenarios.

METHODOLOGY

Through a relationship proposed by Koschmieder (2), visual range for black objects can be related to the extinction coefficient b defined in the Beer-Lambert Law. The relationship also has been found to be a good approximation for calculating the visual range of dark topographical features, such as forest-covered mountains. The relationship for the "limiting distinguishable contrast" of a black object against the horizon sky for a standard observer is given by

$$V = \frac{3.912}{b} \tag{1}$$

where V is the "meteorological visual range." The value used for the limiting distinguishable contrast is 0.02 for the standard observer (3).

The extinction coefficient can be represented by the sum of the following components (4): the light-scattering coefficient caused by atmospheric particles, b_{sp}; the light-scattering coefficient caused by air molecules (Raleigh Scattering), b_{sg}; the light absorption coefficient caused by gases such as NO_2, b_{ag}; and the light absorption caused by particles such as carbon black, b_{ap}. The relationship can be represented as follows:

$$b = b_{sp} + b_{sg} + b_{ag} + b_{ap} \quad . \tag{2}$$

For nonurban areas where the Class I areas of interest are located, the anthropogenically related light extinction primarily is due to light-scattering by particles. The concentration of NO_2 tends to be too low for light absorption by gases to be important and the light absorption coefficient caused by particles tends to be an order of magnitude less than the light-scattering coefficient caused by particles.

The light-scattering coefficient for a particulate species can be calculated as a product of its scattering-to-mass ratio, b_s/m, and its mass concentration, m. Particles 0.1 to 1 μm in radius, comparable to the wavelength of light, are the most effective light scatterers per unit mass (5). Particles in the accumulation mode, formed by the chemical reaction and condensation of gases, are found in this size range (6). The bulk of the particulates in the accumulation mode are composed of sulfate, nitrate, and organics, with particulate sulfate typically comprising the largest mass fraction. Submicron particles in the nuclei mode and particles greater than 1 μm in the coarse particle mode, are not effective light scatterers per unit mass.

Hygroscopic particles will grow in size with increasing relative humidity. This growth will cause the scattering-to-mass ratio to increase and can be mathematically expressed (7-9) as follows:

$$(b_s/m)_{RH} = (b_s/m)_o \bullet [1 - (RH/100)]^n \quad , \qquad (3)$$

where

$(b_s/m)_{RH}$ = the scattering-to-mass ratio at relative humidity RH and

$(b_s/m)_o$ = the scattering-to-mass ratio at zero relative humidity.

Cass (7) has estimated values of n for Los Angeles that range from -0.53 to -1.09 for particulate sulfates and nitrates. Trijonis and Yuan (10,11) used an n value of -1 in their regression analyses. Based upon a multiple regression analysis, Nochumson (7,8) found that n = -1/2 gave the best fit to observed data for nonurban locations. An inspection of plots of the scattering coefficient vs relative humidity based on measurements in Seattle, Washington; Denver, Colorado; and Altadena, California (12), indicate that an n value of -1/2 fits the data better than a value of n = -1.

The light-extinction coefficient calculated from visual range observations has been significantly correlated with the concentrations of particulate sulfates (SO_4), with particulate nitrates (NO_3), with total suspended particulates (TSP) excluding the former two components, and with relative humidity. Regression relationships with these pollutants as independent variables have been fitted to light-extinction coefficient data (7-11). The regression coefficients can be interpreted as the light-extinction coefficient per unit mass for each pollutant.

The use of the light-extinction coefficient to characterize visual air quality has been recommended by Charlson, et al. (13). The limitations of the light-extinction coefficient for characterizing visual air quality are that: it does not account for coloration effects; the visual perception of a scene is affected by the position of the observer and the sun; it does not completely characterize the visual range of colored objects; it is assumed that atmospheric concentrations of light-scattering aerosols are spatially homogeneous; and it does not characterize the case of plume blight. The major advantages of the light-extinction coefficient are that: it is simple to calculate and interpret, the scattering portion can be directly measured, its components are additive, and its relationship to visual range has been successfully tested.

The light extinction for AQCRs is calculated as the sum of: (1) light-scattering caused by gas molecules, b_{sg}; (2) light extinction caused by the sources and light-scattering aerosols accounted for in long-range transport calculations; and (3) the remaining light extinction caused by the sources and light-scattering aerosols not accounted for in long-range transport calculations. The visibility methodology was found to perform well for the case when the light-scattering aerosol concentrations were known. The equation used for the calculation of the total light extinction coefficient, b_T, is

$$b_T = b_{sg} + \sum_{ij} b_{sij} + b_r \qquad (4)$$

where

b_{sg} = light-scattering caused by air molecules. (At sea level, $b_{sg} = 1.2 \times 10^{-4}\ m^{-1}$),

b_{sij} = light-scattering caused by aerosol i and source j in $10^{-4}\ m^{-1}$, and

b_r = the remaining light extinction caused by the sources and light-scattering aerosols that were not accounted for in the long-range transport calculations in $10^{-4}\ m^{-1}$.

The light-scattering coefficient, b_{sij}, is calculated using Eq. (3). Mass concentrations of two aerosols, particulate sulfates and primary fine particulates (PFM), are from long-range transport predictions that were supplied by Brookhaven National Laboratory (BNL) and Pacific Northwest Laboratory (PNL), respectively. The scattering-to-mass ratio for particulate sulfates typical of nonurban locations and at zero relative humidity was estimated from nonurban data using multiple linear regression analysis (8,9). A scattering-to-mass ratio reported in the literature (13) based on the review of numerous field measurement studies and engineering calculations was used as an estimate of the scattering-to-mass ratio for fine particulate matter at zero relative humidity.

The significant anthropogenic air pollutant emission sources were accounted for in the study. Particulate nitrates were not accounted for by the long-range transport analysis and were estimated from Eq. (5). Light-extinction budget calculations indicated that the major share of the light extinction in nonurban areas is due to particulate sulfates, which are formed by the oxidation in the atmosphere of sulfur dioxide (8,9,13). These results are supported by the light-extinction budgets for the Southwest and Northeast prepared by Trijonis and Yuan (10,11). The greatest uncertainty in the calculation of the total light-extinction coefficient is the contribution of natural sources. Except for meteorologic impairment of visual air quality, visual air quality as indicated by airport observations of visual range are strongly related to regional anthropogenic air pollutant emissions (14).

The remaining light extinction was calculated from the baseline light-extinction (14) and factors that account for excluded emission sources (oxides of nitrogen emission sources) and light-scattering aerosols (particulate nitrates). The baseline light extinction was calculated from median visual range statistics, using the Koschmieder relationship. The remaining light extinction was calculated as follows

$$b_r = (b_b - b_{sg}) \cdot (\sum_i F_{1ik} \cdot F_{2ik} \cdot F_{3ik}) \qquad (5)$$

where

b_b = baseline extinction in $10^{-4}\ m^{-1}$,

F_{1ik} = the fraction of the total emissions of pollutant i in region k for sources not accounted for in the long-range transport analysis (it was calculated from the air pollutant emission data for the baseline year 1975),

F_{2ik} = the fraction of the total light extinction caused by aerosol i in region k for the baseline year, and

F_{3ik} = the relative growth of total emissions of pollutant i in region k from the baseline year to a future scenario year.

A methodology has been developed to calculate upper percentile visual range from the calculated median visual range based on the analysis of historical airport visibility data (15). The upper percentile visual range was calculated as follows

$$V_p = f(p) \bullet V_{50} \tag{6}$$

where

p = percent of the time that the visual range is greater than V_p,

V_p = visual range for percentile p in miles,

V_{50} = median visual range calculated from Eqs. (1-5) in miles, and

$f(P) = V_p/V_{50}$.

Functional relationships and tabulations have been developed for f(p) by Gins, Nochumson, and Trijonis (15).

RESULTS

The results from the visibility analysis are summarized in national maps and in tables. A US map of calculated median visual range by AQCR for the baseline year of 1975 is shown in Fig. 1. The best regional visibilities are shown for large areas of the West where population and air pollutant emission densities are generally low. Visual range declines as one approaches the Pacific coast where relative humidity increases and large population centers exist. A large portion of the eastern half of the US, where population, air pollutant emission densities, and relative humidity are generally high, has a median visual range of less than 15 miles. Florida and New England are estimated to have the best regonal visibility in the eastern half of the US.

US maps of calculated percentage change in visual range from the baseline year of 1975 to the scenario years of 1990 and 2000 are shown in Figs. 2 and 3. Improvements in regional visibility are calculated in areas where a reduction in copper smelter SO_x emissions are projected (western US or areas where reductions in SO_x emissions are projected for existing coal-fired power plants to comply with SIP air pollution control requirements). Deterioration of regional visibility is projected in areas: where there is a projected switch from cleaner fossil fuels such as oil and gas to coal (Gulf Coast States); where there is a projected large growth in population such as in the Western States; and where there is a projected large increase in coal burning stationary combustion sources (southeastern and western US in particular).

The visibility calculations were summarized by Federal Region. Some of the results for Federal Regions I and VI are summarized in this paper. Federal Region I is comprised of the New England States of Maine, New Hampshire, Vermont, Massachusetts, Rhode Island, and Connecticut. This region generally has better visibility than other regions in the Northeast although it has poorer visibility than the western regions of the US. The calculated median visual range for the baseline year of 1975 varied from about 10 mi in Massachusetts

NEP II HIGH SCENARIO

MEDIAN VISUAL RANGE – 1975

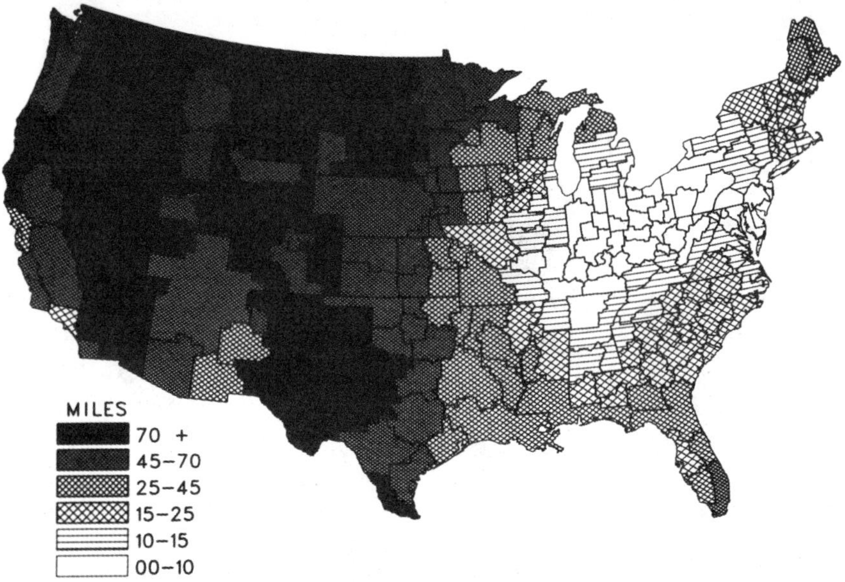

NEP II HIGH SCENARIO
PERCENT CHANGE IN VISUAL RANGE
FROM 1975 TO 1990

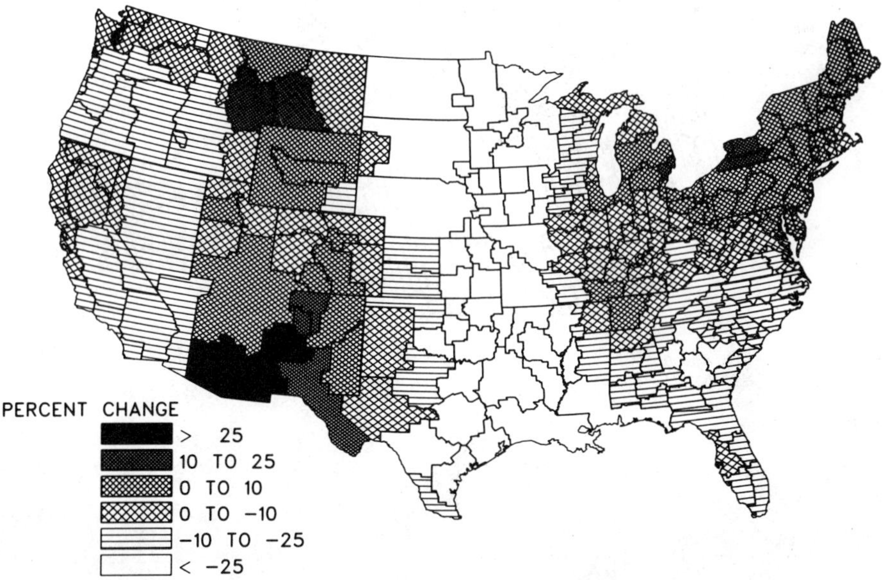

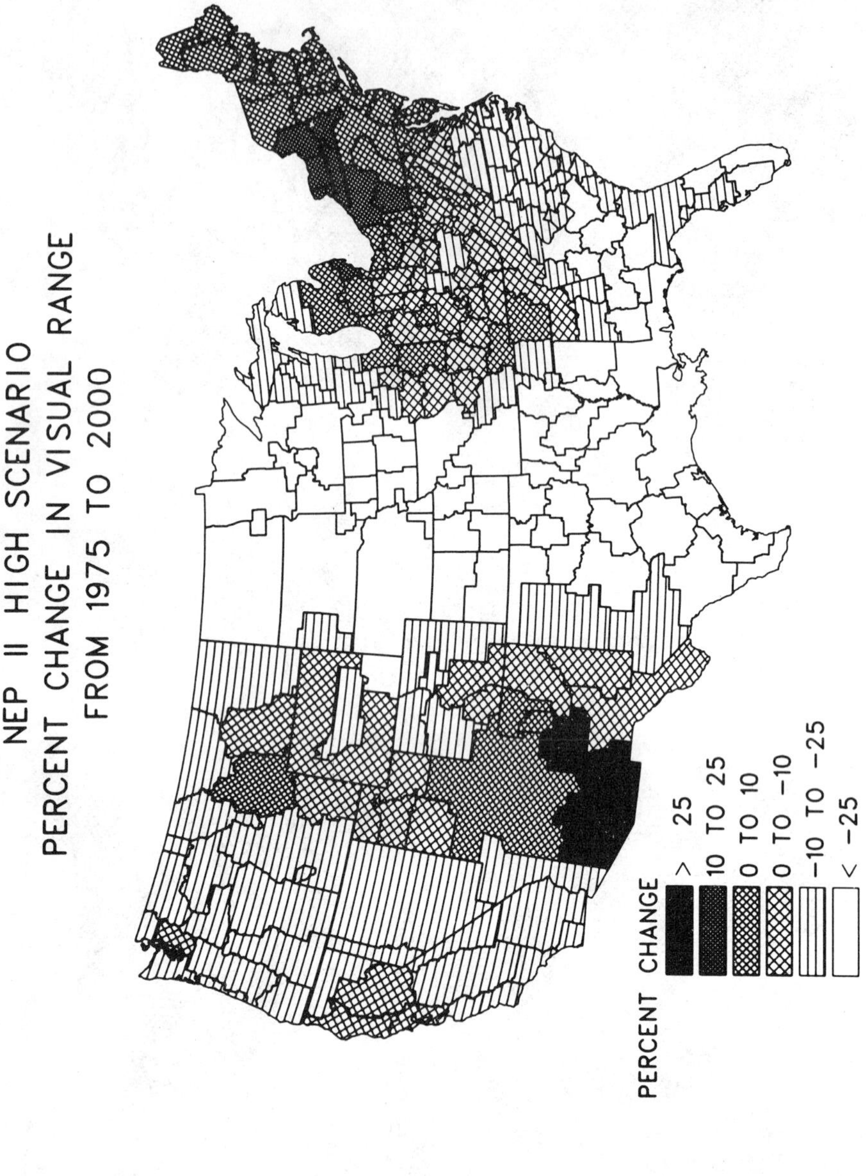
NEP II HIGH SCENARIO
PERCENT CHANGE IN VISUAL RANGE
FROM 1975 TO 2000
PERCENT CHANGE
> 25
10 TO 25
0 TO 10
0 TO −10
−10 TO −25
< −25

to about 35 mi in northern Maine. Poor visibility, which occurs 10% of the time or less (calculated for the baseline year) varied from 4 to 15 mi, again with a north-south regional gradient. The visibility averaged over the region is projected to improve 1% by 1990 and 6% by the year 2000. These projected improvements are related to the scenario's assumption that the state implementation plans in the Northeast will require a reduction in SO_x emissions from existing coal-fired power plants, particularly if the numerous Ohio River Basin plants reduce their emissions. The projected improvements in regional visibility are least in the southern portion of the region.

Table I shows that for Federal Region I coal- and oil-burning stationary fuel combustion sources and transportation sources are major contributors to the anthropogenic aerosols (sulfate, nitrates, and primary fine particulates) that impair atmospheric visibility. Stationary combustion sources that burn coal are estimated to be the largest and a growing contributor to the regional anthropogenic light-scattering aerosol in almost all regions of the US. The contribution from oil combustion to the regional light-extinction budget is higher for Federal Region I than for other regions. Except for the projected reduced contribution to the light-scattering aerosol by oil combustion and residential/commercial sources, there is no significant change in the major source categories for the region. The reduced contribution by oil combustion sources is also projected for almost all other regions. For the baseline and scenario years and for almost all regions of the US, sulfates were calculated to be the major contributor to the anthropogenic component of the regional light-scattering aerosol (followed by primary fine particulates and nitrates).

Federal Region VI is comprised of New Mexico, Oklahoma, Arkansas, Texas, and Louisiana. Visibility in the region varies from fair to excellent. The calculated median visual range for the baseline year of 1975 ranges from over 70 mi in the western portion of the region to just over 17 mi in the eastern portion of the region. Poor visibilites that occur 10% of the time or less (calculated for the baseline year) varied from just over 7 mi to nearly 37 mi with the same west-east spatial gradient as for the calculated median visual range. The visibility averaged over the region is projected to decline 21% by 1990 and 30.9% by 2000. Deterioration of regional visibility of at least 50% is projected for some AQCRs located in the hazier eastern portion. Because of the projected cleanup of the SO_x emissions from the copper smelters located in southern Arizona and New Mexico, a dramatic improvement in regional visibility is projected for most of New Mexico, particularly in the vicinity of the copper smelters. The dramatic deterioration of regional visibility projected for the eastern portion of the region is related to the projected switching from cleaner fossil fuels, oil and gas, for stationary fuel combustion, to coal, a more polluting fossil fuel.

Table II shows that for the baseline year in Federal Region VI, utility sources, copper smelters, gas combustion, transportation, and other sources are calculated to be major contributors to the anthropogenic aerosols. Although the contribution by copper smelters and gas combustion is much higher for this region than for most other regions, the contribution by utility sources and coal combustion is much lower. Growth is projected in the relative contribution to the light-scattering aerosol for utility and industrial sources and coal combustion, and a decline is projected for transportation sources, gas combustion, and copper smelters. For the baseline and scenario years, sulfates are calculated to be the major contributor to the anthropogenic component of the light-scattering aerosol, followed by nitrates and primary fine particulates. Nitrates comprise a larger fraction of the budget of extra light extinction than in most other regions. Sulfates are projected for

TABLE I
NEP II HIGH SCENARIO FOR FEDERAL REGION I
LIGHT-EXTINCTION BUDGET BY SOURCE CATEGORY AND FUEL
(% of Total Extra Light Extinction)

	Sulfates			Nitrates			Primary Fine Particulates			All Pollutants		
Source Category	1975	1990	2000	1975	1990	2000	1975	1990	2000	1975	1990	2000
Utility	59.6	62.8	61.6	2.7	3.3	3.0	1.8	3.0	2.7	64.2	69.1	67.3
Industrial	8.2	10.3	9.0	0.4	0.8	0.8	0.6	0.6	0.2	9.2	11.7	10.0
Residential/Commercial	7.0	3.4	2.7	1.0	0.7	0.6	2.3	1.5	1.3	10.4	5.6	4.6
Transportation	1.1	1.9	2.8	3.2	2.2	2.8	3.1	3.3	4.9	7.3	7.4	10.5
Fuels Extraction/ Processing/ Distribution	0.0	0.0	0.0	0.0	0.0	0.0	0.3	0.2	0.2	0.3	0.3	0.3
Copper Smelters	0.1	0.1	0.1	0.0	0.0	0.0	0.0	0.0	0.0	0.1	0.1	0.1
Industrial Process/ Stone and Clay	0.0	0.0	0.0	0.0	0.0	0.0	2.0	1.1	1.5	2.0	1.1	1.5
Industrial Process/Other	0.0	0.0	0.0	0.0	0.0	0.0	3.3	1.5	2.0	3.3	1.5	2.0
Other	3.1	3.0	3.6	0.0	0.0	0.1	0.1	0.1	0.1	3.2	3.2	3.8
ALL	79.2	81.6	79.8	7.4	7.0	7.2	13.5	11.5	12.9	100.0	100.0	100.0
Stationary Fuel Combustion Sources/Fuel Type												
Coal	61.0	60.3	61.5	0.6	1.5	1.6	1.4	0.8	0.9	62.9	62.6	64.0
Oil	13.9	16.1	10.7	3.4	2.7	1.9	3.2	4.0	2.8	20.4	22.8	15.3
Gas	0.0	0.0	0.0	0.2	0.4	0.5	0.2	0.3	0.5	0.4	0.7	1.0
ALL	74.9	76.5	73.3	4.1	4.7	4.4	4.7	5.1	4.2	83.8	86.4	81.9

TABLE II
NEP II HIGH SCENARIO FOR FEDERAL REGION VI
LIGHT-EXTINCTION BUDGET BY SOURCE CATEGORY AND FUEL
(% of Total Extra Extinction)

	Sulfates			Nitrates			Primary Fine Particulates			All Pollutants		
Source Category	1975	1990	2000	1975	1990	2000	1975	1990	2000	1975	1990	2000
Utility	18.2	29.5	27.5	9.5	10.2	9.2	1.2	2.2	1.9	28.9	41.9	38.6
Industrial	3.5	16.4	20.8	4.0	6.7	7.4	0.6	0.6	0.3	8.1	23.7	28.5
Residential/Commercial	1.8	0.7	0.4	0.6	0.4	0.4	0.7	0.3	0.2	3.2	1.4	1.0
Transportation	1.5	1.8	2.1	8.2	3.8	3.7	3.4	2.5	2.9	13.1	8.1	8.8
Fuels Extraction/ Processing/ Distribution	0.0	0.0	0.0	3.4	2.1	1.4	1.6	1.0	0.8	4.9	3.1	2.3
Copper Smelters	22.9	7.2	6.8	0.0	0.0	0.0	0.0	0.0	0.0	22.9	7.2	6.8
Industrial Process/ Stone and Clay	0.0	0.0	0.0	0.0	0.0	0.0	2.5	1.0	1.1	2.5	1.0	1.1
Industrial Process/Other	0.0	0.0	0.0	0.0	0.0	0.0	3.8	4.2	4.9	3.8	4.2	4.9
Other	11.8	8.7	7.3	0.4	0.5	0.7	0.4	0.2	0.1	12.6	9.4	8.1
ALL	59.8	64.3	64.9	26.0	23.7	22.8	14.1	12.0	12.3	100.0	100.0	100.0
Stationary Fuel Combustion Sources/Fuel Type												
Coal	18.2	41.0	44.2	1.4	9.2	12.0	0.9	1.5	1.4	20.6	51.7	57.6
Oil	5.2	5.5	2.9	1.0	2.2	1.4	1.0	1.2	0.7	7.1	8.9	5.0
Gas	0.2	0.1	0.0	11.7	5.9	3.5	0.7	0.4	0.3	12.5	6.4	3.8
ALL	23.6	46.5	48.7	14.1	17.3	16.9	2.5	3.1	2.5	40.2	66.9	68.1

the scenario to increase in importance in the regional light-extinction budget by 1990 and 2000, whereas nitrates and primary fine particulates are projected to decrease in importance.

Two of the 156 mandatory Class I Federal Areas that are well known, Grand Canyon National Park and Yellowstone National Park (which are given protection against visibility impairment by Congress under the 1977 Clean Air Act Amendments), were examined for potential scenario-related regional visibility impacts. In EPA's visibility rules (1), "visibility impairment" is defined as

> "any humanly perceptible change in visibility (visual range, contrast, coloration) from that which would have existed under natural conditions" and "significant impairment" is defined as "visibility impairment which, in the judgement of the Administrator, interferes with the management, protection, preservation or enjoyment of the visitor's visual experience of the mandatory Class I area. This determination must be made on a case-by-case basis taking into account the geographic extent, intensity, duration, frequency and time of the visibility impairment, and how these factors correlate with (1) times of visitor use of the mandatory Class I Federal area, and (2) the frequency and timing of natural conditions that reduce visibility."

A great deal of uncertainty remains concerning the quantitative criteria for determining visibility impairment.

In a recent study, Trijonis (16) attempted to give estimates of thresholds of perception for contrast based on information from a literature review. The estimates were given for Case A, where the observer is in a plume or haze layer and for Case B, where the observer is outside of the plume or haze layer and viewing the haze against a cleaner background. The plume or haze is expected to be more perceptible in Case B where there is a simultaneous comparison of the plume or haze against the background. Trijonis also related the change in contrast (Δc) to a corresponding change in light extinction and visual range based on the Koschmieder relationship (16). The derived relationship is dependent on the distance of the observer to the target relative to the visual range. The estimates of the contrast thresholds and the corresponding estimated perceptible change in visual range are presented in Table III. Two values of the latter are given, one at a point of maximal impact and a second for an average contrast change integrated over the visual range. It should be emphasized that there is no consensus concerning acceptable thresholds of perception and that those discussed are inadequate for addressing the discoloration problem.

The contrast threshold approach is used to examine perceptible changes in regional visual range from the baseline year to the scenario years of 1990 and 2000 for the Grand Canyon and Yellowstone. Comparisons are made under the assumption that the shape of the frequency distribution of visibility observations for the two Class I areas does not change over time. The Grand Canyon has a calculated baseline median visual range of 65.0 mi and calculated percent changes of +8.4 and +3.4 from the baseline year to the scenario years, respectively. Therefore, there is a projected scenario-related improvement in visibility relative to the scenario year. The +8.4% change would be perceptible near the point of maximal impact for Case B and for the probably perceptible criteria. The projected improvement could change into a deterioration if the SO_x emissions from the copper smelters located in southren Arizona and New Mexico are not effectively controlled by 1990 as assumed in the scenario. Yellowstone has a calculated baseline median visual range of 70.6 mi and calculated percent changes of +1.1 and -9.9 from the baseline year to the scenario years,

TABLE III
THRESHOLDS OF PERCEPTION

	Case A Observer In Plume or Haze			Case B Observer Outside Plume or Haze		
		Percent Change In Visual Range			Percent Change In Visual Range	
	ΔC	Point of Maximal Impact	Case of Average Contract Change	ΔC	Point of Maximal Impact	Case of Average Contrast Change
Possibly Perceptible	0.02	5.4	8.7	0.01	2.7	4.3
Probably Perceptible	0.04	10.9	17.3	0.02	5.4	8.7
Definitely Perceptible	0.06	16.3	26.0	0.04	10.9	17.3

respectively. For the probably perceptible criteria, the -9.9% change would be perceptible for the Case B observer but not for the Case A observer.

By using the contrast threshold approach with the same assumptions, one could identify the natural visibility levels that the estimated 1975 baseline median visual ranges for the Grand Canyon and Yellowstone would be perceptibly different from the natural levels. Under the probably perceptible criteria at the point of maximal impact, the baseline levels would be perceptibly different at natural median levels of 72.1 and 68.5 mi for Case A and 78.3 and 74.4 mi for Case B for the Grand Canyon and Yellowstone National Parks, respectively. Although there is no consensus concerning what the existing visibility levels or the natural levels are for the two national parks, it is likely that the calculated baseline median visual ranges are perceptibly different from natural levels.

SUMMARY AND CONCLUSIONS

NEP II has been evaluated for potential impacts on regional visibility. The major anthropogenic air pollutant emission source categories have been accounted for in this study. The synergistic effect of relative humidity on the enhancement of light-scattering by aerosols has been accounted for in the calculations. The estimates of potential future regional visibility impacts are not predictions of future effects but are projections based on the assumptions of the scenario, regulatory requirements, and the methodology. Primary and secondary light-scattering aerosols transported over long distances can substantially contribute to regional visibility degradation that appears as large-scale regional hazes.

The best regional visibilities are found in the West where population and air pollutant emission densities are generally low. Visual range declines as one approaches the Pacific Coast where relative humidity increases and large population centers exist. A large portion of the eastern half of the US, where population, air pollutant emission densities, and relative humidity are generally high, has a median visual range of less than 15 mi. Florida and the New England States are estimated to have the best regional visibility in the eastern half of the US. Improvements in regional visibility are calculated in areas where copper smelters are projected to reduce their SO_x emissions (the western US) or in areas where existing coal-fired power plants are projected to reduce their SO_x emissions to comply with SIP air pollution control requirements. Deterioration of regional visibility is projected in areas: where fuel switching from the cleaner fossil fuels, oil and gas, to coal is projected (Gulf Coast States); where a large growth in population is projected (in the Western States, for example); and where a large increase in coal burning stationary combustion sources is projected (the southeastern and western US in particular).

Fuel combustion sources, particularly coal combustion sources, are calculated to be a major contributor to the anthropogenic aerosols that can impair atmospheric visibility. Coal combustion sources are projected to become a more important contributor in the future. Copper smelters are calculated to be a major contributor to the anthropogenic regional light-extinction budgets in the West. Oil combustion sources are estimated to be a major contributor but are projected to decline in relative importance in the future. Natural gas combustion sources are estimated to be an important contributor to the particulate nitrate light-scattering aerosol in the Gulf Coast States but are also projected to decline in relative importance in the future. Transportation sources are calculated to be relatively significant contributors to regional light-extinction budgets in several regions of the US. Particulate sulfates are calculated to be a major contributor in all regions to the anthropogenic component of the regional light-scattering aerosols. Primary fine particulates and particulate nitrates are calcuated to be smaller but significant contributors.

The integral vistas of Class I Federal areas were given protection against visibility impairment in the 1977 Clean Air Act Amendments. There is no consensus on quantitative criteria for determining visibility impairment for Class I Federal areas. The contrast threshold approach may serve as an operationally useful approach for defining visibility impairment although this approach is inadequate for addressing the discoloration problem. This approach has been used to evaluate perceptible differences in calculated visibility for Grand Canyon and Yellowstone National Parks. In the EPA visibility rules, visibility impairment for Class I Federal areas was defined relative to natural levels. Because neither existing nor natural visibility levels are currently well defined for Class I Federal areas, this uncertainty introduces significant additional uncertainty in determining cases of visibility impairment in these areas. The EPA and the National Park Service currently have a joint program (Visibility Investigative Experiment in the West) to measure visibilities in parks and wilderness areas in the West.

ACKNOWLEDGMENTS

This paper is based on work performed under contract for the US Department of Energy, Office of the Assistant Secretary for the Environment, Office of Environmental Assessments, Regional Impacts Division. I would like to thank Katie Greiner and Flavio Gurule for generating the national visibility maps, Michelle Jewell and Janice Stelzer for typing the manuscript, and Lidia Morales for editing the manuscript.

REFERENCES

1. Federal Register, Vol. 45, No. 233 (Tuesday, December 2, 1980).

2. W. E. K. Middleton, Vision Through the Atmosphere (University of Toronto Press, 1952), p. 61.

3. W. E. K. Middleton, Vision Through the Atmosphere, p. 105.

4. R. J. Charlson, "Atmospheric Visibility Related to Aerosol Mass Concentration--A Review," Environ. Sci. Technol. 3, 913 (1969).

5. W. H. White and P. T. Roberts, "On the Nature and Origins of Visibility Reducing Aerosols in the Los Angeles Air Basin," Atmos. Environ. 11, 803 (1977).

6. S. K. Friedlander, Smoke, Dust, and Haze (John Wiley and Sons, New York, 1977).

7. G. R. Cass, "The Relationship Between Sulfate Air Quality and Visibility in Los Angeles", Cal Tech Environmental Quality Laboratory Memorandum No. 18, Pasadena, California (1976).

8. D. H. Nochumson, M. Wecksung, and F. Gurule, "Potential Future Impacts on Visual Air Quality for Class I Areas," Proc. National Conference on Applied Techniques for Analysis and Management of the Visual Resource, Incline Village, Nevada, April 23-25, 1979, LA-UR-79-1141.

9. D. H. Nochumson, M. Wecksung, and F. Gurule, "National Impacts on Visual Air Quality from a Future Energy Scenario," Proc. View on Visibility--Regulatory and Scientific, Air Pollution Control Association, Denver, Colorado, November 26-28, 1979, LA-UR-79-3214.

10. J. Trijonis and K. Yuan, "Visibility in the Southwest: An Exploration of the Historical Data Base," Technology Service Corporation, Santa Monica, California (1977).

11. J. Trijonis and K. Yuan "Visibility in the Northeast: Visibility Trends and Visibility/Pollutant Relationships," Technology Service Corporation, Santa Monica, California (1978).

12. D. S. Covert, R. J. Charlson, and N. C. Ahlquist, "A Study of the Relationship of Chemical Composition and Humidity to Light Scattering Aerosols", J. Appl. Meteorol. 11, 968-976 (1972).

13. R. J. Charlson, A. P. Waggoner, and J. F. Thielke, "Visibility Protection for Class I Areas: The Technical Basis," prepared for the Council on Environmental Quality, Washington, DC (1978).

14. J. Trijonis and D. Shapland, "Existing Visibility Levels in the US," Technology Service Corporation, Santa Fe, New Mexico (1978).

15. J. Gins, D. H. Nochumson, and J. Trijonis, "Statistical Relationship Between Median Visibility and Conditions of Worst Case Man-Made Impact on Visibility," Proc. Symp. Plumes and Visibility--Measurements and Model Components, Grand Canyon, Arizona, November 10-14, 1980, LA-UR-80-1605.

16. J. Trijonis, "Thresholds of Visibility Impairment," prepared by Santa Fe Research for Los Alamos National Laboratory, unpublished data, August 1980.

ACID RAIN--UNCERTAINTIES IN SCIENTIFIC KNOWLEDGE

Ralph M. Perhac
Director, Environmental Assessment Department
Electric Power Research Institute
Palo Alto, California

March, 1980

ABSTRACT

Data from the eastern U.S.A. on changes in precipitation acidity over the past few decades are conflicting. Some data sets indicate that rain is becoming more acidic; others show no change. In particular, data from stations which have been in continuous operation for a number of years in New Hampshire, New York, Pennsylvania and the Tennessee Valley give no indication of change. We still do not know what changes may have taken place over the past 20 to 30 years.

We know that acidic rain is essentially a dilute solution of sulfuric and nitric acids; however, we do not understand the reaction mechanisms by which rain drops get their acidity. Nor do we understand the chemical reactions which convert primary emissions from coal-burning power plants into secondary sulfates and nitrates. Finally, we cannot assess the importance of atmospheric long-range transport in defining ambient levels of pollutants at hundreds of kilometers from a source. Because of the many gaps in scientific knowledge, no reliable models have been developed which relate the acidity of rain to emissions, hence we cannot judge what effect an increase or decrease in emissions from coal-burning facilities will have on the deposition of acidic material from the atmosphere. That coal burning by power plants is a major cause of acid rain is a hypothesis to be tested.

The present concern over acid rain focuses on three claims: (1) that acid rain, at least in the eastern United States, is increasing--both in acidity and in area being affected; (2) that acidic precipitation is causing severe damage to ecosystems; and (3) that coal burning, particularly by utility companies, is the major cause of increasing acidic precipitation. These three claims should be viewed as just that--claims. They are not facts; they are hypotheses to be tested. This paper will concern itself with only the first and last of the three claims. It will not consider the biological aspects of the subject, only the physical and chemical.

Although acid rain is falling in the East (Fig. 1), we cannot be certain about what changes may have occurred over the past few decades. Specifically, we do not know the extent to which precipitation has become more acidic. The most reliable way of establishing changes in acidity is to examine precipitation data which were collected over a number of years. Such data as are available, however, are confusing and conflicting.

Most of the claims for increasing acidity are based on maps published by Likens (1976) and Cogbill and Likens (1974). For three time periods (1955/56, 1965/66, 1972/73), these maps show, by use of pH contours, both an increase in the acidity of the rain and in the area being affected by such rain. Drawing such conclusions, however, involves comparing data from one set of stations for one time period with data from a different set of stations for a different time period. The reliability of making comparisons when different networks are involved is subject to question. Using all the data from all the maps, one must draw the conclusion that acidity is increasing. If, however, one compares data only from stations which are common to different time periods, the conclusion is not so obvious.

The original data which were the basis for the Likens (1976) and Cogbill and Likens (1974) studies also involve comparing different sampling procedures (e.g., type of collector used) and different techniques for determining acidity. The 1955/56 study, for example, arrives at pH by a calculated technique, i.e., calculated (H^+) from a chemical analysis of the rain. In contrast, the 1972/73 data are based on measured, not calculated pH values. Differences in collection technique and in acidity determinations can yield data which are not comparable (Miller and Everett, 1979). For example, a study of 32 World Meteorologial Organization samples of rain, for which both calculated and measured pHs have been published, show an average difference between the two techniques of about 0.9 ± 0.6 pH units (Tyree, 1980), with no consistency in direction.

Obviously, a sound way of establishing trends is to have data over a number of years, collected at the same sites(s) with the same techniques. A few such data exist. Likens and others (1977) have maintained a station at Hubbard Brook, New Hampshire (Fig. 2). For at least a 10-year period (1964-1974), no overall trend in rain acidity was statistically significant (Likens et al., 1977, p. 36). And the U.S. Geological Survey has operated nine stations in the Northeast since 1965 (Fig. 3). The data from the stations show no significant change in acidity (Fig. 4-6). Finally, the Tennessee Valley Authority (TVA) has collected precipitation at a number of sites in its operating territory. Similarly, the TVA data show no consistent change in acidity for the 9-year period 1971-1980 (TVA, 1980).

Taken as a whole, the Likens and Coghill maps (1974, 1976) show an increase in acidity. Data from stations where measurements have been made continuously over a number of years, however, do not. The data are confusing and conflicting. We have no firm scientific basis for defining rain acidity trends over the past few decades. Even EPA

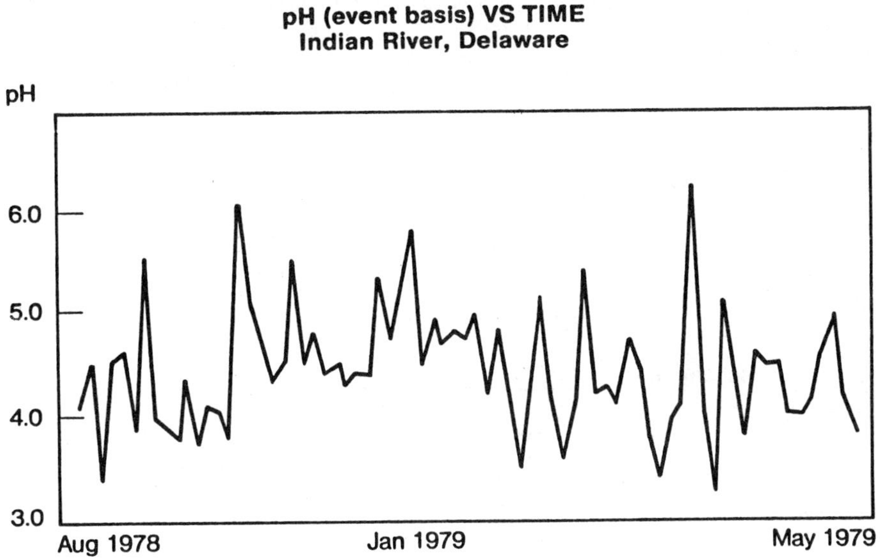

FIGURE 1

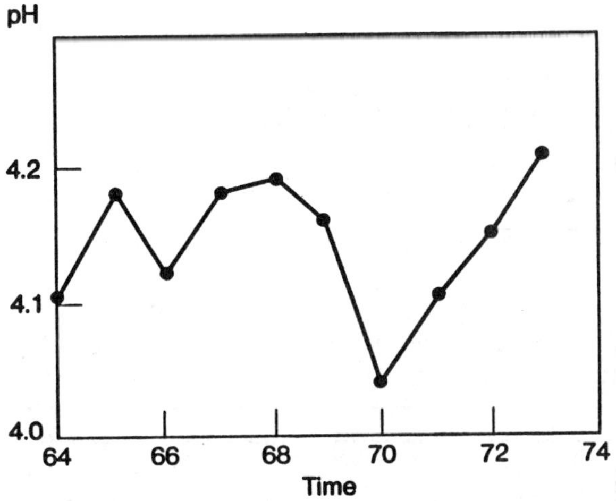

"There has been a slight upward trend in . . . concentration of hydrogen ion between 1964–1965 and 1970–1971, followed by a downward trend until 1973–1974; overall (1964–1974), however, no trend in concentration is statistically significant."

Liken, G. E. et al., 1977, *Biogeochemistry of a Forested Ecosystem*, Spring-Verlag, New York, p. 36.

FIGURE 2

U.S. GEOLOGICAL SURVEY: PRECIPITATION MONITORING STATIONS

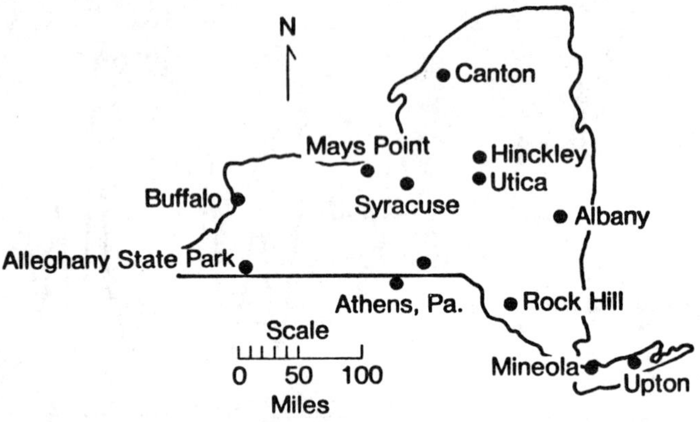

FIGURE 3

pH OF PRECIPITATION: U.S. GEOLOGICAL SURVEY STATIONS

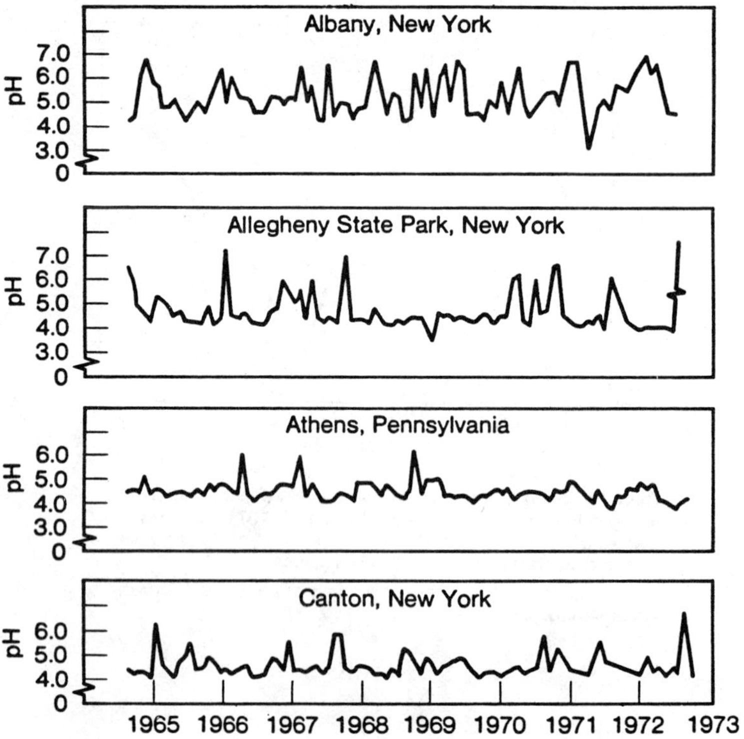

FIGURE 4

pH OF PRECIPITATION: U.S. GEOLOGICAL SURVEY STATIONS

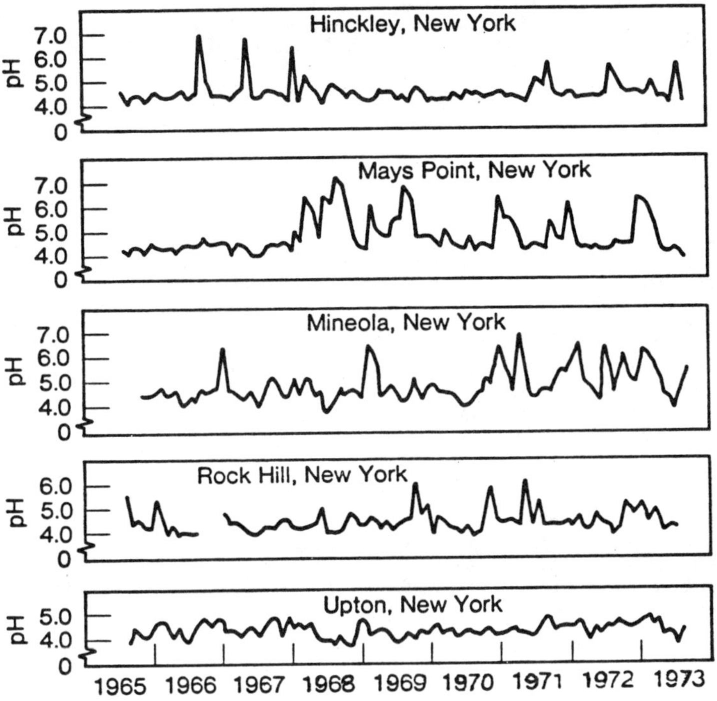

FIGURE 5

U.S. GEOLOGICAL SURVEY STATIONS

pH OF ACID DEPOSITION

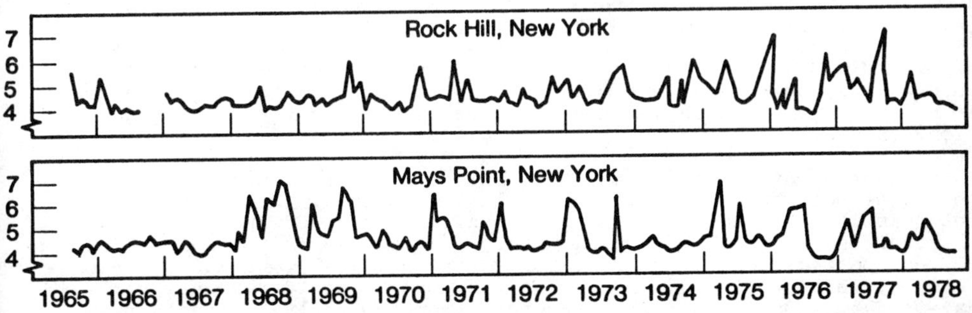

FIGURE 6

cautions that "...the existing trend data were assembled from fragmentary information..." (EPA, 1980a, p.1). EPA further feels that, "The absence of a precipitation monitoring network throughout the United States in the past makes determination of trends in pH extremely difficult and a controversial topic." (EPA, 1980b, chapt. 8). If we are uncertain, therefore, about changes in acidic precipitation, we are in no position to judge the effect of man's changing activities on the question of acid rain. That important point should not be overlooked.

The recent increase (since the 1940s) in both SO_2 and NO_x emissions has been cited as indirect evidence that the acidity of rain is increasing. Admittedly, emissions of these gases have increased but it is highly questionable if the increase is great enough to acount for the reported increases in rain acidity. They need not be, however, if another hypothesis is valid, i.e., the tall-stack hypothesis. According to that hypothesis, the introduction of tall stacks since the 1950s has resulted in emissions of SO_2/NO_x into a higher level of the atmosphere. As a result, the gases remain suspended longer thus allowing opportunity for more conversion to sulfates and nitrates and for transport over longer distances. At best, this is a hypothesis which needs testing.

The hypothesis does have some weaknesses. For one thing, much of the reported increase in rain acidity occurred before extensive installation of tall stacks. Secondly, the conversion/long-range transport segment of the hypothesis would be a valid consideration only for emissions getting above the mixing layer, and we have data to show that a large percentage of power plant emissions stay below the mixing layer (EPRI, 1980).

Let us now examine the other claim--that coal burning by power plants is the principal cause of the increase in acid rain. In examining this claim, of course, we have to be assuming that the rain is becoming more acidic and, as mentioned above, we cannot be certain of that.

In assessing the role of coal burning, we are interested in knowing the relation between stack emissions and what comes down as precipitation. In other words, we would like to quantify the following functional relation:

$$\text{acid rain} = f\ (\text{emissions})$$

Unfortunately, the relation is not simple. It is a complex one comprising at least six components: (1) emissions, (2) chemical conversion of primary emitted compounds (e.g., SO_2) into secondary products (e.g., sulfates), (3) atmospheric transport, (4) scavenging efficiency of clouds, (5) in-cloud processes which result in the acidity of the rain drop, and (6) changes in rain drop chemistry during fall through the atmosphere. We must be able to quantify these components if we are to put the above functional relation into a useable equation, one which defines the utility role in acid rain.

Unfortunately, we can quantify only the first of the six components; we have excellent data on utility emissions. Not only can we not quantify the remaining five, we really do not yet understand the physical and chemical phenomena involved in them. As a result of an EPRI-supported regional sulfate study (Perhac, 1978), we know, for example, that utility SO_2 emissions vary only 5 to 10 percent from day-to-day or month-to-month. Regional, ambient sulfate concentrations, on the other hand, can vary by hundreds of percent, and these variations are related to meteorological phenomena, not changes in emissions (Fig. 7). Our ability to predict ambient sulfate levels in terms of changes in emissions is poor.

AMBIENT SULFATE CONCENTRATIONS AT DUNCAN FALLS, OHIO

SO_2 EMISSIONS FOR EASTERN UNITED STATES

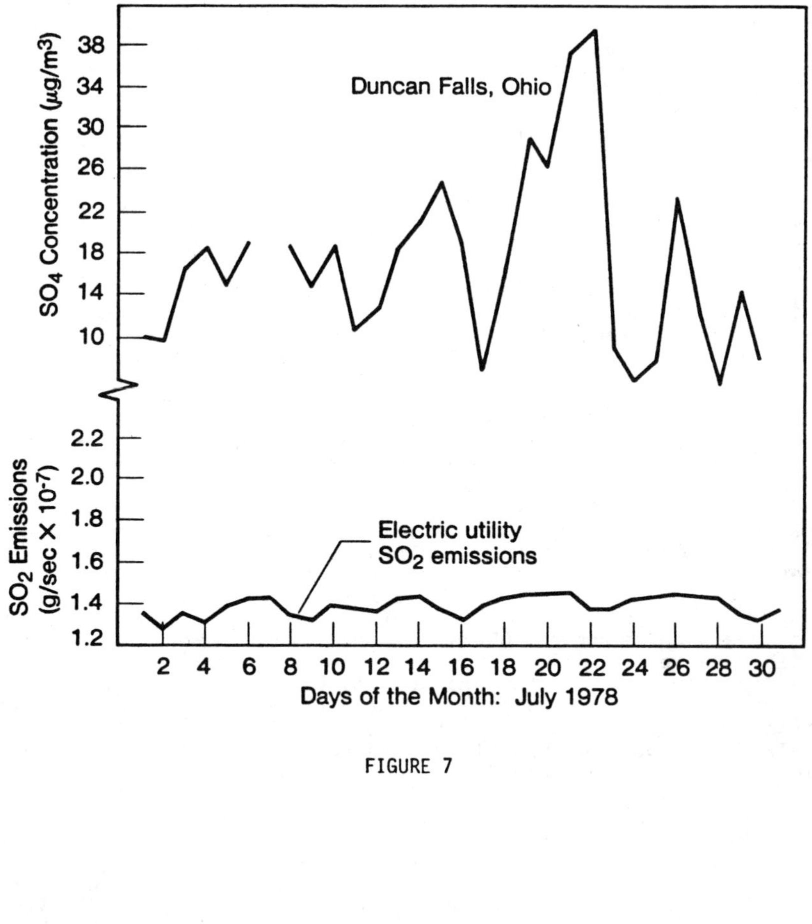

FIGURE 7

Many cities have experienced marked decreases in SO_2 levels without commensurate decreases in sulfate concentrations (Altshuler, 1976). This discrepancy has been explained by many by recourse to long-range transport. In other words, a city may reduce its SO_2 levels simply by burning cleaner fuels but the sulfate levels might not be affected if the sulfates are a result of material being transported in from a long distance. Similarly, acid rain in New England has been attributed to SO_2 emissions from the West coupled with conversion to and long-range transport of sulfates. Unfortunately, our ability to model, or even to recognize, long-range transport is still in a nascent stage. As a result of two years of detailed study in the East (EPRI, 1980), we feel that we can recognize long-range transport of pollutants (over 200-300 km) only a few times a year, a few days at a time. Typically, we cannot distinguish transported material from the local background of pollutants. If we cannot recognize the transported material, we cannot model the relation between emissions and ambient concentrations of pollutants at some distant point, hence we cannot judge what sort of change in emissions would effect what sort of change in ambient pollutant levels. We agree with Costle (1980, p. 65-66) who states, "We have enough difficulty right now, frankly, with our efforts to model the impact of Stack A on hill B 20 miles away. I am not optimistic at all about the prospect of developing the kinds of regional models that are really legally adequate and scientifically adequate to trace impacts of individual stacks 500, 600, 800 miles away."

The problem of relating emissions to acid rain becomes even more difficult when considering in-cloud processes. We cannot quantify the scavenging efficiency of clouds. Identical concentrations of pollutants in the atmosphere can result in markedly different concentrations within the cloud itself. And we know practically nothing about the physico-chemical processes that cause the cloud water to become acidic. Finally, the scientific community has barely started to investigate changes in rain chemistry during fall of the drops. Even if we could predict what concentration of ambient pollutants (in the vicinity of a cloud) would result from a particular emissions level, we could not predict the acidity of rain that would result from that ambient pollutant concentration. We know for example that the hydrogen ion content of rain can change drastically from one storm to the next--by factors of more than 100 (Fig. 1)--whereas the ambient concentration of pollutants is not changing nearly that much.

Until we understand the chemical conversion of emissions, the role of long-range transport, and cloud processes, we cannot model the relation between emissions and acidity of precipitation. "...no model exists today that is capable of predicting the acidic(sic) (H^+ ion content) of matter deposited from the atmosphere." (Smith, 1980). If that modeling capability does not exist, we cannot assess the role of utility operation in the occurrence or distribution of acid rain. We do not know, for example, what effect a 20 percent increase or decrease in emissions will have on the acidity of rain. Too many uncertainties exist.

In view of these uncertainties, regulatory action is difficult. The regulator faces the problem of trying to design a control strategy without knowing quantitatively (or even qualitatively) how effective that strategy will be. For acid rain, it is easy to propose a reduction in emissions from power plants (because it will do no harm and it may do some good). But reducing emissions is not a simple matter. Reducing sulfur dioxide emissions, for example, may involve the installation of scrubbers at a cost of $100 million per plant--a cost which is ultimately borne by the consumer.

The regulator is in a difficult position. Should he propose an emissions control strategy which may be costly to the consumer but without the knowledge of how effective that strategy will be in correcting a problem whose severity is still not well established? The issues are complex and simple solutions are not readily apparent.

WORKS CITED

Cogbill, C. V. and G. E. Likens, 1974, Acid precipitation in the northeastern United States: Water Resources Res., v. 10, p. 1133-1137.

Costle, D. M., 1980, Acid rain: Hearing before the Subcommittee on Oversight and Investigations, House of Representatives, Feb. 27, 1980, p. 65-66.

EPA, 1980a, Acid rain: U.S.E.P.A., Office of Res. and Develop., 1980, v. 1.

EPA, 1980b, External review draft No. 1, Air quality criteria document for particulate matter and sulfur oxides: U.S.E.P.A., 1980, Chapt. 8.

EPRI, 1980, SURE - Sulfate regional experiment: in press.

Likens, G. E., 1976, Acid precipitation: Chem. & Engr. News, Nov. 22, 1976, p. 29-44.

________, F. H. Borman, R. S. Pierce, J. S. Eaton, and N. M. Johnson, 1977, Biogeochemistry of a forested ecosystem: Springer-Verlag, New York, 146 p.

Miller, M. L. and A. G. Everett, 1979, History and trends of atmospheric nitrate deposition in the eastern U.S.A.: Am. Chem. Soc., paper presented before Div. Environ. Chem., Washington, DC, Sept. 1979.

Perhac, R. M., 1978, Sulfate regional experiment in northeastern United States: the "SURE" program: Atmos. Environ., v. 12, p. 641-647.

Smith, Lowell, 1980, Identification of regional air quality models and efforts to validate them: U.S.E.P.A. memo, June 23, 1980.

TVA, 1980, What's wrong with rain: Impact, Tenn. Vy. Authority, v. 3, no. 4, p. 2-10.

Tyree, S. Y., 1980, Rainfall acidity problems: Atmos. Environ., in press.

WHAT ENVIRONMENTAL PRICE DO WE PAY TO CONSERVE ENERGY?

Howard Brandston

Howard Brandston Lighting Design, Inc.

It is a pleasure and a privilege to be here today with an opportunity to present a most urgent opinion on lighting and energy; but first an anecdote which appeared in the Los Angeles Times:

> "The worker's job was simple. He was merely required to move bricks from the top of a two-story building to the ground, but it proved to be more hazardous than it sounded. His full report follows:
>
> 'Thinking I could save time, I rigged a beam with a pulley at the top of the house and rope leading to the ground. I tied an empty barrel on one end of the rope, pulled it to the top of the house, and fastened the other end of the rope to a tree. Going up to the top of the house, I filled the barrel with bricks.
>
> Then I went back down and unfastened the rope to let the barrel down. Unfortunately, the barrel of bricks was heavier than I, and before I knew what was happening the barrel jerked me up into the air. I hung onto the rope and halfway up I met the barrel coming down, receiving a severe blow on the left shoulder. I then continued on up to the top, banging my head on the beam and jamming my fingers in the pulley.
>
> When the barrel hit the ground, the bottom burst, spilling the bricks. As I was now heavier than the barrel, I started to come down at high speed. Halfway down, I met the barrel coming up, receiving severe lacerations to my shins. When I hit the ground, I landed on the bricks.

At this point I must have become confused because I let go of the rope. The barrel came down, striking me on the head, and I woke up in the hospital. I respectfully request sick leave.'"

I believe there is a great deal of similarity between that worker's confusion and the handling of our urgent energy problems in America.

The group of topics gathered here for our consideration under the heading, "Lighting and Energy" seem to be narrowly concerned with "means" without their necessarily associated "ends". Yet, without the "ends" there is no need for the "means". The "END" is the need which generates the method; gives it direction, and measures its' success. To consider "means" without also considering the "ends" to which those "means" are meant to lead, is a bankrupt exercise.

The point is simple. The MAIN thing we should consider today is WHAT IS GOOD LIGHTING DESIGN? I submit the only thing being considered now in the United States is HOW CAN WE LIMIT THE POWER ALLOCATED FOR LIGHTING?

The goals or "ends" of lighting are the results of the design itself - what it is we want to see, or HOW we want to see the spaces we are designing. The "means" are the engineering technology and methodology used to fulfill these design "ENDS".

The engineering approach to lighting puts the "means" first. That is, we would apply technology and methodology to discover results or "ENDS".

Remember, the design approach would set goals or "ENDS" first, and then use engineering expertise to fulfill them.

The difference in the methods usually results in the following scenario: ENGINEERING APPROACH. We observe the room with a light meter, to see if the predictions we made were correct. Whereas in the design approach, we would see if the place looked the way we conceived it.

The engineer looks for engineering results - the designer looks for pleasantness and comfort as well as function. Remember - we are designing and engineering for people - not light meters. Light meters do not occupy or work in the places we design.

During these years of feeling the energy pinch, we are constantly being bombarded with statistics. We are now told that proven resources will be exhausted in the 1990's. We are also told that if projected annual consumption could be decreased by 15% per year, we could delay this exhaustion for only two years. This massive a reduction, we are told on the other hand, would increase unemployment, curtail growth and ruin our economy.

Somehow, mysterious as it is to me, lighting, which only uses 5 to 6% of our total energy, is somehow going to be regulated to make a significant contribution.

If you turned off all the lights, you could gain back this 5 to 6% - but we know that is ridiculous.

It is, however, important to cut back; but we must measure cut backs in terms of gains. We must look at the "ends" first, then the "means" for reducing lighting energy consumption properly.

How can the professionals in practice know what to do to make this contribution? They can turn to the great body of research in lighting, and analyze what approach might be most successful.

Right?? No! - Wrong!! There is no great body of research. There has never been enough money to accumulate a significant body of research. All the lighting research done in modern times has cost less than a single military fighter plane and its operations. And the military runs mostly on petroleum - yet lighting must make this "significant" contribution.

This lack of research funds, political controls which kept energy virtually free, and other economic and psychological factors lead to the over-prescription of illumination in the U.S. This over-prescription and that portion only, is all that can be given back. If we give back more than that, we will be creating gloom and darkness pierced by shafts of brightness worthy of a Wagnerian opera production, justified in the fashionable name of energy conservation.

Some of you may think the situation is not quite as bad or dramatic a problem as I make it out to be. I say it is that bad. Give anyone - especially a federal bureaucrat - an assignment - "Cut down energy consumption in lighting." He measures his "end" result in how much he cut down consumption. When a new man is added to the staff, he has to cut out some too, or how will he be measured? This process is still continuing and each year new people are stepping up to take a slice. When are we going to stop slicing and start to evaluate the impact of these efforts on our environment?

It is now time, to get in situs,new buildings - buildings of the late '70's and early '80's - collect a data base and some post occupancy evaluations before we slice so thin that we are mandating bad lighting.

In my opinion, the current IES document EMS-1 is as stringent as we should get with limiting power for lighting at this point in time. I would like to call a moratorium NOW on further limits to power allocation and to use this IES standard as the model document.

During this moratorium, while data is being gathered and the evaluations are being documented, we can tackle our real national problem in lighting - EDUCATION. Our universities today are turning out no one trained in lighting design.

There are few designers or engineers in practice today that know anything about lighting for people or space they occupy. Yes, they can calculate radiant flux transfer and direct and interreflected components. They can look up in tables how many footcandles are recommended for lighting any task. But few of them know what those numbers mean; not even the people on the committees that put them in the reference material in the first place.

Consider this: if they really knew what human task requirements and footcandles are - why are we such a grossly overlighted country? Why are the same experts suddenly lowering their lighting level prescriptions this year. This new directional thrust to lower lighting levels has little to do with rationality, creativity or humanity. Criteria for lighting are still responding to the politics of energy. There seems to be a political energy law that states: Lighting consumes energy, therefore, when energy is abundant, use may increase, and when energy is scarce, use must be rationed. Please notice, human visual requirements are not a part of this "Political Energy Law". Human visual requirements go beyond task lighting.

Since few know anything about lighting - I repeat we must properly educate a new generation of lighting designers, not just Illuminating Engineers.

As lighting designers, and as members of the design team, what this new breed will decide is how space is seen and the specific contribution that the lighting will make to what is seen. Not making elaborate or convincing drawings, not writing specifications, not performing calculations, not juggling watts per square foot, or following codes. All of those things are done as a matter of course.

Lighting is truly designed only when it is decided in advance the effect that light ought to make and how it will be seen. According to the Oxford English Dictionary, to design is to "conceive and arrange in the mind; to originate mentally." And a design is "a plan or scheme conceived in the mind for subsequent execution; the thing aimed at; the adaptation of means to ends." I believe that what the OED has to say about design is very close to a truth that we must remenber. The new lighting designer must be concerned with what is seen. His job is to decide what should be seen, within the context of lighting, and in relation to every other visual aspect of a project; to decide how it should be seen; and the means to produce this exact visual effect in advance, and in that order.

Because we have so few trained people in lighting design, who is writing our Lighting Energy Regulations? Perhaps illuminating engineers? The real question is - how do people who do not work specifically in this field; who would have trouble seeing good lighting if it were pointed out to them, have the chutzpah to ration it?

How many people rationing lighting today have had their work scrutinized to prove that they have done good lighting; that they know good lighting, and what is requred if good lighting is to be practiced before they got annointed as rationers? None of them. I know this for a fact. Therefore, the energy allocations determined by them will allow a just acceptable level of mediocrity. Is this what we want?

So. ladies and gentlemen, we need this Moratorium on Future Energy Cutbacks in Lighting - we need a time to build and evaluate with the standards we currently have and we must start a war - a war on ignorance in lighting.

We must establish our end or goal to be good lighting practice - this is the only true path to effective conservation.

ESTABLISHING AND CARRYING OUT A SOCIOECONOMIC IMPACT MITIGATION COMMITTEE

David Myhra
Myhra Associates
Reston, Virginia

Introduction

The construction and operation of large energy facilities have definite impacts on the social and economic (socioeconomic) conditions of the surrounding areas, varying from primarily beneficial to generally detrimental. Rural areas with otherwise little industrial development are particularly susceptible to these impacts. Advance planning is the key element that has made the difference in those situations where the surrounding communities have benefitted across a broad spectrum of social and economic conditions from the construction and operation of large industrial complexes. The absence of advanced planning has been the primary cause of detrimental effects of such activities.

Through advanced planning on the part of the sponsoring energy company and the communities surrounding the proposed energy facility, the project can result in long-term benefits to a broad cross-section of the population. At the same time, planning will reduce or control what has been termed the "boomtown" effects of construction.

It has been proposed and widely accepted that the most effective way to plan ahead to handle the influx of construction workers, their families and secondary development needed to provide support services for energy development is to organize representatives of the surrounding communities into a committee to examine the various impacts the project will have. This committee, usually called the "Impact Committee" or "Project Management Committee," develops the plans to deal with the socioeconomic impacts of a sponsor's energy development. The purpose of this report is to present the objectives and organization of the typical Impact Committee and how to establish one.

Objectives

The objectives of a community-oriented, locally based socioeconomic impact organization comprised of representatives of the site area's county government, municipalities, organizations, and private citizens are to accomplish the following goals:

- Identify and evaluate the likely socioeconomic impacts of the energy development project on the citizens and economy, particularly in and around the site area, as a result of an influx of construction workers.
- Translate the impacts of construction worker influx into dollars, i.e., the costs of new facilities, equipment, personnel, and services, and determine the net fiscal deficit for the site area.
- Identify sources of funds and money to purchase the required items necessary in order to mitigate the socioeconomic impacts on a timely basis during construction.
- Develop an impact mitigation implementation plan, and assist the county and municipal governments, agencies, districts and other organizations in implementing plans and programs to minimize the adverse impacts during the construction of the project.

The Impact Committee and How It Will Work

The Impact Committee should consist of a representative cross-section of site area people who, working together, will seek to arrive at an agreement on how socioeconomic costs stemming from the construction of a steam electric plant are to be handled. At issue is the question of the likely impacts anticipated and the level of assistance, form, timing, etc., that should be provided by State and Federal grants-in-aid, the county, the site area cities and towns, and the energy sponsor.

In essence, the Impact Committee, made up of representatives of the various interest groups in the site area, will work in conjunction with the sponsor's planners to identify the impacts of the project, define their costs, and arrive at an equitable mitigation plan. The committee will start by reviewing those projected impacts and estimated costs and then seek to work out an overall agreement on how much, when, and in what form the energy sponsor's assistance should be in order to minimize impacts. Later in the program the committee will develop an implementation plan, describing how all mitigation efforts should be handled. The committee will also make recommendations to the appropriate governing body of changes to existing zoning and comprehensive plans and the like that, if enacted, will serve to influence favorably the impacts of the project.

Carrying Out the Impact Committee's Goals and Objectives

To get an energy facility constructed requires sometimes up to ten years of planning and construction effort. It is the Impact Committee's job to identify and minimize all reasonably anticipated adverse socioeconomic impacts and, by working with the site area surrounding the proposed energy project, help to implement the impact planning at the appropriate times during the project's licensing and construction phases. To achieve these objectives, the following twelve steps as illustrated in Figure 1 should be carefully followed:

Step 1: Recognition of Need and Support for Impact Planning

The construction of an energy facility has the potential to create diverse socioeconomic changes to the site area although, with effective planning, few adverse impacts may actually occur. Even though the energy sponsor may have had a number of studies prepared which predict little or no anticipated adverse change, such studies have often over or underestimated likely consequences. To achieve public acceptance, the energy sponsor should initiate a desire for a more complete assessment of the socioeconomic impacts. This should be announced right after the energy sponsor makes known its intentions to construct an energy facility. However, it is never too late, at least up until the time the sponsor receives its construction permit. But substantial opportunities for generating public good will may be missed if the permitting agencies and the site area residents demand that such an action take place prior to issuance of any construction permits. The longer the sponsor waits the more difficult it becomes to carry out successfully a well-planned program, because the sponsor and the site area people have to play "catch up" planning, and, in the ensuing rush, invariably some impacts will not receive adequate attention.

Once the site area community understands that an energy project is in the offing and that a complete assessment of likely socioeconomic and environmental impacts is desirable, the establishment of a citizen's Impact Committee can be suggested as the best vehicle for accomplishing this assessment with real community involvement.

The purpose of a citizen's Impact Committee is not to develop community support for a proposed energy project. Instead, its purpose is to initiate and support early planning (consistent with the status of project permit approvals), to anticipate and avoid or mitigate adverse socioeconomic impacts, should the project be completed as planned. It is absolutely imperative that this distinction be made clearly if the entire concept is to be understood and supported by the general public.

Supporters of the citizen's committee concept should seek to impress upon the site area the importance of the Impact Committee as a mechanism for predicting, examining, and mitigating or coping with change. Change, they will say, happens to almost every city, county or area due to normal shifts in population. Most area populations are slowly expanding. Expansion could occur at a faster rate should the proposed energy project be constructed. While some residents might welcome the changes its construction and operation would bring to the area, others would be expected to oppose them. Yet there would be many residents and interests who will be undecided at the onset.

Gaining support for the citizen's Impact Committee requires the advocates to suggest that, whether people support or discourage the proposed energy project or are neutral on the issue, everyone can agree that the best position to take is "pro community." Often a "pro community" position is absent in the polarized debate that can occur over the socioeconomic benefits and costs of the proposed facility to the community. Frequently some individuals, area officials, public interest groups, etc., think that no socioeconomic impact planning is necessary in the community because, they believe, the project will not be allowed to be built. But no one knows this for certain; this uncertainty is all the more reason for the residents of the community to start thinking "pro community" and develop a socioeconomic impact mitigation plan, just in case the project is constructed as planned.

In developing this concept for the site area, it should be stated quite strongly that in no way does the acceptance of socioeconomic impact planning acknowledge or approve the energy project's construction. Instead, having a growth management plan is about the same as a family having an insurance policy on its home. No one wants or expects a storm to cause damage to one's home. Yet, should a storm occur and cause considerable damage, people want to know that they are protected against any loss.

The analogy just mentioned applies equally well to almost all communities and energy projects. Just as it is too late to take out an insurance plan after the storm has struck, without an impact mitigation plan, a site area may be unprepared to handle effectively a large influx of new people that will come if the proposed energy project is built. When the formation of a citizen's Impact Committee is put in these terms, people agree that having an impact plan is "pro community" because it protects the residents against a storm of potential adverse impacts.

Step 2: Organizing The Impact Committee

The various steps required to establish an Impact Committee are shown in Table 1. Frequently the energy sponsor and/or area citizens have already contacted site area individuals, county and city officials, superintendents of schools, and others regarding their participation in a proposed Impact Committee. Response is usually completely favorable. The energy sponsor and interested citizens continue to contact potential committee members and interested people, agencies and groups. They describe the committee's concept, its goals and objectives, and the amount of time members will have to spend on committee matters. A steering committee will recommend members. This steering committee will be composed of about twelve individuals from throughout the site area. Typical steering group members are county commissioners, mayors of the several site area cities and towns, and other people. The purpose of the steering group is to recommend individuals who should be contacted to serve with the large committee when it is formed.

Selection of Members

A proposed membership for the Impact Committee is as follows:

Group		Minimum Number
County Commissioners		1
County Planner		1
School Districts (school superintendent)		1
Municipal Officials		4
County/City Planning Commission		10
County Offices		10
The Energy Sponsor		1
County Council of Governments		1
Other County/Municipal groups & individuals		4
Planner(s) from state's Office of Community Planning (ex officio member)		1
Planner(s) from area's Federal Regional Council Office (ex officio member)		1
General Public:		
Citizens	5	
Farmers and Ranchers	5	
Business	5	
Civic Groups	5	
Others	5	25
Estimated Total Members		60 to 70

Leadership and Organization

The committee overall is led by a chairperson, a vice chairperson, secretary and treasurer, elected through a vote of the entire membership.

Once the leadership has been selected, the first major work of the committee begins: learning all about the socioeconomic conditions in the site area. This inventory should be made as comprehensive as possible, so all socioeconomic concerns and issues can be placed into a manageable number of catgories or impact groups. In most areas the Impact Committee can reduce all socioeconomic impacts to about twelve categories. A separate subcommittee is established for each of the twelve groups. This arrangement is shown in Figure 2. The categories include:

1. Housing
2. Education
3. Social Services
4. Roads & Streets
5. Recreation
6. Medical Services
7. Public Safety
8. Utilities
9. Solid waste
10. Planning
11. Sewer & Water
12. Fiscal & Taxes

The primary purpose of the committee is to write an impact mitigation plan. But this occurs only after its subcommittees have inventoried the community in terms of its quality of life, education, social services, public safety, transportation, and other services and facilities. It determines how well the community can or cannot cope with socioeconomic change. If the entire Impact Committee has between 60 to 70 members and there are twelve subcommittees, then each subcommittee is able to have about five or six members to do the investigations and community-wide socioeconomic inventory.

Role of the Energy Sponsor

The energy sponsor's planners will work with the Impact Committee to help it achieve its goals and objectives. The energy sponsor will:

- Assign staff support to develop materials for the committee
- Establish agendas, meeting places, times, etc.
- Maintain a steady menu of action items for the committee and subcommittees to consider, review and act on.
- Provide guidance and direction for the committee.
- Assist the chairperson and members to reach their goals and objectives.
- Provide committee organizational assistance: This will consist of person-to-person contacts with community representatives along with the committee's temporary chairperson and vice chairperson who are likely candidates for the committee membership. Particular attention will be given to the identification of individuals who, because of their stature in the community, interest, and time availability will assume a leadership role.
- Present to the committee in a clear, consistent manner a statement of the existing socioeconomic conditions. The work the energy sponsor has done to prepare this input for the Environmental Assessment (EA) will be the basis for this imput to the committee.
- Define the expected impacts in terms of a qualitative assessment of the effects on various socioeconomic aspects of the communities.
- Apply dollar values to the expected impacts and determine the distribution of those costs to the various towns and communities affected.
- Make recommendations as to the source of money to mitigate the adverse effects. The objective here is to arrive at a fair distribution of costs for the mitigation plan. The availability of other sources of mitigation money will also be included in the committee's evaluation and recommendation.
- Prepare a draft mitigation plan for distribution to the committee. This draft plan will consist of the committee member's preliminary views on the content of the mitigation plan. The draft will then serve as the basis of further discussions.
- Assist the committee in drafting mitigation agreements.

In addition to working closely with the committee, the energy sponsor's planners will conduct an active program of public information, presentation and participation outreach activities with the committee. These activities will be designed to respond to public requests for information, to provide for a reliable understanding of the public's concerns, and to convey information to the public regarding how its concerns are being addressed in committee. The public information and participation outreach activities that the committee will conduct with the community include:

- Furnishing periodic press releases to local newspapers and other media
- Developing an information fact sheet on the committee's activities
- Conducting formal public meetings during the committee's activities

- Establishing and maintaining a speaker's bureau, local exhibits and a mailing list of interested citizens, groups, interests and agencies.

Ultimately, the committee will serve as a key mechanism for informing and educating the public about its activities and the mitigation program. Experience with these committees indicates that the general public will show considerable interest and the committee can be a highly credible and effective information resource. Through small meetings with local civic groups, citizens, farmers, and others, the committee can provide valuable assistance to increase public understanding of its activities and program. Generally, the committee will be responsible for issuing press releases about its progress, assembling mailing lists, preparing fact sheets, organizing workshops, etc. All materials developed will be prepared in conjunction with the committee and cleared with committee members before dissemination.

Role of The County

It will be the County's responsibility to see that all concerns and issues are addressed by the committee and satisfactory mitigating measures decided upon. In addition to accepting this outcome of the committee's efforts, it may be up to the County to implement many of the mitigating measures.

Role of State Planners

It is desirable for the state planning office of state government to have a representative on the Impact Committee. Ultimately the committee will be looking to the state planners and the Federal Regional Council (FRC) for assistance in identifying and obtaining grants-in-aid. However, usually neither the state planners nor the FRC's provide additional input into the committee other than having representatives and planners in attendance from time to time monitor workings and decision-making processes of the committee.

Table 2 summarizes the roles and relationships of the various groups.

Step 2 highlights include:

- A sixty to seventy member citizen's committee appears to be a highly workable size.
- Elect a chairperson, vice chairperson, and subcommittee leaders, etc. who are good organizers, articulate speakers, and hard workers.
- Group all socioeconomic impacts into about twelve categories such as:

1. Planning	7. Roads
2. Fiscal & Taxes	8. Medical
3. Education	9. Public Safety
4. Solid Waste	10. Social Services
5. Housing	11. Sewer and Water
6. Recreation	12. Government

- Have the entire committee meet at least once a month.
- Have each subcommittee meet once a month also.
- The energy sponsor should provide technical assistance and planners throughout the committee's existence.
- Provide financial assistance to the committee to allow field trips, office space and equipment, coffee, and outside consultants if needed and desired.

Step 3: Conducting the Inventory

One of the first tasks to get started on after developing the basic goals and objectives, in short the impact management plan, is the making of an inventory to gather socioeconomic information. Ultimately, the citizen's committee will want to translate the cost of minimizing adverse impacts from rapid populationgrowth into dollars. To do this, the committee through its subcommittees will have to conduct an inventory of area resoures to determine their ability to accommodate added population growth. The data to be obtained includes:

1. A description of the proposed energy facility and its planning sequence.

2. A tabulation of the impact information which has been or can be furnished by the energy sponsor and the site area data sources. This includes information on:

- education
- health care
- police protection
- water systems
- employment
- taxes
- utilities
- energy resources
- natural resources
- land
- sewer systems
- solid waste
- revenue
- planning
- fire protection
- housing
- labor pool
- recreation
- economy
- transportation
- social services

3. An inventory of area resources that is currently unused, such as an excess sewer disposal capacity, unused or underused school classrooms, surplus hospital beds, etc.

4. A description of the likely impacts of the energy facility on the site area's several political jurisdictions (nearby counties and towns, local government districts and state governments); and

5. An assessment of the ability of the project site area to furnish information which would be needed to design and carry out the planning program, and its ability to obtain financial assistance, including:

   - institutional abilities
   - organization structure
   - planning capacity and experience
   - technical abilities

Step 4: Determining the Influx of Construction Workers

Constructing an energy facility requires large numbers of skilled craftsmen such as electricians, pipefitters, boilermakers, carpenters, and iron workers, as well as substantial numbers of unskilled workers. The number of man-years required to build an energy project depends not only upon the project's technology (cooling system, air pollution control system, etc.) but also upon the efficiency of the construction process itself.

Generally energy development projects draw the majority of their construction workers from well outside the site area. The more remote the site is, that is, the further away from urban population centers, the larger the percentage of nonlocal labor the project will require. It is these workers and their dependents who create the need for additional housing, services, and facilities throughout the site community. Since adverse socioeconomic impacts tend to vary in proportion to the number of new workers moving into the site area, this portion of the growth management plan is designed to project the construction worker influx. To estimate construction worker influx one must:

Develop estimated construction labor requirements for the project by craft and job skill over the construction phase of the project;

- Calculate the peak labor demand by craft by year;
- Obtain the number of workers by craft within 60 minutes driving time from area unions (workers are seldom interested in spending more than one hour commuting each way);
- Note any special features about the labor force within a radius of 60 miles, such as working habits, competition and other things such as year-round weather conditions, highway conditions, etc., which might discourage or encourage daily commuting;
- Estimate the number of daily commuters expected by labor craft;
- Estimate the number of movers (non-resident construction workers);
- Calculate the number of dependents who have to be housed, educated, provided with services and facilities may be significant. Where there is insufficient capacity, facilities, staff and resources, adverse impacts can occur. These areas have to be identified.

Step 5: Forecasting the Likely Socioeconomic Impacts

Once the actual worker influx has been estimated the citizens impact committee is equipped to go to the next impact assessment step: forecasting the likely socioeconomic changes in each community and town throughout the site area.

The committee will need to learn more about where the workers are likely to live after moving into the area. The amount of direct and secondary non-project employment in local stores, businesses, government, etc., needs to be identified also. In each case, population information is required including the number of children per household, spending patterns, and other characteristics such as recreational habits, preferred dwelling units - single-family house, apartment, or mobile home, and the cities where workers will probably want to live. Along with this and other characteristics of the work force, estimates will have to be made of each city's increased use of community services, facilities, and utilities, and potential net fiscal deficits, if any, that are likely to occur.

Step 6: Translating Adverse Impacts into Impact Mitigation Needs and Assigning Net Fiscal Costs

In this step, the committee lists the improvements needed for each of the community's services, facilities, and infrastructure in order to accommodate the peak labor influx. This list is usually presented in terms of additional pieces of equipment, buildings, new roads, installation of new sewer and water systems, expansions to parks and playgrounds, additional staff and equipment for the police and fire departments, educational services, social services, and other community-wide activities. The committee categories all reasonably anticipated adverse socioeconomic impacts into six groups:

- Private and public housing
- Human service needs
- Health and social service needs
- Municipal capital facility requirements
- Public service needs
- Educational facility requirements

Within each of the six categories, the committee draws up a "needs" list of items, buildings, equipment, staff, etc., that would have to be provided sometime during the five- to six-year construction period of an energy facility. Each of these items on the "needs" list would have to be filled in order for the site area to maintain its social and economic activities during the rapid influx of population. Included in this list of items will be such things as new equipment for the police and fire departments, additional staff for the school, new roads, and expanded sewer and water facilities so that new housing could be built.

The list of needs associated with providing for the accommodation of new workers in the several site area communities is translated into net fiscal deficits, generally based on a combination of planning ratios and previous community experiences. Some communities have taken the per capita performance level and standards that exist for nearly every urban service and infrastructure such as the desirable number of hospital beds per thousand people, etc. At the same time, they also rely heavily on their own past performance in meeting the needs of their citizens. The important point is that peak labor influx occurs early in the project's construction, generally between the fourth and fifth year of its construction and then falls off sharply as the project nears completion. The objective of the committee in estimating demand and adequacy of services, etc., is to develop a method whereby it can determine the peak demand and then determine how well "in place" services, facilities, and infrastructure are working. At that point committee members can estimate what additions, modifications, financing, etc. are needed in the provision of services for the influx during peak labor demand throughout the site area.

An approach several impact committees have found useful is to make a simple inventory. Basically, the committee identifies the current status of the infrastructure, its present level of performance, its funding, staffing, unused capacity, and so on. This information is obtained basically through Step 3 activities. The committee determines from activities taken in Step 5 the characteristics of the incoming population. This includes its total size, density, income, age, sex, marital status, family members, and so on. At this time the committee looks at the infrastructure which is currently in place and seeks to apply some performance level standards to the service. Wherever shortages or gaps occur is where the site area needs to "beef up" in preparation for the construction worker influx.

At the conclusion of Step 6, some preliminary alternatives for impact mitigation will have been developed. Points to consider include:

- Existing communities may have adequate services and facilities although they may not approach national planning ratios;

- Provision of expanded services should not be based entirely on planning ratios, but on the community's own experience in terms of what has worked well for its citizens over time; and

- In planning additional services and facilities, the site area should use a modified approach. This is based on established ratios nationwide coupled to the site area's own previous experience (citizen satisfaction). Then the community should seek to maintain that level of service or ratio for the new residents.

Step 7: Public Reaction and Feedback

By this time a number of technically sound alternatives are being developed and evaluated in terms of feasibility and effectiveness. The committee initiates forums for obtaining the comments and preferences of the general public. By stimulating and encouraging public dialogue, a consensus is reached on appropriate mitigation strategies and priorities.

Step 8: Development of Final Impact Mitigation Plan

After receiving the public's comments, criticism, ideas and other feedback, the committee prepared a final mitigation strategy. Since the impacts identified are really only estimates, and actual impacts may differ somewhat from those forecasted earlier as the construction of the energy project begins and workers arrive, the final mitigation strategy must not only identify and quantify basic needs but also be flexible enough to allow change as population growth exceeds or fails to reach estimates. Things to consider include:

- Mitigation strategies develop out of the recognition that specific negative impacts will likely occur;
- To reach this state (mitigation) , sufficient decision-making has gone into the identification of what public officials and citizens have acknowledged to be a reasonable expectation of need during rapid growth;
- Population growth may or may not occur as estimated and consequently, mitigation requirements may change also;
- Mitigation strategy involves both professional, public and private judgement which includes the local citizenry's expectations and perceptions of a desired quality of life;
- A list of all area needing assistance such as education, housing, public services, and facilities, etc., should be developed in the form of an impact mitigation planning matrix; and
- An "open planning" process which brings together the energy sponsor, public officials and the general public should be maintained while defining final mitigation strategy.

The energy sponsor's planners will assist the committee in developing practical financial agreements. For example, it is anticipated that each of the school districts in the site area will want to establish some form of impact mitigation agreement with the energy sponsor. The committee will be instrumental in drafting the agreements so that impact assistance will accurately reflect the likely impacts anticipated. Similar agreements will be reached with the several municipalities in the site area, and the committee will be instrumental in drafting these agreements as well. Likewise, it is anticipated that financial assistance will be requested by the site area's county agencies and other constituencies. These agreements will be presented to state officials as required evidence that the energy sponsor and the site area have worked out a satisfactory arrangement for effectively handling all reasonably anticipated socioeconomic impacts. Typical written agreements which the site area may expect the energy sponsor to complete are explained in detail in this paper's Appendix.

A variety of financing mechanisms will be used and include:

- Prepayment of the energy sponsor's property taxes
- State and federal assistance programs
- Loan guarantees
- Technical assistance
- Low-interest loans
- Direct grants from the energy sponsor and other project participants
- Bond guarantees from the energy sponsor and other project participants
- Other forms and types of assistance and aid

After the public review period has been completed and a final consensus has been reached on the mitigation plan, the Impact Committee will present the final plan to the public. At this presentation, the committee could release a final report on its work and the arrangements reached. Summary copies of this report will be made available for public distribution, and complete copies could be placed in special locations (e.g., county libraries and schools for public review).

Step 9: Implementing the Growth Management Program

Implementation of the energy impact growth management program is the responsibility of the site area jurisdictions such as cities, towns, communities, and the county. The citizen's committee seldom implements its own recommendations and would only participate if the site area cities and towns had no planning capabilities or if they only had part-time administrations. Again, the Impact Committee's purpose is to make recommendations that the jurisdictions implement.

One of the more important elements in implementing mitigation measures is timing. This is important because of the lead-times involved; that is, a certain number of months or even years, may be required between deciding to do something in terms of expanding or improving services of facilities and the actual achievements of the planning improvements. Effective coordination within the site area's political jurisdictions, timely decision-making and financing to assure adequate lead-time requirements are all developed in the overall growth management plan.

Step 10: Socioeconomic Impact Monitoring

The primary purpose of a monitoring program is to establish a measurement of actual impacts as they occur and then evaluate the effectiveness of the existing mitigation program. If the mitigation program is not working as planned, or if impacts are occurring in greater intensity than forecasted, then the monitoring program would identify any aggravated conditions that need additional attention. It will provide for the flow and assessment of data in a time frame sufficient to allow appropriate utility and public officials to react to any negative situation not anticipated earlier.

Monitoring programs require collections of data throughout the entire construction period of a project. Wide variations exist among the successful ones. Several citizen impact committees collect data on a monthly basis. Others rely on a quarterly collection approach, while still others feel that data collected on an annual basis is adequate. The geographical areas covered in the monitoring program vary considerably and usually are based on local boundaries, political jurisdictions and other factors. Things to consider include:

- Monitoring allows a measurement of actual impacts as they occur;
- It flags and identifies any aggravated condition, sort of an early warning system;
- It provides for the flow of data and assessment to allow public officials to take appropriate mitigating action or redirect priorities;
- It establishes a communications link with local citizens to keep them informed about socioeconomic change and the effectiveness of impact mitigation;
- Usually industry's role is to provide data while the site area administers the program; and
- Four basic functions of an impact monitoring program, cyclical in nature, are: data collection, analysis, reporting, assessment, and then a return to the data collection and so on.

Step 11: Redirect the Allocation of Impact Assistance

During the construction of the energy project, things may occur to change forecasts of anticipated impacts. Outside factors such as the expected provision of housing by local entrepreneurs may not materialize. Certain functions, once thought to be adequate to accommodate additional use, may in fact not work out as planned. Internally, energy project management may shift construction schedules, bringing in a large number of workers than initially anticipated. Then again, the construction work force may be reduced and the construction schedule slowed. In any event, whenever conditions change, plans must be reviewed and, if necessary, revised to accommodate the changes.

Redirection of impact assistance should not take place too quickly within the committee. It should be carried out carefully. The need for redirection comes first through the monitoring function which has consisted of an overseeing activity. While conducting the monitoring activities, if things are not working out as anticipated, the reasons for this will have to be determined. Maybe the unsatisfactory performance is only temporary and consequently self-correcting; or, the problem may be more long-term and very slow to respond. In any event, before any changes are made in the original plan, an effort will have to be made to determine what might happen if resources are taken from an activity and reassigned. In other words, some form of cost-benefit analysis should be conducted so that the entire task force or committee is aware and can weigh the anticipated benefits of a reallocation of impact assistance. Things to consider include:

- Be prepared to cope with unexpected change;
- Before making changes to established mitigation plans, carefully analyze the need for modification and what other things might occur if a change is made; and
- A well-designed mitigation program may need only minor modifications from time to time.

Step 12: Post-Construction Readjustment

One of the most overlooked aspects of mitigation is the "opportunity" for a site area to influence its post-construction growth in a more manageable manner. Maybe it would like to bring about a more balanced economy through diversification. This readjustment is possible during post-construction activities of the committee.

All too often an overemphasis is placed on the negative elements of an energy project's construction without weighing the positive benefits. One of the positive elements is the impact mitigation process itself. Perhaps at no other time in the site's history do its officials, business leaders, and the general population know more about their socioeconomic environment, its strengths and weaknesses. This is because the site area will have just had one of the most thorough self-inventories ever taken. All the data collected will have been fresh, having been gathered mainly through direct personal interviews with local officials by the committee members. This is a "golden opportunity" for a site area, community or town, and so on to ask itself what direction it should take in the post-construction period. Several alternatives are available. One is to forget about the construction period as quickly as possible and return to a life style in existence prior to the energy project's construction. A second alternative is to encourage expansion to the area's economy in the form of additional jobs, income and diversification. If the choice is to create more and better job opportunities, then all the information gathered during the many months the committee has been in existence in terms of social, economic, fiscal, and infrastructure can be used to aid the site area in very definite and positive ways for many years to come.

Appendix

The following is an agreement for Construction Impact Payments between the Washington Public Power Supply System and Montesano, Washington.

Washington Public Power Supply System
Richland, Washington

AGREEMENT

THIS AGREEMENT is made and entered into in this _____ day of _____, 1977, by and between WASHINGTON PUBLIC POWER SUPPLY SYSTEM, 3000 George Washington Way, Richland, Washington (herein referred to as "the Supply System"), and the city of Montesano, Washington (herein referred to as "The City").

WITNESSETH:

WHEREAS, the Supply System, in compliance with provisions set forth in RCW 54.36.070 and in the "Site Certification Agreement for the Supply System's Nuclear Projects Nos. 3 and 5 (WNP 3 AND 5) between the State of Washington and the Washington Public Power Supply System", dated October 27, 1976, has agreed to pay any valid claims filed against it by any city arising out of a clearly anticipated net financial burden caused by the construction of WNP 3 and 5 (herein referred to as "The Project"), and

WHEREAS discussions between the Supply System and authorized representatives of The City will be obliged to provide as a result of construction activities on The Project and which will result in a net financial burden to The City, and

WHEREAS the cities of Elma, Oakville and McCleary, Washington expect to enter similar agreements,

NOW, THEREFORE, it is hereby agreed as follows:

Article I. Identification of Compensable Impacts

A. The Items Covered by the Agreement

The following list of compensable impacts are claimed by The City, and the amounts listed for each item are the maximum amount of compensation to be paid by the Supply System for each item, and the total amount of compensation to be paid for all items shall not exceed $624,000:

Police services	$218,000
Emergency services	146,000
*City Hall remodeling and general government capital	160,000
*Water tank and line	50,000
*Emergency service equipment	50,000
TOTAL	$624,000

*Capital Facilities and Equipment Items

B. Amendments to List of Compensable Impacts

The parties to this Agreement have exerted their best efforts towards analyzing the probable impact of the construction of the project on The City and it is their combined judgement that the agreement provides fair and equitable compensation for the full period of construction. However, in recognition of the fact that the future course of events is not completely predictable, The City may submit to the Supply System not more often than annually and within thirty (30) days after the anniversary date of the Agreement, any proposed revision to the list, including the total amount, and the Supply System agrees to review the list and to amend the list if the city demonstrates that an amendment is requested because of substantially altered conditions.

C. Future Legislative Changes

In the event that future amendments to state tax laws substantially change the amounts and distribution of tax revenues from The Project to The City, any claims not yet submitted to the Supply System may be suspended pending renegotiation of the list and amounts of compensable impacts. In any such renegotiation, the Supply System shall take under consideration expenses incurred by The City in anticipation of future reimbursement by the Supply System under the terms of this Agreement. Any payments already made shall be non-refundable, but shall be credited in negotiating a revised list.

Article II. Procedures for Payment

A. Annually Recurring Items

Payments for police services, emergency services, and general government services are scheduled to provide for an increased annual level of these services for a period of six (6) years. After this period of time it is anticipated that any increased costs would be borne from other sources or that any increased costs would be reduced. Annual payments shall be made by the Supply System within sixty (60) days of the submission of a claim by The City during its annual budget process; provided however, that the amount claimed shall not exceed the lesser of the amount budgeted for that item for the budget year, or the balance allocated for that item under Article I.A. of this Agreement.

B. Capital Facilities Items

For items which are capital purchases, the Supply System will make payment within sixty (60) days after presentation of evidence of The City to the Supply System that the items for which payment is due according to this Agreement have been invoiced to The City. The amounts paid for capital items shall be the lesser of the amounts invoiced to The City or the value listed in Article I.A.

C. Documentation and Auditability

The City agrees to maintain books, records and documents pertaining to the costs, data and expenses of items compensated for by the provisions of this Agreement. All such records of costs, data and expenses shall be subject to reasonable inspection, audit and reproduction by the Supply System. Unless otherwise agreed upon by The City and the Supply System, The City shall preserve and make available all such records for a period of three years after expiration of this Agreement.

D. Substitution of Items

The payments are all earmarked for specific purposes and cannot be used for other purposes without amendment of this Agreement to reflect such substitution.

Article III. General Conditions

A. Modification

This Agreement can be modified by mutual agreement by the parties.

B. Disputes

All parties shall use best efforts to administer this Agreement in the spirit of cooperation. However, in the event of an unresolvable disagreement between the parties, it is mutually agreed that a settlement shall be determined by a hearing before the Site Evaluation Council.

C. Communications

All communications to the Supply System hereunder shall be in writing and mailed or delivered to the following:

Washington Public Power Supply System,
3000 George Washington Way
P.O. Box 968
Richland, Washington 99352

Attention: Manager of the Technical Division

D. Term

The term of this Agreement shall be for a period of six (6) years from the date of execution; provided however, that the record and audit provisions shall apply for three (3) additional years.

E. Special Condition

Except for a one (1) year payment for police and emergency services, no payments shall be required under this contract until after the Supply System has received a Construction Permit for The Project from the United States Nuclear Regulatory Commission.

IN WITNESS WHEREOF, two (2) identical counterparts of this Agreement, each of which shall for all purposes be deemed an original thereof, have been duly executed by the parties hereinabove named on the day and year first above written.

WASHINGTON PUBLIC POWER SUPPLY SYSTEM

By: ______________________________
Managing Director

CONTRACT DOCUMENTS APPROVED
AS TO FORM:

____
Attorney

CITY OF MONTESANO, WASHINGTON

By: ______________________________

Title: ______________________________

FIVE DESIGN APPLICATIONS OF A LARGE PLANT/SMALL UNIT POWERPLANT CONFIGURATION: RESEARCH AND FINDINGS

Christopher H. Schroeder
Lyna L. Wiggins
Daniel T. Wormhoudt
MAP, Inc.

INTRODUCTION

In 1978 the Berkeley Energy Facility Study group (BEFS) established detailed theoretical justification for a policy proposal which has come to be called "flex-big." The flex-big proposal to site a large amount of coal-fired electric generating capacity at a single site in small (500 MW or less) increments is intended to ease siting and permitting difficulties for coal facilities without compromising environmental goals. The central points presented in the 1978 policy proposal concerned establishing the potential engineering, economic, environmental and social economies of scale to be expected from such a large plant/small unit configuration. The incremental, unit-by-unit phasing plan at a few, ultimate-capacity guaranteed sites provides opportunities for innovative, streamlined regulation; for flexibility in meeting changing demand forecasts; for increased certainty in land use planning; and for the implementation of more sophisticated impact assessment methods.(1)

The second phase of the flex-big study, completed in 1980, worked out and evaluated five design applications of the flex-big concept. The study addressed the question: what would be required to develop conventional direct coal combustion in specific high-potential coal utilization regions in a flex-big configuration, rather than in the haphazard clusters of plants that are already beginning to appear in some regions?

The 1980 research design consisted of three steps: first, the identification of five high-potential coal utilization regions; second, determination of the approximate holding capacity of a representative site within each region based on air quality as the most limiting constraint; and third, identification of region-specific environmental and regulatory issues which might enhance or prohibit the implementation of a flex-big facility.(2)

HIGH-POTENTIAL COAL-UTILIZATION REGIONS

The flex-big design concept requires a large amount of capacity at a single site and a corresponding large consumption of resources, particularly coal, water, limestone, and air quality increments. It was decided that the five design applications, consistent with current siting trends, would be mine mouth facilities. The availability of coal was considered to be a primary locational constraint. Large areas were immediately eliminated due to unfavorable mining or geologic conditions. In the interest of insuring long-run supply security, the search was limited to areas with indicated coal reserves of 500 million to 1 billion tons. Representative information of coal quality in sufficient detail for the estimation of unit system engineering information had to be available before a specific area could be included for preliminary siting consideration. Furthermore, otherwise favorable coal reserves were removed from consideration if conditions indicated that the bulk of the coal was dedicated to other consumers.

Another major physical constraint in location selection was the availability of sufficient water for process and cooling purposes. For this study, only surface water sources were considered. The scope of the water availability survey was limited to known rivers and waterways and existing water impoundments. Although an attempt was made to avoid inclusion in this study of water that was already allocated for other purposes, some of the water considered available may not in fact prove to be available for the flex-big facility. Furthermore, no attempt was made to estimate the required size or specific location of storage reservoirs which would be dedicated primarily to the proposed facility.

During review of areas with favorable coal and water resources consideration was given to the avoidance of potential environmental conflicts. The procedure consisted of noting air quality nonattainment areas, national parks, monuments, wildlife areas, forests, and major population concentrations. The potential problems of siting a flex-big facility in the proximity of such areas were considered a negative factor in the evaluation of a given location and jurisdictional area. The major purpose of the above process was merely to anticipate, and thereby avoid, some of the major conflicts that could arise later.

Five location and jurisdictional areas (LJAs) were identified through this process:

1. South Central LJA. Four adjacent counties in Texas and Louisiana with anticipated coal supply from the Wilcox lignite formation and the Toledo Bend Reservoir as the primary water source.

2. Southeast LJA. Three counties in southwest Alabama with coal supply from local lignite and water from the Alabama or Tombigbee Rivers.

3. Powder River Basin LJA. Two counties in the Powder River Basin of Wyoming using the high-quality, low sulfur, subbituminous coal of the Basin and a potential long-distance water source used in conjunction with dry cooling.

4. Eastern Interior LJA. Three counties in southern Indiana using the underground coal of the region and the Wabash River as the primary water source.

5. North Appalachia LJA. Three adjacent counties in Pennsylvania and West Virginia in an area with wide distribution of high quality coal and the Monongahela and Cheat Rivers as primary water sources.

The design analysis for each of these location and jurisdicitonal areas was necessarily sui generis because of the great variation in types of coal, environmental parameters, history and socio-economic factors, and regulatory and permitting processes.

DETERMINATION OF APPROXIMATE ULTIMATE SITE CAPACITIES

Since the five areas were selected on the basis of excellent resource availability, it was assumed that the provisions of the Clean Air Act would provide the limiting constraints to ultimate site capacities.(3) A representative site in each location and jurisdictional area was selected for air quality analysis. In each case, two potential site layouts were considered. The first had 500 MW units with separate stacks, and the other configuration had groups of three units sharing a common stack. In all cases it was assumed that the units were equipped with state-of-the-art wet limestone flue gas desulfurization (FGD) systems. Because of the design coal variation across LJAs, the operating parameters for each 500 MW (net) unit differed between the five design applications.

In each case both New Source Performance Standards (NSPS), National Ambient Air Quality Standards (NAAQS), and Prevention of Significant Deterioration (PSD) regulations were considered.

All five location and jurisdictional areas were in remote, rural areas where any increase in ambient concentrations due to the proposed source would have to exceed the PSD class II increments before it would approach the NAAQS. Therefore, in each case, the PSD class II increments would be the most constraining air quality regulation. It was assumed that at this time the entire class II PSD increment was available for the proposed source at the site for which the air quality modelling was carried out. The specific increment which would provide the limiting factor was determined through computer dispersion models. EPA specified, second-phase, detailed screening procedures were carried out using a variety of models: PTMAX, CRSTER, VALLEY, and PTDIS. These results were expected to give conservative estimates of potential holding capacities for a representative site within each LJA. The models were used to calculate maximum ground-level concentrations for stability classes A through E and various wind speeds. In no case (except perhaps the Powder River Basin LJA) were the particulate class II increments the most constraining factors; in each case it was the SO_2 increments which provided the limiting constraint. Since the common stack configuration increases the gas volume flow through the top of the stack and increases the height of the plume above the ground, this configuration increases the calculated ultimate site capacity over individual stacks based on these conservative modelling results with a combined stack configuration. Ultimate site capacities were estimated as:

1. South Central LJA - 5400 MW.

2. Southeastern LJA - 2700 MW.

3. Powder River Basin LJA - 19230 MW.

4. Eastern Interior LJA - 2470 MW.

5. North Appalachia LJA - 2400 MW.

Based on the air quality analysis alone, the study team concluded that each of the five location and jurisdictional areas could (conservatively) provide sites for large conventional coal-fired power plants; two of the five could provide opportunities for very large flex-big facilities.

IDENTIFICATION OF REGION-SPECIFIC ENVIRONMENTAL AND REGULATORY ISSUES

The third step of the research design was to identify region-specific regulatory and permitting, socio-economic, environmental, and transmission and generation issues which might enhance or prohibit the implementation of a flex-big facility. Some major conclusions from these analyses are summarized here; details may be found in reference (2).

Regulatory and Permitting.(4) Most new generating capacity is being developed in rural regions of the country. In rural areas, the influence exerted by local government over location decisions, the pre-planning, and the pre-construction phases of the actual project can be considerable, yet predicting what those influences will be is extremely difficult. These regions of the country have little or no zoning, nor much comprehensive planning. There are few bodies of administrative code to help a utility planner understand the pre-existing constraints that local government will impose on new powerplant construction. In some areas of the country, for example in the parishes in Louisiana investigated in this study, it was not even indisputable as a matter of statutory authority that approval by the police jury of the parish was required. Yet in every instance it was clear that local governments can play potentially decisive roles in the formulation of specific plans to build. Oftentimes that role will be accomplished more through negotiation, informal bargaining, and horsetrading than through well structured public hearings and administrative review procedures.

Those interested in streamlining and making more predictable the permitting process for energy facilities might think the informal, unstructured, unpredictable role played by local government in rural areas is a candidate for improvement through formalization and structuring. We do not think so. Although the levers are not always apparent, there are many ways in which local approval can impair or facilitate the construction of a new facility. That approval must be gained through a process resulting in local acceptance of the project. Our discussions with local leaders and others suggested strongly that such acceptance is best gained through an informal feeling-out process. The concessions ultimately exacted from the project applicant are frequently small in comparison with the capital costs of the project, while the gains achieved through such local acceptance are harder to measure, but certainly significant.

Local governments frequently defer to state and federal agencies with respect to questions of environmental quality that require technical skills to resolve. This generally means that the main disputes over the environmental acceptability of the project will occur in relatively more formal state and federal proceedings. Because of the shared responsibility of state and federal agencies in administering many of the major environmental laws, it was again difficult to generalize about the structure of these aspects of the plant approval process. Conditions varied from those in Texas and Louisiana, where the EPA regional office has direct responsibility for the major water and air related permits; to Wyoming, where the state has always sought to retain permitting authority to the maximum extent feasible under the statutes, and has generally succeeded.

The diversity of conditions prevailing in the location and jurisdicitonal areas suggests that one should approach efforts at structural improvements with a certain humility, and a recognition that plans drawn to improve certain permitting patterns in a region of the country may not always transplant to other regions.

The 1980 research has indicated that some of the regions studied are capable, from the perspective of resource availability, of supporting a flexibility-in-scale plan. Several aspects of the permitting regime present obstacles to such a plan, however. EPA's current restrictions on the anticipatory acquisition of PSD approval appears to prohibit a plan that would assure a utility or a utility consortium of sufficient air quality increments to support subsequent additions of capacity at the same site, unless those units were definitely planned and committed at the time of application for PSD approval. Such a commitment to actual units would run counter to the flexibility-in-scale concept, which seeks to preserve the planning flexibility to defer future unit additions until they are truly needed. Water acquisition, in those parts of the country in which actually putting the water to a beneficial use is an increasingly significant requirement, could confront the same sort of permitting obstacle.

Socioeconomic. The five location and jurisdictional areas have development problems peculiar to their rural settings. Socioeconomic activity in these regions has been marked by periods of rapid growth and decline, in each case resulting from heavy dependence of the local economy on a predominant, cyclical industry. No sustained growth or prosperity has occurred in any of the five regions. Energy resource development has been a primary cause of the "boom/bust" phenomenon in two of the location and jurisdiction areas, and all five regions are soon expected to expand their economic bases due to increased mining activity.

Outmigration between 1970 and 1978 was evident in one of the LJAs, and four out of five areas had lower 1977 per capita incomes than the census division in which they are located.(5) The Powder River Basin LJA is the sole exception due to the energy development growth in the area since 1970. All five areas have high percentages of sub-standard housing, increasing use of mobile homes, and shortcomings in the provision of social services and public facilities. All five areas would have difficulty accommodating sudden population growth due to increased mining activity or large facility construction. The Powder River Basin LJA is already experiencing difficulty coping with the current growth rate. The South Central and Southeastern LJAs would be especially vulnerable to housing and service impacts and currently have extremely limited planning capability. The Eastern Interior and North Appalachia LJAs would require aid in upgrading current infrastructure and services.

Each of these areas would be better able to cope with the staged, incremental growth due to a flex-big facility than a conventional plant. With a more gradual buildup of construction and mining activity, and an extended, slower construction phase (perhaps extending 20 years) impacted localities would be better able to plan necessary improvements and expansions of physical infrastructure, and economic and social services.

Environmental. The analysis of probable land use and general environmental impacts of flex-big developments in the five LJAs produces generally the same conclusion as that for socioeconomic impacts. There are broad similarities among the areas, but also, specific and significant differences.

In terms of the standard classification systems used by environmental planners, the five LJAs exhibit considerable variety. For example:

1. Of the ten soil orders used in the revised Soil Conservation Service Classification system, six are represented in the location and jurisdictional areas selected for this study.

2. Of Kuchler's potential natural vegetation types, there are two grassland associations, four broadleaf forests, two broadleaf and needleleaf forests and one grassland and forest combination represented in the location and jurisdictional areas chosen for this study. (The actual vegetation is much more diverse than even this range indicates.)

3. Of the twenty-three physiographic provinces employed in Fenneman's classification system (1931;1938), six are represented in the location and jurisdictional areas selected for this study. Of the twenty-one land surface form types used in Hammond's classification system (1963), seven are represented in the areas chosen for this study.

However, and perhaps somewhat surprisingly, this variety in classification types does not generate a corresponding diversity in the environmental and land use impacts that may be expected as the consequences of flex-big projects. In all cases, air quality effects are the largest concern. As noted above, sizing of each facility was based upon the assumption that the full PSD increments for both particulates and sulfur dioxide (SO_2) were available at a representative site within the LJA for which the air quality modelling was done. Ultimate capacity designations, therefore, are a function of the large differences in design coal characteristics,meteorological conditions and topography across the five LJAs. Even the smallest facility, the 2400 MW plant in the North Appalachia LJA would produce SO_2 emissions on the order of 1000 grams/second and particulate emissions at a rate of about 155 grams/second. These impacts are environmentally significant, even though they are no greater than would be the total air quality effects produced by smaller, dispersed plants with the same aggregate capacity. A significant difference between the air quality impacts associated with a flex-big configuration and plants constructed in the more conventional scattered pattern is that the flex-big plants utilize enhanced plume rise to attain acceptable local ground-level concentrations of pollutants. The remainders would be deposited at distances greater than 50 kilometers downwind and the conclusion is unavoidable that flex-big facilities are more likely than conventional plants built in the same LJA to contribute to the problems associated with long-distance SO_2 transport, such as acid rains.

In terms of other environmental impacts, there are further commonalities. In all cases, flex-big projects would occupy less land per unit of output than conventional plants, due to the utilization of common transportation facilities, common coal-handling and storage facilities,and common cooling and waste stream treatment equipment. Furthermore, it is likely that savings in consumptive water use would be achieved in flex-big plants due to scale economies. Once-through cooling employing a large on-site impoundment with no return to the original source becomes economically feasible at this scale and once-through cooling is much less water consumptive than the now almost standard (wet) evaporative cooling towers. Similarly, common high-efficiency demineralizers would be cost-effective and would increase the cycles of concentration in the stream cycle and thereby minimize boiler blowdown. And, too, generating units in the 500 MW range have favorable heat rates to begin with (roughly, 10,000 Btu/kWh).

Beyond these broad, shared resource consumption and residual production characteristics, flex-big developments in the five LJAs would have some other similarities. In only one of the five areas, the Eastern Interior LJA, are commercial crops grown for which there are demonstrated SO_2 damage functions. In two of the LJAs, timber, mostly pine and oak, is the only vegetation of commercial value. In the remaining two LJAs, cultivated crops are not economically significant, although alfalfa is grown in the Powder River Basin LJA and has been shown in some studies to react adversely to doses of sulfur compounds. All five LJAs include portions of ranges of endangered plants and animals, though evidence concerning specific habitats is everywhere scarce.

There are also, however, extremely important environmental differences among the areas. In the Powder River Basin LJA, it would almost certainly be impossible to obtain sufficient cooling water to operate conventional induced draft evaporative cooling towers and plant cooling would require construction of a wet/dry system, relying on water imported from outside the immediate region. The remaining LJAs have abundant water supplies and few or no present institutional impediments to withdrawals by utilities, although in the South Central LJA future demand for allocations from the Sabine River for use in the Dallas-Fort Worth area may create conflicts. Topography creates a major issue only in the North Appalachian LJA, where mountainous terrain and extremely high local relief variations would produce unacceptable ground-level pollutant concentrations in most locations. Seasonal flooding and unstable soils in the South Central LJA --- and to some extent in the Southeastern LJA --- require substantial investment in construction and maintenance of transportation infrastructure if fuel and building materials are to be moved safely to a flex-big site. In one parish in the South Central LJA, approximately 70 percent of the Parish-maintained bridges are down or otherwise unusable and most of the roadbeds will not carry large loads.

Transmission and Generation. Because of the proposed large size of flex-big facilities, demand forecasts for electric powere were analyzed at the level of the Reliability Council. For each of the five location and jurisdictional areas, the corresponding Reliability Councils have recently been predicting much slower growth in demand, and a good deal of projected generating capacity has been cancelled or delayed.(6) However, in each of the regions, there remains a considerable amount of projected coal capacity. Because of the size of a "flex-big" facility it is most likely that the plant will be jointly owned by a group of utilities within a council, and there is growing experience within the industry in joint ownership arrangements. In each of the five areas there is a likely group of utilities with current projected coal capacity who might be encouraged to combine interests in a flex-big plant.

Although each Reliability Council indicates decreasing demand forecasts, this may be an advantage for the flex-big proposal. Since the flex-big concept relies on phased, incremental construction of small units, with each unit's approval predicated on need, considerable demand response flexibility is built in. And, since a flex-big site would be guaranteed for an ultimate site capacity, additional units could be added quickly without additional siting procedures should demand forecasts prove too low.

CONCLUSIONS

Five high-potential coal utilization regions were identified for use in evaluating the potential of a large plant/small unit "flex-big" configuration. Based on air quality analyses, all five could provide sites for large conventional coal-fired power plants; two of the five could support very large flex-big facilities. In the various location and jurisdictional areas, a great deal of variety in permitting requirements was evident. The largest regulatory obstacles to a flex-big facility may be the inability to acquire in advance air increments and water rights for the entire ultimate site capacity. Planning aid and funds from outside sources for infrastructure improvements would be required in every case, although the flex-big phasing would ameliorate some of the socioeconomic impacts. The most significant negative environmental impact associated with all the flex-big facilities is air quality degradation, particularly long-distance sulfur dioxide transport and deposition. A likely group of joint owners with current projected coal capacity could be identified in each region and flexibility of response to demand forecast changes is an added advantage of the flex-big proposal.

REFERENCES

(1) See Christopher H. Schroeder, Lyna L. Wiggins, and Daniel T. Wormhoudt, Flexibility of Scale in Large Conventional Coal-fired Power Plants, Policy Analysis Division, U.S. Department of Energy, December 1978. Also forthcoming in Energy Policy, June 1981.

(2) Detailed results of the second phase study may be found in Christopher H. Schroeder, Lyna L. Wiggins, and Daniel T. Wormhoudt, with Melissa A. Eisenberg, Julia A. Gould and Dee Winton Alexander Wallis, Analysis of Multi-Jurisdictional Permit Procedures, prepared under Contract DE-AC03-80EV10425, Regulatory Analysis Division, U.S. Department of Energy, forthcoming 1981.

(3) The air quality analysis described in this section was conducted by Burns & McDonnell Engineers of Kansas City. See Reference (2) for details.

(4) The research for the regulatory and permitting sections of Reference (2), with primary data collected from field work, includes detailed descriptions of each required federal, state, and local permit for power plant siting in each study area state.

(5) Comparisons are based on most recently published U.S. Census data.

(6) Recent years of the Reliability Council annual submissions to the Economic Regulatory Commission of Coordinated Bulk Power Supply Programs were examined and compared in this analysis.

INDUSTRIAL ENERGY PLANNING AND PUBLIC POLICY

L. W. Sessions
Vice President, Energy and Chemicals Technology
American Can Company

This afternoon, I want to talk to you about energy conservation and public policy. A few days ago, I read a startling statement by Robert Kirby, Chairman of Westinghouse Electric Corporation which focussed the seriousness of this relationship. He said, "There are a number of different ways to measure our national dependence on foreign oil. For me, one of the most startling ways of putting the issue is this: At the current rate of import, in the next five years the United States will pay over to OPEC a sum of money equal to half the value of all the companies listed on the New York Stock Exchange." Mr. Kirby also reminds us that this enormous transfer of wealth, with all its consequences on gross national product, economic growth rate, and living standards, is still secondary to the national security aspects of continued dependence on volatile governments in the Persian Gulf.

Specifically, I want to talk about a paradox in industrial energy planning, and then suggest public policy measures to help resolve it. Basically, the paradox is this: industrial energy conservation represents one of the most cost effective and immediately realizable opportunities to stretch our energy supply and reduce the nation's appetite for foreign oil. Why, then, isn't more being done to achieve these worthwhile objectives?

It is true that impressive results in improving the nation's energy productivity already have occurred. As you can see from Figure I, the total energy productivity of the United States, measured against GNP, has improved by about 8%-9% since 1972. Meanwhile, some of our major industries, based on reports to the Department of Energy, have improved their energy efficiency by as much as 20%. Still, a key point is that we could be doing much more. Robert Stobaugh and Daniel Yergin, in their book, "Energy Future", make the point as follows:

> If the United States were to make a serious commitment to conservation, it might well consume 30%-40% less energy than it now does and still enjoy the same, or an even higher standard of living. That savings would not hinge on a major technological breakthrough, and it would require only modest adjustments in

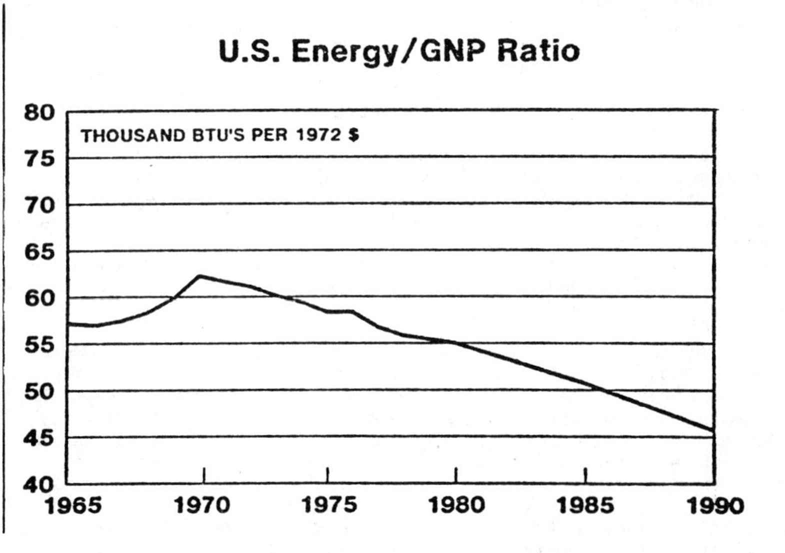
U.S. Energy/GNP Ratio
THOUSAND BTU'S PER 1972 $
80
75
70
65
60
55
50
45
40
1965
1970
1975
1980
1985
1990

FIGURE I

the way people live. Moreover, the cost of energy conservation is very competitive with other energy sources. The possible savings would be the equivalent of all imported oil (8MM barrels a day), and then some.

I expect that most of you would agree with the general thrust of Stobaugh and Yergin's argument when you assess the energy productivity improvement potential in your own companies. The difficulty in getting energy projects approved can be very frustrating to an energy planner because, generally, there are reasons why such projects should be especially attractive. Let us consider some of them for a moment, using Figure II as an outline.

First of all, energy savings represent a dollar-for-dollar carrydown to operating profit. We all know that business general managers are partial to investments which expand the business, or protect market share. These investments are directed to develop sales, and for each dollar of new sales only 5¢-10¢ is typically brought down to profit. But a dollar saved in energy cost is carried straight down as a dollar of increased operating profit. In your energy project presentations, you may find it useful, as we do in the American Can Company, to develop charts plotting energy costs as a percent of operating profit. One such chart for two of our own divisions is shown in Figure III. In this chart, "A" Division represents a relatively energy-intensive business and "B" Division one less so. In "B" Division, energy costs, historically, have amounted to less than 2% of sales. From that standpoint, energy costs might not seem very important. But when these same costs are measured against profits, they will loom much larger, and are likely to command more intensive management attention.

Investments to improve energy productivity usually involve low risk and provide near-term payout. Since most of the existing industrial plants in the United States were built during an era of very cheap energy, the investment to retrofit, or replace, usually is not very exotic. Typically, we are talking about insulation, heat recovery, use of waste fuels, energy load management, and perhaps cogeneration. Technology is well-known and proven. Start-up and run-in is likely to be straightforward. In a recent review of more than fifty proposed energy productivity improvement projects in my company, none involved unusual technology/risk factors.

Energy investments are likely to provide increasing returns in real dollars over time. Rare indeed is the energy planner who had the nerve to forecast the rate of increase in unit energy costs which actually occurred over the past several years. Rarer still is the general management who responded early to optimize plants to protect against these increased costs. Figure IV shows the escalation in costs per BTU of purchased energy for one of our American Can divisions since 1973. There are many elements in the future energy supply picture which argue for a continuing escalation in energy costs outpacing the general inflation rate. OPEC certainly can be expected to index its pricing to inflation, plus an increment. Meanwhile, in the United States, natural gas is grossly underpriced relative to oil, and likely will be "catching up" during the next few years. Energy from coal and nuclear power, seemingly so promising a few years ago, is becoming ever more costly in response to environmental and other regulatory constraints.

Planners, at least in our company, have trouble projecting costs increasing at the rate of 20%-25% per year. Perhaps this is because the United States enjoyed such a long history of relative price and cost stability with an inflation rate of less than 5%. For energy, at least, such days appear to be long gone. In our own energy capital projects, we are now using a projected CPI deflator, plus an energy "kicker."

My last two observations on energy productivity relating to low capital investment and timeliness may be considered together. I am not talking here about the absolute dollar levels of specific energy projects. Rather, I refer to the capital required to save energy, or to produce an equivalent barrel of oil. It is a fact that industrial energy conservation in general is far and away the most cost effective means of stretching our energy supply. Some of the cost relationships to illustrate this are shown in Figure V.

It has been estimated that viable industrial energy conservation projects in 1979 would require around $15 M-$20M of capital investment to save the energy equivalent of 1 barrel of oil per day, or approximately 2 billion BTU's per year. For my own company, the number is closer to $15M for the projects which have a change of being approved.

THE ATTRACTIVENESS OF ENERGY PRODUCTIVITY INVESTMENT SAVINGS

- Dollar for Dollar Carry Down to Operating Profit
- Low Risk
- Near Term Payout
- Real Dollar Increase Over Time
- Often Low Capital
- In Tune with the Times

FIGURE II

ENERGY COSTS VS. OPERATING INCOME

(1972-1980)

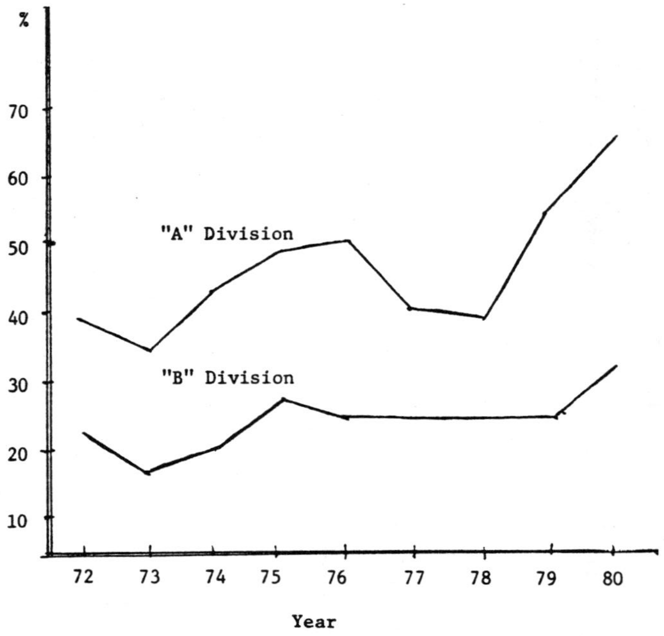

FIGURE III

COMPOSITE PURCHASED ENERGY COST

($/MM BTU)

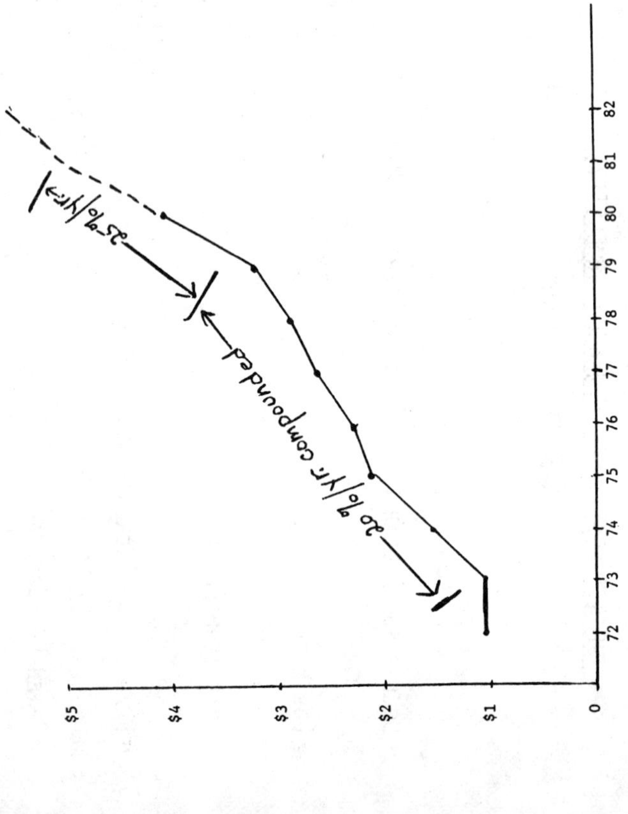

FIGURE IV

INVESTMENT REQUIRED TO

PRODUCE 1 BARREL OF

OIL PER DAY
(1979)

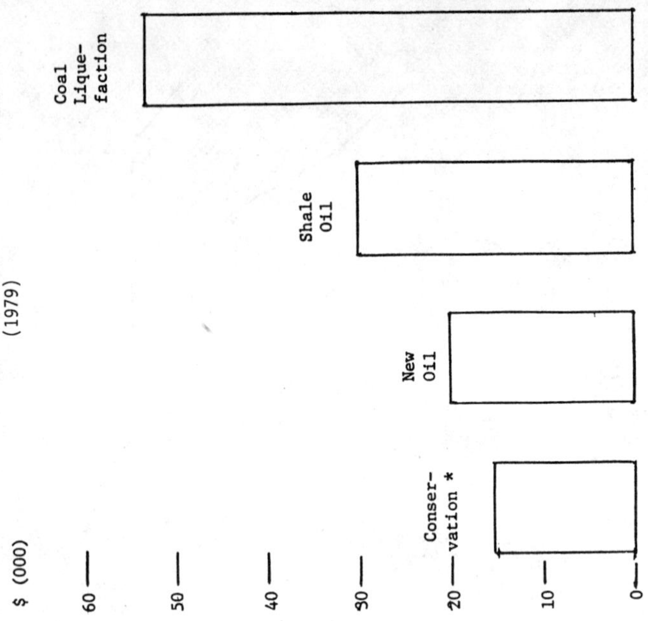

* Assumes 2 year payback

FIGURE V

But to generate the same added oil equivalent through new oil field investment, or synfuels, is likely to be far more expensive. Costs in bringing in new oil and gas discoveries in the United States are now running \$20M-\$30M per barrel of new oil per day. Shale oil investment is estimated at upwards of \$30M and coal liquefaction, according to Exxon, is in excess of \$55M.

Now to return to our paradox. If industrial energy conservation represents such a high yield opportunity, viewed both from the standpoint of individual firms and from public policy, why then do so many projects languish? And what might be done through public policy to move them forward in approval priority in the planning decision process?

Your company experience is probably similar to mine; the problem basically is cash flow. There is an intense competition among capital projects. There are far more urgent priorities than there is cash to fund them all. Minimum hurdle rates for project approval continue to rise. Energy projects, though attractive, tend to stay on the books, "pending approval", or "deferred." It may be argued that there is nothing wrong with this. Energy projects should be expected to compete with all other capital projects — and in the long run, those which meet the priorities of the company will be funded. I confess a personal sympathy toward this view, but it ignores the fact that there is no such thing as a perfectly competitive marketplace. The marketing relationships and management decisions affecting them are conditioned by a myriad of outside influences, many of which fall into the category of public policy.

If the essence of national energy policy is to contain foreign oil imports, an essential element of this policy <u>must</u> be the encouragement of industrial energy conservation. There are many ways this might be done. Recent Administrations have flirted with some of them, such as mandatory conservation targets, or efficiency standards for certain industrial equipment like pumps and motors. Such initiatives were anathema to industry.

The present Administration and Congress seem more inclined toward using the carrot approach rather than the stick. They are proposing to encourage conservation through revision of depreciation rules and tax policy. Congress is now beginning consideration of President Reagan's initial tax proposals, which include a revision of depreciation accounting rules for all industry designed to accelerate new investment, and thus rejuvenate the aging U.S. industrial capital base. Another bill to be introduced in both Houses shortly will propose additional investment tax incentives targeted for energy productivity capital investments. This bill is the Industrial Energy Security Tax Incentives Act of 1981, also known as The Wallop/Heftel Bill.

The rationale of the Wallop/Heftel approach includes the following concepts:

1. The U.S. remains vulnerable to significant energy supply interruptions. This vulnerability should be protected against through a balanced program of increased domestic energy production and improved efficiency of energy use.

2. Improvements in energy efficiency are the most economic, least risky, and most environmentally sound methods of reducing our dependence on foreign energy.

3. Improvements in energy efficiency and conversions from oil and gas to alternate fuels will require extensive capital expenditure.

4. Investment tax credits have a demonstrated ability to spur investment.

The bill pursues this rationale by revising and extending the business investment tax credits first enacted as part of the Energy Tax Act of 1978, which would otherwise expire on December 31, 1982. It also adds a more generalized incentive for major innovative process improvement to produce significant energy savings. Basically, the bill would increase existing energy tax credits from 10% to 20% and would considerably expand the definitions of energy property qualifying for the tax credit treatment.

A bill like this, if enacted, should have considerable impact on energy investments. Figure VI is a plot we prepared in the American Can Company showing the effect

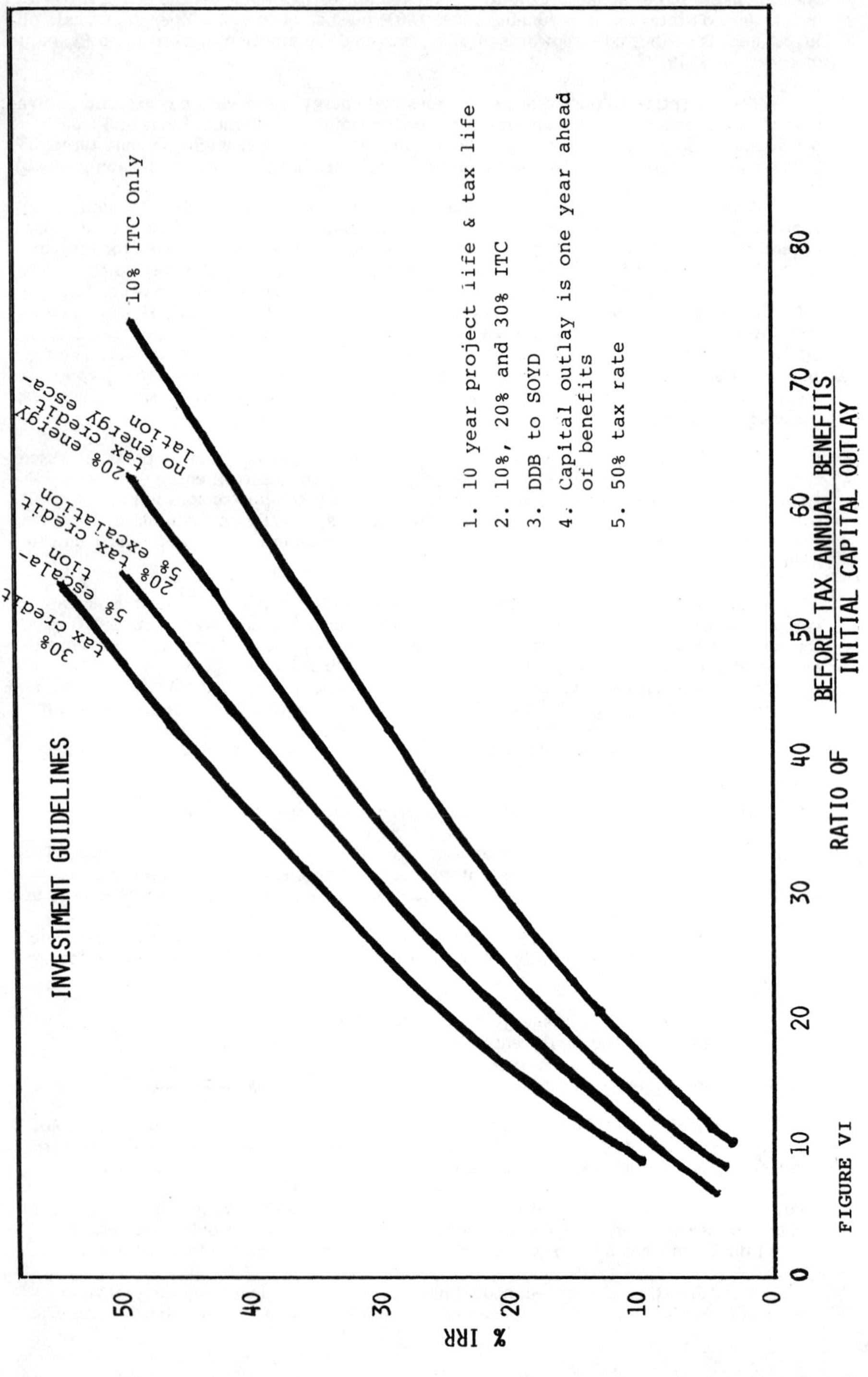
INVESTMENT GUIDELINES
% IRR
0
10
20
30
40
50
0
10
20
30
40
50
60
70
80
RATIO OF BEFORE TAX ANNUAL BENEFITS / INITIAL CAPITAL OUTLAY
30% tax credit
5% escala-tion
20% tax credit
5% excalation
20% energy tax credit no energy esca-lation
10% ITC Only
1. 10 year project life & tax life
2. 10%, 20% and 30% ITC
3. DDB to SOYD
4. Capital outlay is one year ahead of benefits
5. 50% tax rate

FIGURE VI

on internal rate of return in one of our divisions for various levels of investment tax credit. The horizontal scale is conventional pre-tax return on investment, and the vertical scale is after-tax internal rate of return. The two lower curves plot the relationships for 10% and 20% investment tax credits. The two upper ones are for 20% and 30%, but with an added 5% inflation factor for energy.

Note that the impact for both investment tax credit and inflation is much greater for projects which are marginal than for those with high return on investments. For example, consider a project with a 20% ROI. A conventional analysis at 10% investment tax credit would yield an internal rate of return of 12% — a clear reject these days. But, at 30% investment tax credit and 5% real energy inflation, the internal rate of return doubles, and begins to be interesting. Needless to say, net present value also jumps and payback period becomes shorter.

There seems to be increasing sentiment among energy planners that some stimulus as represented by the Wallop/Heftel Bill is needed to move energy projects up the scale of priority toward actual funding. There clearly is increasing sentiment in Congress for this. In fact, some consternation has been expressed concerning the less than expected response resulting from the tax incentives incorporated in the Energy Tax Act of 1978. In the two years after enactment of those credits, apparently only about $300MM had been claimed, compared with a conference committee expectation of more than double this amount. The hesitant industry response, however, probably was caused first by the long delay from the IRS in issuing regulations under the Act, and then the extreme restrictiveness of the regulations themselves.

The Wallop/Heftel proposals would attempt to reassert the original intent of Congress, and extend the program beyond the current 1982 expiration date. The proposals will include several co-sponsors from each House. Broad Congressional support is expected. But Congressional follow-through will depend, of course, on active support from the industrial sector, and communication of this support to Congress. Active industry endorsements seems appropriate to stimulate funding to upgrade energy productivity consistent with the needs of national energy policy.

THE ROLE OF PUBLIC UTILITY COMMISSION IN PROMOTING APPLICATION OF RESEARCH AND DEVELOPMENT RESULTS -- AN INDIANA VIEW

LARRY J. WALLACE

CHAIRMAN, PUBLIC SERVICE COMMISSION OF INDIANA

Each utility must have its own company specific research and development program suited and tailored to the specific needs of that utility. Similarly, because of the vast differences in budget and staff considerations as well as regulatory philosophies, each state regulatory commission with jurisdiction over electric utilities will have its own commission specific approach to the promotion and application of research and development to be performed by or for the utilities subject to its jurisdiction. Some utility Commissions have staff resources allocated just to research and development. Other regulatory commissions may have some degree of authority over state sponsored research and development programs. Perhaps most, however, are more like the Indiana Commission which does not have a research and development staff nor do we oversee or have any authority over a centrally directed statewide entity having responsibility for research and development. Our primary impact on research and development by the electric utilities subject to our jurisdiction is through the day to day regulatory ratemaking process. To that extent the "Indiana View" may well be illustrative of that of most, if not all, state regulatory commissions.

The greatest contribution that any utility commission can make toward the ultimate use of research and development results, regardless of size or sophistication of the commission staff, is the maintenance of a realistic regulatory climate. That climate must recognize that rates should be so constructed that the legitimate operating expense -- research and development -- can be included therein. There must also be a realization that research and development will hold many of the keys to solving the problems of future energy needs within the twin restraints of limited primary energy supply and higher environmental standards.

It is unrealistic to expect commissions to spend all of their time on the energy-utility problems when they are charged by law with regulating a whole spectrum of utilities and transportation companies from billion-dollar giants to very modest entities. But they should evince sufficient interest to learn of the problems, both present and emergent, that threaten the energy utilities. They must take advantage of the information flow that comes from the energy industry. Good examples of the information that is available are the regular journals and reports from the Electric Power Research Institute.

The utility commissions must accept the concept that R&D expenses are legitimately borne by the utility customers. R&D expenses are a portion of virtually every product and service that we buy in today's economy. Every industry that expects to remain in business to the end of this decade is engaged in some type of R&D. The question of whether utility customers should pay for R&D expenses has been laid to rest in large segments of the utility industry where the customers own their own companies, such as TVA, BPA, many municipal systems and public power authorities, and rural electric cooperatives.

Another legitimate area of concern for regulators is the question whether or not the utilities' R&D programs are adequate in light of the energy picture. This type of inquiry will give encouragement to utility managements to face up to their R&D responsibilities. A company's efforts may be compared with other utilities in the state, in the region, or of comparable size. Further inquiry could establish whether they are better or worse off than other utilities and need R&D assistance more or less than other utilities. Geographic locations, demographic changes, available transportation, present generating mix, etc., will all play some part in determining their relative position. Finally, there is the equity question for ratepayers within a given state. Is it fair that only a portion of the ratepayers are assessed R&D costs which will ultimately be of benefit to the entire state? Should there be a state policy directing some uniformity of R&D expenditures? Or should a state support the "free lunch" theory for some or all of its utilities?

Having said all of the above, regulators should not abdicate their position of questioning most severely every expenditure made by a utility. In examining a company's R&D program, the regulator should ask: In light of the conditions in your service territory, your state, and your region, are you pursuing the right path of R&D? How do you choose your program? Will the technologies ultimately yield to the ratepayers of this state a sufficient return on their investment? What influence do you have on decision making within the utilities' central R&D facility, the Electric Power Research Institute? Are you managing your EPRI contribution to the extent that you are participating in the Advisory Committee structure?

During the past few years when the utility industry was organizing its R&D effort and increasing its R&D expenditures, we regulators tended not to ask hard questions about results, because those of us with some familiarity with the R&D process realized that results are oftentimes delayed for years or decades. This highlights one dilemma facing individual regulators. With the average tenure of a regulator now at about four and three-quarter years, the regulator cannot expect to be around when the R&D program approved finally bears fruit. The process seems to me to be sufficiently mature that we can begin to ask our companies: What results are being achieved as returns on your R&D investment? Which of these emerging results are you using at the present time or in the process of implementing? What progress is being made toward commercially feasible developments for short-term application (within the next 5-10 years)? What benefits do you see emerging in the long-run from present R&D activities?

Finally, the regulators can play an important role in the application of R&D results in the certification process and siting approvals. For instance, it would do no good for fuel cell technology to make downtown generating plants feasible if commissions refuse to allow their installation. Also, the general rate structure must yield sufficient income for capital accumulation to pay for investment in new technologies, and novel approaches to capitalizing new technologies should not be rejected out of hand. This is not to infer that investment dollars are necessary for all R&D benefits. Many of the technologies are in the area of items regularly expensed in the operation of a utility, and many of them are techniques and applications to improve the operation and use of present facilities.

A decade ago state regulators could safely sit back in an adjudicative posture merely reacting to proposals of the utility industry. They should not have done so but they could do so and -- I am told -- more often than not that's what they did. In the present state of changing energy conditions, the regulators must keep up with, and in many cases, lead utility managements in applying research and development for the solution of problems to the ultimate benefit of the ratepayers. When I first became a utility regulator in 1974 the utilities in my state were being criticized for not spending enough money on research and development. This was when EPRI was just getting started and there was no centrally coordinated R&D effort within the electric utility industry. Now in some states I understand that electric utilities, and the regulatory commissions, are being criticized for spending too much or not spending the right amounts on the right things. Nevertheless, there does not seem to be any lessening in demand to improve our energy independence and develop better and new ways to use renewable resources for the generation of electricity. Thus no regulators should find themselves in a situation in which effective promotion and application of research and development responsive to the needs of the ratepayers is not demonstrably in the public interest. Our prime responsibility is to provide a regulatory climate in which R&D can flourish and be channeled in beneficial directions in a publicly acceptable and accountable way.

ROLE OF PUBLIC UTILITY COMMISSIONS IN PROMOTING APPLICATION OF RESEARCH AND DEVELOPMENT RESULTS: A NEW YORK VIEW

Dr. David P. Stricos
New York State Department of Public Service

I. INTRODUCTION

The need to accelerate the development and application of new energy technologies has presented, and continues to present a number of challenges to the electric utility industry and its regulators. Clearly, the industry is obliged to involve itself, to some extent, in research programs that might affect its future; but the nature of that involvement is not always self-evident. What should be the level of utility support for R&D? How should individual utilities approach the planning and management of their internal R&D programs? How does one best promote the introduction of advanced technologies into the existing utility system?

Questions such as these have occupied utility companies and regulators for some time, and I am pleased to have this opportunity to describe how regulators in New York State have responded. I will do so by describing very briefly what I believe are the more important actions taken by the New York State Public Service Commission in dealing with research and development issues. I would add that the views expressed here are my own as a member of staff and may not necessarily reflect those of the Commission.

II. LEVEL OF UTILITY SPENDING ON R&D

The new york Commission first expressed itself on the subject of R&D spending levels in a 1971 electric rate case decision.(1) The utilities at that time were applying about 0.2% of revenues to R&D. The Commission, in consideration of the growing technical

(1)Case 26088 - Niagara Mohawk Power Corporation, December 22, 1971.

demands on the company, stated that at least 0.25% of revenues should be devoted to research. In another rate case shortly thereafter,(2) the Commission concluded that "...a research program of adequate dimensions for the electric power industry as a whole should probably involve expenditures of about 1% of revenues..." As an initial step toward that goal, the Commission stated that all utilities under its jurisdiction should increase their R&D spending to at least 0.25% of revenues.

As electric utility R&D expenditures moved toward 1% of revenues, the Commission periodically reexamined the propriety of the 1% guideline. In 1978, following extended hearings on the utility R&D programs, the Commission found that the need to maintain strong internal R&D programs, while at the same time providing support to external research groups such as the Electric Power Research Institute (EPRI), required a modification of its 1% guideline. Accordingly, the Commission stated(3) "...we expect that valid R&D programs will begin to exceed 1% of revenues, perhaps amounting to as much as 2%. But we need not decide here that some figure above 1% is our new target. We need only state that we are prepared to consider R&D expenditures in excess of 1% of revenues, so long as the programs being funded are shown to be sound and in the best interests of the consumers. Other large industries are putting a much greater portion of their revenues into R&D. The record in this proceeding demonstrates that the 1% figure is a reasonable minimum target and that expenditures of between 1% and 2% of revenues are not, per se, unreasonable."

More recently, the Chairman of our Commission stressed the point that our R&D guideline does not represent a mandatory spending level. He said,(4) "We believe that the R&D planner should have some flexibility in assembling an R&D program to meet his or her company's individual needs. ...Requiring a selected spending figure is likely to lead to some wasteful spending, merely to reach the target, and undue concentration on reaching the minimum rather than assuring that the content of the R&D program is sound."

In my opinion, these guidelines have served the very useful purpose of helping to maintain an adequate level of utility support for R&D through periods when the companies might otherwise have looked to reduce expenditures. The guidelines would have been of dubious value, however, were it not for a number of administrative actions accompanying their implementation.

III. THE ADMINISTRATION OF UTILITY R&D PROGRAMS

One of the first steps taken by the New York Commission, following its call for greater R&D funding, was the issuance of guidelines(5) prescribing R&D reporting procedures, defining valid research and discussing research priorities. By this action, the

(2) Case 26105 - Consolidated Edison Company of New York, Inc., March 29, 1972.

(3) Case 27154 - Long-Range Electric Plans, Opinion 78-3, March 6, 1978.

(4) Statement of Charles A. Zielinski before the New York State Legislative Commission on Science and Technology, May 29, 1980.

(5) Research and Development Guidelines for New York State Utility Companies, September 28, 1973.

Commission required annual reports from the companies documenting their research activities. The reports were to include an analysis of research goals, a multi-year research plan and descriptive material on each R&D project. The reports are reviewed by our staff as part of our regulatory review of utility research.

In reviewing current research programs and projects, we try to assure that each program is properly directed toward a reasonable set of research goals and that each research project is consistent with one or more high-priority research objectives. Research goals will differ, of course, from one area of the country to another and from one utility to another. In New York State, for example, the need to reduce our dependence on imported oil leads to such goals as the enhanced ability to burn eastern coal in an environmentally acceptable manner. We have several large ongoing projects relating to that goal. Similarly, electrical transmission problems unique to the Consolidated Edison Company lead to some projects aimed at reducing the cost of AC/DC conversion equipment and to others aimed at developing new ways to generate electricity within New York City. The point I would make here is that there is a role for regulators in helping to make sure that each utility company maintains a proper level of R&D activity prudently directed to high-priority research needs on behalf of the electric consumer. Technical reviews of the research programs make it possible to fulfill this role by helping to identify potential problems as well as opportunities for improvement.

In pursuing these activities, the New York Commission has attempted to provide guidance and direction to research programs that are and ought to be planned and carried out by the companies themselves, not by the Commission. On occasion, however, we have found it necessary or desirable to promote specific research undertakings. In 1975, for example, our Commission issued a statement(6) calling on the utilities to launch a concentrated effort to demonstrate a reliable regenerative scrubber system for the removal of sulfur from the stack gases of coal-fired power plants. The result, due at least in part to this action, was a major scrubber demonstration project on the Niagara Mohawk system.

I might mention that the New York Commission has not attempted to tell the utilities how to develop and manage their internal research programs. Each company develops its own R&D management plan to fit its own needs. Several of the larger New York utilities have established separate research and development departments; some of our companies have set up internal R&D task forces to involve various operating departments more directly in the formulation and conduct of research projects. In addition, the seven major electric utility companies in the State have joined in supporting the Empire State Electric Energy Research Corporation (ESEERCO), which has the responsibility to develop and manage a Statewide research program on behalf of all its members. Our Commission monitors all of these activities very closely and tries, through various regulatory actions, to assist and encourage responsible research.

IV. REGULATORY ACTIONS TO PROMOTE RESEARCH AND ITS APPLICATION

Having established R&D funding guidelines, reporting procedures and regulatory review processes, the Commission turned its attention to ways of promoting useful research and its application. One of

---

(6) Statement of Policy on the Control of Sulfur Emissions from Public Utility Boilers, January 21, 1975.

the Commission's more recent actions in this regard was the institution of deferral accounting of R&D expenditures.(7) Under this accounting procedure, revenues collected for R&D and revenues expended on R&D are recorded in separate subaccounts. The procedure offers several advantages. It assures that funds collected for R&D are used for that purpose only, and it frees the R&D planner from strict adherence to an annual R&D budget and permits temporary distortion of R&D budgets to accommodate a large research project. Any overspending or underspending is simply recorded in the subaccounts, permitting corrective steps to be taken at a later date.

This accounting approach has also been helpful in our efforts to promote the multi-year planning and funding of R&D. Our Commission has required each company to submit a multi-year research plan along with multi-year research funding projections. Where the material is presented in sufficient detail, the Commission can approve the program plan, thereby giving the company added assurance that its long-range research program will have the continued support of the Commission.

As the products of research move closer to commercial application, utility companies and regulators must also find ways to promote needed technologies that are either unproven for utility use or not yet economically competitive. Such technologies as the fuel cell and large-scale solar or wind generating systems are examples of technologies that appear to have much to offer but have not yet demonstrated their reliability to utility standards and would, in any event, prove more costly in the near-term than more conventional electric generation alternatives. Ways must be found to share the technical risks and the added costs imposed by the introduction of such new technologies.

We have made a modest beginning in New York by attempting to establish ground rules for accommodating these kinds of near-commercial undertakings in research budgets. It has been determined, for example, that some research costs may be capitalized rather than being expensed against current revenues, if the intent of the R&D project is to build a commercially useful item of plant. This would help minimize the distortion of R&D budgets that would otherwise result from inclusion of a costly demonstration facility. The Commission also determined(8) that "...a research effort may result in an operable item of plant which is not economical, either because of high initial cost or because of high operating cost over its useful life. In such situations, the difference between the total cost of the item and the prudent cost of an item having similar productive capacity should be considered the R&D effort..." This action paves the way for inclusion in an R&D budget of at least some of the cost of pioneering facilities that are technically acceptable but still more costly than conventional facilities.

It seems to me, however, that the more significant breakthroughs in accelerating the commercial introduction of large new technologies will come from cooperative efforts among utilities, regulators, the government and the investment community to develop acceptable means of sharing the costs and the risks of the emerging technology. As I have already indicated, fuel cells, for all of the potential benefits

(7) Technical Release No. 16, approved by the Public Service Commission, February 6, 1980.

(8) Case 27154 - Long-Range Electric Plans, Opinion 78-3, March 6, 1978.

they offer, will not represent a cost-effective alternative to electric utility companies for a number of years--and then only at a total "commercialization" cost nationwide of hundreds of millions of dollars. Similarly, large-scale wind energy conversion systems hold the promise of low lifetime operating costs, but the initial units will be very expensive and of unproven reliability. It is just possible also that the total investment in a major facility involving technologies such as these might have to be written off as a loss. If we truly intend to accelerate the application of large new technologies, we must recognize that possibility and provide the kinds of incentives needed to promote investment in spite of the risks involved.

V. SUMMARY

The New York State Public Service Commission, beginning in 1972, called on electric utility companies under its jurisdiction to increase R&D expenditures to about 1% of revenues. Shortly thereafter, the Commission issued guidelines for the conduct, reporting and review of utility research activities. These steps helped assure a larger, better planned involvement in R&D by the companies.

The Commission continued its efforts to promote useful R&D by helping to identify needed research, by supporting multi-year research planning and funding, and by instituting such procedural changes as the deferral accounting of R&D expenses. The Commission also has begun to address the challenge of incorporating costly new technologies into the existing system by examining methods by which the costs, or a portion of those costs, can be treated as R&D.

The industry and its regulators have come a long way in seeing to the establishment of a mature and sophisticated utility research capability. The further task of accelerating the application of research results poses new and difficult challenges that we are only beginning to appreciate and overcome. It is clear, however, that the regulator is going to have an important role to play in promoting effective strategies for bringing evolving technologies to full commercial status.

UTILITY RATE DESIGN
OPTIONS & OPPORTUNITIES

René H. Malès
Energy Analysis and Environment Division
Electric Power Research Institute

Introduction

The decade of the 1970's was a time of slowly building credulousness and comprehension that we had entered a new energy era. The prior decade was one in which energy resources were inexpensive and appeared plentiful. The realization which became accepted was that energy would be more expensive and that demand for some forms of energy exceeded supply.

This awareness was built on many events: for example, the control of new gas attachments, the formation of OPEC, the oil embargo, the Iran-Iraq war. In the electricity sector, the change in energy economics was mirrored in the utilities' rising costs of doing business which in turn was reflected in frequent and large requests for rate increases.

When the nation perceives a crisis, it usually spawns an avalanche of responses, some helpful and some irrelevant. It has been no different in case of the perceived energy crisis. Studies have been produced to overload library shelves. A Department of Energy has been created to manage a multibillion dollar budget. Tinkerers and thinkers have invented new and sometimes better devices to produce or utilize fuels. This conference by its sheer magnitude is a testament to the plethora of ideas.

The Electric Rate Design Answer

It was in this environment that was born the idea that electric rates, and more generally load management, could be an answer to some of the challenges of controlling the increasing costs of electricity. A look at history is instructive.

Average electricity prices increased in real terms an average of less than 3 percent per year during the 1970s. But inflation made the actual price increase appear much larger, more than 10 percent per year. Moreover, the costs of producing electricity have risen even more than the price, such that prospective further price increases are assured for some time. For most utilities the increase in cost, however, has not been the same during peak periods and off-peak periods. The cost of peaking power has risen much more sharply, in many cases, than the costs of base load service.

Many have suggested the idea that by shifting customer usage from peak to off-peak periods there could be some reduction in the rate of increase in the utilities' costs. This shifting could be initiated by the direct control of loads by the utility or the customer. Alternatively, it could be achieved by differential pricing with time-of-use rates. Others have suggested that a different basis for deriving rates, namely pricing based on marginal cost, could provide signals to consumers for more efficient consumption of energy, whenever it is used. This would allow the customer to decide for himself if he wants to buy electricity at the incremental cost to the utility.

These suggestions have been studied by EPRI in response to the request of the National Association of Regulatory Utility Commissioners. I have asked Robert Malko, who has been the Project Director of the Rate Design Study, to tell you about the findings of the Study. Bob will also review the options that exist from a rate-making point of view and some of the issues associated with these approaches to rate-making.

The New York State Public Service Commission was one of the earliest to investigate these suggestions for holding down utility costs. Through extensive hearings and innovative actions, they are among the leaders in understanding the problems and opportunities in marginal cost applications and TOU rates. Chairman Charles Zielinski has now heard extensive arguments from utilities, customers, and experts of all varieties. I have asked Chuck to tell you about opportunities he sees to meet the objectives of customers, utilities, and regulatory commissions with utility rate designs.

The objectives I have had in putting together this program is to allow you to get a glimpse at this complex and arcane field of rate design. This is not intended to make you rate design experts. Nor is it intended to prove to you that load management will save you money. Only a specific customer-by-customer, plant-by-plant analysis can accomplish this.

The electric utility industry is facing severe problems of increasing costs. If we approach the problem with the techniques of the past, you are assured of continued increased costs per kWh. This program is intended to demonstrate that one way of reducing the rate of increased costs for many utilities is to get customers to manage loads. A successful load management program could lower costs per kWh over what they otherwise would be. This program is intended to show also that there are opportunities for customers to benefit individually from load management.

LOAD MANAGEMENT OPTIONS

J. Robert Malko and E. Victor Niemeyer
Public Service Commission of Wisconsin
and
Electric Power Research Institute

INTRODUCTION

The significant increase in electricity prices throughout the 1970s prompted consumers, regulators, and utilities to seek means for slowing down the rise in prices. This increase in prices was substantial, relative to the declining trend in electricity prices during the previous two decades. Increases in the costs that electric utilities paid for fuel, for new equipment, and for financial capital during the 1970s were primary factors contributing to rising prices.

One expression of concern on the part of regulators was the passage of Resolution Number 9 by the National Association of Regulatory Utility Commissioners (NARUC), dated December 5, 1974. The resolution called for "a study of the technology and cost of time-of-day metering and electronic methods of controlling peak-period usage of electricity, and also a study of the feasibility and cost of shifting various types of usage from peak to off-peak periods."

In response to this resolution, the Electric Power Research Institute (EPRI), the Edison Electric Institute (EEI), the American Public Power Association (APPA) and the National Rural Electric Cooperative Association (NRECA) joined in sponsoring the Rate Design Study (RDS). Effective participation in the Rate Design Study by the utility industry and its regulators was obtained by the formation of active committees, task forces, and advisory groups, comprised of several hundred persons.

A primary goal of the Study is to improve the ability of regulators and utility managers to assess the desirability of load management. Load Management includes direct load controls and time-of-use rates. The Rate Design Study is divided into three phases. The

Study's first phase (Phase I, 1975-1977) of research was organized into ten topics related to the potential feasibility and value of load management. Results from research on these ten topics are presented in a series of approximately 50 grey covered reports and are summarized in Rate Design and Load Control: Issues and Directions, distributed in November 1977 (1).

In July 1976, NARUC requested an extension of the research activities which were translated into a second phase (Phase II, 1978-1980) of research. During this second phase, the Rate Design Study has developed additional methodologies, data bases, and information on six topics pertaining to the assessment of load management. Results from the research on these six topics are presented in a series of approximately 25 grey covered reports, and summaries of research in each of these areas are being published as six topic papers. In addition, there will be a final report presenting the overall conclusions of the Rate Design Study, Rate Design and Load Control: Final Report and Conclusions.

A third phase (Phase III, 1980-81) of the Study was organized and is being implemented for the primary purpose of transferring the research findings. A series of regional conferences was held during May 1980 to present an overview of preliminary research findings on load management to regulators and utility executives. Workshops on costs and rates and on cost-benefit analysis are scheduled during 1981 to transfer more specific research findings to the technical staffs of utilities and regulatory commissions.

In addition to the research activities of the Study, utilities and state and federal agencies have examined and implemented load management options during the 1970s (2). Moreover, the Public Utility Regulatory Policies Act of 1978 (PURPA) specifies that time of use rates and other ratemaking standards be considered by state regulatory commissions and utilities (3). The methods developed by the Rate Design Study and the experience gained from implementing time-of-use rates should help state regulatory commissions and electric utilities address the ratemaking standards of PURPA.

PRINCIPAL FINDINGS OF PHASE II RESEARCH

The major findings of the Phase II Rate Design Study research are presented in eight general conclusions. The essence of the eight conclusions can be summarized in the following principal findings (4):

- Time-of-use rates and direct load controls can induce customers to lower their peak demand and to shift their electric energy usage from peak to off-peak periods.
- These reductions and/or shifts in electricity use can allow utilities' costs to be lower than they otherwise would be.
- The results of load management programs will depend on utilities' operating characteristics and service territories.
- The specific situations in which particular load management programs are cost effective can be identified with varying degrees of confidence.

The logical policy implications of these findings are:

- Where such situations can be identified, sound utility and regulatory policy calls for implementing the load management programs.

- o Regulators should recognize the large uncertainties associated with load management's costs and benefits, and should adopt policies to mitigate these uncertainties.

The eight conclusions from which the foregoing statements are derived are discussed in detail below.

CONCLUSIONS

The objective of the Rate Design Study was to determine whether load management, defined as both time-of-use rates and load control, could help in holding down costs. It became apparent that utility rate and service practices involve a multitude of other objectives which are in conflict with each other as well as with the objective of controlling peak period usage.

The difficult tradeoffs necessitated by industry as well as national economic and political considerations, coupled with the variations among utilities' operating conditions, require clear specification of objectives prior to any determination regarding appropriate load management strategies. Therefore, based upon its research as well as analyses of historical economic and political trends, the Rate Design Study has concluded that:

CONCLUSION #1: THE OBJECTIVE OF CONTROLLING PEAK-PERIOD USAGE OF ELECTRICITY AND SHIFTING USAGE FROM PEAK TO OFF-PEAK PERIODS SHOULD BE PURSUED WITH REGARD FOR OTHER RATEMAKING OBJECTIVES. THOSE OBJECTIVES VARY BY CUSTOMER CLASS, UTILITY SYSTEM, STATE, AND REGION. THUS THE IMPLEMENTATION OF TIME-OF-USE PRICING AND DIRECT LOAD CONTROL PROGRAMS MUST BE BASED UPON A CLEAR SPECIFICATION OF THE OBJECTIVES TO BE ACHIEVED BY A PARTICULAR APPLICATION.

There is a tradition that costs should be the basis for rates. One question which needed to be examined was whether costs varied by time of use (TOU). In all the cases examined by the RDS, the utilities' costs did vary by time of use. However, this is not enough to conclude that such variation should be reflected in rates. In addition it is conceivable that costs for certain utilities do not vary significantly by time of day.

In those cases where TOU rates are desired, the question becomes which costs are appropriate for ratemaking. This was the heart of the great rate debate between the traditional costing method which we termed "accounting costs" and the newer approach which we termed "marginal costs". The RDS found that in a regulated environment both costs are important. Accounting costs are important because they are used to limit utility revenues to the recorded book costs. Marginal costs are important because they provide signals for economic efficient actions for consumers.

The Rate Design Study research has provided both accounting and marginal cost methods for structuring rates to reflect differences in costs by time of use. The Study also set out the strengths and weaknesses of each method, and relate those to specific rate-making objectives. However, professional opinions differ on the degree to which a given costing method meets such objectives. What the Rate Design Study offers to utilities and regulators are examples of applications of both accounting and marginal cost concepts and experts' comments on each. Further, the Rate Design Study provides a framework for making the difficult judgmental choices among methods as well as guidelines for applying utility-specific data to any particular method.

Thus research on costing and ratemaking has enabled the Rate Design Study to conclude that:

CONCLUSION #2: COST-BASED RATES ACCEPTABLE FOR PRICING ELECTRICITY BY TIME OF USE AND FOR USE IN CONJUNCTION WITH DIRECT LOAD CONTROL DEVICES CAN BE DEVELOPED. SUCH RATES MUST REFLECT THE COMPROMISE MADE IN SATISFYING OBJECTIVES.

The next question was whether equipment needed to implement load management was commercially available. Use of TOU rates requires meters that can track customers' electricity consumption by rate period. Equipment needed for direct control of customer loads can include one- and two-way communication systems and end use energy storage devices.

As experience is gained through the use of such hardware in the field, more equipment is reaching commercial maturity. For example, research and development is being conducted on two-way communications systems as well as on better and less costly load control devices and energy storage equipment. Significant improvements in these technologies are technically feasible, but further research is needed.

Therefore, the Rate Design Study has determined that:

CONCLUSION #3: TECHNOLOGY EXISTS FOR TIME-OF-USE METERING AND FOR DIRECT CONTROL OF CERTAIN TYPES OF END USE LOADS. BOTH HAVE BEEN USED IN COST-EFFECTIVE LOAD MANAGEMENT SYSTEMS AND SUCH EQUIPMENT IS INCREASINGLY AVAILABLE. WHILE RELIABLE, REMOTE ONE-WAY COMMUNICATIONS SYSTEMS EXIST, TWO-WAY SYSTEMS ARE STILL IN THE RESEARCH AND DEVELOPMENT STAGE, AS IS TECHNOLOGY FOR CONTROL OF OTHER TYPES OF END-USE LOADS.

Customer attitudes and degree of understanding and acceptance of load management programs can be important in determining load management results. Recognizing a need for information on these consumer-related issues, the Rate Design Study has developed a framework for conducting and evaluating research on customer acceptance. Through this type of research it has found that:

CONCLUSION #4: A GOOD POTENTIAL EXISTS FOR CUSTOMERS TO UNDERSTAND TIME-OF-USE PRICING AND DIRECT LOAD CONTROL TECHNIQUES. FURTHER, CUSTOMERS TENTATIVELY ACCEPT THE IDEA OF TIME-OF-USE RATES IN GENERAL, BUT NOT NECESSARILY AS THE MOST PREFERRED RATE DESIGN. SUCCESSFUL IMPLEMENTATION OF LOAD MANAGEMENT PROGRAMS WILL REQUIRE A COMMITMENT BY BOTH UTILITIES AND REGULATORS TO AN EDUCATIONAL PROGRAM THAT WILL INFORM CUSTOMERS ABOUT THE CONCEPTS, DETAILS AND POTENTIAL ADVANTAGES OF LOAD MANAGEMENT.

One of the requirements in planning a successful load management program is an understanding of how customers use electricity through load research. Load information is needed for allocation of costs among customer classes and rate periods, selecting rate periods, and in estimating how much load can be directly controlled. Thus the RDS found that:

CONCLUSION #5: THE QUALITY AND ADEQUACY OF LOAD RESEARCH WILL AFFECT DIRECTLY THE SOUNDNESS OF DECISIONS CONCERNING THE PLANNING AND IMPLEMENTATION OF TIME-OF-USE RATES AND DIRECT LOAD CONTROL PROGRAMS.

For many smaller utilities the cost of a sufficient load research program will be more than they can afford. Therefore, combined programs among utilities and understanding how to use data developed by others will be necessary. Transferability of load research data is a question needing further research.

Looking at the results of a number of TOU and load control experiments, the Rate Design Study was able to conclude:

CONCLUSION #6: IN SPECIFIC SITUATIONS, TIME-OF-USE RATES AND DIRECT LOAD CONTROLS CAN INDUCE CUSTOMERS TO LOWER THEIR PEAK-PERIOD USAGE OR SHIFT THEIR USAGE FROM PEAK TO OFF-PEAK PERIODS. ESTIMATING THE MAGNITUDE OF SUCH REDUCTIONS AND/OR SHIFTS IS UNCERTAIN. SOME OF THIS UNCERTAINTY CAN BE REDUCED BY FURTHER EXPERIENCE.

Whether such load shifts are cost-effective is a very complex question. It depends not only on the cost effects on the utility but on customers as well. Also, there may be effects on the society as a whole. The Rate Design Study developed a number of ways to approach this analytical challenge and concluded:

CONCLUSION #7: LOAD MANAGEMENT PROGRAMS CAN BE COST EFFECTIVE IN CERTAIN CIRCUMSTANCES. THE IMPACTS OF TIME OF USE PRICING AND DIRECT LOAD CONTROL PROGRAMS VARY BY CUSTOMER CLASS, UTILITY SYSTEM, STATE, AND REGION. THUS THE RATE DESIGN STUDY HAS DEVELOPED BASIC COST/BENEFIT METHODOLOGIES FOR EVALUATING SPECIFIC LOAD MANAGEMENT PROGRAMS TO DETERMINE THE LIKELIHOOD OF THEIR COST EFFECTIVENESS.

Finally, we cannot be sure of consumer responses, of the savings to the utility, of the cost of equipment, or of many other facets of load management programs. This argues for a degree of caution in implementing such programs. The Rate Design Study suggests:

CONCLUSION #8: REGULATORS SHOULD RECOGNIZE THE UNCERTAINTIES ASSOCIATED WITH LOAD MANAGEMENT'S COSTS AND BENEFITS AND SHOULD ADOPT POLICIES TO MITIGATE THESE UNCERTAINTIES, SO THAT UTILITIES AND CUSTOMERS CAN REALIZE THE POTENTIAL BENEFITS OF LOAD MANAGEMENT PROGRAMS. THESE UNCERTAINTIES ALSO NECESSITATE PERIODIC MONITORING OF LOAD MANAGEMENT PROGRAMS TO ASSURE CONTINUED COST EFFECTIVENESS.

SUMMARY

This paper summarizes the results and conclusions of the research activities conducted for NARUC by the Rate Design Study. The Study's five years of research will assist state regulators in their comprehensive analysis of various aspects of time-of-use pricing and direct load controls as required by PURPA. NARUC's timely request of December 1974 and its continuing support of research has produced information that state regulators will be able to use in the early 1980s to consider and evaluate load management alternatives. Ratepayers will benefit in the decade ahead through a careful application of the lessons learned in load management activities.

FOOTNOTES

(1) For a complete listing of RDS reports, see "Guide to Electric Utility Rate Design Study Reports," Electric Utility Rate Design Study, EPRI, Palo Alto, California, October 1980.

(2) Lessons in Load Management, prepared by National Economic Research Associates, Inc., Electric Utility Rate Design Study, Report No. 91, EPRI, Palo Alto, California, June 25, 1980.

(3) Reference Manual and Procedures for Implementing PURPA, Electric Utility Rate Design Study, Report No. 82, EPRI, Palo Alto, California, March 1980.

(4) These findings are discussed in much greater detail in seven topic papers, some of which are already published and the remainder will be available in early 1981. Consult the following Rate Design Study reports: #84, 85, 86, 87, 88, 89, and 90.

ELECTRIC RATE DESIGN: REGULATORY OBJECTIVES AND ECONOMIC OPPORTUNITIES

by

CHARLES A. ZIELINSKI, CHAIRMAN
NEW YORK PUBLIC SERVICE COMMISSION

On February 19, 1981, a United States District Court Judge in the Southern District of Mississippi issued a six-page Judgment holding the Public Utility Regulatory Policies Act of 1978, commonly known as PURPA, an unconstitutional infringement on the power of states to regulate electric utility companies. (1) In this extraordinary opinion, the District Court Judge found, among other things, that "the jurisdiction of...[the Mississippi] Commission has been accepted and approved by profound lawyers and courts alike since memory of man runneth not to the contrary"; and, the "sovereign state of Mississippi is not a robot, or lackey which may be shuttled back and forth to suit the whim and caprice of the federal government...." The Judge does not, however, even recite what PURPA requires of state public utility commissions.

In fact, PURPA imposes only a requirement that state public utility regulatory commissions consider adoption of certain regulatory practices and ratemaking standards, (2) in light of the conservation, economic, efficiency and equity objectives. Since this

---

(1)The State of Mississippi v. Federal Energy Regulatory Commission, Civil Action No. J79-0212(c), So. Dist. Miss., judgment of District Court Judge Harold Cox, issued February 19, 1981.

(2)Standards relate to: Cost of Service, Declining Block Rates, Time-of-Day Rates, Seasonal Rates, Interruptible Rates, Load Management Techniques, Master Metering, Automatic Adjustment Clauses, Information to Consumers, Termination of Service, and Advertising.

represents a requirement to consider actions in furtherence of goals that most state regulators embraced even before enactment of PURPA, it is highly unlikely that the statute's demise -- assuming, of course, that the Supreme Court upholds the extraordinary judgment of the Mississippi District Court -- would change what state regulatory commissions are now doing. In particular, it is unlikely that state regulators will stop their pursuit of change in electric utility rate structures not only for the purpose of encouraging conservation, economic efficiency and equity, but also to try to hold down the electric utility cost increases that have caused a continuing stream of requests to increase electric rates throughout the country. (3)

In the 1950's and '60's -- the era of cheap energy -- when inflation was modest and more than outweighed in its effect on the cost of electric utility service by the scale economies of expanding production, state regulatory commissions, for the most part, left the design of electric rates to the industry. The electric companies were only rarely requesting rate increases and, when they did, commissions could often cut them sharply. Indeed, they could even negotiate decreases, with a confident expectation that continuing expansion of demand and declining unit costs would produce more than satisfactory company profits and generally good service.

The high inflation rates of the 1970's, in general, and the dramatic increases in the price of oil (used by many electric utilities as a boiler fuel), in particular, brought an abrupt halt to these halcyon days. The cost of providing electricity began to rise dramatically, and requests to increase rates were visited upon regulatory commissions with increasing frequency. And this trend has, thus far, continued into the 1980's.

In these circumstances, even without guidance from a statute like PURPA, regulators look to initiatives like the redesign of electric rates to reflect, as closely as possible, marginal costs rather than average costs, as a way of controlling rising costs.This means redesigning rates so that they differ in accordance with the time of peak demand on an electric system, when a company is, in the short run, incurring higher operating costs to serve each Kwhr demanded because it must run its very inefficient generating equipment, probably fueled by oil, or purchase more expensive power. In the longer run, increases in peak demand also require the company to build new capacity -- unless it has very substantial capacity reserves already -- and the costs of it also should be accounted for in designing rates for service at peak times.

Rates designed in this manner are, of course, economically efficient, because they reflect the differing cost to society of satisfying demand for electricity at peak and off-peak times. But our hope is that rates reflecting the higher cost of peak demand will cause people to take less electricity at the peak, thus conserving the high cost of the greater amounts of fuel that are needed to satisfy such demand in the short run, and the cost of additional capacity in the long run. And, since these rates impose the proper cost of providing electricity on those who demand it, they also promote equity among consumers.

---

(3)A compilation of recent activity of this nature by most state commissions may be found in "Report on State Progress Under the Public Utility Regulatory Policies Act," conducted by the National Association of Regulatory Utility Commissioners (NARUC) Ad Hoc Committee on the National Energy Act (of which I am Chairman), issued December, 1980.

In short, prices based on marginal costs not only serve the goals of economic efficiency, conservation, and equity, but also hold the potential for helping to control electric utility cost, and general rate increases. For these reasons, state regulators see, and are likely to continue to see attractive opportunities in the redesign of electric rates to reflect more closely marginal costs.

The form in which this regulatory interest in rate design is expressed will depend, of course, on the specific conditions on an electric utility system and the state of technology. For example, in principle, rates that vary by time-of-day ought to be charged to all customers to reflect the higher cost of satisfying demand during daily peak periods,and the lower cost of serving demand during night-time off-peak periods. But time-of-day rates require the use of time-of-day meters, and these devices may cost $200 or more each (although that price appears to be declining.) A responsible regulatory body must therefore determine whether the benefits that could be realized from the introduction of time-of-day meters -- in terms of greater economic efficiency, conservation inducement, equity, and system cost savings -- would exceed the cost of investing in time-of-day meters.

In making this calculation in New York, we have not thus far found mandatory time-of-day rates for the residential and small commercial customer classes, whose per customer consumption is, for the most part, relatively modest, to be justified. Instead, we have directed companies to operate experiments with groups of customers, and to offer customers who volunteer a time-of-day rate if they are willing to cover the cost of the meter.

In addition, we have imposed on all of these customers seasonal rate differentials when the company involved experiences a pronounced summer peak demand which is the case with, for example, Consolidated Edison and Long Island Lighting Company. These higher rates for all consumption during the summer period, and lower rates for all consumption during other parts of the year are not a perfect reflection of the marginal consumption costs, but they are a better approximation of marginal costs than rates without seasonal variation, and they do not require investment in time-of-day meters. (4)

Mandatory time-of-day rates for all customers consuming fairly large amounts of electricity, however, are generally a sensible regulatory initiative, for at least two reasons. First, many of these customers already have their consumption recorded on meters that could be used for time-of-day rates, so no further metering investment cost need be incurred. Second, the amount of consumption per customer being influenced here by time-of-day rates makes for a much higher probability that cost savings induced by the time-of-day rates will offset the costs of implementing them.

---

(4)Some people fear that seasonal rate differentials will cause a decline in summer consumption mainly through decreases in daily off-peak consumption, without reducing daily peaks. In these circumstances, engineering efficiency could worsen as load factors deteriorated and existing peaks become needle peaks. But we have had seasonal differentials in New York for a number of years now, and this fear has not been realized.

In New York, this reasoning has led the Commission to mandate time-of-day rates for large commercial and industrial customers throughout the State. On Consolidated Edison's system, there are 200 customers, representing about 14% of the company's total sales, that are now subject to these rates. On Long Island Lighting Company's system, there are also 200 customers representing around 11% of its total sales. On the Central Hudson and Orange & Rockland systems, there are 80 customers equally divided between the two, representing approximately 20% and 15%, respectively, of their total sales. Niagara Mohawk Power Corporation now has 125 large customers subject to time-of-day rates, representing about 25% of its total sales. And the Commission, most recently, ruled that 135 large customers of Rochester Gas & Electric Company, representing 32% of its total sales, should be charged time-of-day rates. (5) In total, these actions represent bringing something like 20 billion kilowatt hours of historical consumption under the influence of time-of-day rates. (6)

The New York Commission has not, however, followed a policy of immediately setting time-of-day rates equal to estimates of <u>full</u> marginal costs, when marginal costs are clearly well above embedded costs. It has set the rates at a somewhat lower level in these circumstances to avoid abrupt, sharp increases, with an intention to increase the rates gradually, over time, toward full marginal costs. This approach serves a number of purposes. First, it gives customers more time to become accustomed to the new rate structure without having to change radically their consumption patterns at the outset to avoid a steep rate increase. Second, it allows the Commission to move toward its goal, while still taking into account future analytical developments in the proper measurement of marginal costs, which is by no means a simple or uncontroversial process. And, third, it properly informs customers of the direction in which their rate structures will continue to go so they can make investments in longer term initiatives that will reduce their consumption at peak times while, perhaps, increasing consumption in off-peak periods. (7)

---

(5)For the specific details of the Commission's reasoning and the time-of-day rates it found reasonable, see: Case 26806, <u>Order Determining Relevance of Marginal Costs to Electric Rate Structures</u>, Opinion No. 76-15, issued August 10, 1976; Case 26887, <u>Long Island Lighting Company</u>, Opinion No. 76-16, issued December 16, 1976; Case 27029, <u>Consolidated Edison Company</u>, Opinion No. 79-6, issued March 15, 1979; Case 27215, <u>Niagara Mohawk Power Corporation</u>, Opinion No. 80-18, issued May 7, 1980; Cases 27344, 27553, <u>Orange & Rockland Utilities</u>, Opinion No. 80-35, issued October 22, 1980; Case 27548, <u>Rochester Gas & Electric</u>, Opinion No. 81- , issued March, 1981.

(6)A number of other state commissions, for example, Wisconsin and California, have also already imposed time-of-day rates on large customers. For a listing of the companies in these and other states already subject to or soon to be subject to time-of-day rates, see the report cited in footnote (3), <u>supra</u>.

(7)When marginal costs are <u>below</u> embedded costs of course, regulators have little reason not to reflect them immediately. In the opposite situation, it may ultimately be necessary to charge rates based on full marginal cost only to customers with elastic demand to avoid producing revenues in excess of embedded cost. This would allow regulators to charge something below marginal cost to customers with relatively inelastic demand, whose consumption is not greatly influenced by price.

This, of course, leads me to the relevance of these electric rate structure changes to a conference and exposition like this one. New electric energy technology should no longer be thought of simply as new developments in the production, transmission and distribution of electricity, or as new developments in the transformation of electric energy into mechanical energy. It now should also comprise technologies that will help industries to restrict and shift their consumption of electricity in response to time-of-day rates. (It should also obviously include developments in metering technology, particularly less expensive time-of-day meters that will make it economically sensible to impose mandatory time-of-day rates on all customers.)

Customers subject to time-of-day rates are clearly becoming a very large market that entrepreneurs should be, and undoubtedly are assessing. These customers will be interested, in particular, in electric storage technologies that would allow them to purchase more electricity in off-peak periods when rates are cheaper, and then to use it at times of peak demand when rates are high in lieu of making purchases from their utility company. They will be interested also in technology that will interrupt automatically less important electric consumption during high cost peak periods -- for example, computer controlled cycling of air conditioning load -- to hold down their operating costs without necessarily interrupting their production. In short, these customers will be increasingly looking for ways to take proper advantage of time-of-day rates in order to maintain or even increase the profitability of their businesses.

Whether this entrepreneurial opportunity for new energy technology will continue depends of course on a continuing regulatory commitment to aligning electric rates with the proper cost of consumption. A growing body of evidence suggests that the commitment is there. And so long as regulators see their ratemaking objectives as economic efficiency, conservation, equity and holding down general increases in electric rates, it is likely that that commitment will continue.

ELECTRIC UTILITY FORECASTING OF CUSTOMER COGENERATION AND THE INFLUENCE OF SPECIAL RATES

Frederick H. Pickel
Utility Systems Group
Energy Laboratory
Massachusetts Institute of Technology

Cogeneration, or the simultaneous production of heat and electric or mechanical power, emerged as one of the main components of the energy conservation strategies in the past decade. Special tax treatment, exemptions from fuel use restrictions, and regulatory policy changes were crafted to encourage its more wide-spread adoption in anticipation of higher energy conversion efficiencies. The expansion of cogeneration still faces a broad spectrum of problems, current and future: environmental restrictions; capital constraints; fuel prices; utility rates and future utility economics; and the difficulties of management.

The most debated issue has been the reform of rates between individual cogenerators and the local electric utility. Many of the major cogeneration studies in the late 1970's urged an analysis of the exact impact from current electric utility rates upon cogeneration project economics (1,2,3). The changes mandated by the Public Utilities Regulatory Policy Act of 1978 (PURPA) are now reaching the final implementation stage and the cogeneration projects of the mid-1970s are nearing completion. To better understand the relationship between utility rates, the economics of cogeneration, and its potential development, the New England Electric System and the Massachusetts Institute of Technology Energy Laboratory Utility Systems Group began a study to refine methods for forecasting cogeneration in a specific utility service area with special attention devoted to the utility rates (4).

This paper surveys the insights gained from this effort, which is now nearing completion. Many of the central issues reflect conditions in New England, but this analysis should provide an approach for examining the question in other regions as well. Since the project has not undergone complete review, however, this paper reflects the opinions of the author alone.

OVERVIEW

A complex network of details governs the operating and resulting economics of a cogeneration plant. For example, as anyone who has tried to read a typical industrial rate schedule knows, a variety of changes in electricity consumption patterns can alter electricity bills for years. Cogeneration plants are based upon a broad range of designs, with substantial differences in the fundemental technologies. Changes in fuel prices and other economic conditions mutually affect utility electricity production costs and the costs of cogenerated power. In New England, both industrial steam needs and the utility's load and operating costs are influenced by weather conditions, especially temperature. Figure 1 illustrates the major interactions between a customer, the utility, and economic conditions.

Cogeneration by an electric utility's customers affects it through several factors: utility revenues are reduced; power system operating costs change; and the overall profile of electricity production by the utility alters. To estimate the potential for local cogeneration and the associated economic conflicts between rates, anticipated electricity and fuel supplies, and cogeneration opportunities, the range of customers who could cogenerate must be assessed. Since it is the individual industrial or large commercial customer who eventually decides whether or not to build a cogeneration plant, this forecasting effort started with a very detailed look at cogeneration from the customer's perspective. The study proceeded in three steps:

1. Conduct a series of detailed on-site visits to about a half dozen plants and examine the cogeneration options available for these plants;

2. Distill the factors in the decision by a customer to build a cogeneration plant and apply this simplified analysis to a prior, extensive survey covering about two thirds of the major industrial customers;

3. Use the information from possible cogeneration development at the visited plants and the surveyed customers to make a system level forecast.

To express the complexity of the economics and the utility rates, the project developed a detailed computer model of the effects of differing cogeneration plant types and sizes upon on a customer's electricity bill, fuel costs, and capital expenditures. This model estimates annual operating costs from a shift-by-shift simulation of plant operation over a multi-year period. The expected minimum annual operating costs together with the financing costs and construction delays from different technologies are used to compute the net present value of the alternatives over the life of the project. Since there are substantial uncertianties in future fuel and electricity prices, the model calculates the discounted worth of the projects under several alternative scenarios allowing for operation to adapt to each year's economic conditions.

The great detail of this model, however, can be an obstacle to interpretation of the problem by the non-specialist. This paper indentifies the major factors influencing customer cogeneration plant choice and discusses them in the context of the number and size of customers who might cogenerate.

THE CUSTOMER'S PERSPECTIVE

When addressing the problems confronting the customer, two special factors are relevant in New England. One, an unanticipated

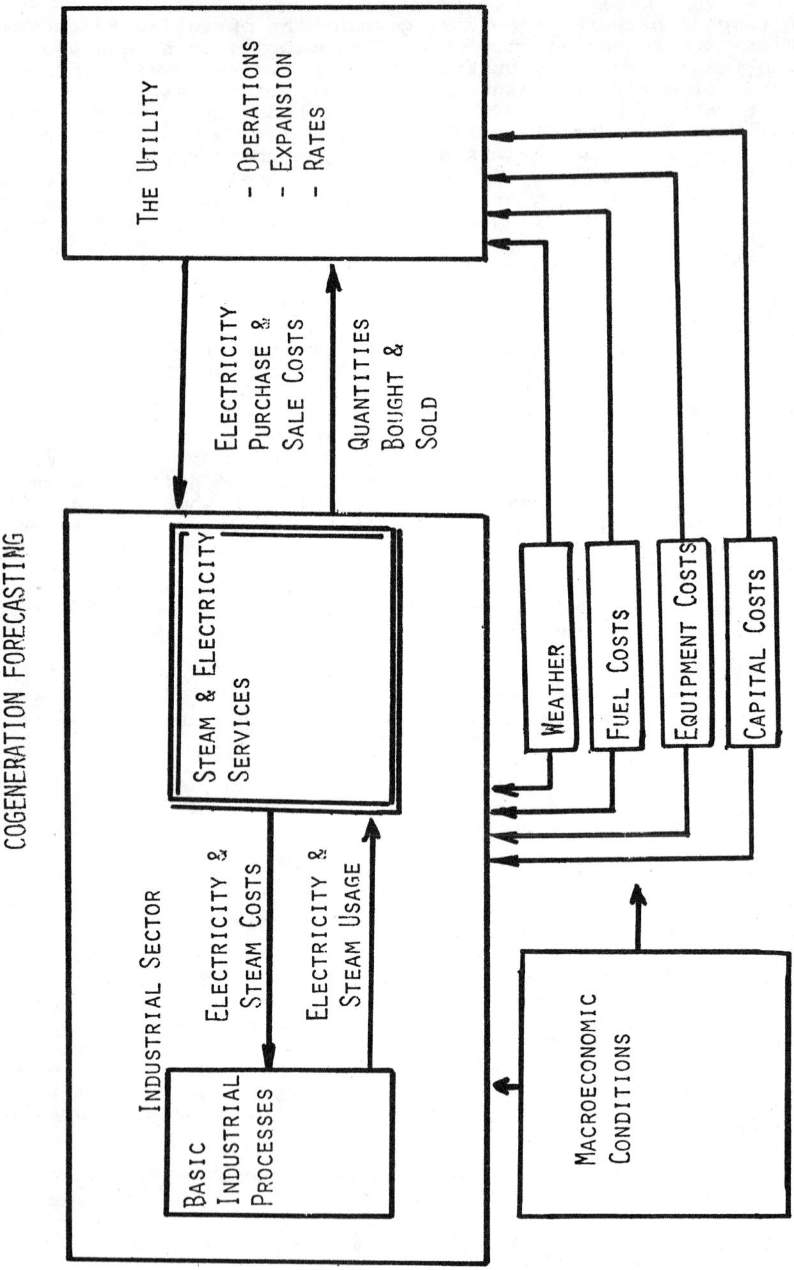
COGENERATION FORECASTING
THE UTILITY
- OPERATIONS
- EXPANSION
- RATES
ELECTRICITY
PURCHASE &
SALE COSTS
QUANTITIES
BOUGHT &
SOLD
INDUSTRIAL SECTOR
STEAM & ELECTRICITY
SERVICES
ELECTRICITY &
STEAM COSTS
ELECTRICITY &
STEAM USAGE
BASIC
INDUSTRIAL
PROCESSES
WEATHER
FUEL COSTS
EQUIPMENT COSTS
CAPITAL COSTS
MACROECONOMIC
CONDITIONS

Figure 1

finding of the on-site visits was the degree of fluctuation in steam usage even in the largest, most energy intensive plants. One site claimed constant steam loads, but examination of the fuel use records showed a winter/summer variation of 30%; this was the lowest variation found. A Major Fuel Burning Installation that already cogenerates had a winter/summer steam load variation of 3:1. Since cogeneration is a capital intensive technology requiring a high capacity factor for economic design, this means that the cogeneration potential should not be estimated from the total annual fuel consumption statistics. A method was needed, then, to calculate the base-load, year-around steam needs at the industrial sites.

Second, since New England is distant from traditional coal sources, oil is the most commonly considered fuel for anticipated cogeneration and boiler facilities. The extent to which New England utilities successfully shift away from oil could render operation of oil-fired cogeneration projects uneconomic although the early project years could produce short-term savings for the project sponsors. For example, as a part of NEESPLAN, New England Electric is planning to reduce its energy production from 78% oil in 1979 to 16% by 1996 through converstion of existing plants to coal and the promotion of conservation and alternate energy projects.

Mindful of these conditions and the large uncertainties in future oil and electicity prices and supplies, the potential cogenerator must ponder the following qusetions:

- In operating the plant,

  -how do you dispose of the power generated by the plant?

  -how would you operate the plant given the electricity usage or sale arrangements?

- In selecting the design,

  -how much do the alternative cogeneration or steam plants cost ,and what uncertainties could make the choice of a given plant regretable?

  -how big should the cogeneration plant be in relation to the steam needs?

  -how should the final size, fuel capability, and technology be determined?

The utility needs to explore which cogenerating customer actions might most significantly affect the stockholders or other customers. This project built several detailed simulation models to tie these and other factors together in analyzing the possible cogeneration plant economics.

Sale of power First, consider the short and long-term disposition of the power produced by the cogeneration plant. The new PURPA rules, in a sense, simplified this problem by allowing the customer to sell all the output of the cogeneration plant directly to the connected electric utility while the customer continues to purchase all its power requirements from the utility (arbitrage); in this case, in-plant electricity usage patterns are of no consequence to the economics of the cogeneration plant. The customer may alternatively sell only the power in excess of its needs to the utility; in this case, the in-plant usage patterns can be important to the cogeneration plant economics. This choice, shown in Table 1, simply depends upon whether the customer can gain more from selling its net needs or by selling in an arbitrage arrangement. The impact on the utility

POWER SALES CHOICE

	Standard Industrial Rate > Avoided Costs	Standard Industrial Rate < Avoided Costs
Customer Choice	Sell power net of needs	Sell all cogenerated power output
Impact on utility and other customers	utility revenue is down more than cost so the rate base is concentrated across reduced sales	unaffected

Table 1

will be discussed later.* As electricity cost and rate conditions change, customers will want to switch from one type of sales arrangement to another.

Operation Next, how the cogeneration plant is operated depends upon the performance of the specific technology, its fuel price, and the price at which the customer can sell the power. Assuming that the cogenerator sells all its power at avoided fuel costs under an arbitrage contract simplifies this discussion, although this only reflects the conditions of an excess capacity utility with rates below the incremental cost of fuel for power generation. Figure 2 illustrates the possible conditions in the relationship between the cogenerator's and the utility's incremental generation heat rates and their fuel costs. For example, if the cogenerator owns a steam turbine back-pressure cogeneration plant with an incremental heat rate of 4500 Btu/kWh running on low sulfur #6 oil, the "avoided cost" utility plants have a heat rate of about 10000 Btu/kWh on coal, and the coal costs 50% per Btu as much as the low sulfur oil, then the cogeneration plant would be in operation (point A). If it was a diesel cogeneration plant with an incremental heat rate of 6600 Btu/kWh, it would be cheaper to shut down the cogeneration plant and get the steam needs from a back-up package boiler also running on the low sulfur #6 oil (point B). If the utility was also running on low sulfur #6 oil as its avoided cost fuel, either type of cogeneration plant design would be economic to operate. If the steam turbine cogeneration plant operated on coal, however, it could always compete with any coal-fired generation by the utility.

This is intended to demonstrate the importance of looking ahead to possible modes for the cogeneration plant operation through-out the projected life of the plant. A plant that is not operating does not help recover its capital costs. On the other hand, a slightly economic plant can become more advantageous if a utility slips from coal as its incremental fuel to oil on the increment. Given the unknowns in coal conversion plans by utilities and in future oil prices, a plant's economic analysis must allow for flexible operating conditions. This can mean the switching from continuous operation to cooperative dispatching with the utility when oil is the incremental

* For simplification, this paper discusses all power purchases in terms of 100% of the utility's avoided fuel costs; in practice, the need for long-term contracts often results in power purchase agreements in which the current year's payment is below from 100% of the current avoided costs. If the utility needs to make no new capacity commitments, then fuel costs reflect the only avoided costs.

ECONOMIC OPERATION OF A COGENERATION PLANT

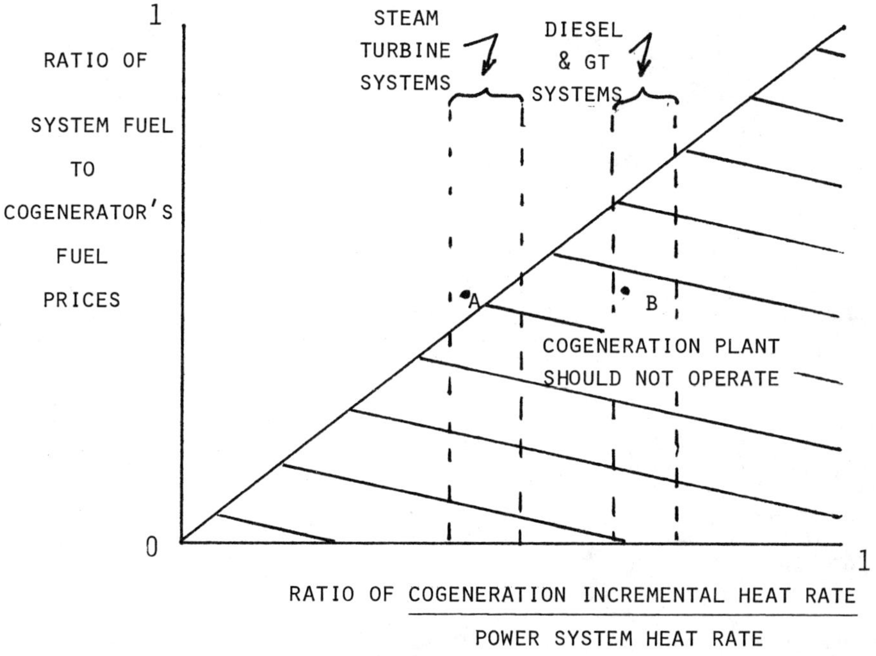

Figure 2

fuel (reference 5 suggests one mechanism for carrying out such dispersed dispatching).

Total plant costs Most approaches to examining the total costs of cogeneration look at the cost in cents per kilowatt-hour. When taking the customer's perspective, however, it is more meaningful to calculate the costs in dollars per million BTUs supplied, although the two approaches are theoretically equivalent. Figure 3a shows the O&M, fuel, capital, cost and power credit components for a #6 oil-fired diesel cogeneration plant on the basis of its design operating time per year; the cost of operating an existing #6 oil-fired boiler is shown at the side. Since the fuel cost and the operating power credit are such a major portion of the total costs, small changes between them heavily influence the final economics. In contrast, the oil-fired back-pressure steam turbine plant cost components shown in Figure 3b have a lower share influenced by the power credit; also a reduction in the plant's capacity factor does not increase the total costs as quickly as in the diesel design. Figure 3c shows the costs for a coal-fired back-pressure steam turbine cogeneration plant. In a region that is burning oil for power generation, if you can meet the environmenal restrictions, it has an overwhelming economic advantage. The size of the different cost components demonstrate how sensitive each plant type is to changes in the underlying factors. In an area that is so uncertain, this gives a quick feeling for the comparative shifts in total plant costs as a function of shifts in the relative prices of fuel, power, O&M, and capital equipment.*

Figure 4a summarizes the total costs for the different plant designs on the basis of 1981 fuel costs in the northeast. The coal-fired system breaks even in cost with an existing oil-fired boiler when it is sized to achieve a very low 35% capacity factor, thus serving most of a typical customer's annual heat load. This extends down to plants significantly smaller than 100 MBtu/hour. If coal is not a viable option, oil-fired cogeneration plants do not appear favorable unless they are designed to operate at high capacity factors; both break even with an existing boiler when sized for about a 75% capacity factor. Even then, they are not a risk-free alternative. Figures 4b and c show two comparative cases for 1985. In the case most advantageous for the cogeneration plants, oil prices have risen by 28% and the local utility still has oil as its incremental fuel 100% of the time. In the worst case, oil prices have dropped 7% while the utility has converted enough of its plants to coal so that oil is the avoided cost fuel only 60% of the year. In this case, it is uneconomic for the diesel plant to operate when coal is the incremental utility fuel, so the diesel plant capital costs must be spread over fewer operating hours. Under these price conditions, the oil and coal steam turbine plants still operate, but the power credit is reduced to reflect the share of coal in the avoided costs.

In summary, uncertainties in fuel and electricity supplies require the examination of possible future operating conditions and not just prices and conditions during the next few years. This demonstrates the flexibility needed in adapting utility rates and cogeneration plant operation. These in turn, show the large uncertainties in cogeneration plant costs; these are risks the customer must recognize. Furthermore, most oil-fired designs are only economic for serving the base steam loads in a retrofit installations.

---

* These costs were derived from representative plant costs listed in the appendix. They reflect differences in tax treatment. Unless noted, the avoided utility fuel is assumed to be oil. The discussion of only diesel and steam turbine cogeneration systems is indended for illustration.

THE COMPONENTS IN THE COST OF COGENERATION SYSTEMS ON A PER UNIT HEAT BASIS

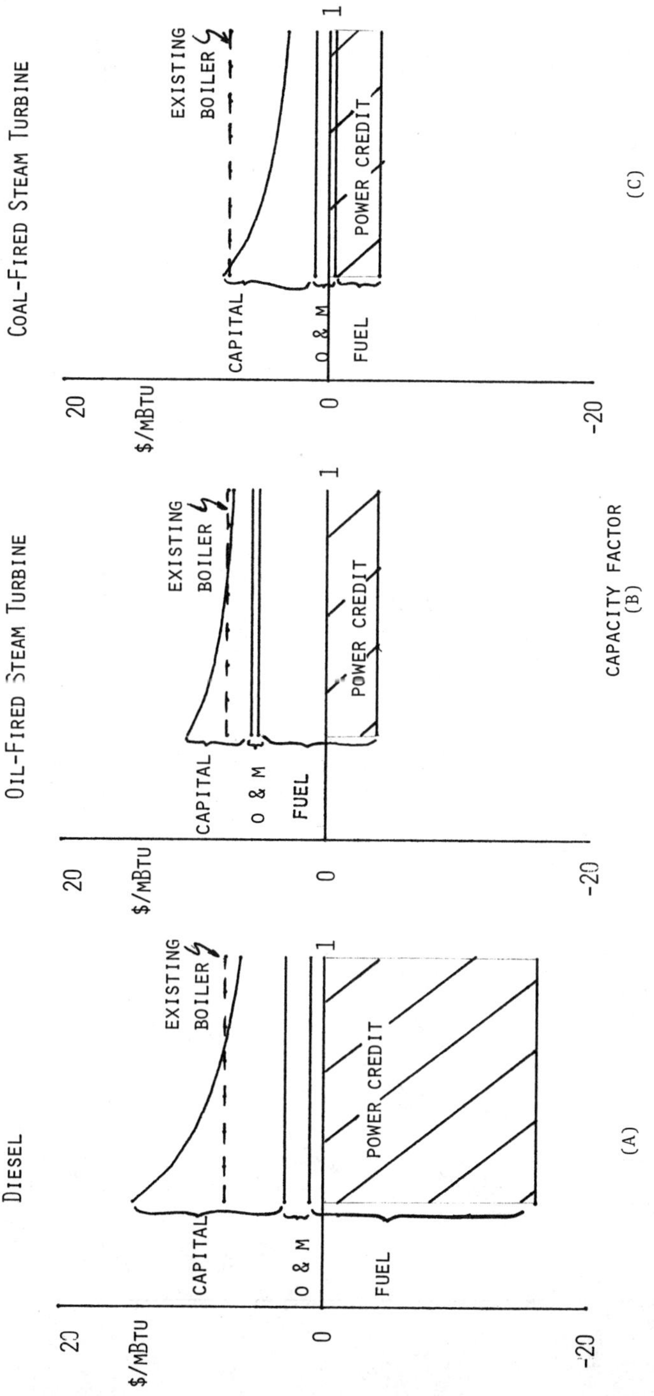

Figure 3

THE COMPARATIVE COSTS OF COGENERATION UNDER DIFFERENT SCENARIOS

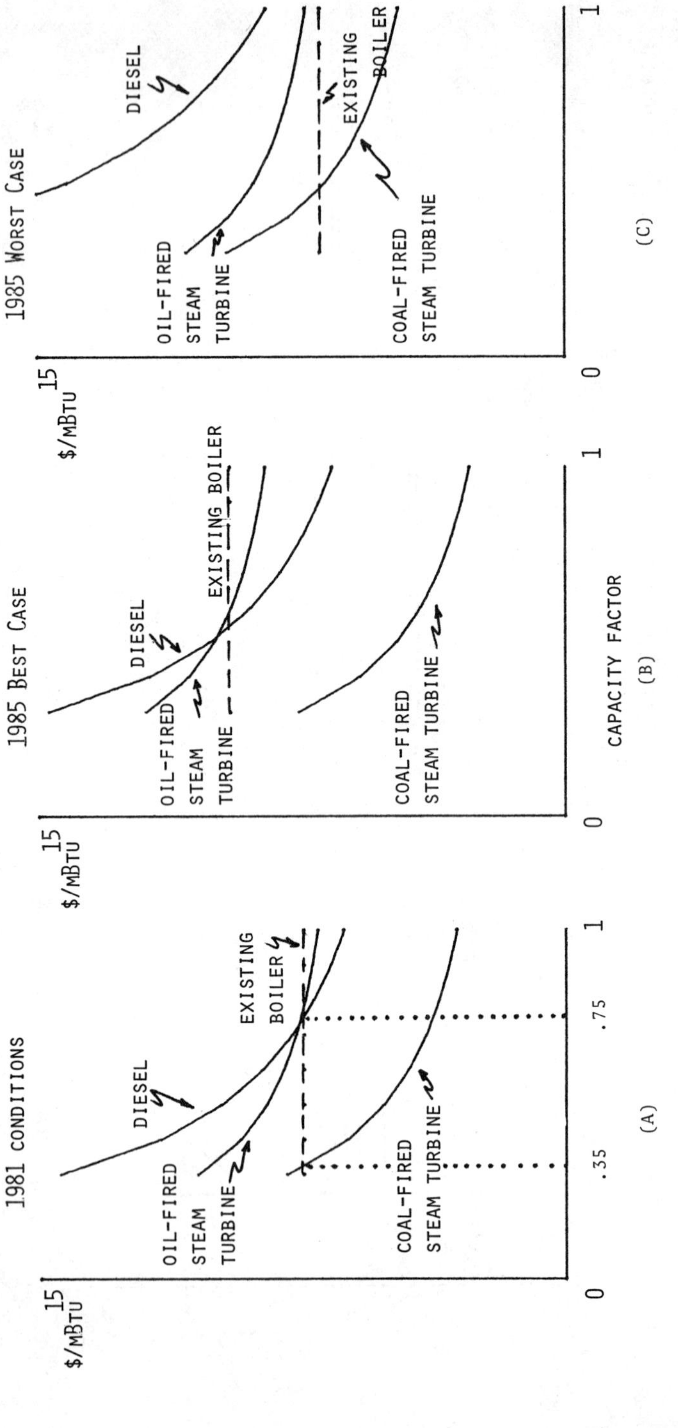

Figure 4

ESTIMATING THE TOTAL IMPACT OF CUSTOMER COGENERATION

A quick, approximate upper limit to the total amount of customer cogeneration can be found by multiplying the total steam load that cogeneration plants could serve and the typical ratio of power to heat output for the corresponding design. Since oil-fired designs are only economic for base load, this requires an estimate of the base heat load, which is difficult to estimate from annual average fuel use; the relationship between the two will vary substantially between even similar plants in the same region. This is the key weakness in many earlier projections.

If coal-fired cogeneration plants can be installed to meet the local environmental requirements, the estimate can be based on the cumulative annual average heat load. The power to heat output ratio for the coal system, however, is much lower than the diesel plant, so the estimate is substantially reduced.

Figure 5a shows data similar to that collected in the 150 customer surveys plotted with the base heat load of each site on the bottom axis and the cumulative hourly base heat load for that and all larger sites plotted on the vertical axis. Each vertical bar represents one customer, demonstrating the concentration of customers in the smaller size range. For example, the total hourly base heat load for all sites greater than 20 MBtu/hour is about 300 MBtu/hour. If all this load was served by steam turbine cogeneration, a total of about 15-20 MW of cogeneration could result; if it was all served by diesel, 90-120 MW could be in service. The disparity in the estimates reflects the significant difference between the technologies in the power output per unit heat load served, as the comparative sizes of the power credits in Figures 3a and b illustrate. This crude approach must be expanded to account for existing cogeneration at these sites; in the actual survey, much more than half of the group with over a 20 MBTU/hour base heat load already had old oil-fired steam turbine cogeneration plants on their sites.

Figure 5b shows cumulative average annual heat loads plotted against each site's average annual per hour heat load. A comparison with Figure 5a demonstrates the low load factors for heat use in New England and also shows the paucity of energy intensive industries. It is this average load chart that can be used for estimating the potential for coal-fired cogeneration -- with the crucial question being the minimum economic size for coal-fired plants.

A careful estimate requires analysis of the existing equipment at each site along with knowledge of the site's heat load characteristics and steam pressure requirements. The survey data then must be expanded to account for the survey coverage of the customer size classes. For example, the 150 customer surveys mentioned above included about 2/3 of all major commercial and industrial electricity customers and half of the major fuel burning installations in the service territory. Potential industrial growth necessitates further adjustment of the forecast, although this increase will be small in relation to cogeneration conversions within New England.

Finally, the financial impact on the utility and its non-cogenerating customers depends on how the cogenerators dispose of their power. If electricity and fuel price conditions encourage customers to sell on an arbitrage basis, there is no negative impact on the system. As shown in Table 1, if conditions encourage the net sale of power to the utility because the industrial rate reflects substantial prior capacity costs imbedded in the rate, the utility's other customers will bear the difference between the industrial rate and the utility's avoided operating costs. A 1¢ difference between the industrial rate and the utility's avoided fuel costs can mean up

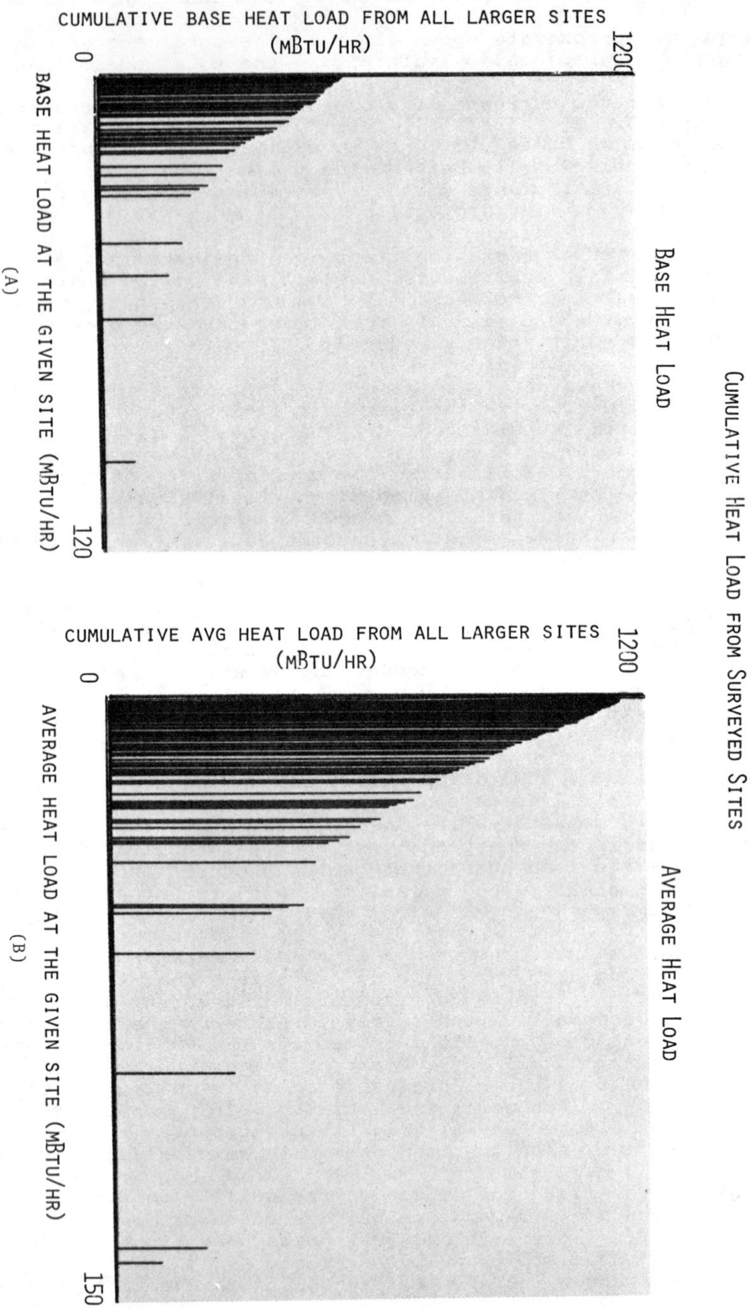
CUMULATIVE HEAT LOAD FROM SURVEYED SITES
BASE HEAT LOAD
CUMULATIVE BASE HEAT LOAD FROM ALL LARGER SITES (MBTU/HR)
0
1200
BASE HEAT LOAD AT THE GIVEN SITE (MBTU/HR)
120
(A)
AVERAGE HEAT LOAD
CUMULATIVE AVG HEAT LOAD FROM ALL LARGER SITES (MBTU/HR)
0
1200
AVERAGE HEAT LOAD AT THE GIVEN SITE (MBTU/HR)
150
(B)

Figure 5

to an additional $87,600 per year gain per 1 MW of capacity for the cogenerator and an identical burden for the other customers.

CONCLUSIONS

This continuing project on utility cogeneration forecasting has focused in detailed on the changing, uncertain cost conditions and electricity rates. The following factors influencing potential customer cogeneration have been indentified:

- Continued oil use by the utilities -- This determines the economic attractiveness of oil-fired cogeneration plants; the degree of utility coal conversion's impact upon the desirability of a given plant type differs substantially between cogeneration technologies.

- Base load designs -- If a cogenerator cannot install a coal-fired plant, and if it can absorb the economic risks associated with an oil-fired plant, the plant will probably serve only the base heat loads with the intermediate and peak steam needs being served by existing oil-fired boilers.

The total impact upon the utility of customer cogeneration depends upon:

- The total amount of cogenerated power -- The first problem in producing an estimate of oil-fired cogeneration is determining the total base-load heat usage in the system. Second, any forecast is very uncertain because of the closeness in the comparative economics and the significant differences in the electrical ouput per unit heat load between cogeneration technologies.

- Payment for the delivery of cogenerated power -- If industrial rates are higher than the avoided fuel costs, the cogenerating customers will first reduce their own bills by internal usage as allowed under the PURPA regulations; this results in a negative impact on the other customers. If the utility's rates are less than or equal to the avoided fuel costs, as is likely under successful utility coal conversion, there will be no adverse impact. In addition, under these conditions it will be to the advantage of oil-fired cogenerators to run only when the utility is also running oil-fired plants as the incremental power source.

The principle source of harm to the utility is the difference between the avoided cost and the standard industrial rate when a cogenerator uses the power internally or on a net sale basis. This difference may diminish two ways: the average costs reflected in the rates falling relative to the avoided costs as the utility reduces its dependence upon oil; or standard rates matching more closely the changing avoided costs, within the limits of the overall revenue requirements.

APPENDIX

The examples in this paper were based upon cogeneration plants providing 15 psig steam to heat loads in the size range of 50 to 100 MBtu/ hr. Base year fuel prices was assumed to be $6.00/MBtu for the low sulfur #6 oil used by the cogeneration plants, $5.00/MBtu for the high sulfur #6 used by the power system, and $2.25/MBtu for coal. The cost figures in Table 2 below represent the combined power and steam related costs divided by the heat output.

PLANT COSTS

Technology	Power/Heat Ratio kWh/MBtu	Incremental Heat Rate Btu/kWh	Total Eff. %	Capital Cost $/kW	O&M Cost $/MBtu
Diesel	290	6600	70	560	1.82
Oil back-pressure steam turbine	70	4110	83	940	.53
Coal back-pressure steam turbine	70	4110	83	1580	1.53
Existing oil boiler	0	0	83	0	.25

Table 2

REFERENCES

(1) Resource Planning Associates, Inc. "The Potential for Cogeneration Development in Six Major Industries by 1985," a report for the U.S. Department of Energy, NTIS, 1977.

(2) Frederick H. Pickel. "Cogeneration in the U.S.: An Economic and Technical Analysis," MIT Energy Laboratory report #MIT-EL78-039. Cambridge, MA: Mass. Inst. of Tech, 1978.

(3) TRW Energy Systems Division and ThermoElectron Corp. "Industrial Cogenration Optimization Program," a report prepared for the U.S. Department of Energy, NTIS, 1979.

(4) Frederick H. Pickel. "Cogeneration and Utility Planning," MIT Energy Laboratory report. Cambridge, MA: Mass. Inst. of Tech., forthcoming 1981.

(5) Fred C. Schweppe et al. "Homeostatic Utility Control," IEEE Transactions on Power Apparatus and Systems, Vol. PAS-99, No. 3, May/June 1980.

PART III

USING ENERGY EFFECTIVELY

Called conservation, efficiency, energy management, cost effectiveness, optimum use, and similar terms, clearly the most important strategy for the near-term and for the indefinite future is to not waste energy. Because of increased prices, government restrictions, and self-preservation instincts, energy is now being used more effectively than ever before in the history of mankind.

This section examines technological, managerial and governmental advances that are contributing to these improvements.

ELECTRIC SUBSTITUTION CONSERVES OIL

Harold B. Finger
Staff Executive
Strategic Planning & Development Operation
Power Systems Sector
General Electric Company

Reducing oil imports continues to be one of the major problem areas facing our nation and one that we have not yet fully addressed.

For the sake of emphasis and to establish the base from which our concern develops, we need only to look at a few data charts.

Our increasing oil imports (Chart 1) have made the difference between a positive trade balance and the deficit position in which we have found ourselves since 1975. In 1980, those imports of oil cost the nation over 75 billion dollars, even though the amount of our imports dropped to just over 5 million barrels a day.

In turn, the impact of rapidly escalating imported oil costs on U.S. energy prices have seriously hurt our efforts to make the productivity gains necessary to compete in world markets.

The concentration of the source of our imports (Chart 2) and the unstable nature of the Middle East raise serious questions about the security of our oil supply. In this highly volatile area, "cutoffs" are becoming the rule rather than the exception and the Soviet shadow looms ever larger.

As a world power, U.S. responsibility must extend beyond our own independence, to help assure that this nation's allies are equally immune to the threat of energy blackmail, for their sake and ours.

These are very serious problems. The first step toward their resolution is to recognize that this country is not energy resource poor. (Chart 3).

Except for petroleum, we have an abundant supply of other energy fuels. Coal and uranium are plentiful.

CONTRIBUTION OF OIL IMPORTS TO NEGATIVE TRADE BALANCE

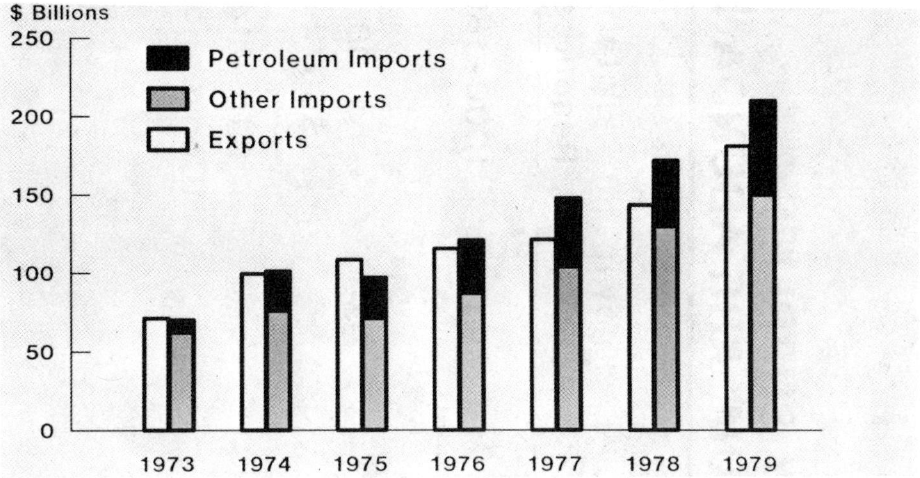

CHART 1

MAJOR SOURCES OF U.S. CRUDE OIL IMPORTS (1979)

(BBL - Thousands/Day)

Saudi Arabia	1,338	Indonesia	376
Nigeria	1,066	Iran	295
Libya	638	Venezuela	292
Algeria	603	United Arab Emirates	280
Mexico	432	Canada	267
Other	891		
Total	6,478		

CHART 2

OUR DOMESTIC ENERGY SUPPLIES OTHER THAN PETROLEUM, ARE ABUNDANT

	Annual Production	Known Reserves	Total Resources
Coal (Billions of Tons)	0.8	218	1000-1800
Uranium (Thousands of Tons)	18	936	1600 ($50/lb.)
Natural Gas (Trillions of Cu. Ft.)	20	194	315-1143 (Conventional)

CHART 3

The picture looks even brighter when you consider that uranium is a potentially extendable resource of fissionable fuel through the application of breeder reactor technology. Of course, that assumes light water reactor orders will resume at a rate which requires us to extend our uranium supplies.

In addition, there are many who have estimated that natural gas supplies are large. The theories of Professor Tom Gold of Cornell University regarding non-biologic sources of gas are particularly exciting; if they are right, they suggest most of the world may have abundant resources of this fuel.

In any listing of energy sources, we must also include the longer-term potential of renewable energies, recognizing however, the requirement for substantial investment of time and dollars before they prove economically competitive.

We are not energy poor.

I believe it's clear that the dangers of continued heavy reliance on oil imports are so great that we must increasingly substitute other fuels that are more readily and reliably available to us.

Although I am here to represent the electrical alternative to imported oil, I want to say that our emphasis must be on using <u>all</u> of our abundant resources with market forces determining which non-petroleum fuel can best perform the needed function.

First I would like to review the electrical substitution for oil concept broadly before getting into the specific element that I would like to discuss in more detail.

I believe that the concept of substituting electricity for oil was broadly, but very effectively presented by Dr. Tom Lee now of MIT, when he was at General Electric in a hearing before the House Committee on Science and Technology. He pointed out that the substitution of electricity for oil and substitution of electricity for natural gas, where that natural gas could then be used to replace oil in other applications, could reduce our imported oil by approximately one-third.

Over the last year, Mr. Herman Hill, Executive Vice President of General Electric emphasized the opportunity for reducing oil use by electrification to a number of organizations including the association of Edison Illuminating Companies, New York State Utilities Executives, Southwest Electric Conference and others, and in a Financier Magazine article.

In addition, the concept of electrification has been expressed very effectively by Dr. Chauncey Starr. In a meeting of the Atomic Industrial Forum last fall, he presented the results of an EPRI analysis that indicated that substantial increases in electricity would be required if the GNP were to grow at a rate that would meet the minimum expectations of our population. He elaborated on the substitution concept further in his introduction to a panel discussion in January at the annual meeting of the National Rural Electric Cooperative Association. His conclusions there were that correction of U.S. energy policy could encourage replacement of oil-fired operating equipment and substitution of electricity and gas for oil in space and water heating plus industrial heat applications, equivalent to 27% of total oil use.

Just to be sure that I'm faithful to my role as a spokesman for the electrical industry, rather than simply as a GE official, I want also to refer to two important presentations made by Westinghouse officials. Mr. John Taylor gave an important presentation before the House Committee on Interstate and Foreign Commerce in December, 1980. His conclusion was that the use of electricity could reduce oil consumption by 3.5 million barrels per day.

And then, at AIF's International Conference 1980, Hunter Chiles described electrical uses in industry that could save substantial amounts of oil. He emphasized three major areas of electrification—pollution control, productivity and conservation/substitution—and then went on to give examples of each.

Electric equipment in pollution control applications will help the primary metals industry comply with EPA regulations, clean up water and power stack gas scrubbers in pulp and paper and reduce dust emissions in the cement industry.

Electronic technologies such as microprocessors and robotics will play an increasingly important role in industrial productivity gains.

Most industries have major opportunities for conservation and substitution.

Continuous casting uses about 20 percent less energy in the final process of steel and iron making and wastes much less crude steel than conventional process.

Petroleum refineries will process heavier crudes and reduce residual oil at the end of the process, but use electricity rather than burning valuable petroleum by products.

Coal-fired industrial boilers will replace oil and natural gas-fired boilers, which will require more electricity in auxiliaries for coal processing and clean up.

Electric heating by electric or induction furnaces provides advantages over alternatives.

The food processing industry is beginning to use a high temperature heat pump called a templifier which utilizes waste process heat for other hot water applications.

These are just some of the industrial opportunities Hunter pointed out in his talk.

I'd now like to focus on a specific subject area—substituting electricity for oil in residential and commercial heating systems.

This is an area of significant opportunity in that some two million barrels of oil a day are being used for space heating by these two sectors of the economy.

An analysis of the cost differential to consumers of oil, gas and electricity (heat pump) in new construction is quite revealing (Chart 4). Today, both the electric heat pump and gas heat, even with air conditioning included, are economically superior to oil heat. As the projections indicate, the economic advantage of electric heat pump over gas becomes apparent within five to six years after construction.

This analysis assumes gas prices will be decontrolled in accordance with the schedule contained in the Natural Gas Policy Act of 1978. If President Reagan's stated intention to speed up deregulation becomes fact, the cross-over point between the electric heat pump and gas heat will occur earlier.

Our analysis also indicates (Chart 5) that there are measurable economic advantages favoring the conversion of existing residential and commercial installations from oil heat to gas or to the electric heat pump.

In the comparative figures, the conversion costs have been factored into both the gas heating and electric heat pump calculations. The conversion or replacement cost payback through fuel savings is two years for gas and three years for the heat pump.

As in the previous case, a speed-up in gas price deregulation will shorten the time during which gas heat has an economic advantage over the electric pump.

We believe, based on these analyses, that the economic merits using the electric heat pump in residential and commercial structures are strong enough to make this the ultimate choice in these sectors.

To determine the potential near-term oil savings (within the next ten years), we have assumed that no new construction will be permitted to use oil heat and that all practical and reasonable conversions will be made to the electric heat pump. Many other conversions will be to gas heat.

NEW CONSTRUCTION (1980) FAVORS HEAT PUMP

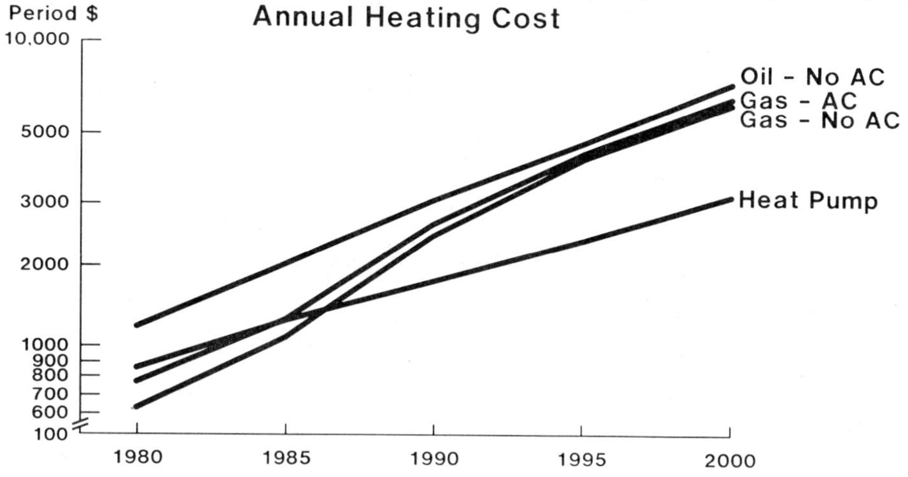

CHART 4

ECONOMICS FAVOR CONVERSION FROM OIL TO GAS OR ELECTRIC HEAT PUMP

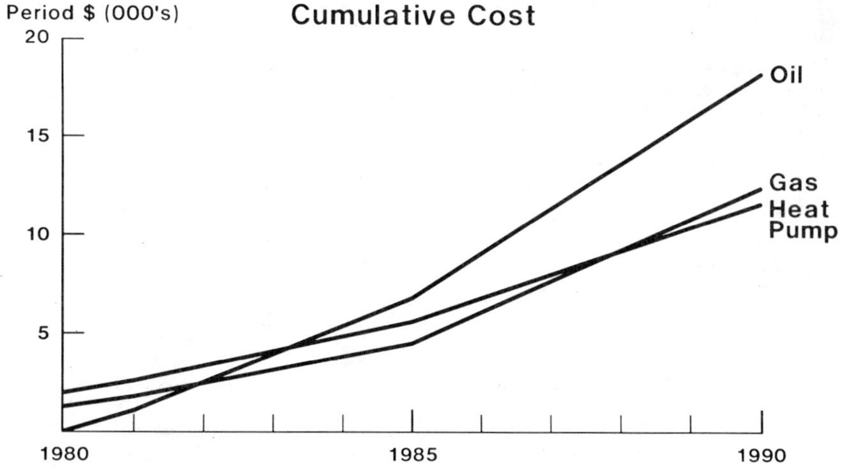

CHART 5

The major assumptions in our analysis include:

- The inclusion of structures in which heat pump conversion is impractical, such as large multi-family dwellings and high-rise office buildings.

- The heat pump conversion of those residential units which have oil-fired warm air or hot water heating systems.

- The heat pump conversion of residential units which have gas-fired heating and central air conditioning, but only at the time of failure of the air conditioning compressor.

- The heat pump conversion of oil heated commercial buildings in low-rise configurations that are less than thirty years old.

- That all heat pump conversions are add-on units which continue to use the existing oil or gas-fired furnaces for low temperature supplemental heat requirements rather than electric resistance heat.

The application of these criteria defines a relatively small universe for conversion to heat pumps — approximately 16 of 80 million housing units, and 4.3 of 32 billion square feet of commercial space.

The projected oil savings resulting from this modest projection is over one million barrels a day in 1990. (Chart 6)

Our calculations of the impact of these oil conversion and diversion actions on the utility load and oil utilization in power generation indicate they are not great. (Chart 7)

In fact, our analysis indicated a reduction in utility oil use of some 22 million barrels per year. Added to the savings generated by the conversion/diversion process, the total is 424 million barrels a year, or 1,162,000 barrels per day by 1990.

In summary:

1. Oil is a special case and is the root of our energy problem. Oil must be reserved for those uses for which there are no reasonable substitutes.

2. Substantial oil savings can be achieved through substitution of electricity, especially in commercial and residential space heating and industrial processes.

3. While the electric generating additions required to meet the conversion/diversion demand are not major, those generating units which already are on order must be completed . . . The units in the field yet to be installed should be put on-line at an early time . . . and new orders must be committed at a rate that will permit the rapid substitution of electricity.

OIL SAVED IN 1990 BY CONVERSION/ DIVERSION TO ELECTRIC HEAT PUMPS

(Residential and Commercial)

Northeast	470,700	Bbls./Day
North Central	349,600	
South	198,800	
West	85,700	
Total	1,104,800	

CHART 6

NET OIL SAVINGS ARE 1,162,000 BARRELS PER DAY BY 1990

(Millions of Barrels Per Year)

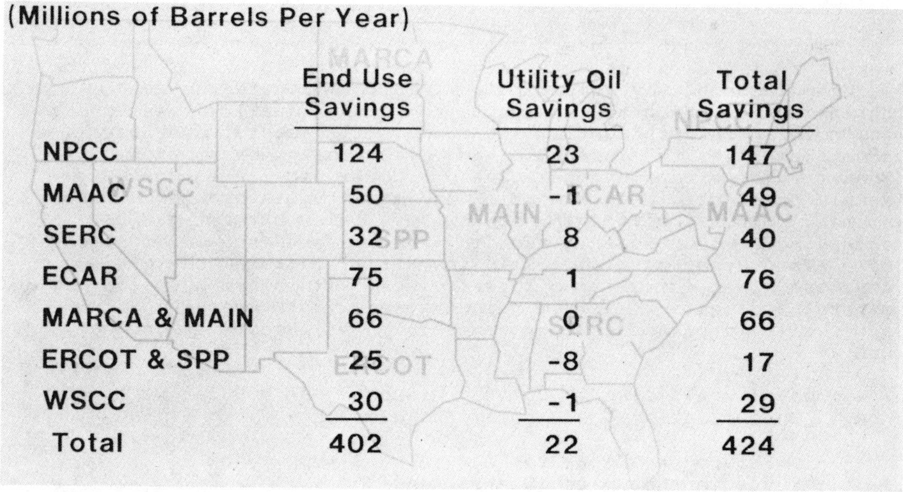

	End Use Savings	Utility Oil Savings	Total Savings
NPCC	124	23	147
MAAC	50	-1	49
SERC	32	8	40
ECAR	75	1	76
MARCA & MAIN	66	0	66
ERCOT & SPP	25	-8	17
WSCC	30	-1	29
Total	402	22	424

CHART 7

IMPLEMENTATION OF THE GAS OPTION

Robert B. Rosenberg
Executive Vice President

Thomas J. Woods
Senior Policy Analyst

Gas Research Institute
8600 West Bryn Mawr Avenue
Chicago, Illinois 60631

INTRODUCTION

Four years ago it was widely believed by energy policy makers that the era of natural gas was coming to an end, and the major purpose of Government policy was to allocate the ever shrinking U.S. natural gas supply to "higher quality" uses such as those in the residential and commercial sectors, and away from "low priority" industrial and electricity consumption. Today, this perception has changed. Confidence in future gas supply prospects, both domestic and imported, is growing. Many states, such as California and New York, view natural gas as occupying a major role in reducing their dependence on imported oil. The use of natural gas in industrial boilers and to generate electricity is no longer viewed with alarm by policy makers; and in fact, is now being viewed with increasing approval. To paraphase the commercial, "We've come a long way, baby."

Today I would like to briefly discuss just how far we have come. Specifically, I would like to discuss -

-- Future prospects for natural gas supply and use in the United States, and

-- The role the Gas Research Institute is playing to maximize the contributions of natural gas at the least cost toward meeting the nation's energy needs.

NATURAL GAS SUPPLY

Prospects for U.S. natural gas supply appeared very bleak in the mid-1970's. The nation was not adding enough reserves every year to replace even 50 percent of its natural gas production. By 1976, U.S. natural gas production had fallen 12 percent from its 1973 peak, and in 1977 the National Energy Plan essentially announced the imminent demise of the U.S. natural gas industry.

But, beginning in 1977 the natural gas "corpse" began to show unmistakable signs of life. Reserve addition rates began to show a small improvement, and production from offshore Gulf of Mexico fields began to surge. As a result, production stabilized and has, in the last two years, even shown modest signs of growth.

The passage of the Natural Gas Production Act (NGPA) in 1978 has added further impetus to the forward movement of natural gas production. Recent exploration activities in the Overthrust Belt have indicated that a region that was once viewed as a modest natural gas producing area when gas prices were very low may well turn out to be one of the richest natural gas producing provinces in the United States at today's higher gas prices. In large part, because of exploratory successes in the Overthrust Belt, U.S. 1980 reserve additions outside of Alaska may exceed U.S. production this year for the first time since 1967.

In addition, the economic incentives of the NGPA have thrown open the door to a vast natural gas resource which had been almost totally overlooked in previous assessments of the U.S. natural gas resource base. These are the natural gas resources found in Western tight sands, Devonian shales, and coal seams.

All recent estimates of the size of the unconventional natural gas resource base agree that it far exceeds the present U.S. proved reserves of conventional natural gas and that resource base limitations will not be a constraint to production from these unconventional sources. However, economic considerations will allow recovery of only a fraction of the resources in-place. Nevertheless, the recoverable resources estimated by GRI are still substantial, as illustrated in Figure 1.

Because current prices for categories of decontrolled natural gas are already on the order of $6.00 per Mcf, one can conclude that the entire 115 Tcf shown in the figure for existing technologies should be considered recoverable. If advanced technologies to produce these resources can be developed and commercialized, the total recoverable resources more than double to 255 Tcf. This is almost 50 percent larger than the 1979 proved natural gas reserves outside of Alaska.

Timely development of these resources can provide the nation with significant amounts of natural gas supplies. As shown in Figure 2, these sources could begin to provide noticeable additional amounts of natural gas in the mid-1980's. Under

Figure 1

GRI RECOVERABLE RESOURCE ESTIMATES FOR TIGHT FORMATIONS AND COALBED METHANE, TCF

	Market Price (1979 $/Mcf)	Existing Technology	Advanced Technology
Western Tight Gas Sands	3.12	30	100
	4.50	45	120
	6.00	60	150
Devonian Shale	3.00	10	20
	4.50	15	30
	6.00	25	45
Coalbed Methane	3.00	10	30
	4.50	15	40
	6.00	30	60

Figure 2

ANNUAL PRODUCTION ESTIMATES FOR COALBED METHANE

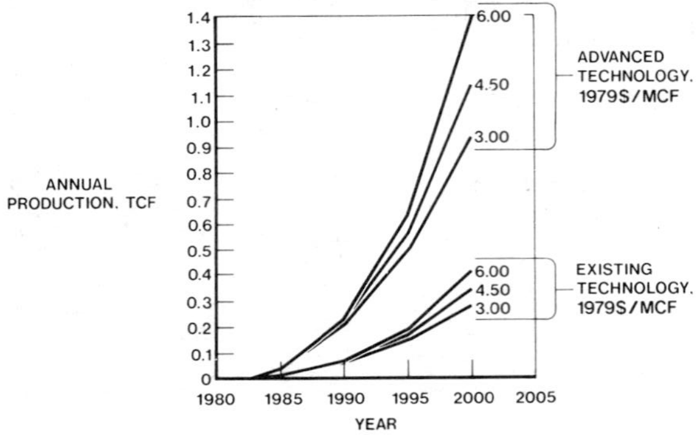

Figure 2

CONSERVATIVE ESTIMATES OF ANNUAL PRODUCTION FROM WESTERN TIGHT GAS SANDS

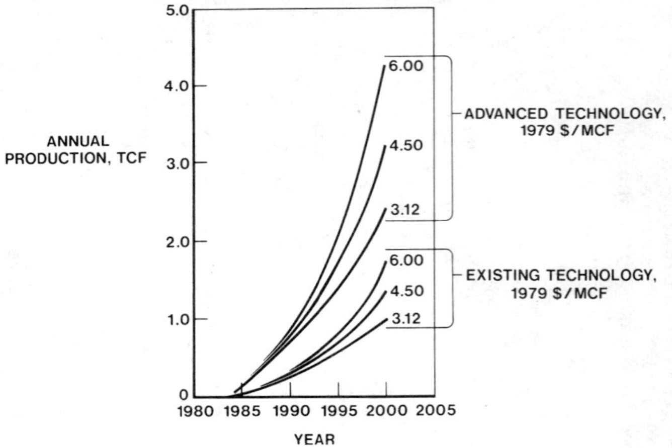

Figure 2

ANNUAL PRODUCTION ESTIMATES FROM DEVONIAN SHALE

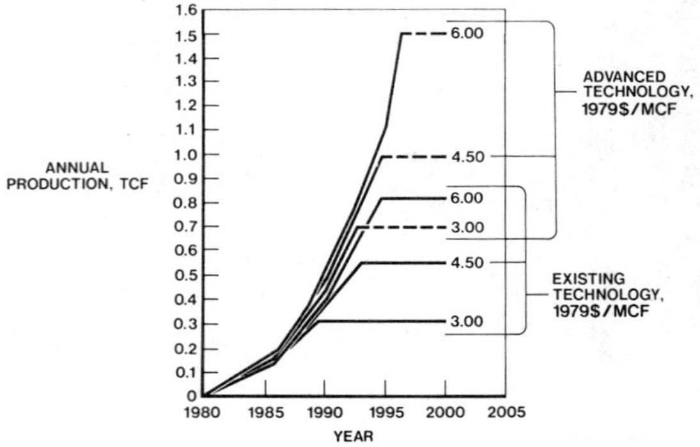

optimistic, but achievable, circumstances production by the end of the century could approach levels equal to about half of the total 1980 U.S. natural gas production levels. (Figure 3).

In addition there is a large potential for natural gas from Alaska and substitute natural gas from solid and liquid fuels such as coal, petroleum, and biomass. Because of the realities affecting Alaskan production, such as lead times approaching 10 years or more, and the state of many of the technologies for substitute natural gas, gas supplied beyond the Prudhoe Bay pipeline, some South Alaska production, and SNG will probably not become significant until the 1990's.

The expected ranges of U.S. gas supply capabilities are shown in Figure 4.

Figure 4 demonstrates very clearly that there is an excellent chance to hold U.S. natural gas supplies reasonably constant through this decade, and that in the 1990's, natural gas supplies could begin to grow at rates approaching or even exceeding 3 percent a year, reaching 25 to 30 Tcf by the end of the century. Under such circumstances the natural gas share of total U.S. energy consumption could approach its peak level of 30 percent reached in 1973.

Figure 4 also shows that, because of the relatively modest supply growth capabilities through the mid-1980's, any displacement of oil imports in the U.S. due to increased consumption would be primarily associated with the increased efficiency in natural gas use that is now being experienced. In the 1990s, however, increasing natural gas supply, coupled with increasing end-use efficiency, could provide for even faster growths in U.S. gas consumption, and thus more displacement of oil imports.

GAS USE

As mentioned earlier, natural gas use in the United States peaked in 1973 at 22 Tcf, accounting for 30 percent of total U.S. energy consumption. By 1979 U.S. gas use was 20.2 Tcf, or 26 percent of total U.S. energy consumption. The major source of the decline in U.S. gas consumption was the 1.8 Tcf decline in industrial sector utilization during the 1974-75 recession. Since that decline, industrial use of natural gas has remained essentially constant at 8.5 Tcf per year. The growth in total industrial energy use has been met totally by increased consumption of electricity and imported oil.

The growing optimism for natural gas supplies since 1977 has resulted in a modest recovery of natural gas use to generate electrcity. Current trends indicate that natural gas use to generate electricity in 1980 should exceed 1973 consumption. Virtually all this increased natural gas use to generate electricity has been reflected in a corresponding decrease in petroleum consumption, and thus in oil imports. Since 1977, oil consumption to generate electricity has declined by about 400,000 bbl/D over this same period. Gas use to produce electricity increased by the equivalent of 250,000 bbl/D.

Figure 3

GRI ESTIMATE OF POTENTIAL GAS SUPPLIES FROM UNCOVENTIONAL SOURCES IN THE YEAR 2000

	TCF
WESTERN TIGHT GAS SANDS	2.0-6.0
EASTERN DEVONIAN SHALE	0.3-1.5
COALBED METHANE	0.3-1.4
METHANE FROM GEOPRESSURED ZONES	0-1

Figure 4

RANGES OF POTENTIAL GAS SUPPLIES

		Tcf/Year	
Source	1979	1985	2000
Natural Gas			
Conventional	19.4	16-18	12-14
Alaskan	0.1*	0.1-0.9	1.5-2.0
Unconventional	-	0.2-0.6	2.6-9.9
Pipeline Imports			
Canadian	1.0	1.0-1.5	1.0-1.5
Mexican	-	0.1-0.5	0.5-1.0
LNG Imports	0.2	0.5-0.8	0.5-1.5
Substitute Natural Gas			
From Petroleum	0.2	0.1-0.5	0.1-0.5
From Fossil Fuels (coal, peat, oil shale)	-	0-0.1	1.5-4.5
From Biomass and Wastes	-	-	0.1-1.0
Total	20.9		
Potential range of supply capability		18-22	25-30

*Quantity consumed in Alaska

The optimism for future natural gas use is further reflected in the growing recognition that the regulations seeking to end or discourage natural gas use in industrial and electrical generating facilities may have been premature. Natural gas use in such situations is now seen as an attractive option to displace petroleum imports. Furthermore, virtually all moratoria on hooking up new customers that were imposed by various state regulatory agencies have been lifted, and the growth in natural gas customers has begun to pick up. Because oil and gas use are interchangeable in many stationary applications, increased natural gas use can provide the nation not only an ability to directly displace petroleum consumption on a permanent basis; it also could provide a back-up capability for the nation in the event of sudden oil supply disruptions. The extent of the overlap between oil and natural gas use can be seen in Figure 5 which shows that up to about 25 percent of total petroleum consumption are in uses which natural gas might be able to displace.

Displacement of oil in stationary applications, particularly distillate oil, would provide the nation with an ability to better meet its transportation needs from domestic sources without having to rely as extensively on imported oil and defer the need to develop expensive synthetic transportation fuel options.

For short-term interruptions, natural gas could play a major role in reducing the negative effects of such a disruption on the nation's economy, particularly if dual natural gas/oil-using capability were more widely spread throughout the nation's energy system. Currently, the natural gas storage system can deliver 38 Bcf per day from underground storage or the equivalent of 6.7 million barrels a day of oil. As can be readily seen in Figure 5, a new delivery capability as large as this could provide enough energy in an emergency to displace all stationary uses of distillate and residual oil. While such an overview is obviously overly simplistic, it does indicate the protection against supply interruptions that natural gas could provide under government policies which had a more general view of strategic energy reserves beyond oil; i.e., the development of a strategic gas reserve (beyond that now in operation) in conjunction with increasing firm-industrial gas loads.

Our assessment of the probable range of long-term trends in gas use in the United States is shown in Figure 6.

The low range is largely representative of a continued unfavorable view by the government toward the electrical generation sector's use of gas and a relatively low economic growth rate. The absolute level of U.S. gas use is projected to decline even though there is a substantial potential for increased supply at economically competitive prices. The high case shows substantial growth in U.S. natural gas use as a result of continued gas use in the electricity generation sector, much higher industrial gas use related to a higher economic growth rate and an aggressive policy of substituting gas for petroleum in stationary applications. Figure 6 indicates very clearly that the major long-term growth prospects for gas use in the United States will be found in the industrial sector, secondarily in the commercial sector, and possibly in electricity generation.

Figure 5

FUEL SWITCHING TO INCREASE TRANSPORT FUELS AND REDUCE OIL IMPORTS

	REFINED PETROLEUM PRODUCT DEMAND (1979), 10^6 bbl/day
GASOLINE	7.07
JET FUEL	1.07
OTHER TRANSPORT FUEL USES	1.63
STATIONARY FUEL USES	
LPG	0.97
DISTILLATE	2.33
RESIDUAL	2.38
RAW MATERIAL AND OTHER USES OF REFINED PETROLEUM PRODUCTS	3.07
TOTAL	18.52

DISPLACE WITH GAS AND COAL- AND NUCLEAR-BASED ELECTRICITY TO MAKE MORE TRANSPORT FUELS

DISPLACE WITH GAS AND COAL

Data Source: EIA Annual Petroleum Statement, 1979 Annual Report to Congress

Figure 6

RANGE OF GAS DEMAND

(Tcf/Year)

	1979	1985	1990	2000
Residential	5.1	4.9	4.9	4.7-4.8
Commercial	2.6	2.7-2.9	3.1-3.4	3.7-4.1
Industrial	8.6	9.1	9.2-9.7	10.6-12.9
Electric Utility	3.6	2.0-3.7	1.2-3.7	1.0-3.7
Transportation	0.5	0.5	0.5-0.6	0.5-0.7
Total	20.2	19.2-21.1	18.9-22.3	20.5-26.2

GRI is currently developing its base scenario of gas-usage growth. The preliminary results support the above conclusions and fall in the mid- to high-band of the range shown in Figure 6.

GRI EFFICIENT UTILIZATION WORK

Although the potential for increased gas supplies is very good, much of this gas will be produced at substantially higher prices than today. If gas is to remain competitive and a viable least-cost option for the consumer, the increased gas prices must be compensated for by increased utilization efficiencies as much as is possible. Otherwise the large potential supplies of gas that the United States could be developing through the rest of this decade will only be realized at relatively modest rates.

Thus, although the availability of gas supplies remains of paramount importance in the long-run, GRI recognizes a more immediate need for the development of high-efficiency appliances and equipment. GRI's Efficient Utilization research is directed at both residential/commercial and industrial applications. In the residential/commercial sector, space conditioning accounts for almost 70 percent of the gas used. GRI's efforts in this area include the development of improved furnace designs as well as a number of gas-fired heat pump concepts that promise significant energy savings for both heating and cooling. In addition, other ongoing projects focus on the improvement of heating-system performance by examining its interaction with the building.

For the industrial market, GRI has identified several processes in which improved gas-fired equipment can decrease product costs, keeping gas the least-cost alternative for providing energy services. These processes include steelmaking, glass melting, and textile drying. GRI is also developing advanced energy utilization concepts such as fuel cells and solar/gas energy systems. These developments should prove beneficial for both residential/commercial and industrial users.

In 1980, several manufacturers committed themselves to the commercial production of new technologies developed with GRI's support. These successes highlight GRI's strategy of involving manufacturers in cooperative efforts so that technical innovations can be developed in the context of practical marketing considerations. GRI's earnings on licenses and patent rights on new developments are used to help underwrite the costs for subsequent developments.

Pulse-Combustion Furnace

A pulse-combustion furnace has been developed with GRI support at the A.G.A. Laboratories. At Lennox Industries, it has demonstrated efficiencies as high as 96 percent in field tests conducted in 1980.

Lennox, a well-known manufacturer in the heating and air-conditioning field, has announced plans to produce and market in mid-1981 eight residential-sized models of the new furnace. The operating cost of the pulse furnace--nearly 50 percent less than that of conventional gas forced-air furnaces--makes the new models attractive as replacements for existing gas furnaces, as well as for the new home construction market. The pulse-combustion furnace achieves its high efficiency by burning gas in a continuous series of rapid pulses within a closed chamber. The pressure of each pulse forces hot combustion gases through a series of heat exchangers that remove most of the available thermal energy. Unlike many conventional gas furnaces, the pulse-combustion model requires no continuous pilot, burners, or vent damper; moreover, the direct venting of the low-temperature flue gas through a small pipe eliminates the need for a chimney.

The introduction of these furnaces into the residential market will not only lower the price of natural gas space heating to consumers; it will also release natural gas for new customers, such as oil users, who could in turn reduce their fuel bills if they were to purchase the pulse combustion furnace. Under such a situation, for every Btu of gas energy saved, about 1.5 Btu of oil energy from an existing oil furnace could be displaced and oil imports correspondingly decreased.

Heat Pumps

Although gas-fired heat pumps are still in the developmental stage, they promise an even higher efficiency than the most advanced gas furnaces. Heat pump designs now being developed by GRI may eventually be able to provide both heating and cooling for a typical cold-climate residence while using less energy than a conventional gas furnace uses for heating alone. When used for heating, heat pumps deliver more thermal energy than is contained in the fuel because they extract heat from outside air, even at low temperatures.

A gas-turbine-driven heat pump developed by AiResearch Manufacturing Company of California (a division of the Garrett Corporation) for commercial-scale application took a large step toward the marketplace in 1980. Bench-scale laboratory tests proved so successful in demonstrating the design's reliability and efficiency that two major manufacturers offered to participate in prototype development and field testing in 1981.

A second important heat-pump design, based on absorption technology developed by Allied Chemical corporation and Phillips Engineering Company, fulfilled performance goals in ongoing tests on six prototype units. Initial test data indicate that the design may offer high efficiencies over a broader range of climate than was earlier expected. Testing of other heat pump concepts is proceeding, with the concepts at various stages of development.

Gas Appliances

GRI promotes the design and development of more efficient ranges, ovens, clothes dryers, and water heaters for residential and commercial markets. GRI-sponsored developments in 1980 have already been accepted by a major manufacturer for use in production-line appliances. Chief among these was a deep-fat fryer incorporating a radiant-burner concept designed by A.G.A. Laboratories. A modification of this design is now being manufactured by Vulcan-Hart Corporation for a major fast-food restaurant chain, which anticipates a 25 percent reduction in related operating costs. This successful development follows the marketing in mid-1980 of a commercial range with a convection oven that consumes 45 percent less gas than existing models.

Development of a residential water heater with an efficiency exceeding 90 percent is a major goal in GRI's improved-gas-appliance efforts. Two new projects were begun in 1980 to achieve this goal: One of them will use the pulse-combustion process, while the other will use power-burner technology. Importantly, the success of a high-efficiency water heater, when coupled with a pulse-combustion furnace, could eliminate the need for a chimney in new homes since both can be direct vented.

Fuel Cells

Fuel cells convert natural gas and other gaseous fuels directly to electricity and heat by means of an electrochemical process--thus avoiding the inefficiencies associated with ordinary combustion. Power plant modules based on fuel cell technology could be used in multi-family dwellings and commercial buildings to supply all required energy services--heating, cooling, hot water, and electricity--using less fuel than is normally used for heating alone. In operation, these advanced cogeneration systems are extremely clean, emitting less than one-tenth as much atmospheric pollution as the cleanest electric power plants now operating.

Successful testing of a 40 kW fuel cell power plant by United Technologies Corporation has demonstrated the technical feasibility and reliability of a prototype unit.

During 1980, an extensive field test program was designed to measure the system's performance and economics in a variety of applications, including apartment buildings, nursing homes, recreation facilities, restaurants, stores, laundromats, and small industries. The fabrication of 45 power plant units is scheduled to begin in 1981. The field test program will involve 20-25 utilities in the U.S. and Japan representing a range of geographic and climatic areas. This entire program is an excellent example of GRI's coordination of funding resources with government, industry, and gas utility sponsors.

Industrial Utilization

While many industrial users could decrease their consumption of gas signficantly through the development of advanced technologies, few can support an effective research program aimed exclusively at gas conservation. GRI's efforts in this area involve improving the efficiency of selected energy-intensive processes and providing more broadly applicable developments, such as industrial heat pumps, burner systems, and energy-recovery systems.

A good example of improved process efficiency is the THERMECON advanced ladle heater developed by Cadre Corporation with GRI support. Previous methods for preheating and curing the ladles that are used to carry molten metals involved simply firing a gas flame into an open ladle. The THERMECON ladle heater uses a compressible ceramic fiber lid to seal the ladle's mouth while extracting hot flue gases through a heat exchanger that supplies preheated combustion air to the unit's power burners. In addition to cutting gas consumption by 50-82%, this new device results in other important advantages, including more uniform ladle temperatures, increased ladle lifetimes, shorter heating cycles, and unattended operation by means of a programmable control unit. Successful field testing of the unit has stimulated widespread interest. Purchase commitments for preliminary production units have already been received by Cadre, as well as requests for the rights to manufacture the advanced ladle heater in several European countries.

GRI initiated two promising industrial heat pump projects in 1980. Industrial heat pumps are used to upgrade steam and other waste-heat sources to temperatures and pressures suited to process applications. One project seeks to develop an open-cycle vapor-compression system driven by an advanced gas turbine; it is ideally suited to large steam-flow capacities. The other project involves a modified Wankel-cycle vapor compressor that could drastically reduce the complexity and high capital costs associated with mechanical vapor-compression machinery.

GRI GAS SUPPLY WORK

Despite all the efforts to improve natural gas utilization, in the end the major factor which will affect the long-term role of natural gas is the extent to which the possibilities for new natural gas supplies discussed previously can be achieved. The increasing attractiveness of natural gas relative to other energy alternatives and the expected high demand for gaseous fuels in the future reaffirm the need to develop all economical gas-supply options to the greatest possible extent.

Western Tight Sands

An excellent example of GRI's program in unconventional natural gas is its work in the area of western tight sands. During 1980, experiments conducted with CER Corporation, in cooperation with DOE, examined the rock characteristics and other production parameters associated with several major areas including the Piceance, Green River, and Arcoma basins. Preliminary data were also obtained on the Raton and Anadarko basins. A two-dimensional seismic array technique developed for GRI by Sandia Corporation demonstrated initial success in locating and delineating potentially productive gas pockets. With Mountain Fuel Resources Company, GRI began a project to determine the commercial potential of the Pinedale, Wyoming, area by drilling a test well for formation stimulation and production research.

Devonian Shale

In the area of eastern Devonian gas shales, GRI continued to collect technical data on the size of this resource, as well as on the effectiveness of various techniques to stimulate production by fracturing the shale formations.

In a joint effort with DOE and Columbia Gas System Service Corporation, GRI supported the drilling and testing of 10 wells in northeastern Ohio. The participation of 30 producers of Devonian shale gas was obtained in a new project for developing a data base on 200 Appalachian wells previously subjected to paced and pulse-loading stimulation techniques. Analysis and compilation of this information will provide industry with a useful tool for evaluating new production technologies.

Coal Seams

Naturally occuring methane trapped in coal-seam fractures and between layers of coal and impermeable rock represents yet another significant but largely undeveloped source of gas. While GRI's activity in this area is at an early stage, the program patterns found in other unconventional resource developments apply here as well: cooperation with DOE and emphasis on technical characterization of the resource and development of improved recovery techniques for use by industry.

Synthetic Natural Gas

Gasification of fossil fuels (primarily coal) offers another source of gaseous fuels for the consumer that will extend conventional and unconventional gas reserves. Several process technologies have been developed and shown to be reliable for use in a commercial plant. GRI's program in this area includes the further refinement of advanced gasification processes to improve their economic advantage over earlier technologies, development of associated technologies to provide needed plant design data and ancillary equipment, and investigation into a means of gasifying unmined coal (in situ gasification).

In the area of coal-gasification processes, two advanced developments continue to receive support from GRI: Westinghouse Corporation's Single-Stage Fluidized Bed and Exxon Corporation's Catalytic Conversion processes. GRI is seeking to further these processes because of their novel design features, which promise greater reliability and on-stream time than existing technologies.

To improve the commercial prospects of gasification, GRI supports the development of several associated technologies. An important research effort is the development of improved catalysts for upgrading the quality of the raw gas produced in the gasifier. A new series of catalysts was developed for GRI in 1980 by the Catalysis Research Corporation that showed improved effectiveness and resistance to sulfur in the gas stream. In tests using simulated raw gas from a commercial reactor, one sulfur-resistant catalyst converted 87.5 percent of the carbon monoxide content of a raw-gas stream to high-Btu methane. When fully developed, this catalyst could be used in a "hot gas" clean-up process that could significantly reduce product-gas costs.

Excellent results have also been obtained from work at the Metal Properties Council. New alloys and materials are now being tested which could greatly improve the reliability and extend the operating lifetime of gasification plants.

SUMMARY

This has been just a brief overview of the future role of natural gas in the United States and what GRI is doing to develop the least-cost components of this strategy for the nation's gas consumer and ultimately the nation's energy system. Clearly, natural gas has a significant potential to displace imported oil and provide least-cost energy services to the consumer. There is every indication that this potential can be increasingly realized in the coming years, and the new technology that GRI is developing to is expected to be an important contribution. The natural gas industry has indeed come a long way in just the last few years; to paraphrase Mark Twain, "Rumors of its demise have been greatly exaggerated."

REDUCING OIL IMPORTS: THE ENERGY CONSERVATION OPTION

by

Grant P. Thompson
Senior Associate
The Conservation Foundation
Washington, D.C.

Energy conservation has been one of the dazzling success stories of the decade, simultaneously proving that economic theory is correct -- people do react to higher prices -- while saving the nation from the ever-increasing burden of oil imports. The figures demonstrating conservation's successes are now commonplace: gasoline consumption peaked in 1979 and started a downward trend that many believe will be permanent; industrial production rose twelve percent during the decade while industrial energy consumption declined more than fifteen percent; electricity growth rates moved sharply down from their historic rate of seven percent per year to less than one percent per year. To be sure, efficiency improvements were not the only reasons for these changes: a slowed economy may have accounted for about half of the industrial change, and curtailments or hardships explain at least a part of the savings in the other sectors. But the long term trend is clear. Even in the relatively short run Americans, once castigated as the world's most profligate energy users, have in fact become significantly more efficient in how they use energy in all forms.

But among energy analysts and policymakers, buffeted by a period in which once reliable predictions have proven wildly wrong, there is a nervousness about this new "source" of energy. Will it continue to play the spectacular role it has so far or has there been a one-time "conservation adjustment" that is now drawing to a close? Put more sharply, how much do we need to worry in the future about increasing production? Do we need to relax environmental regulations dramatically to permit more electricity production from coal and nuclear power? Do western federal lands need to be "unlocked" to permit more production? Must the taxpayer and the ratepayer be asked to subsidize new expensive energy forms either directly through the Synthetic Fuels Corporation or indirectly through rolled-in rates and all-events tariffs?

In answering these questions, it is important that we avoid the silly tone and language that has characterized debate on conservation over the past decade. At one pole, many have spoken of it with the glowing terms that once were reserved for motherhood, flag, and apple pie. "Cheapest," "fastest," "cleanest," and "most democratic" were favorite terms. Seemingly effortless, conservation was promised to provide the country with a return to the good life with no sacrifice. Quite a different picture was painted by others, who referred to conservation as "freezing in the dark," "return to poverty," "washing clothes on rocks by the stream," or even "unAmerican." Conservation was a temporary hair shirt to be endured until technology or increased production or a newly benevolent Middle Eastern regime could solve our problem. Pursued as a conscious policy, so it was said, conservation would be the handmaiden of economic ruin.

But neither of these viewpoints helps us understand conservation or its limitations. By conservation, we should mean only investments and changes in behavior that promote economic efficiency in the use of inputs. The Ford Foundation sponsored study, Energy: The Next Twenty Years (Landsberg, ed., Cambridge, MA: Ballinger Publishing Co., 1979, p. 115) defined the term more neutrally:

> We adopt quite a different viewpoint about energy conservation, a viewpoint guided by the economics of self-interest rather than either hardship or moralistic appeals. We mean by conservation those energy-saving investments, operating decisions, and changes in the goods and services that we buy and use that save money over the life of energy-consuming products. Money can be saved by substituting intelligence, prudence, maintenance, better equipment, or different equipment for purchased energy; the substitution should be made up to the point where the cost of not using the energy is equal to the cost of the energy saved.

This definition provides a foundation for answering the demand questions posed above, but it does not make the task any easier. Conservation is almost uniquely difficult to analyze because of two characteristics. First, it is a source that works in the negative: it is a description of kilowatt-hours and therms that consumers did not use rather than ones that can be measured through meters. More subtly, it can be a measure of fundamental changes in underlying desires, so that substitutions of houses that offer a southern face to the sun for more conventional saltbox designs must be counted as conservation, although they show up statistically only as ordinary housing starts. In more analytic terms, the very bases of econometric modeling for predictions are either inherently unknowable or exceedingly difficult to estimate. There is literally no way econometricians can guess what the future world oil prices will be, and our data on price elasticities of demand in times of rapidly changing prices are poor or non existent (although knowable in principle).

The second reason conservation is so difficult to analyze is that it depends on the actions of millions of individuals, reacting to a variety of stimuli and economic or social conditions. Unlike power production, where the location, amount, and fuels to be consumed can be specified in advance and where the planning and execution are under the control of a relatively small number of economic units, efficiency improvements are decentralized in their decision making level, usually quite small individually (thus depending on aggregation to make them significant for analytic purposes), and subject to fashion, economic conditions for capital formation and a whole host of other hard to predict forces. Difficult though government policy or corporate planning are to predict, prediction of more general behavior is a rudimentary science at best.

Even with all of these difficulties, there are important reasons to believe that conservation will continue to play a dominant role in the energy picture in this country over the next twenty years. The most important reason is, of course, that almost all observers believe that energy prices to the ultimate consumers will continue to rise in real terms (although perhaps not smoothly) over the next two decades. As we have seen, short term adjustments in capital stock can occur quite quickly in response to these prices. What we have only just begun to see are the longer term responses to higher prices (either experienced or expected). Domestic and foreign automobile makers appear to be reacting to an expected continued rise in gasoline prices by producing fleets that easily beat the federally mandated fleet average standards; there are reports that Volkswagen and perhaps the Japanese makers are testing models that get 80 miles per gallon of diesel fuel. In the buildings sector, pulse combustion furnaces that have seasonal efficiencies of ninety-two percent are about to be marketed. The possibilities, in theory at least, for the heat pump might promise coefficients of performance well in excess of the range of two to three that can be purchased now. More fundamentally, designs that include more room for insulation and that capture and store the sun's energy by means of the structure of the building itself promise comfortable homes that cost little more than now-conventional designs yet use eighty to ninety percent less energy than conventional homes. Social and economic trends in society that have little to do with energy, such as smaller homes, fewer children, and two income families may ultimately affect living patterns of cities, making them more compact and contiguous, a design that helps mass transit planners, permits economic district heating, and uses less energy to heat smaller spaces. Energy conservation is truly a magic cup that can keep refilling as necessary, given the impetus of higher prices, the ingenuity of inventors, and the willingness of government to play its part in the process.

What are the ingredients for making this process occur? What, in a basically market-oriented society, should public policy encourage? In the balance of this paper, I address the questions of what the government should do to assist the market to work and the role of the government even where the market is permitted to operate.

The Government and Energy Prices

Decontrol of oil and gas prices is certainly an important element in encouraging energy conservation, but it is far from being the complete solution that some of the members of the public and the federal government believe. This can be illustrated in a variety of ways. In the case of natural gas, immediate decontrol would affect only around half of the gas flowing in the pipelines; the remainder is "old gas" under the Natural Gas Policy Act. Even if the old gas were decontrolled through Congressional action -- possibly with a windfall profit tax similar to that imposed for oil -- there is a great uncertainty about the effect of long term contracts on the price of gas sold into the pipelines. Many of the contracts are silent or ambiguous about the effect of such a decontrol of prices and would have to be litigated or renegotiated. In any case, a large, though uncertain, amount of gas would be rolled in at the lower price. Thus, whether gas decontrol is speeded up or even if old gas is decontrolled, consumers will still for a period of time be paying costs that do not reflect the full replacement cost of gas. Consumers will not engage in a sufficient amount of energy conservation because it will be cheaper to buy a marginal additional amount of natural gas rather than a marginal improvement in efficiency.

In just the same way, electricity prices are set at average costs, not at marginal costs. The price is the average of the cost of generating electricity at older, less expensive plants together with the cost of generating at newer, more expensive plants. To make matters worse, in the absence of meters that are time-actuated, consumers using electricity on-peak (when it is expensive to generate) are subsidized by the users who use it off-peak, when it is less expensive to make.

Research and development that should be performed by private industry is a form of subsidy for a number of energy sources. Arguably much of the nuclear R&D and a great deal of the photovoltaic R&D are in this category.

The capital investment decisions in the country are also skewed by tax incentives for production. Construction investments, drilling expenses, and other uses for capital receive special tax advantages that are not available to those who would seek to make conservation investments. On a day to day operating basis, energy consumed may be deducted from income immediately while investments in energy efficient conservation devices must be depreciated over a number of years against income.

In view of all of these distortions in the market, if the government is really serious about relying on market pressures for encouraging an optimal amount of energy conservation, a program far more pervasive than the ones proposed so far should be pursued. As an example of the pervasive program of "letting the market work," we might propose a program that included, for example, these items:

- Natural gas should be immediately decontrolled and the decontrol should include old gas.
- Pressures should be brought to bear on holders of long term natural gas contracts to renegotiate those contracts to the new, decontrolled price.
- The federal government should be given power to issue mandatory guidelines for state public utility commissions to follow to bring electricity and natural gas prices to consumers to their full marginal cost.
- Eliminate tax subsidies for exploration, drilling, accelerated depreciation, and other production incentives.
- Either eliminate expensing of energy purchases or devise a system that gives equivalent treatment for energy saved.
- Eliminate R&D for solar energy, for nuclear power production, for synthetic fuels, and for other energy forms.
- Eliminate the duties of the United States Geologic Survey that have reference to mapping for fuel resources.
- Charge the full marginal cost of any part of the nuclear fuel cycle that the federal government is involved in, including any disposal costs.
- Disband the U.S. Synthetic Fuels Corporation and

suspend any subsidies under the Defense Production Act for fuels.

- Eliminate expenditures for the Strategic Petroleum Reserve, leaving stockpiling to the consumers of oil.

- Transform natural gas pipelines into common carriers, not purchasers and resellers of gas.

- Eliminate the Solar and Conservation Bank, the various tax credits for alternative fuels such as gasohol, and the tax credits for renewable or conservation investments.

- Forbid rolled-in pricing for natural gas projects.

- Eliminate Corps of Engineers involvement in construction or maintenance of port or canel projects that are used principally for energy shipments.

This list -- far from complete -- gives some indication of the range of actions that would have to be taken to come closer to a free market in energy, a market that would call forth conservation in an optimal amount. The strong shift in economic policies that enacting such a list would involve also gives some clue as to the likelihood that such a program would either be proposed or enacted. But until such a comprehensive program is part of the fabric of the country's energy policy, it is a sham to talk of the energy market as being "free".

Role of Government In Conservation

There are many places in which the market working alone -- even a market far more free from distortions than the Administration is seeking -- cannot and will not bring forth the proper amount of energy conservation. There are three illustrative examples that I would like to deal with. These are (1) Research and Development; (2) Information Programs; and (3) Programs Where Market Structure Itself Prevents Action.

Research and Development. The role of government in research and development has a long history in the energy field. Virtually all of the nation's nuclear energy development has been undertaken by government; synthetic fuels also apparently will benefit from government funding; and solar electricity production traces its development to government financed R&D. But the needs for research and development on conservation are just as compelling as they are on the production side, and indeed, the arguments for government involvement are much stronger. Conservation is not a monolithic solution to the energy problem; it is a diffused system of many small and medium scale contributions to greater efficiency. By the nature of the scale of these solutions, it is hard for single firms to capture the benefits of the research and development. Often these developments do not clearly return the reward to the single entrepreneur. Because of their scale, they are hard to obtain private funding to accomplish. Consider for example the research that needs to take place in the buildings sector. Important parts of the conservation potential in buildings comes not from products but from improved designs of buildings, designs that take into account the heat flows within a building and the contribution that site planning and placement can make for passive heating and cooling. Yet such innovations cannot be protected through the patent system; the industry is fragmented with no mechanism for research and development; and firms are undercapitalized and fear innovation that is not proven. Nor is the government's beneficial effect on private R&D efforts to be ignored when it acts as a competitor. Consider, for example, the Department of Energy's success in stimulating research and product commercialization for the humble light bulb. The light bulb industry is heavily concentrated; DOE's work in supporting the development of the so-called LITEK high-efficiency light bulb hastened the commercial introduction of more efficient bulbs by the regular manufacturers.

There is a great deal of hardware development work that must take place in energy conservation. As noted earlier, conservation can continue to pay dividends if the products exist. Although most of this development should and will go on in the private sector, there is room for at least some publicly supported conservation hardware research. If it makes sense to use hardware R&D to support nuclear, coal, and other supply side options that produce energy at above the world price of oil, it certainly makes sense to support conservation hardware that may save energy at far less than oil prices. Of course, R&D for energy efficiency should be measured by the same strict standards that apply to R&D on the supply side. It should not be conducted when a private firm could do it instead. Private sector involvement should be heavy, and should include cost-sharing arrangements if possible. Government sponsorship should end well before the marketing phase, leaving that to the private sector that does it better.

Information Programs. It is almost a truism of economics that a free market depends upon consumers who have access to information in making their own decisions. Yet energy information is notoriously difficult to obtain, even with interested consumers with technical expertise. Consumers Reports and other non-governmental entities have a valuable role to play in providing energy information to consumers, but the individualized and specific information that consumers require in their major energy decisions simply cannot be supplied by a magazine. Nor can we trust the makers of specific brands of products to give consumers energy information in a form that is directly comparable to information given by other manufacturers. Instead, consumers frequently need detailed advice that speaks to them about what to do to make their own homes more weathertight, to make their own chosen lifestyle more energy efficient, and to make their own purchasing patterns more informed.

Unfortunately, the free market cannot provide such information reliably. Energy information is elusive, technical, and hard to quantify for the individual. Even when it can be calculated accurately by the consumer, comparing similar products requires that assumptions be made by the consumer that are realistic and consistent, a difficult task for a person unfamiliar with how he or she might use a new purchase. Regretably, it will always be in the interest of those manufacturers or builders whose products perform most poorly to obscure and confuse the information provided to consumers, a practice that dilutes the value of reliable and consistent information provided by the majority.

We have many examples of the beneficial effects of accurate, consistent information on consumer behavior. For example, the effect of the miles per gallon labels on automobiles has been impressive. Although far from accurate, consumers at least know that they provide a consistent basis for comparison. Likewise, the nicotine and "tar" labels on cigarettes permit smokers to make choices among the competing brands according to their own preferences.

Thus it is important that programs that provide direct, relevant information to consumers be supported by the government and that the data gathering and research base that underlies an accurate presentation of information be continued as well.

To be sure, information programs should not escape a careful scrutiny to make certain that they are not poorly managed, duplicative of private efforts, or poorly targeted. But those programs that have been evaluated have shown themselves to be so cost effective, so facilitating of informed market behavior, and so inexpensive that it is surprising to see them treated so poorly in recent weeks.

Programs Where Market Structure Itself Prevents Action. Within the building sector, there are a number of important cases in which the relationship between the parties guarantees that market pressures are not able to induce conservation, no matter how high energy prices rise. A few examples make this clear:

> Residential rental units account for almost one-third of American dwellings. Where the landlord pays the energy bills, the tenant has no incentive to conserve energy by, for example, closing windows in the winter or turning off air conditioning in the summer. Where the tenant pays the bills, the landlord has no incentive to make capital improvements that will conserve energy.

> In commercial office space built on speculation, none of the parties to the design of the structure have any great incentives to invest in design or construction practices that will save energy. Each party looks to factors such as low first cost, speed of costruction, conventional design, or other non energy matters. No party to the transaction at the construction phase has the interest of the future tenants or of the nation in mind.

In these cases and others like them, if the government at some level doesn't act, the structure of the transaction prevents economic conservation behavior. Because we have so little experience concerning how to manage programs for the landlord/tenant area, support to state and local governments experimenting with different programs is a wise investment. Without government intervention, the market is unable to help.

Conclusion

The litany of actions that government should take is by now familiar to most students of energy conservation: remove as many subsidies and distortions as possible from the energy market so that conservation can come closer to a fair comparison with supply options; provide information through as many channels as possible in order to influence capital purchases and behavior patterns that have energy saving potential; conduct research and development both on hardware and on patterns of energy consuming or saving behavior; regulate, provide subsidies, or take other actions necessary to correct existing subsidies that cannot be removed otherwise or to reach areas in which the market cannot work; and work cooperatively with states and localities to fashion "delivery systems" for energy conservation that are as efficient as possible. The question over the next few years is whether any of these actions will be taken. Although it is too early to be certain about the shape of the final budget for the coming year, let alone for years in the future, the government appears to have opted out of the business of treating conservation seriously. Yet "depending on the market" will not capture the potential. It may be several years before the value of conservation to the nation becomes clear as it has, eventually, to every other Administration in the past.

LOAD MANAGEMENT: STRATEGY AND CONTROL

Patricia H. Harris

Tennessee Valley Authority

ABSTRACT

This paper provides a general overview of the conservation, renewable resource, and load management activities of the Tennessee Valley Authority. Specific emphasis is placed on load management programs with detailed discussion of TVA's overall strategy in developing and carrying out its programs, as well as discussion of the operating strategies associated with the particular techniques.

INTRODUCTION

The Tennessee Valley Authority (TVA) is a corporation of the Federal Government established May 18, 1933. Its power program is carried out in support of TVA's broad purposes, which include providing an adequate and reliable supply of electricity at the lowest possible rates, enhancing environmental qualities, and developing, demonstrating, and promoting the most effective and efficient utilization of electricity and other energy resources.

In carrying out its power supply responsibilities, TVA sells electricity to 160 distributors, 50 directly served industries, and several Federal installations. Through its power distributors, TVA provides electricity to 2.5 million residential consumers. In fiscal year 1980, the average residential use was 15,130 kWh, almost twice the national average. Much of this consumption is for space conditioning and electric water heating: about 45 percent of the residential consumers have electric heating, 75 percent have electric water heating, and about 65 percent of the homes are either partially or completely air-conditioned.

Although TVA's residential rates are still considerably lower than most areas of the country and lower than the national average, they have nonetheless increased 175 percent between 1970 and 1980. This increase has served to emphasize the need for more aggressive development and implementation of options which reduce capacity requirements and operating costs, while directly assisting consumers in their efforts to reduce or control energy costs. Within TVA, the Division of Energy Conservation and Rates is charged with the responsibility of designing and developing such options including its conservation, renewable resource, and load management programs. Programs are developed to provide both capacity and operating savings to the system, although near-term savings are operating savings. In addition, they offer a measure of insurance against changes in the power supply situation and against the uncertainties of load forecasting.

The following information briefly summarizes some of TVA's conservation and renewable resource programs. More detailed discussion of load management activities is provided later.

OVERVIEW OF CONSERVATION ACTIVITIES

TVA's Home Insulation Program (HIP) was initiated in August 1977. The program is conducted in cooperation with the distributors of TVA power and provides onsite inspections of residential dwellings and no-interest loans to finance weatherization measures. As of December 1980, over 300,000 surveys have been completed and about 138,000 loans made. A similar program designed to assist commercial and industrial customers has resulted in over 3,000 surveys.

As an adjunct to HIP, the Heat Pump Financing Plan was developed to encourage those consumers who heat electrically to install energy-efficient heat pump systems. The program offers financing assistance and installation inspections. As of December 1980, about 14,000 surveys and 6,000 loans have been made.

Other efforts include the New Home Conservation Program which is designed to aid in the development, adoption, and implementation of building standards for energy-efficient new homes. The program offers inspection of new homes on a voluntary basis and provides certification and labeling of those that meet the energy-efficiency equivalent of TVA's standards. Education and technical assistance is provided to encourage builders, lending institutions, and realtors to undertake or support the construction of energy-efficient homes.

In the area of renewable resources, TVA has actively pursued the development and implementation of solar alternatives within the region. Solar Memphis, TVA's first major solar program, was begun in October 1978 to demonstrate the design, installation, certification, and low-interest financing of 1,000 solar water heaters. TVA is now into the second major phase with program expansion to the Nashville area. A limited demonstration of solar water heater projects for commercial establishments has also been recently authorized. Other major activities include the design, construction, and monitoring of solar homes and residential and commercial and industrial biomass projects.

Now we turn to load management in order to round out the review of TVA's overall approach to its conservation efforts.

LOAD MANAGEMENT DEFINITION

For the purposes of this discussion, the term "load management" will refer to those techniques which utilize devices or mechanical systems to reduce peak demand at the point of use. The term embodies those techniques which optimize the pattern of demand on the electric system in order to provide more economical system operation and reduce the need for future generating capacity. The emphasis is to shift energy use to a time favorable to the electric system while still providing the consumer with the benefits of electric service when it is needed. Within the TVA region, such techniques are offered on a voluntary basis.

SYSTEM CHARACTERISTICS

Since the optimization of the system demand pattern is the primary purpose of load management, it seems essential to review the TVA power system load and supply characteristics before proceeding with further discussion of current activities. TVA is basically a winter peaking utility with the system peak normally occurring between 8 a.m. and 9 a.m. with a secondary peak occurring at 6 p.m. The peak and system load shape are strongly influenced by the residential class which uses about one-third of the energy sold while representing about one-half of the coincident peak demand. Thus, much of the swing in the system load curve can be attributed to the residential load pattern which is influenced by heating requirements and daily life styles. (See Figure 1.)

Although the TVA system has historically peaked in the winter, it may be thought of as a dual peaking system because its summer peak load is only slightly less than the winter peak load requirement. For example, in 1980, the system peaked in February at 20,745 MW, but the peak summer load was recorded in July at 20,250 MW. In fact, recent load forecasts predict that the system will become summer peaking in the next several years. It should be noted that the summer load profile is strongly influenced by cooling requirements with load building through the day and generally peaking between 4 p.m. and 6 p.m. (See Figure 2). The summer load curve also differs significantly from the winter curve in that there is a larger variation between the peak and offpeak load requiring a large swing in output from the generating units. This swing is estimated to be about 9,000 MW.

Total installed generating capacity on the TVA system is 29,839 MW. Coal-fired units represent almost 60 percent of total capacity. Hydro, nuclear, and combustion turbines make up approximately 16, 12, and 8 percent of total capacity, respectively, with the remainder coming from ALCOA and Corps of Engineers dams operated by TVA. In addition, about 12,000 MW of capacity is under construction with planned commercial operating dates staggered between 1981 and the late 1980's. Base load requirements are provided by coal-fired and nuclear plants with hydro, combustion turbines, and purchases supplying peak load requirements.

END-USE CHARACTERISTICS

The formulation and establishment of comprehensive, cost-effective load management programs is highly dependent on adequate information concerning diversified end-use load patterns. This information is required in conjunction with system data to assess load management potential as well as define appropriate operating strategies for various load management techniques. The basic source for much of TVA's load profile data is the Water Heating/Space Conditioning Control Test (WH/SC Test).

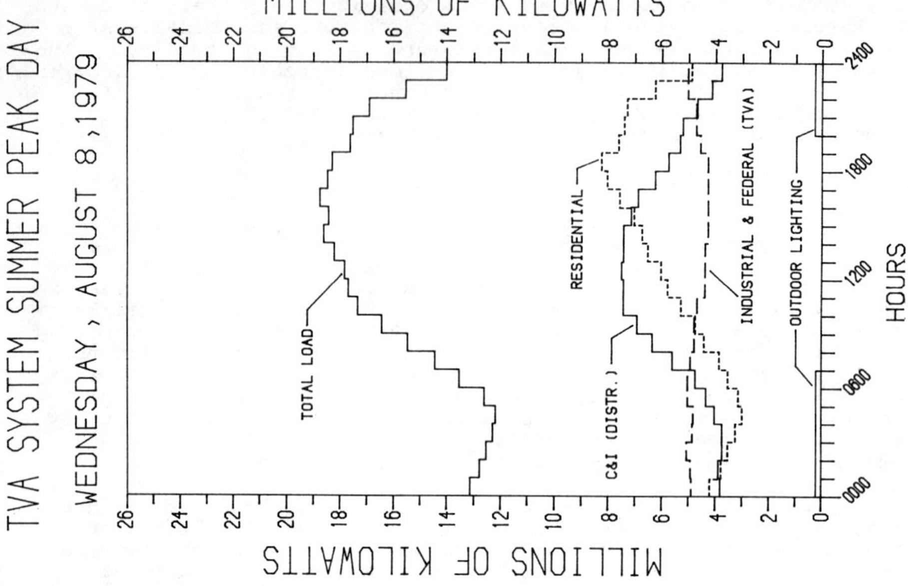

FIGURE 2

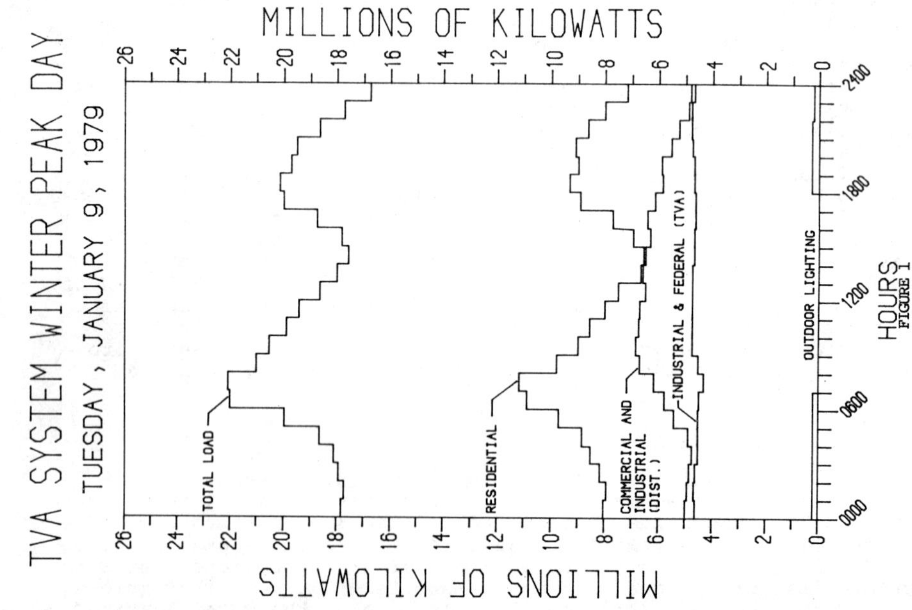

FIGURE 1

The WH/SC Test was TVA's first large-scale attempt to evaluate load management techniques. It was designed to gather and analyze data pertaining to a number of related topics, including the costs and benefits derived from managing residential water heating, the ramifications of extending management to residential space conditioning, customer acceptance of load management techniques, and further research of the various components contained within residential demand for electricity. The test began on July 2, 1979.

Four power distributors representing a broad range of climatic and geographical variations as well as two time zones were selected to participate in the project. Then 457 residential consumers from the distributorships were selected as test subjects. The majority of consumers (358) were in the portion of the test designed to investigate the direct management of standard electric water heaters while the remaining consumers (99) were involved in the assessment of large capacity storage water heaters operated offpeak. Of the 457 participants, 27 also had their electric heating systems cycled and 131 had their cooling systems cycled. A number of other major electrical appliances were monitored.

The test was designed without a separate control group for statistical comparisons. Instead, each customer serves as his own control with load management implemented in alternating weeks and baseline load information collected during the intervening weeks. By comparing properly adjusted baseline loads and managed loads, the appropriate deferral achieved through load management can be quantified.

A sizeable data base for analyzing the potential load reductions achievable from management of water heating and space conditioning has been gathered through this project. The managing of water heaters (through either direct control or use of large capacity offpeak storage water heaters) has the potential for reducing consumer load by roughly 1.3 kW during winter system peak and 0.6 kW during summer peak. The managing of air-conditioning systems promises load reductions of over 0.6 kW. Present estimates provide for reductions in excess of 1 kW on hot days. The test provided load reduction/payback information for varying control strategies for both space conditioning and water heating.

Customer acceptance in the test was positive with over 95 percent of the participants who responded to a mid-test questionnaire indicating that they felt consumers would allow control of appliances. Consumer inconvenience problems encountered in the test could be directly related to water heater malfunctions or undersized storage capacity in relation to family size and habits. Information gathered through followup efforts, together with water and energy usage data, has been used to develop water heater sizing criteria for future water heater management efforts.

Utilizing information gathered from this test along with other available data and analyses, TVA has taken major steps toward the expanded demonstration and implementation of key load management techniques. Its first major activity was the Air-Conditioning Cycling Program (Cycle & Save).

AIR-CONDITIONING CYCLING PROGRAM (CYCLE & SAVE)

As discussed earlier, although the TVA system is presently a winter peaking system, it is becoming summer peaking. In addition, the large swing between peak and offpeak load indicates that any energy shift from peak to offpeak hours could help alleviate potential energy balance problems caused by excess energy from non-cycling generating units during minimum load periods. For these reasons, it was determined that TVA's first major load management efforts would be targeted for the summer period.

The Air-Conditioning Cycling Program (Cycle & Save) provides for the installation of control equipment and the cycling of 50,000 central air-conditioners by the end of 1983. Since TVA is a wholesaler of electricity, the project is being carried out in cooperation with about 12 distributorships through contractual arrangements covering the details of the program. Participating consumers will receive monthly bill credits of $5 per month from June through September.

In addition to providing operating savings, this first phase of a possible system-wide effort is designed to provide needed operating experience regarding the incorporation of air-conditioner cycling into TVA's normal and emergency operating procedures. It will allow the assessment of the acceptability of the control equipment and cycling strategies, emergency curtailment arrangements, customer incentive plans, and other contractual arrangements prior to full implementation.

The control system will be operated by power system dispatchers with the control signal originating from one of five existing area dispatch control centers. The program provides for both normal and emergency operation. Under normal cycling arrangements, electric service to the compressor can be interrupted for approximately 7-1/2 minutes in each 30-minute period from 10 a.m. to 9 p.m., Monday through Friday, from June through September. Such interruptions are initiated by power system dispatchers and may be implemented for economic reasons to modify the daily load shape in order to better utilize less expensive generating units during peak hours.

Emergency curtailment may be instituted for longer periods in accordance with the provisions of TVA's Emergency Load Curtailment Plan. The plan has five basic steps numbered 10, 20, 30, 40, and 50.

Step 10 - Inhouse loads reduced by curtailing the use of non-essential lighting and heating for air-conditioning.

Step 20 - Appeal to the public for voluntary curtailments.

Step 30 - Reduce the voltage 5 percent at certain TVA and distributor substations.

Step 40 - Curtail firm, industrial, and interchange loads.

Step 50 - Interrupt residential firm load on a rotating basis.

The emergency provisions of direct control load management programs will be implemented immediately after Step 30, voltage reductions, with periods of electric service interruption lengthened beyond the 7-1/2 minute interval. Prolonged interruptions of several hours will not be part of the interruptible provision for air-conditioning cycling.

WATER HEATER CYCLING PROGRAM (CYCLE & SAVE)

The Air-Conditioning Cycling Program (Cycle & Save) has now been expanded to incorporate controls on 50,000 electric water heater installations by 1983. The program is being offered to the same distributors as in the initial program. Heavy emphasis will be placed on enrolling participants for both air-conditioning and water heating control in order to maximize the number of dual-function receivers used and thus greatly reduce cost. Consumers participating in the water heater portion of the program will receive a $2 bill credit each month of the year.

There are several differences between the water heater program and the air-conditioning program. One significant difference is that the water heater program has more stringent participation criteria which are provided to ensure that participants are not inconvenienced. As mentioned earlier, the Water Heating/Space Conditioning Control Test revealed that there are a number of potential participants who have storage tanks only marginally adequate to supply enough hot water under non-control conditions. Therefore, water heater sizing criteria are to be applied in the large-scale program to advise consumers about participation. The sizing quidelines to be applied consider variables such as family size, number of bathrooms, ownership of a dishwasher, along with maximum possible outage time to determine the appropriate water tank size.

In this program, TVA has the option to cycle at times between 6 a.m. and 12 p.m. seven days a week. Under normal operating conditions the maximum length of time that the heater can be disabled in any one day is two 4-hour periods separated by at least 4 hours of no control. During emergency conditions, longer interruptions may be implemented following initiation of Step 30 in the Emergency Load Curtailment Plan.

It should also be noted that direct load control devices are being installed in conjunction with TVA's expanded solar water heater efforts to ensure that the systems do not contribute to peak requirements. The same provisions of the standard water heater cycling program apply.

DISTRIBUTOR COOPERATION

As referenced earlier, direct load control programs implemented by TVA must be executed through cooperative arrangements with distributors. Careful assessment of control strategies is made to prevent possible negative impacts of control actions on the participating distributor's wholesale bill, since one element of the determination of the monthly wholesale bill is the distributor's (not TVA's) monthly peak. Strategies applied by TVA will most often have a positive effect on the distributor's load shape. In addition, although TVA maintains the control function, distributors may request load reduction at times critical to the distributor system as long as the customer's agreement is not violated. It should also be noted that possible upcoming restructuring of the wholesale rate may lead to a form of coincident demand billing which will enhance future cooperative efforts.

It is also worth noting that we are now entering an interesting cooperative arrangement with a power distributor who has installed an independent control system. Through cooperative agreement, the distributor will maintain control of the system but will allow TVA to install a direct interface for initiating load control as needed on the TVA system. In return, the distributor's consumers are eligible for bill credits consistent with the Cycle & Save Program. This arrangement should prove very beneficial to all parties involved.

STORAGE OPTIONS

Although discussion has been limited to cycling options, storage options such as storage water heating and storage heating and cooling are potentially very cost-effective options for incorporation into our load management plans. Storage concepts offer significant energy shifts from onpeak to offpeak periods. These large shifts not only offer opportunities for significant load reductions and large energy shifts to reduce generating costs but also can alleviate energy balance problems such as the large swing present in the summer load profile and the midday offpeak period in the winter season. By filling these valleys, operating costs can be reduced by better utilization of non-cycling base-load plants rather than by using more expensive intermediate and peaking capacity to meet the drastic swing in load.

The WH/SC Test described earlier included a test of 99 storage water heaters to gather load data, test offpeak operation, and determine customer acceptance. The test was very successful. Based on test results regarding load reduction and customer acceptance, detailed system benefit analysis has been performed which indicates that such systems operated in conjunction with time-of-day rates are cost beneficial. The program details have been prepared but program implementation is contingent on the adoption of a voluntary time-of-day rate for residential consumers. Adoption of such a rate is under consideration in accordance with the final determination of the time-of-day standard under the Public Utility Regulatory Policies Act (PURPA).

Heat storage is also a very feasible option, but more questions about the various storage technologies remain unanswered. Therefore, TVA has implemented a major thermal storage project in order to address some of the issues. The project is being carried out in cooperation with eight participating power distributors and involves the installation of 75 central heat storage systems (25 brick systems, 25 pressurized water systems, and 25 eutectic salt systems). Some units have been operational for two winter seasons. About 60 are installed. A detailed presentation of the results of this project is being made in another session at this conference.

OTHER ACTIVITIES

A number of other activities are underway which help provide a comprehensive load management program from the standpoint of implementation, demonstration, and research and development. Among these activities are a field test to gather detailed information on the effects of programmable thermostats; a demonstration project to determine the impacts and effectiveness of energy use display meters with/without cycling features, utilizing the standard rate versus a time-of-day rate; demonstration of heat and/or cool storage systems in commercial and industrial establishments; demonstration of residential heat/cool systems; and demonstration of room-sized heat storage systems in multi-unit apartments.

Although the programs discussed previously deal predominately with residential consumers, they are generally available and applicable to small commercial customers (less than 50 kW) as well. Load management opportunities for larger customers are also under consideration with some limited demonstrations underway. In general, load management opportunities will be identified and recommended in conjunction with the ongoing commercial and industrial audit program. As with storage water heating, the availability of a voluntary time-of-day rate is a key item for future successful implementation.

SUMMARY

TVA is continuing to evaluate and assess its load management and conservation alternatives in order to offer those options which can provide both near-term and long-term savings to the power system and the consumer. Programs are in place and under development which can provide operating savings in the near-term and also form a basis for reducing capacity requirements in the years to come. Significant peak reductions, energy savings, and energy shifts can be achieved at very reasonable costs. The programs now in place form a sound foundation from which to evaluate future needs and meet these needs in a timely, economical fashion.

LOAD MANAGEMENT CONSERVES CAPITAL

Donald H. Denton, Jr.
Duke Power Company

Good afternoon. It is indeed a pleasure to be here to discuss a subject which is very important to the electric utility industry and its customers. The conservation of capital through load management must be a primary concern of the electric industry because of its impact on the stockholders and the rate payers. In order to properly describe the impact of growth on a company, I believe it would be beneficial to discuss briefly the history of our industry and place into proper perspective the past, the present, and the future.

Most electric companies were founded about the turn of the century and from that time until the early 1970's, almost 70 years, the companies prospered and grew. This growth was brought about by the normal development activities within our service areas, the expanding industrial climate and the improvements in technology made available to all sectors that utilized electricity - residential, commercial, and industrial.

In this period, we saw automation in the industrial sector; we saw air conditioning growth in all sectors; we saw new and improved appliances to ease the burden of the housewife and to allow a better lifestyle. Through this entire period, the growth in the companies afforded an adequate, even abundant supply of electricity at an ever decreasing cost due to the economies of scale and the improvements in efficiency that we were able to achieve. In addition, it afforded a growth in the value of the companies that improved the value of our stock.

During the first three quarters of this century, the consumer was pleased with his ability to increase his use of electricity and in his perception received greater value than he was having to pay. The stockholder was also pleased because value of his property was increasing and his long-term expectation was continually improving. Unfortunately, the decade of the 70's brought a very abrupt reversal to this mutually agreeable and satisfying relationship between the company stockholders and the company's consumers. In the early 70's the industry was faced with double digit inflation, escalating interest rates on debt and rapidly increasing fuel costs.

In addition, an oil embargo due to the OPEC nations created a concern on the part of the federal government, and a national energy policy was established. In the same manner that growth was beneficial in history, reduction in growth will be beneficial to the stockholder and the rate payer in the future. From the stockholder's perspective, a reduction in the growth of the peak will allow a reduction in the construction program. This reduction in the construction program will mean that fewer shares of common stock will have to be sold in the future. At the present time, the company's stocks are selling 60 to 75% of book value. The selling of stock in this manner will dilute the equity portion on a per share basis of all the existing stockholders. As we all know, it is not good business practice to sell stock below book value. In a completely free enterprise, good business managers would refuse to conduct their business in this manner. However, in a regulated environment, there are the questions of franchise rights and obligations to serve. But, I would hasten to add that the continuation of the sale of stock below book value will ultimately result in bankruptcy of any corporation, regulated or unregulated.

From the standpoint of the rate payer, the reduction of peak load will similarly cause a reduction in generating facilities. If generating facilities can be eliminated, then the company's rate base will be less than they otherwise would have been and the need to recover the costs or revenue requirements of the company will also be less than they otherwise would have been. Therefore, the amount that the consumer would have to pay for his electricity would be less.

Let's then, for a moment, look at the present situation and what the customers of one utility, Duke Power Company, currently are doing to reduce their power bill. Obviously, when an individual or a corporation is looking to reduce its consumption of a product, it will first select those options that carry with it the least impact, the things that are easy to do but also the things that achieve the objective and, in this instance, the objective is reduction of the power bill.

As I pointed out earlier, historically, the company and all utilities were declining cost industries and increased consumption was good. Therefore, in the electric industry rates had traditionally been designed to encourage growth or, said in another way, designed such that growth was not discouraged. Because that is the structure of the rate, at the present time, it works to the detriment of certain concepts.

If a consumer is attempting to maximize his savings at a minimum adverse impact upon himself or his business, he searches for those options that are available to him. Some examples that customers selected did just this. They reduced the power bill through reduction in energy sales but, in many instances, there was no reduction in future demand. Air conditioning can serve to demonstrate this. A residential customer on the Duke Power system can save kilowatt-hours by shutting off his air conditioner when he would normally have used it. Historically, a customer would use the air conditioner when the outdoor temperature was in the 80-82 degree range. Now, he turns it off during those periods and keeps it off for many hours, thereby building up a reserve of funds that he otherwise would have paid. However, at the time of the company's peak when it is very, very hot outside, the customer can turn his air conditioning on for a few hours a day or even a few days during a very hot week and not pay an extreme penalty for those hours of operation because he has more than offset this expenditure in a savings at a time that it was not as inconvenient for him.

Another example would be industry moving its utilization of electricity between the billing demand periods to minimize the payments and yet consume the same amount of electricity. This is known as "peak splitting" and is not an uncommon practice with load controllers in large facilities. In commercial establishments, the use of night setback thermostats or actually turning off heating equipment at night has become very common. This practice saves great amounts of energy and, therefore, has a significant contribution in savings on a commercial customer's power bill if he is utilizing electric heating. However, when he warms up his building in the early morning, he establishes peak demands that contribute significantly to the company's peak. We certainly cannot fault the customer for taking these steps in order to reduce his power bill. We, in the electric industry, must recognize that the consumer will react and will make changes that are convenient to him and will also minimize his power bill.

Let's then, for a moment, take a look at what has happened to Duke Power Company over the last 10 years in the area of kilowatts and kilowatt-hours. This will demonstrate what I have said. In the electric industry, the relationship between the total energy sales for a year and the maximum demand for the system is referred in load factor. If you have a maximum demand of 10,000 KW and you use that full demand for 8760 hours a year, you would have 100% load factor. However, we all fully realize that the load on our system varies from hour to hour, week to week, and season to season. Therefore, the energy for a year and the annual demand always results in a load factor which is less than one.

On the Duke Power system in 1970, we experienced a load factor of 66.6%. Because of the activities of the consumers of the company in conserving their use of energy to a greater degree than their use of demand, that load factor deteriorated. That is to say, the use of energy went down at a faster rate than did the demand. In 1979, we experienced a load factor of 62.3%. In 1980, we experienced an even lower figure - 60.7%. What, then, does this mean for the future?

First of all, it means that we in the electric industry must be creative and innovative so that rate forms are placed into operation which will provide the proper pricing signals to the consumers. The objective will be to provide a clear signal which will make the rewards for their actions quite apparent. In order to accomplish this, at Duke Power Company we have an overall program which we began in August of 1974 called Load Management. It has 43 elements and is directed at the reduction in the growth rate of the peak demand. We have made estimates of our anticipated successes in these 43 programs. The programs will cause a reduction in our peak load in 1994 of 4508 MW in the summertime and 5635 MW in the wintertime. This very aggressive program is, as I said, aimed at reduction in the growth rate of the peak. It has been subtracted from the corporate forecasts of anticipated loads and our construction program is designed to provide capacity on our system to meet the forecasts after load management has been deducted.

Although I have attempted to describe our plans and objectives as well as the underlying reasons, there are always some skeptics or there are always questions about the real world value of the kinds of programs that I have been describing.

Therefore, let me attempt to put into perspective what our program means to our stockholders and our rate payers. From 1904 (the beginning of our company) until 1980, the total all-time investment in Duke Power Company amounted to just over 6 billion dollars. Needless to say, this is a tremendous amount of money. If Duke Power Company's Load Management Program did not exist, our forecasted need for generation to meet the growth in our service area would require our raising 23.8 billion dollars in the next 10 years (1980-1990). This is approximately four times the capital that we have raised in the last 76 years, an unbelieveable amount of money and an impossible undertaking. It is beyond our capabilities to finance this sort of a construction program. The Load Management Program is an attempt at reducing this obligation.

The goal is a winter reduction of 5635 MW and when you add 20% reserves, that would be 6762 MW or the equivalent of six of our largest nuclear plants. Even considering the amount of money it will take for us to be successful in our overall Load Management Program, this aggressive activity results in reducing the 23.8 billion dollars by 10.4 billion dollars.

This will mean that our stockholders will not have to sell 235,000 shares of common stock below book value and further dilute all of the existing stockholders. It will, also, mean that our rate payers will not have to pay for 10.4 billion dollars of capital plant. We feel this 10.4 billion dollars can be eliminated and, perhaps, even more.

UTILITY SYSTEM IMPACT OF HEAT STORAGE

Michael J. Cahill
Niagara Mohawk Power Corporation

There has been a tremendous upsurge of interest by utilities in implementing load management schemes where thermal energy storage(TES) is the major tool. This is not surprising when you consider that the basic principle underlying the load management concept is that some portion of an electric system's energy requirement can be shifted from the normal peak load period. TES will allow large blocks of energy to be stored during low load periods for later use during system peaks. This reshaping of load profiles should increase the load factor and presumably will lead to lower costs by reducing the installed capacity and dependence on premium fuels required to supply the electrical loads.

The impact that load management, employing TES can have, requires that a thorough investigation of all impacts on the generation transmission and distribution systems be made. This investigation should consider the following factors:

- Capacity reduction
- Fuel shifting
- Fuel costs
- Maintenance
- Seasonal and daily reserve capacity
- Distribution constraints
- Supply and delivery management

Different levels of TES penetration must be considered for their impact on generation, transmission and distribution system profiles and corresponding facility requirements. Any number of customer devices in use today which have the potential for greater penetrations in the future, or devices which have not yet surfaced could also have a tremendous effect on the utility system profile. Obviously then, this impact analysis must also consider:

- What today's loads are and how they are being trended.
- What loads can be shaped to an advantage.

- What loads are controllable and for what duration are they controllable.

It becomes necessary to match up what is wanted in the way of impacts to the system and then seek customer loads which can be used to shape the system in that form.

Consideration must also be given to the most appropriate control strategy and rate incentives that can be employed. It may well be that the control strategy selected in the initial implementation may be substantially different than that which is felt is the ultimate control strategy to be used. For example, it may be decided that in the initial implementation, time of day metering is most suitable, but then in a later phase, direct control from the utility may have the most beneficial overall impact to the consumer by minimizing both utility and customer costs. During this discussion, an attempt will be made to identify some of the concerns that must be dealt with in order to make these decisions.

The use of TES in a load management strategy must respond to some clearly defined objective. For example, the utility might choose any one or more of the following objectives:

- To minimize energy usage for heating
- To minimize system or feeder peak demand
- To minimize overall cost to the utility and customer
- To minimixe utility control
- To maximize load shed benefits to the utility
- To maximize customer benefits

Substantially different types of customer devices and control strategies may be required to respond to the individual objectives above. For example, minimizing peak demand requires that storage devices be discharged during peak periods and charging be done during the valley period. If the control strategy selected was intended to respond to a time of day metering arrangement, the equipment selected to respond to this objective would require substantial storage capability. Figure 1 shows a conventional heating profile overlaid with a TES profile responding to time clock control. It can be clearly seen that the total heat energy requirement must be met by a heavy concentration of energy in a relatively tight charging window. If the customer equipment selected was going to respond to direct control, a much lesser storage capacity would be required. Figure 2 again shows the conventional heating profile, this time overlaid with a TES profile responding to direct control. On the day shown only two load shedding signals are depicted. This could of course be expanded to show any number of signals, dependent on the sophistication of the control strategy. It is easy to see that the total heat energy requirement can be met by using a much wider charging window and a much narrower discharge period. This of course requires a much reduced storage capability. In this example, the decision must be made up front as to where the dollars should be invested. Should they be invested in the customer equipment, or should they be invested in the utility control equipment? Should the utility minimize control, or should they minimize overall consumer costs? There are both investment and production cost trade-offs.

It becomes readily apparent that it is difficult, if not impossible, to simultaneously respond to all objectives that were stated earlier. It also becomes apparent that the selection of customer equipment, in a scheme where initial implementation is done through one control strategy and then at a later date changed to some other type of control strategy, must have the built in flexibility to

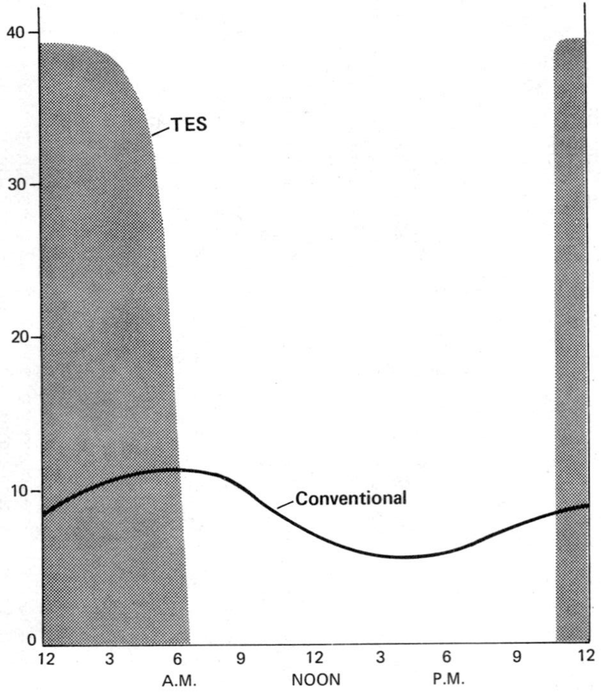

Figure 1

TES is charged as needed in shaded area and is available for discharge through the remainder of the day. With conventional heating, the energy requirement would track the heating requirement.

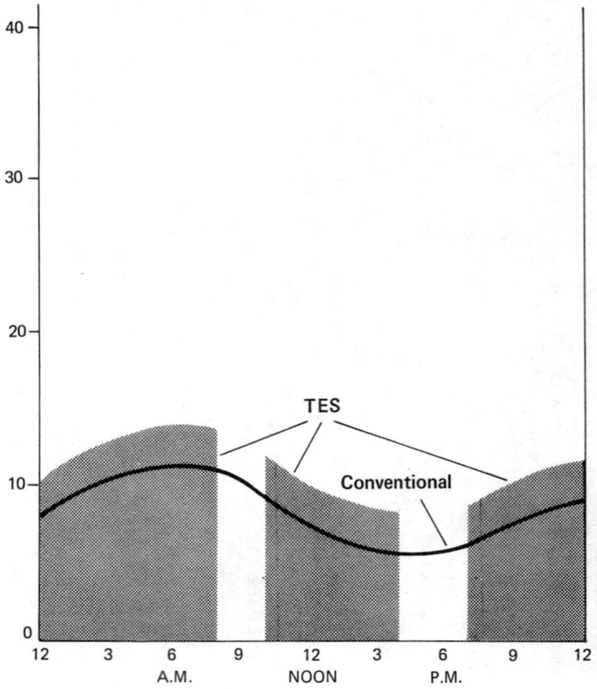

Figure 2

TES is charged as needed in shaded area and is available for discharge as shown. Charging requirement is much less than in Figure 1.

operate over a wide range of control. For example, equipment selected first to respond to time control and then later to direct control must have the capability and the flexibility of accomodating either at the time of installation. Otherwise, the customer will face an unfair economic penalty later. It may have to have the initial economic attractiveness to the customer which requires fairly rapid payback because of the possibility of unused capacity of storage occurring later under the different control strategy. Figure 3 shows a system profile without TES and the effect on it by two different levels of TES penetration. The first level reshapes the profile in a favorable way. If the utility could refuse any additional TES installations at this point everything would be fine, however, this may not be acceptable from a customer relations or regulatory standpoint. If additional TES was allowed on the system responding to the original incentives, the effects would be decidely negative to the utility. It may well be that direct control would become a better strategy at this point, both for existing customers and new ones. Figure 4 shows what the effect may then be with the same penetration as shown in the second level of Figure 3. The magnitude of charge does not have to be nearly as great and all of the load does not have to be shed together. Much finer tuning in line with system requirements is possible.

Ideally, a utility who ultimately wanted a direct control system could install it at the start in order to minimize this storage capacity requirement. Unfortunately, this may not be practical. It is doubtful that enough controllable load would exist at the start and unless there are other advantages to the utility such as distribution automation, etc., it is doubtful that the utility could justify the expense.

Another possibility that must be considered in the implementation of this type of program is the desired penetration of TES devices that will successfully affect the profile. Under or overshooting the mark may create some unwanted consequences such as new peaks, economic penalties, overloaded feeders, etc. The chance of this possibility can be minimized if the utility takes extra care in evaluating the demographics of the area that they plan on affecting. For example, a utility which is experiencing a normal growth of new customers and whose existing system contains relatively new housing with fairly modern heating systems, must recognize that an incentive just large enough to attract the new customers will not be large enough to attract conversions from the existing ones. The incentive must be fairly large if the utility wants to attract all of the new customers in addition to a percentage of the conversions from the existing customers whose present systems are operating well. There is also the problem that, as the number of customers taking advantage of the rate increases, the incentive may have to be reduced.

A different requirement exists, if a utility assesses their system and finds again that normal growth prevails, however, a large portion of their system consists of older and, in many cases, inadequate heating systems. It may well be that there is a market for new heating systems. This case may show that a potential exists for large numbers of TES conversions over a period of years and that the incentive that need be offered may be quite a bit less than that required in the earlier case. It may be necessary just to be attractive enough to select TES over other options available. One big concern here, however, is that many times where you have older and antiquated heating systems, you also may have older and somewhat antiquated distribution systems. There may not be the available capacity built into the distribution system to take on these new blocks of additional load from non-electric heating systems, that would be required by the TES installations. Consequently,

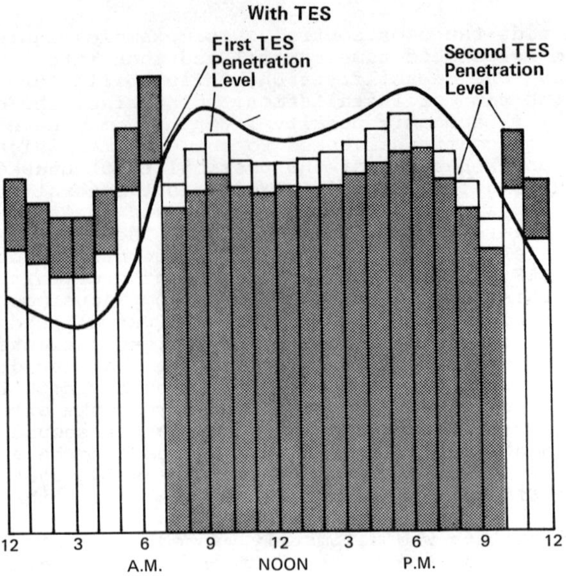

Figure 3

Two levels of TES (under time clock control) are indicated with resulting effect on peaks. System profile without TES is shown for reference.

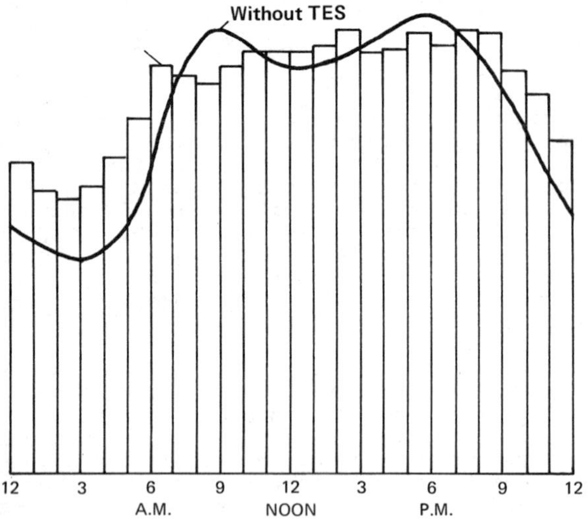

Figure 4

Direct control of TES allows for preferred shaping of profile without the need for large individual storage banks.

it may well be that the distribution system is the constraint which limits the utility in allowing TES installations. It may well be that the cost to upgrade the distribution system to respond to the TES installations may more than offset the other benefits from system improvements which are sought.

As stated earlier, incentive offered to the customer can have a major bearing on the penetration levels that are attained by TES devices. Obviously, the greater the incentive, the greater the penetration will be. One important point that must not be overlooked in the development of these incentives is the proper assessment of the benefits that will be derived by the utility. These must balance.

In the development of an effective program, the ideal strategy would predefine benefit/cost assessment at various levels of penetration. Once a utility has a good fix on these benefit levels, they could then design what incentive could be offered to their customers in order to attract them to put equipment of the desired types onto the system. This incentive must be integrated with a control strategy which is compatible with the TES equipment. Cost obviously plays a major role in this decision.

The implementation of a program which would deal with no control or time control would certainly be much less expensive to implement for the utility than one relating to a direct utility control strategy. However, the cost to the customer for the equipment necessary to respond to the former control strategy would be much higher than it would be for the direct control strategy. The capacity requirements for a time control system are usually much greater than those required for a direct control system. The reason of course being that with a time control scheme usually the system is designed around the most extreme conditions. This leads to shorter charge periods and longer discharge periods. Whereas, with direct control it is possible to tailor the response directly to the real time needs of the system. This usually means allowing a charge to build up prior to an expected peak and then shutting the system down when the peak occurs for a period of only two or three hours. Thereby, much less capacity is needed.

Obviously, any decision to employ TES devices in a load management strategy must consider many different conditions:

- The existing equipment
- The condition of the system
- What are the objectives
- What controls should be used
- What economic advantages need to be gained
- What changes will occur on the system
- How these will be affected by TES applications

Unfortunately, all the pieces of the puzzle are not yet assembled. Consequently, it is imperative that in any program which contemplates implementing TES devices in a load management setting, flexibility must be built into this equipment and the control equipment. Customer equipment and control equipment must have the capability of responding to different strategies which may develop in future years. Otherwise severe economic penalties may be faced by either the customer or the utility (which ultimately affects the customer). It may well be that an incentive offered ten years down the road may be reflective of real benefits which are entirely different than those anticipated during early implementation. More and more information is being generated in this regard and the way is becoming clearer and will continue to do so in ensuing years.

Much study is going on in this area at the present time. The Department of Energy has a massive Customer Load Management Demonstration Program involving several utilities, each with TES systems or cool storage systems. Each of these utilities is including in their projects the assessment of impacts on their systems. The results of this program are intended to give utilities across the country the data base that will assist them in making decisions on whether TES has application for them or not.

Niagara Mohawk designed and manages one project in this program located near Lake Placid, New York. The site, which provided athletes housing for the 1980 Winter Olympics, has now become the Raybrook Correctional Facility. The host for this experimental program is the U. S. Bureau of Prisons. In addition to Niagara Mohawk and DOE, other participants include the Electric Power Research Institute, The New York State Research and Development Authority and the New York State Department of Public Service.

The major emphasis of the research to take place at the Olympic Village location entails evaluating load management strategies utilizing TES technology, or the storage of heat by various methods, for release into building areas as customers require it, with electricity provided at the best times for the utility company. As a result of this two and a half year study, researchers expect to develop a data base which will enable electric utilities to apply this technology to electrical supply systems in the future.

The dormitory buildings will demonstrate different heat storage, electric demand reduction and control concepts. Each building will be considered a separate "customer" with its own metering and control scheme. In order to apply comparative sampling techniques to the study, one of the five buildings will be using customary (resistant) heating only, employing no load management or heat storing methods. In addition, each of the other four buildings also is equipped with the originally specified standard resistant heating, so in effect there will be two methods or controls, which may be used in checking the operation of the TES buildings:

- The one dormitory equipped only with resistant heating
- Switching to the resistant heating in each of the other buildings whenever desired

The Olympic Village setting provides an excellent opportunity to study TES because of the identical physical characteristics and occupancy characteristics of the buildings.

One building is equipped with a slab storage system supplied by Peak Supervision Control, Laval, Quebec, Canada. It relies upon electrified (24V) structural reinforcing mesh located in the first floor concrete floor slab. A transformer, thermostat and limit controls have replaced the "as designed" electric resistance furnaces on the first floor only. Electrical switching is arranged so either the mesh or the "as designed" system can be used for heating purposes.

Another building has a heating system comprised of ceramic bricks coupled with air handlers. It features an electric block storage heater with face and bypass dampers plugged into the inlet side of the "as designed" furnace system. The standard resistant heaters will remain inoperative during the period of the experiment. The ceramic brick type of heating will be in use only in what will be the residence area of the building, with the other areas, namely the administration core area, recreation, entrances and vestibules relying on the systems originally designed for the buildings. The components for the ceramic brick TES facility were provided by Control Electric Corp., Burlington, VT. This system will charge fully in seven hours and discharge over 24 hours.

A third building has a water storage with air handler's system supplied by Megatherm of East Providence, RI. This replaces the "as designed" system in the resistance area with high temperature liquid storage and hot water heating coils, control valves and stats. In the Megatherm system water is heated to 290°F. before air passes over the coil. The water storage takes a full charge in two hours and will discharge in five hours. The remaining parts of the building will be heated with the "as designed" system.

The fourth dormitory building making up the Olympic Village will be equipped with Carrier Corp. heat pumps. In each building wing a 15-ton heat pump unit will replace two 22.4 KW fan coil units as originally designed. The multi-mode heat pump building has the capability of operating in the following modes: air to air, water to water, water to air and air to water. A microprocessor in the system determines the mode. This heating system has a utility intervention point which can be used to interrupt the system and modify its operation to the benefit of the utility system while still providing the required heat. There are four heat pump units with storage tanks located outside. The fifth and final building relies upon the "as designed" furnace.

The actual collection of the data emanating from the various types of TES will be accomplished by means of probes located throughout the buildings under study. With 40 probes in each building, comprehensive records will be kept on temperature, energy use and the rate of air flow. All data are gathered by a remote terminal in each building. Once it is gathered, the information then is sent by leased telephone to a control computer located in Palo Alto, CA., the location of the project consultants, Systems Controls Inc., who will do the monitoring, compilation and analysis of the data.

The results from this experiment, coupled with those from others in this DOE Program, should provide a valuable data base for utility regulators and planners throughout the country to better understand the interrelationship of impacts, both to customers and utilities, when Heat Storage is applied to the utility system.

THE SEASONAL THERMAL ENERGY STORAGE TECHNOLOGY INFORMATION SYSTEM

L.S. Prater, J.R. Eliason

Pacific Northwest Laboratory

V.E. Hampel

Lawrence Livermore National Laboratory

ABSTRACT

A comprehensive information system is being developed to support the Seasonal Thermal Energy Storage (STES) Program, which is managed by the Pacific Northwest Laboratory (PNL) for the U.S. Department of Energy (DOE). The STES Technology Information System (STES-TIS) is being developed to provide a centralized, on-line source of information which will assist with the development of thermal energy storage technologies and will also aid in the transfer of these technologies to the commercial sector. The STES-TIS is a subset of the Technology Information System (TIS) developed at the Lawrence Livermore National Laboratory (LLNL) under the sponsorship of DOE's Office of Advanced Conservation Technologies.

The STES-TIS has been designed to fill the informational needs of the STES Program by providing the capability to store and manipulate various types of data files and numerical models. In general, the STES-TIS contains four types of resources: 1) administrative, 2) bibliographic, 3) computational, and 4) communications. The administrative resources include files containing task descriptions, weekly highlights, a quarterly newsletter, and conference information. The bibliographic resources consist of various data files related to the STES Library. The computational resources provide automated access to various computers on which the economic and hydrologic models developed for the STES Program can be run. The communications resources include electronic mail, conferencing, and the capability to link with or write to the terminal of another user.

More information will be added to the STES-TIS as the Program progresses. The information system will enhance future commercialization of seasonal thermal energy storage technologies by providing a useful, centralized source of information.

INTRODUCTION

A comprehensive information system is being developed to support the Seasonal Thermal Energy Storage (STES) Program, which is managed by the Pacific Northwest Laboratory (PNL) for the U.S. Department of Energy (DOE). The STES Technology Information System (STES-TIS) is being developed to provide a centralized, on-line source of information which will assist with the development of thermal energy storage technologies and will also aid in the transfer of these technologies to the commercial sector.

The STES Program is a large, multi-year program designed to demonstrate the storage of thermal energy on a seasonal basis, using surplus heat or cold. The STES Program is managed by PNL, but it also involves many industrial, university and government contractors who are distributed throughout the United States (Figure 1). Effective management of this large program with its geographically-distributed components depends upon the availability of accurate and timely information. The STES-TIS is being developed to fill this need. The system provides an efficient means for program components to exchange information.

The STES-TIS is a subset of the Technology Information System (TIS) developed at the Lawrence Livermore National Laboratory (LLNL) under the sponsorship of DOE's Office of Advanced Conservation Technologies. It is designed to provide a central source of data and models for use in the development of energy storage systems. The TIS is a computer-based integrated information system, which combines in one mini-computer the information and resources needed to support energy storage research and development, and which makes this information available to the energy storage community. Unlike other information systems that offer only bibliographic, financial, numeric, or computational services, TIS combines all of these resources into one powerful information system. The TIS provides access to interactive models, electronic mail and conferencing, distributed computer resources, and data bases related to energy storage. This information system is a user-oriented, self-guiding system that can be used by program managers, scientists, engineers, and support staff without previous computer experience (Hampel and Rabe, 1980).

The TIS can be accessed from remote computer terminals over commercial phone lines and national computer networks. The system provides user-to-user communications capabilities and acts as an automated gateway for connection to other computers and information centers.

The four major capabilities of the TIS are data base management, interactive modeling, electronic communications, and distributed networking. These capabilities allow the TIS to provide a sound base upon which to build a centralized information and communications system for the STES Program.

Development of the STES-TIS was initiated in FY 1980. PNL contracted with LLNL to provide assistance in the development of the STES-TIS. Considerable progress has been made on this project, although there are still many improvements that could be implemented. Progress to date and status of the project are discussed in the following sections.

PHYSICAL ASPECTS OF SYSTEM DEVELOPMENT

Development of the STES-TIS began with procurement of remote terminal equipment for use at PNL. Project staff at LLNL assisted PNL in the selection and procurement of equipment that would meet the needs of the program and that would be compatible with TIS. These remote terminal systems have been installed and are now serving the

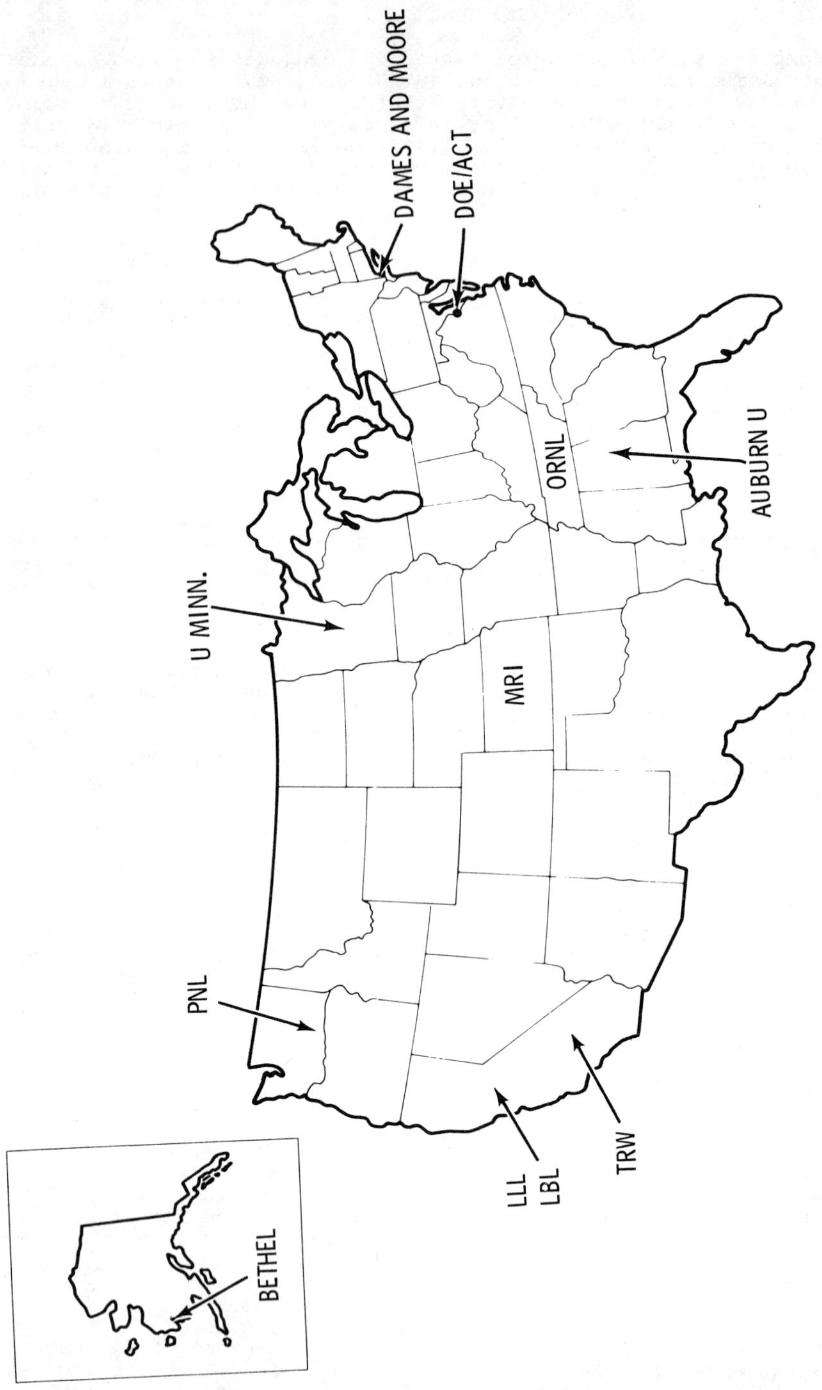

Figure 1. Location of STES Program Components

STES Program office at PNL. In addition, communication capabilities were added to the Qyx word processors used by the STES Program secretaries. These communications capabilities enable the word processors to operate as remote terminals. The secretaries can transfer documents from the memory of the word processors directly into the STES-TIS and can then send the documents to other system users via the electronic mail. This capability has vastly improved communications between the STES Program office and the contractors.

While the remote terminal equipment was being ordered and installed, communications options were being investigated. It soon became apparent that the success and viability of the STES-TIS would depend largely on the ability to establish reliable communications lines between LLNL and PNL. Communications options available to PNL included establishing a dedicated line, linking to the ARPANET, and using voice-grade phone lines. A link to the ARPANET could not be achieved, and the cost of a dedicated line between PNL and LLNL was found to be prohibitive. Therefore, the only option left was to attempt to establish reliable communications over phone lines, which can be somewhat noisy. Error controlling units designed to eliminate transmission of stray characters were installed on both ends of the line. These units have allowed reliable communications at 1200 baud rate. This rate is sufficient for most of the work currently being conducted at PNL. Most of the STES contractors use phone lines to access the STES-TIS.

ORGANIZATIONAL STRUCTURE OF THE STES-TIS

The STES-TIS was designed to fill the informational needs of the STES Program by providing the capability to store and manipulate various types of data files and numerical models. The STES-TIS was organized in a hierarchical manner to reflect the organization of the STES Program. Each level of the hierarchy represents a task or element of the STES Program. Each level may contain numerical models or various types of data files generated by that particular part of the program. Each level of the hierarchy also contains a description of the corresponding program element. Thus, the information system contains a complete, on-line description of the activities being carried out and the data being generated under the STES Program.

In general, the STES-TIS contains four types of resources: 1) administrative, 2) bibliographic, 3) computational, and 4) communications. The administrative resources include files containing task descriptions, weekly highlights, a quarterly newsletter, and conference information. The bibliographic resources consist of various data files related to the STES Library. The computational resources provide access to computers at PNL and at the Solar Energy Research Institute. The communications resources include electronic mail, conferencing, and the capability to link with or write to the terminal of another user.

When a user enters the STES-TIS, he is automatically placed at the top level of the hierarchy, and the resources available at that level are listed. The highest level of the hierarchy contains the following resources:

1 STES Administrative Information
2 Aquifer Demonstration Program
3 Seasonal Storage Technology Program
4 STES Library & Bibliography
5 Integrated Computer Resources
6 News
7 Electronic Mail

The user may select any of these options. If, for example, a user selects option number 1, he will be dropped into the next level of the hierarchy, and the following directory will be printed:

1 STES Administrative Information

- 1.1 STES Program Description
- 1.2 Conference Agenda
- 1.3 Mailing Directories
- 1.4 STES Program Management

The user may select any of these options and continue down the hierarchy, or he may return to the top of the hierarchy and may pursue another branch.

The hierarchical structure of the STES-TIS is very logical and orderly, and it allows inexperienced users to obtain the desired information without in-depth knowledge of the data base structure or the operating system's command language. The hierarchy provides a guide by which a user can navigate through the available data files. Methods that can be used to find and extract information from the STES-TIS are described in the STES Database Tutorial (Gallo, 1980).

INCORPORATION OF DATA FILES AND MODELS

Construction of the hierarchy provided the framework necessary to incorporate various data files and models. As mentioned previously, four types of resources (administrative, bibliographic, computational, and communications) have been incorporated. The project status with respect to each of these resource types is discussed in the following paragraphs.

ADMINISTRATIVE

Several administrative data files have been entered into the STES-TIS. One such data file is the conference data file, which contains information about various STES-related conferences. The file contains details about each conference, names of STES staff members who attended or presented papers, and information about the papers presented at the conference. The conference file is a useful managerial tool which is used by the STES managers to keep a running account of staff members' activities. Another administrative data file contains two types of news: 1) weekly highlights, which are short reports describing the important events of each week, and 2) newsletter articles, which are compiled, printed, and distributed on a quarterly basis.

BIBLIOGRAPHIC

The STES-TIS contains bibliographic data from the Seasonal Thermal Energy Storage Library. These bibliographic data were transferred into the computer from a text-processing system. The data were checked for validity and were converted to the proper format. The information contained in the data file was then checked by the STES librarian to ensure that it was complete and correct, and that it accurately represented the documents contained in the STES Library. The library data file now contains over 1700 citations and is upgraded regularly as new library articles are received.

The STES-TIS provides the capability to conduct customized, online searches of the bibliography. A user may conduct a search on any field of the citation, including author, title, publisher, publication date, keywords, etc. An example of a customized search is shown in

Figure 2. Methods for searching the bibliographic data file have been documented by Kawin (1980).

The citations identified as the result of a search can be printed out in a variety of formats. The entire bibliographic data file can also be printed out on a high-quality printer to produce a camera-ready copy of the STES Bibliography for publication.

Each citation contains fields that can be used by the STES librarian to keep a record of the status of the document that the citation represents. For example, each time someone checks out an article, the librarian can enter the name of the person and the date. Later, this information can be recovered to show which staff members have overdue articles (Figure 3). The ability to keep an on-line record of library transactions eliminates the need for maintaining voluminous hard-copy files.

COMPUTATIONAL

In addition to data files, the STES-TIS can also accomodate numerical models. The STES Program is developing several models that could potentially be put into the system, including both economic and hydrologic models. Some of these models are too long and complex to reside on the small computer which houses the STES-TIS. However, in these cases, the STES-TIS can still be used as an automated gateway to other computer resources which can more easily accomodate them. This capability has already been used to implement one of the complex hydrologic codes on a large computer at the Solar Energy Research Institute, to which the STES-TIS can provide access. The connection to other computer resources is made automatically, with the STES-TIS acting as an interface so that the user does not have to know the telephone number, password, or protocol of the other computer. Some of the computer resources available through the STES-TIS are shown in Figure 4.

Implementation of the STES models on the information system makes them available to a wider community of users, thereby encouraging transfer of energy storage information to the commercial sector. The project staff at LLNL can also assist in converting the models from batch-mode to interactive, allowing users to run the models without having intimate knowledge of the codes. None of the STES models have yet been converted to interactive mode, although this task may be undertaken at some time in the future.

COMMUNICATIONS

The STES-TIS offers a number of communications capabilities. The electronic mail program can be used to leave a message in the "mailbox" of another system user, who will be notified about the message the next time that he enters the system. The STES-TIS also has a conferencing capability, which allows a user to solicit responses to a message from a number of other users. Terminal to terminal communications are also possible, using the "write" and "link" commands. The communication capabilities of the STES-TIS are described more fully by Hampel and Schriebman (1980).

These communication capabilities have proved invaluable to the STES Program, because they have allowed the program office at PNL to keep in close contact with all of the contractors. Use of the communication capabilities of the STES-TIS allows instantaneous transmission of messages and eliminates the delays caused when a letter is sent through the U.S. mail. Rapid transmission of information has led to more efficient management of the STES Program.

```
 Type an Option Number, a Command, or "stop".
*report stes-titles where title contains "aquifer" and title contains "storage"
and date = 1980
                                  (1)

                              STES LIBRARY
                            -CURRENT LISTING-


 1445   A Descriptive Analysis of Aquifer Thermal Energy Storage Systems. 1980
          Reilly, R.W.
          Pacific Northwest Laboratory, Battelle, Richland, WA. 71 pp.
          PNL-3298.

 1481   Aquifer Thermal Energy Storage Reference Manual.                  1980
          Prater, L.S.
          Pacific Northwest Laboratory, Battelle, Richland, WA. PNL-3471
          UC-94e.

 1569   Geohydrologic Aspects of Aquifer Thermal Energy Storage.          1980
          Raymond, J.R.
          Pacific Northwest Laboratory, Battelle, Richland, WA. 21 pp.
          PNL-SA-8834.

 1510   Guidelines for Conceptual Design and Evaluation of Aquifer        1980
        Thermal Energy Storage Systems.
          Meyer, C.F. and W. Hausz.
          GE Tempo, Santa Barbara, CA. GE80TMP-44. PNL-3581.

 1583   Legal and regulatory issues affecting the aquifer thermal energy  1980
        storage concept.
          Hendrickson, P.L.
          Pacific Northwest Laboratory, Battelle, Richland, WA. 53 pp.
          PNL-3437.

 1607   Preliminary Survey and Evaluation of Nonaquifer Thermal Energy    1980
        Storage Concepts for Seasonal Storage.
          Blahnik, D.E.
          Pacific Northwest Laboratory, Battelle, Richland, WA. PNL-3625
 Type an Option Number, a Command, or "stop".
*
```

Figure 2. Example of a Customized Literature Search.

```
                              STES LIBRARY CATALOG

*report stes-cat where out > "80-12-30"

     NAME                DATE OUT     ID       AUTHOR
     -----------------------------------------------------------------------

     Blahnik, D.E.       81-01-08     1645     Szabo, B.S.
     Blahnik, D.E.       81-01-13     1648     Williams, G.T., C.R. Attwater and F.C.
                                               Hooper.
     Blahnik, D.E.       81-01-14     1650     Wettermark, G., B. Carlsson and H.
                                               Stymne.
     Blahnik, D.E.       81-01-14     1651     Moriarty, M.P.
     Blahnik, D.E.       81-02-02     1680     Cha, B.K., D.W. Connor and R.O. Mueller.

     Blahnik, D.E.       81-02-02     1715     Barnes, P.R. and H.G. Shapira.
     Hendrickson, P.     81-02-02     1678     Harbridge House, Inc.
     Kincaid, C.T.       81-01-29     1674     Pollock, D.W.
     Prater, L.S.        81-01-21     1660     Pelka, W.
     Reilly, R.W.        81-01-08     1644     Gozeinik, E.F.
     Reilly, R.W.        81-02-02     1679     Davitian, H. and R.W. Leigh.
     Reilly, R.W.        81-02-02     1716     Metz, P.D.
     Reilly, R.W.        81-02-02     1718     Noll, S.

 Type an Option Number, a Command, or "stop".
*
```

Figure 3. List of Articles Checked Out Since December 30, 1980.

REFERENCES

Gallo, Laurie E. 1980. "STES Database Tutorial." Supplement No. 1 to M-0112 (Draft), Lawrence Livermore Laboratory, Livermore, California.

Hampel, Viktor E. and William G. Rabe. 1980. "STES-TIS Technical Information System." UCRL-85064, Lawrence Livermore Laboratory, Livermore, California.

Hampel, Viktor E. and Jeffrey A. Schriebman. 1980. "A Personal System for Electronic Mail." M-115 (Draft), Rev. 1, Lawrence Livermore Laboratory, Livermore, California.

Kawin, Rick. 1980. "Seasonal Thermal Energy Storage Program On-Line Library System." Supplement No. 2 to M-0112 (Draft), Lawrence Livermore Laboratory, Livermore, California.

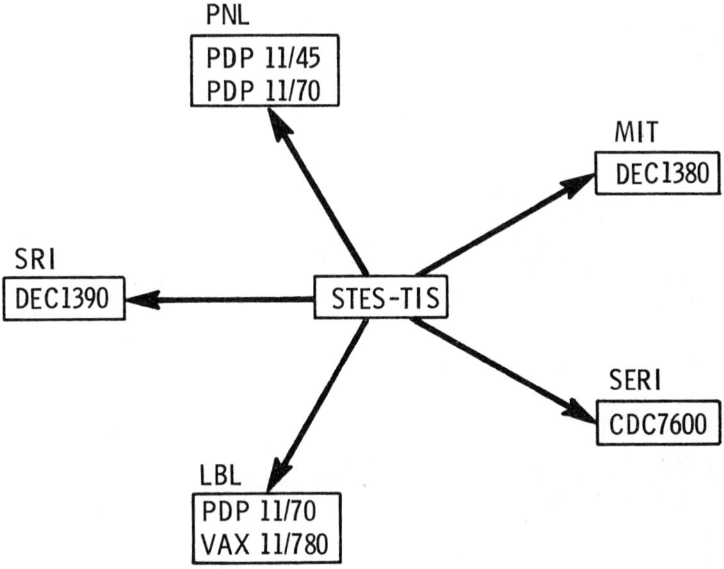

Figure 4. Computer Resources Available Through the STES-TIS

DOE/EPRI/PEPCO STUDY OF ENERGY STORAGE

PETER E. SCHAUB
POTOMAC ELECTRIC POWER COMPANY

ABSTRACT

This paper discusses recently completed preliminary designs of Compressed Air Energy Storage (CAES) and Underground Pumped Hydro (UPH). These designs were prepared by Acres American, Inc. for specific criteria of the Potomac Electric Power Company (PEPCO) under the sponsorship of the Department of Energy (DOE), the Electric Power Research Institute (EPRI) and PEPCO. The results are applicable to other utilities located where the geology includes competent rock suitable for excavation of large caverns between 2500 and 5000 feet underground.

CAES and UPH are technologies available for implementation now, to displace oil-fired peaking capacity in the 1990's. No major technology advances are necessary prior to design and construction. The relatively low environmental impacts greatly increase siting opportunities over those for conventional pumped storage. As developed for PEPCO, CAES or UPH would be flexible in operation to provide load following, frequency regulation, synchronous condensing and spinning reserve capabilities.

INTRODUCTION

The study was begun in 1977 with the objective: to develop estimates of cost, schedule and risk upon which a decision to construct a CAES or UPH plant could be based. The project is now complete. EPRI is in the process of publishing the Final Report, which consists of thirteen volumes and ten appendices (1-23). This paper will present the construction costs and schedules developed and comment on the practicability of the two technologies. Technical discussions may be found, along with full detail on costs and schedules, in the EPRI publications.

ENERGY STORAGE

Electric Energy Storage is the process by which electricity generated in excess of demand at off-peak periods is made available to supplement supply during peak periods. The automobile battery with its charge-discharge service, illustrates the concept. At the utility level, hydroelectric pumped-storage plants are being used. These are operated as conventional hydro facilities to generate electricity. During off-peak periods, when base-load generation is available for the purpose, electricity is used to pump water to the upper reservoir.

Energy Storage saves oil and lowers overall production costs. During off-peak periods, even coal-fired and nuclear plants may be throttled to inefficient load levels. The power used in charging an Energy Storage System could keep these base load plants near full load, and maximum efficiency, thereby lowering production costs. Peak loads are typically met with gas turbines and other fast-response, oil-fired units. Displacing peak period generation from these units, Energy Storage saves oil. Production costs are further lowered since the coal and nuclear-generated storage energy is less costly than the oil-fired energy it replaces.

Siting of conventional pumped-storage facilities can be difficult in view of competition for the required land and water resources. Where extensive transmission to load centers must be built, the economics and environmental impact may simply preclude the option. A number of alternative storage concepts are under development by DOE and EPRI (24), but only two would appear ready for central station commercialization: CAES and UPH.

CAES

CAES operates on the open Brayton cycle. Unlike gas turbines, however, compression and expansion are separate in time. The air is compressed by a motor-driven compressor, cooled, and stored in caverns excavated in rock underground. During the generation period, the air is mixed with natural gas or distillate oil and expanded through a turbine-generator. The benefit is fuel savings. In this cycle, compression work is about 60% of turbine work. With the turbine not driving the compressor, as it does in a combustion turbine, it can deliver two to three times the electrical output for the same fuel input.

CAES has been postulated feasible with three kinds of underground storage. Air might be stored at pressure in caverns leached from a salt formation (25, 26), in caverns excavated in rock, or as a "bubble" within an aquifer. The PEPCO design is for rock cavern storage. To minimize the amount of excavation, constant pressure storage is utilized. A water shaft connected to a surface reservoir applies compensation to the air in storage as the air volume is reduced during generation. Figure 1 is an artist's view of the plant configuration.

Table 1 gives the summary cost estimate for the 4 unit, 924 MW, 10 hour storage capacity CAES plant. The overall estimate was built up from unit costs within subheadings of the FPC (Federal Power Commission) accounts. The estimates were prepared from vendor supplied costs, specialist contractor cost estimates for the underground works, and data from similar work on conventional power plants.

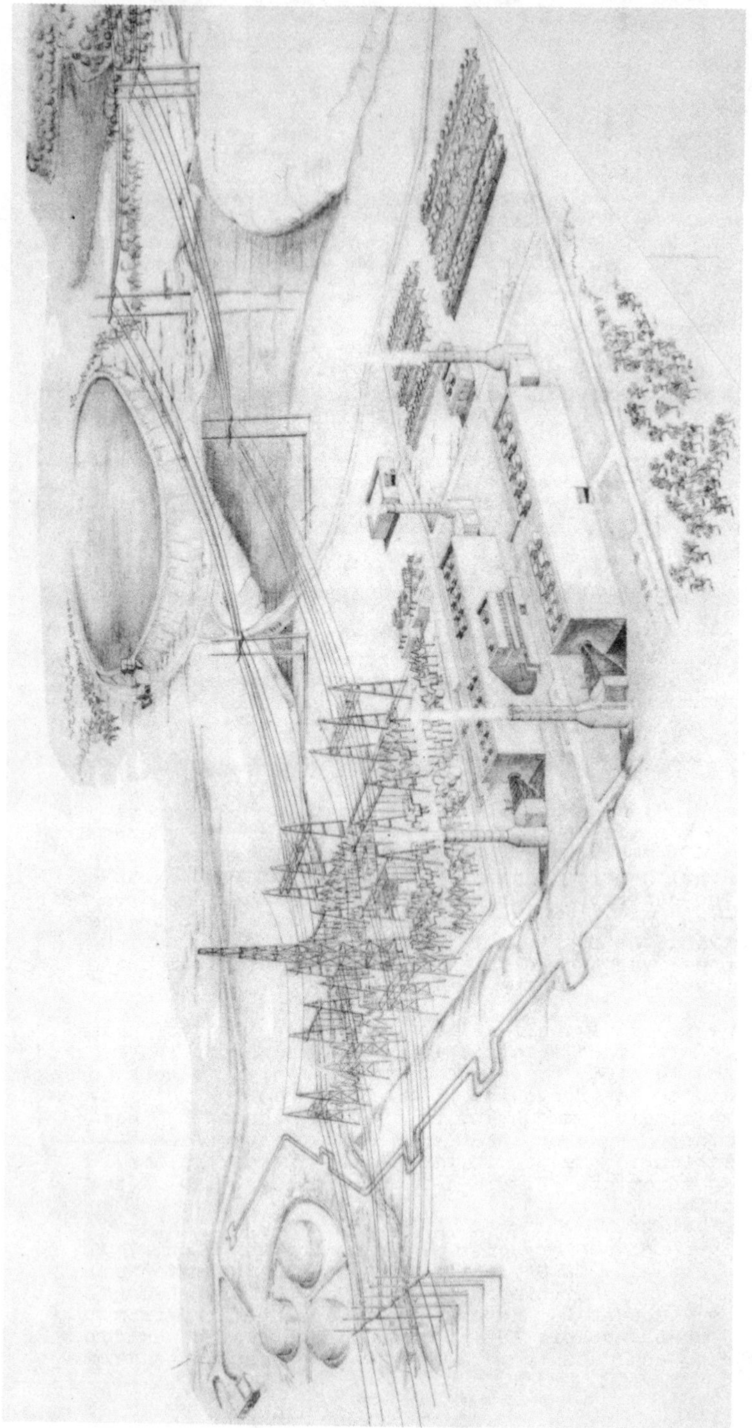

FIGURE 1 - Artist's view of the PEPCO/Acres CAES plant (22). The 3 fuel oil tanks provide a 45-day supply. The switchyard connects the plant to a 500 KV transmission loop. The four units arranged in opposed pairs are provided compression cooling by dry towers located to the right. An impounded site stream forms a pond from which evaporative losses of the compensating reservoir are made up. The reservoir is constructed of excavated rock, and supplies the water column keeping a constant pressure on the air caverns.

All costs are in 1979 $. Neither escalation nor interest during construction have been included, nor have they been accounted for in the contingency. The 45% factor for indirect costs and contingency was provided by PEPCO from previous experience. References (14) & (22) discuss the development of the cost estimate.

The construction schedule shown in Figure 2 is based on the 4 Unit, 924 MW, 10 hour storage capacity plant. The activities shown would place 2 of the 4 units in service and all of the common facilities. The final 2 units could be scheduled in any five year period, overlapping or after the commercial operation of the initial development. Common facilities included in the intial development are the underground caverns, water supply system, compensating reservoir, fuel oil storage system, control and administration building, air and water shafts, and major air piping, Naturally, the economics of various alternatives need to be considered for each project. However, it is deemed unsafe to excavate additional storage caverns at the same site with caverns in operation at 70 atmospheres. References (14) and (22) contain more detail about schedule considerations.

TABLE 1
SUMMARY OF CAPITAL COST ESTIMATE
CAES

Item	Amount Mid-1979 $ x 10^6	
Land, Site Access and Mobilization	8.7	
Surface Facilities	23.5	
Storage System	55.9	
Generator/Compressor System	172.2	
Balance of Mechanical Plant	31.8	
Switchyard	32.5	
Electrical Plant	22.4	
TOTAL DIRECT COSTS	347.0	($375.5/kW)
PEPCO Costs (15% of Direct Costs)	52.1	
Engineering Costs (5% of Direct Costs)	17.4	
Construction Management Costs (10% of Direct Costs)	34.7	
Contingencies (15% of Direct Costs)	52.1	
TOTAL	503.3	($544.7/kW)

NOTE: No provisions have been made in this Capital Cost Estimate for Escalation or Interest During Construction.

No technical breakthrough is required for CAES. It is a practicable technology. Certain features for an economic plant require further engineering, and EPRI is seeing to this. One project is developing the exhaust gas recuperator postulated in the PEPCO/Acres design. No heat exchanger is now in service in similar conditions - 1000 F gas to 120 F air, 660 lb/sec air flow, two thermal cycles per day. Another is examining the underground air/water interface to ensure that the air does not escape through the compensating water column via a solution/dissolution phenomenon. Other projects are underway to reduce the generation fuel consumption by storing the heat produced during compression.

FIGURE 2 - CAES LICENSING & CONSTRUCTION SCHEDULE: Critical path items are shown for placing in-service 2 units and all common facilities of a 4 unit plant. The schedule for the last 2 units would duplicate the Turbomachinery line, and could, in fact, be coincident.

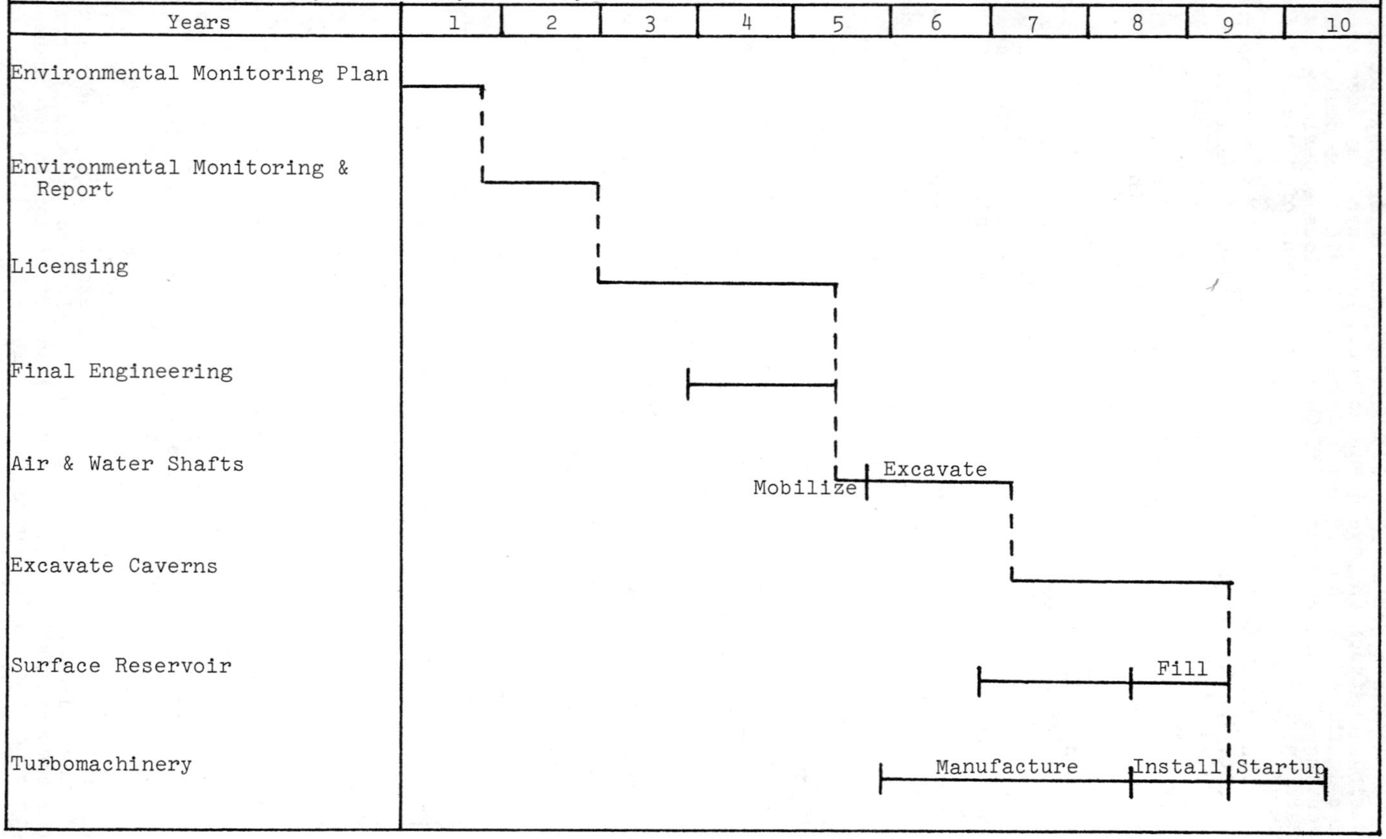

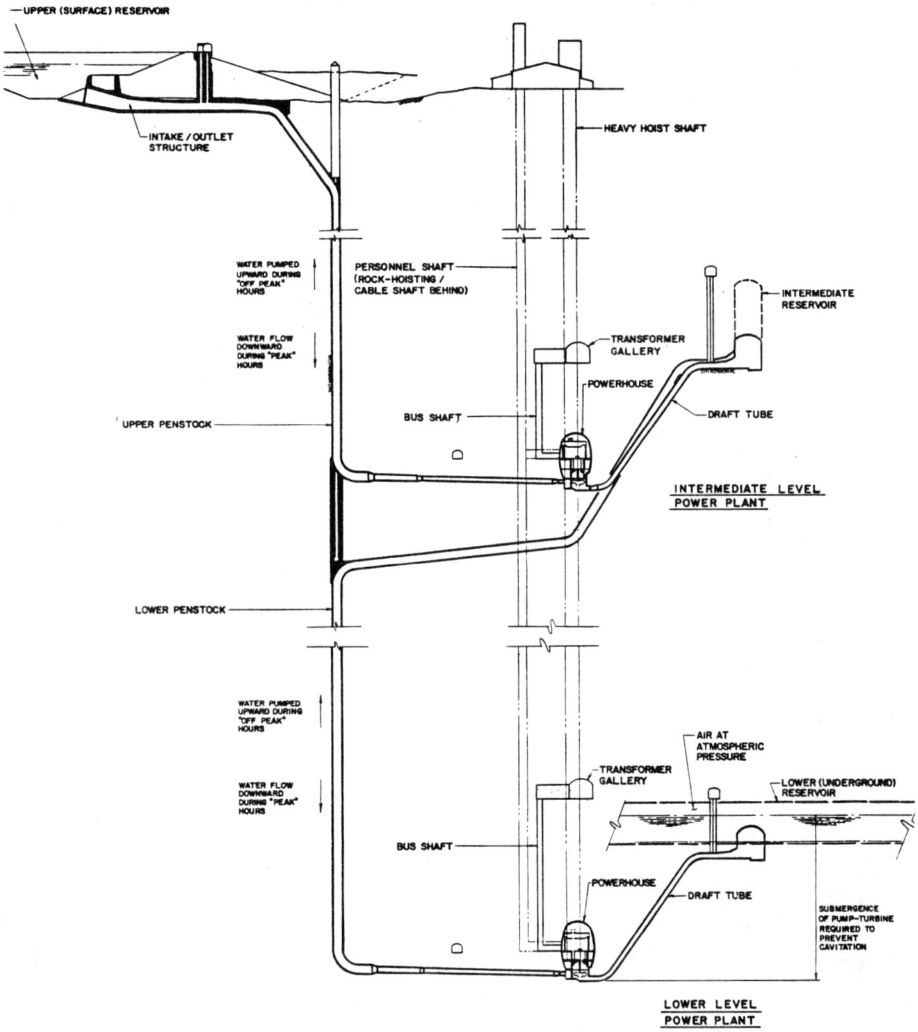

FIGURE 3 - Simplified elevation view of the tandem cascade UPH arrangement (1). Water flows from the upper reservoir to the intermediate powerhouse and into the intermediate (surge) reservoir; then on down to the lower powerhouse and lower reservoir. Pumping results in the reverse path. Pump-turbines would operate in pairs - one from each level working together - and be provided with trim controls to maintain equal flows into/from upper and lower reservoirs.

UPH

UPH is pumped storage with the lower reservoir and powerhouse excavated from hard rock, up to 5000 feet underground. The upper reservoir can be constructed from excavated rock, on level terrain, and be self-enclosed. Since the largest single plant cost is the excavation of the lower reservoir, the size of this excavation should be minimized. The power from such a facilitiy is a function of head (depth) and flow (water volume). Thus, for a fixed power output, minimum overall cost is achieved by increasing the head, and thereby decreasing the flow volume.

Reversible pump-turbine hydrogenerators limit the achievable depth. Current single-stage machinery can achieve heads of 2000 feet, and this may be extended to 2500 feet. By operating pairs of pump-turbines in tandem cascade, with an intermediate surge reservoir, a total depth of 5000 feet can be attained. Figure 3 illustrates the arrangement. The very high head also gives the advantage of a small surface reservoir, requiring modest land and water resources.

Table 2 gives the summary cost estimate for a 6 unit, 2000 MW, 10 hour storage capacity UPH plant arranged in tandem cascade to 5000 feet. Summary costs were built from unit costs within sub-headings of FPC accounts. The estimates were prepared from vendor supplied costs, specialist contractor cost estimates for the underground works, and data from similar work.

All figures are in 1979 dollars. Neither escalation nor interest during construction have been included, nor have they been accounted for in the contingency. The 45% factor for indirect costs and contingency was determined from PEPCO experience with conventional power plant construction. References (8) and (21) discuss the development of the cost estimate.

TABLE 2
SUMMARY OF CAPITAL COST ESTIMATES
UPH

Item	Amount Mid-1979 $ x 10^6	
Land and Site Access	6.2	
Surface Structures	7.1	
Upper Reservoir & Intake	31.2	
Intermediate Reservoir	27.4	
Lower Reservoir	243.7	
Shafts	134.5	
Miscellaneous Tunnels & Galleries	63.6	
Powerhouse Civil Works	46.4	
Pump-Turbines & Valves	56.8	
Motor-Generators	80.9	
Transformers & Electrical Equipment	62.3	
Auxiliary Mechanical Equipment & Hoists	42.3	
Switchyard & Transmission	29.1	
TOTAL DIRECT COSTS	831.5	($415.7/kW)
PEPCO Costs (15% of Direct Costs)	124.7	
Engineering Costs (5% of Direct Costs)	41.6	
Construction Management Costs (10% of Direct Costs)	83.1	
Contingencies (15% of Direct Costs)	124.7	
TOTAL	1205.6	($602.8/kW)

NOTE: No provisions have been made in this Capital Cost Estimate for Escalation or Interest During Construction.

FIGURE 4 - UPH LICENSING & CONSTRUCTION SCHEDULE: Phase I includes 4 units, 2/3 of lower reservoir and all common facilities. Phase II completes the 6 unit plant. The Phase II critical path is entirely machinery-related. Congestion underground prohibits overlap of Phases.

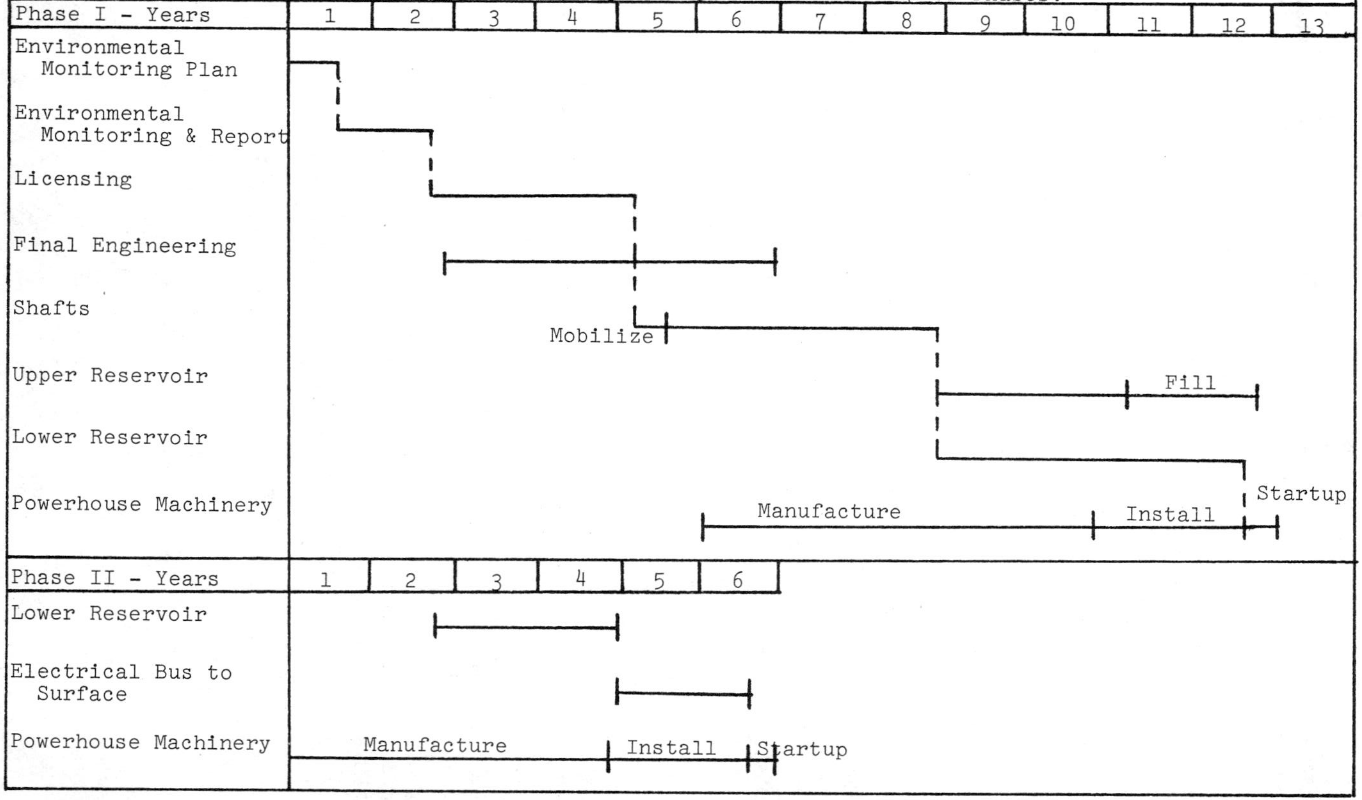

A simplified version of the construction schedule is shown in Figure 4. For the purpose of study, it was assumed that two units in each powerhouse (1330 MW) would be brought on line initially, followed by the remaining two units (670 MW) seven years later. All common facilities are developed in Phase I. Only cavern excavation, final powerhouse concrete, machinery erection, and electrical connections are included in Phase II. The schedule presented for Phase II is flexible to the extent that none of the work should overlap corresponding work in Phase I. The congestion of forces underground and the access limitation of vertical shafts tend to require sequential, vice parallel activity. Further detail is included in References (8) and (21).

No technical barriers need be overcome before implementation of UPH. Comparable facilities, albeit not with such quantity of deep underground excavation, have been built (27, 28). Risks, beyond those normally associated with powerplant construction, are essentially limited to the ability to construct underground works in a specific host rock. Geological emphasis in the siting process and an extensive sub-surface investigation program before construction are the ways to reduce this risk.

Figure 3 indicates the complexity of the tandem cascade arrangement. To reduce this complexity, and lower overall construction costs, EPRI and DOE are pursuing projects to develop two-stage pump-turbines which are suitable for 5000 foot heads, and which meet utility operating criteria.

CONCLUSION

The DOE/EPRI/PEPCO work has determined estimates of cost, schedule and (technical) risk upon which a decision to construct can be based. The costs and construction schedules are comparable to steam plants. The technology is new, but practicable.

Environmental studies (20) have shown both CAES and UPH to be benign in comparison to alternatives. Licensing process evaluations (23) have uncovered no obstacles to obtaining all permits, although a 4 year process is required for both UPH and CAES.

PEPCO planning studies were carried out as part of the project (4, 29). These indicated an economic advantage for a 670 MW, 10 hour capacity storage plant during the 1990's. Both CAES and UPH had a $1 billion advantage, expressed as cumulative present worth of minimum revenue requirements, over combustion turbines. In addition, each produced a net system fuel oil savings of 1 million barrels per year.

Need, cost incentive, and oil savings will be unique for each utility, and variable within the scenarios studied. Nonetheless, the DOE/EPRI/PEPCO work shows both CAES and UPH to be currently viable options for meeting peaking needs in the future.

REFERENCES

1. "Preliminary Design Study of Underground Pumped Hydro and Compressed Air Energy Storage in Hard Rock," by Potomac Electric Power Company and Acres American, Inc. for the U. S. Dept. of Energy and Electric Power Research Institute, EPRI Report EM-1589, 1980. Volume I - Executive Summary.

2. Ibid, Volume II - UPH Design Criteria

3. Ibid, Volume III - CAES Design Criteria

4. Ibid, Volume IV - System Studies

5. Ibid, Volume V - Site Selection

6. Ibid, Volume VI - Site Investigation, Shallow Drilling

7. Ibid, Volume VII - Site Investigation, Deep Drilling

8. Ibid, Volume VIII - UPH Design Approaches

9. Ibid, Volume VIII, Appendix A - Upper Reservoir

10. Ibid, Volume VIII, Appendix B - Shafts

11. Ibid, Volume VIII, Appendix C - Heavy Hoist

12. Ibid, Volume VIII, Appendix D - Power Facilities

13. Ibid, Volume VIII, Appendix E - Lower Reservoir

14. Ibid, Volume IX, CAES Design Approaches

15. Ibid, Volume IX, Appendix A - Air Storage System

16. Ibid, Volume IX, Appendix B - Champagne Effect

17. Ibid, Volume IX, Appendix C - Major Mechanical Equipment

18. Ibid, Volume IX, Appendix D - Mechanical Systems

19. Ibid, Volume IX, Appendix E - Electrical Systems

20. Ibid, Volume X - Environmental Studies

21. Ibid, Volume XI - UPH Plant Design

22. Ibid, Volume XII - CAES Plant Design

23. Ibid, Volume XIII - Preliminary Licensing Documentation

24. Kalhammer, F. R., "Energy Storage Systems", Scientific American, Vol. 241, No. 6, December 1979, pp. 56-65.

25. Herbst, H. and Stys, Z. S., "Huntorf 290 MW The World's First Air Storage System Energy Transfer (ASSET) Plant: Construction and Commissioning", Paper, American Power Conference, 1978.

26. Maass, P. and Stys, Z. S., "Operation Experience with Huntorf", Paper, American Power Conference, 1980.

27. Boivin, R. D. and Anderson, J., "Construction of the Churchill Falls Hydroelectric Development", Proceedings of the American Power Conference, Vol. 34, 1972, pp. 771-783.

28. Vasilescu, M. S. et al., "Blenheim-Gilboa Pumped-Storage Project Design and Construction", Proceedings of the American Power Conference, Vol. 38, 1976, pp. 1006-1024.

29. Driggs, C. L. and Robb, P. W., "Energy Storage for PEPCO - The Economic Feasibility of Compressed Air and Underground Pumped Hydroelectric Energy Storage on the PEPCO System", ASME Winter Annual Meeting, 1980, included in the publication G00195 - "Peaking Power Generation".

BATTERY ENERGY STORAGE FOR UTILITY FUEL SUBSTITUTION

MA McDermott
JA Asher

There is not an energy shortage in this country. We have coal reserves that will meet our current energy demand for over five hundred years. With the breeder reactor our uranium resources would meet our current energy needs for thousands of years. The so-called energy crisis is not a crisis. It is a shortage of liquid petroleum. The crisis exists because with current technology these premium fuels are the exclusive energy sources for transportation and our national effort must be to preserve these fuels for these uses where other fuels cannot be substituted.

I will discuss how the use of energy storage by electric utilities can reduce the consumption of premium fuels in this country, and the potential role that batteries can provide in electric utility energy storage. To better understand how premium fuels are used in this country, the following information is presented. In 1979 46% (37 quads*) of the energy consumed in the United States was from liquid petroleum(1) (see Figure 1). Of this premium energy 9.5% (3.5 quads) was used to generate electricity and 53% (19.2 quads) was used for transportation. During the same time period the US imported 16.6 quads of energy in the form of liquid petroleum products. The liquid petroleum used to generate electricity in 1979 represented 21% of the US oil imports. If the definition of premium fuels is extended to include natural gas, then the combined oil and natural gas used to generate electricity is the equivalent of 40% of US imports (Figure 2).

First, let us examine how the electric utilities currently use premium fuels. The ability of the electric utility industry to reliably provide electric electric power to meet the cyclical demands of industry and home requires a variety of generating equipment. If we examine a daily load cycle for a typical utility we can see how the load dispatcher uses the generating equipment at his disposal to

Figure 1

U.S. OIL CONSUMPTION, 1979

	Quads	Percentage Total	Percentage Imports*
Electrical Generation	3.5	9.5	21
Transportation	19.2	53.0	116
Residential and Commercial	6.9	18.5	41
Other	7.4	19.0	46
	37.0	100.0	224

* 16.6 quads imported in 1979.

Figure 2

U.S. OIL & GAS CONSUMPTION, 1979

	Quads	Percentage Total	Percentage Imports*
Electrical Generation	7.1	12.5	40
Transportation	19.7	34.5	110
Residential and Commercial	14.8	26.0	83
Other	15.2	27.0	85
	56.8	100.0	318

* 17.8 quads imported in 1979.

meet the load variations. Figure 3 shows such a daily cycle and how the various generation equipment is used. Typically, large coal and nuclear plants operating at near peak capacity provide the baseload. These plants are characterized by high capital costs, relatively high efficiency, low operating (fuel) costs and, because of their size, slow response to changing power demands. To meet power demands during peak periods smaller intermediate and peaking plants must be brought on line. These peaking generators are normally oil- or gas-fired steam or oil-fired gas turbine driven generators. Currently, oil- and gas-fired units represent about 35% of the generating capacity of the utilities, and they produce approximately 30% of the power generated by the utility. This includes base generation units as well as peaking and intermediate units.

Energy storage can provide direct substitution of coal and nuclear energy for part of the energy from premium fuels used by the electrical utilities. By increasing the baseload capacity and providing stored energy that can be used during peak periods, the use of premium fuels for electrical power generation can be significantly reduced. Figure 4 shows how this can be accomplished. With sufficient storage capacity and the use of non-premium fuels for baseload generation almost all utility power could be generated from non-premium fuels.

We will now look at the role various energy storage options could have in reducing the consumption of premium fuels. First, to examine the role energy storage could have in reducing premium fuel consumption it is necessary to establish an estimate of the total potential fuel savings from the use of energy storage. This will be examined for a current scenario and projected to the year 2000.

Currently about 25 quads of energy are used per year in the US to produce electricity, including 7.1 quads of oil and gas. The oil and gas consumption includes existing base as well as intermediate and peaking generation units. Therefore, only a fraction of this consumption could be replaced with energy storage. It is estimated (1)(2) that between 1.5 and 5% of electrical energy could be supplied through energy storage. Using an average value of 3.2%, 0.8 quads of oil consumption could be replaced by energy storage. This is equivalent to 4.8% of our current oil imports. Although much uncertainty exists on the growth rate of electric power generation in this country, current estimates indicate an approximate 50% growth between 1980 and 2000, and a corresponding increase in premium fuel savings (1.2 quads) by energy storage is assumed in 2000.

This brings us to an examination of the energy storage alternatives. There are really two categories of energy storage - load management or storage by the user, and storage by the supplier. Load management is accomplished by time-of-day metering which encourages the user to perform high energy consumption operations at periods of low demand. The most common application is the user stores energy, usually in the form of heat, at his site during periods of low demand and uses this stored energy to supplement his needs during periods of peak demand. This is currently used with success in European countries. Little can be predicted about the ultimate use of load management in this country. This involves not only economic but political and emotional overtones. We will assume that load management could reduce peaking requirements by 25%. This would account for 0.2 of the 0.8 quads of premium fuel that could be saved today by energy storage and 0.3 of the 1.2 quads in 2000.

* 1 quad = 10^{15} BTU.

FIGURE 3 ELECTRIC UTILITY DAILY LOAD CYCLE AND GENERATION MIX

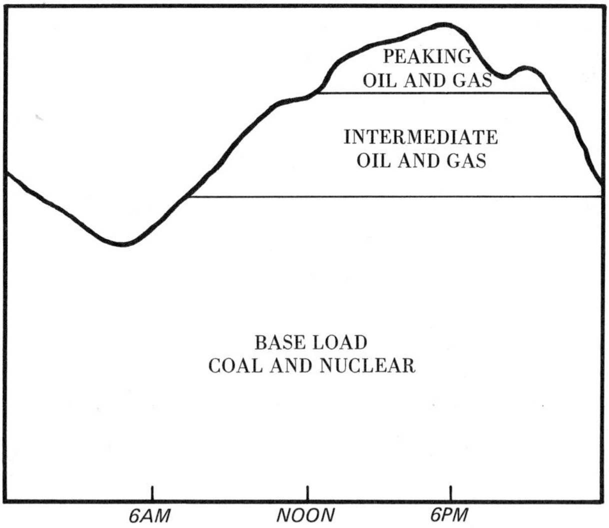

FIGURE 4 USE OF ENERGY STORAGE FOR PEAKING POWER GENERATION

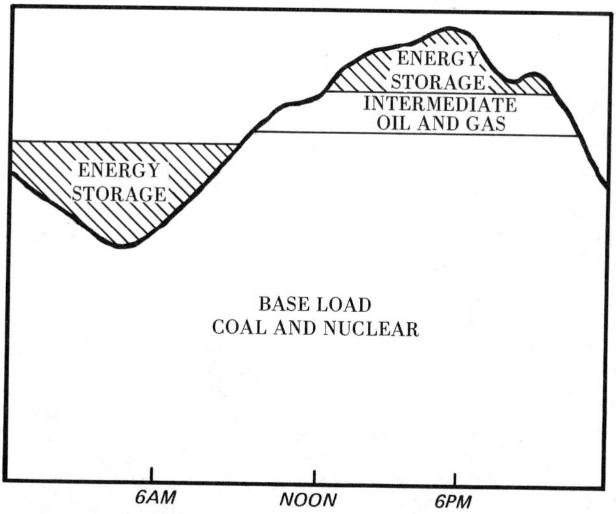

The leading candidates for energy storage by the supplier between now and the year 2000 are pumped hydro, compressed gas and batteries. Pumped hydro plants existing or under construction in the US could provide up to 25 GW of power.(2) This is the equivalent of 20-30 million barrels of premium fuel saved per year (equivalent to about 0.15 quads per year). Construction of additional pumped hydro plants is possible, but meets with environmental constraints and limitations on the numbers of available sites. Use of underground pumped hydro could overcome these obstacles and increase capacity to an estimated 30 GW or 0.2 quads by 2000.

Compressed air storage has a wider choice of geological formations, greater compactness, and a smaller minimum capacity. A 290 MW plant has been operating in Huntorf, West Germany since 1978. This plant uses two 5-million ft(3) capacity salt caverns. No plants are operating in the US but the technology is available. Such systems are under study and commercial plants could be operating before 2000. However, compressed air is not expected to make a significant contribution to energy storage before 2000.

Of the 0.8 quads of energy from premium fuels that can be saved today by energy storage, 0.2 could be saved by load management, 0.15 by pumped hydro and the remaining 0.45 quads require alternate methods of energy storage. This leaves a large energy storage capacity that could be filled by batteries if they are technically and economically ready to enter the market. Recognizing this, the Department of Energy (DoE) and the Electric Power Research Institute (EPRI) are sponsoring programs to develop advanced batteries for electric utility energy storage. Batteries have the advantages of easy siting, flexibility of size and capacity, and short lead times for installation. They could provide up to 50% (or greater) of the utility storage needs if they can be economically developed. Figure 5 shows a projected energy storage scenario for 2000.

At General Electric, with EPRI support, the beta battery is being developed for utility energy storage.(4) In this advanced battery the electrodes are liquid sodium and sulfur. They are separated by a solid beta-alumina ceramic electrolyte (see Figure 6). Features of this battery are:

- battery efficiency improves with temperature and the battery is normally operated at 300 to 350°C,

- there is no self-discharge, resulting in high (~ 80%) electrical efficiency and indefinite shelf life.

- all of the working components are made from readily available and potentially low cost raw materials.

- the energy density is approximately five times that of a lead acid battery.

Figure 6 shows a schematic of a 700 watt-hour beta battery cell. This is a building block for larger battery systems. Because of their modular structure storage batteries could be sized to meet the megawatt sizes required by utilities. Figure 7 is an artist's concept of a 100 MWH battery located at a utility substation. This silent, environmentally benign system occupies approximately one-third of an acre and could produce 20 MW of power for a 5-hour period. The life of the battery will be at least ten years. Although technical challenges remain, such batteries are expected to be available and operating after 1985. These batteries could save at least 0.4 quads of premium fuels per year if they were available today and projected to 2000, could produce savings of 0.8 quads of oil per year.

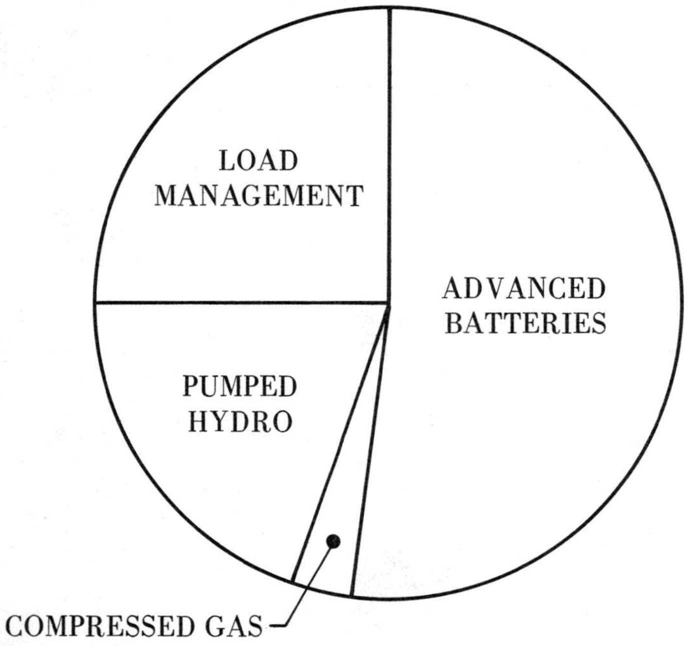

FIGURE 5 PROJECTED UTILITY ENERGY STORAGE DISTRIBUTION BY 2000

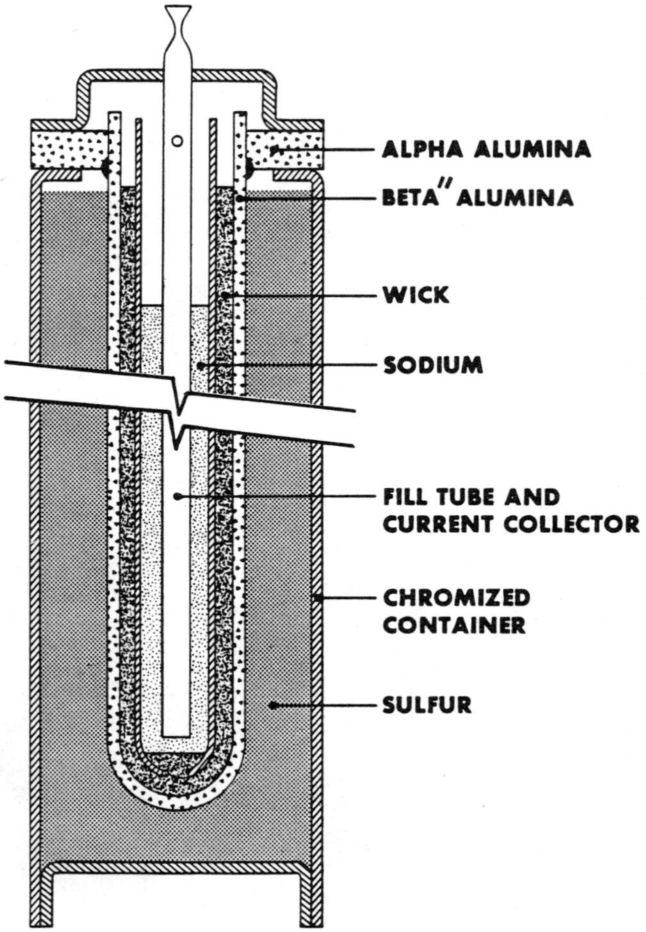

FIGURE 6 GENERAL ELECTRIC ADVANCED BATTERY CELL FOR ELECTRIC UTILITY APPLICATION

FIGURE 7 ARTIST'S CONCEPT OF 100 MWH BATTERY ELECTRIC UTILITY ENERGY STORAGE SYSTEM

In summary, advanced energy storage batteries are expected to have a major role in reducing electrical utility dependence on premium fuels between 1985 and the year 2000. These batteries, combined with other forms of energy storage and load management, will allow utilities to meet daily load demands while using primarily energy from coal and nuclear plants and minimizing the use of gas- and oil-fired peaking plants. Such batteries are under development in the US and are expected to be ready for the market in the mid 1980s.

References

(1) DoE Monthly Energy Review, March 1980.

(2) "Putting Baseload to Work on the Night Shift," EPRI Journal, Vol. 5, No. 3, April 1980.

(3) "Reducing Oil Requirements in the Electrical Utility Industry: The Need for Energy Storage," JR Birk and JW Pepper, Electrochemical Society Proceedings, Vol. 77-4, pp 61-78, 1977.

(4) "Advanced Battery Development at General Electric," JA Asher, JA Bast, FN Mazandarany, 15th Intersociety Energy Conversion Conference, Seattle, Washington, August 18-22, 1980.

LARGE URBAN RESIDENTIAL/COMMERCIAL APPLICATIONS FOR COGENERATION

RICHARD STONE, PRESIDENT
NATIONAL URBAN ENERGY COOPERATIVE FUNDING CORP.,

The focal point of this paper will be the benefit of on-site cogeneration in the production of electric power and thermal energy as applied to large urban residential and commercial applications.

On-site cogeneration can be best described as the simultaneous production of electric power at the point of use along with the production and utilization of the heat energy by-products of the electric generation process.

On-site cogeneration finds its economic attraction in the fact that the practice is far more efficient than utility produced power. The national average for system efficiency of utility produced power is 30%, whereas on-site cogeneration under ideal conditions can achieve a system efficiency in excess of 90%.

As early as 1967, there were some 500 successful co-generation systems in the United States, many operating on natural gas. In New York City, there are a number of cogeneration systems in service, some for many years.

NEW YORK CITY COGENERATION PLANTS:

INSTALLATION	LOCATION	APPLICATION
Rochdale Village	Queens	Residential
Warbasse Houses	Brooklyn	Residential
Proctor & Gamble Mfg. Co.,	Staten Is.	Industrial
Kings Plaza Mall	Brooklyn	Shopping Ctr.
Starrett City Houses	Brooklyn	Residential
Kings County Hospital	Brooklyn	Institutional
11 West 42nd Street	N. Y. C.	Office Bldg.
Seal Kap Corp.,	Queens	Industrial
N. Y. Telephone Co.,	N. Y. C.	Office Bldg.
Elmhurst Dairy, Inc.	Queens	Industrial
Ronzoni Mfg. Co.,	Queens	Industrial
Mangolia Plastics Corp.	Brooklyn	Industrial
Admiral Div. of APL Corp.,	Brooklyn	Industrial
Big Six Towers, Inc.	Queens	Residential
Industrial Scrap Corp.	Bronx	Industrial

The residential plants are among the oldest, The 5,000 family Rochdale Village and the 2600 family Amalgamated Warbasse Houses are both approximately 15 years old and are cooperative housing complexes.

Most of these plants are fired by liquid fuels, No. 2, 4 or 6 oil and natural gas. Some of these plants, predominantly those servicing housing complexes, are conventional fossil fuel fired utility plants relying on boilers to produce steam and steam turbines to generate electricity. Some, like the Kings Plaza Mall have large low speed stationary diesel engines driving generators, whereas the majority of the plants utilize high and medium speed mass produced engine generator sets to generate power.

ENERGY SOURCES:

ELECTRICITY -

Electricity is the most expensive energy source being utilized by man but in many ways the most versatile.

Of all the utilities, Con Edison supplies the highest cost energy. All utilities are subject to rigid regulations by State Utility Regulatory Commissions and Federal Agencies.

Most power is generated by these utilities in utility owned plants, whereas some is purchased from other utilities through grid ties for peaking purposes, or on long term contracts from hydro sources with excess capacity like Hydro-Quebec in Canada and transported via the Power Authority of New York State to be distributed as far south as Con Ed's district.

All utilities rely in varying degrees on hydro, coal, oil, natural gas and nuclear plants. Those customers of utilities in heavy reliance on fuel oil as their prime source of energy suffer under the burden on the highest energy prices.

Attempts to reduce operating costs by conversion to cheaper coal, for example, are presently hamstrung by environmental problems and by the inability of utilities to raise capital as well as the high costs of capital.

Most utilities have the help of a monthly fuel cost increase pass-a-long in the form of a fuel adjustment without the need of the Regulatory Commissions' approval.

Con Edison is oversupplied with generating capacity either by design or miscalculation. The Utility presently has approximately 50% or more capacity, much of it old and inefficient. The retirement of the old inefficient oil fired capacity might be accelerated if it were not for the fact that the Utilities' return is predicated on capital invested and these plants are not yet depreciated.

As regulated monopolies, utilities are guaranteed a rate of return on their investment, thus there is an incentive to increase investment by constructing centralized generating facilities, particularly nuclear, because these facilities are the most capital intensive. Rate of return technology, not only inhibits utilities from developing alternative energy technologies, but encourages them to overdevelop conventional plants. This encourages the upward spiral of utility rates.

Since the utilities are dependent on sustained growth, by continued expansion of their capital investment, they can avoid paying taxes and many do. The growth rate in electric consumption in New York City has been dropping drastically since 1974 and with the advent of a recession and peak shaving efforts by users might well become negative for the next several years, having the tendency to increase rates.

Con Edison estimates that its customers pay $3.8 million in taxes for each 100 MW of electricity sold, and conclude that tax revenue would decrease by $3.8 million for each 100 MW of cogeneration developed. This figure is based on Con Edison's estimate that total sales and gross receipts taxes add 15.4% to a customer's bill. The logical conclusion of this analysis is that all conservation measures are undesirable because they will result in a reduction in municipal tax revenues.

It is questionable, however, whether the discrepancy in the amount of taxes paid by the utility and by the cogenerator is as great as it seems. Just like a utility, a housing project must pay sales tax on the fuel, real estate tax on the value of the property and equipment used for generation, and gross receipts tax on the natural gas bought from the utility. In addition, the property taxes paid to New York City by Con Edison would decrease in any event as Con Edison phases out its in-city capacity.

Con Edison has argued that the proliferation of cogeneration facilities would decrease the amount of electricity demanded by its customers but it would not decrease its fixed costs. Rising costs would have to be spread over a diminishing customer pool. Con Edison estimates that for every 140 Megowatts of generation lost to cogeneration, rates must increase 1%.

Actually, Con Edison's costs are not as fixed as they seem. The loss of revenues from decreasing sales of electricity may be partially offset by increasing sales of natural gas, a prime fuel for cogeneration in an urban area.

Utility arguments against cogeneration are based on the assumption that fossil fuel decentralized energy systems will pose serious environmental hazards. In this context, two considerations should be kept in mind. First, one should measure the impact of the pollutants not in units of fuel consumed, but in units of work achieved. Energy is produced to do work. Thus the pollutants derived from oil-fired cogeneration may be greater per barrel of oil burned, but less oil needs to be burned to achieve the same amount of energy,(or work). Second, oil-fired cogeneration will probably be subordinated to natural gas fired cogeneration and natural gas fired cogeneration will be virtually pollution free.

Con Edison argues that oil-fired cogeneration would decrease use of oil for electricity by only 30% as it replaces centralized oil-fired generation. The limits imposed on the potential for cogeneration are greatly exaggerated by the assumption that the new cogenerator is oil-fired. There is clearly a fourth fuel option: natural and synthetic gas. Both cogeneration and the conventional generating plant can be gas fueled. The alternative which will save more oil, is the one which can be brought on line quickest and is the most efficient in the long run,i.e., cogeneration.

METHODS OF GENERATING ELECTRICITY ON-SITE:

In order to better understand the process of generating electricity on-site, a review of the various methods available follows:

CYCLE	FUEL	TOTAL PLANT INSTALLED COST($/kw)
Gas turbine & waste heat boiler	Gas, #2 oil treated residual SNG (low Btu)	$350 - 400
Diesel engine & waste heat boiler	Gas, #2 oil treated residual Methane	$350 - 500
Steam boiler & turbine	Nuclear - any oil,coal wastes	$500 - 600
Combined cycle & waste heat boiler	Gas, #2 oil SNG	$350 - 450
Steam bottoming	Waste heat	$400 - 600
Organic bottoming	Waste heat	$400 - 700

Cogeneration, using the energy where it is generated, produces greater efficiency by eliminating distribution and transmission losses. Its efficiency is increased still further by allowing energy to be used as near as possible to its original form reducing energy conversion losses.

Another important advantage of the cogeneration concept is the ability to match the size of the plant to the local need. In this manner, capacity and demand are automatically tied together, making it possible to avoid inefficiencies that occur whenever a large power plant is built to meet demand that increases gradually in small steps over a number of years. In cogeneration, capacity can be added to, or withdrawn from a system in much smaller, less costly increments. Not to be overlooked in today's marketplace is the high cost of capital that carries with it an ever increasing impact on the cost of energy. The low capital costs of cogeneration make it an attractive technology.

NATURAL GAS

Natural gas, a major cogeneration fuel, is presently being used in dual fired steam boiler plants like those found in Starrett City or Warbasse Houses, in dual fired engines like those found in Kings Plaza Mall, or in gas engines like those found at Big Six Towers. In most cases cited, gas is being sold by Brooklyn Union Gas under a special low cost rate governed by outdoor temperature.

METHANE

Methane is the major constituent of natural gas mined from underground entrapment by conventional gas well drilling methods. The most attractive source of methane that is presently being explored is found in municipal landfill.

In at least two major housing and Shopping complexes, 15,000 family Co-op City in the Bronx and 5,000 family Starrett City in Brooklyn, Methane from adjacent landfill and sewage treatment facilities can be directly incorporated as a fuel source into existing power houses, a heating plant for Co-op City and a cogeneration plant at Starrett.

OIL

Oil is the only energy source not supplied by a regulated utility but rather through private oil dealers. The bulk of the oil consumed is in residential buildings from low to high density buildings. Many buildings are presently converting to natural gas since price and availability make it more attractive than oil.

WASTE HEAT RECOVERY

Essential to the viability of cogeneration is the need to recover and utilize the heat by-product produced during the process of generation of electricity. Waste heat recovery reduces the amount of fuel otherwise needed for steam, hot water and air conditioning and is particularly valuable in housing and commercial application where fuel costs have exceeded economically reasonable and affordable levels.

UTILIZATION OF WASTE HEAT:

1. DOMESTIC HOT WATER

Domestic hot water is supplied at temperatures of between 130°- 180°F. hence it becomes the most versatile use of waste heat since it can be supplied from low and high pressure steam and low and high temperature water.

2. SPACE HEATING -

Space heating equipment like domestic water is generally designed to operate at temperatures above 130°F preferably at 160°F and over. Space heating equipment can operate on either water or steam waste steam.

3. ABSORPTION COOLING -

Until recently absorption chillers, a process where refrigerated water is produced at 45° by a device that obtains its motive energy from a heating source, required 250°F steam or hot water (or in some cases direct fired gas flame). Since the energy crunch, however, manufacturers have been producing a unit that can be fired by low temperature 160°F hot water in order to make them applicable to use with solar collectors. This modification has placed the absorber in a position to supply refrigeration by utilizing relatively low temperature jacket water from internal combustion engines. Absorbers find application predominately in air conditioning. Other uses like Process heating, process cooling and mechanical conversion find industrial application.

CHARACTERISTICS OF ON-SITE COGENERATION PLANTS:

Most On-site Power generation for residential and commercial facilities will be of the internal combustion engine generation variety for the following reasons:

1.Steam plants, because of the scale required to be cost effective, are relegated to large industrial or large housing complexes in the 5,000 apartment or more category of which there are few and because of financial conditions, less likely to be built in the future.

2.Most small to medium sized residential and commercial facilities that fit cogeneration criteria in the Metropolitan New York area require relatively large electrical supply but limited thermal energy needs on an annual basis, dictating the use of internal combustion engines because of their high electrical to thermal ratio.

3.I.C. engines offer the simplest technology finding wider acceptance than any other technology available.

4.Gas and diesel engine generator sets are the lowest cost available in the sizes that would be considered in New York City.

5.Maintenance backup is most abundant for these types of prime movers.

ENERGY USERS:

In Metropolitan New York, energy users fall into a number of categories: one and two family homes, multiple dwelling residential buildings of the garden and high rise type, office structures, institutional type buildings like schools and hospitals, stores and shopping centers, light industrial type buildings with a sprinkling of buildings used by heavy industry. Of all the occupancies in New York City probably the most energy intensive are some manufacturing facilities, hotels and hospitals.

POTENTIAL FOR COGENERATION:

Numerous estimates have been made of the amount of cogeneration that can be put on-line in the near future. They vary greatly depending upon the assumptions made and the time frame covered.

Con Edison has estimated that there is an economic potential of 100 MW (peak demand) of cogeneration in New York City. Brooklyn Union Gas estimates that at most, 2 MW of cogeneration will come on line, each year in Brooklyn Union's service territory. (Brooklyn Union's territory is primarily residential and is half Con Edison's electric territory). The State Energy Office, however, estimates that only 600 MW of cogeneration statewide is being developed, and that very little of it is in New York City.

Actually the possibilities are far greater. Because not previously counted are the well over 1,000 large governmentally aided housing developments in New York in which energy costs, fuel and electric are the largest and most inflationary operating budget items. Most of these houses, not to mention other types of public and private housing and office and shopping facilities, would require approximately 5 megawatt installations.

In addition, in many parts of the City, in houses of 4 and 5 stories, private rentals, apartment buildings and in commercial and industrial communities, individual operators or small clusters of buildings or businesses can combine to create on-site utility cooperatives with all participants having a share in proportion to their share of consumption. These installations can be built right into existing buildings or into the multitude of vacant or abandoned parcels in many areas of this City. The technique can add dramatically to the cogeneration potential.

TYPICAL INSTALLATION:

GENERAL -

It is difficult to determine a so-called typical cogeneration installation, since in every category a wide divergence exists in size and application complexity. There are not many large cogeneration installations in New York City, though existing or potential, of the size larger than 5-10 megawatts. In the 10KW to 10MW range many applications exist, e.g.:

Residential Sector -	One & Two Family Homes	- 5 - 15 KW
	Multi-Family Dwellings	20 - 5,000 KW
Commercial Sector -	Office Buildings	- 200 - 10,000KW
	Local Shopping Centers-	100 - 250 KW
	Distribution Centers -	250 - 2500 KW
	Regional Shopping Centers-	5,000-15,000KW

The following is a capsule description of 5 typical existing residential and commercial installations in New York City:

FACILITY	TOTAL CAPACITY	PEAKING DEMAND	EXCESS CAPACITY	FUEL	NOTES
Rochdale Village	18,000KW	9,800KW	8,200 KW	dual	5860 units Shopping centers w/o grid connection.
Starrett City	18,000KW	11,200KW	6,800 KW	dual	5000 units + stores w/o grid connection.
Warbasse Houses	12,500KW	4,200KW	8,300 KW	dual	2585 units w/o grid connection
Kings Plaza	11,000KW	7,500KW	3,500KW	dual	1 mil.sq.ft. Shopping center w/o grid
Big Six Towers	4,000KW	2,400KW	1,600KW	dual	1,000 units + stores w/o grid connection.

A good example of a typical installation is that at Big Six Towers.

The Big Six Towers complex stands on more than 12 acres just a few miles from Manhattan. The complex, made up of seven towers, each 18 stories tall, and a small shopping center is a co-op. Each tenant owns stock proportionate with the size of his apartment. One thousand families live in the complex - named after Local Six of the International Typographical Union - which originally financed the 17-year-old complex.

On May 14, 1980, the apartment complex switched to its own newly installed self-generation system, becoming the first occupied residential retrofit.

After the 1977 blackout, the board and the management of the complex became concerned about utility cost and reliability and began looking seriously into the self-generating system.

The initial investment was about $2 Million for equipment capable of handling the electrical needs of the 4,000 resident complex. To further promote the organization and financing of co-generation, Big Six has organized the National Urban Energy Corporation to support energy conservation projects like the co-generation system at Big Six Towers.

This organization has been able to develop an awareness on the part of private investors and lenders of the credit worthiness of these energy-saving installations and thus to provide a means to build and finance these energy conservation programs that would otherwise be difficult, if not impossible, to finance and build completely by conventional means.

The Big Six project took two years to develop and complete. A lot of careful planning and hard work went into the development of the system. The co-generation system has a capacity of 4,000 kilowatts. It is made up of five Caterpillar generators -four 850-kw units and one 600-kw unit. The generators are presently fueled by oil and natural gas. The generators are housed in a concrete block building recently erected on the Big Six Towers grounds.

Before the changeover, the property had an annual electric bill of $550,000. under a master meter plan. Costs with the utility would currently be closer to $700,000. Based on current and anticipated costs, with heat recovery included, Big Six expects to save $300,000.

The profitability and hence, potential for cogeneration will increase still further when equitable buy-back rates are established for excess cogenerated power. Most Cogeneration installations have substantial excess capacity, usually at least, 25% excess even during peak demand periods, considerably more during their off peak periods.

A typical 4 megawatt installation like that at Big Six Towers can generate approximately 30 million KWH a year. It uses only 7,600,000 KWH leaving an annual excess of 22,400,000 KWH.

One of the most crucial incentives for the development of these energy conserving and inflation reducing cogeneration installations is the added energy conservation and financial benefit to be derived from the equitable establishment by the Public Service Commission of the price utilities must pay for the purchase of the cogenerator's excess capacity.

At Big Six Towers, the full purchase of all of its excess power could result in the elimination of the entire cost of electricity to the Housing complex.

In actuality, it is not likely that Big Six, or any other complex, would sell all of its excess capacity. All developments would rather sell the excess power when they can best coordinate the power sale with the maximum use of the thermal benefit produced by the generation of that power. This would work well for all parties in economically helping to meet power requirements of the utilities and still provide very substantial sources of revenue for the cogenerators.

The sensible integration of cogeneration into the greater utility network can only redound to the greater good. Not only would there be conservation and economy but additional emergency standby power, in the event of local or system-wide utility company blackouts.

Decentralized energy producers or cogenerators, especially in large independent housing and commercial communities have an enormous potential. These facilities ought to be encouraged not only because of their conservation value but because they introduce the possibility of competition into an industry once considered a natural monopoly that has grown costly and inefficient as a result.

IMPLEMENTATION OF AN ENERGY MANAGEMENT PROGRAM PRACTICAL VS. THEORETICAL CONSIDERATIONS

WILLIAM C. POPPERT
DIRECTOR OF ENERGY MANAGEMENT
CARLSON COMPANIES, INC.

In recent years there has been much good technical material generated in the field of energy management. The implementation of this information, however, has been lagging. Now, the primary challenge for the corporate energy manager is to implement and maintain energy management programs dealing with the technical, human and external factors influencing energy usage. Once the ample technical and theoretical information has been sorted out, the actual value of energy management will be in the implementation of these ideas.

It is now generally recognized that today's energy situation is not a "crisis" or short-term aberration but a long-term condition of supply and cost. Energy costs have simply become a more important factor in doing business, both in the United States and the rest of the world. Energy management, therefore, is in the process of becoming an established management discipline. There are many ways of approaching energy management programs, however, it is clear that it is an interdisciplinary science. The energy manager is required to have engineering knowledge, management abilities, and to function as an investment analyst and salesman to both management and operations personnel. In some ways, I would liken the embryonic position of energy management departments to that of corporate data processing ten years ago.

ENERGY MANAGEMENT OBJECTIVES

The first step for any organization is to determine their overall energy management objectives and how they can best be met. I use the term "energy management" rather than simply "energy conservation" to emphasize the broad scope of this field. An analysis of the correct role for energy management in the organization must be made. Sufficient data should be analyzed to determine the importance of energy consumption, both financially and in terms of potential business losses due to direct energy supply interruptions. Indirect energy supply interruptions can also present problems in the form of supplier's energy shortages or the impact of energy prices and availability on the consumer's buying habits.

Once management has determined the importance of energy management in the organization, a series of long-term or strategic objectives should be developed. For example:

LONG-TERM ENERGY MANAGEMENT OBJECTIVES

1. Conserve energy and lower effective costs.

2. Ensure adequate supplies and favorable rates for new and existing properties.

3. Contribute to America's energy independence — ensure a favorable business climate.

4. Deal with codes, regulations and government interface.

5. Obtain favorable public opinion for corporate efforts.

6. Identify sales and marketing opportunities in the energy field.

After strategic objectives are established, they must be translated to specific mid-term objectives. For example, in the case of the energy conservation, measurable goals should be set, i.e.; a five-year program with a yearly goal of 3-5% annual reduction in energy consumption.

Consideration must be given to the reactive and pro-active elements of an energy management program. The reactive elements involve reacting to external or internal problems. This includes supply problems, regulations, and troubleshooting or "putting out fires" as they arise. Sufficient effort must also be directed to pro-active tasks; the portion of the program which has been planned and initiated by those responsible for energy management. Whether due to poor planning or lack of manpower, the energy manager must not make the mistake of becoming so caught up in tactical problems that the long-term direction of the program is lost.

Finally, the program goals must be re-evaluated periodically by both the energy manager and corporate management. For example, an organization in the fourth or fifth year of a 5% per year conservation program may find that the goals have changed and the emphasis should now be on maintaining the existing energy efficiencies already won, while looking toward more esoteric technical solutions for further savings.

ORGANIZATIONAL CONSIDERATIONS

Once management has determined the objectives of energy management within its' organization, the administration and organization of the energy program must be established and the specific tasks of the program defined. Unfortunately, in many instances, an energy management program grows out of the existing organizational structure without a great deal of planning. Often, this is not all bad but growing pains and problems with the definition of responsibility frequently arise. Some organizational structures adjust more readily than others to energy management. Ultimately, however, the growing importance of energy management is forcing reorganization and a realignment of priorities.

In most large companies, the first consideration in organizing an energy management program is usually whether it will be handled by a central department or delegated to divisions and/or individual building's management. Often, within organizations we find a hybrid of both types of programs, with some tasks being performed by a central staff and others handled solely on the divisional level. In any case, the key to the success of the energy management program is the competency and responsibility of the building and plant managers involved, whether they are solely responsible or whether they report to a central energy management department.

This raises the second question, that of authority for managing the program. More and more companies are discovering that a central energy management program offers certain economies of scale and specialization. However, it must be clearly defined whether this department is to be a consultative department relying on the authority of operations personnel or whether the department will be a part of the line management structure of the organization. There are pros and cons to both approaches and the only real pitfall would be to not correctly define the division of authority and responsibility for the program.

A central energy management program can be developed in several ways:

ORGANIZATION OF CENTRAL ENERGY MANAGEMENT PROGRAM

- It can become part of an **existing corporate design or engineering department.** This is very often a logical outgrowth of a central engineering department.

- It can become part of an **existing operations/maintenance department**. In a close-knit corporate structure, working through these existing channels may provide the most expedient way to implement the program.

- A **separate energy management department** can be organized. This permits the most autonomy and flexibility for the program. But a staff consultative department must rely on good interface and communications with operations personnel in order to achieve successes in the program. An individual should be identified in each division as responsible for their part of the energy program. This approach works well for a relatively diversified corporate structure.

In many cases existing energy management programs have matured to the point where it is vital for management to take a second look at the organizational considerations of the program and streamline it for better efficiency and more immediate results.

Another organizational consideration involves determining the manpower that should be committed to the program, both on the central department level and in terms of time budgeted for operations personnel. It is also vital that the personnel responsible for the program be trained and capable in the technical skills necessary for success.

Funding the energy management program also must be addressed. Questions must be answered concerning the program's budget and how it will be funded. This can be accomplished through interdepartmental billing, from savings generated, or out of general corporate funds. Likewise, a procedure must be established for inputting energy rate increases and energy conservation retrofit projects to the corporate budgeting process.

Once these organizational decisions are made, it must be determined what specific tactical objectives will be met by an energy management program. These specific tasks could take a form similar to the following:

TACTICAL ENERGY MANAGEMENT OBJECTIVES

- Perform consumption data collection and evaluation.

- Identify conservation and supply priorities
  - Data evaluation
  - On-site surveys
  - Evaluate alternate sources

- Encourage a commitment to conservation
  - Provide manuals and seminars
  - Work to develop follow-through programs in each division
  - Educate managers and operations people

- Provide technical assistance
  - Technical and equipment data
  - Retrofit design and financial analysis
  - Answer inquiries

- Provide information and assistance with government regulations

- Keep up with the state-of-the-art in equipment and procedures

- Continually update goals and priorities and communicate with management.

Lastly, and perhaps most important, there must be visible management support for the program. Without visible support, energy management programs are destined to failure or, at best, obscurity within the organizational structure.

ENERGY DATA COLLECTION AND ANALYSIS

An energy management program **must** be based on a reliable system of energy consumption data collection and analysis. Each building or plant's sources and uses of energy must be tracked and the information made available to the energy management staff and building personnel. Again, we run into a centralized vs. de-centralized question: central data collection and evaluation versus that done on the building level. Experience has shown that **both** the central energy manager and the building or plant staff must be equally aware and involved in the data collection process. Without adequate data neither can evaluate the successes or failures of their conservation strategies and plan accordingly.

For most energy programs, unless they are very highly developed or involved with vast numbers of buildings, a manual data collection program is initially the most effective. This program should utilize forms allowing the period-to-period comparison of consumption and cost. Emphasis must be put on training the accounting or engineering staff for accuracy in filling out the forms, whether using data from meters or utility bills. To avoid errors, ensure that billing periods are as consistent as possible from year to year and source to source. When large amounts of data must be processed or extensive analysis performed, a computerized data system may prove worthwhile. Here, especially, accuracy is required at each stage in the collection and input process.

Energy conservation and the interpretation of energy data has been described as approachable from three views:

THREE VIEWS OF ENERGY MANAGEMENT

- **The engineering approach,** in which efficiency is the most important criteria. BTU analysis is of primary importance and improvements in technology and operating efficiency are given first consideration.

- **The economic approach,** in which costs and cost reduction are the prime considerations. Whatever tactics provide the quickest payback on capital or labor investments are followed. This approach directs efforts to areas where costs are the highest or energy rates are greatest. Ultimately, a business entity can be thought of as consuming dollars as much as BTU's. This approach should, perhaps, be the final determinant for a corporate energy program.

- **The moral approach,** which looks at conservation more as a moral imperative and considers such "big picture" goals as America's future energy self-sufficiency and its' effect on our long-range business climate and world resources.

In reality, energy data must be looked at with a balanced view toward each of the three approaches. The most common forms of energy consumption data analysis are as follows:

METHODS OF ENERGY DATA ANALYSIS

- **Period-to-period comparison** involves looking at weekly, monthly, quarterly or yearly energy consumption in comparison to that same period in previous years. Overall costs, rates and energy units consumed must be analyzed. This is the most basic form of energy data analysis and assists in identifying problems, tracking savings, monitoring rate increases and developing operating personnels' responsibilities. This type of report can also be used to generate deviation reports when a building is over its' energy unit or energy cost budget.

- **BTU per square foot** is an "engineering approach" index enabling the comparison of energy used in a building to a specific standard or to other buildings with similar operations. It identifies a building's comparative energy efficiency but does not take cost into consideration.

- **Cost per square foot** is an "economic approach" index allowing a comparative analysis of buildings, accounting for both energy costs and consumption.

- **Energy cost as a percent of sales or profits** is an extremely important index allowing one to determine the "importance" of energy to a specific building, division or production process.

- **Production indexes** normally take the form of BTU's per unit of production or dollars per unit of production. In a manufacturing process, production indexes go hand-in-hand with BTU and dollar per square foot indexes, allowing a determination of plant efficiency as compared to production rate.

Energy costs and consumption are always affected by external factors. Variables such as weather, business activity and energy prices affect costs and consumption just as much as the basic efficiency of equipment or operations personnel. Although this becomes a complicated area, there have been some inroads in attempting to quantify the effect of these factors on building or process energy consumption. Computer programs are available allowing the factoring of the effect of some of these variables on energy consumption. Proper sub-metering of energy use helps immeasurably in this type of evaluation.

Finally, the bottom line is measuring the effectiveness of the program. The first comparison should be the **percentage change in total energy consumed** or in energy production indexes. This can then be translated into a **dollar savings or "cost avoidance"** figure. This is determined by taking the period-to-period savings (or increase) in energy units and applying the current energy rate. In other words, this figure indicates how many more dollars would have been spent on energy if a reduction in consumption had not taken place. Other, less tangible measures of a program's success involve: fewer supply and business interruptions, better maintained and operated facilities, and public relations benefits. In many cases, a successful energy management program also results in an increase in employee morale.

TRAINING AND MOTIVATION — DISSEMINATION OF TECHNICAL INFORMATION

One of the primary tasks of an energy manager is to ensure that all "no cost or low cost" conservation techniques are implemented. This requires becoming involved with the proper operations and maintenance of buildings and equipment. It is here that the energy manager takes on the role of salesman or motivational expert.

Training of proper operations and maintenance techniques can be provided by manuals which are either written in-house or selected from the large number of existing manuals available. The program also can utilize slide shows, seminars or periodic mailings and reminders. Perhaps most important are on-site surveys or building energy audits. Emphasis on energy management training must also be built in to new facilities and incorporated in new employee training programs.

To obtain the best performance from staff members directly involved with energy use, some form of incentives should be provided. The most basic form of incentive is to include energy performance as part of their job description and one of the criteria for salary review or bonuses. In addition, a recognition awards program should be provided for employees, divisions or buildings that have met their goal or shown outstanding performance. These awards must be very visible within the company and to the peer group. As an additional option, more sophisticated incentive programs are available where employees directly, or indirectly, share in the benefits of their energy conservation programs. These can involve monetary awards or prizes, with each division or individual's incentive tied directly to the amount of energy saved.

IMPLEMENTATION OF RETROFITS AND CAPITAL EXPENDITURES

An energy conservation program seems to develop naturally in "stages". Often various buildings or divisions are at different stages at different points in time. This depends on the technical complexity of the property and the efforts of operating personnel:

STAGES OF AN ENERGY MANAGEMENT PROGRAM

1. **"Easy savings"** are the operations and maintenance savings resulting when operating inefficiencies are corrected.

2. **Low cost retrofits** involve the first capital expenditures of the energy management program. This includes modifications to buildings or equipment with a quick return on investment.

3. **More sophisticated retrofits** involve additional engineering input and capital expenditures. These become increasingly important as the energy management program "matures".

It is important to note that operations and maintenance training is an on-going part of an energy management program and effort must be expended there at the same time that more sophisticated capital expenditure programs are implemented.

The question of technical or engineering assistance must be addressed. The approach varies widely according to the structure of a given energy management program but the options are basically:

TECHNICAL AND DESIGN INPUT

- **In-house** corporate staff handle design for retrofit projects.
- **Outside design consultants** are engaged by the corporate energy staff with bids let for the installation of equipment.
- **Installation contractors** provide design engineering as well as installation.

There are certain values to each of the above approaches and it is a matter of using the best procedures for each task or organizational structure. A small central energy management program may at first rely more heavily on outside consultants but as the program develops and proves cost-effective through savings, more in-house engineering staff may be acquired.

The energy manager must also become a financial analyst, at least insofar as energy equipment retrofits are concerned. It is often difficult to put precise estimates on the savings potential of a specific project; very often a range of savings has to be estimated. These estimates can then be compared to retrofit costs by a number of types of financial analysis. The two most common are:

1. **Return on investment** (ROI) - is expressed as a percentage rate-of-return of annual savings to the capital expenditures involved.

2. **Simple payback** - a technique which compares annual savings to a project's capital expenditure. This shows the amount of time, in years, required to payback the investment from savings.

Although the above are very popular techniques for comparison of alternative investments, one must be aware that the time value of money or potential escalation of energy rates should be factored in for a more complete analysis.

In analyzing financial alternatives, tax considerations must also be accounted for. The energy manager should work with corporate accounting and tax personnel to make them aware that, in addition to the 10% investment tax credit on new equipment, there is a 10% energy tax credit available on many items of retrofit hardware. Further tax benefits and grants are available from the federal government and many states for alternate energy source investments such as solar, wind generation, bio-mass, etc. Leasing or renting equipment may also provide additional tax benefits.

Armed with this financial data, the energy manager must then be prepared to impact budgeting and capital expenditures within his company. This involves becoming aware of the corporate financial structure and budgeting process for each property or division. It also includes determination of a corporate criteria for a minimum return on investment and then insuring that proposals are communicated within the organization.

SUMMARY

Energy management is a new discipline growing from many roots. Its' importance is escalating rapidly as costs increase and supply problems arise. It is now imperative to look closely at a more formal structure for our programs. The technology for improving end-use efficiency and increasing supplies is available. We must meet the challenge of implementing our theoretical ideas or we will lose whatever time advantage America still has in dealing with the energy situation.

CARLSON COMPANIES INC., CASE HISTORY

I offer the following summary of some of our experiences in implementing a program:

Carlson Companies, Inc. is a diversified organization with $1.6 billion in annual sales. Carlson Companies, Inc. encompasses Radisson Hotels, Carlson Marketing Group (Premiums & Incentives), Country Kitchen and TGI Friday's restaurants, Ardan and Naum retail stores, Gold Bond stamps, and Carlson Properties (real estate). Corporate energy sources are: electricity 40%, natural gas 45%, fuel oil 5%, municipal steam 5% and miscellaneous 5%.

Energy costs and efficiency were used increasingly as a criteria for decisions throughout the early 1970's. A formal energy department was established in late 1977 as a corporate department to work with operations people in the various divisions. We primarily operate as a consultive department with building energy conservations or cost avoidance as our main thrust.

We currently have a three person central staff with indirect authority for energy management over building operations and engineering staff. With over 500 buildings, of varying sizes in the organization, the department relies heavily on each division's operations manager to follow through on recommendations as well as to assist in data gathering and troubleshooting of consumption problems.

The ongoing programs can be divided into two types of tasks, reactive and proactive. Some components of our program are as follows:

REACTIVE

1. Review proposals and answer inquiries from the divisional level.
2. Troubleshoot consumption problems.
3. React to government regulations.
4. React to energy price increases.

PROACTIVE

1. Keep central energy records.
2. Issue recommendations for procedures and hardware.
3. Assist in developing operations manuals and training programs.
4. Issue criteria for building remodeling and new construction.
5. Perform on-site energy surveys.
6. Follow through on previous recommendations.
7. Continually reevaluate goals and priorities.

Our mid-term goal set at the department's inception is to achieve a 5% yearly reduction in corporate energy consumption for a five-year period. This goal was achieved in 1978, 1979 and 1980 year-to-date by the participating divisions.

We have devoted the most time to items that eliminate waste and are now starting to get into more programs that require significant capital expenditures. At this stage, however, most capital expenditures under consideration have a simple payback of less than three years.

Management feedback is obtained in periodic meetings, with an annual presentation to the President and Finance Committee.

The most important organizational issues that have developed in the past year include the following:

1. Central department expansion and stronger backup on a divisional level.

2. Interdepartmental billing for energy department services.

Our long range strategy will address itself more to items related to energy supply and the effect of energy issues on marketing strategy in areas such as location of retail facilities and the marketing of energy management services.

TOTAL ENERGY MANAGEMENT WITH MINI-COMPUTER SYSTEMS

N. Chatterjee, D. J. Hersh, J. R. Couch
Air Products and Chemicals, Inc.

INTRODUCTION

The Process Systems Group of Air Products and Chemicals is responsible for the operation of over a hundred industrial gas and liquid producing facilities. The products include large tonnage quantities of oxygen, nitrogen, argon, hydrogen, carbon monoxide, etc., in both gaseous and liquid form. From an operational perspective these plants are characterized by:

- being extremely energy intensive, and
- being minimally staffed at the production site.

The energy intensive characteristic of the process is dictated either by the necessity of separation at low temperatures as for gaseous oxygen, nitrogen, hydrogen, carbon monoxide, etc.; or because the product itself is at a low temperature, i.e., as liquid. One way of illustrating the energy intensive nature of cryogenic processes is to tabulate the theoretical (or ideal) work required to produce some of these products as liquid:

Product	Boiling Point (^{o}R)	Ideal Work (BTU/lb)
H_2	36.7	5,002
N_2	139.2	321
A	157.1	201
O_2	162.4	265
CH_4	201.0	453
NH_3	431.6	100

Another way of illustrating the importance of power is that for a plant built in 1979, the cost of energy to produce 1,000 tons per day of oxygen at 1200 psig is between 60-70% of the total cost; the rest is related to capital and others including labor and maintenance.

The large number of plants mentioned earlier are spread out all over the United States. In order to insure the maximum effective operation at minimum cost, the facilities are minimally staffed at site and are supported by a highly experienced and trained staff in Central Operations located at Air Products' headquarters in Allentown. Thus some plants are totally unattended, some have one single operator per shift, while some of the large and complex facilities may have two. As an example, the facility at LaPorte, Texas, which includes three 1,000 tons per day O_2 plants, a 300 ton per day liquid plant, an argon purification plant, and a liquid product loading terminal, is manned by one panel operator and one field operator per shift.

Prior to the oil embargo of 1973, the emphasis in plant operation was primarily on safe and reliable supply of product to the customer with normal attention to operating costs. With the post-1973 rapid escalation in energy costs another requirement was imposed on the plant operator; in addition to operating the plant reliably and safely, he (or she) was now required to operate the plant considerably more efficiently. To meet the added requirement and still maintain minimal staffing at site, it was necessary to provide the operator with an additional tool: a mini-computer system.

MINI-COMPUTERS IN PROCESS PLANTS

OBJECTIVE: The primary objective of the mini-computer system was to improve the performance of the plant. Thus at fixed production requirements it was to minimize the consumption of power, fuel and raw materials; and at fixed power (or fuel) requirements it was to maximize production within the prescribed constraints. Since most of the existing facilities had pneumatic instrumentation, extensive retrofitting with electronic instrumentation was necessary. Hence the requirements for any project to install mini-computers at operating plants were:

- It had to be cost-effective. Thus the benefits in power savings, or increased production, or a combination of both, had to justify the cost of the computer system and associated applications software, the cost of maintenance, and, most significantly, the cost of retrofitting the plant with electronic instruments.

- It had to be reliable. Although operating the plant at maximum efficiency usually involved operating close to a "brink", the reliability of the plant could not be compromised.

- It had to be acceptable. Since the mini-computer was to be a tool used by the operator it had to satisfy the operator's perceived requirements for better operation. Otherwise it would quickly become "a piece of junk" gathering dust in a corner.

- It had to be standardized. In keeping with the philosophy of a relatively small technical staff supporting a large number of mini-computer systems, it was necessary to keep the hardware, the instrumentation, the operating systems and the applications software philosophy the same from plant to plant.

HARDWARE CONFIGURATION: Based on the above requirements the following configuration of the mini-computer system evolved. As can be seen it satisfied all of the above requirements while keeping most of the additional knowledge acquired "in-house" for further improvements in design or operation.

- Central Processing Unit: In all the plants the equipment is purchased from a single vendor.
- Process Input/Output Equipment: In all plants this is also purchased from a single vendor.
- Back-up Analog Controller: In order to insure safe and reliable operation, each "loop" controlled by the computer has a back-up analog controller. All such controllers are also purchased from a single vendor.
- Electronic Instrumentation: The vendors for the various pressure and flow transmitters, the pressure-to-current and current-to-pressure transducers, etc. are all the same from plant to plant.

This sole-source philosophy for hardware merits a comment. It was recognized that the total cost of hardware is made up of the initial purchase and installation cost plus the cost of calibration, maintenance, spare parts, training, and trouble-shooting. It is easy to see that in keeping with the philosophy of a centralized Operations' support groups the optimum cost leans towards standardization.

CONTROL STRATEGY: In order to optimize the efficiency or performance of a plant it is necessary

- to have available instantaneously all data dealing with the plants' production and the actual performance of all critical pieces of equipment,
- to have all process and equipment knowledge to determine the ideal operating condition to achieve optimum performance,
- to have the capability of adjusting the plant conditions in small but frequent steps from the actual to the ideal conditions in the shortest time interval possible, and
- to have the capability of efficiently accommodating changes in production requirements, product splits, ambient conditions, etc.

The above requirements reflect a "best operator philosophy" and are incorporated into the applications program that is stored in the computer memory. The chief functions of this program can be divided into:

- Monitoring: All important signals are tied into the computer to determine the plant's
  - Production of all products
  - Consumption of all resources, power, fuel, feedstocks, etc.
  - Performance of the over-all plant and all critical pieces of equipment.

One significant aid to the operator through computer monitoring was the development of performance standards also stored in the computer memory. These standards are based on extensive performance testing aided by process simulation and are available to the operator at request. This enabled the operator to monitor the plant and the computer system performance.

- Control: All critical variables that enable the plant to operate at "rock steady" conditions are controlled by the computer through direct digital control. The controls may be

    - Single DDC loop

    - Single or multiple cascade control

  All signal conditioning, limit checking, and appropriate defaulting to safe analog mode is included in the controls software.

- Optimization: At a given production requirement and ambient condition there is one optimum operating condition. The optimization routines enable the plant to operate at that condition and appropriately adjust for changes in production rates, product splits, ambient conditions, etc. Usually the plants operate under either of the following two constraints:

    - Fixed Production - the program minimizes consumption

    - Fixed Consumption - the program maximizes production

  Both the options are included in the optimization program.

BENEFITS: As a result of the implementation of mini-computer control and monitoring systems in operating plants significant improvements in plant operating efficiency was achieved. In some of the best operating plants the improvements noted were 3-5%; in facilities which were large and complex or had frequent changes in production rates the benefits were even higher. Since all cryogenic processes are energy intensive the small percentage savings were significant enough to justify a computer retrofit program that has, to-date, installed mini-computer systems in all the medium and large plants at a cost of over $10,000,000. These plants account for over 85% of the Process Systems Group's energy consumption. In addition, all new facilities are provided with such mini-computer systems, and a program to install microprocessors in the small unattended plants is also under way.

There were several contributors that led to the improvements achieved. They were:

- Improved quality of data. The retrofitting of higher quality instruments and the capability of integration of flows significantly improved the quality of the data. Thus one was no longer limited to taking instantaneous data and hoping the plant had been steady. Bu running precisely controlled tests over a period of several hours and even several days, one was able to evaluate changes in performance to within 1% or less! This helped considerably in determining optimum operating conditions in the field.

- Instantaneous availability of performance parameters. This enabled very early detection of any small deviation from optimum operating condition.

- Improved control. By enabling the closure of many manual loops and the implementation of cascade controls the facility was able to operate close to the desired optimum conditions.

- Improved response to production changes. The use of feed forward control and ramping of several process variables from one set of pre-determined set-points to another significantly increased the capability to accommodate production rate changes and reduce inefficiency.

- Improved documentation. Daily and shift logs of all the production and performance data provided operators with ready access to historical data for reference or improvement. The incorporation of performance standards was also of benefit.

- Improved reliability. A special feature of the computer system was the checks that were inherent in the program. In some cases if the plant condition approached a limit the control system would automatically move it away from the constraint. Appropriate software initiated action due to instrument drift or failure was also beneficial.

It soon became apparent that the monitoring capability of the plant computer systems could be used for something more than just save energy in an operating plant. It could easily be incorporated into an over-all energy management or optimization program for the entire group of plants operated by the PSG Operations Department.

CENTRAL PLANT MONITORING

In addition to improving the performance of an operating plant, the mini-computer system can be effectively used for additional benefits. These benefits result from the system's ability to provide extremely accurate data that can be stored for long-term trends or references. As an example, the scope of these computer systems included the capability to carry out

- daily comparisons of the plants' overall performance to the "best expected" performance and flag unwarranted deviations.

- comparison between the actual performance of all critical pieces of equipment and their "best expected" performance in order to quickly identify degradation in performance.

- evaluation of penalties incurred in the form of lost production, lost efficiency, unproductive power, etc., due to poor performance of plant and equipment.

- better planning and scheduling of maintenance and outages through early detection of equipment degradation and instantaneous quantification of the costs.

- better training of new operators since historical data and optimum performance standards are available for reference.

Since all support of the operating plants are provided by a Central Operations Department, it was essential that all such plant information be available to Central Operations. This required the tieing in of all the plant mini-computers to a central "host" computer with extensive storage capability and is referred to as the Central Plant Monitoring System.

SCOPE: The Central Monitoring System, which included the installation of a host computer system and the appropriate modifications in the plant mini-computers had to satisfy the following size and reliability requirements:

- Have the capability of tieing into over forty computer systems in various locations in the United States and Canada and have the capability of expanding as needed.
- Have a very high on-stream time so as to insure maximum accessibility of the plants' information and knowledge to Central Operations and vice versa.
- Have all the security and protection necessary to insure the confidential and sensitive nature of the data.
- Have adequate memory for "on-line" storage of at least one year's worth of data from each plant, plus off-line storage beyond one year.

The operating functions of the Central Monitoring Computer included:

- Obtaining from the plant mini-computer systems the daily production and performance logs together with the specific standards. This had to be done on a daily basis.
- Obtaining from the plant all significant measured and calculated data, also on a daily basis.
- Providing the plot of historical data on any measured or calculated variable to identify trends in performance.
- Making available to the Central Operations terminals live data from any of the plants tied into the system.
- Having the capability of accessing the plant computer system for loading and unloading programs from the host computer to the plant computer.

BENEFITS: The benefits that result from implementation of such a central monitoring system are:

- Higher plant efficiency through higher on-stream time for the plant computer system. This is achieved primarily through faster response to hardware and software problems which enables the plant to operate under computer control for longer periods of time.
- Lower operating cost. This is due to timely identification, evaluation, and correction of equipment degradation and other problems.
- More effective use of Central Operations staff. Trouble-shooting or advising plant personnel can be carried out more effectively from Central Operations instead of having to go out to the field. This also reduces the travel and living expenses.

- Optimum allocation of individual plant production. In instances where a single facility produces multiple product for multiple customers, or where multiple facilities distribute one or more product over an entire region or country, maximizing revenue does not necessarily imply that every plant has to run at its maximum efficient operating point. The data obtained from central monitoring together with cost data has been useful in deciding on the optimum energy or product allocations to individual plants. This has led to a true overall production and energy management.

CONCLUSION

The rapidly escalating energy prices since 1973 made it essential that all plants perform at the optimum achievable efficiency at all times. Because of the energy intensive nature of the processes involved, efficient design has always been the practice. Implementation of mini-computer monitoring and control systems was found to be cost-effective in both new and, more significantly, in existing pneumatically instrumented facilities. Hence Air Products invested more than $10,000,000 in mini-computer retrofitting programs that have had extremely attractive payouts. Furthermore, the necessity of effective overall production and energy management of the large number of computerized plants in the country resulted in the implementation of a Central Plant Monitoring System. Only with such a system in place, with all facilities operating under mini-computer control, and with an effective organization in Engineering and Operations capable of designing, developing, implementing and maintaining such systems effectively, is it possible to insure that all facilities operate at the optimum efficiency achieveable in a world of high energy prices.

INTERNATIONAL HARVESTER FUEL STRATEGY

THOMAS T. DINGO
INTERNATIONAL HARVESTER COMPANY

Energy in the form of natural gas, oil, electricity, and coal is the lifeblood of industry. Industry, as a whole, has been faced with skyrocketing fuel costs, increased supply uncertainties and increased government intervention on these essential commodities. At International Harvester, we have mounted a comprehensive long term program to alleviate, and to avoid the impact of future fuel problems.

Because of the above, energy management for industry has become a complex challenge. A company cannot afford to direct its efforts to only today's energy problems and expect to perform in a cost effective manner in the future.

International Harvester has committed itself to a long term energy program. The program is directed to three main segments designed to complement each other and be responsive to our view of the energy situation. These areas are energy awareness, conservation, and supply and strategy. Although none of the segments will do the job alone, all are important ingredients of a successful program.

Awareness keeps the concept of energy management highly visible throughout the organization. At International Harvester, this segment includes a commitment by top management to a specific conservation goal of 3% per year through 1984 and the establishment of a special capital fund of $16 million for energy projects. Also included are the placement or designation of energy coordinators at every plant; monthly reporting of energy costs, consumptions and efficiencies; and the company-wide exchange of energy savings ideas.

Conservation provides an economical way to addressing today's energy supply problems. At International Harvester, major improvements have been made in the energy efficiency of our manufacturing plants. The overall effectiveness of IH's conservation program is best demonstrated by comparing its performance for 1979 against 1972. After adjustments for production and temperature effects, IH shows over 16% energy savings. By 1984, we hope to achieve and have set a corporate goal of another 15% improvement. In 1980, because of a prolonged strike and downturn in the economy, IH's performance slipped to a 13% improvement. In 1981, with normal levels of production, IH expects to recover the lost ground and achieve a 22% reduction. Elements of the conservation program include facility audits, detailed plant surveys, and the formation by each plant of a five year conservation plan that includes energy saving technologies that can provide significant energy savings. These projects are:

- Electronic Energy Management Systems
- Solid Waste Boilers
- Burner Controls and Process Modifications
- Submetering
- Heat Recovery
- Cogeneration

Supply and strategy, the third segment of the International Harvester energy program, is designed to improve IH's energy position relative to fuel desirability, thus increasing the probability of continued operation and providing energy capabilities for expanded use of productive facilities in the future. Although coal is the most available and desirable fuel, costs to retrofit for its use in our present and projected size of facilities with current operating practices are prohibitive. This paper addresses the expected fuel availability and cost, and technical developments that International Harvester assessed in the development of a fuel strategy into the 1990's.

IH FUEL CONSUMPTION

Before the details of the IH fuel strategy are covered, a quick overview of its operations is helpful to put the problem in perspective. International Harvester is a capital goods manufacturing company producing trucks, agricultural equipment, turbomachinery, and construction and industrial equipment. IH has 48 manufacturing plants worldwide with 18 plants in the U.S. During its fiscal year of 1979, IH's plants consumed about 19.4 trillion BTU's at a cost of $85.8 million. At 1981 levels of cost, every percent reduction in domestic consumption will mean almost $900,000 in savings.

Individually, no single International Harvester facility can be classified as a large industrial consumer of energy. In fact, we are not really considered an energy-intensive industry compared to high energy users like the chemical or metal producing industries. But collectively in 1979, our plants consumed about 7.2 BCF of natural gas, 94 thousand tons of coal, almost 1 billion kWh of electricity, and over 15 million gallons of fuel oils. On a BTU basis, IH depends on oil for only 19% of its energy needs, but depends on natural gas for 44% of our requirements.

The average IH facility consumes about 900 billion BTU's annually. The largest, the Ft. Wayne Plant, consumes 3.5 trillion BTU's and the smallest, Shadyside Stamping, less than 100 billion BTU's.

Thus, any strategy that was developed must include recommendations for both large and small facilities where economics can vary drastically.

FUEL AVAILABILITY AND COST

In developing a fuel strategy, projections of the availability and cost of the various fuels were made over different time spans. Throughout the 1980's, the same supply and cost problems which industry faced in the 1970's can be expected.

The availability of both natural gas and oil will continue to deteriorate. Entrance of eastern bloc nations into the world energy market and the restrictive oil production policies of foreign governments will cause increased pressure on supplies. During the decade, minor disruptions and curtailments of IH production operations can be expected if extreme weather conditions are experienced as in 1977 or if the Middle East countries experience domestic upheavals as occurred in Iran in 1978 and 1979.

As the end of the decade approaches, increasing uncertainty of both oil and gas supplies will become critical. At that time, coal will become an alternative to alleviate IH's energy supply uncertainties in a cost effective manner.

Projections of the fuel cost have been made by IH's Resource and Technology Forecasting. They projected in May, 1980, gas and oil to have significant rates of increase when compared to coal and electricity. Over the next 10 years, IH projects that coal costs will increase at an average 10%/yr; electricity at an average 11%/yr; oil by 14%/yr; and natural gas by a whopping 18%/yr.

The reasons for the large expected increases in gas and oil prices are continued deregulation and government manipulation of prices through the Natural Gas Policy Act and the Public Utility Regulatory Policy Act. The net effect of those policies will force industry to bear increased fuel costs.

For the short term, these higher prices will cause large industrial and utility users of fuel to voluntarily convert to coal, freeing up both natural gas and oil for the majority of the IH plants. Additionally, the prices will accelerate voluntary conservation programs with all users, further lowering the demand for both gas and oil. Thus, the average industrial user should enjoy an adequate supply of gaseous and liquid fuels but at significantly higher costs. In summary, for the next decade, the uncertainty of reliable supply will increase, and the rise in cost will continue to be drastic. The challenge for IH is to implement a strategy that will allow us to realize the benefits of coal.

NEW TECHNOLOGIES

Again, before developing a strategy, an assessment of new and emerging technologies was made to determine if there was a possible fit into IH-type facilities. Unfortunately, over the next decade, no breakthroughs that will significantly affect the cost of using coal can be projected. In the assessment, fluidized bed boilers, coal gasification, coal-oil mixtures, hybrid and synthetic fuels were evaluated as to commercial timing and cost of implementation. None of these technologies would be applicable to our facilities until the 1990's.

THE STRATEGY - COAL

After assessing the future cost and availability of fuels, upcoming technologies, and the consumption patterns of the various plants and future installations, a plan was set to facilitate the economical use of coal at IH plants. The plan addressed different time periods to take advantage of the economics.

For the next five years, the IH fuel policy will be to conserve and not convert. Conservation should be stressed because it can provide the same margin of safety as coal conversions of existing plants while giving superior return on investments.

During this time period, in spite of the rapid escalations in real fuel cost, conversion to coal based on economics cannot be justified. This is best shown by analyzing the cost and savings associated with making a conversion at the Ft. Wayne Plant, IH's largest consumer of oil and gas in boilers. Using a discounted cash flow analysis, it shows that a conversion at Ft. Wayne in the early 1980's would take almost 20 years to break even.

The economics of coal conversion is further dramatized by showing the effects of converting all the IH facilities to coal on the projected domestic 1990 fuel bill. If IH converted its eight largest gas/oil powerhouses to coal, the fuel mix within the domestic operations would change so that coal would supply about 40% of its total energy needs.

The projected 1990 corporate cost of fuels (1980 dollars) with that mix would be lowered only to $12.15/MM BTU or $194,000,000, a reduction of $30,000,000 (1980 dollars). The investment, however, to achieve this savings would be almost $140,000,000, yielding only a $215,000 annual return for every $1,000,000 invested.

For the short term, the IH strategy will be a concerted conservation effort. Corporate Facility Energy Management has projected that $16,000,000 investment in equipment modifications and energy management systems will achieve a 18% reduction in energy consumption. By 1985, this would approximate an annual cost savings of $17,500,000, yielding $1,100,000 annual return for every $1,000,000 invested.

Although the economics of coal conversion are poor, coal provides a security of supply. Past 1985, the IH strategy will be to take advantage of this availability in a more economical manner. This can be done by increasing our utilization of steam. Currently, coal conversions of our powerhouses are uneconomical because of the load factors (the ratio of the average hourly load to our installed hourly capacity). Our load factor is approximately 22% because steam is used mainly in winter months to heat our facilities. Very little steam is used for process applications.

The long term goal for both our large existing and new plants is to increase the load factor on our boilers to 60%. Load factors can be improved by using steam to heat ovens and/or spray booths, generate electricity and/or mechanical power, make chilled water and provide air conditioning. At that level, the economics of coal begin to look more favorable. More important, it is at these high levels that coal is really providing the desired protection from supply cutoff. Unless steam is used in processes, coal conversions do not protect our productive capacity. For instance, plants using coal, like Indianapolis, were affected by the gas curtailments during the winter of 1977. Indianapolis' coal base did nothing to reduce dependence on gas for processes.

The economic effect of increasing the steam load factor on a typical facility has been analyzed. Looking to the future, most IH facilities will be 1,000,000 square feet and will require about 100,000 pph in installed steam capacity to meet peak load heating requirements. In the analysis, a new coal-fired installation was compared to the costs for a new gas/oil-fired installation. It assumed that the installation will take place in 1990 and that there will be no real rate of fuel cost increase after 1990. The analysis showed that if the load factor could be increased to 60%, the years to break even improved from 15 to 5 years.

APPLICATION AND SUMMARY

As described, the long-term strategy for IH is to increase steam utilization to justify economically the use of coal in large facilities. Small plants, however, will rely on conventional and synthetic fuels. Coal cannot be justified at the small plant because of high capital investment requirements for coal handling equipment and emission control devices.

The implementation of the strategy at the specific plant locations has been done. It is intended to outline a means by which a specific plant can reduce, economically, the risk of production curtailments because of energy cutoffs. The goal for the 1990's is to use steam as our primary energy source for both process and heating applications and generate that steam with coal.

To achieve that goal, the IH strategy will be phased as follows:

- Short-Term (1980-1985)

  The IH fuel policy for the short-term should be to conserve and not convert. Conservation should be stressed because it can provide the same margin of safety as coal conversions while giving superior return on investments.

- Mid-Term (1986-1990)

  The mid-term strategy at IH facilities is a function of size and boiler load factor. Facilities less than 1,000,000 square feet should concentrate on conservation and develop fuel standby capabilities such as propane and fuel oil storage since, again, coal conversion at those plants is uneconomical.

  Facilities above 1,000,000 square feet should start to modify processes and heating and ventilation equipment to use steam. Such modifications will allow taking advantage of the next period's economics. For instance, operations should consider providing the ability to use steam in the proposed paint system and any modifications of existing paint systems. The ability to utilize steam as part of future and planned HVAC systems should be planned.

- Long-Term (1990+)

  The long-term strategy is to use coal in facilities above 1,000,000 square feet to generate steam that will cover process, heating and ventilation energy requirements. This expanded use of steam is critical to the overall economics of coal conversions. It can mean the difference of a 15 year payback with current steam practices to one of 5 years with expanded usage.

In facilities below 1,000,000 square feet, direct coal use will remain impractical. Synthetic fuels, however, which are coal-based liquids and gasses, and other unconventional sources of gas will be on the market and will provide the needed safety of supply.

An expanded steam based energy system in our large facilities will protect us from curtailments while providing IH with an economical method to use coal in the future.

AST ENERGY CONSERVATION STORY

Martin A. Mozzo, Jr.
Director, Energy Management
American Standard Inc.

ABSTRACT

American-Standard is a major worldwide manufacturer in four diversified areas: Transportation Products, Building Products, Security and Graphic Products, and Construction and Mining Equipment. Energy supply and costs play a dominant role in our manufacturing process. In 1975, the Corporate Energy Conservation Program was begun in the U.S. only. This was expanded in 1979 to include all American-Standard facilities worldwide in recognition of the fact that energy is a global problem and the corporation is a global manufacturer.

The initial energy problem was one of supply. Curtailments of certain forms of energy, especially natural gas, interrupted production. The Energy Conservation Program initially was aimed at reducing consumption, thereby meeting production requirements with allocated supplies. Today, energy supplies to industry for the most part are adequate, and the energy problem has swung to one of high prices. Our Energy Conservation Program is now geared towards reducing energy consumption and thus energy costs to meet this new problem.

The Energy Conservation Program reviews facilities, equipment, processes, and products to reduce energy consumption and costs. Many major energy projects have been accomplished worldwide since 1975. 1980 results, the best year yet, showed a savings of over $5½ million (annual) from actions taken in the year, about 8 percent of the almost $70 million spent in 1980 on energy.

INTRODUCTION

Like most industries, the energy problems of the 1970's had significant impacts on our manufacturing operations worldwide. We have had supply interruptions which have curtailed production or at least reduced it. The natural gas shortage of 1977 in the U.S. caused several of our Eastern and Midwestern plants to shut down. One Brazilian facility was subject to a reduced fuel oil allocation beginning in 1979 which necessitated new equipment purchases converting from oil to electricity. The rapid rise in energy prices has also affected operating income as now energy has become a significantly higher portion of cost-of-goods sold. Government involvement in rate-structuring, such as incremental pricing in the U.S., has had a significant impact worldwide in our energy costs, not consumption.

In 1975, American-Standard responded to the energy problem by establishing a Corporate Energy Conservation Program. This Program is now 5 years old and has been quite successful. This paper reviews the energy conservation organization utilized, the impact of the energy problem on our operations, the program undertaken to reduce energy consumption, and the results achieved.

THE COMPANY

American-Standard is a major worldwide manufacturer in four diversified product areas, each area represented by a global business group. The Building Products Group manufactures plumbing fixtures and fittings, steel doors, pre-fabricated fireplaces, and conference room folding doors. The Security and Graphic Products Business Group manufactures safes, vault doors, safety deposit boxes, and prints bank checks, business forms, and commerical printing publications. Our Transportation and Industrial Products Group manufactures automotive parts such as truck braking systems, railroad braking systems, signalling and switching systems for railroads and mass transit systems, and a line of heat transfer and fluid power components. The fourth business group, Construction and Mining, makes large off-the-road trucks used for coal and mineral mining operations. All business groups have facilities located worldwide. The importance of energy supplies and prices vary within the four groups; however, all groups and facilities are expected to be full participants in the Energy Conservation Program.

ORGANIZATION

In 1975, a Corporate Energy Conservation Program was started in the U.S. only. Initially energy supply was the key problem for American-Standard. A number of U.S. plants were experiencing natural gas curtailments, and several plants on occasion were shut down. Every U.S. group was asked to participate and to develop specific energy cost reduction projects at plant level. In 1979, recognizing that American-Standard was a global corporation, and that the energy problem was a global one, the Energy Conservation Program was expanded to cover all American-Standard groups and facilities.

Figure 1 shows the organization structure used by American-Standard to carry out energy conservation goals. At all plants worldwide, a plant energy coordinator has been appointed. His responsibilities for his facility are: (1) develop and enforce an Energy Awareness Program, (2) prepare and monitor energy budgets, consumption, and costs, (3) plan energy conservation projects and insure they are budgeted, and (4) implement approved capital energy projects.

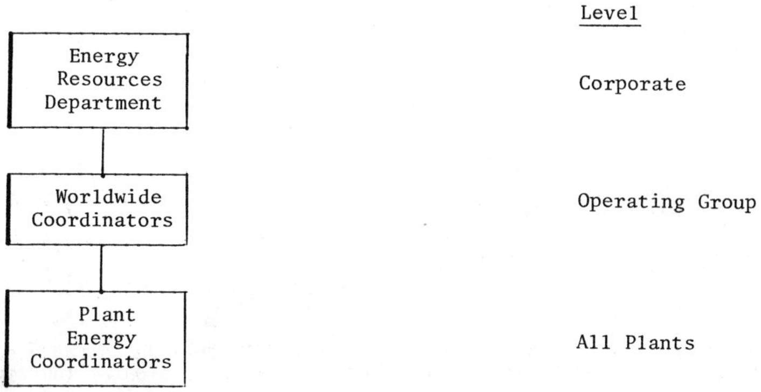

Figure 1. Energy Organization, American Standard Inc.

The plant energy coordinator has a functional reporting responsibility to an energy coordinator at the operating group, a suborganization to the global business group. We currently have 19 operating group energy coordinators located worldwide. The operating group energy coordinator's responsibilities include: (1) monitor plant energy programs to insure that energy conservation is practiced, (2) insure compliance with corporate energy policies and guidelines, (3) coordinate all plant efforts and provide focal point for exchange of ideas within the group, and (4) review Operating and Strategy Plans to insure inclusion of major energy projects which require budgeting of capital funds.

At the corporate level we have an Energy Resources Department to provide policy, guidance and engineering to establish and accomplish energy programs. The Energy Conservation Program has two individuals in the Department, a Director, Energy Management and an Energy Engineer. Responsibilities for these individuals include: (1) reviewing status of Energy Conservation Programs at operating group and plant levels, (2) collecting and analyzing energy consumption and cost data, (3) forecasting future energy costs, (4) review impacts of regulations, rate structures, and so forth, and plan corporate strategies accordingly, (5) review Operating and Strategy Plans for inclusion of sufficient funds to accomplish major energy programs, (6) provide in-house engineering consultant capabilities, and (7) review energy impacts of planned corporate moves, such as new facilities or major equipment.

THE PROBLEM

As stated previously, American-Standard formally started an Energy Conservation Program in 1975, and expanded it worldwide in 1979. The initial energy problem was one of inadequate supplies to run our facilities without disruption. Today, energy supplies appear to be adequate, and the shift of the energy problem has been one from inadequate supply to one of high prices adversely affecting manufacturing costs.

Table 1

Energy Costs and Conservation Savings ($000's)

Year	Global Energy Costs	Annual Savings	Percentage Savings
1976	37,982	2,651	7.0%
1977	45,501	3,156	6.9
1978	48,028	3,266	6.8
1979	54,471	4,109	7.5
1980	67,065	5,709	8.5
1981	82,300	6,842	8.3
1982	104,333	8,034	7.7
1983	119,569	8,968	7.5
1984	131,940	9,500	7.2
1985	145,150	10,160	7.0

Our Energy Conservation Program is now 5 years old. Table 1 shows the global energy costs and annual savings for the corporation during the period of 1976-1985. The annual savings shown is for new energy conservation projects completed during the year, and it does not reflect any carry-over from year to year. 1980, the first full year of our global operation, has been the best year to date with an annual savings of $5.7 million from new projects. As indicated in Table 1, our plan is to continue the Energy Conservation Program, with the goal of achieving a minimum of 7 percent per year conservation savings from new projects accomplished each year.

Our energy costs have almost doubled since 1976, and will more than double again in 1985. We expect a relatively constant consumption of natural gas, electricity and fuel oils in developing these costs, with conservation efforts almost offset by increased sales, productivity, EPA, and OSHA requirements. Thus the "problem" that we now face in our Energy Conservation Program is primarily one of price. In forecasting our energy costs, we have assumed that global crude oil prices will rise to $60/barrel by 1985, and that natural gas will be priced the same as oil on a BTU basis by 1984. Some experts forecast an even higher price for oil, a statement which if true, will further accentuate our energy problem and further necessitate the Energy Conservation Program.

THE PROGRAM

In order to meet the annual savings goals set out in Table 1, all elements of organization in American-Standard's Energy Conservation Program must constantly develop plans to become more energy efficient. The Energy Conservation Program reviews facilities, equipment, processes, and products to reduce energy consumption and costs. Table 2 lists some of the considerations all operating groups evaluate under each of these four headings.

During the past five years, several capital energy projects have been accomplished to reduce energy costs and consumption. Some of these projects are:

Facilities

- Replace single-pane windows with double-pane or other type insulating panels
- Added insulation to walls and roofs of existing structures
- Build new, energy efficient buildings, replace older existing facilities

Equipment

- Replace old boilers with new, higher efficient units
- Rebuild enameling furnaces with ceramic insulation
- Install recuperators on heat treat furnaces
- Install recuperator preheater on foundry cupola
- Install waste heat recovery systems on sanitary ware kilns
- Install better control equipment and burners on boilers, furnaces, and kilns
- Replace existing energy-inefficient kilns with modern, direct-firing kilns

Processes

- "No-Pickle" enamel steel tub processing
- Single coat enamel steel sanitary ware
- Low mass kiln car for sanitary ware

Products

- Machined product versus foundry casted product

THE RESULTS

The objective of our Energy Conservation Program is to insure adequate energy supplies to our facilities while reducing energy consumption and costs. The goal is to maintain at least 7 percent annual savings per year from new projects. We have achieved through the first five years at least 6½ percent annual savings per year.

One very desirable result from the Program is the cumulative effect of annual energy savings. Most energy conservation projects continue to save energy year af-

Table 2

Programs

Facilities

- Design new facilities for low energy use
- Update existing facilities during renovations to improve energy efficiency
- Consider energy costs in the site selection process
- Design efficient distribution systems of all utilities
- Use new efficient lights and control intensity
- Install meters to regulate energy usage
- Provide adequate storage facilities for primary and alternate fuel supplies
- Conduct regular audits to review energy management procedures

Equipment

- Specify energy efficient equipment in new CEARs
- Consider replacement of old, less efficient equipment
- Develop preventative maintenance programs for existing equipment
- Increase efficiency of existing equipment
- Install controls on equipment to insure energy efficient usage
- Provide dual-fuel equipment where economically justified
- Install heat recovery equipment on present equipment
- Investigate new equipment technologies

Processes

- Utilize good process flow layouts to minimize required areas
- Maximize equipment utilization rates
- Reduce BTUs required to produce chinaware
- Use lower temperature finishes
- Change pickling specifications
- Investigate new processing technologies
- Reduce rework and scrap

Product

- Design for simplified manufacturing
- Change material specifications to increase energy efficiency
- Reduce weight and mass

AST Energy Costs
With and Without Conservation

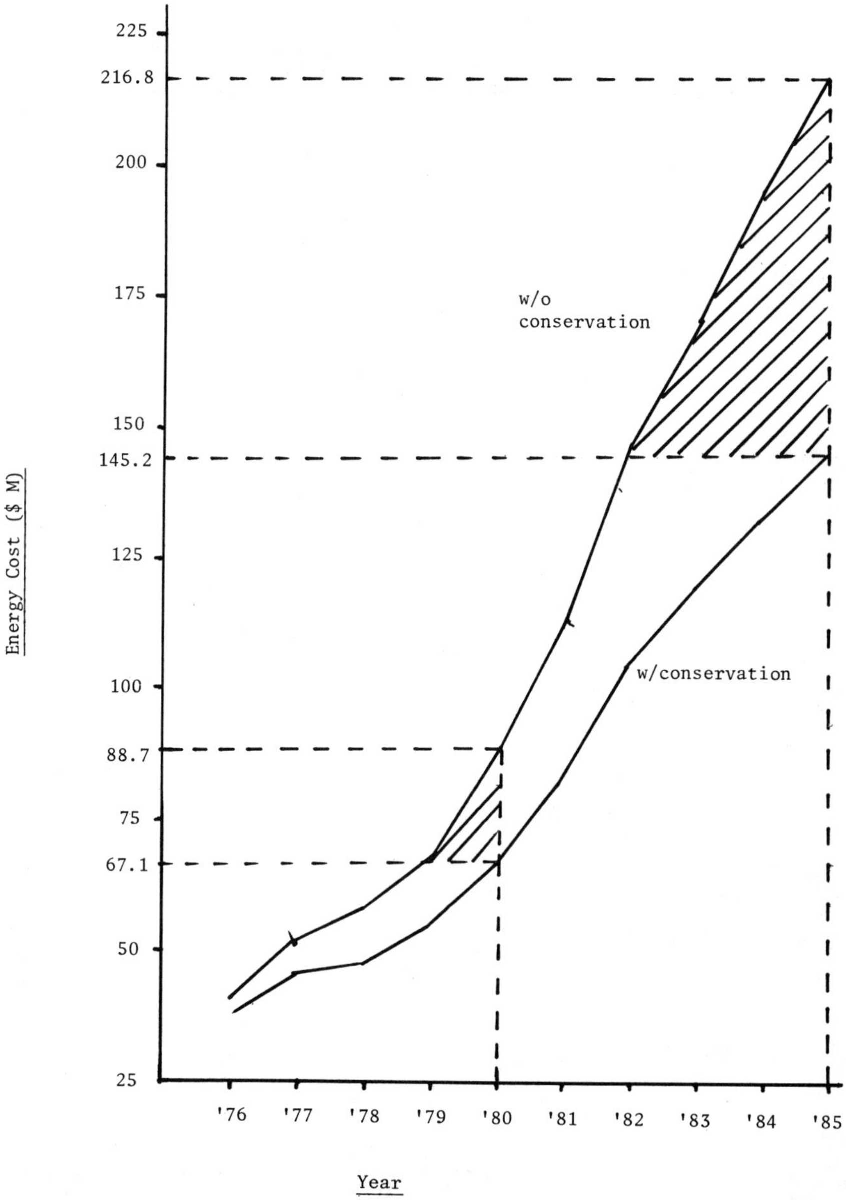

Figure 2

ter year. Without savings from the Energy Conservation Program during the past 5 years, the 1980 global energy costs would have been $89 million as compared to $67 million actual (see Figure 2). Forecasted costs for 1985 is $145 million with conservation versus $216 million if no Conservation Program is in place, a difference of $71 million.

CONCLUSION

American-Standard has had a very active Energy Conservation Program for the past 5 years, and it has plans to continue the Program in the future. The Program involves all operating groups worldwide and is primarily geared to making facilities, equipment, processes, and products more energy efficient. The continuing reduction in energy costs and consumption that we have achieved in the past 5 years is the prime reason for management acceptance and support of the strategy reflected for the next five years of the American-Standard Energy Conservation Program.

THE VALUE OF AN ENERGY PROGRAM

CARL E. SALAS, P.E.
VICE PRESIDENT
O'BRIEN & ASSOCIATES, INC.
335 SOUTH 11TH STREET
SAN JOSE, CALIFORNIA 95112

Plant engineers, consulting engineers, government officials and building owners all seem to have developed an independent definition of the term Energy Audit. In this era, we can all take security in the fuzzy definition that has evolved. It must be noted however, that results of an Energy Program in terms of actual dollars savings are as significant as the savings resulting from a highly capitalized energy improvement program.

It now becomes important that a generic definition comes into focus so that 1) energy savings in both Btu's and dollars are maximized for all facilities, 2) responsible personnel can develop the "tools" necessary to monitor those savings and 3) responsible personnel can take credit for these savings.

THE CONTRAST

It is rare for me as a consultant to walk into a hospital or industrial facility, and talk to a facility engineer who doesn't say "We've already implemented all energy-related fixes." Further investigation generally reveals that "all" means that the plant has purchased an economizer for their boiler or a chilled water temperature controller, or worst of all...an energy monitoring and control system (ECMS). Rarely have I found the real ground work has been laid so that there is historical justification for these projects and so that the real savings from these projects can be monitored.

I've had similar experiences with facility managers who have had an Energy Audit developed at their facility. "We've already developed an audit, look at the size of this report!". Sometimes, it's as if size or capital investment automatically means that energy utilization has been optimized at the facility. As pointed out in reference 1, there are many types of energy audits, all of which are useful. However, it is only the Energy Audit that actually accounts for long-term facility energy savings that can be called an effective energy program.

DEFINITIONS

Energy Audit

No matter what you've been told, forget it. The most practical definition of an Energy Audit is...

> "An energy audit is only a DEVELOPING UNDERSTANDING of the specific energy using patterns of a given facility."

Energy Program

An Energy Program naturally follows the definition of an energy audit if the audit definition is broken up into 9 areas, whereby questions can be asked and knowledge can be developed.

Each area of understanding allows us to better control the situation yet that same area opens up new questions leading to new projects and new potential for saving increased quantities of energy.

To reiterate, an energy audit is a <u>developing understanding</u> of the specific energy using patterns of a given facility. In that regard, the energy program is a planned and organized energy audit which ensures that energy utilization is optimized. It is comprised of 9 different areas:

*HISTORY	*IMPROVEMENTS	*BUDGETING
*INVENTORY	*MONITORING	*PERSONALIZING
*OPTIMIZATION	*PLANNING	*MANAGING

THE FIRST LEVEL, <u>History</u>

In order to eventually conquer your energy problems, the primary question to ask is: "What has this facility been doing with respect to energy use over the last 5 years?" Until this pattern is established, there is little sense in rushing to install even the most simple energy saving devies. Sure they'll save energy but in the meantime, they will skew your baseline to the point where you cannot use your history as a guideline to plan a comprehensive energy program.

Comparative Indexes of use or cost are important parameters of the facility and become important here. A discussion of these indexes and how to calculate them is included as Reference 2.

In the event that time or available records are such that a 5 year history is unavailable, it is still mandatory that the most recent yearly history be observed. Energy use, energy cost and Energy Utiliation Indexes (EUI's) must be observed on a month by month basis over the past year. Is energy usage increasing? Why?

Why is the history so important?

1) It provides a baseline from which to evaluate future energy savings

2) It allows monthly and yearly trends in energy use to be evaluated (and predicted)

It should also be noted that historical energy use information is generally available through the local utility representative.

A striking example of the value of a simple energy audit (one comprised of just EUI's) may be observed from the Energy Summary Analysis of the Cupertino Union School District. By ranking the school buildings by EUI and ECUI (energy use and energy cost respectively) a natural prioritization of District energy problems becomes apparent. From this ranking, major decisions concerning facility replacement, facility improvement and facility abandonment are being made. This energy summary is also included as Appendix A.

THE SECOND LEVEL, <u>Inventory</u>

Once again even though one thinks that he knows exactly how a facility is made up, "energy understanding" follows only when an inventory of ALL energy using equipment is completed in an orgainized manner. Seemingly major pieces of equipment become very low energy users while others that seem unlikely turn out to be high energy users. This level marks the most tedious portion of the energy audit but, as in life, the rewards from the organized effort are significant.

The inventory need not be overly complex. It is more important to make it all inclusive than it is to make it complex. For example, even equipment nameplate information is not mandatory.

It is very important that equipment be grouped into categories such as:

* Furnaces	* Fans	* Boilers
* Presses	* Pumps	* Hot water heaters
* Elevators	* Chillers	* Lighting

Also estimates of the following information should be obtained for each category:

* KW demand/horsepower (for lighting, motors)
* Rated energy usage or production (for chillers, boilers)
* Hours of operation (is equipment on timeclock control?)

The benefits of a completed inventory along with an energy history are enormous. It makes you smart!! No outside engineer can "snow" you and no salesman can trick you when you know exactly how much and where your energy is being used.

Specific benefits of the inventory are:

1) It provides a convenient source of information when projects (energy and non-energy related) are being evaluated

2) Provides a basis for deciding which categories have the highest payback potential for energy savings

The end product of the best inventory is a "pie" chart energy summary. This chart takes each energy using category as a percentage of the entire facility energy use by multiplying kilowatt (KW) demand by total yearly operating hours. A pie chart of a "not-so-typical" campus is attached as Appendix B. This campus was thought to use a very low portion of heating energy. The equipment inventory along with a heater monitoring program proved that heating energy was the most significant energy problem at the campus. The important point here is not a pretty pie, but rather a real idea of where energy dollars are going.

THE THIRD LEVEL, Optimization

Now we must work with the existing equipment. Is a constant volume reheat system really that bad? Maybe yes and maybe no... but before replacement is called for, it is vitally important to the overall energy plan to take that system and "tweek" it so that it is working at its peak efficiency.

So called Operation and Maintenance changes (O&M's) are identified here and are of tremendous significance. Savings of 10-20% are common. This percent is scoffed at by plant at by plant engineers who don't yet understand their facility but is always realized by those who do! A good maintenance/calibration program goes hand in hand with O&M modifications. At the Golden West College, for instance, the heating and air conditioning systems were continuously fighting each other. Costs for this lack of calibration were on the order of $10,000 per month.

The savings realized from your O&M changes not only will add to your confidence but will also provide you with funds which can be used to implement larger projects.

Certainly there are thousands of written pages of information on each piece of equipment. Variable speed motors or resheaving are just a few of many the different schemes for more closely matching your equipment to your situation. You don't have to be an authority on all your equipment, rather it is your position to ask the questions. "What has been done, what is being done and what must be done for improvement? When does the equipment have to be used, and for how long? . . . What is the real load and what happens when it is shut off?" These questions need answers for every piece of energy-using equipment including the following:

* Lighting
* Heating, Ventilating, Air-conditioning
* Special Processes
  (large energy-using pieces of industrial equipment)

At this point in your energy audit, you now may have a real interest to call in a specialist who can answer your questions. The key result of this audit process is that it makes you knowledgeable enough to ask the pertinent questions of a specialist who can help you in the area of his expertise.

THE FOURTH LEVEL, Improvement

Now and only after all of the above, it is time to consider capital improvement projects. What really makes sense and what really pays back. It seems to me that as I become more familiar with the energy audit approach I must continuously go back and review the first three levels each time I consider a new project. Many different solutions appear, some of which give me an energy payback, but surprisingly enough do not fit into the overall program plan. This can occur for 2 reasons...

1) Some of the projects, although providing significant energy savings on their own, will save lesser amounts of energy when combined with other porjects in the overall program plan. If this is not kept in mind, (through comparisons of projected energy savings with previous usage as recorded in level one of the audit) projected energy savings may be greater than the energy usage.

For example chiller reset controls may not provide significant energy savings if economizer controls result in reduced chiller operation at low loads.

2) Even though payback is high, maintenance is also increased. For example, an EMCS provides ultimate control of many facilities, however, additional help is required by programmers and software experts.

There is an enormous number of energy saving devices on the market. Because energy conservation is a newly developing field, many of these products are unproven. We are still discovering local vendors whose products make good sense. We are also discovering large, "name" firms who are marketing production grade "conservation" equipment which is not particularly conducive to saving energy in many situations.

Therefore, look at all the products, schemes and equipment on the market. Call faclities that have installed them. Find out what problems have been encountered and what real savings have resulted from the installation. Then begin laying out your long range energy plan. What projects pay back the fastest? What projects look good but require further development? Most importantly, where are you going and what will be your premier project?

THE FIFTH LEVEL, Monitoring

Monitoring of the ever evolving energy history is an important part of an energy conservation program. Changes in operations of the facility may provide significant insights into energy saving patterns for future projects. At the same time, results of energy savings projects and changes to date are easily quantified and can be used to justify future projects, awareness and interest in energy. Another advantage of a consistant monitoring program is that it allows early detection of a "systems" problem. A sharp rise in energy usage in a given month may be due to control mis-calibration.

Unfortunately, it is not easy to justify a monitoring program when most maintenance departments are understaffed. Of importance here is that you, the responsible energy engineer, are aware of the benefits of the monitoring program. This is the main area where you can justify your existance more than any other position in the maintenance department. You, the energy coordinator now have a payback.

A monitoring program can be sophisticated but a sophisticated one is unnecessary until such a time as you are closely monitoring energy cost and energy use on a monthly basis. As in the energy audit itself, the process of program development is more important than having the most sophisticated, state-of-the-art monitoring program available.

THE SIXTH LEVEL, Planning

The growth and change of businesses that are on the move and continuously expanding, is the mainstay of our economy. Each new facility change must be evaluated for long term duration by the RESPONSIBLE PLANT PERSONNEL who will be the working with the equipment and paying the energy bill.

A prime example of poor energy planning in a facility expansion program is evidenced by the San Jose/Santa Clara Water Pollution Plant. With the addition of tertiary recovery, the facility energy costs jumped from 1.0 million to 2.5 million in one year as a result of a decision to go to electric blower motors with no heat recovery. A preliminary evaluation of the planned expansion, with respect to energy use, could have saved from $10,000 to $100,000 per year in energy costs.

THE SEVENTH LEVEL, Budgeting

Identifying a real location and magnitude of a budget figure for energy conservation in the master facility annual budget could be the single most important step in ensuring that long range energy utilization will be achieved at the facility. I have generally found that utility costs are usually grouped togwether somewhere in the maintenance department's operating budget and are, therefore, considered as overhead items by the facility administration.

The strategic value of an energy program is not evident to administrative level personnel because they are primarily concerned with profits. However, by breaking out a line item for energy conservation activities and placing it adjacent to energy related utilities, the upper management can begin to observe a relationship between monies invested and utility costs.

Because of the difficulty in establishing a first year figure for this budget, it is recommended that a budget estimate of 5% to 10% of the annual energy costs be listed. It is the experience of this consultant that savings of 10% are easily achieved and, as a minimum, the program pays for itself in the first year.

THE EIGHTH LEVEL, Humanizing

Possibly the most important, yet illusive, area of energy conservation potential is related to the human element. Personnel at any one facility can affect total energy consumption by 15% or more.

To actually define a methodology for capturing savings associated with this human element would be to pretend that there was one. Savings in energy cannot be calculated and cannot be controlled in this critical area. However, SAVINGS CAN VERY EASILY BE FACILITATED WITH JUST A SMALL EFFORT.

Incentive programs in each division or department have worked quite effectively. Promotion, in the form of posters and projects related to energy conservation have also been effective. Perhaps the single most effective method with the best results has been to simply report the monthly bills (gas, electric, etc) to all of the employees. Once the people using the energy develop a sense for the magnitude of the costs, they become much more responsible for using it.

The above are suggestions based on this consultant's experience. It is up to the facility energy manager to implement and experiment with these pro-humanizing programs which work best at his facility.

THE NINTH LEVEL, Managing

Managing an energy program or an energy audit is no different than managing a project or a department. Planning of your program is of utmost importance. How much energy do you want to save? How fast do you want to save it? How many dollars do you have to invest in the program? How large a utility bill is projected for the upcoming years? Once these goals are set, the management process has begun.

The other key ingredient which naturally compliments the planning pricess is PATIENCE! It must be realized by the energy manager that, in order to control the energy utilization at his facility, he must have patience. As in life, the nicest things come very slowly. I've put a graph of the energy consumption pattern of an old client of ours as Appendix A. Note that the turn around came quickly but that now, two years later, the capital has been provided so that many of the projects suggested in the original audit have begun. The program is now in effect in its entiret and rates have more than doubled since it's inception.

THE VALUE OF AN ENERGY PROGRAM

Conquering your facility energy problems is impossible. Even in situations where all energy is generated internally through the uses of waste-burning incinerators, maintenance is increased ten-fold and a continuing process of energy optimization is required. Conquering is therefore more a matter of understanding, respecting and planning for energy.

In my experience, I've seen better managed facilities who have an "energy coordinator" who just looks at his bills once a month than those who have the most sophisticated energy management and control system computer. Sure the EMCS can give you a "payback" in energy savings. The computer, however, is limited while you are limited only by your ability to understand and to become responsible for energy at your facility.

Your importance lies in the advantage of being able to provide your management with an overall picture of your facilities energy usage and providing an energy consultant with the type of backup information he needs in order to suggest practical projects to fit your real needs.

How much is it really worth? Think for a second. Do you know how much your present energy bill is. Do you know what it was last month...last year? Is 1% yearly savings on your energy bill acceptable when you can save 10%? With the audit in hand you can do it. If no other benefit is derived from this paper, I beseech you to spend one entire day evaluating your facility with respect to each of the 9 categories. After obtaining an understanding of what today's usage is, set a simple goal for next years energy use. Even 5% of a $500,000 utility bill can justify a new position, or better yet, bonuses for you and all of your staff.

5% percent savings is easy. 10-20% savings can be obtained with little capital investment. So ask yourself the question... "How much am I worth?" If you have the guts to pose that question, then you'll very rapidly find out the real value of an energy program.

BIBLIOGRAPHY

1. "Energy Audit Management",
Published by Carl E. Salas for the
1981 ENERGY AUDIT SYMPOSIUM,
February 1981, San Francisco, CA

2. "The Value of an Energy Audit",
Published by Carl E. Salas for the
1980 AIPE WESTERN PLANT ENGINEERING CONFERENCE,
July 1980, Anaheim, CA

APPENDIX A

Energy Utilization Ranking of a School District

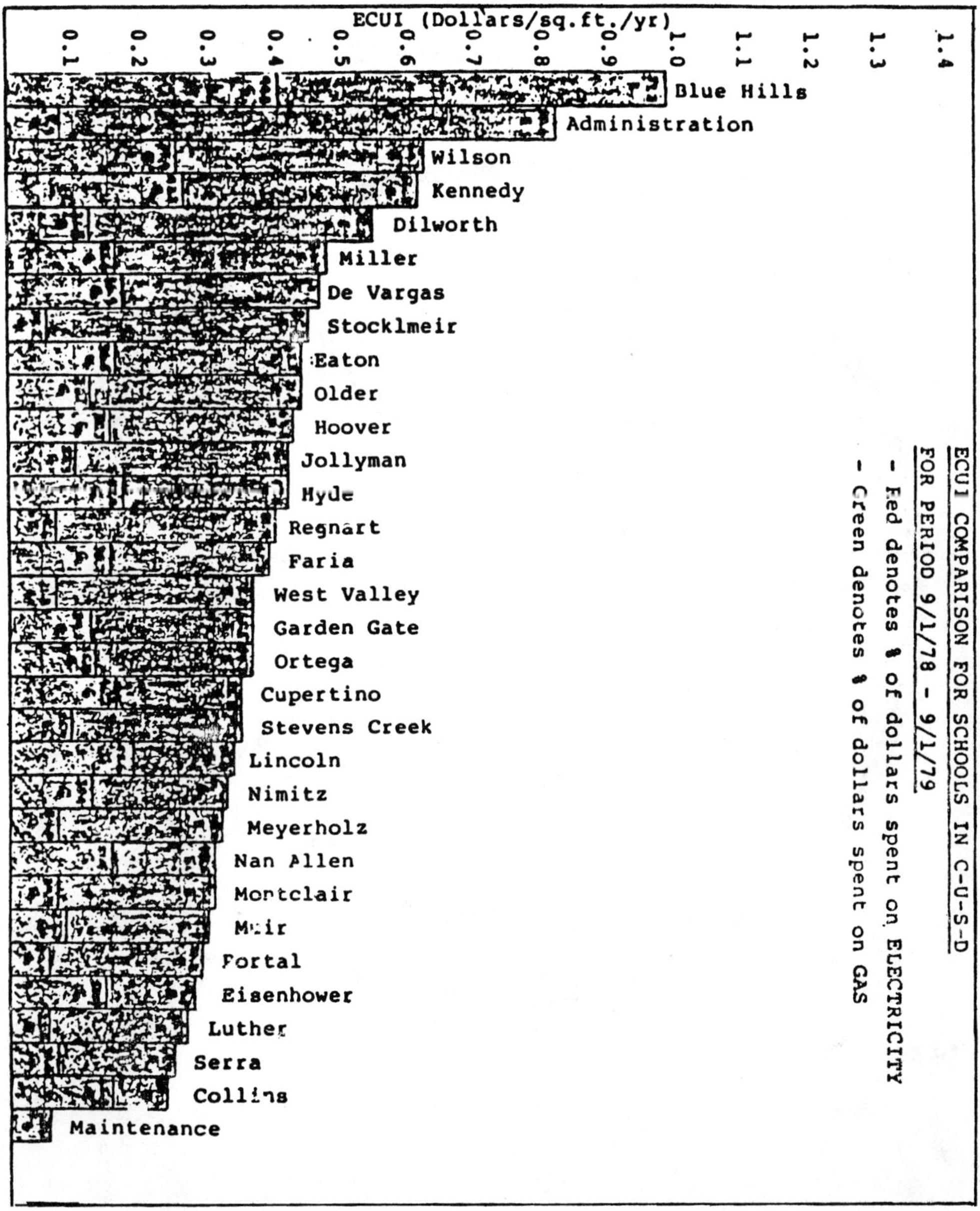

APPENDIX B

Standard Facility Energy Use Profile

GOLDEN WEST COMMUNITY COLLEGE

ENERGY USE PROFILE

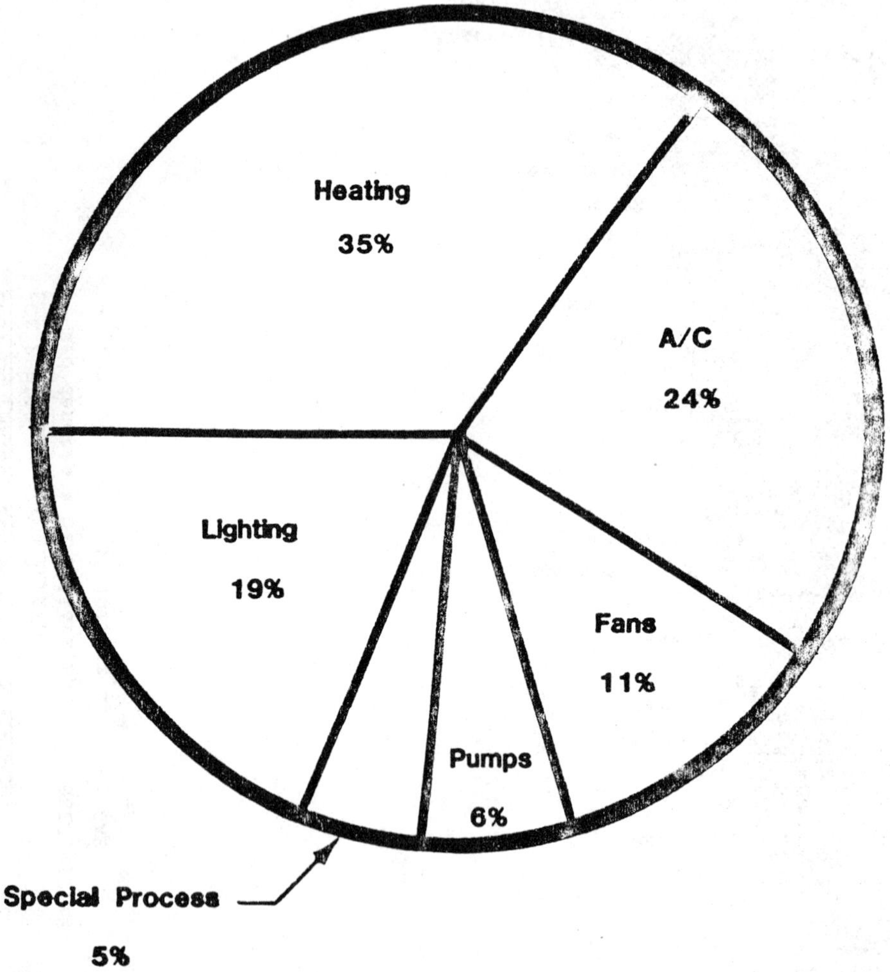

APPENDIX C

Standard Historical and Present Energy Use Comparison

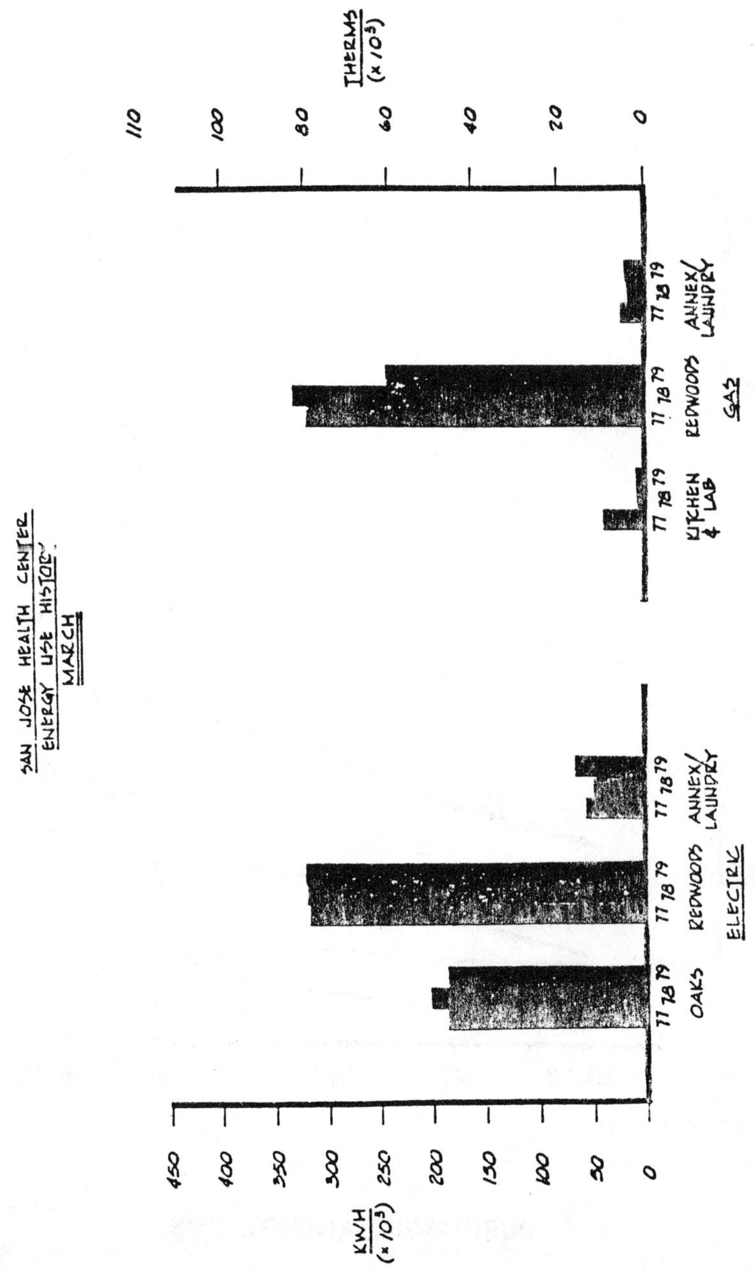

APPENDIX D

Energy Cost History at the

San Jose/Santa Clara Water Pollution Control Plant

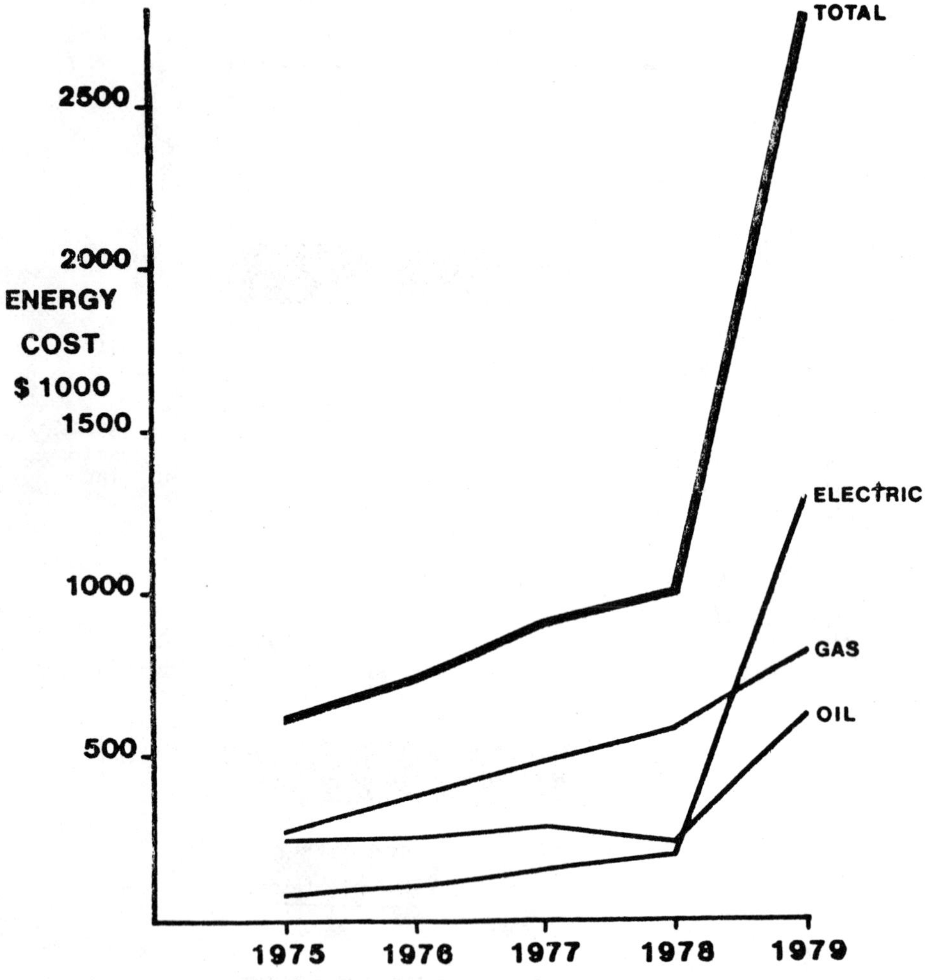

PURCHASED ENERGY COST

ENERGY-USE DOCUMENTATION

Robert E. Fite, P.E.
North Carolina State University at Raleigh

Prior to 1973, the term "Energy Conservation" was only a background expression for most Americans, however, the energy crunch of that year placed these words in the day-to-day vocabulary of essentially everyone.

The conservation efforts that emerged created a new need -- that of documenting the energy savings realized. This discussion will address a method of documenting energy savings, taking into consideration major factors affecting energy use and also utilizing energy-use profiles derived from historical data as measured at the output of the energy source, whether it is a boiler or an electrical substation.

North Carolina State University's steam energy use documentation will be described. North Carolina State University has an enrollment in excess of 21,000 students -- 5,600 in residence -- and a physical facility totaling roughly 5 million sq. ft. of floor space and approximately 100 buildings. (Floor area has not increased appreciably since 1972). Most buildings receive heat from a centrally located steam generating plant on campus and electricity from a centrally located sub-station and accompanying cable distribution network. Last year (1979-80) more than 350 million lbs. of steam and approximately 75 million kilowatt-hours of electricity were used at a total cost of about $3.5 million.

The simplest and perhaps the most widely accepted and used method of documentation is the degree day method. The energy used during a period of time is divided by the number of degree days in that period. Degree Days represent the average outdoor temperature in any one day as compared to 65°.

Following is a tabulation (Fig. 1) showing the comparison of steam energy use per degree day at North Carolina State University for the fiscal years 1971-72 through 1979-80.

ANNUAL STEAM LOAD

	Steam Production Pounds	Heating Degree Days	Pounds Per Degree Day
1971-72	446,894,688	3,112	143,604
1972-73	468,361,340	3,450	135,757
1973-74	377,331,040	2,800	134,761
1974-75	390,627,100	3,597	108,598
1975-76	393,413,182	3,005	130,920
1976-77	399,460,150	4,266	93,638
1977-78	403,096,560	4,052	99,481
1978-79	360,733,787	3,327	108,426
1979-80	352,214,291	3,564	98,826

Fig. 1

There appears to be an overall gradual decrease in steam use, but at the same time there are sharp differences from one year to another. These inconsistencies prompted a closer look at documenting steam use, taking into consideration other factors affecting steam use.

Factors affecting steam energy use on the NCSU campus include:

1. Weather Conditions - Certainly the outdoor dry-bulb temperature is the most influential of the weather factors; however, wind and solar conditions play an important role, especially if there are high wind conditions during cold weather.

2. Heating Comfort Level - Lowering the thermostat settings conserves energy, and this is one of the main targets for documentation.

3. Internal Heat Contribution - Electricity and people are suppliers of heat energy in buildings and should be considered in documenting total energy use.

4. Work Schedules - Changes in work schedules must be taken into account when comparing energy use in like periods. The interaction of work schedules with weather conditions becomes most significant when comparing one month's energy use with the same month in another year. Periods of similar weather conditions can produce completely different energy use patterns. Mild weather during weekends and holidays and cold waves during high energy use periods produce a completely different energy-use picture from that produced when conditions are reversed. Yet, both conditions may produce the same number of degree days.

5. Calendar Days - Generally, similar months in consecutive years have different combinations of weekdays, Saturdays, and Sundays. Weekdays can vary as much as 10% in any month from one year to another.

6. Distribution System Capabilities - Distribution system capabilities, when inadequate to maintain desired comfort levels during extremely cold weather, act as a restriction to steam flow; therefore, steam production is lower and out of proportion with the number of degree days.

7. Base Load - This is defined as that energy which is not weather affected and is the average minimum energy load. Included are such things as energy for heating domestic hot water, line losses, process energy, and the like.

8. Energy Conservation Efforts - This is the bottom line item that needs to be documented and includes all those actions taken to reduce energy used in the base load, the comfort level maintenance, and the reduction of waste in any segment of the energy system.

Development of energy use profiles was selected as the first step in attacking the problem.

The next chart (Fig. 2) shows instantaneous boiler output plotted against current outdoor temperatures at 10 a.m., 1 p.m., and 4 p.m. on workdays, January through May, 1977. No outstanding patterns are evident. Knowing that there are other factors involved, points on this chart were moved horizontally and vertically by trial and error to produce the best fit of a reasonable pattern.

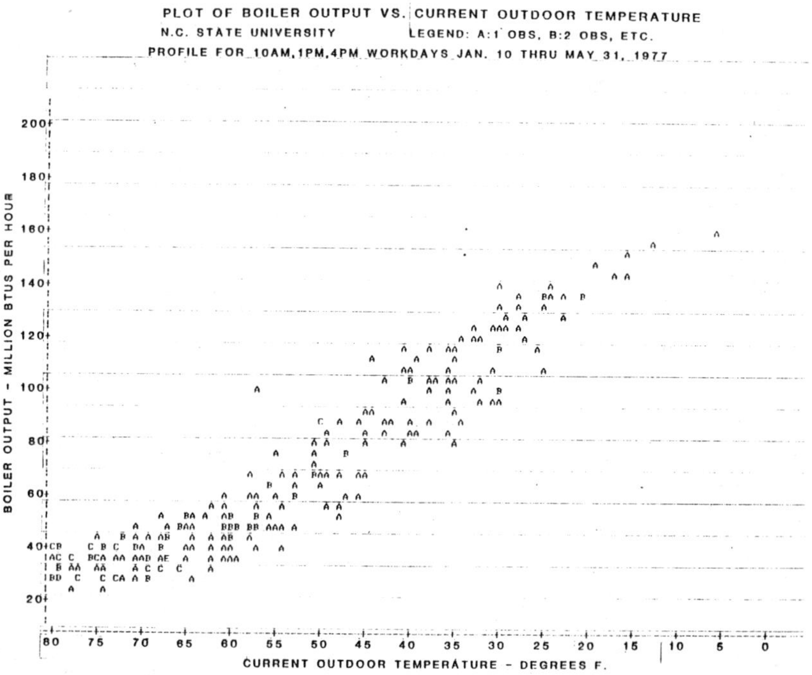

Fig. 2

Outdoor temperatures have been converted to a composite temperature, using on-the-hour-temperatures for eighteen previous hours. This was done to compensate for the heat storage qualities of the building walls, etc. This moves the points in Figure 2 horizontally toward the colder temperatures on a rising thermometer and toward the warmer temperatures on a falling thermometer.

Compensating for the wind moves the points down in a vertical direction.

Compensating for the solar contribution moves the points up in a vertical direction.

Compensating for the effects humidity has on the steam load is very slight and is in a downward direction.

Compensating for the internal load contribution to maintaining the minimum comfort level moves the points upward.

So compensating for these five factors -- composite outdoor temperature, wind, solar, humidity, and internal load -- produces a heating requirement profile at no wind, no sun, no humidity, and no internal load conditions and is based on outdoor composite temperature only.

The next chart (Fig. 3) shows these same points after compensations are made. Now heating requirement is plotted against composite temperature, and certain patterns become legible. To illustrate the identity of the patterns better, the higher end of the temperature range is enlarged.

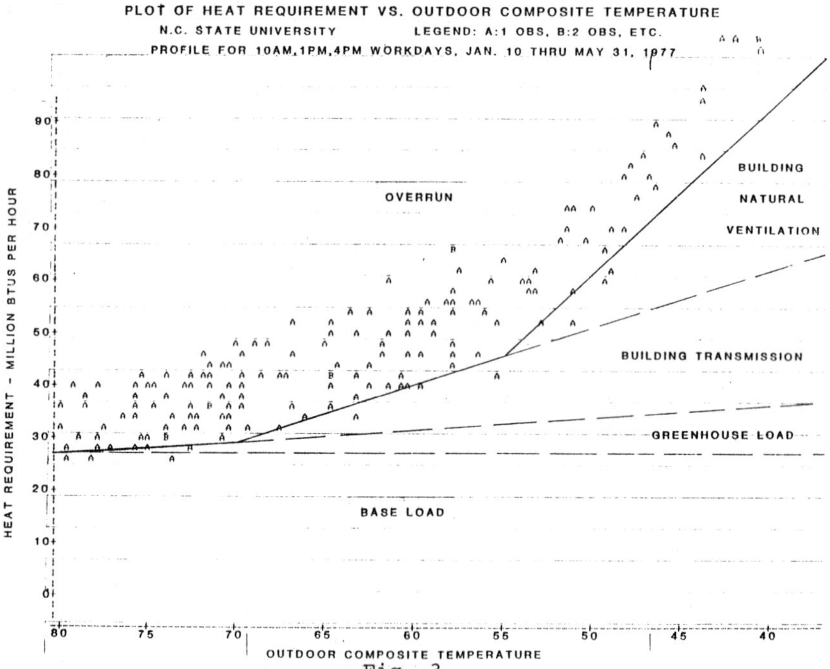

Fig. 3

First of all, the base load is established at about 27 million BTUS/HR.

Secondly, there appears to be a weather-affected heat requirement between 80 composite degrees and 70 composite degrees. This is the campus greenhouse load where temperatures are maintained as high as 85° F. (The Greenhouse load is about 8% of the total campus steam load).

Thirdly, at about 70° F. and below, one would expect buildings to require heat, and a straight line appears between 70 composite degrees and about 55 composite degrees. This is designated as the building transmission load line.

The last pattern line begins at about 55 composite degrees and extends into the colder temperatures and is designated as the natural ventilation load line.

The profile lines are drawn at the bottom of the comfort range. Points above the line signify overrun, much of which is unneeded energy use and could be conserved. Fluctuations in the base load also appear in the overrun. The profile line, or "minimum comfort line" is so designated because of the absence of points falling below this profile line or out of the comfort range. (People will not allow themselves to become uncomfortably cold for extended periods of time).

Figure 4 shows the profile lines at the expanded scale of composite temperatures, 80° to 0°.

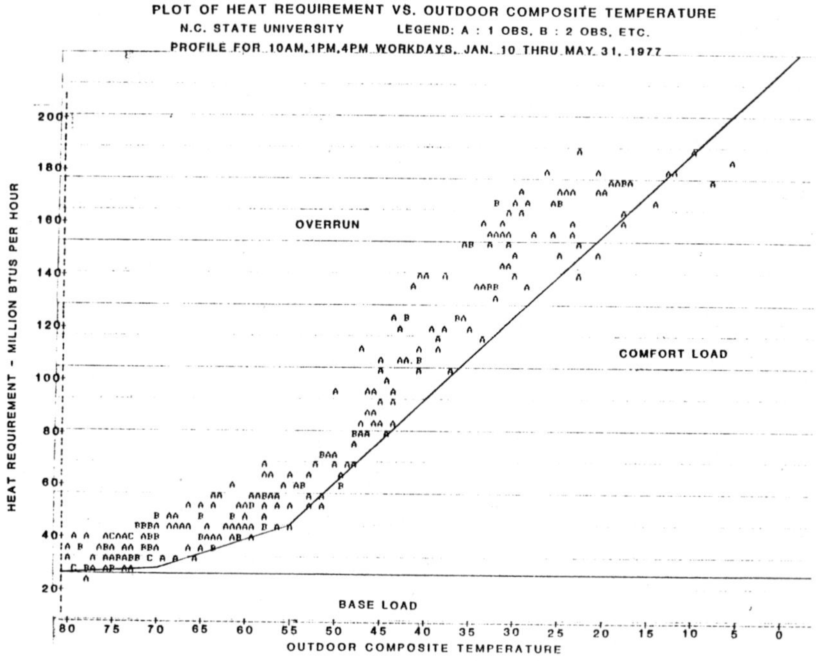

Fig. 4

The three main components of energy use from this chart become:

1. Base load
2. Comfort Load
3. Overrun

To illustrate the accuracy of the profiles, Fig. 5 shows the same plot of boiler output vs. current outdoor temperature as in Fig. 2 except all days occurring before February 2, 1977, are shown by circles. There is a significance to the February 2 date. During January, severe weather (for North Carolina) occurred which aggravated an oil shortage in our State, and an emergency condition was declared by the Governor on February 1, 1977. On that date he issued a general order to lower thermostats to 62° F. during daytime and 55° F. at night. Figure 5 illustrates the difficulty in distinguishing between the energy use levels before and after February 1 (the points are mixed).

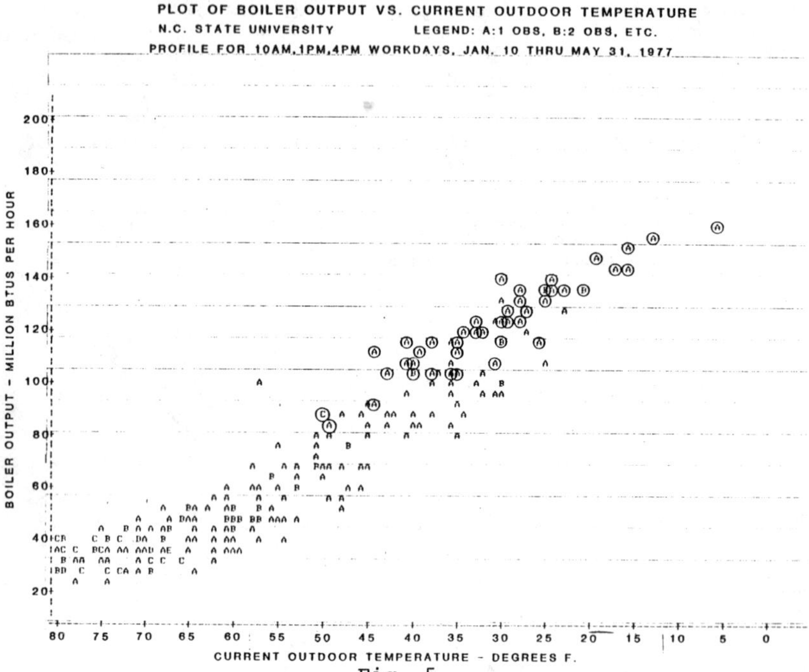

<u>Fig. 5</u>

However, in Figure 6, the compensated values are shown with the same workdays prior to February 2 circled. A clear division between the two energy-use levels appears and the University can factually state that energy use was lowered by approximately 15%.

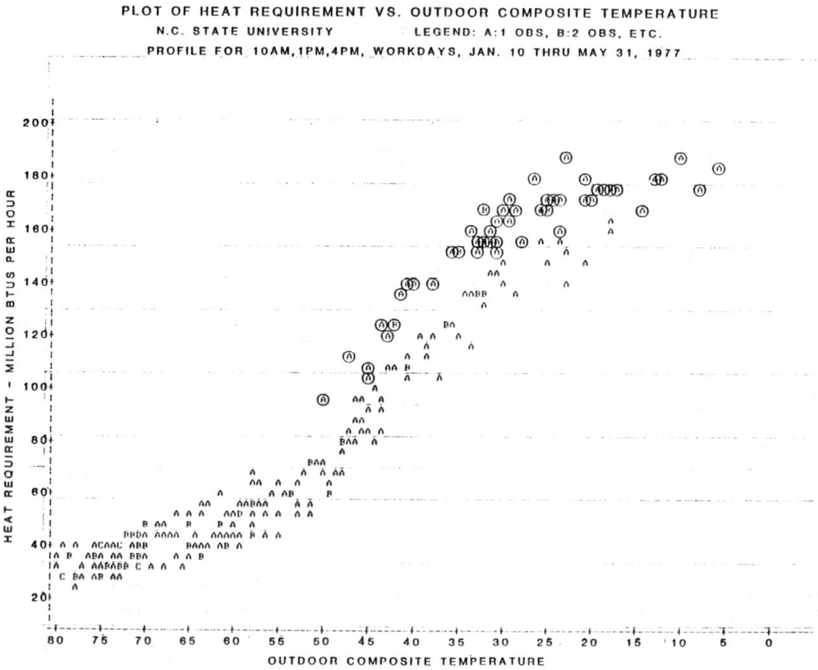

Fig. 6

So far, compensations have been made for weather conditions and for internal load conditions, and the use profiles describe historical energy use at no sun, no wind, and at no internal load conditions. To compensate for heat cutback within a 24-hour period, profiles are constructed for each hour.

To compensate for the changes in energy use between different types of days (work schedule and calendar days), sets of 24 hourly profiles are developed for each type of day. For the North Carolina State University operation, seven different types of days have been selected as shown in Figure 7. Also shown are the number of the various type days occurring in a typical 12-month period.

ACADEMIC CALENDAR DAY CATEGORIES

1. Workdays	- Students on Campus -	217
2. Saturdays	- Students on Campus -	42
3. Sundays	- Students on Campus -	41
4. Workdays	- Students Away -	33
5. Holiday	- Between beginning and ending of student holidays -	21
6. Holiday	- Day before Students resume classes -	5
7. Christmas Holidays -		6
		365

Fig. 7

A total of 24 profiles for seven different kinds of days or 168 profiles are used at North Carolina State. For each hourly profile the base load is determined. Compensations are thereby made for the base load.

Reviewing then, compensations for weather conditions, internal load contribution, work schedule, calendar days, and base load variations, have been applied to historical energy use data. Energy-use profiles result for each hour of each type of day in the academic year.

USE OF PROFILES

The first step in documenting energy use has been completed; that is, profiles have been developed for the base year to be used (in this case, the fiscal year 1971-72). The next step is to develop an energy summary for the base year. This is accomplished by again adjusting the boiler output for weather and internal load conditions and by comparing these quantities to the appropriate base year profile as measured at the composite dry-bulb temperature. The base load, comfort load, and overrun are documented for each hour and summarized daily, monthly, and for the total year (Fig. 8). This is called a "summary" and is described as comparing a set of weather and energy-flow data with a set of energy-use profiles.

As the current year begins and progresses, a similar process using current year's data and the base year profiles yields hourly, daily, monthly, and finally the total current year summaries. As each hour is summarized, the overrun of the current year is subtracted from the overrun of that hour in the base year, giving the overall energy savings, as shown in Figure 8. As a new month begins and progresses, hourly summary amount comparisons are available at the end of each day, showing the differences in the overrun amounts (savings) compared to the same month in the base year. At the end of the month, summaries are terminated and the process begins again for the new month.

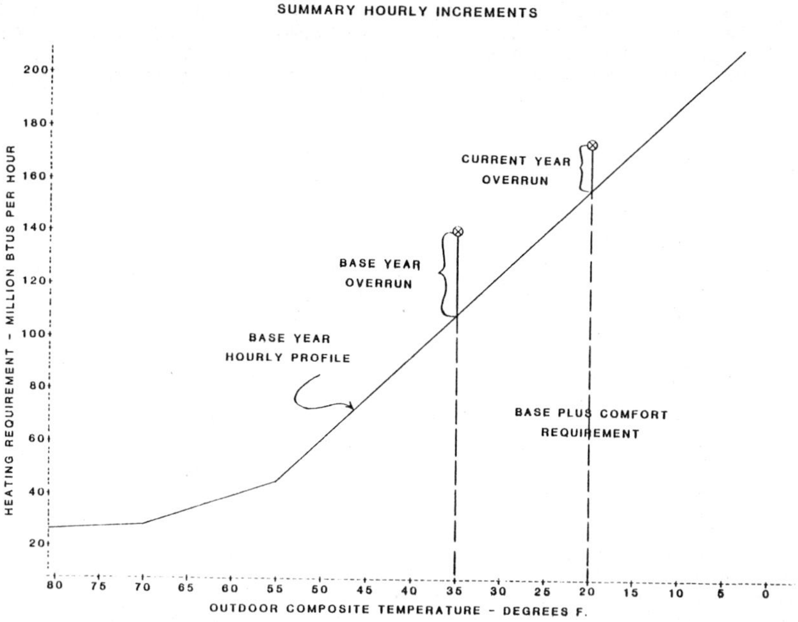

Fig. 8

Monthly and annual overall savings summaries are shown in Figure 9. Cost avoidance is based on prevailing fuel costs. "Percent energy savings" is the savings divided by the sum of the steam produced and the savings. Savings in the current year 1980-81 thru January are running about 4% behind last year.

ENERGY SAVINGS COMPARED TO 1971-72

	1978-79			1979-80			1980-81		
	Energy Savings MBTUS	% Engy Svgs	Dollar Cost Avoidnc	Energy Savings MBTUS	% Engy Svgs	Dollar Cost Avoidnc	Energy Svgs MBTUS	% Engy Svgs	Dollar Cost Avdnc
July	9,974	41.0	27,528	5,829	21.7	18,769	6,989	27.3	30,262
August	7,685	30.0	20,135	8,066	28.5	22,827	8,893	33.8	35,216
Sept.	6,374	24.4	18,867	8,538	29.0	32,188	8,019	30.4	32,477
Oct.	7,363	19.4	22,384	13,349	32.6	47,522	11,241	28.7	58,903
Nov.	8,538	20.2	21,857	14,989	31.1	54,110	8,863	17.8	49,456
Dec.	13,453	22.6	34,843	20,460	31.7	68,950	18,471	27.8	69,820
Jan.	6,652	10.1	16,630	14,294	21.5	54,031	11,789	15.9	77,925
Feb.	10,047	15.0	29,237	15,924	23.6	44,428			
March	12,041	23.5	39,374	15,784	26.7	64,714			
April	8,363	23.7	28,351	8,257	24.2	32,202			
May	6,671	23.4	21,347	8,326	31.0	34,386			
June	4,647	18.3	14,685	6,517	28.6	30,043			
Totals	101,808	20.8	295,238	140,333	27.2	504,170			

Fig. 9

As stated earlier, the differences in the overruns are the overall energy savings of one period compared to another. This is done by comparing summaries derived by a common denominator -- for overall savings, the same set of profiles. To break down the overall savings into its components -- savings in the base load, comfort load, and overrun -- the current year's energy use profiles must be constructed. The summary derived from the comparison of the current year data with the current year profiles is compared with the summary of the current year data with the base year profiles. The common denominator in this case is the current year data. This comparison yields the difference (as savings) in the two base loads and in the two comfort loads. The remaining savings is that of the overrun. Summaries are listed as follows:

(1) Base Year Data with Base Year Profiles
(2) Current Year Data with Base Year Profiles
(3) Current Year Data with Current Year Profiles

Meaningful comparisons result when matching (1) to (2), yielding overall savings, Base Year Profiles being the common denominator and when matching (2) with (3), yielding the breakdown of overall savings, Current Year Data being the common denominator. The third combination, matching (1) with (3) has little meaning; there is no common denominator.

The breakdown of the overall energy savings is shown in Figure 10. The increase in the savings in the comfort load and the overrun of 1979-80 over 1978-79 is attributed to the fact that the University implemented an automated control system program to limit steam use in the campus residence halls (approximately 1 million sq. ft.). Not only was the comfort level lowered for the overall campus but also the runaway use of energy appearing as overrun in the comfort range was reduced.

BREAKDOWN OF OVERALL ENERGY SAVINGS
Compared to the Base Year 1971-72
(Energy Units in Million BTU's)

	1978-79				1979-80			
	Base	Comfort	Overrun	Total	Base	Comfort	Overrun	Total
July	6,293	12	3,669	9,974	6,289	41	-501	5,829
August	6,167	171	1,347	7,685	6,179	92	1,795	8,066
Sept.	6,088	39	247	6,374	6,062	360	2,116	8,538
Oct.	6,333	1,304	-274	7,363	6,281	2,668	4,400	13,349
Nov.	6,052	2,310	176	8,538	6,046	4,877	4,066	14,989
Dec.	6,367	5,524	1,562	13,453	6,091	9,066	5,303	20,460
Jan.	6,147	4,517	-4,012	6,652	6,168	10,220	-2,094	14,294
Feb.	6,428	3,845	-226	10,047	5,907	9,433	584	15,924
March	6,177	2,990	2,874	12,041	6,192	8,475	1,117	15,784
April	6,073	1,166	1,124	8,363	6,083	2,050	124	8,257
May	6,055	1,080	-464	6,671	6,033	666	1,627	8,326
June	6,153	192	-1,698	4,647	6,161	35	321	6,517
Totals	74,333	23,150	4,325	101,808	73,492	47,983	18,858	140,333
% Svgs	15.2	4.7	.9	20.8	14.2	9.3	3.7	27.2

Fig. 10

The breakdown of percent savings for 1979-80 (as compared to 1971-72) according to the different calendar days is shown in Fig. 11. Almost 62% of the steam use occurred on the 217 full workdays with students on board and a corresponding 60.5% of the savings occurred on these days.

1979-80 Percent Savings
Compared to 1971-72

	Base Load	Cmft Load	Over-run	Total	No. Days	% Total Steam Used	% Total Savings
Workdays	9.00	4.99	2.46	16.45	217	61.7	60.5
Saturdays	2.42	.49	.94	3.85	41	10.0	14.2
Sundays	2.04	1.08	.50	3.62	42	10.2	13.2
Workdays/ Stu. Away	1.41	1.01	- .02	2.39	32	8.7	8.8
Holidays	- .63	1.74	- .22	.89	34	9.4	3.3
Totals	14.2	9.3	3.7	27.2	366	100.0	100.0

Fig. 11

The 1979-80 savings for the 217 workdays are broken down further in Fig. 12. These graphs show the percent savings according to the hours of the day. Base Load Savings occurred fairly uniformly across the 24-hour period except for the high hot water use in the early morning. The comfort load graph shows good savings during the night cut-back hours. The savings during the high use daytime hours was moderate which indicates thermostats have not been lowered to a great degree. However, the savings in the overrun during the high use hours indicates the faculty and staff have had better control over the run-away overheating of building spaces. The bottom graph, which is the total of the top three, indicates that some of the energy saved by night cut-back is lost back in the early morning warm-up. However, the comfort load graph does not bear this out as being true. The dip in the savings at about seven and eight o'clock is accented by the build-up of savings in the overrun and base load during the daytime hours.

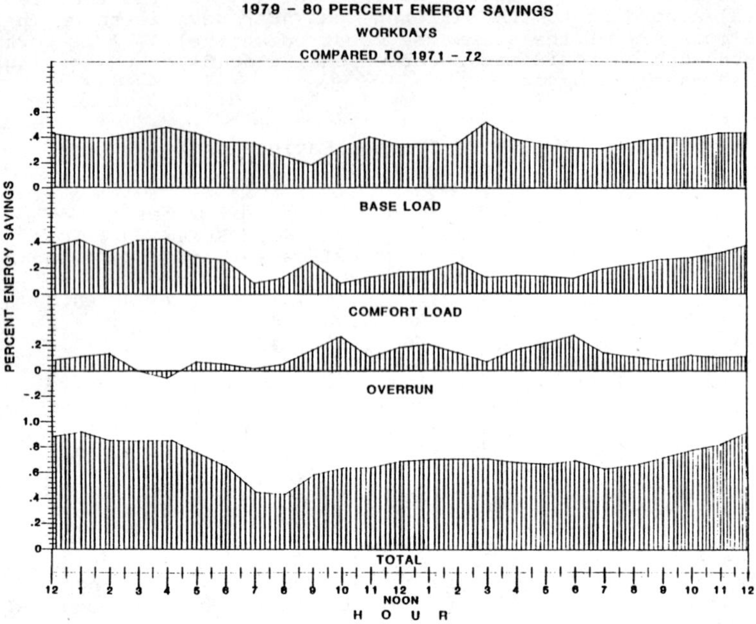

Fig. 12

Figure 13 illustrates the dangers involved in documenting by using degree days, and shows the comparison of percent savings derived by dividing the pounds of steam produced by the number of degree days in the period with the percent savings as derived by the documentation program. Significant are the differences noted in November, the opposite trends in January and February, and the differences in March, April, and May.

STEAM ENERGY SAVINGS

	1978-79 Pounds Per Degree Day	1979-80 Pounds Per Degree Day	Percent Savings Pounds Per Degree Day	Percent Savings Documentation Program
July	---	---	---	19.3
August	---	---	---	1.5
September	---	---	---	4.6
October	166,300	140,700	15.4	13.2
November	115,800	84,300	27.2	10.9
December	73,300	66,800	8.9	9.1
January	74,900	69,400	7.3	11.4
February	72,100	62,800	12.9	8.6
March	98,500	77,000	21.8	3.2
April	147,100	198,700	-35.1	0.5
May	507,700	562,300	-10.8	7.6
June	---	---	---	10.3
Total	108,426	98,826	8.9	6.4

Fig. 13

STEAM ENERGY MANAGEMENT

The steam documentation program with the use of hourly profiles can be used to conserve heat energy in much the same way as the conventional electrical power management computer program limits electrical use. This can be done by cycling off heat to buildings, thereby limiting overall heat flow to as near the hourly profile line as possible. About 15% of the steam used at NCSU is comprised of the overrun, the major portion of which is believed to be wasted heat in overheated spaces. Such a steam management system would be self-documenting and would also have the capability of lowering comfort levels by a prescribed amount on command.

CONCLUSION

The documentation process that has been described is one that is in daily use at North Carolina State University, one that does in fact conserve energy, and one that lets the University know where it stands in its energy consumption and conservation on a daily basis. This enables the University to evaluate the energy conservation program and take corrective action as it happens rather than after it has happened.

ENERGY MANAGEMENT SYSTEMS . . . A SPECIFIER'S GUIDE

DICK FOLEY
CONNTROL ENERGY INC.

"... Several issues will have to be examined before money can be committed..How much energy will it save?What will it cost? Will the investment be worthwhile?Are there other alternatives that would provide a higher energy savings, while requiring a lower capital investment..?"

As energy management programs have evolved the past few years, it has become apparent that if they are to be successful, a heavy emphasis must be placed on financial analysis.

Thus, energy conservation decisions involving "automation" must be made primarily on the economic merits of the measure... not solely on how much energy is saved.

Although an economic analysis has an obvious advantage, it has not always been applied to energy management programs. Some energy managers have had little choice but to implement measures, regardless of their economic basis, in order to avoid a forced shutdown of the building.

For other managers, an economic analysis was not necessary, since projects were low cost and offered a rapid payback.

The situation has changed somewhat today. Many of the easy, low cost projects have already been implemented while the need for conservation grows daily. There are numerous other projects that could be implemented, but they require a capital investment.

Several issues will have to be examined before money can be committed to a particular energy management system (EMS). How much energy will it save? What will it cost? Will the investment be worthwhile? Are there other alternatives that would provide a higher energy savings, while requiring a lower capital investment?

Without answers to these questions, management cannot select the optimum alternatives and make the most of the energy conservation dollar. To answer these questions - we need tools. In a moment we will discuss these tools, but first, we must recognize that four essential ingredients are needed for a successful energy conservation program. The key ingredients are : Top Management Commitment; Team Effort; Accurate Measurements; and Incentives.

Top management's attention has to be gained to establish a plan which defines the intended objectives, establishes required policies, and sets definitive procedures which tie the whole process together. Important procedures to be set for energy management system evaluation and selection are energy audit criteria and realistic technical standards.

The following is a systematic approach so you can minimize your risk and avoid major causes for past failures of energy management systems:

- Incorrect Assumptions
- Neglect of Viable Alternatives
- Lack of Commitment
- Neglecting All Available Resources

Illustration of each of these problems can reinforce their importance. First, using nameplate data for load size can conceivably cause errors of savings estimates by the degree to which the motors are oversized. Second, viable alternatives like preventative maintenance, recalibrating controls and modifying operational schedules, have to be considered as first priorities. Usually management has difficulty realizing that preventative maintenance is important, and therefore there is no mechanism for assigning an engineer adequate time to research the problem. An obvious approach to solving this problem is to assign suitable people, avoid risking slow results, and keeping reliable records. Accurate measurements, not "hand waving" during the energy management survey, should be used to prepare bid specifications, savings calculations and financial analysis. Many people feel unqualified to make an EMS selection, and therefore, depend on the vendor for recommendations. However, the problem is that in most cases, vendors do not have the time to do your homework. Hence, a company's commitment for success is limited. I must also interject that this area is well suited for objective recommendations from consultants who specialize on a day-to-day basis in evaluating needs and understand energy management systems capabilities.

RESOURCES CHECKLIST

Finally, a checklist of all resources should be made up front, and each utilized throughout your systematic program. Resources like utilities, vendors, in-house staff, financial people, instrumentation and consultants, are just a few that are available. The energy survey allows you to determine "where you are" and how much you have to pay for conservation prior to investing in an automation system. An electrical facility profile or "fingerprint" analysis of your building is another tool which is invaluable for identifying such items as abnormal peak demands, start-up and shut-down operating procedures, plus base load conditions. These problem areas must be identified when assessing needs for peak demand reduction, operational schedule changes, and elimination of unnecessary equipment operation. Facility profiles can be obtained from the utilities or by commercially-available instrumentation.

CONTROLLABLE LOADS

Controllable loads are those loads which you can control without reducing your productivity or degrading environmental conditions. Control of these loads requires load control equipment. Whether simple or complex, load control works to reduce equipment operating time, and thus saves money and energy. It can do so by any of several techniques: demand limit load shedding; fixed-start-stop schedules; optimization of equipment running time and air sources; and load cycling. Utilities spend a great amount of money to construct generating plants. They must build them to meet increased demand for electricity. To recover these added costs, the utilities monitor the average rate of energy use (the demand) at 15 or 30 minute intervals. They charge at the maximum demand level, and part of what you pay for electricity reflects your peak rate of use. This part of your payment is your peak demand charge. This peak demand charge could represent 25 to 40 percent of your electric bill, particularly if the utility charges "time-of-use" rates. Demand limit techniques can eliminate abnormal peak demand. They reduce your peak demand by shedding loads for short intervals at the time of peak. You also can reschedule the loads. Both will reduce your peak demand, but not your total electricity consumption. Demand limit reductions of ten percent are typical. You can use the same amount of electricity at less cost. Fixed start-stop schedules automatically turn on your equipment in the morning, and shut it off again at night. You and your employees no longer have to walk around the building throwing switches. These fixed schedules generally insure desired comfort levels even under extreme weather conditions. You also are assured of shutting off equipment when you don't need it.

Techniques for optimization of equipment running time should monitor environmental conditions. For instance, they can reset fixed start-stop schedules to further delay their start and to accelerate their shutdown based on actual daily weather changes. Thus, they further reduce operation time. They provide comfort when other than design-day conditions exist. Optimized air source techniques can position air dampers to draw upon cool outside air. When they do so, you avoid using the excessive power required for mechanical cooling. Pre-cooling, for instance, could avoid excessive energy use during the time you would pay most for it: the time of peak demand. Optimized air source techniques also can monitor other than dry bulb temperatures. They measure humidity or wet bulb temperatures. They calculate enthalpy. They reposition dampers to again minimize energy use.

Load cycling periodically shuts off fans for short periods throughout their normal time of operation. Each load has its own on-off schedule . . . its duty cycle. Load cycling should monitor space temperatures so as to ensure your comfort by resetting the off time to avoid excessive room temperatures. However, load cycling should be considered as a last resort. Alternative methods such as motor speed reduction can save equal energy and potentially minimize tenant discomfort due to air stratifications and excessive wear and tear on motors due to load cycling techniques. Load control equipment for commerce and industry ranges from the expensive and complex to inexpensive and simple. You may find you need to install elaborate mini-computer or micro-processor systems. On the other hand, you may find that programmable controllers or dispersed black box systems serve you just as well and for far less cost.

Load Control Equipment is available in five basic forms: central mini-computers, central microprocessors, programmable controllers, dispersed black boxes and distributed systems.

CENTRAL MINI-COMPUTER

The central mini-computer systems work best in large facilities or multi-building complexes. They provide all the energy-related features, plus process control and other non-energy related functions. These systems can be well suited to meet government "EMCS" specifications. As computers have evolved, the magnetic core in the central memory of the digital computer has gradually been replaced by a semiconductor memory. This simplifies access circuitry and permits low cost and high component density, which reduces physical size. The speed of input-output devices, such as printers and card readers, has also increased steadily. Peripheral storage media, particularly magnetic tapes and disks, have been improved by more reliable mechanical functioning and increased recording densities. The appearance of the floppy disk and the adaptation of the familiar audiotape cassette for storing data, provide low-cost versions of the versatile disk and tape media.

CENTRAL MICROPROCESSOR SYSTEMS

Central microprocessor systems lack the general program features of the mini-computers. They do not provide business data processing or control of industrial production. They also have smaller memory capacities than do mini-computers. They perform only the energy management functions of mini-computers. As energy managers, they are controlled through an easy-to-use keyboard that has its own simple language. They provide fixed schedule start-stop, load cycling, demand limit load shedding, and all optimization techniques. A microprocessor chip can also be configured into a stand-alone microcomputer, with the addition of a memory chip and an imput/output (I/O) interface. In a microcomputer, the program instructions are entered through a CRT or some other peripheral device connected to the I/O interface. The program is stored on a random-access-memory (RAM) chip until it is ready for execution. Sometimes a program is stored permanently on the read-only-memory (ROM) chip. The ROM and RAM chips serve the same purpose as the central memory in a mainframe or minicomputer. When a program is being executed, the CPU reads the instructions on the RAM or ROM and decodes them. As in bigger computers, data can be input or output through the I/O interface at any time during program execution. Microprocessors (Microcomputers) have advantages other than size, cost, and computation speed. Their available memory can be expanded simply by adding more memory chips to the microcomputer board. Also, special versions of high-level languages are available for some microprocessors, and thus, their programming is easier, especially for someone with little previous programming experience. The interfacing and implementing of a microprocessor-controlled system could become a major effort, especially if there is no previous experience in the particular application for which it is being considered.

PROGRAMMABLE CONTROLLERS

Programmable controllers are suited for any size facility (typically greater than $1000/month bill). They provide a limited number of EM features such as fixed schedule stop-start, demand limit shed and load cycling. These devices typically provide relay contact outputs and are more sophisticated than simple time clocks. Usually microprocessors with small memory capacity are used and EM controls are limited. Also, they usually only provide environmental override based on extreme conditions. Interface is through pushbutton keyboard or thumbnail switches.

DISPERSED BLACK BOXES

Dispersed black boxes are suited for any size facility. Time clocks may be utilized for lighting control; however, if once the time clock has turned the lights off, and the lights are then manually turned on by cleaning crews, after-hour workers, etc., they may remain on. The computer, on the other hand, may be programmed to turn the lights off, and then, during the off period, repeat the command on a 30-minute basis...ensuring that the lights will remain off during unoccupied hours. Time clocks cannot easily be reprogrammed. It is not necessarily true that the start and stop times for various pieces of equipment will be the same every day. With the use of central control center, the on and off times can be varied to accommodate changing requirements. Time clocks are usually not updated at daylight savings time. Therefore, during many months equipment may be operating on an invalid schedule. Building operators then often resort to manual control, resulting in 24 hours a day equipment operations. Economizers, like computers, measure differences between outside and return air temperatures. By controlling outside and return air dampers, they enable cooling equipment to make good use of outside air. Electro-mechanical pneumatic devices are available to optimize chiller operations as well. Some demand-limiting black boxes can provide demand-limit load shedding, but not effectively, because they require demand interval data. Most utilities do not provide such data. They frequently vary the timing of demand intervals to avoid interval splitting. Many commercial and industrial firms find the simplicity and low cost of dispersed black box systems initially attractive. Those who go on to purchase such systems, however, generally will find them cost effective in smaller facilities. In larger facilities, the dispersed black box system often fails to coordinate the operation of various equipment. Without such coordination, true energy management becomes impossible. Consumers should not view their energy problems from so narrow a perspective as to overlook the limitations.

DISTRIBUTED SYSTEMS

Distributed systems incorporate a central mini-computer and remote microprocessors. Their purpose is primarily twofold: eliminate catastrophic failure of control when central system fails, and maintain real time control when a large number of loads have to be controlled. Usually, these systems are very expensive and require extensive evaluation. It is difficult to substantiate the dollars saved through the ability to monitor various energy related parameters. It is a proven fact, however, that having the ability to obtain instantaneous information related to equipment efficiencies, energy consumption profiles, and reports of equipment malfunctions, can only increase the potential savings available in the area of energy conservation. An additional benefit of computerized optimization, is the ease in which the system may be re-programmed to perform different functions. As an example, let's look at the changes that have effected electric demand control. Initially, demand billing was based on fixed intervals, (15,30, or 60 minutes). Recently, several utilities have changed to a method commonly referred to as the "Sliding Window". If a hardware demand controller was installed and operating based on the fixed demand interval, this device would have to be replaced in order to incorporate the correct control strategy required for "Sliding Window" methods of demand billing.

UTILITY"S "PEAK PERIOD"

Additionally, many utilities have adopted demand billing according to a time-of-day schedule. This method of billing has initially been based on the user's electric peak. Already underway is a entirely different method of demand billing which is based on the utility system peak. At this point, the "black box" demand controller approach would become completely ineffective and an entirely different system, which would monitor the system peak, and compare it to the user peak, would have to be devised. Software programs for central monitoring and control systems are already available for the reduction of energy costs relative to the utility system peak. On the other hand, "black box" state-of-the-art is not sufficiently advanced to inexpensively change for "Sliding Window" modifications. An area of centralized control that is related to energy conservation is that of computerized maintenance management. It is only natural that if the central monitoring system has the ability to totalize equipment running time, calculate equipment efficiencies, and alarm equipment malfunction, then this information should become part of a comprehensive maintenance management program.

TASK INFORMATION

The function of maintenance management is to schedule and output task information. These tasks are put in the form of work orders and are scheduled based on calendar time, equipment running time, and/or performance index. The work orders not only specify the task that must be performed, but give a description of the skills required, hours or labor required, materials required, and tools necessary to perform that task. Computerized maintenance management provides savings in the following areas:

1. Longer equipment life.
2. Lower maintenance costs.
3. Less manpower required due to regulated manpower loading.
4. Reduction in inventory.
5. Forecasting of required materials.
6. Cost forecasting for budget planning.

Let us look again at the areas in which central control and monitoring systems increase the return on investment. The "black box" approach to energy conservation only provides savings in a few areas (mainly thoseinvolved in the reduction of equipment operating times). They can increase, rather than reduce , manpower requirements and, without the ability to monitor abnormal conditions, they offer nothing toward the increase of equipment life. Central control and monitoring systems may be expanded to include functions in the area of life/safety and security. Central control and monitoring systems can control building access, detect fires, and limit liabilities.

Only through total building automation can one expect to reach the acceptable level which approaches optimum building operation. If you do decide to buy load control hardware, you face a problem that is at least complex and, at best, frustrating. The EMS project must undergo financial scrutiny. A simple payback analysis will not be sufficient. Without belaboring the details, our own method of financial analysis incorporates (NPV) Net Present Value, which measures all the economic consequences of the project by finding the net gain or loss at a given discount rate, usually specified by the client. NPV is the sum of discounted cash flows (net cash flow=cash inflow - cash outflow). We use this to measure the amount by which the benefits exceed all costs. If the NPV is positive, the project is acceptable and shows the degree of desirability. To find the Projects Internal Rate of return, we determine what discounted rate of return makes the NPV equal to zero. This process is calculated for all vendor proposals to objectively compare each marginal value to the project, and most important, definitively select the best solution for the client. Hardware alternatives can be compared by checklists, although not all inclusive, they itemize important points for evaluation. A final selection depends on weighing all requirements. In the final analysis, the energy consultant could be the bridge between client needs and product capabilities . . . day - to-day involvement is critical.

ENERGY MONITORING AND CONTROL SYSTEMS (EMCS) USER SATISFACTION AND RELIABILITY STUDY

Gary W. Dickinson
National Bureau of Standards*

ABSTRACT

A study of 86 energy monitoring and control systems (EMCS) was made to determine users' satisfaction and system reliability. Many of these systems reportedly were not meeting users' expectations for energy, dollar, or manpower savings. This study revealed that nearly one-third of the EMCS users were not satisfied with their system's performance. Undependability of the manufacturer and proprietary hardware and software were the two most common problems found. Factors found to have no apparent affect on reliability were: the type of facility the EMCS is installed in, which manufacturer supplied the system, and whether the system was owned by the Federal Government or was non-federally owned.

The data suggest that reduced user dependence on the system manufacturer, EMCS maintenance training, and in-house maintenance capability may increase system reliability.

1. INTRODUCTION

This interim report summarizes the findings of an initial study to determine failures experienced with energy monitoring and control systems (EMCS). A related study is scheduled to begin in March 1981 and it will be summarized in a future report.

The goals of this initial study were to determine user satisfaction with the reliability of their EMCS and to identify common problems and possible cures, through interviews with system users.

1.1 EMCS IN PERSPECTIVE

Design specifications vary for each EMCS installation, but a common goal of all EMC systems is to reduce the energy costs associated with heating, ventilating, and air-conditioning (HVAC) (1), while meeting a preselected set of temperature and humidity goals. EMCS included in this study have been designed for use in applications ranging from small buildings with as few as 50 monitor and control points, to large multi-building complexes with as many as 15,000 points. These systems have environmental and control sensor data brought together at a central control area where the starting and stopping of HVAC devices can be performed.

The more advanced EMC systems also include one or more computers that turn energy consuming devices on or off, control dampers, decide which source of heating or cooling is most efficient for present conditions, and shed electrical loads to keep within peak electric demand limits.

1.2 UNRELIABILITY IN PERSPECTIVE

In this report unreliability is a broad term which includes not only the frequency of system failures, but also user dissatisfaction with EMCS performance. System failures include degraded performance of hardware or software systems as well as breakdowns and "catastrophic" failures. Many users' complaints result from frustration in trying to deal with the new EMCS technology.

Information relating to unreliability is not based on precise failure data; it contains a degree of uncertainty due to user subjectivity. Two users of the same EMC system may perceive its performance differently, for example. Another factor that could affect integrity of the data is the possible reluctance of users to admit failures that some people may construe as inadequate operator skills.

Previous studies about general EMCS reliability have not been found. However, two guidelines [1,2] that provide detailed approaches to planning, specification, and installation of EMC systems are available. The U. S. Navy guideline [1], referred to as the inter-agency guide spec., is designed to assist all government agencies with their EMCS procurements. The Veterans Administration guideline [2] is a research report that focuses on V.A. hospital EMC systems. Although neither guideline's primary effort is the discussion of EMCS problems, they may, by their comprehensive approach, provide a basis for improving reliability of these systems.

Among the approaches recommended by these guidelines is the decentralizing of system control. This "distributed processing" technique will reduce the load on a system's main computer and may resolve transmission link problems that this study uncovered. The "smart" field interface devices (FID's) recommended are able to continue functioning when the main computer fails, thus improving reliability. Some of these new distributed type systems were coming into use as this initial study was in progress; but any improvement in reliability of the new systems must be determined after they have been operating for some time. Results of studying these new systems will be included in a final EMCS reliability report.

2. APPROACH

A study was conducted to determine failure experiences in a large number of EMC systems installed in diverse facilities. By contacting users close to the system operation, major problems causing unreliability were uncovered. Discussions with EMCS users centered around the following five questions.

Are you satisfied with your EMCS performance? What are or have been the major problem areas? Are you using all of the functions purchased with the system?

(1) A glossary of EMCS terms is located at the end of this report.

Is EMCS maintenance done in-house or by contract? Has your maintenance team had EMCS maintenance training?

Responses to these questions were pursued in detail. For example, a complaint about hardware was pursued further to localize it to computer, peripheral device, or sensor failure. Because of sometimes conflicting responses careful analysis of data was necessary. "Yes, I'm satisfied with my system's performance", may have come from the same user who responded, "We are expanding our system with manufacturer B's equipment due to all the problems we had with manufacturer A".

3. SYSTEM COMPARISONS

Two differently structured EMCS installations were observed to compare user experiences and system configurations. Both systems are described below and are considered reliable by the system operators.

The first system (2) was a relatively small system configured like many that do not have optimizing functions. It has remote monitoring and data logging of control points, and is used as a manual start/stop center for single building HVAC equipment. The automatic mode is not used because operators anticipate heating or cooling load changes and adjust the system manually. They are performing functions like optimum start/stop and load shedding.

One factor contributing to this system's reliability is in-house maintenance. Many replacement parts are kept on hand to assist in repairing by replacing suspect components. The system manufacturer is only occasionally called for repair assistance, while the system operators play the major role in making this system reliable and effective. It should be noted, however, that this system does not operate near its potential for automatic control with the existing equipment.

The second system (3), in contrast to the first, was one of the largest in the country and is used as a multiple building HVAC control system. It has many optimizing functions including optimum start/stop, load shedding, chiller/boiler optimization, and enthalpy control. Due to a high electric demand rate structure, load shedding is used extensively. A complete back-up computer and trend log are available to insure reliable operation.

Some of the factors that users say contribute to this system's high reliability are: personnel training, English language man-machine interfaces at control consoles, electric surge protection to minimize lightning problems, as-built drawings and documentation, and maintenance management software to keep track of preventive maintenance schedules.

Maintenance is done jointly by the Ft. Eustis team and the system manufacturer. Five full-time people keep the system running on a three shift basis. Personnel commitment, capability of using optimizing functions, and backup systems are all factors in this system's high reliability.

This comparison shows that perceived satisfaction or reliability does not depend on system size or the use of automatic controls.

4. PROBLEMS IDENTIFIED

Figure 1 shows the problems that were identified during the survey and the number of times each problem was reported. Since the complaints are not completely independent, the following examples may help differentiate the categories.

(2) Washington, D. C. Bulk Mail Center (U. S. Postal Service)

(3) Ft. Eustis, VA. (U. S. Army)

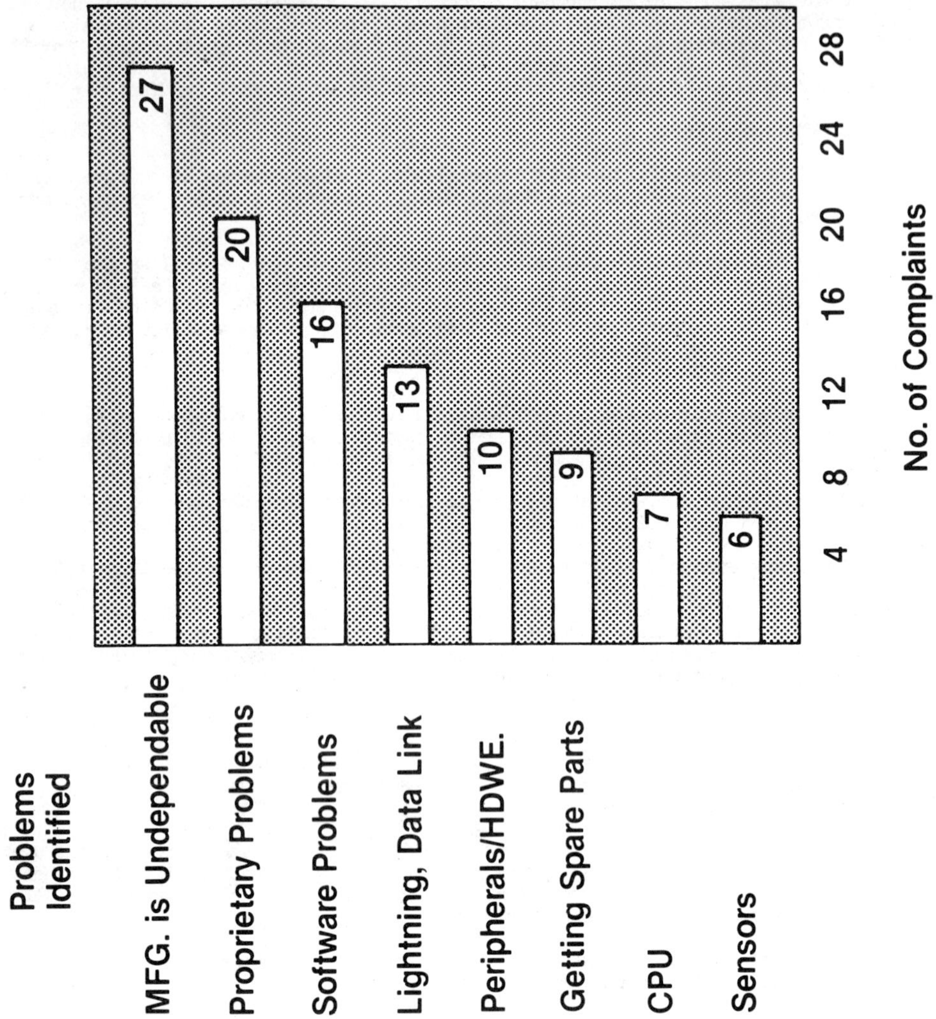
Problems Identified
MFG. is Undependable
Proprietary Problems
Software Problems
Lightning, Data Link
Peripherals/HDWE.
Getting Spare Parts
CPU
Sensors
27
20
16
13
10
9
7
6
4
8
12
16
20
24
28
No. of Complaints

FIGURE 1

The most common problem was undependability of the manufacturer. Most of the complaints concerning this problem centered around the inability of the manufacturer to keep a system operational. Inadequate documentation, poorly trained service people, and slow response on service calls were frequent complaints. Other complaints included slow turn-around for software changes and repeated calls to the manufacturer to get repairs made.

Proprietary problems were reported as the most frustrating. Skilled technicians at EMCS sites were unable to repair hardware or to make software changes because the manufacturer refused to supply adequate documentation. In some cases users could not obtain hardware or software maintenance training from the system supplier because such training would reveal proprietary information.

Users mentioned only a few specific software problems. Among them was the inability of the software to perform simple data gathering or averaging routines. Other software problems were fairly general, such as intermittent failures in system programs causing excessive downtime, or the inability to bring optimizing programs up to an acceptable performance level.

Lightning, in several cases, destroyed major electronic assemblies in EMC systems. Most other related complaints were about noise effects on the data transmission link which resulted in significant data losses.

Peripheral device and other hardware problems were broadly distributed. There were slightly more complaints about printers, but tape and disc drives, loop remotes, and data gathering panels all had about the same number of reports of failures.

Obtaining spare parts was also reported as a frustrating problem. Some users encountered delays of several weeks in getting parts, while others encountered delays exceeding six months before getting replacement parts for their systems.

Most central processing unit (CPU) problems were intermittent and users were not specific about the cause, but, like intermittent software problems, the source of these problems was hard to locate and that caused much system downtime. Only a few users reported specific CPU problems such as temperature sensitivity.

Complaints about sensor problems were very consistent: all were about calibration drift.

5. CHARACTERISTICS OF RELIABLE AND UNRELIABLE SYSTEMS

Data indicate that there are major differences between reliable and unreliable systems in three areas. The data in Figure 2 show reliability may be related to the three characteristics. Many more users who have had EMCS maintenance training reported that their systems were reliable than unreliable. Also, more than twice as many users who have in-house maintenance capability judged their systems' performance to be reliable than unreliable.

These two maintenance-related characteristics may be associated with the third, which is the use of optimizing functions. These data show that for the systems that have optimizing capability, more than twice as many users report that their systems are reliable than unreliable.

6. SCOPE: HOW EXTENSIVE IS THE UNRELIABILITY PROBLEM?

Of the 86 EMC systems surveyed, 29% of the users were not satisfied with their system's performance. While making a comparison of reliable and unreliable systems data, several areas were investigated. These are: type of facility, system age, manufacturer, and user group. Analysis of these data follow.

<u>Unreliability vs. Types of Facilities - Figure 3</u> - The data presented on this graph show that facilities other than hospitals have a larger unreliability problem than 29%. If hospital data were not included with the other survey data,

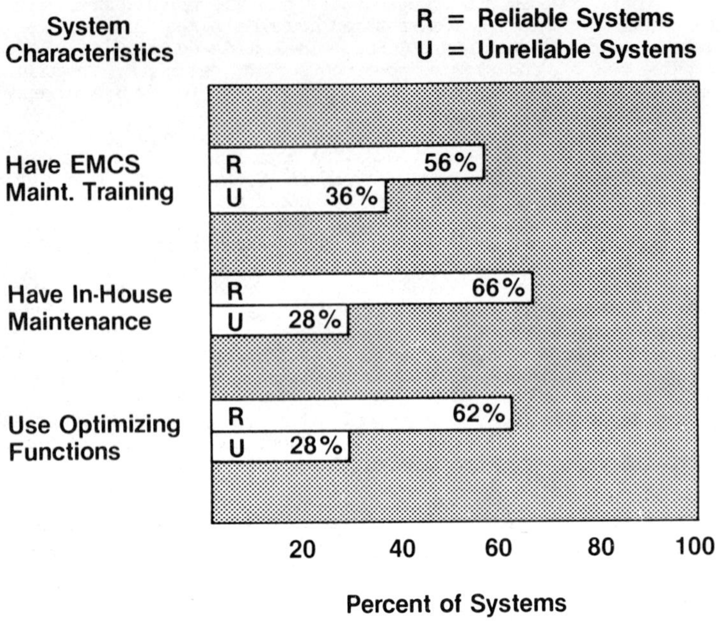
System Characteristics
R = Reliable Systems
U = Unreliable Systems
Have EMCS Maint. Training
R 56%
U 36%
Have In-House Maintenance
R 66%
U 28%
Use Optimizing Functions
R 62%
U 28%
20 40 60 80 100
Percent of Systems

FIGURE 2

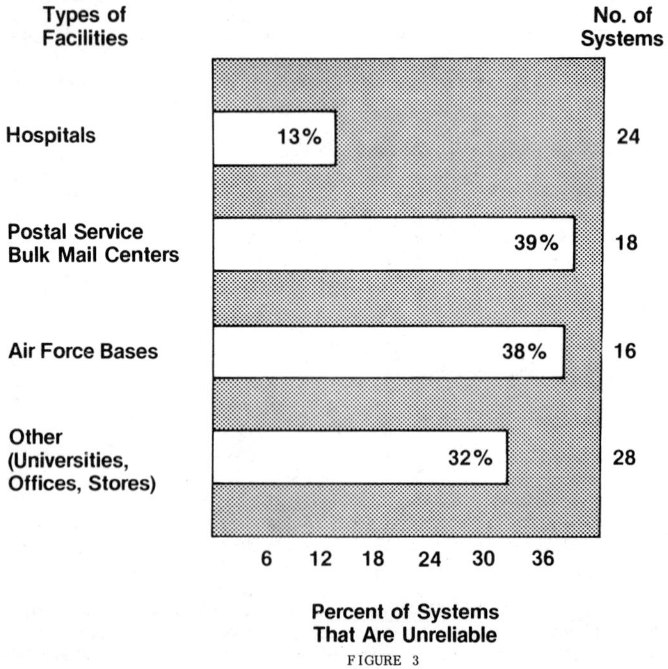
Types of Facilities
No. of Systems
Hospitals
13%
24
Postal Service Bulk Mail Centers
39%
18
Air Force Bases
38%
16
Other (Universities, Offices, Stores)
32%
28
6 12 18 24 30 36
Percent of Systems That Are Unreliable

FIGURE 3

for example, the general unreliability problem would increase from 29% to more than 35% of systems.

Why do the hospitals included have systems which appear more reliable than others? Studying age profile, manufacturer, and system characteristics does not account for this performance difference. One factor that helped hospital reliability is the V.A. "Guidelines" [2]. By increasing user awareness of latent problems and providing an effective approach for handling EMCS technology in the hospitals improved reliability resulted. Since V.A. hospitals make up two-thirds of the total hospital sample, the effect of their improved reliability affects results for the entire group.

Unreliability vs. System Age - Figure 4 - The overall trend of this graph shows a general increase in unreliability with increasing age. This seems reasonable since failures in mechanical parts are often due to wear and fatigue which can increase with age. No reason has been found to explain the break in trend shown by the highly unreliable 4 year-old systems (1976 vintage). Also, because the data for systems that are less than two years old is based on a very small sample, those results could be misleading.

Unreliability vs. Manufacturers - The study data indicate that all manufacturers' systems have a similar reliability record.

Unreliability vs. User Groups - To compare reliability information, the system data were separated by user group into federal military (25 systems), federal non-military (41 systems), and non-federal (20 systems). Comparison of these data show that each user group had similar reliability experiences.

7. SUMMARY

The data show that many problems do exist with EMC systems; about one-third of the system users in this initial study reported trouble keeping their system operational. Although the study revealed no causative factors for the identified problems some associative factors have been identified. Since data suggest that about half of all identified problems are related to dependence on the system manufacturer, reducing that dependence could be a key factor in improving EMCS reliability.

Analysis of system characteristics for reliable and unreliable systems revealed that maintenance training and in-house maintenance capabilities were found much more often with reliable systems. Also, there are more than twice as many reports of reliable systems than unreliable systems in the group of users whose systems have optimizing capability.

System age was found to affect reliability in a logical way; that is, unreliability increases with increasing age. To minimize old age effects on reliability some users recommend that a preventive maintenance program be implemented.

From other system data three factors were found to have no apparent affect on reliability: the type of facility the system is installed in, which manufacturer supplied the system, and whether or not the system was owned by the Federal Government.

ACKNOWLEDGMENT

This is one of a series of reports documenting National Bureau of Standards (NBS) research and analysis efforts in developing energy and cost data. The data support the Building Energy Conservation Criteria Program sponsored by the Office of Buildings and Community Systems, U. S. Department of Energy (DOE). The work described in this report was supported by DOE/NBS Task Order No. A 008-BCS under Interagency Agreement No. EA 77 A 01 6010.

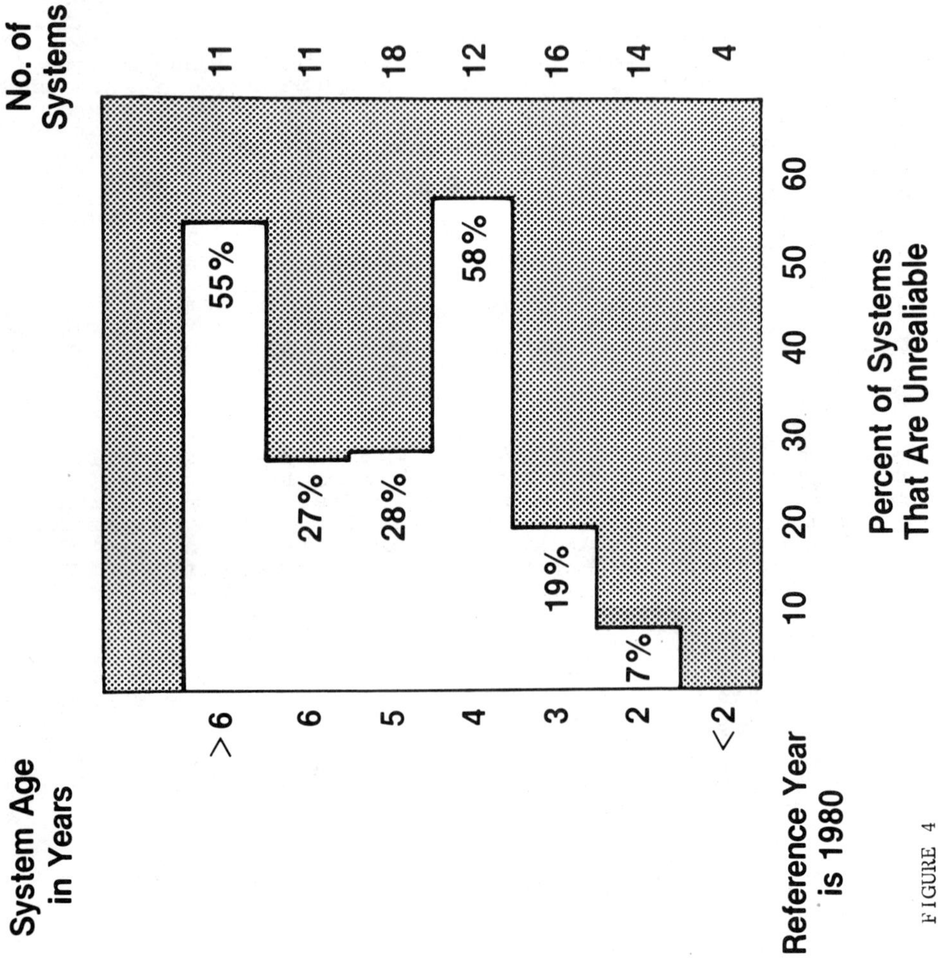
System Age in Years
No. of Systems
>6
6
5
4
3
2
<2
Reference Year is 1980
55%
27%
28%
58%
19%
7%
11
11
18
12
16
14
4
10
20
30
40
50
60
Percent of Systems That Are Unrealiable

FIGURE 4

8. REFERENCES

[1] Guide Specification for Energy Monitoring and Control Systems (U.S. Navy Ref. # TS-13941 thru 50) March 1980.

[2] Guidelines for Energy Management and Control Systems for V.A. Facilities (Veterans Administration Project #99R082) June 1979.

APPENDIX I

GLOSSARY OF EMCS TERMS

Control points - switches through which remote control functions are accomplished.

Distributed processing - a technique which uses automatic controllers at remote sites to collect and store sensor data until called for by the central computer.

Electric demand limit - a predetermined rate of electricity use beyond which large additional fees are added to a customer's electric bill.

Enthalpy control - an optimizing function that allows use of outside air for cooling when total heat content of the outside air is less than that of the inside air.

HVAC - heating, ventilating, and air-conditioning.

Intermittent hardware/software problems - those problems that occur randomly and cannot be initiated by any known sequence of events.

Load shedding - an otpimizing function that cycles electrical loads off and on to keep energy use below the electric demand limit.

Optimizing functions - those which improve efficiency of the EMCS by automatic, real-time adjustment of system equipment.

Optimum start/stop - an optimizing function that saves energy by starting or stopping systems after evaluating weather conditions, thermal inertia of the buildings, and hours of occupancy.

Trend log - a display of system parameters that is updated frequently to establish temperature and energy use trends.

Unreliability - a broad term which includes the frequency of system failures and user dissatisfaction with EMCS performance.

ENERGY AUDITS OF PUBLIC BUILDINGS IN ILLINOIS

BY

DONALD S. GLICKMAN
EXECUTIVE DIRECTOR
ILLINOIS CAPITAL DEVELOPMENT BOARD

AND

DONALD L. TERRY
SUPERVISOR, TECHNICAL RESEARCH UNIT
ILLINOIS CAPITAL DEVELOPMENT BOARD

INTRODUCTION

The Illinois Capital Development Board is a State governmental board which manages the construction of $200-400 million of public buildings and related facilities per year.

This paper will discuss the technical experience gained in the Capital Development Board's energy audit program wherein 327 public buildings have been audited and recommendations made to change operating and maintenance procedures and make construction changes. When implemented, the changes, estimated to cost $29 million, will save approximately $241 million over a 15 year period and enough energy to heat over 4,000 typical Illinois homes every year.

Types of buildings audited to date include dormitories, schools, dietary facilities, hospitals, offices, steam generation plants, and general and mechanical stores. Approximately 14 million gross square feet have been audited. The largest building audited was 444,000 GSF, the smallest 1500 GSF.

The cost of performing the audits is 3.5¢/sq. ft. and is, on the average, repaid in two years by the savings realized from implementation of recommended operating and maintenance changes alone. This program was praised by Region V DOE Headquarters, Chicago, Illinois.

A brief review of the coal conversion program in State-owned building complexes will also be made in this paper.

HISTORY OF AUDIT UNIT

The State of Illinois owns and manages the operation of approximately 3000 code agency and University buildings, plus some 5000 elementary and secondary school buildings. A survey conducted in early 1977 by the Illinos Capital Development Board (CDB) revealed energy usage in some buildings as high as 470,000 BTU per square foot per year. The potential for better energy utilization was thus determined to be substantial. In addition, the general deterioration of many structures, some approaching 100 years of age, was well known.

The Capital Development Board's Energy Management Unit was organized in 1977 for the purpose of fulfilling CDB's obligations under Plan 16, "Energy Management for State Buildings and Facilities", of the Illinois Energy Conservation Plan funded by the Federal Department of Energy (Public Law 94-163) via the Illinois Institute of Natural Resources. This federally funded unit was authorized to audit state owned buildings larger than 50,000 gross square feet (GSF) with a staff of seven persons. Plan 16 required this seven person federally funded staff to conduct energy audits of 164 code agency buildings comprising 11.8 million GSF by the end of calendar year 1980. Plan 16 was completed on schedule.

The early audits done under the federal audit program confirmed CDB's suspicions of a large energy savings potential, not only in the buildings covered by Plan 16, but also in code agency buildings of less than 50,000 GSF and in central heating plants and utility systems. Since the federally funded unit was not authorized to investigate conservation opportunities in these other locations, CDB added a state funded audit staff of nine persons. Attachment "A" shows the number and size distribution of all significant state code agency and university buildings. This unit will also be examining central heating and cooling plants and utility distribution systems. To date, the State funded unit has audited and prepared reports on 163 buildings. At the completion of the federally funded audit effort the federal team was incorporated into the state team.

All of the energy audit teams follow up the energy audit report recommendations by revisiting the audited buildings and meeting with maintenance and administrative staff to explain and assist in implementation of operations and maintenance recommendations listed in the audit report. The Unit also presented a review of the most commonly recommended operations and maintenance changes at the 1979 Annual Illinois Chief Engineers Short Course held in Allerton Park, Illinois. Copies of CDB's "Operating and Maintenance Practices for Energy Conservation in Buildings" were distributed to the approximately 120 state building chief engineers present at the conference.

The energy auditors are mechanical or electrical engineers by degree and/or experience. All engineers in the Unit have extensive experience with buildings, their construction, systems, operation, and controls.

Attachment A

User Agencies	NUMBER OF BUILDINGS					
	0-10,000 GSF	10,000-20,000 GSF	20,000-30,000 GSF	30,000-40,000 GSF	40,000-50,000 GSF	TOTAL
Code Agencies						
Conservation	5	0	0	0	0	
Mental Health/DD	486	169	123	56	22	
Corrections	443	94	42	12	8	
Transportation	338	33	14	5	5	
State Fair	80	10	1	1	0	
Child. & Fam. Svcs.	52	15	9	5	1	
Law Enforcement	51	1	1	0	0	
Military & Naval	45	20	18	8	7	
Administrative Svcs.	10	11	0	2	4	
Historical Library	9	1	0	0	0	
Secretary of State	0	3	3	2	1	
Public Health	1	0	0	0	0	
Agriculture	3	1	0	0	0	
Courts of Illinois	0	2	0	1	0	
Registration & Educ.	8	1	0	0	0	
Veterans Affairs	27	6	4	5	0	
Subtotal	1,558	367	215	97	48	2,285
Higher Education						
Univ. of Illinois	123	37	20	11	17	
So. Illinois Univ.	231	21	6	6	4	
Board of Governors	31	9	6	8	3	
Board of Regents	99	5	6	7	3	
Subtotal	484	72	38	32	27	653
TOTAL	2,042	439	253	129	75	2,938

AUDIT PROCEDURES

All audit teams use the same methodology in conducting energy audits and report preparation. Initially, contact is made with the building manager or chief engineer to arrange for a mutually agreeable time for the site visit and also to request that as-built drawings of the building and historical energy consumption data for the building be made available to the audit team. The drawings are reviewed by the team so as to familiarize themselves with the building design and construction including the mechanical and electrical systems and their operation and control. If possible, control drawings are obtained from the manufacturer of the building temperature control system. The historical energy consumption data is analyzed to determine the Energy Utilization Index (EUI) which is the British Thermal Units (BTU) per gross square foot of building area per year. A comparison of this index with recommended levels categorized by building type is a valuable indicator of the energy efficiency of the building.

After this preliminary analysis is completed, a site visit is performed to determine the extent of building variances from the construction drawings; the condition and probable efficiency of energy consuming systems; maintenance, occupancy and operational schedules; and the present programmatic uses of the building and their effects on the energy consumption of the building. The chief engineer or building manager is asked to accompany the audit team on the site visit so that questions on the facility may be answered by a person with intimate knowledge of the building systems and their operation. Building personnel are also asked to contribute any ideas of their own for energy conservation and also to respond to audit team suggestions for possible operational changes, such as night temperature setbacks. Any additional energy consumption records not already obtained by the audit team are also requested. These might include hot water and steam consumption at a building located in a facility with a central boiler plant. Site visits also include the taking of pictures to document and illustrate points of inefficient energy use, and the taking of quantitative measurements such as light levels, air flow, fluid flow and heat transmission through the building envelope.

After the site visit is completed, a mathematical model of the building is constructed with the aid of a computer which simulates the following: heating, cooling, ventilating and lighting systems and their operation and control; the building envelope, its insulating values including glass areas, and all fenestrations and their contribution to infiltration; operational and occupancy schedules. Once the energy analysis model has been constructed actual historical energy consumption data is used to verify and refine it so that it agrees with the actual energy performance of the building. With a reliable energy analysis model of the building, the analysis of possible energy conservation opportunities (ECO) requires only that the input data of the model pertinent to the proposed modification be altered and the model be run through the computer again. The difference between the original building energy consumption and the reduced consumption of the altered model is the energy saving potential of the proposed modification. Typical ECO's for a building include increased roof insulation, reduced lighting levels, reduced domestic hot water temperatures, conversion of multizone heating, ventilating and air conditioning systems to variable air volume systems, or the installation of a solar heating system. In the case of operational or maintenance changes which can be implemented by building staff for little or no cost, a recommendation for immediate implementation is made to the building staff and these changes are also reiterated later in the published report. In the case of retrofit recommendations requiring capital expenditures, a complete economic analysis is performed on the promising energy saving ideas. The economic analysis performed is as follows: Given the CDB bond interest rate, plant and O & M savings, fuel savings for the first full year of operation, the investment cost of the retrofit project, and a 10% discount rate as recommended by the Federal Department of Energy, a computer program computes the payback period. Annual cash flows for each year throughout a 25 year period are determined by the computer program. The annual cash flow is equal to the algebraic sum of the annual bond retirement payment (which is equal to the investment divided by 25 year bond life plus annual bond interest) plus fuel savings (gas, coal, electricity, oil propane, etc.). The program compares the present value of the cumulative savings for each year with the present value of the cumulative outstanding debt. When the present value of the cumulative savings equals the present value of all of the remaining outstanding debt service payments, payback is said to have occurred.

The program takes into account the estimated escalations of the various fuels. The source for these escalation factors is the "Federal Methodology and Procedures for Life Cycle Cost Analysis" as published in the January 23, 1980 Federal Register and other research done at the University of Illinois. At the present time, the federal guidelines recommend a discount rate of 10% and that rate is being used. The CDB bond interest rate is currently 6.67% and that rate is automatically programmed into the computer. In many cases, there are negative fuel savings which also must be allowed for and these are entered into the program simply as negative numbers. The program also computes the 25 year present value life cycle cost to permit comparison of alternative schemes. CDB is currently not recommending funding projects of more than five year payback period. Projects of five to ten year payback are, however, listed in audit reports for possible future funding. The fuel escalation factors now in use are as follows:

	ANNUAL PERCENTAGE INCREASE		
FUEL	1980-'85	1986-'90	AFTER 1990
Natural Gas	16.93%	16.93%	12.71%
Distillates	11.78%	13.54%	12.29%
Coal	11.14%	11.14%	11.9%
Electricity	10.6%	10.45%	10.53%

Analysis by CDB of fuel consumption records at state facilities for the last eight years showed fuel escalation rates somewhat different than those above. Therefore CDB is recommending the use of the following rates for payback analysis on energy conserving retrofit projects in the future:

	ANNUAL PERCENTAGE INCREASE		
FUEL	1980-'85	1986-'90	AFTER 1990
Natural Gas	23.15%	16.93%	12.71%
Distillates	30.0%	13.54%	12.29%
Coal	17.4%	11.14%	11.9%
Electricity	13.2%	10.45%	10.53%

These numbers are all based upon an assumed overall inflation rate of 10% per year throughout the balance of this century.

With a favorable economic analysis indicating an investment payback of less than five years, a retrofit recommendation is included in the energy audit report. This report contains a description of the existing facility including its configuration, envelope construction, and condition, and the type and use of all existing energy consuming systems in the building. It also contains a complete energy use analysis after which all operations and maintenance and retrofit recommendations are delineated. The retrofit recommendations section includes the economic data used to evaluate its viability. These include the estimated project cost, the projected first year cumulative savings, and the calculated payback period for each recommendation.

The following are examples of recommended retrofit projects which resulted from energy audits:

PROJECT	ESTIMATED COST	ESTIMATED FIRST YEAR SAVINGS	ESTIMATED FIFTEEN YEAR SAVINGS	ESTIMATED PAYBACK PERIOD
Kankakee Community College (Kankakee, Illinois)				
VAV fan system conversion	$440,000	$76,000	$1.6 million	4.15 yrs.
Energy Management System Installation	$ 86,000	$19,000	$383,000	3.32 yrs.
Lewis and Clark Community College (Godfrey, Illinois)				
Energy Management System Installation	$ 48,000	$ 3,000	$ 60,000	8.71 yrs.
HVAC System Modifications	$ 35,000	$15,000	$397,000	1.73 yrs.
Adlai Stevenson Hospital (Dixon, Illinois)				
Recirculating Air System Installation	$ 58,000	$41,000	$969,000	1.2 yrs.
Electric Chiller Installation to Replace a Steam Absorber	$167,000	$13,000	$353,000	7.4 yrs.
Lakeland Community College (Mattoon, Illinois)				
Energy Management System Installation	$ 91,000	$25,000	$463,000	2.66 yrs.
Solar Assisted Heating and Cooling System Installation	$306,000	$ 9,000	$240,000	13.78 yrs.
Mt. Vernon Regional Office Building (Mt. Vernon, Illinois)				
Ventilation Air Reduction	$ 27,000	$ 12,000	$967,766	1.85 yrs.

HISTORICAL STATISTICS

In its three year history, CDB's Energy Management Unit has audited 327 buildings, making up approximately 14 million square feet of space. The audits have resulted in 289 measurable recommendations, such as lowering nighttime heating temperatures, that can be implemented by the maintenance staff at almost no cost. The unit also has recommended 527 energy saving remodeling construction projects.

PRESENT ACTIVITIES

Currently, design or construction of retrofit projects is underway at 24 facilities. These projects will cost the state $2 million, but result in 15-year cumulative energy savings of $34.9 million. The unit also has identified $2 million in energy-conserving remodeling projects included in the Governor's Fiscal Year 1982 budget.

FUTURE ACTIVITIES

For the future, approximately 2,000 state buildings remain to be audited. Data from audits already completed show an average estimated energy savings of 55% of present consumption for all audited buildings, an average 15 year estimated cumulative savings of $19 per square foot versus an estimated retrofit cost of $1.65 per square foot of retrofitted buildings.

TECHNIAL RESEARCH UNIT

Energy savings projects are being identified by the Energy Management Unit while another CDB Unit is researching alternate sources of energy to power state facilities. This unit, the Technical Research Unit, serves as the research arm of the Capital Development Board, keeping the Board staff current in building design and construction technology by conducting research and intra-agency technology transfer sessions with other Board staff, primarily, but not exclusively in the three primary areas of energy (coal combustion, coal conversion, pollution control regulation and technology; energy efficient building mechanical and electrical systems; and alternate and renewable energy sources), technical research (design criteria, life cycle costing, higher education co-op projects, and product material research); and building and energy codes (CDB and ANSI Accessibility Standards, code organization activities, and a proposed State Building Code).

Specifically, Unit staff perform the following duties:

A. Energy

1. Study informational documents, attend workshops, seminars, etc., and invite industry experts to CDB for lectures, prepare technical reports, and work with other CDB staff and User agencies to include desirable new technology in future projects.

2. Develop funding proposals for desirable alternate energy and energy conservation projects for presentation to the State Legislature for funding.

3. Boiler plant upgrading by conducting studies of existing central heating and cooling plants for determining feasibility and cost of new coal stoker techniques and coal and ash handling techniques, all of which enable the State to get longer and more efficient life from existing plants.

4. Conduct feasibility studies for conversion of existing central heating and cooling plants from gas/oil firing to coal firing. The bulk of the coal conversion effort now underway at CDB results from the recommendations of the Governor's Coal Conversion and Energy Conservation Task Force. That report recommended the FY 81 construction of one conversion project, the FY 81 design of three other conversion projects, and the FY 81 feasibility study for twelve other sites. It is assumed that some of the FY 81 feasibility studies will result in design and construction of coal conversion projects in future years. In addition to all this effort, a conversion study is currently being conducted for the Stateville Department of Corrections Center as a part of a steam distribution study. I will take these sites one at a time and discuss briefly the status of each as follows:

a. Stateville Correctional Center near Joliet, Illinois. Two conversion schemes are being studied. One scheme consists of reconversion of existing 80,000 #/hr. boilers that are on gas and were formerly on coal. The second scheme consists of the installation of new fluidized bed boilers in lieu of conversion of existing boilers. The overall study is approximately fifty percent complete; however, no firm cost estimates for either scheme are available today. We expect completion of this study in approximately one month.

b. The University of Illinois at Champaign-Urbana, Illinois. $6.9 million dollars has been provided for funding of design and construction of a project which would include reconversion of three 150,000 #/hr. gas/oil fired boilers to coal firing. It was decided to prebid the pollution control equipment as early as possible so as to get this equipment ordered and fabrication begun while design on the balance of the project was underway. The delivery time for the pollution control equipment is quite long and this fast-tracking method will save money by allowing the pre-purchase of the pollution control equipment. The pollution control equipment bid package will be going out for bids in about one month. It is expected that the design of the balance of the project may require eight months. Construction would begin approximately two months after completion of the design.

c. Eastern Illinois University, Charleston, Illinois. FY 81 funds are available for design of the conversion project for this location which will include the startup of three existing coal fired boilers which are already in place, plus the installation of particulate control equipment to make them comply with EPA regulations. These boilers have not been used for approximately twelve years. Upon completion of startup of these units the two existing 100,000 lbs. per hour gas fired boilers would no longer be used. The project cost is estimated to be $4 million dollars, the payback period less than 4 years.

d. Western Illinois University, Macomb, Illinois. The feasibility study for reconversion of the Western Illinois University Heating Plant coal firing has been completed and it is estimated that $4.6 million dollars will be required to construct this project. Funds for preparation of bid documents is available, and A/E has been hired, and the design work is approximately forty percent complete. We expect to bid this project in the late Summer of 1981. This project will consist of reconverting three 45,000 #/hr. boilers that were formerly on coal, are now on gas, back to coal. New stokers, new coal handling equipment, and new ash handling equipment will be required, as well as pollution control equipment.

e. Illinois State University, Normal, Illinois. The project will consist of reconversion of four 28,000 #/hr. gas fired boilers to coal firing. These four boilers were formerly on coal and are now on natural gas. The feasibility study is only approximately twenty percent complete. Design money is now available for FY 81 design, and design will begin as soon as final project cost and cost effectiveness are identified, which should take place in two to three months.

f. Feasibility studies are underway at thirteen other sites throughout Illinois to determine project cost, and project cost effectiveness, of conversion to coal firing in lieu of gas or gas oil firing in the facility heating plants. The location of these sites are as follows:

Northern Illinois University, DeKalb
Southern Illinois University, Carbondale
Southern Illinois University, Edwardsville
University of Illinois Circle Campus, Chicago
University of Illinois Medical Center, Chicago
Dixon Developmental Center, Dixon
Alton Mental Health Center, Alton
Shapiro Mental Health Center, Kankakee
Manteno Mental Health Center, Manteno
Tinley Park Mental Health Center, Tinley Park
Chicago Read Mental Health Center, Chicago
East Moline Correctional Center, East Moline
Pontiac Correctional Center, Pontiac

In most of these locations two conversion schemes are being examined. At the Medical Center/Circle Campus, Chicago, the A/E is also looking at the desirability of installing a new steam transmission interconnect between the two existing heating plants (which are located 1.3 miles apart) to allow production of all steam at one location (the Med. Center plant). Scheme I is for reconversion to coal, from oil or natural gas, either by reconverting boilers that were formerly on coal and subsequently converted to gas/oil or by installation of conventional coal fired boilers in place of existing boilers that are now on gas or oil. Scheme II is for the installation of fluidized bed combustion boilers to replace existing gas or gas fired boilers. As you probably know, the advantage in using fluidized bed boiler equipment is that expensive and difficult to maintain flue gas scrubbers are not required for flu gas desulfurization. The desulfurization is inherent in the fluidized bed combustion process. The feasibility studies on these thirteen locations are approximately fifty percent complete, with the final cost estimates and payback periods computed for each due June 30, 1981.

5. Research new and emerging air and water pollution technology to permit the increased use of Illinois coal as an energy source and to identify new methods and products for improvement of the Illinois environment.

6. Identify and categorize those building mechanical and electrical system types and energy management system types which are most energy efficient; and keep Board staff informed of same to facilitate their incorporation into CDB projects.

7. Prepare a passive solar energy design manual for issuance to A/E firms for their use on CDB designed projects to promote energy conservation in State owned buildings.

8. Select, for funding, three cost effective active solar projects, three wind power or biomass projects, and three total energy or co-generation projects for FY 82 funding. This will require consultation with recognized authorities on the national level, at universities, the Illinois Institute of Natural Resources, and attendance at regional conferences and seminars.

9. Select and evaluate heat recovery equipment capable of use on CDB projects and keep Board staff informed of the desirability of its use on future projects.

B. Technical Research

1. Update and maintain CDB's design criteria contractual documents and establish design standards for User agencies, by building occupancy type.

2. Evaluate material and project quality, performance, user acceptance, and energy efficiency and develop a feedback mechanism to improve the design process.

3. Conduct research to develop appropriate fuel cost escalation information for use by the Energy Management Unit and Technical Services Section in their life cycle costing activities.

4. Coordinate agency and the University of Illinois programs in architecture.

C. Building and Energy Codes

1. Involve Unit staff in recognized building code agency activities (BOCA, Uniform Building Code, National Building Code, ANSI, NIBS, NEC, NFPA, ASHRAE, etc.) for purposes of monitoring their energy related activities and extracting information useful to CDB.

2. Identify the need and justification for a State Building Code and purpose the enactment of one into law for the purpose of promoting more efficient and less costly construction in Illinois.

3. Monitor the process of revision of ASHRAE 90-75 and 100-P Series of energy conservation standards and recommend to the Board those standards and/or revisions found to be desirable and beneficial to CDB's activities.

4. Study the ramifications of the implementation of the federal Building Energy Performance Standards (BEPS) upon CDB so as to minimize any adverse impact in the event of BEPS being mandated into law.

Through these and many other diverse projects, the Illinois Capital Development Board has assumed a leadership role in the State's ongoing efforts to conserve diminishing energy resources.

NEW FUEL AREAS IN THE 80's
BARRIERS AND INCENTIVES TO BUILDING CONSERVATION

DR. MAXINE SAVITZ
DEPUTY ASSISTANT SECRETARY FOR CONSERVATION
DEPARTMENT OF ENERGY
WASHINGTON, D. C.

Conservation has a major role to play in the energy situation by reducing vulnerability to oil disruption, alleviating pressures on conventional supplies and facilitating the transition to new domestic sources. Conservation is one of the least expensive sources of energy available to the United States in the 1980's.

Conservation means developing energy use patterns that reflect the true cost of additional energy production and use. This includes improving the efficiency of energy use as well as some shifting to less energy-intensive goods and services. Higher energy costs necessarily imply some loss of economic welfare; cost-effective energy conservation is part of the response that minimizes that loss. Thus, by its very definition, conservation reinforces the fundamental national goal of long-run economic efficiency. Conservation is not tantamount to sacrifice, and should be clearly distinguished from arbitrary curtailments. Conservation should connote energy efficiency and energy productivity.

Actions to achieve higher energy productivity -- more efficient use of energy -- should be based on a strategy of minimum total cost. Many studies have amply documented the fact that even at current prices there are huge opportunities for cost-effective capital investment and operational changes to get our residential, commercial, industrial, and transportation infrastructure more energy efficient. Larger potential savings are available by improving the productivity with which we use energy. Weatherizing and improving the space conditioning systems in residential and commercial buildings is estimated to provide the equivalent of 2.5 million barrels of oil per day at a cost of $10 or less per barrel. Energy efficiency does not imply economic stagnation. Energy consumption and economic growth can be decoupled. In many cases investments can be made today in higher energy productivity (e.g., insulation in buildings; furnace modifications; micro-processor controls in autos, buildings, and industrial processes) that are deflationary (by providing the same amenity at a lower total cost to own and operate). There are many examples of companies which have increased production while reducing their energy use during the past 10 years.

The buildings sector which accounts for 38 percent of the total U.S. energy consumption offers significant opportunity for more efficient energy use. With most of the 76 million residential units and 27 billion square feet of non-residential buildings constructed in a time of inexpensive and abundant energy, there is substantial potential for energy savings through improvements in the energy efficiencies of existing structures. This is especially compelling when it is realized that certain portions of the building stock will not change dramatically between now and the year 2000. It is estimated that in the year 2000, 60 percent of all existing commercial and almost 85 percent of existing residential buildings will still be in use.

The Mellon Institute's Least-Cost Strategy Study identified the nation's building stock as the largest single opportunity in the economy to reduce the cost of energy services. The Mellon analysis revealed that investing $148 billion in improving existing buildings could save $22.8 billion per year and the equivalent or 4.6 million barrels of oil per year. The largest single category of investment in buildings was in existing residential structures, which would account for $88 billion of investment. Single family homes make up 75 percent of these structures and the retrofit opportunities are technically similar and exist today. Very little of this potential has been realized.

The barriers to energy efficiency in this sector are not the lack of technology but range from market forces (inaccurate price signals); insufficient information (cost and benefit of energy efficient methods/delivery systems), to regulation (codes and standards).

<u>Market Forces</u>

A major factor in discouraging capital investments in energy efficiency has been the inability to allow energy prices to rise to the level where they reflect the cost of obtaining new supply. Price controls, use of average costs and rolled-in pricing discourage energy investments which might be undertaken if energy were priced at its true marginal replacement cost. Low prices encourage continued energy intensive and energy wasteful investment and discourage energy efficient investments. Consider the case of a Washington condominium apartment currently heating with oil. The condominium association receives two estimates for replacing its furnace -- one for solar heating, and one for conversion of the oil burner to natural gas. Natural gas, which is currently subject to some price controls, is the more cost-effective option. Solar heating could not compete with the gas price, but it is competitive when compared with oil. Natural gas is selected. This is one case, but is repeated all over the country. It means that thousands of conservation measures are not being pursued because the right price trade-offs are not being made.

Price patterns are beginning to change. President Reagan accelerated the decontrol of oil in January. Even with correct market signals, this sector (particularly the individual homeowner) has limited access to capital markets. There is a high cost of capital, and it is more expensive to retrofit with energy efficiency than to incorporate energy efficiency into new buildings. The Federal Government currently offers a variety of financial assistance programs such as tax credits and grants to accelerate investments in energy efficiency in existing buildings.

A residential energy tax credit - 15 percent, with a limit of $300 for conservation; 40 percent, limit of $4,000 for renewable resources and business investment energy credit of 10 percent for renewables in addition to 10 percent business tax credit is available. From 1977-1979, over 10,000,000 returns were filed using the residential tax credit provision, representing over $1 billion in credits. However, less than 3 percent of households with incomes of $10,000 to $15,000 claimed a tax credit compared with 20 percent of those with incomes $25,000 to $30,000. The amount being invested is about $750 per household instead of the optimum $1500-$2000 investment. About 15,000 businesses have used the credit for $340,000,000 of investment. In addition, 44 states have some form of tax credit or tax incentive for conservation and renewables. The question remains, "Why aren't more taxpayers taking advantage of the assistance?"

Several programs are available to low income people. It is estimated that approximately 14 million dwellings are occupied by low-income families and that those families are spending at least 21 percent of their income on home energy needs. There is the low-income energy assistance program (LIEAP) which is providing this winter $1.85 billion to help low income families pay their fuel bills. It is administered by Health and Human Services. The Department of Energy administers a program which provides funds for weatherization activities which include insulation, caulking, weather stripping, storm windows and furnace modifications. Expenditures are limited to $1600 per dwelling unit. Production and expenditures have increased. In FY 1980, 265,000 homes were weatherized with expenditures of $182,000,000. This is in comparison to FY 1979 with 94,000 homes weatherized with expenditures of $40,000,000. Recent studies in the States of Indiana, Minnesota, and Pennsylvania indicate a range in energy savings of between 16 to 35 percent per weatherized home. This range would represent for each weatherized home, a savings of between 2.8 and 6.2 barrels of oil annually. Thus, for an estimated $700 per home there is a $100 to $200 annual savings, based on $34 per barrel.

DOE is currently conducting and evaluating a pilot project incorporating furnace efficiency improvements for low-income families. The pilot project is being done to demonstrate the feasibility of using windfall profits tax (e.g., LIEAP) to effect long-term energy conservation measures in low-income homes. Retrofitting oil burning furnaces can save 15 to 25 percent of the oil consumption, which is comparable to improving the shell. The pilot project involves 30 fuel oil dealers in Philadelphia who have received training in upgrading furnaces. These include:

- correcting any unsafe conditions of the heating system;
- thoroughly cleaning all flue passages;
- replacing the existing burner with new "flame retention" oil burners - primary element of retrofit;
- constructing an instrumented furnace tuneup to optimize firing rates and overall performance; and
- installing thermostats.

The retrofit cost is $500 and there can be a payback in 2-2.5 years for a 1000 gallon a year user.

Schools and hospitals have had rapidly escalating fuel bills and often have not been able to make energy efficient improvements. The National Energy Conservation Policy Act (NECPA, PL 95-619), authorized DOE to provide 50 percent cost sharing grants to schools and hospitals for preliminary audits, energy audits, technical assistance (TA) and actual implementation of identified energy conservation measures (ECM). To date $260 million has been obligated, resulting in more than 8,000 grants. More than 58,000 buildings have been audited. Applications for TA/ECM have been received for over 60,000 buildings. During the last cycle, DOE funding was provided for only one-third of the applicants (20,000 buildings). The average payback period for most of the investments is approximately 3 years. While the recipient institutions for TA/ECM grants are required to provide matching funds, some states have programs to assist the recipient institution in providing their match (Maryland). California and Massachusetts are providing funds for those institutions turned down. In addition, health organizations such as Blue Cross/ Blue Shield and State Health Associations and Departments of Health provide funds.

Insufficient Information

The average consumer, in contrast to the typical businessman, does not routinely calculate the life-cycle cost of equipment when purchasing a refrigerator, a water heater or a furnace. Equipment is often sold through different markets. For new construction, the builder will buy the equipment -- often the same size space conditioning equipment for a 1,200 square foot house as for an 1,800 square foot house. Replacement appliances, such as dishwashers, refrigerators, are purchased by the consumer; but a furnace that stops functioning in the middle of winter is often replaced by a heating contractor the next day. Thus, equipment to upgrade the efficiency of a home is not sold as a package; consumers lack the expertise needed to decide which combination of measures to buy. Even though some information exists, the consumer is not sure which to believe.

There are several information programs to encourage energy efficient purchases - residential conservation services, fuel oil marketing, appliance labeling and realtor programs.

For example, the purpose of the Residential Conservation Service Program (RCS) is to provide on-site energy audits and other energy conservation information to residential customers of large gas and electric utilities. The program covers both energy conservation measures such as ceiling insulation or storm windows and renewable resource measures such as solar hot water systems. The program is implemented by utilities.

Plans have been received from States and non-regulated utilities submitted June 4, 1980 onward; plans from 47 states plus Puerto Rico and the District of Columbia, and 34 non-regulated utilities. As of February 16, 1981, 35 states and 30 non-regulated utility plans had been approved by DOE. An additional 4 state plans have been disapproved. We are now completing review and final negotiations on the remaining plans.

RCS programs have been initiated in several states already but will not begin in most areas until 6 months following plan approval. When the program is in full operation, it will reach 90 percent of the single to four-family homes in the country. In addition to offering energy audits, the utilities will also provide lists of installers, suppliers, and lenders willing to assist homeowners to get measures installed, and will offer to arrange jobs or loans with the listed firms. It does not provide grants or loan guarantees for weatherization, but it provides for energy audits at no more than a $15 charge. The balance will be paid for by allowing utilities to recover their costs. Utilities can provide financing for the retrofitting and several have had zero interest or low interest loan programs for several years.

A national survey performed in September indicated only 5 percent of the surveyed population had had an audit. Seventy-two percent of those who had the audit took some retrofit action. Early indications are that in areas with high fuel bills, audits will be in demand. For example, the Connecticut RCS program called Conn Serve was implemented by the Connecticut Department of Energy in October 1980. Since its inception, over 12,000 audits have been conducted (averaging 4000 monthly) and there remains a 20,000 audit request backlog. To date there has been no let up in requests and the State expects 60,000 audits to be conducted the first year. There are 150 auditors and another 100 people in training in Connecticut.

Programs such as RCS and appliance labeling provide information for the consumer already in the home, but what about the person purchasing a home. His contact is often a realtor. Realtors and real estate appraisers need information concerning energy use. Rising energy costs and the uncertainty of some energy supplies have made prospective homeowners increasingly more conscious of the energy efficiency of homes. A recent survey showed that 83 percent of homeowners surveyed believed that energy saving features added to the resale value of a property. Thus, realtors have become extremely conscious of the value of energy saving features.

To encourage residential energy conservation, DOE and the National Association of Realtors (NAR) have developed a training program to introduce realtors to using energy to sell homes. A "Realtors' Guide to Residential Energy Efficiency" has been developed as part of a training seminar. This program will train 100,000 realtors in energy efficiency by December 1982. The partnership between DOE and NAR (676,000 membership) has DOE developing the materials and NAR disseminating the information and performing the training. DOE has invested $80,000 in this program which results in a cost of $.80 per realtor for training.

Training is being completed for all instructors. In addition, a "Cooling Supplement" to the Guide is being completed to highlight technologies which are particularly relevant for the warmer parts of the United States.

In spite of the information programs, tax and market incentives, retrofitting is not occurring at the rate one would expect and the opportunity which exists. It appears that several barriers still inhibit large segments of the residential market from taking action. These include a lack of technical knowledge to sort through the barrage of conservation product claims, a fear of being cheated, and a lack of convenient and advantageous mechanisms for financing and installation of retrofits.

These are problems of the market for delivery of services, rather than problems of financial attractiveness. Organizations which offer service delivery mechanisms which are complete and convenient could receive a much larger market response and would be attractive business ventures. Several firms have entered the energy conservation delivery service business: Ebasco Services, Inc.; Scallop and Servidyne. These firms currently in this business believe that the market is huge and has substantial profit potential. One of the firms believes that their market could be more quickly penetrated if they could offer loans with terms of 7 to 10 years so that conservation investments produce positive cash flows. They would like to demonstrate to a bank that such loans would be profitable to them. Other firms such as Honeywell, General Electric, and smaller companies have analyzed the residential conservation market, but appear to be holding back because of the lack of proven methods of obtaining the large market penetration necessary to justify the substantial overheat costs of auditing, quality control, and customer interaction.

Similar barriers are believed to inhibit retrofit investments by owners of commercial and multi-family buildings, although building owners in these sectors tend to be somewhat more knowledgeable. In addition, retrofit or commercial and multi-family buildings is inhibited by the division of interests between owners and occupants, by cretain tax code provisions which limit the period of ownership, and by the competition for capital within companies which often favors investments in market share improvements regardless of the attractiveness of conservation investments.

A few firms have begun offering energy conservation services in a way which avoids many of these problems for some owners. The most common is the energy services company, which guarantees a certain level of service (temperature, lighting, etc.) at a fee specified for a term such as 5 years. In return, the service company invests in operating and capital improvements in the building and retains the savings as its revenues. The existing service firms appear to be offering a limited range of conservation improvements and to be working in those market segments where savings are easiest. There is a significant profit potential in commercial conservation services, but there are also potential risks in entering this market of diverse buildings and owners and making substantial capital improvements in the face of rapid energy price increases.

By conducting limited demostrations, one should be able, at relatively little cost, to accomplish the following:

-- To implement fully and test the most promising new and potentially transferrable mechanisms for delivering conservation services to existing buildings;

-- To develop credible information for businesses and others concerning the operations, market penetration, and factors affecting viability of those delivery systems which prove most promising. A limited number of demonstrations will be performed this year.

Regulations

Codes and standards which were developed prior to higher priced and scarce energy can discourage energy efficiency. With regard to new construction, 46 states, the District of Columbia and 4 U.S. territories have adopted some form of energy conservation code or standards. Most of the state codes are based on the technical requirements of ASHRAE 90-75. These requirements have been incorporated into the national model building code. There does need to be training of code officials for them to become familiar with energy aspects of design and construction and model programs have been developed.

Ventilation is often one component of codes. Until the 1970's, ventilation requirements were established with little scientific basis. The goal of the DOE ventilation program is to determine minimum ventilation requirements for safety, comfort, and health, and to develop energy efficient, economical ventilation systems which meet those minimum requirements.

For the past 4 years DOE has funded a research, development and demonstration effort in the area of ventilation and indoor air quality. This effort has served as the principal conduit by which the problem of indoor air quality (IAQ) versus energy conservation has been raised. Accomplishments include: (1) the development of new instrumentation for formaldehyde and radon, (2) field measurement in energy efficient homes, offices, schools and hospitals of indoor air pollutant levels and ventilation rates; and (3) the development through the voluntary consensus process of a standard entitled "Ventilation Required for Acceptable Indoor Air Quality." This standard is now proposed, and expected to be final by June, 1981, by ASHRAE. This standard must be updated as new information is collected in research programs (next update is scheduled for about 1985). The program is also analyzing and demonstrating low cost residential ventilation systems now used in Sweden and Japan. These incorporate air-to-air heat exchangers.

The opportunity exists for conservation -- this new source efficient use of energy in buildings -- to be found and delivered in the 1980's. It is an opportunity to be exploited by the private sector -- with the government sector removing any barriers, and not simultaneously creating new barriers.

Reference

1. "Residential Energy Conservation," Office of Technology Assessment, 1979.

2. "Creating New Choices, Innovative Approach to Cut Home Heating and Cooling Costs," Mellon Institute, 1980.

3. Energy Future, Ed by R. Stobaugh and D. Yergin, 1979.

4. National Academy of Sciences, Energy in Transition 1985-2010: Final Report of the Committee on Nuclear and Alternative Energy Systems, (CONAES), National Research Council, 1979.

5. "Drilling for Oil and Gas in Buildings," M. Ross and R. Williams, Center for Energy and Environmental Studies, Princeton University, Report No. 87, July 1979.

FINANCING ENERGY CONSERVATION

REMARKS OF HAROLD OLIN
Director, Architectural & Construction Research
U. S. League of Savings Associations

If I were to summarize the core of my message to you today it is that those of us here who believe in solar energy have a big job to do. That job is to educate our communities to understand and appreciate the benefits of energy conservation and alternate renewable energy sources. I also want you to know that savings and loan people are eager to learn, and many are learning about solar first-hand by building and working in solar offices.

Based on a 1980 League survey of 447 lenders (Table 1) it appears that nearly 2/3 (64%)* of savings and loan associations consider energy features in appraisals, and more than one-third (39%)* consider these facts in underwriting. The reason those numbers aren't nearly 100% -- as we would like them to be -- is that the typical lender has been deluged with tons of information from me and you, manufacturers and solar advocates, government agencies and consumer groups, telling him to be a better citizen and finance energy conservation. Except the lender hasn't heard much from one group he knows he has to listen most closely to -- that is, the home buyer.

*Original survey statistics have been increased in accordance with response 5(1) indicating 16% more energy-conscious associations by year-end 1980.

June 1980
Energy Workshops

Table 1

U.S. League of Savings Assns.
Federal Home Loan Bank Board

ENERGY QUESTIONNAIRE

1. Your City and State: ____________

2. Institution Asset Size: Up to $50 mil (1) 18% 50-100 mil (2) 25%
100-500 mil (3) 42% 500-1000 mil (4) 8% Over 1000 mil (5) 6%

3. Your institution type: a. Federal Charter (1) 64% State Charter (2) 32%
b. Stock Company (1) 7% Mutual Company (2) 15%

4. Does your institution now consider energy in the mortgage loan evaluation process?
a. In Appraisals: NO (1) 50% YES (2) 48%
b. In Underwriting: NO (1) 67% YES (2) 23%

5. If you do not now consider energy features, are you likely to do so in the future?
NR
Likely by year-end (1) 16% Likely in 1981 or later (2) 43%

6. If appraisers or loan officers consider energy, which items most likely to be noted:
Adequacy of insulation (a) 62%
Double/Triple glazing (c) 30%
Solar hot water (e) 18%
Passive solar space heating (g) 15%
Other (i) 2%
Storm doors and vestibules (b) 47%
Efficient furnace/furnace modifications (d) 36%
Active solar space heating (f) 13%
Low cost items (stripping, caulking, etc.) (h) 26%

7. Does your organization offer home improvement loans for energy conserving items?
a. NO (1) 27% YES (2) 70%
b. If yes, would you indicate the annual number of loans funded? Under 50 (1) 48%
50 to 250 (2) 17% Over 250 (3) 2%

8. Does your institution offer preferred financing for energy-efficient installations?
NO (1) 90% YES (2) 6%
If yes, check in table below:

	Higher LTV Ratio	Lower Int. Rates	Reduced Service Charge	Longer Term	Higher Income Ratio
Energy Efficient Homes	(a) 1%	(b) 2%	(c) 0	(d) 0	(e) 1%
Energy Conserving Items	(f) 1%	(g) 3%	(h) 0	(i) 0	(j) 1%
Solar Energy Systems	(k) 0	(l) 3%	(m) 0	(n) 0	(o) 1%

9. On existing houses, do you counsel loan applicants and offer to finance the extra cost of energy items? NO (1) 76% YES (2) 21%

Table 1 Continued

If yes, this is done by:

Increasing the first mortgage by the cost of the improvement:	(a)	7 %
Increasing mortgage based upon improvement's contribution to appraised value:	(b)	12 %
Offering Personal/Unsecured Loans:	(c)	1 %
Offering Secondary Financing:	(d)	3 %
Offering Home Improvement Loans:	(e)	15 %
Other: (describe) ______	(f)	0

10. Do you advertise or publicize your energy related lending activities?
NO (1) 78 % YES (2) 15 %

11. Do you find the public response: Significant (1) 3 % Mixed (2) 12 %
Negligible (3) 20 %

12. Have you identified energy as a cuase of delinquencies or foreclosures?
Frequently (1) 1 % Occasionally (2) 11 % Rarely (3) 27 % Never (4) 49 %

13. If delinquency was caused by energy inefficiency what steps did you take to assist the borrower?
None (a) 13 % Forebearance (b) 9 % Grant additional financing to cure the energy deficiency (c) 4 % Loan Modification (d) 1 %
Intensive Counseling (e) 5 % Other (f) 2 %.

14. Can you suggest actions the U.S. League should take to help you?

More technical information:	(a)	24 %	Name specific topics:	______
Periodic workshops:	(b)	34 %	State how often:	______
More consumer information:	(c)	34 %	Describe type:	______

15. Please comment upon any other of your energy related programs or policies.

______

______

16. Can you suggest any regulatory actions the Federal Home Loan Bank Board could take to assist mortgage lenders with regard to energy? (Use other side if necessary)

______

______

17. Can you suggest any legislative actions, the U.S. Congress could take?

______

______

Optional but helpful: Your name and title: ______
Name of Institution: ______
Address of Institution: ______

RETURN QUESTIONNAIRE TO: Harold B. Olin
U.S. League of Savings Associations
111 E. Wacker Drive
Chicago, IL 60601

In the survey referred to above, we also asked if associations publicized their energy-related efforts, and only 15% said yes. Let's see why the percentage was so small: Replying to another question about energy advertising, 20% said the public response was negligible and 15% mixed; only 3% said the public response was significant.

I suppose we might reason then that the home lender should promote energy conservation merely to protect his security interest. After all, don't excessive energy bills cause thousands of delinquencies or foreclosures yearly? Apparently not, for only 1% of respondents identified energy as a frequent cause of delinquencies or foreclosures; 38% said energy was occasionally or rarely the cause; and 49% said energy was never the cause.

With that kind of background let me repeat my thesis: lenders are not more keenly interested in energy conservation financing because most of what they know about energy has not come from "grass roots" public interest, but was "force fed" so-to-speak from above. To put it another way -- when large numbers of consumers begin to believe that energy saving homes are either a financial necessity or status goal, lenders will become informed and responsive to the demand for solar, wind and other renewable energy systems.

Let me try to document what I'm saying with findings from another recent survey of lenders (Table 2), by a University of Chicago student, Earl Andrews, who polled a representative sample of 36 commercial banks and savings institutions in Illinois. Of this survey sample, 55.6% had never been asked for a solar loan; 27.8% had only one or two applications. Out of a special group of 10 financial institutions who were on the State Energy Office's list of solar lenders, 3 did not respond to the survey at all and 3 indicated that they had never been asked for a solar loan.

Can you believe that? Here's a list of institutions which must have indicated their willingness to make solar loans and the State spreads the list around, and 3 institutions don't receive a single loan application! Why? Not because lenders aren't interested, but because the homeowning and home buying public isn't!

So that's why I say that you and I have a big educational job to do -- both on the public and on the lenders. Let's now examine the primary concerns about solar systems as expressed by lenders. In a 1975 survey (Table 3) of savings and loan executives, commercial bankers and mortgage bankers Prof. Ronald Melicher of the University of Colorado established that these lenders shared (in order of importance) these concerns about solar energy systems: (1)

Table 2

Number of Solar Loan Applications*

APPLICATIONS	PERCENT	(number)
None	55.6%	(20)
"Some" (1 or 2)	27.8%	(10)
"More" (3 or 4)	5.6%	(2)
"Many" (5 or 6)	11.1%	(4)
	100.0%	(36)

SPECIFIC LOANS	NUMBER
New home construction using:	
passive designs	6
active systems	17
New home purchase using active system	1
Home improvement using active systems	11
Total Loans	35

*Earl R. Andrews, Financing The Solar Home In Illinois (Chicago: University of Chicago, 1980).

Table 3

DEGREE OF CONCERN FOR SOLAR SYSTEM FACTORS+

	Average of Responses*			
	Total Responses (130)	Savings and Loan Associations (79)	Other Financiers (52)	T-Values
Expected Life of Solar Equipment	3.02	2.94	3.16	-1.52
Fuel Cost Savings	2.88	2.89	2.88	.03
Maintenance Expense	3.07	3.05	3.10	- .35
Warranty Coverage on Solar Equipment	2.77	2.77	2.76	.05
Reliability of Solar System	3.63	3.64	3.61	.31
Damage Due to Water Leaks	2.64	2.51	2.84	-1.82
Effect on Salability of Home	3.33	3.17	3.59	-2.76**
Insurability of Home	2.92	2.65	3.33	-3.43**

+Ronald W. Melicher, Lending Institution Attitudes Toward Solar Heating and Cooling of Residences (Colorado: University of Colorado, 1975).

*Responses could range from 1 (little concern) to 4 (great concern). The T-Value tests for possible significant differences between the savings and loan association and other financier responses.

**Statistically significant at the .01 level.

reliability, (2) salability, (3) maintainability, (4) longevity, (5) insurability, and (6) accuracy of energy savings projections.

Please remember that loan officers and appraisers -- no matter how informed and energy-conscious -- must by training and established procedure reflect in their professional judgments the public opinion of the community. The loan officer who judges the degree of risk and establishes lending terms takes into account all of the concerns which the marketplace holds. So does the appraiser, but with a greater emphasis on salability and market value. Since the market place is the community at large, and the homeowner or home buyer in particular, we must convince first of all these folks that solar systems are RELIABLE, SALABLE, INSURABLE, LONG-LIVED, MAINTAINABLE AND COST-EFFECTIVE.

While this effort is a continuous, ongoing process for those who are interested in spreading the solar gospel, I want to address this issue from the viewpoint of what information leading solar lenders require from solar loan applicants and what information they would find useful. On the matter of reliability, leading solar lenders require complete plans and specs of the solar and backup systems, heat load calculations and estimates of the solar contribution. Many lenders charge the loan applicant for an independent evaluation regardless whether the loan is consummated or not. They may also look for 3 year minimum warranties on materials and installation, collector efficiency certification, maintenance contracts, and insurance coverage. Any additional information that can be provided by the applicant to demonstrate a successful track record for the solar system, for collector manufacturer or system installer would be welcomed by the lender.

The matter of future solar access -- that is the continuing ability of the solar system to receive sunshine across neighboring properties -- is of paramount importance to solar lenders. To finance a property without assurance of solar access would be like making a loan for a new "synmobile" without knowing whether the owner can get enough synfuel to run his new vehicle. Lenders will generally wnat the assurance of properly drafted solar easements or restrictive covenants to run with the land.

On the subject of salability, this information would be useful: (1) about the selling price of comparable solar structures or systems, (2) recent sales of similarly-equipped older building and (3) evidence that the subject property is not drastically different in appearance from those that seem to be selling well in that marketplace. One of the concerns lenders frequently have about solar equipped homes is that the solar system not constitute on over-improvement. A leading solar lender, Midland Federal Savings of Denver, reports most solar systems it finances cost 8 to 12% of the home's total value.

On the matter of insurability there seems to be no real problem for most systems and these are insured routinely by major carriers.

On the question of system life, there must be real concern in the mind of many lenders about how soon collectors, pumps or phase change storage materials would have to be changed and at what expense to the owner (and risk to property value). Any positive information which can be provided on these questions to reassure the lender would certainly be welcome.

No matter how good the system's original product quality and installation adequacy, the system will surely require periodic servicing. This is especially true with active systems and will be most needed during the first year's shakedown period. So a maintenance contract with a reliable, reputable dealer is extremely valuable, both to the owner and to the lender's peace of mind.

Finally, we come to the question of cost-effectiveness. This question is often responded to by citing life-cycle values of solar systems. While life-cycle valuation can demonstrate the economics of higher first-cost solar systems compared to lower cost conventional heating system, I don't believe the life-cycle approach is a good way of convincing a potential homeowner or home lender that a solar system is cost-effective. Its horizon of 20 years is too distant. Number of years to positive cash flow, return on investment or even simple payback, are much more comprehensible to most lenders.

As far as the underwriting decision, much has been written in solar publications about the importance of adding E (energy) to the PITI (principal + interest + taxes + insurance) housing expense-to-income

$$\text{formula:} \quad \frac{\text{Housing Expense (PITI)}}{\text{Gross Monthly Income}} = \text{not more than } 28\%$$

The problem with this approach is that there is no consensus within the financial community either which household energy costs to include and how much above 28% to increase the allowable ratio, when energy is included.

A different approach to the historic expense/income formula is being used by the VA office in Denver. There, a house with a high energy-efficient rating is assumed to save on utility bills, therefore releasing extra funds towards principal and interest for a larger mortgage. This is how such an approach could be made to work for a solar system, anywhere in the U. S., and without benefit of an energy-efficiency rating system.

Let us assume a proposed solar system is capable of delivering 70,000,000 BTU's of solar heat a season. At a local fuel cost of $15.00/MMBTU's, the system saves $1050 a year. At 15% for 20 years, we can finance approximately $7860 of additional first cost, thus covering some or all of the solar system costs.

Yet another approach to the problem would be to simply define "the energy-efficient house" in some absolute way and add 1% or more to whatever the underwriter determines is a reasonable income ratio. The 1% extra on the income ratio generally adds about $1500 to $3500 of additional loan value (Table 4). The definition of the energy-efficient house can be related to a locally operating program such as Econergy in Denver, EE-7 in Baltimore, or perhaps some national voluntary consensus on Building Energy Performance Standards.

Much of the solar promotional activity by savings and loans to date depended on some type of interest rate discount. It was generally understood that financial institutions could only engage in this type of public subsidy for a short time, to "prime the solar pump", so-to-speak. In the present economic environment, they can't even prime the pump for a few minutes.

Although the current administration has no great commitment to the solar alternative its accelerated deregulation campaign may be just what the solar industry needed. As deregulated domestic oil and natural gas proceeds to its inevitable rendezvous with escalating oil prices and the cost of electric power escalates likewise, the time is not distant when solar power can compete unaided with other energy sources. Then the issue becomes merely one of availability of mortgage funds for solar installations.

One encouraging example of private sector creativeness comes from the West Coast. In San Francisco, Continental Savings & Loan is attracting substantial funds from savers who want their money to work for solar energy. Such earmarked dollars go into a special fund designated to be used for solar loans. While neither the savers nor the borrowers get any special financial consideration, this marketing approach channels funds not otherwise available into solar systems.

Besides supporting solar and alternate energy systems with special financing programs, at least 25 of our members have also built solar offices (Table 5). They are doing this in order to learn first-hand about the benefits -- and sometimes, the problems of solar heating.

I now want to share with you some pictures and facts about our members' solar office building efforts. These slides are part of a slide presentation available form the U.S. League titled HOME LENDERS LET THE SUN SHINE IN.

Effect of Additional Energy Efficiency Expenditures of $1 and $2 Per Square Foot on Home Buyer's Housing Expense/Income Ratio

Table 4A

Assumptions: Based on the median new house price ($57,600), mortgage rate (9.18%), and income ($27,000) for 1979 as reported in the NAHB 1979 Survey of New Home Owners, and assumptions about the size of the house and total housing expenditures,[1] the effects of additional construction costs on the home buyer's housing expense/income ratio for three different downpayments are as follows:

	(a) As is	(b) $1/Sq. ft. Extra	(c) $2/Sq. ft. Extra
10% Down			
Price	$57,600	$59,254	$60,908
Payment	423.85	436.02	448.19
Tot. hous. exp.	632.61	644.78	656.95
Median income	2,250	2,250	2,250
Expense ratio	28.1%	28.7%	29.2%
15% Down			
Price	$57,600	$59,254	$60,908
Payment	400.30	411.80	423.29
Tot. hous. exp.	597.46	608.96	620.45
Median income	2,250	2,250	2,250
Expense ratio	26.6%	27.1%	27.6%
20% Down			
Price	$57,600	$59,254	$60,908
Payment	376.75	387.57	398.39
Tot. hous. exp.	562.31	573.13	583.95
Median income	2,250	2,250	2,250
Expense ratio	25.0%	25.5%	26.0%

[1]The size of the median house (1,654 sq. ft.) was calculated based on the ration of the NAHB's median price figure to the median price of an FHA 203B home for the same period (1.21) times the FHA median house size. The total housing expense was based on the assumption that the mortgage payment was about two-thirds of the total expense in 1979, using the NAHB's figures for mortgage rates and total housing expense.

Source: U.S. League of Savings Associations, Economics Department.

Conclusions: Approximately one-half of one percentage point is added to the monthly expense/income ratio for each additional $1 per sq. ft. spent. In the 10%-20% range, the size of the downpayment does not significantly change this relationship.

Table 4B

Assumptions: The January 1980 median house price ($64,400) was obtained from the NAHB Economic News Notes. Assuming a 20% downpayment, the same home size and total housing expense/mortgage payment relationship as before, and using a 14.0% mortgage rate, three incomes where chosen to yield a 20%, a 25%, and a 30% housing expense/income ration for an "as is" house. The effects on these ratios of additional construction expenditures are as follows:

	(a) As is	(b) $1/Sq. ft. Extra	(c) $2/Sq. ft. Extra
Price	$64,400	$66,054	$67,708
Payment	610.45	626.12	641.80
Total expense	911.12	926.79	942.47
Income A	4555.60	4555.60	4555.60
Expense ratio	20.0%	20.3%	20.7%
Income B	3644.48	3644.48	3644.48
Expense ratio	25.0%	25.4%	25.9%
Income C	3037.07	3037.07	3037.07
Expense ratio	30.0%	30.5%	31.0%

Source: U.S. League of Savings Associations, Economic Department.

Conclusions: The 20% ratio provided by the largest income increases to 20.7% with the addition of $2.00 per sq. ft. to construction costs, the 25% ratio increases to 25.9%, and the 30% ratio increases to 31%. Thus, neither the price of the house nor the mortgage rate has much effect on the increase in the expense/income relationship of .5% per $1.00/sq. ft., but the size of the initial ratio does; the smaller the initial ratio, the less it will increase.

Table 5

THRIFT INSTITUTIONS WITH SOLAR OFFICES

- CONTINENTAL SAVINGS & LOAN ASSN., San Francisco, CA (Main Office)
- CITIZENS SAVINGS & LOAN ASSOCIATION OF SAN FRANCISCO, Sacramento, CA (Branch Office)
- HOME FEDERAL SAVINGS & LOAN ASSN., San Diego, CA (3 Branch Offices)
- HEART FEDERAL SAVINGS & LOAN ASSN., Auburn, CA (Main & Branch Offices)
- COLUMBIA SAVINGS & LOAN ASSN.— PIONEER DIVISION, Montrose, CO (Main Office)
- GOLDEN SAVINGS & LOAN ASSOCIATION, Golden, CO (Branch Office)
- MIDLAND FEDERAL SAVINGS & LOAN ASSN., Denver, CO (Branch Office)
- UNITED FEDERAL SAVINGS & LOAN ASSN., Des Moines, IA (4 Branch Offices)
- CITIZENS MUTUAL SAVINGS & LOAN ASSN., Leavenworth, KS (Branch Office)
- MID-KANSAS FEDERAL SAVINGS & LOAN ASSN., Wichita, KS (Branch Office)
- KEYSTONE SAVINGS & LOAN ASSN., Westminster, CA (Branch Office)
- ARLINGTON COOPERATIVE BANK, Arlington, MA (Branch Office)
- GREAT FALLS FEDERAL SAVINGS & LOAN ASSN., Great Falls, MT (Branch Office)
- FIDELITY FEDERAL SAVINGS & LOAN ASSN., Hickory, NC (Branch Office)
- WESTCHESTER FEDERAL SAVINGS & LOAN ASSN., New Rochelle, NY (Branch Office)
- PRUDENTIAL SAVINGS & LOAN ASSN., Philadelphia, PA (Main Office)
- FIRST FEDERAL SAVINGS & LOAN ASSN., Perkasie, PA (Branch Office)
- FRIENDSHIP FEDERAL SAVINGS & LOAN ASSN., Pittsburgh, PA (3 Branch Offices)
- TEXAS SAVINGS & LOAN ASSOCIATION, San Antonio, TX (Branch Office)
- NEWPORT NEWS SAVINGS & LOAN ASSN., Newport News, VA (Main Office)
- PORT ANGELES SAVINGS & LOAN ASSN., Sequim, WA (Branch Office)
- EASTERN LIBERTY FEDERAL SAVINGS & LOAN ASSN., Washington, DC (Main Office)
- HOME SAVINGS & LOAN ASSOCIATION, Madison, WI (Branch Office)
- PIONEER SAVINGS & LOAN ASSOCIATION, Racine, WI (Main Office)
- PROVIDENT FEDERAL SAVINGS & LOAN, Casper, WY (Branch Office)

Home lenders have long ago recognized the threat of high energy costs to their customers and engaged in various educational campaigns. Midland Financial Savings of Des Moines, Iowa was one of the first to issue a "call to action" and offered special financing terms for energy-related improvements to both new and existing customers.

That association took full-page newspaper ads, rented TV and radio time, and sent mail stuffers to their customers urging "Fight Back", Defend Yourself."

They also installed a lobby exhibit prepared by Iowa Power & Light showing by heat sensitive aerial photography which houses were conserving energy and which were wasting heat through under or uninsulated attics.

A permanent, staffed, Energy Information Center was set up in the lobby to answer questions of those poeple who wanted to do something about reducing their home energy waste and cost.

Another association which pioneered energy-efficiency programs is Home Federal Savings of Rockford, Illinois. They promoted their Energy Efficient Home Program to borrowers by stressing that energy-efficiency (1) saves money by saving fuel, (2) reduces interest rates on loans obtained at Home Federal and (3) enhances marketability to future home buyers.

The association developed energy-efficiency specifications which were widely publicized to builders, realtors and prospective buyers. These specifications formed the basis of an energy building code adopted by the City of Rockford early in 1981.

At least 35 savings and loan offices have either solar space heating, solar water heating or both. Several have solar air conditioning as well. An examination of the geographic distribution of these solar offices will show that contrary to popular opinion, solar <u>space</u> heating installations are most common in the frost belt states of the Northeast, Midwest and Rocky Mountains rather than the sunbelt states of the South.

One of the most solar-minded associations -- with three solar offices to its credit -- is Friendship Federal Savings of Pittsburgh. President Dick Knapp sums up their many years of experience with solar offices thus: "Our feeling that the added cost of the solar system would be offset by community goodwill more than met our expectations. In light of recent energy shortages more and more people are looking to the sun as the long-term answer to our energy needs."

The first solar branch office was opened the summer of '76 in Greensburg, a suburb of Pittsburgh. The system cost $10,500 and has been providing about 2/3 of the space and water heating needs of the office. The estimated payback period is 15 years. This is an active system, meaning that the roof mounted collectors require pumps and energy to circulate the water from collectors to storage tank or heaters.

The second office in Ingomar, another Pittsburgh suburb, was opened in the spring of '77. Here the system cost $20,000, but because the system heats an area twice that of the Greensburg office, with the solar component still at 2/3 of the total heat needed, this system calculates out at a 15 year payback. The dark gray roof areas are the active solar collectors which provide heated water either to a heat pump, to a storage tank for later use, or to the water heater. Because of the publicity, this office has experienced 40% higher savings flows than a comparable office without a solar system.

The third Friendship office in Butler, Pennsylvania, is the most interesting and innovative. Completed in the spring of '79, it employs a passive solar approach to deliver the same 2/3 solar fraction as the other two offices. A passive solar system uses natural phenomena such as radiation from warm surfaces or hot air rising to distribute the heat; active systems on the other hand require man-made energy to drive pumps or fans to move the heat from collector to storage to user. While the cost of the system was $30,000, the payback is still about 15 years, because of the higher energy costs in '79.

John Dowling, President of Westchester Federal Savings of New Rochelle, New York says of his new office: "Why solar? The cost of conventional energy in the N.E. United States has risen to the point where we are losing industry to the warmer southern states. The obvious salvation for the homeowner and businessman is the development of cheap and unlimited solar energy."

The Westchester Federal office in New Rochelle from the street front looks like any other well designed contemporary building. However, when viewed from a high point nearby, the rooftop collectors can be plainly seen. They not only collect heat, but generate electric power. An array of 8 units tracks the sun and collects enough heat to provide 1/4 of the space heating needs, as well as electricity to run several computers and chandeliers. The association advertises that "You never have to be in the dark about your savings, because our computers are run by the sun."

In this interior view, note the sign, "Solar Powered" below the chandelier and the sign "Solar Energy Heat" under the air vent to the right. The total system cost was $15,000. Deducting from this the federal tax credit of $2200 we have a net cost of $12,800. With annual savings of $1500 the payback is expected to be 8.5 years or less.

Jim Bradley of Home Savings of Madison, Wisconsin says this of sunpower: "Solar energy is a big thing these days; from a publicity viewpoint you can't beat it. But it really is our responsibility as lenders to set an example and get people and builders thinking about energy-efficient homes. Our home-like office is proving that 2/3 of a home's heating energy can be saved."

It is hard to say whether it is the solar system or the basic rustic appeal of the building which is responsible for savings flows in the first year doubled those anticipated. This consideration overshadowed the outstanding performance of the $18,000 solar system. The building interior echoes the rustic, homey Norwegian styling of the exterior. There is nothing here that identifies this as a solar building.

Lambert Mills, President of Citizens Mutual Savings of Leavenworth, Kansas has this to say of his new solar office: "The publicity generated by our solar building has been trememdous. We are now having school groups tour our building. The public is very interested in our building and solar energy in general."

The entire south facing roof was covered with active solar collectors of the water cooled type. The huge system is intended to provide 90% of the space heating, 70% of the space cooling and 94% of the water heating. The actual solar contribution has been somewhat lower.

The huge collector, measuring over 2500 sq. ft., dominates the south facade. It produces water at nearly 200°F for cooling and lower temperatures for heating. The total system cost nearly $190,000 of which the U. S. Department of Energy paid $103,000 as a demonstration grant.

Bob McChessney, Vice President of Heart Federal Savings says: "Our Auburn solar system has given us a tremendous amount of both national and international attention. Due to conventional energy shortages, we expect to be using solar equipment a great deal in the future. In fact, we are building another office right now, which will be heated and cooled by solar power."

Here Heart President Wendel Robie proudly poses on the roof of his Susanville office. The 2000 sq. ft. collector can deliver more than 3/4 of the space heating, nearly 2/3 of the air conditioning and 100% of the water heating needs. Total system cost was $133,000 of which $54,000 was for the solar portion. Calculated payback is 10 years.

Rollin Barnard, President of Midland Federal Savings, says: "We at Midland Federal understand the finite nature of fossil fuels and are concerned by escalating energy prices. We have adopted the solar approach in our new facilities both to save energy and to focus public attention on the solar potential."

This passive solar office in Arapahoe County just outside Denver uses sunspaces, or greenhouses, to collect solar heat. The heated air is then distributed by fans and ducts all over the building.

The sunlit bright areas adjacent to the greenhouses accomodate loan counselors and savings counselors. Interior window coverings will control heat gain in the summer. Excessive heat will be exhausted through the rooftop cupola.

The rock filled ledge of the greenhouse will also tend to moderate air temperatures, by absorbing and storing excess heat. Extra heat in winter can also be pumped to a crawlspace rock storage area under the concrete slab.

One of the most ardent supporters of solar energy comes from the chilliest part of the country: Ray Wirth, President of Great Falls Federal Savings believes that, "The inescapable fact about energy is that the problem is real and the clock is ticking fast. Cheap energy is a thing of the past. We of Great Falls Federal are deeply concerned, and for us, the search for energy is serious."

The attractive contemporary building in Conrad, Montana is equipped with vertical solar air collectors in the south wall. The solar portion of the heating system cost $16,500 and contributes approximately 2/3 of the space and water heating needs. The calculated payback is 14 years.

This fireplace and a warm air gas-fired furnace provide the back-up to the solar system. Note the tempered glass fireplace doors to prevent drawing heated interior air up the chimney.

Groups of school children frequently visit the office and learn about the operation of the solar system first-hand from permanent displays and personal examination of the equipment.

The slide program concludes with the suggestion that lenders, who from firsthand experience become convinced of the conservation potential of solar systems, should not only lenders but educators be.

To protect the security of their portfolios and to serve their customers, they hopefully will ask, "Will the next home my association finances be energy-efficient?" It should be and it can be -- if you and I do our proper jobs in educating lenders and the general public in your community and mine.

PROGRAMMING AND IMPLEMENTING A LODGING INDUSTRY ENERGY CONSUMPTION RESEARCH PROJECT

Eugene A. Scales
Honeywell Inc., Technology Strategy Center

INTRODUCTION/OBJECTIVES

The key to developing meaningful energy conservation strategies for the hotel/motel industry is to obtain an understanding of where energy is used, when it is required and how much is used for various hotel functions such as guest rooms, kitchen and laundry. Therefore the Hotel/Motel Conservation Demonstration program (Figure 1) was initiated to instrument six properties for collection of actual energy consumption data, to analyze that data with respect to hotel/motel business activity, and to develop strategies to reduce energy consumption. One phase of this program, the planning and implementation of the instrumentation and data collection, is described in the following paragraphs.

SITE SELECTION

Six sites were selected as representative of existing hotels/motels in terms of:

- Type of property--transient, convention, resort;
- Size--number of guest rooms;
- Geographic location--heating/cooling;
- Type of HVAC system--central, through the wall;
- Service areas within building--kitchens, laundries;
- Energy used--electric, gas, oil, solar.

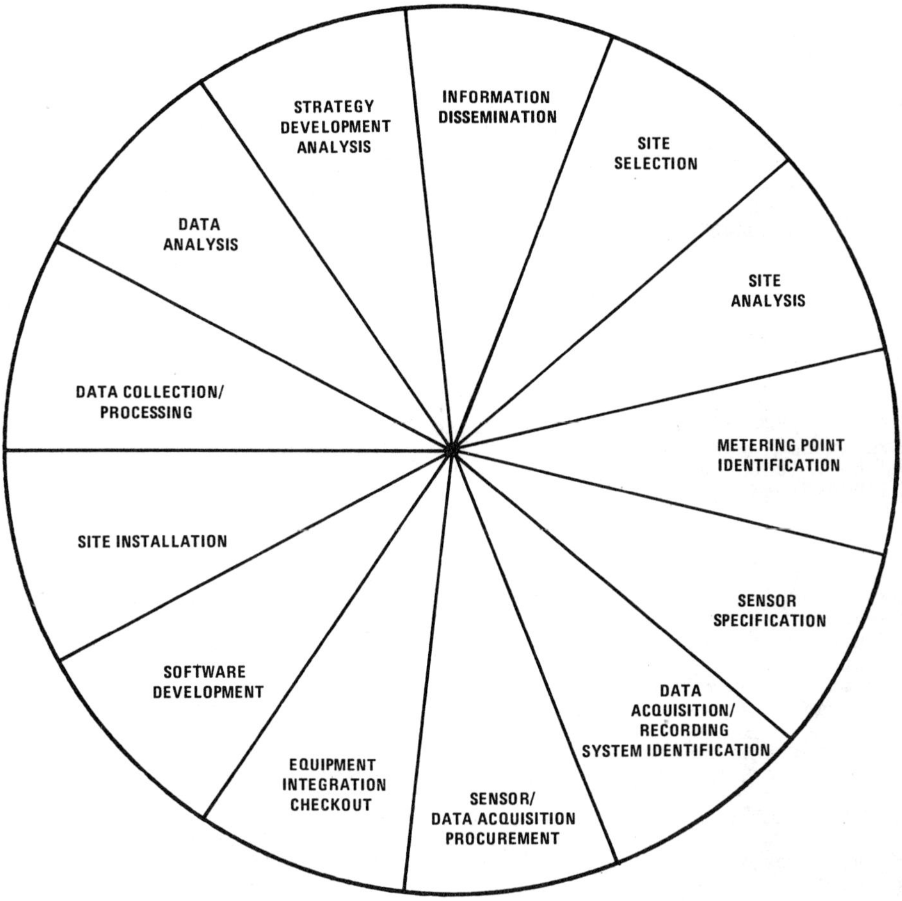

Figure 1. Hotel/Motel Conservation Demonstration Program

The six properties are described in Table 1. The sites were selected from a list of volunteers, and each participating site owner agreed to share the cost of the project in terms of coordinating the installation, monitoring the metering equipment and collecting business activity data (i.e., number of guests, meal covers).

SITE ANALYSIS

Site analysis is the foundation of the instrumentation effort. It involves examining each site in detail to determine how energy is used, what types of energy are used, and where the energy is used within the building. Based on this important information, metering points are identified and instrumentation is selected and purchased.

On the Hotel/Motel Conservation Demonstration program, the basic approach consisted of obtaining building plans, examining them in detail, and developing a preliminary analysis of where the energy is used within the building and what points should and could be monitored. After this initial step, each site was visited and the initial analysis and metering plan revised.

The latter step proved beneficial from a number of perspectives. First, it is difficult to become familiar with a building merely by examining drawings. Second, many buildings have been modified in the intervening years since they were constructed. Last, it afforded the opportunity to establish direct liaison with building maintenance personnel, who eventually would coordinate installation and be responsible for collecting business activity data within the buildings.

MONITORING POINT IDENTIFICATION

Once a complete profile of the building's energy consumption is obtained, the monitoring points can be selected. In this study, the basic rationale for selecting monitoring points was to:

- Limit the number of points per site to 48 because of budget constraints.
- Divide the properties into functional areas (i.e., guest rooms, laundry, kitchen). The energy consumed for space heating, cooling, domestic hot water, lighting and appliances was to be monitored for all functional areas possible. Because of piping and electrical layouts at certain properties, it was not always possible to isolate a functional area completely.

Figure 2 illustrates this approach for the instrumentation of the domestic hot water and heating/cooling system at Woodstock Inn, Woodstock, Vermont. Additional points were selected for monitoring electricity consumption but are not shown in this illustration.

SENSOR/METER SPECIFICATION

Once the basic monitoring points were identified, it was necessary to characterize each point such that purchase specifications could be developed.

The specification of most sensor types (temperature, electricity and gas) is relatively straightforward because it is possible to

Table 1. Representative Sites

SITE/OWNER • SIZE • TYPE	SPACE CONDITIONING SYSTEMS		DOMESTIC HOT WATER SYSTEMS	HEATING DEGREE-DAYS	COOLING DEGREE-DAYS
	PUBLIC/SERVICE AREAS	GUEST ROOMS			
HIBBING, MN/KAHLER • 126 ROOMS • TRANSIENT	FOUR-PIPE SYSTEM • A/C—ELECTRIC CHILLERS • HEATING—BOILERS (NATURAL GAS/NO. 2 FUEL OIL)	A/C AND HEATING—FAN COIL UNITS	NATURAL-GAS-FIRED BOILERS	9396	122
WOODSTOCK, VT/ROCK RESORTS • 121 ROOMS • RESORT	FOUR-PIPE SYSTEM • A/C—ELECTRIC CHILLERS • HEATING—BOILERS (NO. 4 FUEL OIL)	A/C AND HEATING—FAN COIL UNITS	NO. 4 AND NO. 2 FUEL OIL AND SOLAR ENERGY	7636	507
ATLANTA, GA/RADISSON • 391 ROOMS • CONVENTION	ROOFTOP UNITS FOR COOLING (ELECTRIC) AND HEATING (NATURAL GAS)	A/C AND HEATING—ELECTRIC THROUGH-THE-WALL UNITS	NATURAL-GAS-FIRED BOILERS	3348	1735
SAN ANTONIO, TX/LaQUINTA • 200 ROOMS • TRANSIENT	BLDG. NO. 1: FOUR-PIPE SYSTEM • A/C—ELECTRIC CHILLERS • HEATING—BOILERS (NATURAL GAS)	BLDG. NO. 1: FAN COIL UNITS	NATURAL-GAS-FIRED BOILERS	1998	2994
	BLDG. NO. 2: NO PUBLIC AREA	BLDG. NO. 2: ELECTRIC THROUGH-THE-WALL UNITS	NATURAL-GAS-FIRED BOILERS		
CORPUS CHRISTI, TX/LaQUINTA • 200 ROOMS • CONVENTION	ROOFTOP UNITS FOR COOLING (ELECTRIC) AND HEATING (NATURAL GAS)	A/C AND HEATING—THROUGH-THE-WALL UNITS, AIR-TO-AIR HEAT PUMPS	NATURAL-GAS-FIRED BOILERS	1456	3691
SCOTTSDALE, AZ/RADISSON • 220 ROOMS • CONVENTION/RESORT	FOUR-PIPE SYSTEM • ELECTRIC CHILLERS • ELECTRIC RESISTANCE HEATING	A/C AND HEATING—FAN COIL UNITS	ELECTRIC RESISTANCE	840	4521

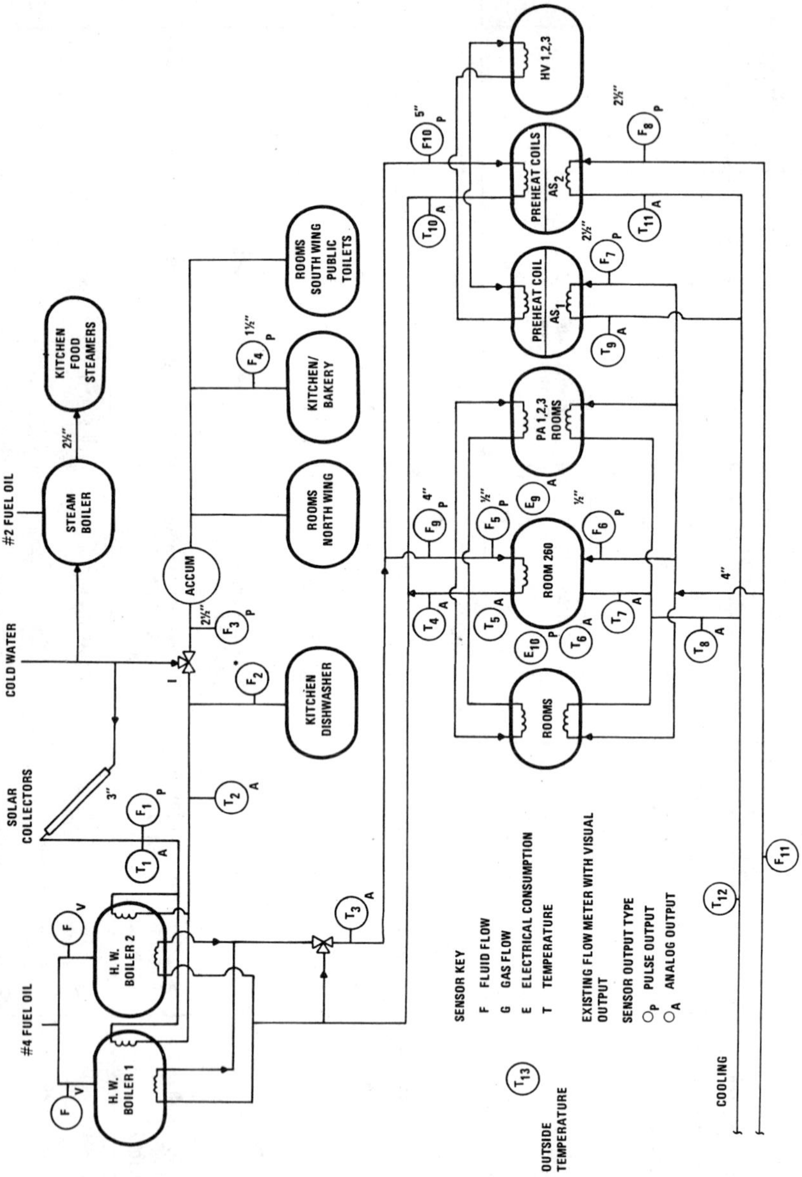

Figure 2. Instrumentation at Woodstock Inn, Woodstock, Vermont

measure and/or visibly gauge the necessary range of the sensor or meter. Fluid flow proved more difficult due to lack of information on flow ranges. In some instances, these had to be estimated based on pipe sizes.

DATA ACQUISITION/RECORDER SYSTEM

Specification of the data acquisition and recording system to be employed was based on the following considerations:

- Reliability--the system must run essentially unattended in the field.
- Sensor Compatibility--the system must be compatible with all sensor inputs, e.g., pulse-initiating consumption meters and analog rate sensors.
- Recording Capacity--the system must provide 2 to 3 weeks of data storage at 15-minute scan intervals.
- Battery Backup--the system should be capable of automatic start-up upon power interruption.

ORDERING EQUIPMENT

Numerous vendors are available that can provide the types of sensors and data acquisition and recording equipment required for the Hotel/Motel Conservation Demonstration program. Two approaches commonly used to purchase equipment are:

- Developing equipment requirement specifications and soliciting bids from all potential vendors. This approach ensures maximum coverage of possible vendors, however it also requires a considerable amount of time and effort.
- Developing equipment specifications and soliciting bids from a prescreened few (three or four). This approach requires a general familiarity with available vendors and products.

The latter approach was employed for the subject program. The number of vendors for each meter/sensor was limited to one to reduce coordination efforts. Those selected are listed in Figure 3.

Main considerations in the selection of equipment and vendors include:

- Reliability,
- Maintainability,
- Product maturity,
- Cost.

Most of the considerations are self-explanatory, with the exception of product maturity, which refers to the length of time a product has been on the market, its operational history, and how many have been produced. The answers to these basic questions often will shed light on the other considerations listed. If there is uncertainty about vendor responses, current users should be contacted.

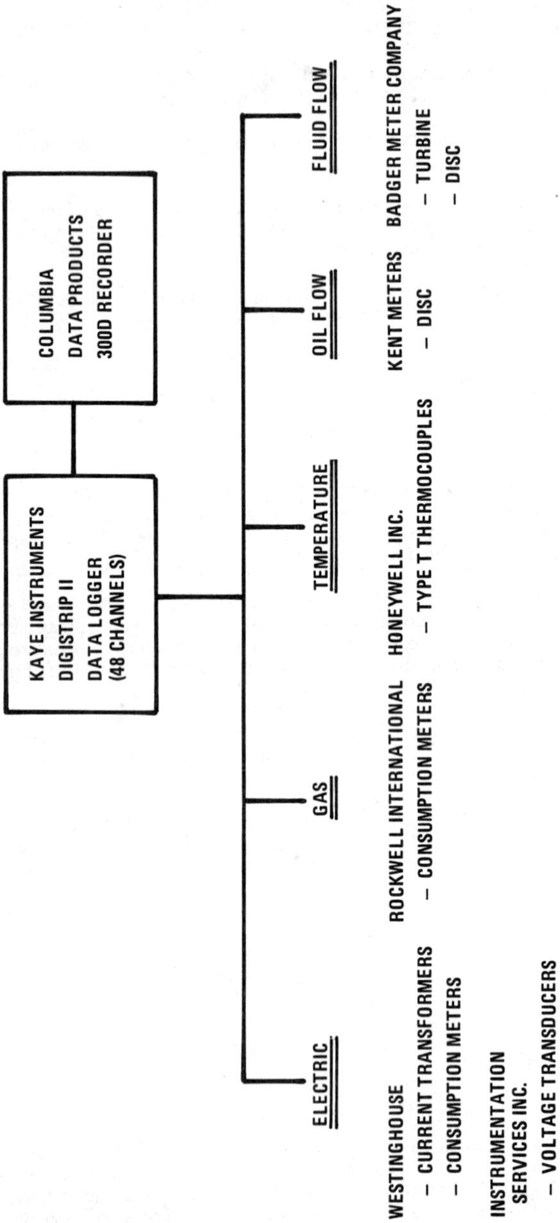

Figure 3. Selected Instrumentation Types/Vendors

The vendors selected and the reasons for selection include:

- Flow Meters--Badger Meter Co., Milwaukee, Wisconsin: Wide range of sizes, especially in hot water meters, competitive costs, visual readout, signal output compatibility, delivery time, reliability.

- Electric Meters - Westinghouse, Inc., Raleigh, North Carolina: Wide range of meters, visual readout, competitive costs, delivery time, signal output compatibility, reliability.

- Temperature - Honeywell Inc., Minneapolis, Minnesota: Competitive costs, delivery time.

- Current Transducers - Instrument Services, Inc., Minneapolis, Minnesota: Competitive costs, delivery time.

- Data Logger Recorder - Kaye Instruments, Bedford, Massachusetts/Columbia Data Products, Columbia, Maryland: Competitive costs, ability to count contact closures, compatibility with in-house equipment.

SITE INSTALLATION

The installation of equipment can pose enormous problems of coordination, especially when the sites are remotely located. How does one get a myriad of sensors and meters properly installed and checked out for a reasonable cost? The answer, while not immediately obvious, is to develop a team approach involving site personnel and vendors that normally service the site. The approach is aimed at early involvement and contribution of participating parties.

In all cases, coordination of the installation was the responsibility of site maintenance and engineering personnel. This had to be the case because the installation of some gas, flow, and electric meters required that portions of the hotels/motels be taken out of service temporarily.

Metering equipment was installed by site personnel and local contractors. At sites where personnel were encouraged to participate in the installation, maintenance staffs were capable of performing some of the work. At others, electrical and plumbing contractors that normally service the property performed the installation. The latter were selected because of their familiarity with the properties and because it was felt that they would have the most incentive for installing meters properly due to their continuous involvement.

A number of steps were taken to facilitate the proper installation of sensors and meters, including:

- Developing a brief manual describing the critical and important installation procedures for each type of sensor.

- Reviewing the installation manual with site contractors and hotel maintenance personnel prior to installation.

- Tagging the location of each sensor in the presence of site contractors and hotel maintenance staff.

- Providing utility contacts and, in some cases, support for the installation of electric and gas meters.

DATA

The Hotel/Motel Conservation Demonstration program was initiated in the fall of 1979. The installation of all instrumentation was completed by November 1980. Data is currently being collected and analyzed.

AHMA - D.O.E. HOTEL/MOTEL CONSERVATION DEMONSTRATION PROGRAM

SAM STEWART, P.E.
SAM STEWART & ASSOCIATES, INC.

Gene Scales has described the kind of instrumentation installed at each of the six demonstration sites. During the next few minutes, I hope to show you how the information collected from each site is being analyzed and also give you an indication of the results of our preliminary analysis.

Before I begin discussing data analysis procedures, let me quickly review the objectives of this program. They are:

OBJECTIVES

(1) Determine energy consumption for hotel/motel functions

(2) Determine energy parameters for hotel/motel functions

(3) Determine conversion efficiencies and standby losses (energy input vs. output)

(4) Develop hourly, daily, weekly and monthly energy profiles (consumption and demand) for specific equipment and systems

(5) Determine hot and cold water consumption for hotel/motel functions

(6) Determine potential of energy saving strategies

(1) DETERMINE ENERGY CONSUMPTION FOR HOTEL/MOTEL FUNCTIONS

The primary objective of the D.O.E./AHMA Demonstration

Program is to determine energy consumption for specific hotel/motel functions. We have identified major hotel/motel functions as:

- √ Guest Rooms
- √ Public Areas
- √ Service Areas
- √ Kitchens
- √ Dining Areas
- √ Laundries
- √ Pools
- √ Elevators
- √ Other

Energy is used within each of the above functions for different purposes. For example, for guest rooms, energy usually is used for space heating, space cooling, domestic water heating, lighting, and HVAC auxiliaries (guest room HVAC auxiliaries are the fans and pumps which require energy for their operation). Or, for kitchens, energy usually is used for the same functions as guest rooms plus for cooking (broilers, steam kettles, fryers, ovens, etc.)

In order to compare the energy used by each function in different hotels and motels, we agreed on the following energy use categories:

- △ Space heating
- △ Space cooling
- △ Domestic water heating
- △ Cooking
- △ Clothes drying
- △ Lighting
- △ HVAC auxiliaries
- △ Elevators
- △ Other

Some categories of energy use -- such as space heating -- are typical of most hotel/motel functions. Others -- such as domestic water heating -- may be typical of some, but not all, functions. Together, these categories should cover every major use of energy in hotels/motels.

We were not able to meter each energy use category for each hotel/motel function at all demonstration sites. However, at most sites we have installed enough instrumentation to understand how the energy is consumed. Data from measured functions at other sites will be extrapolated so that we have a complete energy profile for each site.

A matrix approach is being taken to analyze the data obtained. The cornerstone of this metering program is accounting for energy use data in a format that can be analyzed readily using uniform data analysis procedures. The matrix has hotel/motel functions on the vertical axis and energy use categories on the horizontal axis. For example:

FIGURE 1.

	Space Heating	Space Cooling	Domestic Water Heating	etc.
Guest Rooms	____Btu	____Btu	____Btu	____Btu
Public Areas	____Btu	____Btu	____Btu	____Btu
Kitchens				
General Use	____Btu	____Btu	____Btu	____Btu
Dishwasher	____Btu	____Btu	____Btu	
Dining Areas	____Btu	____Btu	____Btu	____Btu
etc.				

Using this matrix, we can easily determine the total energy requirement for a hotel/motel function or for an energy use category. This information can help hotel/motel operators make more educated energy conservation investment decisions. To obtain the total energy required for a specific function, add the Btu consumption identified on the lines to the right of the function. Hotel/motel operators often prefer to undertake energy measures which do not directly compromise guest comfort. This matrix can identify the total energy used outside of the guest room function. If nearly all of the energy is used by guest rooms, perhaps energy conservation aimed at other hotel/motel functions won't be cost-effective.

To determine the total energy used in a specific category, add all of the figures in the column and enter the total at the bottom. By comparing the totals at the bottoms of all the columns, you can identify where most of the energy in the property is used, and, therefore, which energy using systems offer the greatest potential for energy savings. For example, if you're trying to decide whether to install new ventilating system controls or new boiler controls, knowing that 50% of the energy is used for space heating and 12% for HVAC auxiliaries might help you choose the boiler controls.

The following figures illustrate the preliminary energy use data being collected at two sites.

FIGURE 2. HEATING ENERGY - WOODSTOCK INN

	DAY 283 GUESTS 247 HDD 20 AVE TEMP 47.1 ^{0}F	DAY 284 GUESTS 246 HDD 23.4 AVE TEMP 40.7 ^{0}F
	BTU x 10^6	BTU x 10^6
HEATING ENERGY INPUT		
Fuel Oil	16.800	19.600
Boiler Air Fan/Oil Pump	0.433	0.433
Circulating Pump	0.945	0.945
Circulating Fan 1	1.166	0.573
Circulating Fan 2	0.890	0.455
Circulating Fan 3	0.975	0.482
Room Fan Units	0.144	0.245
Fans/Pumps (%)	21.3	13.7
HEATING ENERGY OUTPUT		
Guest Rooms	5.461	4.714
Public Areas	6.264	7.636
Hot Water	4.10	4.216

FIGURE 3. KITCHEN ENERGY USAGE - RADISSON SCOTTSDALE

KITCHEN AREA	MEAL COVERS - 242 THURSDAY, OCTOBER 16			MEAL COVERS - 498 FRIDAY, OCTOBER 17		
	GALLONS	kwh	Btu x 10^6	GALLONS	kwh	Btu x 10^6
HOT WATER						
Sinks	107		.038	128		.046
Dishwasher	1161		.416	1303		.468
DISHWASHER						
Booster Heater		61.8	.211		74.5	.245
Heater		9.7	.033		16.4	.056
Pump		31.5	.108		40.6	.138
EVAPORATIVE COOLING		97.1	.331		96.1	.328
KITCHEN HOODS		71.4	.244		71.1	.243
LIGHTING		192.2	.656		185.2	.632
APPLIANCES		577.6	1.972		567.2	1.936
TOTALS	1268	1041.3	4.008	1431	1051.1	4.101

FIGURE 4. KITCHEN ENERGY USAGE - SCOTTSDALE, THURSDAY, OCTOBER 16

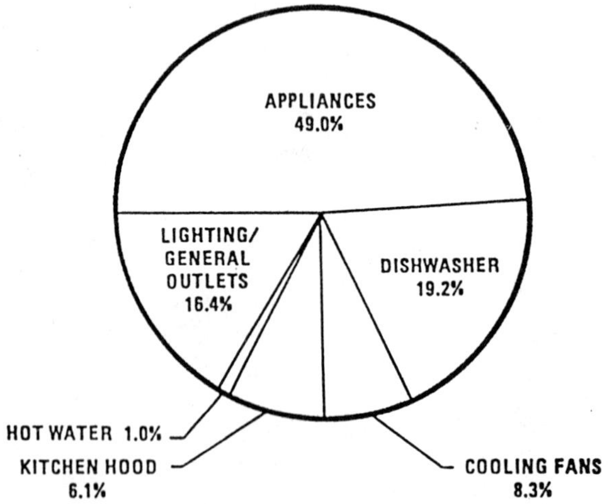

(2) DETERMINE ENERGY PARAMETERS FOR HOTEL/MOTEL FUNCTIONS

All of us recognize that absolute energy consumption data determined for a specific property -- as illustrated by Figures 2 through 4 -- is really meaningful only to that property. Absolute energy consumption data becomes useful to other properties when it is the basis of an energy parameter. In this program an energy parameter is defined as a reference point that makes comparisons possible for the same hotel/motel function in different properties. Using an energy parameter, one can see how well a function is performing both compared to itself and to the same function in other hotels and motels.

One of the first tasks we undertook was to identify energy parameters common to the lodging industry. After evaluating all hotel/motel functions and energy use categories, we concluded that the following five parameters best describe energy use in hotels/motels:

- Btu/square foot/heating degree day
- Btu/square foot/cooling degree day
- Btu/guest day
- Btu/meal served
- Btu/square foot of floor area

We are still considering several different approaches to incorporating occupancy rate into the Btu/square foot/heating degree day and Btu/square foot/cooling degree day parameters. Btu/guest day, Btu/meal served, and Btu/square foot parameters are used in more than one of the hotel/motel functions to identify levels of consumption. For example, Btu/guest day can be applied to guest room domestic water heating and to laundry hot water heating.

In determining and using parameters, categorizing energy information as either output or input is extremely important to assure a uniform approach. You can't compare Btu <u>input</u> to a water heater at one site to Btu <u>output</u> from a water heater at another site. Conversion and standby losses might make input appear twice as high as output even at the same water heater. We decided to use only output data when making calculations for the following reasons:

- much of the instrumentation installed at the sites (electric meters, Btu meters, etc.) measure output data

- actual conversion efficiencies can vary from 30 percent to 70 percent so no valid default value can be established

- output data probably is more useful to the industry because conversion efficiencies and standby losses can be applied that are appropriate for equipment at each site

In order to determine and use parameters such at Btu/guest day, in addition to energy consumption data we have had to collect weather data, business activity data, and functional use area information.

The intent of this project is to show how parameters vary according to occupancy, climate and business activity. For each of the properties instrumented in this program, we are now calculating parameters where absolute energy use data is available. Energy parameters for each site are being determined for time periods ranging from a day to an entire heating or cooling season.

Domestic water heating provides a good example of how these parameters can be used. The figure (5) below shows the kinds of parameters that could be useful to evaluate energy consumed to heat water.

FIGURE 5. EXAMPLE ENERGY USE PARAMETERS

Hotel/Motel Functions	Energy Use Category: Domestic Hot Water Heating
Guest Rooms	
- Guest Days	Btu/Guest Day
Kitchens	
- Meals Served	Btu/Meal Served
Laundries	
- Guest Days	Btu/Guest Day
- Meals Served	Btu/Meal Served
Pools	
- Size (gallons)	Btu/1000 gallons/HDD

This example assumes that we know hot water is used in the guest rooms, the kitchen, the laundry, and for pool heating. If we also know the Btu's consumed for each of these hotel/motel functions over a certain time period, we then can calculate energy parameters which relate energy consumption to business activity. For example, energy required to heat water for guest rooms can be expressed in Btu/guest day. Energy to heat water for the kitchen can be expressed in Btu/meal served, or this parameter can be further refined and shown as Btu/meal served for general kitchen use and Btu/meal served for the dishwasher. Energy use parameters can be developed in similar fashion for all energy use categories. The intent of this program is to fill in a matrix with parameters showing energy consumption ranges for all business activities.

Energy parameters can help hotel/motel operators save energy by:

(1) providing a basis for comparing similar properties. If a property uses twice as many Btu/square foot/heating degree day as the average, reducing energy use to the average Btu/square foot/HDD could result in significant energy savings.

(2) providing a means of setting priorities for energy saving efforts. A property may have a Btu/square foot/<u>heating</u> degree day parameter well within the average. However, the Btu/square foot/<u>cooling</u> degree day parameter may be 50% higher than the average. First priority could be given to energy conservation in the area where the property uses significantly more energy than the average.

(3) establishing energy use guidelines for existing facilities. It is our intent to show a range for each parameter so you can compare your property to a similar property using the most energy efficient parameters. Using your own building floor areas, local weather data, and business

activity data, you can calculate parameters. If your parameters don't fall within the "energy efficient" range, you can use the project parameters as guidelines for developing an optimum-use energy budget for your property.

(4) designing new buildings and systems that are more energy efficient. The parameters -- Btu/square foot/heating degree day, for example -- can serve as a requirement for the performance of new buildings and systems.

You probably are curious about how your property is performing compared to the test properties. Figures 6 and 7 provide an indication of the magnitude of preliminary energy use parameters determined in this demonstration project.

FIGURE 6. WOODSTOCK INN (GUEST ROOM 258) OCTOBER 1980

DAY	HEATING (BTU)	FAN (BTU)	LIGHTS (BTU)	BTU/HDD	BTU/OCC.	BTU PER FT^2	NUMBER OF OCCUPANTS
283	20,581	1,194	9,858	1,023	15,824	90	2
284	35,291	2,023	5,753	1,504	21,533	123	2

FIGURE 7. LAQUINTA - SAN ANTONIO KITCHEN ENERGY USAGE

DATE	DOMESTIC HOT WATER (BTU)	APPLIANCES AND LIGHTS (BTU)	TOTAL* (BTU)	MEAL COVERS	BTUS PER MEAL COVER
SEPT. 28	0.9 10^6	3.3 10^6	4.2 10^6	150	28K
SEPT. 29	0.9 10^6	3.4 10^6	4.3 10^6	135	32K

*EXCLUDES ENERGY FOR SPACE COOLING

(3) DETERMINE CONVERSION EFFICIENCIES AND STANDBY LOSSES

Using the format discussed earlier for energy consumption, we have recorded functional use energy output on one matrix and energy input on another matrix. This procedure makes it easy to determine the overall conversion efficiency and standby loss when we have measured output and input energy use data. For example, we have both measured output and input data where a gas meter records energy input to a hot water heater and a flow meter and temperature sensors provide information to determine the Btu output from the water heater.

(4) DEVELOP HOURLY, DAILY, WEEKLY AND MONTHLY ENERGY PROFILES

Hourly, daily, weekly and monthly load profiles are essential to understanding what causes peak load conditions and which energy saving options should be considered to reduce peak demand.

The following three figures (8,9 and 10) illustrate the kind of load profile information being generated by this program.

FIGURE 8. WOODSTOCK INN (ROOM 258) THURSDAY, OCTOBER 23, 1980

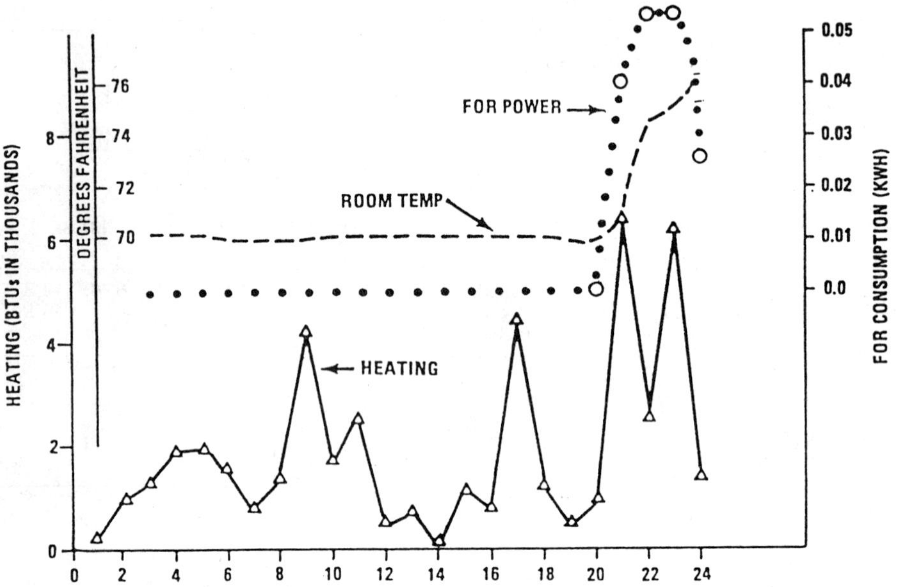

FIGURE 9. HOT WATER CONSUMPTION WOODSTOCK INN THURSDAY, OCTOBER 23, 1980

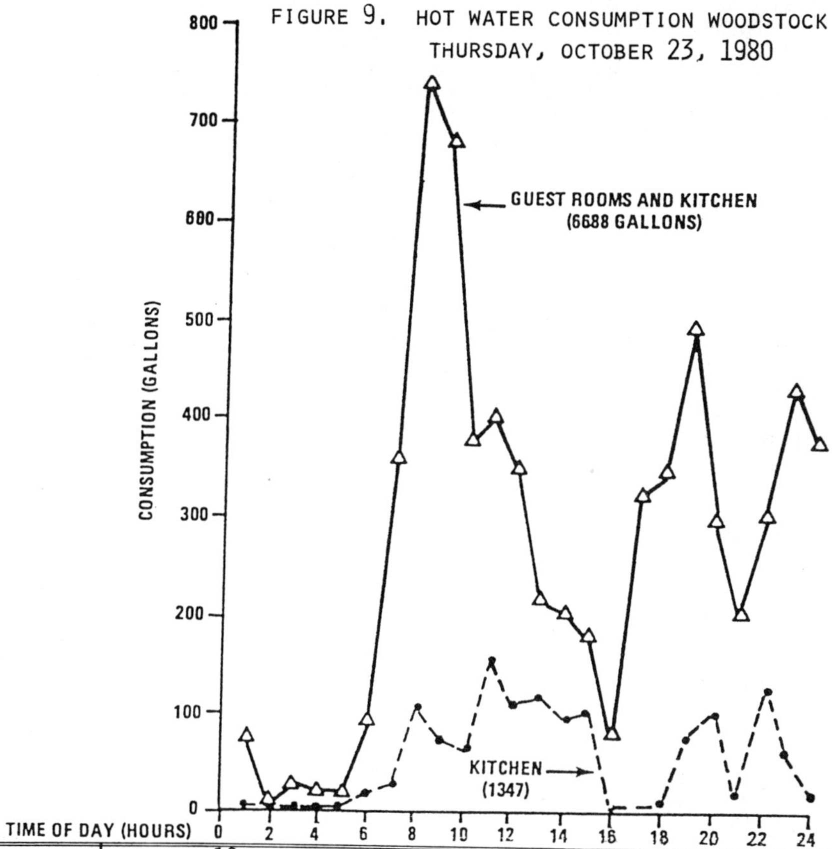

FIGURE 10. KITCHEN ENERGY CONSUMPTION PROFILE - SCOTTSDALE

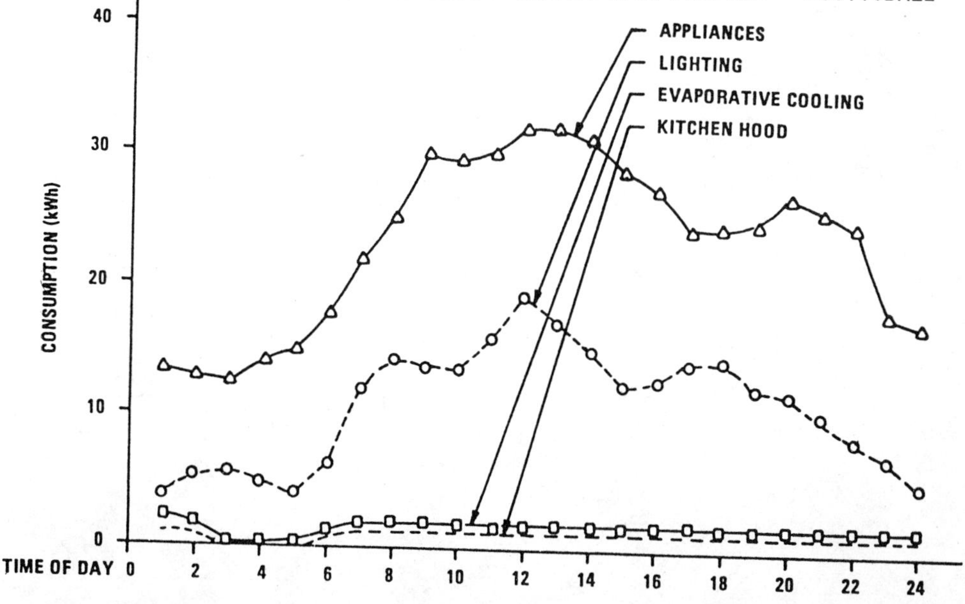

(5) DETERMINE HOT AND COLD WATER CONSUMPTION FOR HOTEL/MOTEL FUNCTIONS

Hot water consumption at all of the test sites has been much higher than anticipated. The following three figures (11, 12, and 13) give a preliminary indication of hot water consumption for guest rooms and kitchens at two sites.

FIGURE 11. HOT WATER CONSUMPTION/DISTRIBUTION WOODSTOCK INN

FUNCTIONAL AREA	DAY OF YEAR			
	283 (THURSDAY)		296 (WEDNESDAY)	
	GALLONS	PERCENT	GALLONS	PERCENT
GUEST ROOMS	5,482	53	5,205	69
KITCHEN SINKS	1,568	15	1,202	16
DISHWASHERS	3,280	32	1,160	15
TOTALS	10,330		7,568	

FIGURE 12. HOT WATER ENERGY CONSUMPTION VERSUS MEAL COVERS

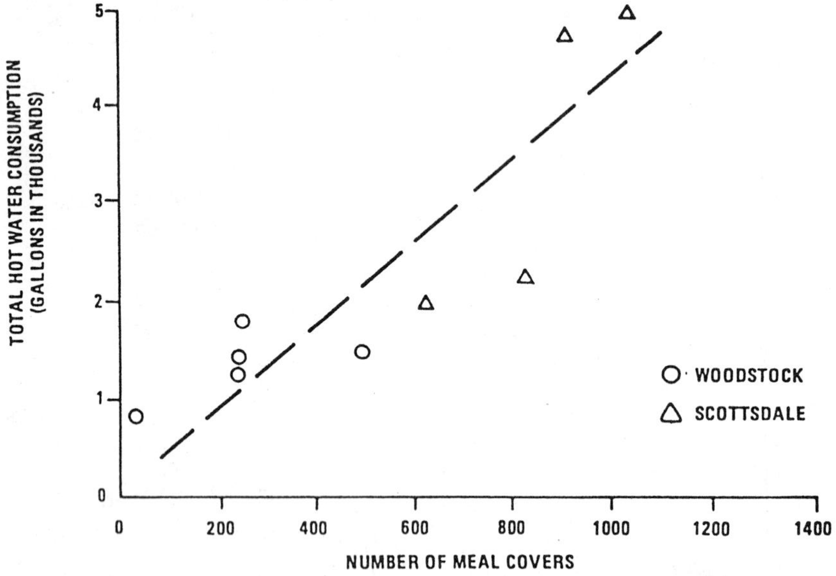

FIGURE 13. HOT WATER CONSUMPTION VERSUS NUMBER OF GUESTS

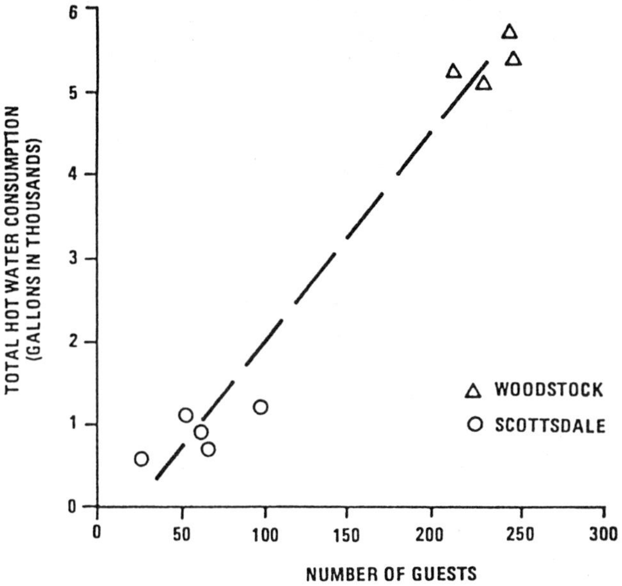

(6) DETERMINE POTENTIAL OF ENERGY SAVING STRATEGIES

Different energy saving measures are currently being considered for implementation at each site. As soon as we have established a pattern for basic energy use, many key measures will be implemented and energy savings will be monitored using the demonstration program instrumentation.

We expect to accurately determine actual Btu savings resulting from each conservation strategy. Among the proposed energy saving strategies are:

- Replacement of incandescent fixtures in guest rooms with fluorescent ones
- Replacement of incandescent lamps with fluorescent adaptors in corridors and stairwells
- Installing a central automation system which controls public area and/or guest room HVAC equipment
- Reducing fan speeds if energy conservation measures have reduced the air conditioning requirement
- Installing new night setback thermostats
- Shutting off domestic hot water to portions of the building not occupied
- Setting up a schedule of preheating times for kitchen appliances
- Installing flow restrictors in faucets
- Using low temperature laundry detergent

- Installing heat recovery equipment to capture condenser waste heat and to preheat domestic hot water

- Adding vestibules

- Installing a new energy efficient "dilution" or 100% recirculated air kitchen hood

- Insulating steam and hot water piping, boilers and hot water storage tanks

- Replacing all existing electric motors with high efficiency motors

"The Design & 1st Year Performance of the California Farm Bureau Federation Energy Conservation System"

Shlomo I. Rosenfeld, PE
Associate Partner & Chief Mechanical Engineer
Syska & Hennessy, Inc., Engineers
575 Mission Street
San Francisco, CA 94105
415/495-7711

INTRODUCTION

In 1978 Syska & Hennessy, Engineers, San Francisco, designed the Heating, Ventilating, Air Conditioning, Plumbing, Fire Protection and Electrical systems for the new headquarters of the California Farm Bureau Federation Building (CFBF) in Sacramento, California. The California Farm Bureau Federation sought a new Federation headquarters that would be symbolic of a new era in Farm Bureau programming. They felt the building, in itself, should reflect the organization's commitment to energy conservation; the energy conserving features had to be cost effective, yet not at the expense of a comfortable office environment. Consequently, economy and energy conservation were prime considerations in the mechanical, electrical, and plumbing design.

The engineers met this challenge through a combination of simple but effective use of efficient and innovative mechanical systems design. This resulted in a three time Energy Conservation Award Winning Building which is air conditioned without the use of a chiller, even when the outside temperautre exceeds 100 degrees F.

PROJECT SUMMARY

The construction cost of this 135,000 sq ft two-story office building was $11,000,000, of which $371,000 was for energy conservation systems for heating, air conditioning, and solar domestic water heating. The building was opened in December 1979 and includes a kitchen, cafeteria, computer room, gymnasium, plus agarage under the first floor. The air conditioning system capacity is 304 TR cooling and 2,160 MBH heating.

Sacramento weather analysis shows cool nights that have the excellent potential of providing cool water by evaporation. The cooling towers, which would have been turned off at night, were converted, during the design, to evaporative coolers and harnessed to perform "free" night cooling. Four 24,000 gal tanks were provided to complete a system that charges its "batteries" at night, storing chilled water for the following day. The natural stratification of warm water above cold water is ineffective under 70°F. Thus, separation of the return water, in chilled water storage, is mandatory to maintain the temperature "energy potential". Utilizing a three full/one empty tank system, with water returned to the empty tank, the desired separation is provided.

The uniqueness of the Heat Recovery System, which recovers heat rejected from the condenser of the computer room air conditioning unit, is its ability to utilize low temperature condenser water for heating. A value analysis resulted in using a conventional technique, the "Four Pipe System", for energy conservation. Using the six row cooling coil for heating provides sufficient heating with only 95°F water. By using low temperature condenser water, the efficiency of the heat recovery is improved. The heat is recovered continuously and is stored in two 24,000 gal tanks for use during morning warm up and during heating seasons.

DESCRIPTION OF THE ENERGY CONSERVATION SYSTEM DESIGN AND OPERATION

A. Night Cooling: Evaluation of the weather data for Sacramento, California, shows that during the summer the daily temperature range is very high. (ASHRAE 2-1/2% design conditions are 97°F DB, 70°F WB, and 33°F daily range.) The following is an example of one summer day and one spring day temperature (°F) profile taken from the Sacramento weather tape:

Hour	2	4	6	8	10	N	14	16	18	20	22	M
Temp. °F			AUGUST									
D.B.	60	58	56	62	75	83	88	92	90	79	70	65
W.B.	56	55	53	57	63	65	66	65	69	63	60	58
Temp °F			APRIL									
D.B.	48	50	50	52	62	71	76	79	79	70	59	53
W.B.	45	43	41	43	49	54	56	57	57	55	52	48

The temperature during the day can be in the 90's, while the night temperature may drop into the 50's. This phenomenon is the basis for the "free" cooling energy conservation system designed for the project. Water is cooled by an evaporative cooler during the night and stored in underground storage tanks to be used for air conditioning the building during the following day; this reduces the energy consumption by minimizing the use of the chiller. In order to maximize the use of the evaporative coolers and the potential cooling energy stored in the tanks, the cooling coils were selected to operate at 55°F entering chilled water temperature, and over 15°F water temperature rise. This is higher than a conventional selection of approximately 45°F entering chilled water temperature and 10°F rise. Summer indoor design conditions, by the State of California Code, are 78°F DB. Dehumidification is normally not required for the Sacramento Valley area.

The following table shows the energy input requirements to generate chilled water by the chillers and associated equipment as compared to the evaporative coolers and its associated equipment. Both cycles are operating with 480 GPM cooled from 70°F to 55°F; output capacity of 3,600 MBH.

CHILLED WATER GENERATION ENERGY INPUT REQUIREMENTS COMPARISON

EQUIPMENT DESCRIPTION	MOTORS KW	INPUT FOR CHILLED WATER GENERATION BY: CHILLERS	EVAPORATIVE COOLERS
CHILLERS	210	210KW	NONE
COND. W. PUMPS	15	15KW	NONE
CH. W. PUMPS	7	7KW	7KW
EVAP. COOLERS	41	41KW	41KW
TOTAL ENERGY INPUT		273 KW (or 928 MBH)	48 KW (or 163 MBH)
C.O.P. = $\frac{\text{output}}{\text{input}}$		$\frac{3600}{928} = 3.9$	$\frac{3600}{163} = 22$

The chillers, when operating during the day, require 273 KW input at a coeficient of performance (C.O.P.) of 3.9. The evaporative coolers cooling the same water at night require only 48 KW at a C.O.P. of 22.

This information points to the conclusion that any time the evaporative coolers can cool water 3 ^{O}F or more $(70^{O} - 55^{O})$F x 3.9/22 = 2.7^{O}F , the evaporative coolers' C.O.P. is superior to the C.O.P. of the chiller system. Therefore the night cooling system should operate only when a minimum of 3°F cooling can be expected. If one assumes a 5°F Wet Bulb approach of the leaving water temperature from the evaporative coolers, the operating conditions of the night cooling cycle will be when the outside Wet Bulb temperature is (5°F approach + 3°F miniumum cooling =) 8°F below the water temperature entering the evaporative coolers.

The night cooling cycles operate in the following sequence and under the following conditions:

1. Night pre-cooling of chilled water storage tanks by the evaporative coolers: Cycle starts at midnight; cycle ends at 3 AM and operates only when outside air wet bulb temperature is 8°F below the water temperature leaving the tanks.

2. Night pre-cooling of chilled water storage tanks by the chillers: Cycle starts at 3 AM; cycle ends at 6 AM and operates only when chilled water temperature in the storage tanks is above 56°F. Note that the performance of the chillers is better at night due to the low Wet Bulb temperature as compared to daytime chiller performance. The result is that each morning the system has charged "batteries," i.e., three full tanks with 55°F water or 72,000 gallons.

3. Night pre-cooling of occupied space by outside air: Cycle starts at 5 AM; cycle ends at 6 AM and operates only when space temperature is above 78°F and outside air temperature is 8°F or more below the space temperature.

B. Air Conditioning: The system was designed for energy conservation but it has the capability to perform demand control as well. The design intent was to evaluate the performance of the system without using the storage of chilled water for electric demand control for the first year, and then to evaluate (presently done by the owner), if demand control has a higher potential of cost savings than energy consumption savings. Therefore, the air conditioning cycles operate in the following sequences during the occupation of the building and under the following conditions:

1. Cooling by the outside air economizer cycle: only when outside air dry bulb temperature is below 63°F.

2. Cooling by chilled water generated by the evaporative coolers directly: only when any of the chilled water valves is partially open, b) only when the outside dry bulb temperature is above 63°F, and c) only when the outside wet bulb temperature is below 52°F.

3. Cooling by the chilled water stored in the storage tanks during the previous night: a) only when any of the chilled water valves is partially open, b) only when the outside dry bulb temperature is above 63°F, c) only when the outside wet bulb temperature is above 52°F, and d) only when the chilled water temperature in the storage tank is below 56°F.

4. Cooling by the chilled water generated by the chillers: a) only when any of the chilled water valves is open, b) only when the outside dry bulb temperature is above 63°F, c) only when the outside wet bulb temperature is above 52°F, and d) only when the chilled water storage tanks are exhausted.

C. Heat Recovery for Space Heating: The building houses an electronic data processing facility, which requires air conditioning 24 hours a day. The "heat of rejection" from the 45 ton refrigeration computer room air conditioning system condenser is recovered and stored in tanks to heat the building. In Sacramento, due to the cool nights, heating is very often needed for warm-up during the morning hours, even during the summer. This heat recovery reduces substantially the natural gas consumption by minimizing the use of the boiler.

In order to maximize the use of the potential heating energy of the condensing water at the lowest possible temperature, a four-pipe system was selected for the cooling/heating coils. The advantages are that the same coil is used for heating and cooling, and since the coil is sized for cooling, the six row coil is "oversized" for heating and will be capable of providing adequate heating capacity when supplied with the hot water temperatures of 95°F available from the computer room air conditioning condenser. Furthermore, the initial cost and air pressure drop of a separate heating coil is completely eliminated. The heat recovery and space heating cycles operate during the following conditions:

1. Heat recovery from the computer room air conditioning system: only when any of the computer room A/C units are on. Hot water from the condenser is pumped to hot water storage tanks.

2. Rejection of excess heat from the computer room air conditioning system: only when water temperature (leaving the hot water storage tanks) to the condenser is above 90°F. Performed by secondary pumping to a dedicated evaporative cooler.

3. Heating by hot water from the hot water storage tanks: a) only when the space temperature is below 70°F and b) only when any of the hot water valves is partially open.

4. Heating by hot water from the boiler: a) only when any of the hot water valves is partially open, and b) only when hot water supply temperature, from storage, drops below the set point 95°F.

D. Chilled Water Storage Tank Operation:

The chilled water storage tanks are charged by the evaporative cooler and chiller during the night-cooling cycle and utilized by the cooling coils during the day air conditioning cycle as described above.

In order to maintain the "energy potential" stored in the tanks, the mixing of supply water and return water in the tanks must be prevented. This is achieved by the chilled water storage system design consisting of four tanks, three of which are full, with the fourth empty. Pumping is always done from one of the three full tanks to the empty tank. This precludes mixing by fully separating the supply and return water.

Each tank has an inlet control valve, outlet control valve, level indicator, low level liquid controller, high level liquid controller, and a temperature sensor to measure the tank water temperature. The pressurized tanks are also used as expansion tanks for both the hot and chilled water systems.

A tank is selected during the night-cooling cycle as follows: If two tanks are partially full, the tank with the warmest water of the two is selected first and the water is pumped through the evaporative cooler, cooled and discharged to the other partially full tank. The low level controller will transfer pumping operation to the next tank. The next tank will be selected on the basis of being the warmest of the remaining tanks. The water from the warmest tank is pumped through the evaporative cooler, cooled and discharged to the empty tank.

In the event that the evaporative coolers cool the water to a temperature above 55°F, the chillers will cool the water in the tanks to 55°F before the building opens. When the chillers are used to supplement cooling of the tanks, the warmest tank will be selected first, second warmest tank next, etc.

During the daytime air conditioning operation, when the tanks are connected to the cooling coils, water is first pumped from the coldest tank, next from the second coldest, etc. It is mandatory that when using storage tanks to heat and cool a building, the heating and cooling coils be equipped with 2 way valve control. This results in maximum use of the stored energy as well as conservation of pumping energy.

E. The Evaporative Cooler Operation: The evaporative cooler consists of 2 sections. In order to achieve a better approach to the Wet Bulb temperature and lower the temperature of the water leaving the evaporative cooler, during the chilled water night-cooling cycle the control valves of evaporative cooler direct the flow (480 gpm) to the two sections of the evaporative cooler in series.

During the operation of the chillers, when the evaporative coolers are cooling condenser water, the control valves at the evaporative cooler direct the flow to the two sections in parallel (400 gpm each).

F. Fan Energy Conservation and Economizer Cycle: The air conditioning system includes an economizer cycle which provides air conditioning the building with outside air when the weather conditions permit. The fan energy conservation feature consists of an air bypass around the coils. When the coils are not in use during the time the economizer cycle operates, the bypass damper will open. In combination with variable air volume control, the bypass damper reduces the electrical energy consumed by the fans by reducing the coil static pressure. The 60°F leaving air temperature of the cooling coils permits the use of the economizer cycle and bypass for longer periods than in a conventional system. A conventional system will use 52°F leaving air temperature. The following is the air by-pass cycle operation:

> Coil bypass damper opens to reduce pressure drop during the following conditions: only when the chilled water control valve and hot water control valves are both in the closed position, - i.e. coil unused.

G. The Perimeter System:

The air distribution system consists of a dual conduit system; variable volume system for the interior, and constant volume variable temperature for the perimeter. The perimeter system was sized to only handle the building skin transmission losses or gains. It is important to remember that the transmission load on the HVAC system is the only load which can change phase; i.e., positive or negative. All other loads on the HVAC system are positive (gain loads) which require cooling. There are many hours in the year during which the outside weather is mild, resulting in negligible transmission gain or loss. Thus, the perimeter system can be turned off when the outside air is between 60 and 85 F, resulting in fan power energy conservation.

H. Garage Ventilation:

The building includes a partially open garage under the first floor. The design intent was to permit natural ventilation to ventilate the garage. Therefore, the mechanical ventilation system is controlled by use pattern and carbon monoxide sensors. This design allows also for the cool night air to cool the first floor slab to provide additional cooling "storage."

I. Solar Energy:

An architecturally integrated solar energy collection system is provided for the domestic hot water heating system which serves the kitchen, showers, and lavatories. The system uses natural gas only when solar energy is not available. It was noted by the building maintenance engineer that the gas consumption was more than double during a month in which fog covered Sacramento for two weeks.

J. Economic Analysis:

The incentive for conserving energy is the savings of operating cost. Thus, energy conservation features should be economically feasible. One way to evaluate the feasibility of energy conservation features is to determine their payback period. The payback period is simply the time in which the savings from a proposed investment will equal the cost of the investment. The investment in this case is the additional initial cost of the Energy Conservation System, and the savings are the cost of the net energy conserved. It is important to remember that the annual energy cost savings are not a uniform recurring cost due to anticipated fuel escalation, and they do not occur at the same time as the initial investment. Therefore, in order to relate the savings to the investment, the present value of the savings must be calculated and the payback period will be the number of years that will make the total present value of all the annual energy cost savings equal to the initial additional investment. Federal tax benefits should also be included in the payback period calculation, where applicable.

Syska & Hennessy Information Systems, a division of Syska & Hennessy utilized the AXCESS Computer program to simulate a full year of weather, the building, and the system performance. This analysis played an important role in evaluating the energy conservation design options considered for the project.

The annual savings resulting from the energy conservation system is a comparison of the estimated energy consumed by the systems included in this project with conventional systems of the same capacity. The table below shows the result of the cmputer simulation.

COMPARISON OF THE ANNUAL ENERGY SAVINGS

DESCRIPTION	FUEL	CONVENTIONAL SYSTEM ENERGY CONSUMPTION	CFBF SYSTEM ENERGY COMSUMPTION	SAVINGS & PERCENT SAVED
SOLAR/DOMESTIC HOT WATER HEATING	Gas	5,330 TH.	1,930 TH.	3,400 TH.
	Elect	---	700 KWH	- 700 KWH
				63.4%
HEATING AND COOLING SYSTEM	Gas	14,830 TH.	600 TH.	14,230 TH.
	Elect	345,700 KWH	187,000 KWH	158,700 KWH
				74%

The figues shown should be used for comparison between the systems only.

NOTE: TH = THERMS = 100,000 BTU; KWH = 3,400 BTU

The first year energy cost saving was estimated to be about $7,000. Note that Sacramento electrical rates are very low--about one cent per KWH. The actual energy savings are higher than that estimate; the building is utilized by two shifts and is frequently occupied on weekends.

The solar system was estimated to save about 63% of the energy consumed by a conventional domestic water heating system resulting in an approximate 14 year payback period. The heating and cooling systems designed was estimated to conserve about 74% of the energy consumed by a conventional system resulting in an approximate 11 year payback period. Although the energy conservation results are impressive, the low cost of electric energy in Sacramento results in moderate payback periods. Comparison of the first year energy bills to referenced energy budget targets established in the construction industry shows the following: (Note that the first year energy bills had to be adjusted to reflect the operation schedule of the building and to include systems which are included in the bases for comparison.)

ACTUAL PERFORMANCE COMPARED TO ESTABLISHED ENERGY BUDGET STANDARDS

ENERGY CONSUMED:	GSA	STATE OF CAL T-24	CFBF
ON SITE BTU/SF/YR	55,000	N.A.	40,000
AT THE SOURCE BTU/SF/YR	N.A.	134,000	100,000

MEETING THE CLIENT'S NEEDS

The building and systems design mesh nicely with the intent of the California Farm Bureau Federation to occupy a building that is energy efficient and cost effective Optimizing the use of major pieces of equipment met the Owner's cost conscious criteria. Whenever possible, equipment served a dual purpose, or had multi applications. To make this effective, the essence of the function of a piece of equipment had to be clearly understood, and applied in heretofore unused manner. Economy and the "use of the available" guided the selections resulting in a building which consumes 25% less energy as compared to current energy budget standards.

TECHNIQUES FOR DEVELOPING BUILDING ENERGY CONSERVATION STRATEGIES

by

Douglas S. Stenhouse, AIA, AICP, Assoc. ASHRAE
President, ENERGY MANAGEMENT CONSULTANTS INC.
Los Angeles California

ABSTRACT

Increases in the cost of energy relative to other expenses is a matter of great concern to building owners. Budgeting of future energy costs as well as the cost of any proposed energy conservation items is, therefore, an important planning task. Various innovative techniques are needed to develop building energy conservation strategies.

The author discusses a number of these techniques which include developing building energy use budgets for construction, plotting of past energy use and costs for existing building(s) over time, performing sensitivity analyses for future energy costs, identifying and comparing various energy use and cost objectives, identifying and comparing alternative business investments, comparing energy use for your building(s) with other buildings and energy conservation performance standards, rank ordering alternative energy conservation opportunities, evaluating the importance of energy conservation as a public relations tool, and comparing building energy use through the entire project delivery process. Examples of projects which have involved some of these techniques are presented and discussed briefly.

The paper promotes comprehensive energy planning and its value as a corporate planning tool. The discussion points to the need for a multidisciplinary approach and coordination of various specialized consultant services.

BUILDING ENERGY USE PROGRAMMING

When an owner plans to build, there are a number of cost and investment considerations. Historically, land and construction costs were the major items. But with the increasing cost of all forms of energy, there is more concern about this latter item as an isolated issue or compared to other operating costs or as a percent of gross sales.

Budgeting future energy costs for new or existing construction is becoming an important planning task each year, and moreso over time as it becomes more difficult to turn over a profit. During hard times, reduction of energy costs is of prime concern to most businesses. The difficult economic conditions we face today provide an excellent incentive for saving energy.

Building owners, as they become more aware of the proportionately large increase in the cost of energy compared to other operating costs, and as they think about how to reduce energy costs, will certainly consider various energy conservation options to improve plant performance and therefore profitability of their business operation. This goes for new as well as existing construction. For example, it is indeed possible that as building owners are able to identify the benefits, they may wish to go well beyond what they are required to do by today's building energy conservation consensus standards.

Generally, the basis for decision-making is and should be based on expected return on investment. Identified capital requirements for energy conservation must compete favorably with other investment alternatives if they are to be implemented.

Before embarking on any building energy conservation program, building owners must first take a hard look at energy costs relative to other operating expenses over time. Once they have done this, they can decide how much they wish to reduce this cost over some specified time frame and how much they can invest to accomplish these objectives. It is then possible to begin identifying various energy conservation strategies, to rank order them, and finally develop a package of proposals that is projected to produce target dollar savings and return on investment.

TECHNIQUES TO DEVELOP BUILDING ENERGY CONSERVATION STRATEGIES

Both in the design of new buildings and the redesign of existing buildings, a number of useful techniques have been developed to assist building owners and professional designers in identifying the most appropriate strategies for energy conservation. Though this emerging specialization is relatively new, some very promising and innovative tools have been developed. The purpose of this discussion is primarily to introduce them and discuss them briefly. As building owners or operators, utility concerns, regulators, and professional designers, you can decide which ones may be most suitable. Certainly many of them are not unique, but their use in comprehensive energy conservation planning context may be:

- Developing building energy use budgets for new construction
- Plotting past energy use and costs for existing building(s) over time
- Performing sensitivity analyses for future energy costs
- Identifying alternative energy sources
- Identifying and comparing various energy use and cost objectives
- Identifying and comparing alternative business investments
- Comparing energy use for your building(s) with other buildings and energy conservation performance standards
- Rank ordering alternative energy conservation opportunities
- Evaluating the importance of energy conservation as a public relations tool
- Comparing building energy use through the entire project delivery process

DEVELOPING BUILDING ENERGY USE BUDGETS

The energy performance of a building is important to most building owners and will become increasingly moreso as the cost of <u>all</u> forms of energy increase. The best way to tell how much energy a new building will use is to simulate its performance. The problem is that all the techniques we might wish to use for this kind of analysis (at various stages in the project delivery process) are not readily available, some have only recently been put into use, and others have yet to be developed.

One such manual technique concentrates on calculations of four basic components of thermal performance or the building load (internal heat gain, solar heat gain, building envelope heat gain/loss, ventilation heat gain/loss) (1). For industrial buildings, process loads must also be factored in. And for all buildings, miscellaneous equipment must be considered. This technique allows one to develop annual energy budgets at very early stages in the building programming effort, even before a specific design has been developed. One must have discrete information about all space uses, the proposed site, and climate. Heating and cooling plants must be selected along with distribution systems and energy source types.

Other techniques such as the "modified bin method" developed by Sizemore (2) and the SASEAP method by SUD Associates (3) can be used to determine a building's annual energy use once the particular building configuration has been identified. The major load programs (DOE-2, TRACE, BLAST, E-CUBE, for example) are suitable for developing energy conservation targets for larger projects and later in the design process. These more elaborate energy analysis programs, however, require time, care, and experience in the formatting of input data and the selection of design assumptions, as well as identifying and compiling relevant totals from detailed output.

Generally much simpler computer programs are developed for in-house use, such as the one described in Reference 4. It has been developed to account for climate and mass of structure in the development of annual energy use projections if the building can be simplified into two or three space heating/cooling zones. Much further work is required in this area to develop simple, non-proprietary energy budget calculation techniques.

These programs are useful in establishing an energy target or goal early in the building planning process.

PLOTTING ENERGY USE OVER TIME

For existing buildings, annual energy use can be obtained from utility bills. These can be plotted over a two year period, for example, to illustrate what the pattern of energy use and cost has been over time.

Computer programs have been developed which allow for graphic presentation and practical communication of data(5). One problem associated with this effort is the difficulty in transposing data for uniform intervals of time. There is often considerable work and delay involved in assembling data. Examples of plots for energy use and cost are illustrated in Exhibits 1 and 2.

PERFORMING SENSITIVITY ANALYSES

Understanding the economic value of building energy conservation investments is crucial in the decision making process. No one really knows what the future cost of energy will be. Government agencies and even private business will often dictate the cost of future energy for the purpose of energy analyses. However, these assumptions tend to be conservative and they often rule out viable options. Failure to consider realistic discounting and other economic factors in life cycle costing will play havoc with the results of such analyses. The use of simple payback (first cost divided by first year savings) may be justified for items that obviously will pay for themselves in the first year or two is justified. But for longer paybacks, life

EXHIBIT 1

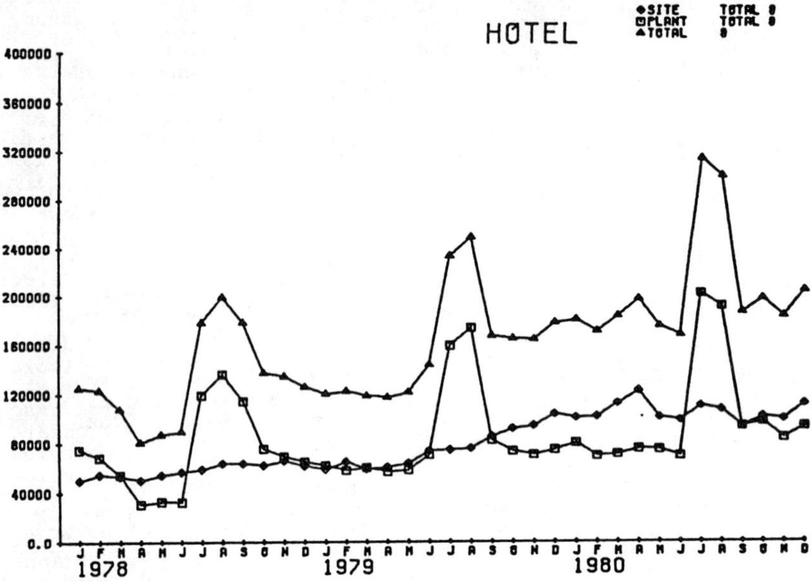

EXHIBIT 2

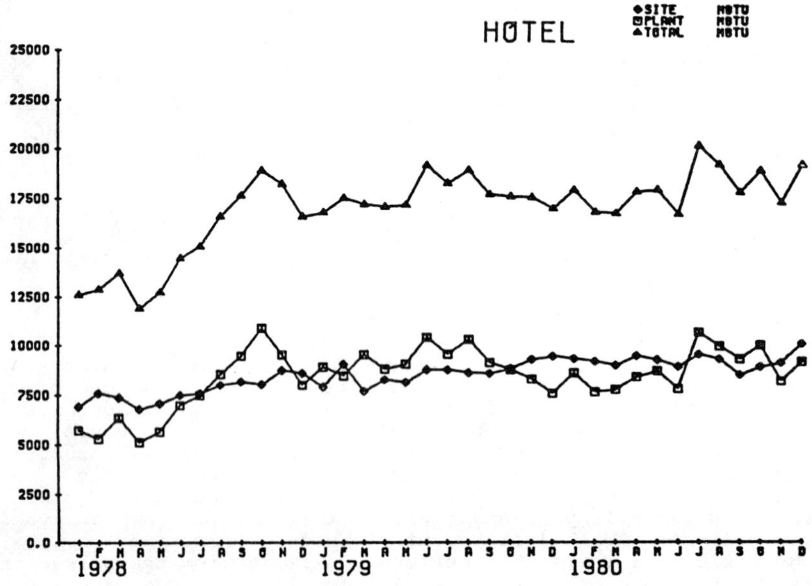

cycle costing techniques should be employed.

We have found that where no future energy costs have been agreed upon or where given assumptions are absurd, a sensitivity analysis illustrating the relative payback is important to illustrate. An example of how this can be done is shown in Exhibit 3. The technique is useful to graphically show the relative effect of various assumptions concerning future cost of energy, the opportunity value of money, etc. Building owners are able through such comparisons to develop a frame of reference for decision-making.

IDENTIFYING ALTERNATIVE ENERGY SOURCES

Conventional energy sources of natural gas, fuel oil, coal, and electricity are generally the only energy sources we consider to heat, cool and light our buildings. But increasingly over time, renewable energy sources or onsite power generation will become viable options.

Cogeneration is economically worthwhile where the cost of electricity is high and where new hook-ups for large customers are limited and where there is a sizable year-round heating load. Active solar energy is already practical for low temperature process water heating in many parts of the US. Passive solar energy is particularly suitable for residential space heating applications. Daylighting offers substantial opportunities for reduction of lighting and associated heat-of-light space cooling for office buildings in temperate climates, particularly where electricity costs are high. Otherwise wasted heat from industrial operations should be considered for lower-tiered heating applications.

An understanding of the relationships between energy use for different functions and building types, for the industry and transportation sectors as well, provides the basis for expanding the scale of our interests in energy conservation. In so doing, we are able to have a more comprehensive understanding of the variety of energy sources that are actually available today. This, of course, is the basis for promoting energy planning not only in a neighborhood or community context, but the larger urban, regional and even national, international context as well.

COMPARING ENERGY COST OBJECTIVES

We have already discussed the value of plotting historical trends of energy use and cost. And we have discussed briefly the need for establishing objectives with regard to the annual cost for energy. Plotting the disaggregated cost of different energy sources which together represent the total cost of energy for an existing or planned structure help us get even a better handle on where to attack the problem. Comparing various objectives with respect to total energy costs constitutes a sensitivity analysis with respect to goals and may identify alternative strategies for the most cost-effective energy conservation plan (See Exhibit 4).

COMPARING ALTERNATIVE INVESTMENTS

Generally a business has certain planned objectives in mind which deal with the commodities or services they promote. Financial resources are used to capitalize plant expansions, additional staffing, acquisition of new equipment, land, material, etc. Return on investment becomes the basis for financing future operating needs.

Because the cost of energy has traditionally been so small a portion of the annual operating budget and/or because the cost for same had escalated less than other line budget items, it has been often overlooked. But times have changed. Now it pays to evaluate

EXHIBIT 3.

Economic Return Probabilities

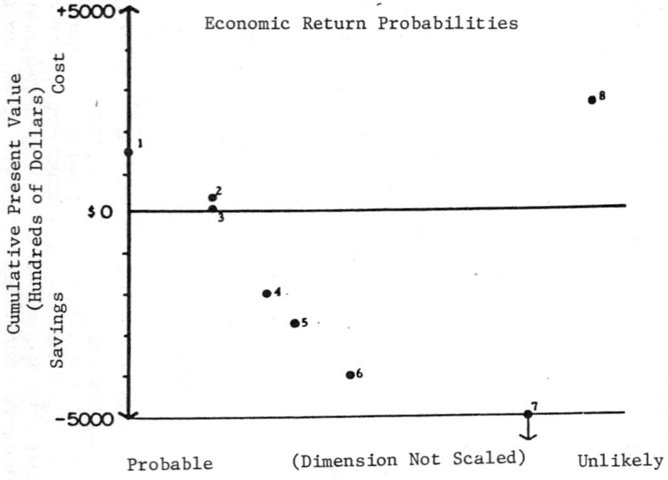

1. Contract given values
2. Discount rate reduced to 5% (average of NBS corporate and private returns)
3. NAVFAC fuel cost inflation, 5% discount rate
4. 0% discount rate
5. Revised fuel cost inflation (NAVFAC), 0% discount rate
6. "High fuel cost inflation, 0% discount rate
7. Reduced system cost, high fuel cost inflation, increased performance, 0% discount rate
8. Increased system cost, low fuel cost inflation, reduced performance, 10% discount rate

EXHIBIT 4

Projected Energy Costs

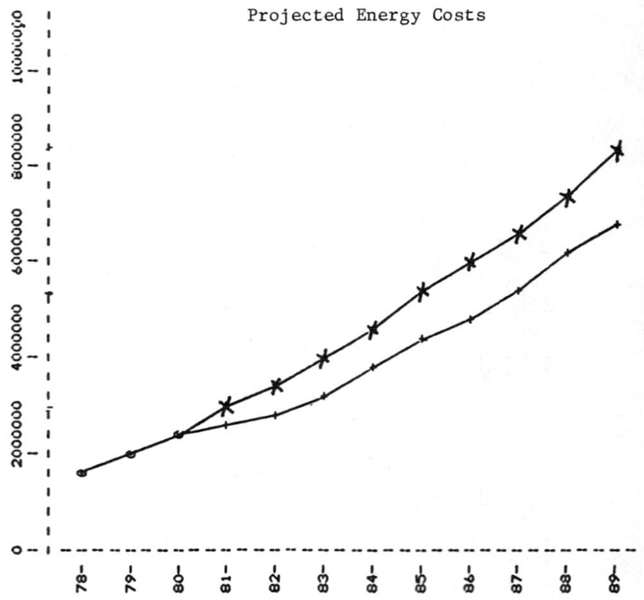

X ANNUAL $ WITHOUT CONSERVATION

+ ANNUAL $ WITH CONSERVATION

these more traditional investments against the opportunities that may accrue from investments in energy conservation. Methods should be developed in the budget planning process to equitably compare the profitability of energy conservation investments both in new and existing construction.

COMPARING BUILDING ENERGY USE

When energy costs are rising, the obvious thing that each building owner wants to know is how well he is doing compared with his or her competition. Various business associations have developed comparative statistics for building energy use. There has been work done to compare the energy efficiency of various manufacturing processes. And, if you are familiar with building energy conservation standards, you know that certain States have developed energy budgets as guidelines or alternatives to prescriptive codes. National standards (BEPS) have been developed for all building types for most cities throughout the US. These are available for your comparison. And regardless when they will go into effect, if at all, they are useful in the planning process to establish norms and projections for future energy use. For one of our corporate clients, from a comparison of the California State energy budgets and the proposed federal BEPS standards, we developed somewhat reduced standards which they plan to put into effect for all new and major rehab construction projects. These standards and proper documentation procedures will become part of other programmatic requirements. The purpose is to not only reduce future energy costs for future construction by 15-20% over what they would otherwise have been, but to stretch out the design capacity/need for expansion of their existing central plant. A current on-going energy conservation program for existing buildings is expected to produce surplus heating and cooling for new construction.

RANK-ORDERING ECOs

In performing building energy audits of existing buildings and for even studying alternative design options for new buildings, techniques have been developed for comparing energy conservation opportunities (ECOs).

The basic procedure is to 1) define the ECO, 2) estimate the initial cost, and 3) calculate the payback. Where the payback is obviously 1 or 2 years, a simple division of cost by savings may be suitable. Where the payback is extended, life cycle costing is required to account for discounting, inflation of the cost of energy, maintenance, replacement, etc. However, once this has been done for each ECO, comparisons can be made to select the best combination of items to achieve a given result.

There are certain precautions that must be taken to avoid duplication of energy savings. Since the process involves considerable number-crunching, computer programs have been developed to assist in rank-ordering individual ECOs (6). This process is an excellent way to sort out options and develop a long-range plan for retrofitting construction. It also obviously can be extremely useful in selecting design options for new construction. An example of how the results look for a recent energy audit we performed is shown in Exhibit 5.

IMPROVING PUBLIC IMAGE

One might say that improving the public image of a business or other organization cannot be claimed to be a technique for conserving energy. However, there are many examples where this is indeed the case and where it may become the final convincing argument for doing something. Whether or not this technique applies will probably depend on an

EXHIBIT 5

ENERGY CONSERVATION/LOAD MANAGEMENT RECOMENDATIONS

	RECOMENDATION	CORR*	ANNUAL KW SAVING	ANNUAL KWH SAVING	ANNUAL ($) COST SAVINGS KW	ANNUAL ($) COST SAVINGS KWH	TOTAL $ SAVINGS	INVESTMENT $	PAYBACK*	NET PRESENT VALUE**
1	SUP-2	1	18	46176	1084	2212	3296	0	0	
2	CS-17	1	88	540360	0	25115	25115	700	0	
3	LOT-1	1	22	57220	1325	2741	4066	7611	1.9	
4	LOT-3	1	154	191569	9276	9192	13469	45000	2.4	
5	CS-2	1	90	0	5454	0	5454	15400	2.8	
6	AD-3	1	0	120000	0	5749	5749	20000	3.5	
7	ENV-6	1	9	19005	0	910	910	5076	5.6	

THE FOLLOWING ITEMS EXCEED PAYBACK LIMITS
AND ARE THEREFORE NOT RECOMENDED:

	RECOMENDATION	CORR*	ANNUAL KW SAVING	ANNUAL KWH SAVING	ANNUAL ($) COST SAVINGS KW	ANNUAL ($) COST SAVINGS KWH	TOTAL $ SAVINGS	INVESTMENT $	PAYBACK*	NET PRESENT VALUE**
8	ENV-7	1	1	1846	0	88	88	1890	NOT COST-EFFECTIVE	
9	ENV-2	1	19	33947	0	1865	1865	55655	NOT COST-EFFECTIVE	

* POST CONSERVATION CORRECTION FACTOR (SAVINGS ESTIMATE BASED ON TOTAL ELECTRIC USE

** LIFE CYCLE COSTING IS ONLY REQUIRED ON SPECIFIC LOCATION & ITEMS

IN ADDITION TO THE AREAS WHERE ELECTRICAL ENERGY USE COULD BE REDUCED, THERE ARE PLACES WHERE NATURAL GAS CONSUMPTION COULD BE DECREASED. ALTHOUGH REDUCING NATURAL GAS CONSUMPTION WAS NOT THE MAIN OBJECTIVE OF THIS SURVEY, THOSE ITEMS WHICH AFFECT NATURAL GAS CONSUMPTION HAVE BEEN NOTED, AND CALCULATIONS PERFORMED ON CERTAIN OF THOSE ITEMS.

SOME OF THE RECOMMENDED ECO'S SHOW REDUCTIONS IN GAS AS WELL AS ELECTRICAL CONSUMPTION. FOR EXAMPLE, BUILDING ENVELOPE IMPROVEMENTS WOULD TEND TO REDUCE COOLING AND HEATING LOADS, ALTHOUGH SAVINGS WERE CALCULATED FOR REDUCTIONS IN ELECTRICAL (COOLING LOAD) CONSUMPTION ONLY. THIS UNDERESTIMATES THE ACTUAL SAVINGS ACHIEVED FROM THE BUILDING OWNER'S INVESTMENT.

THE ECO'S INCLUDED IN THIS REPORT ARE INTENDED TO IDENTIFY AREAS FOR FUTURE CONSIDERATION AS THEY SHOW POTENTIAL FOR REDUCTION IN GAS CONSUMPTION.

* See Appendix A-7 for Life Cycle Cost Analysis

independent, objective evaluation of the relevance of improved public image to business success. The subject, however, merits consideration.

We have been employed to accomplish a specific reduction in energy use, which done for no other reason, ostensibly, than to demonstrate to the public that the proposed facility had some redeeming value. This was then successfully used to exact specific land development tradeoffs that were to the distinct financial benefit of the building owner.

ENERGY CONSERVATION THROUGH THE ENTIRE DESIGN PROCESS

Understanding how to integrate energy conservation into the total design process offers the potential for not only creating wholesome buildings but a new sense of beauty and order around us.

The steps involved in integrating energy into this process are illustrated in Exhibit 6. A detailed discussion of the techniques and procedures to design, build, and operate energy conserving structures is covered in a brand new AIA publication which we authored(7).

CONCLUSION

The traditional process for designing buildings is no longer a valid one. Preconceptions of how buildings should look, how spaces should be planned, and what systems should be used must change. Professional designers, building owners, and investors must be more innovative, attempting to discover new roles, new patterns for involvement, new strategies if they expect to continue to make a profit, render a service, and thus stay in business.

It is our thesis that comprehensive energy planning has great value as a corporate planning tool. It signals a need for doing things differently than we have in the past. It calls for teamwork and a multidisciplinary approach to problem solving. It offers the prospect, however, for better buildings and, we feel, increased profitability for those that take the time, trouble, and expense to do things right.

Exhibit 6

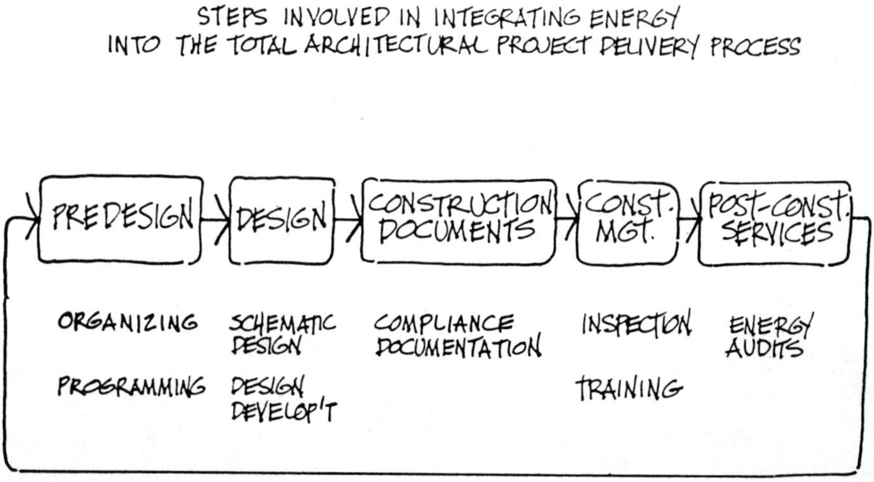

NOTES AND REFERENCES

1. Hart, Kurtz & Whiddon; Energy Graphics: A New Approach to Energy-Conscious Design; Booz.Allen & Hamilton, Inc.; Washington, D.C.; August 1979.

2. Sizemore et al; Energy Planning for Buildings; The American Institute of Architects; Washington, D.C.; 1979.

3. Booz•Allen & Hamilton, Inc.; Assessing The Energy Conservation Benefits of Historic Preservation: Methods and Examples; Prepared for The Advisory Council on Historic Preservation; Washington, D.C.; January 1979.

4. The "PASS-ONE" Program written by Energy Management Consultants, Inc., 672 S. LaFayette Park Place #38, Los Angeles, CA 90057. (A listing of the program is available in fortran for the IMSAI 8080 microcomputer).

5. The "MIPLOT" Program written by Energy Management Consultants, Inc. has been developed to interface with a plotter. A listing of the program is available in fortran for the Watanabe plotter).

6. The "ECON" Program written by EMC.

7. Energy Management Consultants, Inc.; Energy Analysis Techniques; American Institute of Architects; Washington, D.C.; 1981.

LEASING ENERGY CONSERVATION AND SOLAR ENERGY EQUIPMENT

Martin Klepper, Esq.
Lane and Edson, P.C., Washington, D.C.

Leasing offers a very attractive means of providing 100% financing for energy conservation and solar energy equipment installed in buildings. Property owners (and long term lessees of buildings) are understandably reluctant to make a significant capital investment in equipment that may be new or untested, such as a solar hot water heater or a computer controlled energy management system. They are also unlikely to be willing to wait five years or longer to recover the full value of their energy conservation investment. Often these property owners lack the necessary capital to acquire the equipment, even if they are inclined to do so. The energy tax credit, even if it is available for the equipment under consideration, is unlikely by itself to provide an adequate incentive to justify the acquisition of the energy saving device. In these circumstances equipment manufacturers, installers and property owners should consider the significant advantages of leasing energy equipment instead of purchasing the equipment outright.

ADVANTAGES TO THE LESSEE

There are many advantages to leasing energy equipment. Leasing provides a built-in source of financing and often permits the lessee to acquire equipment he cannot finance elsewhere. For example, assume that an owner has just conducted an energy audit of his office building. The audit suggests that he could reduce utility costs by $50,000 a year if he installs a computer-controlled energy management system (the system). The auditor advises him that the system will cost $250,000; in other words, the system will pay for itself in five years, if utility costs and energy use remain constant during that period.(1) The "simple payback" period will be even shorter if utility costs and/or energy use increase during the period.

By leasing the system, the owner avoids the initial cash outlay necessary to purchase the system. Instead of making a capital investment of $250,000, he may be able to obtain the full use of the system under a five-year lease with monthly payments of approximately $4,000.(2) Total lease payments would be $2,000 a year less than the owner's projected utility cost savings. By leasing instead of purchasing, the owner pays for the system without increasing his normal operating expenses. The owner may, of course, be the beneficiary of even larger energy savings if utility costs increase during the lease term.

Leasing provides the owner with 100 percent financing and makes it possible for him to avoid the necessity of borrowing funds. Avoiding additional bank loans may be particularly advantageous if the mortgage on the building prohibits additional secured financing without the mortgagee's approval. The owner encumbered by such a mortgage would be unlikely even to consider the system except on a lease basis.

The owner's lease payments may be lower than his comparable monthly cost for financing the system with a bank loan. The lessor receives tax benefits from accelerated depreciation of the system and from applicable tax credits. In addition, the lessor retains the right to the residual value of the system after the lease terminates. The lessor should pass some of these benefits on to the lessee through lower lease payments. The lessee can deduct his lease payments as an ordinary business expense.

The owner may also discover nonfinancial advantages from leasing the system. A lease can shift to the lessor part of the risk that the equipment will not provide the projected energy savings. If the system does not provide the expected results, the lessee may have the right to cancel or renegotiate the lease. The lease could contain an adjustment clause that would decrease or increase lease payments according to the energy cost savings that the system produces. This provision would protect the lessee if the system performed below expectations. It would reward the lessor if the system performed better than expected. If the owner had purchased the system and its performance fell below his expectations, his only recourse would be the manufacturer's warranty.

---

(1) This is a hypothetical example. Actual cost and projected savings will vary depending on the building and type of energy equipment. A more detailed life-cycle cost analysis of the proposed investment would evaluate financing costs, maintenance and service costs, and tax and cash flow aspects of the proposed acquisitions, along with calculating the present value of projected future energy savings.

(2) A $4,000 monthly payment would cover debt service on the lessor's assumed $175,000 loan (70 percent of the cost of the equipment) at 14 percent interest plus a small cash return to the lessor on his equity investment of $75,000.

Another advantage is that the lease can be terminated after the initial lease term. Usually the lessee will not have paid the full cost of the system during the initial lease term. Hence, he will have had the benefit of the system for five years while only paying a portion of its cost.

Leasing can be used to shift to the lessor part of the risk of obsolescence. If, after five years, a newer, more efficient, less costly energy system exists, the lessee can terminate the lease and acquire (or lease) the newer system.

Leasing may be most attractive to a net lessee of a building, who generally has little incentive to acquire equipment outright, particularly if the remaining term of the building lease is less than ten years. Property managers may also find leasing energy equipment desirable because they can lower operating costs, thereby increasing the return to the building's owner, without asking the owner to make any additional capital investment.

ADVANTAGES TO LESSOR

Leasing energy equipment also offers significant advantages to the owner/lessor of the equipment. The lessor benefits from the following:

1. The right to use accelerated depreciation to amortize the equipment's cost;

2. A tax deduction for interest paid on any loans that he obtains to purchase the equipment;

3. The 10 percent investment tax credit (and the 10 percent energy tax credit if the equipment qualifies for these credits); and

4. The equipment's residual value. (The equipment's residual value would increase if oil and gas prices continue to increase because its value will be closely tied to the energy savings that it produces.)

Leasing can be used by an equipment manufacturer in two ways. A manufacturer can lease equipment directly to property owners and keep the tax credit for itself. The manufacturer would use its own credit to borrow necessary funds from a bank or other financial institution. Alternatively, a manufacturer can sell equipment to a syndicated limited partnership which then leases the equipment to the manufacturer's customers. The manufacturer could agree to service the equipment in return for a fee from the lessor partnership. The limited partnership would consist of investors who contribute capital to the partnership- capital that is used to purchase the equipment from the manufacturer.

Energy equipment leasing can provide attractive yields to investors seeking tax benefits, particularly if they purchase the equipment with debt financing. Using the earlier example, if the investor purchased the system for 30 percent equity ($75,000) and a 70 percent loan ($175,000), he would be entitled to a 10 percent ($25,000) tax credit that would return one third of his investment. Lease rental of $4,000 a month would pay the debt service on the $175,000 equipment loan and provide him with an annual cash return. This cash return would probably be partially sheltered from taxation during the early years of the lease because of the benefits of accelerated depreciation available to the system and the deduction of interest on the borrowed funds.

TAX ADVANTAGES OF OWNING ENERGY EQUIPMENT FOR LEASE

There are two different federal tax credits that might be available to the owner/lessor of energy equipment. The Energy Tax Policy Act of 1978, as amended by the Crude Oil Windfall Profit Tax Act, provides a tax credit of 10 to 15 percent (depending on the type of equipment) for qualified investment in certain specifically defined categories of energy conservation and solar energy equipment. Such equipment qualifies for the "energy credit" even if it is deemed to be a structural component of a building.

Energy equipment may also qualify for the 10 percent regular investment tax credit if it satisfies the provisions of Section 48 of the Internal Revenue Code. However, property that is deemed to be a structural component of a building does not qualify for the regular 10 percent investment tax credit nor does property installed in buildings with permanent residents, such as apartment buildings.

State business energy tax credits can provide additional incentives for leasing transactions. California, for example, provides a 25 percent tax credit for businesses that purchase solar energy equipment.

A separate analysis must be made in each situation to determine which federal and state tax credits apply to specific equipment that is to be leased. The identity of the lessee is important because the investment tax credits are not available if the equipment is used by tax-exempt organizations and government units. In addition, final energy tax credit regulations adopted by the Internal Revenue Service on January 21, 1981 provide that specially defined energy property, such as a computer controlled energy system, does not qualify for the energy tax credit if it is used to heat or cool an office building, retail store or shopping center. A computer controlled energy management system leased to a food processing plant would probably qualify for a 20 percent tax credit; but a computer installed in an office building only qualifies for the regular 10 percent tax credit.

Investment tax credits are available to individuals and partnerships that are equipment lessors, provided the lease is not a "net lease" as defined in Section 46(e)(3)(B) of the Internal Revenue Code. The term of the lease must be for less than 50 percent of the equipment's useful life, and certain deductions taken during the first twelve months of use of the property must exceed 15 percent of the rental income from the equipment, or the lessor must be the equipment manufacturer. Thus, manufacturers can lease equipment to property owners and retain the benefits of the energy tax credit.

THE LEASE MUST QUALIFY AS A LEASE FOR TAX PURPOSES

To obtain the tax benefits described above, the lease must be structured to qualify as a lease, and not as a conditional sales contract (or mere financing transaction) for federal tax purposes. The Supreme Court recently held that the circumstances of a sale-leaseback qualified as a true lease (3). A detailed analysis of the IRS lease requirements is beyond the scope of this article. (4)

---

(3) See Lyon v. United States, 441 U.S. 910 (1978); Rev. Rul. 55-540, 1955-2 C.B. 39.

(4) For Internal Revenue Service advance ruling purposes, a lease must meet the requirements of (1) Rev. Proc. 75-21, 1975-1 C.B. 715; (2) Rev. Proc. 75-28, 1975-1 C.B. 752; and (3) Rev. Proc. 76-30; 1976-37 I.R.B. 25.

However, the Internal Revenue Service generally requires the lease transaction to satisfy the following conditions, among others:

1. A 20 percent equity investment in the equipment must remain "at risk" throughout the lease;

2. Rent must provide the lessor with a projected positive cash flow;

3. Rent that the lessor charges during any period after the fixed term should be the fair rental value of the equipment;

4. Any right of the lessee to purchase the equipment should require that he pay the fair market value of the equipment; and

5. At the end of the lease term, the equipment must have an estimated value of not less than 20 percent of its cost, and it must be commercially feasible for some party other than the lessee to use the property at the end of the lease term. (5) (This last requirement will be satisfied if the cost of removing the equipment and reconstructing it at a new location does not exceed the value of the equipment.)

If the lease fails to qualify as a "lease" for tax purposes, the lessee will be deemed the owner of the equipment and only a portion of his lease payments will be deductible for tax purposes. More importantly, the lessor would lose the right to the federal tax credits and could not depreciate the equipment.

A recent IRS private letter ruling (6) held that a lessor of a solar hot water system was entitled to the 15 percent solar energy business tax credit where the lease to a residential homeowner was for less than one-third of the useful life of the equipment and there were no options to renew the lease. If the homeowner elected to purchase the system at any time during the lease term, thirty percent of his rental payments would be applied to the purchase price.

LEASING TO A MUNICIPALITY - TAX-EXEMPT LEASING

Equipment leased to a municipality does not qualify for federal tax credits. However, a tax credit may exist for equipment that is used to provide an energy service - such as heat or air conditioning. Alternatively, a lease to a municipality could be structured in a manner that permits the lessor to treat part of the lease payment as tax-exempt interest. Tax-exempt leasing by municipalities is an accepted method of avoiding debt ceilings in many communities.

CONCLUSION

Equipment manufacturers and installers should not overlook the significant marketing advantages that exist by establishing a program that permits leasing of energy saving equipment. Lease financing is an established method of selling equipment on terms that may be more attractive than debt financing. The ability to lease equipment will enable owners to acquire conservation equipment in many situations where they would not otherwise do so.

---

(5) Id.

(6) I.R.S. Letter Ruling 8017082, January, 1980.

INCREASING FUEL EFFICIENCY THROUGH DIRECT FLAME WATER HEATING SYSTEM

Dr. Yen-Hsiung Kiang, P.E.

Trane Thermal Company
Conshohocken, Pa. 19428

The world shortage of energy and continual increase in fuel prices requires a radical and innovative approach for energy saving. Traditional water heating systems have a thermal efficiency around eight percent. Since the efficiency of a heating system is dependent on the temperature of the stack flue gases, the lower the stack temperature, the more efficient the heating systems are. To reduce stack temperature, and thus increasing overall heating system efficiency, the direct flame submerged exhausted combustion water heater can be used. Thermal efficiency of a typical direct flame water heating system is over ninety percent. A comparison of thermal efficiency between traditional and direct flame water heater is illustrated in Table 1. The efficiency of direct flame water heater is three percent higher than traditional heater when economizer is not used, and nine percent higher when economizer is used.

Table 1 Thermal Efficiency Comparison

Heater Type	Without Economizer	With Economizer
Traditional[2]	81%	86%
Direct Flame[3]	84%	95%

Note: 1) Fuel: Natural Gas, 25% Excess Air, and 140°F water temperature.

2) Stack Temperature - 400°F without economizer
200°F with economizer

3) Stack Temperature - 140°F without economizer
100°F with economizer

To be presented in the following sections are the description of the equipment, basic principles, control and safety, and applications.

EQUIPMENT DESCRIPTION

The direct flame water heating system makes use of submerged exhaust combustion principle. The basic equipment is illustrated in Figure 1; it consists of a burner, a downcomer, a weir, and a tank. The burner, complete with its integral combustion chamber, is located at the top of the downcomer tube above the water level in the tank. The hot products of combustion from the burner are conveyed through the downcomer tube into the water. The weir tube, concentric with and outside the downcomer tube, forms an annulus that contains the escaping combustion gases and forces a gas lift to occur, raising the water to the spill over level of the weir. This weir action increases the gas bubble travel distance and improves heat transfer. It promotes better circulation within the tank, thus producing a more uniform temperature and it keeps the upper portion of the downcomer tube cool and free from scaling and corrosion.

The heart of the direct flame water heating system is the burner. In conventional burner, the design is usually based on heat release of 50,000 to 200,000 Btu/hr./cu.ft. Under this condition, a long flame usually results. This type of burner is not applicable to the direct flame water heating system. The quenching effect of the downcomer cools the flame and causes incomplete combustion, which is the cause of soot contamination of the water being heated, and the loss of thermal efficiency. A special designed burner, developed by Trane Thermal Company, is used in the direct flame heater. This burner, Figure 2, differs widely from the conventional burners. The heat release, which occurs entirely within the integral combustion chamber of the burner, is on the order of 10 Million Btu/hr./cu.ft. of combustion space. Thus, the cool downcomer tube has no effect on the combustion efficiency of the burner.

Another feature of this burner is that it is of the nozzle mixed type. Premixed type burner is not recommended for direct flame water heating application because of the existence of back pressure. As the back pressure increases, the flame propagation rate also increases, improving the possibilities for flashback in the premix system design.

Usually, natural gas and propane are the primary fuel for direct flame water heater. Sometimes, distillate type fuels are required as the primary fuel. Proper combustion of a liquid fuel depends not only upon fine atomization of the liquid prior to combustion, but also upon complete and rapid mixing with combustion air. Even this does not insure a short flame, which is essential in direct flame water heating system. In this case, the Thermal HV combination gas-oil burner, Figure 2, can be used. This recirculating design provides for internal vaporization of the fine liquid droplets produced by the burner nozzle, so that the air and fuel mixture approach the combustion zone as a vapor, rather than as air mixed with liquid droplets. On distillate type fuels, both the flame characteristics and completeness of combustion of a Thermal HV combo burner are equivalent to that of the gaseous fuels. This makes it possible for the Thermal HV combo burner to be used in a direct flame water heating application.

The downcomer is a cylindrical tube which is used to direct and exhaust the hot combustion gases below the water level. The downcomer tube must be cooled from outside to avoid overheating. The passage between the downcomer and weir tubes are used for heat transfer.

To maximize the efficiency of the direct flame water heating system, a heat recovery unit, the economizer, can be used (Figure 3). The economizer is usually in the form of a packed tower. This packed tower could be part of the tank and could consist of a wide variety of standard tower packings. The feedwater, under this design, would be brought in at the top of the packed tower section, just beneath the entrainment separator and flows down through the packing into the tank. The saturated products of combustion from the tank would pass up through the packing and out the stack. This provides a counterflow arrangement

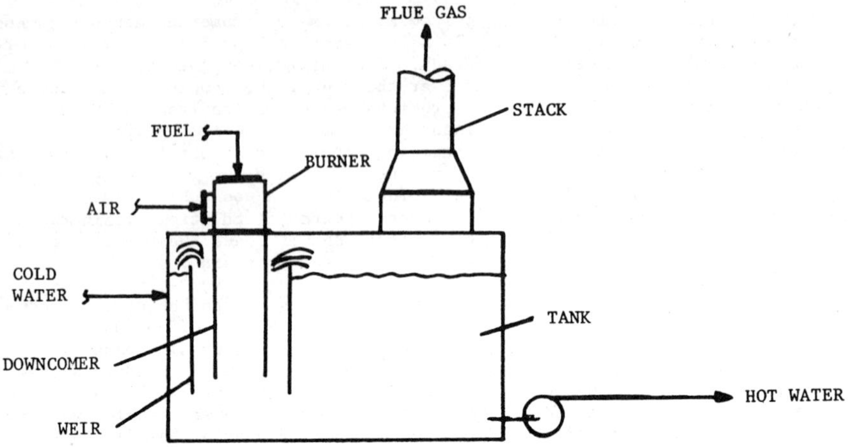

Figure 1. Direct Flame Water Heater

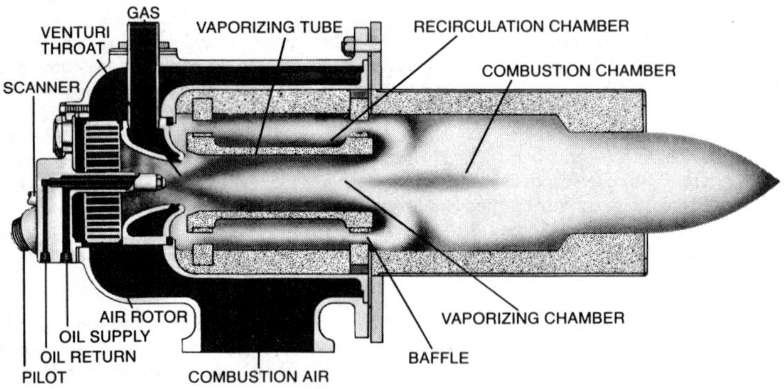

Figure 2. Trane Thermal HV Burner

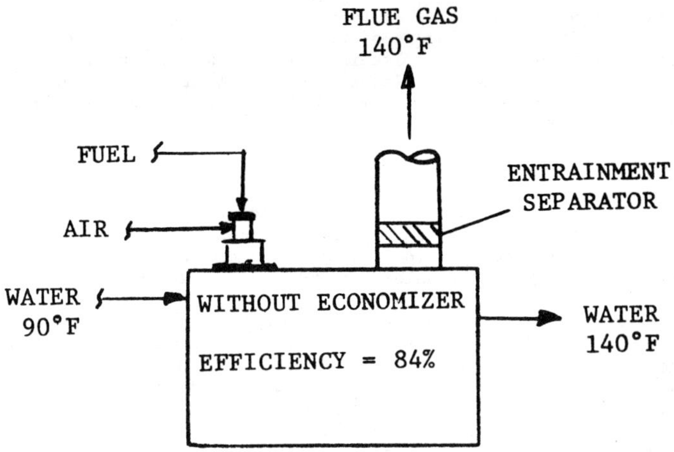

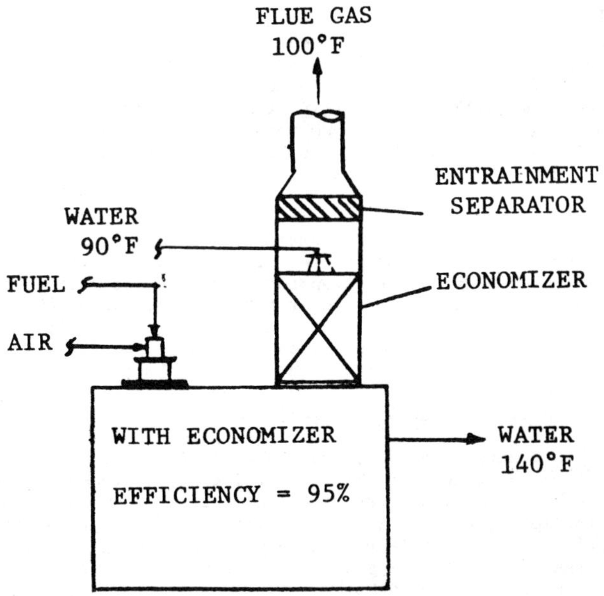

Figure 3. Comparison - Heaters With and Without Economizer

where the gases can exit at the inlet water temperature. Thus, the efficiency of this system is at the final temperature of the gases leaving the stack; rather than the temperature of the gases leaving the tank which usually is the same as the water temperature. For example, refer to Table 1, the overall thermal efficiency could increase by nine percent if the exhaust gas temperature was cooled to 100°F, instead of 140°F.

BASIC PRINCIPLES

The direct flame waterheaters make use of the concept of submerged exhaust combustion principle. There are specific phenomena connected with the submerged exhaust combustion principle, which involves the contacting of the hot products of combustion with the water below the water level in the tank.

In submerged exhaust combustion, the large volume of gases generated by the combustion of the fuel at high temperature must be brought in contact with the water. This is accomplished by using an orifice plate to distribute the gases equally around the end of the downcomer tube. When these hot gases come in contact with the water, they form many thousands of small bubbles (Figure 4). The number and size of bubbles formed depend upon a wide variety of factors, such as the diameter of the orifice, the relative shape of the orifice, the angle at which the orifice allows the gas to enter the water bath, the velocity of the hot gases entering the water bath and the depth of the submergence of the exit ports. It has been estimated that one cubic foot of gas when broken into bubbles, 1/1000 of an inch in diameter, represent an area of 72,000 square feet of heat transfer surface. The water bath acts as a vast heat sink, capable of absorbing the heat from the combustion products almost as rapidly as it is delivered. The bubbles will rise directly from the orifice exit to the surface of the bath, possibly combining and recombining with other bubbles. The gas leaving the liquid is practically at the same temperature as the water itself, or, essentially a zero approach is accomplished.

The flue gas leaving the water bath may evaporate or condense water, depending on the bath temperature. For example, for natural gas (as 100% methane), the saturation temperature of the flue gas is 140°F (Figure 5). If the bath temperature is above 140°F, evaporation occurs and thermal efficiency decreases. At 140°F, the saturation temperature of methane, an exact balance occurs. The amount of water added to the bath is exactly equal to the amount of water leaving in the flue gas. When the water bath temperature is below the saturation temperature, the water generated by the combustion will condense and give up its latent heat, maximal thermal efficiency is achieved. Water balance curve for propane is illustrated in Figure 6.

Illustrated in Figures 7 and 8 is the thermal efficiency for natural gas (as 100% methane) and propane. The thermal efficiency decreases as the bath temperature increases. Thus, for water heating, direct flame application is very efficient for low water temperature requirement; as the temperature requirement goes higher, the efficiency decreases. To combat this drop in efficiency for high water bath temperature requirement, the heat recovery system described in Figure 3 can be used.

Another distinct feature of the direct flame water heater is the boiling point of water. Referring to Figure 7, all of the heat in the fuel has been utilized at approximately 190°F. This is the highest temperature a direct flame heater can achieve at sea level (14.7 psia). Thus, the normal boiling point for a direct flame water heater application is 190°F. The boiling point depression of roughly 22°F is due to the fact that the inert products of combustion exert their own partial pressure on the system. The vapor pressure of water at 190°F is 9.3 psia. The difference between 14.7 psig and 9.3 psia is the partial pressure of the inert gases in the system. The theoretical boiling point of water can be reduced by the addition of excess air or can be increased by the increasing of system back pressure.

The products of combustion leave the surface of the water bath saturated with water and also carry entrained water droplets. It is necessary to remove

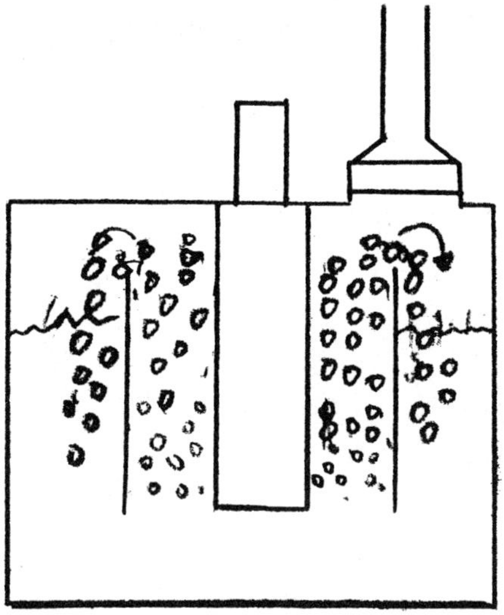

Figure 4. Submerged Exhaust Combustion Principle

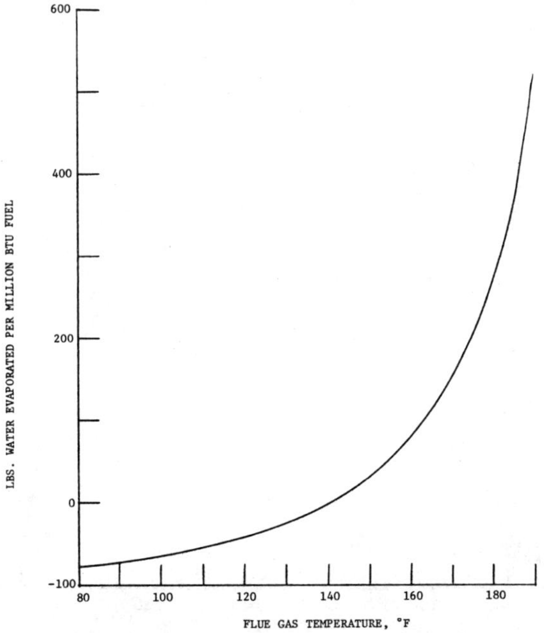

Figure 5. Water Evaporation Rate of Methane

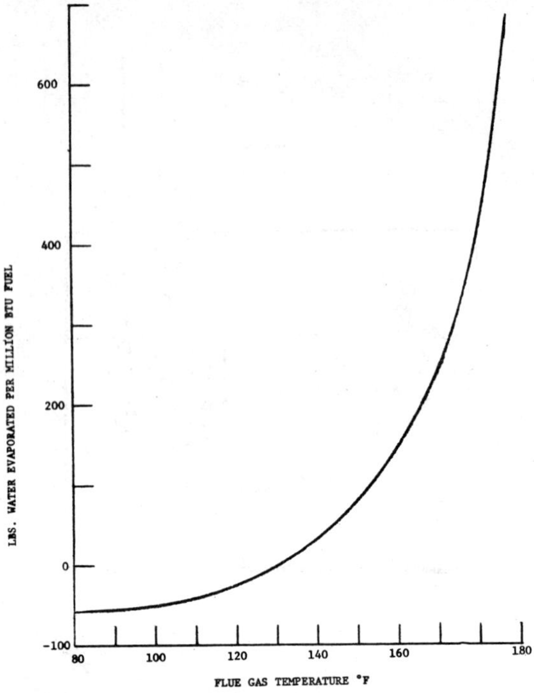

Figure 6. Water Evaporation Rate of Propane

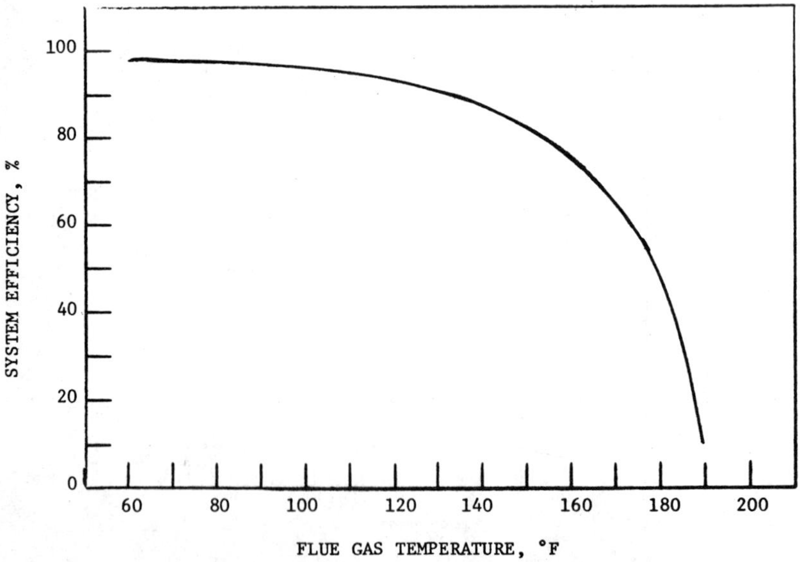

Figure 7. Thermal Efficiency for Methane

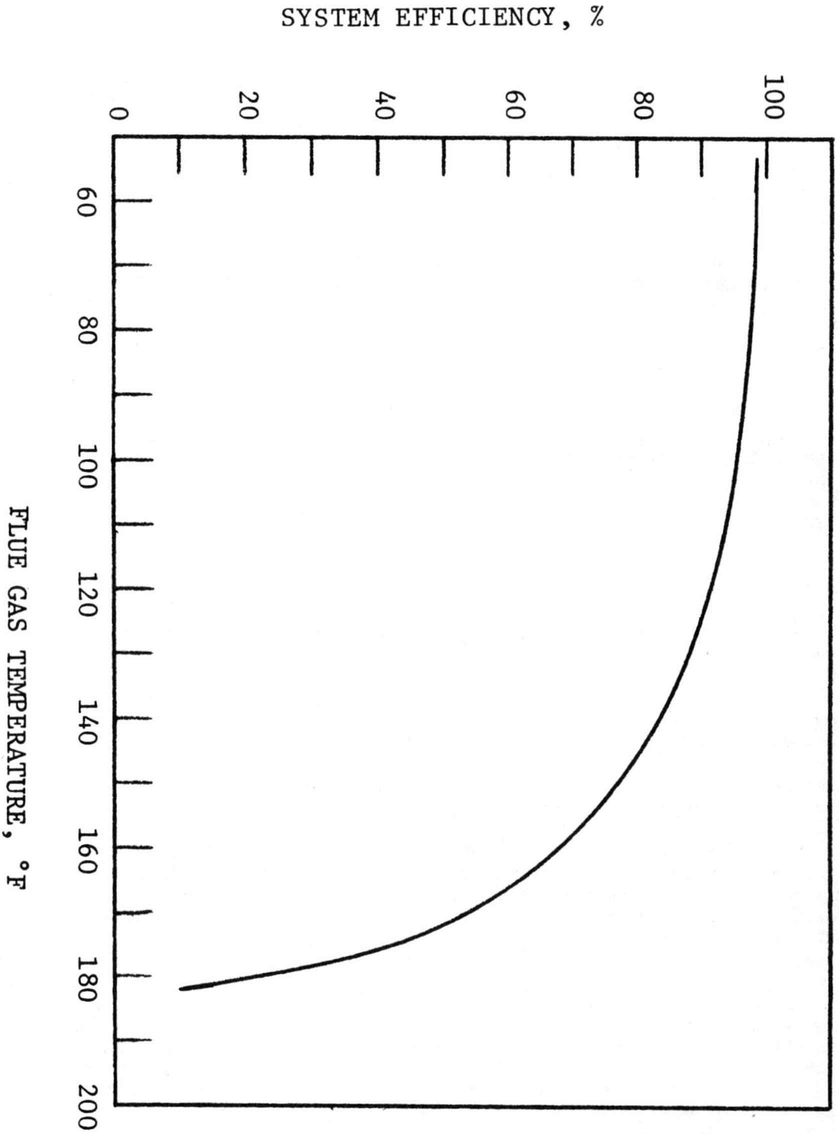

Figure 8. Thermal Efficiency for Propane

this entrained water. The exit plume from a direct flame water heater is normally visible. The plume is the result of the condensation of the saturated water vapor in the flue gases.

In direct flame heating system, there is a possibility of acid build-up. This problem is not critical if the system is a once through system, i.e., water is heated in the heater, used, and then discharged. However, if the water is recycled in the system, a critical look at the acid build-up is necessary.

There are two mechanisms for the formation of acids in the direct flame water heater:

a) the formation of weak acids - carbonic acids.

b) the formation of permanent acids - sulfuric and nitric acids.

The principle of acid formation is the same, Figure 9, for both weak and permanent acids. Carbon dioxide, sulfur dioxide or nitrogen dioxide are absorbed in the water and then dissociate to form acids. The maximal pH value of water in contact with a typical combustion flue gas (12% CO_2) is illustrated in Figure 10. The final pH value is a function of temperature and carbon dioxide concentration in the flue gases.

The acid build-up problem in direct flame water heaters is the main restriction for its application. Sometimes, it can be solved by neutralization. In general, it is necessary to study the acid build-up problem and its solution before each specific system design is made.

CONTROL AND SAFETY ACCESSORIES

The basis of control for a submerged combustion heating system is usually the final temperature desired. Therefore, some type of temperature control device measuring and controlling the outlet temperature from the heating vessel by modulation of the burner output is the best solution. It should be kept in mind also that a constant level in the heating tank is imperative and that some type of level control is necessary. This can often be achieved by merely an overflow nozzle located at the proper level height in the process vessel. In some cases, however, it is desirable to use a level controller and to throttle the feedwater by means of a level control valve. This level control valve can also be placed in the outlet line from the unit. The modulation range of the submerged combustion burner is usually about 4 to 1, from maximum to minimum. If greater turndown is required, multiple burners should be considered.

The safety aspects of most submerged combustion heating systems are controlled by insurance requirements usually associated with the combustion of gas or fuel oil. Some type of flame failure protection at the burner is usually required to shut down the entire system if the flame in the burner should go out. Pressure switches to prevent excessively high or low fuel pressure are usually a must in such a system. A pressure switch is utilized in the combustion air line to shut the unit down in the event of some mechanical failure in the combustion air blower. A high level and low level alarm are often utilized to shut the unit down in the event of some interruption in the water feed or draw off rate from the vessel. An excess temperature switch is usually located in the stack. This would be indicative of a low level condition or a burnout condition of the downcomer tube which would allow the hot gases to bypass the water bath and pass into the stack at very high temperatures. All of these safeties and interlocks are tied into the fuel system through a safety shut-off solenoid type valve in the fuel line which automatically closes in the event of any failure or irregularity in the unit.

A typical control diagram is shown in Figure 11. Ignition of the burner is generally accomplished by means of a high voltage ignition transformer and a spark plug. Occasionally, however, a gas pilot is used which must also be monitored by the flame failure device. Time delay relays are incorporated to

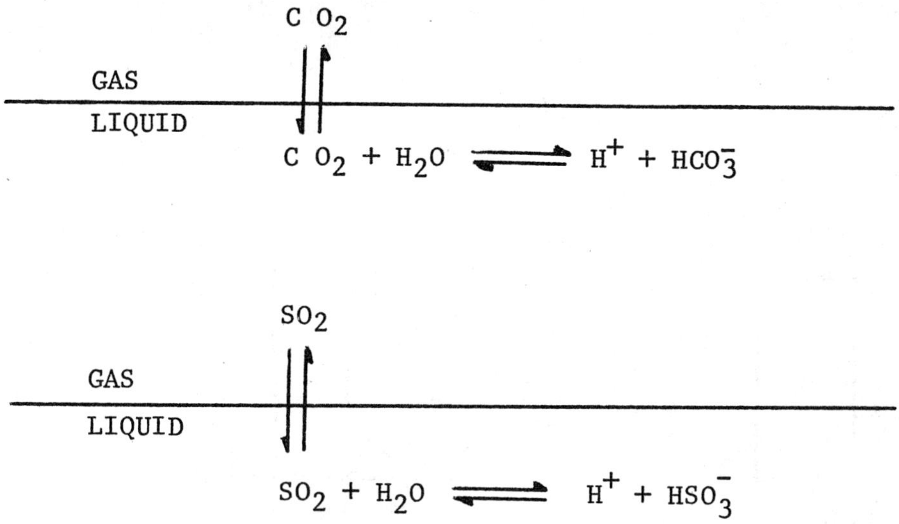

Figure 9. The Formation of Acids

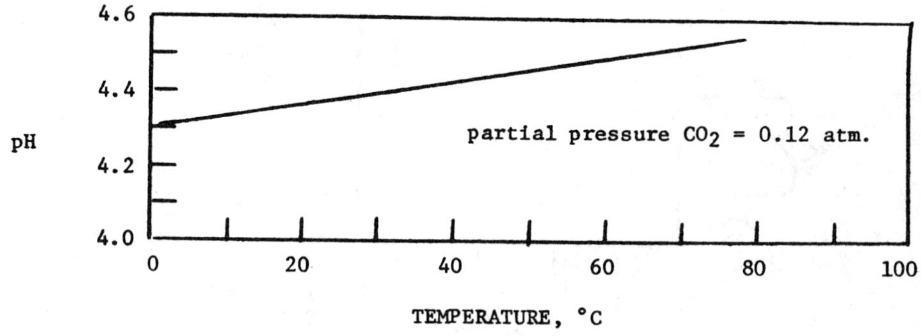

Figure 10. Equilibrium pH for Carbon Dioxide-Water System

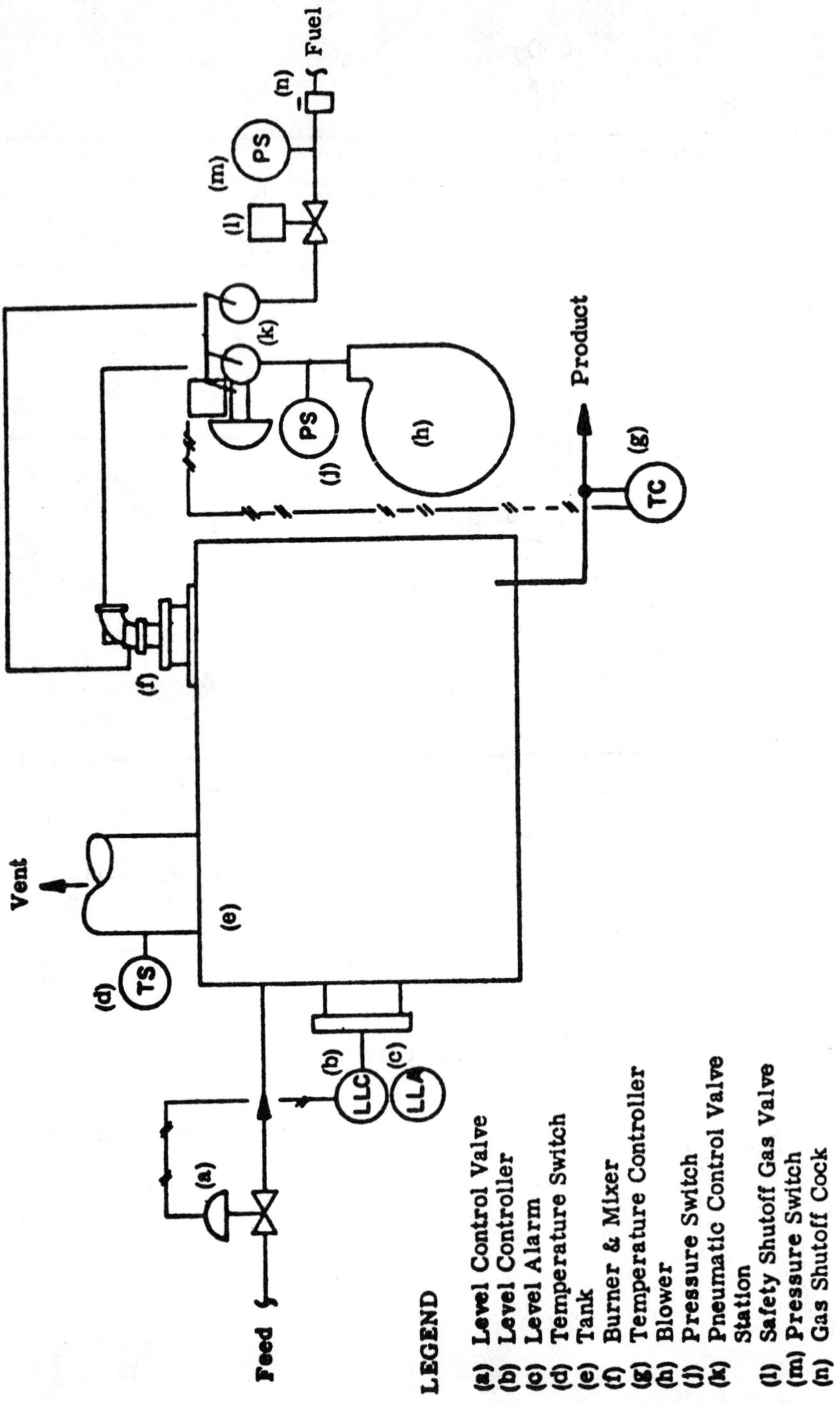

Figure 11. Schematic Control Diagram-Direct Flame Heater

program the combustion system as prepurging and postpurging of the burner occurs upon start-up and shutdown so that no concentration of combustible materials can develop in the downcomer tube. Usually, a low fire light-off safety switch is incorporated in the main valve system preventing the operator from starting the unit at any point above the minimum turndown of the system.

The proper proportion of air and fuel is usually maintained by mechanically linked air and fuel valves which are designed to provide the proper proportion of air and fuel throughout the entire operating range. These valves can be operated by either a pneumatic or an electric valve operator which receives its signal from the temperature controller. Separately operated pneumatic valves on each line have also been used under certain conditions, but the mechanical linkage arrangement is the more usual system.

The flame failure systems can be either the infra-red radiation type, scanning the flame with a lead sulfide cell, or the ultraviolet type.

APPLICATIONS

Direct flame water heaters are usually used for open loop, once through water heating applications. For example, direct flame heaters have been used for laundry water heating application, to heat water for gold leaching in gold mining process, and as pulping water heater in pulp and paper industries.

Very early in the development of direct flame water heaters, the application to a closed loop central heating system of a building was suggested. The idea was never considered entirely practical because of the problems incurred with acid build-up. However, a combination domestic hot water and central heating system may offer the solution to the acid build-up problem. In the combination system, Figure 12, the water discharge from the central heating system is the make-up water for the domestic hot water system. The water discharge rate for the central heating system will be determined to maintain proper pH value in the system. Thus, a balance between the domestic hot water requirement and the water discharge requirement of central heating system has to be maintained. Other approaches, such as using the discharge water as feedwater preheater, can also be used.

REFERENCES

1. Trane Thermal Company, "Thermal Manual of Submerged Combustion", Conshohocken, Pa., 1981.

2. Kiang, Y., "Waste Energy Utilization Technology", Marcel Dekker, N.Y., 1981.

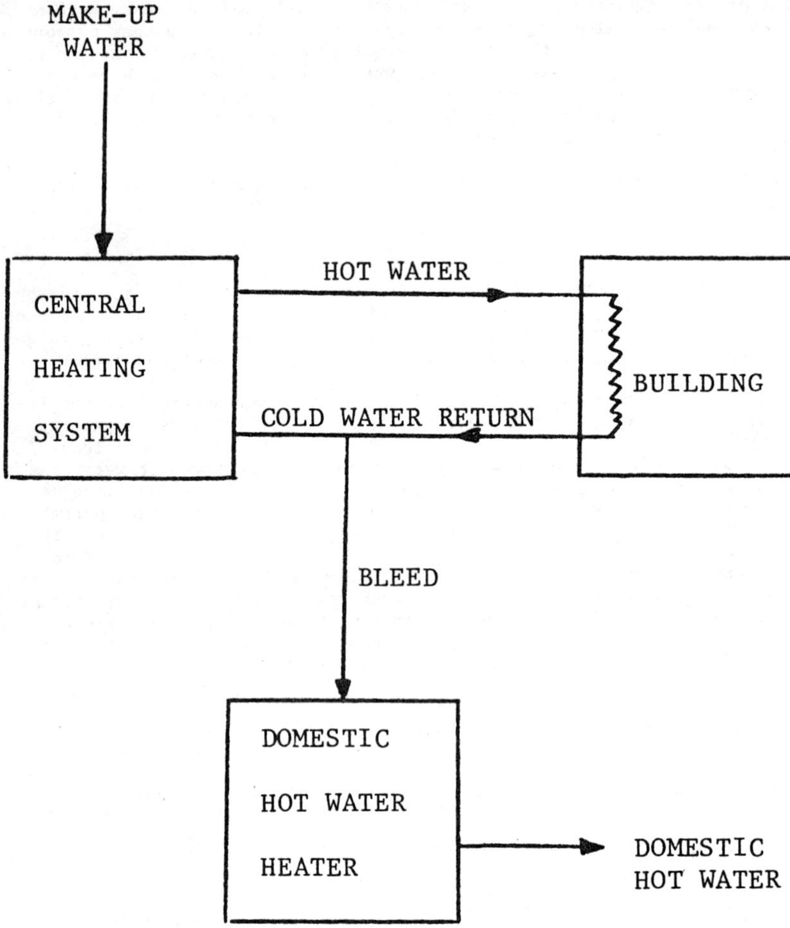

Figure 12. Combination Domestic Hot Water and Central Heating System

DISTRICT HEATING IN URBAN CENTERS

William C. Hanselman
Ira J. Joseph

Resource Development Associates
5060 Wadsworth Road
Dayton, Ohio
45356

1. FRAGMENTED ENERGY EFFORTS

It is a reasonably safe statement that energy awareness is a topic of much discussion and effort. The last few years have seen a proliferation of ways by which energy users can meet their energy requirements.

Another safe statement would be that for the most part these evaluations of energy use, and the proposal of new approaches to energy consumptio, and development of new energy supplies are all looked at in a fragmented manner. Particularly with regard to energy supply is this condition found to be true. There is no interrelation between synfuels and solar power, or between resource recovery and geothermal energy, to name just a few. Part of this problem is due to the geographical limitations of various potential sources. Geothermal energy is not available everywhere. Likewise, hydro applications are restrained by localized water supply.

If the United States were an agrarian society, largely decentralized, this issue would be of no concern. The fact is, however, that we are an urbanized nation with large concentrations of population in metropolitan areas. The dynamics of urban centers do not lend themselves to energy solutions which are one-dimensional. Instead, energy solutions must be able to flexibly utilize the myriad of available energy sytsems and also take advantage of the inherent physical, functional and spatial characteristics of urban areas.

2. IMPORTANT URBAN CHARACTERISTICS

What are the characteristics of urban areas which should be considered in the formulation of energy supply solutions? Perhaps the most important is the density of development and intensity of land use. In terms of energy's common denominator, BTU's, high density and intensity of use often translate into a high concentration of BTU's per unit of land area. Additionally, urban areas exhibit a diversity of land uses, many of which have concentrated energy loads. The central business district of an urban area is often the spatial segment with the greatest intensity of use. This can be seen in an evaluation of the CBD's floor area ratio (F.A.R.), which compares the number of square feet of usable floor space with the number of square feet of land area. In most urban areas the amount of available floor area is at least double the available land area. In fact, many of the newer types of development in urban areas have dramatically high floor area ratios. For example the Charles Center project in Baltimore has an F. A.R. of 5.3. Peachtree Center in Atlanta contains more than 6 million square feet of built area on a 10 acre tract, for an F.A.R. of 13.8. Illinois Center in downtown Chicago, as of 1975, had over 8 million square feet of built area on 10 acres, for an F.A.R. of 18.8.

Clearly the major development efforts in cities are based on the intensity of land use permitted in the central city. It can be expected that this trend will continue to be the norm for major development and redevelopment activity in the nation's cities. What this means in terms of energy demand is that the evolution of the mixed-use development as an urban building form lends itself to an energy system particularly attuned to cities.

Aside from these new mixed-use developments which combine retail, residential and office uses, there are the more traditional separate uses: high-rise apartment and condominium structures, large department stores and office buildings, and numerous institutional uses such as schools, and hospitals, and cultural attractions such as museums.

Cities are also frequently the home for major industries, many of which use large amounts of energy in their mechanical processes. Many also waste or reject substantial amounts of thermal energy.

Finally, a good number of urban centers have power plants located in close proximity to the central business district. This placement is not restricted to the northeastern U.S. cities, although the great majority of these cities exhibit this characteristic. Other cities in other parts of the country have power plants for the production of electricity near the downtown areas. Why are power plants important in this discussion? They use only about 1/3 of the energy that goes into the plant to produce power. The rest, nearly 70 percent, is wasted, either by discharge into a nearby river or evaporation through a cooling tower or pond.

As most of us are aware, however, the condition of our nation's cities is not all peaches and cream. Our cities face major problems, particularly the outmigration of workers from core areas, the decline of the urban infrastructure, and the movement of businesses and industries to locations offering easy access and large tracts of land. Northern cities, with longer heating seasons, must address the exodus of firms to areas of the nation with cheaper labor and energy costs. Solid waste disposal is also a pressing concern; the use of landfills for disposal is creeping toward the point where it is no longer a viable alternative due to cost, public sentiment and land availability.

By way of summary, it appears advisable that a centralized energy supply system for urban areas take into consideration the following opportunities and constraints presented by cities:

- o High density of development & intensity of use
- o Diversity of land uses with concentrated energy loads

  - o high density residential
  - o commercial and office space
  - o mixed-use developments
  - o industrial developments
  - o power plants
  - o Declining infrastructure
  - o Outmigration of labor
  - o Disinvestment by business
  - o Solid waste disposal concerns

3. DISTRICT HEATING'S APPLICABILITY

The energy system with the potential for integrating the energy dynamics of cities is an old technology; it is called district heating. When first developed more than 100 years ago, the system transported steam from a central plant to various buildings in a city for heating purposes. The systems were reasonably reliable--while New York was paralyzed by a blizzard in 1888, the city's steam system performed without a problem.

In the 1900's, however, the increasing use of oil and natural gas decreased the desirability of the steam heat system. In addition, the steam systems were not highly efficient and were thus limited in the land area they could service. Many cities with district steam systems have let the networks deteriorate, and the cost advantages of other fuels have served to further diminish the systems' markets. Today, there are still more than 50 cities in the U.S. with district steam systems. Their customers, however, are declining in number and the systems are in need of revitalization.

Why then is district heating such an attractive energy system for an urban center? To understand what appears to be a paradox, it is necessary to consider three issues: 1) modern district heating technology; 2) the change in the price of fossil fuels; and 3) the characteristics of urban areas described earlier.

The first consideration is the present state of district heating technology. While steam may have been the medium by which thermal energy was transported 100 years ago, it is not the case today. Hot water is a superior thermal transport medium and it can carry energy for greater distances with minimal loss of BTU's. Also, piping system technology is far superior to that employed in the older steam systems. Finally, heat exchangers for end-user energy extraction are now modern and efficient.

The cities of Scandanavia and Europe offer ample evidence of the reliability of modern hot water district heating systems. Most have been built in the years since 1945 as a result of those countries' declining energy resources. They are now a basic infrastructure component in many cities.

The second reason to reconsider district heating lies in the energy marketplace. With the recent decontrol of oil, and the iminent deregulation of natural gas, two major fuels used in urban areas, the price disadvantage which precipitated the decline of the old district steam systems may in fact be removed.

The third consideration in an assessment of district heating for urban areas centers on the characteristics of cities described previously. Cities can be viewed as a combination of heat islands--those land uses or configurations which are either large users or suppliers of energy. Certainly the CBD, with its relatively high floor area ratios, contains large users. High density residential neighborhoods can be seen in a similar light, as can hospitals and schools. Industries can be both users and suppliers, depending upon how much thermal energy is left over from the manufacturing processes. It has been noted that power plants waste roughly 2/3 of the energy produced in the generating process; this, then, is a substantial energy source for cities with power plants.

4. THERMAL LOOP INPUTS

The primary reason that district heating can be applied in cities is because of the flexibility of the distribution system, or "thermal loop." It is important to keep in mind that it is not oil or gas that is transported by the thermal loop; rather, thermal energy is transported in water to individual structures, where it is extracted by the heat exchangers. Such a system enables the energy already being wasted or underutilized at one location to be transported for use at another location. In its simplest sense, then, implementation of a thermal loop allows the community to match its local energy requirements with its own resources.

The ability of district heating to help a community recover and reuse energy presently being wasted might appear to be an adequate justification for system development. However, there is an additional positive aspect to be considered and this concerns the integration of new energy technologies into the thermal loop.

This additional flexibility can be seen in a consideration of several "alternative energy" technologies: geothermal, solar and resource recovery. Geothermal energy can be utilized in much the same manner as excess energy from a power plant. The BTU's can be fed into a thermal loop for distribution, answering the question of what to do with geothermal energy once it is out of the ground. Solar energy has only limited application in urban areas because of space constraints. However, it is feasible to tap solar collectors located outside of an urban area for integration into the distribution system. Resource recovery does not become economically feasible unless there is a market for the thermal energy produced as a result of waste incineration. A thermal loop system is capable of providing the necessary distribution mechanisms for selling thermal energy. (It is interesting to note that of the more than 330 resource recovery plants in Europe, all but a handful are connected to district heating systems.)

The three alternative energy technologies mentioned above are not the only potential inputs into a thermal loop system. In actuality the flexibility of such a system is great, and the use of hot water as a heat carrying medium is far superior to the older steam systems with their limited service areas. This potential can be seen in the plans to use the waste heat from a nuclear power plant located 60 miles from Stockholm, Sweden to heat water for distribution to district heating customers in Stockholm.

5. IS DISTRICT HEATING GEOGRAPHICALLY LIMITED?

Throughout this paper the term "district heating" has been used. This could lead certain readers to conclude that such a system is appropriate only in colder climates. One could also point to the geographic location of European district heating systems to support such a contention.

This geographical limitation does exist if a "heat only" system is considered. If this configuration is desired it is necessary to use heating degree days as the primary criterion, and 4000 heating degree days becomes the relevant yardstick.

Looking at system application in this light, however, often bypasses a characteristic of urban centers previously mentioned: the high density commercial and office space in downtown areas. These large buildings often have significant cooling loads and are frequently serviced by absorption chiller systems inside the buildings. These systems not only run on hot water, they also are among the least expensive systems to retrofit for district heating. In certain circumstances this situation can lead to heating and cooling loads which are roughly equivalent.

Despite these potentials, it is important that the load factor, actual demand for thermal energy, be the key barometer guiding system development. This consideration will, in the long run, determine the economic success of the system.

6. CONCLUSION

This paper has been decidedly nontechnical. There are two reasons why this approach has been taken. First, it is not necessary to prove the technical feasibility of district heating systems. A modest amount of checking will reveal that the technology is both implementable and reliable. The second reason for the nontechnical approach is the key point of this paper: cities, by the nature of their form and development characteristics, have excellent potential for district heating development. Hopefully, this paper has helped to reveal this opportunity.

DISTRICT HEATING, JOB CREATION AND ECONOMIC REVITALIZATION

Gerald S. Leighton
U.S. Department of Energy

In the post-embargo years, as the Nation struggled to adjust to a new energy reality of periodic supply interruptions and steady price escalations, it became obvious that forging a new energy base might well turn out to be one of the most complex national undertakings in U.S. history. There are numerous reasons why this should be so. But surely, the complexity of the energy problem is due above all to the pervasive and diffuse character of energy production, delivery, and use. It is precisely because of the astronomical number of actors involved, each moved by often conflicting interests, that reaching a consensus on how to resolve the energy problem has been so difficult.

Paradoxically, the source of this difficulty may in some cases be turned to advantage. By definition, almost our entire energy-intensive society is directly or indirectly influenced by the price and availability of energy. It follows that energy choices can have a marked impact on economic and social developments. To borrow a term from the economists, energy has a "multiplier" effect.

This paper explores this multiplier effect. Specifically, it develops the argument that the choice of a particular energy system--in this case, district heating--allows one to accomplish several apparently unrelated tasks. The primary objective of a district heating system is to conserve energy and provide a community with a reliable and reasonably priced source of energy to heat and cool its buildings. But, as we shall see, district heating delivers more than energy: it creates jobs where they are most needed; it retains large sums of money within a community; it attracts new and helps to hold old industries; and, in general, it can serve as a powerful agent of economic revitalization. The focus of this paper is precisely this leverage effect of district heating systems.

Development and Decline in the U.S.

District heating is not a new concept. The use of a central power plant to produce--and a network of underground pipes to distribute--thermal energy to buildings, it was first applied in the U.S. in Lockport, New York, as long ago as 1877. By 1962, more than fifty cities had district heating systems, but by then district heating was very clearly on the wane.

Three developments contributed to this reversal. One was the growing popularity of inexpensive, clean and plentiful oil and gas for space heating; another was the technological progress achieved in the production and transmission of electricity which made it possible to build large steam-electric power plants remote from urban areas. And a third element was the exodus of people from cities to the suburbs, where the scattered nature of homes and businesses made district heating uneconomical.

The end result of these three converging developments is that today district heating supplies only about 1 percent of U.S. heating demand, while 20 percent of all oil consumption is for space and water heating.

Energy Benefits of District Heating

Just as the advent of cheap oil and gas helped to reduce the attractiveness of district heating, so, too, the sharp rise in the price of the two fuels since the winter of 1973-74 has once again drawn attention to the energy benefits of these systems. Estimates for the total energy conservation yield of district heating range from 2 to more than 5 quads,[1] and the Department of Energy has predicted that by the end of this century district heating could realistically displace 2,500,000 barrels a day of imported oil.[2] A closer look at how district heating systems can function reveals why they yield such large overall savings in general and an opportunity to displace oil in particular.

District heating is an extraordinarily flexible system: the central plant can be fueled by coal, nuclear, solid waste, geothermal, or solar resources. Basically, these systems may be compared to a street. Once built, the street is not restricted to small cars, trucks, or anything in between. And what this flexibility allows one to do is to substitute a variety of fuels for increasingly more expensive and insecure petroleum.

The energy savings potential of district heating systems is due among other things to their use of cogeneration techniques and industrial waste heat. Expected to be a major source of heat for district heating in the U.S., cogeneration power plants produce both electricity and useful thermal energy. This dual function yields a remarkable improvement in efficiency. "In an electric-only

---

1 J. Karkheck, J. Powerl, E. Beardsworth, "Prospects for District Heating in the United States," Science Vol. 195 (March 1977). "Potential for Scarce Fuel Savings in the Residential/Commercial Sector Through the Application of District Heating Schemes," Working paper, Argonne National Laboratory (1977). "A Case Study Application of Venture Analysis: The Integrated Energy Utility," Report 347-117-03 by the Future Group and General Electric - TEMPO (1978). C. L. McDonald, "An Evaluation of the Potential for District Heating in the United States," presented at the Converence on Alternative Energy Sources, Miami, Florida (November 1977).

2 Gerald S. Leighton, "Energy Policy and the District Heating Strategy," Modern Power Systems, Vol. 1, No. 1 (January 1981), Page 23.

plant, at most 30-40 percent of the thermal energy produced is converted to electric energy; the rest is rejected to the environment through stack-gas losses and the plant's cooling system. By utilizing most of the rejected heat, a cogeneration plant can operate at a fuel utilization efficiency of more than 85 percent. This requires some reduction in energy output, but for each unit of electricity sacrificed, five to ten units of thermal energy are made available."[3]

Cogeneration is only one element in a broader effort to recover and use industrial waste heat--a vast potential source of useful energy. And district heating systems are especially well suited to use this energy resource, thus again yielding both improved efficiencies and fuel substitution.

Environmental Benefits

In addition to the purely energy benefits of district heating, they also improve air quality. In a district heating system, the burning of fuels in individual buildings is replaced by piped-in heat. Consequently, emissions from many uncontrolled sources of pollution are replaced by emissions from a central plant. Such a plant is more likely to be fueled by non-polluting resources such as solar or geothermal than individual buildings. Thus a recent study of the effect of district heating with cogeneration on sulfur dioxide concentrations in the atmosphere in the Minneapolis-St. Paul area indicated that district heating can decrease pollutant concentrations in ambient air. Measurements in Sweden and elsewhere in Europe have shown an actual improvement in air quality in cities with district heating systems.[4]

Energy Choices and the Economy

In recent years, as the manifold consequences of energy choices became a matter of concern, a number of investigations focused on the relationship between energy policy and the economy. Indeed, the Joint Economic Committee of the Congress felt the subject was important enough to warrant two days of hearings; the published proceedings of these hearings (held March 15 and 16, 1978) bring together a great deal of data and now constitute an essential starting point for any serious discussion of the topic.[5]

As just one example of the impact of energy choices on local economic life, one can point to the results of a year-long study in the Long Island, New York, region comparing the economic implications of a nuclear power plant to a package of conservation/solar measures yielding an equivalent amount of energy. The major findings of this study may be summarized as follows:

1. Over a thirty-year period, the solar/conservation package would create 178,000 jobs, compared to 72,000 jobs created by the nuclear option.

2. The types of jobs created by the two options are also dramatically different. In the nuclear case, most of the construction work is carried out over a seven-year period, much of it is fairly specialized, and many workers are imported.

---

[3] "National District Heating Program Strategy," Internal Department of Energy Working Paper (October 1980), Page IV.

[4] Ibid., Page 4.

[5] U.S. Congress, Joint Economic Committee, Creating Jobs Through Energy Policy: Hearings, 95 Congress, 2nd Session.

"Conservation and solar require small businesses in the regional economy to supply not only the material but also the manpower to install and maintain this equipment . . . We are examining types of labor where it is needed the most, and that is in the building trades where the highest unemployment exists now; for carpenters, plumbers, sheet metal workers . . . There is a fortuitous linking of more jobs being created plus being created where jobs are most needed."[6]

3. The total cost of the two options is $6 billion for the conservation/solar choice and $7 billion for the nuclear choice.

4. "The same package which costs $6 billion for conservation and solar will result in an overall energy savings of 6 cents per kilowatt hour, or $32 billion over a 30-year period. The $7 billion investment in nuclear powerplants, paid for at the 6 cents per kilowatt hour results in a cost to the households of $17 billion.

   One of the most important aspects we are looking at in more detail is the effect of these households paying $17 billion to utility companies. The money then flows into large financial institutions where it is reinvested in other projects in society, usually very large ones.

   In comparison, with the conservation and solar scenarios, there is a savings in energy per household which result in disposable income being increased by a net amount of $26 billion. That means those homeowners make the decision on how to spend that additional $26 billion which has been released through conservation and solar energy, rather than paying out $17 billion to the utility company. This reinforces the argument we are hearing more often about, democracy--about democratic considerations related to the difference of energy supply technologies, to social equity. Who is making the decisions and how is the money spent? Much of this $26 billion will be spent by households in the local economy. It seems only logical that local, small businesses and the regional economy itself is going to be stimulated much more directly than having the $17 billion flow out of the regional economy through the utilities, to pay for nuclear powerplants."[7]

The important point about this example is not the specific comparison between the two options. There may in fact be times when noneconomic considerations--national security, for example--dictate that a more capital-intensive choice best serves the national interest. What is instructive rather is the relationship the example exposes between an energy choice and the life of a local economy. As one looks at the potential markets for district heating, it immediately becomes apparent that it is best suited to exactly the locale where it's multiplier effect will do the most good.

District Heating, Job Creation, and Economic Revitalization

District heating can serve four types of markets: (1) densely populated urban areas, (2) single-family residential areas, (3) industries requiring process heat at low temperatures, and (4) institutional complexes such as universities and shopping centers. Previous experience, as well as a number of studies, support the view that district heating systems are most likely to be economically viable in the first of these markets.

[6] Creating Jobs, Page 29.

[7] loc cit.

To date, the best analysis of the economic implications of district heating has been carried out by Argonne National Laboratory for the Department of Energy. As Figure 1 shows, Argonne estimates indicate that even when compared to other relatively labor-intensive energy options, district heating will create more jobs per dollar of construction investment, with substantially more low-skill construction jobs created.

As depicted in Figure 2, compared to new oil and gas supply investments, most of district heating's investment will take place in and near its market area--the urban centers of the Northeast and midwest. What this means, of course, is that district heating not only provides more jobs than its competitors, but it also does so where workers are available, where jobs are needed, and where there is a great need to revitalize the economy.

Figure 1: An indication of construction jobs created for each dollar of investment in energy system capacity

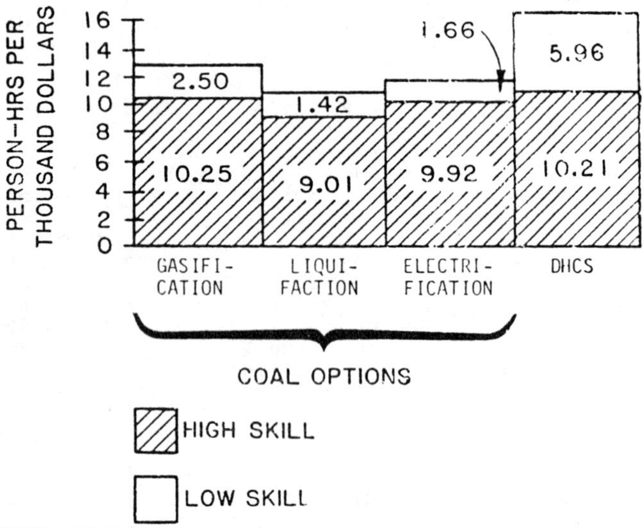

Figure 2: Percentage of construction estimated to occur in the center city by various heating methods

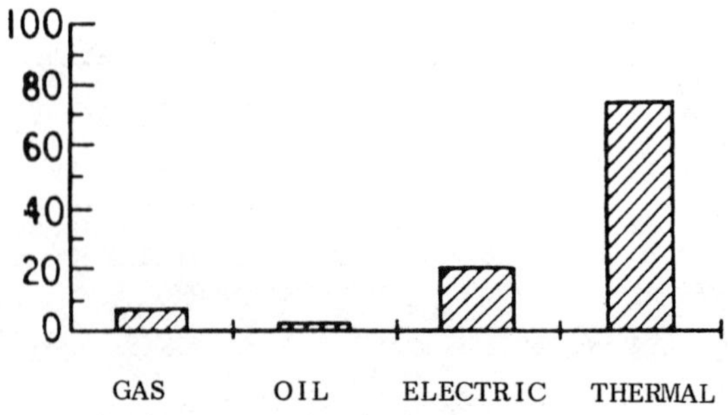

The way in which district heating systems will most likely be built will also have a beneficial impact on jobs. Because they constitute large capital investments, district heating systems will ordinarily be built in increments--each built section creating revenues making it possible to build the next segment. One of the consequences of this procedure is that it helps to avoid the cyclical, boom-bust character of construction trades.

Two additional economic consequences deserve mention. Because they are energy-conserving, district heating systems will same consumers money; dollars ordinarily leaving the community can thus be kept at home to invest in the economic well being of one's immediate environment. Another beneficial result of installing these systems is that insofar as they supply reliable and fairly priced energy they can both retain old and attract new industries.

In all these ways, then, district heating offers a community multiple and not originarily linked benefits: it saves energy and money; it creates jobs where they are badly needed; it nourishes economic development by holding money in the community; and it can draw industry to those cities where America's industrial revolution began and where so many millions of her people continue to live.

HEAT RECOVERY SYSTEMS FOR THE STEEL INDUSTRY

AUTHOR - FRANCIS H. BRICMONT, PRESIDENT
SPEAKER - PRAKASH M. SHETE, VICE PRESIDENT

BRICMONT & ASSOCIATES, INC.

Energy efficiency has become a top priority objective in most industries. The steel industry, dependent on energy for the majority of its operations, is no exception. Concerned with optimizing energy use in its furnace systems, many parts of the industry have adopted heat recovery as an accepted strategy for meeting this objective.

We are convinced that heat recovery merits such attention, but encourage companies to attempt it only with a thorough understanding of what it implies. For instance, heat recovery efforts ought to be launched after the basic nature of a BTU is acknowledged. This fact is central to the successful heat recovery effort: Once generated, the BTU cannot be destroyed. Essentially, this means that heat recovery becomes a tracking strategy. We must follow the BTU in the system where it is used, carefully studying and evaluating where it accomplishes the most and where it accomplishes the least. Our goal is to ensure that we get as much mileage as possible from that unit of energy.

Most assuredly, tracking means we must examine the entire heating system. Through applying this strategy, we have found that the best way to improve heat recovery is to transfer greater percentages of available heat to the product during the heating process. Thus, the most energy efficient furnace will be one that has been designed to minimize all other forms of heat consumption.

Until recently, this "systematic" energy tracking strategy was not acknowledged widely. Although virtually everyone has been concerned with energy conservation, few realize that efficiency is unattainable until inefficiency has been identified and analyzed. In effect, we have to account for the multiple causes of heat loss in a system before we take action to minimize them.

Without a systematic approach to heat recovery, improvements to a system often fall short of expectations. In the steel industry, the lack of a systems approach to this system-based problem has encouraged some companies to expect significant solutions through recuperation and waste gas boilers alone. It has encouraged others to reject proposals that would have lowered the temperature of waste gas as it leaves the furnace, because this reduces the temperature of preheated air. Still others have abandoned new, fuel conscious furnace designs in favor of less costly, but demonstrably less efficient -- modifications to old designs.

Thus, although everyone wants to improve fuel efficiency, not everyone is aware that optimum efficiency necessitates an understanding of the entire furnace system, including furnace design and its impact on energy consumption. When maximum heat recovery is the objective, all options for improvement ought to be investigated. Under certain circumstances, the best investment may be modifications to an existing furnace, or a new design rather than an old one for a new furnace. In other situations, a systematic approach may show that economic considerations are crucial: The upgrading of existing furnaces and furnace designs may be the more feasible of available alternatives. Regardless of which solutions are accepted, the systems approach will have yielded far better choices than the short-sighted "one problem, one piece of equipment" causal approach.

The systems approach is a new way of thinking about heating systems. It allows us to develop perspectives on furnaces and furnace designs. For example, a survey of the industry will show that most systems in operation today were designed and installed prior to the escalation of fuel costs and the frequent occurrence of fuel shortages. Although these furnaces may have met the objectives for which they were designed, we should not expect them to meet today's criteria.

Given the contemporary energy situation, the maximization of fuel efficiency will remain a primary objective for the steel industry. Undoubtedly, heat recovery will play a major role in reaching this objective. Central to successful heat recovery is a recognition of the complexity of a furnace system and the multiple factors that are responsible for inefficient heating in such a system. The systems approach will prove to be an invaluable tool in isolating these factors. It also will lead to proposals that facilitate proper heat recovery, because it addresses solutions through the careful study, evaluation and analysis of an entire heating system. On the other hand, if we exclude this approach, from an energy standpoint the impact of resulting changes may be minimal.

In the following paragraphs, we will examine a case study that shows how we helped one client improve fuel efficiency through applying the systems approach. The case study is drawn from our experience with the client's furnace needs over many years. It illustrates the evolution of the bar mill furnace in a mini-mill operation. We have chosen this example because the mini-mill is a particularly enterprising operation. Due to market competition and to the relatively small size of the mills involved, operators are extremely sensitive to increasing fuel costs. Consequently, they are among the first in the steel industry to find and implement devices through which operating costs can be minimized. Not surprisingly, the mini-mill operators are quite receptive to strategies that will increase the thermal efficiency of their reheat furnaces.

Following a description of the three-zone pusher furnace that existed at the mill prior to our involvement, we will examine two separate furnace projects and their results. The object of both projects was to improve the heating efficiency of our client's furnaces. Our first project investigated and subsequently improved, through the addition of skid pipe insulation and a recuperator, an existing three-zone pusher furnace. Our second project was an investigation and subsequent installation of a new, sweeparound, two-zone pusher furnace that included a recuperator and a waste gas boiler. For both projects, the product heated remained the same: 5" square billets.

The existing bar mill furnace was designed and implemented at a time when the cost of fuel was approximately $0.25 per million BTUs, prior to the first major fuel crisis. The furnace has been chosen by the mill from among several alternatives because it required the minimum capital investment. It was a conventional over-and-under fired, three-zone, continuous reheat furnace (Figure 1). In such a furnace, the product is supported on water-cooled "skid" pipes.

No provisions for recovering heat from the waste gases were included in the furnace's original design. In fact, its combustion system was designed for cold air in most elements, including burners, ducts and controls. When the system was installed, however, the cold air combustion system was quite acceptable because the price of fuel was too low to justify the cost of preheated air.

The existing furnace produced about 75 tons of the 5" square billets per hour. Its bar heating characteristics included a waste gas temperature of 1850°F at maximum production levels, a significantly high temperature by today's standards (Figure 2). The theoretical fuel rate on this furnace was 2.06 MM BTU/ton at 100% operating capacity. The actual fuel rate, however, was quite sensitive to the furnace's operating level (Figure 3). Since production at 100% furnace capacity was not possible at all times, the actual monthly fuel rate was higher, about 3.00 MM BTU/ton.

With the passage of time, fuel costs went up and fuel shortages became more frequent. Consequently, our client asked us to study and improve the existing furnace's heating efficiency. Our analysis showed that heat distribution in a typical furnace can be attributed to a number of sources (Figure 4). Among these are: Heat transferred to the product, heat lost to water-cooled parts, heat lost through furnace walls and doors, and heat lost through the waste gas. A percentage distribution of sources of heat loss in the existing furnace revealed that significant losses could be attributed to the water-cooled parts and waste gases (Figure 5).

One modification we considered as the result of our analysis was conversion of the existing furnace to a hot air combustion system. The cost of such a conversion was quite reasonable when the recuperator cost was considered alone. When the costs for modifying the burners, ducts and controls were added, however, the hot air combustion system became an economically unattractive alternative. Recuperation, reliable as a means to reduce energy consumption, would have increased the mill's operating costs when retrofit to the existing furnace. At that point in the sixties, fuel costs remained low relative to the installed costs for a new recuperator and hot air combustion.

The other modification that we considered, skid pipe insulation, proved to be more practical under the circumstances and was implemented. According to our analysis, this insulation would increase fuel efficiency through reducing heat lost to water-cooled parts by about 50%. In contrast to the modification we had first considered, the cost of installing skid pipe insulation would be off-set by the savings generated by the insulation's favorable impact on fuel consumption. The furnace operation realized an important reduction in its fuel rate as a result of the skid pipe insulation: From 2.06 MM BTU/ton to approximately 1.75 MM BTU/ton (Figure 6).

Fuel costs continued to rise, however, and pressure mounted to further reduce the furnace's fuel rate. As fuel costs increased and fuel shortages became more common occurrences, recuperation became a more economically attractive alternative. Consequently, a recuperator capable of producing 750°F preheated air was installed on the existing system. The fuel rate on the furnace dropped from 1.75 MM BTU/ton to 1.39 MM BTU/ton (Figure 7).

At this point, our on-going analysis of the furnace's efficiency began to show a decline in benefits realized from the skid pipe insulation. Further study revealed a tendency among the operators to neglect maintenance on the skid pipe insulation. The correlation between deteriorating insulation and a noticeable rise in the fuel rate was evaluated. Operators were appraised of the importance of maintenance in this area (Figure 8).

By this time, the minimization of fuel use had become a top priority item in the industry. The client, satisfied with the improvements we had made to

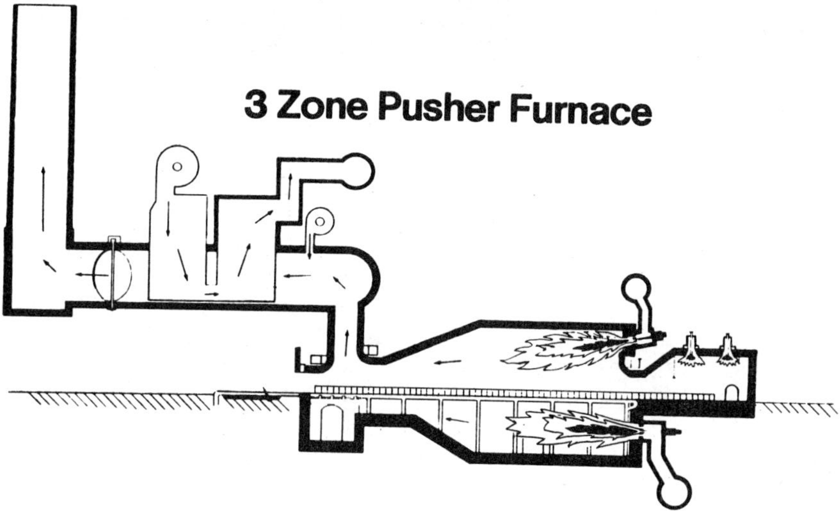

FIGURE 1

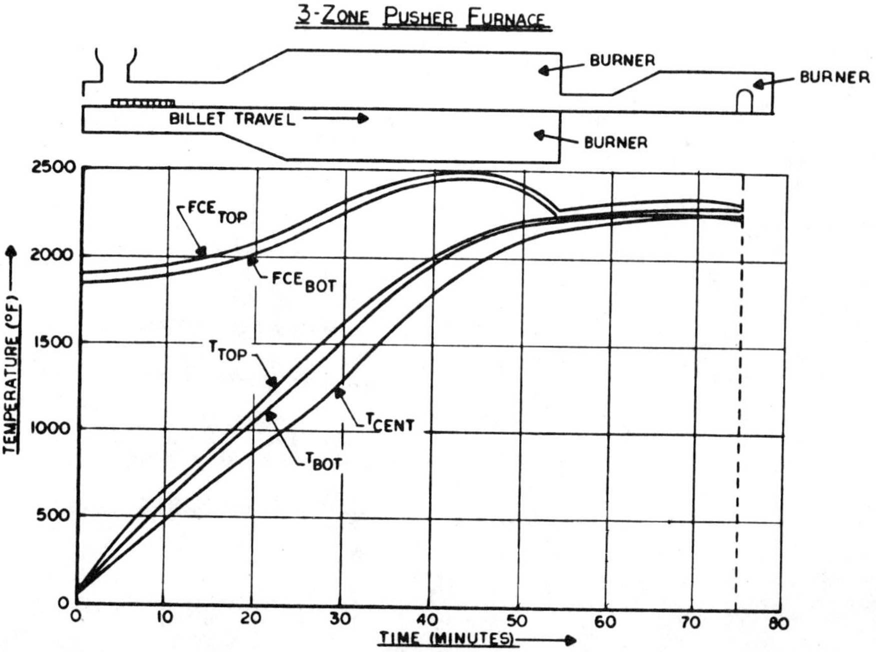

FIGURE 2

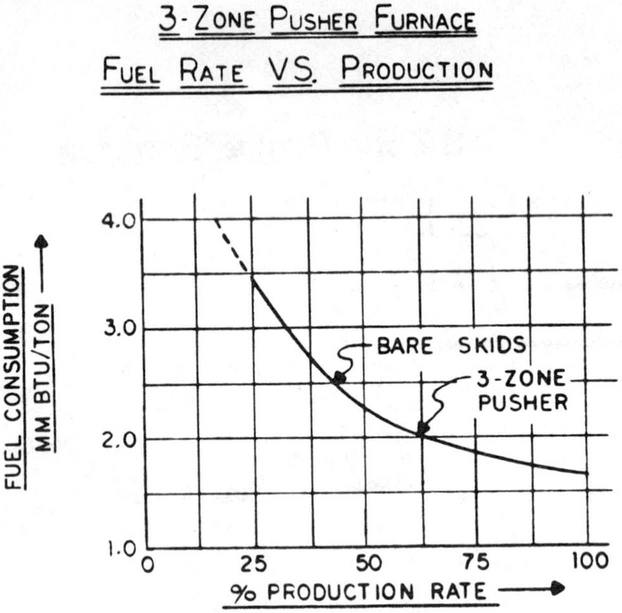

FIGURE 3

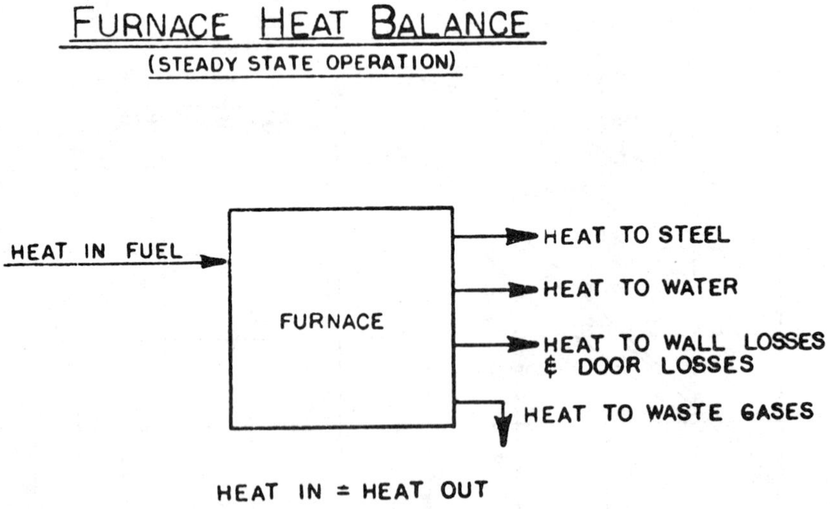

FIGURE 4

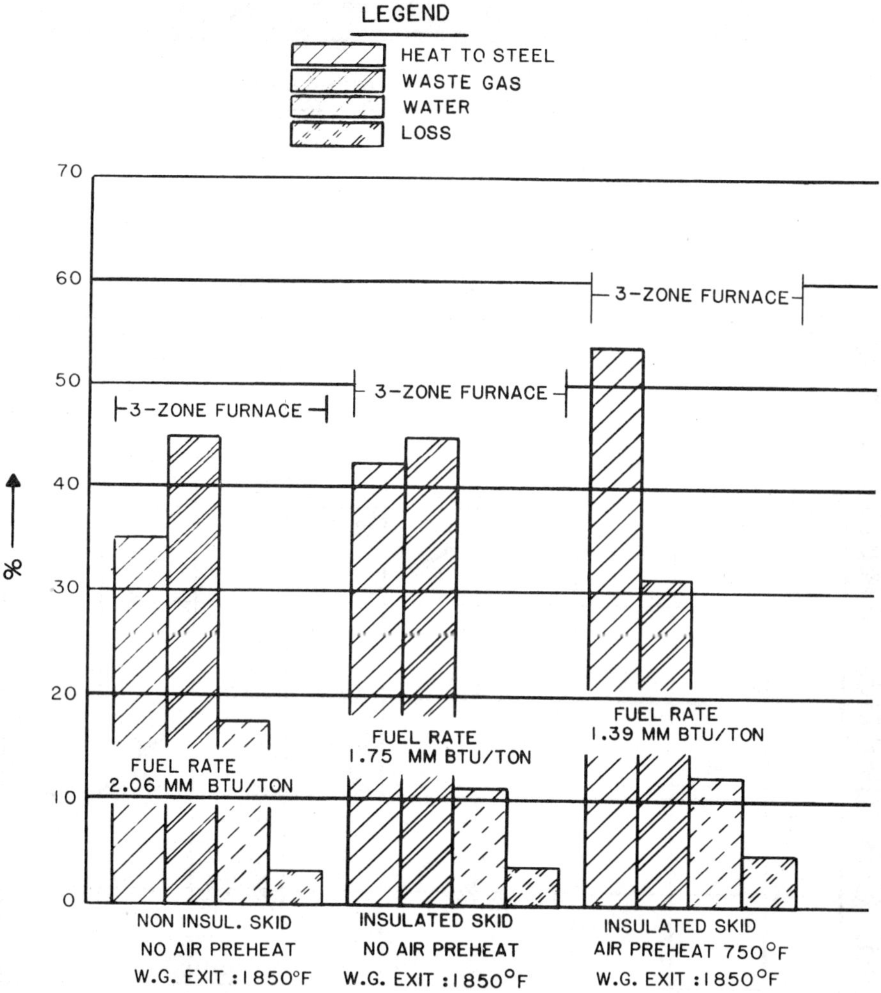
LEGEND
HEAT TO STEEL
WASTE GAS
WATER
LOSS
70
60
50
40
30
20
10
0
%
3-ZONE FURNACE
3-ZONE FURNACE
3-ZONE FURNACE
FUEL RATE
2.06 MM BTU/TON
FUEL RATE
1.75 MM BTU/TON
FUEL RATE
1.39 MM BTU/TON
NON INSUL. SKID
NO AIR PREHEAT
W.G. EXIT :1850°F
INSULATED SKID
NO AIR PREHEAT
W.G. EXIT :1850°F
INSULATED SKID
AIR PREHEAT 750°F
W.G. EXIT :1850°F

FIGURE 5 FIGURE 6 FIGURE 7

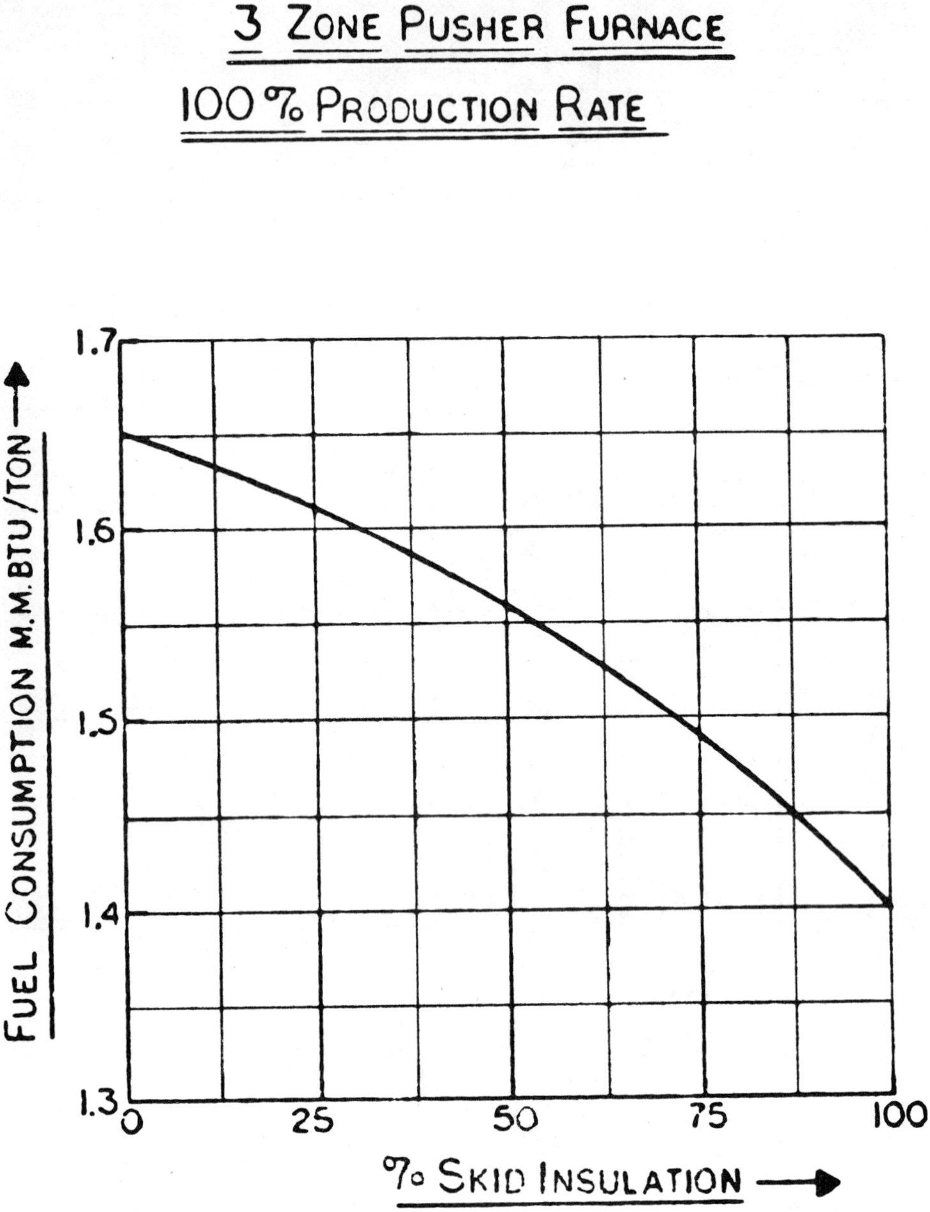
3 ZONE PUSHER FURNACE
100 % PRODUCTION RATE
FUEL CONSUMPTION M.M.BTU/TON
1.7
1.6
1.5
1.4
1.3
0
25
50
75
100
% SKID INSULATION

FIGURE 8

fuel use on the existing three-zone furnace, requested that we propose a design for a new system that was to be installed in another part of the bar mill operation. The new furnace was to have a capacity and product identical to that of the existing furnace.

Design criterion for the new furnace identified a critical need to minimize fuel consumption. Recalling our previous analysis, it seemed apparent that heat lost to water-cooled parts and through the waste gases would have to be reduced. Both objectives could be accomplished if the water cooling could be eliminated or minimized and if the waste gases could be discharged from the furnace at a lower temperature.

In the existing furnace, water-cooling occurred primarily on the skid pipes. To reduce the amount of heat lost through this cooling process, the product would have to be heated on a solid hearth. To adequately heat this 5" square billet, however, the hearth would have to be extremely long. Moreover, the waste gas temperature in a system with a solid hearth of this length would be almost identical to that of the existing three-zone furnace.

To circumvent these problems, the design for a two-zone, sweeparound pusher furnace was investigated (Figure 9). The sweeparound design allowed gases from the top zone to circulate around and under the product during part of the heating process. As a result, the billets could be heated sufficiently on a relatively short hearth. The new two-zone, sweeparound furnace also would improve the amount of heat transferred to the product by providing the benefits of two-sided heating through part of the heating cycle.

The two-zone furnace design that we considered had yet another advantage. Since it would require only minimal water cooling, the amount of heat lost to water-cooled parts would decline dramatically from levels typical of the three-zone furnace. Furthermore, the temperature of the waste gas as it leaves the sweeparound section would be 1500°F, 350°F lower than the exit temperature experienced on the three-zone furnace (Figure 10). The new design was approved and we installed the two-zone sweeparound furnace in our client's mill.

Even without a recuperator, the fuel rate for the two-zone sweeparound furnace was about 1.29 MM BTU/ton (Figure 11). This rate was about 10% less than that of the three-zone furnace with both insulated skid pipes and recuperated hot air. To optimize heating efficiency in the proposed furnace, the final design included a recuperator capable of providing 750°F preheated air. The fuel rate for the sweeparound furnace is 1.05 MM BTU/ton with the recuperator (Figure 12). This is much lower than the 1.40 MM BTU/ton rate experienced after both skid pipe insulation and recuperation were added to the existing, three-zone pusher furnace. At maximum production, the fuel rate for the sweeparound furnace is about half that of the client's original furnace.

The design of this furnace also featured improved turndown (Figure 13). Compared with the existing three-zone system, the sweeparound, two-zone furnace's fuel rate is insensitive to fluctuating production levels. This can be attributed primarily to the fact that more heat is transferred to the product in the two-zone, sweeparound furnace than in the three-zone furnace.

After the recuperator was installed on the sweeparound furnace, a downstream waste gas boiler was added to the system. This addition was proposed primarily because it could be used as a source of steam for atomizing No. 6 fuel oil, which at times had become the only available fuel. When the boiler is operating, the actual fuel rate is .85 MM BTU/ton, only 41% of the fuel rate on the original, three-zone pusher furnace.

A comparison of fuel rates reveals the systematic reduction of the fuel rate which resulted from our step-by-step systems analysis (Figure 14). Designed specifically to maximize fuel efficiency, the two-zone sweeparound clearly outperforms the mill's three-zone furnace.

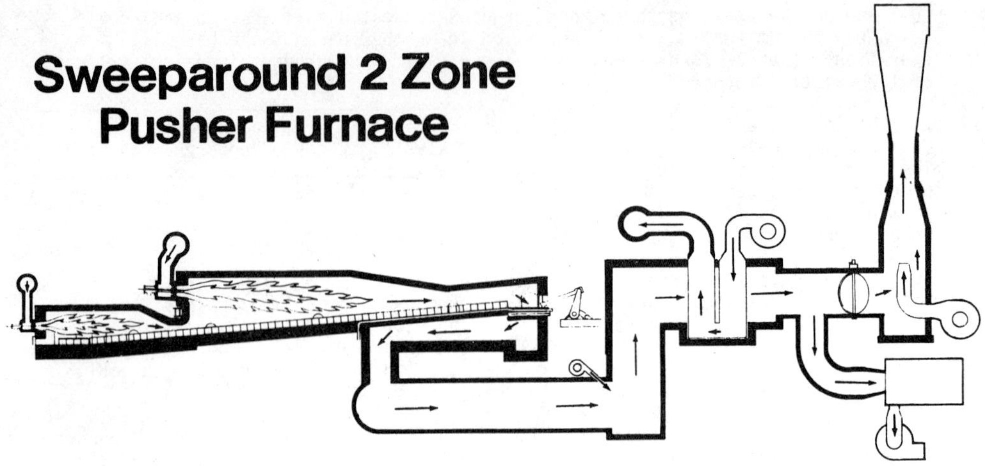

FIGURE 9

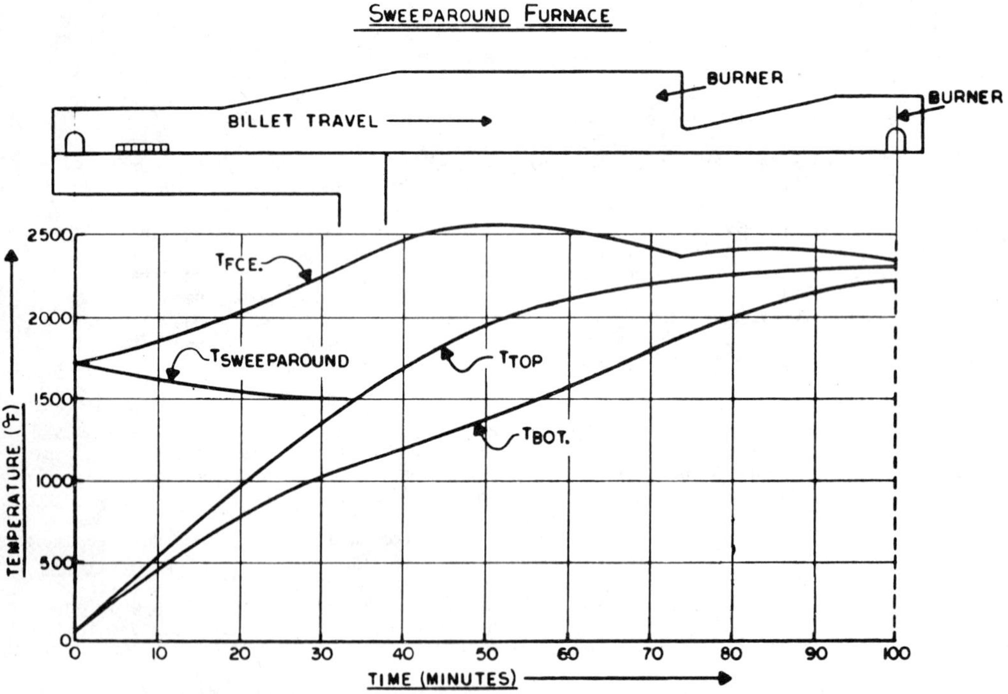

FIGURE 10

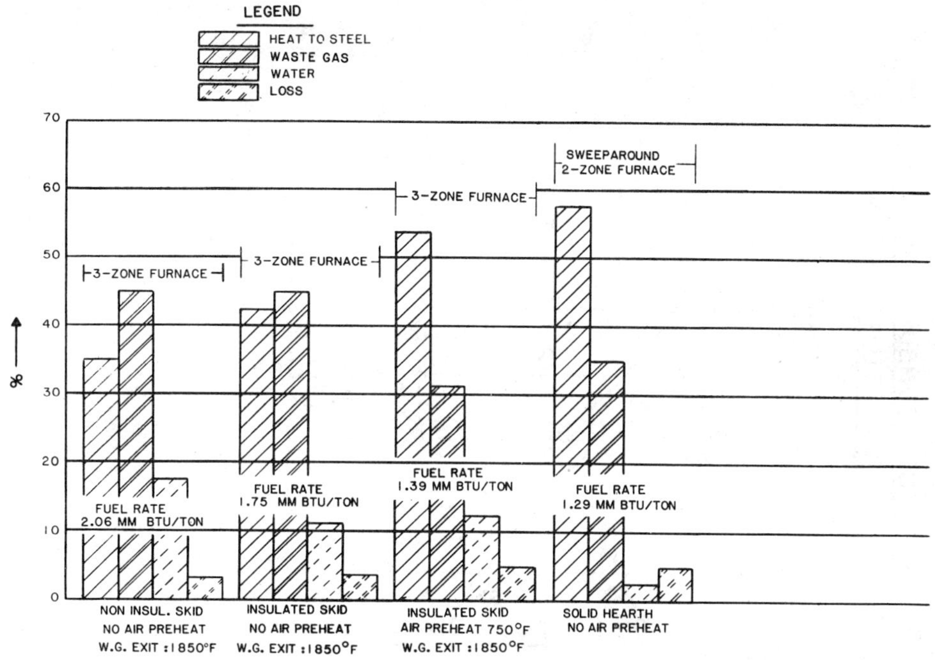

FIGURE 11

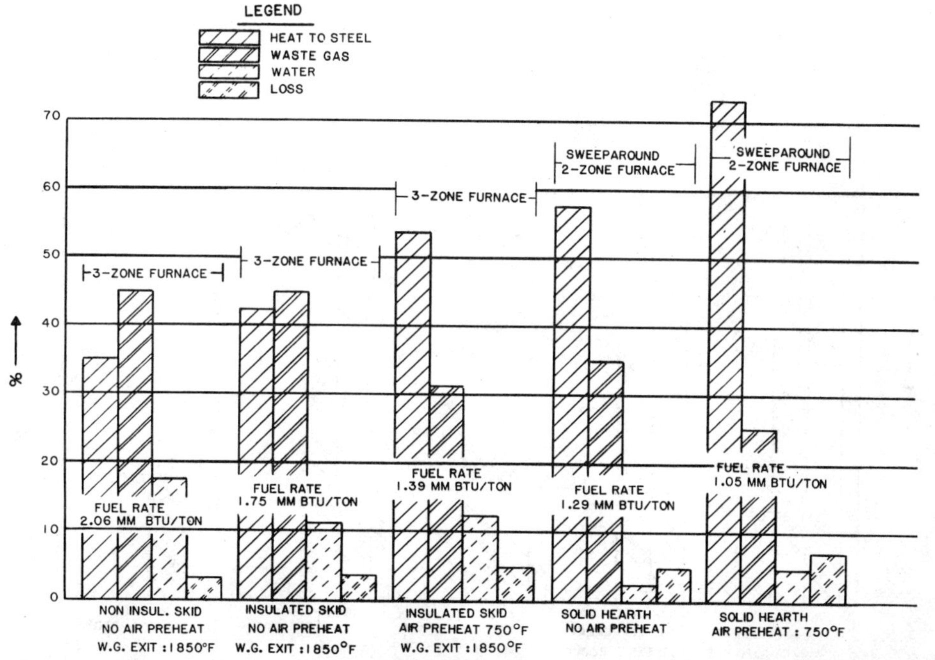

FIGURE 12

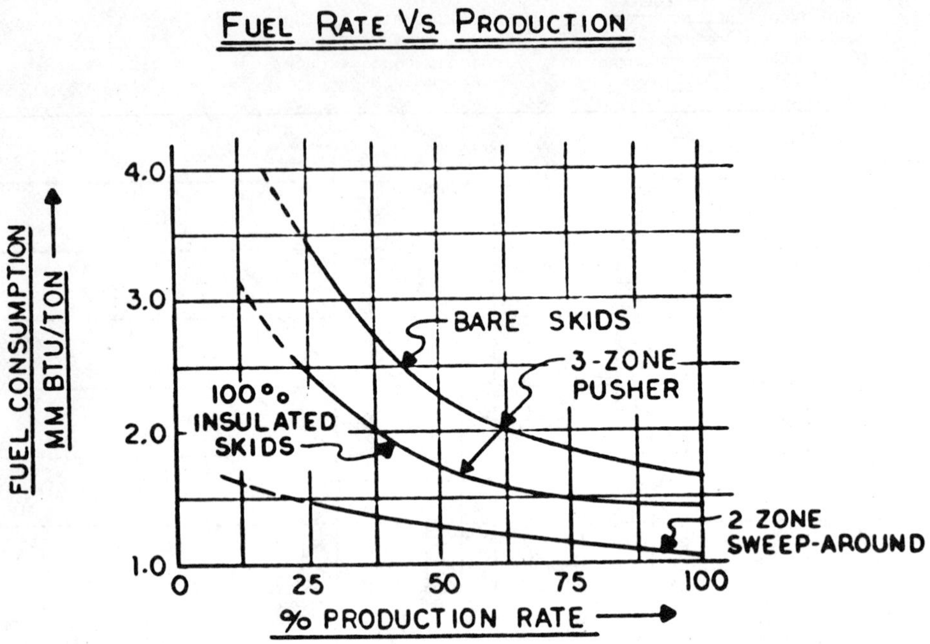

FIGURE 13

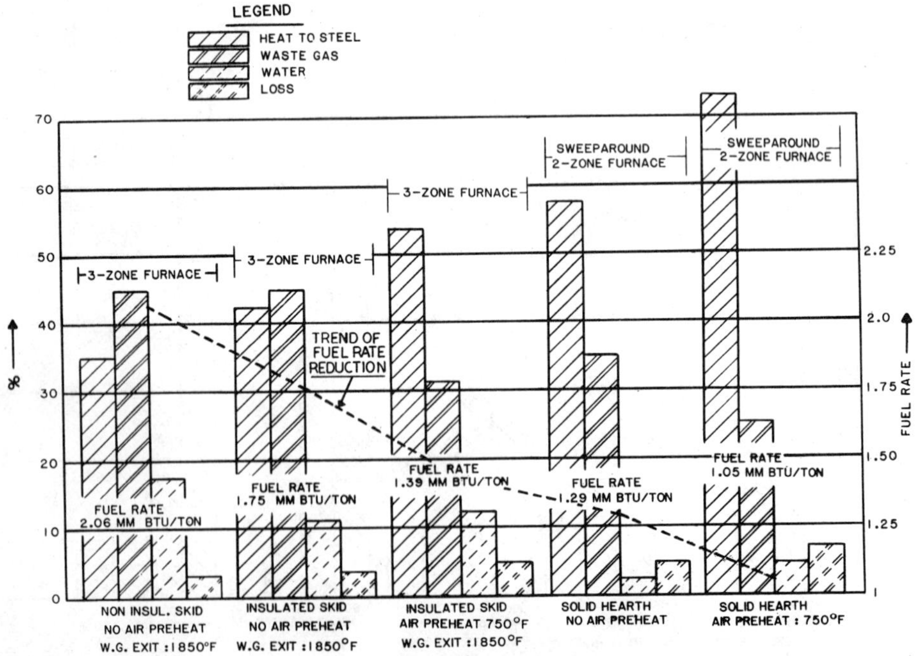

FIGURE 14

Our experience with the mini-mill illustrates the benefits of designing a furnace with fuel efficiency as a primary consideration. As demonstrated in the evolution of the client's bar mill furnaces, heat recovery can be maximized only when a systems approach to fuel conservation is adopted. Prior to implementing modifications and installing additional equipment, we must analyze carefully the entire furnace system, isolate elements that contribute to heat loss and remedy each of these within the system.

Arbitrary modifications may improve the fuel rate, but will prove incapable of ensuring optimum energy efficiency. For instance, if we had attempted to modify the existing furnace to match the efficiency of the new furnace, a preheated air temperature greater than 1700°F would have been required. This preheat temperature is a practical impossibility. In fact, even with the skid pipe insulation, the existing three-zone furnace would have required a 1300°F air preheat to match the fuel rate of the new sweeparound, two-zone furnace. This preheat temperature is economically almost impossible to achieve.

The waste heat boiler on the sweeparound system was installed after the recuperator. It was added because no other dependable equipment was available to consume the steam produced. This point should always be considered carefully when new furnace systems are being investigated. When justifying recovery systems, the system's interaction with other equipment can be quite significant.

To maximize heat recovery in reheat furnaces, all factors contributing to the inefficiencies must be evaluated. In our experience, we sometimes have discovered that the installation of a furnace with a new design may be preferable to extensively revising an older design. In the mini-mill case, the recuperated sweeparound will generate about $1,200,000 fuel savings on an annual basis. The projected savings is based on an annual production of 300,000 tons and a fuel cost of $4.00 per million BTU. The payback time for the costs of the additional features on the new sweeparound furnace, without adjustment for continuously rising fuel prices, was less than one year.

The existance of older furnace systems certainly is not unusual in the steel industry. To improve the efficiency of reheat furnace systems, we must remember that such furnaces were built to meet an entirely different set of design criteria. We should not expect one piece of new equipment, whether it is a recuperator, waste heat boiler or control, to transform an old furnace into a modern, energy efficient heating system.

If we are to improve significantly the energy efficiency of our furnaces, we must adopt the systems approach and isolate the sources of inefficient heating. After considering relevant economic and operating conditions, we can determine how to best modify existing furnaces or incorporate changes in new furnace designs. To effectively minimize heat losses, it is essential that we account for each source individually and then understand it within the context of the overall furnace system. As we have found, the system itself can be the primary source of inefficient heating, since it may have been designed to meet quite different criteria during an earlier time when fuel was both inexpensive and plentiful.

REVIEW OF RECUPERATION
IN THE AUTOMOTIVE AND METALS INDUSTRIES

Robert W. Meyer

HOLCROFT A DIVISION OF THERMO ELECTRON CORPORATION

In discussing the application of recuperators in the automotive and metals industry we will restrict our attention to "Modular" recuperators. Modular is a term used to distinguish recuperators applied to small combustion sources from larger units first developed for the primary metals industry. In general, modular recuperators, in a cylindrical configuration, will vary from 6 inches to a maximum of 3 feet on the inside diameter and support firing rates from 30,000 BTUH to 10 million BTUH. Although the initial use of recuperators; primarily to achieve higher flame temperatures, dates back to the turn of the century, their application to relatively small combustion sources has only occurred in recent years. Hence we are still on the learning curve with respect to their design and application. We will concentrate our review on the most significant aspects of this learning curve.

Before beginning our review a quick examination of the basic concept of recuperation is beneficial. The Sanke diagrams of figures 1 and 2 illustrate this concept. Energy normally exhausted in hot waste gasses is in part returned to the system in the form of preheated combustion air.

A more generalized concept of recuperation must include criteria for a systematic review of the systems components functions which are affected by recuperation, such as, process temperatures, combustion equipment, and mode of operation.

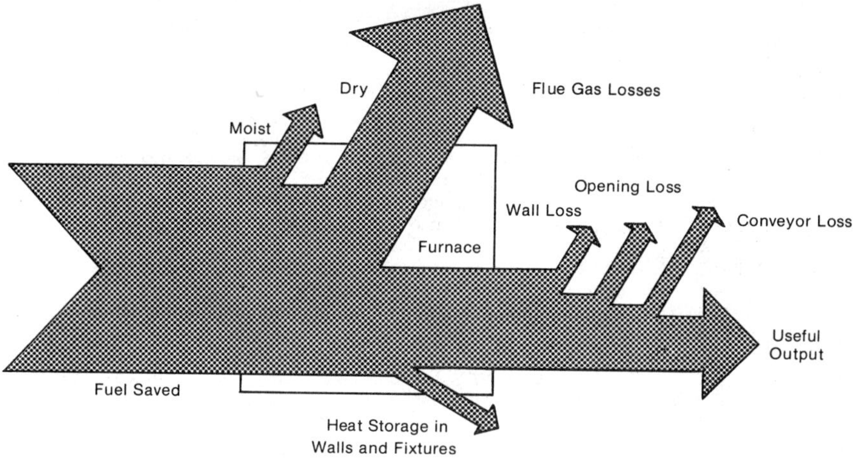

FIGURE 1

**SANKE DIAGRAM
UNRECUPERATED FURNACE**

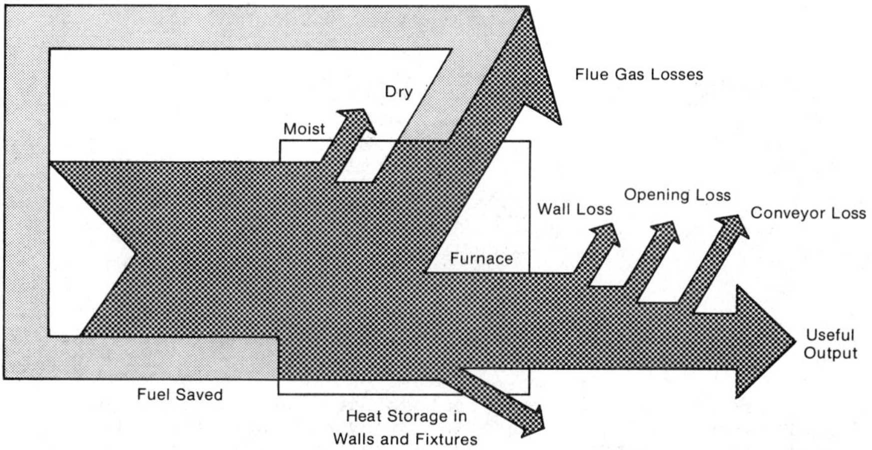

FIGURE 2

**SANKE DIAGRAM
UNRECUPERATED FURNACE**

Expanding this generalized concept to an application sequence we identify the following steps:

1) Determine expected preheat, and associated savings.
2) Establish recuperated flue to burner relationship, generally by zone of control.
3) Select recuperator size and calculate associated flow and pressure drop through combustion system.
4) Establish suitability of burner for hot air application.
5) Select new combustion system components as required.
6) Review system for consistency with today's insurance standards.

We will touch briefly on each of these steps in relation to their impact on the application of modular recuperators.

As would be expected, preheat and associated savings vary considerably from application to application even for seemingly identical processes. The concept of loss of preheat due to extensive piping and variation in insulation is well understood. Not as frequently considered is the variation of performance with fuel and air/fuel ratio. Table 1 illustrates the variations possible.

Fuel	Percent Excess Air		
	10	15	20
	Percent Savings		
Natural Gas	21.82	23.80	24.37
Propane/Air	20.73	21.94	23.18
Propane/Natural	21.14	22.36	23.50
Water Gas	19.03	20.11	21.20
Coal Gas	20.23	21.39	22.57
Blast Furnace Gas	22.39	23.70	25.03

800°F Preheat 1600°F Flue

Saving from Preheated Air

Table 1

As is seen in this table relatively small variations in fuel and air/fuel ratio can result in a 25 percent variation in realized savings.

While few users of modular recuperators have noticed significant saving effects from fuel composition, many have seen the effects of excess air when conversion to recuperation has led to closer maintenance of air/fuel ratio and much higher savings than can be attributed to preheated air alone. In some cases savings as high as 50 percent have been accredited to recuperation, where in fact only half of that savings is attributable to preheated combustion air. Other contributing factors are better heat transfer and better air/fuel ratio control.

A specific example of this effect can be given for the case of a conversion of a radiant tube furnace utilizing atmospheric burners with on/off control. Conversion to recuperation required sealed head burners. The use of sealed head burners greatly improves the capability of maintaining air/fuel ratio and this accounted for a 7 percent savings. During the burners off cycle a limited draft cooling effect occurs. Elimination of this effect coupled with improved heat transfer characteristics resulted in a further 11 percent savings. The preheated air gave a 22 percent savings for a net savings of 40 percent.

In many cases of retrofit recuperation careful consideration must be given to furnace performance prior to the installation of recuperation. For example, let us consider a direct fired two zone furnace heating to forging temperature. If the furnace has been operating with the second zone firing at maximum capacity a counterflow heating effect from that zone may reduce the required input in zone 1. Operating in this manner has little effect in a non-recuperated heating environment. However, upon retrofit to recuperation significant flue flow from zone 2 into zone 1 and out zone 1's flues may result in overheating and failure of metallic recuperators due to a mismatch of flue flow and combustion air being preheated. With a properly balanced zone set-up metallic recuperators have been very successful in such an environment.

As implied by our example the application of recuperation can be considered as special case with special concerns when dealing with high temperature (above 1800°F) processes. Generally, the special concerns will fall in three categories; material selection, flue contaminates, and thermal expansion. Thermal expansion is generally the least troublesome of these concerns in that an appropriate general design eliminates potential problems. Material selection and contamination combined to require evaluation of specific installations. One of the key criteria used in selecting either metallic or ceramic recuperators should be the temperature of the heat transfer surface. If designed so that surface temperature is well below flue gas temperature and the surface is shielded from direct radiation from the furnace, metallic recuperators have given good service in high temperature applications. Of more concern than simply temperature is possible contaminants that will attack recuperators at elevated temperatures. High temperature alloys have been used successfully in high temperature applications where flue gas has been clean. Stress corrosion alloys have been used with some success where contamination is a factor.

In some cases of application to reclamation processes carry over of charge contaminants has not been as detrimental to recuperators as might be expected. Although it is very difficult to project prior to installation, carry through deposits have been observed to stabilize at a relatively small thickness soon after installation. A decreased rate of heat transfer is seen but further build-up and rapid deterioration does not occur. Possible reasons for this are a degree of limited protection by initial build-up and thermal cycling limits extensive thickness of build-up. It would be very risky to project this for any general case of contaminants, however in some cases the potential benefits may warrant the risk of a trial installation.

There have been conflicting views on the appropriate combustion train conversions that are required when using recuperation. These have related primarily to burners usage and mass flow control. The conflicts relate not as much to specific application as the extent to which general guidelines are valid for a variety of applications.

There has been a tendency to associate a burner change with retrofit recuperation. Generally, this has evidenced itself by a tendency to derate burners or to select a larger capacity burners. Caution should be exercised here to avoid the potential of unecessary cost addition in retrofit of recuperation. It has been found a successful approach is to consider the burners as a simple orifice. That is, first consider increased pressure drop for a given mass flow through the burner. If adequate flow potential exists frequently minor changes, such as conversion to an alloy gas tube, can make a burner acceptable for hot air use if it is not initially. Still many burners are usable with hot air up to 700-800°F without modification. However, it is generally true that an existing blower will not provide sufficient flow at the required pressure drop with hot air. Hence the most frequent combustion system change is that of the blower rather than burner. In some cases, it has been cost effective to step up to a larger recuperator with less pressure drop and retain all combustion system components. In general, however, this will not provide the best economic payback.

The use of mass flow control is dependent on furnace operation, in particular temperature and load cycling. Few continuous furnaces in the automotive industry today can achieve a payback of mass flow control within the time frame of recuperation payback. This situation is changing as fuel cost increase faster than mass flow control device cost and a wider variety of control devices become available. Today a very rough rule of thumb is that a control zone of 7 MMBTUH with a 50/50 heat/hold cycle will pay back mass flow control at the same rate as recuperation.

For modular recuperators we may separate applications into three areas: (1) Application to radiant tubes. (2) Clean direct fired applications, and (3) Direct fired applications with flue contaminants. Although these applications overlap somewhat the technologies of application are distinctly different. The trend of usage of recuperation is shown in figures 3 and 4 by types of application. Figure 3 presents estimated total units and figure 4 presents fuel savings. As is shown in these figures the application of recuperators to radiant tubes is at a fairly stable usage level, while fuel cost increases continue to increase dollar savings. The application of recuperation to relatively small combustion sources for direct fired applications is still strongly in its growth curve. The application of modular recuperation in corrosive/high temperature flue environments has just begun.

In comparing figure 2 and 3 we see a proportionally much higher savings associated with direct fired units as compared to radiant tube. This is due simply to a higher average firing rate for burners

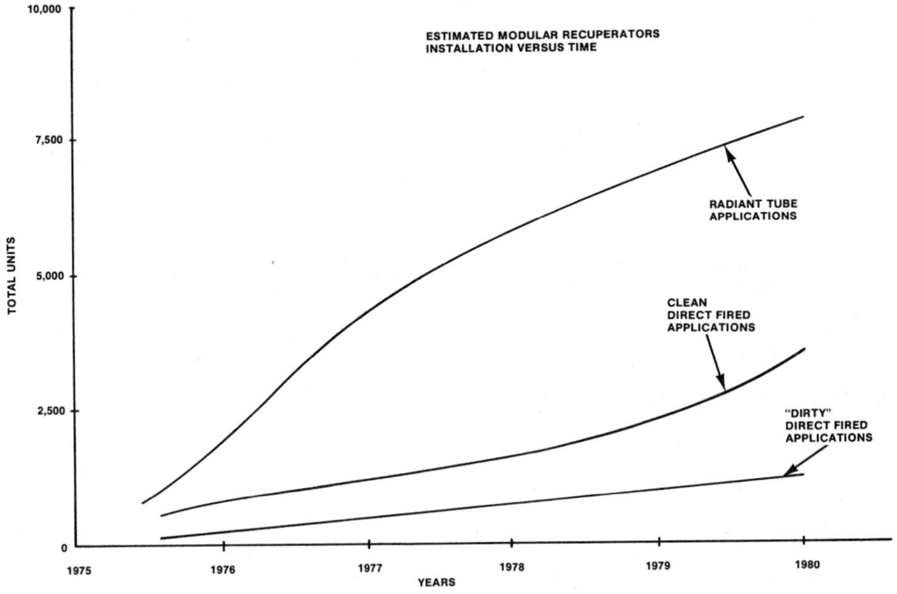

FIGURE 3

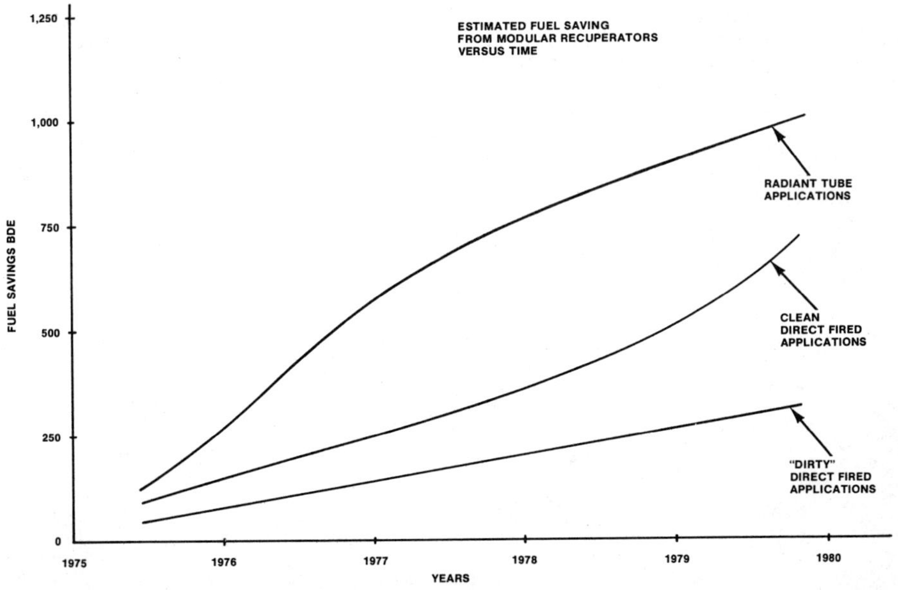

FIGURE 4

in direct fired applications. However, the figures serve to illustrate the increasingly more significant effect on fuel savings as the application of modular recuperators continues to rise in this category.

Considerable work is being carried out today to improve the adaptability of recuperators to adverse environment. The limits of metallic recuperators are close to being defined. Work continues on ceramic recuperators with their unique advantages and disadvantages; high temperature application and greater corrosion resistance as advantages, with sealing and low shock resistance as general disadvantages.

The fuel savings achieved by the application of modular recuperators in the automotive and metals industry is presented in Table 2.

Modular Recuperators Installed	8,500	units
Average Firing Rate Supported	325	SCFH
Average Savings	23	%
Average Yearly Operation	5,500	hours
Estimated Yearly Savings	3.5×10^{12}	BTU
	5.57×10^{5}	BBL Crude
	1,721	BDE
Fuel Cost Based On Natural Gas	3.25	/KKBTU
Estimated Yearly Cost Savings	$11,375,000	

Table 2

While these savings are significant, it is estimated that they represent only 10 to 15 percent of the potential applications in this industry segment for which a proven technology exists.

WASTE HEAT RECOVERY FOR STEAM GENERATION

W. H. MARSH

OPENING REMARKS

In order to put this case history in perspective, I am going to tell you about the Company, the manufacturing processes, and the feasibility study which resulted in construction of a waste heat boiler. Then we will look at the entire system and go through the component parts step by step.

At the risk of giving away the plot, I want to call your attention to three significant aspects of this project.

1) The availability of high temperature clean flue gas in large volumes.

2) A control system which was designed to allow a very flexible operation without disturbing the furnace providing the heat source.

3) The first known use, on a waste heat application, of refractory lined duct with lining installed at the factory. Very minor refractory work was required on site.

I. INTRODUCTION

A. Copperweld Corporation has executive offices in Pittsburgh, Pennsylvania with major manufacturing facilities of:

Copperweld Steel Company - Warren, OH

Ohio Steel Tube Company - Shelby, OH

Regal Tube Company - Chicago, IL

Alumoweld Products Division - Glassport, PA

Flexo Wire Division - Oswego, NY

Copperweld Southern Division - Fayetteville, TN

B. Ohio Steel Tube Company at Shelby, Ohio is on a 106 acre tract and has a floor space of 805,000 square feet.

C. Raw materials processed at this plant are hot rolled round steel billets and hot rolled steel coils.

D. These raw materials are converted into seamless steel tubing and electric weld steel tubing. Total production capacity is about 50,000 tons of seamless and 90,000 tons of electric welded tubing annually.

II. MANUFACTURING PROCESSES

A. Seamless

1. Billets are heated to 2200° F in a rotary hearth furnace.
2. Pierced or center-punched.
3. Rolled through a piercing mill to make a tube shell.
4. Elongated by rolling on Assel Mill to nearly finished size.

B. Welded

1. Coil steel stock is fed into an electric weld mill where tube is formed;
2. Welded;
3. Cut to length.

C. Processing of tubes manufactured by either process.

1. Tubes are annealed in a roller hearth furnace.
2. Pickled with hot sulfuric acid to remove scale from hot rolling or incoming hot-rolled strip.
3. Cold drawn to finished size.
4. Stress-relieved.
5. Followed by pickling, straightening, cutting, testing and packing.
6. Heat required for solution heating of pickling acid tanks is presently supplied by fuel-fired boilers.

III. UTILITIES - BOILER FUEL IS EITHER:

A. Natural gas - 80,000 to 110,000 MCF per month;

B. Propane-air - 374 M gallons stand-by storage; or

C. Fuel oil - 1,242 M gallons stand-by storage.

IV. ENERGY CONSERVATION

A. In 1976 and 1977, Ohio Steel Tube Company faced drastic curtailment (up to 96%) of natural gas supply.

B. The rotary hearth furnace, used to heat billets for seamless tube production, was found to consume 33% of the total Plant energy.

C. Boilers use 15% of the total fuel to produce steam for heating pickling facilities.

V. FEASIBILITY STUDY FOR ENERGY CONSERVATION

A. The rotary hearth furnace was identified as the most likely candidate for energy conservation since it:

1. Consumed the major portion of fuel;
2. Has long hours of operation;
3. Has direct relationship with pickling requirements;
4. Has large volume of high temperature (2400°F), clean flue gas.

B. Heat balance of furnace

1. Description of furnace - Inner hearth diameter is 22 feet and the outer hearth diameter is 44 feet. There are three heating zones, with a total burner input of 71 million BTU/hour. The maximum firing rate corresponds to a through-put of 26 tons per hour (TPH) of steel billets. Analysis shows 25% of the heat goes into steel, 2.6% is lost through furnace walls, 0.2% into water seal, 24% through charging doors, and 48% out with flue gases.

C. There are three 14 square foot flue openings which discharge into the building. At 26 TPH production rate, 34 million BTU/hour is discharged through flues. A conservative 9 TPH (slightly below average production rate) was used for justification; and at this production rate, 12 million BTU/hour is contained in the flue gases.

D. It was at this point that the engineering firm of W. Vance Middough and Associates was commissioned to conduct a feasibllity and cost study. A conservative preliminary report indicated that total costs of $725,000 would provide a waste heat recovery installation generating 280 boiler/horsepower at a minimal 9 TPH billet heating rate. Maximum steam loads in the range of 500 boiler

V. FEASIBILITY STUDY FOR ENERGY CONSERVATION (Cont'd)

D. horsepower could be generated at a furnace heating rate of 16 TPH. The economic justification considered production at 9 TPH, natural gas cost of \$2.30 per MCF, and payback in approximately five years. (Present cost of natural gas has risen from \$2.30 to \$3.00 per MCF).

E. Operation and maintenance was planned to be handled by the existing labor force. In order to generate steam with this waste heat, the flue gases must be collected, conducted through a boiler, and furnace pressure control maintained (0.05" wg). Also, there must be provisions for positive shut-off of flue gas to the boiler for maintenance purposes.

VI. WASTE HEAT RECOVERY SYSTEM

A. Boiler feed-water comes from the existing system, since steam generated by the waste heat boiler is directly replacing present boilers. Piping treated water from the old boiler house to the waste heat boiler was more economical than new installation. It should be noted that there is no condensate return in this plant. All steam used is directly sparged into pickling, caustic, or rinse tanks.

B. The spray type deaerator was sized to handle 150% of maximum boiler throughput with 100% make-up water.

C. The waste heat boiler is a three-pass design, fire tube type, built for 150 psig steam working pressure. Nominal specifications call for the boiler to reduce hot gases from 2100° F inlet temperature to 460° F outlet temperature. Large gas passages hold pressure drop to a maximum of 12 inches water column.

The boiler was factory insulated and skid mounted, provided with refractory lined inlet.

In addition to normal water column, blow-off valves, and safety valves, there were thermocouples inserted in the inlet and outlet gas streams. Controls furnished with the boiler monitor pressure, water level, and gas temperatures. Automatic shutdown, with indication of malfunction, is built into the system.

The waste heat boiler and auxiliaries, as a system, are rated at 600 boiler horsepower or 20×10^6 BTU/hour; which is well in excess of the present steam requirements. All plant steam is used for process solution heating.

VI. WASTE HEAT RECOVERY SYSTEM (Cont'd)

D. The exhaust flues were extended vertically with furnace pressure control lids relocated to the top of extensions. Flue additions incorporate a flanged outlet for breeching which conducts the hot gases to the boiler. The breeching was shop fabricated with castable refractory centrifugally cast into 1/4 inch thick A.S.T.M. A-36 steel jackets. Twelve foot lengths were assembled on site with tees, elbows, and expansion joints. Isolation dampers were installed immediately downstream of furnace flues to shut off the hot gas stream. These butterfly dampers are fabricated of Inconel 601 with air cooled shafts. They are opened and closed by hydraulic actuators, no modulation. Maximum gas flow of 18,450 ACFM at 3100 fpm and 2200°F.

E. Gases are pulled through the breeching system and boiler by a high temperature induced draft fan. An inlet damper, hydraulically modulated, controls flow through the system. Gases are discharged through a steel duct into a 50 foot tall refractory lined stack. Expansion joints are provided on both sides of the fan. Specifications are:

Cold (70°F) 18,820 CFM 18.92" S.P.
Hot (450°F) 18,820 CFM 10.97" S.P.

Fan is driven by a 75 hp motor.

F. The damper control system is hydraulically operated and consists of three parts:

1. Three 33 inch diameter butterfly dampers in breeching near existing flues, open-close operation.
2. Induced draft fan inlet damper, modulating control.
3. Transfer valve to block existing flue lid dampers hydraulic controls.

By manual control, the butterfly dampers are closed, isolating the waste heat boiler. All products of combustion discharge through the rotary hearth furnace flues. Furnace pressure is controlled by modulation of the lid dampers (original operation).

In the waste heat mode of operation, the butterfly dampers are opened and furnace pressure is controlled by modulation of the induced draft fan damper. When steam demand is greater than the output of waste heat boiler, package boilers are also on line. The transfer valve mentioned earlier has blocked the existing hydraulic controls and closed the lid dampers.

An override control on the I.D. fan inlet damper is activated at high steam pressure. This occurs when the waste heat boiler can supply more steam

VI. WASTE HEAT RECOVERY SYSTEM (Cont'd)

F. than required. The flow of hot gas is reduced, and control of furnace pressure is turned back to prior existing system. Excess flue gases are vented through flue dampers, while the I.D. fan inlet is throttled according to pressure requirements.

G. The boiler deareator, I.D. fan and controls are housed in a new building addition immediately adjacent to the rotary hearth furnace building.

CLOSING REMARKS

The entire project went quite smoothly, from inception through start-up in December, 1979. Costs ran about 1½% above the appropriation made 18 months prior to completion.

However, there is one problem which prevents the project from being an overwhelming success. Upon application for an operating permit or license, we found that an agency of the State of Ohio requires a licensed boiler operator to be on duty while generating steam. Fuel savings of about $230,000.00 per year would therefore be partially offset by additional labor cost of $140,000.00 per year. The $90,000.00 per year net savings does not justify the project.

A legal action is pending with the State of Ohio to obtain relief from a law which dictates that a boiler operator must observe the operation of this system in addition to the furnace-tender who controls the firing of the furnace. Due to a misunderstanding or misinterpretation of regulations, we believed that no additional personnel would be required. I caution you to fully investigate this area if considering the installation of a waste heat boiler.

OPERATING FLEXIBILITY OF FOUR HEAT STORAGE SYSTEMS

Michael J. Cahill
Niagara Mohawk Power Corporation

In recent years, several U.S. utilities have offered rates which attract customers to installing thermal energy storage (TES) devices. These utilities are confident that such devices can provide them customer loads which can be controlled either in real time or by clock mechanisms to reshape their system profile, while still having the ability to perform the utilization function sought by the customer. For a winter peaking company, these potential swings in load can be significant. This application is not a new one, with fairly widespread European experience, especially in Germany and England. The question often arises as to when does the concept have application for American utilities, and would it gain customer acceptance.

To help answer this question, I think it is important first to examine key items that today's customer expects from a home heating system.

- Comfort - In European countries that have had success with TES devices, the systems that they replaced usually have been room space heaters, or fireplaces and the like. American homeowners have been used to central heating systems with balanced heating throughout the home. They demand this same type of stable heating condition in any heating system offered for their consideration. Wide fluctuations of temperature would not be acceptable.

- Economical operation - The heating systems in the American homes were designed primarily for comfort and secondarily for economical operation due to the abundance of cheap energy. The rapidly escalating cost of energy in recent years has given the impetus for more effort to be devoted to the improvement of efficiency in the heating devices. Conventional heat pumps, add-on heat pumps, pulse combustion furnaces, furnace modifications, etc., have all played a part in maximizing the BTU output per unit of energy input. TES devices coming into the marketplace must measure up against these new standards of performance.

- Low maintenance costs - The American customer, over the years, has been conditioned to taking his heating system for granted. The common, central, forced-air system requires little more than filter changes and an occasional lubrication. Several years may go by before anything beyond that is called for. Hydronic systems may have occasional valve failures, but otherwise operate trouble free year after year. Electric baseboard may cut out on thermal protection or have a line thermostat fail, but again, will operate trouble free year after year. Heat pumps, because of their more complex nature, usually require a little more maintenance than those other devices mentioned, however, it is still fairly minimal. In order to gain acceptance in the American marketplace, TES devices must offer this same degree of reliability.

- Safety - The American public has become more and more safety conscious and demand high safety standards for any device that they place in their homes. Here again, the same foolproof safety measures must be available for TES devices as are available for conventional devices.

There are several TES devices available in the marketplace which may have the ability to respond to these expectations:

- Ceramic Brick Systems - These are available either in room units or in central block furnaces which utilize a high density brick made from magnesite or some similar material. This brick can be heated in excess of 1400°F. By encasing these bricks in a heavily insulated jacket, it is possible to blow air across them in a controlled fashion and draw off the necessary heat required for comfort heating over a period of time. This is usually done through a series of face and bypass dampers which mix the system air with room air. (See Figure I)

- Hot Water Storage Systems - There are several configurations utilizing water as the storage medium. The most common of these utilizes a high pressure storage vessel where water can be heated to 280°F or higher. A secondary water loop is passed through this container and then circulated directly into a standard hydronic system or is circulated into a water coil which is located in the air stream of a forced air system. (See Figure II) Not as common as this, but attracting more and more attention, are water storage systems (usually at normal pressures) that utilize heat pumps. In this application, several different strategies are being tried, including the use of the heat pump to raise the heat of the water storage in off peak periods, and then using the stored water as a source during the on peak period, or prioritizing the operating mode dependent on a signal from the utility. (See Figure III)

- Floor or Under Floor Heating - Systems employing the floor slab or a storage bank underneath the floor slab are also commonly available. This type of application obviously does not have the dynamics associated with the systems discussed earlier, however, the simplicity of installation is attracting more and more attention. It may well be that there is a place for this type of system in a TES setting, especially when used with direct control.

In addition to these, there are also several approaches being experimented with which employ phase change materials, such as eutectic salts, paraffins, etc.

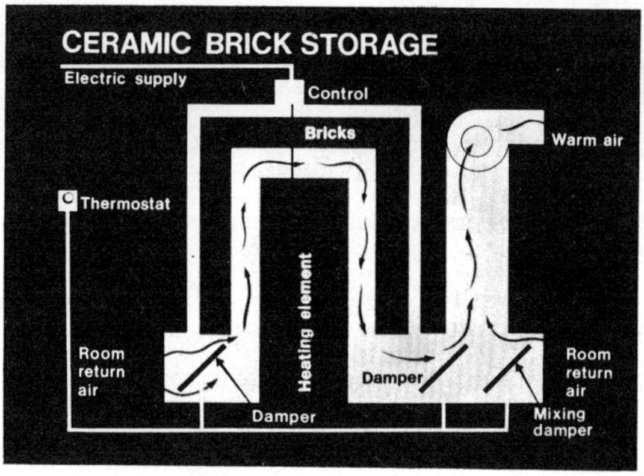

Figure I.

Schematic of the ceramic brick storage system.

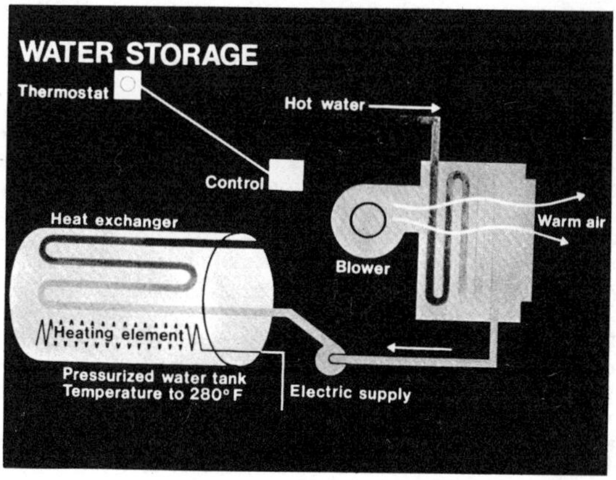

Figure II.

Schematic of the water storage system.

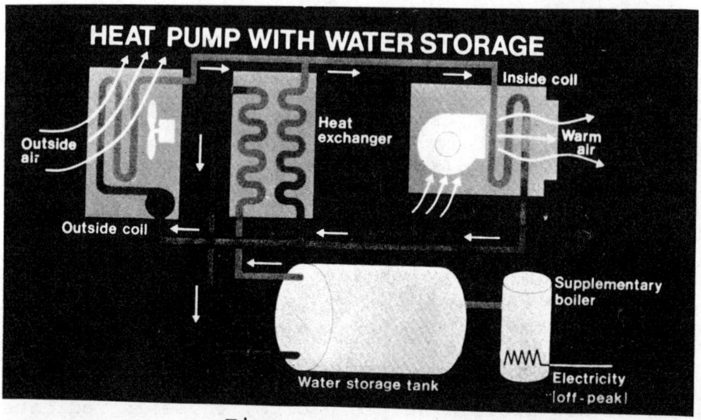

Figure III.

Schematic of the heat pump with water storage system.

Whether all of these or any of these will work, varies from one utility application to the next. An examination must be made of the device to determine its ability to respond:

1. To controllability - Can the device respond to the control strategy that the utility wishes to employ today?

2. Can the device respond to control startegies which may be employed in the future?

3. Can the device satisfy the customer's heating requirements based on today's control strategy?

4. Can the device satisfy the customer's needs with control strategies that may be employed in the future?

Projecting the growth of electric loads and where they will occur has become increasingly difficult due to the instability in the energy picture. Such things as the widespread use of electric vehicles, conversion from critical fuels, or any number of industrial processes can have dramatic affects on a utility's load profile. As these load profile changes occur, it may be preferrable to utilize these controllable loads in an entirely different fashion so that they can track or offset these perturbations that effect the load profile.

The utility company at first might decide on a control strategy utilizing time clocks which may well be the best answer for their immediate needs. They usually will have a very well defined off peak period that will allow large blocks of load to be shifted into the valley, gaining significant benefits on their generation system. The problem that may arise from this, however, is that as the TES load increases on particular feeders, it may well be that they become overloaded and the benefits gained from the affects on generation will be more than offset by the negative affects on the distribution system. This basic realization has convinced more and more utilities that a more effective control strategy that could be utilized with load management is one which gives them direct control over the customer's loads. This latter approach also has the added benefit that the customer loads that need to be available for this direct control do not have to be of the magnitude of those that are utilized in a time control method.

Unfortunately, it may not be practical from an economic standpoint for a utility company to put in a direct control system up front because of the immediate lack of controllable loads available on its system. Consequently, many utilities opt to put in time control first with the thought that later on they can switch to direct control when it becomes of an advantage to do so.

The incentives offered to customers can also play a large part in shaping this profile. An assessment of today's marketplace may dictate that a utility select a particular incentive which would be designed to attract the preferred number of customers towards installing TES devices. The objective would be to have a favorable effect on their generation profile. It may well be, however, that after a short period of years, the rapidly changing market conditions may require an entirely different incentive.

There is a distinct balance that needs to be maintained between the type of equipment that is utilized, the controllability of this equipment and the incentives that are offered to attract the customer. As new profile shapes evolve, these ingredients may get out of balance.

Obviously, it becomes highly desirable that the TES systems installed should have the flexibility to respond to a wide range of control strategies and incentives. This flexibility ideally should be built into the TES device at a cost to the customer which, when coupled with the initial incentive, can still be attractive to the customer. Products coming into the marketplace should be evaluated on their ability to meet this criteria.

There are systems available which can respond to fixed Time of Use rates that will charge during the off peak periods, and then discharge over the remainder of the day. This discharge period, many times, will exceed 16 hours a day. These systems usually will also have the flexibility of being able to respond to a contract demand rate at a future date, however, the customer may be required to invest in storage that he may not need later. The system requirements may call for discharging over a two or three hour period at intermittent spots during the day. Unfortunately, the first cost associated with these high capacity storage systems is usually quite high when compared to a conventional system. It may well be that a customer will expect a fairly long commitment from a utility assuring that an attractive rate will be available to cover the lifetime costs of the system. If it is expected that a much lower capacity system will do the job as the load management program evolves, then the utility may well be faced with the responsibility of keeping the customer connected on the line at a rate which is no longer responsive to real costs, but must be maintained in order to satisfy customer relations. If a balance is sought between today's needs and tomorrow's needs by splitting the difference, it may well be that the undersized systems that would result would respond to the utility needs only at moderate times. When the utility system needed controllable loads the most (during the coldest periods assuming that this was coincident with their peak) the customers would be overriding the controls and putting unusually high peaks on the system at a time when the utility could least afford it.

Charging strategies can differ greatly, depending on what results a utility is seeking. Ideally, from an energy use standpoint, a TES system would charge immediately preceding the discharge period to the exact level required to handle the full discharge period. This would keep stand-by losses to a minimum and would maximize the utilization of the device. This may not be practical, however, not only because of the lack of sophistication in available control hardware, but also because the utility system impacts may not be those sought. Utilities may very well seek to stagger the charging of the TES devices over a fairly wide charging window to minimize these impacts, especially those felt on the distribution system.

The ability to respond to these different charging conditions place decidely different requirements on the TES device. For example a TES device designed to accept full charge at the start of the charging period must be insulated well enough to retain that charge until it is called to discharge. If, on the other hand, the charge period immediately precedes the discharge period, the ability to retain the thermal energy over an extended period is eliminated, greatly reducing the insulating requirement. This consideration becomes that much more critical if weekend/weekday charge cycles are considered.

Not only is it important to store the heat appropriately for later use, it is also very important to consider the ability to extract heat from the storage medium. The rate of delivering heat can be critical for many applications, especially where the charge is built up over an extended period and then discharged very rapidly.

The uncertainty of these future requirements make it essential that a utility look for flexibility in whatever TES devices that they are trying to encourage coming onto their system.

Niagara Mohawk is currently managing a thermal energy storage demonstration at the Raybrook Correctional Facility located near Lake Placid, New York. The site was formerly used as the Olympic Village during the conduct of 1980 Winter Olympics. Sponsors in this program include the Department of Energy, Electric Power Research Institute, the New York State Energy Research and Development Authority and the New York State Department of Public Service. The host for the experiment is the Federal Bureau of Prisons. The main thrust of this program is to determine impacts that can result on a utility system from the use of TES devices when employed in a load management concept. In addition to this objective, one other very important part of this program will deal with the flexibility of TES devices. An attempt will be made to find out what the operating parameters of these various systems really are, and under what control modes that they can be the most effective. The program will determine the ability of each of the devices to respond to different control strategies under minimal charging conditions, as well as maximum charging conditions.

It is expected that the results of this program will provide a valuable data base to utility companies and customers in determining which TES devices may have application for their purposes.

OPERATING EXPERIENCE WITH OFFPEAK BRICK FURNACES, PRESSURIZED WATER SYSTEMS, AND EUTECTIC SALT SYSTEMS

Harold G. Womble
Anthony B. Dahmus
E. Tylor Claggett, Jr.
Edward L. Colston

Tennessee Valley Authority

ABSTRACT

This paper describes the residential thermal storage efforts of the Tennessee Valley Authority (TVA). Emphasis is placed on TVA's operating and installation experiences with three particular offpeak heat storage technologies and a discussion of the analysis planned for the data collected during the test.

INTRODUCTION

TVA is a corporation of the United States Government serving approximately 2.5 million residential consumers in seven Southern States. The power program within TVA is completely self financing. TVA sells wholesale electric power to approximately 160 independent retail power distributors. The 1980 winter peak demand established on TVA's system was 20,745 megawatts at 8 a.m., February 1.

TVA's interest in load management intensified during the late 1970's as a possible means of mitigating the impacts of a daily load swing of several thousand megawatts. The benefit of managing the daily swings includes the deferring or delaying future capacity construction plans and the moderating of future rate increases for TVA consumers.

The first load management experiments with offpeak thermal energy storage in end-use applications were tests of the then available (in limited quantities) central ceramic brick storage systems and a test of two prototype heat/cool storage units. The test of ceramic brick systems demonstrated that 85 to 90 percent of the total home energy could be shifted offpeak, while the heat/cool storage systems were a failure. The heat/cool storage systems test was terminated after 9 months of operation because of poor system performance, maintenance problems, and consumer complaints.

Encouraged by the preliminary success of the ceramic brick systems, TVA decided to expand its thermal storage efforts through further field testing. This proposal became known as the Residential Thermal Storage Field Test.

TEST OBJECTIVES AND EXPERIMENT DESIGN

The heat storage units are being placed in the homes of volunteers in eight Valley distributor areas. The actual selection of test homes is a function of the individual consumer's willingness to accept the unit, physical considerations of the storage units, and the availability of qualified installers. The thermal storage test has been designed to permit:

1. The determination of the effects of thermal storage units (in conjunction with time-of-day electric rates) on the TVA winter load profile and the consumer's heating costs.

2. The determination of the thermal storage system's reliability and efficiency.

3. The determination of cost/benefit information for alternative thermal storage systems.

4. The determination of consumer acceptance of thermal storage systems with time-of-day rates.(1)

ANALYSIS AND DATA REQUIREMENTS

The ideal tool for evaluating the feasibility of thermal storage would be a comparison of load profiles from the same test residence, under identical conditions, with and without such units. Without such an ideal, the two basic evaluation strategies that follow remain.

---

1. The experimental time-of-day rate has the following characteristics:

   Customer charge:

   $5.10 monthly

   Energy Charge:

   All onpeak kWh per month at 6.8 cents/kWh
   All offpeak kWh per month at 2.2 cents/kWh

   Onpeak Hours:

   During the months of January through April and November through December from 7 a.m. to 11 a.m. and from 5 p.m. to 10 p.m., excluding Saturdays and Sundays. During the months of May through October from 11 a.m. to 10 p.m., excluding Saturdays and Sundays.

   Offpeak Hours:

   All hours not onpeak shall be offpeak.

1. The individual heating load profiles observed during the test with thermal storage will be compared to proxy heating load profiles which would have occurred without thermal storage.(2)

2. An alternative method is a direct comparison of billing data from previous years without thermal storage to billing data with thermal storage and time-of-day rates. This method has the advantages of being cheaper and independent of modeling difficulties although it is analytically less rigorous.

Descriptive home construction data will be required to utilize the proposed models to their best advantage. This and much necessary demographic, behavorial, and attitudinal data will be gained through the use of one or two pretest surveys. In addition to the home data, hourly weather information is also required. Representative weather variables include: (1) dry-bulb temperature, (2) wet-bulb temperature, (3) wind speed, (4) cloud cover, and (5) ground temperatures. This list is not all inclusive; the idea is to gather as much weather data as possible.

Most of the test homes are located close enough to TVA power plants to make use of weather stations located at those sites. For those locations distant from TVA facilities, arrangements have been made with other Government agencies to obtain the required meteorological data.

Ambient conditions inside the test home will be collected by temperature/humidity recorders.(3)

Standard electrical service submetering data is a test data requirement. The total home load, water heating load, and space heating load are being collected by magnetic tape demand recorders.(4) While the test is underway, detailed complaint, failure, and maintenance logs, along with complete installation and other cost records, are being kept by power distributors and TVA field engineers.

The data collected from submetering packages, modeling efforts, and weather observations will be used to estimate kW and kWh shifts from onpeak to offpeak hours. These estimates can then be used to determine potential effects on the TVA winter load profile assuming the widespread use of thermal storage. The same shifting estimates, coupled with rate data, will then be used to determine the effects on residential heating costs.

---

2. The proxy home load profiles will be generated by computer models designed for this purpose. The two probable models are the TRNSYS residential models developed by the University of Wisconsin and the Department of Energy's model DOE 2.1.

3. When indoor conditions are not recorded, a suitable substitute is the test home occupant's answers on test surveys as to what his thermostat settings are during the various periods of the day.

4. Metering consists of both time-of-day watthour meters and standard watthour metering equipment equipped with pulse initiators. Pulse data from these meters is recorded on a 4-channel magnetic tape recorder for load profile analysis.

   All homes are equipped with GE IR-70 time-of-day meters. These three register meters are field programmed to register onpeak,

The data will also be used to compare differing technologies. Coupled with failure and maintenance records, system reliability and efficiencies will be determined. The electrical submetering data and meteorological data will be stored in computer files for future TVA heat storage implementation efforts.

An appraisal of consumer acceptance of thermal storage will result from analyzing the data taken from the post-test questionnaire and complaint log.

EQUIPMENT SELECTION AND INSTALLATION

After receiving a formal project authorization from the TVA Board of Directors on August 30, 1979, the engineering staff of TVA's Load Management Branch conducted a review of the recent thermal storage activities of other utilities. Simultaneously, a review of existing technologies was completed to determine the available heat storage equipment. Three manufacturers were selected to test brick heat storage, eutectic salts heat storage, and pressurized water heat storage.

BRICK HEAT STORAGE - TENNESSEE PLASTICS, INCORPORATED (TPI)

The TPI offpeak thermal storage unit is a central warm air system having a ceramic brick heat storage medium. The bricks are charged using resistance heating elements embedded in the bricks. All controls, wiring, blowers, etc., are furnished with the TPI systems. The ratings of the units may be varied according to the number of heating elements and storage bricks. The 30-kW units have a storage capacity of 200 kWh, while the 21-kW units have a capacity of 140 kWh.

The 25 TPI units installed by TVA have operated reliably so far with only minor maintenance required. Installation of the units was relatively straightforward with excellent step-by-step instructions supplied by the manufacturer. Component failures have been limited to two logic boards, one transformer, two thermocouples, and one hydraulic core limit.

TVA purchased 20 of the 21-kW models and 5 30-kW models. The 30-kW units strained the circuit breaker capacity of the 200-amp service homes because these large units were required in larger

---

offpeak, and total electrical consumption of the home as well as control the times when the storage units may recharge.

The magnetic tape channel allocations for all TPI and Megatherm installations are as follows:

Channel A - Water Heater
Channel B - Storage Unit
Channel C - Total Home Load

The OEM installations are metered as follows:

Channel A - Heat Battery Elements
Channel B - Heat Pump Load
Channel C - Total Home Load

The OEM 120-gallon storage water heater load is monitored with a watthour meter, but not recorded on the tape due to the constraints of the recorder. In all cases, the fourth channel records time.

homes with proportionately higher appliance and lighting loads. In addition, these larger homes often had 4 tons or more of air-conditioning which required a greater air flow than the TPI blower could provide. Even homes sized for the 21-kW units experienced problems in the warmer areas of the Valley where cooling loads require larger blower capacity. Condensate problems were noted on all units installed with the cooling coil in the return side of the system. The moisture condensed on the uninsulated lower plenum of the TPI unit and this was solved by the addition of external insulation.

Units were located in areas suitable to the utilization of their standby heat losses. Basements and semiconditioned space were actively sought and attached garages were avoided. Units of 30 kW were installed in homes with calculated heat losses of up to 75,000 Btu/hr and 21-kW units in homes with up to 50,000 Btu/hr. There are no known instances of storage capacity being exceeded. The homes equipped with TPI units have averaged offpeak consumptions of 92 percent with an average energy consumption of 3,600 kWh per month during November through January.

EUTECTIC SALTS HEAT STORAGE - OEM, INCORPORATED

The OEM Heat Battery (TM)/heat pump system is an American Air Filters water-to-air heat pump which is supplied 90°F source water by the OEM Heat Battery (TM) (Model 3672). The Heat Battery (TM) is a tank containing hydrated sodium sulfate (phase change material) as the heat storage medium. The Heat Battery (TM) is charged with water heated by conventional resistance elements. The water is circulated through a heat exchanger submerged in the tank. A second pump circulates warmed oil from the top part of the tank through a distributor system in the tank. The circulation continues until a temperature sensor is satisfied in the tank, indicating the salt solution is melted and therefore at full charge. The heat is extracted from the tank via the warm water loop between the Heat Battery (TM)/heat pump. The tank storage capacity is 100 kWh.

The OEM units were the first units delivered to TVA; therefore, the first installation was an OEM system. Although the unit was accompanied by comprehensive installation instructions, the complexity of the unit often intimidated the contractors during initial installations. This necessitated the presence of manufacturer's personnel at initial startups. After assistance on the first installation, contractors were able to handle subsequent installations with relative ease. However, the system's complexity continues to make troubleshooting difficult for local contractors.

Some of the more frequent causes of unit malfunctions are:

1. Leaking and/or sticking control, solenoid, and expansion valves,

2. Failure of the logic board controlling air flow,

3. Unbalanced condenser fan motors,

4. Improper internal wiring,

5. Pump failures and leaks,

6. Heating element failures, and

7. Heat pump freon leaks and other freon system problems, such as charging difficulties.

In an effort to solve problems as they occurred, OEM made several design changes after installations began.

The major problem with the OEM system resulted from a chain of circumstances involving the pump and element failures. The element failures resulted from the interruption of water flow accompanying the malfunction of the water pumps. Only one of the three elements in the boiler assembly usually burned out during a pump malfunction thus making detection of the trouble difficult. In addition, the element continued to draw electrical current, setting up an electrolytic reaction between the copper heat exchanger submerged in the salt and the steel tank of the Heat Battery (TM). Rapid degradation of the tank followed. Four OEM tanks developed holes which prompted costly and messy leaks. Subsequently, a flow switch control has been designed to prevent future element failure due to loss of water flow.

Although excessive maintenance problems have caused a reduction from 25 to 15 units on the OEM test, the basic operating principle of the unit is sound and offers significant potential for adaptation to multiple uses. The low storage temperature of the OEM unit (90°F) makes it readily compatible with solar heat sources. With the addition of collectors and a more efficient heat pump, the system coefficient of performance could be increased from the present value of 1.0 to 6.0 or better.

Energy consumption for homes equipped with OEM heat storage units have averaged 83 percent offpeak. This percentage is encouraging in light of the fact that many of the units were operating on backup resistance heat over 50 percent of the time. Steady-state, reliable operation of the units would probably increase the offpeak percentage to over 90 percent.

PRESSURIZED WATER HEAT STORAGE - MEGATHERM

The Megatherm thermal storage unit is an ASME steel tank using pressurized water as the storage medium. The water is heated using resistance elements inserted into the tank. Heat is extracted from the tank through a water-to-water heat exchanger located in the tank. (Recent models of this unit have eliminated the heat exchanger.) Warm water is pumped to a hydronic coil in the warm air system. Only the heat storage tank and associated controls are supplied by Megatherm. All other coils, pipe, valves, and pumps must be purchased locally. The Megatherm (Model RH-30) has a storage capacity of 90 kWh.

The only serious difficulties encountered with the Megatherm units occurred during the installation and startup phase. Twenty RH-30 Megatherms and five RHA-30 models have been installed. The RH-30's are heat exchanger-less models designed to improve efficiency and simplify installation.

One universal complaint of the Megatherm system is that although the units are relatively simple, the additional pumps, valves, and expansion tanks necessary for installation were not easily obtained by local contractors. These components contribute considerable additional expense to the heat storage unit price.

A lack of comprehensive installation instructions hampered initial installations. The heat exchanger-less models reduced the problem, but detailed instructions would have been helpful.

Reports of frequent tripping of the unit's internal circuit breakers led to the replacement of the original breakers by the factory. High temperatures in the central electrical panel areas resulted in the breakers tripping. The replacements supplied by Megatherm were calibrated for higher operating temperatures. The breaker problem and the need to disassemble and clean several mixing valves are the only factory-related operational problem encountered.

TVA is utilizing only three of the available five 6-kW elements in the Megatherm units for a total of 18-kW input. The units installed in homes having up to 40,000-Btu/hr heat losses have supplied adequate heat. Consumer reaction to the units has been positive and reliability has been excellent. The only known instance of inadequate capacity was in a home with a calculated heat loss of 47,000 Btu/hr at 15°F. The homeowner knew the unit might be undersized, but specifically requested to try the unit. Storage capacity was exceeded during a period of extremely cold weather when the temperature dropped below 10°F.

Megatherm-equipped homes have averaged 92 percent offpeak consumption with an average energy consumption of 3,600 kWh per month during the period of November 1980 through January 1981. High standby losses have been noted on some units installed in unconditioned space. Because all the units purchased were indoor units, extra insulation was recently added in some cases to reduce excessive standby losses; however, it is too soon to ascertain the effects.

STORAGE SIZING TECHNIQUES

The three thermal storage systems have different kWh storage capacities, different connected loads, and different operating characteristics. Therefore, the sizing requirements for each system were calculated using different methods. Each system has a unique single limiting factor which must be considered when matching its capacity to the heat loss of a home. The OEM unit is the simplest of the three systems to size because the only considerations necessary are the limits of the heat pump. The heat pump used in this project delivers a maximum of 43,500 Btu/hr in the heating mode while its cooling capacity is rated at 26,000 Btu/hr. The Heat Battery (TM) has a storage capacity equivalent to 12 hours of continous operation with this heat pump, which is more than ample for TVA's time-of-day schedule which incorporates a recharge period. A complete line of heat pumps with different capacities are available for this system to adapt to nearly any given heat loss.

TPI recommends sizing their units based on the heat loss of a home at the design temperature and then installing a unit with a connected load of three times this heat loss for a typical 8-hour charging period.

This results in an extremely oversized unit since solar gain, thermal lags, daily temperature savings, and heat content of home mass and furnishings are not taken into consideration. Since these units have large storage capacities when compared to TVA's relatively short daily onpeak periods of 4 to 5 hours, respectively, sizing of these units is a function of kW input or total connected load as TPI suggests but not with their degree of conservatism. With offpeak charging periods of 9 hours and 6 hours, we settled on a minimum sizing factor of 1.3 (as compared to a sizing factor of 3 by the manufacturer) for these units and have experienced no shortage of stored heat in any unit to date.

The Megatherm units do not have the plentiful storage capacity of the TPI units (90 kWh as compared to 140 kWh); therefore, storage capacity had to be a consideration in sizing these units along with home heat loss and the charging time. Because of their storage limitations, the sizing factor for the Megatherm units was set at 1.5 for the 18-kW units which were installed in this project. This translates into a maximum home heat loss of 41,000 Btu/hr for which they could be safely sized to provide heat.

STORAGE WATER HEATING

Each home in the test is also furnished with an A. O. Smith, Model EES-120, water heater. Storage water heaters were included in the test to permit maximum utilization of the experimental time-of-day rate. These 120-gallon water heaters meet or exceed the ASHRAE 90-75 heat loss standard of 4-watts/ft^2/hr. The units have two 4,500-watt elements with nonsimultaneous thermostat operation.

The storage water heaters are controlled by the same control circuit as the heat storage units. The GE IR-70 meter allows the units to charge only during offpeak hours by means of a dry contact output from the meter.

INSTALLATION COSTS

The installation costs for all three types of units included extensive electrical submetering costs as well as contractor training manhours. The portion of these costs which can be directly attributed to submetering and training alone could not be exactly determined from the installation costs received but appear to constitute a major share of these costs. The total average installation costs for each of the three types of storage units are as follows:

TPI	\$2,730
OEM	\$3,290
Megatherm	\$4,025

The TPI unit was the simplest unit to install, as reflected in its average cost, with a complete and self-explanatory instruction booklet to aid the contractor. The OEM units were by far the most complex system to install, requiring manufacturer's assistance for the initial installations and resulting in slightly higher installation costs. The Megatherm systems were relatively simple to install except for the extensive plumbing work involved which is included in their average installation costs. These installation costs do not include the cost of the thermal storage units.

PRELIMINARY PERFORMANCE

Preliminary performance results of these units are very encouraging. The effects of such units on TVA's capacity situation will not be available until a full winter's data has been collected; however, preliminary studies indicate load reductions of up to 6.0 kW (at the time of TVA's system peak) per thermal storage system installation.

Average offpeak energy consumption levels for November 1980 through January 1981 range from 83 percent for the OEM systems to 92 percent for the Megatherm and TPI systems. This translates to a customer savings of approximately \$25 per month for the OEM units and \$45 per month for the TPI and Megatherm units based on the consumer's average conventional consumption levels for the same 3 months.

CONSUMER REACTIONS

Customer satisfaction and comfort levels for both the Megatherm and TPI systems has generally been very favorable, especially since the temperature of the supply air for both units is variable and can be adjusted to the homeowners' satisfaction. Also, these units employ blowers commonly found in conventional central systems so velocity of these units is compatible with most normally sized existing central systems. Customer reaction to the OEM units has been mixed because of the excessive maintenance problems experienced with these units which often resulted in supply temperatures below 90°F in some homes. When operating properly, the OEM system supplies air at 100°F+ to the home as compared to the 95°F supply temperature of a conventional heat pump.

SUMMARY

The installation phase of the Thermal Storage Field Test is almost completed. Some installations have been beset by problems from equipment malfunction, contractor's inexperience, and the simple logistical considerations of installing units in eight different locations. Overall, the systems have performed well and preliminary indications are that thermal storage heating may be a viable solution in the future for TVA's problems with load swing and capacity planning and offers rate relief for Valley consumers. The remaining 2 years of the test should provide valuable data to respond to some of these concerns.

HEAT PIPES FOR HIGH TEMPERATURE INDUSTRIAL WASTE HEAT RECOVERY

Michael A. Merrigan
Los Alamos National Laboratory, Los Alamos, NM

INTRODUCTION

With increasing cost and decreasing availability of fuels has come increased interest in energy conservation in industrial heating operations. This has resulted in reductions in fuel consumption through the use of improved insulation materials and tighter operational controls, particularly in control of excess air in furnaces. However, even with good insulation and close control of excess air, the energy loss in exhaust gas will be more than 50% of the total furnace input for flue gas temperatures over 1800°F. Figure 1 indicates the percentage energy loss in flue gas as a function of the exhaust temperature for unrecuperated furnaces. The figure shows that significant fuel use reductions are achievable through recovery of energy from exhaust gases by means of high temperature recuperation. As over 25% of the total fuel consumption in industrial applications is in installations having flue gas temperatures above 1800°F, the potential for increased energy savings is large.

MATERIALS LIMITATIONS

The most significant impediment to more widespread use of high temperature recuperation is the limitation of the materials used for heat exchange surfaces. Heat exchangers made of stainless steels are limited to surface temperatures of ∿1300°F while alloys such as Incaloy 800 are useful to only about 1700°F. In the higher portion of their operating range, all the metallic alloys are reduced in strength and therefore limited in application to comparatively low pressure differences and well supported tube runs. Therefore, high effectiveness recuperators designed to operate above ∿1600°F have to be made of ceramic materials. The practical operating limits for heat exchanger surfaces of various materials are shown in Fig. 2.

High temperature metallic recuperators, such as radiation recuperators, in use in the higher temperature ranges are limited in effectiveness by surface temperature restrictions. The influence of recuperator effectiveness on fuel

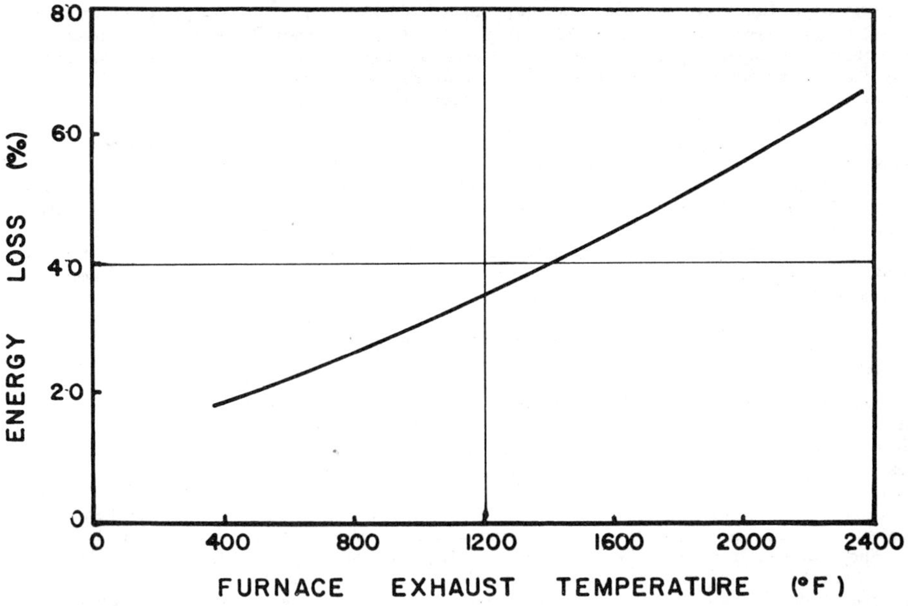

Fig. 1. Energy loss in unrecuperated furnace with 10% excess in air

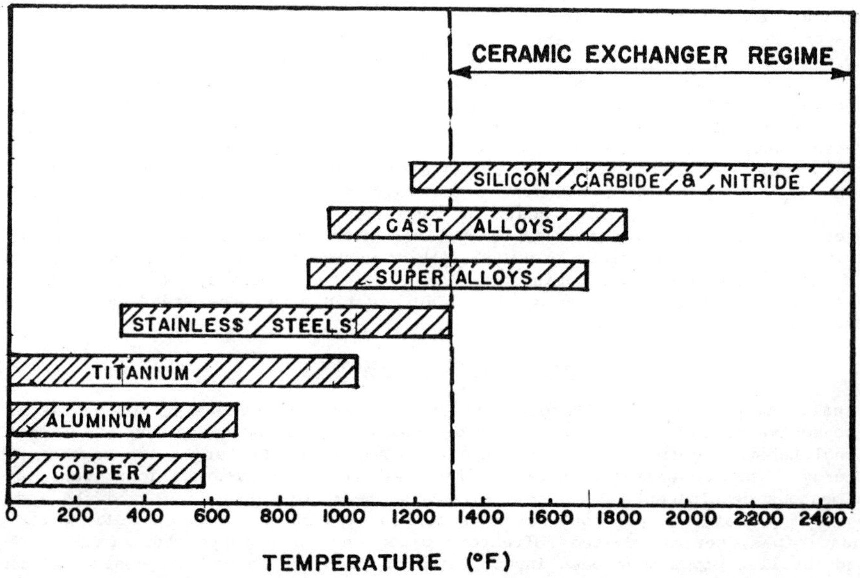

Fig. 2. Operating limits for heat exchanger materials.

savings as a function of flue gas temperature is shown in Fig. 3. A reduction of exhaust gas temperature from 2300 F to 750 F, corresponding to a recuperator effectiveness of about 75%, lowers fuel use by about 50%. Heat exchange surfaces capable of use at these temperatures traditionally have been fabricated of ceramic materials in the form of cemented matrices of tubes and bricks. This type of ceramic exchanger is no longer in common useage because of the inherent leakage problems caused by the multiplicity of joints. As newer ceramic materials having good conductivity and strength have become available in recent years there has been a renewal of interest in ceramic heat exchangers in configurations more closely resembling conventional tube and shell metallic exchangers. These material advances have resulted in the present availability of ceramics such as silicon carbide, having good high temperature strength, thermal shock resistance, conductivity, and density; which have made possible the development of ceramic heat pipes and their use in recuperators.

THE CERAMIC HEAT PIPE RECUPERATOR

The characteristics of the ceramic heat pipe heat exchanger are illustrated in Fig. 4. Each tube is supported at the flow divider at a single point so that no sliding seals are required. Loss of an individual heat pipe does not compromise the operation of the heat exchanger as a whole, a decided advantage when using brittle materials. The nature of heat pipes allows the design of a counterflow heat exchanger, an inherently efficient design. The extremely high internal conductance of the heat pipes ensures that the heat transfer from the pipe to the surrounding gas is the controlling factor in overall heat transfer. Therefore, the use of extended heat transfer surfaces on the tube exterior will improve heat exchanger performance. The tubes are accessible for cleaning by removing side covers of the duct without the need for disassembly of any tube seals.

The heart of these high temperature recuperator designs is the ceramic heat pipe illustrated in Fig. 5. This device consists of an evacuated enclosure lined with a wicking material and containing a limited amount of working fluid. In operation, the fluid is distributed in the porous wick structure with the internal volume of the pipe occupied by the vapor phase at the saturation pressure corresponding to the average temperature of the pipe. When a local heat source is applied to the heat pipe, the fluid increases in temperature and vapor pressure and local evaporation occurs. The vapor flows through the pipe to lower temperature areas and condenses, giving up its heat of vaporization to the pipe wall. The operating fluid used in the heat pipe must be selected to match the operating temperature range of interest. For high-temperature industrial heat recuperation, the normal working fluids are liquid metals such as potassium, sodium, or lithium. These alkaline metals are excellent heat transfer fluids, having good conductivity, high heats of vaporization, and good surface tension. However, in general, they are not compatible with the candidate ceramic materials. The pipe is therefore provided with a protective inner liner matched to the ceramic expansion characteristics. The liner materials may be comparatively thin films of chemical vapor deposited (CVD) coatings or free standing metallic tubes.

CERAMIC HEAT PIPE DEVELOPMENT

The realization of the high temperature ceramic recuperator design advantages is the objective of the ceramic heat pipe program being conducted by the Los Alamos National Laboratory for the Pittsburgh Energy Technology Center of the Department of Energy. The Los Alamos program has involved material investigations, fabrication methods development, compatibility tests, heat pipe operation, and the modeling of application conditions based on current industrial useage. Solid ceramic heat pipes, ceramic coated refractory pipes, and high-temperature oxide protected metallic pipes have been investigated as part of the general consideration of high temperature recuperation requirements. Economic studies of the use of heat pipe based recuperators in industrial furnaces have been conducted and payback periods determined as a function of material, fabrication, and installation costs.

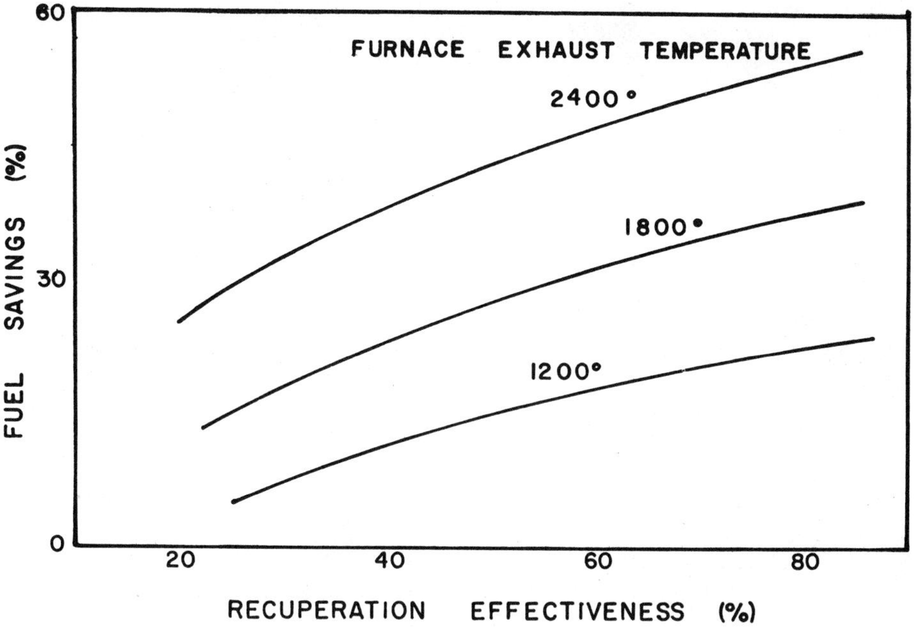

Fig. 3. Fuel savings due to exhaust gas recuperation as a function of recuperator effectiveness with 10% excess air.

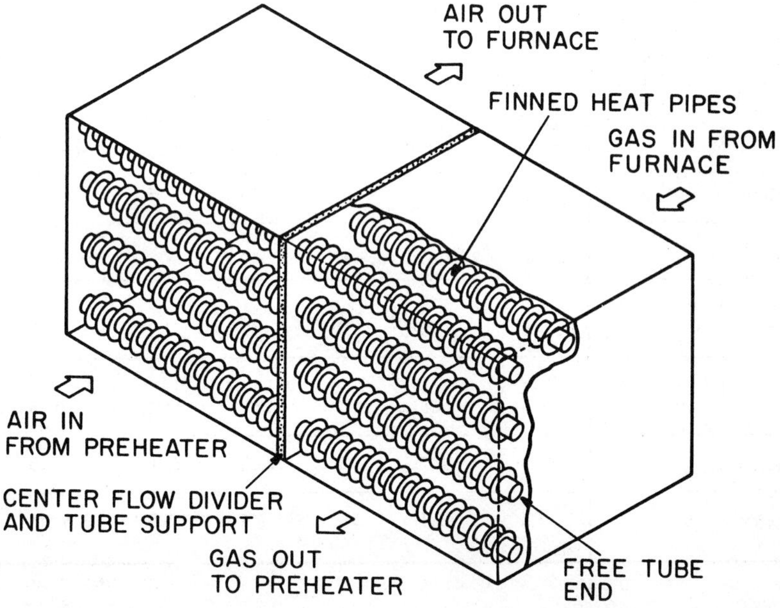

Fig. 4. Ceramic heat pipe recuperator arrangement.

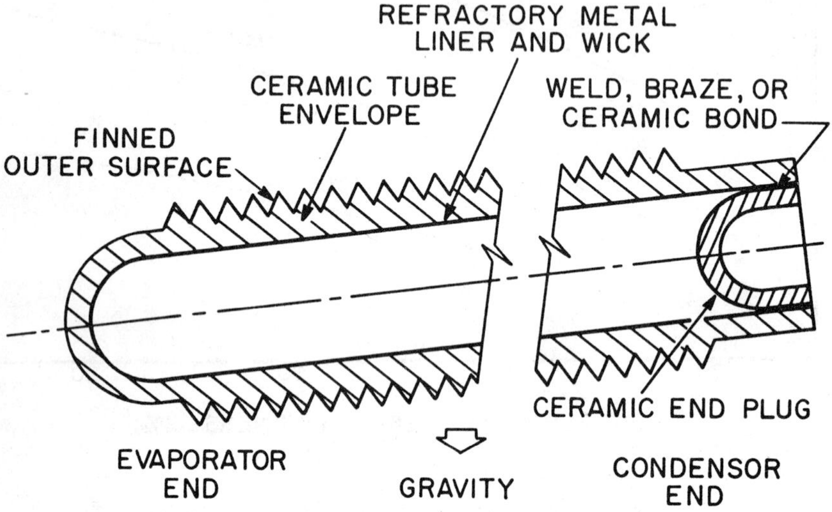

Fig. 5. Cermic heat pipe configuration.

	Si_3N_4 REACTION BONDED	SiC Si BONDED	SiC SOLID ALPHA	SiC CVD	Al_2O_3	SiC-C COMPOSITE
FAILURE STRENGTH KSI AT 2000°F	35	45	50	45	55	45
THERMAL EXPANSION IN./IN./°F	$1.7x10^{-6}$	$2.6x10^{-6}$	$2.5x10^{-6}$	$3.2x10^{-6}$	$7.0x20^{-6}$	3.0 TO $5.0x10^{-6}$
THERMAL CONDUCTIVITY AT 2000°F BTU/H-°F-FT	6	20	11	10	2.2	20
MAX. USE TEMP. FOR EXTENDED TIME, °F	2800	2450	2450	3000	3000	2400 TO 3000
FABRICABILITY	GOOD	GOOD	GOOD	GOOD	GOOD	GOOD
DENSITY, g/cm^3	2.8	3.1	3.2	3.2	3.9	2.5 TO 3.5

Fig. 6. Candidate ceramic materials for high temperature heat pipes.

The program began with a survey of commercially available materials and the conduct of compatability tests on ceramic materials and working fluids. Ceramic materials evaluated for use on the program are listed in Fig. 6. Initial compatibility tests verified the need for use of metallic liners with alkaline metal fluids except for high-purity alumina material. Based on review of the material properties of metallic liner materials and ceramics, alumina-niobium-lithium, alumina-sodium, and silicon carbide-tungsten-sodium combinations were selected for tests.

Initial experimental development was concentrated on the alumina-sodium heat pipes because of the favorable economics of the material, the possibility of achieving reliable operation without the use of a metallic liner, and because a successful high temperature ceramic seal had been developed for closure using 25% Y_2O_3-75%Al_2O_3 reaction bonding. However, none of the alumina heat pipes that were assembled survived the severe thermal shock imposed by heat pipe start-up and development of high temperature alumina heat pipes was discontinued. A low temperature alumina pipe intended for use in corrosive atmospheres such as wet sulfur dioxide scrubber environments was subsequently assembled and successfully operated. This low temperature (300°F) ceramic heat pipe used toluene as a working fluid and consequently did not experience the severe thermal shock environment encountered with the liquid metal working fluids.

The first successful operation of high temperature ceramic heat pipes was achieved using CVD silicon carbide as a container material and CVD tungsten as a protective liner. Controlled deposition of the tungsten was used to give a textured surface for wicking. The working fluid used was sodium with the heat pipe closure accomplished by a tungsten-to-tungsten braze using a palladium-cobalt brazing alloy. These heat pipes were operated in air and in combustion gases at temperatures to about 1700°F. Peak measured heat transfer through the 0.75 inch diameter heat pipe was about 2 kilowatts with operational limits estimated to be higher. Figure 7 shows one of these heat pipes operating in air with a gas fired heat source. After about 100 hours of operation and 30 cold starts one of these pipes was sectioned and the tungsten-silicon carbide interface examined for evidence of deterioration or reaction. The reaction zone thickness was found to be in agreement with predictive models developed on the program. However, some evidence of liner-shell bond deterioration was observed prompting a continuing investigation of improved fabrication methods.

The brazed plug closures used in these heat pipes required close fitting of the tube and plug and limited maximum operating temperature to below braze reflow temperature. Continuing investigation was focused on the development of welded enclosures. Two different approaches to the development of an all welded enclosure were pursued. In the first a complete, operational heat pipe was assembled of molybdenum tubing using a molybdenum screen wick, hafnium gettering, and electron beam welded molybdenum end caps. After verifying operation this heat pipe was overcoated with an interface layer of CVD tungsten, approximately 0.005 inch thick, and then with CVD silicon carbide, approximately 0.010 inch thick. This layered configuration was developed in thermal shock tests of coated tube samples. Figure 8 shows two of the molybdenum core heat pipes prior to ceramic coating while Figure 9 shows the completed heat pipe in test in air. Initial test data for this heat pipe is given in Fig. 10. In continued testing there has been some local deterioration of the ceramic coating on this heat pipe resulting in oxidation of the core molybdenum end cap at the hemispherical end closure, but the remainder of the coating has been unaffected by the limited testing conducted to date. Developmental work on this concept is continuing in parallel with solid ceramic tubing based designs.

In a second approach to a welded closure development an investigation of joining of the tungsten liner material to a tungsten end cap by electron beam welding was initiated. The silicon carbide shell was first removed from the weld area to leave a protruding shell of tungsten which was end welded to the cup shaped cap. This technique imposes minimal mechanical constraint on the weld area and is intended to minimize weld cracking problems. In application the tungsten end cap and joint would be covered with a ceramic cap for oxidation protection. Work on this closure method is continuing.

Fig. 7. Close-up view of heat reject end of tungsten-lined SiC/sodium heat pipe. Heat pipe temperature 1560°F.

Fig. 8. Molybdenum core heat pipes before CVD-SiC coating.

Fig. 9. Molybdenum core heat pipe operating in air.

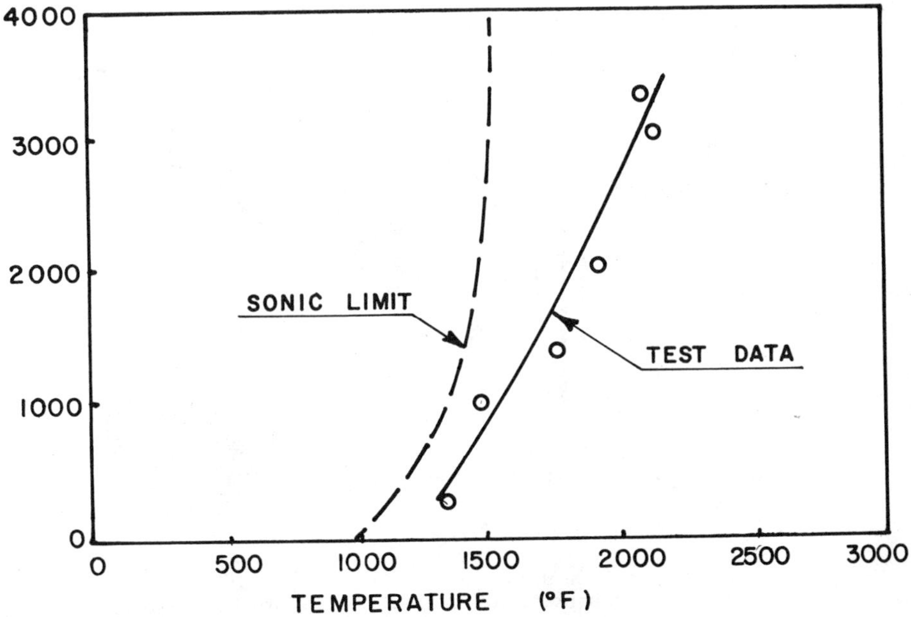

Fig. 10. SiC-W-Mo heat pipe performance.

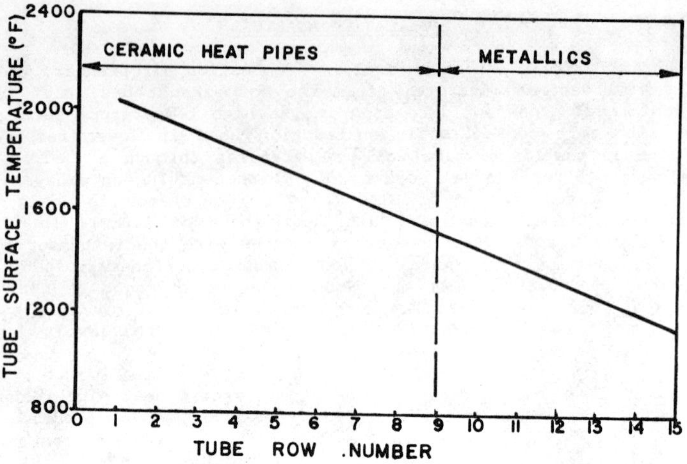

Fig. 11. Tube surface temperatures through a 15 row counterflow recuperator.

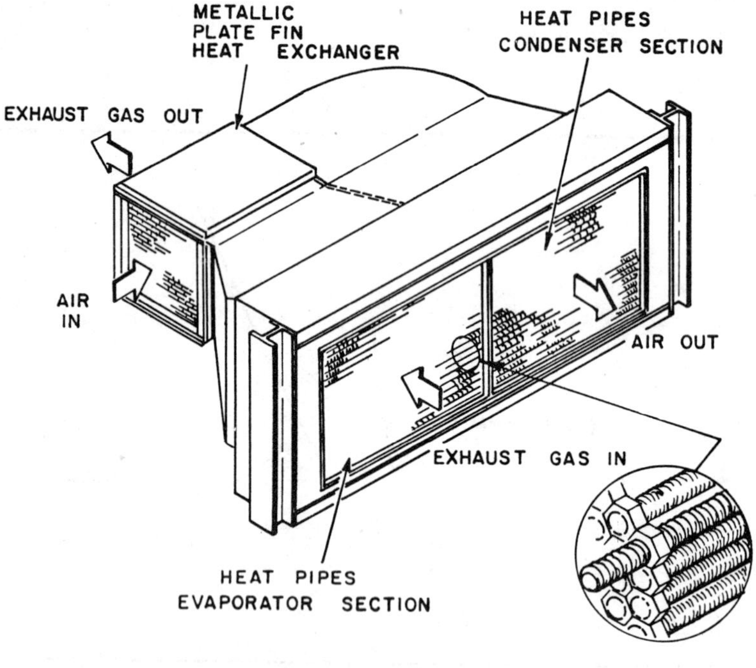

Fig. 12. A ceramic heat pipe exhaust gas recuperator used in conjunction with a stainless steel plate fin preheater.

APPLICATION ANALYSIS

In a counterflow heat exchanger used for combustion air preheat, the tube row temperatures will decrease uniformly from the hot exhaust gas inlet to the air inlet. Ceramic heat pipes would be used in the high temperature sections of the recuperator with more conventional construction used in lower temperature regions. Figure 11 gives the tube wall temperatures through a 15-row heat pipe recuperator designed for a steel soaking pit furnace having an exhaust mass flow of 5.75 lb/s at a temperature of 2150°F. The recuperator is designed for an effectiveness of 70% with an inlet air temperature of 100°F. In this design example, the first 9 rows would use ceramic tubes with the remainder of the exchanger using high-temperature metallic heat exchange surfaces. Depending on the overall size of the recuperator, the metallic sections might be heat pipes or other surfaces. The maximum axial power density predicted for this ceramic recuperator design is about 500,000 BTU/h ft^2, well below the levels demonstrated in ceramic heat pipe tests.

Figure 12 shows a configuration employing a ceramic heat pipe recuperator in conjunction with a stainless steel plate fin heat exchanger as a preheater. This design has been the subject of a cost payback analysis conducted by Garrett-AiResearch for Los Alamos that indicated payback periods of less than six months at fuel costs of $3.35/million BTU.

CONCLUSION

Work on the Ceramic Heat Pipe Program has demonstrated the feasibility of the use of ceramic heat pipe recuperators in high temperature industrial waste heat recuperation. Realistic heat transfer rates have been achieved at temperatures in the range of interest. Economic analysis has indicated favorable payback periods at current fuel costs. Continuing work is intended to provide improved fabrication methods and to demonstrate long term suitability of the technology.

REFERENCES

1. W. A. Ranken, "Ceramic Heat Pipe Heat Exchanger," Los Alamos National Laboratory report LA-6514-MS (September 1976).

2. W. R. Pierson and W. T. Hanna, "Applications for Ceramic Heat Exchangers in Process Heat Industries," Battelle, Columbus Laboratories, Columbus, Ohio, January 9, 1979.

3. W. A. Ranken and L. B. Lundberg, "High Temperature Heat Pipes for Terrestrial Application," 3rd Inter. Heat Pipe Conf., Palo Alto, CA, May 22-25, 1978.

4. L. B. Lundberg, "Silicon-Carbide-Tungsten Reaction Kinetics," American Chemical Society Spring Meeting, April 6, 1979, Honolulu, Hawaii.

5. E. S. Keddy and W. A. Ranken, "Ceramic Heat Pipes for High Temperature Heat Removal," 18th National Heat Trans. Conf. (AICHe-ASME), Aug. 5, 1979, San Diego, CA.

6. L. B. Lundberg, "Silicon Carbide-Tungsten Heat Pipes for High-Temperature Service," IEC Product Research and Development, Vol. 19, p 241, June 1980.

7. H. J. Strumpf and W. S. Miller, Garrett-AiResearch Report 79-16480, 1980.

8. M. A. Merrigan and E. S. Keddy, "High Temperature Heat Pipes for Waste Heat Recovery," AIAA 15th Thermophysics Conf., July 14-16 1980, Snowmass, CO; LA-UR-79-3437.

9. M. A. Merrigan, "Economics of High Temperature Recuperation Using Ceramic Heat Pipes," Amer. Soc. for Metals Conf., 1980, Pittsburgh, PA; LA-UR-80-1985.

10. M. A. Merrigan, "A Heat Pipe Heat Exchanger Model for High Temperature Recuperators," CUBE Symposium, Oct. 22-24, 1980, Lawrence Livermore National Laboratory, LA-UR-80-1985.

EXPERIENCES WITH RERADIANT RECUPERATION FOR ALUMINUM MELTING

Dennis H. Larson
Institute of Gas Technology

P. K. Ellis
Reynolds Metals Company

INTRODUCTION

Industrial furnace waste-heat recovery has been an established practice for over 20 years. In the past, waste-heat recovery in the form of preheating combustion air was practiced to increase the combustion temperature in order to increase production capacity. The low cost of fuels at the time made waste-heat recovery solely to save energy uneconomical. Consequently, the market size for waste-heat recovery equipment remained relatively small for a long period of time. This tended to discourage equipment manufacturers from developing new, more efficient designs or less costly equipment approaches. The soaring cost of fuel in the last few years together with projections of continuing price increases, however, have dramatically changed this picture.

Waste-heat recovery methods using recuperators are now being found economically attractive for a large number of furnace types solely on the basis of reducing cost by reducing fuel consumption. This has significantly increased the size of the market for recuperators and encouraged equipment manufacturers to seek improvements in their products. The most sought after improvements are:

- Applicability of the generic design to a wide variety of furnace types in different industries,
- Decreased capital and operating costs per unit of energy saved,
- Increased waste-heat recovery efficiency in order to increase net energy savings, and
- Increased service life.

In many instances, the Federal Government, user industries, and R&D organizations have teamed with the recuperator manufacturer to achieve these goals. This paper describes one such team effort and its results.

The purpose of the project was to field test an engineering prototype of a so-called reradiant recuperator whose design promised better performance than conventional stack type recuperators. The test program was cooperatively undertaken by the Reynolds Metals Company and the U.S. Department of Energy. The reradiant recuperator was developed and tested by the Institute of Gas Technology and built by the Thermal Transfer Corporation. Earlier experimental development work was supported by Southern California Gas Co. and Consolidated Natural Gas Co.

The test work and performance evaluations were conducted at Reynolds' Alabama Reclamation Plant near Sheffield, Alabama. The test consisted of installing and operating two reradiant recuperators; one unit showing lower capital cost per unit of energy saved and a second unit showing higher recovery efficiencies.

The first recuperator installed at the site suffered extensive heat and corrosion damage after approximately 30 days of full operation. It was later found that a faulty linkage controlling the fuel valve allowed part of the fuel to burn within the recuperator at fuel-rich conditions. The failure occurred within 2 hours of the leakage. Because its operating life was short and spanned Christmas-New Year season, no accurate data were collected on this recuperator.

The second recuperator was installed at about the same time as the failure of the first unit. The second recuperator was operated over two 3-month periods; each followed by a short evaluation of furnace performance without recuperator. Following the first period of operation, the lower one-third of the radiant insert had deteriorated. This insert section was replaced with a different type alloy, and a corrosion study was initiated for the second test period.

SYSTEM DESCRIPTION

The furnaces on which the recuperators were installed are nominal 95,000-pound reverberatory aluminum melters (Figure 1) of Reynolds' own design. The furnaces had 12,000,000 Btu/hr dual-fuel burners capable of burning natural gas or #2 fuel oil. The furnaces used oil for most of the recuperator trials. Cold scrap in various forms is charged into an external charging well. A salt flux in the well removes impurities from the metal and serves as a cover for the well. Submerged arches separate the main hearth area from the charging well.

The recuperator is a stack-type system (Figure 2). The main difference between the reradiant recuperator and a conventional design is the reradiant element consisting of thin metal leaves suspended within the flue-gas passage (Figure 3). A four-leaf configuration was used for the evaluation trials as it represents a good compromise between increased performance and additional cost. Added features of the reradiant recuperator are a) small floor space requirements and b) resistance to plugging by solid matter build-up.

This type of recuperator operates by having the hot flue product gases pass through the center and the colder combustion air flow through an annular passage surrounding the flue passage. Heat is transferred from the hot flue products to the steel alloy separation wall. The separation wall, in turn, transfers the heat to the combustion air. Heat transfer on the flue-gas side occurs primarily by radiation with some convection (Figure 4).

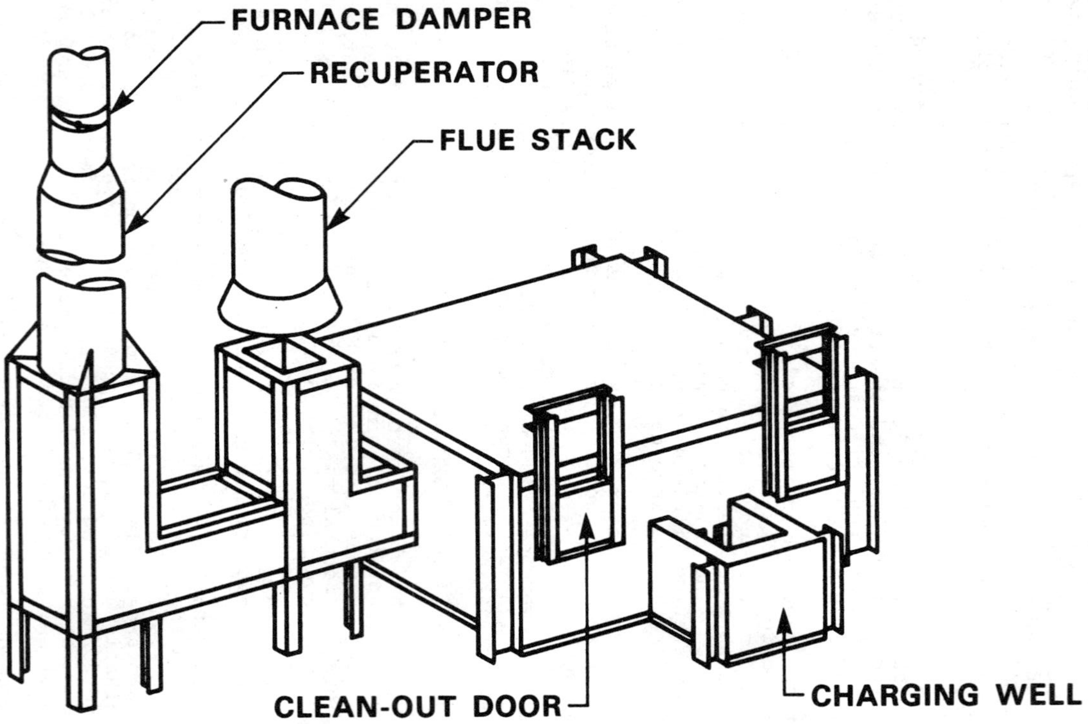

Figure 1. WET HEARTH REVERBERATORY ALUMINUM MELTER

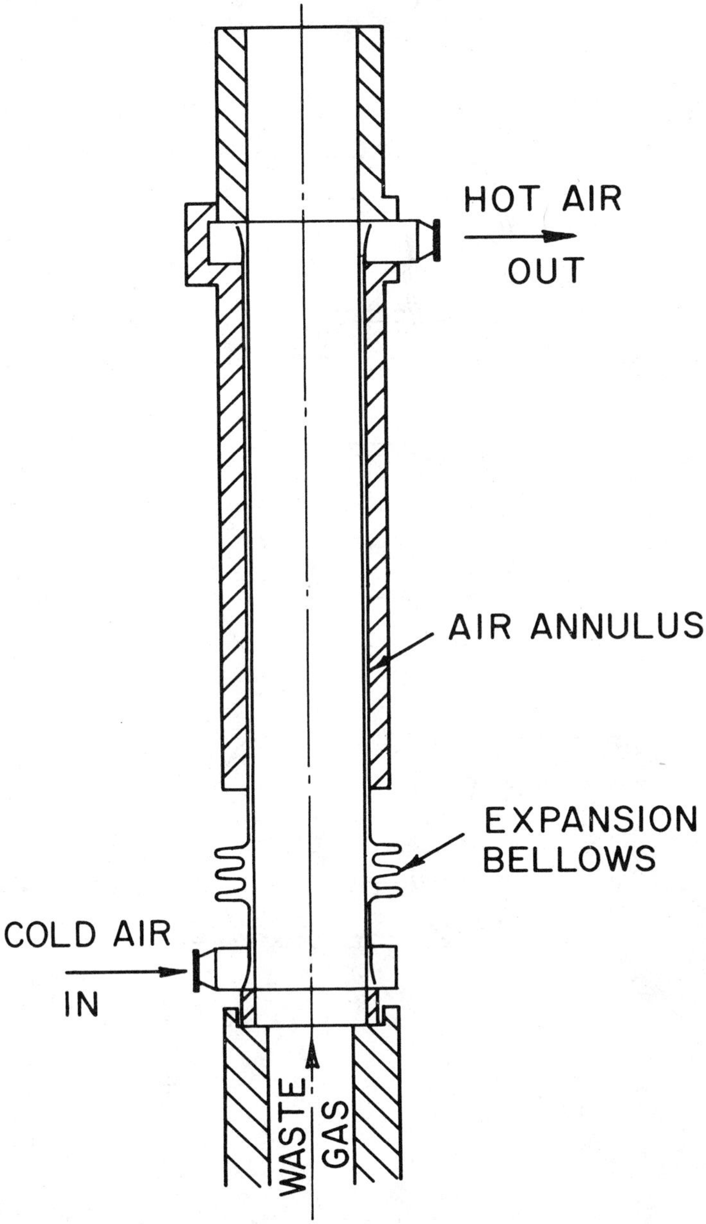

Figure 2. STANDARD RECUPERATOR

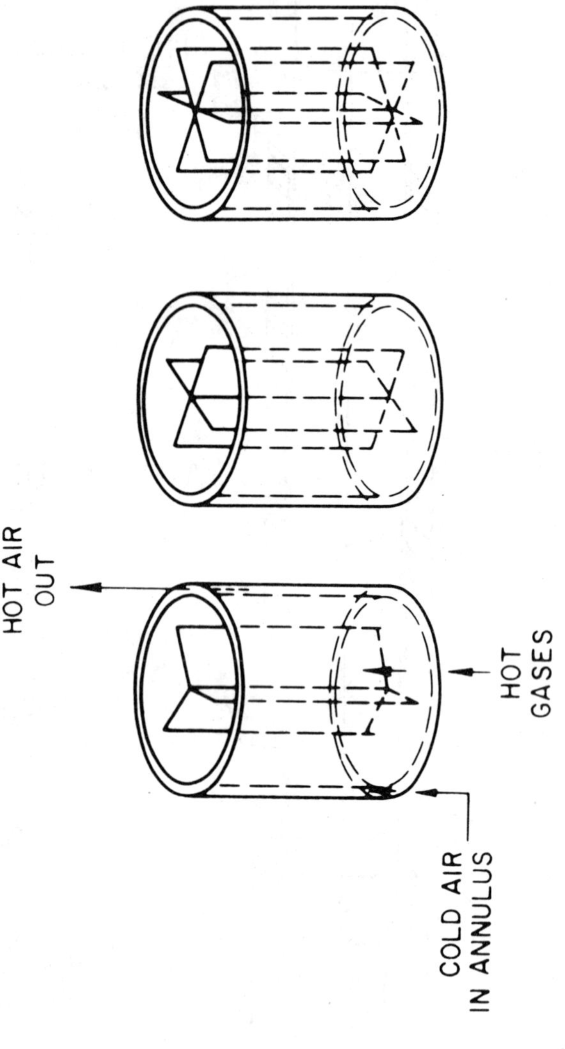

Figure 3. EXAMPLES OF RERADIANT RECUPERATOR INSERTS

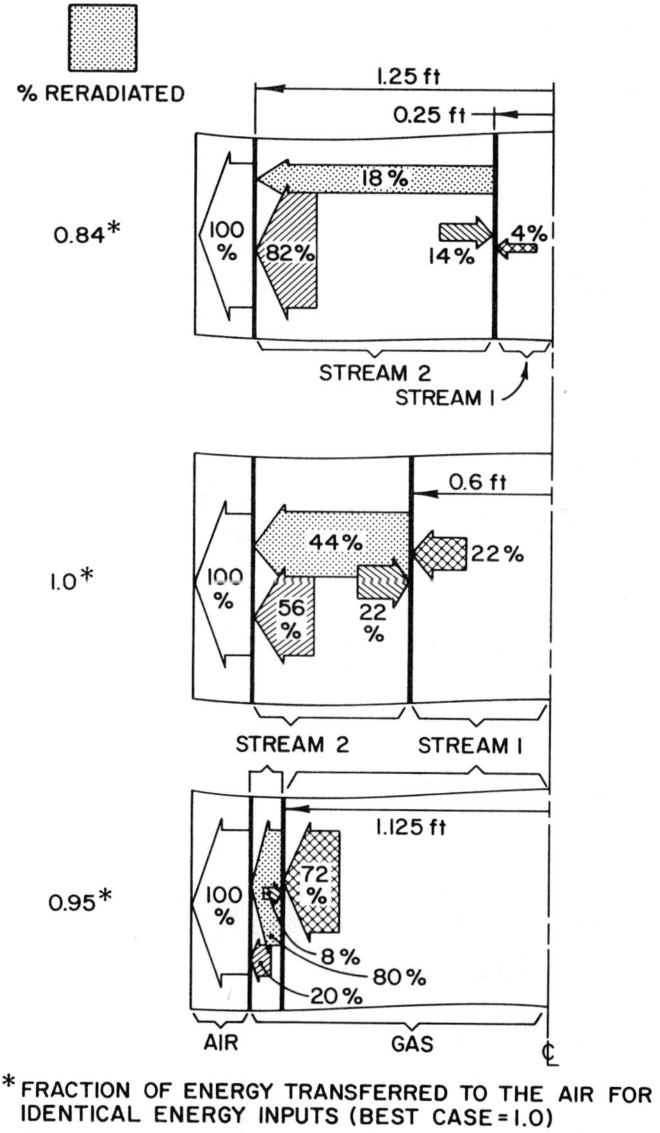

Figure 4. HEAT FLUXES AT 30 FEET FROM ENTRY IN A RECUPERATOR WITH RERADIATORS OF VARIOUS SIZES AND SEPARATED STREAMS

The reradiant insert increases the radiation component of the overall heat-transfer mechanism by increasing the "effective" surface area of the annular separation wall. The insert accepts heat by convection on its surface and radiates this energy to the separation wall. This reradiated thermal energy is in addition to the heat transferred directly from the hot gases to the separation wall. This added heat transfer increases the heat recovery efficiency and fuel savings.

The reradiant insert is not to be confused with the "core-buster" type insert. The primary purpose of the core-buster is to induce turbulence in the hot gas stream and, thereby, increase convective heat transfer. The reradiant insert is intentionally designed to minimize turbulence in order to minimize any increases in flue-gas pressure drop.

The recuperator installation also included a) furnace pressure control, b) mass-flow fuel/air ratio control, and c) various furnace and recuperator over-temperature safety controls. These controls, as shown in Figure 5, were found to also contribute to the overall furnace efficiency improvement.

The burners were controlled by furnace roof temperature. This is accomplished by controlling the flow of combustion air. The flow of combustion air signals the ratio controller to adjust the fuel flow. Feedback for comparison comes from differential pressure across an orifice for natural gas. For oil, it is done indirectly by measuring an air flow signal through an orifice in parallel to the actual oil flow. This system was developed by Bloom Engineering, Inc., primarily for use with heavy oil.

In addition to the air/fuel ratio controller, the instrumentation consists of temperature controllers on the furnace roof temperature, the metal temperature, the combustion air temperature, and the flue-gas temperature near the base of the recuperator. There is also a comparator to monitor the fuel flow and air flow as a backup to the ratio controller. This device sounds an alarm if the air/fuel ratio is greatly out of balance.

The controls and instrumentation required more attention than on an unrecuperated furnace. More instrumentation is needed for safe operation of the equipment and the need for the instrumentation to be at peak performance is greater. The possibility for extensive damage to the recuperator is very high in the case of a malfunctioning instrument. An over-temperature condition of even a few minutes duration could cause buckling of the inner shell of the recuperator. An off ratio burner in a fuel-rich condition while burning oil will cause rapid corrosion of the reradiant insert. Both of these conditions were experienced on the first recuperator.

Other maintenance requirements did not differ significantly from those on an unrecuperated furnace.

PERFORMANCE RESULTS

During the initial operating period of approximately 3 months, the performance of the recuperator system was intesively monitored for 51 consecutive shifts. When compared with a similar 59 shift periods of unrecuperated operation, fuel savings in excess of 30% based on a Btu per pound basis were realized, as shown in Figure 6.

In addition, the average production rate of the furnace increased by about 13% for the recuperated period as compared with the non-recuperative period (Figure 7). However, the data collected are insufficient to determine if the increase was due to recuperation or operating practice.

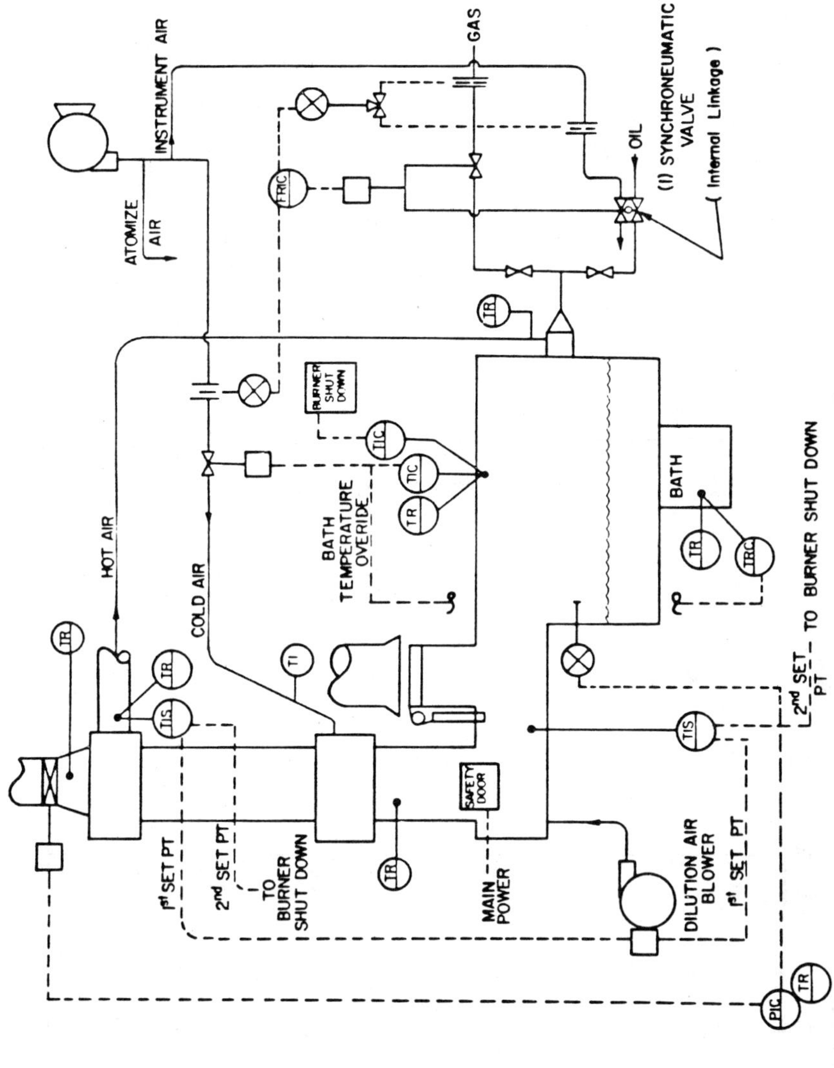

Figure 5. SCHEMATIC OF THE FURNACE/RECUPERATOR CONTROL SYSTEM

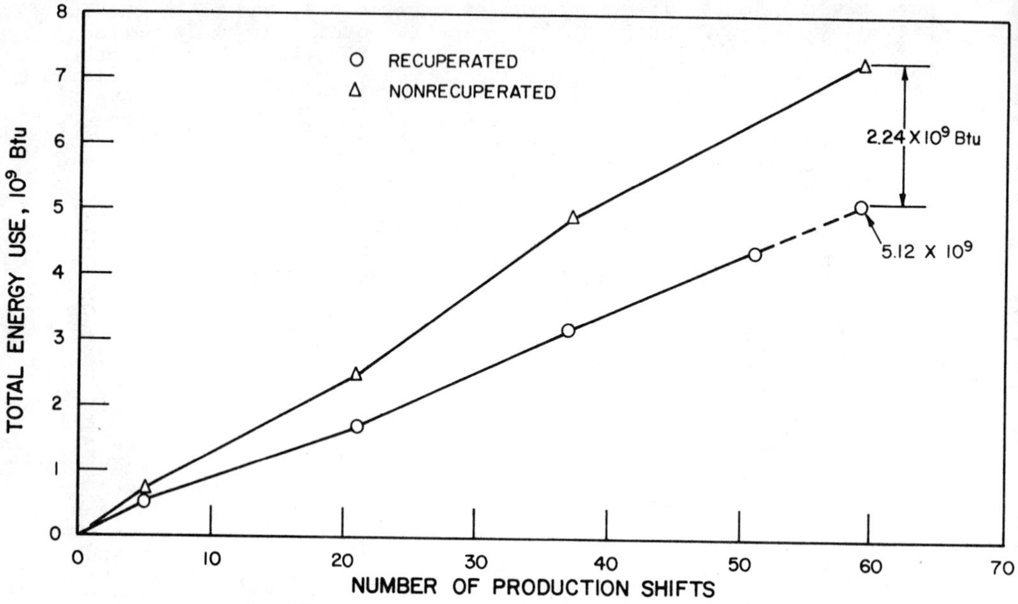

Figure 6. COMPARISON OF A RECUPERATED AND NONRECUPERATED ENERGY USE

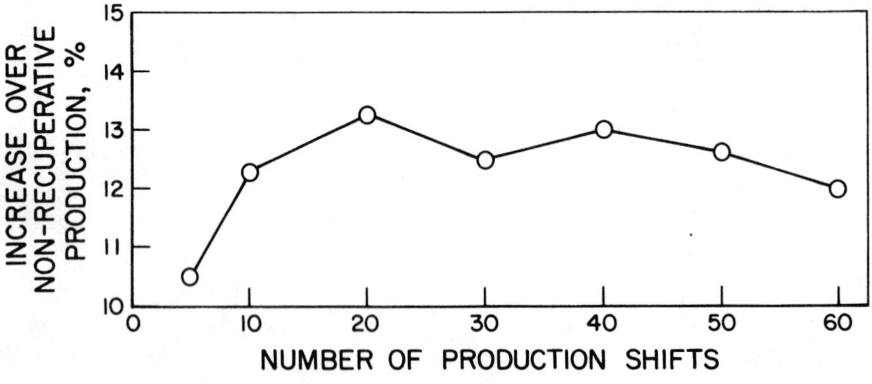

Figure 7. COMPARISON OF PRODUCTION WITH AND WITHOUT RECUPERATION

We fully realize that the fuel efficiency of the furnace is strongly related to its operating conditions and, hence, to the fuel savings realized by recuperation. The two operating shift periods for this example were chosen because of the similarity of all other operating conditions. It is possible, however, to compare non-similar shift operating periods through the use of data collected in the test program. The principal operating conditions that affect furnace efficiency, aside from heat recovery, are —

- Type of scrap charged
- Scrap charging rate
- Furnace pressure as it relates to air infiltration or flue product exfiltration.

Furnace pressure had the most pronounced effect on furnace fuel efficiency. When the furnace operates in a non-recuperative mode, there is no pressurization system on the furnace. Figure 8 shows measured combustion air preheat temperatures for variations in furnace pressure with all other parameters constant. The high air temperature and, hence, maximum fuel savings occurred at a furnace pressure of about ± 0.05-in. H_2O. Lower furnace pressures allows outside air to enter through furnace openings and cool the flue products, which reduced the recuperator recovery efficiency. Higher furnace pressures caused some of the flue gases to leave the furnace through openings, which reduced the mass flow through the recuperator and, hence, reduced its recovery efficiency. The change in the recuperator heat recovery performance or effectiveness, as expressed in Equation 1, can be directly related to heated air temperature, as shown in Figure 9.

$$E = \frac{T_{HA} - T_o}{T_{FG} - T_o}(100) \qquad \text{(Eq. 1)}$$

where —

T_{HA} = temperature of combustion air, ^oF
T_{FG} = temperature of flue gases, ^oF
T_o = temperature of ambient air, ^oF

The scrap charging or overall production rate had the next largest effect on fuel consumed per pound of production, which we have defined as E, work efficiency. Figure 10 shows measured data on work efficiency versus production rate expressed as a percentage of the maximum design production rate.

The comparison of energy requirements for a recuperative and non-recuperative operating case was made for periods of about 100% of the maximum design production rate. As can be seen in Figure 10, the work efficiency is about 2000 Btu/lb for the non-recuperative case and 1400 Btu/lb for the recuperative case.

The type of scrap charged showed a measurable but small effect on the fuel efficiency of the furnace. Three types of scrap were commonly charged. First was the charging of "briquettes." This is usually can stock compressed into 18-inch cubes weighing approximately 30 pounds each. These briquettes are palletized into a 4-foot cube and charged as is. Another type of scrap is "bales."

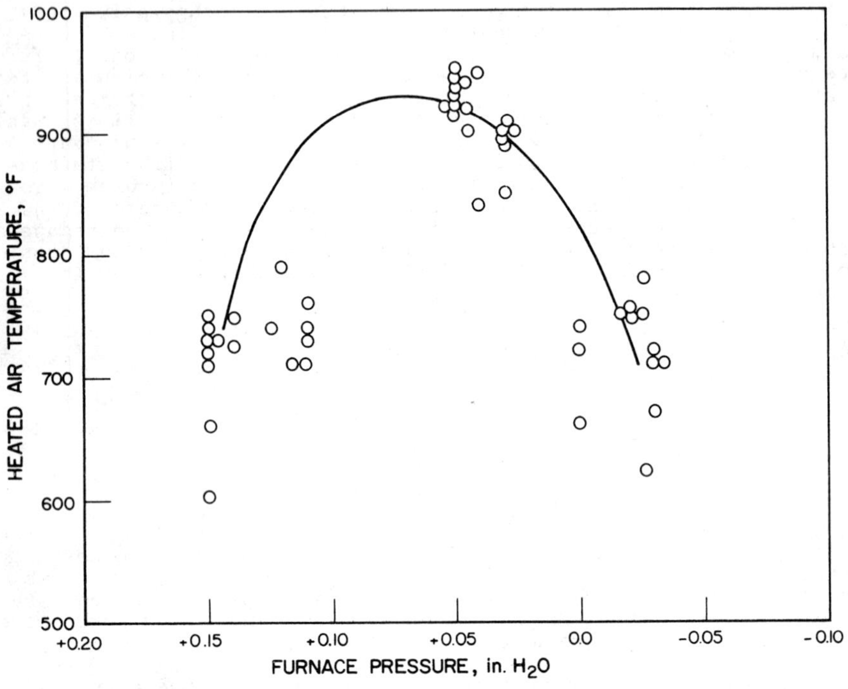

Figure 8. EFFECT OF FURNACE PRESSURE ON RECUPERATOR PERFORMANCE

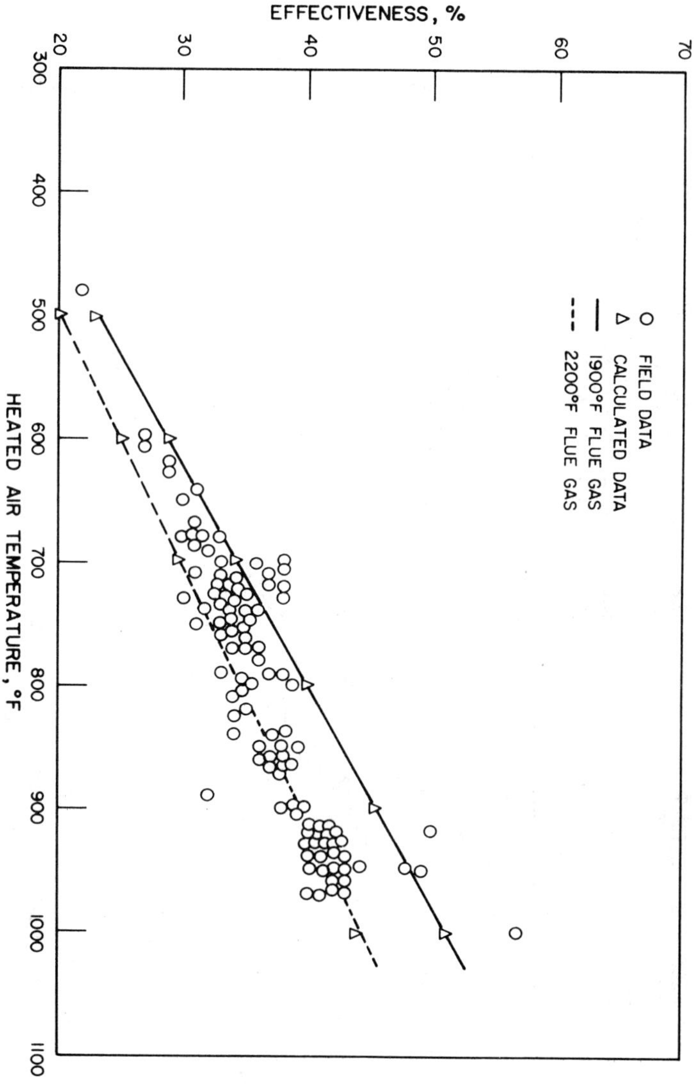

Figure 9. EFFECT OF HEATED AIR TEMPERATURE ON RECUPERATOR PERFORMANCE

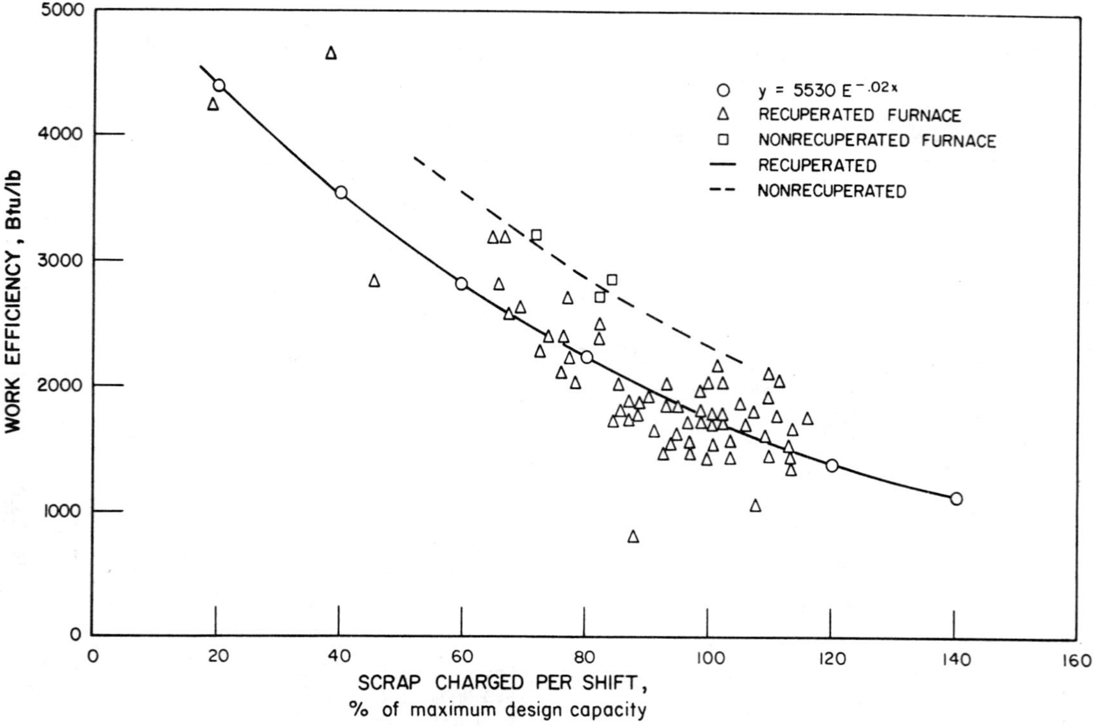

Figure 10. RECUPERATED AND NONRECUPERATED FURNACE WORK EFFICIENCY

This is also compressed scrap but of larger size than briquettes. Each bale weighs about 500 pounds and is approximately 3 feet wide and deep by 6 feet long. A pallet of two bales is fed to the furnace.

The most diverse type is called "loose" scrap. This scrap comes in many forms: loose trimmings, dippings and tailings, denser materials such as can lids, or smaller pieces obtained from shredding.

Figure 11 shows the furnace work efficiency as a function of percentage of loose scrap charged. As expected, the more loose scrap fed, the lower the work efficiency, except when the loose scrap is of a denser quality.

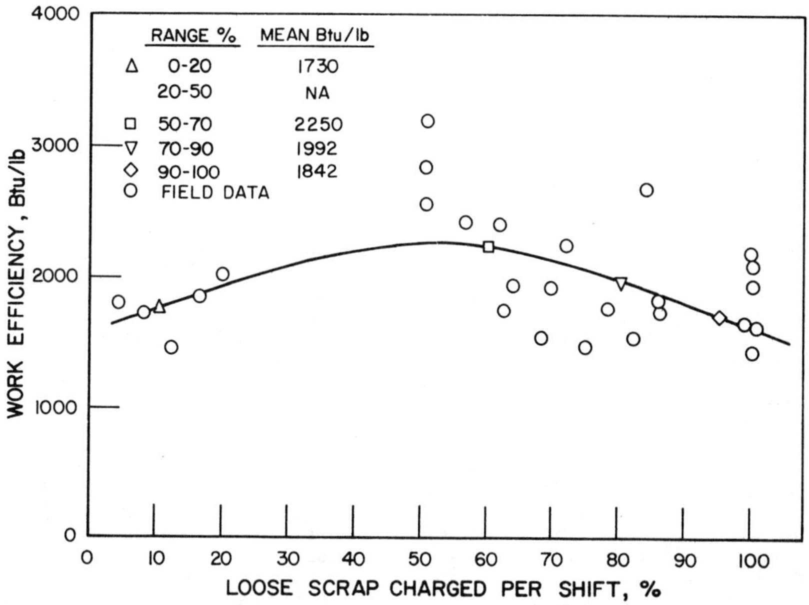

Figure 11. EFFECT OF LOOSE SCRAP ON WORK EFFICIENCY

SURVEY OF ADVANCED HEAT PUMP DEVELOPMENTS FOR SPACE CONDITIONING*

Phillip D. Fairchild
Energy Division
Oak Ridge National Laboratory
Oak Ridge, Tennessee

ABSTRACT

Because of the heat pump's appeal as an energy conservation tool and its potential impacts on electric and gas utilities, a great deal of recent attention has been focused on this type of space conditioning equipment by national energy research organizations. This paper presents a survey of heat pump RD&D projects with special emphasis on those supported by the Department of Energy, the Electric Power Research Institute, and the Gas Research Institute.

INTRODUCTION

Looking ahead to the residential and commercial space conditioning needs of the future, it is apparent that heat pumps will play an important role in this energy use sector. This is emphasized by the remarkable growth in sales of electric air-to-air heat pumps during the past few years and the extensive heat pump research and development activities under way in the United States and abroad.

At Oak Ridge National Laboratory (ORNL), we have been conducting in-house research on heat pumps since 1976, and since 1978 we have assisted in managing a contracted RD&D program for the Department of Energy (DOE) on residential and commercial heat pumps. This assignment has given us an opportunity to interface with many other organizations supporting and/or conducting similar R&D work. It is the intent of this paper, then, to present a survey of these projects (whether recently completed, ongoing, or about to start).

---

*Research sponsored by the U.S. Department of Energy under contract W-7405-eng-26 with the Union Carbide Corporation.

With regard to scope, this survey paper covers conventional electric as well as experimental gas-fired heat pumps and focuses on unitary equipment for residential and commercial space-conditioning applications. It does not cover solar-assisted heat pumps or chemical heat pumps, both of which may be appropriate technologies for this application and which are receiving R&D support from DOE and elsewhere. This omission was not for technical reasons, but because of lack of specific project information and time limitations for obtaining it.

HISTORICAL NOTES ON HEAT PUMP DEVELOPMENT

The only type of heat pump being manufactured today for application in residential and light commercial buildings is the electric heat pump — predominantly the air-to-air heat pump. The electric air-to-air heat pump evolved directly from the electric air conditioner (which may be characterized as a cooling-only heat pump). It was introduced on the commercial market in the 1950s and for several years was a relatively successful new product. Then, as a result of some severe equipment reliability problems, the product's survival was threatened. Many manufacturers dropped heat pumps from their product lines. Others restricted sales by region to the more southern climates. Also, during this period a ban was imposed on further installation of heat pumps in military-base housing due to the poor reliability record (1). But the industry's heat pump advocates persevered.

Thanks to the diligence of several major manufacturers during the 1960s and early 1970s, the heat pump was not abandoned and, gradually, real progress was made toward developing improved designs and components that could withstand the more severe stresses of heat pump service (2). Credit should also be given to programs sponsored by the Edison Electric Institute and certain individual electric utilities such as American Electric Power and Alabama Power aimed at improving the quality of the heat pump "as installed" (3,4,5). These utility programs focused attention on installation and servicing requirements, as well as providing detailed failure data which was fed back to manufacturers to help them determine specific design weaknesses. Developments over that 15-year period yielded proven solutions to the reliability problems by some (but not all) manufacturers (1,5).

As a result, a reliable and relatively efficient heat pump product was available when the demand for it occurred in the mid-1970s due to shortages and curtailments of natural gas and heating oil in many areas of the country. For many home buyers and builders, the heat pump became the only reasonable home heating alternative during that period of moratoriums on new gas connections in many regions and rapidly escalating prices for fuel oil and electric power. The combination of these events led to a period of dramatic growth in heat pump sales.

MARKET AND EQUIPMENT TRENDS

As illustrated in Fig. 1, heat pump sales have grown markedly in the United States since 1970, when shipments totaled less than 100,000 units per year, to approximately 500,000 to 600,000 units per year during the last three years. The heat pump, which was installed in only 4% of single-family homes in 1973, captured a 25% market share last year (6). This wide acceptance in the marketplace indicates that, in the purchasers' minds at least, the mechanical and reliability problems of the past have been overcome and, further, that the current heat pump product's initial costs are acceptable in return for decreased energy consumption and lower monthly utility bills.

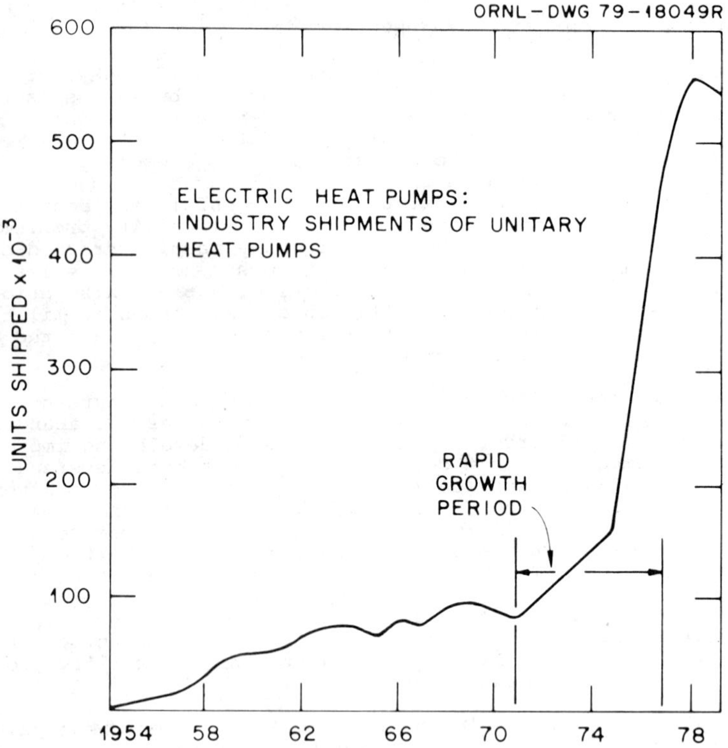

Fig. 1. Growth of heat pumps sales in the United States since 1954.

Source of 1954 to 1977 data: Gordian Associates, Inc., Heat Pump Technology: A Survey of Technical Development Prospects, U.S. Department of Energy, HCP/M2121-01 (June 1978); source of 1977 to 1979 data: Air Conditioning, Heating and Refrigeration News Statistical Panorama, Apr. 7, 1980.

Heat pump manufacturers usually offer several lines, ranging from an economy builder's model to a top-of-the-line, high-efficiency model. The following tabulation characterizes the efficiency of current heat pumps in terms of Air-Conditioning and Refrigeration Institute (ARI) steady-state ratings (7).

	Heating COP* at 47°F	Heating COP at 17°F	Cooling EER†
High-efficiency model	3.2	2.1	8.8
Typical model	2.5	1.8	8.0
Small, economy model	2.0	1.3	5.9

At present, the higher efficiencies are obtained by the use of high-efficiency compressors, larger heat exchangers, and improve controls.

Just as reliability was the key issue for heat pumps in the 1960s and early 1970s, performance and efficiency have been the key target for improvement since the mid-1970s. Because of the heat pump's appeal as an energy conservation tool and its potential impacts (both positive and negative) on electric and gas utilities, much of the impetus for performance improvement has come from outside the heating, ventilating and air-conditioning (HVAC) equipment manufacturers.

Research programs supported by DOE (and its predecessor organizations, the Federal Energy Administration and the Energy Research and Development Administration) began in that period, including the work at the National Bureau of Standards (NBS) and at ORNL. Those programs, together with special contracted studies on heat pump technology (8) and input from organizations such as EPRI and, later, GRI were influential in formulation of the DOE R&D programs that will be described in subsequent sections of this paper.

However, it is doubtful that those in the HVAC industry would rate the increase in heat pump R&D as the highlight of the last five years in the heat pump business. Rather, it is more likely that they would identify the federal energy conservation legislation that resulted in establishment of new government test procedures and proposed minimum efficiency regulations for their heat pump and air conditioner products (9,10).

Much of the manufacturer's engineering resources are and will likely continue to be involved, near-term, either in responding directly to those new testing procedures or in striving to obtain maximum available rating credit under the new procedures. And although the chief effect of the efficiency standards may be to eliminate the lower efficiency, "economy" models now being sold, there should be some effect on advancing the state-of-the-art. Near-term trends in heat pump performance improvement are therefore likely to be highly influenced by regulations, as well as by outside R&D support and market forces.

*COP, Coefficient of Performance, is defined as the ratio of the heating capacity in Btuh to the electrical power input in Btuh (Watts × 3.412 Btuh/Watt) at a given rating condition and is the standard efficiency measure for heat pumps in heating mode.

†EER, Energy Efficiency Ratio, is defined as the ratio of the cooling capacity in Btuh to the electrical power input in Watts at a given rating condition (expressed in Btuh/Watt).

RECENT ELECTRIC HEAT PUMP RESEARCH

Research work initiated by NBS during the 1974-75 winter season in a test house (11) and later augmented by their own laboratory investigations (12) and those of Carrier Corporation (13) was influential in focusing R&D attention on two effects that significantly degrade heat pump performance under actual operating conditions: (1) "on-off" cycling losses resulting from the need to reestablish steady-state operating conditions in the system after each "off" cycle and (2) effect of frost buildup on the outdoor coil under certain weather conditions and the associated energy required for a periodic defrost operation.

During the same period, research was undertaken at ORNL to determine the potential for improving steady-state efficiency or COP utilizing state-of-the-art computer modeling techniques for heat pump design and component matching. This work (which is summarized on Table 1) has involved experimental studies to assemble a detailed data base of system and component performance for heat pumps in addition to development of the computer simulation models.

Table 1. DOE/ORNL heat pump technology studies

Project	Objective
Heat pump performance data base	Detailed component and system performance/model validation (14,15)
Heat pump computer model	Explore efficiency improvements/ tool for industry use (16,17)
Heat pump optimization	Define efficiency limits, best cost-effective system (18)
Cycling and frosting tests	Characterize loss mechanisms (15)

ORNL also developed and demonstrated an innovative heat pump/thermal storage concept called the Annual Cycle Energy System (ACES) during the late 1970s. The system employs a unidirectional water-source heat pump to heat the house and simultaneously to make ice during the heating season; this ice is stored in a large, insulated tank to provide a source of cooling in the following summer (air conditioning) season. Work has continued on this system, both in-house and under subcontracts, to evaluate and improve the system's economic competitiveness against more conventional systems (19,20).

EPRI began to support contracted research on electric heat pumps in the mid-1970s and, over the past 5-6 years, has sponsored several systems analysis and field testing projects aimed at evaluating seasonal performance of air-source heat pumps and potential effects on electric utilities. Those projects, listed in Table 2, have focused primarily on the heating seasonal performance factor (SPF)* and its improvement in northern climates and on utility concerns about maximum electric power demand and time-of-day load profiles.

---

*SPF, Seasonal Performance Factor, is defined as the total heating or cooling effect delivered during the season in Btu's divided by the total electrical energy input in Btu's (Watt-hours × 3.412) during the same period (an integrated COP value).

Table 2. Recent EPRI-sponsored heat pump research projects

Project No.	Contractor	Title/description
RP-432[a]	Westinghouse	Load and use characteristics of heat pumps in single-family residences/Field tests of heating operation at 120 houses in 12 regions (21)
RP-1100	Gordian Associates	Impacts of residential heat pump technologies on utility system loads/Analysis of data developed in RP-432 to assess effect on utilities
RP-544	Westinghouse	Investigation of methods to improve heat pump performance and reliability in northern climate/Comprehensive systems analysis and methodology development (22)
RP-789[b]	Carrier Corporation	Investigation of heat pump performance in northern cliamtes through field monitoring and computer simulation (23)
RP-1201-6	Science Applications Inc., (SAI)	State-of-the-art assessment of hybrid electric heat pump system with oil or gas supplementary heat (in lieu of electric resistance) (24)

[a]Jointly sponsored with the Association of Edison Illuminating Companies with participation of 12 electric utilities.

[b]Cofunded with Niagara Mohawk Power Corporation and Carrier Corporation.

DEVELOPMENT OF ADVANCED ELECTRIC HEAT PUMPS

Most of the research work in the late 1970s led to the same general conclusions about the key technical issues in improving electric heat pumps for the future:

- improve part-load performance (reduce system "cycling" losses),
- improve low-temperature performance (reduce supplementary electric resistance heat),
- improve system steady-state performance (optimize system design and improve component efficiency),
- reduce frosting/defrosting losses (reduce frost buildup, improve defrost method/controls),
- continue to maintain/improve reliability.

Development of cost-effective variable-capacity compressors addresses two of those issues, namely reduction of cycling losses and reduction of supplementary resistance heat requirements at low ambient temperatures. Accordingly, development work on capacity modulation (variable-capacity compressors together with appropriate advancements and modifications in other components and controls to take full advantage of variable capacity on a systems basis) of the electric air-source heat pump is considered a high development priority.

A second need is for innovative defrost methods to replace the current refrigerant flow reversing scheme. Currently, losses attributed to frost buildup and those associated with the defrost cycle are considered second only to cycling losses in most locations (climates), especially when it is considered that a significant fraction of the electric resistance heat is used to "temper" the air supplied to the house when the system is switched over to the cooling mode. In addition, the effect of the defrost transient on the compressor is seen to represent the most severe "shock" that the compressor experiences; therefore, its elimination may represent one of the primary means of improving reliability.

Component improvement and design optimization is a third research and development area, first for determining the limits of heat pump steady-state efficiency (maximum reasonable COP without cost constraints), then for determining the highest COP which is possible in a cost-effective configuration heat pump, and finally for determining the relationship between dynamic (seasonal) and steady-state efficiencies by integrating dynamic losses (cycling, defrosting, etc.) into the optimization process.

In the subcontracted RD&D program that ORNL manages for DOE, system development or "product development" projects provide a way of addressing such technical issues on an integrated or "systems" basis. This approach is also consistent with the program's commercialization thrust in that developments can be more readily implemented if developed by or for a present or potential heat pump manufacturer in a manner similar to that used for developing a new product line.

Development work has started on two such residential-sized advanced electric heat pump projects (see Table 3) selected in 1979 and 1980 via a request for proposals (RFP). Like all major DOE/ORNL product development projects, these projects involve cost sharing by the contractor. They employ different approaches to capacity modulation and related advancements to achieving their seasonal efficiency targets.

Table 3. Electric heat pump system development (residential application/size)

System	Developer	Target SPF[a] improvement
Advanced electric heat pump I (systems approach, featuring 2-step capacity modulation of compressor, high-efficiency fans/motor, new defrost, microprocessor control, diagnostics)	Westinghouse	20% min. (northern climate SPF 2.5)
Advanced electric heat pump II (fully variable modulating compressor with linear motor)	MTI	25-35%

[a]SPF is defined as the total heating or cooling effect delivered during the season divided by the total electric energy input during the season.

Much of the current hardware development in the Westinghouse project is based on the system analysis and methodology developed under EPRI Project No. RP-544 (see Table 2).

As mentioned previously, market and regulatory pressures will also result in new heat pump models and product features designed to improve energy-efficiency ratings. Variable-capacity compressors, for example, have recently been introduced by some manufacturers, including both two-speed and continuously speed-modulated compressors and a dual-compressor configuration. Field and laboratory studies are planned as part of both the DOE/ORNL program and the continuing EPRI-supported research activities to test and evaluate these different variable-capacity designs, their seasonal efficiency potential, and their cost-effectiveness.

In addition to the field testing and systems analysis activities sponsored by EPRI, we understand that some "product development" work similar to the DOE/ORNL-sponsored projects is in the planning stage at EPRI. EPRI has advised us that they plan to report on the ongoing and planned EPRI-sponsored heat pump R&D projects at the upcoming 16th IECEC in Atlanta (August 9-14, 1981) (25).

WELL WATER AND GROUND-COUPLED HEAT PUMPS

Another way to improve heating performance at low ambient temperatures is to use alternative heat sources such as well water or ground-coil heat pump arrangements. These relatively constant-temperature heat sources offer considerable potential for improved capacity and COP during extreme winter weather when ground (and groundwater) temperatures are much higher than outdoor air temperatures. Also, because the outdoor coil is replaced by a refrigerant-to-water heat exchanger, frosting/defrosting losses will not be a factor in performance degradation. A notable achievement in this area during the past year was completion of a DOE-sponsored comprehensive study on well water heat pumps (by the National Water Well Association), including assessment of groundwater availability and quality on a state-by-state basis, investigation of environmental and legal problems, and an analytical evaluation of energy use and economics (26). Due to the analytically assessed energy conservation potential of this currently available technology (energy savings of 10-25% on an annual heating and cooling basis and 20-60% for heating only relative to an air-source heat pump), field performance projects have been initiated by DOE/ORNL to establish energy savings of well water heat pumps under actual residential service conditions. One of these is a cooperative project with EPRI, and a second is being cosponsored with New York State Energy Research and Development Authority (NYSERDA) and Empire State Electric Energy Research Corporation (ESEERCO).

Other projects are under way to assess ground-coil heat pump technology. Both EPRI and DOE/ORNL are supporting research in this area, although the effort is relatively small at present, pending favorable indications of energy conservation potential and commercial viability.

HEAT-ACTUATED HEAT PUMP DEVELOPMENT

A major portion of the DOE/ORNL contracted R&D resources are allocated toward development of heat-actuated heat pumps. Conventional combustion heating equipment (gas or oil furnace, for example) currently achieves about 60% efficiency and, when developed to its ultimate potential, can approach but not exceed 100% efficiency. Heat-actuated heat pumps, of which the gas-fired heat pump is the predominant example, offer potential heating COPs exceeding 1.0, that is, delivering a heating effect greater than the heating value of the input fuel. Heat-actuated heat pumps are a product concept not presently manufactured or marketed in the United States.

Such a product could be an important option for improved efficiency in residential and/or commercial space heating in the mid- to late-1980s if its viability in the marketplace can be established. A gas heat pump currently envisioned as a typical 1990 residential product should provide both heating and cooling service using 30 to 50% less gas, depending on climate, than a conventional gas furnace uses for heating alone. About 55% of the nation's homes (over 40 million homes) are heated using natural gas. Furnaces and other gas heating appliances generally have a 15-20 year life expectancy, which creates a continuing replacement market of about a million units per year. With addition of the new home market, the total residential gas heating market is estimated at approximately 1.5 million units annually. With the cooperative support of the gas utility industry, the gas heat pump is now emerging as a leading candidate for providing more energy efficient heating for these consumers in the mid-1980s and beyond (27).

There are two basic types of gas-fired heat pumps. The first uses an absorption cycle and the second replaces the electric motor of the conventional heat pump with a gas-fueled prime mover (engine-driven). Of course, one of the key advantages of the heat-actuated system is that in the heating mode, waste heat can be recovered to augment the heating effect produced by the heat pump cycle, thus increasing the heating COP and capacity significantly.

Development efforts are under way on four residential and two commercial gas-fired heat pumps. As summarized in Table 4, two of the residential heat pump projects are of the absorption type and two are Stirling-engine-driven. Both commercial-sized heat pump systems are engine-driven, as described in Table 5. In most cases, systems developed for use with natural gas can also be oil-fired with appropriate modifications to the combustion system.

For the heat-actuated (gas- or oil-fired) heat pumps, the priority issues and technical risks seem to focus on three elements:

1. seasonal efficiency (which involves the same potential losses due to cycling and defrosting as the electric heat pump faces but with different design options to alleviate some of those losses);

2. endurance, reliability, and reasonable maintenance (e.g., seal life in engine-driven systems); and

3. initial cost/selling price.

The final risk facing any heat-actuated heat pump development venture is attractiveness to potential manufacturers as a business opportunity. With the substantial investment in development and manufacturing facility costs required, manufacturers will need to be assured of a substantial rate of return on investment before accepting the risk. Outside support can reduce the development costs but not the tooling and facility costs.

At the present time, absorption heat pump technology is in the more advanced stage as a result of prior work in developing absorption chillers and air conditioning equipment. Therefore, the lower performance potential for the absorption heat pump is offset by the reduced technical risk associated with its development, as compared to the higher risk Stirling engine heat pumps. Thus,

Table 4. Heat-actuated heat pump system development (residential application/size)

System	Developer (Major subcontractor)	Assessed risks (technical and business)
Free-piston Stirling[a] (spring-mass resonating engine/compressor)	General Electric[b]	High
Diaphragm free-piston Stirling[a]	Consolidated Natural Gas (with MTI)	High
Organic fluid absorption[c] (R133a/ETFE)	Allied Chemical[b] (with Phillips Engineering)	Moderate to high
Ammonia-water absorption[c]	ArkLa Industries	Moderate

[a]Target fuel COP of 1.6 (heating) and 0.85 (cooling), where fueld COP is defined as the heating or cooling effect divided by the energy value of the fuel used. Target values are steady-state ratings (excluding parasitics) for prototype hardware.

[b]Cosponsored with GRI.

[c]Target fuel COP of 1.25 (heating).

Table 5. Heat-actuated heat pump system development (commercial application/size)

System	Developer (Major subcontractor)	Assessed risks (technical and business)
Gas turbine driven[a]	Garrett/AiResearch	Moderate to high
Braun linear engine driven[b] (free piston, internal combustion engine)	Honeywell/Techtonics	High

[a]Target fuel COP of 1.5 (heating) and 1.3 (cooling), where COP is defined as the heating or cooling effect divided by the energy value of the fuel used. Target values are steady-state ratings (excluding parasitics for prototype hardware.

[b]Proof of concept.

the general strategy is one of supporting different concepts with correspondingly different levels of risk and ultimate performance potential. Because no single heat pump concept is likely to be optimum for all climatic regions, then more than one option could be developed and commercialized on a competitive basis.

Two of the residential and one of the commercial gas-fired heat pump projects are continuing efforts cosponsored by the GRI with cost sharing by the respective contractors (General Electric, Allied Chemical, and Garrett/AiResearch). The other three were selected via competitive proposals solicited by RFP in 1979 and also involve contractor cost sharing.

In a following talk one of the ongoing projects cosponsored by DOE/ORNL and GRI will be described in some detail. It involves development of a commercial-sized Brayton/Rankine gas-fired heat pump system by AiResearch.

In addition to the gas-fired heat pump projects being cosponsored by GRI and DOE, GRI is supporting two other heat pump projects (27). Mechanical Technology, Inc., (MTI) is conducting one of those development projects, which involves a Rankine/Rankine system in a novel configuration employing rotating heat exchangers. That project is cost-shared by MTI, DuPont, and NYSERDA. The other gas-fired heat pump project involves a Phase I feasibility study of a Stirling/Stirling system. That project is being conducted by Sunpower, Inc.

INTERNATIONAL INTEREST IN HEAT PUMPS

As this survey indicates, there is much R&D activity under way in this country to develop advanced heat pumps for the future. Similar efforts are under way in other countries and, in fact, the R&D programs in Europe may be even more extensive than in the United States. DOE and ORNL have been involved in a cooperative program on advanced heat pumps through the International Energy Agency (IEA) since 1977. The final report on the first phase of that work provides a fairly complete technology survey in the 13 participating countries (28). In addition, EPRI (jointly with RWE, a German electric utility) and Carrier have sponsored international conferences, the proceedings of which provide additional information on the varying energy situations in other countries and the role that heat pumps may play in their energy conservation efforts (29,30).

CONCLUSIONS

This paper represents an attempt to describe not only the heat pump R&D projects with which we at ORNL are involved as part of our responsibilities to DOE, but also the work being conducted and/or sponsored by other national research organizations. As this survey illustrates, the private sector HVAC equipment manufacturers continue to play the major role in developing advanced heat pump systems, and the general objective of contracted R&D programs is to accelerate and complement the internally supported efforts of industry. As can be inferred from the variety of projects described, there is no consensus as to the "best" space-conditioning system. There are too many unpredictable factors, such as fuel availability and electric-to-gas price ratios, to accurately judge the ultimate commercial viability of some of these developmental systems. Therefore, much of the impetus for the development is to increase the available space-conditioning options for the future by ensuring that advanced energy-efficient heat pump products are available to meet the needs of various market sectors.

REFERENCES

1. R. D. Heap, Heat Pumps, E.&F.N. Spon Ltd., London, 1979.

2. "Heat-Pump Reliability Shows Big Gains," Electr World, 78-80 (Aug. 1, 1973).

3. Nance C. Lovvorn, "Heat Pump Compressor Reliability as Demonstrated in Alabama Service Program," Air Conditioning, Heating & Refrigeration News (Jan. 27, 1975).

4. Dennis Orange, "Service, Product Quality Keys to Continued Heat Pump Success," Air-cond Refrig Bus, 112-17 (January 1977).

5. E. F. Gorzelnik, "Heat Pumps get High Grades Up North," Electr World, 142-44ff (Sept. 15, 1978).

6. "One-family Housing Dipped for First 9 Months of 1980," Air Conditioning, Heating & Refrigeration News (Jan. 5, 1981).

7. Air-Conditioning and Refrigeration Institute, Directory of Certified Unitary Air-Conditioners, Air-Source Unitary Heat Pumps, Sound-Rated Outdoor Unitary Equipment, Central System Humidifiers Effective July 1-December 31, 1980, Air-Conditioning and Refrigeration Institute, Arlington, Va.; 1980.

8. Gordian Associates, Inc., Heat Pump Technology: A Survey of Technical Development Market Prospects, U.S. Department of Energy, HCP/M2121-01 (June 1978).

9. U.S. Department of Energy, "Test Procedures for Central Air Conditioners, Including Heat Pumps," Fed. Regist. 44(249): 76700 (Dec. 27, 1979).

10. U.S. Department of Energy, "Advanced Notice of Proposed Rulemaking Regarding Energy Efficiency Standards for Heat Pumps," Fed. Regist. 45(16): 5602 (Jan. 23, 1980).

11. G. E. Kelly and J. Bean, Dynamic Performance of a Residential Air-to-Air Heat Pump, NBS Building Science Series 93, March 1977.

12. W. H. Parken, Jr., R. W. Beausoliel, and G. E. Kelley, "Factors Affecting the Performance of a Residential Air-to-Air Heat Pump," ASHRAE Trans. 83(1): 839-50 (1977).

13. C. E. Bullock and W. R. Reedy, "Heat Pump Cyclic Performance and Its Influence on Seasonal Operation," Proceedings, Third Annual Heat Pump Technology Conference, Oklahoma State University, Stillwater, 1978.

14. A. A. Domingorena and S. J. Ball, Performance Evaluation of a Selected Three-Ton Air-to-Air Heat Pump in the Heating Mode, ORNL/CON-34 (January 1980).

15. W. A. Miller, Performance Evaluation of a High-Efficiency Air-to-Air Heat Pump in the Heating Mode, ORNL/CON-69 (to be published).

16. R. D. Ellison, F. A. Creswick, C. K. Rice, W. L. Jackson, and S. K. Fischer, "Heat Pump Modeling: A Progress Report," Proceedings of the Fourth Annual Heat Pump Technology Conference, Oklahoma State University, Stillwater, Apr. 9–10, 1979.

17. R. D. Ellison, F. A. Creswick, S. K. Fischer, and W. L. Jackson, "A Computer Model for Air-Cooled Refrigerant Condensers with Specified Refrigerant Circuiting," ASHRAE Trans., Vol. 87, part 1.

18. C. K. Rice, S. K. Fischer, R. D. Ellison, and W. L. Jackson, "Design Optimization of Conventional Heat Pumps: Application to Steady-State Heating Efficiency," ASHRAE Trans., Vol. 87, part 1.

19. R. E. Minturn et al., ACES 1979 Capabilities and Potential, ORNL/CON-48 (June 1980).

20. L. A. Abbatiello, "The Annual Cycle Energy System Characteristics and Performance," Proceedings of the International Congress on Building Energy Management, May 12-16, 1980, Povoa De Varzim, Portugal.

21. Paul J. Blake and William C. Gernert, Load and Use Characteristics of Electric Heat Pumps in Single-Family Residences, Westinghouse Electric Corporation, EA-793, Vol. 1 (June 1978).

22. H. S. Kirschbaum and S. E. Veyo, An Investigation of Methods to Improve Heat Pump Performance and Reliability in a Northern Climate, Westinghouse Corporation, EPRI EM-319, Vol. 1 (January 1977).

23. Gerald C. Groff and Wayne R. Reedy, "Investigation of Heat Pump Performance in the Northern Climate Through Field Monitoring and Computer Simulation," ASHRAE Trans., Vol. 84, part 1, pp. 767–85 (1978).

24. M. H. Blatt and R. C. Erickson, State-of-the-Art Assessment of Hybrid Heat Pumps, Science Applications, Inc., La Jolla, Calif., EM-1261 (December 1979).

25. Personal communication between Arvo Lannus, EPRI, and P. D. Fairchild, Oak Ridge National Laboratory, Dec. 23, 1980.

26. Dana M. Armitage et al., Ground-Water Heat Pumps: An Examination of Hydrogeologic, Environmental, Legal, and Economic Factors Affecting Their Use, National Water Well Association, DOE/CS/20060-5120 (November 1980).

27. Gas Research Institute, GRI 1981-1985 Five-Year R&D Plan and 1981 R&D Program.

28. D. L. Hodgett et al., International Energy Agency Programme of Research and Development on Advanced Heat Pump Systems, Annex 1: Common Study on Advanced Heat Pump Systems, Final Report, Vol. 1, Battelle-Institut e.V. Frankfurt (July 1980).

29. Rheinisch-Westfälisches Elecktrizitätswerk Aktiengesellschaft and Electric Power Research Institute, The Electric Heat Pump, Papers of the German/American Conference on Technology and Applications, June 18-20, 1980, Düsseldorf.

30. Carrier Corporation, "Heat Pump and Space Conditioning Systems for the 1990's" Proceedings of an International Symposium, February 1–2, 1979.

LIGHT COMMERCIAL BRAYTON/RANKINE SPACE CONDITIONING SYSTEM

David Friedman, Kenneth Trester, Wiley Cowell

AiResearch Manufacturing Company
2525 W. 190th Street, Torrance, California 90509

Updated February, 1981

SUMMARY

A heat-actuated space conditioning system that provides more efficient use of natural gas has been developed for 26.4- to 87.8-kW (7.5- to 25-ton) commercial applications. The system consists of a subatmospheric, natural-gas-fired, Brayton-cycle engine that drives a Rankine-cycle heat pump. A centrifugal Freon compressor is driven directly from the Brayton engine rotating group through a permanent-magnet coupling. Unique features that offer high life-cycle performance include a hermetically sealed magnetic coupling, air foil bearings, an atmospheric in-line combuster, and a high-temperature recuperator. Predicted overall engine efficiency is 27 percent and predicted overall coefficient of performance (COP) at the energy source is 1.0 in in cooling and 1.2 in heating.

INTRODUCTION

The increasing energy shortage has resulted in new developments in energy-saving devices. In the space conditioning field, heat pumps driven by off-site electric motors have been successfully marketed as energy-saving systems. However, on-site, gas-powered generation systems can significantly improve upon the performance ratings of these existing systems. As shown in Figure 1, with a gas-powered engine directly driving the heat pump, the heat loss from the power cycle is not wasted. The heat can be recovered and put directly into the building. With the electric-motor-driven system, however, 71 percent of the energy required never reaches the building because it is lost at the power plant and during transmission.

AiResearch Manufacturing Company, a division of The Garrett Corporation, and Dunham-Bush have been investigating the critical technology for a gas-fired Brayton/Rankine 35-kW (10-ton) space conditioning system.* This program, first sponsored by the American Gas Association, is now sponsored by the Gas Research Institute and

*Garrett and Dunham-Bush are members of The Signal Companies.

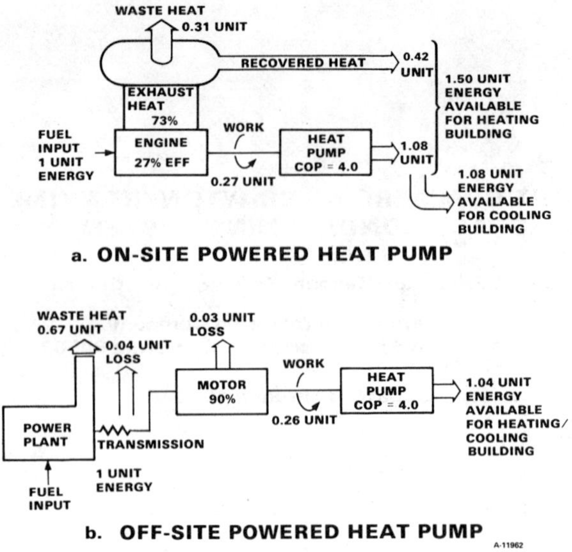

Figure 1. Off-Site vs On-Site Heat Pump Cycles

the Department of Energy. The system offers high performance potential by utilizing the most advantageous working fluid in each cycle: air is used in the heat engine where high turbine inlet temperatures provide high thermal efficiency, and Freon 12 is used in the heat pump because of its desirable thermodynamic characteristics. The total system is slightly larger than a 35-kW (10-ton) electric rooftop heat pump and measures 1.83 m high by 1.68 m wide by 2.29 m long (6 ft by 5.5 ft by 7.5 ft). Figure 2 shows a scale model of the space conditioning system.

The principal technical achievements during this program have been the development of (1) the Brayton/Rankine rotating group, including a magnetic coupling that provides efficient torque transmission at the high speeds associated with the gas turbine engine, and (2) an engine cycle that provides high thermal efficiency while burning natural gas at low pressure.

CYCLE DESCRIPTION

The basic system, shown in Figure 3, consists of two thermodynamic loops. The first is a conventional vapor compression heat pump system that uses a high-speed centrifugal vapor compressor instead of the usual positive displacement type. The second loop is the Brayton-cycle engine, which runs the centrifugal compressor and supplies waste heat energy to the building during the heating mode of operation.

The actual Brayton cycle is a semi-open subatmospheric-pressure cycle consisting of a centrifugal compressor, a radial inflow turbine, a recuperator, a sink heat exchanger, and an in-line atmospheric combustor. Ambient air is drawn through the recuperator, where it is preheated before being introduced into the atmospheric-pressure, natural-gas-fired combustor in stoichiometric proportion. Compressor discharge gas is also cycled through the recuperator and used as diluent to provide added flow and the desired turbine inlet temperature. Expansion takes place through the turbine component, from which sufficient power is extracted to drive both the Brayton and Freon compressors. The turbine discharge, which is at subatmospheric pressure, is processed through the recuperator, where it preheats combustor inlet air. The temperature of the low-pressure gas is further reduced by using the sink

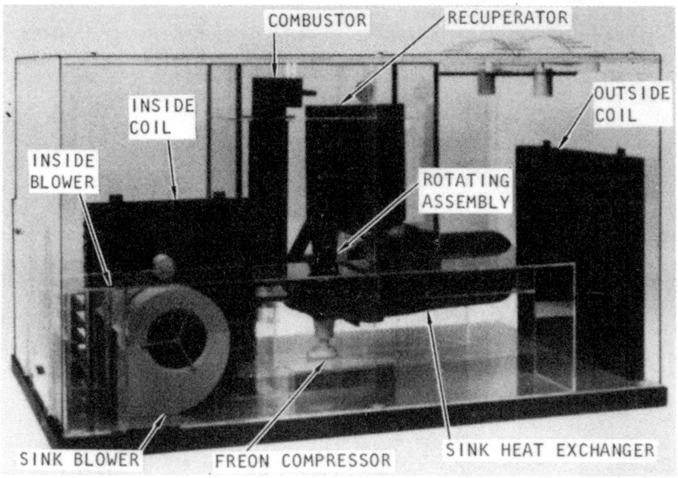

Figure 2. Model of Brayton/Rankine Space Conditioning System

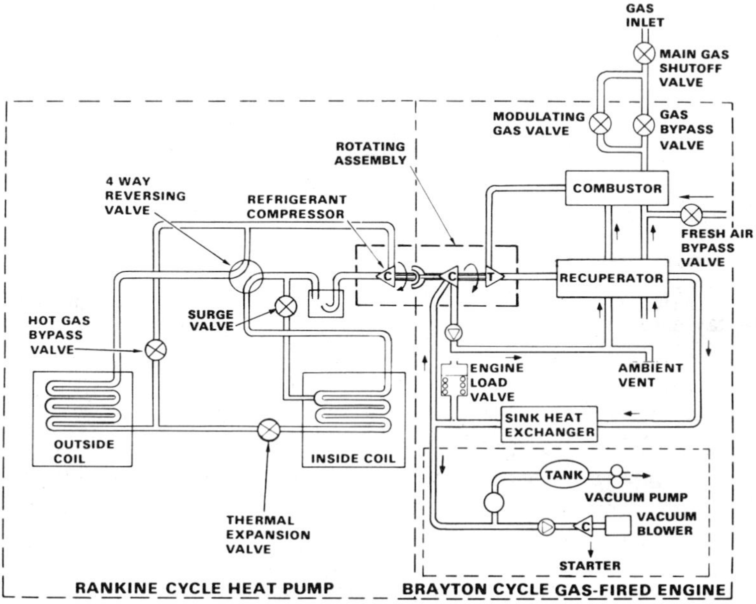

Figure 3. Basic Brayton/Rankine System Concept

heat exchanger to reduce compressor power consumption. The compressor pumps the gas back to atmospheric pressure and a small portion is exhausted; the remainder is recycled as diluent. The sink heat exchanger cooling air, which now has a temperature greater than 37.8°C (100°F) is combined with building recirculation air in the heating mode, supplementing the vapor cycle unit heating capacity.

SYSTEM PERFORMANCE PREDICTIONS

Figure 4 presents engine performance characteristics for a 35°C (95°F) ambient day. Cycle efficiency and output shaft power are plotted as functions of speed for varying turbine inlet temperatures. At the design point of 75,000 rpm and turbine inlet temperature of 843°C (1550°F), the predicted cycle efficiency is nearly 30 percent and the delivered power is 7.65 kW (10.2 hp), which is sufficient to drive a 35-kW (10-ton) refrigeration system that provides 35 kW of cooling for the 35°C (95°F) ambient day.

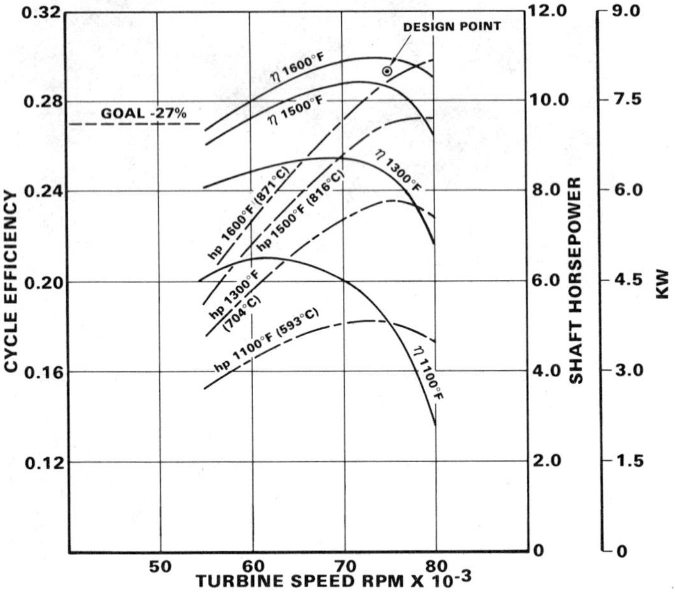

Figure 4. Brayton Engine Performance

The combined performance characteristics for the heat pump system for a 35°C ambient day cooling mode of operation show a system COP of 1.3. (With all parasitic losses included, the overall COP is reduced to 1.0.) This is equivalent to an electrically driven heat pump energy efficiency ratio (EER) of 15.3 (11.8 after inclusion of parasitic losses). A similar performance curve was generated for an 8.3°C (47°F) ambient heating day. With the system operating to match predicted heat load--turbine inlet temperature at 621°C (1150°F)--the system COP is 1.5 (1.2 with all parasitic losses included). This is equivalent to a heating COP of 5.17 for an electrically driven heat pump system (4.14 after inclusion of parasitic losses). In Figure 5 the Brayton/Rankine system COP is plotted as a function of ambient temperature and compared against existing equipment. As shown, Brayton/Rankine system performance surpasses that of existing conventional systems at all conditions.

ENERGY SAVINGS AND PAYBACK COST ANALYSIS

An energy consumption and energy cost analysis was made to compare the Brayton/Rankine heat pump system with a variety of competitive year-round space conditioning systems. The following competitive systems were used in the analysis:

(a) Present conventional electric heat pump (HP) with EER of approximately 8.0 at 35°C (95°F)

(b) Present conventional year-round air conditioning (YAC) with gas heating furnace efficiency of 60 percent overall and electric cooling EER of 8.0 at 95 percent

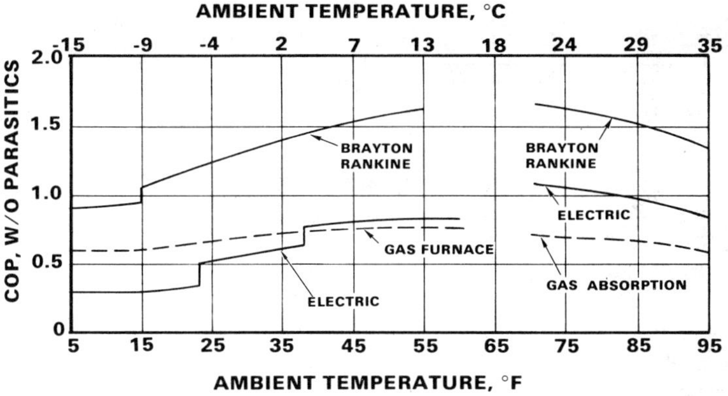

Figure 5. Comparative Performance

(c) Future all-electric heat pump with predicted EER of 11.5 at 35°C (95°F)

(d) Future year-round air conditioner incorporating pulse-combustion gas furnace with overall efficiency of 90 percent, and electric cooling EER of 11.5 at 35°C (95°F)

(e) Future two-speed heat pump with high efficiency at maximum design condition, while providing lower cyclic losses at off-design conditions

Three cities were chosen to represent climatic conditions for three different areas of the United States: Canton, Ohio; Birmingham, Alabama; and Phoenix, Arizona. Air Force weather data were used for each city to establish the time for each 2.9°C (5°F) temperature bin. A representative load line for a commercial 35-kW (10-ton) system was generated for each city, and a computer program was designed to determine the energy consumption for each system. Included within the analysis were degradation factors for on-off cycling losses, where appropriate, and for defrost. The degradation factors for cyclic losses were utilized for heat pumps, air conditioners, and the Brayton/Rankine system.* The pulse-combustion furnace and standard furnace were assumed to have 90-percent and 60-percent efficiency, respectively, at all conditions, with no degradation factors.

Figure 6 shows the results of the computer energy consumption analysis for the three cities and each of the five systems analyzed. The figure shows the annual seasonal energy efficiency ratios (SEER's) for each system. To compare gas systems with electrical systems, the gas source energy numbers were divided by a conversion factor of 0.29 to arrive at a comparable SEER. The data include all parasitic losses for each system, including electrical fans for indoor and outdoor coils, furnaces, and sink heat exchangers. As shown, the Brayton/Rankine system performance exceeds that of an advanced year-round conditioning system or a heat pump, even with those systems showing a performance improvement of approximately 40 percent over their current counterparts in the marketplace.

Figure 7 shows the energy savings converted to dollar savings, with each system being compared with the Brayton/Rankine space conditioning system. The figure reflects three different energy rates for both gas and electricity for each of the three cities analyzed. As shown by the figure, annual savings over future YAC's or HP's are greater than $600, with nominal energy costs of $4 per million Btu for gas and 7 cents per kW-hr for electricity. With high energy costs, the savings against today's systems could be as great as $2400 per year.

*Didion, D.A., and G.E. Kelly, "New Testing and Rating Procedures for Seasonal Performance of Heat Pumps," ASHRAE Journal, September 1979, Vol. 21, No. 9, pp. 40-44.

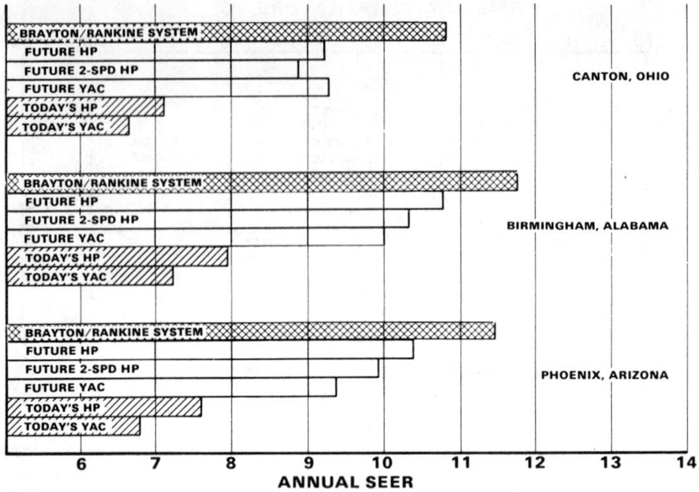

Figure 6. Annual Seasonal Energy Efficiency Ratios

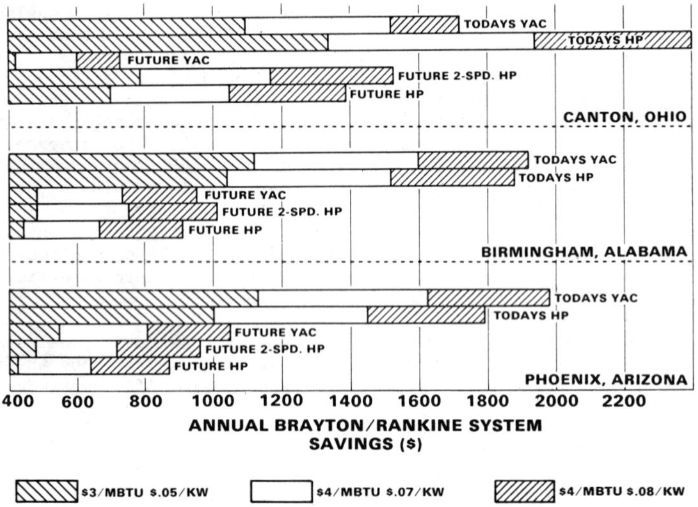

Figure 7. Comparative Annual Savings

Figure 8 shows the total cost savings to the user over the expected life of the equipment. The Brayton/Rankine system is compared with present and future YAC's for the Canton, Ohio, test case. Energy rates of $4 per million Btu for gas and 7 cents per kW-hr for electricity were assumed. The equipment cost was included in the analysis and was assumed to be financed for 10 years at a 10-percent interest rate. Energy costs were compounded annually at 8 percent. Installation costs and maintenance costs were assumed to be the same for all systems. The equipment was assumed to have no salvage value at the end of 10 years, and depreciation was ignored in the analysis.

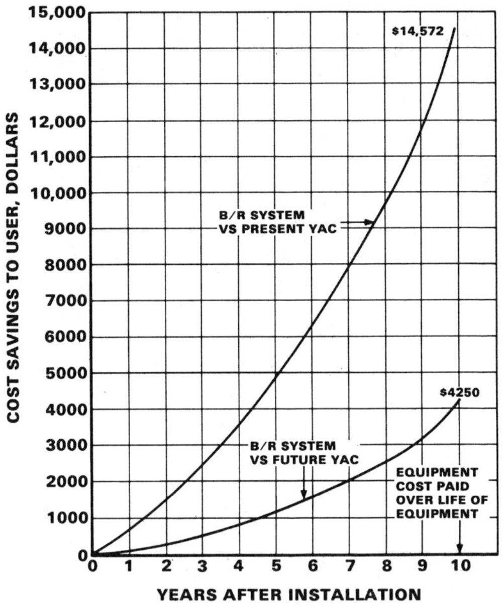

Figure 8. Cost Savings to User

Estimated equipment costs were based on expected 1984 prices. The future YAC cost was assumed to be 14 percent greater than that of the baseline conventional YAC, while the Brayton/Rankine system was assumed to have a 42.5 percent greater selling price than the baseline conventional YAC.

As shown, even with the higher first cost, the Brayton/Rankine system shows positive cash flow in the first year when compared against either YAC under this type of financing arrangement. The savings to the user over the life of the equipment is shown to be $14,572 when compared with the conventional system, and $4,250 when compared with the future YAC.

It should be noted that the savings would be even greater if the Brayton/Rankine system were compared with an electric heat pump in the northern climate of Canton, Ohio.

The results of this analysis should encourage prospective purchasers if those purchasers are the actual users of the equipment, rather than contract installers who do not pay the monthly energy bills.

HARDWARE DEVELOPMENT

Freon Compressor and Magnetic Coupling

The first development effort was the design, fabrication, and test of the Freon-12 compressor and magnetic coupling. The Freon compressor unit shown in Figure 9 is basically very simple; its one moving part comprises a single-stage, shrouded impeller at one end of the shaft, and the driven magnet cartridge at the other end. The shaft is supported on two standard, precision ball bearings. The compressor wheel is made of aluminum alloy. A stationary cover seal encloses the driven magnet and forms a part of the Freon compressor unit. The magnets are samarium-cobalt. The primary requirement of the magnetic coupling is to run at rated speed without slipping.

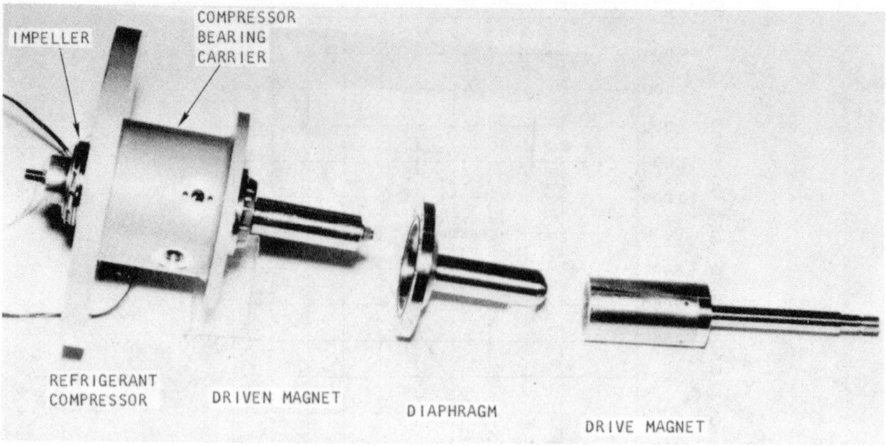

Figure 9. Refrigerant Compressor and Magnetic Coupling

Brayton Rotating Group

The last phase in the program was the development and test of the Brayton rotating group shown in Figures 10 and 11. The Brayton rotating assembly is connected to the Freon compressor via the magnetic coupling. The assembly consists of a cast, single-stage, radial turbine and a single-stage, radial compressor; these are bolted back-to-back to the shaft to form an integrated rotating unit. The shaft is supported on long-life, maintenance-free, foil air bearings. The foil bearings use the loop air as both coolant and lubricant and are entirely self-acting.

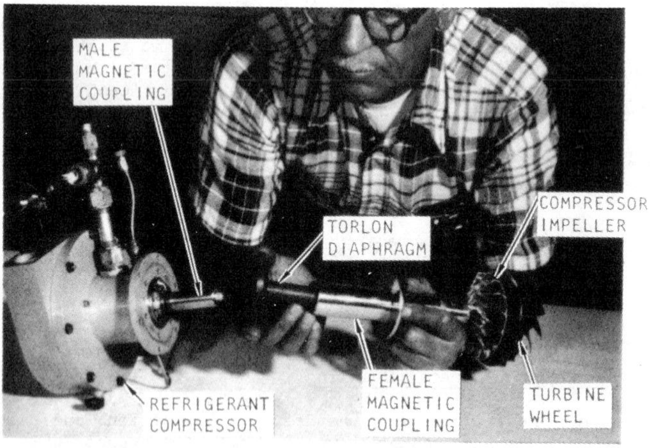

Figure 10. Brayton/Rankine Rotating Group Details

The turbine wheel is of high-temperature cast alloy, and the compressor wheel is cast steel. Cast turbine and compressor scrolls, a turbine nozzle, a compressor diffuser, a thermal shield, and a compressor inlet housing complete the drive design.

Figure 11. Assembled Brayton/Rankine Rotating Group

The design arrangement draws cooling air over the bearings into the compressor inlet housing. A labyrinth seal keeps the bearings essentially at ambient atmospheric pressure. A thermal shield between the turbine and compressor wheels, plus other thermal design features, minimize thermal leakage.

The unit shown in Figures 10 and 11 is compact, having overall dimensions of 0.74 m (29 in.) in length, including the attached Freon compressor, and 0.36 m (14 in.) in diameter at its widest point.

Recuperator

The high-temperature recuperator is constructed from a formed tube sheet of nitride-dispersion-strengthened, 400-series, stainless steel. Its counterflow configuration provides an effectiveness of 0.90.

The core consists of alternate layers of gas and air fins. The development unit is shown in Figure 12. Air enters a plenum and is distributed by turning fins into the central counterflow section of the core, where most of the recuperation takes place. After leaving the counterflow section, air flows through additional turning fins and out the return air plenums. Gas enters at the return air end and passes through the core in a counterflow direction. High-performance rectangular strip fins are used in the counterflow section in both the air and gas passages.

Both the air and gas fins are brazed to the tube sheets with a high-temperature, nickel-base braze alloy. Brazing maximizes heat transfer and makes the entire core an integral structural assembly capable of withstanding high pressures and temperatures without external support. This type of design is used by AiResearch in its large industrial regenerators.

Combustor

The use of a subatmospheric gas turbine leads to a distinct advantage in design of the combustor, which is a low-cost, in-line, atmospheric design that utilizes natural gas at only a few inches of water pressure. The in-line design, in which all combustion products are introduced into the flow stream, results in a very efficient combustor that provides 100-percent combustion efficiency with only 10-percent excess air above stoichiometric proportion. The diluent and quenching gasses are

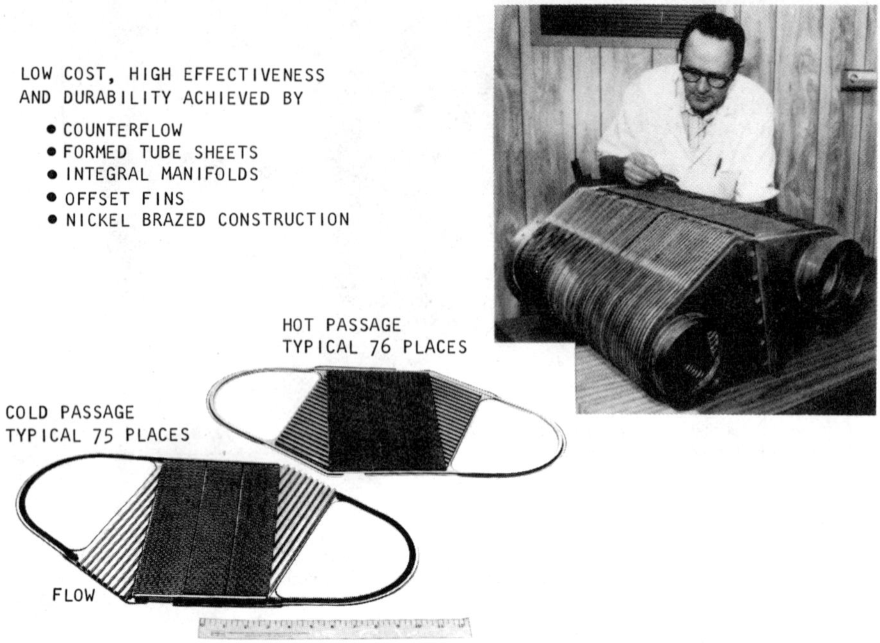

Figure 12. Recuperator Core Construction

recirculated combustion products that provide a reasonably low combustion chamber temperature. The unit is a cylinder having a diameter of approximately 0.18 m (7 in.) and a length of approximately 0.75 m (30 in.).

PROGRAM STATUS

Development

The development program schedule shown in Figure 13 has proceeded basically as depicted, with most of the major technical challenges already met. It is expected that the development effort will be completed in early 1981, and a full demonstration prototype system will be fabricated and assembled, ready for test, by mid-1981.

The introduction of this system into the marketplace is still a substantial effort. A demonstration unit is yet to be fabricated; the field test program, scheduled for 1982, is still to be accomplished; and the economics of the approach has yet to proceed past the preliminary analysis stage. Assuming these tasks are completed successfully, full production is anticipated in the 1984 time frame.

Program Expansion

The program is being expanded to allow a second heat pump manufacturer to participate in the development effort, and also to survey the international market. In this regard, assistance in development efforts is being negotiated with foreign gas utilities.

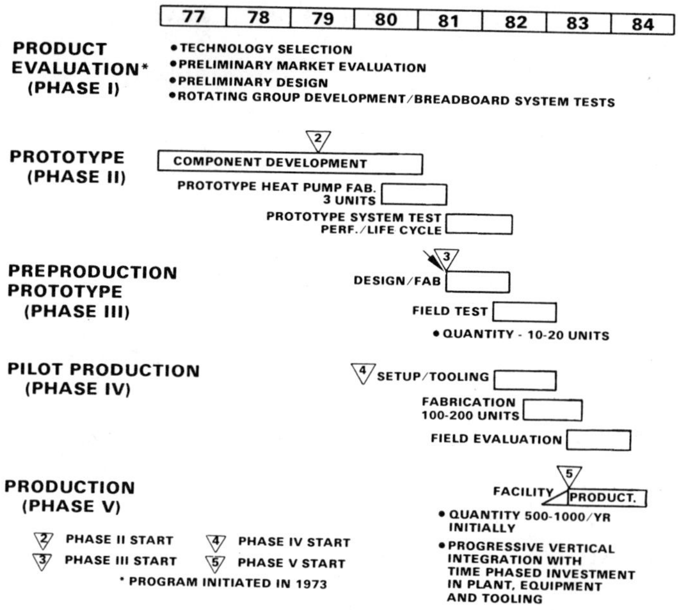

Figure 13. Brayton/Rankine Conditioning System Master Program Schedule

AiResearch is also investigating the application of common hardware for use in power generator sets, including solar power generation. To this end, a study contract has been awarded to AiResearch by GRI to investigate the application of this hardware for use in a solar-assisted, gas-fired heat pump. It is planned to utilize support hardware similar to that needed for a solar-dispersed power generating system.

Figure 14 shows the overall arrangement of the solar-assisted, gas-fired heat pump. Basically, the system would incorporate the subatmospheric Brayton engine and the 10-ton Rankine heat pump under development. The system will require repackaging for integration with a solar parabolic concentrator and a thermal receiver. The Brayton engine will be mounted at the focal point of the concentrator behind the receiver, and the heat pump will be on the ground. A simplified system schematic is shown in Figure 15.

Solar equipment used in the system has the highest efficiency of any such hardware now under development. The high temperatures attainable from the concentrator/receiver correspond to high potential power system efficiencies, which will result in minimum collector area requirements and minimum installation cost.

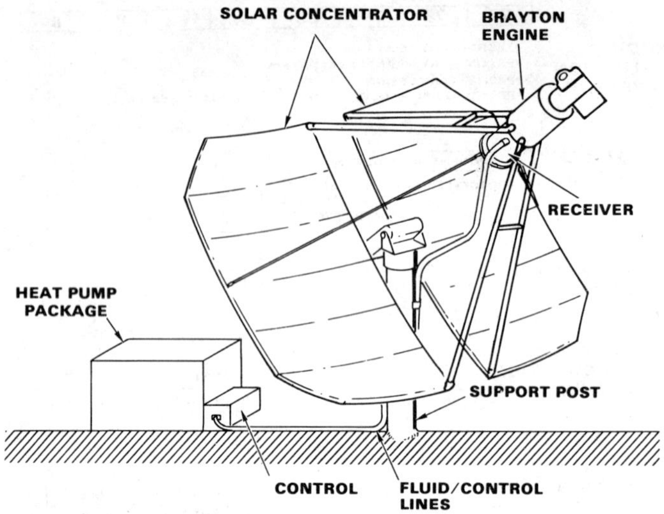

Figure 14. Overall Arrangement of Solar-Assisted, Gas-Fired Heat Pump System

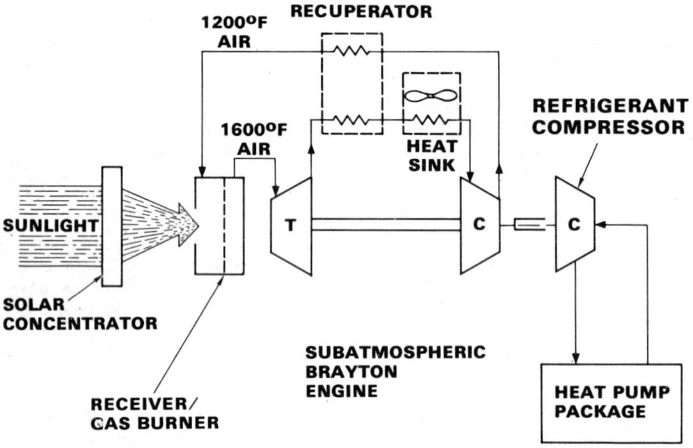

Figure 15. Schematic Diagram of Solar-Assisted, Gas-Fired Heat Pump System

WASTE HEAT RECOVERY WITH OPEN WATER VAPOR COMPRESSION

Heinz Jaster
General Electric Company
Schenectady, New York 12301

Rapidly increasing cost of energy as well as uncertainty about steady energy supplies have resulted in considerable engineering activity to define alternative approaches to industrial energy conservation. Industrial process heat pumps, which are capable of raising the temperature of waste heat, represent one approach for substantial industrial energy savings. A competing approach is the use of open water vapor compression.

This paper presents a discussion of the relative performance advantages of water vapor compression versus closed cycle heat pumps. Due to the absence of heat exchangers and their associated thermodynamic irreversibilities, open cycle systems provide for significant efficiency improvements over closed cycle heat pumps. Alternative cycle configurations will be described, the effect of intercooling will be discussed. Applications are for the use of warm industrial waste water for process steam generation.

Closed cycle heat pumps for thermal upgrading of industrial waste heat have now been commercially available for a number of years. These are closed cycle vapor compression systems which use fluorocarbon refrigerants with higher temperature capability than those used for space air conditioning. Similar compressors, heat exchangers and thermodynamic cycle configurations are employed for space conditioning and process heat pumping applications. Open cycle systems, in the form of vapor recompression applications, are finding increasing use in the chemical process industry.

Increased and increasing energy prices and relative unreliability of future oil and gas availability to industry, as well as governmental regulation concerning limitations on the use of oil and gas for new industrial process heat generation capacity, have begun to make industrial process heat pumps (IPHP) economically attractive – despite substantially higher first costs than for oil or gas fired heat generators. Heat pumps can, in some cases, successfully compete against coal fired systems on the basis of first cost.

Closed cycle IPHP, because of the presence of heat exchangers and associated temperature differences, operate at lower efficiency than can potentially be achieved with open cycle systems. And because currently used refrigerants are non-perfluorinated fluorocarbons which have thermal stability limits which are low, currently available IPHP are incapable of producing process heat at temperatures greater than approximately 220 $^{\circ}$F.

This paper points out differences and similarities between closed and open cycle vapor compression devices, the effect of alternative cycle configurations and shows the approximate efficiency of open cycle systems.

A general description of a mechanical energy activated heat pump is: a device which operates between two unequal temperatures and causes, with the addition of mechanical energy, heat to be drawn from the region of lower temperature and heat to be rejected to the region of higher temperature (see Figure 1). For the use of upgrading of waste heat, the useful output is the heat rejected. The heat pump's efficiency is measured by the coefficient of performance (COP) which is defined to be the non-dimensional ratio of the rate of heat rejected to the mechanical power drawn:

$$COP = \frac{\dot{Q}_H}{\dot{W}} \tag{1}$$

In the steady state,

$$\dot{Q}_H = \dot{Q}_L + \dot{W} \tag{2}$$

Where $\dot{Q}_L$ is the rate of heat drawn from the low temperature region. For the case for which the temperatures are a sink and a source respectively, it can be shown that the maximum possible COP is the COP for a Carnot heat pump:

$$COP_{Carnot} = \frac{T_H}{T_H - T_L} \tag{3}$$

Where T_H and T_L refer to the low and high temperature reservoirs respectively.

Figure 2 shows the essential components for a simple closed cycle vapor compression heat pump. A refrigerant vapor is compressed from state 1 (low pressure and temperature) to state 2 (high pressure and temperature). Between states 2 and 3, the refrigerant gives off heat at approximately constant pressure. The resulting liquid is irreversibly expanded from state 3 to state 4 to result in a low temperature, low pressure mixture of liquid and vapor. The refrigerant takes on heat between states 4 and 1, resulting in vaporization of the liquid fraction. The source of heat is assumed to be some warm fluid stream (transferred by the evaporator), the sink for heat is assumed to be some fluid stream which when heated serves to heat a process (transferred by the condenser).

The open cycle analog to the closed vapor compression cycle is shown in Figure 3. Here a warm liquid is isenthalpically expanded to state 2 (a mixture of liquid and vapor at low temperature and pressure). The phases are separated in a tank, the vapor is compressed

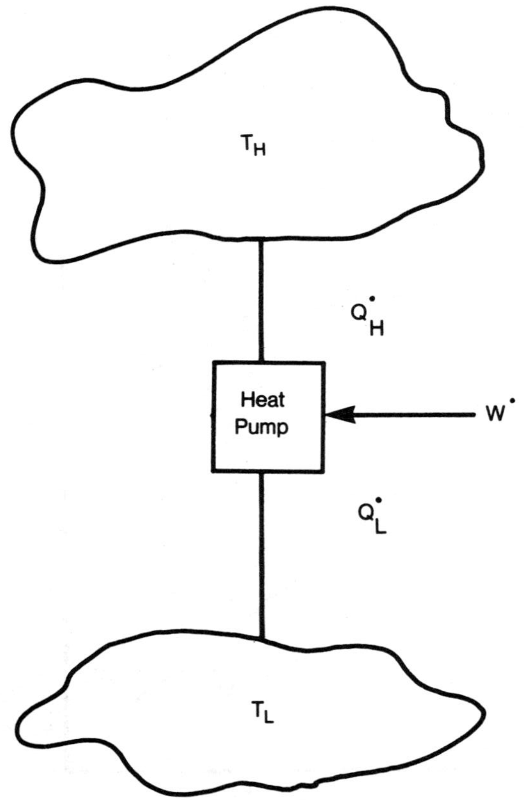

Figure 1. The heat pumping process - general

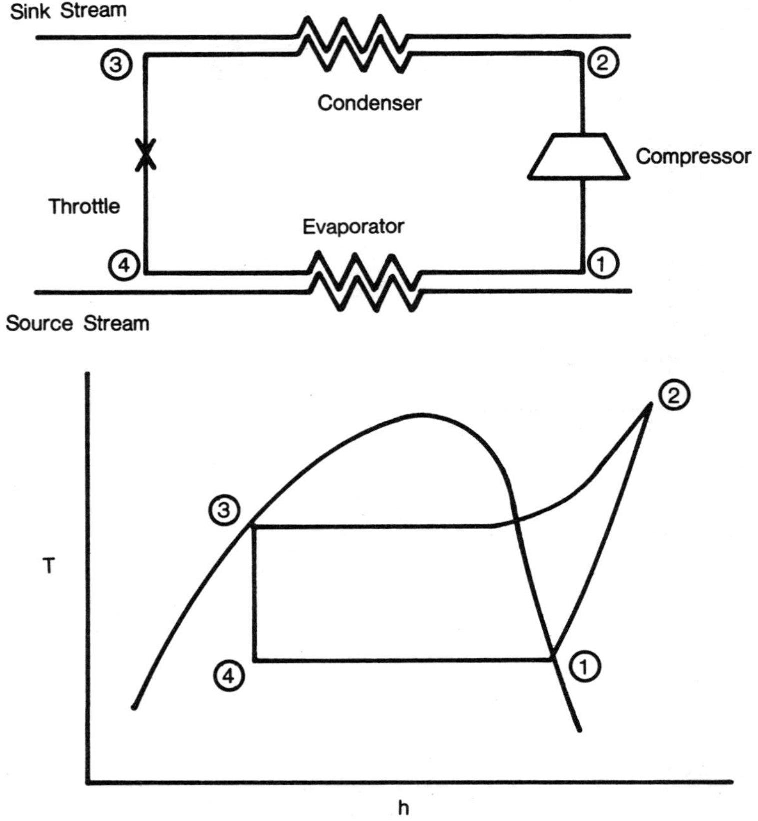

Figure 2. Closed vapor compression cycle

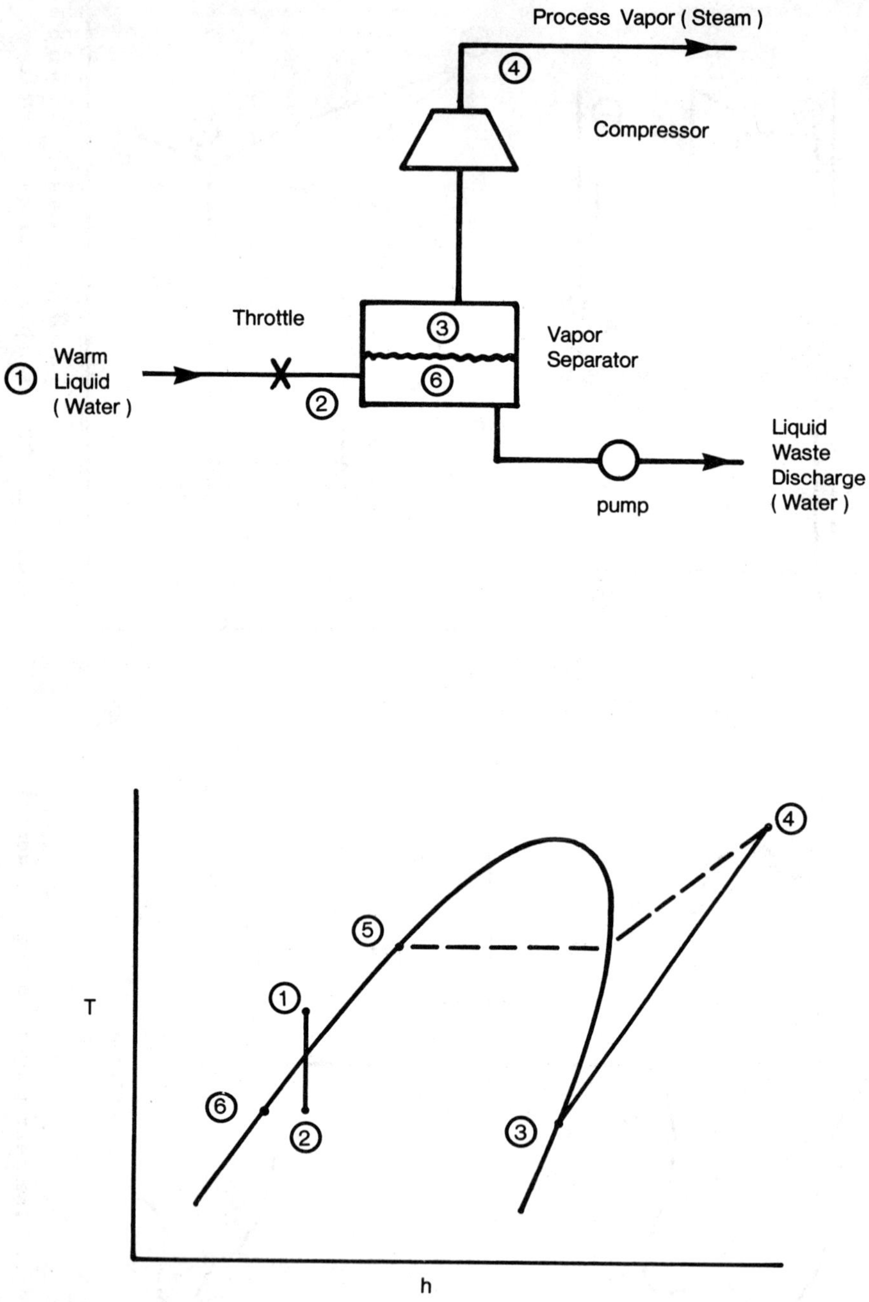

Figure 3. Open vapor compression cycle

to state 4, the liquid fraction is discharged via a pump. This cycle (a misnomer) derives its heat by converting sensible heat of the liquid into latent heat of vapor. Because the cycle contains no heat exchangers, that is, the transfer fluid is identically also the refrigerant, refrigerant saturation temperatures are equal to source and sink fluid temperatures respectively. This is in sharp contrast to closed cycle systems and this contrast is responsible for a considerable efficiency advantage of open cycle systems. Figure 4 shows the thermodynamic difference. For given source and sink saturation temperatures, the closed cycle needs to produce a temperature lift of ($T_b' - T_a'$), the open cycle a lift of ($T_b - T_a$). The temperature differences ($T_b' - T_b$) and ($T_a - T_a'$) for the closed cycle are necessary to transfer heat across heat exchangers. Equation (3) shows the effect, because the required temperature lift for the closed cycle is greater, the COP is lower. While, for moderately and economically sized heat exchangers, each heat exchanger temperature difference may only be 15 ^oF, such a difference accounts for an increased lift of 30 ^oF. Heat pump total lifts in excess of approximately 150 ^oF are uneconomical at current prices of energy, so that the heat exchangers in potentially viable systems account for a likely efficiency penalty of 20% or more. In fact, it is thought that most commercially economical lifts are less than 100 ^oF, here then, the open cycle will improve efficiency over its closed cycle counterpart by more than 30%.

All this, of course, depends on the thermodynamic suitability of the open cycle fluid. If, as is bound to be the case, most suitable industrial waste streams are water, then the efficiency advantage is greater yet, since water makes, from a thermodynamic point of view, an excellent refrigerant.

The simple cycles displayed as Figures 2 and 3 do not provide for optimum performance. Each can be modified to achieve increased efficiency. The modifications are similar for closed and open systems. I will discuss two types of modification: multistaging and multistepping.

A closed two stage vapor compression cycle is shown on Figure 5, the modification is addition of a second expansion and a second compression stage. The thermodynamic advantage is due to two effects. For one, not all refrigerant is irreversibly throttled from the highest to the lowest pressure, some of the refrigerant is bled off out intermediate pressure (as vapor) and recompressed. The other effect is due to intercooling the vapor before it reaches the second compression stage. Since, for a given pressure ratio, required compression work is approximately proportional to compressor inlet temperature, the second stage operates at lower work input since its temperature is decreased due to desuperheating.

Multistaging for the open cycle system is shown in Figure 6. The advantageous effects are the same, the vapor discharged from the first compression stage is desuperheated at intermediate pressure. Staging also results in dramatically lower system discharge temperatures, thereby making possible the use of less expensive compressor materials.

Figure 7 shows a closed and an open system configured for source stepping. In each case, instead of providing the entire temperature lift in a single step, only the last of the two steps must do so. This assures that the first step generates heat at a higher COP (the temperature lift is smaller). Since some of the heat is generated more efficiently, system efficiency, or COP, is raised. While only source stepping is shown, sink stepping is also possible if the desired heat output is in the form of sensible heat.

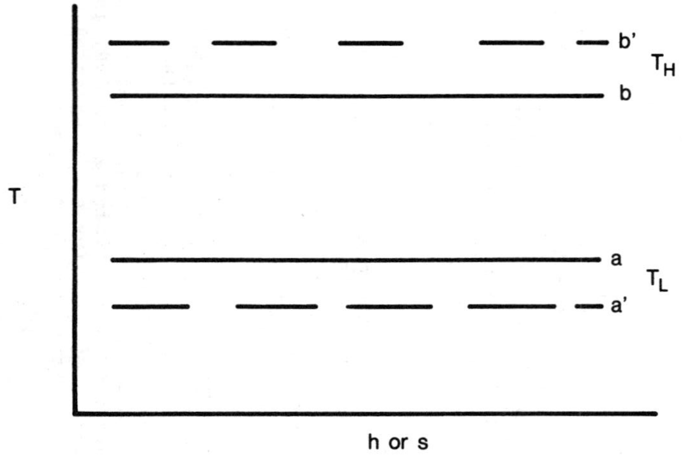

Figure 4. Saturation temperatures of closed and open cycles.

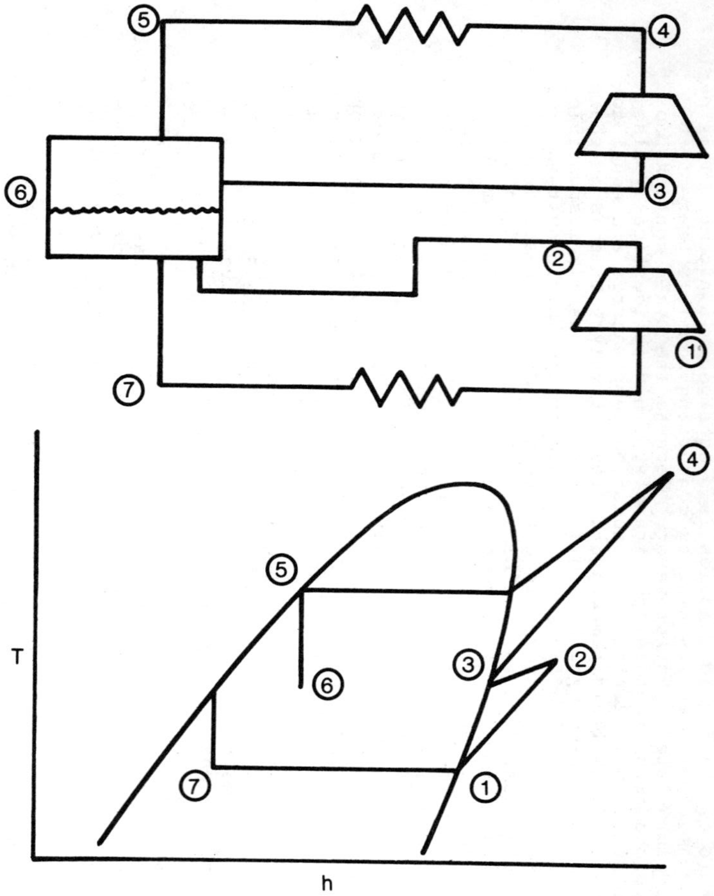

Figure 5. Two stage closed vapor compression cycle.

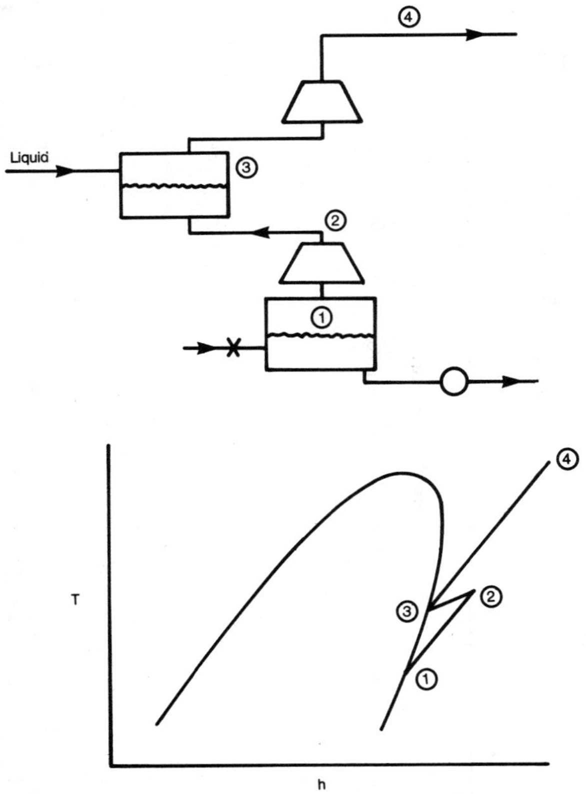

Figure 6. Two stage open vapor compression cycle

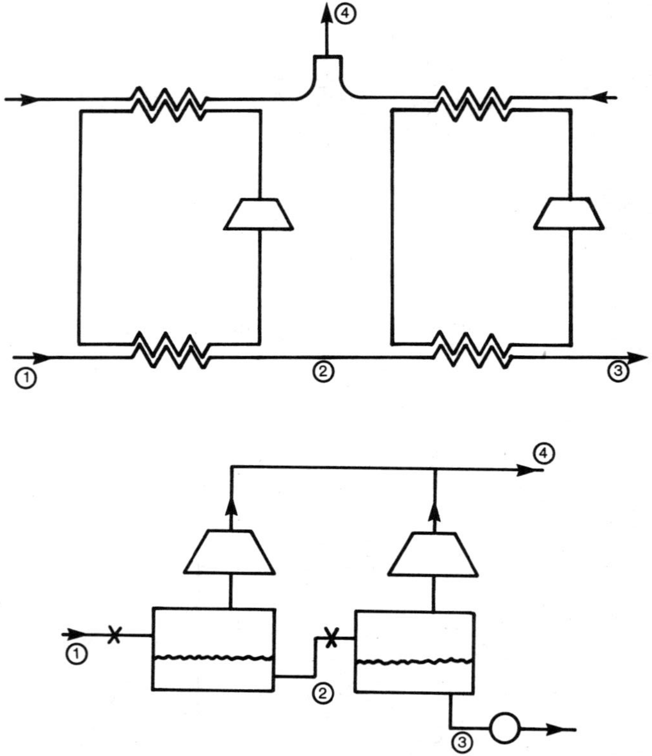

Figure 7. Source stepping with closed and open cycle systems

Approximate maps of COP as a function of source and sink temperature can be generated if refrigerant, compressor efficiency and heat exchanger effectiveness are known. For the case of a water based open cycle, estimated performance is shown on Figure 8. The source temperature is here taken to be the temperature of the discharged waste water, the sink temperature is the saturation temperature of the discharged steam.

The improved efficiency of water based open cycle systems over closed cycle analogs does not come without cost. Water vapor at, for example, 160 F has a density which is 1/300 as great as that for R-114. That is not to say that 300 times the compressor inlet volumetric flow is required for the same rate of heat delivery. The latent heat of vaporization is 20 times as large for water as for R-114. It still requires, for this example, that water vapor volumetric inlet flow must be approximately 15 times as great as for a R-114 charged closed cycle system. The open system saves the cost of heat exchangers but requires more expensive compressors. The water based open cycle system attains significant efficiency improvements and its temperature limitations are not those which unstable refrigerants prescribe, but are only constrained by water's critical temperature (705 F).

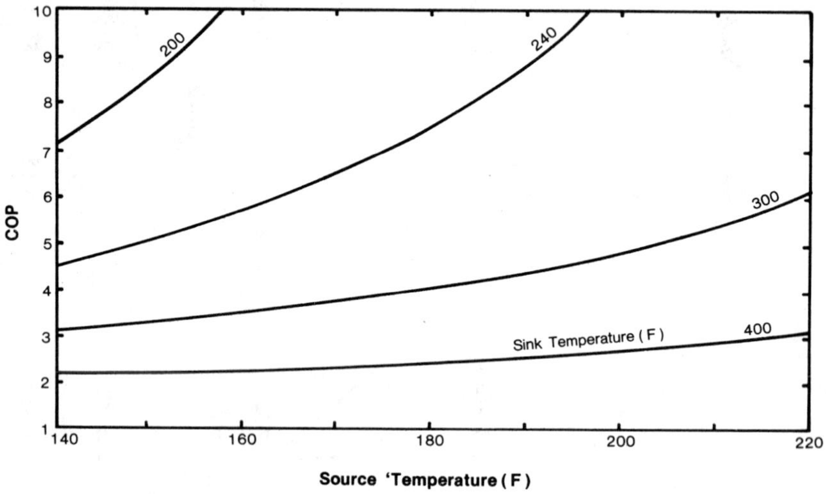

Figure 8. Estimated heating COP for open cycle water vapor compression.

ENERGY SAVINGS POTENTIAL IN CURRENT INDUSTRIAL INSULATION FOR MECHANICAL SYSTEMS APPLICATIONS

GARY ACINAPURA

CERTAINTEED CORPORATION

Business and industry consume 62% of our nation's fuel requirements if you include power generation. This is more than home heating and transportation combined and 1/3 is wasted.

The bill for all this waste -- is 8½ billion dollars per year.

We're paying for this -- you and I -- in the form of higher prices for nearly everything we buy.

But we don't have to. We could save half of the waste which could mean lower prices or bigger profits or both...

...which could also lead to more capital available to invest in more efficient production equipment and facilities...

...which could contribute to lower costs, lower prices and a more competitive position in world markets.

...which would call for expansion with more efficient equipment, more jobs and less energy consumption per unit of output.

And then, let's add to this cycle, a tax credit for company's who invest in fuel saving devices and equipment.

Which, in turn, speeds up the whole cycle -- makes more capital available -- to more firms, faster.

And the more there are, the less reliance we have on fuel imports and a better balance of payments.

...which helps strengthen the dollar. And on and on. This is not a fairy tale or conjecture. It can become a reality.

Let's take only one energy-use sector of our industrial universe and see how a simple upgrading of insulation can save 2.3 billion dollars a year which can go a long way toward achieving these goals.

We found these potential dollar savings through TIMA. Our firm, CertainTeed is a member. TIMA is an association of insulation manufacturers serving the non-residential building community.

TIMA observed only a one quarter inch average thickness increase in pipe insulation shipments from 1975 thru 1980. During that time fuel prices quadrupled and home insulation thicknesses almost doubled.

Something was wrong, so we set out to determine how much insulation industry was using and how much they could save if they had the right amount and what would be their return on investment for the added insulation.

We isolated 15 SIC classifications which use 85% of the total energy consumed by the industrial sector. This was determined from an independent research firm's energy-use data base compiled for DOE.

Next, we determined that steam process piping was the largest single category of energy-use, representing 42 percent of the total.

This was then researched using a two phase approach which combined the use of the extremely sophisticated DOE energy-use data base and a direct survey of 500 actual plants across the country.

In the first phase of the study we isolated the amount of energy used in creating the steam of the chosen SIC's and the amount of linear feet of pipe used to transport that steam. We did this by researching each one of the SIC's as well as developing information on over 400,000 actual plants across the country contained in the Dun and Bradstreet plant file.

The second phase of the study analyzed the insulation practices of 500 actual plants which were statistically stratified to be representative of the entire country as well as each one of the industrial classifications.

The survey incorporated an indoctrination letter followed by a telephone survey to the energy manager of each one of the 500 plants.

We determined the average operating temperatures of the steam lines in each of the plants, as well as how much and what type of insulation existed on those pipe lines. In addition, we further validated the information on amount of energy used and the amount, in length, of the steam lines which we estimated in the first phase of the study.

The survey results were then put into the computer and 192 economic thickness calculations were made...

...covering many input variables and pipe surfaces.

The conclusions were quite startling and clearly indicate a considerable amount of waste energy, not only through the existing insulation but from bare, uninsulated hot steam lines. There's alot of steam process piping out there.

...over 325 million lineal feet. That's enough to circle the globe two and one half times.

Twent two percent of this total, or one half the circumference of the earth 13.5 thousand miles -- has no insulation at all.

Heat loss from this piping costs 265 thousand barrels of oil equivalent per day.

The rest is underinsulated with only 1.7" thickness on large sized piping and 1" on small. It's been estimated that there are 32 million underinsulated houses in America, now we know, there are 325 million feet of underinsulated pipe.

For the piping, add the two together -- the uninsulated and underinsulated and industry is losing $6.8 million dollars per day or 2.5 billion dollars every year. This wasted energy is more than the whole plastics industry uses. More than used by the textile and lumber industries combined.

So, if we insulate the uninsulated lines and the underinsulated lines up to economic thicknesses, industry could not only save all this money, but 305 thousand barrells of oil equivalent every day-- just from steam process piping alone.

Then we asked the computer to add up all these savings facts and average them out for the 15 SIC categories and calculate economic thicknesses for the large piping and the small using this equation.

ETI thicknesses came out to 3" for large piping above 2" IPS and 2" thick for small sizes below 2". This compares with existing thicknesses of 1.7" on large pipe and 1" on small.

Besides having computer data on averages for all SIC's we have them for individual SIC classifications. Here, for example, is the one for class 28 chemicals and allied products. We have the ETI thicknesses broken down by SIC's too.

For instance in SIC 28 there are 66 million 400 thousand feet of steam pipe in the chemical and allied products category. Average temperature is 379 degrees. 50% of pipe is outdoors and length by pipe size is shown.

19 percent of the small size pipe in SIC 28 is uninsulated -- no insulation at all. And the insulated pipe has 1.1 inches on it. The mix, but insulation type, shows a preponderance of fiberglass.

Large sized pipe is better insulated, with 12% reported as bare and 1.7 inch thickness on the insulated. A greater percentage of the pipe is insulated with calcium silicate than the smaller pipe.

The computer analyzes many variables to calculate the ETI -- or economic thicknesses -- and then projects the savings in quantities of fuel and in dollars. Shown here are savings by insulating the bare pipe with ETI thicknesses in the chemical industry. These savings are over 25 thousand barrels of oil equivalent per day or over 667 thousand dollars.

Add the upgrade in thickness, from existing insulated lines to ETI thicknesses, and the total savings in dollars is almost 900 thousand per day. This is just for SIC 28.

Thickness recommendations are as shown which are calculated to provide the most return for the dollar cost based on the averaged input data for this SIC category.

The computer program will work for individual companies with specific energy savings goals, too, all we need is your input data on this form...

...it will then calculate just the right thickness to do the job -- not too much and not too little. That's why it's called ETI, Economic Thickness Insulation. It works for all hot or cold surfaces -- piping, equipment tanks -- flat, round or irregular. It is not for building walls or roofs.

It can tell you how much insulation you need on your next job.

It can tell you how good a job your present insulation is doing.

It can tell you how much to retrofit over existing insulation and what that additional amount will save.

It gives you an estimate of what insulation will cost and calculates the payback.

ETI costs you nothing except your time, and if you have a computer and want the program yourself, it's for sale for one hundred dollars, in Fortran IV language on cards or paper tape.

Any TIMA member company can run an ETI analysis of your particular insulation needs.

They may call it something different then ETI but it will be the same. There's no charge.

Let's get back to the scenario I opened this message with, where I referred to tax credits for energy saving investments.

During the last session of Congress there were 2 parallel bills introduced in the House and the Senate calling for a 20% investment tax credit for these kinds of investment. They will be re-introduced in this session and stand a good chance of becoming law. This means you can save 20% of the cost of an insulation job by subtracting that amount from the bottom line on the IRS form.

With this kind of inducement, coupled with the substantial savings and very attractive ROI insulation can bring, insulation could very well be the best investment your company could make this year and the years ahead.

Fuel costs are not going to go down, now would be a good time to bring your insulation up to modern energy saving thicknesses.

Contact a TIMA manufacturer representative and start saving money.

INDUSTRIAL/COMMERCIAL INSULATION FOR MECHANICAL SYSTEMS APPLICATIONS

Frank J. Powell
Manager, Thermal Insulation Program
Office of Energy Programs
National Engineering Laboratory
National Bureau of Standards
Washington, D.C. 20234

For the past 10 years, the main thrust for energy savings from the use of thermal insulation came from applications in residential housing. Much has been done and yet a great deal still needs to be done with new and existing houses.

Forecasts for the 1980's to improve energy utilization efficiency via the use of thermal insulation are emerging. Executive reports and outlook articles indicate that a steady recovery of residential building activity will continue in 1981 and beyond with increasing use of insulation. Also, the industrial/commercial market in 1981 will at first decline but then start to rebound in late 1981. One executive sees 6.5 percent overall decline in the industrial/commercial insulation market during 1981. However, with 32 million uninsulated houses in America and 325 million feet of underinsulated steam pipe and given tax credits to provide the necessary incentive for industry to invest in insulation - a tremendous demand for thermal insulation products could occur (1). Similarly, in an outlook for the eighties (2), an assessment reveals a recurring theme of heat recovery and energy efficiency via a restructuring of tax laws to promote reindustrialization and capital development with payback periods of 2-5 years.

What then is the potential for impacting the industrial and commercial sectors, where over half of the nations energy use occurs, through improved thermal insulation of mechanical equipment and systems? Past estimates need to be updated and reevaluated. In 1976, it was estimated for the six largest energy consuming

---

(1)"Time is Right for Recognition of Insulations Role," Rene Goutte, President, Insulation Group, CertainTeed Corporation, Heating/Piping/Air Conditioning, January 1981, page 156.

(2)American Society for Heating Refrigerating and Air Conditioning Engineers (ASHRAE) Journal, January 1981, pages 41-46.

industries, approximately 1.5 x 10^{15} Btu (1.5 quads; 250 x 10^6 barrels of oil-equivalent) could be saved annually through more effective use of thermal insulation (3). About half of these savings could result from increased insulation and maintenance of steam distribution systems alone. Recently, a study on the current use of insulation on steam process pipe lines for 15 Standard Industrial Classification (SIC) categories representing 85 percent of industrial energy use has been completed by the Thermal Insulation Manufacturers Association (TIMA) (4). The study shows that 71.8 million feet of pipe are not insulated at all out of a total of 325 million feet. All pipe is underinsulated by today's economic thickness standards. Bringing pipes up to today's standards would produce potential savings of 104 million barrels of oil equivalent per year. To do this would cost $6.2 billion (40 percent material and 60 percent labor) but the payback time would be a short 30 months.

The above quantities of energy (1.5 quads) can be saved by insulating mechanical equipment and systems, such as pipes, ducts, tanks, vessels, boilers, furnaces, and surfaces at which heat is transferred over the operating temperature range of -300°F to +2800°F. Further, a variety of insulating materials currently exists and enough reliable technical data and installation techniques can be found that if collected, evaluated, published, widely disseminated and used, the energy savings envisioned would be accomplished. Where lack of enough information in the current data base, test methods, standards, codes, specifications and installation guidelines exists, a research, development and demonstration program should be initiated.

Planning and commitment are necessary. The following items are suggested for action in both the public and private sectors for both new systems as well as the retrofitting of existing systems.

1. Prepare an updated forecast of the level of energy savings possible by optimum utilization of available insulation materials and techniques for reducing heat gain or loss in industrial and commercial equipment and systems. Include detailed specific energy conservation measures such as insulating bare steam pipes or increasing the thickness of insulation on inadequately insulated equipment.

2. Describe the kinds of information on materials, systems, test methods, standards, specifications, codes, cost/benefit analysis, installation guidelines and operations and maintenance practices, based on current technology, that must be collected, evaluated, published, disseminated and used. Indicate how this information could be used to optimize immediate energy conservation benefits from improved insulation practices of industrial and commercial equipment and systems. List the sources of this information and the nature of cooperative efforts that must be carried out amongst the source organizations.

3. Organize and outline the kinds of reports, manuals, evaluations and assessments, tabulations of test methods, codes, guidelines and other materials that must be prepared and made available for education, training and workshop programs. Describe the education and training programs required and the procedures needed for executing training programs to diagnose, calculate, and audit the energy conservation potential from insulation in new and existing systems.

---

(3)"Industrial Thermal Insulation - An Assessment," TID-27120, Energy Research and Development Administration, August 1976, page xiv.

(4)"Assessment of Industrial Process Steam Piping Survey," General Energy Associates, Drexell University, 1980.

4. Conduct a study of effective incentive concepts that could be implemented by manufacturers, contractors, owners, consumers, and governmental bodies to facilitate and promote these methods of energy conservation including an indication of the nature of new incentives and/or disincentives. Usually the motivation to translate a need into action is based on either a legal requirement or a financial incentive. Legal requirements are usually set forth in building codes and regulations and generally apply only to new buildings and systems. Financial incentives in the United States and in Europe are in a number of forms such as direct grants up to a specified maximum, the use of loans sometimes combined with low interest rates and extended repayment periods, and tax allowances based on the cost of installation or the interest paid on loans.

5. Identify gaps and inadequacies in available technical data, test methods, standards, codes, materials and systems, installation and maintenance guidelines, including durability/reliability, and recommend types of programs and organizations that should be used in research, development, and demonstration efforts to advance the technology in a longer range context.

Implementation of the above items and the subsequent application of appropriate insulation to industrial systems should significantly reduce energy use and reduce dependence on foreign fuel sources. Suggestions for incentives to insulate such as; tax breaks, subsidies, loans or other similar ideas should be developed. Ongoing efforts such as new diagnostic ideas and techniques should be exploited. New selection/decision criteria should be developed, i.e., the balance between economic payback and energy conservation. New or revised standards, codes, workshops, and other educational means should be rapidly developed and provided.

These items should be developed jointly by the public and private sectors with opportunity for all to participate and/or review the results of the work. Input should be actively solicited from committees and organizations in this field such as the ASHRAE Technical Committee 4.4, Thermal Insulation and Moisture Barriers, the Thermal Insulation Committee of the National Institute of Building Sciences (NIBS), the American Society for Testing and Materials (ASTM), Committee C-16 on Thermal and Cryogenic Insulation, the International Standards Organization (ISO) Committee 163 on Thermal Insulation, relevant Commissions of the International Institute of Refrigeration (IIR), National Model Code groups, trade associations, and Federal/State agencies. Public workshops should be offered and followed by endorsement, voluntary adoption by Federal and State government agencies, and by a large number of relevant public and private organizations for cooperative implementation.

NEW ENERGY SAVING LIGHTING PRODUCTS

KENNETH L. SCHMIEDER

FLACK + KURTZ CONSULTING ENGINEERS

The increased emphasis on energy conservation and the rising cost of electrical energy have spurred development and research into new lighting system products. These products offer the potential to maintain satisfactory artificial illumination levels, with considerably less expenditures of energy. Energy for lighting currently consumes approximately 50 percent of the total energy requirements of a typical commercial office building. By careful system design and application of the new technological advances made in lighting system components, energy consumption can be reduced by more than one fourth. The dollar saving through avoided energy costs is significant. It would not be unusual to save $100,000 per year in a 500,000 square foot New York City office building. The cost for accomplishing this: six months to two year's savings.

Recent developments have been made in fixtures, ballasts, lamps, and the approach to lighting system design. By proper application of these products, energy savings can be achieved in a cost effective manner.

Let's look at what's happening in fluorescent fixtures. The fluorescent fixture is certainly one of the most basic components of interior illumination systems. A conventional lensed fixture is roughly 55 percent efficient. Although many improvements have been made to the conventional fixture, such as increased reflectivity of internal surfaces and more efficient shielding mediums, the maximum efficiency of a lensed luminaire is still only about 60 percent. Most parabolic fixtures easily achieve 60 percent efficiency and benefit from an improved maintenance factor since there is no lens for dirt to accumulate on. Some parabolics are available with vastly increased efficiencies of over 75 percent. That is 25 percent more efficient than standard. Examples of super-efficient fixtures are LPI - X - Parabolic and Edison Price's parabolic fixture.

In addition to how much light comes out, how the light comes out is also important.

Some fixtures will produce a "batwing" distribution. This distribution pattern derives its name from the shape of the candlepower curve. A batwing fixture will develop peak candlepower or light output between 25° and 55° from vertical in a plane perpendicular to the lamp axis. The importance of this feature can best be appreciated by developing an awareness of illumination system quality.

The quality of an illumination system is an extremely complex concept, the details of which are beyond the scope of this discussion. It includes consideration of concepts which are not entirely quantifiable, such as visual focus, relative luminance of surfaces, and visual sparkle. Quantifiable concepts of illumination quality are quantity of light, direct glare coming from the fixture to the eye, and indirect glare coming from the fixture and reflected off the task surface into the eye. These concepts are measured in terms of footcandles, visual comfort probability and equivalent sphere illumination respectively. Equivalent Sphere Illumination (ESI) levels account for contrast, fixture location, observer location, distribution pattern of light emitted by the fixture, and raw footcandle levels. Use of a batwing candlepower distribution is a very efficient means of producing high levels of ESI from available raw footcandles.

Batwing distribution may be achieved via one of two methods. An open bottom or louvered fixture with a reflector of specific design to produce the batwing effect, or a lensed fixture where the lens is patterned to create the desired batwing candlepower distribution. The LPI X - Parabolic is an example of the open-bottom type, while the Edison Price Parabolic is an example of the louvered type, both dependent upon reflector design to achieve the desired effect. The radial lens by Carrolite exemplifies the lensed approach to producing the batwing pattern. By locating batwing fixtures so that the light will usually strike the task from the side (remember placing the reading lamp over your left shoulder), glare is reduced and visual acuity is improved with less light and less energy.

That's what is happening in fluorescent fixtures. Now let's discuss fluorescent ballasts, that essential component of a fluorescent lighting system. The ballast is necessary to obtain the required starting and operating conditions of voltage and current for the lamp. Conventional electro-magnetic fluorescent ballasts consume an additional 15 percent of the lamp wattage for that purpose. Recent development of the "low heat ballast" has reduced that to as low as 5 percent; however, more is yet to come. It is fairly well known that fluorescent lamp efficacy can be significantly improved by operation at higher frequencies, above 10,000 Hertz. Since transmission of electrical energy at this frequency is not practical due to existing equipment which is designed for 60 HZ operation and the tremendous line losses which would be associated with distribution at that frequency, an alternate solution was sought. That solution has recently arrived on the market place as the "static ballast"

The static ballast is a solid state device which converts the applied electrical energy to high frequency energy for operation of the lamp. This results in energy savings of over 35 percent while maintaining 100 percent light output as compared to conventional electro-magnetic ballasted systems. These ballasts are currently marketed by Electronics & Energy Technology, Thomas Industries and soon by GTE.

Due to the operating frequency of these devices, it is possible to create interference with some radio bands. It is my understanding that the FCC is promulgating new, more stringent standards to be met by these devices. This is simply a matter of proper shielding and filtering and I do not foresee any difficulty in widespread application of these devices.

The next area where significant advances have been and will continue to be made are in lamp technology. Starting with fluorescent sources which have seen such recent improvements as:

1. Low energy tubes, which can be used with standard ballasts or new low heat ballasts to obtain nearly the same light output while reducing energy requirements by 12-20 percent. The G.E. Maxi-Misa System is one such product.

2. Increased output fluorescent lamps utilizing improved phosphors and cathode shields to reduce end blackening such as G.E.'s "Super Bright" Fluorescent Lamp which provides 10 percent more light with no increase in energy consumption.

3. A miniaturized self-ballasted fluorescent, marketed by Norelco, which directly replaces incandescent lamps, will be available mid-1981. This will simplify the direct replacement of incandescent lamps with the more efficient fluorescent source.

I feel the major achievement in fluorescent lamp technology will be the use of glass fibers within the tube to increase effective arc length and thereby significantly increase luminous flux without the associated decrease in lamp efficacy. This same technique will enable small lamp envelopes to maintain a relatively high lumens-per-watt ratio.

One of the major drawbacks to employing fluorescent sources where incandescents are presently used has been the sharp drop-off in lamp efficacy as the arc length is decreased; this obstacle should be overcome shortly by the introduction of the glass fibers within the lamp envelope.

The addition of glass fibers in a conventional fluorescent tube would not necessarily improve lamp efficacy; however, it would mean the use of fewer lamps since the lamp will deliver more lumens per inch of tube. This will allow better focusing control which, when combined with an efficient fixture design capable of utilizing this feature, will improve the efficiency of the overall illumination system. Norelco is doing considerable research in this area.

Improvements have also been realized in incandescent lamp technologies. For example, Elliptical Reflector Lamps to minimize light lost within certain fixtures have application when re-lamping existing systems and also, Krypton Gas Lamps, which have improved lamp efficacies over conventional lamps. The most signficant development is likely to result from work done by Duro-Test, in the use of Dichroic Filters integrated with the spherical bulb. A dichroic element permits transmission of light within the visible spectrum but reflects infra-red energy. The potential for this development is recognized when we review how the incandescent lamp functions. A filament is heated by an electric current, the heated filament radiates energy in the visible and invisible spectrum due to thermal excitation of its molecules. The dichoric filter, by reflecting the infra-red, which is heat energy energy, back into the filament, will maintain the filament at its operating temperature while reducing the electrical energy required for incandescence. Huge savings in residential lighting is likely from this development.

In the High Intensity Discharge (HID) area, specifically Metal Halide and High Pressure Sodium sources, the major advances will be improved color and availability of lower wattages. Westinghouse has recently marketed an improved color high pressure sodium source called Ceramalux 4. This lamp is currently available in 250, 200 and 150 Watts with a Color Rendering Index (CRI) of 65. Westinghouse anticipates marketing a 100 Watt Ceramalux 4 by the end of this year, and a 50 and 70 Watt model within two years.

Metal halides are also experiencing improved color characteristics and are available with CRI's between 65 and 70. This lamp is comparable to cool white fluorescent in its affects on colored surfaces. In addition, their output is being punched up to nearly match the High Pressure Sodium Lamps.

Availability of lower wattage HID's currently as small as 35 Watts for conventional High Pressure Sodium and 175 Watts for Metal Halide will enable their application to lobby environments. However, their efficacy is currently challenged by fluorescent sources once the HID lamp size is reduced to 35 Watts.

This concludes our discussion of lamps, now, let's examine another source which is receiving renewed interest; the sun. Indirect sunlight, reflected from the sky, is the most beneficial source. Many factors affect the illumination level from skylight such as building orientation, type of glazing, cloud cover, time of day, time of year, geographic location, and exterior surface reflectances. Generally, however, on cloudy days you may anticipate one footcandle per square foot of clear glass, fifteen feet from the perimeter.

Direct sunlight, generally of little value to interior illumination due to glare and difficulty of predicting its changing effects, is also being studied with renewed interest. The current thinking is to utilize direct sunlight by reflecting it into the space. The initial reflection would take place outside the building in the summer, to reduce the undesirable heating effects, and possibly inside the structure in the winter when the heating effects of direct sunlight could be beneficial. By reflecting the sunlight into the space via lightwells and windows, it may be possible to have the sunlight provide the major portion of the ambient light, in a task ambient illumination system.

In order to realize benefits from a system employing daylighting, adequate controls must be provided to enable the artificial illumination system to respond to the varying contribution from daylight. The control system should employ a photocell to sense available light and automatically adjust the input energy by relay control or preferably by dimming control of the artificial illumination system.

Now let's examine some of the dimming systems available to us. Photocell Controlled Dimmer Systems such as that manufactured by Widelight for HID Sources and Lutron and others for fluorescent sources, offer energy savings irrespective of available daylight. All artificial illumination systems depreciate in light output over time, due to dirt accumulation on fixtures, walls, ceilings, and lumen output depreciation of the source. Because of this, systems are designed to produce a specified illumination level at some specific time after the initial installation, typically 8,000 hours for fluorescent systems. This results in illumination levels and energy consumption, greater than what is actually required until the design time has been reached. A means of reducing the illumination level to that actually required and thereby reducing energy consumption is to employ a Photocell Controlled Dimmer System.

Another dimming system available is the Integral Prism Type. The Integral Prism System consists of a daylight sensor (photocell) connected by a light pipe directly to an electronic dimming controller which serves four forty Watt fluorescent lamps. This packaged system costs about $35 and will pay for itself in approximately two years in New York City. Its only disadvantage is that when full light output is required due to aging of the tubes or loss of available daylight, the system consumes slightly more energy than a conventional system.

Bulk Type Dimming Systems which require no special ballasts are available from Lutron and Deerfield and have major application to building retrofit programs. The Lutron System controls one lighting circuit with up to ninety 40 Watt lamps per device and is suitable for interface with photocell control. The Deerfield System is designed to affect an entire lighting system, and is capable of controlling in excess of 4,000 lamps. The Deerfield System is not suitable for photocell control or when existing local control has been provided. The Deerfield System allows for starting the lamps at normal voltage and then by use of a tapped auto transformer, reduces the input voltage to the illumination system reducing lighting output and energy consumption.

Other opportunities for energy conservation are available with some of the more sophisticated control systems such as General Electric's Programmable Lighting Controller. This System is capable of ten independent program functions for up to 8,000 independent points. Lights may be turned on and off in accordance with occupancy schedules and building cleaning crew schedules. Overtime use of the illumination system in a specific area can be arranged by simply initiating telephone instructions to the micro processor which serves as the basis of the system. In addition, records of overtime use can be accurately kept for billing purposes, a valuable tool in today's multi-tenanted commercial office buildings.

In closing, let me add that unless someone finds a means of transforming water into oil, the technologies briefly reviewed here will continue to evolve and receive wider and wider application due to national policy and economic self interest. It is important for all of us to follow developments in the lighting field and be fully aware of their potential applications.

- -

CANADIAN AND OVERSEAS ACTIVITIES AND RESEARCH REGARDING METHANE POWERED VEHICLES

BY Richard E. Wyman
Director of Research, Canadian Hunter Exploration, Ltd.
and
Vice-President, CNG Fuel Systems, Ltd.

The 1980's will become known as the decade of mass conversion to alternate fuels. Methane powered vehicles will almost certainly lead this conversion effort in many countries around the world. The arguments for converting to methane (especially compressed natural gas known as CNG) are simply too compelling:

- Supply is plentiful in short and long term
- Practical technology exists today
- Costs are lower because of plentiful supplies, inexpensive processing, and easy transportation
- Fuel is non-toxic and safe
- Engine conversion is simple
- Environmental damage is significantly reduced
- Methane burns more efficiently with higher octane (130) than gasoline
- Engine life is extended
- Engine operation is smoother
- Vehicle maintenance is reduced
- Easier starting in cold or hot weather
- Fuel may be easily switched from natural gas to gasoline or back again while driving in dual fuelled vehicles.

Most of these advantages have been well documented[1,2,3,4] and will not be rediscussed in this paper. A review of world wide activities with emphasis on the supply and transport of natural gas will be discussed here. Special attention will be given to the emerging interest of methane powered vehicles in Canada and a review of research and development around the world.

SUPPLY OF NATURAL GAS

Before we can seriously discuss any alternate fuel that might have wide spread application, we need to understand if the potential supply is adequate. In the case of natural gas the world wide supply appears overwhelmingly promising. This is the main reason (along with its lower cost) that methane is considered more attractive as an alternate fuel for mass conversions than other candidates, such as propane or butane (LPG). As with crude oil, certain parts of the world are blessed with more natural gas than others; but the technology of moving natural gas from one part of the world to another is also rapidly improving.

Table I shows those countries that might be in a particularly favorable position to consider methane powered vehicles (as we will discuss later, many of them already have). All but two of these countries had proven reserves of at least 5 trillion cubic feet (TCF) at the end of 1979.[5,6] Since the amount used, compared to the reserve, is also important, any country with reserves of at least 3 TCF and a reserve to production ratio (R/P) of over 25 is also included. This does not mean, for instance, that in a country with an R/P of 15, there are only 15 years of production left; historically most, if not all, of the gas produced during a year is replaced with new reserves through continued exploration and development. In fact, of the 37 countries listed in Table I, 67% of them have actually increased their reserves from those recorded in 1977. To put this in context with transportation fuel, approximately 1 million vehicles could be powered for only 0.2 TCF per year in a high consumption country such as the United States or Canada. Most countries use far less gasoline per capita than North America.

But Table I represents only the tip of the potential supply iceberg. As price and extraction technology for a resource increase, the amount available for extraction improves markedly. The concept of the resource triangle as it applies to natural gas was outlined by Masters[7] in 1979 (See Figure 1). He demonstrated that while a large amount of gas is available in so called conventional reservoirs, an enormous amount is available in lower quality formations such as tight sands. As economics warrant, other gigantic sources of methane can be tapped including coal seams, organic fractured shale, ultra deep basins (over 20,000 feet deep), geopressured water and hydrates. The amount of methane in these resources is unquestionably huge with estimates of world supply running to thousands of years. Then on top of all this are renewable sources such as recovery from waste or biomass conversion. For instance, studies at the Institute of Gas Technology, of methane recovery from kelp and other rapid growth water plants are pointing the way to potential large scale conversion systems. By farming marine biomass from deep water off the shores of North America, it is estimated that several times the current natural gas production in North America could be generated.

TABLE I

NATURAL GAS
(1979)

COUNTRY	PROVEN RESERVES (TCF)	RESERVES-TO-PRODUCTION RATIO (R/P)
Algeria	131	348
Argentina	22	50
Australia	30	105
Bahrain	9	107
Bangladesh	8	*
Bolivia	6	49
Canada	87	21
China	30	9
Columbia	8	41
Ecuador	4	25
Egypt	3	33
France	6	23
Germany, West	6	9
India	12	64
Indonesia	24	24
Iran	485	548
Iraq	28	570
Italy	8	18
Kuwait	31	152
Libya	24	149
Malaysia/Brunei	26	78
Mexico	65	33
Netherlands	62	19
New Zealand	6	152
Nigeria	41	115
Norway	43	32
Pakistan	15	65
Qatar	60	1476
Saudi Arabia	110	450
Thailand	8	v. large*
Trinidad/Tobago	12	49
Tunisia	6	*
United Arab Emirates	22	340
United Kingdom	25	21
United States	191	10
USSR	920	63
Venezuela	42	101

* Data not available

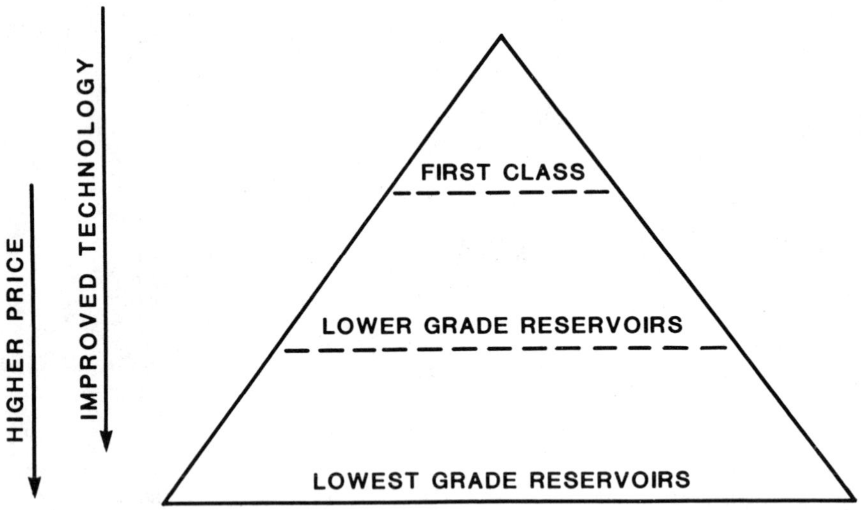

Figure 1. Resource triangle

Of course, the key to all these unconventional sources of natural gas is economics and technology, but many of these sources are expected to be economically competitive with conventional gas during the coming decade. The main point is that there is no need to be concerned about a shortage of methane as long as the free market can determine the source.

TRANSPORT OF NATURAL GAS

It used to be thought that widespread use of natural gas was limited to those areas near the source. The technology of transporting natural gas has grown rapidly over the past decade and will certainly continue to expand during the next decade. Very long pipelines overland or underwater are now commonplace. Huge LNG (liquified natural gas) tankers regularly deliver natural gas from surplus gas areas of the world such as the Middle East and Southeast Asia.

A couple of recent examples of how new pipelines will drastically change the energy use in countries around the world, is the offshore Thailand development and the trans-Mediterranean lines. The Thailand line is the longest submarine pipeline in the world with 264 miles underwater. This line will be capable of delivering 500 million cubic feet per day from the Gulf of Thailand to the doorstep of Bangkok. Bangkok's streets are now choked with smog creating vehicles burning imported liquids. Here is just one example where methane powered vehicles could significantly improve the enviroment and economy at the same time.

Another record breaking line which spans the Mediterranean from Tunisia through Sicily across the Strait of Messina, to a point near Bologna, will deliver over 1.2 billion cubic feet per day from the giant Hassi R'Mel field in Algeria to Italian markets (See Figure 2). This gas will supplement that from the Po Valley and other imports for the use in methane powered vehicles in Italy. It should give a significant boost to the security of supply for over 250,000 methane powered vehicles now operating in the central and northern part of Italy.

Meanwhile, intercontinental movement of natural gas has increased many times over the past decade by LNG tankers. These tankers are presently moving large amounts of natural gas each year. Further increase of this traffic is expected when satisfactory long term pricing negotiations are completed. A typical LNG tanker with 125,000 cubic meters of LNG could fuel 10,000 to 20,000 vehicles for a year.

In brief, world wide transportation systems are now in place and improving each year to economically distribute surplus gas. Because of this, many more places in the world will now be able to convert to a stable natural gas economy as crude oil supplies become scarce, expensive and insecure.

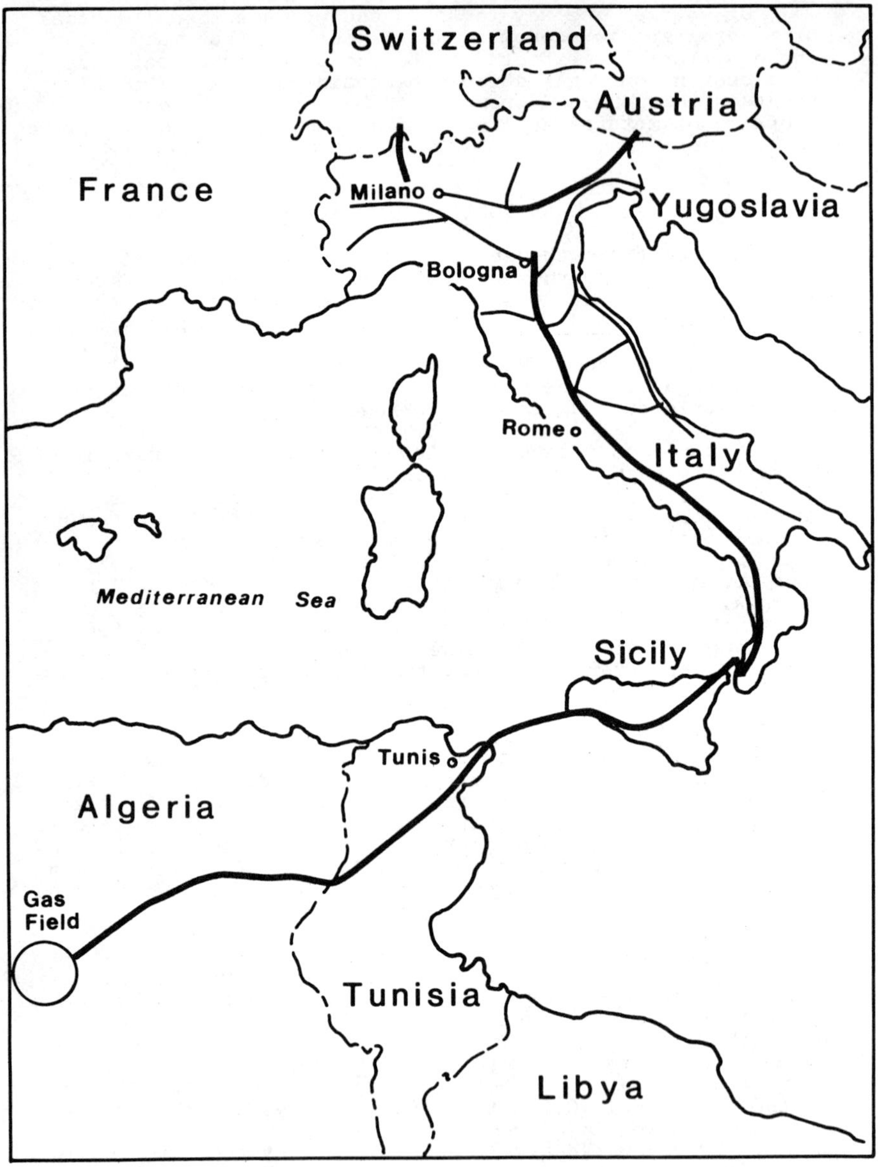

Figure 2. Trans-Mediterranean pipeline

ACTIVITIES AROUND THE WORLD

AUSTRALIA

Australia, with its large surplus of natural gas has been investigating the use of methane powered vehicles since 1937. It now has about 70 CNG vehicles and four refueling stations in different parts of the country. Officially the government of Australia expects there will be over 400 vehicles within the next two to three years. Unofficially there are those close to the CNG scene in Australia that believe there will be several thousand in this period of time. West and Brown[2] have documented the history of CNG in Australia in their excellent report on Compressed Natural Gas.

CANADA

Canada has been experimenting with methane powered vehicles for over ten years. It now has only two CNG refueling stations (Vancouver and Calgary) and a few experimental vehicles. However, a rapid increase is expected later this year in the number of refueling stations and CNG vehicles as the country embarks on a serious effort to convert to alternate fuels. A more detailed discussion of activities in Canada follows this section.

COLOMBIA

There are no methane powered vehicles in operation at the present time in Colombia, but the government expects considerable activity in this area within the next five years.

EGYPT

There are no methane powered vehicles in Egypt at the present time. However, studies by the government of Egypt and the World Bank are being conducted to determine the feasibility of widespread conversion of gasoline and diesel engines to CNG. Estimates have been made that 100,000 vehicles could be converted within ten years. There is strong economic incentive to do this for the displaced crude oil would be available for additional export revenue.

FRANCE

France has been using CNG in vehicles since 1941, reaching a peak usage of 20,000 to 25,000 vehicles and 30 to 40 refueling stations in the late 1950's. Most of these vehicles were privately owned. The use of methane has declined considerably since that time, due to massive increases in taxation on CNG. The government is currently investigating the substitution of alcohol for gasoline. The government's aim is to see that 25 to 50% of the motor fuel used in France is of national origin by 1990.

IRAN

Within the last three years, Iran installed two self contained CNG refueling stations (assembled by Dual-Fuel Systems Inc.) and operated about 1,000 taxicabs on CNG. Recent correspondence indicates their interest remains high. Reduction of air pollution along with conservation of refined products, provides the chief motivation. As seen in Table I, Iran has enormous reserves of natural gas.

ITALY

To Italy goes the honor of being the leading country in the world in the use of CNG. They have been using CNG since the early 1920's; now there are about 260,000 vehicles operating on CNG. These are mostly private vehicles, refueling from about 250 stations in Northern and Central Italy around the Po Valley.

Some of these stations are quite large and can accommodate as many as 40 vehicles at a time, although the average station has about 6 refueling ports. An Italian refueling station using hydraulically actuated compressors is shown in Figure 3.

A variety of vehicle conversion systems, storage cylinders, and compressors are manufactured in Italy specifically for CNG operation. This early start in the development of technology for CNG has led to a rapidly increasing world wide export market for Italy.

Italy has an enviable safety record with CNG. During the past 30 years, there have been no accidents or explosions attributable to CNG components. Although CNG vehicles were involved in their fair share of accidents during hundreds of millions of miles of travel, no injuries or deaths were caused by CNG.

An outstanding report on Italian technology, safety and activities regarding CNG was prepared by Abrams, Titchener, and West[3] for the Liquid Fuels Trust Board of New Zealand.

MALAYSIA

Although Malaysia does not have methane powered vehicles in operation at this time, it is a country with large reserves of natural gas and a shortage of crude oil. Exxon recently discovered a very significant gas deposit in their contract area off Trengganu. In January of this year, Datuk Leo Moggie, Minister of Energy, Telecommunications and Posts, made headlines when he announced the government was looking into the use of compressed natural gas to power vehicles.

Figure 3. Large public CNG refueling station in Italy showing five Idromeccanica compressors

MEXICO

Mexico is known to be interested in the potential use of CNG and is actively investigating its use for vehicles.

NEW ZEALAND

In mid 1979, the government of New Zealand announced ambitious plan to convert 150,000 vehicles to run on natural gas by the end of 1985. The incentives for New Zealand to convert to natural gas are quite evident when one realizes that they must import virtually all their crude oil while a great surplus of natural gas is within easy access to the North Island. The giant offshore Maui gas field and the Kapuni field shown in Figure 4, together contain about 6 TCF of proven reserves.

As of January, 1981, New Zealand has 24 refueling stations with an additional 15 under construction. To date, about 4,000 vehicles have been converted with more being added every week. Training courses have certified over 1,000 mechanics. These trained mechanics along with new government incentives should accelerate the rate of conversions in the future.

The government of New Zealand has decreed that all appropriate Government-run vehicles, including over 2,500 Post Office vehicles, are to be equipped for CNG. In addition the government is offering cash grants (eg. 25% on cost of refueling equipment and $200 per motorist for conversion) and accelerated tax writeoffs.

PAKISTAN

Pakistan is another country with a large surplus of natural gas, most of it being produced from the Sui field in Boluchistan. Active exploration is expected to find even more gas during the 1980's. The government is anxious to see natural gas used in the transportation sector which currently is a heavy energy drain on the country. Although there are no vehicles running on CNG at the moment, they are expected to have about 200 vehicles and two refueling stations in operation later in 1981.

THAILAND

At the present time, Thailand does not have any vehicles operating on CNG. They do have approximately 9,000 small vehicles, mostly taxis, using LPG. The government policy at this time, is to use LPG when the gas conversion plant is completed for the offshore gas mentioned earlier. Some investigations are now being made regarding the use of CNG. The attractive economics and safety of CNG may well change this policy when their pipeline comes on stream in late 1981.

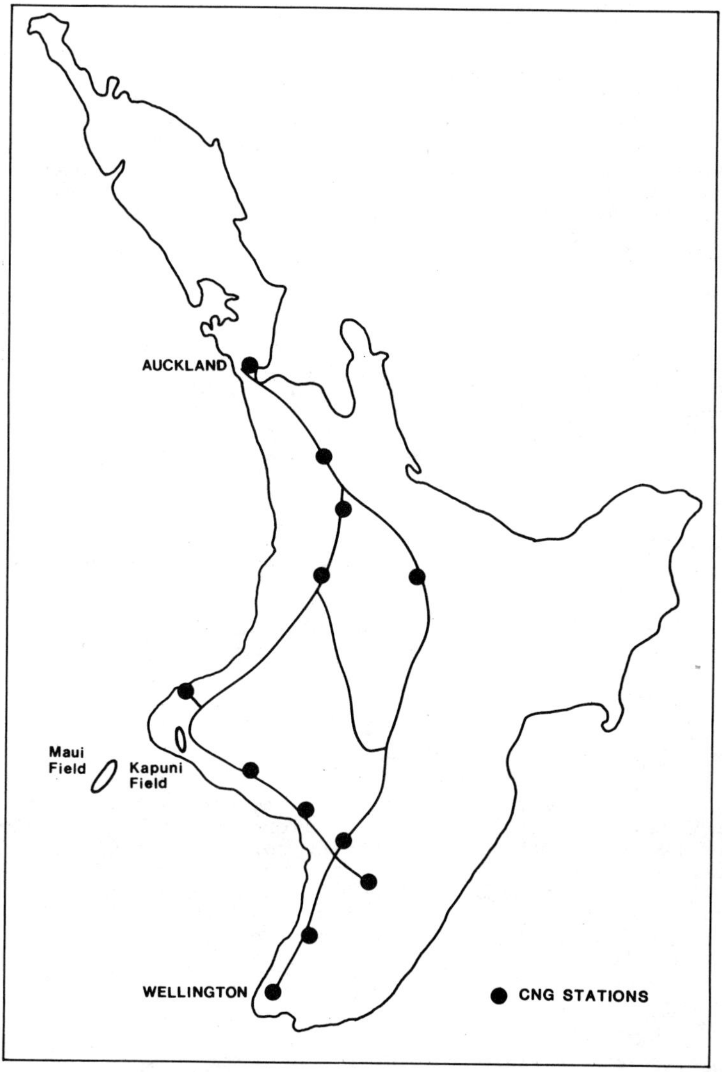

Figure 4. New Zealand gas fields and CNG service centres

U.S.S.R

According to a press release from the TASS news agency in late 1979, Moscow is planning to convert all busses to liquified natural gas by 1983. All taxi cabs are to be converted to LNG by 1987.

UNITED KINGDOM

CNG probably will not be used in signifcant quantities in Great Britain. A low level of experimentation has been going on since 1971. The major supplies of natural gas are committed to industrial purposes and domestic heating. One or two vehicles have been converted to run on methane from manure and sewage on an experimental basis.

UNITED STATES

It is estimated there are approximately 20,000 methane powered vehicles in the U.S.; practically all of these are fleet vehicles. All but a few of these have been converted to run on CNG. Although it appeared there was a shortage of natural gas only a few years ago, this was more a result of artificial price restraints than actual shortages. Since the easing of price regulations of natural gas, the reserve additions have been climbing each year. With deregulation, additions are expected to be close to production in the coming years. In the United States the security of supply is as important as the favorable economics for conversion to CNG.

As in Italy, the safety record in the U.S. for CNG has been excellent.[8] A more detailed discussion of current U.S. activities is to be presented by John Wright at the 8th World Energy Conference.[9]

WEST GERMANY

There are fewer than 10 methane powered vehicles in West Germany. Most of these are powered by LNG. The main incentive for investigating methane powered vehicles is to reduce emissions. Experiments with both CNG and LNG on diesel engines and gasoline engines have been conducted with public transportation buses during the past decade.[10,11] Most of the natural gas used by West Germany must be imported and is primarily used for heating purposes.

ACTIVITIES IN CANADA

The importance of CNG to the future of Canada was given official Federal recognition in the National Energy Program of 1980.[12] The government stated that it was interested in the potential to use CNG as a motor fuel in Canada and anxious to encourage the development of this newer option.

Since that time, it has been moving aggressively with private industry and provincial governments to insure the speedy and safe implementation of this alternate fuel. The prime motivation for this is that Canada is presently importing above 600,000 barrels per day of crude oil while surplus natural gas is shut in. The annual cost of this in 1980 was over 8.0 billion dollars. This figure is expected to be far greater by 1985.

Canada has been adding over 5 TCF per year to its proven reserves for the past five years. In 1980 only 1.6 TCF was used in Canada while slightly over 1 TCF was exported. If markets and reasonable prices prevail in the future, there is every reason to believe the industry can continue to add at least this amount each year. In fact, a more realistic figure by 1990 would be an annual addition of 8 TCF with contributions from the Frontier areas. By 1990 the National Energy Board (NEB) estimates less than 3 TCF will be needed within Canada even with expanded domestic markets (eg. as to the Maritime Provinces) to be served by the proposed pipelines indicated in Figure 5. If 500,000 high mileage vehicles were converted to run on CNG, they would require only about 0.1 TCF per year. Yet nearly 200,000 barrels per day of crude oil are required to provide gasoline for this number of vehicles. This is equivalent to two thirds of Canada's net imports. As shown in Figure 6, such a projection would still leave Canada with an annual surplus of over 5 TCF per year which could be available for export in 1990; this, of course, is dependent upon deliverability. Further substitutions of natural gas for fuel oil, gasoline and diesel fuel could bring Canada significantly closer to energy self sufficiency within a decade.

The National Energy Program outlined certain incentives including support of demonstration fleets and the necessary fuel facilities through the existing Federal-Provincial Energy Demonstration Agreements, and, as necessary, directly by the Government of Canada. After standards for Canadian CNG Systems are approved it is anticipated that incentives will be provided for CNG along the lines of the propane conversion incentives. The 1980 program provides for taxable grants of up to $400 for each vehicle of a commercial fleet converted to propane.

In addition to these incentives, individual provinces are actively seeking ways to encourage the use of alternate fuels. Ontario, for instance has removed the retail sales

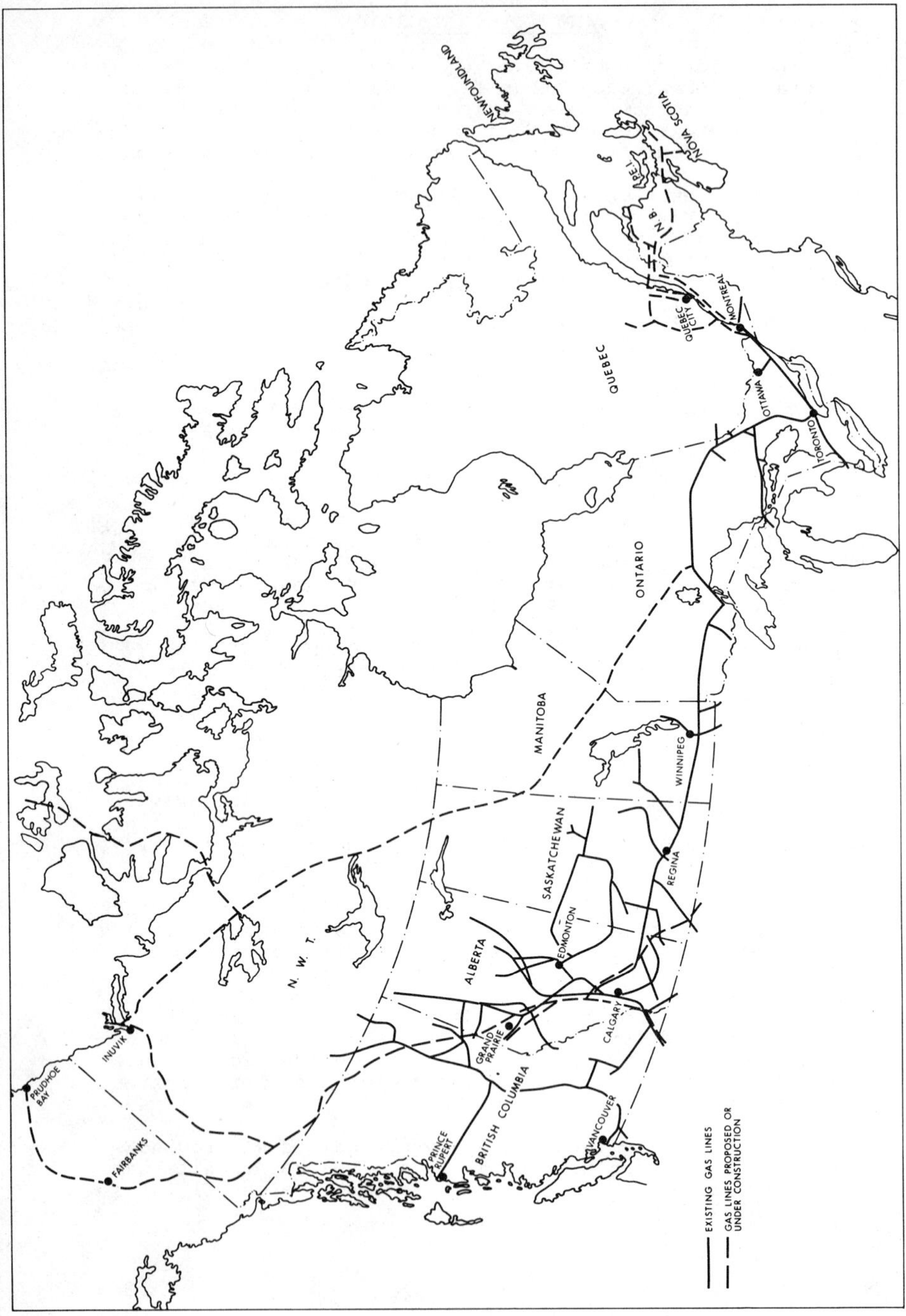

Figure 5. Natural gas pipelines in Canada

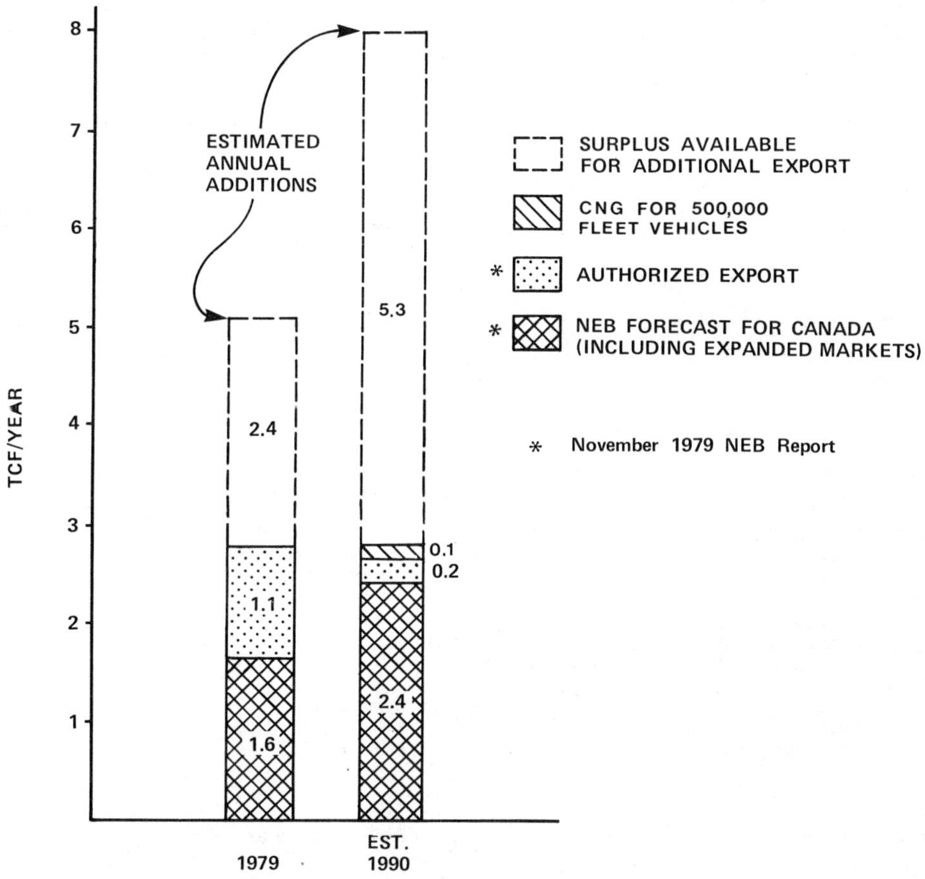

Figure 6. Natural gas annual additions to reserve and demand in Canada

tax on vehicles powered exclusively by alternate fuels which include natural gas. Also, Ontario has eliminated the gasoline sales tax on alternate fuels. The province of British Columbia is actively promoting interest in methane powered vehicles and providing research grants and helping to educate the public. Alberta has also expressed a great deal of interest and is expected to announce their program to encourage use of CNG shortly.

In the first quarter of 1981, a joint delegation of Federal, Provincial and Industry representatives visited Northern Italy. The purpose of this visit was to become more familiar with the widespread, long term use of CNG in Italy. A report on this visit is now being prepared. It is expected that Canada will utilize this experience to help emplace their safety codes and regulations and to accelerate the movement in Canada to convert to CNG.

A committee has now been established under the sponsorship of the Canadian Gas Association to recommend safety codes for CNG conversions and refueling stations. Representatives from various levels of government and industry are working together to ensure that safe and reasonable codes will guide the conversion efforts. Past experience in Canada as well as other countries, especially Italy, United States, and New Zealand, are being utilized in this effort.

Actually, experiments with methane powered vehicles have been in progress in Canada for over ten years. The Canadian Western Natural Gas Company installed a combination slow fill and quick fill CNG station in Calgary that can serve over 100 vehicles (See Figure 7). This station is still in operation (in fact, it has been a source of CNG for my car for the past year). They expect to upgrade their experiment shortly with a new compressor and a pilot fleet equipped with the latest conversion equipment.

Other experiments have been tried in Canada with CNG vehicles including trials at Consummers Gas in Toronto, and The Edmonton City Council. B.C. Hydro in British Columbia experimented with a couple of LNG vehicles, but now propose to install their own CNG refueling station. B.C. Research has a compressor station and four experimental methane powered vehicles at this time.

The first commercial conversion centre and refueling station in Canada is scheduled to open in Calgary in the second quarter of 1981 (See Figure 8). CNG Fuel Systems Ltd. indicated they expect to open additional centres in Toronto, Vancouver and Edmonton within the next year.

Some of the earlier experiments lost momentum primarily because of the unsatisfactory cost/benefit ratios with the artificially low prices of gasoline. Other reasons cited included lack of refueling stations, loss of power in older vehicles, maintenance problems with the older compressors and high taxes on natural gas (these taxes have now been removed in Ontario).

Figure 7. Canadian Western Natural Gas CNG refueling station (installed 1970)

Figure 8. First commercial CNG installation centre in Canada. Calgary, Alberta

Most, if not all, of these drawbacks are rapidly dissolving as the price of gasoline continues to increase, and tax relief and incentives are instituted. In addition, the newer technology in compressors and vehicle conversion are helping to overcome previous problems of maintenance and loss of power. The "chicken and egg" cycle of number of methane powered vehicles and refueling stations now appears to be started with the commitment of commercial interests and the concurrent emerging provincial and federal encouragement.

One of the concerns for the widespread use of CNG as an alternate fuel in Canada is the lack of refueling stations. While this is true at the present, it is interesting to note that in reality the infrastructure is already in place. Like many other countries, Canada is already served by a widespread network of natural gas pipelines as shown in Figure 5. So with no major additional manufacturing plants, truck fleets or storage depots, this alternate fuel is available now throughout the country; all that is needed is to hook up compressor stations at the desired locations to deliver this surplus alternate fuel. Certainly, New Zealand has recognized a similar situation with their comprehensive national plan to convert to CNG (See Figure 4).

The sensitivity of the payback to the price of gasoline is illustrated in Figure 9. This graph (based on Alberta prices) shows the increasing savings that result from using CNG as the price of gasoline goes up even though the price of natural gas and operating costs of the refueling station are assumed to increase proportionately. In this example a taxi cab that uses 30,000 litres (7,925 U.S. gallons) per year could save $5,600 each year if the price of gasoline is $0.25 /litre ($.95 per U.S. gallon). As the price of gasoline doubles, the savings increase to $11,358 per year even though the price of natural gas has also increased. This example assumes that the price of natural gas is $2.00 per 1,000 cubic feet when the price of gasoline is $0.25 per litre.

In summary, Canada has strong national incentives for converting to methane powered vehicles. Furthermore, the Federal and Provincial governments are co-operating with private industry to implement an aggressive program of using alternate fuels during the coming decade.

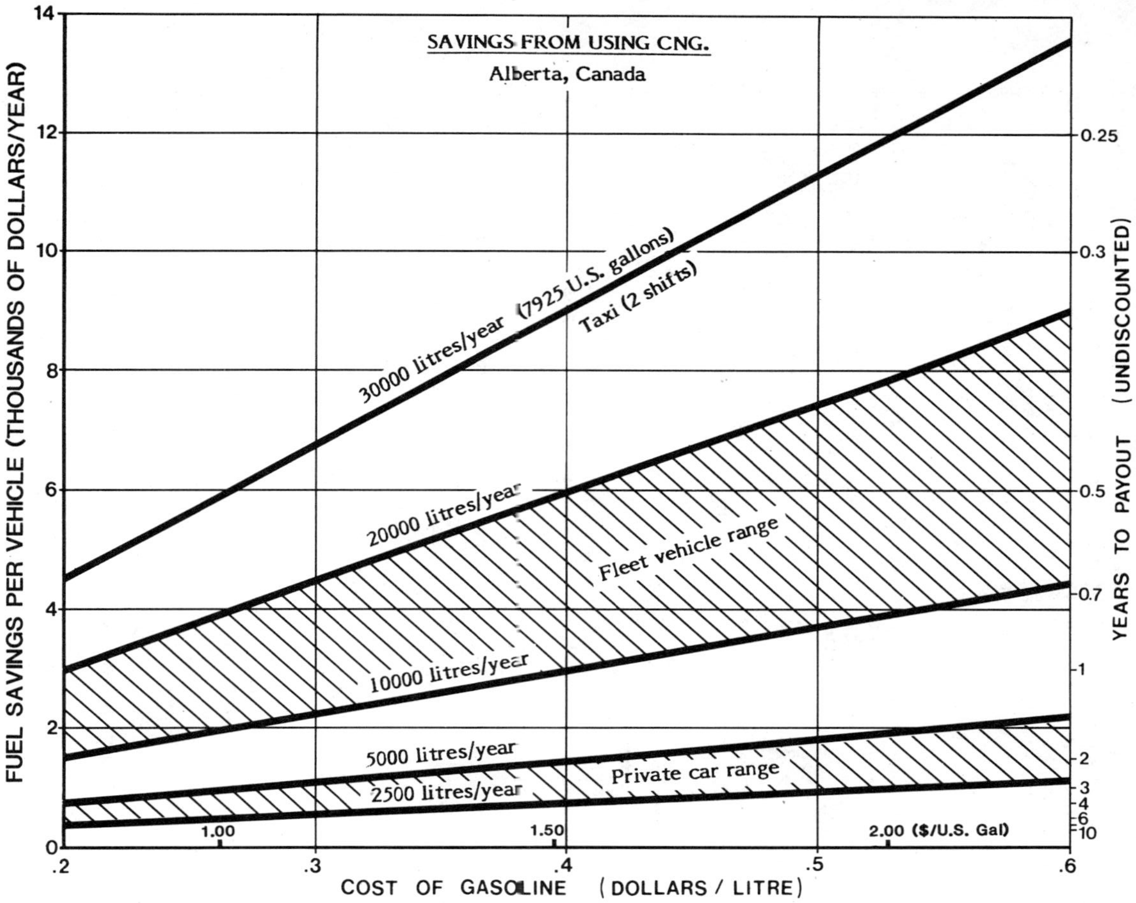

Figure 9. Estimated annual savings from CNG conversion is strongly dependent upon price of gasoline (Courtesy of CNG Fuel Systems, Ltd.)

RESEARCH AND DEVELOPMENT

The basic technology for methane powered vehicles is well established and practical conversions are being made daily around the world. However, as with any relatively new technology, there is considerable room for further research and development.

There are some investigations in the use of LNG in vehicles in the United States, West Germany and Russia; however, the preponderance of use and research around the world now appears directed towards CNG. The primary areas of investigation for CNG include:

- Refueling stations
- Fuel storage
- Spark ignition engines
- Diesel engines
- Chemistry and physics of gaseous fuels and combustion

REFUELING STATIONS

The heart of any refueling station is the compressor. Compressors suitable for high pressure air (eg. 3600 psi) are not necessarily suitable for compression of methane at high pressures. Therefore modified or completely redesigned eccentric rod compressors have started to appear on the market within the last decade. The need to refuel vehicles with natural gas at these relatively high pressures has also led to radically new designs such as the hydraulically actuated compressors shown in Figure 3. These compressors, manufactured in Italy, have proven very reliable over the past decade with relatively low operating and maintenance costs.

Although improvements in design of individual parts and packaging of the peripheral equipment are being investigated by most manufacturers, there are no major research projects that have been made public. One area that is being studied in Canada is the feasibility of small refueling stations suitable for the home or apartment complex. Special criteria that would apply to a home refueling station include safety, reliability and quietness at a reasonable price. Security and convenience of refueling at home has enormous appeal.

As public refueling stations become more common the desire to have a single metering device increases. The present convention in Italy and New Zealand is to rely upon measurements of tank capacity, pressure and temperature. Research on a reliable metering device has taken place in the U.S., New Zealand and England. Ultrasonic and electronic devices are being investigated, but no reasonably priced, accurate meter has yet appeared on the market.

FUEL STORAGE

The desire to have more onboard storage of CNG without punitive increases in weight has led to investigations of other materials for cylinders. New metal alloys with improved manufacturing techniques are leading to stronger, lighter weight cylinders. Another promising technical advance comes from the areospace industry, where lightweight metals, such as aluminum or titanium are wrapped in a cocoon of light-weight synthetic fibers such as Kevlar. However, cost projections are still too high to be practical for mass usage. B.C. Research, located next to the University of British Columbia, is focusing on some of the crucial problems of storage.

SPARK IGNITION ENGINES

Considerable testing has been done around the world on the performance of conventional spark ignition engines running on methane.[1,2] The variations of power, torque and emissions of standard vehicle engines running on methane versus RPM, spark setting and inlet gas pressure are well documented. Less research has been done on design of engines to run specifically on CNG. It is understood that major auto manufacturers are currently doing research and testing on CNG vehicles. Fundamental understandings on cyclic pressure variations of engines using CNG are being developed at the University of Calgary. There is opportunity for improved engine performance by taking advantage of the 130 octane methane; vehicles using only CNG may have compression ratios in the order of 14 to 1.

The government of British Columbia is funding $500,000 to B.C Research this year to study critical problems in methane combustion. They are investigating ignition systems, timing, and air/fuel ratio control devices.

The Mechanical Engineering Department at the University of Calgary has been studying engine combustion with methane as well as mixtures of other fuels for a number of years.

Dr. Michael Seal, Director of the Vehicle Research Institute at Western Washington University, is doing significant development work on methane powered engines. Recent inventions on new methane injection techniques and ignition systems have improved the power of the CNG engines.

DIESEL ENGINES

The operation of diesel engines on natural gas is far from new but most of the applications have been with stationary engines. Diesel engines must have a pilot stream of diesel fuel when operating with methane. If the methane is injected into the cylinders under pressure, this pilot stream may be less than 5%; however, this requires specially manufactured engines or major modifications of conventional diesel engines. A far simpler conversion can be achieved by bringing CNG through the induction manifold. Under such conditions, a 25 to 35% diesel stream is normally required. An appreciable amount of work along this line has been carried out in Italy and other European countries such as Germany. The European communities have been doing research and development work with methane powered tractors. Also considerable research has been done by Dr. Karim at the University of Calgary in the use of a high percentage of methane in diesel engines.[14]

Lloyd Brown, general manager of the Wellington Gas Company, of New Zealand, with the assistance of the Tessari Company in Italy has invented an electronic controlled governor that allows the practical use of CNG in diesel powered vehicles (see Figure 10). This device allows a conventional diesel engine to run on 85% CNG and results in a 10% increase in torque with no increase in fuel. It also runs smoother, produces significantly less pollution and operates at lower temperatures.

CHEMISTRY AND PHYSICS OF GASEOUS FUELS AND COMBUSTION

Fundamental studies of gaseous fuels and combustion have been going on for some time.[15] But with renewed world wide interest in methane powered vehicles the research along this line will be increased.

Propane-methane mixtures and their suitability for internal combustion engines are being studied in Calgary, Canada.

CONCLUSIONS

There is very strong interest around the world in methane powered vehicles. Compressed natural gas technology has been proven economical, safe and environmentally clean by about 300,000 vehicles currently in use. Although Italy, United States, and New Zealand are leading the world in conversion to compressed natural gas for vehicles at this time, there is a strong interest in many other countries. Canada, in particular, appears on the threshold of moving ahead aggressively to start replacing gasoline with their surplus natural gas.

It is becoming abundantly clear that while there are huge proven reserves in the world, there are far greater potential methane resources yet to be developed. Besides the conventional natural gas reservoirs, there are sources from tight sands, coal seams, fractured shales, ultra deep basins, geopressured waters and hydrates. An increasing amount of research is also being done on renewable methane resources.

Research by private industry, government, and universities are expected to accelerate during the early 1980's for CNG. Experimentation with LNG is also continuing in several countries; however, the application to vehicles has been limited to date.

It is expected that CNG usage will increase very rapidly during the 1980's. There will be a great challenge to meet this demand of new equipment and trained personnel to engineer, install and service conversions to methane powered vehicles.

Figure 10 CNG 6 cylinder Ford diesel engine equipped with electronic controlled governor (Tessari–Welgas)

REFERENCES

1. Porter, W.J., Grosser, M., Prospects for Using Natural Gas in Light Transportation Vehicles, American Gas Association, Arlington, Virginia, December, 1978.

2. West, J.P., and Brown, L.G., Compressed Natural Gas, New Zealand Energy Research and Development Committee, University of Auckland, Auckland, New Zealand, April 1979.

3. Abram, R.N., Titchener, A.L., and West, J.P., Report of Overseas Visit to Investigate Compressed Natural Gas in Italy, Liquid Fuels Trust Board, New Zealand, November 1979.

4. Brown, L.G., "CNG and Vehicles", The Australian Gas Journal, March, 1979.

5. World Energy Industry, 1979 Annual Issue, Business Information Display, Inc., Volume I, No. 2, 1980.

6. Oil and Gas Journal, "World Wide Issue", December 29, 1980.

7. Masters, J.A., "Deep Basin Gas Trap, Western Canada", Bulletin, AAPG, (February, 1979).

8. Porter, W.J., Preliminary Analysis of the Safety History of Natural Gas-Fueled Transportation Vehicles, Report prepared for the American Gas Association, December, 1979.

9. Wright, J., "U.S. Activities; Technology and Economics Regarding MPV's", 8th World Energy Conference, Washington, D.C., March, 1981.

10. Pischinger, T.F., Chair for Applied Thermodynamic, Rheinisch-Westfalische Technical University in Aachen, Private Correspondence, 1981.

11. Van Dipl.-Ing. H. Hardenberg und Dipl.-Ing. V. Rubi Stuttgart, "Methangasmotoren fur Nutzfahrzeuge und Omnibusse", VDI-Berichte Nr. 224, 1974.

12. Energy, Mines and Resources Canada, The National Energy Program, 1980.

13. Karim, G.A., Sarpal, G., "A Statistical Analysis of Engine Cyclic Pressure Variations Using a Data Acquisition System", Published as a Technical Note in the Transactions of CSME, 4, No. 4, 1977, p. 234-236.

14. Karim, G.A., "A Review of Combustion Processes in the Dual Fuel Engine-The Gas Deisel Engine", Prog. Energy Combust. Sci., Vol 6, Pergamon Press Ltd., Great Britain (1980).

15. Karim, G.A., Klat, S.R., "The Knock and Autoignition Characteristics of Some Gaseous Fuels and their Mixtures", Journal of Institute of Fuel, March 1966, 39, p. 109-119.

COMPRESSED NATURAL GAS (CNG)
A "HERE TODAY" SUBSTITUTE FOR GASOLINE

JOHN E. WRIGHT
GAS SERVICE ENERGY CORPORATION
CNG DIVISION
2023 WASHINGTON
KANSAS CITY, MISSOURI 64108

Gas Service Energy Corporation, a wholly owned subsidiary of The Gas Service Company, is actively and aggressively marketing all of the components that will enable an operator of a fleet of vehicles to use natural gas as an engine fuel. The components consist of vehicle conversion kits, natural gas compressors, fast fill reservoir assemblies and control panels and slow fill post assemblies.

In early 1970, The Gas Service Company began using Compressed Natural Gas (CNG) in its fleet of vehicles. By the end of 1981 it will have nearly 500 vehicles operating on CNG. Since 1970, the Company has enjoyed the numerous benefits CNG provides, and through Gas Service Energy Corporation wishes to make the equipment and concept available to its customers and others across the country.

Needless to say, over this period of time improvements have been made in equipment and technology. Today, the concept is proven, our experience in the field is broad, our desire to share the CNG concept sincere, and foremost, our interest is in serving our customer well to his satisfaction.

Call us today to discuss CNG for your fleet of vehicles!

WHY COMPRESSED NATURAL GAS (CNG) AS A FUEL FOR YOUR FLEET OF VEHICLES?

CNG PROVIDES:

Considerably Lower Fuel Cost

A Second Fuel System - You Retain The Use of Conventional Gasoline

An Option To The Effects of Gasoline Rationing

Extends the Range of Your Vehicles Due To Addition of CNG

A Cleaner Burning Engine Thereby Extending Engine Life

Savings By Reducing Spark Plug Replacements

Savings By Increasing Time Between Engine Tuneups

Fewer Oil Changes

Reduced Vehicle Exhaust Emissions Extending Life of Exhaust System

A System That May Be Transferred From One Vehicle To Another With Minor Expense

A Reduction of Fuel Loss From Pilferage

A System Where The Fuel Is Paid For After It Is Used, Not Before, As With Gasoline

ADDITIONAL AREAS FOR CONSIDERATION

Space Must Be Allocated For Fuel Cylinders

Engine Power May Be Slightly Reduced

A Complete Natural Gas Fueling System Must Be Established and Maintained

SUBJECT: COMPRESSED NATURAL GAS (CNG) VEHICLES

The subject of gasoline is currently one of the most hotly debated, least understood and emotional issues facing Americans. The price has escalated dramatically in recent years and its availability in any amount is questionable. For many years we have looked at the problem of motor vehicle fuel and believe that the use of natural gas in motor vehicles represents an ideal solution at this time. The information which follows provides some generalized technical information on how we believe natural gas can be used for motor fuel purposes.

In addition to answering some technical questions, one must consider the practical, political and economic effects of a massive switch to natural gas powered vehicles.

Natural Gas Supply - A considerable quantity of natural gas is available today, and the outlook for the next eight to ten years is excellent. More importantly, as residential and business conservation increases, more gas will be available.

Technology - Compressed natural gas has been used in motor vehicles in the United States for over a decade and has worked well. In other countries natural gas has been used in motor vehicles for over forty years. As with all technology it can be improved, but the fact is it is a proven concept.

Environment - Natural gas is clean burning and substantially reduces auto emission pollution far exceeding present EPA quideline objectives.

Imported Oil - Assuming that a vehicle uses 1,000 gallons of gasoline a year (a low estimate), then for each 100,000 vehicles there can be a reduction of approximately 15,000 bbl./day of imported oil. If as few as one quarter of all U.S. vehicles are converted, the potential savings in oil is nearly 3,750,000 bbl./day. We believe this would significantly help to control world oil prices and domestic inflation.

A number of questions have been asked whenever we discuss the subject of CNG vehicles, and we hope to answer some of these questions in the next few paragraphs.

How safe is the system? Very safe! Several characteristics of natural gas make it a safe vehicular fuel. It is lighter than air, and thus will dissipate into the atmosphere if a leak should occur, as opposed to liquid fuels which could "puddle" on the ground in the event of a collision and present a potential fire hazard. The ignition temperature of natural gas is approximately 1,200 degrees F, or approximately 400 to 500 degrees F higher than that of gasoline or diesel fuels. Our parent company has been using cars and trucks fueled with CNG for ten years. By the end of 1981 it expects to have nearly 500 vehicles operating on CNG. We believe that these vehicles are as safe, if not safer, than the conventional gasoline powered vehicles. The possibility exists that in a severe collision leakage of natural gas could occur in the valves or fuel lines. From all we are able to determine, we do not believe that this is a significant factor. In fact, we are aware of some tests which have been run which show that significant damage to a vehicle does not cause a corresponding damage to the CNG system.

How far can a vehicle travel on one filling of CNG? One hundred twenty cubic feet of natural gas contains about the same energy as one gallon of gasoline. The average cylinder, when fully pressurized, has a capacity of approximately 380 cubic feet of CNG. Each cylinder contains approximately 3.2 gallons of gasoline equivalent, and therefore the vehicle will travel the same distance that it will go using 3.2 gallons of gasoline. If we assume that the vehicle has two CNG cylinders and averages 12 miles per gallon of gasoline, we can further assume the vehicle will get about 75 miles of travel between fillings, 6.4 gallons of fuel equivalent is sufficient for one day's usage in the majority of vehicles.

How do we fuel a vehicle? The vehicles may be fueled overnight (slow fill) or in three to five minutes (fast fill). The filling system consists of a fully automatic high pressure compressor, a high pressure piping system, slow fill post assemblies (if slow fill is desired) and/or cascade reservoir assemblies (if fast fill is desired) and temperature compensated regulators. Please note, a system may consist of slow fill only, fast fill only, or both. The slow fill system consists of a temperature compensated regulator, a high pressure piping system, and post assemblies that have hoses and valves for two or four filling positions. In the slow filling process, the compressor pumps regulated natural gas through the high pressure piping, to the post assemblies whose hoses are connected to a filling connection on the vehicles. Natural gas passes through the filling connection directly to the cylinders permanently mounted in the vehicle. The fast fill system consists of a high pressure piping system, a control panel and a cascade reservoir assembly. Natural gas is pumped to the reservoir and is stored until a vehicle is connected to a hose attached to the cascade control panel. Once connected, valves are turned to the "on" position and a regulated supply of natural gas is passed from the cascade reservoir directly to the cylinders mounted permanently in the vehicle. A vehicle may be fast filled in approximately three to five minutes.

Can only fleets of vehicles use CNG? At the present time, the only compressors commercially available are for fleets of vehicles. However, we are having a small compressor developed that ultimately can be used at home to supply CNG for one or two vehicles on a slow fill basis.

<u>What is the potential for use of CNG?</u> We envision two markets that will use CNG. They are commuter vehicles and fleet operated vehicles. On the average, a commuter goes between 10 and 30 miles one way to work, or uses less than five gallons of gasoline a day. Fleet vehicles, on the other hand, travel between 70 and 100 miles per day and would use between 6 and 9 gallons of gasoline each day. Initially, we envision the use of CNG would be by fleet operators. Once the fueling system is established at a plant location, the plant owner may also find that it is worthwhile to provide to employees for their personal or commuter vehicles as a fringe benefit or to ensure they have fuel in order to report for work.

<u>How much natural gas would a vehicle use in a year?</u> Our estimate is that each vehicle using CNG will use approximately the same amount of natural gas in a year that a house heating customer does; that is, something between 125 Mcf and 200 Mcf.

<u>What is the cost of CNG as compared to an equivalent gallon of gasoline?</u> Presently, the cost of CNG per gallon as compared to a gallon of gasoline can range from $.30/gallon dependent upon the applicable cost of natural gas, electricity, and the road use tax. This equivalent cost included the cost of natural gas, the cost of electricity to drive the compressor, the cost of maintenance of the compressor, and a road use tax similar to that paid on gasoline.

<u>What is the cost to convert a vehicle?</u> Currently, costs are ranging from $2,300 to over $3,000 per vehicle dependent upon driving range and type of refueling system required. These costs include the conversion kit installed on the vehicle and the cost of equipment and installation of the refueling station (compressor, slow fill apparatus and fast fill apparatus). In any event, upon considering these costs and the fuel savings, the CNG system can show a simple payout in the range of from 1.5 to 3.5 years, which, in most instances, is an attractive investment. It is important to note that nearly all of this equipment is reusable in that it may be transferred from one vehicle to another. Therefore, the real savings to a company will show up after the initial write-off period. An additional factor is that the price of gasoline will most likely track the price of crude oil and it is not unreasonable to expect that the price of gasoline will very shortly be $1.50 per gallon and quite possibly, even as high as $2.00 per gallon within the next year or two. We may also see a shortage of gasoline. Consequently, while the natural gas vehicle would be attractive at the cost shown, it will be significantly more attractive simply because the natural gas to fuel the vehicles will be available. Finally, this simple economic scenario gives no consideration to special tax provision now being considered for the use.

<u>How many CNG vehicles are in operation today?</u> There are approximately 20,000 CNG vehicles in the U.S.A. at the present time. Worldwide there are an estimated 300,000 to 400,000 motor vehicles burning CNG as an motor fuel.

<u>How many miles have CNG vehicles driven?</u> According to the AGA report entitled "Preliminary Analysis of the Safety History of Natural Gas-Fired Transportation Vehicles", December, 1979, fleet vehicles have traveled in excess of 175 million miles.

Also mentioned in the report was the information that:

No deaths and only one injury occurred in which natural gas was a contributing factor.

In an estimated 1,360 collisions involving natural gas fueled vehicles there were no failures or fires involving the natural gas system.

Over 180 of these collisions were rear-end collisions involving CNG vehicles. Although the natural gas tanks are stored in the car trunk, none of these collisions caused a failure of the natural gas system.

HOW DO CNG VEHICLES COMPARE WITH ELECTRIC VEHICLES?

Item	CNG Vehicle	Electric Vehicle
Cost of conversion	When the system is added to a gasoline vehicle $1,000/vehicle plus installation costs.	Not feasible to convert an existing gasoline vehicle without a major retrofit.
Range	CNG - 60-100 miles Gasoline -Unlimited	40 - 60 miles on a charge.
Cost of operation per mile	At 55¢/gallon and 12 miles/gallon - 4.6¢/mile	4 - 5¢ per mile.
Alternate fuel capability	Gasoline	None
Recharge time	Fast fill - 5 minutes Slow fill - Overnight	Fast charge 30 min. Slow charge - 8 - 10 hours.
Can the fuel system be transferred to a new vehicle?	Yes, with a minor expense involved.	Not readily since it is an inherent part of the vehicle.
Fuel storage system and replacement time table	Cylinders - some similar types of cylinders have been in use over 70 years. Hydrostatic tests cost $15 every 5 years.	16-20 six volt batteries-they will need to be replaced every 2-4 years at a cost of $60/battery for a total cost of $960-$1,200.
Simple payout period	3 - 4 years. Sometimes as low as 6 months on high usage vehicles.	Difficult to predict due to $12,500/vehicle cost when handmade. In volume perhaps $6,000/vehicle.
Additional weight added to the vehicle	Approximately 130 pounds/cylinder in rear of vehicle. This is of great help on slick streets in winter.	Batteries weigh 1,260 lbs.
Safety of fuel system in event of a collision	Natural gas is lighter than air and if leakage should occur it dissipated into the	Dependent upon protective construction.

	atmosphere.	
Vehicle performance characteristics.	Very similar to gasoline.	Very similar to gasoline.
Passenger	Any size gasoline vehicle can be converted to CNG.	2 - 5 passengers.

COMPONENTS OF A COMPRESSED NATURAL GAS (CNG) REFUELING SYSTEM

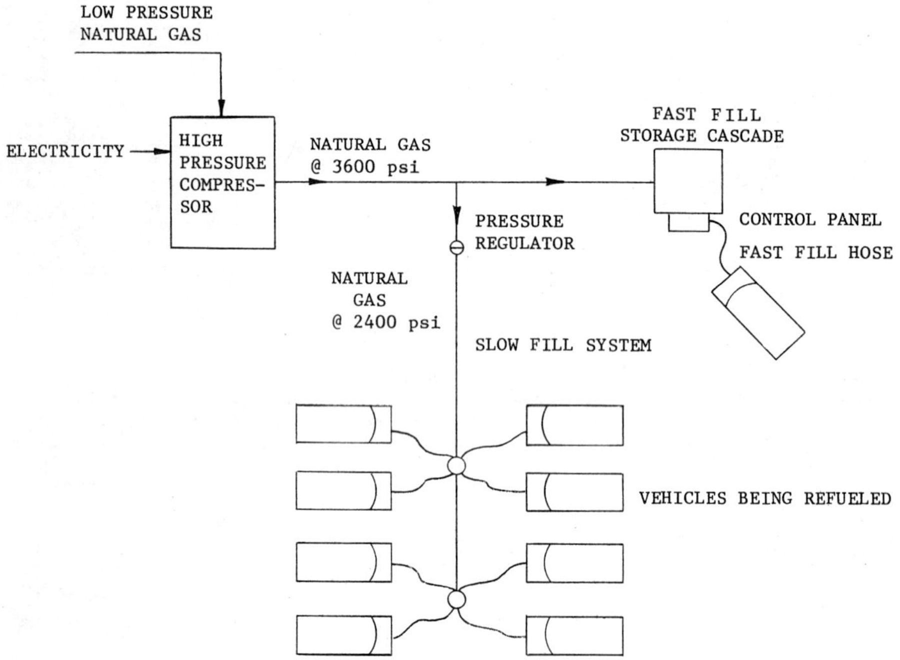

COMPONENTS OF A CNG VEHICLE SYSTEM

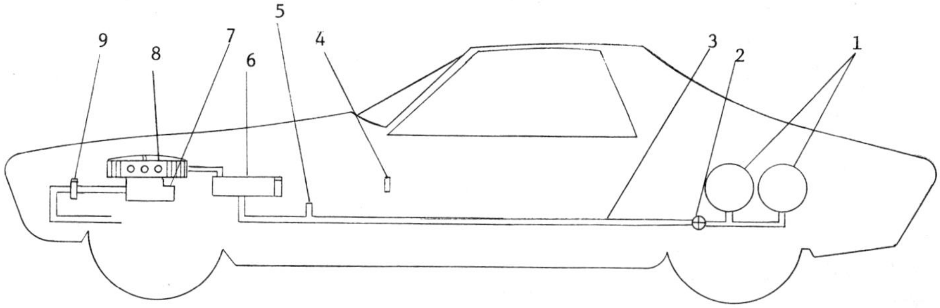

(1) NATURAL GAS CYLINDER
(2) MANUAL SHUT-OFF VALVE
(3) HIGH PRESSURE FUEL LINE
(4) FUEL SELECTOR SWITCH AND GAUGE
(5) NATURAL GAS FILL VALVE
(6) PRESSURE REDUCER AND NATURAL GAS SOLENOID VALVE
(7) ORIGINAL EQUIPMENT GASOLINE CARBURETOR
(8) NATURAL GAS MIXER
(9) GASOLINE SOLENOID VALVE

SYSTEM OPERATION OF A CNG VEHICLE

Natural gas is compressed to 2400 psi by a special compressor and is placed aboard the vehicle through the natural gas fill valve (5) where it is transferred to high pressure cylinders permanently installed in the rear of the vehicle. When natural gas is required by the engine, it leaves the cylinders (1), passes through the master manual shut-off valve (2), the high pressure fuel line (3), and enters the engine area. At this point the natural gas enters the pressure reducer (6). Here the pressure is reduced from 2400 psi to atmospheric pressure. The solenoid valve is used to shut off the natural gas supply when the engine is not running or when it is being fueled by gasoline. After the natural gas solenoid valve has opened, the natural gas passes through the pressure reducer into a specially designed natural gas mixer (8) where it is properly mixed with air, flow down through the gasoline carburetor (7) and enters the engine's combustion chambers. The gasoline supply to the gasoline carburetor is controlled by the gasoline solenoid valve (9). A fuel selector switch (4) is provided so that the operator of the vehicle can select fuel from either the CNG fuel system or the gasoline system.

EQUIVALENT COST
NATURAL GAS PER GALLON GASOLINE

Data

Gasoline = 115,000 btu/Gallon
Natural Gas = 1000 Btu/Cubic Foot
Cost of Natural Gas = \$3.00/1000 cf
Cost of Electricity = \$.05/KWH

A. Natural Gas Cost Per Gallon Equivalent

$$\frac{115{,}000 \text{ Btu/Gal Gasoline} \times \$3.00/\text{Mcf}}{1000 \text{ Btu/cf} \times 1000 \text{ cf/Mcf}} = \$.3450/\text{Gallon}$$

B. Electric Cost Equivalent Per Gallon
(To Power the Compressor)

11KW x \$.05/KWH x 1 Hr = \$.55/Hr

$$\frac{17 \text{ cfm} \times 60 \text{ m/hr}}{115 \text{ cf/gal.}} = 8.86 \text{ Gal/Hr}$$

$$\frac{\$.55/\text{HR}}{8.86 \text{ Gal/Hr}} = \$.0620 \qquad \$.0620/\text{Gallon}$$

C. Maintenance Cost Equivalent Per Gallon

$$\frac{\$2000/\text{YR}}{8.86 \text{ Gal/HR} \times 6000 \text{ HRS/YR}} = \qquad .0376/\text{Gallon}$$

D. Road Use Tax* .0700/Gallon

Total	$.5146/Gallon
Total less Road Use Tax	$.4446/Gallon

* Not Applicable to Most Municipal Governments

COMPRESSED NATURAL GAS (CNG) EUIPMENT MAY BE LEASED
AND
PAID FOR OUT OF AN EXISTING OPERATING BUDGET
IN A RELATIVELY SHORT PERIOD OF TIME

Typical Example:

Cost for Conversion and Refuleing Systems Per Veh.:	$3000
Typical Gasoline Usage Per Month Per Veh.:	250 Gallons
Approximate Cost of Gasoline:	$1.10/Gallon
Cost of Natural Gas Equivalent:	$.45/Gallon
Lease-Purchase Cost Per $1000 Per Month:	$50/1000 (24 mos.)
	$35/1000 (36 mos.)
	$30/1000 (48 mos.)

	24 Month Lease	36 Month Lease	48 Month Lease
Lease Purchase Cost Per Gallon:	$.60/Gal.		
(24 Month Lease): $\frac{\$50 \times 3}{250}$			
(36 Month Lease): $\frac{\$35 \times 3}{250}$		$.42/Gal.	
(48 Month Lease): $\frac{\$30 \times 3}{250}$			$.36/Gal.
Cost of Natural Gas Per Equiv.: Gallon of Gasoline	$.45/Gal.	$.45/Gal.	$.45/Gal.
Total Cost Per Gallon:	$1.05	$.87	$.81
Cost of Gasoline:	$1.10	$1.10	$1.10
Savings Per Gallon:	$.05	$.23	$.29

CERAMICS FOR THE AGT101 AUTOMOTIVE GAS TURBINE

D. M. Kreiner and J. M. Wimmer

INTRODUCTION

An Advanced Gas Turbine (AGT) powertrain for automotive application is being developed by the Garrett Turbine Engine Company and Ford Motor Company under DOE/NASA Contract No. DEN 3-167. This effort represents a significant portion of the Gas Turbine Vehicle Systems Program established by the Department of Energy, with NASA Lewis Research Center delegated to provide project management and administration. The goals of the program are:

- Fuel consumption of 42.8 miles per gallon on No. 2 diesel fuel in a 3000-pound car (Sea level, 59°F, Combined Federal Driving Cycle)
- Substantially the same overall vehicle driveability and performance as obtained with a conventional spark ignition internal combustion engine
- Low emissions (1984 EPA goals)
- Ability to use a variety of fuels (5 maximum)
- Reliability and life comparable to current powertrains
- Competitive initial and life cycle cost
- Acceleration suitable for safety and consumer expectation
- Noise and safety characteristics meeting 1984 Federal standards

AGT101 POWER TRAIN

The AGT101 powertrain, as shown in Figure 1, consists of three major modules: a power section, gearbox, and transmission. The power section is a single-shaft, regenerated gas turbine configuration utilizing ceramic components in the high temperature portion of the engine. The 4-speed AOD transmission currently in production at Ford will be used with minor modifications to the mounting and controls. The gearbox module matches the output of the single-shaft turbine with the speed characteristics of the 4-speed transmission. This is accomplished by utilizing a split-path gearbox with a torque converter. Gearbox pads are arranged to mount standard automotive accessories.

The powertrain is 60.8 inches long and has a maximum diameter of 21.5 inches at the power section flange. This package is designed to be installed in the rear wheel drive Ford Fairmont without modification to the engine compartment structure.

The predicted fuel consumption on the Combined Federal Driving Cycle is 42.8 miles per gallon on DF-2 in a 3000-pound car at sea level on a 59°F day. Although diesel fuel is used in the mileage rating, the turbine engine will utilize a variety of fuels, and the combustor is being designed with this versatility in mind. The entire range of operational fuels has not been finalized, but the leading candidates are gasoline, Jet A, a shale oil derivative, and methanol. As the effort to develop other alternative fuels in the U.S. expands, other candidates will be considered.

AGT101 PROJECT SCHEDULE

The AGT101 Project was started October 1, 1979, and the tasks are proceeding according to the schedule shown in Figure 2. The initial program milestone, the Preliminary Design Review (PDR) of the Reference Powertrain Design (RPD), was successfully completed January 30, 1980. The RPD represents the ultimate design to be available at the end of the program; a preliminary, MOD I powertrain will be run initially, and will be updated as the program progresses. A second preliminary design review of the MOD I powertrain was completed April 30, 1980. The two design reviews were accomplished on schedule and completed the first contract milestone.

The MOD I, Build 1 engine has all metallic components except the ceramic regenerator, and has an aerodynamic flow path identical to the MOD II or RPD. As the program progresses, ceramic components will be incorporated into the MOD I engine to replace hot section metallic components, (with the exception of the metallic rotor). In the MOD II engine, the rotor also will be ceramic.

Component and powertrain development currently are underway to support subsequent program milestones. This includes efforts at the ceramic subcontractors (Carborundum, Ford, AiResearch Casting Company, Pure Carbon, and Corning Glass Works) in developing processes to fabricate the ceramic components for the power section.

The design and analysis conducted thus far support the initial engine concept, as no significant changes to the design have been required. Test data obtained during this period also support the design projections. Individual testing of all major components is scheduled to be complete prior to the first engine build.

AGT POWER SECTION

The AGT MOD II power section is shown in Figure 3. This gas turbine is flat rated at 100 horsepower at a maximum operating speed of 100,000 rpm. The single-shaft rotating group is composed of a turbine, compressor, and output gear supported by an air-lubricated, foil

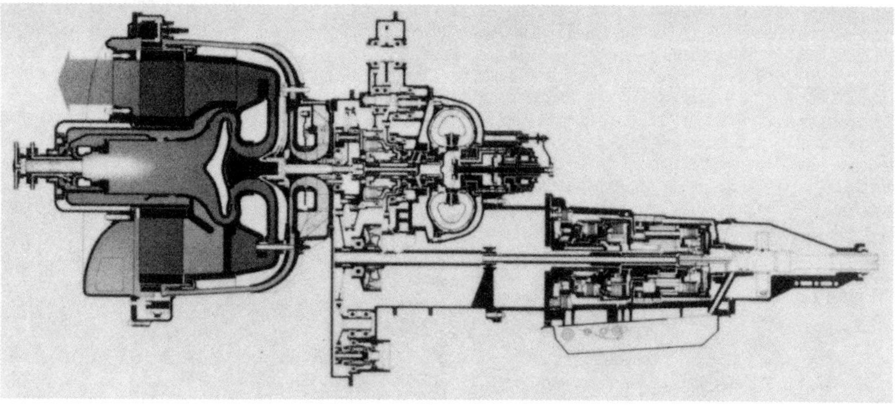

Figure 1. Garrett/Ford AGT101 powertrain.

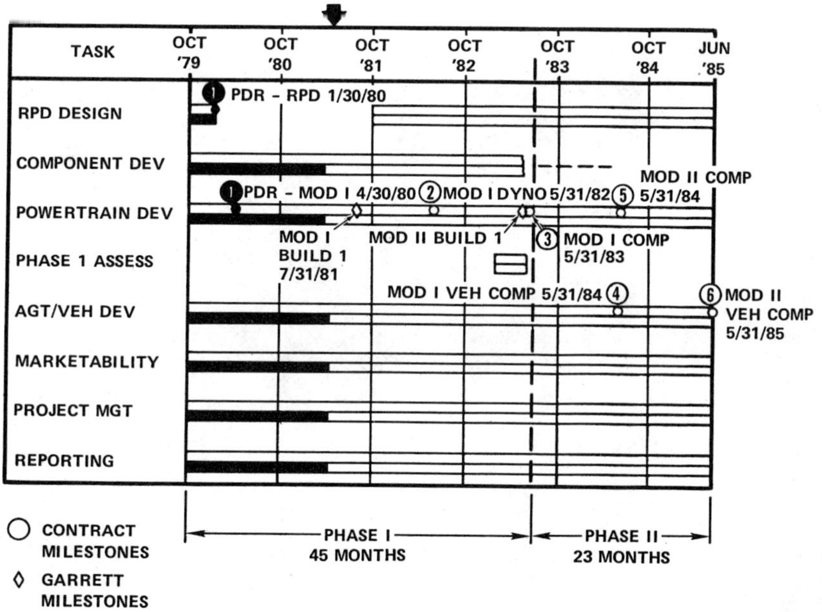

Figure 2. AGT101 project schedule.

Figure 3. AGT101 automotive gas turbine engine.

journal bearing and an oil-lubricated ball bearing. Air enters the engine through variable inlet guide vanes, passes through a single-stage compressor (pressure ratio 5:1), around the perimeter of the engine and through the high pressure half of the ceramic regenerator face area. The partially heated air then flows to the combustor. Variable geometry is used in the combustor to limit the primary zone maximum temperature to 3000°F, and thus suppress NO_x formation. The combustor design also provides a prevaporizing/premixing zone to limit particulates and unburned hydrocarbons. A high level of mixing, along with minimal wall quenching, assures low CO formation. Hot gas exits the combustor at a maximum temperature of 2500°F and expands through the turbine stators and radial rotor. It is then routed through the low pressure half of the rotating regenerator and out the exhaust.

The individual ceramic components that comprise the hot flow path of the engine are shown in Figure 4. Design complexity ranges from a simple cylindrical shape for the regenerator shield to closely controlled flow path contours on the stator vanes and turbine shroud. Dimensions vary from a 1.5-inch diameter duct spacer to a 15-inch diameter inner diffuser housing. The turbine rotor is expected to be one of the most difficult components to develop due to the large mass, high hub stress, and complex blade geometry. Several material and fabrication approaches are being taken to develop the rotor, and the various static components. Prior to describing the specific approaches, it is appropriate to discuss the general design approach used for ceramics, since it differs significantly from the approaches used for metals.

CERAMIC DESIGN APPROACH-GENERAL

The major difference between ceramic and metal behavior is the mode of failure. A metal part locally stressed beyond its yield strength will plastically deform to accommodate the stress. In contrast, when the strength of a ceramic part is exceeded, it responds by crack propagation which often results in complete fracture. Fracture initiates at imperfections (flaws) either on the surface or in the interior of the part; the general relationship is described by,

$$\sigma_f = \frac{1}{Y} \frac{K_C}{c^{1/2}} \quad (1)$$

where σ_f is the fracture stress, K_C is the fracture toughness of the material, c is the flaw size and Y is a constant related to flaw geometry. Typical flaws consist of interior voids and inclusions which occur during component fabrication, and surface scratches and minute notches which result from surface machining. Since these flaws cannot be completely eliminated, the objective is to optimize the ceramic processing and machining steps to achieve as small a flaw size as possible.

Although flaws of a particular type will have an average size, there will be a distribution of sizes around this average, i.e., there will be flaws larger and smaller than the average size. This leads to the observation that the strength of a ceramic part is dependent on its own size, since the probability of containing one of the larger flaws in the distribution increases with component size. The fact that there is a flaw size distribution also means that only the probability of failure can be specified at a given stress level. At low stresses, the probability of failure will be small but finite.

Current ceramic design practice uses Weibull statistics to calculate the reliability, as described by:

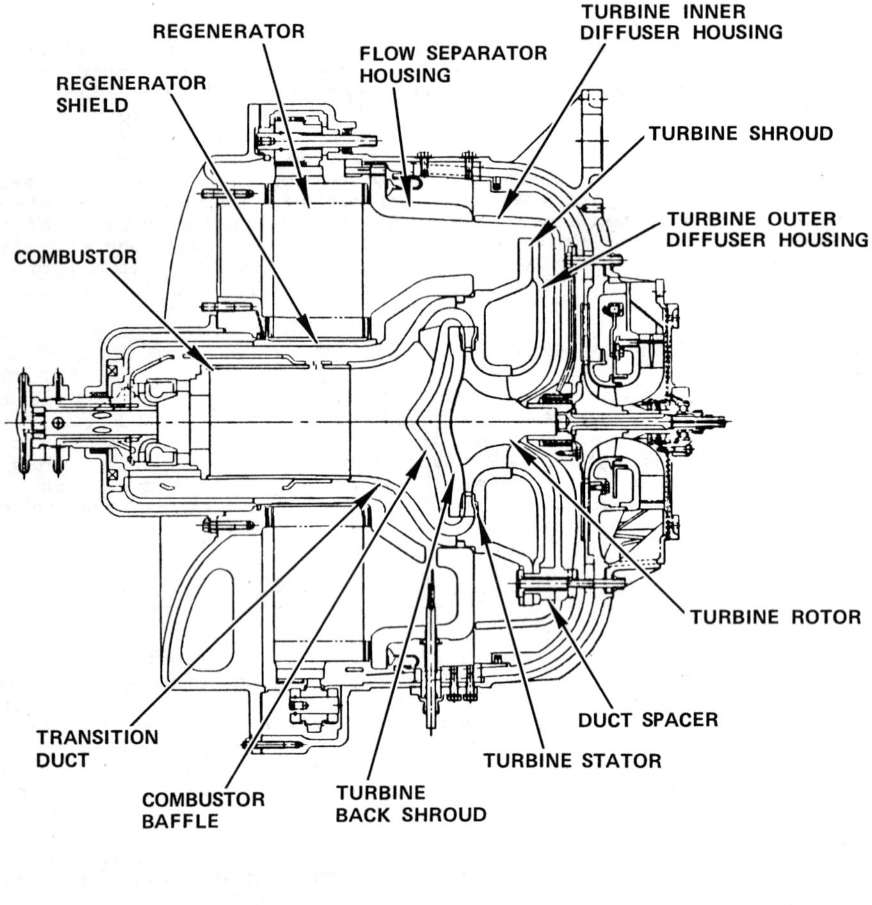

Figure 4. AGT101 ceramic components.

$$1 - F = \exp - \left[\frac{A}{A_t} \left(\frac{\sigma}{\sigma_\theta} \right)^m \right] \qquad (2)$$

where, F = probability of failure

σ = applied stress

A = component area

σ_θ = characteristic strength

m = Weibull modulus

The characteristic strength and modulus are referred to as the Weibull parameters and are determined on test specimens of known area (A_t), fabricated from the same material as the components, and having the same surface condition. Equation (2) applies to a uniformly stressed area. Since a finite element analysis generally is used to calculate both the temperature and stress distribution in a component, these same elements are used in the statistical analysis. Equation (2) is used to calculate the reliability (1-F) of each finite element and the total component reliability is then calculated as the product of the element reliabilities. A similar approach is used to predict failure due to a volume flaw distribution.

Although the procedure outlined above is straight-forward, its success depends on being able to design specimen tests in which fracture occurs from the same type of flaws which can lead to component failures. Differences in the stress distribution and material properties between test specimens and components often make this a difficult task.

AGT101 CERAMIC ROTOR DESIGN APPROACH

Figures 5 and 6 illustrate the results of the above procedure as applied to the rotor design. Since material property improvements were anticipated during the program, rotor reliability goals of 0.95, 0.99, and 0.999 were established for the first through third years. Based on a finite element and Weibull analysis, the corresponding characteristic strengths and Weibull moduli were calculated. The results are shown in Figure 5.

A material with a low Weibull modulus exhibits a large variation in strength, and therefore must have a high characteristic strength to meet the reliability requirements.

Although some failures are predicted at low stresses, i.e., well below design speed, these do not need to occur in service and proof testing will be used to insure rotor reliability. Since transient thermal stresses add to the rotational stresses, overload proof testing will be used. In the case illustrated in Figure 6 a proof test of 115 percent of design speed is utilized. Using first year goals this results in a loss of 65 percent of the rotors during proof testing, a number considered acceptable in a development program. Using third year goals, only 2-3 percent would be lost during proof testing.

CERAMIC COMPONENT MATERIALS

Silicon nitride (Si_3N_4) and silicon carbide (SiC) both have the combination of strength and thermal shock resistance which make them prime candidates for use as turbine components. All AGT101 components are being fabricated from various forms of these two materials, with the exception of the flow separator housing and the regenerator, which are being fabricated from a low expansion aluminum silicate.

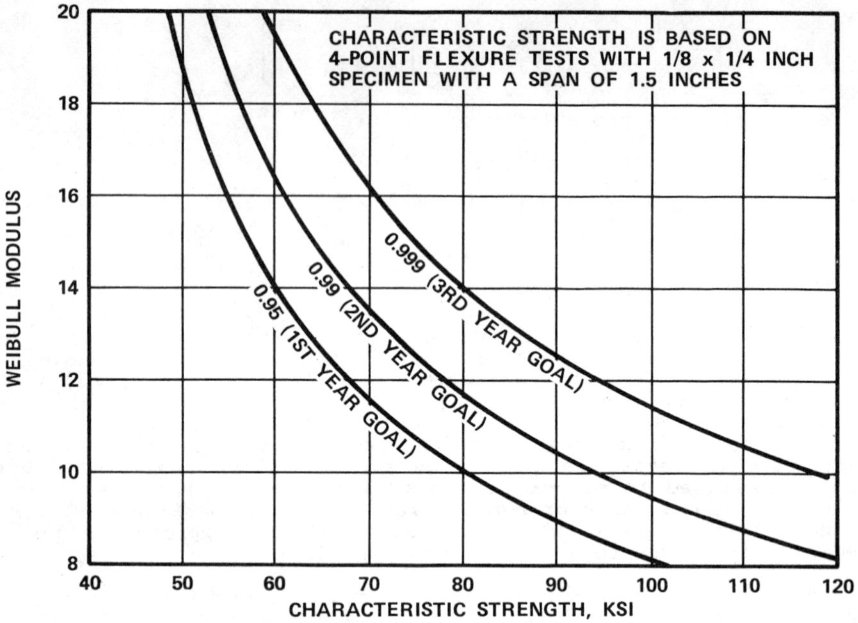

Figure 5. Corresponding reliability and material property goals for the AGT101 turbine rotor at 100K rpm.

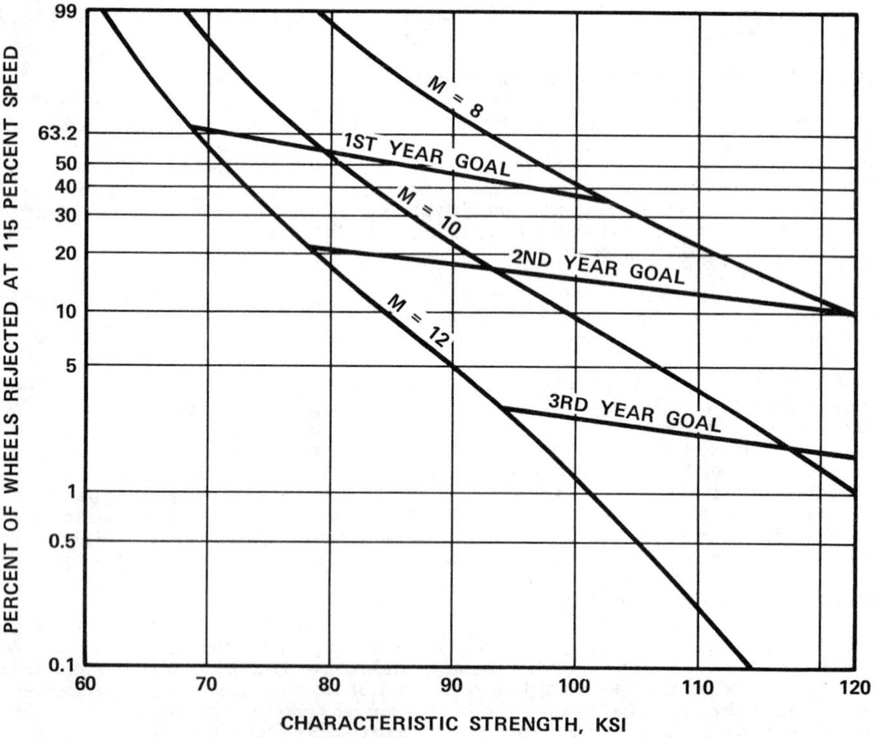

Figure 6. Predicted failures during overload proof testing for various material properties.

Both reaction bonded silicon nitride (RBSN) and sintered alpha silicon carbide (SASC) are being considered for static components, while sintered silicon nitride (SSN), reaction sintered silicon carbide (RSSiC), and SASC are candidates for the radial rotor. Typical strength characteristics of these candidates are shown as a function of temperature in Figure 7. The strength of both RBSN and SASC appears to be adequate for static components. However, the rotor candidate materials require improvement: SSN at high temperature, and SASC and RSSiC over the whole temperature range. Compositional and processing improvements are being pursued to improve material properties.

CERAMIC COMPONENT FABRICATION

Ceramic component fabrication consists of four major steps: green shape forming, firing, final machining, and inspection. The primary approaches to green shape forming for AGT components are green machining, slip casting, and injection molding. Table 1 indicates the current material and fabrication approach for the primary AGT101 components. Green machining consists of machining the component from a soft (green) block of isopressed powder. In slip casting, an aqueous suspension of powder is poured into a mold and the water is drawn out of the suspension through a plaster surface. The component is further dried after removal from the mold. For injection molded components, powder is mixed with a plasticizer, heated, and injected into a metal die. This approach is particularly useful for complex shapes. However, there is currently a thickness limitation on injection molded parts due to the difficulty of removing the plastic binder from the molded shape without damaging the fragile part. In the case of silicon carbide, this has led to a two-piece approach to rotor fabrication in which an injection molded ring containing the blades is bonded to a core made in a separate process.

During firing, close control of temperature is mandatory. RBSN parts start out as elemental silicon powder shapes which are then nitrided. Although there is negligible shrinkage during firing, the nitridation rate must be slow enough to prevent overheating due to the exothermic nature of the reaction. For sintered components (which shrink up to 20 percent during firing) uniform temperatures must be maintained to prevent non-uniform shrinkage and cracking.

An efficient turbine engine requires close dimensional control of the flow-path components to achieve aerodynamic performance. As a result, all mating surfaces will be final machined to achieve the required tolerances. Although this is currently a slow and expensive operation requiring diamond impregnated tools, alternate machining approaches are being pursued. In addition, machining costs are estimated to be low in automotive production quantities.

After fabrication, all components must be inspected to assure that flaws are below detectable limits. This usually consists of visual, x-ray, ultrasonic, and fluorescent dye penetrant inspection. New techniques such as scanning laser acoustic microscopy (SLAM) and scanning photo-acoustic microscopy (SPAM), as well as higher frequency ultrasonic techniques are also being developed to aid in the detection of the small flaw sizes relevant to ceramic failure. These inspection techniques coupled with component proof testing will be conducted to assure reliable engine operation.

CONCLUSION

Although many challenges remain, the basic ceramic design methodology, materials, and component fabrication approaches exist to develop a viable ceramic gas turbine engine. This powertrain has been designed to meet the AGT101 program goals of high mileage, competitive

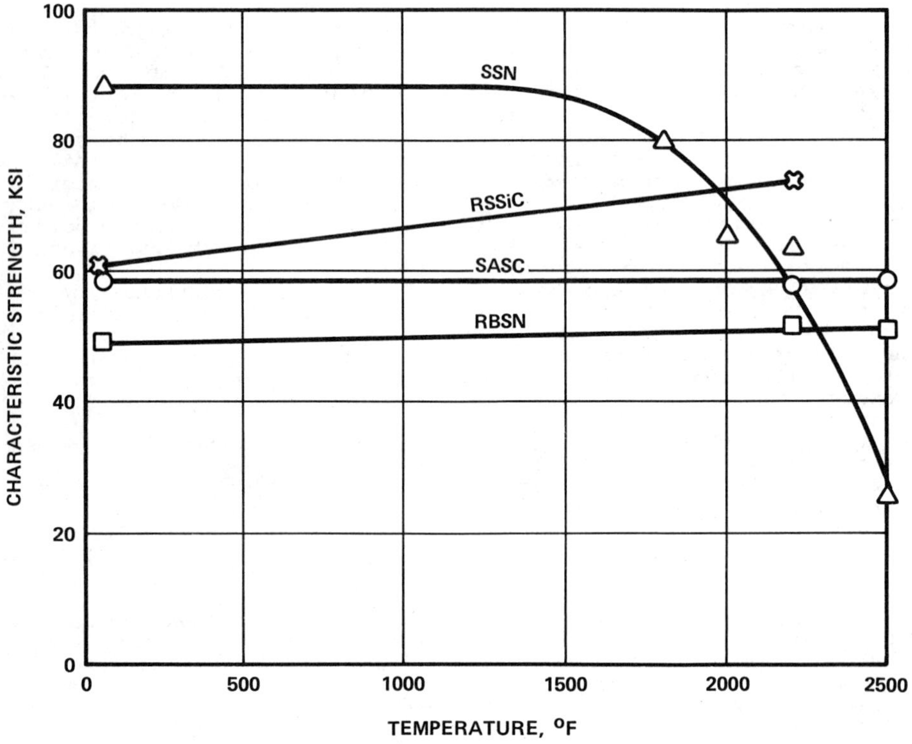

Figure 7. Temperature dependence of candidate material strength RBSN = reaction bonded Si_3N_4, SASC = sintered alpha SiC, SSN = sintered silicon nitride, RSSiC = reaction sintered silicon carbide.

cost, low emissions, and multiple fuel capability in a reliable and safe automobile. In meeting these objectives, the AGT powertrain is projected to be competitive with any other alternative powertrain.

TABLE 1. AGT101 CERAMIC COMPONENT MATERIALS AND FABRICATION APPROACH

Component	Material	Fabrication Approach
Transition Duct	SASC RBSN	Slip Cast Slip Cast
Combustor Baffle	SASC	Slip Cast
Turbine Back Shroud	SASC	Green Machine
Turbine Stator	SASC RBSN	Injection Molded Injection Molded
Turbine Rotor	SSN SSN SASC RSSiC	Injection Molded Slip Cast Injection Molded Injection Molded
Turbine Shroud	SASC RBSN	Injection Molded Slip Cast
Turbine Outer Diffuser Housing	RBSN	Slip Cast
Turbine Inner Diffuser Housing	RBSN	Slip Cast
Duct Spacer	RBSN SASC	Slip Cast Green Machine
Flow Separator Housing	AS	Slip Cast
Regenerator	AS	Tape Forming
Regenerator Shield	SASC	Green Machine

UPS BROAD FUEL TOLERANCE STRATIFIED

CHARGE ENGINE - PROGRESS REPORT

J.M. Lewis
Automotive Engineer
United Parcel Service, Inc.

ABSTRACT

United Parcel Service, the operator of the world's largest commercial vehicle fleet is developing a broad fuel tolerance stratified charge engine to meet the fleet requirements of the 1980's.

Under contract with UPS, Texaco, assisted by Ricardo Consulting Engineers, has applied the Texaco Controlled-Combustion System (TCCS) to an existing GMC 292 (292 in^3) six cylinder gasoline engine.

The converted engine has been designated as the UPS 292 SC and has successfully completed initial prototype evaluation. The projected benefits of increased fuel economy, low emissions, low noise level, reasonable cost, ease of retrofit and broad fuel tolerance appear to be entirely viable at this stage of the development.

The program has been expanded to include ten additional prototypes for further development work followed by five hundred pre-production engines for field test. The ten prototype engines are nearing completion, but work on the five hundred field test engines is held up pending clarification of the Tampering Section of the Clean Air Act.

INTRODUCTION

Thirty-four months ago, UPS entered into an engine development contract with Texaco, assisted by Ricardo Consulting Engineers as sub-contractor, to apply Texaco's TCCS (Texaco Controlled Combustion Technology) to power parcel delivery vehicles.

This is an unusual endeavor for a parcel delivery company whose sole activity is the use of motor vehicles rather than their manufacture.

Many factors prompted this effort. A study of the available technology and the potential conditions facing UPS as well as others revealed three main problems:

1. Emission control requirements will become more stringent.

2. Vehicles will have to become more fuel efficient, and there will be strong government pressure to introduce diesel engines in light and medium trucks.

3. Fuel availability will be questionable both as to quantity and type of fuel, which may include synthetics, alcohols, and shale derivatives.

No engine is currently available nor is being developed to meet these requirements and still satisfy cost criteria and other practical considerations.

In addition to the problems stated, UPS has a unique fleet characteristic that further complicated the company's search for a solution - the long life of its vehicles.

Any decisions in regard to future power plants had to include the possibility of retrofitting this large fleet of over 35,000 package delivery vans that will be operating through the 1980's. (See Figure 1 and Figure 2.)

UPS DIESEL EXPERIENCE

At first glance, the diesel engine would appear to offer the greatest promise in meeting the problems and requirements of the UPS fleet.

Careful review of experience with light and medium duty diesels tested in the UPS fleet since the 1960's revealed the following problem areas:

Cold weather starting
Exhaust fumes and odor
Short starter life
Increased noise level
Loss of payload due to increased weight
Parts and service availability
High cost of engine and driveline maintenance
Costly modification for retrofit
High initial cost

Despite the relative fuel economy advantages of diesel engines, the overall UPS experience does not justify use of existing light and medium duty diesels in low annual mileage, class 5, package delivery vans which are currently powered by high volume, durable, and cost-effective gasoline engines.

Other alternatives such as Sterling engines, gas turbines, Wankel engines, electrics, hybrids, and others offer no near-term practical solution for package delivery operations.

TCCS TECHNOLOGY

The major feature of the TCCS technology as applied to the UPS 292 SC engine is the controlled combustion cycle which eliminates knock. At the beginning of the cycle, (see Figure 3) air swirls into the cylinder through specially designed swirl ports in the

P-800 P-600 P-400 P-200

FIG. 1

UPS PACKAGE CAR FLEET DESCRIPTION

No. of Veh.	UPS Des.	Net H.P.	Avg. MPD	Avg. MPG	Truck Class
2300	P-200	118	221	12.0	2
860	P-300	126	134	10.8	2
7300	P-400	126-130	117	9.5	5
4500	P-500	126-130	103	8.0	5
13800	P-600	126-130	66	7.2	5
6000	P-800	126-130	42	6.1	5
224	P-1100	180	29	4.0	6
160	Van	180	42	4.1	7

FIG.2

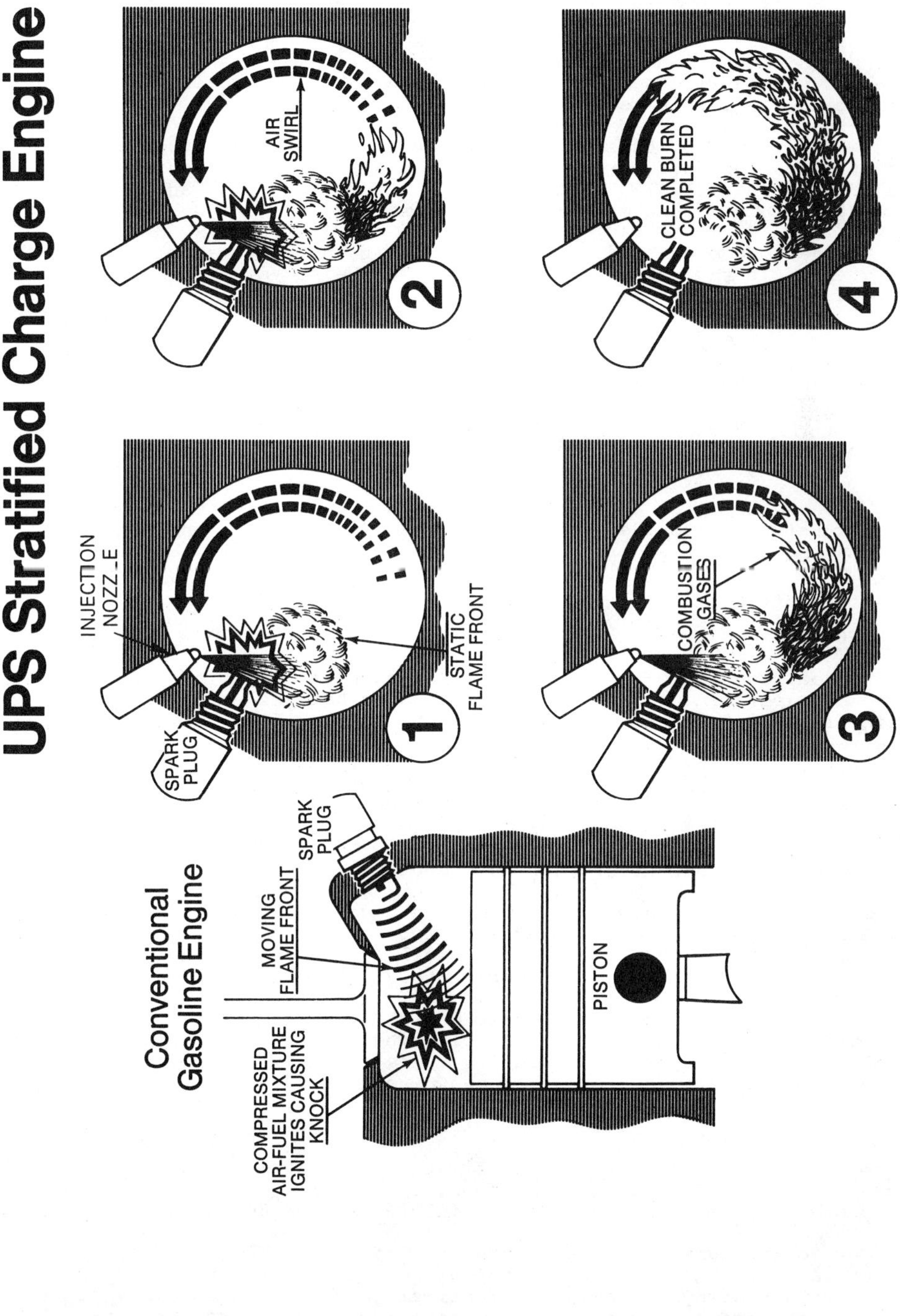
UPS Stratified Charge Engine
1
SPARK PLUG
INJECTION NOZZ_E
STATIC FLAME FRONT
2
AIR SWIRL
3
COMBUSTION GASES
4
CLEAN BURN COMPLETED
Conventional Gasoline Engine
COMPRESSED AIR-FUEL MIXTURE IGNITES CAUSING KNOCK
MOVING FLAME FRONT
SPARK PLUG
PISTON

FIG. 3

cylinder head. The cup in the piston crown enhances the swirl during the compression stroke. As the piston nears top dead center, fuel is injected through a nozzle toward a special spark plug. Ignition is coincident with fuel injection and continues longer than in a conventional gas engine. A flame front is established downstream of the spark plug. As the cycle progresses, swirling air supports combustion while the flame front remains static. Combustion gasses are carried around the cylinder by the air swirl until fuel injection is terminated.

The result of this process is a clean burn of the precisely controlled fuel/air mixture which minimizes exhaust emissions and allows the use of all types of automotive fuels which would include methanol, ethanol, diesel fuel, gasoline and synthetics.

The Texaco stratified charge engine technology was chosen as a viable alternative to the typical gasoline and diesel medium duty truck engines shown on Figure 4.

An engine designed with this technology is insensitive to octane or cetane ratings, and can operate on a wide boiling range fuel that can be more efficiently producted at a refinery.

It therefore offers greater flexibility in fuels that can be used and a substantial fuel savings at the refinery as well as on the road.

These apparant advantages over existing engines and the opportunity to provide a demonstration of a practical engine technology that will provide more miles from a barrel of crude - prompted UPS to enter into a development contract with Texaco in April of 1978.

CONVERSION

Modification of a manufacturer's standard engine to increase speed or performance is commonplace in the hot-rod and racing fraternity. In these applications, the engine block and components serve as a foundation to which performance components are fitted to obtain the desired results. This is the approach being used by UPS in the development of a stratified engine conversion.

Selection of the GMC 292 engine for conversion came after a thorough feasibility study by Texaco to determine if it was possible and practical to convert the basic engine.

The GMC 292 converted to TCCS has been designed as the UPS 292 SC and serves as the foundation for the new components which are added to achieve the conversion as well as to provide engine reliability, durability, and performance. (See Figure 5.)

The main change to the basic GMC 292 engine included the substitution or addition of the following new components and systems: cylinder head, manifold, camshaft, valve train, valve cover, fuel pump, water pump, ignition system and fuel injection system.

A single cylinder development engine was built by Texaco to evaluate piston design, injector design, air swirl pattern, fuel pump requirement, spark plug configuration, ignition system characteristics, clearance problems, operation on various fuels, and other parameters important to multi-cylinder design.

Ricardo Consulting Engineers, utilizing the TCCS know-how provided by Texaco, designed the engine conversion components. Patterns, castings, machining, fabrication and final assembly of the prototype engine components were handled by Ricardo as well as initial engine assembly and brief run-in.

Engine Comparison

Item	Gasoline Engine	Diesel Engine	UPS 292 SC
Fuel Economy	Fair	Excellent	Excellent On All Fuels
Fuel Requirement	Octane	Cetane	No Octane Or Cetane
Initial Cost	Low	High	Moderate
Performance	Good	Fair	Equal to Gas Engine Better Than Diesel Engine
Noise Level	Low	High	Higher Than Gas Engine
Vibration	Minor	Heavy	Minor
Exhaust Odor	Low	High	Low on Gas & Broad Cut High on Diesel
Starting Problems	None	Cold Weather	None
Retrofitability	—	Difficult	Easy
Emission Control	Moderate	Moderate	Moderate
Projected Maintenance Cost	Moderate	High	Low

FIG. 4

FIG. 5

The first prototype engine was shipped to Texaco for performance testing in Feb., 1980. The second engine was shipped to Texaco in June, 1980 to replace #1 engine in the test stand. The #1 engine was installed in a UPS P-600 parcel delivery van in August, 1980.

TEST AND EVALUATION

The proto-type UPS 292 SC stratified charge engine after undergoing initial performance testing was installed in a UPS P-600 package van. The van chosen was a 1976 GMC that had previously been powered by a standard GMC 292 engine. The stratified charge engine installation was straightforwarded with only a few modifications to the chassis required.

All fuel economy and performance evaluations of the UPS 292 SC engine have been made in a direct comparison with a 1979 GMC 292 gasoline engine installed in an identical 1976 UPS P-600 package van. This vehicle serves as the control vehicle for a "two truck test".

The engine has operated very satisfactorily over a full speed range from idle to 3200 rpm without misfire or knock. An upper limit of 3200 rpm was imposed on the engine operation to assure prototype engine life with the newly developed components.

Although the majority of the preliminary road tests were conducted with gasoline as the primary fuel, a simulated wide boiling range fuel (75% gasoline, 25% diesel) also was tested. The engine fuel and ignition system settings were not modified to accommodate the wide boiling range fuel during the tests. No significant difference in starting performance of fuel economy was noted. Increased exhaust smoke occurred during acceleration and under certain load conditions. Some of the smoke could be eliminated by fine tuning for the wide boiling range fuel, but the development of an injection pump load control will be needed to eliminate smoke with all fuels.

Acceleration, driveability, and response of the UPS 292 SC powered P-600 is almost identical to that of the standard GMC 292 powered P-600.

The engine idles smoothly without "hunting" at about 700 rpm. A throttle body was originally installed on the intake manifold to aid idle operation and provide vacuum at idle, but was found to be unnecessary because of the excellent idle characteristics and the sufficient output of the vacuum pump.

Hot and cold ($100^{o}F$, $10^{o}F$ ambient temperatures) starting characteristics have excellent with the standard Delco starter motor.

The vehicle can be immediately accelerated to highway speeds without stumble or misfire. No false starts have been noted. Further cold start evaluations will be made. Past TCCS engine conversions have exhibited excellent startability in extremely cold weather. The engine is equipped with a no-leak thermostat with controlled flow characteristics and with an electric temperature demand fan to closely control engine temperatures.

Idle emission checks were conducted on both the UPS 292 SC and the GMC 292 powered package vans utilizing NDIR state inspection type equipment. The UPS 292 SC showed <u>115 ppm HC</u> while the standard 1979 GMC engine showed <u>30 ppm HC</u>. When field tested for CO, the UPS 292 SC showed <u>.05% CO</u> and the standard GMC 292 showed <u>.15% CO</u>.

Exterior and interior noise levels of both the standard and stratified charge powered package vans were taken according to

BMCS stationary run-up test procedures. The exterior noise level of the UPS 292 SC powered van was equal to that of GMC 292 powered van for all practical purposes and well under the maximum BMCS decibel level. The interior noise level of the stratified charge powered van just meets the BMCS maximum with the present rear engine cover. The addition of accoustical material to the rear engine cover has reduced the interior noise level to 88 dba.

Fuel economy improvement was the predominant reason for UPS's decision to enter into this engine development. Therefore, it was extremely important to obtain valid fuel economy and performance comparisons in order to provide management with the necessary data to decide on the future of the program.

The test vehicles were identical in test load, transmission, rear axle ratio, tire size and body design. Factors affecting fuel economy and performance such as lubricants, wheel bearing adjustment, wheel alignment, tire pressure, etc. were continually monitored.

Flow meters reading in hundreths of a gallon increments were installed on each vehicle. Odometer and speedometer accuracy were checked and corrected to assure valid results.

Each vehicle was loaded to a 12,000 # GVW which represents a 4,000# payload.

Early fuel consumption test work has been concentrated on an "On Area Delivery Test Cycle" specified by the UPS Industrial Engineering Department as representative of 1/4 the average P-600 daily delivery area driving activity below 40 mph. This test is indicative of idling, stop and go, and the less than highway road speeds associated with a package delivery operation. It should be noted that UPS vehicles are shut off at each delivery stop - therefore idling in the sequence represents stoplights and traffic delays.

The test results were as follows:

	UPS 292 SC	GMC 292	% Improvement
Idle	.31 g/hr.	.75 g/hr.	142
City	9.4 mpg	7.2 mpg	30.6
On Area	8.7 mpg	6.3 mpg	38.1
Highway Rolling Country	10.3 mpg	8.6 mpg	19.8

Although results achieved have exceeded expectations, the repeatability of the data has provided a high confidence in the excellent fuel economy characteristics of the UPS 292 SC engine.

PROGRAM STATUS AND EXPANSION

As engineering work has progressed over the past years, contacts have been made with various component suppliers and machine shops to establish that engines beyond the original prototypes could be produced at reasonable cost.

A major breakthrough was the interest shown by a Michigan machine shop facility with production type machinery. Utilizing this facility, plus the outstanding support shown by other component suppliers, it is now possible to build 10 additional prototypes within the next four months.

Initial component, tooling, machining pricing, indicate that engines for a 500 engine field test, can be built for less than the cost of a comparable diesel engine of the same horsepower.

Emission regulations remain the largest cloud over this development program. The EPA Tampering Enforcement Section regards this conversion, even though directed to retrofit, as being in violation of the Tampering Amendment to the Clean Air Act which now includes fleets.

UPS has obtained an exemption from the EPA to continue test and development work on 10 engines, but the engines must be destroyed or returned to their original form at the end of two years.

A ray of hope of clearing the way for the planned extended field test of the UPS 292 SC beyond the ten engines, was obtained at a recent House of Representatives Power and Energy Sub-Committee hearing.

The sub-committee expressed concern that a development project of this type, which offered promise in saving fuel, using alternative fuels, and serving as a "bridge engine" was impeded rather than encouraged by government agencies.

Clarification of this question will be necessary before any extended field test can be conducted.

Based on the experience to date, UPS management has approved the expansion of the program to include:

1. 10 more prototype, beyond the 4 in existance, to be built within the next four months for further development work, durability testing, emissions testing and road tests.

2. The procurement of patterns, castings, tooling, machining, etc. for a 500 engine field test evaluation in the UPS fleet.

IN CONCLUSION

UPS is engaged in a major effort to develop and test an engine that appears to have the greatest promise in meeting the problems facing the fleet in the 1980's.

In addition, this project provides a demonstration of a practical technology that has the prospect of saving fuel both on the road and at the refinery while meeting performance, noise, emissions, cost and operational criteria.

This is an opportunity and a challenge UPS cannot afford to pass up in its quest for more efficient vehicles.

BIBLIOGRAPHY

1. A Stratified Charge Multifuel Military Engine - A Progress Report - E. Mitchell, M. Alperstein, J.M. Cobb, and C.H. Faist, Paper 720051 Society of Automotive Engineers, Automotive Engineering Congress, January 10-14, 1972.

2. Texaco's Stratified Charge Engine - Multifuel, Efficient, Clean, and Practical, M. Alperstein, G.H. Schafer, and F.J. Villforth III, Paper 740563, Society of Automotive Engineers, Southern California Section, May 14, 1974.

3. The Texaco Controlled-Combustion System - A Stratified Charge Engine Concept - Review and Current Status - W.T. Tierney, E. Mitchell and M. Alperstein - The Institution of Mechanical Engineers "Power Plants and Future Fuels" Conference, London, England, January 1975.

4. Optimization of the Vehicle - Fuel - Refinery System - R.F. Wilson and W.T. Tierney, Preprint No. 16-76, American Petroleum Institute, 41st Mid-year Meeting, May 12, 1976.

5. Stratified Charge Engine Development with Broad Fuel Tolerance - J.M. Lewis and W.T. Tierney, Society of Automotive Engineers, National Truck Meeting, Fort Wayne, Indiana, November 8-9 1979.

6. United Parcel Service Applies Texaco Stratified Charge Engine Technology to Power Parcel Delivery Vans - Progress Report, National Truck Meeting, King of Prussia, Pennsylvania, November 10-13, 1980.

ELECTRIC VEHICLES OFFER NEW LIFESTYLES TO URBAN CONSUMERS

George Tesar
University of Wisconsin-Whitewater

INTRODUCTION

Modern electric vehicles emerged in the United States in the mid-1960s in response to market pressures created by public interest groups concerned with increasing air pollution problems in major American cities. By the late 1960s all major automobile companies in the United States were experimenting with electric vehicle prototypes. General Motors Corporation introduced the Electrovair and the Electrovan; Ford Motor Company introduced the little Comuta; and the American Motors Corporation built the Amitron.

The development of electric vehicle prototypes moved from the automobile companies to the battery manufacturers and component developers. ESB, Incorporated hired McKee Engineering Corporation of Palatine, Illinois to build the Sundancer. By the early 1970s the battery manufacturers served as major change agents in the development of the electric vehicle industry.

In the early stages of electric vehicle development there were approximately 16 developers and builders of electric vehicles. Through these developers and builders electric vehicles evolved in three different directions. Initially, a number of the firms simply redesigned or converted electric golf carts into on-the-road vehicles; others concentrated on conversions of conventional automobiles; and a relatively small number of the firms designed electric vehicles from the ground up.

The critical point for the electric vehicle industry in the United States came in the fall of 1973. The energy crisis provided a new direction for the electric vehicle industry. However, due to the lack of capital, the lack of general entrepreneurial skills, and most of all a lack of understanding of the consumer and industrial markets, most of these firms have phased out of the emerging electric vehicle industry.

The second impetus for the electric vehicle industry came in September 1976 with the passing of the Electric and Hybrid Vehicle Research, Development, and Demonstration Act of 1976 (Public Law 94-413). Within the scope of this law many government leaders see the potential of electric vehicles to help reduce air pollution and the dependency on foreign supplies of oil, along with other economic benefits. Selective demonstration and commercialization projects are being funded by the United States Department of Energy.

In the semi-private sector the Electric Power Research Institute is sponsoring a two-year cooperative program with the Tennessee Valley Authority to identify research and development areas needed to technologically improve existing electric vehicles.

A recent study conducted by the Electric Vehicle Council indicates that there are 15 electric vehicle manufacturers in the United States today. Ten additional firms have specific programs to develop electric vehicles, and another three firms intend to produce electric vehicles in the near future. The 15 firms responsible for today's production are primarily small manufacturers with limited investment capital and a narrow understanding of the actual and potential markets for their products.

The firms that have special programs to develop electric vehicles tend to be large firms with considerable market experience in the automobile industry or in automotive- and/or transportation-related industries. Their approach to the development of electric vehicles and participation in the electric vehicle industry is more systematic and follows a specialized technological path. They are more interested in battery, component, drive-train, motor, and instrumentation development. These efforts are not market-oriented, but instead, are research- and development-oriented.

Systematic evaluation of the existing and emerging electric vehicle industry in the United States indicates that the greatest inefficiency of the industry is its lack of consumer orientation. The entire electric vehicle industry needs to develop a comprehensive understanding of consumers' needs for electric vehicles and the behavior that is associated with these vehicles in those markets that are positively predisposed to adopt electric vehicles. Urban consumers constitute one market segment that is positively predisposed towards electric vehicles.

NEED FOR CONSUMER ORIENTATION

Today's consumers operate in a complex system of technological, economic, lifestyle, political, and legal forces. These forces shape the day-to-day lives of consumers. These forces also directly influence the type of consumption pattern individual consumers will develop as a function of their current income. Some consumers may feel that they are better off than others even though their income levels might be identical. This is because individual consumers can select among available purchase alternatives in such a way that they carefully develop the type of lifestyle that they find most comfortable.

Since the 1973 energy crisis, the number of environmental forces has increased and some of the forces have intensified so that they influence a greater number of consumers than ever before. The combination of inflationary pressures, disruptions in delivery of gasoline, deterioration of the physical environment, and unfavorable consumer sentiments regarding the purchase and consumption of durable goods in general superimposed on reduced speed limits, increased sales of small automobiles, limited sales of gasoline, and increasing cost of gasoline are drastically altering the personal mobility of individual consumers.

The emerging electric vehicle industry and, more specifically, the electric vehicles themselves, promise new personal transportation options for a broad spectrum of consumers. Potential consumers perceive electric vehicles in a number of ways. To some market segments, electric vehicles provide a more economical and stable mode of transportation within a given set of constraints and transportation missions. Other market segments are seeking new levels of independence from gasoline-powered automobiles. And finally, a significant market segment is looking for the opportunity to create a new lifestyle around the concept of an electric vehicle. Within the context of energy conservation and energy management, the introduction of new lifestyles is the most important preoccupation of consumers today. The electric vehicle industry must develop a sound understanding of the anticipated lifestyles and lifestyle alternatives that are sought by potential electric vehicle buyers.

Contemporary electric vehicles project to potential consumers a set of complex and highly interrelated attributes. Consumers perceive electric vehicles favorably, cautiously, and with an element of curiosity. Perceptions, attitudes, and preferences are the most important factors in positioning a product on the market. The electric vehicle industry must realize that these perceptions, attitudes, and preferences will quickly dissipate if the industry attempts to market inadequate, inferior, or unrealistic products. Also, false or misleading promotion, insufficient consumer services, unprofessional sales techniques on the retail level, or insufficient service facilities will produce strong negative behavioral factors for both electric vehicles and the entire electric vehicle industry.

The industry's need to understand the behavior of potential consumers for electric vehicles goes even further. Electric vehicles are durable products. Marketers are well aware of the fact that consumers do not buy durable products based only on their functional attributes. Consumers purchase a variety of services that the product will perform over a period of time. In the minds of consumers, the manufacturer's obligation to consumers extends beyond the product; it continues over the reasonable life span of the product. In order for the manufacturer to understand the services that consumers expect from the electric vehicle, the manufacturer must maintain a close relationship with existing and potential consumers.

This relationship can be developed in two ways. First, the manufacturer must communicate with existing and potential consumers through the standard channel of distribution or through the infrastructure. Second, the manufacturer must periodically conduct marketing research studies to determine not only the changes in the external environment that shapes the consumption process of individual consumers, but also the specific changes in consumers' perceptions, attitudes, and preferences. Careful management of both types of information will assist the manufacturer in the design of a better electric vehicle and placement of that vehicle in optimal market segments.

LIFESTYLES OF URBAN CONSUMERS

In socio-economic terms electric vehicles constitute a new technological product. It is a product that is socially desirable and theoretically can be successfully integrated into the social structure of society. Yet, from an examination of the adoption process for this type of product, it is apparent that the product needs a major stimulus before it is partially accepted by the adoptive group categories.

Introduction of technologically-new products frequently requires identification of the most compatible adoptive group. By matching the lifestyle attributes of the adoptive group with the attributes of the product, the product can be introduced in a relatively concentrated effort to produce high market visibility. In the case of electric vehicles, this requires a sound understanding of consumers' lifestyles, the constituency that formulates each lifestyle, and the physical environment in which each lifestyle is preferred.

Lifestyle is generally referred to as the distinctive mode of orientation an individual or a group has towards consumption, work, and play. Lifestyle segmentation is becoming increasingly more important for marketers; they are targeting their products to lifestyle groups and are studying new product opportunities arising out of lifestyle analysis. In the field of automotive marketing, Volkswagen introduced lifestyled automobiles: a car for a "good citizen" emphasizing economy, safety, and ecology; a car for a "car freak" emphasizing handling, maneuverability, and sportiness.

The constituency that formulates the lifestyle refers to the socio-economic orientation that the group represents in its aggregate form. The professional suburban communities, the rural farming communities, or young, highly educated urban condominium owners are such groups.

The physical environment in which a life-style is implemented is represented by the traffic patterns, availability of parking, and availability of public transportation among other factors such as the quality of air and climatic zones which contribute to the specification of a lifestyle.

The identification, specification, and interpretation of lifestyles, the constituency, and the physical environment can be accomplished through marketing and consumer research studies using the concept of lifestyle segmentation. These studies can be supplemented by environmental analysis and environmental impact studies. A number of studies conducted by both the private and the public sectors indicate that the urban consumer does represent a specific lifestyle.

A number of studies clearly indicate that the major cities in the United States are growing. The lifestyle patterns associated with transportation, commuting, and shopping are similar in each large metropolitan area. Each metropolitan area can be segmented according to socio-economic factors. Segments have different transportation requirements depending on their distance from the center of the metropolitan area. For example, the Washington, D. C. metropolitan area can be clearly segmented into several transportation preference zones. The inner zone includes a population segment that consists extensively of young professionals that tend to live in apartment buildings or condominiums and who rely heavily on the public transportation system. The ownership of private automobiles in this segment is significantly lower than in the second zone.

The second zone includes most of the suburban areas consisting of professional families having relatively higher incomes than the inner segment, but relying on automobiles for commuting, primarily due to the prestige factor of being assigned a parking space. This segment, in fact, practices two-car automobile ownership. One automobile is used for regularly scheduled commuting and the other is used for shopping, taking children to school, and other daily chores.

The third zone, the outside zone, is a segment of less affluent younger professionals that prefer to live in a detached house on the periphery of the metropolitan area and out of necessity are dependent on public transportation. Although many of these families own two automobiles, one automobile tends to be underutilized.

The three segments described above are found in a majority of the major metropolitan areas such as St. Louis, Chicago, New York, and Cleveland among others. These patterns tend to be somewhat different in California, but other factors contribute to the stimulation of specific lifestyles regarding electric vehicles.

IDENTIFICATION OF MISSIONS FOR ELECTRIC VEHICLES

Within the context of the first segment identified above, the following implications can be seen for electric vehicles. Since the lifestyles that exist in this segment rely more on public transportation systems than on a private automobile, the electric vehicle can be positioned as a limited-use share-ownership vehicle. This implies that large apartment complexes, condominiums, or cooperatives own a fleet of electric vehicles. The residents of these buildings then can rent these vehicles on a short-term basis to drive to the suburban shopping centers, drive to a party on the weekend, or other non-routine trips. This gives the residents the flexibility they lack in public transportation systems, but it also gives them independence from a fixed investment in a conventional automobile.

In the second segment the missions associated with the ownership of electric vehicles are different. The commuting trip performed by the individual members of this segment are routinely characterized by fixed distances, speeds, and time periods. Each driver covers a specific distance in a given time period daily. However, each driver is faced with inefficiencies produced by the conventional automobile while the automobile is standing in traffic, waiting at a traffic light, or waiting to leave the parking lot. In this lifestyle electric vehicles can reduce commuting costs by eliminating the inefficiencies of the internal combustion engine, and can improve the physical environment by not polluting the air.

This segment provides new opportunities for the electric vehicle industry to position the electric vehicle within the commuting public; it also provides an opportunity to develop a new concept of the electric vehicle service and recharging industry. In this segment the electric vehicles can be serviced, recharged, and maintained while they are parked in the on-the-job parking lot during the day. A special fleet of service trucks could be designed and equipped to service electric vehicles. These services could be offered by the local utilities, independent garages, or by electric vehicle dealers.

This concept would increase the efficiency of servicing electric vehicles in more concentrated locations, and at the same time would increase the efficiency of operations for the owner of the electric vehicle. The owner would not have to rely on service in his or her neighborhood.

In the third segment the mission defined for the electric vehicle is generated by the concept of replacing the second vehicle that tends to be underutilized. In this segment the second automobile in the family performs the routine functions of shopping, pick-up and deliveries, and other short-run driving responsibilities. A majority of these trips are inefficient and detrimental to the conventional internal combustion engine. Specially-designed electric vehicles that could successfully function in the suburban environment, i.e., vehicles that could carry two adults and two children with some capacity for grocery bags, could serve effectively.

The essential problem that faces the marketing specialist in all three market segments described above is the fact that these missions can be identified and systematically analyzed using simple demographic data. Secondary socio-economic data can also be useful in developing some understanding of the make-up of the family units and its preferred consumption function. Neither the demographic nor the socio-economic data can explain the fundamental behavioral factors that are needed to explain the actual motivations or personal reasons why the individual or the family unit would purchase an electric vehicle. This information has to be collected for each segment. In-depth consumer behavior studies are needed to develop a sound understanding of the individual segments.

INFORMATION NEEDS OF THE ELECTRIC VEHICLE INDUSTRY

The type of consumer information the electric vehicle industry needs can be classified into the following categories. The first category consists of information that helps develop the fundamental understanding of the environmental forces shaping the consumers' day-to-day operations. This includes purchasing behavior, driving habits, and consumers' understanding of the environmental changes that are taking place. This information can be collected and analyzed by information specialists that deal with broad aspects of societal changes.

The second type of information that is needed deals directly with identification of the different socio-economic and social groups that can be identified in various metropolitan areas. Once these groups are identified the next step is to identify the individual lifestyles that exist within these groups. There must be a clear understanding of the perception and the product fit of the electric vehicle into each individual lifestyle.

The third type of information that is needed concerns the individual members of the groups. In-depth study of the individual consumer's perceptions, attitudes, and preferences of electric vehicles must be undertaken.

Contemporary marketing research techniques offer techniques, procedures, and tools that can be used to generate all three types of information discussed above. Consumer panels, shopping center-based surveys, and mail and telephone surveys can be used to generate this type of information. Statistical techniques including factor analysis, discriminant analysis, conjoint analysis, and other multivariate techniques can generate the type of information that the electric vehicle industry needs to penetrate these markets.

AN OPTIMAL POTENTIAL MARKET FOR ELECTRIC VEHICLES

Based on trends both in the United States and abroad, it appears that the optimal market for electric vehicles is in large metropolitan areas. The market consists of two independent segments. One segment represents changes in the growth and restructuring of the cities. Many large cities are closing their downtown areas to traffic. These areas are being converted into shopping districts and malls. At the same time, it is becoming fashionable to live in the center of the city. Many old and historical buildings are being converted into expensive condominiums or apartment buildings. Residents of these buildings demand quiet streets free of commercial traffic. The merchants servicing these areas find it difficult to have merchandise delivered to their stores and therefore rely on small delivery vans to bring the merchandise from nearby warehouses.

The small delivery vans currently have small gasoline-powered or diesel-powered engines. These engines are noisy and produce pollution. In cities such as Munich, Copenhagen, Paris, and London attempts are being made to completely replace these vehicles with electric vehicles. This market will grow and create new opportunities for the entire electric vehicle industry.

The second segment of this market consists of the individuals who live in these areas of the city and do not have sufficient need for private large automobiles, but who do occasionally travel to other parts of the city. These downtown areas can be served by special firms that rent electric vehicles exclusively to the residents of these areas.

The concentration of electric vehicles in one central area in the center of the city would not only provide the opportunity to study the adoption process of electric vehicles, but would also provide the visibility that the electric vehicle industry needs today.

BIBLIOGRAPHY

The enclosed bibliography is designed to include material and sources that will help the electric vehicle industry better understand marketing; it also selectively highlights some of the environmental trends and developments that facilitate the integration of electric vehicles into the appropriate lifestyles of urban consumers.

Drucker, Peter F., "Technological Trends in the Twentieth Century," in Technology Management and Society Essays by Peter F. Drucker (New York: Harper & Row, Publishers, 1970), 55-73.

Effects in Washington, D. C. Area of 1979 Gasoline Shortages: Supplies Less Than National Average; Price Increases Comparable, Report to Senator John W. Warner of the United States by the Comptroller General, EMD-80-70, June 24, 1980.

Energy Conservation: An Expanding Program Needing More Direction, Report to the Secretary of Energy by the U.S. General Accounting Office, EMD-80-82, July 24, 1980.

"EV Territory: Downtown," Electric Vehicle Progress, 1, April 1, 1979, 2.

Frank, Ronald E., William F. Massy, and Yoram Wind, Market Segmentation (Englewood Cliffs, NJ: Prentice-Hall, Inc., 1972).

Hannay, N. Bruce and Robert E. McGinn, "The Anatomy of Modern Technology: Procegomenon to an Improved Public Policy for the Social Management of Technology," Daedalus, 109 (Winter, 1980), 25-53.

Hisrich, Robert D. and Michael P. Peters, Marketing A New Product: Its Planning, Development, and Control (Menlo Park, CA: The Benjamin/Cummings Publishing Company, Inc., 1978).

Kollat, David T., Roger D. Blackwell, and James F. Robeson, Strategic Marketing, (New York: Holt, Rinehart and Winston, Inc., 1972).

Korff, Walter H., Designing Tomorrow's Cars (Burbank, CA: M-C Publications, 1980).

Kotler, Philip, Marketing Management: Analysis, Planning and Control, 4th Ed. (Englewood Cliffs, NJ: Prentice-Hall, Inc., 1980).

Linder, S.B., The Harried Leisure Class (New York: Columbia University Press, 1970).

Lovelock, Christopher H., "A Market Segmentation Approach to Transit Planning, Modeling, and Management," in Marketing in Private and Public Nonprofit Organizations, Edited by Ralph M. Gaedeke (Santa Monica, CA: Goodyear Publishing Company, Inc., 1977), 125-142.

Mahajan, Vijay and Eitan Muller, "Innovation Diffusion and New Product Growth Models in Marketing," Journal of Marketing, 43 (Fall, 1979), 55-68.

Merklein, H.A. and William P. Murchison, Jr., Those Gasoline Lines and How They Got There (Dallas: The Fisher Institute).

Naidu, G.M., George Tesar, and Gerald G. Udell, The Electric Car: An Alternative to the Internal Combustion Engine (Acton, MA: Publishing Sciences Group, Inc., 1974).

"Research and Development of Electric Vehicles in Japan," Agency of Industrial Science and Technology, Ministry of International Trade and Industry in cooperation with the Society of Automobile Engineers of Japan, Inc., 1977.

Safdie, Moshe, "Beyond the City Limits," Saturday Review/World, August 24, 1974, 54-57ff.

Sassin, Wolfgang, "Energy," Scientific American, 243 (September, 1980), 118-132.

Special Issue on Cities, see Scientific American, 213 (September, 1965).

St. John, Harry M., "The Energy Market for High Technology Companies," Journal of Marketing, 42 (October, 1978), 46-53.

Tesar, George, "Marketing/Buyer Profile," EVC Expo 80 Conference Proceedings, Electric Vehicle Council, 1980, EVC No. 8009.

Tesar, George, Paul Hugstad, Harvey J. Iglarsh, "The Role of Empirical Research in Government Decision-Making and Public Policy Formulation," Proceedings and Abstracts, American Institute for Decision Sciences, 7th Annual Meeting, Western Regional Conference, San Diego, March 2-3, 1978, 289-91.

Wind, Yoram, "Issues and Advances in Segmentation Research," Journal of Marketing Research, XV (August, 1978), 317-37.

NEAR-TERM BATTERIES FOR ELECTRIC VEHICLES

C. C. Christianson, N. P. Yao, F. Hornstra
Argonne National Laboratory

INTRODUCTION

Electric vehicles offer the promise of reducing the nations petroleum consumption since they use electrical energy and more than 85% of electrical energy generated by utilities is based on non-petroleum energy sources such as coal, nuclear, or hydroelectric. In recognition of this fact Congress passed Public Law 94-413 (Electric and Hybrid Vehicle Research, Development and Demonstration Act) in 1976, later amended by Public Law 95-238 in 1978. The objective of this law is to stimulate introduction of electric vehicles into the nations vehicle fleet. The law provides for the research and development of improved batteries, propulsion systems and electric vehicles; demonstration of electric vehicles in appropriate applications; and economic incentives to stimulate electric vehicle commercialization. The law is being implemented by the Department of Energy (DOE) through the Electric and Hybrid Vehicle (EHV) Division of the Transportation Program Office.

One of the five projects of the EHV Division is the Electric Vehicle Project which has a goal of stimulating industry commercialization in 1986 of competitive electric commuter vehicle capable of a 100 mile urban range in stop-and-go driving. In contrast, the urban range of presently available state-of-the-art electric vehicles is limited to about 50 miles. Improved batteries will be required in order to meet this goal. In order to meet this need, DOE assigned to Argonne National Laboratory the project management responsibility for the Near-Term Electric Vehicle Battery Project. This paper presents the battery performance and cost goals associated with achieving a competitive 100 mile urban range electric commuter vehicle and discusses the project approach and the results obtained to date.

NEAR-TERM ELECTRIC VEHICLE (EV) BATTERY GOALS

Improvements in battery, propulsion system and vehicle design are required to achieve the 100 mile range goal. Under contract from DOE, General Electric and Chrysler recently designed and fabricated an improved four-passenger Electric Test Vehicle (ETV-1) with low road and aerodynamic energy losses and an efficient propulsion system. The power and energy requirements of this vehicle have been used to establish the performance and cost goals for the batteries being developed under the Near-Term EV Battery Project. Table 1 presents the key vehicle performance and cost goals and the associated battery goals. State-of-the-art lead-acid batteries presently used in electric vehicles display a specific energy of 30 watt-hours/kg, a specific power for 30 seconds of 70 watts/kg (at 50% depth of discharge), a cycle life of about 250 cycles, and an OEM cost of about $55/KWH. Thus, the near-term EV battery goals require major performance improvements, particularly in specific energy and life.

Table 1

Key Performance and Cost Development Goals for Near-Term Electric Vehicle Batteries

Battery Parameters	Development Goals	Expected Vehicle Performance*
Specific Energy	56 W-hr/kg (C/3 Rate)	100 miles/charge (SAE 227aD Urban Driving)
Specific Power (at 50% DOD)	104 W/kg (30 sec. peak)	Accleration 0 to 30 mph in 8 sec.
Cycle Life	800 cycles (to 80% DOD)	>60,000 miles (life)
OEM Price	70 $/KW-hr	~$2000 battery cost

*Based on DOE/ETV-1 Vehicle with Regenerative Braking.

NEAR-TERM BATTERY PROJECT STRATEGY AND APPROACH

The project strategy is to support industry-based research and development of those battery types and contractor development approaches which have a clear potential of meeting the project goals. This is accomplished through cost-sharing R&D contracts with private battery developers. Initial R&D contracts are approximately three years in duration. Support after this initial period will be provided to only those contractors demonstrating the most promising results. Most R&D contracts were initiated in 1978 and thus are nearing the end of the initial technology development phase.

The three battery candidates receiving major R&D emphasis by ANL during this initial technology development phase are lead-acid, nickel/iron and nickel/zinc types. Eight R&D contracts are presently underway. These are listed in table 2 along with the contract periods and contract values. A ninth contract, also shown in table 2, was terminated in 1979 by mutual agreement between ANL and the battery developer.

Cells/modules and batteries developed by the contractors are periodically delivered to the ANL National Battery Test Laboratory (NBTL) for technology verification testing. This modern computer controlled facility performs a broad spectrum of standarized performance and life test thus accomplishing an independent and equitable evaluation of each contractors battery technology.

Table 2

Near-Term Electric Vehicle Battery R&D Contractors

Battery	Contractors	Contract Period	Contract Value[a] ($000)
Lead-Acid	Eltra	04/78-05/81	2,580
	Exide	02/78-01/82	3,386
	Globe	03/78-09/81	4,633
Nickel/Iron	Eagle-Picher/SU	03/78-09/81	3,945
	Westinghouse	12/77-09/81	3,022
Nickel/Zinc	Exide	04/79-01/82	1,197
	Energy Research Corp.	02/78-09/81	3,608
	Gould	03/78-09/81	8,105
	Yardney	07/77-05/79[b]	1,000

a - Includes Contractors Share
b - Terminated by Mutual Agreement

LEAD-ACID BATTERY DEVELOPMENT APPROACHES

In the case of lead-acid batteries the greatest performance improvement is needed in specific energy although improvement is required in all of the performance areas. Athough the contractual goals are the same, each of the three lead-acid battery R&D contractors is utilizing a different technical approach to achieve the common contractual goals. Key aspects of each contractors approach are discussed in the following paragraphs.

Exide is using factorial tests to evaluate various combinations of design variables in order to determine which combination best provides the desired performance and life. These variables include positive electrodes of both the flat plate and tubular plate design as well as other cell construction and electrolyte concentration variations. The use of lead-plastic and lead-copper composite grids for the negative plate are also being investigated in order to reduce lead weight.

Eltra is stressing a low cost approach based upon using an expanded metal grid structure rather than cast grids for both the positive and negative grids. They have installed a pilot line facility for producing the expanded metal grids and manufacturing the positive and negative plates. Eltra is also evaluating use of lead-calcium alloy grid material in addition to the conventional lead-antimony alloy. Lead-calcuim alloys display lower self-discharge rates, reduced water consumption and no emission of the toxic stibine and arsine gasses during charging.

Globe's approach emphasizes use of a sophisticated proprietary computer program to optimize the plate size and aspect ratio, number of plates, grid configuration, and acid concentration to achieve the desired performance. In addition, they have developed an innovative intracell electrolyte mixing pump and system which increases both the specific energy and cycle life. They also are developing a lead-plastic composite negative grid and the associated manufacturing process equipment in order to reduce the amount of lead required.

NICKEL/IRON BATTERY DEVELOPMENT APPROACHES

Development emphasis in the case of nickel/iron batteries is on cost reduction with lesser emphasis given to improvements in specific energy. In this case also the two contractors are using different technical approaches to achieve the same goals. Key aspects of each contractors approach are presented in the following paragraphs.

Eagle-Picher is stressing a high performance approach which utilizes sintered nickel positive electrodes, and negative iron electrodes developed by the Swedish National Development Company. Cost reduction is being achieved through a use of fewer but thicker electrodes. Eagle-Picher has developed techniques for making nickel electrodes up to 3 millimeters thick using the continuous wet-slurry method. Previously, electrodes made by this method were limited to a thickness of 1 millimeter or less. The Swedish National iron electrode utilizes a sintered iron plaque. It is a high performance electrode which overcomes the poor low temperature performance previously characteristic of the nickel/iron battery.

Westinghouse is emphasizing the development of a battery having low initial cost. Thus, their objective is to limit the amount of nickel to essentially that required for the active material. Consequently, Westinghouse is using sintered steel wool substrates for both the positive and negative electrodes. In the case of the positive electrode, the steel wool is nickel plated to avoid undesirable side reactions. In addition, Westinghouse is using a central electrolyte management system. This system utilizes a common electrolyte reservoir for all cells and circulates the electrolyte through the cells. The system also provides for the central collection and ventilation of the hydrogen gas produced by the battery and thus minimizes the associated safety hazard.

NICKEL/ZINC BATTERY DEVELOPMENT APPROACHES

In the case of the nickel/zinc batteries, the major development emphasis is being placed on improving the life cycle cost. In this case also, the three contractors involved are utilizing different technical approaches to achieve the same goals. Key aspects of each of the contractors approaches is presented in the following paragraphs.

Gould is stressing a high-performance approach which utilizing sintered nickel electrode substrates and sintered zinc electrodes. Much emphasis is being given to the control of zinc electrode shape change and densification and zinc dendrite growth since these lead to limited cycle life. Development efforts are concentrating on the development of improved co-polymer membrane-type separators having the desired performance and life characteristics plus the use of additives which alter the characteristics of zinc dissolution and deposition in such a way as to suppress dendrite formation and shape change.

The Energy Research Corporation is emphasizing a low-cost approach directed at minimizing the total nickel requirement. Thus, low-cost, plastic bonded nickel electrodes are being developed which utilize graphite as a current collector. They are also developing improved separators and investigating the use of additives to reduce zinc electrode shape change and dendrite growth.

Exide is surmounting the life difficulties by use of a unique cell design where the zinc electrode is vibrated during charging. This eliminates the zinc electrode shape change and dendrite problems and provides a cell having a long life, but at the sacrifice of reduced specific energy and volumetric energy density. Thus, Exide is concentrating on the development of design changes which will increase the specific energy of their battery. These design changes include a reduction of the inter-electrode spacing and the use of electrodes of minimum thickness.

PRESENT TECHNICAL STATUS

Considerable technical progress has been made since the inception of the R&D programs in 1978. The present status of the technology is illustrated in figure 1 along with the project goals. The data presented represents performance of three to six cell modules. In all cases the data presented in figure 1 are the results of tests at NBTL. The variation between contractors is shown by the range indicated on each bar. In the case of cycle life, contractor testing significantly leads the NBTL testing and thus contractor life results are even better than those shown in Figure 1. For example, a cycle life of 800 cycles has been measured by one lead-acid battery contractor and greater than 600 cycles has been obtained by one nickel/iron battery contractor. Comparable life tests are underway at NBTL and preliminary indications are that the contractor data will be verified. Figure 1 also shows the urban range expected from an ETV-1 vehicle powered by 488 Kg of the indicated batteries. These range predictions are based on tests of the batteries in NBTL using the simulated power profile of the ETV-1 vehicle with regenerative braking.

As indicated previously, the state-of-the-art lead-acid battery presently being utilized in electric vehicles has a specific energy of 30 wh/kg, a specific power of about 70 w/kg, a cycle life of about 250 cycles, and would provide a range of about 51 miles in the ETV-1 vehicle. Thus, a considerable performance and life improvement has been achieved by the near-term batteries with the exception of the cycle life of nickel/zinc batteries which presently does not exceed 200 cycles for full-size modules. However, much progress has been made in the area of nickel/zinc battery cycle life in small laboratory cells, and it is expected that these improvements will be demonstrated in full-size modules in the near future.

PLANNED FUTURE PROJECT ACTIVITIES

Battery testing in the future is being expanded to include tests of full-scale batteries in the laboratory and in electric vehicles. The purpose of such tests will be to demonstrate the capability of

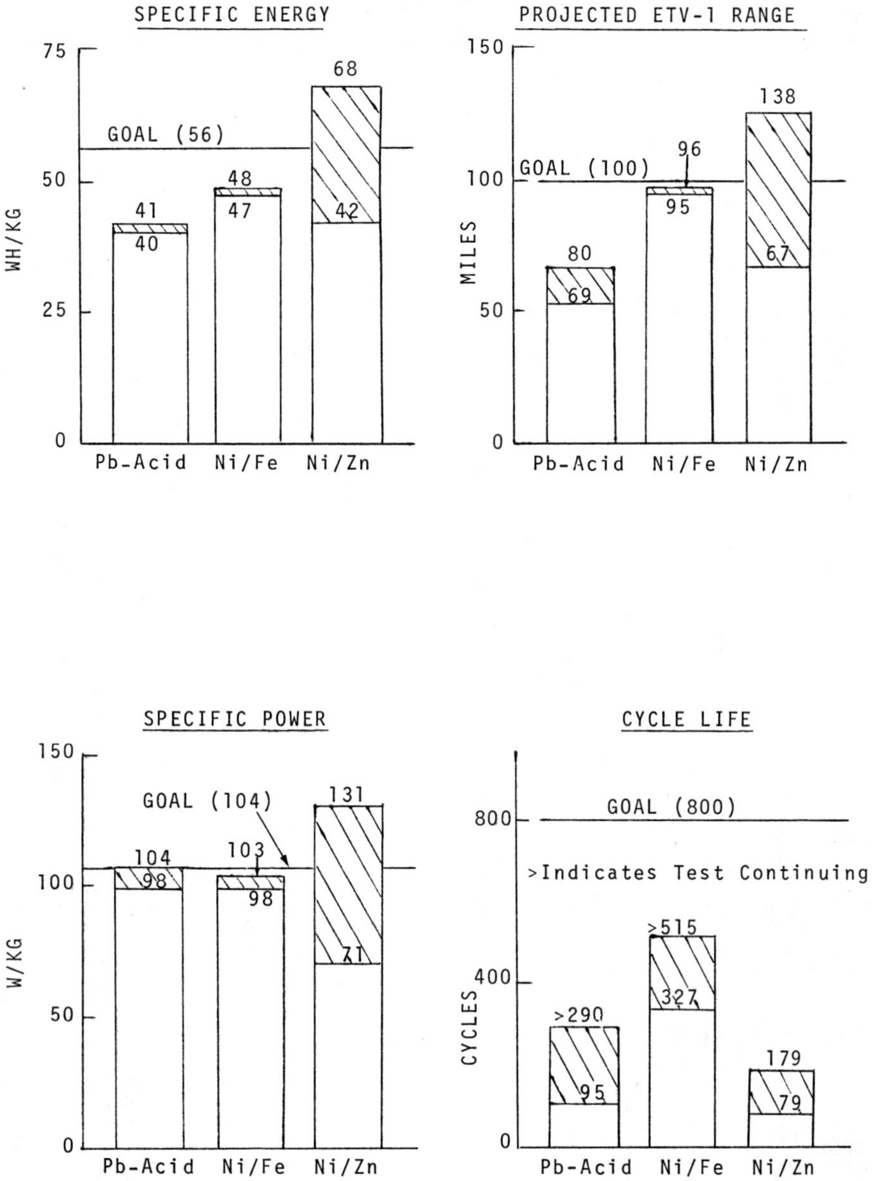

FIGURE 1: PERFORMANCE AND LIFE OF IMPROVED ELECTRIC VEHICLE BATTERIES AS MEASURED AT THE ANL NATIONAL BATTERY TEST LABORATORY AS OF JANUARY 31, 1981

these improved batteries in electric vehicles and to determine any design aspects needing further improvements before the battery is placed into full-scale production.

In September of 1981 the number of contractors per battery type will be reduced. This will permit the limited available funds to be concentrated with those contractors demonstrating the most promising technology and commercialization prospects. Also, the future program will give more emphasis to demonstrating life and reducing cost. Pilot production processes will be developed and batteries will be produced for field tests.

SUMMARY

Major progress has been achieved in the lead-acid, nickel/iron and nickel/zinc battery technology development since the initiation of the Near-Term EV Battery Project in 1978. Against the specific energy goal of 56 wh/kg the demonstrated specific energies are 41 wh/kg for the improved lead-acid batteries, 48 wh/kg for the improved nickel/iron batteries, and 68 wh/kg for the improved nickel/zinc batteries. These specific energy values would allow an ETV-1 vehicle to have an urban range of 80 miles in the case of the improved lead-acid batteries, 96 miles for the improved nickel/zinc batteries, and 138 miles for the improved lead-acid batteries. All represent a significant improvement over the state-of-the-art lead-acid battery capability of about 30 wh/kg with approximately a 51 mile urban range for the ETV-1 vehicle. The project goal for specific power of 104 w/kg for 30 seconds at a 50% depth of discharge has been achieved for all of the technologies with the improved lead-acid demonstrating 111 w/kg, the improved nickel/iron demonstrating 103 w/kg and the improved nickel/zinc demonstrating 131 w/kg. Again this is a significant improvement over the state-of-the-art lead-acid battery capability of 70 w/kg. Substantial progress has been made against the life cycle goal of 800 cycles as evidenced by the demonstrated lead-acid battery achievement of >295 cycles in ongoing tests, the nickel/iron demonstrated capability of >515 cycles in ongoing tests, and the nickel/zinc demonstrated capability of 179 cycles. Except for the nickel/zinc batteries, the demonstrated cycle life is better than the state-of-the-art lead-acid battery cycle life of about 250 cycles. Future program emphases will be on improving cycle life and further reductions in cost.

Based on the continued technical progress expected in the future, it can be expected that these near-term batteries will play a major role in stimulating the broad application of electric vehicles in the nations vehicle fleet.

ACKNOWLEDGEMENT

This work was supported by the Department of Energy, Division of Electrochemical Energy Storage Systems and the Division of Electric and Hybrid Vehicles. DOE's program manager is Mr. William Webster (DOE/EESS).

REFERENCES

1. N. P. Yao, C. C. Christianson, R. C. Elliott, T. S. Lee, and J. F. Miller, "DOE's Near-Term Electric Vehicle Battery Program--Status of Improved Lead-Acid, Nickel/Iron and Nickel/Zinc Battery Developments", Proc. EVC Expo '80 Conf., St. Louis, May 20-22, 1980, Paper EVC No. 8029, Electric Vehicle Council, Washington, DC, 1980.

2. F. Hornstra, et al., "Results of Simulated Driving Profiles in the Testing of Near-Term Electric Vehicle Batteries", Proc. EVC Expo

'80 Conf., St. Louis, May 20-22, 1980, Paper EVC No. 8034, Electric Vehicle Council, Washington, DC, 1980.

3. F. Hornstra, et al., "Test Results of Computer-Generated Simulated Driving Profiles Applied to Near-Term Batteries", Proc. 30th Annual Conf., IEEE Vehicular Technology Soc. (Convergence 80), Dearborn, MI, September 15-17, 1980, Paper No. C-4 2, IEEE, New York, 1980.

4. Office for Electrochemical Project Management, Argonne National Laboratory, "Development of Near-Term Batteries for Electric Vehicles Summary Report, October 1977-September 1979", Argonne Report ANL/OEPM-80-5, June 1980.

5. Eltra - C&D Batteries Division, "Annual Report for 1979 on Research, Development and Demonstration of Lead-Acid Batteries for Electric Vehicle Propulsion", Argonne Report ANL/OEPM-79-7, June, 1980.

6. Exide Technology and Management Co., "Annual Report for 1979 on Research, Development and Demonstration of Lead-Acid Batteries for Electric Vehicle Propulsion", Argonne Report ANL/OEPM-79-8, June, 1980.

7. Globe-Union Inc., "Annual Report for 1979 on Research, Development and Demonstration of Lead-Acid Batteries for Electric Vehicle Propulsion", Argonne Report ANL/OEPM-79-9, June 1980.

8. Energy Research Corporation, "Annual Report for 1979 on Research Development and Demonstration of Nickel/Zinc Batteries for Electric Vehicle Propulsion", Argonne Report ANL/OEPM-79-10, June, 1980.

9. Gould Inc., "Annual Report for 1979 on Research, Development and Demonstration of Nickel/Zinc Batteries for Electric Vehicle Propulsion", Argonne Report ANL/OEPM-79-11, June, 1980.

10. Exide Technology and Management Co., "Annual Report for 1979 on Research, Development and Demonstration of Nickel/Zinc Batteries for Electric Vehicle Propulsion", Argonne Report ANL/OEPM-79-12, June, 1980.

11. Eagle-Picher Industries, Inc., "Annual Report for 1979 on Research, Development and Demonstration of Nickel/Iron Batteries for Electric Vehicle Propulsion", Argonne Report ANL/OEPM-79-13, June, 1980.

12. Westinghouse Electric Corporation, "Annual Report for 1979 on Research, Development and Demonstration of Nickel/Iron Batteries for Electric Vehicle Propulsion", Argonne Report ANL/OEPM-79-14, June, 1980.

HISTORY AND OVERVIEW OF STIRLING ENGINES

Worth H. Percival
United Stirling, Inc.

INTRODUCTION

THE STIRLING ENGINE

The Stirling engine is an external combustion heat engine operating on a closed, thermally regenerative, cycle consisting, ideally, of two constant volume and two constant temperature processes, shown in figure 1. It was invented by the Stirling brothers in Scotland in 1816, and often termed the hot-air or hot-gas engine. The historical background is presented in the last section. The modern Stirling engine was introduced by the Philips Research Laboratories of Eindhoven, The Netherlands, in their publications in 1946, the work having started in the late 1930's. NV Philips has licensed several companies since 1958 to join the Stirling development program, to share technical information and refine the computer based cycle analyses. These companies have included General Motors, Ford Motor, MAN-MWM, and United Stirling. United Stirling of Sweden is now the most active licensee.

The major incentives for developing the Stirling engine include its high thermal efficiency -- 35 to 40 percent -- combined with very low exhaust emissions; low noise and vibration level -- less than 85 dB(A) at one meter; wide heat source flexibility -- gas, liquid and solid fuel capability in addition to solar energy; zero oil consumption; a wide speed range and fast response.

Today's Stirling engines are external combustion, closed cycle, piston machines; but they do not operate on the ideal cycle because of continuous piston motion and because compression and expansion involves a blending of adiabatic and constant temperature processes. This results in a real P-V diagram shown in figure 2. Like all heat engines, the Stirling produces power by compressing the working

Figure 1.

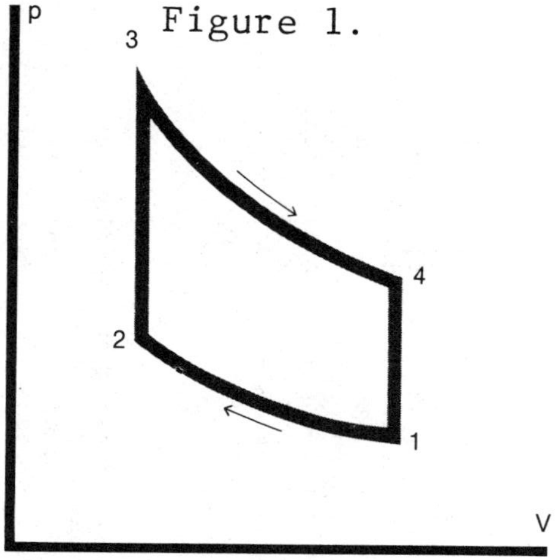

Theoretical pressure/volume curve

1—2 Compression
Work is supplied by compressing the working gas on the cold side; the gas is cooled at low pressure.

2—3 Displacement
The gas is moved from the cold to the hot side at constant volume. The regenerator gives off stored heat. Pressure increases.

3—4 Expansion
Work is performed when the working gas expands on the hot side while it is heated at high pressure.

4—1 Displacement
The gas moves from the hot to the cold side at constant volume. Heat is stored in the regenerator. Pressure declines.

Figure 2.

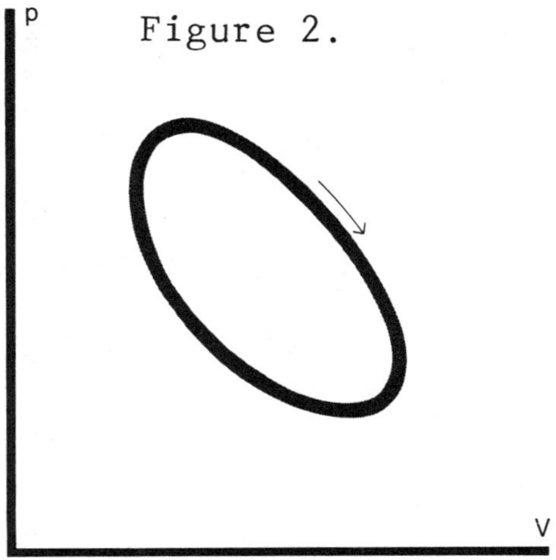

Actual pressure/volume curve

fluid, a gas, at low temperature and expanding it at high temperature. During each cycle, heat is supplied at a nearly constant high temperature and rejected at a nearly constant low temperature. Gas temperature is changed during the cycle by displacement of the gas from a hot space through a regenerative heat exchanger to a cold space, and vice-versa. An important distinction between the Stirling and other historic hot air engines is the method of controlling gas flow by volume changes (displacement) rather than by valves. Engines having valves to control flow were usually termed Ericsson engines. Modern Stirling engines developed by Philips and United Stirling operate without valve gear. Ideal cycle efficiency equals the Carnot.

The Stirling is different from the familiar gasoline or diesel engine in two respects; first, it uses the same working fluid over and over again, and second, the heating and cooling takes place by heat transfer through metal walls, and from a continuous rather than an intermittent combustion process. The Stirling differs from a steam (Rankine cycle) engine in that the working fluid does not change phase; it is always a gas, usually hydrogen or helium.

The mechanical arrangement of modern Stirling engines can be divided into two classes, single-acting (displacer) and double-acting machines.* The displacer type requires 2 pistons for each power unit, as in the historic engines; the double-acting type requires only 1 piston per cylinder, which functions as a compressor, expander and displacer by interaction between neighboring cylinders. From 3 to 7 cylinders are necessary for one power unit, the most common being 4 arranged either in-line or axially in a square.

For most applications, the engine can be considered to have two distinct fluid systems. The internal system is filled with working gas at an elevated pressure, and is comprised of two variable volumes and three heat exchangers, called the heater, cooler and regenerator. The external system supplies heat to the engine heater, usually by continuous combustion of liquid fuel, and recovers most of the heat in the exhaust products by means of an air preheater.

The Stirling thermodynamic process can operate from any source of sufficient heat at the proper temperature. Normally, heat from the combustion of gasoline or a distillate fuel is used; but gas, liquid or solid fuels as well as solar heat, stored heat, isotope heat or a nuclear reactor heat source might be used. Because of the rapid rise in the price of liquid fuels, there is a growing interest in operating the Stirling engine on solid fuels -- coal, forest and agricultural residues, municipal and industrial refuse, etc.. However, practical experience with such fuels is limited. The Stirling engine has received much attention recently from government agencies and private organizations for its near term cost-effectiveness as a solar energy converter.

UNITED STIRLING

United Stirling (USS) is a Swedish research and development company, established in 1968 for the purpose of developing the Stirling cycle engine -- to realize its potential as a reliable, economical and environmentally acceptable means of converting energy.

*References 1 and 2 thoroughly describe the numerous design variants of Stirling engines.

United Stirling employees number 135, 90% of whom are engineers and technicians.

In 1979, a wholly owned subsidiary of USS was formed, United Stirling, Inc., now located in Alexandria, Va., for the purpose of aiding the Stirling technology exchange between Sweden and the United States.

United Stirling's large, modern headquarters, comprising over 5,500 square meters of laboratory space, is located at the Kockums Shipyard, Malmo, Sweden. It is equipped with advanced technical facilities including 18 test cells, a cold room, electronics laboratory, assembly rooms and a work shop. The design philosophy at USS is to develop components and accessories in separate test rigs to a certain confidence level, followed by integration of them with the basic engine for testing and evaluation. Much field experience has been gained over the years by testing engines in two boats, a city bus, a truck, 5 passenger cars, a generator set and a heat pump installation. A solar thermal engine, described later, is expected to be tested in a parabolic concentrator this coming summer.

In 1972, the decision was made to concentrate on double-acting rather than single-acting (displacer) engines. The double-acting principle has resulted in a lighter, more compact engine of reduced complexity and lower cost. The drive mechanism consisted of a single crankshaft V-4 arrangement. Figures 3 and 4 illustrate the double-acting arrangement and the V design.

The first V-4 engine successfully ran in December, 1971. Over the next two years eleven engines of this type, with progressive improvements, were designed, built and tested. Accumulated operating time totaled about 6000 hours.

THE USS 4-95 ENGINE

In 1975, a new double-acting 40 kW engine was designed and first tested in 1976. It was originally termed the P40 but more recently designated the 4-95, having a displacement of 95 cc/cylinder. The design objective was to achieve a reliable experimental engine for the development of specific components such as the heater head (the high temperature heat exchanger receiving heat from an external source), piston rod seals, piston rings and control systems. This called for a concept with parallel cylinders placed in a square, a heater head with rotational symmetry, and a twin crank shaft drive unit. This configuration was considered as having greater potential for manufacturing than the V design. The 4-95 cross-section is shown in figure 5, and the engine on a dynamometer is shown in figure 6.

The engine is structurally built up from three main assemblies, the drive unit, block and heater head. It is possible to split the engine between block and drive unit without disassembling the heater head. This option reduces the time needed for assembling and disassembling in conjunction with modifications and servicing.

Twenty one 4-95 engines have been built for in-house use as well as for special testing by government agencies and private organizations in the United States, Britain and West Germany. The 4-95 is playing a key role as a baseline engine in the DOE/NASA Automotive Stirling Engine program. Three passenger cars, so far, have been operating with the 4-95 engine.

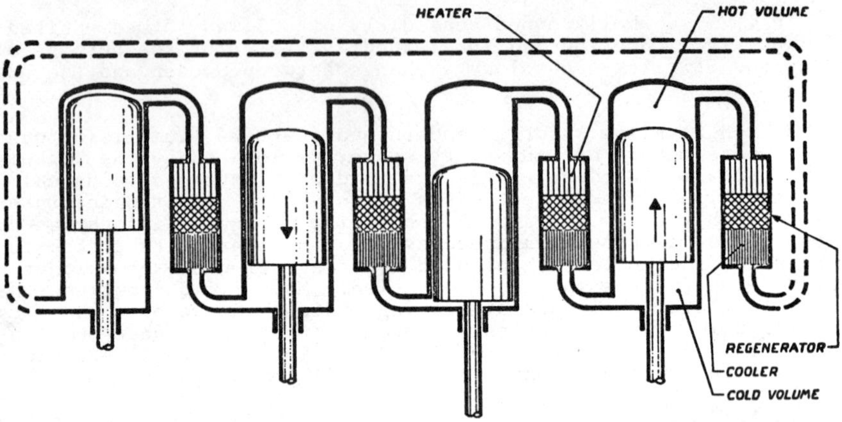

Figure 3. Double acting Stirling engine.

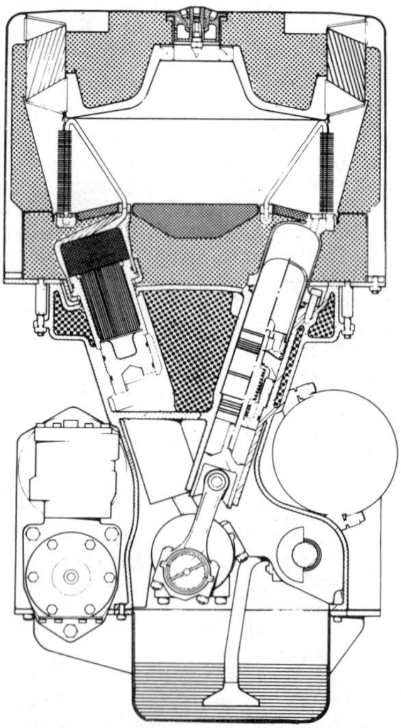

Figure 4. Four cylinder V configuration.

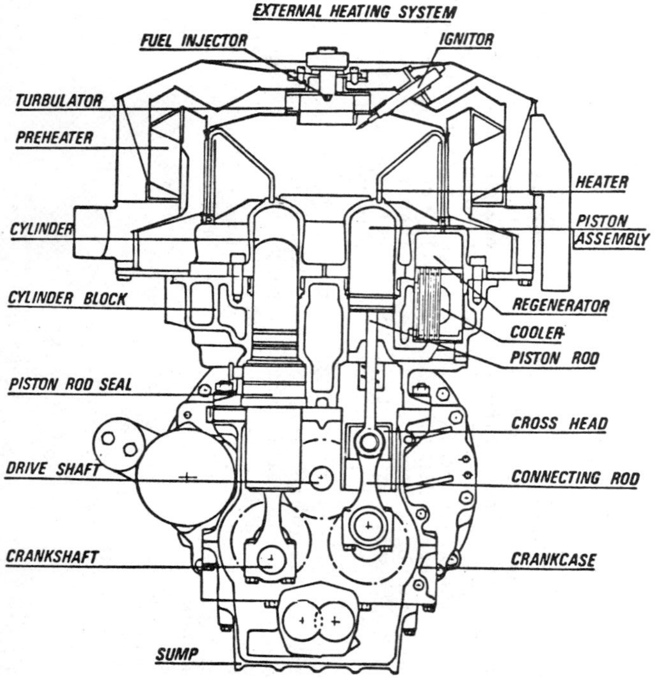

Figure 5. Four cylinder twin-crank model 4-95 engine.

Figure 6. Model 4-95 engine on dynamometer.

Several conceptual and design features give the engine a potential for long life between overhauls. Such unique features include:

- absence of sharp pressure impulses within cylinders;
- inherent low linear and torsional vibration;
- absence of valve gear;
- lubrication system operates in non-contaminating atmosphere;
- piston rings and seals operate in cool region; and
- cross head design eliminates side forces on piston assembly.

As of December 1980, total test time for all 4-95 engines on dynamometers and in demonstration programs exceeded 13,000 hours. One engine operating on a special high temperature (820 °C) endurance cycle has been running over 5600 hours. The critical piston rod seal, known as the new PL design, has achieved approximately 120,000 hours of successful running on all seal units, with one seal exceeding 7000 hours without failure. Additionally, about 150,000 hours of separate component and accessory testing contribute to overall reliability of the 4-95.

The ultimate objective of the United Stirling development program is to achieve volume production of Stirling engines. The program emphasis has two phases -- the DOE Stirling automotive development which is aimed at achieving a 30% improvement in economy over the spark ignition engine by 1984 -- and the internal United Stirling commercial development program. The latter is aimed at solar thermal power systems, gas fired heat pump and cogeneration systems, biomass combustion systems, medium duty truck and bus applications, and underwater propulsion systems.

SOLAR AND BIOMASS EXPERIENCE

During 1980 United Stirling has been under contract from the Jet Propulsion Laboratory, for the modification of one of their test engines, for operating as a solar power plant in a parabolic dish concentrator. The engine with its receiver (solar heat exchanger), alternator and control system is to be installed on a concentrator located at the JPL Parabolic Dish Test Site at Edwards, California, in June 1981. The engine will incorporate a solar, direct-coupled hybrid receiver, shown in figure 7. It will permit the engine to operate at constant load on either a "solar only" mode, or with a natural gas burner in a "combustion mode" during cloud cover or at night.

The objective of the program is to demonstrate that the Stirling engine is a practical, efficient and reliable energy converter when integrated with a parabolic dish concentrator, and that it has the potential of being cost competitive with fossil fueled electric generating systems of today. Predicted performance of the solar engine is shown in Table I.

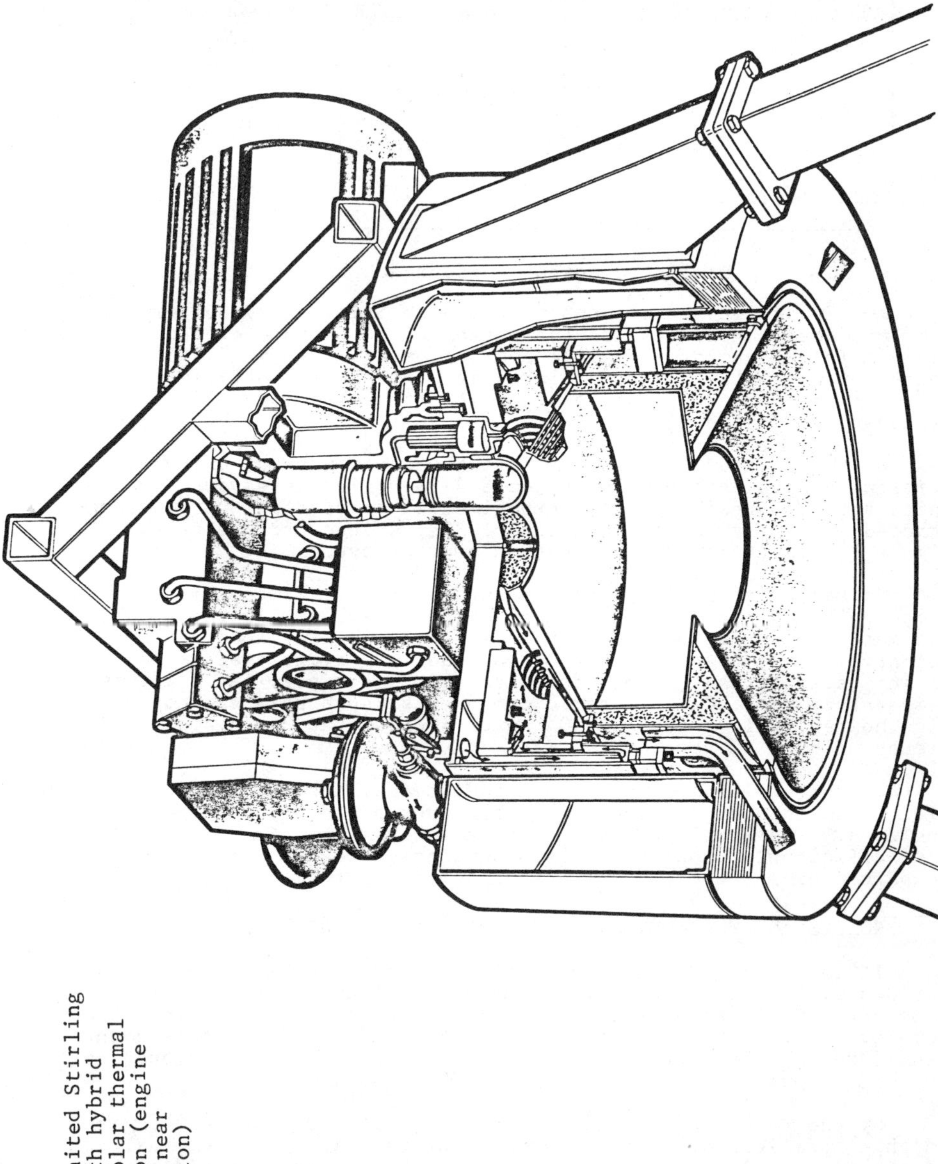

Figure 7. A United Stirling 4-95 engine with hybrid receiver for solar thermal power generation (engine operates in or near inverted position)

COOLANT TEMP. 50°C MEAN PRESSURE 15 MPa	NOMINAL OUTER TUBE WALL TEMP. 710°C		NOMINAL OUTER TUBE WALL TEMP. 810°C	
	He	H_2	He	H_2
Max. power, kW	24	26	27	28
Max. efficiency, %	36	38	40	41

TABLE I. Predicted engine performance in a solar application.

Recent studies have shown that a Dish/Stirling system employing mass produced components has the potential to produce electricity for 50-70 mils/kWh and at a capital cost of under $1000/kW (2)(3)(4). Contributing to this is the relatively high thermal efficiency of the Stirling, in excess of 35%, and its projected low selling price of under $60 per kW of peak power.

In the area of biomass and solid fuel combustion, United Stirling has successfully operated a 4-95 engine on sawdust, wood chips and corn residues. The experimental combustion system demonstrated extremely low content of CO, NO_x, HC and particulates. Figure 8 illustrates the combustion control system.

United Stirling has a continuing program for improvement of components and accessories for all engine designs. In the area of new materials, the introduction of a ceramic heater head has the potential for substantially reducing the life cycle cost of the engine, as well as the need for strategic materials. At a working temperature of 1100 °C, ceramic components, such as a silicon carbide heater, will produce a 50% power increase and a thermal efficiency of over 45 percent.

In some solar applications a sodium cooled solar receiver/heater will be advantageous, especially when thermal energy storage is included. The Stirling engine with a sodium heater head operates more efficiently (approximately 3 percentage points) since the heater tubes can be shorter and temperatures more uniform.

AIR ENGINES AND THERMODYNAMICS IN THE 19th CENTURY

It is believed that 19th century inventors were motivated to build air engines because of a wide spread fear of steam boiler explosions. Early statistics are lacking, but from 1862 to 1879 there were over 10,000 boiler explosions in England, while in this country from 1880 to 1919 there were 14,281 recorded boiler explosions, resulting in over 10,000 deaths and 17,000 injuries.

By the term "air engine" is meant a reciprocating heat engine which compresses cool air and expands heated air. The internal combustion engine is excluded because historic air engines did not employ combustion directly in the cylinder. Gases other than air were sometimes used, but the important distinction is that the gases did not change phase, as in steam powerplants. Air engines, generally, have been divided into three types: Type I were open cycle machines

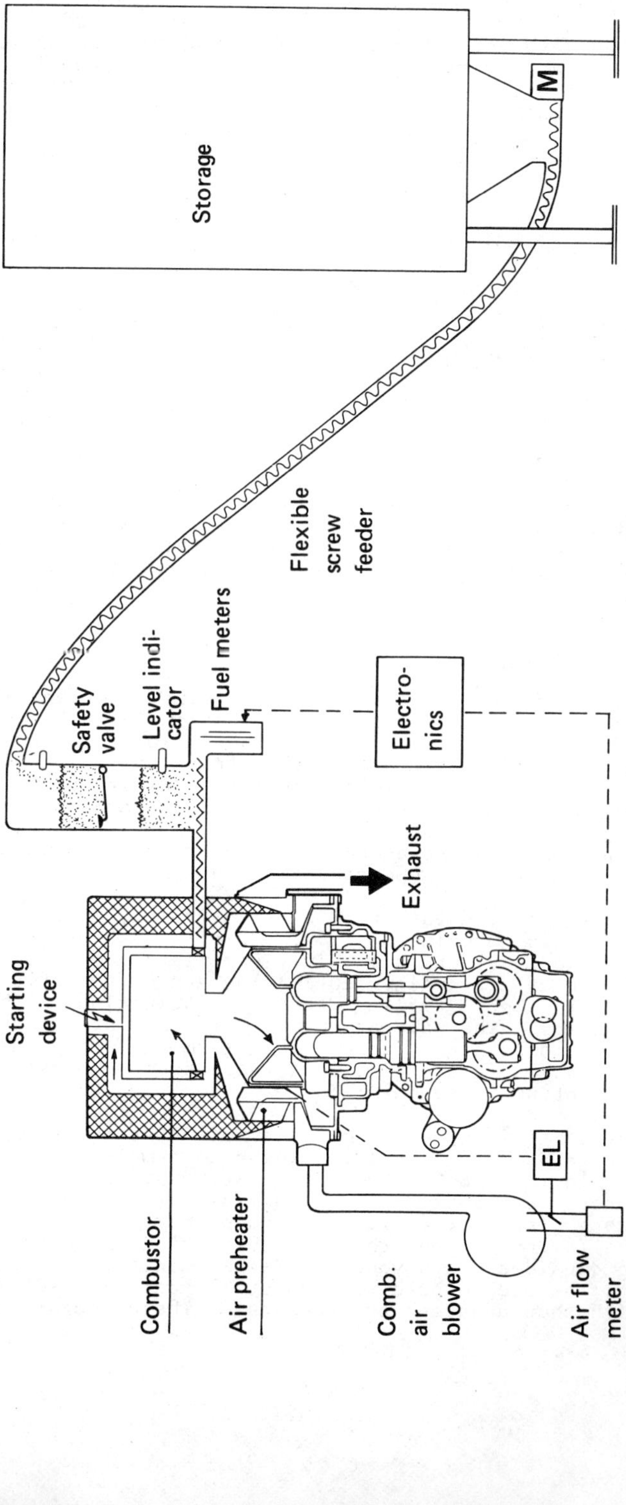

Figure 8. Biomass combustion control system with 4-95 engine.

in which the products of combustion from a closed coal-burning chamber mixed with the working air and were admitted to the power cylinder through a valve. They are usually referred to as furnace gas engines, and approximated the Brayton cycle. Type II were closed cycle, externally heated machines, and are called Stirling or constant volume engines. Type III were open cycle, externally heated machines, and generally known as Ericsson or constant pressure engines. Only type II, Stirling engines, will be discussed here.

SUPERSTITION AND THERMODYNAMICS

It is difficult to review the history of the air engine without including the history of thermodynamics. The failure of the air engine to achieve the success of the steam engine during the 19th century can be ascribed to a misunderstanding of the nature of heat. Although Carnot wrote his <u>Reflections on the Motive Power of Heat</u>, in 1824, acceptance of the kinetic theory of heat was not general until after 1860. The first experimental measurements of the mechanical equivalent of heat were published by Joule in England in 1843. Until this time, the <u>caloric</u> theory of heat was prevalent. Heat or caloric was considered a substance -- heat absorbed in the expansion of a gas became latent, but remained in the gas and was again evolved on compressing the gas. The caloric theory gave no explanation of the source of work produced by gas expanding against a piston. These errors resulted in many attempts to build perpetual motion heat engines during the 19th century. While the caloric theory also held back progress of steam power, it was easier for inventors to rationalize a steam-boiler-engine system than the more elusive air engine.

THE STIRLING INVENTION

Robert Stirling was born in Perthshire, Scotland, in 1790. He was licensed by the Presbytery of Dumbarton in 1815, and was ordained September 19, 1816. Less than two months later, he applied for a patent for a regenerator and an air engine which included the regenerator. The title was, "Improvements for Diminishing the Consumption of Fuel, and in Particular an Engine Capable of Being Applied to the Moving of Machinery on a Principle Entirely New". Figure 9 is a view of the engine as shown in the patent drawings. An engine of this type was first operated in 1818.

Stirling's grandfather, Michael Stirling, invented the first rotary threshing machine in 1756. Stirling's brother, James, was a member of the Institution of Civil Engineers and cooperated in the construction of air engines during the years from 1818 to 1847. Four sons of Robert Stirling attained distinction in the profession of engineering. The obituary notice of James stated that they were, "descended from a highly respectable farming family, nearly every member of which, females included, possessed mechanical talent almost amounting to genius". (6)

Robert Stirling became minister of Galston Parish Church, in southwestern Scotland, in 1824, where he remained until his death in 1878. In 1842 the University of St. Andrews conferred upon him the honorary degree of Doctor of Divinity. Apparently he was esteemed as a minister, a classical scholar, as well as an inventor. Reference (6) points out the difficulty of concluding which activity Stirling considered his vocation -- the ministry or engineering. Apparently he picked both, for about 25 years, by setting up a blacksmith shop adjacent to the manse. There, he and James experi-

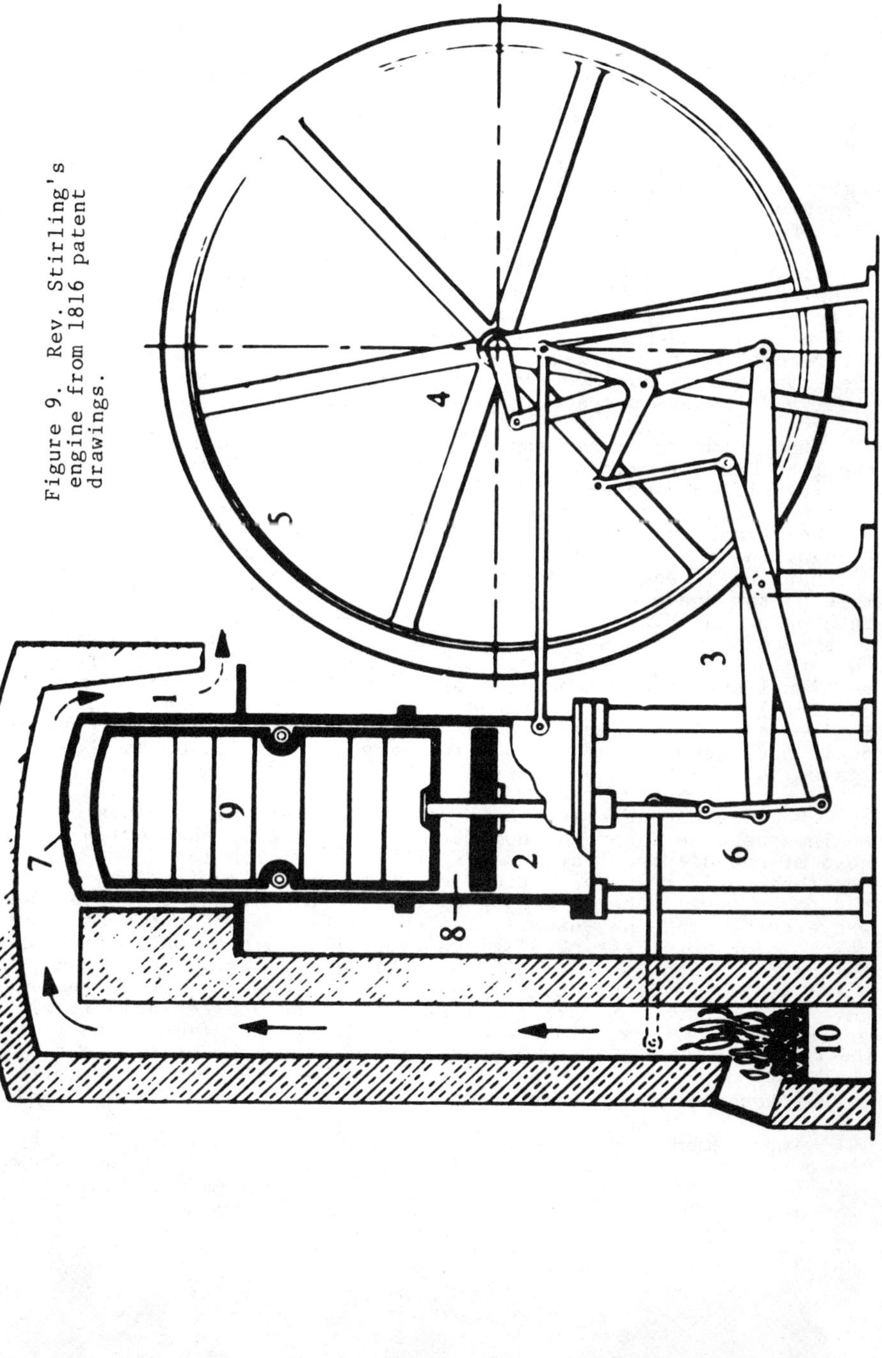

Figure 9. Rev. Stirling's engine from 1816 patent drawings.

mented with air engines, and as was reported, "also constructed with great neatness and dexterity many optical and other scientific instruments".

Although an explanation of the Stirling cycle has been given earlier , it is of interest to note several aspects of the original invention which were unique compared to other engines. The Stirling engine required no valves nor ports controlled by piston position. The power piston was located in the cool zone, thus assuring that it remained at a low temperature throughout the cyclic process from idle to full load. Operating on a closed cycle permitted the minimum cyclic pressure to be raised above atmospheric, which increased the specific output proportionally. This feature was credited to James Stirling, and the result was the brothers' most successful engine, which they installed at a Dundee foundry in 1843. The mean effective pressure and thermal efficiency of this engine exceeded that of any subsequent air engine for the next 95 years.

In their patent of 1827 the brothers anticipated the advantages of using a working fluid other than air and apparently had tested other gases. They stated, "and though we generally work the engine with common atmospheric, yet whenever a supply can be conveniently obtained of nitrogen, carbonic acid (CO_2) or any permanent gas which does not readily corrode iron nor cause explosion, we use it with equal and in some respects greater advantage."

Many commercial air engines produced during the latter half of the 19th Century eliminated the regenerator, or "economizer" as Robert Stirling preferred to call it. Apparently, this was done through misunderstanding its function or to simplify the design, but it sacrificed the high potential efficiency of the cycle. Although recognizing the vital contribution of the regenerator, the Stirling brothers believed the exterior heating surface was only required to make up the unavoidable thermal and friction losses. Robert stated in a paper in 1852, "the greatest enemy of the economizing principle, is the continual passage of caloric from the hot to cold parts of the engine, by radiation, conduction, and etc., which requires a continual supply of caloric to maintain the proper temperature of each."(7) In a paper presented before the Institution of Civil Engineers in 1845, James also expressed the belief that the engine could have been a "perfect" engine except for friction and heat losses.(8)

Because of these erroneous views, the heating surfaces of their engines were designed to include only the area of the hemispherical head of the displacer cylinder, with no provision for fins or projections to increase heat transfer area. With insufficient area for the power level, the safe working temperature of the cast iron heads was exceeded and they usually failed within less than a year of operation. The son, Patrick Stirling, reported to the Institution of Engineers of Scotland in 1860 that the heaters were, "kept red hot". (9) After four such heaters had cracked, the Dundee foundry gave up the use of Stirling's last air engine and replaced it with a steam engine. Materials were available at that time which might have enabled the Stirling brothers to increase the heater area by means of tubes, as they had already successfully done with their rather sophisticated cooler designs.

Improvements on the original engine were patented jointly by the brothers in 1827 and again in 1840. Their final engine, shown in figure 10, ran the lathes and other machinery at the Dundee foundry from March 1843 to January, 1847. The power cylinder inside diameter was 16 inches and piston stroke was four feet. Operating speed was 28 rpm. There were two displacer cylinders connected to the double-acting power piston. Each of the four foot diameter displacer cylinders contained an annular regenerator composed of thousands of 1/50th inch thick iron sheets, each measuring 38 by 1.75 inches, with a heat transfer area said to equal 3200 square feet. The minimum cyclic pressure at full load was about 150 psi and the maximum pressure about 240 psi. A water cooled tube type heat exchanger appears to be located above the regenerator, much as in modern Stirling engine designs.

By means of a prony brake an average of 21 brake horse power was measured. The engine was said to consume 50 pounds of coal an hour. The specific consumption of 2.4 pounds per BHP hour compared to about 4 pounds for the best steam engines of the day. According to Stirling, the coal was of "inferior quality"; but if one assumes a heating value of 12,000 BTU/pound, the brake thermal efficiency is calculated to be 8.9 percent. If one then assumes a reasonable mechanical efficiency of 75%, and from additional data furnished by James Stirling on the heat to the cooling water, the indicated thermal efficiency appears to have been about 21% and the furnace efficiency about 57 percent. The temperature limits of the internal air were said to be 150°F and 650°F, but no measurements were known to have been made, except of the cooler metal temperature. The Carnot efficiency for this temperature range is 45 percent. Steam locomotives of the 1930's used from 3.5 to 6 pounds of coal per BHP hour.

REGENERATOR DEBATES

Engineers were divided into three groups in regard to the utility of the regenerator, and it was a subject repeatedly debated in the British Institution of Civil Engineers. One group denied that the regenerator had any effect whatever. A second group, which included the Stirling brothers, concluded that a perfect regenerator would result in an engine with practically no fuel consumption. The third group took the middle course: the regenerator was a useful device, but as to exactly how it worked they were not sure, although they were convinced it could not produce a perpetual motion machine. The first published scientific explanations of the device were by Siemens in 1853 and Rankine in 1854.

At an engineering meeting in December 1860, at which Patrick Stirling gave a paper on his father's engine developments, it is evident from the comments of eleven discussers, that only one, Professor Rankine, understood that the engine had to reject the difference between heat input and work output. (9) Even in 1875 there was still disagreement over how the cycle operated. It was generally believed that the loss of efficiency was chiefly due to keeping the engine heaters constantly hot throughout the cycle, which was not true, and to imperfections of the regenerator. Rankine estimated the effectiveness of Stirling's 1845 regenerator at between 90 to 95 percent, which indicates that the major problem was not the regenerator. Rankine said of the Stirling cycle that, "this was the best air engine that had been put in operation and it was a pity that it had been laid aside".

Towards the end of his life, in 1876, Dr. Stirling, in

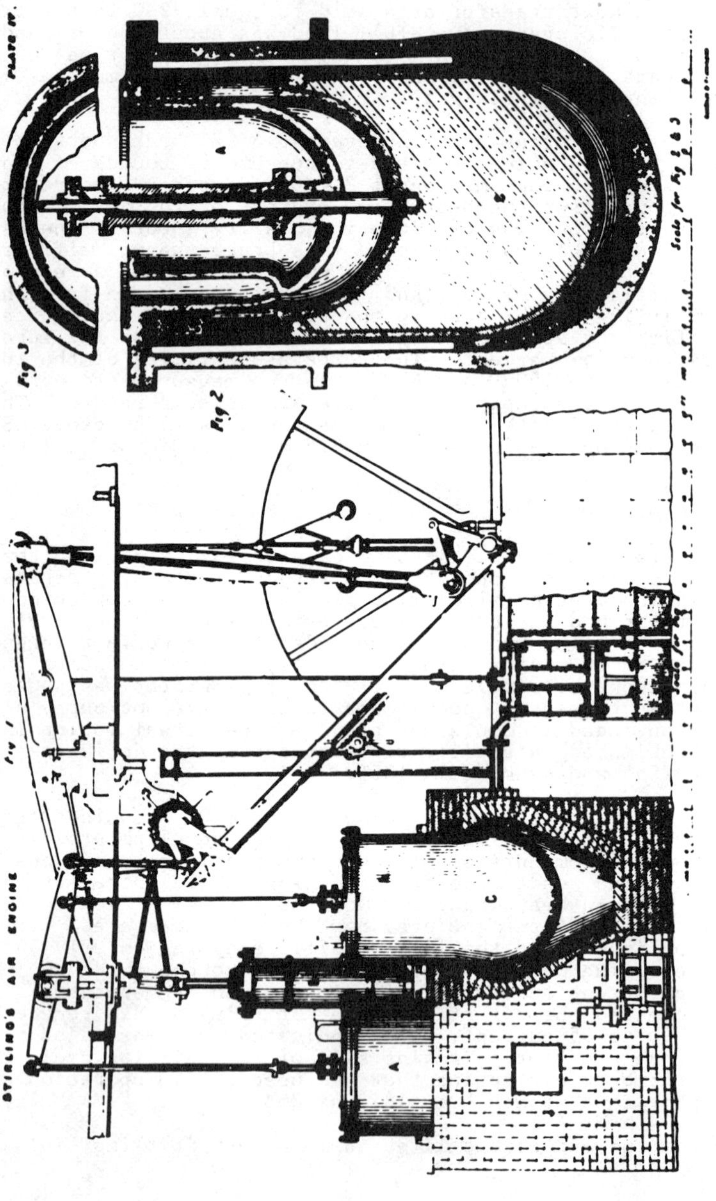

Figure 10. Stirling Brothers' 1845 Dundee foundry engine.

reminiscing said: "....It remains for some skilled and ambitious mechanist in a future age to repeat it under more favourable circumstances and with complete success." (6)

STIRLING ENGINES IN PRODUCTION -- 1870-1910

During the last quarter of the 19th Century several companies started to manufacture air engines which operated on a quasi-Stirling cycle. Actual production figures are unknown but it is estimated that at least 3000 were sold for pumping water. Perhaps the best known builder was the Rider-Ericsson Engine Co. of New York, successor to the Rider Engine Co., established in 1870. John Ericsson, the famous Swedish inventor noted for his armored warship Monitor, the screw propeller and his "caloric" engines, joined Rider in 1898. Their engines, operating silently on coal, kerosene or gas, were widely used for domestic water service in homes, buildings and industry as well as for railroad tanks and irrigation. No operator's license was required, in contrast to the requirements for steam boiler operators.

Figure 11 is a cross section of the Rider engine from the 1906 catalog. It has twin, parallel cylinders, with "opposed" pistons rather than a displacer, and it included a regenerator. Mean effective pressure was less than 4 psi, and the price was $540. The Ericsson engine, shown in figure 12, was a displacer type, and operated without a regenerator. From the catalog data, its brake thermal efficiency was slightly more than half of the Rider engine which was just over 2%, based on anthricite coal with a HHV of 13,000 BTU/lb. The Rider engine had a specific weight of about 4000 pounds/hp and a specific volume of about 120 cu ft/hp. Modern Stirling engines have reduced the weight and volume by three orders of magnitude while increasing the efficiency to 40 percent.

A list of some of the persons who owned Rider-Ericsson engines reads like a page from Who's Who, as can be seen in the catalog, figure 13. The Rider catalog for 1885 lists 231 water pumpers installed in homes, churches and businesses, in addition to 123 engines installed in apartments, all in the New York City area. Also listed were 297 engines owned by 93 domestic and 7 foreign railroad companies. Foreign references included 15 countries.

In spite of their enormous bulk and weight, there was little competition from other silent power plants (steam engines were somewhat lighter, but no less bulky), and until electricity was better distributed and more central water systems installed, these engines met a definite need. Compared to Robert Stirling's designs, however, they represented a step backwards in specific power and efficiency. In the authors opinion this could not be charged to a lack of suitable materials for construction; it was a lack of understanding of: 1) Robert Stirling's experimental work; 2)thermodynamics; and 3) how the Stirling cycle operated.

20th CENTURY REBIRTH

Revival of the Stirling cycle took place at the Philips Research Laboratories in The Netherlands in the late 1930's. It was a result of their search for a portable, silent heat engine to power small generator sets for operating Philips' radio equipment in remote regions. It is of interest to note their unbiased approach to the subject. Authors of powerplant and thermodynamic texts generally

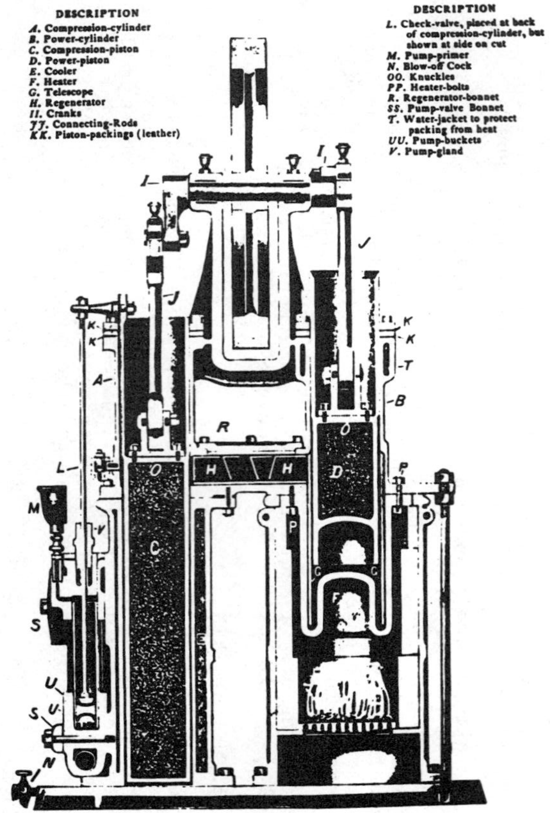

Figure 11. Rider engine from 1906 Catalog.

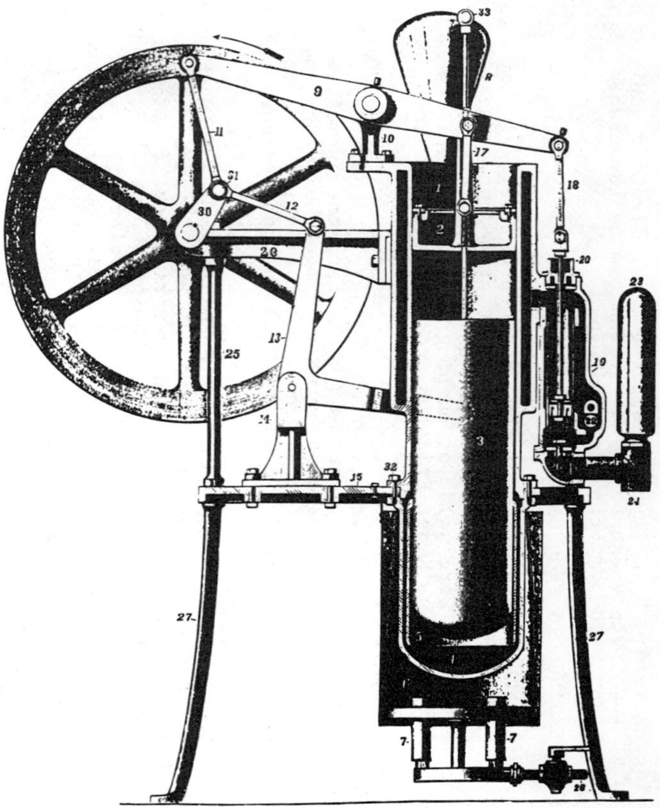

Figure 12. Ericcson engine from 1906 Catalog.

THE FOLLOWING names are selected from our testimonials as representing discriminating persons of national importance, who would under any circumstances buy the best article that could be obtained for any purpose.

United States Government
Dominion Government, Ottawa, Canada
New York City Government
J. Pierpont Morgan, New York
Khedive of Egypt
Whitelaw Reid, New York
W. K. Vanderbilt, New York
George Vanderbilt, New York
Frederick W. Vanderbilt, New York
Chauncey M. Depew, New York
Charles Lanier, New York
H. H. Westinghouse, Pittsburg
King Edward VII
Ex-Mayor Seth Low, New York
Andrew Carnegie, New York
H. O. Havemeyer, New York
Montague Allan, Montreal
Sir William Van Horne, Montreal
H. M. Flagler, New York
Colgate Hoyt, New York
Sultan of Turkey
Ex-Gov. Frank Black, New York
Harvard University
Michigan University
Cornell University
University of California
Ex-Senator D. B. Hill, Albany
Valentine Blatz, Milwaukee, Wis.

This does not mean that the Hot-air Pump is an expensive luxury. The man of moderate means will find it the cheapest sort of household investment—furnishing all the comforts and luxuries which accompany an abundant and constant supply of water.

RIDER-ERICSSON ENGINE COMPANY

New York Boston Chicago Philadelphia

Sydney, Australia

Figure 13. Testimonial names from 1906 Catalog.

concluded that since all attempts to build practical air engines had failed, therefore the Stirling cycle was not worth considering. Philips investigators took an opposite approach. They noted the large difference between the brake thermal efficiency of historic engines (which they tested) and their potential Carnot efficiency, and correctly concluded that it was due to basic mistakes in the hardware design, arising from ignorance of the Stirling cycle, heat transfer and thermodynamic laws.

After the interruption of WW-II, Philips first revealed their Stirling research in 1946. (10) The invention of the transistor nearly stopped the work since the need for generator sets was eliminated. But Philips persisted, and by 1950 they had manufactured over a hundred small single-cylinder displacer engines, hoping to sell them in the military market. Also developed was a reverse cycle machine for producing liquid air, which they began to market in 1955.

A fresh approach to the hot gas engine was begun in 1954, aimed at a higher power level by operating above 50 atmospheres, on hydrogen, and with improved regenerators and tubular heat exchangers. By 1958 Philips had tested a one cylinder 30 hp displacer engine and had constructed a four cylinder in-line engine capable of developing 350 bhp, both of which incorporated the Philips invention known as the rhombic drive. Measured thermal efficiency, without auxiliaries, ranged from 30 to 36 percent.

Perhaps the greatest contribution of Philips was their development of the Stirling cycle analyses. Effective mathematical modeling of the cycle to match the experimental results and, especially, developing programs to optimize engine design required a tremendous effort and a unique scientific approach. The author estimates that Philips and their licensees have devoted over 150 man years to this work since 1945. Without the electronic computer the analysis might be only marginally better than that of Schmidt in 1871 (11), which had become the only acceptable analysis prior to Philips. The basis of Philips' computer program is the so-called "semi-adiabatic engine", which allows for the fact that gas and heat transfer are not ideal, and for the influence of compressibility and finite heat transfer in heat exchangers and cylinders.

In 1958 a cooperative R&D program was started between Philips and General Motors. Numerous GM papers were published and some 330 internal reports were written, many of which have recently been released through NASA Lewis. A summary of GM's work from 1960 to 1970 is given in reference (12). Reference (13) summarizes the work at Ford Motor Co. from 1973 to 1978, and references (14) and (15) describe early work at MAN-MWM in Germany. The major developments at United Stirling in Sweden are covered in selected references, numbers (16) through (20). A practically complete bibliography on the subject of Stirling theory and hardware has been prepared by Martini (21).

CONCLUSIONS

The Stirling engine is a prime example of the stringent limitations imposed on engineers and technicians of the 19th Century in their search for safer and more efficient heat engines -- essentially as a result of ignorance of the laws of thermodynamics. It is also evident that the pioneering work of the Stirling brothers, whether through inspiration, intuition or genius, led to a more efficient air engine than did all the subsequent developments for the next 95 years.

The rebirth of the Stirling concept by Philips required a strong need, a scientific approach and good detective work. Two major inventions, the electronic computer and Teflon, have contributed greatly to the engineering success of the modern Stirling engine -- the former to understanding and optimization of the cycle, the latter to improved piston rings and seals which are critical for a long life engine.

The work of United Stirling in Sweden during the 1970's, and in particular the past 4 years, has produced the greatest thrust towards a commercial design of a Stirling engine. More than any predecessor, USS has made thorough marketing studies as well as hardware demonstration programs covering a variety of special applications for vehicles, surface vessels and submarines, solar thermal power, biomass combustion, heat pumps, generator sets and cogeneration systems. A solar Stirling engine at the focal point of a parabolic concentrator appears to be one of the most cost competitive applications for the future.

Finally, the Stirling engine is also a prime example of international technology transfer. Beginning with Scotland in 1816 it has expanded to major developments in at least 9 countries and smaller development studies in many others. The manner in which this transfer has taken place, in 5 steps, is shown in Table 2.

REFERENCES

1. Walker, G., Stirling-Cycle Machines, Clarendon Press-Oxford, 1973

2. Kolin, Ivo, The Evolution of the Heat Engine, Longman Group, Ltd., London, 1972.

3. Pons, R.L., "The Performance of Solar Thermal Electric Power Systems Employing Small Heat Engines," ASME Paper 80-Pet-25, November 1979.

4. Stearns, J.W., et al, "Solar Stirling System Development," A.I.A.A. Terrestrial Energy Systems Conference, Paper No. 79-1009, June 1979.

5. Truscello, V.C., Williams, A.N., "The JPL Parabolic Dish Project," 15th Intersociety Energy Conversion Engineering Conference, Paper No. 809346, August 1980.

6. Edelman, H., "In Search of Stirling," The Announcer, Vol. 23, No. 1, pp. 6-12, February 1969, Publ. Press Dept., N.V. Philips, Eindhoven.

7. Proceedings of the Institution of Civil Engineers (England), Vol. 12, 1852.

8. Proceedings of the Institution of Civil Engineers (England), Vol. 4, 1845, pp. 348-361.

9. Proceedings of the Institution of Engineers of Scotland, Vol. 4, 1860, pp. 13-15, 40-50, 141-158.

Table 2

THE STIRLING ENGINE

A PRIME EXAMPLE OF INTERNATIONAL TECHNOLOGY TRANSFER

1818----1847	1860----1910	1945----1969	1969----1978	1978----198X
DISCOVERY AND INVENTION	LOW PRODUCTION ERA AND A STEP BACKWARDS	REVIVAL AND RESEARCH	COMMERCIAL DEVELOPMENT	APPLICATIONS REFINEMENTS PRODUCTION
*SCOTLAND	*UNITED STATES GERMANY	*HOLLAND AND U.S.A.	*SWEDEN, HOLLAND, GERMANY, USA	*SWEDEN, USA, GERMANY, JAPAN, U.K., FRANCE, CANADA, USSR, CHINA
*STIRLING BROS.	*RIDER-ERICSSON LEHMANN	*PHILIPS GENERAL MOTORS	*UNITED STIRLING(USS) PHILIPS MAN-MWM FORD	*ENGINE DESIGNS AIMED AT AUTOMOTIVE, SOLAR, BIOMASS, COGENERATION, MARINE, HEAT PUMP
*SEVERAL ENGINES BUILT, COAL FIRED	*THOUSANDS OF WATER PUMPERS SOLD IN 15 COUNTRIES	*SERIOUS RESEARCH AND COOPERATION; CRYOGENERATORS ARE COMMERCIALIZED	*BREAKTHROUGH IN STIRLING ENGINE DESIGN AND COMPONENT DEVELOPMENT	*NASA/USS/MTI COOP FOR TECHNOLOGY TRANSFER TO UNITED STATES
*THERMODYNAMIC IGNORANCE	*POOR UNDERSTANDING OF HEAT	*MARINE APPLICATIONS DISPLACER ENGINES	*TRUCK,BUS,MARINE DOUBLE-ACTING ENGINES	*INITIATION OF PRODUCTION PLANNED BY UNITED STIRLING BY 1985
*EFFICIENCY 8%	*EFFICIENCY 2%	*ELECTRONIC COMPUTER TEFLON	*COMPUTOR OPTIMIZATION OF DESIGN	*EFFICIENCY POTENTIAL OVER 55%
		*EFFICIENCY 33%	*EFFICIENCY 40%	

10. Rinia, H., DuPré, F.K., "Air Engines," Philips Technical Review, Vol. 8, 1946, pp. 129-136.

11. Schmidt, G., Theore der Lehmannschen Calorischen Maschine, Z. Ver dt. Ing. 15, No. 1, 1871.

12. Percival, W.H., "Historical Review of Stirling Engine Development in the United States from 1960 to 1970," NASA CR-121097, July 1974.

13. Kitzner, E.W., "Automotive Stirling Engine Development Program," (Ford Motor Company), NASA CR-159836, May 1980.

14. Feurer, B., "Degrees of Freedom in the Layout of Stirling Engines", Von Karman Inst. for Fluid Dynamics, Lecture Series # 53, February 1973.

15. Zacharias, F.A., "Advanced Development on Stirling Engines at MWM", 2nd Symp. on Low Pollution Power Syst. Development, Dusseldorf, September 1974.

16. Hallare, B. and Rosenqvist K., "The Development of 40-150 kW Stirling Engines in Sweden and their Application in Mining Equipment, Total Energy Systems and Road Vehicles," Proc. 4th Int. Symp. on Automotive Propulsion Systems, Vol. 3, Sess. 8/9, Washington, D.C., March/April 1977.

17. Hoagland, L.C., Percival, W.H., "Potential of the Stirling Engine for Stationary Power Applications in the 500-2000 hp Range," Proceedings of the 13th I.E.C.E.C., Paper # 789355, August 1978.

18. Ortegren, L.G., Hansson, B.L., "Biomass Fueled Stirling Engines," Bio-Energy '80 Congress, Atlanta, Georgia, April 1980.

19. Rosenqvist, K., Haland, Y., "United Stirling's P40 Engine - Three Years Experience of Testing, Evaluation and Improvements," 5th International Automotive Propulsion Systems Symposium, April 1980.

20. Holgersson, S.,et al, "The Stirling Engine - A Ready Candidate for Solar Thermal Power Systems", SAE International Congress, paper # 810456, February 1981.

21. Martini, W., "Index to the Stirling Engine Literature--1300 references", Martini Engineering, Richland, WA., May 1980.

AUTOMOTIVE STIRLING ENGINE DEVELOPMENT: PROGRAM STATUS

PAUL LOMBARDI
PATRICK SUTTON

DEPARTMENT OF ENERGY
CONSERVATION AND RENEWABLE ENERGY
OFFICE OF TRANSPORTATION PROGRAMS
AUTOMOTIVE TECHNOLOGY DEVELOPMENT DIVISION

INTRODUCTION

The configuration of today's typical automobile has changed greatly since the early 1970's. The passage of stringent auto emissions standards, the 1973 oil embargo, subsequent fuel economy standards, the impact of foreign imports, and rising fuel costs continue to have significant effects on the types of vehicles produced by the auto industry.

The early attempts by auto manufacturers to comply with exhaust emissions standards resulted in reduced fuel economy. In order to meet the requirements of better fuel economy and lower emissions levels, in conjunction with noise and safety requirements, industry has emphasized smaller vehicles and materials substitution using spark ignition and diesel engines.

The conventional internal combustion engine-powered automobile is in the mature stages of its technology. Further improvements are likely to come only in small increments. These improvements may be initially expensive and difficult to maintain. In economic terms, conventional engine improvements have probably reached the point of diminishing returns on investment. The use of alternate automobile propulsion systems offers an option to realize much higher fuel economy improvements, especially if adherence to environmental and safety regulations is to be maintained.

Stirling engines were evaluated in the early 1970's as being one of the engines offering the best potential for improved performance in fuel economy and emissions among alternate power systems. The findings of analyses at that time rated Stirling technology one of the best candidates for development and use in automobiles.(1)

In 1977, under a DOE contract, the Ford Motor Company estimated possible improvements in the Stirling engine fuel economy of 38 percent to 81 percent over a spark ignition baseline, depending on degrees of risk in development.(2) After the fuel economy assessment task was completed, Ford stopped research and development on Stirling engines due to the increasing demand for their manpower and other resources. These resources were needed to satisfy near term requirements for fuel efficient and environmentally acceptable automobiles that the market demands. This problem, underscored by the Ford Motor Company's decision to terminate the contract, illuminates the underlying rationale for government involvement in Stirling engine development. It was recognized that Stirling technology was basically in its infancy and that many technological and institutional barriers would have to be solved prior to the acceptance of this revolutionary type of engine in transportation. The government funded R&D on alternative power systems is, by its nature, directed toward relatively long-term goals. Long-term R&D is expensive, and the risks of recovering the R&D investment can be high. In addition, since the early 1970's the auto industry has been working against legislative deadlines to meet fuel economy, emissions, and other standards. The immediacy of that need precluded investments in alternative design systems. Instead, improvements in the conventional engines were actively pursued by the industry.

DOE PROGRAM

The Federal government first began supporting alternate heat engine development in 1971. In recent years, DOE's policy has been guided by the Automotive Propulsion Research and Development Act of 1978 (Title III of Public Law 95-238). The DOE-sponsored efforts are designed to supplement and enhance private sector research and development in alternative heat engine systems. Programs are presently underway to develop both the Stirling engine and the gas turbine engine. These engines were the only types to survive the development criteria established by the DOE program.

The program has specific objectives for both fuel economy and emissions performance by September 1984. When installed in a typical late model vehicle, the Automotive Stirling Engine (ASE) system must demonstrate at least a 30 percent better combined fuel economy over a comparable production vehicle. The driveability and performance of the ASE and the spark ignition system must be substantially the same. It should be noted, however, that the government program is a technology development program rather than a specific engine development for a specific vehicle class. The purpose of demonstration in a vehicle is to provide realistic configuration control as well as on-road experience. In demonstrating this level of technology, it is anticipated that final engine HP, vehicle design, preproduction and production engineering would be the direct province of industry. The government's role is

---

(1) Should We Have a New Engine? An Automobile Power Systems Evaluation, Volume I, Summary, Jet Propulsion Laboratory, California Institute of Technology, August 1975, p. 76.

(2) Automotive Stirling Development Program, Ernest W. Kitzner, Ford Motor Company, prepared for NASA Lewis Research Center under Contract EC-77-C-02-4396, for U.S. Department of Energy, Conservation and Solar Energy, Office of Transportation Programs, March 1980, p. xxix.

to underwrite the higher risk elements of the development.

The ASE demonstration vehicle will be expected to show the potential for gaseous emissions and particulates levels less than the following:

NOX	=	.4 g/mile
HC	=	.41 g/mile
CO	=	3.4 g/mile
particulates	=	.2 g/mile after 50,000 miles.

In addition to the above quantitative objectives, five design objectives are considered major goals of the program.

- o The ability of the ASE to use a broad range of liquid fuels from many sources (including coal and shale oil)
- o Noise and safety characteristics to meet projected Federal standards for 1984
- o Reliability and life comparable to 1985 vehicles
- o Initial and life cycle costs competitive with 1985 vehicles
- o Acceleration capability suitable for safety and maneuverability.

PROGRAM ORGANIZATION

The Automotive Stirling Engine Development Program is being carried out within the Division of Automotive Technology Development (ATD), Office of Transportation Programs (OTP) at DOE. ATD, as Program Manager, has the responsibility for formulating and managing automotive research and development projects. DOE and the National Aeronautics and Space Administration (NASA) are currently working together toward development of the automotive Stirling engine under an Interagency agreement. Figure 1 describes this functional relationship.

The NASA Lewis Research Center is the technical project manager for the project. Their Stirling Engine Project Office implements the every day technology development tasks associated with the project. These include implementation of a strong quality assurance program and feedback program. NASA Lewis is responsible for accomplishing the project's technical objectives. The NASA Lewis team interfaces directly with its technical counterparts in the auto industry, and helps resolve any technical or contractual problems that arise during development.

In March 1978, Mechanical Technology Incorporated (MTI) became the prime contractor in the development of the advanced automotive Stirling engine designed to meet the program goals. Since most of the state-of-the-art technology resides in Europe, MTI is also responsible for technology transfer to the United States. MTI heads a team composed of United Stirling of Sweden and AM General--a wholly-owned subsidiary of the American Motors Corporation. MTI is also responsible for systems engineering, program and business management, mechanical component and systems development, engine and vehicle testing and evaluation, and computer code development.

United Stirling of Sweden (USS) is MTI's major subcontractor. USS is responsible for developing, designing, building, and delivering automotive Stirling engines to the United States. They are presently carrying out developments on components and systems,

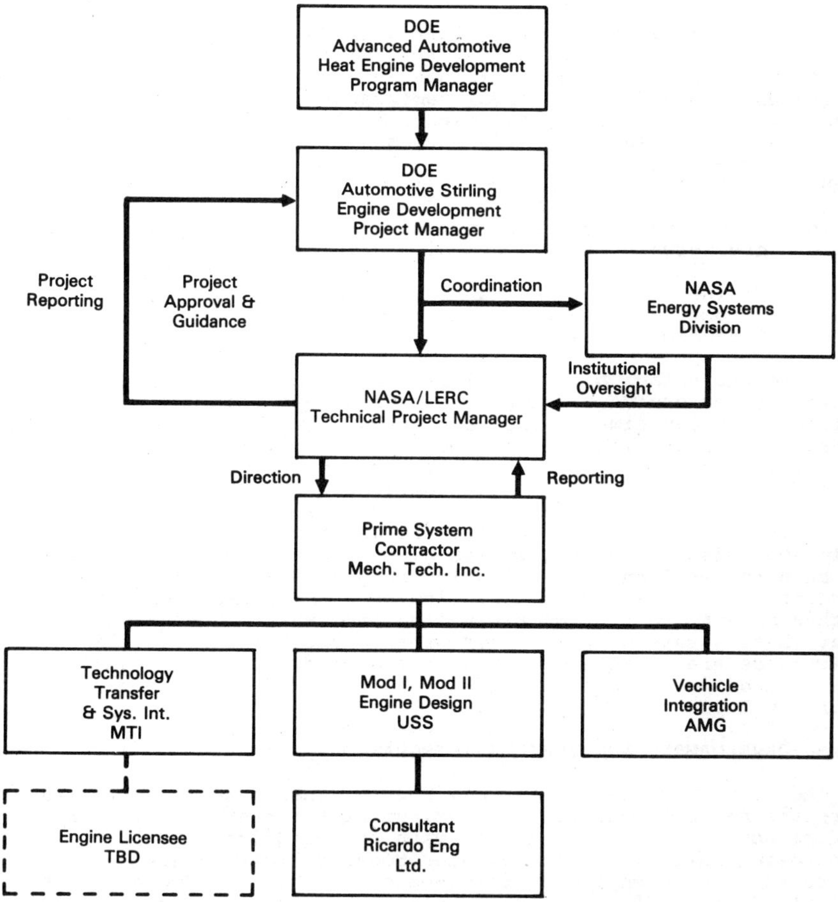

Figure 1: Program Organization

installing engines in vehicles, and conducting engine and vehicle systems evaluations.

AM General is MTI's subcontractor responsible for vehicle design, integration of Stirling engines into American passenger cars, market and trend forecasting, and testing and evaluation of engine and vehicle systems.

In addition to technical evaluations, DOE is sponsoring investigations into other factors that will affect the acceptance of automotive Stirling engines. Issues involving economic, environmental, social, and legal factors are under investigation. Through this research, these non-technical factors can be identified and dealt with to assist in addressing any inhibiting influences.

TECHNICAL APPROACH

There are three major elements of the program. Figure 2 shows the work breakdown structure associated with these elements.

- o Engine development and vehicle integration
- o Supporting research and technology
- o Technology transfer and market integration.

The basic technology being developed in the current program will be applicable for a wide variety of engine sizes--from small to large vehicles. Design simulations have shown that the engine could be as attractive for use in a compact car as in a five- or six-passenger car. The strategy of the program is to develop the baseline Stirling engine through two consecutive generations called MOD I and MOD II. These prototype engines will be used to test the technology and prove the Stirling engine's performance and feasibility. The early emphasis in the program was placed on delivering and testing the baseline P-40 engine, an engine developed entirely by United Stirling. This testing is now going on at MTI, NASA, and AM General in this country to provide a baseline from which improvements can be made. Assessments of engine performance and overall vehicle performance are an integral part of the plan. In addition, the P-40 engine functions as a test rig for engine components and subsystems under development.

ENGINE DEVELOPMENT AND VEHICLE INTEGRATION

The program currently has six P-40 engines undergoing testing. There are several different objectives of the tests. Baseline performance and emissions levels with engine driven auxiliaries are being determined, data for computer model validation are being generated, and in-engine performance of advanced components is being evaluated. In addition, NASA is gaining operating experience in an engine dynamometer which will aid in its independent evaluation of future contractor furnished engines.

The following table (Table 1) shows where these engines are being tested, and the total hours testing time experienced to date. The NASA and MTI engines are on conventional dynamometer tests at NASA and MTI respectively. The USS high temperature engine is on an engine dynomometer endurance rig. The USS annular is an engine dynamometer test to study a regenerator arrangement where each cylinder is wrapped with its annular regenerator. The Spirit and Concord engines are installed in AMC vehicles which will be discussed later.

The major problems with the engines have been in the areas of static seal leakage, heater head manifold cracks, and other

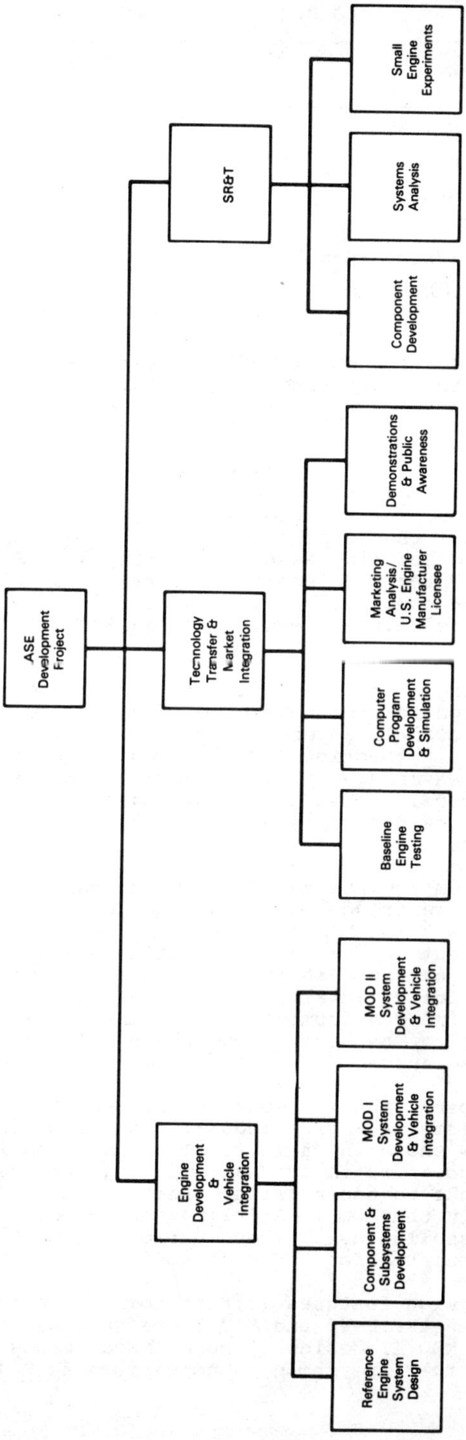

Figure 2: Work Breakdown Structure

TABLE 1 P-40 TESTING

Engine	Operation Time
ASE 40-1 (NASA)	204.2
ASE 40-7 (MTI)	202.4
ASE 40-8 (SPIRIT)	245.4
ASE 40-12 (CONCORD)	67.3
ASE 40-4 (USS-High-Temp)	5620.7
ASE 40-13 (USS-Annular)	140.6

structural failures. The MOD I reflects this baseline knowledge and has been designed to incorporate the appropriate fixes to relieve these problems.

Computer models of the engine are concurrently being developed under the automotive Stirling engine program. An early version of these codes has been verified and delivered. These new codes will eventually allow U.S. industry to simulate Stirling engine parameter sensitivity and will then supplement and replace computer programs developed by United Stirling Sweden. Testing with Stirling engines in this country is providing empirical data for computer program verification. Good computer simulation capability is considered necessary for future Stirling engine design work to continue in this country.

The Stirling engine program utilizes a reference engine system design. This design is a projection of the best MOD II engine design using projected 1985 technology. As our knowledge and experience with the automotive Stirling engine increase, the reference design can change to embody newly recognized, tested and verified capabilities. In addition to projecting the future engine design, the reference design is used as a guide for component and subsystem development.

The advance from the present P-40 engine capabilities to the realization of the reference engine system design is planned to occur in two stages. The MOD I engine is presently being tested. This is the first automotive Stirling engine design utilizing the experience gained working with the P-40 baseline technology. This experience is considered invaluable in lowering the risks associated with the final engine development. It is the essence of the program that MOD II technology be based upon proven MOD I technology achievements and advances.

The expected performance comparison between the baseline P-40 engine, MOD I, Reference Engine (MOD II), and a typical 1984 spark ignition engine is shown in Table 2. The MOD II engine will have the highest horsepower rating of the Stirling engines, with the lowest engine weight to power ratio. Peak thermal efficiency will improve to 43.3% in the MOD II Stirling, and composite MPG will rise to 42.1 mpg for gasoline use. These parameters meet or exceed the initial goals formulated for the program.

Seven basic design features (either components or systems) will be improved between the P-40 and MOD I design, and then again between MOD I and MOD II design. These improvements are summarized in Table 3, where they are shown in comparison to P-40 features.

TABLE 2
PERFORMANCE COMPARISON

	Baseline-P40	Mod I - 1981	Mod II - 1984	S.I. Engine 1984 4-Cy. (2.5L)
Engine Power (HP)	54	72.4	90	90
Demo Vehicle Wt (lb)	3750	3375	3170	3125
Engine Weight/Power	13.4	8.1	4.5	3.8
Max. Thermal Efficiency (%)	29	36	43.3	31
Composite MPG (Gasoline)	20.8	27.5	42.1	27.0
Composite MPG (Diesel)	23.8	30.5	48.5	—
Exhaust Emissions	Meets 1985 Federal Standards	Better Than Standards	Better Than Standards	Meets 1985 Federal Standards
0-60 Acceleration (sec)	34	22	15	15
Vehicle	1979 AMC Spirit	1981 AMC Spirit	1984 Pontiac Phoenix	1984 Pontiac Phoenix

SUPPORTING RESEARCH AND TECHNOLOGY

In addition to the main MTI contract, supporting research and technology activities are being conducted to support the automotive Stirling engine development effort. The activities are managed by NASA and consist of in-house, industry, and university research and development directed towards component and subsystem development. The areas under investigation include:

- Seals - development of piston rod seals to retain the high-pressure hydrogen working gas and to prevent oil contamination
- Materials - emphasis on hydrogen containment and low-cost alloys
- Controls - achievement of more efficient and reliable engine power controls
- Small Engine Experiments - conduct of component development and validation of computer code simulation models
- Combustors - development of low-emissions combustion technology through more uniform temperature and fuel distributions
- Systems Analysis - verification of NASA-generated Stirling engine computer models, and assessment of the fuel economy potential of Stirling vehicles.

TECHNOLOGY TRANSFER AND MARKET INTEGRATION

One of the principal program objectives is to accelerate the ultimate large scale use of the Stirling engine technology into the transportation sector by the 1990's. However, Stirling engines have not been universally accepted and manufactured, basically due to the internal combustion engine's marriage with automobile mass production. As such, the Stirling technology, however significant in meeting future national conservation goals, will find difficulty in supplanting the well-established track record of the internal combustion engine. This is compounded by the fact that most of the recent active research associated with Stirling engines has been pursued in Europe rather than in the United States. The market integration approach has been designed to build up U.S. infrastructure acceptance as well as take more traditional approaches associated with new product development.

In this light, technology transfer has become a key element of the development program. The baseline P-40 Stirling engine has been the workhorse for the technology transfer activities with engine characterization testing at both MTI and NASA, and both vehicle chassis dynamometer and on-road testing by AM General. In addition, the P-40 engines all serve as test beds for advanced component designs at MTI and for the supporting research and technology activities at NASA.

Work is underway to identify the market sectors associated with production. Market penetration rates, target sectors, supplier capabilities, engine manufacturing costs, capital investment and after market consideratons are but some of the areas being explored. A licensing strategy is being developed to meet the overall program objectives, including technology transfer at the engineering level. The establishment of a viable U.S. engine manufacturer is of primary importance and appropriate market research has been initiated to secure such a licensee.

MILESTONES AND ACCOMPLISHMENTS

The Automotive Stirling Engine Development Program plan uses schedule guidelines which trace the program's progress through milestones, or decision points. These are either major achievements or efforts which must be accomplished during the life of the program. Figure 3 shows this schedule for the present government funded phase of the program.

The first two milestones have been achieved on schedule. The MOD I basic Stirling engine design freeze occurred in FY80 and the first build of the MOD I engine began dynamometer testing in FY81 as planned. Figure 4 depicts this USS Mod I installation. USS has a very well-equipped test laboratory with a real-time display of computer process test data. Results of early MOD I engine testing are in good agreement with the engine design.

The Reference Engine System Design was approved on schedule. As was shown in Table 2, the MOD II engine installed in a 1984 Pontiac Phoenix should be a very competitive system. Figure 5 shows a cross-section drawing of the MOD I engine. The changes in engine size between the P-40 and the MOD II engine designs are indicated.

The P-40 engine has been installed in several automobiles in order to obtain test data on the performance of the automotive Stirling engine in a vehicle system. A 1977 Opel housed a P-40, installed in Sweden (see Figure 6). This vehicle provided credibility to the idea of an automotive Stirling engine application and provided an initial definition of the baseline for the program.

Component/System	ASE MOD I	ASE MOD II
External Heat System	– lower matrix position – CGR burner type – good flow pattern – rather wide – thicker insulation – improved seals – removable matrix	– lower matrix position – CGR burner type – good flow pattern – compact design – air cooled thin insulation – designed for ease of fabrication – simplified assembly – changed gas flow – improved seals – removeable matrix Alternative: regenerative system
Heater Head	– same tube geometry – constant tube gaps in each row – good inside flow to housings – 8 instead of 12 housings – simpler pressure vessel castings (separate manifolds) – longer regenerator manifolds – increased cylinder dome length – axisymmetric vessels (improved flange arrangement) – better gas seal cooling	– same tube geometry – constant tube gaps in each row – good inside flow to housing – 8 instead of 12 housings (annular version possible) – simpler pressure vessel castings (separate manifolds) – long split regenerator manifolds – less increase of cylinder dome length – axisymmetric vessels – requires improved material (820°C) (stress level assumed) – better gas seal cooling – smaller regenerator diameter – extended regenerator housing – big thread connection (cooler regenerator housing) – compact quadrants – smaller heat exchanger Alternative: annular concept (4 vessels) to be studied
Regenerator	– 4 instead of 8 – scaled size, moderately reduced from experienced volumes – improved matrix cover	– 4 instead of 8 – possibly rolled matrix – very small diameter (optimized part-load) – improved matrix cover – lower filling factor – low cost potential

Table 3: Salient Basic Engine Design Features as Quantatively Compared to Baseline Engine (P-40)

Gas Cooler	– conventional type, possible Al – improved cooling flow distribution	– conventional type, possible Al – improved cooling water flow distribution – reduced size – integral duct plug – different attachment – smaller o-rings
Cylinder Block	– split in water jacket, duct plates, etc – light alloy jacket – two parallel water passages – equal flow distribution – separate cyl liners – separate duct plates – less casting exposed to working gas – all pieces clamped by reg/cyl bolts – relatively compact – separate but improved check valve housings – all gas o-rings radially mounted – cooler better kept in place – easier serviced	– split in water jacket, duct plates, etc – light alloy jacket – two parallel water passages – equal flow distribution – separate cyl liners – duct plates integral with crosshead liner (Possibly Al) – less casting exposed to working gas – only clamped by cylinder bolts – very compact – check-valve housings integral with duct plates – less o-ring length – cyl liner clamping seal housing cover – all gas o-rings radially mounted
Piston Rod Seal System (and Piston Rings)	– Principally P40-system but refined successively – housing and cyl liner floating in duct-plate – provision for pressure-connection from piston ring volume – integral piston rod and cross-head	– the same system but extremely pushed into piston – housing floating – pressure-connection from piston rings through liner Alternative: flexible membrane system
Drive Unit	– Aluminum structure – integral front cover – integral rear end cover/water pump housing and drive – compact lub oil system – rearranged balance weights – central synchronization (chain links) feasibility	– aluminum structure – integral front cover – compact lub oil system – rearranged balance weights – central synchronization option Alternative: transverse engine design to be studied, possibly with power take-off from crankshaft

Table 3: (continued)

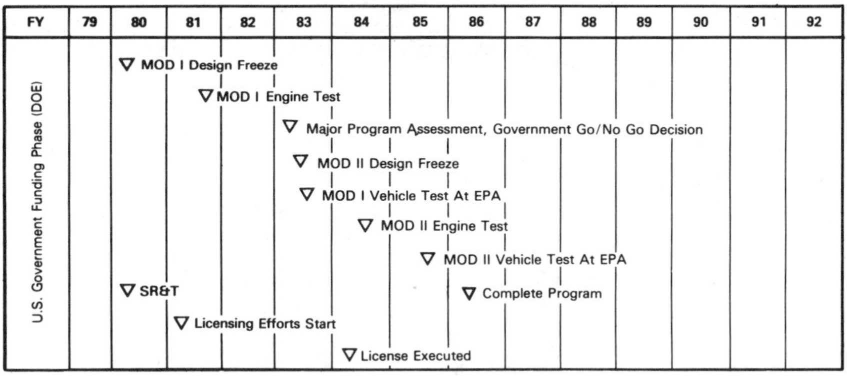

Figure 3: Program Schedule

Figure 4: USS MOD I Installation

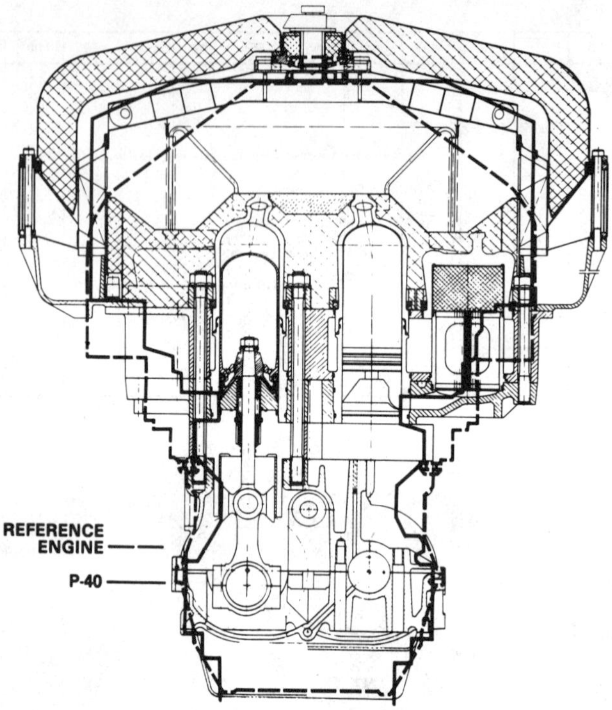

Figure 5: MOD I Size vs Reference Engine an P-40

Figure 6: Stirling Opel

A 1979 AMC Spirit was fitted with a P-40. The Stirling Spirit (Figure 7) was made to provide vehicle experience to AM General and MTI. The Spirit has been shown at demonstrations and is being used for vehicle testing. The Stirling Spirit engine installation is shown in Figure 8. This is typical of all three vehicles. The 1980 AMC Concord is the third vehicle to be fitted with a P-40 engine (see Figure 9). It is being used to help inform the public of what a Stirling engine is and of our program to develop it as a viable alternative for use in automobile engines. The Stirling Concord is designed to emphasize the fuel flexibility of the automotive Stirling engine. The Concord fuel tank is divided into three compartments, each of which can be filled with a different fuel. While the car is driven, the fuel in use can be switched. No difference in performance is apparent to the driver as the fuels are switched. Today the tanks are filled with unleaded gasoline, gasohol, and diesel fuel. Plans are to run the vehicle on other fuels in the future. Other technical accomplishments include:

- Over 4,000 hours of engine testing experience was accumulated at the 820° Centigrade temperature for MOD II heaterheads. The feasibility of increasing temperatures in the MOD I engine was established.

- A pumping Leningrader (P.L.) seal has been successfully operated in an engine for 3,500 hours. Thus, the durability of one critical piece of automotive Stirling engine equipment was proven.

- Over 70,000 hours total experience has been accumulated on the P.L. seal.

- The transfer of Stirling technology in the United States was aided by the beginning of operation of a test engine, a seal rig, and a regenerator rig at MTI.

SUMMARY

The Automotive Stirling Engine Development Program is based on cooperation between government and private industry. The government role emerges from economic fundamentals. In an environment of limited resources, the automobile industry is being forced to concentrate its efforts on work leading to the near-term changes mandated by regulation. Research on Stirling engines for automotive applications, although promising, could not be privately justified or supported. The kind of long-term research and development effort needed is being supported by DOE in an effort to ensure this nation's personal mobility in the future.

A highly skilled team has been assembled to do the work for the program. Continuing progress is being made on schedule. The MOD I engine now being tested is evidence of the capability to meet program goals. We at DOE are pleased with the program's progress to date and we will continue to keep you informed of the status of the Automotive Stirling Engine Development Program.

Figure 7: Stirling Spirit

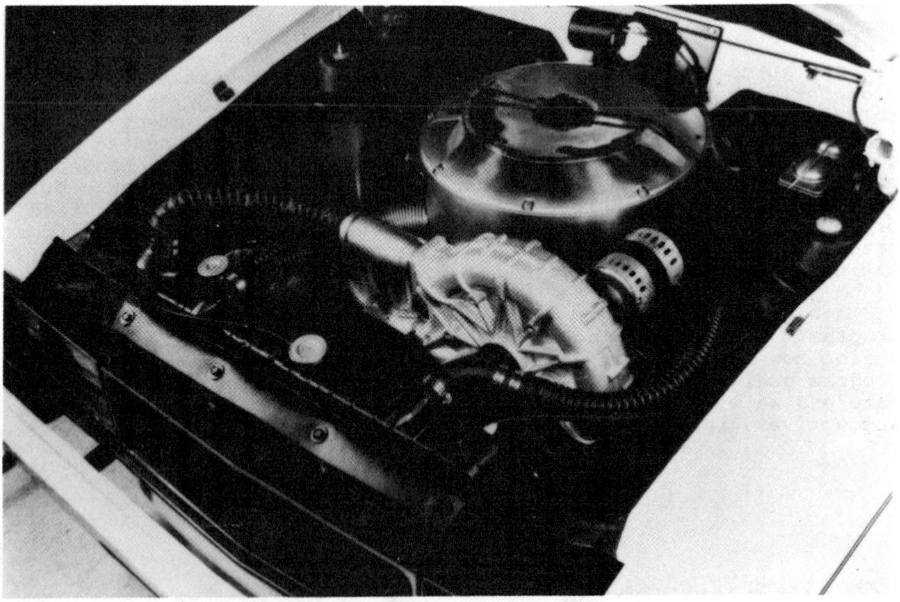

Figure 8: Stirling Spirit Engine Installation

Figure 9: Stirling Concord

STIRLING ENGINES FOR STATIONARY APPLICATIONS

Bruce Goldwater
Mechanical Technology Incorporated

INTRODUCTION

Unstable petroleum markets and spiraling energy costs require a reduction of the United States' dependence on imported oil. A concentrated effort to develop higher efficiency equipment and to provide increased supplies of natural and synthetic fuels must be supported. Critical to the achievement of these objectives is the reduction of fuel consumption in heat engines, a dominant utilization segment. The development and introduction of engines with the flexibility to burn newly developed coal-derived fuels will reduce the economic impact of supply shortages. Stirling engines, with the capacity to achieve improvements in fuel efficiency while preserving the environmental quality demanded by public opinion, can provide the necessary technological hardware to meet the energy needs of the United States.

The subject of this paper is limited to a discussion of the stationary engine applications that can be served by the Stirling engine. Although the transportation sector remains an important factor in the total market, and perhaps the market pull that is needed to support the additional applications that are less dramatic in impact, their discussion is covered in other papers within this session. It should be noted, however, that development of the Stirling engine for the automotive application is already underway. Extensive funds and resources have been committed by the Department of Energy, Division of Conservation and Solar, to meet this objective. Although many of the technical risk areas for the automotive engine will be addressed during this program, the stationary engine application is sufficiently different in size requirements, operating requirements, and application needs, that direct substitution of the prime mover is valid only in a few areas and, in general, is not an appropriate approach. For the automotive application, high efficiency at part load conditions and high specific power at a sacrifice in operating life define the design criteria, while tradeoffs in power density with initial cost and increased life are prime considerations in the design of stationary engines.

A broad spectrum of stationary engine applications as potential Stirling engine markets have been studied. In this paper, the discussion will concentrate on the most important market areas, i.e., residential and commercial heat pumps, and electric power generation. Consideration of engines to satisfy these applications requires a power range from 1 to 1000 kW in order to address the entire market. As a result of this wide range, it is projected that two different Stirling engine configurations are required to cover the power needs. In the small size, less than 15 kW, the hermetically sealed free-piston Stirling engine concept seems to dominate, while above 15 kW, the kinematic direct-shaft output configuration is prominent. Both engines are discussed within the context of the various applications.

GENERAL DESCRIPTION OF STIRLING ENGINES

CYCLE DESCRIPTION

The Stirling engine can be described as a closed-cycle piston engine with cyclic recirculation of the working fluid. Power is produced in the conventional manner; that is, compression of the working gas at low temperature, heating the gas, expanding it at high temperature and, as a result, cooling it again. A major difference between the Stirling engine and conventional piston machines is the method of adding heat. In the internal-combustion engine, output work is achieved when a fuel-air mixture is compressed at low temperature, the mixture is rapidly heated by combustion, and the gases are allowed to expand at a decreasing temperature. In the Stirling engine, the required heat is added continually during expansion to the working gas inside the engine through a heat exchanger wall. Since this wall has a high heat capacity, it is not possible to rapidly heat and cool the wall; therefore, by one means or another, the working gas is alternately shuttled between two stationary chamber volumes called the hot space and cold space (see Figure 1).

Since the alternating heating and cooling of the shuttle gas would inherently waste large quantities of heat, a matrix type regenerator is placed between the hot and cold sources. Heat is then stored in the regenerator as the gas moves toward the cold space (compression space), and then released as the working fluid returns from the cold space to the hot space (expansion space).

The ideal Stirling engine cycle is shown in the pressure-volume (P-V) and temperature-entropy (T-S) diagrams of Figure 2.

a) Process 1-2 is an isothermal compression where heat is transferred from the working fluid at the minimum cycle temperature to the external sink (heat rejection from the engine).

b) Process 2-3 is at constant volume where heat is transferred to the working fluid from the regenerator matrix (heat released).

c) Process 3-4 is an isothermal expansion where heat is transferred to the working fluid at a maximum cycle temperature from the external source (heat added to the engine).

d) Process 4-1 is at constant volume where heat is transferred from the working fluid to the regenerator matrix (heat absorbed).

If the heat that is transferred to the working fluid from the regenerator matrix is the same as that transferred from the fluid to the matrix, then only the external heat transfer processes remain, and the efficiency is consistent with the Carnot cycle efficiency. The advantage of the Stirling engine cycle is that the two isentropic processes of the Carnot cycle are replaced by two constant volume processes, thereby increasing the area under the P-V diagram, reflecting in higher specific work output levels without resorting to very high pressures and high swept volumes.

The ideal cycle described assumes that the processes are thermodynamically reversible; that is, the expansion and compression processes are isothermal, and that infinite heat rates exist in addition to infinite heat capacities. In the ideal analysis, the effects of regenerator matrix voids, clearance spaces, and cylinder volumes were neglected. In addition, the pistons were assumed to move in a discon-

Figure 1

Engine Configuration

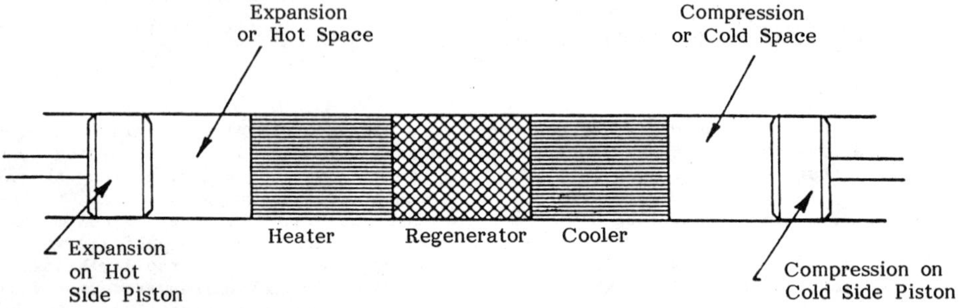

Stirling Thermodynamic Cycle Vs. Carnot Cycle

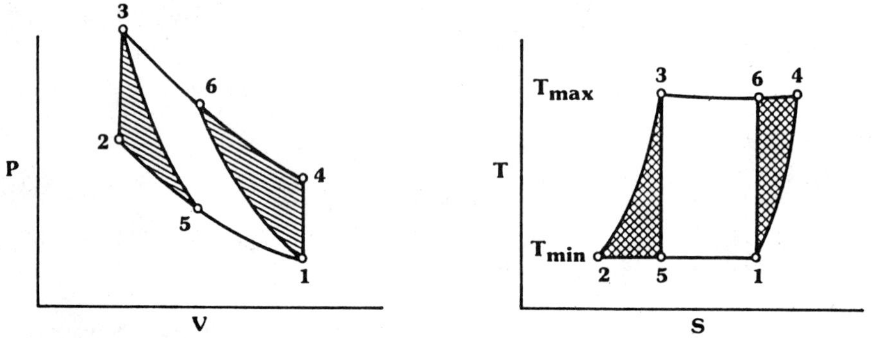

1-2 Isothermal Compression Heat Transfer From Working Fluid at Tmin to External Dump

2-3 Constant Volume — Heat Transfer to Working Fluid From Regenerator

3-4 Isothermal Expansion — Heat Transfer to the Working Fluid at Tmax From Eternal Source

4-1 Constant Volume — Heat Transfer From Working Fluid to Regenerator Matrix

Additional Work Made Available by Substituting Constant Volume Processes For Isotropic Processes

Increase in the Quantity of Heat Supplied And Rejected in the Same Proportion as the Available Work

Figure 2

tinuous manner, whereas in reality, the motion is a smooth continuous path; therefore, in reality, the theoretical P-V and T-S diagrams are rounded off. Aerodynamic and mechanical losses were also neglected. Inclusion of these losses, of course, results in a lower net cycle output power and lower efficiency.

The addition of the heater and cooler components changes the real heat transfer to a more adiabatic situation rather than the assumed isothermal processes. Penalties in additional aerodynamic flow losses and increased dead volume result. The use of practical equipment imposes on additional reality — that the fluid not only is heated as it flows to the expansion space, but also as it flows in the reverse direction from the expansion space to the cooler, thereby introducing additional losses.

STIRLING ENGINE DESCRIPTION

As mentioned previously, at the present time, extensive Stirling cycle machinery development is being directed at two different engine approaches designed to achieve the cycle. For the past five years, a high level of concentrated work on the development of a unique Stirling cycle engine concept in the class of machinery termed free-piston has been ongoing. This approach is considered a breakthrough in the technology of small, highly efficient, long life, low cost, and fuel flexible heat engines. Combined with the development of linear generators, resonant piston compressors, process fluid gas bearings, and hermetic package design, this technology yields an advanced system that can now address both the current small stationary engine market and the potential new markets anticipated as an outgrowth of the availability of this engine.

The free-piston Stirling engine priciple, although a simple application of a mass-spring system, requires extensive knowledge of Stirling cycle thermodynamics, as well as dynamic machinery theory. The basic thermodynamic cycle of the engine is exactly the same as the kinematic engine designs; however, the method of delivering output power is quite different. Kinematic engines are direct-shaft output systems which achieve the thermodynamic cycle requirements by phasing the pistons mechanically by use of a conventional crankshaft, connecting rod, and crosshead arrangement. Figure 3 is a schematic cross-section of the first generation automotive Stirling engine, a representative kinematic design.

Free-piston Stirling engines, shown pictorially in Figure 4 and schematically in Figure 5, on the other hand, achieve the thermodynamic cycle requirements by phasing the pistons via proper design of the gas springs and piston mass selection. The engine is, therefore, essentially a resonant system whose frequency is largely dependent on the charge pressure of the working fluid. Power output from the engine can be commonly achieved via electric generators contained within the hermetic cylinder, shown in the figure, or via gas compression devices, as would be required in a heat pump system.

STATIONARY STIRLING APPLICATIONS

Potential use of the Stirling engine in the United States is seen as a possible substitution for conventional equipment in four major areas, as shown in Figure 6. Space conditioning and hot water energy use account for 40% of the total United States energy consumption, while an additional 35% is attributable to energy usage in manufacturing processing and general residential and commercial electric power needs. Estimates of the market segment that may be served by Stirling engines equated to energy consumption is overlaid on the figure, indicating areas of potential penetration for both free-piston and kinematic engines.

RESIDENTIAL AND COMMERCIAL HEAT PUMPS

The space conditioning category of energy consumption represents a major market potential in which penetration requires the availability of a highly efficient and reliable prime mover that, when combined with conventional refrigeration components, would provide a heat-activated heat pump product. Both residential and commercial markets could be addressed and, as shown, on an energy basis, a nearly equal use of

Schematic of ASE Mod I

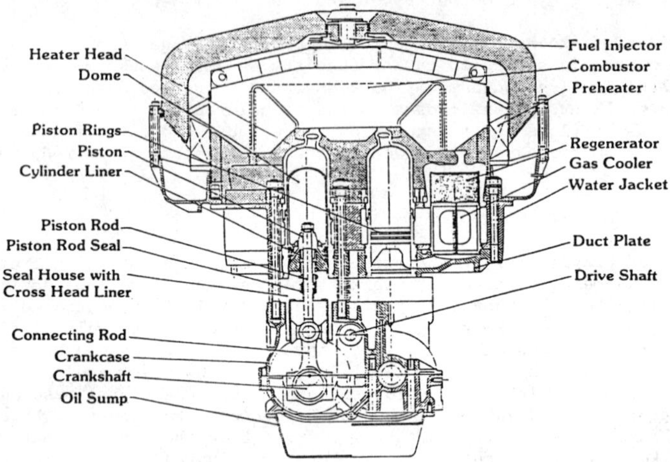

Figure 3

Typical Stirling Engine

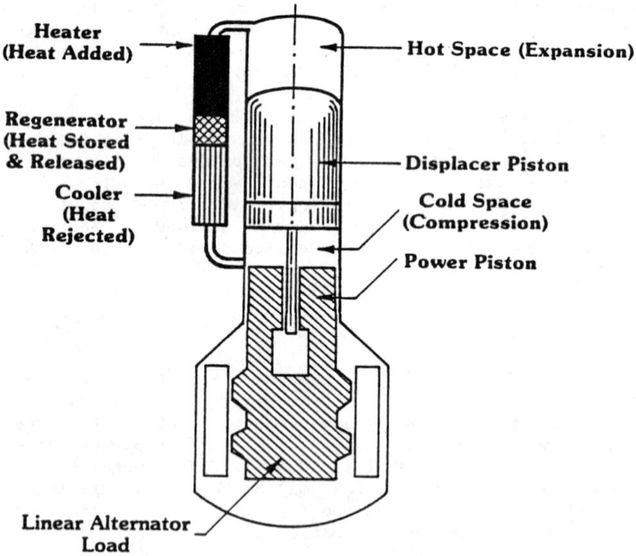

Free-Piston Type Stirling <50kW

Figure 4

MTI/DOE Free-Piston Stirling Engine Demonstrator

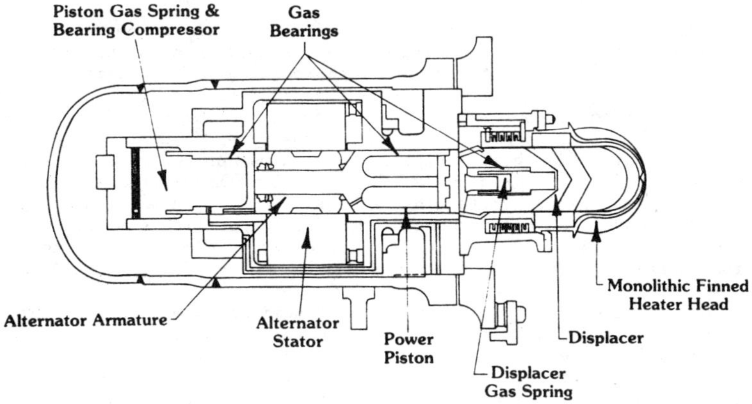

Figure 5

Potential Use of Stirling Engines in the U.S.

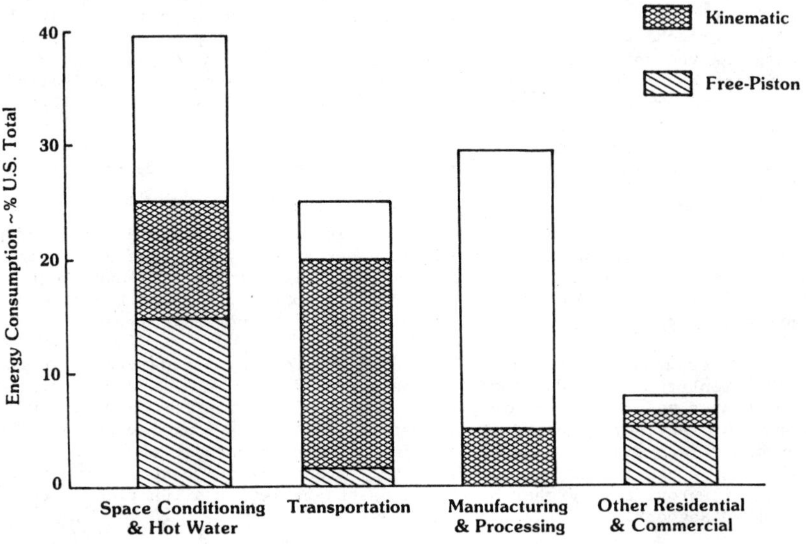

Figure 6

both large and small engines could be anticipated.

Residential heat-activated heat pumps, defined as on-site heat-engine-driven refrigeration systems, a schematic of which is shown in Figure 7, operate as follows:

Fuel is burned in a conventional atmospheric combustion system which consists of a combustion blower, air-fuel control, and recuperative exhaust gas/inlet air preheater. Engine power is used to drive a reciprocating refrigeration compressor in a typical heat pump loop. Rejected engine heat is recovered in a closed loop heat exchanger system to further increase the seasonal performance factor (SPF) in the heating mode operation. Conventional heat exchangers and heat pump controls are used throughout. Residential heat-activated heat pumps can be shown to be 50% more efficient than advanced fuel-fired furnaces, and 75-100% more efficient than electric heat pump systems when operating in the heating mode. This is due to the system's ability to recover the on-site engine heat rejection with simple heat exchanger components. Figures 8 and 9 show the difference between an electric heat pump and fuel-fired heat pump operation, exemplifying the reason for the improvements.

Natural gas-fired equipment is presently the dominant fuel for residential and commercial heating systems, and projections indicate that this will continue. In addition, the demand for cooling has not subsided and, consequently, the market for combined systems is projected to grow substantially. Recent incentives have also improved the pipeline supply of natural gas to the point where storage facilities in the off-season are limiting, and the need to use the excess is again evident. Gas-fired cooling satisfies this need. It is also projected that in order to justify SNG on an economic plane, utilization equipment must be available which is of an efficiency to be able to support the increased costs to the consumer, and yet be cost-effective on a life cycle basis and, more importantly, on a first cost payback basis. Free-piston Stirling-engine-driven heat pumps satisfy this equipment requirement. Typical annual energy costs for various heat pump systems are shown in Figure 10 for both natural and coal-derived fuel costs, including auxiliary electric power costs.

Market projections and penetration of that market with various first cost penalty assumptions for the new housing, replacement, and conversion market have been estimated and are shown in Figure 11. Significant levels of production are anticipated, assuming the optimum product features are achieved. Although the first cost of the Stirling heat-activated heat pump is projected to be higher than that of competitive advanced equipment, overriding performance advantages and generally lower maintenance costs will reflect in an acceptable short-term payback situation for the buyer. Most importantly, the Stirling engine provides the consumer relative independence of the fuel situation due to the engine's ability to burn virtually any fuel.

Commercial heat pump systems, as indicated in Figure 6, represent a major portion of the space conditioning market that can be served by the kinematic Stirling engine. This market represents the commercial and light industrial equipment needs more applicable in size and operating requirements to the design of the kinematic engine system. A typical system is shown schematically in Figure 12.

Market assessments performed by W. E. Hill and Company indicate that in the size range of interest (20-200 tons) for the nonresidential HVAC systems, a potential annual sales volume in 1981 is projected at $360 million for over 900,000 installed tons of refrigeration. Unitary systems, defined as a system whereby the major components are sized, packaged at a factory and shipped to the site, account for nearly 50% of the market, and are generally standardized up to 125 ton packages. Tables I and II provide a further breakdown of the market as a function of building type or market segment.

Penetration of this market with the proposed system is estimated to depend on a number of critical factors:

1. Life cycle costs of the proposed system are less than those of competitive equipment on the market today.

2. Payback on the first cost of this equipment is less than three years.

Stirling Engine Heat Pump System

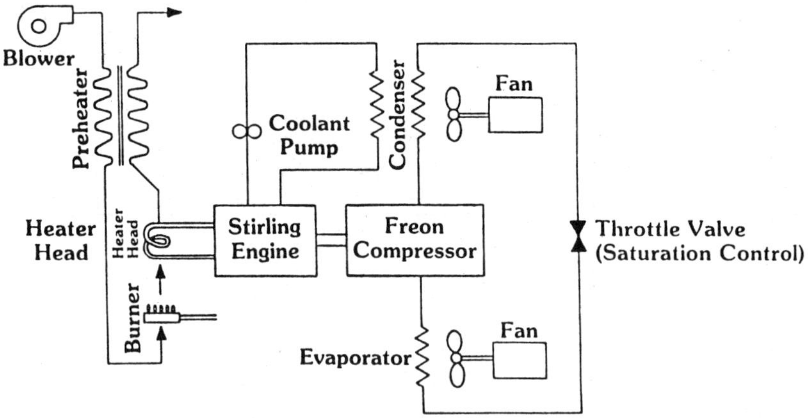

Figure 7

How An Electric Heat Pump Works

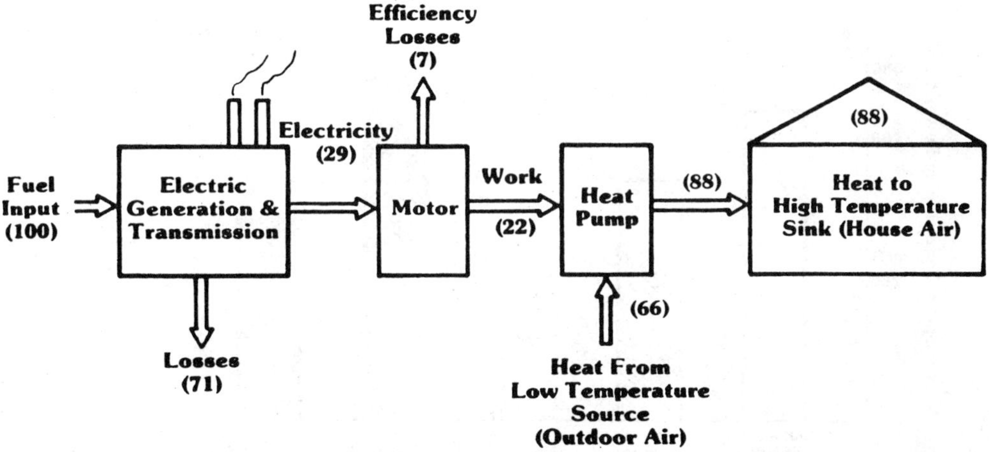

Figure 8

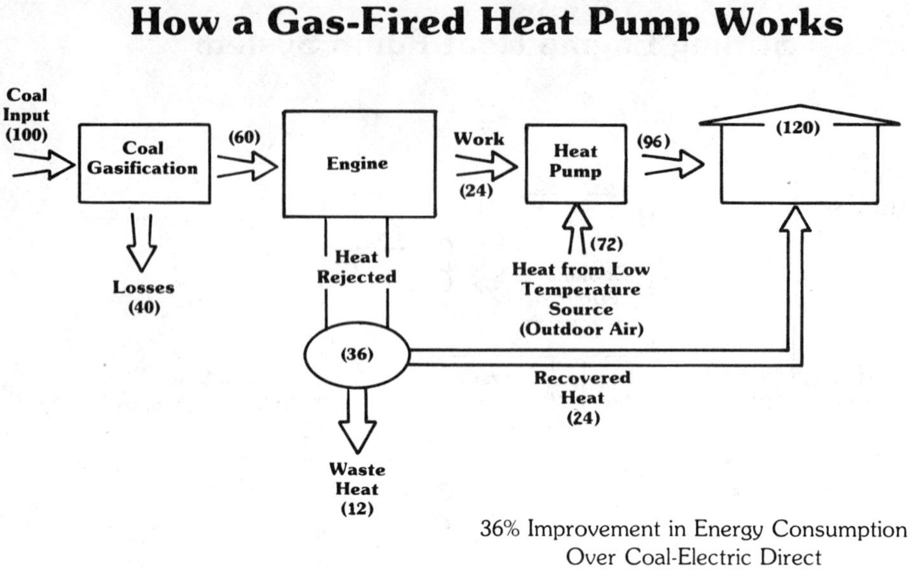
How a Gas-Fired Heat Pump Works
Coal Input (100)
Coal Gasification
(60)
Engine
Work
(24)
Heat Pump
(96)
(120)
Losses (40)
Heat Rejected
(36)
(72)
Heat from Low Temperature Source (Outdoor Air)
Recovered Heat (24)
Waste Heat (12)
36% Improvement in Energy Consumption Over Coal-Electric Direct

Figure 9

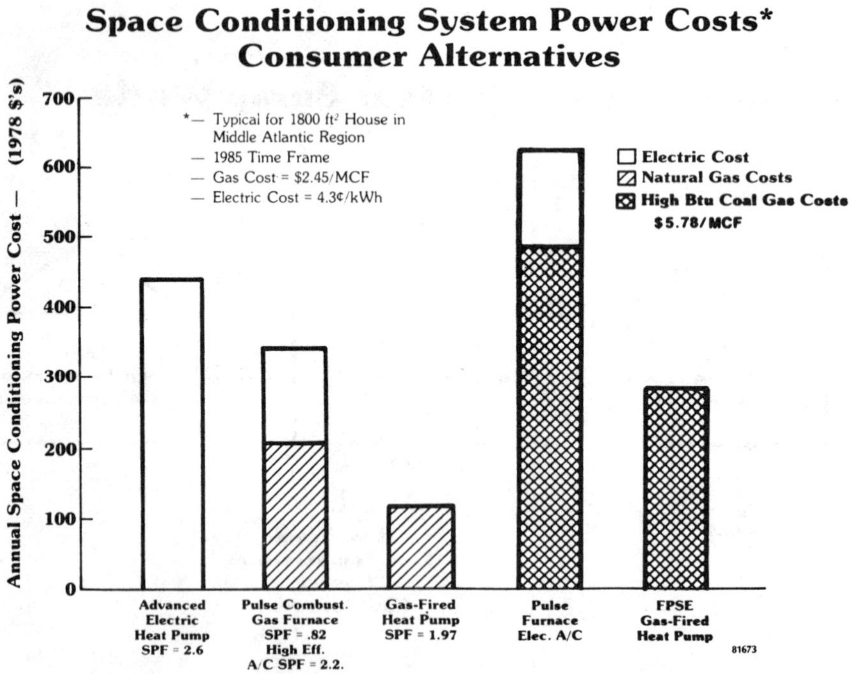
Space Conditioning System Power Costs*
Consumer Alternatives
Annual Space Conditioning Power Cost — (1978 $'s)
700
600
500
400
300
200
100
0
*— Typical for 1800 ft² House in Middle Atlantic Region
— 1985 Time Frame
— Gas Cost = $2.45/MCF
— Electric Cost = 4.3¢/kWh
Electric Cost
Natural Gas Costs
High Btu Coal Gas Costs
$5.78/MCF
Advanced Electric Heat Pump SPF = 2.6
Pulse Combust. Gas Furnace SPF = .82 High Eff. A/C SPF = 2.2.
Gas-Fired Heat Pump SPF = 1.97
Pulse Furnace Elec. A/C
FPSE Gas-Fired Heat Pump
81673

Figure 10

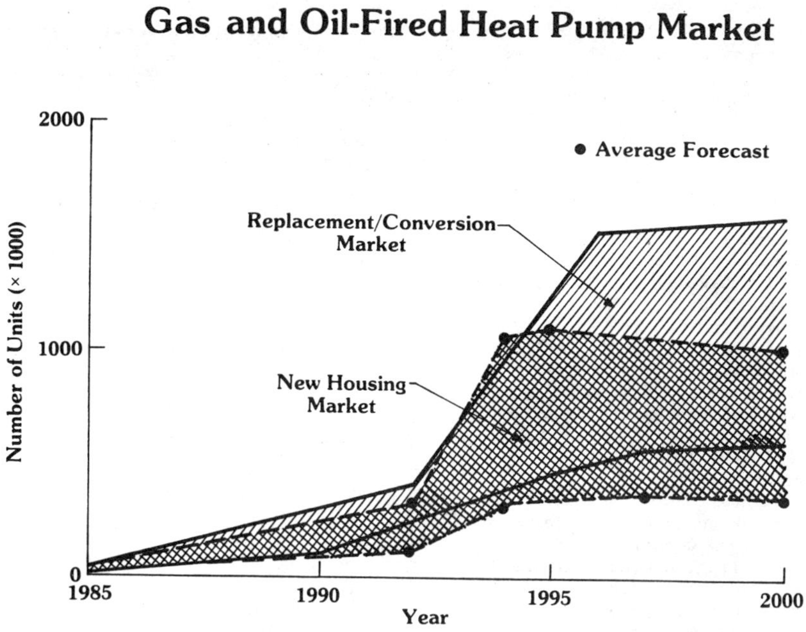

Figure 11

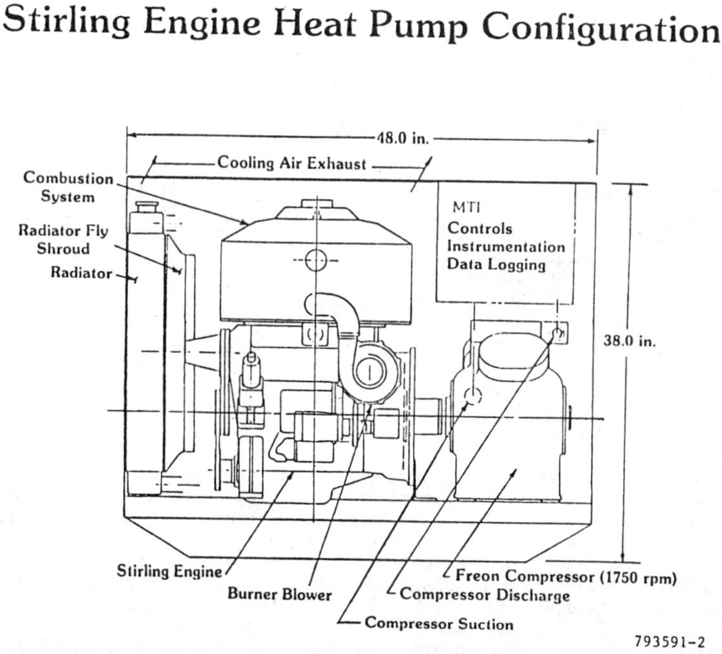

Figure 12

3. Maintenance schedules for this new equipment are comparable to service requirements for conventional HVAC systems.

4. The prime mover is environmentally acceptable.

Table I - Projected National Market for 7-1/2- to 50-Ton Unitary Equipment in 1981

Building Type or Market Segment	1981 Market (Thousands of Tons)
Commercial	
Banks and Offices	
High-Rise	0.0
Low-Rise	27.4
Stores and Others	73.0
Industrial	42.5
Institutional	
Education and Science	22.3
Hospital and Health Treatment	11.8
Public	6.9
Religious	1.1
Other	16.3
Apartments	0.0
Nonhousekeeping Residential	5.9
Total New Construction	207.2
Remodeling, Replacement, & Retrofit	105.0
Total Available Market — North Central Region	312.2
Total U. S. Market	1449.7

Table II - Projected National Market for 50- to 400-Ton Equipment in 1981

Building Type or Market Segment	1981 Market (Thousands of Tons)		
	Unitary	Built-Up	Total
Commercial			
Banks and Offices			
High-Rise	0.0	84.0	84.0
Low-Rise	29.1	113.3	142.4
Stores and Others	60.4	102.0	162.4
Industrial	28.5	90.6	119.1
Institutional			
Education and Science	29.9	65.9	75.8
Hospital and Health Treatment	11.2	110.1	121.3
Public	7.2	35.8	43.0
Religious	1.9	9.0	10.9
Other	8.4	42.6	51.0
Apartments	2.1	108.4	110.5
Nonhousekeeping Residential	1.0	32.1	33.1
Total Available Market	179.7	793.8	973.5

The plan to achieve marketable Stirling engine heat pump systems for the nonresidential market is to leverage the technical work being performed in the ASE development program, modifying certain identified components to extend life. In addition, anticipated production volume for the automotive engine will enhance the business potential due to the "piggyback" effect. The planned use of standard heat pump components will also be advantageous since conventional maintenance and service practices can be applied and, therefore, no new technology is needed.

A comparison between the automotive engine design and that required for stationary applications reveals that the need to extend life to the 50,000 hour range is presently not covered in the ASE development program. In support of this need, a study was conducted to evaluate the potential of successfully modifying the ASE engine to achieve the required 50,000 hour life. The conclusion of this study indicated that derating the advanced ASE engine from rated output to 50 hp by reducing engine operating temperature, engine mean pressure, and operating speed would result in a design life of 50,000 hours. It should be noted that the study concentrated on the main structural limitations of the engine, and did not review in any significant detail the other life-limiting components such as the drive, pistons, piston rod seals, or preheater.

ENGINE-DRIVEN POWER GENERATORS

Stationary electric generator markets, normally termed APU's (Auxiliary Power Units), are currently supplied by the internal-combustion engine and diesel engine manufacturers. If today's market is projected to what could be expected in the mid/late 1980's, approximately 600,000 engines per year will be manuractured in power ranges up to 1000 kW. Most of this market (over 70%) resides in the very small gasoline engine applications, leaving a likely market of about 150,000 engines per year that could conceivably be penetrated by a Stirling engine in the early production years. Looking more carefully at the market data reveals that the need for engines divides conveniently into two major groups — below 15 kW (60% of the market volume) and above 15 kW (40% of the market volume), with the major need in the large engine market between 100 and 400 kW. Obviously, when the market dollar sales volume is examined, then it is seen that the large engine market is four times that of the small engine field; however, a substantial dollar value is still associated with the small engine market.

SMALL ELECTRIC POWER GENERATORS

In the small heat engine field, neglecting possible long-term transportation applications (hybrid electric vehicles), the features of fuel flexibility, high efficiency, and environmental considerations were matched with market needs and application areas. Table 3 shows that a wide range of conventional applications for both government and commercial use benefit highly from this technology. In addition, the engine's inherent features now permit more wide-reaching market brainstorming in the areas of solar, agriculture, and coal-derived fuel applications. Figure 13 presents the life cycle cost comparison of various advanced energy conversion systems in ¢/kW-hr., compared with the free-piston Stirling engine. The effect of projected life cycle costs in the solar application can be related directly to the cost of the concentrator component of the system. In Figure 14, thermal efficiency (a parameter that is factored into the life cycle cost analysis) for small (<15 kW) engines is shown together with the relative concentrator acquisition cost as a function of system efficiency. The well-known conclusion is that engine efficiency is critical in determining collector size and, in fact, may be the determining factor in making dispersed (Figure 15) thermal solar systems a reality.

The same effect can be shown when biomass energy power conversion systems are examined. It is believed that a substantial market in agricultural water pumping applications in the United States and, most importantly, in developing countries, will be matured. Figure 16 shows the effect that system efficiency has on the amount of biomass needed in varying climates to produce a given quantity of power for the pumping application. As shown, only the Stirling engine with its associated high system efficiency would provide a positive economic position in the arid climates where, or course, water pumping for irrigation is most needed.

Advanced free-piston generator designs already executed through the preliminary phase, based on experimental hardware test efforts, have shown that these engines are realistic and cost-effective. Design studies have shown that a complete, self-contained power system, shown in Figure 17, can be packaged in a volume of 13.5 cubic feet for a military application, and weigh less than 275 pounds, meeting the stringent military requirements in Figure 18.

If one studies the life cycle cost data and future projections for this type of engine configuration, it can be seen that the free-piston Stirling generator unit is clearly the leader over present conventional systems. A comparison with the competitive systems indicates, as shown in Figure 19, that on a life cycle cost basis, the free-piston Stirling generator system outperforms the diesel by over a three-to-one margin. Only the fuel cell is competitive in the long-term; however, a signif-

TABLE III - FREE-PISTON STIRLING ENGINE GENERATOR APPLICATIONS

FREE-PISTON STIRLING GEN. FEATURE — FUNCTION	MARKET BENEFIT	APPLICATIONS	
		GOVERNMENT	COMMERCIAL
• LOW NOISE	— LOW DETECTION PROBABILITY — QUIET LEISURE ENVIRONMENT — QUIET WORKING CONDITIONS	— REMOTE SITES — FIELD OPERATIONS POWER	— RV'S/CABINS/BOATS — CONSTRUCTION SITES
• LOW EMISSIONS • EXTERNAL COMBUSTION • MULTIPLE FUEL CAPABILITY	— CLOSED ENVIRON. OPERATIONS — RELIABLE STARTING — LONG LIFE POTENTIAL — STRATEGIC ADVANTAGE	— UNDERGROUND SHELTERS — ARCTIC OPERATIONS — REMOTE SITES, SPACE — FUEL SOURCE INDEPENDENT POWER	- MINING OPERATIONS - COLD WEATHER CONST. - SOLAR ELEC. CONVERS. - COAL ELEC. CONVERS. - AGRICULTURAL PRODUCT ELEC. CONVERS.
• HIGH CYCLE EFFICIENCY	— FUEL CONSERVATION — REDUCED OPERATING COSTS — REDUCED SYSTEM COSTS	— SPACE NUCLEAR — REMOTE SITES	— CONSTRUCTION POWER — TOTAL ENERGY SYSTEMS — DISPERSED SOLAR
• HERMETICALLY SEALED	— LOW MAINTENANCE	— READY FIELD POWER	— REMOTE POWER IN DEVELOPING COUNTRIES
• NO OIL CONSUMPTION • GAS BEARINGS	— LONG LIFE	— NAVIGATION EQPT. — SPACE POWER	
• LIGHT WEIGHT • HIGH SPECIFIC POWER	— PORTABLE/TRANSPORTABLE	— MOBILE POWER UNITS	— ELECTRIC HYBRID VEHICLES

Figure 13

Small Electric Generator Life Cycle Cost

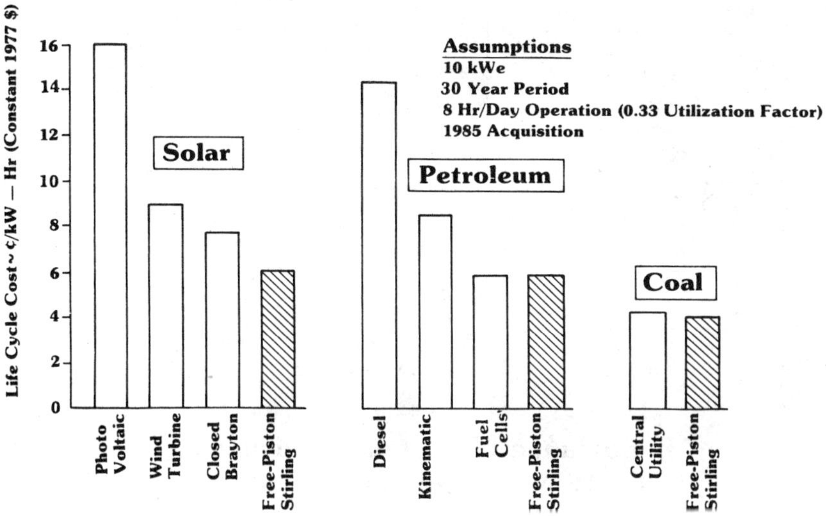

Figure 14

Impact of Power Plant Efficiency on Concentrator Cost

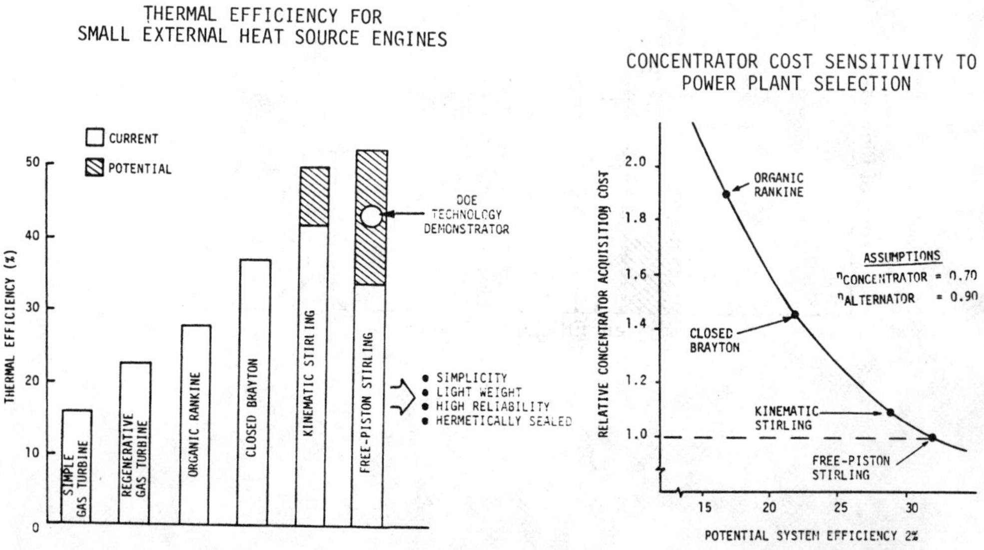

DISTRIBUTED SOLAR POWER SYSTEMS

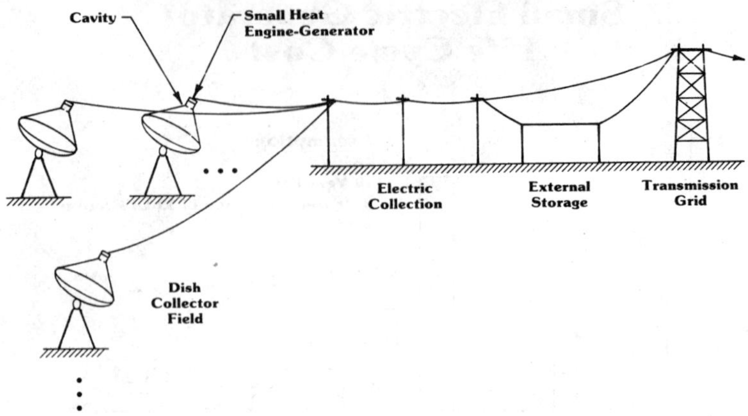

Figure 15

Importance of System Efficiency for Biomass Energy Systems

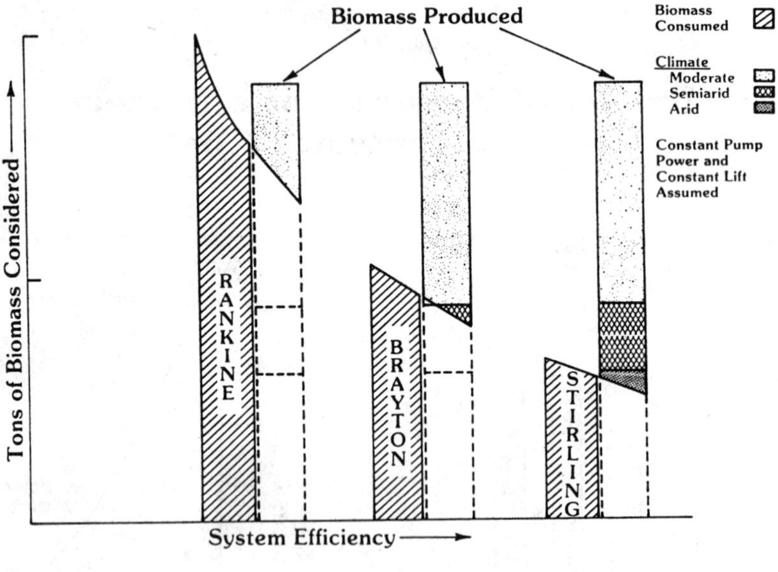

Figure 16

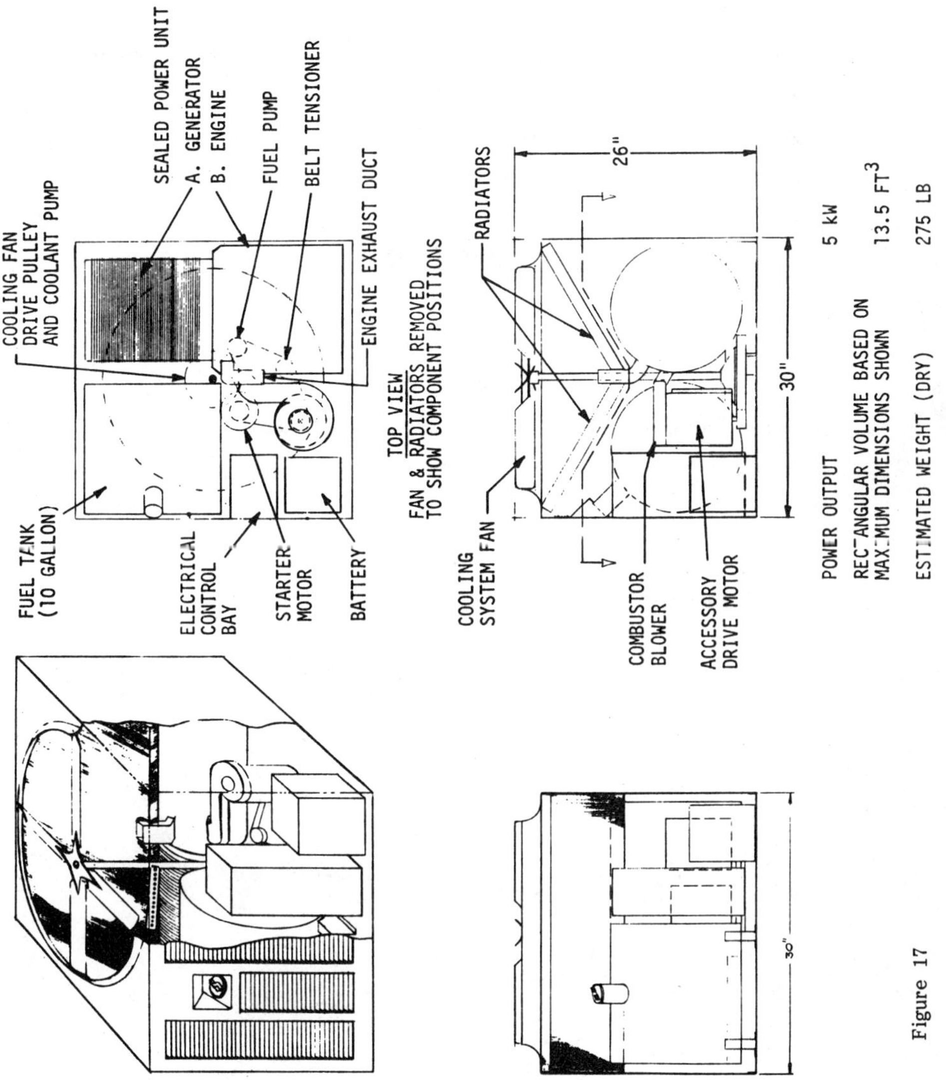
COOLING FAN DRIVE PULLEY AND COOLANT PUMP
SEALED POWER UNIT
A. GENERATOR
B. ENGINE
FUEL PUMP
BELT TENSIONER
ENGINE EXHAUST DUCT
FUEL TANK (10 GALLON)
ELECTRICAL CONTROL BAY
STARTER MOTOR
BATTERY
TOP VIEW
FAN & RADIATORS REMOVED TO SHOW COMPONENT POSITIONS
RADIATORS
26"
30"
COOLING SYSTEM FAN
COMBUSTOR BLOWER
ACCESSORY DRIVE MOTOR
POWER OUTPUT 5 kW
RECTANGULAR VOLUME BASED ON MAXIMUM DIMENSIONS SHOWN 13.5 FT³
ESTIMATED WEIGHT (DRY) 275 LB
30"

Figure 17

Figure 18

Military Power System Characteristics

Power — 5 kWe, 60 Hz, 120/240 Volts, Single Phase

SFC — 0.54 lbm/kW — hr (Diesel 2)

Weight — 300 lbm Max (Less Fuel)

Volume — 15 ft³ Max

Fuel — Multi-Fuel Capability
- Diesel 1 and 2
- Combat Gas
- JP4 - JP5
- Navy Distillates

Noise — Below NC 60 at 10 ft

Exhaust Temperature — 25°F Above Ambient Max

Emissions — Below 1985 EPA Standards

Response, **Wave Form**, **Frequency Control**, **Voltage Control** } Equivalent to Rotary Generators

Figure 19

Fuel/Electric System Total Cost

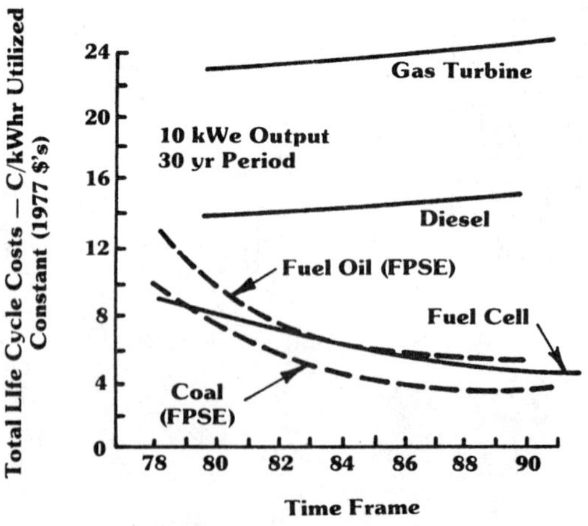

icant advantage in multifuel capability and lower weight makes the free-piston Stirling engine package a more desirable system from an overall standpoint. The comparison is made for a 30 year user period and, therefore, accounts for projected life of the engine configuration, as well as the cost to maintain and operate the system.

LARGE ELECTRIC POWER GENERATORS

The large Stirling engine application can be considered to be divided into two market segments. The first relates to those applications (perhaps less than 100 kW) similar to the commercial heat pump application where the engine can be derived from the technology being developed in the Automotive Stirling Engine Development Program. In the case of electric power generators, this holds as well since with proper upgrading of the design, increased life for the stationary application can be achieved. Increasing the power output of the engine to cover those markets where the sales volumes are attractive requires much additional equipment in the combustion section of the engine, while minimum redesign in the basic engine would be expected. Figure 20 shows, in block form, a possible multifuel, large stationary power system to address the large engine application.

The discussion that follows pertains most closely to the large Stirling engine application; however, the fuel and emission considerations apply as well to the small free-piston Stirling engine.

Although the Stirling engine has the potential for burning a very wide range of fuels much like the steam boiler, the size, cost, and complexity of the combustion system for large-scale applications can vary widely with the types of fuel to be used and the emissions' requirements. The simplest systems are those for burning clean-distillate fuels from petroleum, shale, or coal sources; however, greater complexity is required for burning heavy coal liquids and petroleum residual fuels containing nitrogen and ash, while the system to burn raw coal would again be different.

Clean-distillate fuels having characteristics similar to No. 2 burner fuel, the basic fuel for gas turbines and diesel engines, can be burned cleanly with acceptable pollutant emissions, and used directly in a compact heat exchanger integrated with the Stirling engine. Much of the technology developed for gas turbine combustion systems would be applicable to engine designs for such fuels. The higher nitrogen content of shale and coal fuels, and the high aromatic hydrocarbon content of fuels, would require special consideration in design of the combustion system to minimize emission of NO_x and smoke.

Heavy liquid fuels, including residual oil from petroleum and the heavier coal-derived liquid fuels, may require a fuel system in which the fuel is heated to reduce the viscosity for pumping and atomization. A combustion system that will accommodate the larger combustion volume required and the higher flame radiation which increases heat transfer to combustion walls, and the use of additives to control corrosion and deposits in the heat exchanger, may also be required. The relatively high ash content of these fuels reduces the attractiveness of compact heat exchanger surfaces, as it will be necessary to provide means of removing ash deposits periodically. This may introduce the need for a secondary heat exchange loop to minimize dead space in the Stirling engine heat exchanger, and maintain an accpetable pressure ratio. The combustion of heavy fuels which may contain much more fuel-bound nitrogen than current fuels will also require special staging techniques to minimize NO_x emission, while at the same time, avoiding soot deposition in the heat exchanger, and visible smoke.

Low BTU gas, because of its very large volume, will require a somewhat larger burner than would be required for medium BTU gas or pipeline (high BTU) gas, and may require a combustion space comparable to, or slightly larger than would be required for heavy liquid fuels. It also has a rather narrow flammability range which must be accommodated in burner design for adequate turndown and variable-load operation. If the gas is cleaned to remove hydrogen sulfide and particulates at the gasifier, no special emission problems would be expected at the engine. It is assumed that low BTU gas would be piped from a central gasifier to the engine site, as a 1000 kW installation would not be large enough to justify a dedicated gasifier.

MULTI—SOLID FUEL STIRLING STATIONARY POWER SYSTEMS

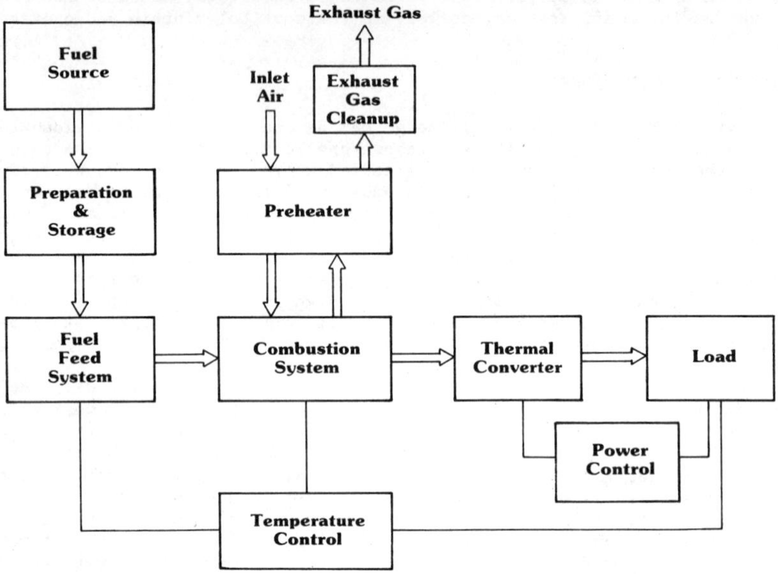

Figure 20

ALTERNATIVE MULTI-FUEL CONVERSION SYSTEMS

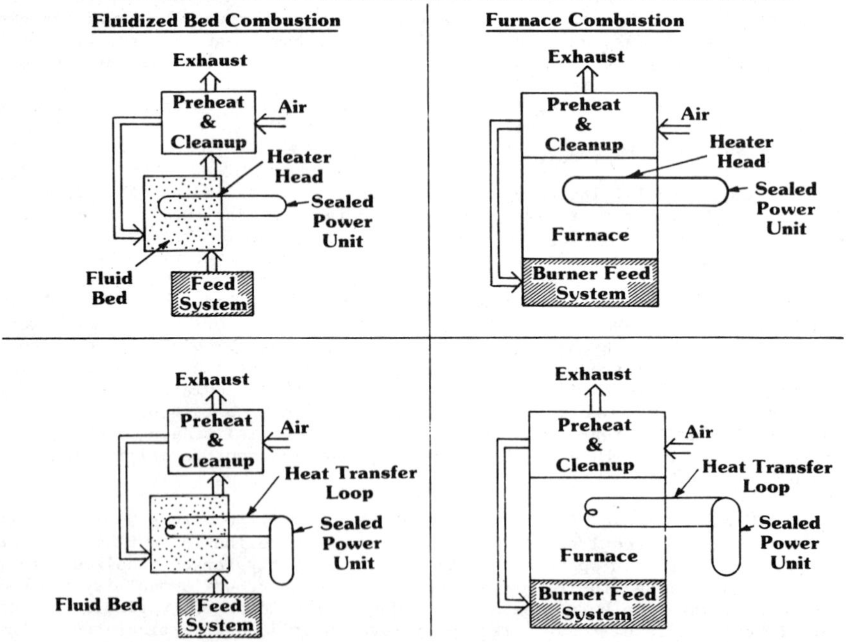

Figure 21

Coal in solid form contains considerable ash and sulfur, leading to potential problems of corrosion and deposits in heat exchangers, and potential emission or particulates and SO_2. It appears to isolate coal combustion products from the Stirling engine by means of a secondary heat-transfer loop to permit use of a small internal volume in the Stirling primary heat exchanger, and use of large, cleanable heat-transfer surfaces in the secondary heat exchanger exposed to coal combustion products. With this type of design, it would be possible to fire coal directly in pulverized form on a stoker, or in a fluidized-bed. Pulverized coal or stoker firing would exhaust coal ash and sulfur dioxide with the stack gas. To meet environmental requirements, it would probably be necessary to include a stack-gas scrubber and particulate collector in the exhaust system. An alternative way of firing coal is in a fluidized-bed containing limestone for sulfur recovery of $CaSO_3$. With the conventional fluidized-bed, up to 80% of sulfur capture appears economic, eliminating the need for a stack-gas scrubber. In addition, the heat-transfer coefficients to tubes immersed in the fluidized-bed are extremely high, minimizing the amount of surface needed. Such a system would appear attractive for heating a secondary heat-exchange loop.

For gaseous and liquid fuels, the technology is established for combustor design using fuel nozzles, primary zone mixing air, and various types of wall cooling. A typical liquid-fuel combustion system used in the MOD I engine, shown previously in Figure 3, uses an air blower to make preheated air through the combustion system where fuel is introduced and ignited. The combustion products are passed over the heater tubes, transferring heat to the engine, and exhausting through the preheater, completing the cycle.

For solid fuels, as mentioned previously, the conventional methods of stoker-fired and pulverized coal combustion are well-established. A cyclone type burner produces a high heat release rate per unit of volume, and may be attractive for Stirling engine application. In addition, recent research and development on fluidized-bed combustion has produced some approaches that could be used for Stirling engines, and that merit further investigation.

For the large Stirling engine application, a remotely coupled engine/combustion system is probably required to provide design flexibility in the heater head component. Liquid-liquid transport loops utilize the sensible heat capacity of a liquid metal or a molten salt to carry thermal energy by convection. The primary design considerations are the choice of the transport medium and the technique for start-up and shutdown, since most potential mediums are solid at ambient temperature. Choice of a suitable transport medium needs to consider: toxicity, flammability, chemical stability, ease of handling, materials compatability, cost, thermal transport properties, and solidification temperature.

Liquid-gas (or heat pipe) transport loops use the latent heat of vaporization of a suitable transport medium to transfer thermal energy from the hotter to colder region of the loop. The primary design considerations are the choice of fluid, the method for ensuring adequate wetting of the evaporator surface, and the technique for start-up and shutdown. Choice of suitable transport medium depends on the same considerations as in a liquid-liquid system, with the additional consideration of the fluid vaporization properties at operating temperature.

Gas-gas transport loops utilize the sensible heat capacity of a gas to carry thermal energy by convection. There are, in general, two design approaches: the hermetic loop and the nonhermetic loop, or air loop. In the latter case, the transport medium is air at ambient temperature. The air is circulated by a suitable high-temperature blower past the receiver heat-transfer surfaces and over the engine heater head tubes. Both the combustor heat-exchanger surface and the engine head tubes would be thermally augmented by fins or other techniques to offset the low heat transfer capability of the air. The amount of surface augmentation required would depend on a trade between the convective temperature drop, the convective pressure drop, and the heat exchanger size. The hermetic system depends on the use of a pressurized gas (possibly air, but more likely helium) to improve the thermal performance of the system. The potential performance improvement must be traded against increased structural and fabrication difficulties, and the addition of another reliability factor — transport loop gas leakage.

The appropriate system for implementing a large Stirling engine for stationary power generation has not been thoroughly investigated; however, the concepts presented, shown schematically in Figure 21, do seem to have technical merit. Whether economic justification can be shown remains an issue t. be resolved.

SUMMARY

Current analysis indicates a sizable stationary engine market representing a target for penetration with advanced Stirling engines. Successful development and marketing will encourage new market opportunities for gas-fired heat pumps for the residential and commercial consumer. As a further application, long-range solar uses of Stirling engine technology appear probable. Of particular interest to the agricultural and third-world market in the near term, is a Stirling engine application using a biomass fuel farm approach, an application which will depend on successful Stirling engine development. The critical factor in future commercialization is cited as development success since inherently the engine has the necessary features in its multifuel capability, high efficiency, and reliability.

From a national interest standpoint, significant petroleum and natural gas reserves could be maintained with the convergence of Stirling engine readiness and synthetic fuel development. It is openly recognized and admitted that substantial funds and resources are required to support this advanced technology to ensure that the potential impact is realized in the appropriate time frame and across the wide range of applications. Development must proceed rapidly so that Stirling engine technology is in place when the economic climate and natural market demand is ready. Continued government participation in funding long-range, high technology programs that are in the national interest must be maintained to ensure availability when the private sector is ready.

ACKNOWLEDGEMENTS

The author would like to express appreciation for the use of certain data in this summary paper to Thomas J. Marusak and Steven J. Piller of Mechanical Technology Incorporated.

REFERENCES

1. "Assessment of the State of Technology of Automotive Stirling Engines," MTI 79ASERE2, DOE/NASA 0032-79-4, or NASA CR-159631.

2. "Design Study of a 15 kWe Free-Piston Stirling Engine Linear-Alternator for Dispersed Solar Electric Power Systems," G. R. Dochat, et. al., DOE/NASA 0056-79/1, or NASA CR-159587.

3. "The Automotive Stirling Engine: Prime Mover for a Nonresidential Heat Pump," B. Goldwater, SAE 1981, paper to be published.

4. "Current Free-Piston Stirling Engine Technology and Applications," B. Goldwater, IECEC 1979, 8412-0513-2/79/0779-246.

5. "A High Seasonal Performance Factor Gas Heat Pump for the North Central United States," Consolidated Natural Gas, MTI, and W. E. Hill Company, 9 volumes, CNG, January, 1980.

Part IV

TECHNOLOGY FOR FOSSIL, NUCLEAR AND GEOTHERMAL RESOURCES

The world and the U.S. have substantial fossil, nuclear and geothermal resources yet untapped. The challenge is to convert these resources to usable energy forms in economic, safe and environmentally sound manners.

This section examines the progress in the further development and use of technologies for using coal, oil, natural gas, oil shale, nuclear energy, and geothermal resources.

REBIRTH OF THE OIL SHALE INDUSTRY IN THE UNITED STATES OF AMERICA

Paul A. Petzrick

Department of Energy

In 1846, the United States was illuminated by whale oil, town gas, and candles. Seven-hundred thirty-six ships were at sea collecting whale oil as the liquid fuel of the day. By 1855, the gradual extermination of the sperm whale led to an astronomical price of $60 a barrel for whale oil. Meanwhile, in France in 1838 and later in Scotland shale oil was being produced. European immigrants, based on this experience, started small shale oil operations in the Ohio Valley and by 1859, fifty-five such plants operating over 100 retorts were in business. In 1859, Titus finished his first oil well. By the following year oil dropped to 10¢ a barrel and the oil shale plants were out of business. Many of them were saved from financial disaster by converting to oil refineries.

With the invention of the automobile there was a sudden expansion in the requirements for liquid fuel and during the period 1915-1920 there was a boom in oil shale interest as domestic production of liquid fuel did not meet our needs. Acting under the Mining Law of 1872, 30,000 oil shale claims were staked. The claims that were ultimately patented have become the private property now owned by several leading oil companies and are the basis for their initial oil shale activity in Colorado at this time. At the end of World War I interest waned and passage of the Mineral Leasing Act of 1920 removed oil shale from the list of locatable minerals and ended the staking of claims. The discovery of large oil fields in east Texas condemned oil shale to second place in the eyes of oil men. Interest began to grow again in the 1930's and the shortages of World War II led to passage of the Synthetic Fuels Act of 1944. Under this act the Bureau of Mines established the experimental facility at Anvil Points and initiated development of the gas combustion retort. The facility was subsequently leased to six oil companies who continued improvement of the gas combustion retort. This activity has provided the basis for many of the present surface retorting technologies. Responding to the oil shortages of the early 70's, the Department of the Interior in 1973 leased four prototype oil shale leases in Colorado and Utah. Activity on these leases has proceeded slowly during the decade of energy uncertainty that we have just completed. Simultaneously, in the early seventies the Department of Defense confirmed that oil shale was the best alternative for the production of middle distillate fuels, the backbone of defense mobility fuels and mass transportation fuels. Thus, even if spark ignition engine switch to methanol, the demand for diesel will remain high for trucks, buses, ships and planes with a special national security need for a secure source of defense mobility fuels.

Each of the booms of interest in oil shale in the United States during this century has provided us with important elements of our current readiness to proceed with large scale activity during the remainder of the century.

Specifically, the claims of the 'teens and leases of the seventies have supplied the resources to the private sector; Government and industry experiments of the '40's and '50's have provided the extraction technology; and the favorable properties of the shale oil itself assure its market in a world that will never again know cheap petroleum.

With every element for a successful venture apparently present, will oil shale finally proceed to commercial production having failed so often to take this final step? A big difference is the presence of the big companies. In any case, the answer is being carved in the marlstone of the Green River Formation right now.

Chevron is there at Red Point at the fork of No Name Creek and Willow Creek preparing for mine development to support operations producing over 100,000 barrels per day.

Superior is there designing a 15,000 barrel per day plant for the Pacific property they now share with Sohio and Cleveland-Cliffs. The plant will be comparable to the traveling grate iron ore operation in La Perla, Mexico.

Union is there having completed construction of their mine bench and access road for their module on the East Fork of Parachute Creek. Mine development is underway for the 10,000 barrel per day module to be expanded to a 50,000 barrel per day plant.

Occidental is there ready to rubblize their 7th and 8th commercial size retorts on the D.A. shale property.

Tenneco is there with Occidental having completed shaft sinking and several hundred feet of adit to develop lease C-b by the modified in situ method or combination methods depending on technical developments.

Exxon is there as a 60% partner in the Colony Project having just let the contract to Brown and Root for construction of the 47,000 barrel per day plant at the head of Parachute Creek. Site construction, road construction, and construction of the community at Battlement Mesa is underway. TOSCO is busy keeping with the pace Exxon has set.

Paraho, although not a big company, has been backed by several of the big companies. Their experimental work at Anvil Points continues and they are designing a 30,000 barrel per day plant for a site near Bonanza, Utah.

Gulf and Standard of Indiana are there developing Lease C-a for a possible combination of modified in situ and surface retorting. Most recently they have filed a modification of their Detailed Development Plan to start a small open pit mine at the northwest corner of the lease to support a Lurgi Module on land they have purchased north of the lease.

Sohio, Sun, and Phillips continue to seek removal of the constraints on developing leases U-a and U-b while looking at production on a larger scale than the originally planned 100,000 barrel per day. Phillips is acting particularly bullish on shale having also purchased significant unpatented mining claims in the Piceance Basin for their Mahogany Zone Shale Project and is actively leasing eastern shale lands.

Quintana Petroleum is a newcomer but has been particularly aggressive in preparation to develop a 30,000 barrel per day operation near Bonanza, Utah. New consortia and partnerships expected to be announced shortly will lead to involvement of virtually all of the major oil companies in western oil shale development.

This activity by the oil shale companies is occuring in a national atmosphere of support for oil shale development. The states and local governments involved recognize that oil shale development must be managed not blocked. Changing policies at the Department of the Interior are expected to lead to a permanent predictable leasing program for oil shale, resolution of state indemnity land selection disputes, and resolution of other public land management issues which have impeded oil shale development.

Most importantly decontrol and current or proposed tax changes are expected to provide the enormous capital required for construction of the plants by large companies or consortia. As the cheapest alternative, oil shale will command support in the free market approach to synthetic fuel development. Chase Manhattan recently reported that synthetic fuel financing would be third only to oil and gas exploration and power plant construction during the next decade. Absent some of the very lucrative incentives for more expensive alternative fuels, oil shale will receive most of this financing. Some companies, particularly middle size and smaller, will still need some form of risk sharing and financial assistance. The Synthetic Fuels Corporation will hopefully meet that need. Support requested under the existing Alternative Fuels Program provides important clues on the readiness of oil shale projects to proceed to commercial production.

Requests under Public Law 96-126 for Feasibility Studies and Cooperative Agreements came from companies traditionally utilizing government support to perfect their technologies and for the feasibility studies of upgrading plants. The Transco project has since withdrawn as Phillips came in to support the Institute of Gas Technology. Union also had problems with government red tape and withdrew its request for assistance with its upgrading facilities.

Requests under Public Law 96-304 included more of the same, but added some important newcomers seeking to determine the feasibility of commercial operations at smaller size and on marginal resources, particularly eastern shales. Funding provided under Public Law 96-304 is the subject of recission under the Program for Economic Recovery.

Request for support approved by Public Law 96-294, under the Defense Production Act and Federal Nonnuclear Energy Research and Development Act requiring early production were dominated by large shale projects seeking risk sharing with the Federal Government to move their projects more rapidly.

Every indicator points to a state of readiness, commitment, and action in oil shale that is unmatched by any alternative for large scale production. The 1980's will finally be the decade in which oil shale moves to the constant growth portion on its curve of evolution.

THE PERSPECTIVE OF A NEW VENTURER

D. ROGER LOPER
CHEVRON SHALE OIL COMPANY

Although my subject emphasizes that Chevron is viewing the oil shale venture as a new entrant in the field this is true only in a narrow sense. In the broad sense, we have a long history of activities in preparation for the development of production of oil from shale. In 1944, Standard Oil Company of California planners began the acquisition of properties in Colorado oil shale country. This was a very early chapter in the modern history of shale oil, the same year in which the government opened a research center for the shale oil industry. Standard of California's activities in acquiring and consolidating oil shale properties went on for about ten years and in that period the Company developed holdings of two mineable properties, both in Garfield County. One of these properties, made up of about 15,000 acres, is located along the western edge of Parachute Creek and the other, some 28,000 acres, lies across Clear Creek and Roan Creek. These lands are held in fee and combined they make Chevron the largest private land holder in the Piceance Basin, which we consider to be the most promising oil shale reserve in the nation.

Because of this large holding, Standard of California has naturally been continually interested in shale oil technology. Periodically over the years the Company has examined numerous processes and projects to determine whether development of these properties could be economical. Over all these years, Chevron's engineers and scientists could not find a way to produce shale oil competitively with petroleum. However, with the Arab embargo in 1973 and the consequent growing recognition of the critical dependence of the United States on imported crude oil, the desirability of recovering oil from shale continued to increase, both economically and politically. It is, of course, by no means clear how the economic contest between petroleum and shale oil will develop. Many of the factors affecting this relationship are outside of the control of the United States. Nevertheless, it appears to us and, apparently to others in the industry, that domestic sources of liquid hydrocarbons may become critically important to our economy in the very near future.

For these reasons, in 1974 Chevron joined with 16 other partners in the Paraho

Project and in this association participated in the expansion of the Anvil Point mine and development of a retort for experimental extraction of shale oil. This first phase of the Paraho Project was completed in 1977 and the next phase has recently been underwritten by industry and government with Chevron continuing to participate. Over the years, the Paraho retort produced over 100,000 barrels of shale oil and the shale and the oil were subjected to considerable study in Chevron's laboratories.

Chevron's most extensive investment in the development of synthetic fuels was the construction of a laboratory dedicated to that activity by Chevron Research Company and built in stages in its Richmond, California research complex beginning in 1975. This laboratory houses several important pilot plants for production of liquid fuels from coal and oil shale. Among them is an oil shale retorting pilot plant which processes one ton of shale per day. Results of the operation of this pilot plant have encouraged Chevron to develop and patent a unique process which holds the promise of lower capital and operating cost, significantly greater energy efficiency, and the production of a more environmentally acceptable form of spent shale. In addition to the one-ton-per-day retorting pilot plant, Chevron Research has constructed and operated a 300 ton-per-day "cold" oil shale retort module at Richmond, California. This model allows scientists to study the flow of crushed shale through a processing sequence modeling the handling of random sized fragments, one of the key technological problems of all shale retorting processes.

In other recent studies into shale oil technology, Chevron's Salt Lake City refinery successfully processed 3500 barrels of shale oil. This experiment was aimed at obtaining understanding of shale oil refining and was made in anticipation of the ultimate need to process shale oil in commercial volumes.

While Chevron recognizes that all this activity is indicative of shale oil technology in a relatively early stage of development, we nonetheless believe that these developments have been sound and that there is a practical and sensible engineering basis for proceeding, first with demonstration and later with commercial applications of this experience and knowledge.

So much for history. Our current activities began in 1979 with an intensive program to put the Company in position to develop the Colorado shale oil properties if and when it becomes economical to do so. We have completed a $21 million program to provide us with a substantial basis for making the conceptual decisions and blueprinting the development of these properties. Included in our present activity is a contract with Morrison-Knudsen and Foster-Wheeler for developing an overall plan leading to commercial production of the properties within the next eight to ten years. This assignment includes technical evaluation of the properties; selection between the two mineable blocks; mine location; mining methods; processing studies including selection of process and the development of economical increments; development of necessary infrastructure, including accommodations for construction and operating personnel and all related requirements for water, power, sewage disposal, treatment and disposal of industrial wastes, transportation, and logistics. Included in this very broad assignment is a study of the socioeconomic effects of our development and alternative strategies to mitigate these effects. These studies are undertaken in the knowledge that they form a part of an overall problem in that others will be developing large industrial facilities in the same area and in the same time frame.

Our present engineering and planning activities include a contract with Environmental Research and Technology of Fort Collins, Colorado to perform environmental baseline studies and to develop plans for the acquisition of the necessary permits to allow our project to go forward.

Our 1979 and 1980 activities included considerable effort at improved definition of the resource base on our Colorado properties. To do this we drilled 64 core holes distributed throughout the properties in order to guide our mining studies and ultimately, assess the overall strategy for development. This year we will drill 16 more to answer specific geological and mine development questions. We also took a sizeable bulk sample of ore from the Clear Creek property, ten thousand tons, which is being used for crushing, beneficiation and retorting tests over the next several months.

The results of all of these activities will put us in position to make the necessary decisions about the future development of our properties. We are now commited to continue and complete this pre-project planning and data acquisition program. We also have the support of Corporate management in advancing the Chevron Research retorting technology which we plan to do by building a small plant -- we call it a semiworks plant -- of 350 tons-per-day capacity as a significant extrapolation of the present pilot plant program. We are pursuing the design and construction of the semiworks plant as a very high priority item. It will not be built on the oil shale property but will be built adjacent to the Chevron USA refinery in Salt Lake City.

Now, a look to the future. Out of all of these studies we begin to perceive the outlines of a large, multibillion dollar development plan for the properties. Depending upon the availability of government incentives to help reduce the risks, we foresee the possibility of developing a mine and building a demonstration module for retorting shale on the property beginning construction perhaps as early as 1983 and coming into production in 1985 or '86. In all likelihood, this demonstration module will use Chevron's retorting technology if successfully proved in the semiworks plant. Alternatively, we might pick the most suitable of alternative retorting technologies available, if some unforeseen problem develops in the Chevron process. Assuming that the Chevron semiworks plant and the demonstration facilities are successful, we visualize expanding the mine and building additional retorting modules in 1986 to 1990 to bring the production to approximately 50,000 barrels per day of shale oil upgraded to a high quality synthetic crude oil.

Again depending on conditions that exist in 1989 and 1990 our plans are laid in such a way that we could readily continue the development of mining and retorts to be able to produce up to 100,000 barrels per day by the mid-1990's.

Later, I will discuss risks we face in proceeding with these major investments. Suffice it to say at this point . . . they cannot be made without assurance of financial assistance from government and without assurance of cooperation from all parties in minimizing delay.

I am aware that many people -- political leadership in Washington and Colorado, our competitors, vendors and contractors, and our neighbors in western Colorado to name a few -- would be interested to know many details of these developing plans. It is premature to try to anticipate these details since our studies will produce them with some reliability between now and the end of the year. Suffice it to say that these outlined plans will require several thousands of construction and operational working people and that in the area where these developments must be concentrated all of these people and all of the facilities to sustain them and all of the facilities to develop the properties must be provided since these are essentially rural areas and remote from centers of transportation and population. Chevron has defined and solved similar problems for housing construction and operating people in remote areas on many projects, both in the United States and abroad. When we can see these requirements for the Shale Oil Project with greater clarity we plan to discuss them first with the western Colorado communities and counties directly affected.

In the present climate of government and public interest in alternatives to imported crude oil, there have been a number of optimistic predictions made about the speed and vigor with which these developing plans for large quantities of shale oil production can be pursued. I would like to discuss some of our concerns which have caused us to take a more deliberate and somewhat cautious position. Basically, we agree that the shale oil reserves,the apparent upcoming shortage of liquid fuels, and the positive position of the embryonic shale oil retorting and processing industry indicate a likelihood that the development of these properties will occur in this next decade. To bring this about, we must address some very sobering realities. The project we perceive for producing shale oil offers grave risks to the private developer, and I wish to express these to you in very strong terms. These risks are all economic in the end, but they can be classified for discussion into specific categories:

Risk of technological failure
Risk of delay
Risk of scale
Risk due to changing cost/price relationships

I have already commented briefly on risks inherent in the developing shale oil technology, but the subject deserves expansion. Retorting of oil shale has intrigued inventors for well over a century and there are several thousand patents on the subject in the current U.S. patent literature, with many hundreds more filed abroad. Yet the fact is that there is no process proven which produces shale oil from rock in anything like commercial quantities. The difficulties of handling many thousands of tons of shale through processing steps requiring heating of the entire mass to 900°F, capturing the vapors, and cooling and disposing of the spent shale are enormous. Efforts to date have handled no more than 10% of the quantities required for commercial production -- most have been far smaller -- and none have been reliable enough to perform continuously day-in and day-out, week-in and week-out as they must do in a profitable commercial venture. These processes all depend upon highly reliable continuous performance of many large pieces of equipment operating in series -- for example, crushers, grading screens, conveyor belts for raw shale, feeders, the retorts themselves with their large attendant fans and compressors, the power plants providing electricity, the cyclones and dust collectors cleaning the gases and the pumps, piping and process vessels for collecting, cleaning and treating the product oil. The plant is the sum of each of these separate parts and failure of one part to function will bring the whole to a stop. What is needed is the skill and foresight of the legendary deacon who built the wonderful one-hoss shay -- each part cunningly contrived and built to perform exactly as long as all the others so that at the end of a long life it simply disintegrated all at one time. I have no doubt that with long experience we will produce oil shale plants which do perform reliably over long continuous runs, as we do now with steel mills, oil refineries, chemical plants and many other compex systems. But we will not likely do it the first time -- or the second or third. It will be a long process, even though we will do our best to do our engineering extremely well and to learn as fast or faster than we have ever done before.

Just one more thought about technological risk -- the shale oil retorting business has the attention of some of the best minds in the world and it will be changing rapidly. Decisions on plants and equipment made in 1980 will be implemented in the field from 1985 onward. Processes and equipment selected now may well be obsolete within a few years after start-up -- or even, possibly, before they ever get into operation.

All of this sounds ominous in the extreme -- and it is; but our Company approaches these technological risks as the least of the several types of risks to the project -- primarily because the technological risks are under our own control -- not fully, of course -- but in greater measure than any of the others.

The second risk to the success of this project is <u>delay</u> -- from whatever source. Technical failures of equipment or systems can produce this delay, of course. But the most evident source of delay is the complex system of permitting required to proceed. A major shale oil project will require more than 200 permits from federal, state and local agencies of many kinds. The process to assemble the required information and follow the established procedures to obtain these permits is inevitably time-consuming. This has been recognized at both Federal and State level, and efforts such as the Colorado Joint Review Process have been made to coordinate, simplify and expedite the process. Even under the Joint Review Process, to gain major permits for a commercial shale oil venture is expected to require about 40 months. While many important projects tasks can be done concurrently with the obtaining of these permits, a prudent project manager will curtail his activities pretty sharply and keep his costs down until he can be fairly sure he can proceed to completion. Consider that one billion dollars invested in savings and loan accounts will produce $295,000 in interest per day. Delay in getting a multi-billion dollar project into production is a risk no Company -- no society -- can lightly accept.

The third great risk in the project we contemplate is the risk of scale. These projects are and must be among the largest ever undertaken by private firms. The Chevron Shale Oil Project appears to be several-fold larger than any project previously tackled by Standard of California. Depending upon the rate of inflation we actually experience in the years ahead, it may cost as much as $5-7 billion dollars, or about one-third of the assets of the Company according to the last annual report. Another measure of scale is the mine -- at the level of 100,000 BPD of production of shale oil -- our target for the early '90's -- we will be mining some 200,000 tons of ore per day and moving perhaps twice that amount in ore, waste, and overburden. This will be bigger than any mine in the world today -- although by that time there will in all likelihood be several as large or larger. A further measure will be the facilities required to house the workers, their families, and the service people needed to sustain the community they will occupy together -- a city or town of several thousand inhabitants. The bigness of such a venture produces risk by its very size alone.

Fourth, the risk of the market place. We can only speculate about what the market place will be like in 1988 and later years. Based on the experience of the last eight years, we might quickly conclude that liquid fuels will continue to increase in real value -- and real cost -- against other commodities in the marketplace. But a longer view of the history of energy costs will take us back to pre-1973 -- into the many long years after World War II when crude oil and petroleum products stayed at about the same price, while the prices of other goods and services increased with steady regularity. In those post-war/pre-Arab embargo years the real cost of crude oil and products declined. We do not expect to see a reversal like that. But we do need to consider possible alternatives to shale oil which may be our real competition in the 1990's. Our studies lead us to believe shale oil will be increasingly valuable in real terms in those future years, and our decision to proceed with our project will depend heavily on this belief. But we recognize a real risk that we may be wrong. Considering the expected competition for oil shale mining and retorting facilities and equipment; and the expected shortage of labor, services, and accommodation in western Colorado; we can readily imagine that the increase in cost of the facilities required for the production of shale oil due to inflation might outstrip the increased price of the product in the marketplace.

Although I have dealt with the risks in some detail, we believe that the needs of the nation for liquid fuels will be such that orderly development of our shale properties is a high priority item for our Company and the nation. It may be possible to accelerate this development by government action either within the present framework of legislation or, more likely, by some additional tax incentives. Even in the presence of these incentives, it is our view that the majority of the risks should still be carried by the industrial organizations forming to develop these facilities. In the case of the Chevron properties, we intend to view each new step against the context of the overall plan to be sure it is appropriate, timely and necessary in the proper development of the facilities to win the oil from the shale. The time frame for reaching our goal of 100,000 barrels per day will depend on our success in each of these early steps and probably could be achieved as early as 1990 and, perhaps, as late as the mid to late '90's. Properly done, we can expect to be able to continue to produce oil from shale on one of these properties for many years, perhaps fifty to seventy years, depending upon the final economic size of the venture. To do this will require a partnership, a partnership made up of the industrial organizations responsible for the development, the local communities affected by it, the state and Federal governments and the ultimate consumer of the product. It is a similar partnership to those we have developed for many years in the various facets of our business but it is sufficiently different to require that we proceed with the best expertise and the most careful technical and management supervision.

PROSPECTS FOR KENTUCKY OIL SHALES
A PROGESS REPORT

Thomas L. Robl, Lee E. Brecher and James E. Funk
Institute for Mining and Minerals Research

INTRODUCTION

During the past two years significant interest and initiatives have been generated in the commercial development of an oil shale industry in the eastern United States. The principal focus of oil shale development since the turn of the century has centered around the shales of the Tertiary Green River Formation of Colorado, Utah and Wyoming (Figure 1). This has been due to the higher yields of oil, as determined by Fischer Assay, compared to other shales.

The causes for the renewed interest in eastern Devonian and Mississippian oil shales are several. The location of the western resources, in a remote semi-arid area, has presented impediments to commercialization. These impediments, including inadequate water resources, inadequate social and industrial bases and institutional problems related to the public ownership of these lands, are not such large factors in the development of an eastern shale industry.

Recent initiatives by the Institute of Gas Technology into the development of retorts which employ hydrogen have demonstrated that yields of oil well in excess of Fischer Assay are feasible from eastern shales (1,2). In addition, symposia (3) and resource studies (4) by that agency have helped to highlight the potential of the eastern resource.

Within recent months, mineral and oil companies began leasing mineral rights to oil shale in Kentucky. These companies include Pyramid Minerals Co. of Olive Hill, Kentucky; Breckinridge Minerals, Inc., a subsidiary of Southern Pacific Petroleum, an Australian based firm, and Phillips Petroleum of Bartlesville, Oklahoma.

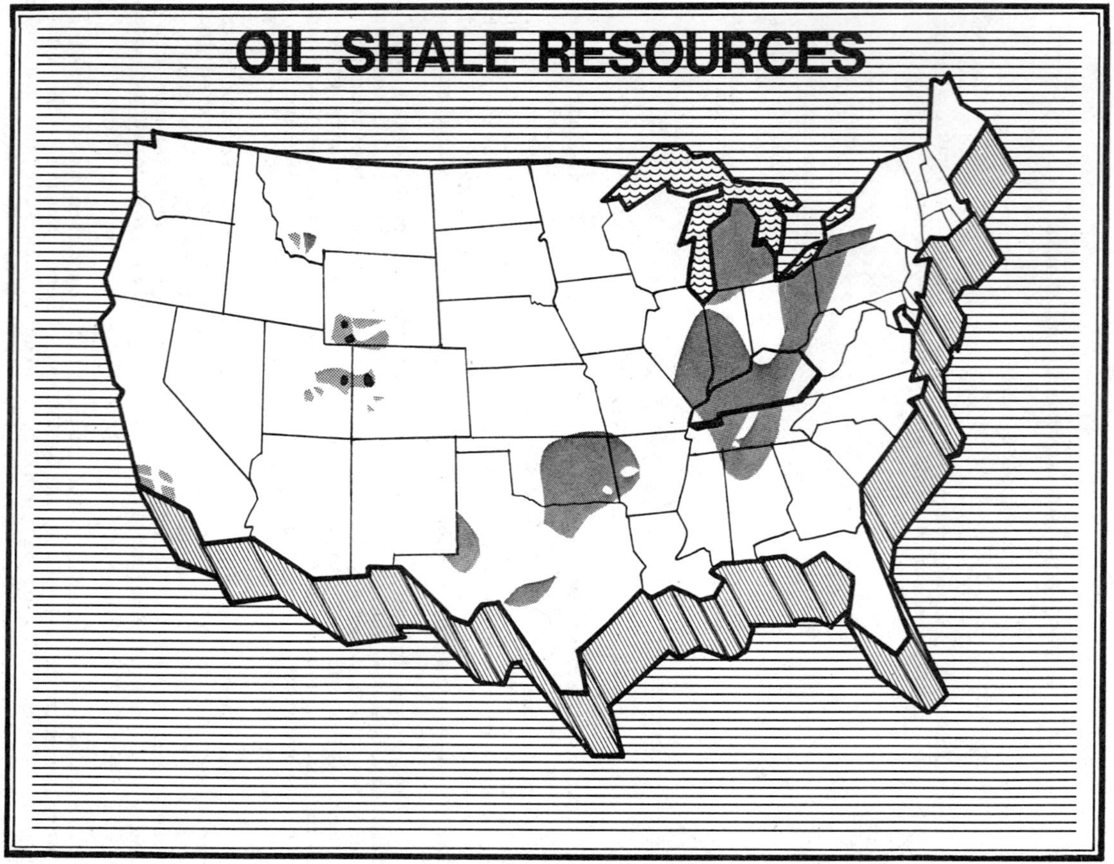

Figure 1. Distribution of oil shale resources in the United States.

The above factors and activities have contributed to the initiation by state and federal agencies of research into the feasibility and nature of an oil shale industry in the east. This paper summarizes recently completed research in Kentucky and briefly outlines the nature of ongoing research.

SYNTHETIC FUELS FROM EASTERN SHALE--THE DAVY MCKEE STUDY

The most recent and extensive study on the feasibility of an eastern shale oil industry was completed in January, 1981 (5,6). This study was funded by the U.S. Department of Energy in the form of a grant to the Buffalo Trace Area Development District of Kentucky (Mr. Terrell Ross, Director). The prime contractor on the project was the Davy McKee Corporation, a large architect and engineering firm located in Cleveland, Ohio. The Davy McKee team, headed by Dr. Kirit C. Vyas was responsible for technology assessment and overall project direction. The Cleveland-Cliffs Iron Co. of Cleveland, Ohio was selected as a subcontractor and, under the direction of Mr. Gary Aho, was responsible for mine planning and evaluation. The Institute for Mining and Minerals Research (IMMR) of the University of Kentucky was selected as the subcontractor responsible for the resource assessment portion of the study. This work was coordinated by Dr. Thomas L. Robl. All of the participants cost-shared extensively in the project.

Resource Assessment

Devonian and Mississippian aged black shales outcrop in Kentucky in a crescent shaped pattern around the central Bluegrass area and less extensively in the Cumberland Saddle area of south central Kentucky (Figure 2). These represent some of the more extensive surface and near-surface deposits of black shales in the eastern United States.

The resource assessment work for the project was confined to Lewis and Fleming Counties, which are located in the northeastern portion of the outcrop area and which are within the jurisdiction of the Buffalo Trace Area Development District.

The resource assessment work included the core drilling, sample analysis, shale characterization, reserve estimation and mine site selection. Ten cores drilled along the oil shale outcrop, 7½-minute geologic quadrangle maps and field notes were used to determine the quality and quantity of oil shales. The pertinent geologic strata consist of (from top to bottom) the Borden Formation (series of siltstones and shales), the Sunbury Shale (oil-bearing black shale), the Berea Sandstone (shaly siltstone), the Bedford Shale (silty shale), the Ohio Shale (oil-bearing black shale), the Bisher Limestone (dolomite), and the Crab Orchard Shale.

The Ohio Shale is divided into (top to bottom) the Cleveland Member (50 to 65 feet thick), the Three Lick Bed (12 to 20 feet thick) and the Huron Member (135 to 208 feet thick). The Ohio ranges from approximately 300 to 200 feet in thickness in the study area, thinning from north to south and from east to west. The Sunbury ranges from 15 to 20 feet in thickness throughout most of the area. These two oil shales are separated by the Berea Sandstone and Bedford Shale, which combined are approximately 100 to 120 feet thick in northern Lewis County, decreasing southward to 30 to 35 feet in Fleming County.

Chemical determinations of the shales and associated strata included C, H, N, S, major element oxide, and trace elements. Carbon concentration in the oil shales averaged 11.42, 9.04 and 5.61% in the Sunbury, Cleveland, and Huron intervals respectively.

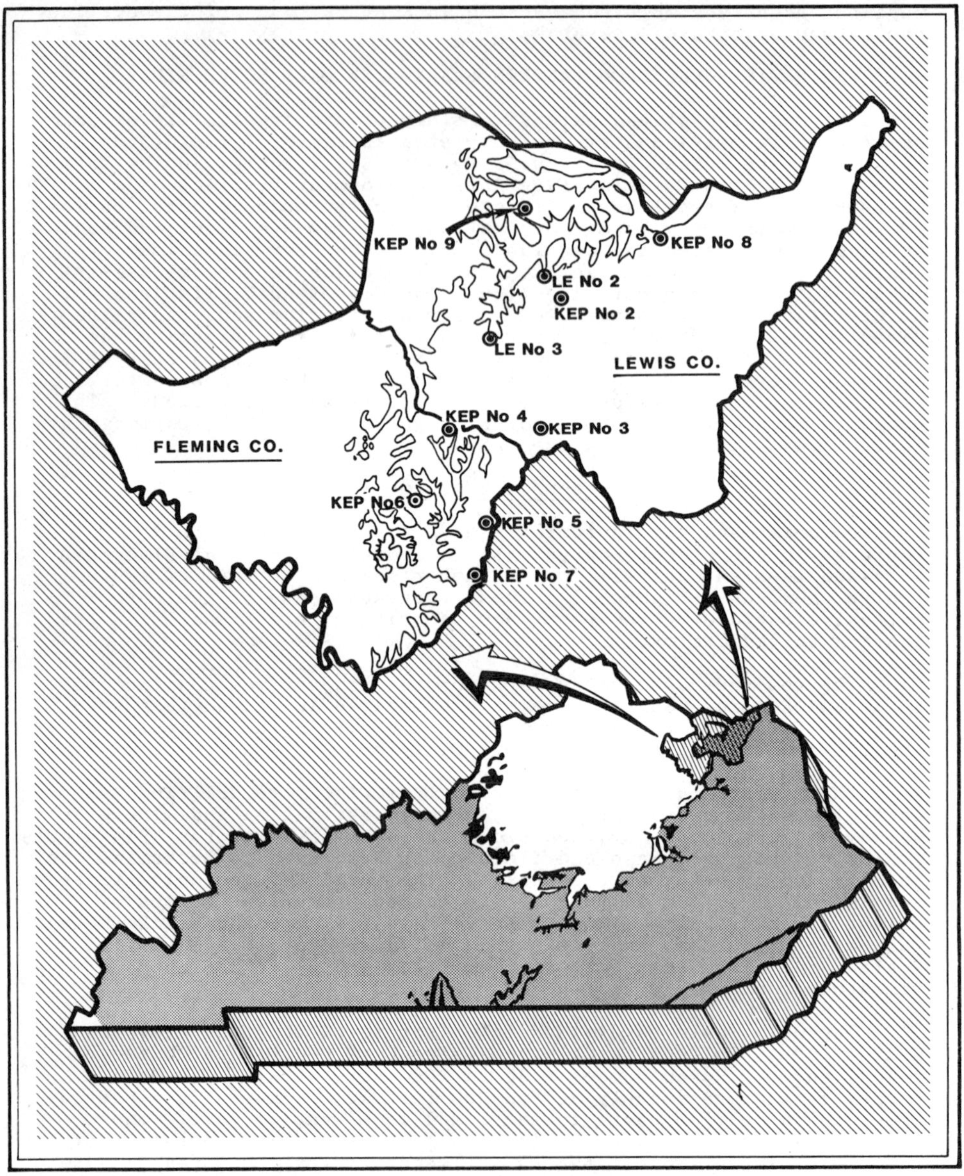

Figure 2. Distribution of Devonian and Mississippian oil shale deposits in Kentucky. Solid black lines indicates outcropping of shale, shaded areas are where shale is present in the subsurface. Insert shows placement of study cores.

The concentration decreased from top to bottom in the Cleveland, and increased from top to bottom in the Huron (Figure 3).

The carbon data were used to define two economic zones: the entire Sunbury Shale and a high grade zone (HGZ) in the Cleveland Member of the Ohio. The Cleveland HGZ extends from the top of the unit to a point where the carbon concentration drops below 8% by weight. The Cleveland HGZ averaged 30.8 feet in thickness for the 10 cores and ranged from 27.9 to 32.8 feet in thickness. Carbon concentration averaged 11.1% in the Cleveland HGZ. Although some intervals with greater than 8% carbon were found in the Huron Member of the Ohio Shale, none of these shales were considered as economic. The principal reason for this was the thickness of lean or barren shales overlying the Huron was too great to maintain a reasonable stripping ratio during their recovery.

Hydrogen averaged 1.2% in the Cleveland HGZ and Sunbury Shale, and was found to increase by 15% on a geographic basis from north to south in the Sunbury shale in the study area. Fischer Assay values for the Sunbury and Cleveland HGZ averaged 10.3 and 11.9 gal/ton for all the cores, and ranged from 9.2 to 11.6 and 10.8 to 13.0 gal/ton. Oil yield was found to increase in a north to south direction. For example, the total oil yield per square foot of surface area per length of defined economic zone was 34.6, 36.5 and 34.1 gal/ft^2 for the three northernmost cores and 41.1, 40.0 and 42.5 gal/ft^2 for the three southernmost. Sulfur content in the shale averaged 3.54% and 2.57% in the Sunbury and Cleveland HGZ respectively and was principally pyritic in nature (Figure 4).

The major element composition of the shale (Table 1) found them to be siliceous (approximately 60% SiO_2, 16% Al_2O_3), low in calcium (< 1.0%), and high in potassium (4.3%), which indicates a significant quartz content, a low carbonate mineral content, and a clay content that is probably illitic in nature. The chemistry of the shales is highly consistent among the cores, with coefficients of variation only slightly greater than analytic variability for many elements.

Several trace elements occurred in interesting, but probably sub-economic concentrations. These are Cu, Co, Mo, Ni, V, U, and Zn (Table 1). Sulfur and trace elements concentrations were low in potential overburden materials, indicating that they would probably not be a source of pollution during mining.

Rock mechanics tests indicate the oil shales to have compressive strengths in the 10,000 to 13,000 psi range. Ash fusion tests indicate the shales to be refractory with a temperature of 1240°C (2260°F) recorded for the initial ash deformation. Bulk densities of the economic oil shale intervals average 2.28 g/cm^3 (142 lb/ft^3). Total volatile hydrocarbon yields from the shales range from 30 to 530 cm^3/kg (1 to 17 ft^3/ton) indicating that gassy conditions may be encountered in underground shale mines.

Shale reserves for the economic shale intervals of the study (i.e. Sunbury and Cleveland HGZ) were calculated by overburden category on a 7½-minute quadrangle map basis for those quadrangles where the shale could be recovered by surface mining. A total reserve of 5.2 million acre-feet or 16.4 billion tons of shale, conservatively representing 4.4 billion barrels of oil, were found. Of this, 6.7 billion tons, representing 1.3 billion barrels of oil, are in a configuration within a stripping ratio of 2.5:1 overall. Most of the shale in the latter category lies in eastern Fleming and southwestern Lewis Counties, in an area of approximately 42,000 acres.

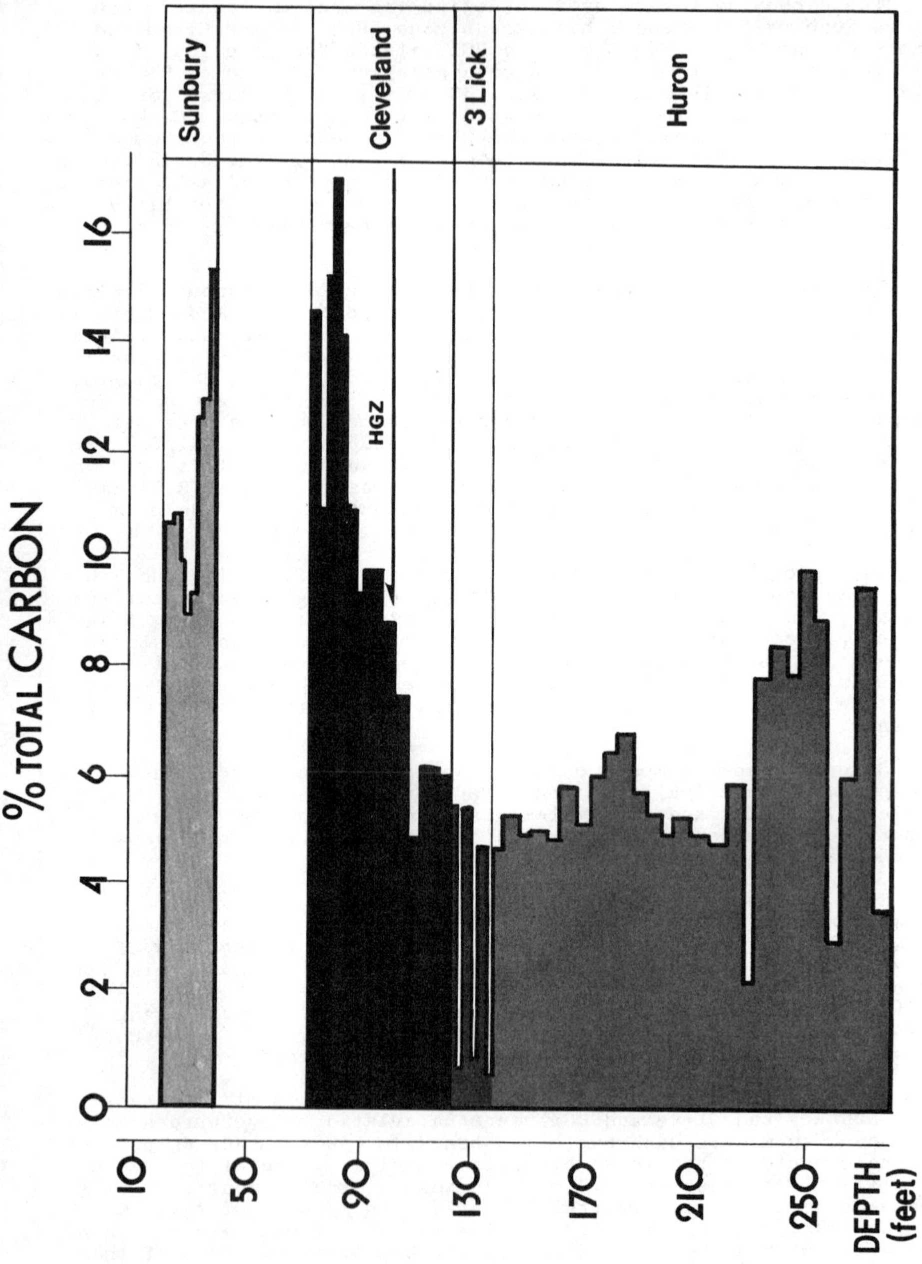

Figure 3. Distribution of carbon with depth for KEP No. 5 core.

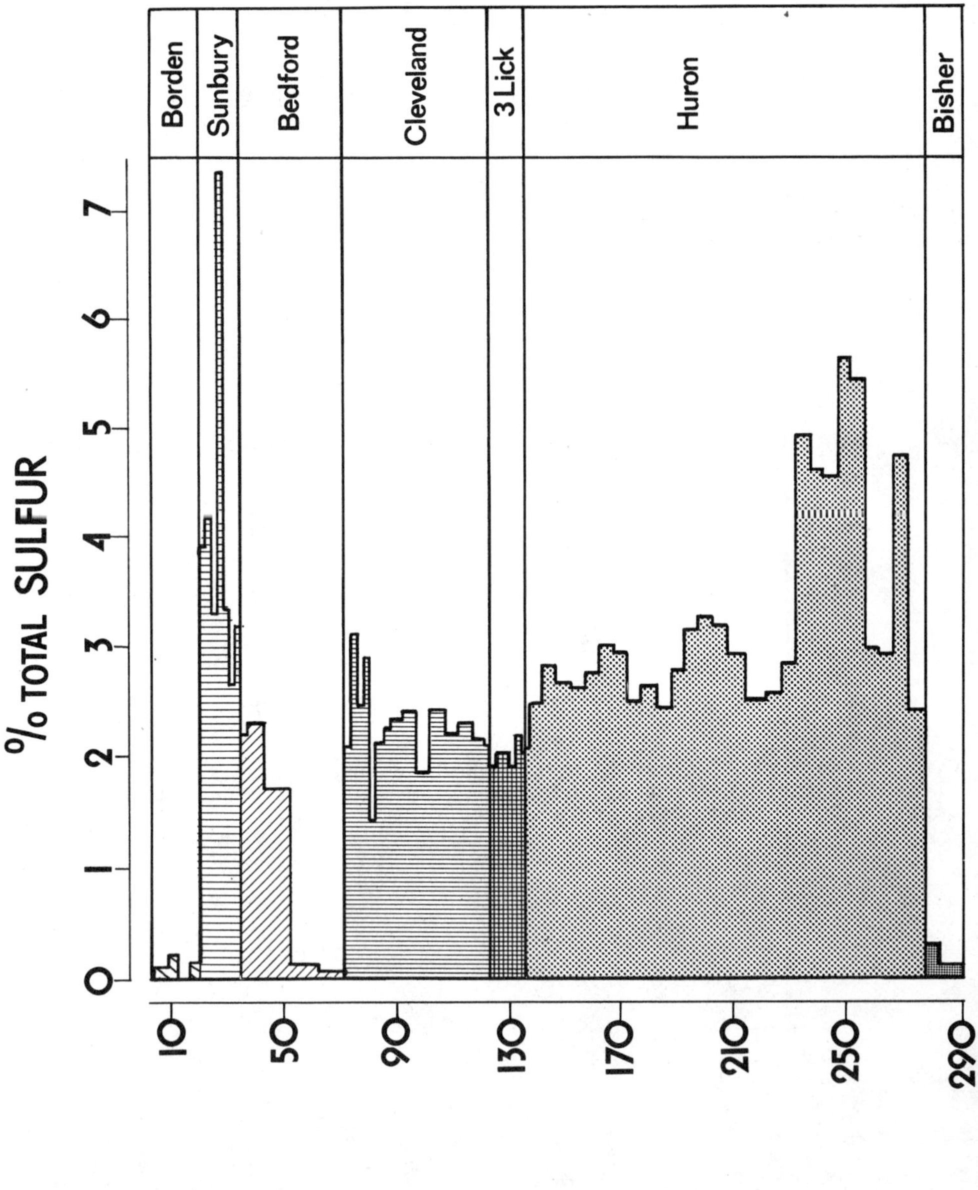

Figure 4. Distribution of sulfur with kepth for KEP No. 5 core.

Table 1. Composition of economic shale units of Lewis and Fleming Counties.

	Sunbury	Cleveland HGZ
C	11.4	11.1
H	1.2	1.2
N	0.4	0.4
S	3.5	2.6
SiO_2	63.7	66.5
Al_2O_3	16.7	16.2
CaO	0.4	0.6
MgO	1.5	1.4
Fe_2O_3	8.7	7.0
K_2O	4.4	4.2
Na_2O	0.4	0.5
TiO_2	0.8	0.8
Ba	647	607
Co	30	22
Cr	218	213
Cu	108	106
Mo	365	124
Ni	282	146
Rb	185	171
Sr	118	122
U	37	20
V	1,533	1,024
Y	57	59
Zr	203	210
Zn	1,126	622

C, H, N and S (all weight %) and U (ppm) are based on raw shale. Elemental oxides (weight %) and trace elements (ppm) reported on the basis of 500°C ash. Data are averages for all cores.

Shale Mine Plan

The mine plan devised by Cleveland-Cliffs was designed to supply sufficient shale for a 10,000 barrel-a-day plant. The mine is a ridge-top removal surface operation site specific to a location in Fleming County. The plan includes the sequential mining of two parallel ridges, in five benches, with operation of 10 years in duration on each ridge. Both the Cleveland HGZ and the Sunbury Shale would be recovered, with the spent oil shale being returned to the ridge top. The overburden and interburden materials would be used as cover with excess materials used as hollow fill. Reclamation plans and sediment control structures were also included in the mine plan.

Stripping ratios of 1.5:1 and 1.9:1 were calculated for the two ridges. A total of 274 million tons of shale would be recovered during the operation of the mine.

The mine plan incorporates conventional mining methods and equipment. Diesel-electric 11 and 12 cubic yard shovels and 85-ton end-dump trucks would be used. Included in the costs are a service building containing offices, warehouses, first aid station, change rooms, shops and maintenance facilities. Also included are road maintenance equipment, a power distribution system, an explosive magazine, fuel docks and a primary crusher facility.

A three-year period was estimated for mine development with preproduction costs calculated ($\pm$ 25%) at \$5.9 million and preproduction capital costs of \$65.1 million (mid-1980 dollars). Operating expenses were estimated at \$0.88/ton for shale and overburden and \$2.19/ton for shale alone. Operating expenses contain no capital associated costs.

Process Evaluation

Two contrasting technologies were used for process evaluation. These are the Paraho Development Corporation's Paraho retort and the HYTORT retort, developed by the Institute of Gas Technology.

The Paraho system is a refractory lined, verticle kiln technology where a moving bed of crushed shale is contacted countercurrently with an upward flow of hot gas to retort the organic constituents of the shale. The system was modified to retort eastern shale by the addition of a second combustion stage to take advantage of the energy left in the form of residual carbon produced by the pyrolysis of these materials. In this modification, the shale is passed through a dynamic seal and goes to a combustion zone, where the leftover carbon is burned, the shale cooled and discharged. The products of combustion are used to heat the recycle retorting gases and then are passed through a boiler to generate steam for in-plant use and power generation.

Both the Paraho and HYTORT plants were designed as grass roots facilities and include mining, material handling, retorting, oil recovery, components to meet environmental standards and all necessary offsites.

The Paraho plant requires a 65 acre site, about 2,000 gpm of makeup water, 1,066 tons/day of crushed limestone and 428,600 scfm of air. The plant produces 8,130 bbl/day shale oil, 2,280 bbl/day light oil, 14,470 scfm of high Btu gas and about 48.3 mw of excess power.

The HYTORT technology is based upon direct, non-catalytic hydrogenation of kerogen at high pressure. The retort is a refractory lined, high pressure, vertical vessel in which a moving flow of shale is reacted with a countercurrent flow of hot gases. Since the vessel operates at 550 psia, high pressure feed and spent shale removal systems are required. The HYTORT process recovers oil at 116% of Fischer Assay, as compared to 90-95% for the Paraho system. The HYTORT system also requires a more complicated process configuration.

The HYTORT plant requires 65 acres, 4,100 gpm makeup water, 565 tons/day oxygen, 45.3 mw electric power and 269,700 scfm of air. The plant produces 10,220 bbls/day of shale oil and 104 tons/day of anhydrous ammonia.

Based upon a detailed process analysis (± 25%) initial plant costs for Paraho and HYTORT of $504 million and $695 million respectively were calculated. Total initial capital costs were $647 million and $890.3 million respectively.

Gross annual operating cost of $63.1 million and $86.1 million were calculated for Paraho and HYTORT respectively, with net costs, after by-product credit adjustments of $10.6 and $74.1 million respectively. Cash flow analysis at 100% equity found a return of 12% at $40/bbl for Paraho and $60/bbl for HYTORT. For the Paraho process, further analysis at 75% debt, 12% interest and 25% equity on a pass-through basis found a return of 24% on equity for $40/bbl.

RESEARCH IN PROGRESS

Since the completion of the McKee study, several new research initiatives into Kentucky oil shales have been started. Funding has been extended by the U.S. Department of Energy to the Kentucky Department of Energy to extend the resource assessment work to the south of Lewis and Fleming Counties. This work, which includes coring, reserve analysis, and geochemistry, focuses on the shales of Rowan, Bath, Montgomery, Powell and Estill Counties of Kentucky. Additional funding in the form of a grant to the Institute for Mining and Minerals Research (IMMR) of the University of Kentucky from the Cleveland-Cliffs Iron Co. will be used to extend the coring to the rest of the shale outcrop areas of Kentucky.

The Kentucky Department of Natural Resources and Environmental Protection has initiated a program into the characteristics of spent oil shale. This project, funded by the Ohio River Basin Commission, will be executed by the IMMR on behalf of the Kentucky Department of Energy. Research will focus on the leaching, fill emplacement and reclamation properties of spent oil shale and associated overburden materials. The Kentucky DNR & EP has also been charged with the development of a set of shale mining and processing regulations for environmental protection.

SUMMARY

The McKee study found a large reserve of shale oil, conservatively estimated at 1.3 billion barrels by Fischer Assay, in a surface-minable configuration in a relatively small area of northeastern Kentucky. The operating costs for a site specific surface mine were calculated at $2.19/ton of shale delivered to the retort. A shale oil selling price competitive in today's market (i.e. $40/bbl) was found to provide a return of about 24% on investment for a 75% debt, 12% interest and 25% equity on a pass-through venture for a grass roots mine and retort based on the Paraho technology.

This recently completed study, in conjunction with mineral leasing activities, indicates a strong prospect for the commercial development of an oil shale industry based in the eastern United States.

Ongoing research efforts in Kentucky, including resource characterization work, studies into the chemical and physical characteristics of spent oil shale materials and the development of a fair and coherent regulatory program will help to delineate the nature and extent of the resource and the environmental practices necessary for its safe utilization.

REFERENCES

1. "The IGT Hytort Process for Hydrogen Retorting of Devonian Oil Shale," Weil, S.A., et al., presented at the Sixth National Conference on Energy and the Environment, Pittsburgh, Pennsylvania, May 21-24, 1979.

2. "Synthetic Fuels from Eastern Devonian Oil Shale," Weil, S.A., et al., Seventh Energy Technology Conference and Exposition, Washington, D.C., March 24-26, 1980, p. 673-681.

3. "Synthetic Fuels from Oil Shale," Symposia sponsored by Institute of Gas Technology, Atlanta, Georgia, December 3-6, 1979.

4. "A Survey of the Major Devonian Oil Shale Resources in the Eastern United States," Institute of Gas Technology, IIT Center, Chicago, Illinois, 1977, 150 p.

5. "Synthetic Fuels from Eastern Oil Shale, Volume 1," Vyas, K.C, G.D. Aho and T.L. Robl, et al., Final Report, Project NC-5477, Federal Grant No. DE-FG-4480R410185, prepared for Buffalo Trace Area Development District, Maysville, Kentucky, January 1981, 346 p.

6. "Synthetic Fuels from Eastern Oil Shale, Volume II, Supporting Data," Robl, T.L., et al., Final Report, Project NC-5477, Federal Grant No. DE-FG-4480R410185, prepared for Buffalo Trace Area Development District, Maysville, Kentucky, December 1980, 248 p.

SASOL - A PROVEN PRESCRIPTION TO CONVERT TONS TO BARRELS!

William A. Samuel, P.E.
Fluor Engineers & Constructors, Inc.

INTRODUCTION

This paper is a report on the current status of the Sasol projects in South Africa, with emphasis on the operating results from Sasol Two. This is especially timely because it appears that many otherwise knowledgeable people are under the misimpression that coal liquefaction is not yet a commercially proven procedure. As I shall describe in this paper, South Africa is currently converting tens of thousands of tons of coal to tens of thousands of barrels of fuel - every day. And more is coming!

Sasol Two was constructed in two phases, beginning from an empty field. Phase One represented about 80 percent of the total construction effort. It included 50 percent of all the units consisting of two or more trains, plus all offsite facilities and utilities as well as critical facilities common to both phases. The construction completion of the first Phase One units was achieved in late 1979, and unit completions continued through the early part of 1980. Commissioning of the Phase One units progressed in accordance with their mechanical completion, and all of the Phase One units are now in operation.

Construction of Phase Two is now also complete, and commissioning of these units is now complete. Full production of the complex should be achieved early this year.

Sasol Three

Sasol Three is essentially a duplicate of Sasol Two. Its construction is now moving on a schedule to bring it into start-up in late 1982, even though its authorization came as recently as early 1979. It is over 50 percent complete. The accelerated schedule is a result of our ability to utilize the existing engineering as done for Sasol Two and to employ our worldwide purchasing capabilities to get new quotes for equipment duplicating Sasol Two for delivery on the Sasol Three program.

OVERALL FLOW SCHEME

Each plant is designed to process about 40,000 metric tons of coal per day. The attached simplified flow sheet, Figure 1, entitled Sasol Synthol Process illustrates the overall process scheme for both Sasol Two and Sasol Three (1). Figure 2 shows physical layout. In the following section each of these units will be discussed in some detail, but a brief general description at this point seems appropriate.

Coal passes from the mine to the coal preparation units, from which the fine coal goes to the power stations and the sized coarse coal to the Lurgi gasifiers. Air enters the air separation system to supply the oxygen, which passes to the Lurgi apparatus along with steam from the power station. In the gasifiers, the reactions occur to produce a raw gas which also includes tars, oil, naphtha, phenols, and ammonia. These liquid by-products are separated in the gas liquor separation system, with the tars going to further refining and the aqueous gas liquor going to units to recover phenols and ammonia. The raw gas proceeds to a Lurgi Rectisol system for removal of the acid gas constituents to produce a pure gas suitable for Fischer-Tropsch reaction.

This purified gas stream goes to the Synthol fluid bed reactor units along with hydrogen and carbon monoxide produced by reforming the methane formed in the processing. The reaction products from the Synthol system go to product recovery units from which the ethylene is separated as a salable chemical. The methane goes to the methane reforming unit, and the other products are sent to a product refinery from which are produced the LPG, gasoline, diesel, fuel oil, jet fuel, and other chemicals as noted on the flow sheet.

UNIT BY UNIT COMMENTS

Coal Mines

Sasol Two and Three will be supplied by the Bosjesspruit Colliery, which is situated in the Highveld coal field. Four shaft systems are planned with combined coal production capabilities of 30 million tons per year. This will supply the design needs of Sasol Two and Sasol Three of 27.5 million tons per year. The mining methods are highly mechanized and include long wall and continuous mining. Mine depth varies between 390 feet and 590 feet with an average of 490 feet. The average seam height is about 10 feet. Working reserves are at least 70 years, with the time being not so much determined by the available coal as by the distance between the ultimate coal front and the plant. The current production rate amounts to over 30,000 metric tons per day. This will be increased beyond this figure in the next few months.

Power Stations

Up to 30 percent of the total coal will go to the power stations which produce steam for process use as well as power generation. Each of the boilers can produce 1.2 million pounds of steam per hour. Sasol Two imports most of its power from outside, but Sasol Three will be almost in balance as far as electric power is concerned.

FIGURE 1

SASOL SYNTHOL PROCESS SIMPLIFIED FLOWSHEET SASOL 2-3

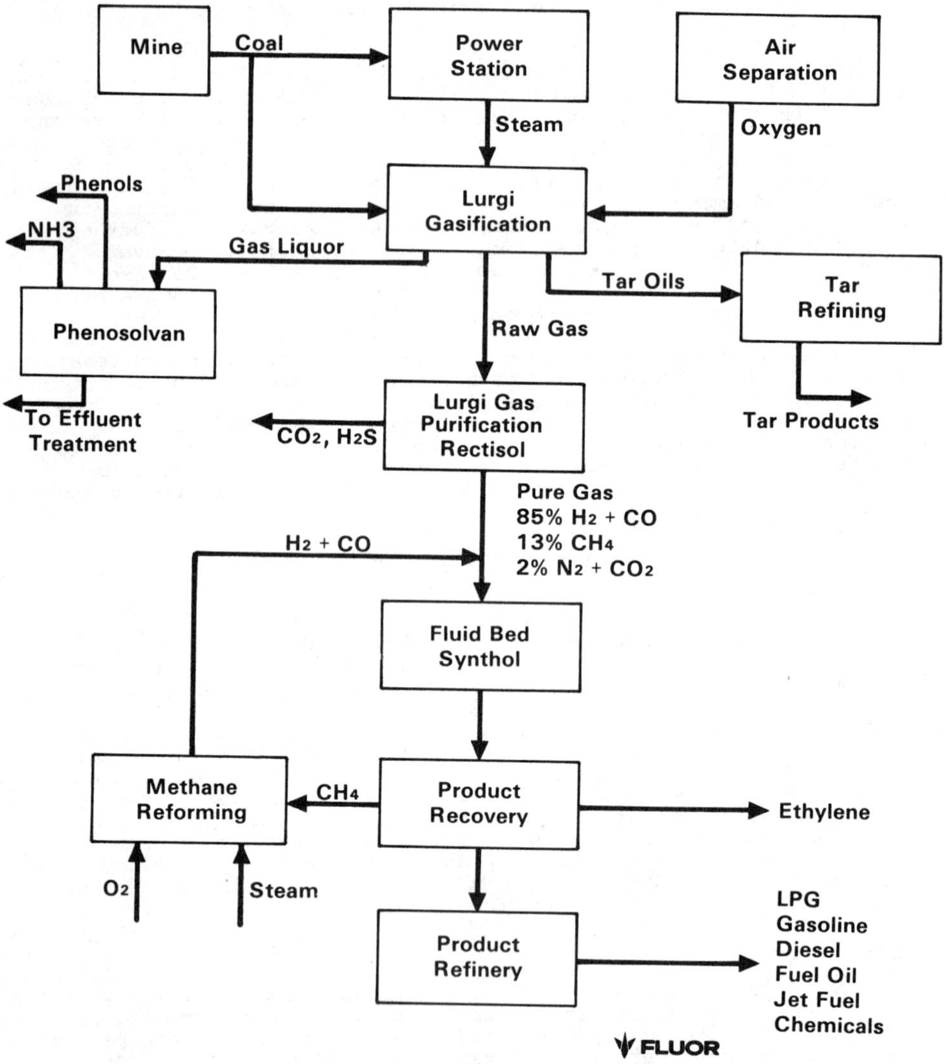

FIGURE 2

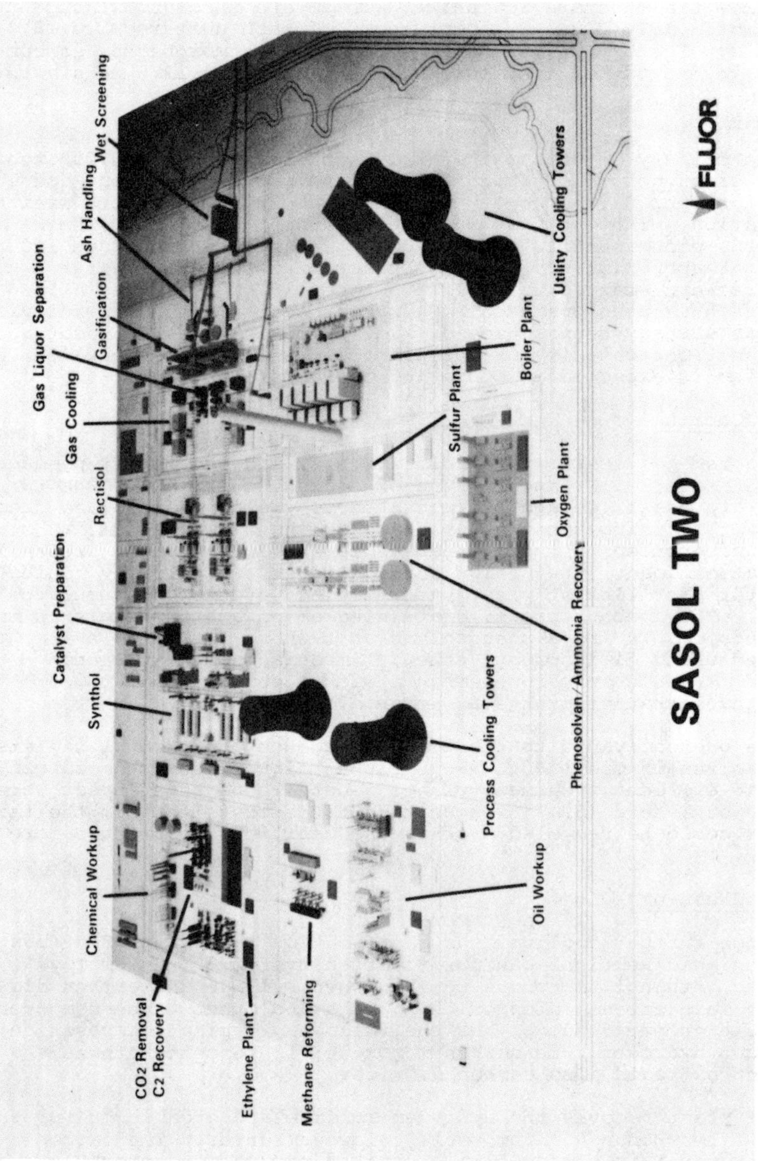

Wet Screening
Ash Handling
Gasification
Gas Liquor Separation
Gas Cooling
Rectisol
Catalyst Preparation
Synthol
Chemical Workup
CO2 Removal
C2 Recovery
Ethylene Plant
Methane Reforming
Oil Workup
Process Cooling Towers
Phenosolvan/Ammonia Recovery
Oxygen Plant
Sulfur Plant
Boiler Plant
Utility Cooling Towers
SASOL TWO
FLUOR

Air Separation System

The biggest energy consumers are the air separation units, in which the oxygen component of air is separated from the nitrogen. Sasol Two has an installed air separation capacity of 13,600 tons a day from 6 air separation units. These are the largest such units in the world (2). The air compressors of each unit require 28.5 megawatts. The oxygen, being required at 35 atmospheres, is compressed in 6 oxygen compressors, each consuming 12.5 megawatts.

Gasifiers

The battery of 36 Lurgi Mark IV gasifiers require 8,000 tons per day of oxygen, along with 28,000 tons per day of steam and 28,000 tons per day of coal. Normally, about 30 gasifiers will be in operation, with the rest on turn-around or stand-by. These gasifiers, each about 13 feet in diameter and weighing some 140 tons, operate at approximately 30 atmospheres pressure. The first gasifiers were started up in late October 1979, and today all 36 gasifiers have been operated. Guaranteed capacity on the individual units has already been achieved. Gas composition is in accordance to design. The capacity of the plant is still being evaluated, but no problem is expected in achieving design production.

Raw Gas Cooling

The raw gas leaves the gasifiers hot, and cooling it produces condensed tars, oils, and a phase of aqueous materials. The tar/oil streams are distilled and hydrogenated to yield gasoline and diesel in the tar refining section, as shown on the flow diagram. The current production of these products is in line with the present gasification load. The design load, however, was already achieved on the tar distillation plant by employing feedstock accumulated earlier. The aqueous phase, containing ammonia and phenols as main components, is fed to a phenosolvan unit where the phenols are extracted using di-isopropyl ether. Ammonia is stripped and purified to yield anhydrous ammonia with a purity equal to, or better than, that of synthetic ammonia.

The phenosolvan with the ammonia recovery unit completed its guarantee run in mid-1980. The stripper liquor from phenosolvan is fed to an effluent treatment plant. [Details of this aspect were recently published (3).] For United States installation, the tar and oils could be recycled back to the gasifiers to produce more syngas.

Raw Gas Clean-up

After further cooling, the gasification raw gas is fed to a Lurgi-designed Rectisol gas purification system. This employs cryogenic methanol to remove the hydrogen sulfide and carbon dioxide from the main stream, along with naphtha and other trace components. The resulting purified gas has the following typical composition: 86 percent hydrogen plus carbon monoxide; 13 percent methane; 1 percent nitrogen plus carbon dioxide.

The plant removes the hydrogen sulfide and organic sulfur to levels of less than 0.1 ppm, well below guarantee. The carbon dioxide removal is achieved to less than one volume percent against a guarantee of 1.5 volume percent maximum.

The carbon dioxide off-gas from the Rectisol, which includes the hydrogen sulfide, will be fed through a Stretford unit where the hydrogen sulfide will be converted into sulfur. The main CO_2 stream will then be released to the atmosphere. The Stretford unit, however, is still under construction, and no operating experience is available as yet.

Synthol Fischer-Tropsch Units

The purified synthesis gas from the Rectisol plant, along with the CO and H_2 from the methane reformer, next pass to the Synthol units - the real heart of a Sasol complex. The Synthol reactors employ a circulating fluid bed with a promoted iron catalyst, as shown in the accompanying simplified schematic diagram, Figure 3. The pressures and temperatures are moderate, and no exotic materials of construction are required.

The Fischer-Tropsch reaction always produces a range of hydrocarbons, along with some oxygenated products. A typical mass distribution of such products has been reported as shown in the table in Figure 4 (4).

By proper selection of reaction conditions, one can influence this distribution to some extent. At Sasol Two and Three, the emphasis is on transport fuels. Thus far all four Phase One Synthol loops have been in operation, and the three Phase Two have run. An eighth unit, with tie lines to Sasol Three, is being added to increase flexibility. No problems were experienced in achieving design feed rate, conversion, and product selectivity.

Note that the contaminant-free feedstock gas from the Rectisol system insures clean products: no sulfur, cyanides, etc.

Product Recovery

The synthesis product is first recovered by air and water cooling of the reactor tail gas. In this manner, most of the synthetic oil (C_5 and heavier fraction) is condensed from the gas.

The residual gas stream is then fed to a low temperature separation unit where first the residual C_3's and heavier hydrocarbons are covered. Then the ethylene and ethane are recovered, and, finally, the hydrogen and methane streams are separated into hydrogen-rich and methane-rich gas streams. The hydrogen-rich stream is returned directly to the synthesis system, whereas the methane-rich stream is reformed using steam and oxygen to produce hydrogen and carbon monoxide for recycle to the synthesis reactors.

The low-temperature plant has been in operation at loads commensurate with the available tail gas from the synthesis unit; that is, $\sim$ 3/4 of design. No specific problems have been experienced during this operation.

The product refinery utilizes conventional refinery unit processes and unit operations to produce the mixture of fuel products: LPG, gasoline, diesel, fuel oil, and jet fuel. Other units process and refine the various chemical products.

As already mentioned above, this area consists of conventional refinery units, so no problems are expected.

FISCHER-TROPSCH FLUID BED REACTOR

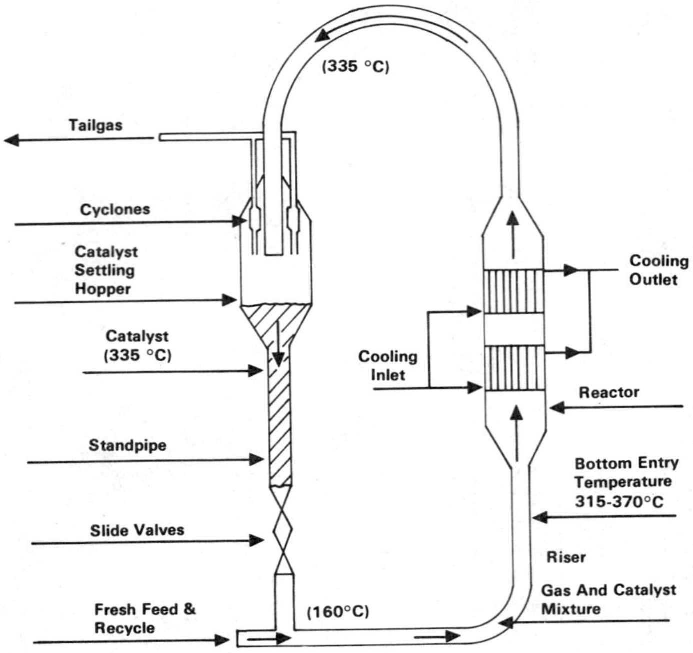

FIGURE 3

FISCHER — TROPSCH
TYPICAL PRODUCT SELECTIVITY (MASS BASIS)

	FIXED-BED	FLUID-BED
Methane	5.0	10.0
Ethane/Ethylene	2.6	10.0
Propane/Propylene	4.8	14.0
Butanes/Butylenes	5.2	9.0
C_5 — 375F	22.5	39.0
375F — 750F	15.0	5.0
750F — 970F	23.0	4.0
Heavier Than 970F	18.0	2.0
Chemicals	3.9	7.0

FIGURE 4

IN SUMMARY

All systems have been commissioned, and all plants are now operating at overall capacity of 75 to 80 percent. Some mechanical improvements are being incorporated in the Fischer-Tropsch Synthesis area, and design capacity is expected to be reached between the first and second quarters of 1981.

On most of the front end units, the ability to produce at or in excess of design has already been demonstrated. On all the other units, no specific problems have been experienced which would prevent them from achieving design capacity. Commissioning exercises have generally yielded fewer problems than expected, compared to more conventional hydrocarbon processing. It is expected that it will still be possible to better the original production schedule.

ACKNOWLEDGEMENTS

The author gratefully acknowledges the help and cooperation of J. C. Hoogendoorn of Sasol, and of R. F. Chew and J. J. Kovach of Fluor's Sasol project group, who have provided data current through the end of 1980.

REFERENCES

1. The Sasol Process; Sasol Public Relations, Sasolburg, South Africa. January 1980, p. 8.

2. Ibid, p. 3.

3. Rolke, D. E.; Treatment of Waste Water from a Lurgi Coal Gasification Plant. Proceedings of "Coal Technology '80," Houston, Texas, Volume 5, p. 148. November 1980.

4. Hoogendoorn, J. C., Jackson, S. B.; Sasol Projects in South Africa. Presented at "Coal Technology '79," Houston, Texas. November 1979.

GRACE/DOE
COAL-TO-METHANOL-TO-GASOLINE PROJECT

JAMES A. DOYLE

W. R. Grace & Co., Agricultural Chemicals Group

The coal-to-methanol-to-gasoline proposed project is the responsibility of Grace's Agricultural Chemicals Group headquartered in Memphis, Tennessee.

In August 1977, Grace and Energy Research and Development Administration (ERDA) signed a three-phase contract to design and consider the construction of a 1,200 ton per day ammonia-from-coal plant at a site recommended by the State of Kentucky on the Green River at Baskett, Kentucky. The Department of Energy (DOE), formerly known as ERDA, and Grace agreed, at the end of Phase I after the conceptual design work was completed, to utilize the technology developed and change the project to a coal-to-methanol-to-gasoline plant and increase the size of the project. As a result, Grace and DOE entered into a Cooperative Agreement October 6, 1980, to design a 50,000 barrel per day coal-to-methanol-to-gasoline plant. The contract is for 24 months and will have as its principal objectives the following:

Contract Highlights

Objectives:

- Utilize existing data from ammonia-from-coal effort to design a commercial gasoline-from-coal plant

- Develop process/mechanical design
- Prepare capital and operating cost estimate
- Perform environmental and economic assessments
- Obtain land options
- Develop preliminary construction and operation planning documents
- Investigate financing alternatives, including provision for early application to the SFC.

Considerable environmental work was completed on the ammonia plant that will apply to the gasoline project and should save from one to two years of environmental monitoring and permitting activity.

Grace has entered into subcontracts, with the Ralph M. Parsons Company as the A-E contractor, with Texaco Development Corporation, Ebasco Services Inc., and Mobil Research & Development Corporation, and will enter into other contracts as needed.

The key production data for the project is shown below:

KEY PRODUCTION DATA

Production	Daily	Annual
Gasoline (Bbl)	50,000	16,500,000
Butane (bbl)	5,000	1,650,000
LP Gas (Bbl)	3,500	1,160,000
Sulfur (ST)	850	280,000
Raw Materials		
Feedstock Coal (ST)	25,000	8,300,000
Fuel Coal	3,900	1,300,000
Total Coal	28,900	9,600,000
Oxygen (ST)	22,000	7,260,000

A single plant producing the quantities that have just been indicated (at first glance) appears to be a very large project. However, if we can put the quantities in the proper perspective, the plant will only supply .4% of the total U. S. petroleum supplies used in 1980, shown below:

COAL-TO-METHANOL-TO-GASOLINE PROJECT
IN PERSPECTIVE

UNITED STATES 1980 PETROLEUM SUPPLIES
(Barrels per day, Crude Oil Equivalent)

Domestic Production	9,700,000
Imports - Crude and Products	6,800,000
Total U. S. Supply	16,500,000

ENERGY SECURITY ACT OBJECTIVES

Crude Oil Equiv. Barrels per Day	Percent of Domestic Prod.	Percent of Imports	Percent of Total U.S.Supply
1987 - 500,000	5.2%	7.3%	3.0%
1992 - 2,000,000	20.6	29.4	12.1

COAL-TO-METHANOL-TO-GASOLINE PROJECT
IN PERSPECTIVE

GRACE OBJECTIVES

Objective Barrels per Day	Percent of Domestic Prod.	Percent of Imports	Percent of Total U.S.Supply
Gasoline 50,000 (Equiv.to 65,000 of crude oil)	0.7%	1.0%	0.4%

TO ELIMINATE oil imports, 100 synfuel plants would be required to produce the equivalent of 50,000 barrels.

DOMESTIC GASOLINE requirements are about 7,000,000 barrels per day, of which the Grace project would provide about 0.7%.

COAL-TO-METHANOL-TO-GASOLINE PROJECT

BLOCK FLOW DIAGRAM

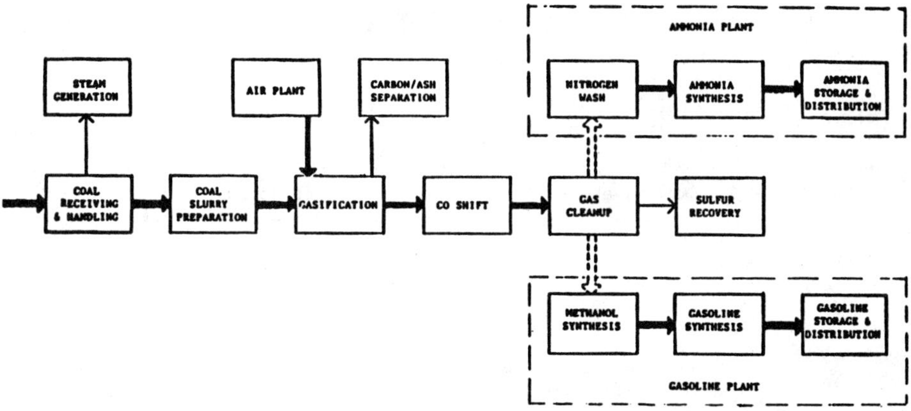

About 70% of the engineering that has been accomplished on the ammonia project can be used in the design of the methanol plant. As you know, methanol and ammonia processes are very similar. In fact, the 1,200 ton per day ammonia plant that was designed is equivalent to a 1,200 ton per day methanol plant.

The project requires production of 16,000 tons per day of methanol for conversion to 50,000 barrels per day of gasoline, and will be accomplished by building multiple units comparable to the 1,200 ton per day ammonia plant. The size of the Texaco gasifiers will be the same (about 14 will be required), and gas purification techniques will be similar all the way through sulfur recovery.

BASIS OF GRACE PARTICIPATION

- Over 25 years' operational experience in Texaco gasification technology

- Extensive experience in design, construction, and operation of major agricultural chemical facilities

- Advanced stage of environmental work and land acquisition at Baskett site

- Substantial engineering work performed applicable to gasification

- Grace expertise in catalysts, oil, gas, and basic research.

TEXACO COAL GASIFICATION

- The Texaco Partial Oxidation technology has been utilized in over 75 plants throughout the world

- The Process will:

  operate at a carbon conversion efficiency in excess of 96%

  accept high-sulfur agglomerating coals

  produce a clean gas excellent for methanol production

  operate at a pressure suitable for economical production of methanol

  produce no phenols, tars or other environmentally unacceptable by-products

  produce a nonhazardous slag

- The Process has been proven by:

  sustained pilot plant run at Texaco's Montebello facility

  the gasification of over 40,000 tons of coal in a 150 ton per day commercial demonstration plant in West Germany.

COAL TO METHANOL TO GASOLINE PROJECT

TEXACO COAL GASIFICATION PROCESS FLOW DIAGRAM

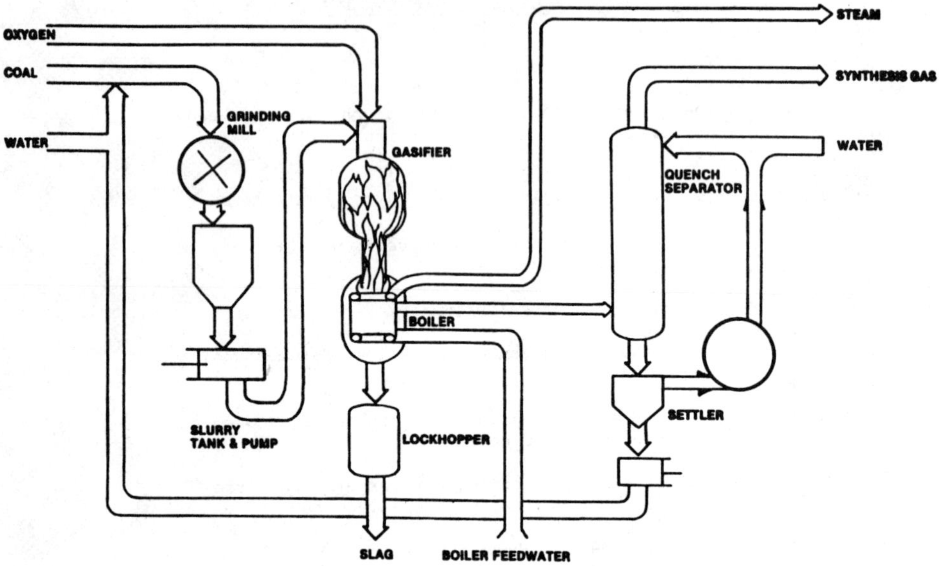

Coal will be wet ground, slurried, and pumped to the Texaco entrained flow slagging gasifier. The gasifier will operate in a temperature range of 2300-2800° F. and will discharge the slag through a lockhopper arrangement. About 4,000 tons per day of granular slag will be produced and, hopefully, sold as a by-product for sandblasting and roofing granules.

MOBIL METHANOL TO GASOLINE PROCESS

- Produces high-octane, sulfur-free, nitrogen-free, unleaded gasoline that is virtually identical to that used in today's cars

- Highly efficient, producing hydrocarbons with 95% efficiency

- Overall process technology is commercial

- The zeolite catalyst utilized in the Mobil process converts methanol directly to gasoline

- The New Zealand Government has recently chosen the Mobil process over the Sasol/Synthol route for a 13,000 barrel per day liquid fuels complex.

PYROLYSIS ROUTE FOR COAL LIQUIDS PRODUCTION

George R. Hill
Chemical Engineering Department
University of Utah

The production of hydrocarbon liquids from coal can be accomplished by three general methods. 1) Essentially total conversion can be accomplished by the direct-route (for example, Bergius, Pott-Broche, Exxon-Donor Solvent, H-Coal, Consolidation Coal, and Solvent-Refined Coal II processes). 2) The Indirect total conversion route involves gasification of the coal to produce carbon monoxide and hydrogen followed by the catalytic recombination to produce hydrocarbons and/or oxygen-containing organic molecules. Examples of the Indirect route are the Fischer-Tropsch and related syntheses, including Methanol and Mobil gasoline syntheses, and the production of motor fuels and chemicals in the Sasol South Africa processes. 3) The third method, which is a primary focal point of this paper, concerns the production, by pyrolysis, of the hydrogen-rich components as a hydrocarbon liquid, and, as a co-product, a char or coke from which the volatile matter has been removed.(1)(2)(3)(4)

The pyrolysis processes utilizing coal prior to World War II had as their prime objective the production of a coke for blast-furnace or other metallurgical operations, or of a char for processes requiring a high-surface area adsorption or reducing agent. The char obtained from the pyrolysis of coal produces a smoke-free combustion product. Elimination of air pollution in England, Europe, and some locations in the United States thus provided the incentive for char production in early pyrolysis plants. A comprehensive listing and description of low-temperature carbonization processes was published as a report of the Utah Conservation and Research Foundation in May 1939.(5) The bibliography in that volume cites 1,412 papers, books, and patents referring directly to low-temperature tars and oils, as well as an additional 17 references to potential carcinogens in the coal tar.

During World War II, the focus shifted from pyrolysis for production of solid fuels to production of liquid fuels. "By 1944, Germany had 15 synfuels plants using the process developed by Friedrich Bergius in 1921...the synfuels plants attained peak production of 340,000 metric tons a month in March 1944."(6)

Table 1 lists the hydrogenation plants in Germany with their location, main raw material, hydrogenation process, pressure, and capacity, as translated from Kroenig.(7) It is most important to note that the primary material hydrogenated to produce the gasoline and diesel fuel was not coal, as has been generally believed. As will be seen in the column "Main Raw Material" in Table 1, the principal feed in most of the plants was not coal, but was a liquid, low-temperature tar, coke-oven tar, or residual oil. Based on the capacity of the individual plants, between 50 and 70 percent of the four million tons a year of liquids produced in these plants had pyrolysis liquids as the feed. I am indebted to G. R. Romovacek of the Koppers Company for calling our attention to these facts.(8)

After World War II specific efforts to improve the pyrolysis process and to broaden the feed base to include caking and semi-caking coal were carried out by Union Carbide Corporation at Institute, West Virginia, in the 1940's and '50's, and by the FMC Corporation, Princeton, New Jersey, during the 1960's and '70's. The Union Carbide effort involved pyrolysis in a hydrogen atmosphere at around 500 lbs/in^2 pressure. The presence of hydrogen under those conditions increased the liquid yield and resulted in a lower-molecular-weight, hydrogen-rich oil than was produced in atmospheric-pressure pyrolysis. (9) The FMC Project COED (Coal/Oil Energy Development) was a staged-pyrolysis operation in which coal was converted to char by sequentially heating to higher temperatures, avoiding the formation of a fluid or a semi fluid coal phase which would stick to the metal surfaces of the equipment. Caking-type bituminous coal could be processed with minimum difficulty using the FMC sequential pyrolysis process.(10)(11)

Current interest in pyrolysis and hydro-pyrolysis of coal has a two-fold basis. The first basis is for the production of a liquid supplement to petroleum feedstocks in oil refineries. The value of hydrocarbon oils which can serve as a petroleum-refinery feedstock has increased from a nominal \$2/barrel value in the 1950 decade to its present value of over \$30/barrel. The second basis is the production of a char product from the pyrolysis. Char can be as good a fuel for electric power generation as coal. In general, the char has essentially the same heating value as the parent, high-volatile bituminous coal. The ash content of the char will be higher, depending upon the percent of initial coal that the char represents. The char can be a single fuel in a properly designed boiler system or can be blended with high-volatile bituminous coal in a conventional pf boiler, in an existing power plant. Tests will be under way shortly to determine the upper limit of char (probably 40 percent) that can be burned satisfactorily in a powdered fuel coal/char mixture.

The renewed interest in producing liquids by pyrolysis and hydro-pyrolysis of coal stems from the fact that the equipment required is much simpler to build and to operate than the equipment required in the total liquefaction processes. The liquid yield from pyrolysis of high-volatile bituminous coal is from about 0.8 barrel to 1.6 barrels of oil plus tar per ton of coal, depending upon the coal used.(12)(13) Swabb, et. al.(14), report that the net yield of liquid products from the SRC-II, H-Coal and EDS processes can range from 35 to 45 percent of the weight of dry coal fed to the plant, depending upon the feed coal, the choice of products to be produced, the operating conditions, and the particular process configuration employed. This would amount to two to three barrels per ton of coal fed to a liquefaction plant, about twice the yield from the pyrolysis route. Table 2 gives the

TABLE 1. HYDROGENATION PLANTS IN GERMANY

Hydrogenation Plant	District	Main Raw Material	Hydrogenation Process	Pressure (atm.) Liquid Phase	Pressure (atm.) Vapor Phase	Capacity (1000's of T/yr.)
Leuna	Central Germany	Brown Coal (brown coal LT* tar)	Liquid + Vapor Phase	200	200	600
Bohlen	Central Germany	Brown Coal LT Tar	Liquid + Vapor Phase	300	300	240
Magdeburg	Central Germany	Brown Coal LT Tar	Liquid + Vapor Phase	300	300	230
Zeitz	Central Germany	Brown Coal LT Tar	TTH + MTH	300	300	300
Wesseling	Lower Rhein	Brown Coal	Liquid + Vapor Phase	700	300	200
Brux	Sudetenland	Brown Coal LT Tar	Liquid + Vapor Phase	300	300	400
Scholwen	Ruhr	Bituminous Coal	Liquid + Vapor Phase	300	300	200
Gelsenberg	Ruhr	Bituminous Coal	Liquid + Vapor Phase	700	300	350
Blechhammer	Upper Silesia	Bituminous Coal	Liquid + Vapor Phase	700	300	500
Welheim	Ruhr	Coke Oven Tar	Liquid + Vapor Phase	700	700	180
Lutzkendorf	Central Germany	Petroleum Residue Coke Oven Tar	Liquid + Gas Phase	700	700	50
Politz	Stettin	Bituminous Coal Coke Oven Tar Crude Oil Residue	Liquid + Gas Phase	700	300	600
Moosbierbaum	Austria	Petroleum Gasoline Hydroforming	HF	---	ca.15	100
Ludwigshafen	Upper Rhein	Hydrogenation Gasoline Hydrogenation Reforming	DHD	---	ca.50	50
TOTAL Hydrogenation Plant Capacity						4000

*Low-Temperature

TABLE 2. PYROLYSIS PROCESS COMPARISONS

Process	Reactor Type	Residence Time	Barrels Oil and Tar Per Ton of Coal	Coals Tested	Maximum Size Tested	Status
Lurgi-Spülgas	Downflow-Moving Bed	1-2 hrs.	0.9	Bituminous, Brown	500 TPD	Commercial
Lurgi-Ruhrgas	Tubular Screw	> 20 sec.	1.1	Bituminous, Lignite, Brown	880 TPD	Commercial (?)
CDC-Pyro (FMC)	Fluidized-Beds	1-4 hrs.	1.1	Bituminous, (Utah) Sub-bituminous	36 TPD	Pilot
Occidental	Entrained-Bed	1-3 sec.	1.6	Bituminous Sub-bituminous	3 TPD (5 lb./hr.)	PDU/Bench
Union-Carbide	Fluidized-Bed	5-12 min.	1.6	Bituminous Sub-bituminous	20 TPD (40 lb./hr.)	Pilot/Bench

liquid tar and oil yield based upon preliminary work using some of the currently available pyrolysis and hydro-pyrolysis processes.

A description of some current residua up-grading processes which should be applicable to low-temperature tar as well as to heavy residues from petroleum has recently been reported by Bridge, et. al.(15) Application of such process schemes would determine the ultimate yield of gasoline, diesel oil, and heavy oil from coal pyrolysis processes. (16)

Among the many hundreds of pyrolysis processes which have been investigated in the past, the following six are among those which have survived through process-development-unit and pilot-plant testing. Some of them are used in commercial plants in operation in Europe. The list includes: Lurgi Spülgas, which was the backbone of the German effort during World War II, and Lurgi Ruhrgas (17); FMC's process, which is based upon the COED effort (18)(19); the Union Carbide process, which has been updated by Union Carbide's Linde Division (20); the Toscoal process, which is based on an adaptation of the aluminum-oxide ball oil shale retorting system (21); and the Occidental Petroleum pyrolysis process, which includes a "Polystop" quench of the primary tar and oil product in a hydrogen-donor solvent.(22)(23)

The electric utility industry has expressed increased interest in pyrolysis and hydro-pyrolysis processing of coal, since the char can be utilized for its heat content either as a blend with coal in existing plants, or as the total feed in fluidized-bed plants or in boiler systems designed for use with low-volatile solid fuel. If the liquid pyrolysis product can be upgraded at a reasonable cost using modern petroleum refinery hydrogenation techniques and provided to an oil refinery as a feedstock, the char would be valued at less than the cost of coal. This would enable the electric utility to reduce the electricity cost to its customers. Some electric utilities are finding it extremely difficult to write long-term contracts for the fuel oil needed to keep existing oil-fired plants in operation. For those companies a captive supply of the hydro-treated coal tar to replace petroleum base fuel oil has great appeal.

It is to be hoped that the petroleum, chemical, and utility industries will give proper consideration to the construction of commercial plants utilizing pyrolysis and/or hydro-pyrolysis processes for the production of petroleum supplements, as well as to supporting the construction of commercial plants utilizing the total liquefaction and the indirect conversion processes.

---

(1)Martin A. Elliott, ed., Chemistry of Coal Utilization, Second Supplementary Volume, March, 1981: Jack B. Howard, "Fundamentals of Pyrolysis," Chapter 12; Leonard Seglin and Sidney A. Bresler, "Low Temperature Pyrolysis Technology," Chapter 13; Michael Perch, "Solid Products of Pyrolysis," Chapter 15; Robert W. Reeve, Harold Shalit, and E. Aristoff, "Low-Temperature Tar," Chapter 16.

(2)Richard F. Hill, Synthetic Fuels Summary, ESCOE Report FE-2468-82 (August, 1980).

(3)Liebowitz, H., and N. Korens. Preliminary Engineering Evaluation of Promising Coal Liquefaction Concepts. Final Report, EPRI AF-884. Palo Alto, California: Electric Power Research Institute. (December 1979).

(4)Mullowney, J. F., and P. F. Mako, "Coal to Transport Fuels and Chemicals, Sasol Two/Sasol Three." 3333 Michelson Drive, Irvine, California 92730: Fluor Engineers and Constructors, Inc.

(5)Utah Conservation and Research Foundation, Low Temperature Carbonization of Utah Coals. Salt Lake City, Utah: Quality Press (May 1939).

(6)Krammer, Arnold, "Fueling the Third Reich," Technology and Culture, 19:394 (July 1978).

(7)Kroenig, Walter, "Catalytic Hydrogenation of Coal Tar and Oil Under Pressure," unpublished transl. of Katalytische Druckhydrierung. Berlin/Gottingen/Heidelberg: Springer-Verlag (1950).

(8)Romovacek, George, Koppers Company, Pittsburgh, Pennsylvania (private communications), February 1980 and January 1981.

(9)Kasbaum, M., and D. W. Linde, "Union Carbide Hydrocarbonization Technology" (private communication), June 1980.

(10)FMC Corporation, Char Oil Energy Development, U.S. Energy Research and Development Administration Report FE-2468-82, Washington, D.C.: (September 1975).

11)FMC Corporation, The Combustion Performance of COED Char, R&D Report No. 73, U.S. ERDA Contract No. 14-32-0001-1212.

(12)Utah Conservation and Research Foundation, ibid.

(13)Youngblood, E. L., et. al., "Hydrocarbonization Research: Completion Report," Oak Ridge, Tennessee: Oak Ridge National Laboratory, January 1981.

(14)Swabb, L. E. Jr., G. K. Vick, and T. Aczel, "The Liquefaction of Solid Carbonaceous Material," Future Sources of Organic Raw Materials--Chemrawn I, Pergamon Press, New York, 1980, pp. 233-255.

(15)Bridge, A. G., G. D. Gould, and J. F. Berkman, "Residua Processes Proven," TECHNOLOGY, Oil and Gas Journal, June 19, 1981, pp. 85-98.

(16)Qader, S. A., and G. R. Hill, "Hydrocracking of Petroleum and Coal Oils," Industrial and Engineering Chemistry Process Research Development, 8 (October, 1969), 462-469.

(17)Rammler, Roland W., "One Process Retorts Three Feeds for Synthetic Crude," Mining Engineering, September 1970, p. 63.

(18)FMC Corporation, Char Oil Energy Development, ibid.

(19)FMC Corporation, The Combustion Performance of COED Char, ibid.

(20)Kasbaum, M., and D. W. Linde, ibid.

(21)Carlson, F. B., L. H. Yardumian, and M. T. Atwood, "Toscoal Process for Low Temperature Pyrolysis of Coal," SME Transactions, 161 (June 1974), 128-131.

(22)Che, S. C., G. E. Dolhear, and J. R. Longanbuch, "Properties of Coal Liquids Produced by the Occidental Flash Pyrolysis Process," presented at the 1976 Coal Chemistry Workshop, Stanford Research Institute, August 1976.

(23)Sass, A., "Garrett's Coal Pyrolysis Process," Chemical Engineering Progress, 70, No. 1 (1974), 72-73.

EXXON DONOR SOLVENT COAL LIQUEFACTION PROCESS: DEVELOPMENT PROGRAM STATUS V

W. R. Epperly
K. W. Plumlee
D. T. Wade

Exxon Research and Engineering Company,
Florham Park, N.J.

INTRODUCTION

This paper describes the distinguishing features of the Exxon Donor Solvent (EDS) coal liquefaction project. Included is an overview of the government-industry cost shared development and a description of Exxon's integrated approach to the development. The status of the laboratory and engineering research and development studies and the status of the 250 T/D large pilot plant operation, test program and a 70 T/D FLEXICOKING* prototype program are presented. The process description includes discussions of coal feed and product flexibility. Potential product utilization schemes are also surveyed. The literature contains past status reports of the development program,[1-12] and discussions of the potential for commercialization,[13] as well as the structure of the EDS Project.[14]

Figure 1 lists the project participants. The U.S. Department of Energy is providing 50% of the funding through a unique government/industry cost sharing arrangement, the Cooperative Agreement[14], with Exxon Research and Engineering Company. The remaining funding for the liquefaction program is provided by Exxon Company, U.S.A., Electric Power Research Institute, Japan Coal Liquefaction Development Company, Phillips Coal Company, ARCO Coal Company, Ruhrkohle AG and AGIP. Private sector support of the FLEXICOKING prototype construction and operation is provided by Exxon Company U.S.A., Japan Coal Liquefaction Development Company, ARCO Coal Company, Ruhrkohle AG and AGIP.

The overall objective of the project is to bring the technology to a state of commercial readiness so that commercial plants can be designed with an acceptable level of risk. The EDS project includes the process blocks of liquefaction, solvent hydrogenation, and bottoms processing; it includes work on hydrogen generation, fuel gas generation, and environmental controls as well.

*"Service Mark"

EXXON DONOR SOLVENT PROCESS
PROJECT
SPONSORED BY
U.S. Department of Energy
Exxon Company, U.S.A.
Electric Power Research Institute
Japan Coal
Liquefaction Development Co., Ltd.
Phillips Coal Company
ARCO Coal Company
Ruhrkohle AG
AGIP
Exxon Research
and Engineering Company
Project Manager

Figure 1

INTEGRATED DEVELOPMENT

To achieve the objective of commercial readiness, the EDS program integrates all phases of process development. Bench scale research, small pilot unit operation, and engineering design and technology studies support operation of a 250 T/D coal liquefaction pilot plant and a 70 T/D FLEXICOKING prototype program. Work is also in progress to evaluate the use of either a bottoms partial oxidation process for generation of hydrogen or fuel gas, or direct combustion of bottoms for fuel.

As shown in Figure 2, the integrated approach involves optimum use of the laboratory and engineering R&D programs, and the 250 T/D liquefaction pilot plant and 70 T/D FLEXICOKING prototype with their associated test programs to obtain the data for a commercial design. The design data for the key development areas, e.g., slurry drying, liquefaction, distillation, solvent hydrogenation, FLEXICOKING, product quality, and environmental control, will be obtained in the most appropriate project area at the minimum development cost. For example, the role of the ECLP (250 T/D) pilot plant is to provide operability and design data in the slurry drying, liquefaction, distillation and bottoms processing areas.

Design data from the large pilot plants will be collected through extensive engineering technology tests. ECLP was planned to include facilities for 115 individual tests which are expected to provide a comprehensive basis for a commercial plant design. Included in this program are test activities such as materials evaluation, erosion and corrosion, coal slurry preheat furnaces, slurry pumping, high pressure letdown valves and environmental monitoring and control. While the details of a similar effort for the FLEXICOKING prototype operation are not finalized, planning has already begun. Critical engineering design areas for FLEXICOKING include fluid bed operation, product quality effects and environmental control. The planned test program will include these areas plus materials evaluation, high pressure vacuum bottoms pumping, and coke particle integrity and gasification activity.

EDS PROCESS BLOCK DIAGRAM

One configuration of the EDS process is shown in Figure 3. Feed coal is crushed and dried by mixing with hot recycle donor solvent. The coal-solvent slurry is fed along with gaseous hydrogen to the liquefaction process block. The liquefaction reactor design is relatively simple, consisting of an upward plug flow reactor with design conditions of 800-900°F and about 2000-2500 psi total pressure. The reactor product is separated via conventional separation and fractionation steps into chemical and light hydrocarbon gases, C_3-1000°F distillate, and vacuum bottoms containing 1000°F+ liquids, unconverted coal, and coal mineral matter.

Part of the 400-800°F fraction of the C_3-1000°F distillate is taken as the recycle hydrogen donor solvent. This spent (dehydrogenated) solvent stream is hydrogenated in a conventional fixed bed hydrotreater using commercially-available hydrotreating catalysts.

Conversion and liquid yield can be increased by recycling liquefaction bottoms to the liquefaction reactor. This has the effect of increasing the residence time of heavy coal molecules.

The light hydrocarbon gases can be fed to a steam reformer to produce process hydrogen. The vacuum bottoms stream can be fed to a FLEXICOKING unit to produce additional liquid products and low BTU fuel gas while concentrating the coal mineral matter for disposal. FLEXICOKING is a commercial petroleum[15] process that employs integrated coking and gasification reactions in circulating fluidized beds. The process is a

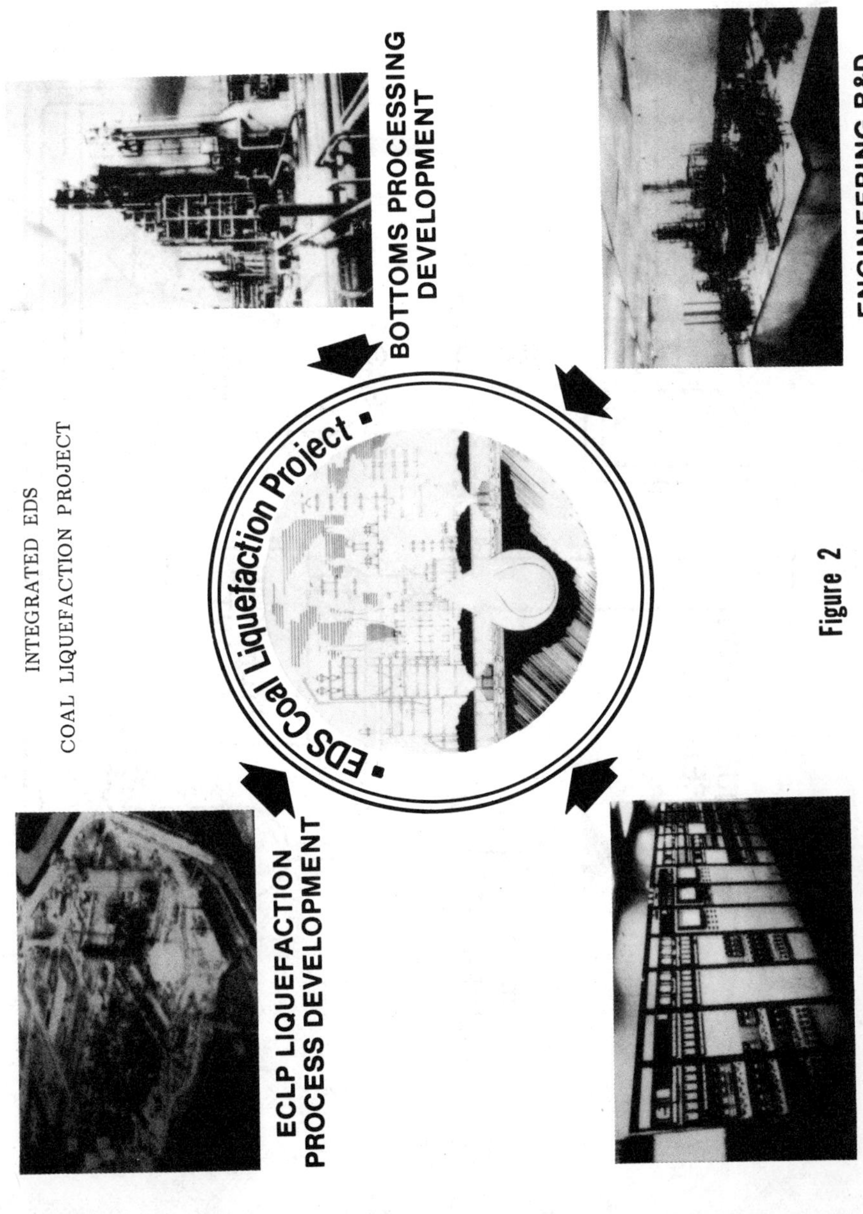

INTEGRATED EDS
COAL LIQUEFACTION PROJECT

Figure 2

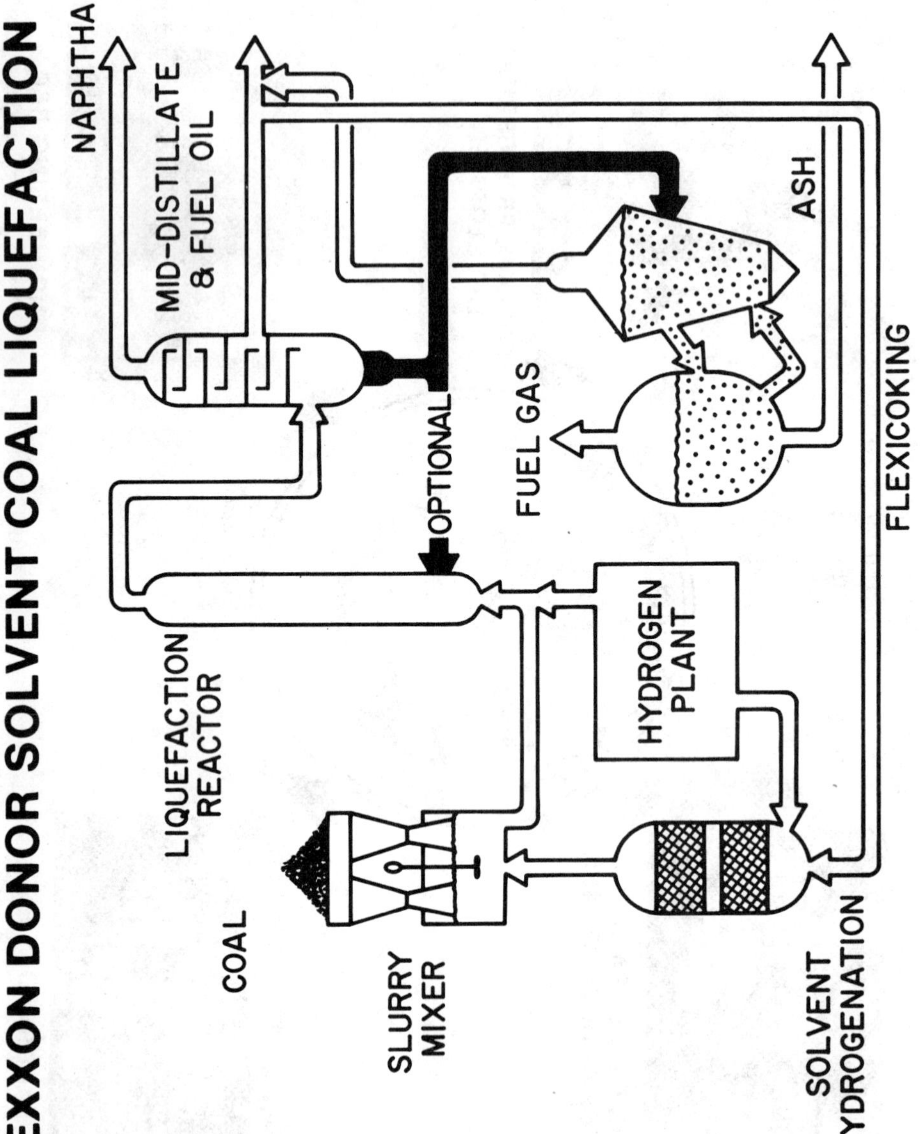
EXXON DONOR SOLVENT COAL LIQUEFACTION
NAPHTHA
MID-DISTILLATE
& FUEL OIL
LIQUEFACTION
REACTOR
COAL
SLURRY
MIXER
OPTIONAL
FUEL GAS
HYDROGEN
PLANT
ASH
FLEXICOKING
SOLVENT
HYDROGENATION

Figure 3

low pressure (<50 psi) and intermediate temperature (900-1200°F in the coker, 1500-1800°F in the gasifier) operation. FLEXICOKING recovers essentially all of the feed carbon as product liquid or fuel gas. A small amount of carbon is purged from the unit with the coal mineral matter.

Partial oxidation of the vacuum bottoms (not shown) can produce either process hydrogen or intermediate BTU fuel gas.(16) This configuration frees the light gas stream for sales or furnace fuel.(17) In high conversion coal liquefaction operating modes, process hydrogen can be generated by partial oxidation of coal.(18) Partial oxidation units typically operate in the 2500-2800°F range and at 400-1000 psi. Partial oxidation is a commercial process employing oxygen to gasify petroleum fractions. The process does not recover additional liquid product but has the potential to consume effectively all of the feed carbon in the production of hydrogen or fuel gas. Direct combustion of vacuum bottoms can be used to provide process heat and raise steam.

DONOR SOLVENT

The EDS process is unique in the use of an externally hydrogenated recycle donor solvent. As shown in Figure 4, conversion is dependent on the availability of solvent of sufficient quality, i.e., a solvent having the required donor hydrogen to insure rapid and efficient capturing of free radicals generated from thermal disruption of the coal matrix. The data in Figure 4 for the Illinois Monterey and Wyoming Wyodak coals illustrate the effects of solvent quality index (SQI) on coal conversion. SQI is a measure of the ability of the solvent to donate hydrogen to the coal. These data show that for the Monterey coal, an SQI of about 3 is required to obtain highest conversion while an SQI of 4 is required for the Wyodak coal. This requirement of higher SQI is consistent with the observed higher reactivity of the Wyodak coal and may be typical of lower-rank coals. Explanation of the degree of fall-off in conversion for the two coals at the lower SQI becomes complicated with possible interactions with the mineral matter present, the molecular hydrogen partial pressure, the presence of H_2S and possible reactant mass transfer limitations.

COAL FEED FLEXIBILITY

Coal is located in many places in world, and differences in deposits can be very important in coal liquefaction. As shown in Figure 5, in the U.S. bituminous coals are found in eastern, mid-western, and southern Rocky Mountain regions. Subbituminous coals are found primarily in the west. Lignites are found in the west and the Gulf Coast. One of the technical challenges is to be able to convert this wide variety of different quality coals into liquids.(19)

The EDS process is suitable for a wide range of coals. Figure 6 shows that bituminous, subbituminous and lignite coals can be liquefied using the process. The liquid yields on a dry, ash-free basis resulting from once-through and bottoms recycle liquefaction are shown on this chart. Bituminous coals studied produce with once thru operations 39-46% liquids, subbituminous coals about 38% liquids, and the lignite about 36% liquids. The liquefaction yield can be increased substantially by recycling vacuum bottoms back to liquefaction. As shown yields for bituminous coals are increased to 55-60% liquids, for subbituminous coals to 44-50% liquids and for the lignite to 47%. Although not shown, additional liquids can be recovered by FLEXICOKING the vacuum bottoms stream. FLEXICOKING yields range from 5 to 10 weight percent additional liquids on coal for the coals shown and correspond to about 25 percent of the ash free vacuum bottoms yield.

In addition to increasing conversion to liquid products, the bottoms recycle process improvement results in improved pilot unit operability. Figure 7 shows the improvement in service factor for lower rank coals

EFFECT OF SOLVENT QUALITY ON CONVERSION

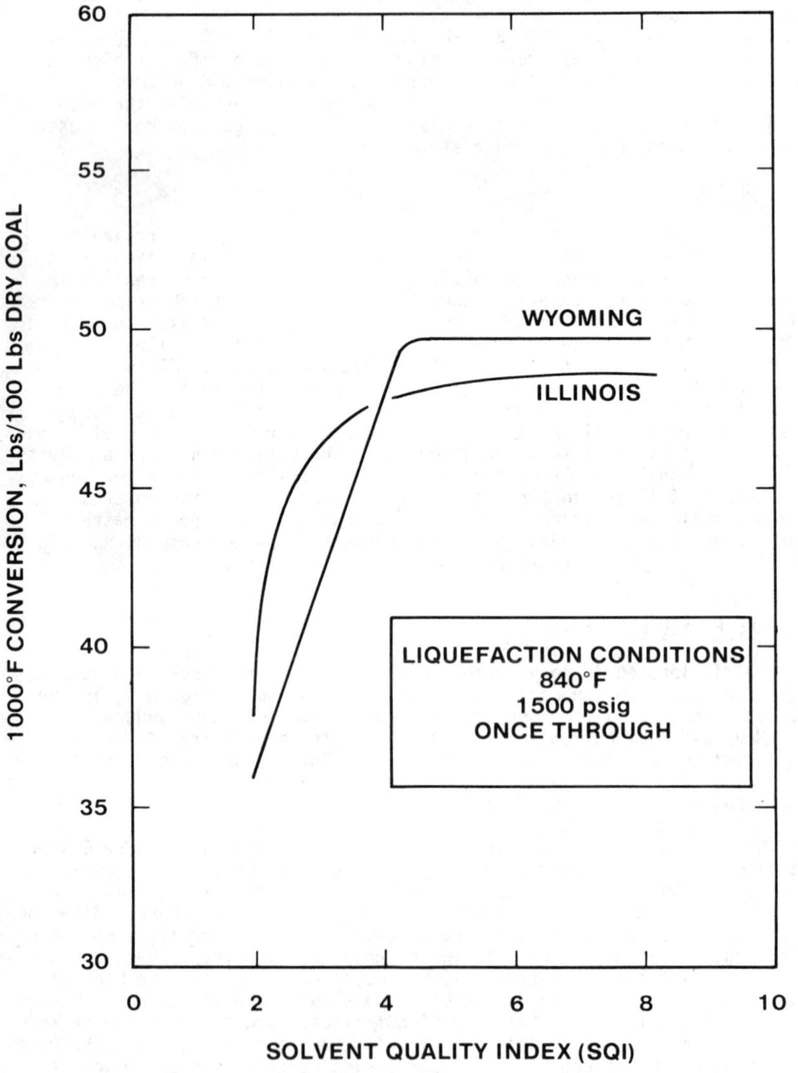

Figure 4

LOCATION OF EDS COALS

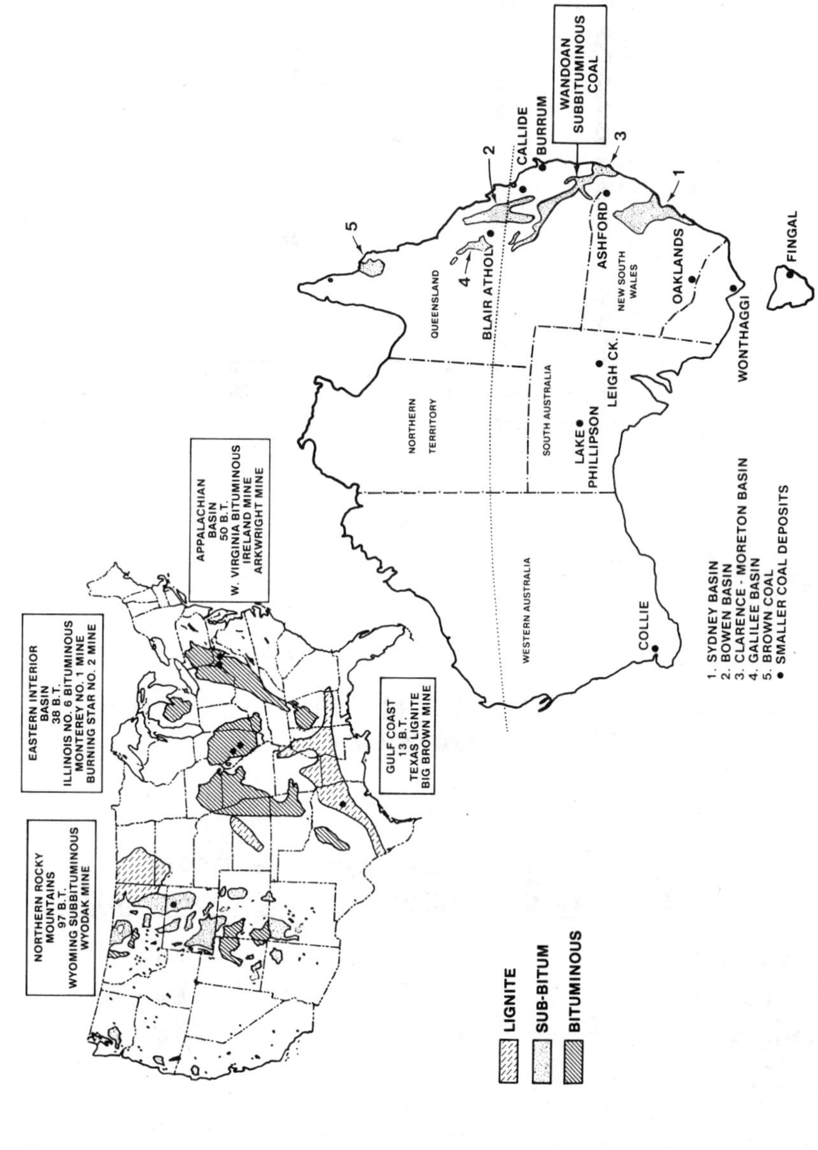

Figure 5

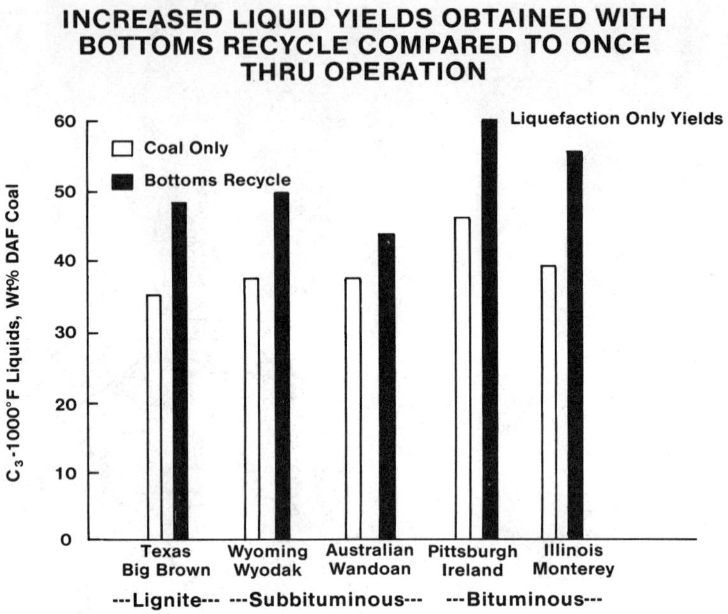

Figure 6

OPERABILITY SUBSTANTIALLY IMPROVED BY BOTTOMS RECYCLE

	RCLU SERVICE FACTOR	
	ONCE THROUGH	BOTTOMS RECYCLE
ILLINOIS	76	73
WYODAK	58	68
BIG BROWN	44	64

Figure 7

observed in the 50 and 100 pounds per day recycle coal liquefaction units (RCLU). The improvement in service factor is felt to result from the reduction in bottoms viscosity which accompanies bottoms recycle operation. The bottoms recycle improvement reduces the viscosity of Wyodak and Big Brown bottoms to a level comparable with Illinois once-through and recycle bottoms, which are equivalent.

PRODUCT FLEXIBILITY, UTILIZATION

The EDS process offers a range of product slates depending on process conditions and configurations. For example, bottoms recycle liquefaction serves to lighten the product slate. Figure 8 shows the product distribution which was projected in the base case design in 1978. The process basis, designated original configuration, included once-through liquefaction, bottoms FLEXICOKING to produce additional liquids and fuel for the plant and hydrogen generation by steam reforming of light gas. Under these conditions, about 35% naphtha and 35% of 400/850^oF distillate were produced while obtaining 25% heavy (850^oF+) fuel oil. As the interest in a lighter product slate grew, emphasis was placed on this area. A number of possibilities were investigated, and bottoms recycle was found to be the most economically attractive.

The improvements in yield distribution shown are obtained using bottoms recycle, recycle FLEXICOKING, and coal POX for hydrogen. Naphtha plus middle distillate yield is unchanged, but naphtha is increased to 46%; fuel oil is eliminated and 21% C_1/C_2 gas is produced. Even higher naphtha yield can be obtained by operating bottoms recycle at very high conversion.

Bottoms recycle is effective in increasing the total yield of coal liquids and the amount of lighter products. The increase in lighter products is approximately equal to the yield of liquids obtained in FLEXICOKING. These higher yield levels increase the number of options for achieving a balanced operation, that is, one without excess coal liquefaction bottoms.

BOTTOMS PROCESSING

Figure 9 shows that the bottoms process selected can also significantly affect the cost of the product. In all of the cases shown, the same product slate has been maintained using external upgrading of the heavy product to adjust for differences in the product slate from the liquefaction plant. The base case from which the improvement or reduction in the RISP (required initial selling price) is determined is shown on the left. It involves steam reforming of light gas to produce hydrogen and FLEXICOKING of bottoms to produce the plant fuel. It was called "original configuration" on the previous figure.

Using coal partial oxidation to produce the hydrogen while maintaining FLEXICOKING for plant fuel (the improved case on Figure 8) has the potential to reduce product cost about 20% if the C_1/C_2 gas product is valued in BTU parity with the liquids produced. The improvement results from the higher efficiency of coal partial oxidation for producing hydrogen and the revenue generated by selling the C_1/C_2 gas instead of producing hydrogen from it.

If, in addition to coal partial oxidation for hydrogen, bottoms partial oxidation is used to produce fuel for the plant, an additional 2% improvement in RISP is possible.

Finally, if a hybrid boiler is employed in which process heat is provided in the convection section and liquefaction bottoms are used as the fuel, an improvement in RISP of about 28% is possible.

SIGNIFICANT PRODUCT FLEXIBILITY DEMONSTRATED

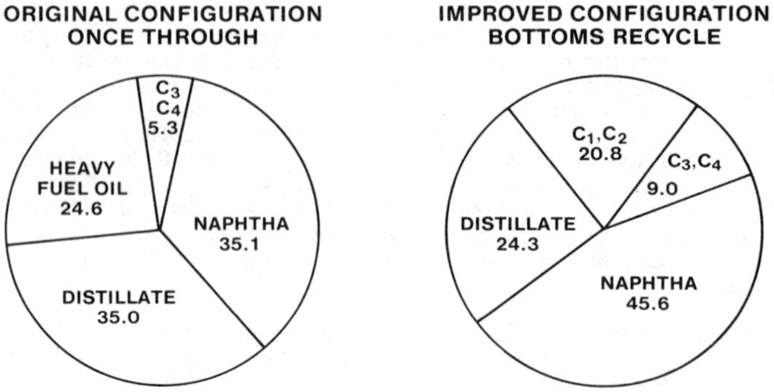

Figure 8

EFFECT OF BOTTOMS PROCESSING CONFIGURATION ON PRODUCT COST

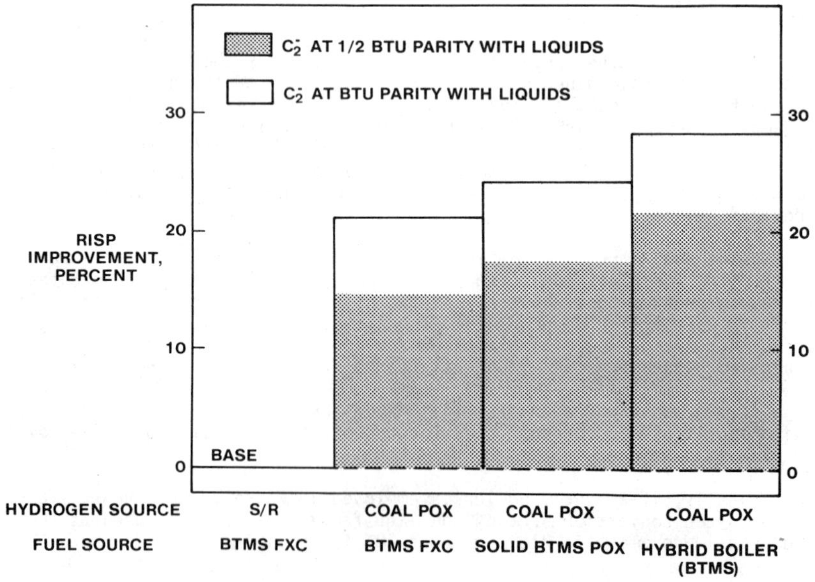

Figure 9

The shaded area shows that valuing the gas at 1/2 BTU parity with liquids lowers the extent of the improvement in RISP but does not change the conclusions in terms of the relative attractiveness of the different bottoms processing options. Thus, there is an indicated incentive to use coal partial oxidation, bottoms partial oxidation and a hybrid boiler fired by liquefaction bottoms.

All of the raw (unhydrotreated) EDS liquids contain significant levels of sulfur and nitrogen. The heteroatom concentration increases with increasing boiling range. Primarily because of heteroatoms, the products tend to degrade during storage.

The EDS project includes an effort to define suitable product uses, but does not include a major upgrading program. Limited hydrotreating studies are included in the project, to define the treatment required to stabilize the products for storage and shipment and to make them suitable for limited direct utilization.

Figure 10 shows several upgrading options and utilization routes for the EDS product streams.[20-25] Under each product the upgrading options shown represent increasingly severe processing. Our studies and the work of others[23] indicate that these are technically feasible upgrading options although their economic viability remains to be determined. The naphtha stream can be hydrotreated to reduce sulfur and nitrogen contaminants and reformed to increase the octane number and recover net hydrogen. The reformate is then ready for motor gasoline blending or aromatics extraction.

The mid-distillate stream can be hydrotreated mildly to reduce sulfur and nitrogen and sent directly to end-use as either a No. 2 fuel oil or stationary turbine fuel. Other options would involve more severe hydrotreating to produce jet or diesel fuel, or hydrocracking to produce a motor gasoline blend stock.

The vacuum gas oil stream also can be hydrogenated mildly to reduce sulfur and nitrogen contents to levels suitable for use as a specialty fuel. At this level of hydrotreating these heavy distillate fuels require special handling due to their higher than normal viscosities and their incompatibility with comparable petroleum fuels. Alternatively, hydrocracking of the vacuum gas oil to produce clean products or transportation fuels appears possible based on stream inspections, extrapolation of petroleum-based correlations, and the work of others.[24]

Figure 11 shows some of the results that UOP[25] has obtained on reforming the hydrotreated naphtha to a gasoline-blending component. It also shows results from an Exxon funded program evaluating reforming of EDS naphtha. The yield, as percent of C_5+ liquid reformate produced per barrel of feed, is plotted against the research octane--clear or unleaded--of the product. By comparison, the yield on naphtha from a Saudi Arabian feed is much lower for any given octane. Hence, the EDS-derived naphtha would be an excellent reformer feed when compared with a major current source of crude in the world market.

ENVIRONMENTAL PROGRAM

The first step in formulating an environmental program was defining those areas expected to be different from petroleum experience.[26] As shown on Figure 12, three general areas were identified. The coal feed is expected to impact the following environmental areas: air as fugitive dust emissions generated during coal handling and crushing, and coal fines disposal; noise generated during coal crushing; and worker health in light of potential dust emissions and noise levels. The products are anticipated to pose a potential health hazard to workers due to their high aromatics

UPGRADING/UTILIZATION OF EDS PRODUCT STREAMS

STREAM	UPGRADING	UTILIZATION
NAPHTHA	HYDROTREATING REFORMING	MOGAS BLENDING
MID-DISTILLATE	HYDROTREATING	NO. 2 FUEL OIL STATIONARY TURBINE FUEL JET FUEL/DIESEL FUEL
	HYDROCRACKING	MOGAS BLENDING
VACUUM GAS OIL	HYDROTREATING	SPECIALTY FUEL OIL
	HYDROCRACKING OR HYDROTREATING/ CATALYTICCRACKING	STATIONARY TURBINE FUEL JET FUEL/DIESEL FUEL MOGAS BLENDING

Figure 10

EDS NAPHTHA IS EXCELLENT REFORMER FEED

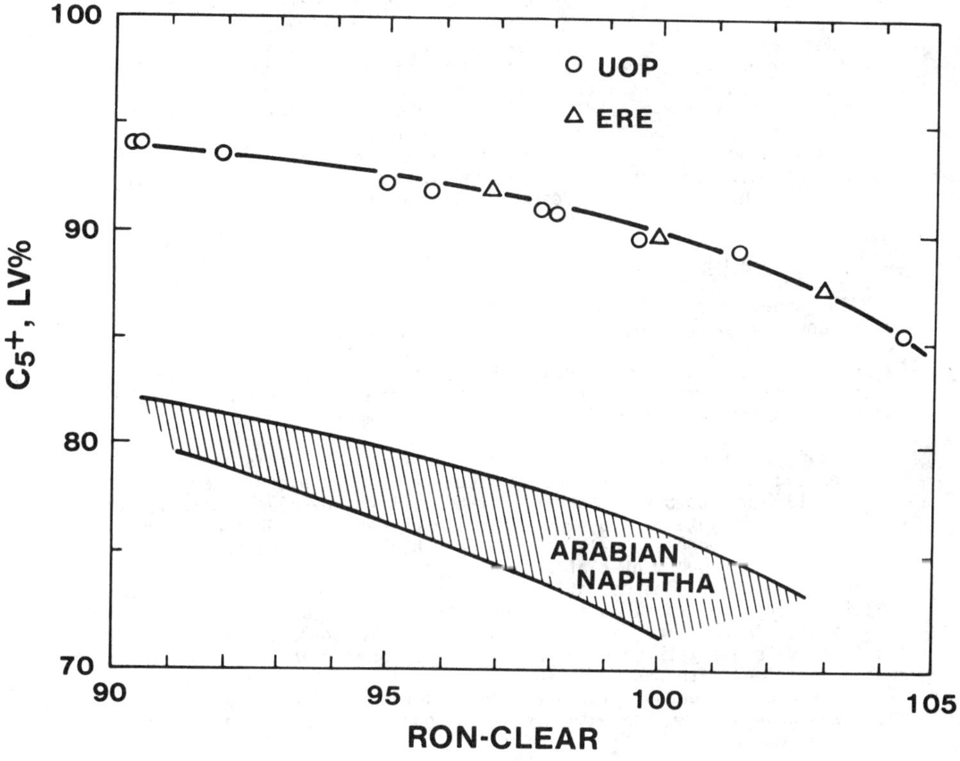

Figure 11

ANTICIPATED ENVIRONMENTAL IMPACT AREAS

SOURCE	AIR	WATER	SOLIDS DISPOSAL	NOISE	OCCUPATIONAL HEALTH
COAL FEED	X		X	X	X
PRODUCTS					X
PLANT DISCHARGES	X	X	X		

Figure 12

content.[27] Plant discharges are expected to impact due to fugitive dust and hydrocarbon emissions to the air, aromatic hydrocarbons and phenols in process and runoff water, and solid waste leaching in landfills.

Figure 13 shows the planned development strategy for addressing these concerns. The general approach is to define the problem using large pilot plant data and engineer solutions based on existing control technology. For example, air quality control measures based on existing petroleum refining and electric power industry control technology are being used in ELCP and the FLEXICOKING prototype unit. Also, all wastewater streams analyzed thus far appear to be treatable using existing refinery technology. This will be verified in bench scale tests. Programs to protect workers from both coal and product-based emissions are based on Exxon experience in coal liquefaction in more than 10 years of research and the experience of others.[28]

EDS PROJECT STATUS

The schedule of the EDS Project is shown on Figure 14. In addition to the continuing laboratory and engineering programs, the schedules for detailed engineering, procurement, construction and operation of the 250 T/D Exxon Coal Liquefaction Pilot Plant (ECLP) and the 70 T/D FLEXICOKING prototype unit are shown. Construction and operation of the two large pilot plants are under the direction of Exxon Company, U.S.A.

Figure 15 shows an overview of the ECLP operations this far. The detailed engineering, procurement, and construction phase of ECLP which began in 1977 was completed in March, 1980. Start-up solvent circulation was initiated in late May, and coal feeding commenced on June 24, 1980. A total of sixteen process goals were established for the shakedown period. During the first four months of shakedown, eleven of the sixteen goals were reached. These included operations on 8 mesh coal, demonstration of the ability to dry coal to 4% moisture, and several fractionation section objectives. In addition, the target of 50% on-stream factor was achieved. The five goals not yet achieved include steady operations at conditions near design coal feed rate and 1.2:1 solvent-to-coal ratio, operation of the reactor solids withdrawal system and operation of the slurry drier. The plant was brought down for a turnaround October 11. Coal feeding was resumed December 30, 1980.

An assessment of the experience gained in the first four months of operation is shown in Figure 16. Essentially all of the initial plant problems were mechanical rather than process oriented.[29] The mechanical problems included erosion of the vacuum tower transfer line, leakage from high-temperature reciprocating pump packing, control of slurry levels and slurry/water interface levels, breakdown of the solids handling systems and plugging of the slurry heat exchangers.

Two process related problems were identified. Pressure drop increase across the lead solvent hydrogenation reactor was observed. Additional care in controlling the recycle solvent endpoint has served to minimize this problem. The preliminary observation of apparently lower coal conversion than expected remains unresolved. Studies are underway to determine the actual ECLP conversion levels and the cause for any differences that may exist.

The slurry letdown valves and centrifugal pumps performed excellently. The slurry letdown valve showed negligible wear on inspection. The centrifugal pumps were judged to have demonstrated wear lives equivalent to nearly one year in commercial service.

The key to successful operation was avoiding solidification of heavy materials and solids plugging. The plant is unforgiving and service factor was a strong function of the time required to unplug the equipment after a coal outage due to solidification based plugging.

ENVIRONMENTAL CONTROLS DEVELOPMENT STRATEGY

- **AIR**
  - — QUANTIFY ECLP/FLEXICOKING PROTOTYPE EMISSIONS
  - — ADAPT CONTROL TECHNOLOGY FROM ELECTRIC POWER/ PETROLEUM REFINING INDUSTRY
- **WATER**
  - — CHARACTERIZE WATER FROM LARGE PILOT PLANTS
  - — SIMULATE TREATING SCHEME USING EXISTING TECHNIQUES
- **SOLIDS**
  - — PERFORM LEACHING/CHARACTERIZATION ON FLEXICOKING SOLIDS
- **NOISE**
  - — IDENTIFY/QUANTIFY SOURCES IN LARGE PILOT PLANTS
  - — ADAPT CONTROLS FROM ELECTRIC POWER/PETROLEUM REFINING
- **OCCUPATIONAL HEALTH**
  - — MONITOR WORKPLACE
  - — ASSESS ADEQUACY OF HEALTH PROGRAMS

Figure 13

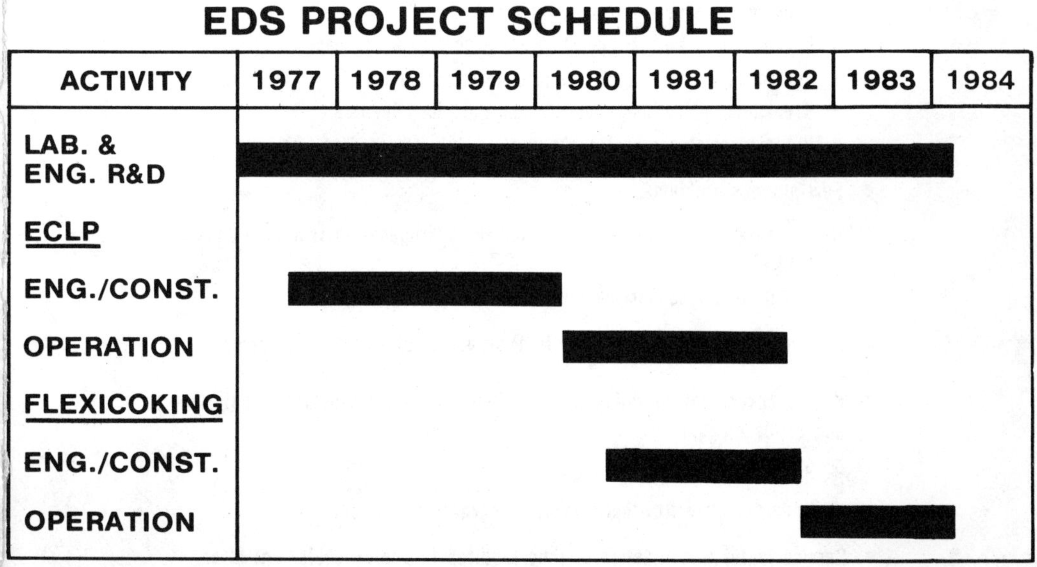

Figure 14

ECLP OPERATIONS OVERVIEW

- Plant mechanically completed March 21, 1980
- Start-up solvent circulation initiated May 28, 1980
- Coal feed commenced June 24, 1980
- Most all objectives achieved during planned four-month shakedown followed by one-month turnaround
  + Shakedown continued until October 11, 1980
  + Met 50-percent service factor objective with 1,245 hours on coal
  + Eleven of sixteen process goals reached
  + Turnaround conducted between October 11 and November 15, 1980
- Restart on coal accomplished December 30, 1980

Figure 15

ECLP OPERATING EXPERIENCE

- Initial problems mechanical rather than process
  + Erosion of vacuum tower transfer line
  + Leakage from high temperature reciprocating pump packing
  + Control of slurry and slurry/water interface levels
  + Breakdown of solids conveyors and gate lock feeders
  + Plugging of slurry heat exchangers
- Few process problems
  + Pressure drop increase in lead solvent hydrogenation reactor—under control
  + Apparent low conversion—unresolved
- Excellent performance of slurry letdown valves and centrifugal pumps
- Key to successful operation was avoiding solidification of heavy materials and solids plugging
  + Plant very unforgiving to these
  + Underdesigned Dowtherm system aggravated problem
- Service factor was function of time required to unplug equipment after coal outage due to solidification

Figure 16

The data collected during the ECLP turnaround is summarized in Figure 17. The significant observations were that corrosion was minimal and erosion while not extensive was significant where it occurred. The vacuum tower transfer line, the liquefaction reactors around the distributors and the receiver downstream from the slurry letdown valve had significant erosion.

During the turnaround, most of the significant equipment deficiencies were corrected, and additional test program facilities were added.

PICTURES OF ECLP/PROTOTYPE FLEXICOKING UNIT

Figure 18 shows an overview of the pilot plant. The relative positions of the administration building, the coal storage and preparation facilities, the process area and the product tankage areas are shown.

Figure 19 is a view of the Prototype FLEXICOKING Unit. The large cylinder is the coke storage silo. The coking reactor is behind the silo and the gasifier is in the structure to the left.

1979 EDS PLANT STUDY DESIGN

The EDS Study Design [30] explores the technical and economic issues which might be faced if an EDS commercial-size plant were to be designed on Illinois coal. It depicts the state of EDS technology in 1978, after approximately ten years of development work, as this technology might be applied in a commercial-size facility. Potential process improvements conceived after the design basis was set in 1978, such a liquefaction bottoms recycle and direct combustion of liquefaction bottoms, are not reflected.

The objectives of the Study Design were to identify the technical issues facing a potential commercial-size EDS plant design, to provide a reliable basis for estimating the environmental impact and the cost of EDS products, and to furnish research guidance to the EDS Project.

The Study Design consists of two distinct studies in which different processing schemes are used to produce the hydrogen and fuel gas required by the plant. These two studies are referred to as the "Base Case" and the "Market Flexibility Sensitivity Case." In the Base Case, hydrogen is generated by steam reforming of the light hydrocarbon gases produced in the plant. Fuel gas is generated by feeding the bottoms stream from the liquefaction section vacuum pipestill to a FLEXICOKING* unit. In the Market Flexibility Sensitivity (MFS) Case, the bottoms stream from the vacuum pipestill is split, and about half is sent to the FLEXICOKING unit for recovery of additional liquid product and production of fuel gas. The remainder of the bottoms stream is converted to hydrogen in a Partial Oxidation unit.

Selected highlights from the 1979 EDS Study Design with respect to coal feed rate, yield, investment, and thermal efficiency are given in Figure 20 for the Base Case and the MFS case. As shown in Figure 20, generating hydrogen from the vacuum bottoms via partial oxidation rather than from the high Btu gas via steam reforming significantly improves the yield and thermal efficiency of the plant. The investments shown do not include the attendant coal mines. The thermal efficiencies shown account for all coal feed to the plant and that used to produce the purchased electricity. It is estimated that the plant would take nine to ten years to complete after project initiation, would require a peak construction work force of 7800 people and would occupy 1300 acres of land. A major portion of the land requirement is associated with solids waste disposal.

ECLP TURNAROUND EXPERIENCE

- Corrosion Minimal
- Erosion not extensive, but significant where it occurred
- Most of significant equipment deficiencies corrected
- Test program facilities added

Figure 17

Figure 18

Figure 19

1979 EDS STUDY DESIGN HIGHLIGHTS

	BASE CASE	MFS
• ILLINOIS COAL FEED RATE kTONS/CD	27.0	28.0
• HYDROGEN SOURCE	STEAM REFORMING	BOTTOMS POX
• FUEL SOURCE	FLEXICOKING	FLEXICOKING
• PRODUCT RATES (kFOEB/CD)		
HIGH BTU GAS	--	10.0
C_3/C_4 LPG	1.6	3.7
NAPHTHA	17.3	16.9
FUEL OIL	32.2	30.8
TOTAL PRODUCT (kFOEB/CD)	51.1	61.1
• INVESTMENT 1980$	3.4	3.6
• THERMAL EFFICIENCY, %	56	63
• CAPACITY FACTOR, %	85	87

Figure 20

Within the EDS Study Design, coal liquids costs were evaluated considering a variety of technical and economic factors.[31] About 90 factors must be specified in order to calculate a specific coal liquids cost. The major factors studied and their impact as a percent of total coal liquids costs are shown in Figure 21.

Major technical factors were stage of development, process configuration, and plant reliability. Stage of development refers to the current status of the technology with respect to commercialization. Since EDS is a developing technology, some allowances must be made for identification of unknown factors during the development which will change the cost. In addition, the site for an EDS plant has not been chosen, therefore, allowances must also be made for those factors relative to a specific site which would impact on the total cost. Within the EDS Study Design, approximately a 35% contingency was added to reflect these uncertainties and this increased the product cost by about 15%. Obviously, these contingencies remain a matter of judgment until a commercial plant has actually been constructed. Process configuration refers to major changes in the liquefaction, fuel and hydrogen process blocks. In the EDS Study Design, this impact is illustrated by comparing the market flexibility sensitivity case with the base case. Moving from the base case to the MFS case changes the product cost by about 30% in this study design. Reliability refers to the actual plant output as a fraction of the ideal plant output. Various factors influence expected plant reliability including the design of the process unit as well as sparing philosophy of the plant designers. In the case of the EDS Study Design, expected plant throughput was about 15% less than ideal throughput. The potential impact of reliability on product cost can be illustrated by arbitrarily assuming ideal throughput with no increase in fixed costs. For this example, product cost would be reduced about 10%.

In general, project specific economic factors had a greater impact on projected product cost than the technical factors. Among the most important economic factors are: assumed inflation rate, tax treatment of investment and products, desired return rate. Within the range of the economic factors considered in the EDS plant Study Design, product cost varied up to 40%.

The principal conclusion from the above is that coal liquid cost can only be accurately described within the context of a consistent evaluation of specific supply alternatives. That is, the principal question facing those interested in building liquefaction plants is what are the most economically attractive ways to provide additional fuel supply. Realistically, this decision can only be made by evaluating the actual investment alternatives for specific technologies, at specific locations, using specific feedstocks, and making specific products within the framework of a consistent set of economic bases.

SUMMARY

EDS project activities are directed toward achieving commercial readiness. Successful operation of ECLP is a high priority activity this year, and a continuing effort on process improvements will be made in an effort to reduce product cost. The project will also develop liquefaction bottoms FLEXICOKING and evaluate other bottoms processing alternatives. Bottoms processing is a critical step in achieving commercial readiness.

1979 EDS PLANT STUDY DESIGN: MAJOR FACTORS AFFECTING COAL LIQUIDS COST

TECHNICAL FACTORS	PERCENT CHANGE IN COST
STAGE OF DEVELOPMENT	15%
PROCESS CONFIGURATION	30%
RELIABILITY	15%
ECONOMIC FACTORS	
INFLATION OUTLOOK; TAX TREATMENT; RETURN RATE	UP TO 40%

Figure 21

REFERENCES

1. Furlong, L. E., Effron, E., Vernon, L. W., and Wilson, E. L., The Exxon Donor Solvent Process, Chemical Engineering Progress, p. 69, (August 1976).

2. EDS Coal Liquefaction Project Technical Progress Reports prepared for the Department of Energy; Interim Report FE-2353-13, EDS Commercial Plant Study Design, February 1978; 1976 Phase IIIA Annual Report FE-2353-9, December 1977; Phase IIIA Final Report, FE-2353-20, March 1978; 1977-1978 Phase IIIB Annual Report, FE-2893-17, September 1978; July-September 1978, Phase IV Quarterly Report, FE-2893-21, November 1978; October-December 1978, Phase IV Quarterly Report, FE-2893-25, March 1979; January-March, 1979 Phase IV Quarterly Report, FE-2893-29, June 1979; 1978-1979, Phase IV Annual Report, FE-2893-35, October 1979; July-September, 1979 Phase IV Quarterly Report, March 1980.

3. Epperly, W. R. and Taunton, J. W., Exxon Coal Liquefaction Process Development, Proceedings of the Thirteenth Intersociety Energy Conversion Engineering Conference, August 20-25, 1978, San Diego, CA.

4. Epperly, W. R. and Taunton, J. W., Status and Outlook of the Exxon Donor Solvent Coal Liquefaction Process Development, Proceedings of the Fifth Energy Technology Conference, February 27 - March 1, 1978, Washington, D.C.

5. Epperly, W. R. and Taunton, J. W., Donor Solvent Coal Liquefaction Process, Coal Processing Technology, Vol. 5 (1979), p. 78.

6. Epperly, W. R. and Taunton, J. W., Exxon Donor Solvent Coal Liquefaction Process Development, ACS Symposium Series No. 110, Coal Conversion Technology (1979), p. 71.

7. Epperly, W. R. and Taunton, J. W., Development of the Exxon Donor Solvent Coal Liquefaction Process, The Annual Fall Conference of the Canadian Society of Chemical Engineering, Sarnia, Ontario, Canada, Sept. 30-Oct. 3, 1979.

8. Epperly, W. R. and Taunton, J. W., Progress in Development of Exxon Donor Solvent Coal Liquefaction Process, 72nd AIChE Annual Meeting, San Francisco, CA, November 28, 1979.

9. Epperly, W. R., Plumlee, K. W., and Wade, D. T., Exxon Donor Solvent Coal Liquefaction Process: Development Program Status, 1980 NPRA Annual Meeting, March 23-25, 1980.

10. Epperly, W. R., Plumlee, K. W., and Wade, D. T., Exxon Donor Solvent Coal Liquefaction Process: Development Program Status, American Mining Congress International Coal Show, May 5-9, 1980.

11. Epperly, W. R., Plumlee, K. W., and Wade, D. T., Exxon Donor Solvent Coal Liquefaction Process: Development Program Status, Seventh Annual International Conference on Coal Gasification, Liquefaction, and Conversion to Electricity, August 5-7, 1980.

12. Epperly, W. R., Plumlee, K. W., and Wade, D. T., Exxon Donor Solvent Coal Liquefaction Process: Development Program Status, Intersociety Energy Conversion Engineering Conference, August 18-22, 1980.

13. Swabb, L. E., Jr., Liquid Fuels from Coal: From R&D to an Industry, Science, 199, 619 (February 10, 1978).

14. Epperly, W. R., Cooperative Agreement--A New Mechanism for Joint Government/Industry Projects, Contract Management, July 1979, p. 4.

15. Blaser, D. E. and Edelman, A. M., FLEXICOKING for Improved Utilization of Hydrocarbon Resources, paper presented at the API 43rd Mid-year Meeting, Toronto, Canada, May 8-11, 1978.

16. Stretzoff, S., Partial Oxidation for Syngas and Fuel, Hydrocarbon Processing, December 1974, p. 79-87.

17. Robin, A. M., Hydrogen Production from Coal Liquefaction Residues, EPRI Final Report AF-233, Research Project 714-1, Texaco, Inc., December 1976.

18. van der Burgt, M. J. and Kraayveld, H. J., Technical and Economic Prospect of the Shell-Koppers Coal Gasification Process, paper presented at the ACS National Meeting, Anaheim, CA, March 16, 1978.

19. Given, P. H., et al, Characterization of Mineral Matter in Coals and Coal Liquefaction Residues, EPRI Annual Report AF-832, Research Project 3361, Pennsylvania State University, December 1978.

20. Quinlan, C. W. and Siegmund, C. W., Combustion Properties of Coal Liquids from the Exxon Donor Solvent Process, paper presented at the ACS National Meeting, Anaheim, CA, March 14, 1978.

21. Fant, B. T. and Barton, W. J., Refining of Coal Liquids, presented at the API 43rd Mid-year Meeting, Toronto, Canada, May 8-11, 1978.

22. Zaczepinski, S., Kamienski, P. W., Toups, H. J., Smith, R. S., and Turner, D. W., Upgrading of Coal Liquids, paper presented at the Mid-Year API Meeting, May 14-17, 1979, San Francisco, CA.

23. Sullivan, R. F., Refining and Upgrading of Synthetic Fuels from Coal and Oil Shale by Advanced Catalytic Processes, FE-2315-37, Chevron Report, Quarterly Report, January-March 1979.

24. Reidl, F. J. and DeRosset, A. J., Hydrocracking of EDS Process Derived Gas Oils, UOP Report, FE-2566-33, November 1979.

25. Pan, G. and DeRosset, A. J., Hydrotreating and Reforming EDS Process Naphtha and Fuel Oil, UOP Report FE-2566-25, February 1979.

26. Green, R. C. "Environmental Controls For The Exxon Donor Solvent Coal Liquefaction Process," paper presented at the Second DOE Environmental Control Symposium, March 19, 1980, Reston, Va.

27. Montgomery, C. H., Coal Liquefaction Occupational Health Program, Presentation at the 65th Annual Meeting of the American Occupational Medical Association, April 21-25, 1980.

28. Sexton, R. J., The Hazards to Health in the Hydrogenation of Coal, Archives of Environmental Health, September 1960, Vol. 1, p. 181-233.

29. Lendvai-Lintner, E., and Sorell, E., Materials Performance, Vol. 19, No. 4, p. 19-25, April 1980.

30. __________, EDS Commercial Plant Study Design, Interim Report, FE-2353-13, February 1978.

31. __________, National Energy Plan II, National Technical Information Service, 1978.

H-COAL® COMMERCIAL PLAN

George R. DeVaux
Hydrocarbon Research, Inc.
Research & Development Division
(A Subsidiary of Dynalectron Corporation)
134 Franklin Corner Road
Lawrenceville, N.J. 08648

INTRODUCTION

Two paths for commercial coal liquefaction using H-Coal® technology are possible. The quantum strategy involves the construction of a large, independent facility, and results in a very high capital investment. The incremental approach deals with stepwise additions of coal to a hydrogenation unit; may involve association with an existing facility; and may result in a substantially smaller investment. HRI's unique and commercially-proven Liquid Phase Hydrogenation systems permit the project owner to select the strategy most suited to his needs. The ultimate goal of coal liquefaction can be reached by either route. The H-Coal® Program supports this goal. Commercial coal liquefaction projects will be undertaken around the world when the value of the products exceeds the costs.

QUANTUM STRATEGY

Because the quantum strategy involves investments, on the order of $2 billion or more, owners will seek to reduce the investments and hedge the risks. One approach to hedging is to design parallel trains and install less than complete capacity. A second approach is to undertake only minimum upgrading at the site. The products are spread out in a large existing refinery system. Other approaches to investment reduction and hedging include purchase of coal or stockpiling vacuum bottoms rather than converting the material.

INCREMENTAL STRATEGY

The incremental strategy is more subtle and has received less public attention. This approach is based on the capability of

Hydrocarbon Research, Inc.'s Liquid Phase Hydrogenation (LPH) systems to convert coals, oils, or mixtures. The H-Oil® process is well known. In the presence of a catalyst, heavy petroleum material is reacted with hydrogen to form lighter, more desirable materials. The H-Coal® process is also well known. A self-produced recycle oil is used as a slurry vehicle for the coal. The slurry is reacted with hydrogen in the presence of a catalyst to form distillate materials. In the COIL process, petroleum oil is used to slurry the coal. The slurry is reacted with hydrogen in the presence of a catalyst.

Following the incremental strategy, a single owner might install a Liquid Phase Hydrogenation system within an existing refinery. If the initial feed is all petroleum, it is an H-Oil® unit. Later, the owner could add coal feeding equipment, more hydrogen capacity and ash withdrawal systems. This is the COIL mode of operation and permits the "cream" to be skimmed from the coal. The ratio of coal and oil can be varied to optimize the refinery.

The final stage of the incremental strategy is installation of equipment to recycle coal-derived oil to slurry the coal. This is the H-Coal® option.

The composition of a vacuum residual oil is contrasted to a coal in the table below:

	Composition, W%	
	Resid	Coal
Carbon	83.8	67.3
Hydrogen	10.3	4.8
Sulfur	5.6	4.9
Nitrogen	0.3	1.0
Oxygen	-	12.1
Ash	-	9.9
TOTAL	100.0	100.0
Atomic H/C ratio	1.47	0.85

Because the oil has a higher hydrogen content and much lower ash and oxygen contents, it is clearly superior to the coal for conversion to liquid products. Therefore, the first step in an incremental strategy will probably be an H-Oil® unit. As the relative value of liquids increases, more resid will be converted. Later, it may be economically desirable to add coal to the system.

Texaco has announced plans for a multi-reactor H-Oil® unit to convert vacuum resid. This project will have about the same effect as converting 7,000 tons-per-day of coal to about 24,000-BPD of liquid. Several other resid conversion projects are in advanced planning stages.

STRATEGIC CHOICE

The choice between a quantum or an incremental strategy will be governed by the needs and capabilities of the project owner.

The incremental strategy using COIL as an intermediate step has several major advantages:

- Low Capital Investment
- Feedstock Flexibility - Oil or Coal or Both
- Very Low Risk
- Economic in Small Volumes
- Process Simplicity

The quantum strategy is also valuable for certain owners because it provides:

- Quantity
- Focus
- Timing
- Economy of Scale

Both strategies for commercializing the H-Coal® process will draw on the technology now being demonstrated.

PROCESS DEVELOPMENT UNIT EXPERIENCE

The H-Coal® Process Development Unit (PDU) has been operated intermittently over the last 16 years to demonstrate scaleup of yield data, to demonstrate equipment operability, and to obtain products for downstream testing. Fourteen PDU runs, typically of about 30 days duration, have been carried out since the demonstration program began. Some of the major accomplishments are summarized below:

- Illinois No. 6, Kentucky No. 11, Wyodak and other coals successfully processed.
- Continuous catalyst addition and withdrawal used to approach equilibrium catalyst conditions.
- Syncrude, fuel oil, and intermediate modes of operation demonstrated.
- Emergency operating procedures for the Pilot Plant were tested while providing operator training.
- Critical operating limits, such as maximum gas velocity, were evaluated.
- Process improvements demonstrated.
- Two-stage slurry letdown system designed for the Pilot Plant demonstrated.
- Improved catalyst tested.
- Irradiated catalyst used to test ebullated-bed mixing and catalyst deactivation.
- Demonstration runs for Commercial Plants.

The PDU is scheduled for further use to demonstrate process improvements coming out of the on-going R&D program and for demonstration runs on specific coals for commercial plants.

Further improvements are anticipated from:

- Better Catalysts
- Recycle Management
- Feedstock Selection
- Feedstock Preconditioning

LARGE PLANT PROJECT

The H-Coal® Process has been thoroughly tested on bench and PDU-size equipment and is being demonstrated in commercial-size equipment at the Catlettsburg, Kentucky Plant. The H-Coal® Plant has several major objectives which are not obtainable on laboratory-scale equipment. These objectives include:

- Demonstration of the mechanical operability of the equipment.
- Provision of products for commercial testing at rates of 100 to 300 tons-per-day.
- Verification of yields in commercial-size equipment.
- Collection of scale-up and engineering data.
- Selection of appropriate materials of construction.
- Establishment of maintenance requirements for key items of equipment.

Two operating configurations have been designed into the Plant. The testing program encompasses syncrude and boiler-fuel mode operations using three different coals. The Plant has successfully operated on several grades of oil, and break-in operations with Kentucky No. 11 coal have been completed. The Plant has been revamped to demonstrate Illinois No. 6 in the syncrude mode. The schedule for the second year calls for boiler fuel operations with those two coals and a return to the syncrude mode using Wyodak coal.

Plans are currently being developed for future operations. Objectives are expected to include several tests for specific commercial projects and a long-term test.

DEVELOPMENT PATH FOR COMMERCIALIZATION

The development path for commercialization of H-Coal® is similar to that used for the commercial H-Coal® residuum and heavy crude hydroconversion process. The H-Oil® reactor system was scaled-up from the bench and the PDU to a large pilot plant demonstration unit, and finally to the commercial-scale plant.

	H-Oil® Reactor Diameter	H-Coal® Reactor Diameter
Bench Unit	3/4"	3/4"
Process Development Unit	8-1/2"	6" & 8-1/2"
Large Pilot Plant	4'-6"	5'
Commercial Plant	13'-6"	10'-13'

Currently operating commercial H-Oil® reactors treat up to 27,500-barrels-per-day of residual oil. An H-Coal® reactor of the same size operated in the boiler fuel mode could produce 17,500-barrels-per-day of product. Based on current practices, reactors as much as 50% larger may be economic.

Commercial plants will have parallel reactors. The number of trains will be governed by economies of scale and the availability of capital. The pioneer plant will have about ten times the throughput per train as the Catlettsburg Demonstration Plant. This will require a reactor diameter scale-up between about 2 and 3, which is not large by standards for successful process development.

COMMERCIAL PLANTS

Commercial H-Coal® plants will be built where there is:

- need
- economic capability
- feedstock.

Several countries have a greater need for coal liquefaction than the United States. We produce about half the liquid fossil fuel used. Many countries have little or no captive production of liquid fossil fuels.

The economic capability needed to undertake a quantum project is difficult to specify. However, if survival is at stake, a $30 billion per year economy is certainly adequate.

Access to a coal feedstock supply with reasonable security, production cost, and transport cost is also required.

Based on these criteria, candidate locations include:

United States	Europe
Canada	Russia
Japan	South Africa
Australia	South America
China	

In order to concentrate HRI's resources on the North American markets and a few other markets, HRI has entered into an optional sale agreement with Ruhrkohle. Under this agreement, Ruhrkohle will have the right to market H-Coal® technology in Europe.

PRODUCT COST AND VALUE

Commercial H-Coal® projects will not be implemented until the value of products exceeds their costs.

The reported cost of coal liquids from hydroliquefaction cover a very wide range and generally fall between $28 and $55 per barrel.

Current market prices for petroleum are within this range. In some situations, the value of coal liquids may already be higher than the cost. This is particularly true when national security, political autonomy, and monetary autonomy are considered.

SUMMARY

A project owner can choose either a quantum or incremental strategy to get a commercial coal liquefaction plant built and running. The potential exists for the first commercial projects in the United States to be achieved by the incremental approach. Owners in other countries may prefer the quantum strategy.

-oOo-

ACKNOWLEDGMENT

The author gratefully acknowledges the contributions of R. M. Eccles, P. H. Kydd, A. G. Comolli, H. H. Stotler, and J. B. MacArthur.

SOLVENT REFINED COAL
SRC-I

W. T. RECTOR
INTERNATIONAL COAL REFINING COMPANY

INTRODUCTION

The Solvent Refined Coal program, designated SRC-I, is a program under the sponsorship of the U.S. Department of Energy to demonstrate the SRC-I direct coal liquefaction technology on a commercial scale. It is anticipated that this demonstration will lead to the commercial application of the technology to produce significant volumes of synthetic fuel by the early 1990's, and thus be a major factor in reducing the U.S. dependence on imported oil. This project report gives a brief overview of the program, the technology employed, the project approach and schedule, and a review of the current status.

PROGRAM

Objective

The SRC-I program objective, as stated by the DOE, is to "demonstrate on a commercial scale, the technical, economic and environmental acceptability of the SRC-I process for converting high sulfur coal to clean burning fuels". To accomplish this objective, the DOE has entered into a contract, or cost-sharing agreement, with the International Coal Refining Company (ICRC) that provides for the design, construction, testing, operation, evaluation and ownership of a plant utilizing the SRC-I technology. (Fig. 1)

Who is ICRC?

ICRC is a partnership of Air Products and Chemicals, Inc., and Wheelabrator-Frye, Inc., formed for the explicit purpose of carrying out this program under the DOE contract. Two additional companies,

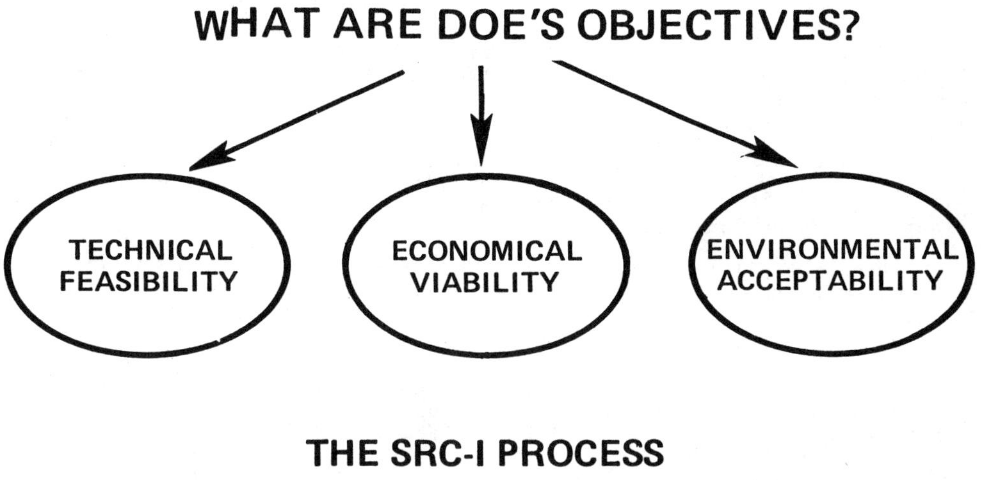
WHAT ARE DOE'S OBJECTIVES?
TECHNICAL FEASIBILITY
ECONOMICAL VIABILITY
ENVIRONMENTAL ACCEPTABILITY
THE SRC-I PROCESS

Figure 1

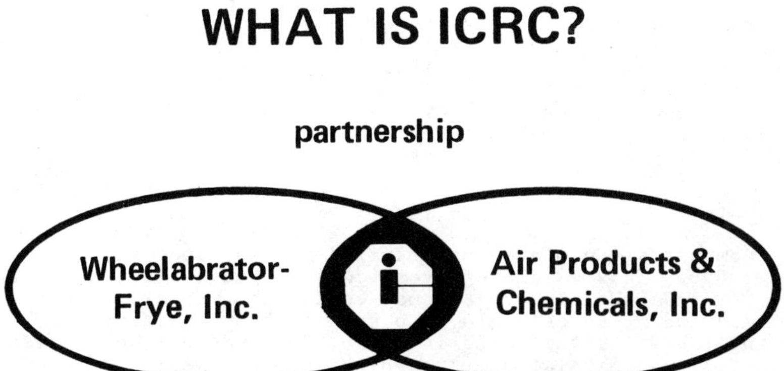
WHAT IS ICRC?
partnership
Wheelabrator-Frye, Inc.
Air Products & Chemicals, Inc.

Figure 2

Alcoa and Cities Service, have been active in coal liquefaction technology and are now negotiating to participate in the ownership of ICRC. It is expected that these negotiations will be completed in the near future. (Fig. 2)

Air Products and Wheelabrator have both participated in the evolution of the SRC-I coal liquefaction technology during the 1970's. Wheelabrator-Frye, through its subsidiary, The Rust Engineering Company, designed and built a fifty ton per day pilot plant employing the process. This plant, in Ft. Lewis, Washington, was started up in 1974.

Air Products and Chemicals, Inc., through its subsidiary company, Catalytic, Inc., designed, built and now operates a six ton per day pilot plant employing the SRC-I process at Wilsonville, Alabama. This plant started up in 1973.

These two pilot plants are currently operating and have served as a major source of data for the design of the demonstration plant.

Commonwealth of Kentucky

In addition to the DOE and ICRC, there is a third active participant in SRC-I. The Commonwealth of Kentucky, through the Kentucky Department of Energy, funded a feasibility study in 1976 to consider building a 2,000 ton per day coal liquefaction plant in western Kentucky, using high sulfur Kentucky coal as a feed stock. The study was carried out by Rust Engineering, with Catalytic serving as a subcontractor for the design of the liquefaction area. This work led to the selection of the proposed site for the SRC-I demonstration plant, and provided background data for the environmental impact statement. This study, along with strong political support within the state for coal liquefaction, has led Kentucky to become an active participant in the demonstration program. (Fig. 3)

Cost Sharing

The contract between DOE and ICRC, signed in August, 1980, calls for ICRC to contribute $90,000,000 toward the cost of the plant, and for the state of Kentucky to contribute $30,000,000. The balance of the cost will be funded by DOE. The contract provides for revenue sharing from product sales, for the ultimate buy-out of the government's interest in the demonstration plant by ICRC, and for sharing of royalties from licensing future plants. It is expected that this will lead to a full recovery of the government's cost and that the total cost of the demonstration program will ultimately be borne by private industry. It is also expected that the plant will be expanded to a commercial plant five times as large as the demonstration plant. (Fig. 4)

Cost Estimate

The demonstration plant is estimated to cost approximately $1,400,000,000. This estimate covers costs through start-up at the end of 1984, and is exclusive of coal costs and product revenues during operation. The estimate is based on the conceptual engineering (Phase 0) completed in late 1979, and includes reasonable contingencies, and escalation through the project period at about 9% per year. As the detailed design (Phase I) has proceeded, these estimates are being revised and are currently being reviewed with the DOE. (Fig. 5)

WHY SRC-I IN KENTUCKY?

- Feasibility Study Funded By State ($5MM)
  - Site Selection and Land Options
  - Environmental Studies
- Financial Support for Project
  - $30MM for Pollution Control Equipment
  - Funded From Coal Severance Tax
- Political Support

Figure 3

COST SHARING AGREEMENT AUGUST 1980

- $90 MILLION ICRC
- $30 MILLION COMMONWEALTH OF KENTUCKY
- BALANCE DOE
- ICRC CAN BUY OUT GOVERNMENT AFTER DEMONSTRATION PERIOD
- EXPAND PLANT TO 30,000 TPD

Figure 4

SRC-I DEMONSTRATION PLANT PROJECT
COST FORECAST - DECEMBER 1979
(MILLIONS OF DOLLARS)

	FY 80	FY 81	FY 82 - 84	TOTAL
PHASE I (DESIGN)[(1)]	34	79	69	182
PHASE II (CONSTRUCTION)[(1)]	12	141	906	1,059
PHASE IIIA (START-UP) [(1) (2)]	- 0 -	- 0 -	160	160
TOTAL	46	220	1,135	1,401

(1) CURRENT YEAR DOLLARS
10% ESCALATION FY 80
9% ESCALATION FY 81 AND BEYOND

(2) PHASE IIIA INCLUDES ONLY START-UP EXCLUSIVE OF REVENUES

Figure 5

PROCESS

The SRC-I process is a development of the coal liquefaction technology developed in Germany prior to World War II. This technology was used to produce synthetic fuels from coal for Germany's military use. Following the war, there was no further development (because of the availability of plentiful and cheap oil and gas) until 1960. Spencer Chemical then initiated research on the process under the sponsorship of the Office of Coal Research. This work accelerated in the late sixties and early seventies with the design and operation of the pilot plants at Ft. Lewis and Wilsonville. The initiation of the Kentucky design study and the evolution into the current demonstration program then followed. (Fig. 6)

Product

The primary product from the SRC-I process is a clean burning solid fuel that looks, handles and stores like coal. During the processing, essentially all of the sulfur and ash have been removed and the BTU content has been upgraded to approximately 16,000 BTU's per pound. The product can be easily pulverized, has a low melting temperature, and as a slurry or melt can be used directly as a replacement for oil in equipment designed to burn fuels such as No. 6 oil. This is a key point which has been clearly demonstrated in recent tests. (Fig. 7)

Product Slate

In addition to solid SRC as the primary product, the process has the flexibility to produce varying quantities of other products, such as naphtha, medium and heavy oils, anode coke and sulfur. It is believed that the ash contained in the coal feed is made inert enough to be used as a landfill or as an aggregate in concrete. The slide shows the product slate and quantities expected at design conditions, but the flexibility does exist to vary this considerably. (Fig. 8)

Process

The flow schematic shows the basic SRC-I process for converting high sulfur, high ash coal to molten SRC. The washed, sized coal is pulverized to -200 mesh and slurried in a coal-derived process solvent. The slurry is pumped to 2,000 psi and mixed with a hot, hydrogen-rich process gas stream. The mixture is then heated in process heat exchangers and fired heaters to over 800°F. Coal dissolution is accomplished in the fired heaters and hydrogenation begins. Additional hydrogen is added and the mixture flows into the dissolvers, where the hydrogenation and desulfurization reactions are completed. The dissolver effluent is flashed to remove the gases, which are purified and recycled, and then distilled to separate the process solvent and lighter liquids. (Figs. 9-10)

The remaining slurry, containing the SRC, ash and some unconverted coal, then flows to a Critical Solvent Deashing Unit, where the ash and unconverted coal are separated from the molten SRC. The ash and coal residue then serves as feed to the gasifiers, where it is reacted with oxygen and steam to produce make-up hydrogen and where the ash is rendered inert. (Fig. 11)

Molten SRC Processing

The molten SRC is further processed through one of three finishing steps. The first is simple chilling with water which produces the solid SRC previously described.

SRC-I CHRONOGRAM

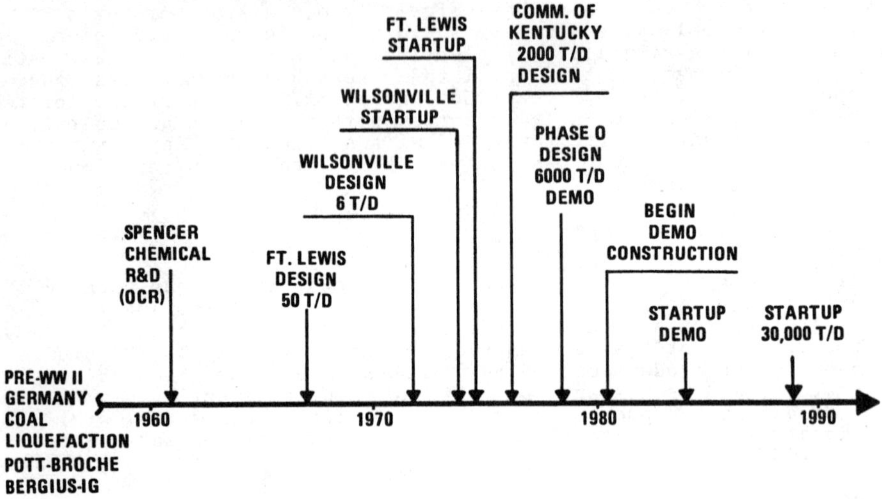

Figure 6

IMPORTANT SRC CHARACTERISTICS

- HANDLES AND STORES LIKE COAL
- LOW SULFUR
- LOW ASH
- HIGH BTU
- EASILY PULVERIZED
- LOW MELTING TEMPERATURE

Figure 7

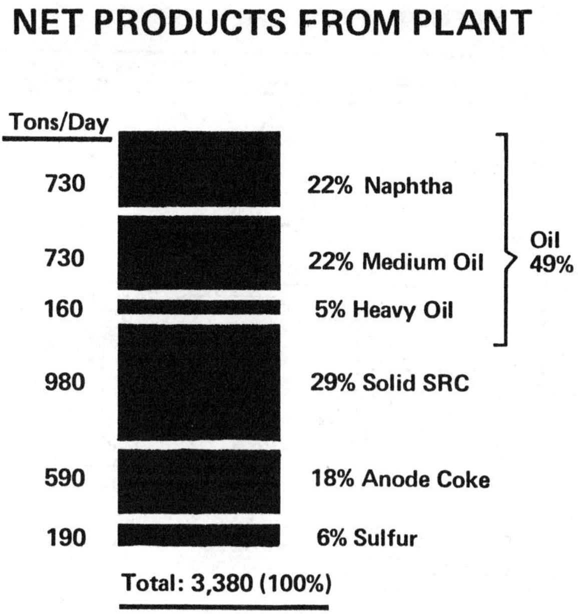

Figure 8

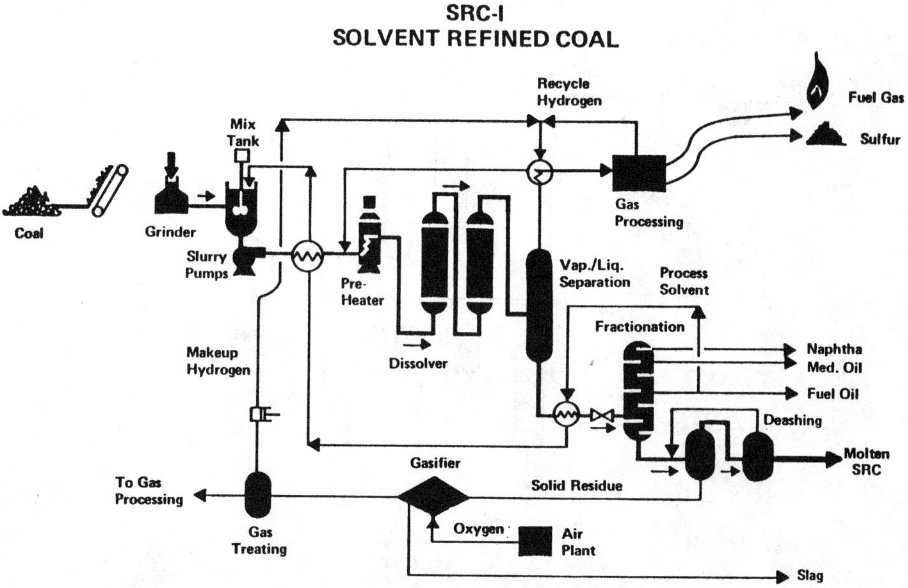

Figure 9

DEMO PLANT SRC PROCESSING

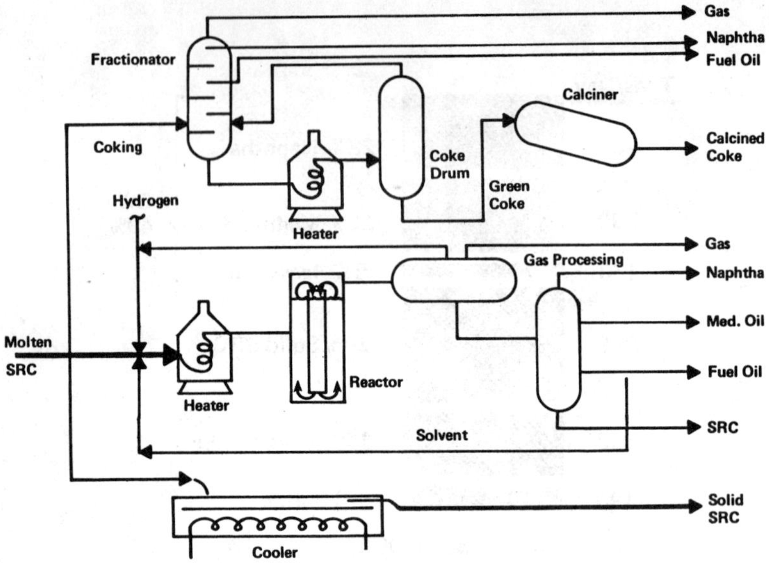

Figure 10

SRC DEASHING SCHEMATIC FLOW DIAGRAM

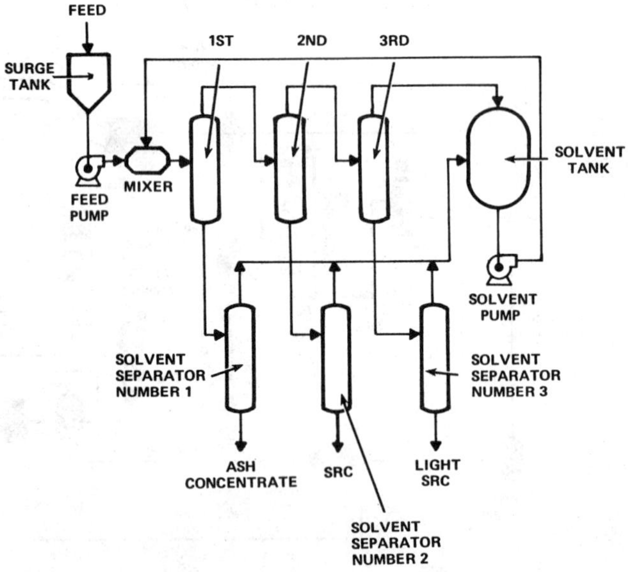

Figure 11

In a second alternate, a portion of the molten SRC is processed through an LC Finer where it further reacts with hydrogen, in the presence of a catalyst, to produce gases, naphtha, medium oils, fuel oils, and molten SRC. This step leads to the terminology of two-stage liquefaction (TSL). The products from the second stage have a lower sulfur content than the equivalent products from the first stage. (Fig.12)

A third alternate for processing is to feed molten SRC to a coker and convert it to coke, which is calcined and further processed into anode coke for the aluminum industry.

Key Process Characteristics

The SRC-I process has two fundamental characteristics that distinguish it from other direct liquefaction processes. The first is that the hydrogenation in the first stage is quite moderate. This improves the energy efficiency, but requires a more difficult ash removal process when compared to other approaches. The use of the Kerr-McGee Critical Solvent Deashing process is the key to the success of this approach, which leads to a high efficiency and a greater flexibility in the product slate.

The second characteristic of SRC-I is the use of the LC Finer to create the Two Stage Liquefaction process. This again improves the energy efficiency and flexibility, and can produce cleaner products when compared to other liquefaction processes.

PROJECT

The plant that is being designed to demonstrate the SRC-I technology on a commercial scale is sized to process 6,000 tons per day of high sulfur, high ash coal, equivalent to Kentucky No. 9. (Fig.13)

Location

The plant will be located on a site in Daviess County in western Kentucky, on the Green River, approximately twelve miles west of Owensboro, Kentucky. The site covers approximately 1500 acres of essentially flat farm land, and is sufficiently large to contain the demonstration plant, and to later be expanded in capacity five-fold to full commercial size. (Figs. 14-15)

Major Contractors

The overall engineering/construction approach to the project is somewhat unique. The overall project management is the responsibility of ICRC. There are six major area contractors, each providing the engineering and major equipment procurement for a defined area of the plant. These contractors are: (1) Air Products and Chemicals, Inc., Allentown, Pennsylvania - oxygen plant and hydrogen purification equipment; (2) Catalytic, Inc., Philadelphia, Pennsylvania - SRC area; (3) C-E Lummus Company, Bloomfield, New Jersey - coking, calcining and LC Finer; (4) The Ralph M. Parsons Company, Pasadena, California - gasification, acid gas treatment, compression and sulfur recovery; (5) The Rust Engineering Company, Birmingham, Alabama - coal preparation, utilities and off-sites; and, (6) Johnson Controls, Inc., Philadelphia, Pennsylvania - instruments and controls. (Fig. 16)

There will be one construction manager for the project, who is yet to be named. This contractor will have the responsibility for procurement of standard equipment and bulk materials, for establishing and managing all construction subcontracts, and for direct line management of all force account labor.

LC-FINING PROCESS

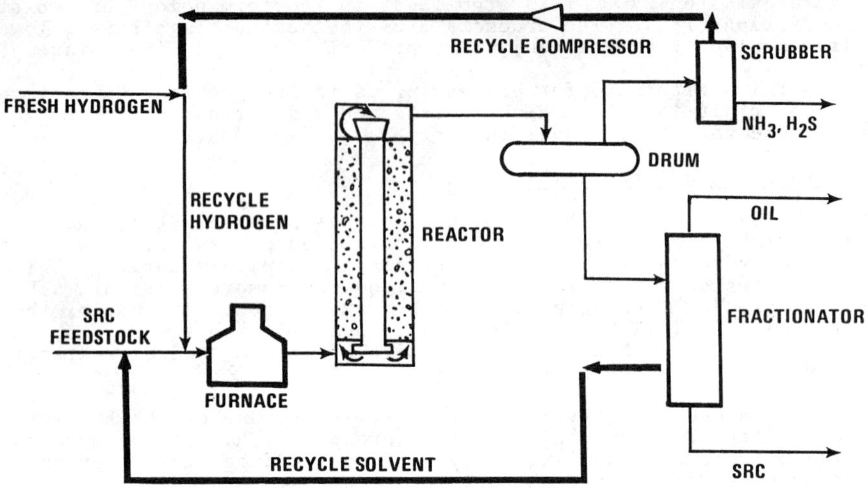

Figure 12

SRC I DEMONSTRATION PLANT

PROJECT OBJECTIVE:

DESIGN, CONSTRUCT, OPERATE AND EVALUATE A DEMONSTRATION PLANT WHICH WILL PROCESS 6000 TONS PER DAY OF COAL FEEDSTOCK AND PRODUCE PRIMARILY A SOLID SRC PRODUCT.

Figure 13

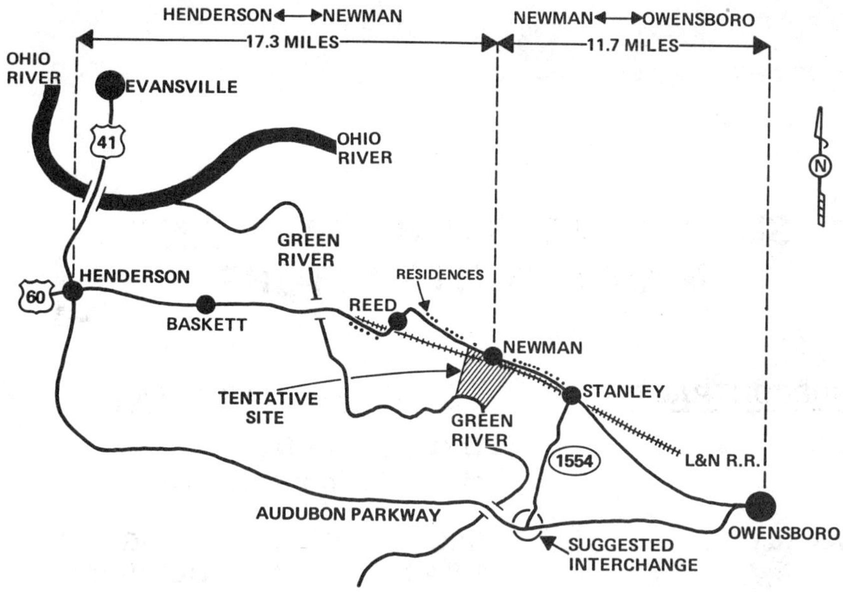

Figure 14

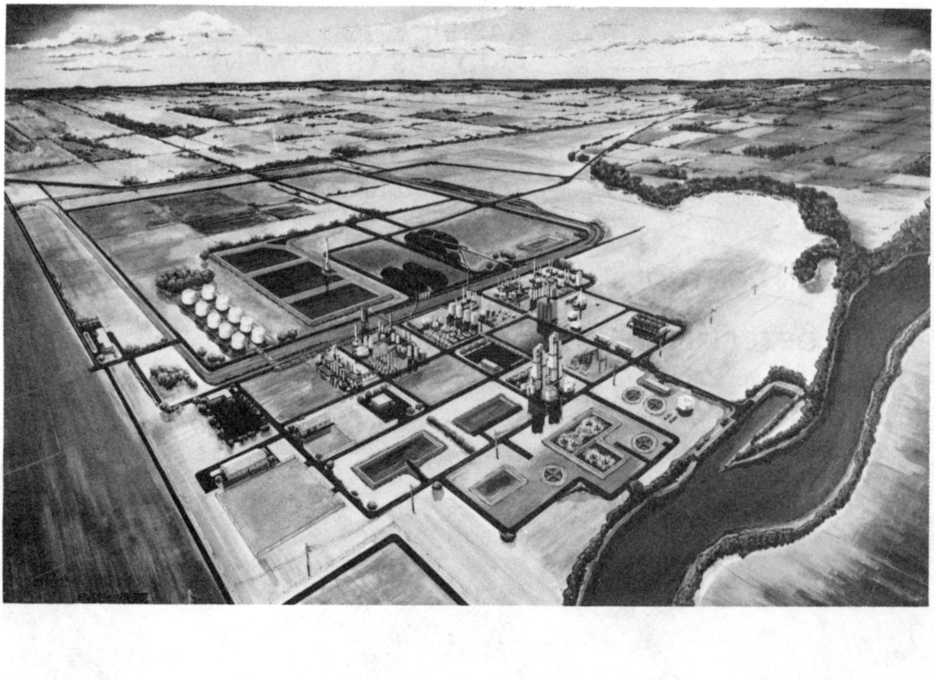

Figure 15

SRC-I DEMONSTRATION PLANT MAJOR CONTRACTORS

SUBCONTRACTOR	PROCESS AREA
APCI	OXYGEN PLANT HYDROGEN PURIFICATION
CATALYTIC	SLURRYING, DISSOLVING, FRACTIONATION, SOLIDIFICATION AND DEASHING
LUMMUS	COKING AND CALCINING LC-FINING
PARSONS	GASIFICATION ACID GAS TREATMENT COMPRESSION SULFUR RECOVERY
RUST	COAL PREPARATION OFFSITES AND UTILITIES
JOHNSON CONTROLS	INSTRUMENT & CONTROL SYSTEMS
TO BE DETERMINED	CONSTRUCTION

FIGURE 16

Schedule

It is expected that the site construction will begin in the fall of 1981 and that the plant will be mechanically complete by the end of 1984. The peak labor force during the most active construction period will be just under 2,500 workers, not including management and engineering staffs. Two separate socio-economic studies are now under way to determine more accurately the impact of this work force on the area. At this time, it appears that the impact is nominal and that the labor will be available as needed. Some special measures to handle the traffic to and from the site during the peak period will be required and commitments have been made by the State of Kentucky to put these measures into effect. (Fig. 17)

Following mechanical completion of the plant, there will be a thirty month period for start-up and operation of the plant in a demonstration mode. Provisions are available to extend the demonstration period by up to three years.

Environmental Impact Statement

The major factor at this time that determines the schedule is the Environmental Impact Statement. This statement is the responsibility of the DOE, as the sponsoring agency for SRC-I. This was released on January 9, 1981, approximately six months behind schedule, and is now open for public comment. Public hearings in the Owensboro/ Henderson area were held February 23rd through February 25th. It is expected that a record of decision date of June 1, 1981, will be achieved. This would lead to the early fall construction start date. Any delays in the record of decision would impact the projected start date. (Fig. 18)

Other Potential Problems

Other potential delay problems to be faced in the near future are: (1) obtaining the construction permits, principally those that are associated with environmental controls; (2) obtaining the final parcels of land, perhaps by condemnation; (3) changing the zoning from agricultural to industrial; and, (4) obtaining sufficient construction labor to maintain progress as projected. At this point in time, we do not expect that any of these will develop into serious obstacles to achieving the progress as projected. (Fig. 19)

PROJECT SCHEDULE

EIS RECORD OF DECISION	JUNE 1981
BEGIN CONSTRUCTION	FALL 1981
MECHANICAL COMPLETION	LATE 1984
COMPLETE DEMONSTRATION	MID 1987

Figure 17

ENVIRONMENTAL IMPACT STATEMENT SCHEDULE

DRAFT EIS TO EPA	9 JANUARY
PUBLISHED IN FEDERAL REGISTER	16 JANUARY
PUBLIC HEARINGS	23–25 FEBRUARY
END PUBLIC COMMENT	2 MARCH
FINAL EIS TO DOE	17 APRIL (Informal)
FINAL EIS TO EPA	24 APRIL
PUBLISHED IN FEDERAL REGISTER	1 MAY
RECORD OF DECISION	1 JUNE

Figure 18

SRC-1
POTENTIAL SCHEDULE PROBLEMS

-- ENVIRONMENTAL IMPACT STATEMENT
-- CONSTRUCTION PERMITS
-- REZONING
-- LAND PROCUREMENT
-- LABOR AVAILABILITY

Figure 19

STATUS OF THE SRC-II (LIQUIDS) PROJECT

Mr. Donald M. Jackson
Mr. Bruce K. Schmid
Dr. Alan S. Schwartz

SRC INTERNATIONAL, INC., P. O. BOX 3396, ENGLEWOOD, COLORADO 80155

Prepared under U. S. Department of Energy Contract No. DE-AC-05-780R03055

ABSTRACT

A 6000 T/D demonstration plant to produce liquid fuels from high-sulfur bituminous coal by the SRC-II process is now being designed. The design, construction, and operation of the plant is being carried out by Solvent Refined Coal International, Inc., a joint venture company consisting of The Pittsburg & Midway Coal Mining Co., Ruhrkohle of Germany, and Mitsui SRC Development Co. of Japan. The total cost of the project, estimated to be about 1.5 billion dollars, will be shared between the United States Department of Energy, the governments of West Germany and Japan, and SRC International. The SRC-II Demonstration Plant will produce fuel oil for use in boilers and turbines, naphtha suitable as reformer feedstock to make high octane gasoline, and by-product light hydrocarbons, from West Virginia coal.

Process design work and preliminary detail design in non-process areas is now underway. The final Environmental Impact Statement has been issued, and a construction management subcontractor (Kellogg-Kaiser Engineers) has been selected. Site preparation work is expected to begin during the summer of 1981. The overall construction work will require about four years. After an initial operating period of two and one half years, the joint venture company will have the option of purchasing the plant. It is expected that such purchase would lead to expansion of the plant to a full-size commercial facility.

INTRODUCTION

The Pittsburg & Midway Coal Mining Co. (P&M), a wholly owned subsidiary of Gulf Oil Corporation has been working on development of the solvent refined coal technology for 18 years, primarily under the sponsorship of the United States Department of Energy and its predecessor agencies. During the last four years, a major part of this effort has involved operation of a 30-ton-per-day pilot plant at Fort Lewis, Washington on an improved version of the process known as SRC-II. The SRC-II process converts high-sulfur coal to distillate liquids, naphtha, and light hydrocarbons. As a result of encouraging results in both the pilot plant program and supporting laboratory work, P&M contracted with the Department of Energy to design, build and operate a 6000 ton-per-day demonstration plant using the SRC-II process. The proposed site for the plant is near Morgantown, West Virginia and it is anticipated that the plant will use high-sulfur West Virginia and other bituminous coals.

Initial work on the demonstration plant contract involved developing a preliminary design for the 6,000 ton-per-day demonstration plant, as well as conceptual designs for a 30,000 ton-per-day commercial plant developed by expanding the demonstration plant and for a 30,000 ton-per-day grass roots commercial plant. In addition, schedules and cost estimates were developed and supporting studies, both technological and market oriented, were conducted. All of this work was completed in what is now known as Phase Zero of the Demonstration Plant Project. Phase Zero was completed in July of 1979 and Phase One was started in October of 1979. This paper discusses the current progress of the Demonstration Plant Project, now in Phase One, and the funding and organizational agreements that have been put in place to carry it out.

PROCESS DESCRIPTION

The SRC-II process converts coal to liquid and gaseous products by first dissolving the coal in a slurry recycled from the process, then hydrogenating and hydrocracking the dissolved coal in the presence of hydrogen at elevated temperature and pressure. The hydrogenation and hydrocracking reactions are enhanced by the catalytic activity of the inorganic matter contained in the recycle slurry and in the feed coal.

A schematic flow diagram for the 6000 ton-per-day SRC-II demonstration plant is shown in Figure 1. Raw coal is pulverized and dried, mixed with hot recycle slurry from the process and pumped, together with hydrogen, through a fired preheater to a reactor maintained at about 860°F and 2000 psig. The reactor effuent flows through a series of vapor-liquid separators, where it is ultimately separated into process gas, light hydrocarbon liquids and slurry. The process gas, consisting primarily of hydrogen and gaseous hydrocarbons, together with minor amounts of hydrogen sulfide and carbon dioxide, is cooled to about 100°F and goes through an acid gas treating step for removal of hydrogen sulfide and carbon dioxide. The treated gas then goes to a cryogenic separation step for removal of the hydrocarbons. The purified hydrogen is recycled to the process, while the recovered hydrocarbons become by-products of the process. The C1 fraction is sent to a methanation unit to convert any remaining CO to methane, then is sold as pipeline gas. The other light hydrocarbon gases are fractionated to produce ethane, propane and butane streams.

All of the hydrocarbon liquid streams collected from the various condensation steps, plus the overhead stream from the vacuum tower, are sent to a fractionator. In the fractionator, the total liquid is separated into naphtha (C5 - 380°F nominal boiling range) and a middle distillate (380-600°F). The product slurry is split, with one portion being recycled to the process for slurrying with the feed coal. The other portion of the product slurry goes to a vacuum tower where the lighter portion of the distillate is removed overhead and sent to the fractionator. A heavy distillate product is removed as a side stream. The residue from the vacuum tower is sent to a high pressure slagging gasifier for production of synthesis gas. The synthesis gas goes through shift conversion and acid gas removal steps to produce pure hydrogen for the process. Any hydrogen produced in excess of that required for the process is burned as plant fuel.

FIGURE 1

FLOW DIAGRAM FOR SRC-II DEMONSTRATION PLANT

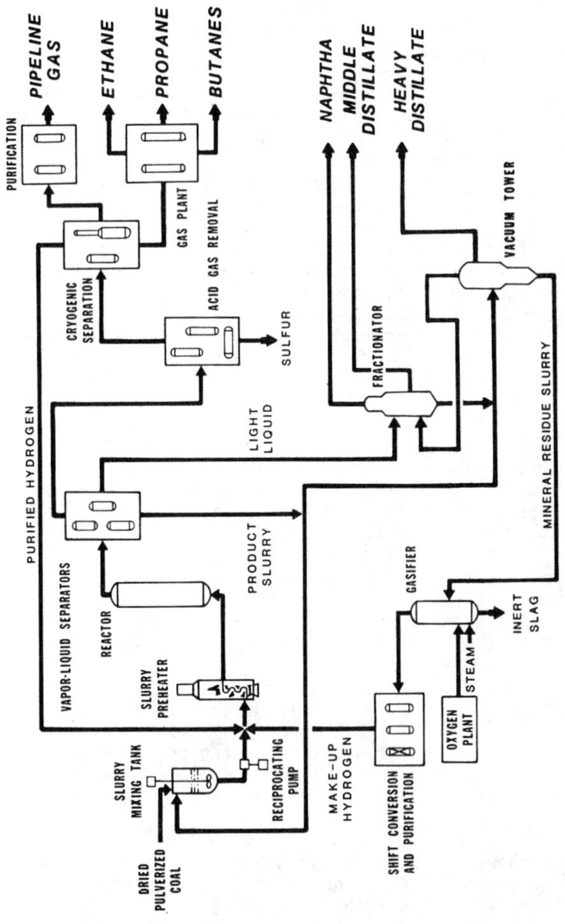

PROGRAM AGREEMENTS

The organization, funding and international participation of the SRC-II project were established with agreements signed in the latter part of 1980. Agreements signed on July 31, 1980, between Gulf and DOE and between the governments of the United States, Germany and Japan provided for funding of the project. Within this framework, Gulf accepted responsibility for carrying out the SRC-II Demonstration Project.

On September 20, 1980, a joint venture company, comprised of The Pittsburg & Midway Coal Mining Co., Ruhrkohle AG of West Germany and a Japanese corporation, Mitsui SRC Development Co, was formed. The name of the joint venture company is Solvent Refined Coal International, Incorporated, but it will be generally known as SRC International. Concurrent with its formation, Gulf assigned the demonstration plant contract to the joint venture company.

Total cost of the project is expected to be approximately $1.5 billion, DOE paying roughly 50%, the foreign governments about 25% each. The contractor's total contribution in cash and kind is $100 million, a $50 million front-end contribution and a $50 million performance fund which would not be drawn until later in the project. The $50 million performace fund, (DOE will contribute a like amount) will be drawn after the completion of mechanical construction. At that time, the contractor will have this level of financial exposure to share equally the costs of any modifications required to make the plant meet its performance goals.

The project will be under DOE's direction from the Oak Ridge operations office. DOE's program managers will in turn be responsible to a 6-person Steering Committee comprised of two appointees each by the governments of Germany, Japan and the United States.

DOE's objective in supporting the SRC-II Demonstration Project is to enable the acceleration of the commercialization of synfuels technology. To this end, it is intended that SRC International will have an option to purchase the plant at the end of an initial period of operation (2½ years) and expand it to commercial size.

SITE ACTIVITY

The demonstration plant site is located on the Monongahela River in Monangalia County, West Virginia, just south of the Pennsylvania state line. This was the one site selected as optimum during an intensive Phase Zero evaluation.

Figure 2 shows the proposed site and identifies the major plant areas. Total acreage is about 2,400, including land required for expansion to a commercial facility. Topographically, most of the site is gently rolling upland. A narrow strip of river terrace, generally about 200 feet wide, parallels the river. At the southern end, this strip is some 600 feet wide and will be used for tankage and rail facilities.

FIGURE 2

PROPOSED SITE FOR SRC-II DEMONSTRATION PLANT

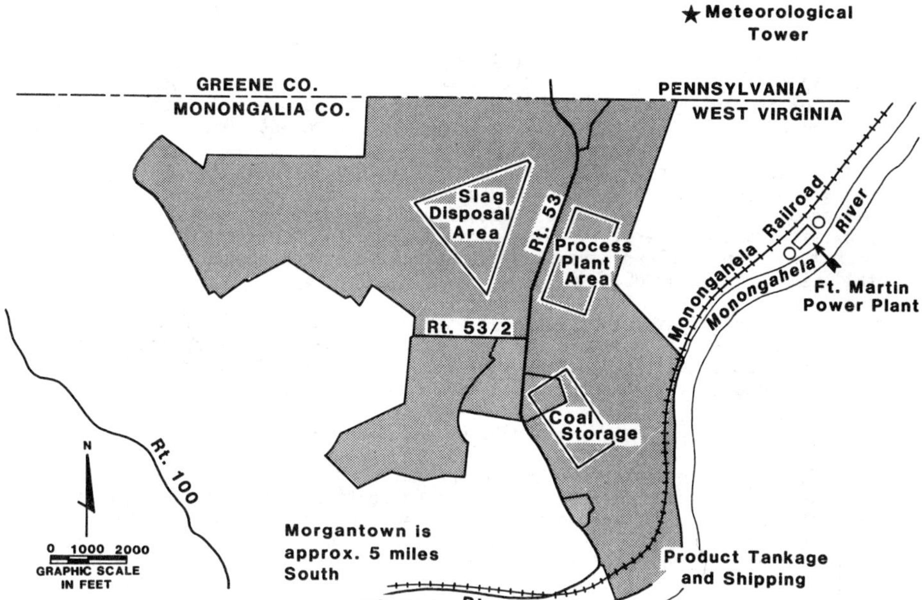

FIGURE 3

AREA OF POTENTIAL COAL SUPPLY FOR THE SRC-II PLANT

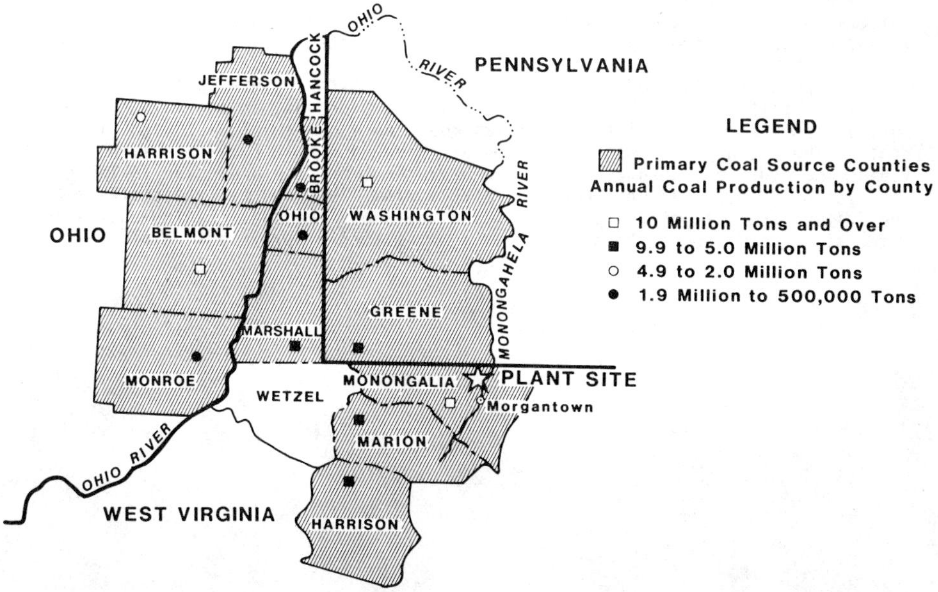

A draft Environmental Impact Statement was issued by DOE on May 30, 1980. In response to comments received from the EPA and several environmental groups, a number of modifications were made. The revised document was publicly issued on January 30, 1981 as the final Environmental Impact Statement for the project.

The acquisition of feed coal in sufficient quantities for an initial operating period of 2½ years is now proceeding. For this initial operating period a dual-source coal supply, based on recent process evaluations and on proximity to the rivers, is being pursued. The two coals are from the Powhatan #6 and Ireland mines. Negotiations have been opened with representatives of both sources.

It is anticipated that a subsequent 2-3 year operating period could employ other bituminous coals (as yet undetermined) after operating performance has been established during the initial demonstration period. Figure 3 shows the counties in the tri-state area considered as possible sources.

PHASE ONE DESIGN

The Phase One design work was begun in October, 1979, upon completion of the Phase Zero work. The process design subcontractor is Badger Engineering, Inc. of Cambridge, Massachusetts, while design of the nonprocess areas, as well as detailed design in all areas, is the responsibility of Stearns-Roger Corporation of Denver, Colorado. The Phase One process design was initiated with a review and evaluation of the Phase Zero design by both Gulf and Badger during the latter part of 1979 and early 1980. This work has led to a somewhat modified design basis for Phase One.

The design priorities for the demonstration plant are different than they might be for a subsequent commercial venture. Environmental protection and plant operability have been assigned the highest level of priority for the project, with yields and thermal efficiency at a somewhat lower priority. In addition, a substantial effort is being made to incorporate additional flexibility into the plant design, to minimize the potentially adverse effects of the uncertainties involved in a first-of-a-kind plant.

The Phase One design is based on a specific West Virginia Coal rather that the hypothetical coal used for the Phase Zero design, as a result of further laboratory and pilot plant experiments on a number of West Virginia coals. The design feed coal properties are given in Table I. The design yields and hydrogen consumption for the hydrogenation step are given in Table II. The C1-C4 hydrocarbons, naphtha, middle distillate and heavy distillate are primary products of the process, while the 900°F residue, insoluble organic material, and ash go to the gasifier as a hot molten mineral residue slurry.

TABLE 1

ANALYSIS OF DESIGN FEED COAL

Coal Analysis (moisture-free basis), Wt. %

Carbon	70.5	
Hydrogen	4.9	
Nitrogen	1.1	
Oxygen	7.9	
Sulfur (total)	3.6	
Pyritic		2.0
Organic		1.6
Ash	12.0	
Total	100.00	

TABLE 2

YIELDS FROM SRC-II HYDROGENATION STEP

	Wt. % of Moisture-Free Coal Charge
Methane	5.8
Ethane	4.4
Propane	4.1
Iso-Butane	4.1
n-Butane	0.2
C5-380°F Naphtha	10.2
380-600°F Middle Distillate	17.9
600-900°F Heavy Distillate	7.9
900°F + Residue	24.0
Pyridine-Insoluble Organic Material (IOM)	4.6
Ash	12.0
Residual S	1.1
Carbon Monoxide	0.1
Carbon Dioxide	1.0
Hydrogen Sulfide	2.5
Ammonia	0.4
Water	6.2
Total	104.4
Hydrogen Consumption	4.4

PROJECT SCHEDULE

As previously discussed, the project is now in Phase I (process and detailed design). The process design work and preliminary detail design in non-process areas is progressing with ongoing evaluation of the scheduled work activities.

Phase II (construction) is scheduled to begin later in 1981, although preliminary work such as land acquisition, site surveys, and selection of a construction management subcontractor for the project have been underway for some time. The construction management subcontractor will be Kellogg-Kaiser Engineers, a joint venture company formed by the M. W. Kellogg Co. and Kaiser Engineers, Inc. Current plans are to start earthwork at the site during the summer of 1981. About four years will be required for construction. In Phase III, the contractor (SRC International) is to operate the demonstration plant and evaluate the commercial viability of the SRC-II process. Although the schedule shows operation beyond 1988, the contractor has an option to purchase the plant before that time.

PRODUCTS

Table 3 shows the expected product slate for the demonstration plant. Excess hydrogen (above that required for the process) butane and methane rich gas (prior to methanation) are assumed used as plant fuel. The pipeline gas, propane, by-product sulfur and ammonia will be produced to industry specifications and marketed accordingly.

TABLE 3

PRELIMINARY ESTIMATE OF PRODUCT SLATE

Net Product Output*	M lb/hr	Bbls/day
(Main Products)		
Pipeline Gas	19.0	1,670
Liquid Propane	17.4	2,320
Naphtha	51.0	3,850
Fuel Oil	120.0	8,350
Total	207.8	16,190
(By-Products)		
Sulfur	16.7	
Ammonia	2.5	
Tar Acids	0.6	

* Products after fuel consumption and power generation

The raw naphtha, comprising about one third of the total liquid product, will be hydrotreated at the demonstration plant to make it suitable as a catalytic reforming feedstock. The hydrotreated naphtha will be distributed to a petroleum refiner.

Properties of raw and hydrotreated SRC-II naphtha are shown in Table 4. The total cycloparaffins and aromatics are much higher than in most petroleum naphthas, indicative of a feedstock which should yield a high octane reformate with relative ease. This has been borne out by subsequent reforming studies. The SRC-II naphtha produces gasoline of octane number equivalent to that obtained from typical petroleum naphthas at substantially lower reforming temperatures and with much improved reformate yields. The reforming studies have shown no problems with catalyst aging.

TABLE 4

PROPERTIES OF RAW AND PRETREATED SRC-II NAPHTHAS

	Raw	Hydrotreated
Gravity: °API	37.7	49.3
Sulfur: PPM	2400	0.08
Nitrogen: PPM	8800	0.2
Hydrocarbon Analysis: % by Vol.		
Paraffins	21.4	23.8
Cycloparaffins	44.8	62.3
Aromatics	33.8	13.9
Octane Ratings		
Research, Clear	--	63.4
Motor, Clear	--	61.7

Properties of the fuel oil, the major product, are given in Table 5. The fuel is an all-distillate product, liquid at ambient temperature, and very low in sulfur. Preliminary studies on the properties of blends of SRC-II fuel oil with conventional products suggest that such blends are stable, and that metal corrosion is unlikely to be a problem. The sulfur content of 0.3% is low enough to meet the stringent sulfur dioxide emission regulations in major metropolitan areas.

TABLE 5

PROPERTIES OF SRC-II FUEL OIL

Gravity: °API	8.3
Viscosity: SUS at 100 °F	40
Flash Point: °F	150
Pour Point: °F	-20
Sulfur: % by wt.	0.3
Nitrogen: % by wt.	1.0
Boiling Range: °F	350-900

A full-scale burning test was conducted in fall of 1978 by Consolidated Edison of New York in a commercial electric utility boiler. The results were very encouraging. Combustion efficiency was comparable to that of petroleum fuels and, while the NOx emissions were somewhat higher than for conventional petroleum fuels, they were much lower than would be expected on the basis of nitrogen content. The emissions not only met all applicable regulations for NOx, but it was found that the emissions could be further reduced by staged combustion.

CONCLUSION

Process design work is proceeding for the 6,000 ton-per-day demonstration plant to convert West Virginia coal to synthetic liquid fuels by the SRC-II process. A construction subcontractor is being selected, and construction is expected to begin early next year. Successful demonstration of the process at this scale could lead to an expansion of the plant to a full-scale commercial facility within the decade. The major products of the SRC-II process are a low-sulfur liquid fuel for the electric utility industry, naphtha which can be converted to high octane gasoline and light hydrocarbon fuels such as methane.

CATALYSTS IN COAL LIQUEFACTION*

Michael G. Thomas
Sandia National Laboratories

INTRODUCTION

Coal is commonly referred to as a hydrogen-deficient solid. This reference follows from the fact that the liquid and gaseous hydrocarbon products most desired from coal contain much higher concentrations of hydrogen (and lower heteroatom concentrations) than the coal itself. A typical eastern high volatile bituminous coal, for example, has an atomic hydrogen-to-carbon ratio of $\approx$ 0.8, whereas fuel oil nominally is 1.8 to 2.0. Assuming that a source of hydrogen is available, we need to efficiently transfer it to the coal or to a coal-derived counterpart. This may be accomplished by direct coal liquefaction, which is a process whereby coal and solvent are heated in the presence of a high partial pressure of hydrogen to product liquid products, see Figure 1.

The direct liquefaction of coal is a catalytically intensive process. The temperatures and long times (800-850°F, 1 to 1.5 hours) required for adequate thermal conversions favor dehydrogenation and repolymerization, not the production of highly desirable hydrogenated liquid products in high yields. Although high hydrogen partial pressures and hydrogenated solvents can counteract the severity of these processing conditions to some degree, the possibility of tailoring product slates through process control, via catalysis, is a necessity and a continuing goal.

* This work supported by the U.S. Department of Energy.

FIGURE 1 GENERIC LIQUEFACTION PROCESS

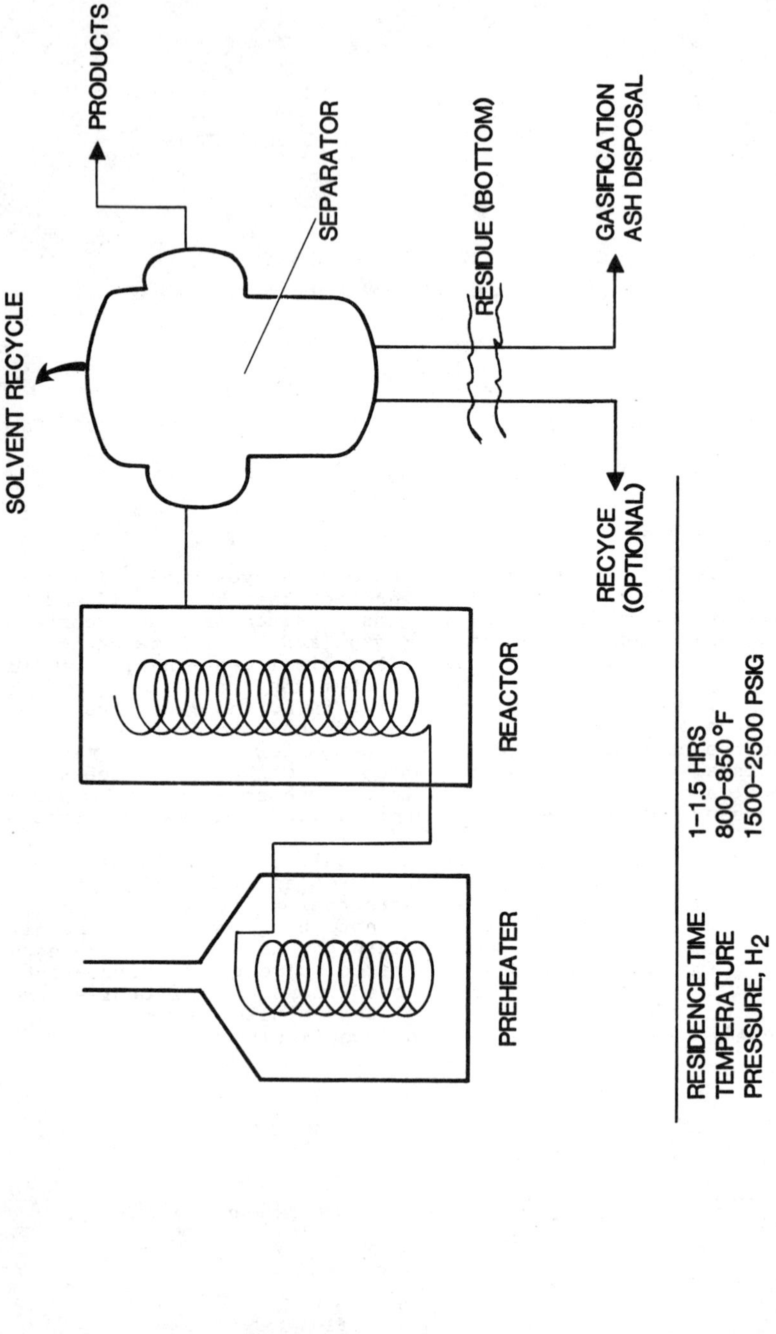

IMPACT OF COAL LIQUEFACTION MECHANISMS

The classical objective of catalysis is to improve the net quality and value of the products, by selectively enhancing the rates of certain reactions. In coal liquefaction, our goals are to produce higher yields of distillate products and to accomplish this with lower consumption of hydrogen. Many current mechanistic descriptions of coal conversion have two common themes:

1. Decomposition of coal is a thermally-induced free radical reaction (1,2); and

2. many parallel paths may be followed (3,4).

The first is of great concern in terms of utilization of catalysts in general because organic free radical reactions are classically non-selective. The second is a matter for debate. From the catalysis viewpoint, the various parallel paths provide an opportunity for selectively catalyzing one or more of the avenues for reaction. Much recent data, however, indicate sequential reaction paths (5,6). With the latter, catalysis can yield overall rate enhancement and allow for process modifications, but not necessarily provide a net compositional improvement at equilibrium.

This rather technical description can be represented as shown in Table 1. Catalysts can not only allow us to improve a product value (selectivity), but they can also reduce capital expense, operating costs, and improve our data base and yield more flexibility to our processes. Many trade-offs exist, and, although a catalyst may be found that increases rates so that temperatures could be lowered, or reduces need for high pressures, these advantages (based upon catalysis) may be alternatively realized by reducing residence times.

Table 1. What Can Catalyst Do in Coal Liquefaction?

Catalytic Effect	Process Advantage			
	Product Value	Operating Expense	Capital Expense	Process Flexibility
Improve Product Slate				
Lower viscosity	Up	Down	?	Up
Lower functionality	Up	Down	?	Up
Improve H/C	Up	Down	?	Up
Lower molecular weight	Up	Down	?	Up
Decrease Hydrogen Consumption	?	Down	?	Up
Decrease Operating Severity				
Lower pressure	?	Down	Down	Up
Lower temperature	?	Down	Down	Up
Reduce residence time	?	Down	Down	Up

A descriptive representation of coal liquefaction is given in Figure 2, and is, conceptually, a series of degradation reactions (7). Coal has properties of a high molecular weight solid. Most of the crystallinity of the reactive part is lost upon heating (8). In the presence of solvent, as is the case with direct

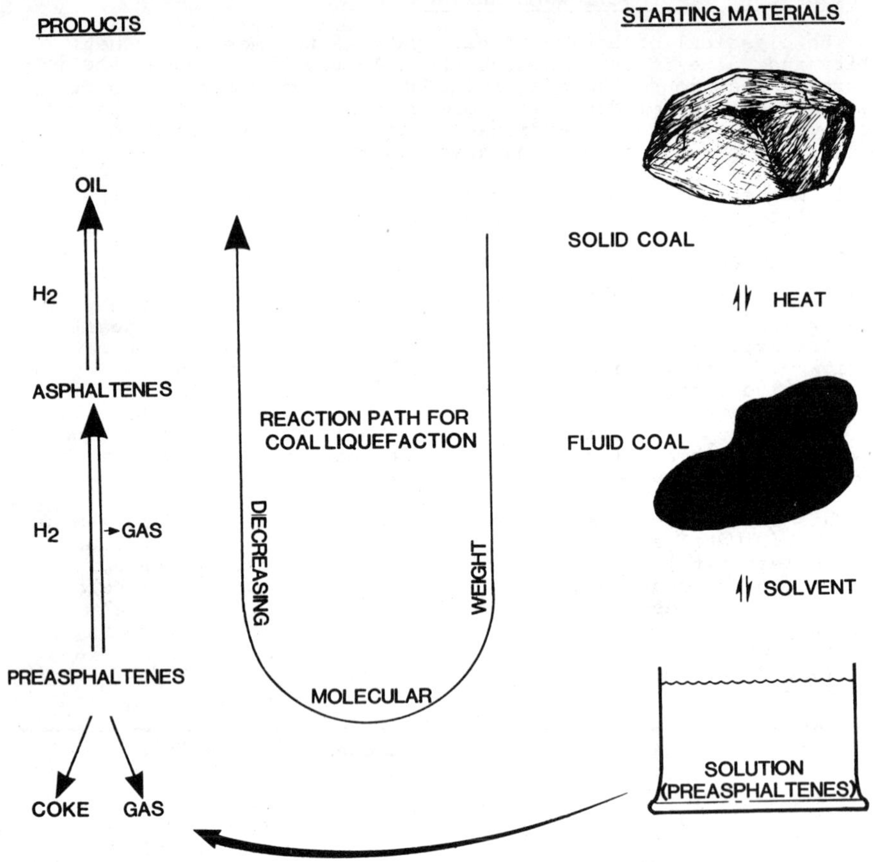

FIGURE 2 REACTION PATH FOR COAL LIQUEFACTION

liquefaction, the fluid-like coal is solvated (in the temperature range of 200-350°C) by what probably is a non-concerted reaction. Thus it is unlikely that coal dissolution could be directly affected by catalysis--at any competitive rate--during the solid-fluid transformation or the solvation process. The solvated species, preasphaltenes, are large, highly functionalized, low H/C solvent-adducts which are the initial product of coal dissolution. At high temperatures, they can polymerize to insoluble material, coke. The preferred reaction is a bond-breaking following by hydrogen transfer from gas or solution phase H_2 or solvent. Though difficult to observe during liquefaction because it is a reactive intermediate, we have shown that isolated preasphaltene reactions clearly are affected by catalysts, as portrayed in Figure 3. Here, conversions are increased by as much as 1.5X, resulting in the production of oil (9). The examples used are FeS_2, the major catalytically active species in coal minerals, MoS_2, the active phase generally used for hydrodesulfurization of fossil fuel feedstocks, and supported (ground) CoMo, a high surface area form of MoS_2 enhanced by the presence of cobalt.

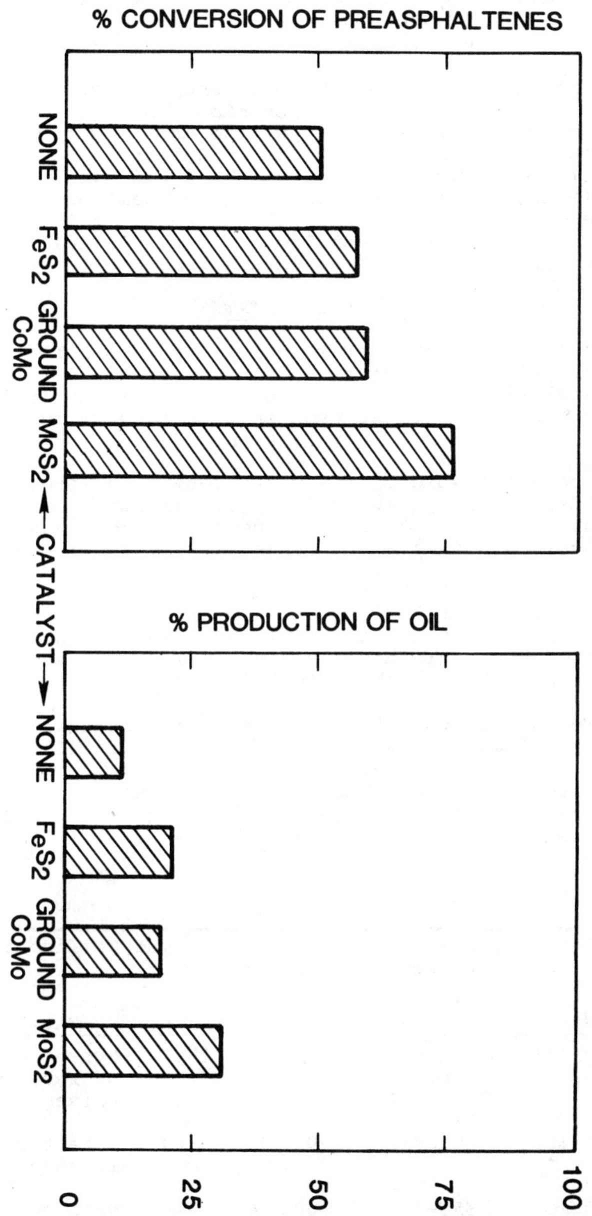

FIGURE 3 CONVERSION OF PREASPHALTENES AND OIL YIELDS ENHANCEMENT BY VARIOUS CATALYSTS

We have also observed enhanced dissolution concurrent with the catalyzed decomposition of preasphaltenes, Figure 4 (10,11). Though this apparently negates our earlier statement, it reinforces data presented earlier by this author that preasphaltene concentration appears to be in steady-state during dissolution. Thus, the catalysis of preasphaltenes drives the dissolution equilibrium.

Coal liquefaction is preasphaltene chemistry (12). The process requires a hydrogen-transfer mechanism. Due to relative concentrations of the hydrogen sources, the majority of the hydrogen does and should come from the solvent. Catalysts can enhance the hydrogen transfer from gas or solution phase hydrogen to solvent and/or preasphaltene, and from solvent to preasphaltene. Because of the advantages of reduced temperature on the overall process, a potential process option with catalysis, it is unlikely that solvent alone can function as well as catalyst and solvent together.

CATALYTIC UPGRADING AND CURRENT STATUS

Once a product free from coal and preasphaltene is obtained, we are more at a point of refining than converting coal. The major use of catalysis in coal liquefaction is this upgrading, and catalysts typical of those used in petroleum refining have been adapted for use in coal liquefaction.

As shown in Table 2, there are four major processes--H-Coal, SRC-I, SRC-II, and EDS--which are all completed or scheduled for completion at the pilot plant or demonstration level (13). Each of these processes utilizes catalysts, except SRC-I, and for very similar reasons, improved final product quality, particularly in terms of high distillate yield and low gas make. In SRC-I, the only one of the four that explicitly uses no catalyst for processing, there is low production of light gases but no real attempt is made to produce distillate (see Table 3). Thus there is no real way to compare SRC-I with the other processes.

If we analyze the other processes for catalyst (i.e., hydrogen efficiency) effectiveness in terms of this distillate yield to gas make selectivity, then H-Coal and EDS are similar. Both H-Coal and EDS have successfully treated both eastern and western coals and the presence of effective catalysis may be the key to the process flexibility.

Table 2. Current Status of Catalyst Utilization in Direct Liquefaction

Status	Process	Size Ton/Day	Date	Catalyst
Pilot	Exxon Donor Solvent	200	1980	Hydrogenation of recycle solvent
Pilot	H-Coal	200-600	1980	Direct coal liquefaction
Demo	SRC-I	6000	1984	None specified
Demo	SRC-II	6000	1984	Mineral matter

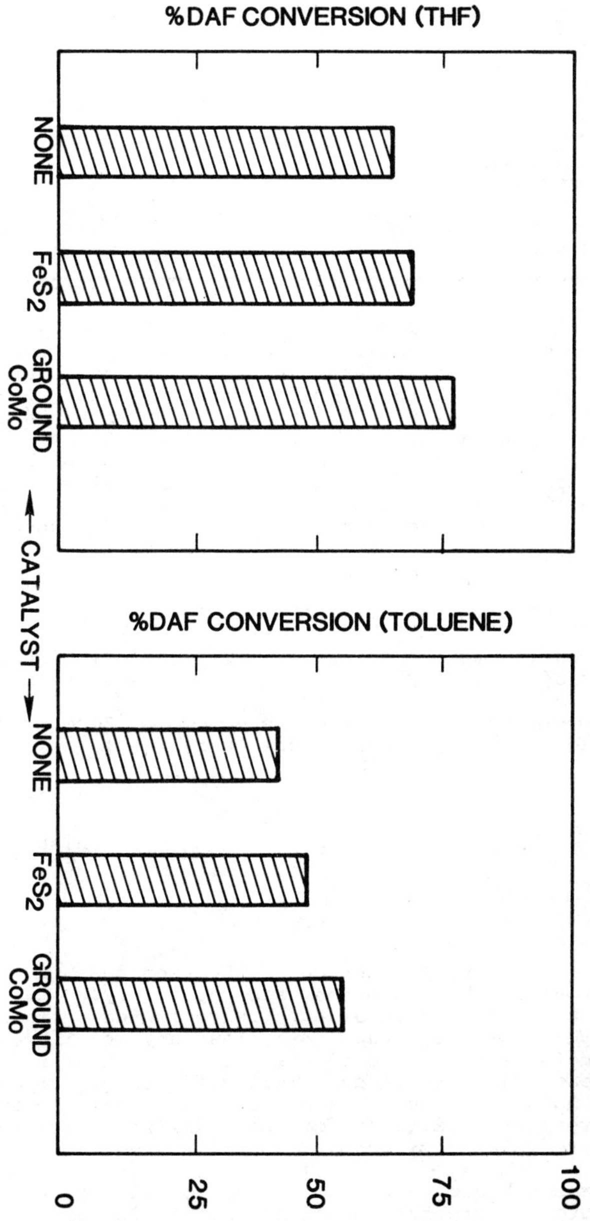

FIGURE 4 DISSOLUTION AND PRODUCT ENHANCEMENT WITH VARIOUS CATALYST

Table 3. Efficiency of Hydrogen Use for the Production of Distillate

	SRC-I	H-Coal	EDS	SRC-II
Distillate Yield	Not Applicable	High	High	Moderate
Gas Make	Low	Moderate	Moderate	High
Hydrogen Efficiency* of Process	-	High	High	Low

* Hydrogen efficiency is a ratio of the distillate yield to gas make (14).

H-Coal has the advantage of design simplicity in that it is a one-step process. However, a complication for H-Coal appears to be catalyst deactivation. Catalyst tagging studies have shown that aging of catalyst in the H-Coal process, exposed to primary and secondary coal products, is rapid. This aging is accompanied by rapid deactivation. For example, in Figure 5, we present a description of an experiment performed in conjunction with the operation of the H-Coal PDU. Radioactively tagged catalyst was added to the bed and ultimately isolated in the withdrawals. These tagged catalysts, of known residence times in the PDU, showed a marked decline, within a very few days, in the ability to enhance the decomposition of preasphaltenes (15).

FUTURE USE OF CATALYSTS IN COAL LIQUEFACTION

Currently, catalysts are utilized in coal liquefaction for upgrading of initially dissolved products and refining.

Coal —(Dissolution)→ Initial Products (preasphaltenes, asphaltenes) —(Upgrading)→ Secondary Products (oils)

Secondary Products (oils) —(Refining)→ Finished Products (gasoline, diesel, fuel oil, etc.)

Emphasis is on the production of the secondary and finished products only. This function, and the catalysts currently used, is the result of many years of experience in the petroleum industry. It is highly unlikely, therefore, that any major breakthrough for coal liquefaction will be made in terms of catalyst development as long as catalysts are used primarily for these purposes. There are, however, major process modifications that could ultimately result if catalysts were tailored for, and utilized prior to, upgrading in a more integrated approach.

Improved hydrogen-transfer functionality is the key to easing temperature and pressure constraints. It is not clear that conventional hydrodesulfurization or hydrodenitrogenation is paramount for the early stages of coal liquefaction, although they are clearly required in upgrading. Other catalysts may be much more effective in transferring hydrogen from the solvent to the coal-derived products.

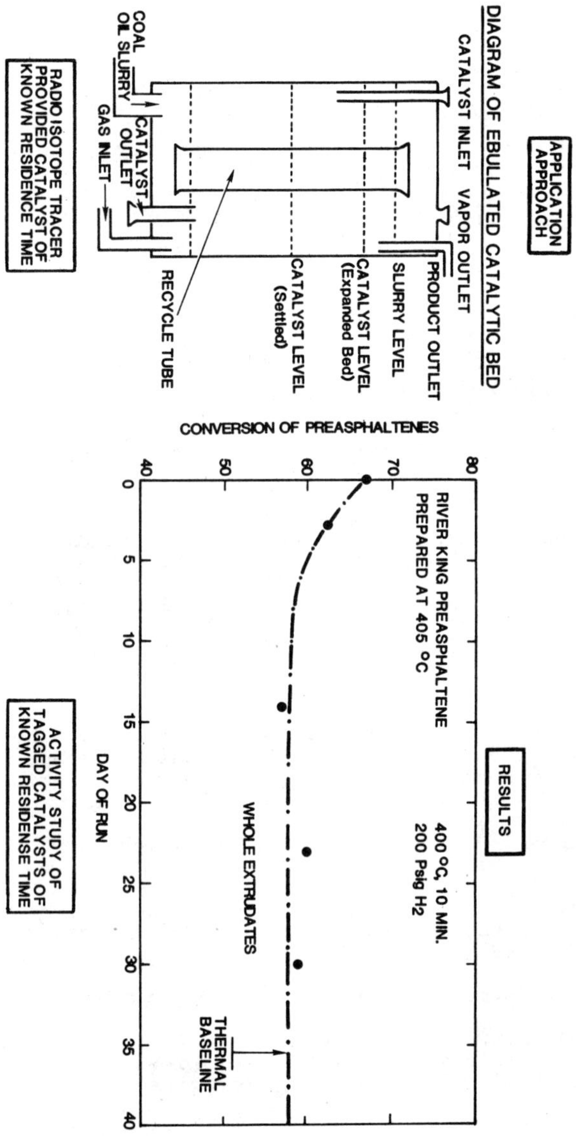

FIGURE 5 DEACTIVATION OF CATALYSTS USED FOR UPGRADING PRIMARY PRODUCTS

In summary, catalysts are used but are not absolutely necessary for direct coal liquefaction. Current technology utilizes catalysts primarily for upgrading and refining, relying on thermal chemistry for most of the initial reaction. Major process modifications due to catalysis are not likely under this constraint. In the absence of this limitation, we could utilize the effects that catalysts have on the initial chemistry of coal liquefaction, and these effects could provide significant process improvements.

REFERENCES

1. Wiser, W. H., Anderson, L. L., Quaged, S. A., and Hill, G. R., J. Appl. Chem. Biotechnol., 21, 82 (1971).

2. Larsen, John W., Ed., Organic Chemistry of Coal, ACS Symposium Series 71, Amer. Chem. Soc. (1978).

3. Shaw, Y. T., Cronauer, D. C., McIlluried, H. G., and Paraskos, J. A., Ind. Eng. Chem. Process Des. Dev., 17 (3) 288 (1978).

4. Neavel, R. C., Fuel, 55, 237 (1976).

5. Thomas, M. G., and Traeger, R. K., Preprint, Amer. Chem. Soc., Div. of Fuel, 24 (3) 224 (1979).

6. Weller, S., Pelipetz, M. G., and Friedman, S., Ind. Eng. Chem., 43, 1572 (1951).

7. Pelofsky, A. H., Coal Conversion Technology, ACS Symposium Series 110, Amer. Chem. Soc. (1979).

8. Walker, P. L., Jr., Spackman, W., Given, P. H., White, E. W., and Jenkins, R. G., "Characterization of Mineral Matter in Coals and Coal Liquefaction Residues", EPRI366-1, Dec. (1975).

9. Unpublished results, Sandia National Laboratories.

10. Bickel, T. C., Granoff, B., Stephens, H. P., and Thomas, M. G., "Coal Liquefaction Process Research Final Report", in process (1981).

11. Thomas, M. G., Bickel, T. C., Granoff, B., "The Effects of Catalysts on Short Contact Time Coal Liquefaction", paper to be presented at the National American Chemical Society Meeting, Div. of Fuel, March (1981).

12. Sternberg, H. W., Raymond, R., Schweighard, F. K., Science, 188, 49 (1975).

13. Moroni, E. C., "DOE Overview - Advanced Direct Coal Liquefaction", Fifth Annual Contractor's Conference, EPRI, Palo Alto, CA, May 7-8 (1980).

14. Private communication, Norman Stewart, EPRI.

15. Thomas, M. G., and Bickel, T. C., "Catalyst Behavior in H-Coal PDU Runs 9 and 10", paper presented at the H-Coal Research Technical Advisory Committee Meeting, January (1981).

SNG PRODUCTION IN THE MEMPHIS DEMONSTRATION PLANT

ROBERT W. GRAY
MEMPHIS LIGHT, GAS AND WATER DIVISION

ABSTRACT

The Memphis Light, Gas and Water Fuel Gas Demonstration Plant concept comprises an installation engineered to generate and distribute industrial fuel gas of approximately 300 Btu/SCF and high Btu pipeline gas. From 3,158 tons of coal per day, the plant will produce 154 million cubic feet per day of industrial fuel gas and approximately 4.5 million cubic feet of pipeline gas. The Institute of Gas Technology's U-GAS^R Process, which utilizes a fluidized bed gasifier, will be used to gasify the coal and make a medium-Btu fuel gas and the Conoco Methanation Process is used to make pipeline quality gas.

The overall project is being conducted in three phases over an 8-year period. The initial 26-month Phase I including process studies, development work, definitive design and project cost estimates has been completed. DOE has approved Phase II Final Design and Construction and Phase III Operations and the contract was signed on May 21, 1980, in Memphis and spans 51 months. Plant operations and testing in Phase III will be conducted over a 20-month period. The plant will then be operated as a commercial unit for 20 years.

The estimated cost of the project is $815 million of which Phase II and III will be cost-shared. This price includes project cost plus escalation and contingencies.

INTRODUCTION

The United States needs synthetic fuels to become energy self-sufficient and reduce imported foreign oil. New and dependable energy sources are especially needed for industrial growth. One of the many new sources that is most plentiful is from coal gasification. Unlike some other substitute fuels, coal gasification has tremendous resources of coal available and is clean burning and environmentally acceptable.

At the present time, coal gas is not being produced in large amounts that contribute much to our energy requirements. However, recently coal gasification has made some tremendous steps. This paper will discuss the U-GAS[R] Process, coal gasification and Conoco Methanation.

Memphis, Tennessee needs energy. The Memphis Light, Gas and Water Division's Coal Gasification Plant is a joint effort on the part of a utility, the U. S. Government and industry to produce medium-Btu gas and pipeline quality gas from coal. The plant will add a new dimension to the future of energy in Memphis.

The Memphis Light, Gas and Water Division has the responsibility of supplying electric, gas, and water services to the customers of Shelby County, Tennessee.

In 1970 Memphis residences and industry used 93.4 billion cubic feet of gas as compared to 70 billion cubic feet in 1980. Today there are ample supplies of natural gas. After deregulation, who knows what supply and prices will be in the market? When natural gas supply becomes unavailable, then supplies are curtailed under Federal guidelines. This proposed Industrial Fuel Gas Demonstration (IFG) Plant is to provide a substitute fuel for these industries and to meet the demands of existing and future commercial and residential customers as shown in Fig. 1.

The plant will make the equivalent of 50 million cubic feet of natural gas per day from 3,158 tons of coal into medium-Btu gas (90%) and SNG pipeline gas (10%). The medium-Btu gas will consist mainly of hydrogen and carbon monoxide, and will have an estimated heating value of 300 Btu/SCF and the pipeline quality will be 965 Btu/SCF.

Construction and operation of the plant will advance the state of the art in the field of industrial fuel gas and SNG production from coal, permit detailed evaluation of the costs and benefits of the expanded technology, and allow identification of environmental and social impacts, regional and national economic benefits.

The Industrial Fuel Gas Plant and distribution system will supply customers in the Shelby County area. The facility would also provide energy for Memphis to offer to new industry---that most other cities in the country do not have to offer. The facility would provide enough fuel to meet the needs of eight more large employers such as Firestone Tire and Rubber Company, which employs 2,800 people.

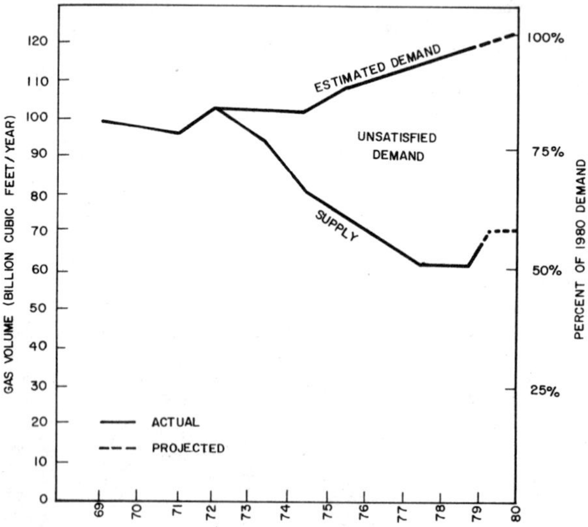

FIGURE 1 MLGW NATURAL GAS SUPPLY AND ESTIMATED DEMAND

PARTICIPANTS AND FUNCTIONS

ORGANIZATION

DOE - The Department of Energy selected MLGW to construct and operate an IFG plant. DOE's objective was to demonstrate this technology and to have a plant operating and producing IFG and SNG in the mid 1980's.

MLGW - The Memphis Light, Gas and Water Division is a municipal utility that distributes electric, gas and water to Shelby County, Tennessee. MLGW is the prime contractor for the IFG Plant to be located in Memphis, Tennessee as shown in the organizational chart in Figure 2.

FWEC - Foster Wheeler Energy Corporation in Livingston, New Jersey is the architect engineer and construction manager. FWEC is well qualified in handling large engineering projects all over the world.

IGT - The Institute of Gas Technology in Chicago, Illinois is the developer of the U-GAS[R] Process. IGT is an independent, not-for-profit educational and research organization. Test results for a design of the plant have been obtained from a pilot plant in Chicago.

DRC - Delta Refining Company in Memphis, Tennessee is an oil refining company that will provide operations and safety experience. DRC will also be a customer of the IFG Plant.

EIA - Energy Impact Associates, Inc. in Pittsburgh, Pennsylvania did the Environmental Report (ER) for the project. EIA is a subcontractor to FWEC. Memphis State University and Ramcon assisted in the collection and analysis of the environmental data.

ORNL - Oak Ridge National Laboratory in Oak Ridge, Tennessee has prepared the final draft FIS and is writing the Final Environmental Impact Statement (FIS) for DOE.

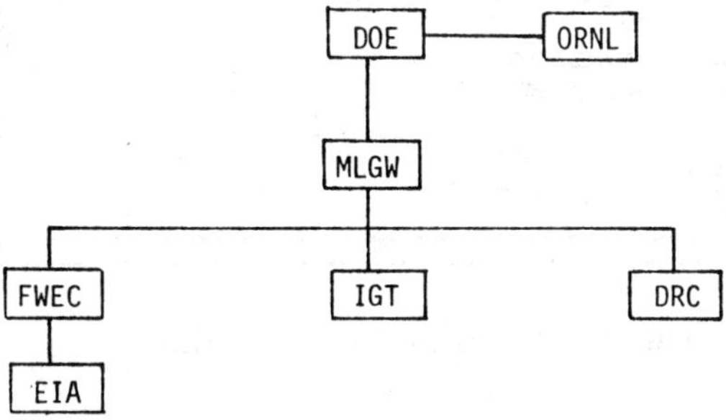

FIG. 2 ORGANIZATION CHART

PROJECT PLAN AND SCHEDULE

The Project is divided into three phases which are shown in Table 1. Phase I Conceptual Design lasted 26 months and the cost was $11.7 million which was 100% funded by DOE. Phase II Final Engineering Design and Construction of the plant is to be completed in 51 months at an estimated cost of $450.0 million. Phase III Operations and Testing will be conducted for 20 months at an estimated cost of $80 million. The total project is scheduled to be completed in 97 months at an estimated cost of $542 million (1979 dollars). Phases II and III will be cost-shared by DOE and MLGW.

TABLE 1 - PHASES OF PROJECT

PHASE	DESCRIPTION	COST MILLIONS $	GOV. FINANCING MILLIONS $	GOV. %	MLGW FIN.	MLGW %	TIME
I	DESIGN	11.7	11.7	100	0	0	26 mo.
II	CONSTRUCTION	450					51 Mo.
				COST-SHARED			
III	OPERATION	80					20 Mo.
		542					97 Mo.

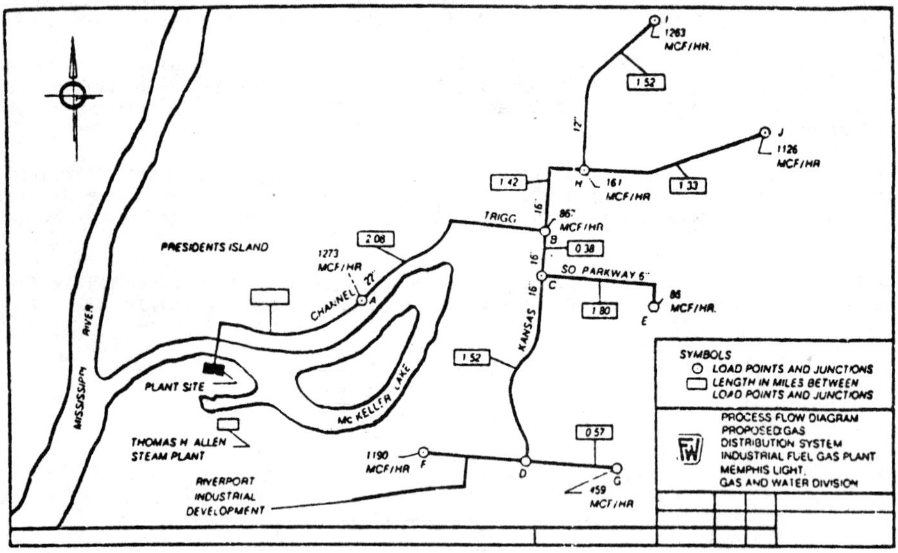

FIG. 3 DISTRIBUTION SYSTEM

PLANT LOCATION

The IFG Plant is to be located in Memphis, Tennessee at a site shown in Figure 3.

The location of the plant will be adjacent to the TVA Allen Steam Generating Plant, south of President's Island and near the Mississippi River port facilities and industrial parks.

Separate IFG gas mains will be installed to economically serve the customers who sign contracts for the fuel gas. The high Btu methanated gas will be put in the Memphis natural gas pipeline system.

PROJECT STATUS

PILOT PLANT

Successful tests have been conducted at IGT's U-GAS[R] Pilot Plant in Chicago. These tests have generated sufficient data to do adequate design work for the commercial plant. Much design data has been obtained for the Demonstration Plant design, but additional tests are being planned to confirm current design. A successful test lasting five days was just completed to confirm design of a prototype cyclone.

DESIGN

Phase I Preliminary Design was completed in December, 1979. Since May, 1980, detail engineering and Project Management activities have been conducted. Approximately 9% of the detail engineering of the Plant has been completed.

SITE

MLGW owns the proposed site and has worked out a satisfactory agreement with TVA on a prior lease agreement. Requests for bids for the site preparation have been advertised and bids for dredging and site preparation are scheduled to be received March, 1981.

ENVIRONMENTAL

MLGW under its contract with the Department of Energy, conducted an Environmental Impact Study and an Environmental Report was issued. Foster Wheeler Energy Corporation subcontracted with Energy Impact Associates, Pittsburgh, PA. The report was submitted to the Department of Energy. MLGW has been advised that the Environmental Impact Statement will be released approximately the last of March, 1981, with a Record of Decision approximately May, 1981. The Environmental Report states, and MLGW believes that the Environmental Impact Statement, will concur that the Coal Gasification Plant, as proposed, will meet all presently applicable federal, state, local, and environmental regulatory requirements.

CONTRACT

MLGW completed Phase I (Preliminary Design) in December, 1979. Phase II (Final Design and Construction) and Phase III (Operations) contract was signed in May, 1980. DOE funding is in a state of confusion. Presently funding on all DOE projects has been deleted from the FY82 budget and the Administration is proposing to recind funding for FY81. Alternate sources of financing are being reviewed at the present time with the best alternative being through the Synthetic Fuels Corporation.

COSTS

Phase I costs were $11.7 million. The Phase II costs since May, 1980, total approximately $11.5 million and approximately $38 million has been made available to Memphis at the present time.

SCHEDULE

The project as of now is on schedule and has made all the required deliverables to DOE. If the present schedule is met, the Memphis IFGDP will be the first major synfuels plant online producing IFG gas in 1985 as shown on the following schedule. At the present time site work is planned to start in June, 1981.

PHASE II AND III SCHEDULE

The following dates are on the proposed schedule.

December 1, 1979	MLGW submitted Phase I to DOE in Chicago.
December 15, 1979	Decision by MLGW Board on Proceeding into Phase II.
January 1, 1980	Submitted Phase I package to DOE Washington with Chicago recommendations.
February 22, 1980	Decision by DOE Washington to proceed.
May 21, 1980	Contract signing and begin Phase II Final Design.
April, 1981	Record of Decision.
June, 1981	Break ground for Phase II construction
October, 1984	Begin testing and shakedown.
November, 1985	Complete operation and tests on Phase III.

PLANT DESIGN

U-GAS^R PROCESS

The U-GAS^R Process has been selected by MLGW and DOE for the demonstration plant program. The U-GAS^R Process has been developed by the Institute of Gas Technology (IGT) to produce a medium-Btu (300 Btu/SCF) fuel gas from coal in environmentally acceptable manner.

In Figure 4 a single-stage, fluidized-bed gasifier accomplishes four important functions. It decakes coal, gasifies coal, and agglomerates and separates ash from char.

Figures 5 and 6 show overall balance and block flow diagram respectively.

METHANATION-CONO-METH Process

The CONO-METH Process is comercially demonstrated. In fact, it is the first and only successful commercial demonstration of the manufacture of SNG from coal. The CONO-METH Methanation Process from Conoco has been licensed for this project.

As shown in Figure 7, the CONO - METH Methanation Process is a catalytic fixed-bed, adiabatic, gas recycle process for the manufacture of SNG from synthesis gas. The CONO-METH Process requires a feed gas having a stoichiometric H_2/CO mol ratio. If the raw gas is deficient in H_2, it must be shifted-converted to provide additional H_2. Before the methanation step, a portion of the CO_2 in the feed gas is usually removed by conventional acid gas processing. Substantially all of the sulfur-bearing components in the synthesis gas must also be removed prior to methanation.

The recycle mode of operation has proven successful in maintaining catalyst activity and suppressing carbon formation in the methanation reactors. Further, the use of multiple bulk methanation reactors in series-parallel reduces recycle requirements by allowing effective inter-reactor heat transfer. Following bulk methanation, the gas is upgraded to SNG quality in a clean-up step.

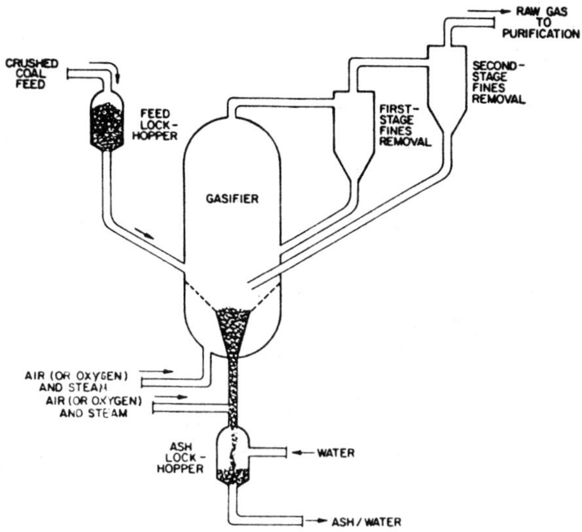

Figure 4. Flow Diagram for U-Gas® Gasifier

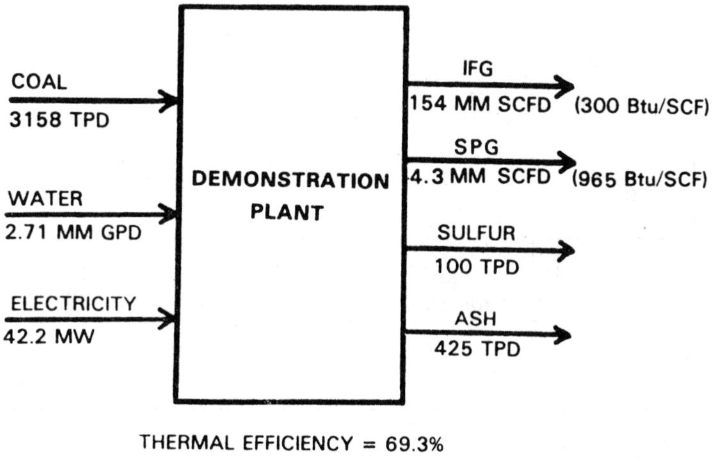

Figure 5. Demonstration Plant Overall Balance

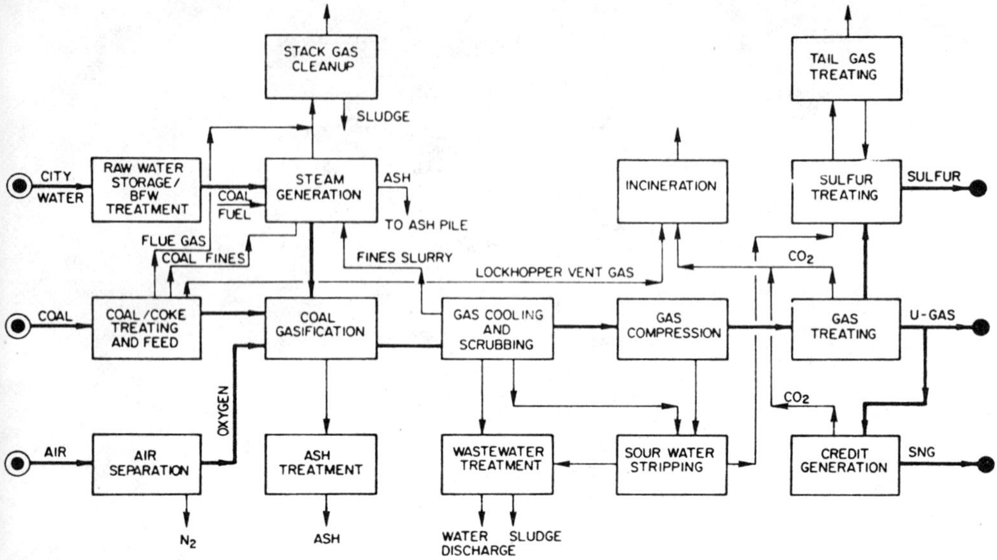

Figure 6 Demonstration Plant Flow Diagram

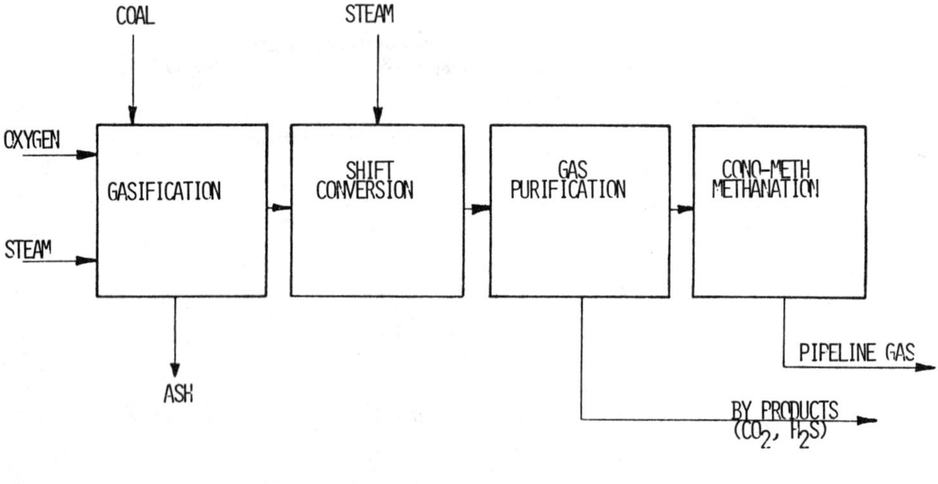

Figure 7 Simplified Block Flow Diagram of a SNG Plant with Cono-Meth Process

MARKETING

STUDIES

A burner study on the combustion ability of medium-Btu gas has been conducted by IGT. No major problems are foreseen based on tests conducted on industrial burners.

A marketing study was conducted by SRI International on medium-Btu gas for MLGW. The study shows that a potential market exists and the utility with a reliable supply and a good load factor can deliver the medium-Btu gas at the most economical price.

Another marketing study "Analysis of Industrial Markets for Low and Medium Btu Coal Gasification" by Booz, Allen and Hamilton, Inc., for the office of Resource Applications, U. S. Department of Energy, was completed in July 1979. The major conclusions that medium-Btu gas plants that have 1) multiple uses, 2) reliable supply, 3) utility financing, 4) less environmental problems, 5) operating experience, 6) coal supply region, 7) natural gas curtailment, and 8) shortage of alternate fuels are the most attractive configurations.

Based on the above items, Memphis can meet all of these conditions and would be an excellent place to build a medium-Btu coal gasification plant. Since a portion of this medium Btu gas can be enriched to SNG and put in the MLGW natural gas system the marketing feasibility is increased tremendously.

RELIABILITY

The main selling point of the Fuel Gas Demonstration Plant is the reliability and the assurance of supply. In order to increase the attractiveness of this fuel gas to potential industrial customers, the reliability of supply must be insured, even during periods of plant shutdown or repair and maintenance. The plant is designed to enhance reliability by the use of modular gasifier trains and several backup systems, but is not cost effective to build complete redundancy in the plant. For the present, reliability is of special concern because only one plant, rather than several independent plants, as would be the case for an already developed system, will be available to produce gas for customers.

CREDIT SYSTEM

The reliability is to be obtained by using the existing natural gas system as backup and establishing a credit system. As the fuel is produced in excess of the system demand, the excess fuel gas can be injected into the natural gas system in small quantities to dilute the natural gas. Another alternative is to consider a propane-air mixture, and a third alternative is to methanate a portion of this excess U-GAS[R].

During fuel gas operation up to 30% of the product gas from the IFGDP can be methanated to natural gas quality and introduced into the existing Memphis natural gas system, thereby accruing "credit" against periods of time when the plant is not operating. During these periods the "credited natural gas" will be withdrawn, diluted with air to the proper medium-Btu heating value and distributed to the industrial customer. These conditions are shown in Figures 8, 9, 10, and 11.

This plant would supply industry the supplemental fuel that is needed and also provide a ready and available fuel for new industry in the area. The industrial fuel gas is expected to be competitive with fuel oil and other alternate forms of energy replacing natural gas.

Also, as part of the marketing effort, surveys on customers' burners, processes, and uses will be conducted. Technical assistance, conversion procedures and estimated costs will be presented to each customer. Distribution pipe sizing metering and operation procedures will be suggested in the proposal to the customer.

FIGURE 8 "CREDIT SYSTEM" - PLANT NORMAL OPERATION

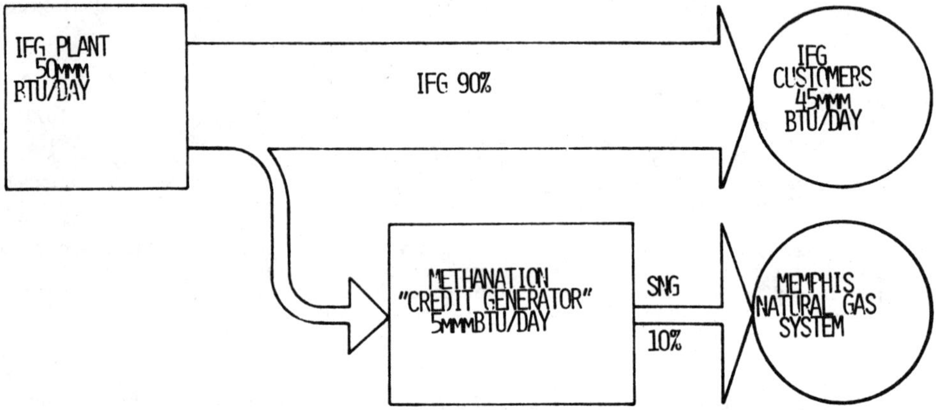

FIGURE 9 "CREDIT SYSTEM" REDUCED DEMAND OPERATION

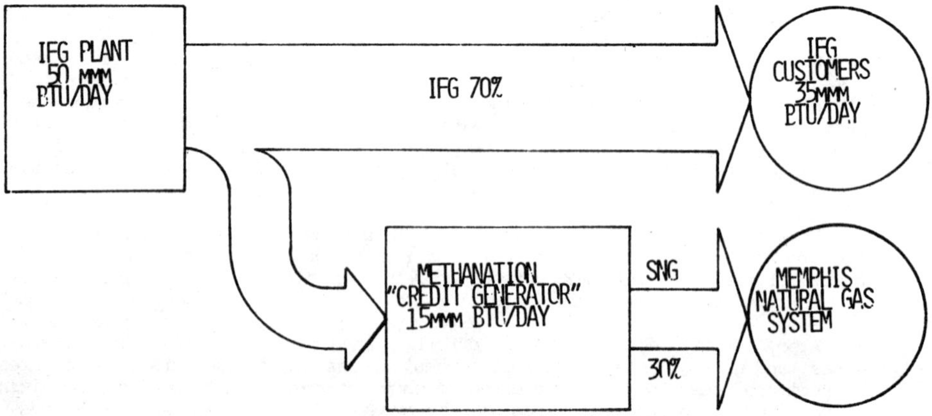

FIGURE 10 "CREDIT SYSTEM" - OPERATION PLANT OUTAGE

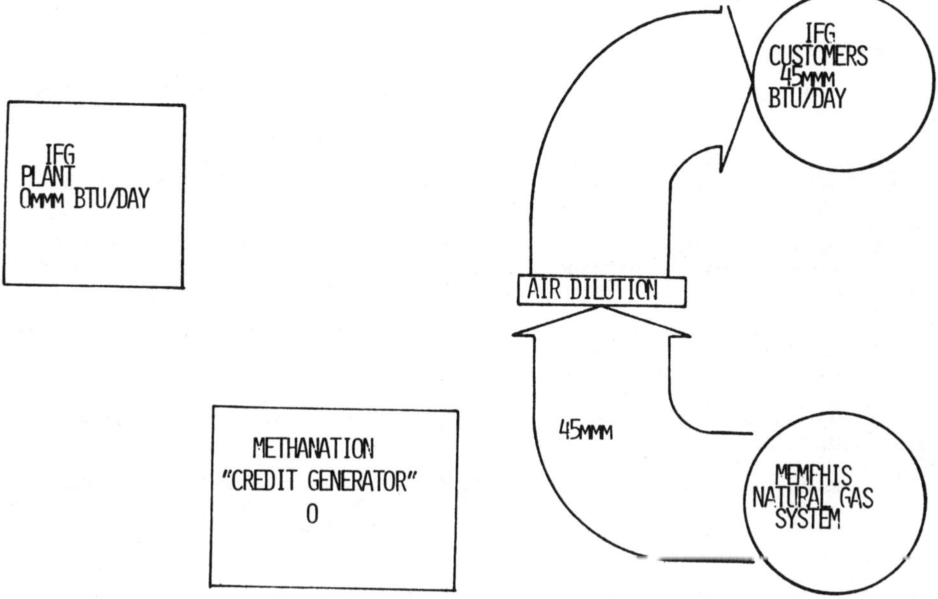

FIGURE 11 NATURAL GAS CREDITS FOR IFG CUSTOMERS

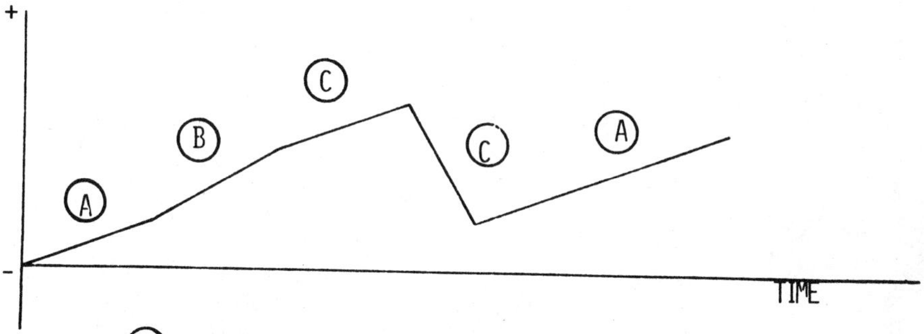

(A) CREDITS ACCRUE DURING PLANT NORMAL OPERATION

(B) CREDITS ACCRUE DURING PLANT REDUCED DEMAND

(C) CREDITS ARE USED DURING PLANT OUTAGE

The market of existing customers consists of 25 to 125 potential customers depending on their usage and distance from the proposed plant. Three large industrial parks are being planned within the proposed industrial fuel gas distribution system. The industrial fuel gas will be a tremendous asset in developing these industrial sites.

REFERENCES

1. Patel, J.G., "Clean Fuel From Coal is Goal of U-GAS[R] Process," Oil and Gas Journal, pp. 51-54, August 1, 1977.

2. Sandrom, W.A. et al., "The Gasification of Coal Chars in a Fluidized Bed Ash Agglomeration Gasifier." Paper presented at the 69th Annual Meeting of AICHE, November 1976, Chicago, Illinois.

3. Mason, David M, J. G. Patel, "Chemistry of Ash Agglomeration in the U-GAS[R] Process." Paper presented at American Chemical Society Annual Meeting, Fuel Chemistry Division, September 9-14, 1979, Washington, D.C.

4. SRI International "Industrial Market Potential for Intermediate BTU GAS from Coal in the United States".

5. Booz, Allen and Hamilton, Inc. "Analysis of Industrial Markets for Low and Medium Btu Coal Gasification" report for Office of Resource Applications, U. S. Department of Energy 30, 1979.

6. Koch, B. J. & Schwartz, M. H. "CONOCO's Methanation Process" Paper presented at Coal Technology '80 3rd. International Coal Utilization Conference and Exhibition, November 18-20, 1980, Houston, Texas.

7. Sudbury, John P., Koch, B. J., & Yoon, H., "Conoco's Capabilities in Methanation of Coal Derived Syn Gases". Paper presented at the First International Gas Research Conference, June 9-12, 1980, Chicago, Illinois.

THE COGAS PROCESS DEMONSTRATION PLANT

Ralph Bloom, Jr., COGAS Development Company

Richard L. McCray, Illinois Coal Gasification Group

From its beginning, one of the objectives of the COGAS Development Company (CDC)* has been to bring the COGAS Process to commercialization in as short a time as possible. The COGAS Process is a versatile process which produces pipeline gas and/or industrial fuel gas plus a substantial quantity of marketable liquid products. The development of this process started with a pilot plant program and a parallel process engineering program. While the pilot plant was being designed and built, small scale gasification tests were conducted. The CDC funded pilot plant work led to the ICGG demonstration plant program which started in mid-1977 and is currently making excellent progress toward construction.

The demonstration plant program is sponsored by the Department of Energy, Fossil Energy section under contract with the Illinois Coal Gasification Group (ICGG)**. CDC is the principal process licensor and a major sub-contractor, as is Dravo Engineers and

---

* A partnership of:

Consolidated Gas Supply Corporation, a subsidiary of Consolidated Natural Gas Company
FMC Corporation
Panhandle Eastern Pipe Line Company
Tennessee Gas Pipeline Company, a division of Tenneco, Inc.

** A partnership of subsidiaries of:

Northern Illinois Gas Company
The Peoples Gas Light and Coke Company
Central Illinois Light Company
Central Illinois Public Service Company
North Shore Gas Company

Constructors who are designing the plant. Process design is complete and detailed engineering design is nearing completion. Commercial venture studies have been initiated.

In this fluidized bed Process, the pyrolysis of the coal permits recovery of gas and oil from the volatile matter and the resultant char is gasified to raw synthesis gas. These coal conversion steps of pyrolysis and gasification are carried out at low pressure to provide high on-stream time reliability.

The Process can handle all types of coals and uses as feed the entire coal, crushed to a fluidized bed size consist. Conceptual commercial designs are completed for midwestern bituminous coal and can be derived for other coals. Illinois bituminous, Western subbituminous and Appalachian bituminous coals are to be run in the demonstration plant to cover the range of coal ranks of grestest interest for application of the Process. The commercial plant designs are based on producing high Btu pipeline gas (SPG), low sulfur No. 2 home heating oil, No. 6 industrial fuel oil and a gasoline reformate grade naphtha. By-products are anhydrous ammonia and sulfuric acid.

Overall material balances for the three coals based on producing approximately 250 billion Btu of synthetic pipeline gas per day are presented in Table 1.

Typical analysis of the SPG product from Pittsburgh seam bituminous coal is shown in Table 2. Analyses of the fuel oils and naphtha are shown in Table 3.

Estimated capital requirements and product prices for Illinois and Pittsburgh seam coals based on utility-type financing in 1980 dollars are presented in Table 4. Included are estimates for 3-train, 250 billion Btu/day of SPG plants for each coal. Estimates are also presented for 1 and 2 train plants for Illinois coal to indicate the effects of staged construction which is the approach coal gasification projects are tending to take.

With the price of incremental supplies of oil rising more rapidly than the plant costs, the COGAS Process economics are benefited when the oil products are priced on the basis of those incremental supplies. The gas prices estimated for 1978, 1979 and 1980, Table 5, decrease because of the increase in oil product values at the average rate of 48 per cent per year. Potential future trends of product prices are shown in Figure 1 for a COGAS Process plant. These data show that with operating costs escalating at an average rate of 8% per year, the average price for the gas and oil products would escalate at an average of only 4.6 per cent over 20 years of operation. This low rate of price escalation occurs in this analysis with capital related costs maintained at a constant value rather than at a reducing rate as would occur with utility financing.

There is a growing interest in the U.S. in the development of energy parks, a cluster of energy-intensive industries around a central energy producing plant. The COGAS Process can fit the needs of such concepts. Using only coal and water as raw materials, a COGAS Process plant can be designed to produce the oils and naphtha, clean medium and high Btu gases and steam. There are advantages to producing both medium and high Btu gases from one plant. The medium Btu gas, a typical analysis of which is shown in Table 2, can be a synthesis gas for a petrochemical plant and/or an industrial fuel gas. The high Btu gas can be used by existing natural gas consumers without requiring investment in conversion of the plant facilities

TABLE 1

COMMERCIAL COGAS PROCESS PLANT MATERIAL BALANCES
250 BILLION BTU/DAY PIPELINE GAS

	PITTSBURGH SEAM	WESTERN SUBBIT.	ILLINOIS NO. 6
Coal Feed Rates, TPD	20,370	31,080	26,000
Product Yields			
Pipeline Gas, MM scfd	265	265	265
No. 2 Oil, bbl/sd	7,100	5,925	12,450
No. 6 Oil, bbl/sd	7,540	940	4,760
Naphtha, bbl/sd	2,010	2,880	3,815
Sulfuric Acid, TPD	1,600	550	2,180
Anhydrous Ammonia, TPD	94	45	105

TABLE 2

COGAS PROCESS GAS COMPOSITIONS
PITTSBURGH SEAM COAL

PIPELINE GAS (SPG)

CH_4, mol %	94.30
CO_2, mol %	0.29
CO, ppmv	10
H_2, mol %	0.61
Inerts, mol %	4.78
H_2O Vapor, lb/MM scf	6.6
Hydrocarbon, dew pt.°C	-40 max.
Total Sulfur	Neg.
Wobbe No.	1263.9
Lifting Index	0.988
Flashback Index	0.981
Yellow Tip Index	0.993
HHV, Dry, Btu/scf	954.3

MEDIUM BTU GAS (MBG)

COMPONENT	MOL %
H_2	56.85
CO	29.62
CO_2	0.20
CH_4	9.51
C_2-C_4	1.91
Sulphides	250 ppm
N_2	1.72
HHV, Btu/scf	415

TABLE 3

COGAS PROCESS
PITTSBURGH SEAM COAL

FUEL OILS

	NO. 2	NO. 6
° API	24.6	7.7
HHV, Btu/lb	18,630	17,830
Nitrogen, wt %	0.3	0.8
Sulfur, wt %	Nil	.09
Carbon/Hydrogen, wt ratio	7.5	8.8

NAPHTHA

Avg. mol wt	90.8
° API	65.9
HHV, Btu/lb	20,440
Nitrogen, ppmv	<1.0
Sulphur, ppmv	<1.0
Carbon/Hydrogen, wt ratio	6.0

TABLE 4

RESULTS OF ECONOMIC ESTIMATES - COMMERCIAL COGAS PLANTS
MID - 1980 $

ILLINOIS NO. 6 SEAM COAL

CASE NO.	I	II	III	PITTSBURGH SEAM COAL
No. Trains	1	2	3	3
10^9 Btu/D	83	167	250	250
Total Plant Investment, $MM	653	1298	1744	1676.6
Total Capital Reqmt., $MM	779	1538	2045	1981.23
Gas Price, $/MM Btu (First Year)	5.65	5.46	4.82	5.16
Coal Price - $1.00/MMBtu				

TABLE 5

COGAS GAS PRICE VARIATION WITH YEAR OF ESTIMATE
ILLINOIS NO. 6 SEAM COAL

YEAR	1978	1979	1980
Avg. Liquids Price, $/bbl	15.98	23.63	35.54
First Year Gas Price, $/MMBtu	5.38	5.19	4.82

Figure I.
EFFECT OF ESCALATION AT 8% / YR. ON PRODUCTS.
COGAS GAS AND OIL, AVERAGE PRICE
BASIS : ILLINOIS NO. 6 SEAM COAL

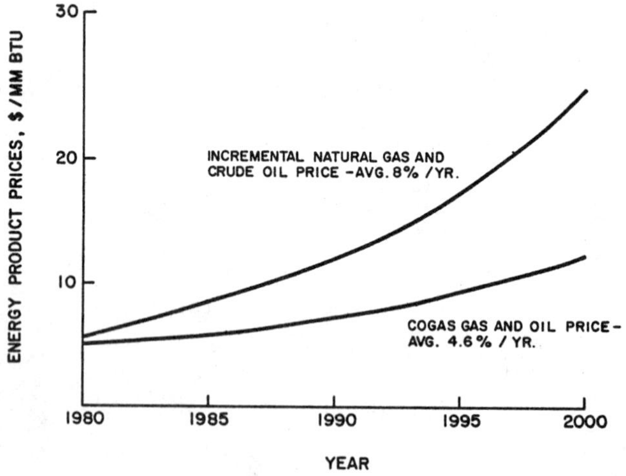

NOTE:
PLANT LEVEL COST ESCALATES
CONSTANT CAPITAL RELATED COSTS

and can be sold to natural gas utilities. Sales of varying volumes of the manufactured high Btu gas to pipelines can provide for this gas to be the swing gas to balance the variations in demand to be expected for the medium Btu gas.

Typical production from such a two-train COGAS Process plant based on Illinois bituminous coal operating an average of 330 days per year is:

No. 2 Fuel Oil, bbls/yr	3,069,000	
No. 6 Fuel Oil, bbls/yr	1,047,420	
Naphtha, bbls/yr	840,000	
405 Btu/scf Gas, 10^9 Btu/yr	18,350	(55.5×10^9 Btu/d)
950 Btu/scf Gas, 10^9 Btu/yr	39,650	(120×10^9 Btu/d)

The ratio of gas production in this example is arbitrarily chosen. The ratio for any venture would be based on market potential.

The process steps of the COGAS Process demonstration plant and commercial plants are shown in Figure 2. This second generation pyrolysis and gasification system is shown schematically in Figure 3. The gasification-combustion system, designed to use air rather than oxygen is shown in Figure 4. Heat production for the gasification reaction is from a slagging combustor which burns char fines collected from the gas streams and rejects ash. Heat is transferred to the gasifier by a circulating stream of bed char. Energy is recovered from the flue gas, after char separation, in the flue gas power recovery system. The other process steps use commercially available technology except for the combined shift-methanation which is a development of the R. M. Parsons Co. Use of this concept of methane synthesis has efficiency and economic advantages.

The latest estimated demonstration plant statistics for the planned three feed coals are shown in Table 6.

In 1975, the demonstration plant construction was estimated to cost $195 MM in 1976 dollars. Due to inflation, escalation and cost growth as the details of the plant design developed, the plant construction is now estimated to cost $450 MM in late 1979 dollars.

If the plant construction starts in mid-1981, completion should be in mid-1985. A 42-month operating program is planned including 12 months on the primary coal, an Illinois No. 5 and No. 6 seam blend, and 6 months each on a Pittsburgh seam coal and a Western subbituminous coal.

The plant site is in Perry County in southwestern Illinois adjacent to the Captain Mine of the Southwestern Illinois Coal Co., the planned supplier of the coal. The site is land reclaimed from earlier surface mining operations. Water for the plant will be drawn from the Mississippi River through a 20-mile pipeline. There will be no wastewater return to the River because the plant is being designed for zero liquid discharge.

This demonstration plant can provide the data base to support decisions to build commercial plants using the advanced COGAS Process. It is not unreasonable to expect commercial production to start in the early 1990's. The participants in the program are looking forward to achieving that goal.

We hear a lot lately about the abundance of natural gas obviating the need for developing coal gasification. It is true that more natural gas reserves are being found now, but deliverability and cost of the gas from newer resources must be considered and the

FIGURE 2

ICGG DEMONSTRATION PLANT SIMPLIFIED BLOCK DIAGRAM

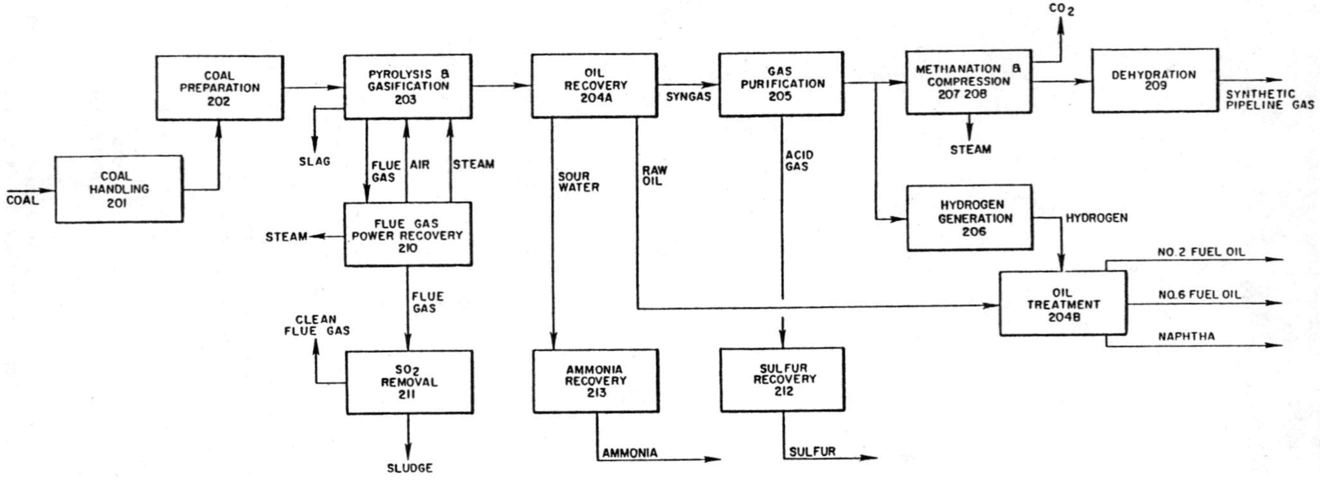

NOT SHOWN:

214 THERMAL OXIDIZER AND FLARE
215 UTILITIES
216 WATER SUPPLY
217 WATER TREATMENT
218 WASTE TREATMENT
219 FIRE PROTECTION
220 FACILITIES (STORAGE)

FIGURE 3

COGAS PYROLYSIS–GASIFICATION

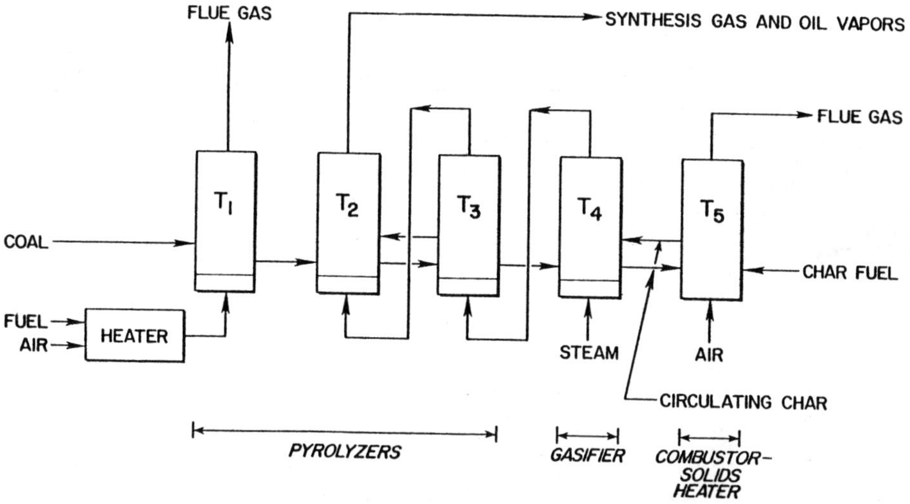

FIGURE 4

COGAS PROCESS GASIFICATION-COMBUSTION

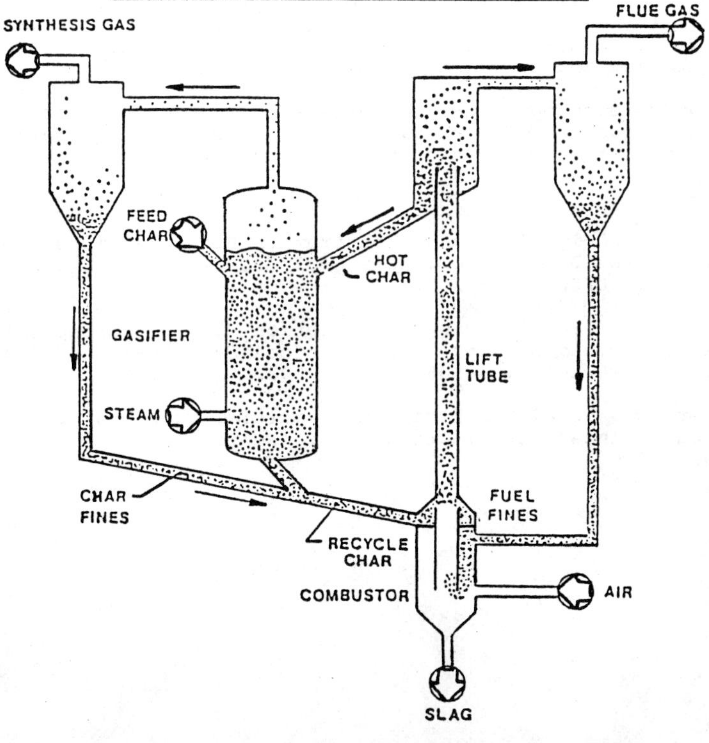

TABLE 6

DEMONSTRATION PLANT PRODUCTION
PRIMARY PRODUCT AND BYPRODUCT PROFILE

	ILLINOIS NOS. 5 & 6 SEAM	PITTSBURGH NO. 8 SEAM	WESTERN SUBBITUMINOUS
Coal Feed, TPD	2210	1927	2757
Water, MM GPD	1.34	1.47	1.26
SPG, MM SCFD	23.1	25.3	25.9
No. 2 Fuel Oil, bbl/d	1117	658	253*
No. 6 Fuel Oil, bbl/d	242*	447*	0*
Naphtha, bbl/d	360	174	210
Anhydrous Ammonia, TPD	3.72	2.33	0.43
Elemental Sulfur, TPD	56.2	47.4	19.4

* Net for export

rate of finding new reserves will probably decrease. The cost of natural gas can be expected to continue to increase over the years whereas the cost of gas from a COGAS Process plant might actually decrease over the same time span. The AGA projections for meeting gas demand the next 20 years include supplying a portion of the gas from coal. Therefore, it is in the best interests of the U.S. to proceed with coal gasification process development and construction of demonstration and commercial plants.

The authors acknowledge the permission of ICGG, CDC and DOE to present this paper and the assistance of members of the staffs of CDC and ICGG in the preparation of data presented herein.

REFERENCES

1. Eby, R. J., "Pipeline Gas Demonstration Plant - Demonstration Plant Process Design", Illinois Coal Gasification Group, Report No. FE/2012/Z/89/09.

2. Eby, R. J., "Pipeline Gas Demonstration Plant - Conceptual Commercial Plant Design", Illinois Coal Gasification Group, Report No. FE/2012/Z/80/06, July 1980.

3. Skamser, R., "Coal Gasification Commercial Concepts Gas Cost Guidelines", C. F. Braun & Co., Report No. FE-1235-1, January 1976.

THE TVA AMMONIA FROM COAL PROJECT - 1981 UPDATE

by

G. W. Alves - Brown & Root Development, Inc.

D. A. Waitzman - Tennessee Valley Authority

The Tennessee Valley Authority has coauthored this paper with Brown and Root Development Inc., Houston, Texas, a contractor on the TVA Ammonia from Coal Project at Muscle Shoals, Alabama. Mention of companies and trade names for processes, equipment, and commercial products does not constitute an endorsement by TVA or the U.S. Government.

THE TVA AMMONIA FROM COAL PROJECT - 1981 UPDATE

Operation of the TVA ammonia from coal facility was scheduled to be started early in March which was too late for operating data to be included in this paper. Design data, rather than operating data, are included herein; operating data, if available, will be presented during the conference.

Project Objective

The TVA Ammonia from Coal Project consists of retrofitting an 8-ton-per-hour (t/h) coal gasification and gas purification facility onto the front end of an existing small, modern natural gas-steam reforming ammonia plant located at the National Fertilizer Development Center, Muscle Shoals, Alabama. The main objective is to provide technical and economic information to the U.S. fertilizer industry for the substitution of coal for natural gas as a feedstock for producing ammonia.

Plant Capacity

The ammonia production capacity of the TVA plant is 225 tons per day (t/d). The plant can be operated at a turndown rate of 60 percent. The coal gasification facility is designed to produce 60 percent of the gas required to operate the ammonia plant at 100 percent rate. Therefore, the ammonia plant can be operated at the design rate with 60 percent of the feed gas supplied from coal and the remaining 40 percent from natural gas; or, the plant can be operated at 60 percent of design rate (135 t/d of ammonia) with all the feed gas supplied from coal. The capability of operating the ammonia plant with 100 percent natural gas feed will be retained. This arrangement makes the greatest use of the existing plant and minimizes the amount and size of new equipment required. Also, the coal gasification facilities can be operated independently from the ammonia plant by burning the carbon monoxide and hydrogen gas in an existing steam boiler.

Plant Design and Construction

The coal gasification unit is based on the Texaco partial oxidation process. The engineering, procurement, and erection of the coal gasification and gas purification facility were performed by Brown and Root Development, Inc. The air separation plant required to provide high-purity oxygen and nitrogen for the process was handled similarly by Air Products and Chemicals, Inc. The engineering, procurement, and construction of the coal handling and preparation area, interconnections to the existing ammonia plant, slag disposal, and services and utilities required for the complex were performed by TVA.

Plant Cost

A percentage breakdown of the costs of the ammonia from coal facilities is shown in table I. The total plant cost is estimated to be about $43.2 million. Since the TVA Ammonia from Coal Project contains developmental and first-time-out design features, this cost should not be scaled-up for costs of commercial plants.

TABLE I

PERCENTAGE BREAKDOWN OF COSTS OF AMMONIA FROM COAL FACILITIES

FACILITIES	PERCENT OF FACILITIES COST
Coal Handling and Preparation	5.3
Wet-Grinding	2.7
Coal Gasification	9.5
Acid Gas Removal, Shift Conversion, Heat Exchange, Sulfur Recovery	37.1
Wastewater Treatment	10.1
Slag Disposal	1.2
Air Separation Plant	13.5
Control Room, Miscellaneous	8.1
Services, Utilities, Ammonia Plant Modifications	5.7
Modifications During Preliminary Operation	5.4
Common Costs, Administration, Etc.	1.4
Total	100.0

Process Description

A simplified block diagram covering the overall facility is shown in figure I. Coal is unloaded from hopper cars in an undertrack unloader system and is conveyed to either a coal storage pile or a hammer mill crusher. The coal may be reclaimed from the coal pile with a front-end loader and fed to the crusher or may flow directly to the crusher and conveyor which transports the crushed coal to a 250-ton live coal storage bin. From the storage bin the coal is conveyed to a feed hopper within the wet-grinding area. From the feed hopper the coal goes through a weigh-feeder to a disc-mill-type wet-grinder where it is ground to the desired particle size distribution. Water is added to the disc mills to obtain the desired coal concentration in the slurry. From the wet-grinders, the slurry goes to one of two mix tanks where the solids content of the slurry is adjusted to the desired level. The slurry is pumped to a 10-hour-capacity feed tank and then metered to the reactor (gasifier) at the process rate of about eight tons of coal per hour. Gaseous oxygen from the air separation plant is fed to the reactor at about eight t/h through a metering system interlocked with the coal slurry feed system.

The gasification process takes place in the reactor at a pressure of about 510 psig and at a temperature in excess of 2200° F. The carbon in the coal is reacted with steam to produce carbon monoxide and hydrogen. Oxygen is injected to burn part of the coal to provide heat for the endothermic reaction. In addition to the gasification reaction and coal combustion to form carbon dioxide (CO_2), sulfur compounds in the coal are gasified in the reducing atmosphere to produce primarily hydrogen sulfide (H_2S) and some carbonyl sulfide (COS). Small quantities of other

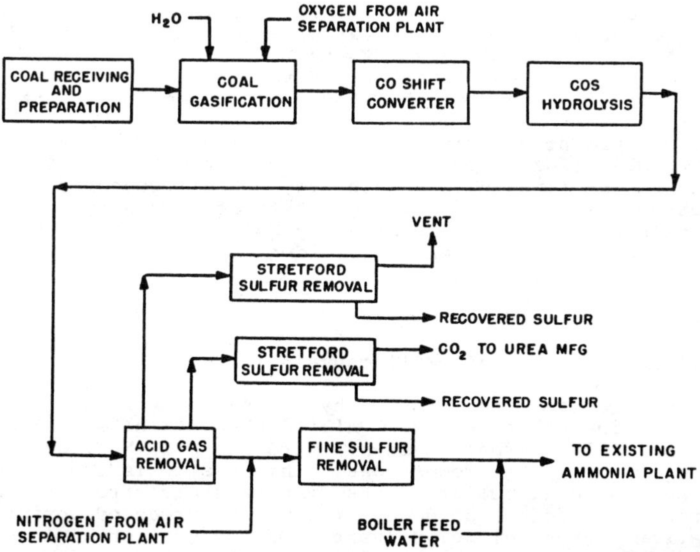

FIGURE 1

FLOW SCHEME FOR TVA'S AMMONIA FROM COAL PROJECT

compounds such as ammonia and methane are also formed. According to Texaco's pilot-plant experience, essentially no long-chain or aromatic hydrocarbons are formed. Slag produced from the ash in the coal is removed from the reactor through a lockhopper system. The slag is glassy in appearance and is very similar to the bottom ash produced in a coal-fired power plant boiler. Initially, trucks transport the solids to a disposal area. A pumping system may be installed later to transport the slag to the disposal area as a slurry. In such a system, the slag would be washed and screened to remove oversize material which is crushed to a size suitable for slurrying and pumping. The gas leaving the reactor is water-quenched and particulate matter (fly ash) is removed in a scrubber. A blowdown is taken from the water recirculating loop and pumped to a wastewater treatment facility, which uses both chemical, physical, and biological treatment processes. In the chemical treatment unit, the wastewater is first treated in a clarifier by addition of ferrous sulfate and hydrated lime to flocculate solids. The liquid fraction from the clarifier is steam-stripped to remove ammonia. The ammonia is recovered and routed to the coal slurry preparation area to neutralize the acidic slurry. The stripped aqueous material containing organic matter, primarily as formates and cyanates, along with water from washdown operations is sent to an equalization-cooling basin for pH control, mixing, and cooling. After aeration the effluent from the equalization-cooling basin flows to the activated sludge unit for biological treatment. The overflow from the activated sludge unit is metered and sampled on its way to discharge. The sludge from biological treatment is combined with flocculated sludge from the clarifier and conditioned with ferric chloride to improve filtration. The conditioned sludge is then pumped to the filter press where the solids are removed for disposal. The filtrate is returned to the wastewater treatment system.

The process gas from the quench scrubber flows to two carbon monoxide (CO) shift converters. The converters are charged with sulfur-tolerant catalyst marketed by Haldor Topsoe. The CO content of the gas entering the converter is about 22 percent (wet basis). After a controlled shift, the CO content is about 2 percent, which matches the CO content of the gas entering the low-temperature shift converter in TVA's existing ammonia plant.

The COS produced during the gasification process is not affected by the sulfur recovery process and, therefore, must either be vented to the atmosphere or sent to another process for further treatment. To decrease the quantity of COS, a hydrolysis unit containing a catalyst marketed by Haldor Topsoe is provided between the CO converters and the acid-gas removal (AGR) system to promote the reaction:

$$COS + H_2O \rightarrow CO_2 + H_2S$$

The effluent from the unit will contain less than 10 ppmv COS.

The process gas from the COS hydrolysis unit flows to the AGR system. The AGR system uses Allied Chemical's Selexol process (a physical absorbent system) to remove CO_2, H_2S, and the remaining COS from the process gas. This system is capable of decreasing the total sulfur in the synthesis gas stream to less than 1 ppm. Two reject acid gas streams are produced during regeneration of the Selexol solvent. One stream containing up to 4 percent H_2S is sent to one train in the Holmes-Stretford sulfur recovery system. The Holmes-Stretford system, furnished by Peabody Process Systems, Inc., uses a proprietary solution containing an oxidized form of vanadium salts. The H_2S is oxidized in the solution to produce elemental sulfur according to the following reaction:

$$2H_2S + O_2 \rightarrow 2S + 2H_2O$$

The reduced metal salt is regenerated by blowing air through the solution. This operation also floats the elemental sulfur to the surface. The sulfur is skimmed off and filtered to produce a wet cake. The tail gas from the Holmes-Stretford system contains less than 160 ppmv H_2S, less than 30 ppmv COS, and less than 500 ppmv CO.

The second stream from the AGR solution regeneration system is relatively pure CO_2. This gas is sent to a separate Holmes-Stretford absorber and then to a sulfur guard containing zinc oxide to decrease the sulfur content to less than 0.5 ppm to meet requirements for urea manufacture. Both Stretford absorbers feed a common regeneration and sulfur skimming system.

Nitrogen from the air separation plant is added to the process gas from the AGR system to produce an H_2:N_2 ratio of 3:1. The gas then flows to the ammonia plant upstream of the low-temperature CO shift converter. The pressure of the gas at the battery limits is about 385 psig.

The composition of the process gas manufactured from coal is very nearly the same as the composition of gas leaving the high-temperature CO shift converter in the ammonia plant.

<u>Plant Instrumentation</u>

The plant is designed as far as practicable for centralized control and operation, with two major exceptions:

- A separate control station is provided for coal unloading and conveying.
- All compressors are started locally.

One central control room serves both the air separation plant and the gasification unit. The control system consists of individual electronic analogue controllers that are coupled to television-type consoles for data display and control. Two consoles are used--one for the gasification unit and another for the air separation plant. An operator at either console, however, can control any loop or take callup alphanumeric or graphical displays for any of the 494 data points in the entire plant. Historical, as well as current, data can be displayed because all points are continuously recorded on 24-hour tape cassettes.

An on-line process computer is an integral part of the system. It is used for data logging, flow totalization, and process calculations such as heat and material balances. The computer is capable of supervisory process control and may be used in this manner later in the project.

On-stream process gas analyzers are used to monitor the composition of gas at various locations throughout the plant area. Table II gives a description of the analyzers, the frequency of analysis, and the analysis performed.

TABLE II

DESCRIPTION OF ON-STREAM PROCESS GAS ANALYZERS USED TO MONITOR COMPOSITION OF GAS

TYPE ANALYZER	# OF ANALYZERS	# OF SAMPLE STREAMS	FREQUENCY OF ANALYSIS	COMPONENT MEASURED
F-C	2	4	Cont.	O_2
Paramag.	1	2	Int.	O_2
T-C	2	2	Cont.	O_2
F-I	1	1	Cont.	H-C
COAL GASIFICATION AND GAS PURIFICATION UNIT				
Chrom.	4	9	Int.	H_2, N_2, H_2O, CH_4, CO, CO_2, H_2S, COS,
I-R	4	4	Cont.	CO_2, CO
Dry F.C.	3	1	Cont.	O_2
Photo-F.	2	2	Cont.	Total S
WORKPLACE SAFETY MONITORS				
Chrom.	1	7	Int.	CO
SS Sensor	5	5	Cont.	H_2S
SS Sensor	8	8	Cont.	Combust.

Operating Personnel

Forty operating and maintenance employees are assigned to the ammonia from coal facilities. A foreman and six operators are used each shift. Two operators are required to operate the air separation plant and four operators are required for the coal gasification, gas purification, and wastewater treatment areas. A supervisor, an auxiliary operator, three coal receiving and handling operators, and seven maintenance craftsmen are assigned on the day shift only.

Because of the experimental nature of the project, an engineer and engineering technician are also assigned to each shift for test and evaluation duties.

Data

Design data for each subsystem are given below.

Coal Preparation and Wet-Grinding--Coal is unloaded at a rate of 100 t/h on day shift only (Monday through Friday). The coal analysis is given in table III. The 1/2-inch by 0-inch coal is fed to the disc mill and pulverized to the desired particle size, and water is added to give a concentration above 60 percent solids.

TABLE III

COAL ANALYSIS

PROXIMATE ANALYSIS (AS RECEIVED)	PERCENT BY WEIGHT
Moisture	7.8
Volatile Matter	37.3
Ash	13.3
Fixed Carbon	49.4
ULTIMATE ANALYSIS (DRY AND ASH FREE)	
Carbon	78.1
Hydrogen	5.5
Nitrogen	1.3
Oxygen	10.9
Chloride	0.1
Total Sulfur	4.3
HEATING VALUE	**BTU/LB**
As Received	11,480
Dry	12,375
Ash And Moisture-Free	14,279

ASH FUSION ANALYSIS

ASH FUSION (IN REDUCING ATMOSPHERE)	TEMPERATURE, °F
Initial Deformation	1980
Softening Point	2100
Fluid Point	2220

Air Separation Plant--Flows, compositions, temperatures, and pressures in and out of the air separation plant are given in table IV. The air compressor is a four-stage centrifugal compressor driven by a 4000-hp motor. The oxygen and nitrogen compressors are three-stage reciprocating compressors driven by 1750 and 1250 hp motors, respectively.

TABLE IV

AIR SEPARATION PLANT

	AIR INLET	OXYGEN OUTLET GAS	NITROGEN GAS	NITROGEN LIQUID
Flow, SCFM*	18,750	3,470	3,390	
", Tons Per Day	1,033	211	180	3
Temperature, °F	96	300	310	-307
Pressure, psig	92	665	475	75
Components, Volume %				
Oxygen	20.95	99.5	0.00	0.00
Nitrogen	78.12	0.0	99.99	99.99

*SCFM at 60°F and 29.92 in Hg

Gasifier--The coal slurry flow to the preheater and the gasifier is about eight tons of coal/hr. Oxygen is admitted to the gasifier at a rate of 7.3 t/h, at a pressure at 660 psig and a temperature of 240°F. Gas, blowdown water, and slag flows and compositions from the gasifier are given in table V.

TABLE V

GASIFIER STREAMS

	EXIT GAS FROM SCRUBBING WATER SEPARATOR	SLAG OUT	BLOWDOWN TO WASTEWATER TREATMENT
Temperature, °F	433	160	140
Pressure, psig	505	0	50
Flow, lbs/hr	83,350	4,230	24,045
Flow, SCFM*	10,957	-	-
Wt. % Solids	-	50	0.5
Component	VO.%	WT.%	WT.%
Hydrogen	10.9	-	-
Nitrogen	0.1	-	-
Carbon Monoxide	14.6	-	-
Carbon Dioxide	5.7	-	-
Methane	0.1	-	-
Argon	0.1	-	-
Hydrogen Chloride	-	-	-
Hydrogen Sulfide	0.4	-	372ppm
Water	68.1	50	99.3
Carbonyl Sulfide	255ppm	-	-
Ash, Grains/SCF	-	-	-
Slag	-	45.5	-
Carbon	-	4.5	-
Ammonia	-	-	0.6
Total	100.0	100.0	100

*SCFM at 60°F and 29.92 in Hg

The slag passes from the gasifier through the lockhopper and screens to the sump. All slag is taken to a disposal area by dump truck. The gasifier outlet gas enters a venturi scrubber for removal of entrained particulate matter and then flows to a scrubbing water separator. The solids in the separator bottoms are sent to the clarifier, and then to the slag sump. The clarified liquid is returned to the venturi. Clean condensate is sprayed in the top of the scrubbing water separator below a demister pad. Particulate loading in the cleaned gas is not more than 0.002 grains/SCF.

Shift Catalyst--After appropriate heat exchange, the gas is admitted to the first two stages of the CO shift converter. The shift converter uses Haldor Topsoe's SSK catalyst. Gas flows, compositions, temperatures, and pressures in and out of the two stages of shift conversion are shown in table VI.

TABLE VI

CO SHIFT CONVERSION

	SHIFT 1 INLET	SHIFT 1 OUTLET	SHIFT 2 INLET	SHIFT 2 OUTLET
Temperature, °F	450	797	605	644
Pressure, psig	483	478	470	465
Flow, SCFM*	17,416	17,395		17,395
Components, Volume %				
Hydrogen	17.0	35.6		37.6
Carbon Monoxide	23.0	4.1		2.1
Carbon Dioxide	9.0	27.7		29.8
Hydrogen Sulfide	0.6	0.7		0.7
Carbonyl Sulfide	400ppm	58ppm		36ppm
Methane	0.1	0.2		0.2
Water Vapor	50.0	31.4		29.3
Nitrogen	0.2	0.2		0.2

* SCFM at 60°F and 29.92 in Hg

COS Hydrolysis--From the shift converter, the gas flows through further heat exchange to a COS hydrolysis unit using Haldor Topsoe's activated alumina catalyst. Flows, compositions, temperatures, and pressures in and out of the hydrolysis unit are given in table VII.

TABLE VII

COS HYDROLYSIS

	INLET	OUTLET
Temperature, °F	401	401
Pressure, psig	462	458
Flow, SCFM*	17,395	17,395
Components, Volume %		
Hydrogen	37.6	37.6
Carbon Monoxide	2.1	2.1
Carbon Dioxide	29.8	29.8
Hydrogen Sulfide	0.7	0.7
Carbonyl Sulfide	36ppm	7ppm
Methane	0.2	0.2
Water Vapor	29.3	29.3

* SCFM at 60°F and 29.92 in Hg

Acid Gas Removal--The gas flowing from the COS hydrolysis unit is admitted into the Selexol acid gas removal system licensed by the Allied Chemical Company. Flows, compositions, temperatures, and pressures in and out of this system are given in table VIII.

TABLE VIII

ACID GAS REMOVAL

	Product Gas		CO_2 TO HOLMES-STRETFORD	
	INLET	OUTLET	VENT	UREA MFG.
Temperature, °F	289[(a)]	100	104	80
Pressure, psig	448	412	7	23
Flow, SCFM*	13,970	8,660	3,097	896
Components, Volume %				
Hydrogen	37.6	75.5	0.1	1.0
Carbon Monoxide	2.1	4.1	163ppm	0.2
Carbon Dioxide	29.8	19.5	85.0	97.9
Hydrogen Sulfide	0.7 }		3.6	0.8
		1ppm		
Carbonyl Sulfide	7ppm }		29ppm	14ppm
Methane	0.2	0.3	61ppm	0.1
Water Vapor	29.3	-	0.2	-

* SCFM at 60°F and 29.92 in Hg

(a) Inlet to absorber is 55°

Sulfur Recovery System--The two reject streams from the acid gas removal system are sent to a Holmes-Stretford sulfur recovery system provided by Peabody Process Systems, Inc. Flows and compositions in and out of the Stretford system are given in table IX.

TABLE IX

SULFUR RECOVERY

	CO_2 TO UREA		CO_2 TO VENT	
	INLET	OUTLET	INLET	OUTLET
Temperature, °F	80	90	104	90
Pressure, psig	23	15	7	3
Flow, SCFM*	896	785	3,118	2,867
Components, Volume %				
Hydrogen	1.0	1.1	0.1	0.1
Carbon Monoxide	0.2	0.2	163ppm	176ppm
Carbon Dioxide	97.9	96.4	85.0	83.3
Hydrogen Sulfide	0.8 }		3.6 }	190ppm
Carbonyl Sulfide	14ppm }	0.5ppm	29ppm }	MAX.
Nitrogen	71ppm	80ppm	11.1	12.7
Methane	0.1	0.1	61ppm	66ppm
Water Vapor	-	2.2	0.2	3.9

Sulfur Recovered

Quantity, lb/hr	587
Moisture Content of Cake, %	48

* SCFM at 60°F and 29.92 in Hg

Wastewater Treatment--A small slipstream of water from the gasification system is sent to a wastewater treatment plant consisting of chemical treatment, ammonia stripping, and biological treatment. The solids removed in the wastewater treatment system are disposed of by recycling to the gasifier. Flows and compositions in and out of the wastewater treatment facility are given in table X.

TABLE X

WASTEWATER TREATMENT

	INLET	OUTLET	SOLIDS TO LANDFILL
Flow, gal./min.	25*	29	-
Solids, lb./hr.	-	-	83
Total Suspended Solids, mg/l	330	30	-
Total Dissolved Solids, mg/l	1842	150	-
Ammonia -mg/l	1600	9	-
Sulfide -mg/l	100	<1	-
Cyanide -mg/l	45	<8	-
Formate -mg/l	350	150	-
Chloride -mg/l	1320	1140	-
Total Organic Compounds -mg/l	760	-	-
Total Inorganic Compounds -mg/l	104	-	-
Bod, ppm	-	30	-
Cod, ppm	-	60	-

*Normal Blowdown from Gasification area. There will be additional flows from miscellaneous sumps.

Product Gas--After the addition of nitrogen and water, the product gas is sent to the ammonia plant. The flow, composition, temperature, and pressure of this stream are given in table XI.

TABLE XI

PRODUCT SYNTHESIS GAS TO AMMONIA PLANT

	WET BASIS	DRY BASIS
Temperature, ^{o}F	595	595
Pressure, psig	385	385
Flow, SCFM*	15,630	10,847
Components, Volume %		
Hydrogen	41.8	60.24
Nitrogen	14.2	20.46
Carbon Monoxide	2.3	3.31
Carbon Dioxide	10.8	15.56
Hydrogen Sulfide / Carbonyl Sulfide	0.1ppm	0.1ppm
Methane	0.2	0.29
Argon	0.1	0.14
Water Vapor	30.6	-
Total	100.00	100.00

* SCFM at 60°F and 29.92 in Hg

Services and Utilities--Steam, air, water, and other service and utilities for the air separation plant and the coal gasification and gas purification unit are given in table XII.

TABLE XII

SERVICES AND UTILITIES

	AIR SEPARATION PLANT	COAL GASIFICATION AND GAS PURIFICATION UNIT
Electricity, kWh/hr.	4,750	2,000
Steam Exported, lb/hr. at 140psig	0	0
Steam Imported, lb/hr. at 140psig	0	0
Cooling Water @ 86°F, gal/min	1,800	2,600
Boiler Feed Water, lb/hr. Treated for 140psig Steam	0	41,000[a]
Instrument Air, scfh	3,000	6,000
Operating Labor Per Shift	2	6[b]

(a) used as O_2 free water in process gas, pump seals, etc.

(b) includes 2 operators on day shift only in coal handling and preparation area.

Conclusion

The data presented herein are the design conditions for the facility and do not necessarily represent operating conditions. After initial runs and imporvements are made operating information will be released. Caution should, therefore, be used in extrapolating these data for drawing conclusions or for making economic analysis and interpretations at this time.

THE HUMBOLDT COAL GASIFICATION PROCESS

P. Paschen, R. Pfeiffer, H.-D. Waldhecker
KHD Humboldt Wedag AG, Cologne,
Federal Republic of Germany

I. INTRODUCTION

KHD Humboldt Wedag AG, a subsidiary of Kloeckner-Humboldt-Deutz AG, Cologne, Federal Republic of Germany, is developing a new coal gasification process.

By now quite a number of coal gasification processes have reached an advanced stage of development or are commercially available. If a company nevertheless decides to develop a new process, this must offer significant advantages with respect to process technology and economy compared with other systems.

The HUMBOLDT coal gasification process uses molten iron for coal gasifying.

This idea is not entirely new.

It was taken up, patented and tested by the Applied Technology Corporation (ATC), Pittsburgh, USA, at the beginning of the seventies.

In laboratory-scale trials, coal and oxygen were introduced through lances into the molten iron bath. These experiments had to be stopped because of technical difficulties in handling the lances (cooling, material problems, lack of mechanical stability).

KHD Humboldt Wedag AG has again taken up the basic idea of coal gasification in a molten iron bath and intensively dealt with feasible alternatives.

Instead of the lance system, bottom tuyeres similar to the Q-BOP technology, proven on large scale by the steel-making industry, were successfully tested.

In 1978 KHD Humboldt Wedag AG acquired the patent rights from Applied Technology Corporation.

II. PROCESS DESCRIPTION

The HUMBOLDT coal gasification process is based on the dissolution of carbon in molten iron.

Schematically described, the following process steps take place when injecting coal (grain size up to 3 mm) in a molten iron bath:

o The volatiles in the coal escape immediately and are cracked due to the high temperatures;

o The carbon is dissolved in the iron;

o The coal ash rises to the surface together with the additives (e. g. lime) and forms a liquid slag;

o The sulphur from the coal reacts with the iron to form iron sulfide (FeS). This FeS reacts with the added lime (CaO) to calcium sulfide (CaS) which enters the slag;

o The carbon dissolved in the iron reacts with the gasification agent (oxygen, air, steam) and with the iron oxide respectively.

Gasification is based on carburizing and simultaneous decarburizing of the iron. Therefor the carbon content of the molten iron remains constant in a continuous operation. Consequently, an equilibrium will be established provided the quantities of coal and oxygen introduced have been adjusted.

By way of stoichiometric operation with respect to the formation of carbon monoxide, a gas is produced which consists essentially of carbon monoxide (CO) and hydrogen (H_2) and is almost free of carbon dioxide (CO_2).

Decomposition of the coal moisture, of the gasifying agent steam and of the volatiles, results in the formation of hydrogen and additional carbon monoxide.

The high process temperature avoids the formation of carbon dioxide corresponding to the Boudouard reaction.

The gas produced is practically free of sulphur. The coal sulphur is discharged almost completely with the slag.

All types of coal can be used, regardless of such characteristics as caking qualities, content of volatiles, ash, sulphur and the like.

In the gasification process, the coal is completely converted. There are no by-products such as tar and other heavy hydrocarbons.

III. TECHNICAL CONCEPT

The gasifier is a refractory lined reactor containing a molten iron bath. Crushed coal with particle size up to 3 mm is injected into the iron bath through cooled tuyeres of special design together with transporting gas (e. g. recycled raw gas), lime and oxygen. At the same time additional gasifying agents such as air, steam, CO_2 etc. can be fed through the tuyeres as and when required.

The iron bath, which does not deplete itself, has a temperature of about 1350 - 1400 °C.

A simplified flow diagram of the HUMBOLDT coal gasification process is shown in Figure 1.

IV. OPERATING RESULTS

The theoretically expected advantages of the HUMBOLDT gasification process could be confirmed during an extensive test program.

The tests have been carried out in an external research institute where a converter (capacity: 6 t of iron) with all required equipments such as feeding and injecting systems, gas scrubber, on-line process control devices, etc. were available. All results have been recorded by computer prints.

The main parameters varied during the tests are listed below:

- o coal throughput
- o ratio of $C:O_2$ injected into the molten iron
- o slag basicity ($CaO/(SiO_2 + Al_2O_3)$)
- o bath geometry
- o iron quantity
- o number and arrangement of tuyeres
- o refractory materials.

In Figure 2 the test facilities are shown.

The crushed coal which had been dried to a residual moisture of about 1.5 % was mixed with lime of the same grain size and pneumatically injected with nitrogen into the iron bath through the tuyeres.

Figure 3 shows the multi-media tuyeres.

The carbon/lime-mixture was injected through the central tuyere pipe whereas oxygen was blown through the adjacent annular gap.

For protecting the high-grade steel tuyeres against excess heat, a protective gas was blown through the outer annular gap.

For instance, propane, methane, CO_2, purified recycled gas or steam can be used as protective gas. Propane was used mainly for the tests made.

Table 1 shows the coal feeding rate as well as the specific consumption figures of oxygen, nitrogen and propane:

TABLE 1: Consumption Figures

Coal	250 - 400 kg/h . t_{Fe}
Oxygen	0.58 m^3/kg of coal
Propane	0.1 m^3/m^3 O_2
Transporting gas (N_2)	0.1 m^3/kg of solid material

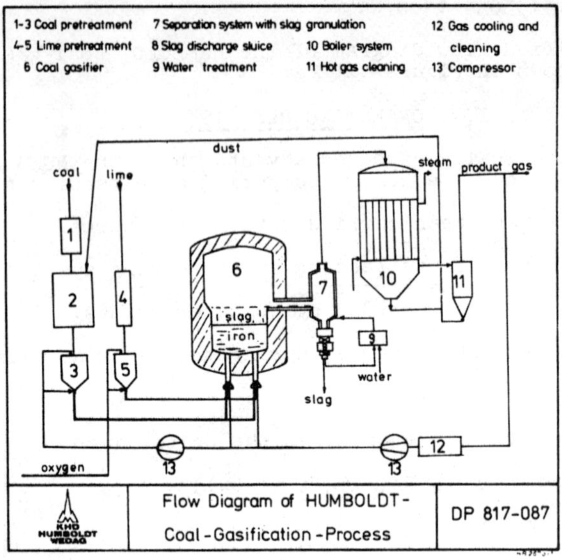

Figure 1: Flow Diagram of HUMBOLDT Coal Gasification Process

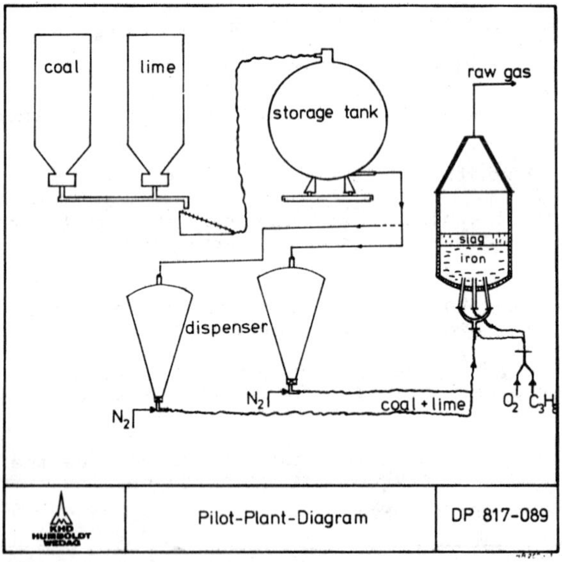

Figure 2: Pilot Plant Diagram (Test Facilities)

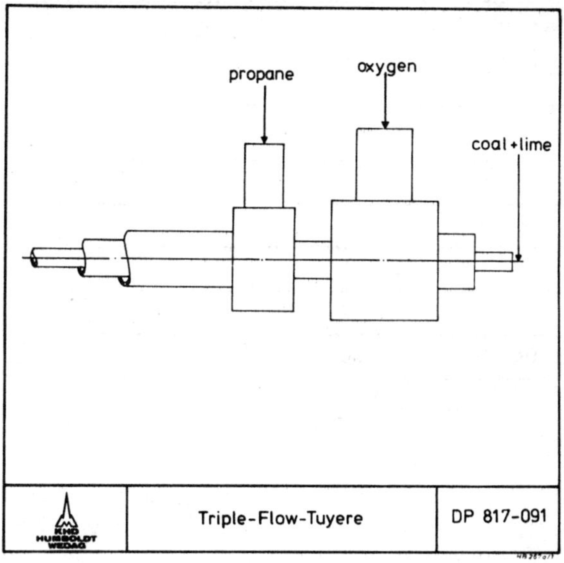

Figure 3: Triple Flow Tuyere

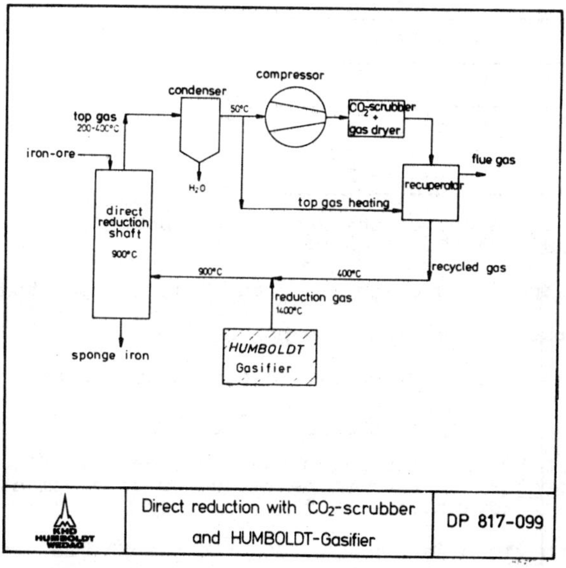

Figure 4: HUMBOLDT Coal Gasification Integration with Direct Iron Ore Reduction

The coal used in the trials had the composition given in Table 2.

TABLE 2: Coal composition

Fixed Carbon	67.5 %
Sulphur	1.0 %
Volatile Matter	22.0 %
Ash	8.0 %
Moisture	1.5 %
Net Heating Value	7,500 kcal/kg (31.4 MJ/kg)

The carbon content of the molten iron bath was kept constant at approximately 3.5 %. The average temperature of the bath was about 1,400 °C.

The coal gas produced had the composition shown in Table 3. The nitrogen amount coming from the transporting gas has been eliminated.

TABLE 3: Raw Gas Composition

CO	65 - 70 %
H_2	25 - 30 %
CO_2	below 0.3 %
S	below 20 ppm
CH_4	below 0.1 %

The heat content of the gas was of the order of 3000 kcal/m^3 corresponding to ~ 12.500 KJ/m^3.

The raw gas production was about 2,100 m^3 per ton of coal.

The carbon conversion rate has been 98 %.

The tests were carried out at atmospheric pressure.

V. PROCESS APPLICATIONS

The HUMBOLDT process is an advanced, new gasification system which, compared with other processes under development, offers the following main advantages:

o a high yield (above 200,000 m^3 of gas/h and gasifier unit)

o an extremely low amount of oxidizing components in the raw gas (CO_2 + H_2O below 1 %)

o an extremely low sulphur content in the raw gas (below 20 ppm).

These advantages make the process well suitable in the fields of

- o metallurgy
- o chemistry
- o heat- and energy market.

It is the only process available at present which, due to the low content of sulphur and oxidizing components (CO_2, H_2O) can be integrated for example in direct reduction plants without the need for interposing an intermediate gas scrubbing system.

Due to the extremely low sulphur content, the gas is especially suitable for a combined cycle gas/steam turbine process to generate electric power. Costly desulphurization facilities are not necessary.

Due to its highly favourable composition (e. g. CO- plus H_2-content > 98 %) the gas is most suitable as synthesis gas for chemical processes.

VI. ECONOMICS

The high carbon conversion rate and the large gasifier throughput means a significant reduction of the specific capital investment. The excellent quality of the coal gas makes complicated, purifying installations unnecessary. This fact will considerably lower the capital and operating costs of the complete plant again. Further, the process is capable to use relatively coarse coal which has a positive effect on the expenses of the coal preparation section.

In a pre-feasibility calculation the costs of the coal gas when using as reduction gas for direct reduction of iron ores, have been considered.

Figure 4 shows the connection of the coal gasification system to a direct reduction plant.

The process flow diagram has been built up in a manner that the gas leaving the shaft is recycled to the process after having been treated in a CO_2-scrubbing step. The recycled gas is mixed with the fresh reducing gas produced by coal gasification. This mixing operation permits adjusting the gas inlet temperature into the shaft. This arrangement reduces in a significant manner the amount of fresh gas from the coal gasification process.

The layout data taken as a basis for the feasibility calculations have been compiled in Table 4.

TABLE 4: Layout data

Plant Capacity	: 450,000 tons of iron/yr
Reducing Gas Demand	: 1,500 m^3/t of Fe - 750 m^3/t of Fe by coal gasification - 750 m^3/t of Fe by gas recycling
Operating Time	: 8,000 h/yr
Gasification Plant Investment	: DM 60 mill.
Capital Charge	: 14 %/yr

The production costs of coal gas have been calculated for different coal prices.

Figure 5 shows the results.

The costs of reducing gas made by the "Steam Reforming" process have been determined by taking different prices per thermal unit of natural gas as a basis.

These calculations result in identical expenditure for gas produced of coal available at a price of about 50 $/t and for reducing gas produced by steam reforming from natural gas having a price per thermal unit of approximately 3,80 $/MMBTU.

VII. OUTLOOK TO FUTURE DEVELOPMENTS

On the basis of extensive research and development work a plant is being built for testing this process on a larger scale wherein 10 t/h of coal can be gasified at a pressure of 10 bar and a temperature of 1350 to 1400 °C. The amount of gas produced is about 20.000 m^3/h. The operation of the plant, going on stream at the end of 1982, is to yield the data for the commercial implementation of the process. This development program is expected to be completed by 1984.

It shall be followed by the construction of a plant including a 200 t-reactor which corresponds to a gas production of at least 200,000 m^3/h. This plan is realistic because individual parts of the system have been under large-scale testing for decades.

The project is sponsored by the Ministry for Economy of the Government of North-Rhine Westphalia at Duesseldorf, Federal Republic of Germany.

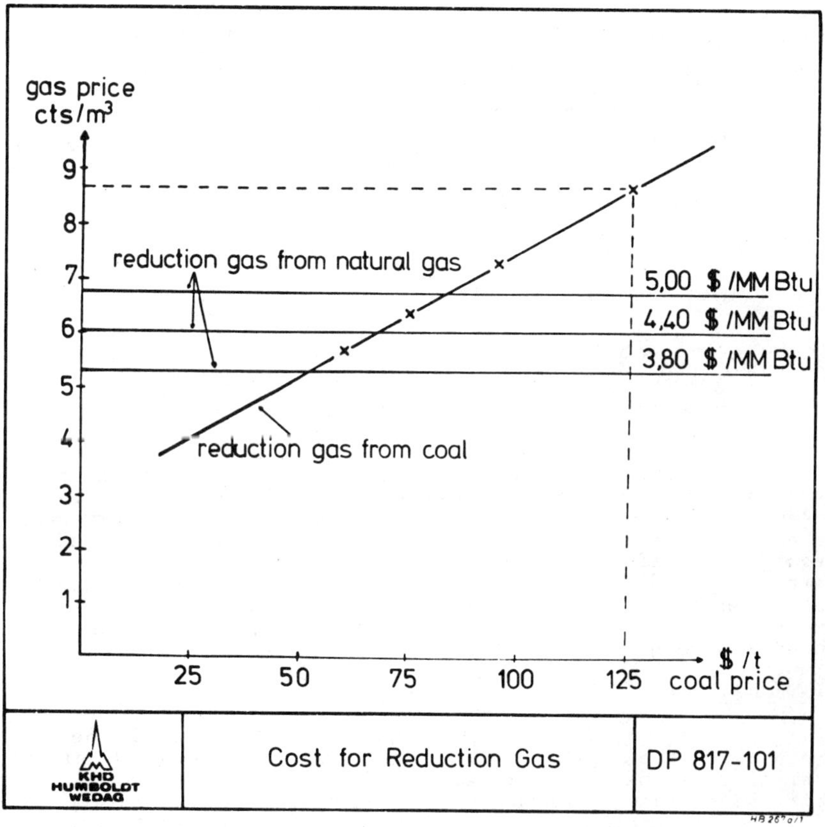

Figure 5: Production Costs for Reduction Gas

COMPONENTS FOR COAL GASIFICATION: THE TRANSITION FROM PILOT TO DEMO PLANT

J.T. McCabe, Mechanical Technology Incorporated
F.D. Freeburn, U.S. DOE, Morgantown Energy Technology Center

ABSTRACT

Data acquired from U.S. coal gasification pilot plants are analyzed to determine potential equipment problems in future demonstration and commercial plants. The major causes of pilot plants' forced shut-downs are discussed. Specific components and mechanical subsystems that may cause maintenance problems in future full-scale plants are analyzed and methods for maximizing performance are outlined. Recommendations are given for establishing and monitoring equipment performance goals. Component and system technology needs are presented in terms of areas where new concepts are needed and areas where existing equipment should be improved. Finally, a recommendation is made regarding the types of test facilities that will be needed to support new, large production plants.

INTRODUCTION

The objectives of this paper are to analyze the performance record of the equipment in coal gasification pilot plants and to relate these data to the component and technology needs for the transition to future plants of the synthetic fuels industry. The gasification of coal or its derivatives will be a key technology in this industry. Gasification is the heart of all substitute natural gas processes; it is the source of synthesis gas for indirect liquefaction processes and it is the proposed source of hydrogen for many direct liquefaction processes.

In 1930, the cost of imported oil caused Germany to accelerate its experimental synthetic fuels work. Lack of adequate equipment with sufficient durability for the process conditions was cited as a major obstacle to carrying out coal conversion processes in scaled-up plants (1)*. More than 50 years later, with vastly superior materials

*Numbers in parentheses indicate references at end of paper.

and components technology, a number of key equipment problems still remain to be solved.

The collection of comprehensive process data for scale-up from the first-of-a-kind U.S. pilot plants has proved difficult. During the study of chemical process parameters operation records indicated that, while equipment damage was incurred when design limits were exceeded deliberately, a number of components failed unexpectedly. Because of the limited duration of the pilot plant program, not all components were selected for long-term performance. Also, some equipment, such as coal mills, are more rugged in larger sizes. Although the mechanical equipment was generally well maintained, the unanticipated failures considerably slowed the collection of data needed to evaluate the processes. Investigation of the causes for poor performance points to a need for establishing equipment performance goals and upgrading specific component technologies.

Three companion reports which assess the mechanical equipment used in the CO_2 Acceptor, SYNTHANE and HYGAS pilot plants (2,3,4) constitute the primary data sources for this summary. Except for the CO_2 Acceptor plant, all data were collected by MTI in conjunction with an on-site DOE representative and a representative assigned by the operating contractor. Data from SYNTHANE and HYGAS were obtained by studying all available maintenance records, interviewing key personnel and observing repair and maintenance procedures, where possible. Data from the CO_2 Acceptor plant were obtained by interviewing personnel stationed at the plant during mothballing and by reviewing some of the maintenance records. Additional related data which has been factored into this paper may be found in References 5 and 6.

SUMMARY AND ANALYSIS

Equipment problems common to two or more of the pilot plants specifically evaluated, and which also have a high probability of occurrence in large-scale plants, are discussed herein. Although most of the data was obtained from plants utilizing fluid-bed gasification processes, the majority of conclusions regarding the equipment are believed to be valid for all types of gasification processes. Not discussed herein is the large group of components that consumed considerable maintenance manpower in the pilot plant but were special items for these facilities and will be replaced by different types of machinery in large plants. For example, some of the low-flow, reciprocating compressors that experienced repeated valve and packing rod problems will be replaced with high-flow centrifugal compressors which have completely different failure modes. Control of the failure modes of the replacement equipment, however, could consume at least the same amount of time as was expended on the pilot plant equipment because many root causes remain and the equipment will be larger.

Plant Forced Shutdown

Plugging of pipes and components caused the majority of forced shutdowns in these pilot plants. All coal conversion facilities contain piping systems carrying multiphase flow streams. Design guidance from analysis of multiphase flow is usually superficial because measured data for confirmation is scarce and the problems are complex.

Although shutdowns due to plugging in pipelines and components will probably diminish as operating procedures are refined in the new demonstration and commercial plants, plugging due to the accumulation of solids in solid/liquid handling equipment may require design modifications. The types of modifications suggested from pilot plant operating experience include improved piping layout on the suction side of pumps, improved solid/liquid separation equipment and the selection of equipment capable of handling the unavoidable fluctuations in solids loading that occur during transient operation. In addition, the frequency of plugging which occurred during off-design point operation suggests that a better analytical understanding of the design and operation of equipment for these periods should be developed.

Equipment Malfunctions

The principal cause of equipment failure can be summarized as a mismatch between component capabilities and actual operating requirements. Generally, the reason for this mismatch was a lack of operating data; the specifications for many key components were based on operating requirements that were calculated or assumed, because the plants were first-of-a-kind. As a consequence of incurred differences, about two-thirds of all major original components were redesigned, rebuilt or replaced.

Data to correct this situation for future plants are limited because the accumulated time under steady-state operating conditions in pilot plants was not adequate for exhaustive analysis. In addition, the data that was collected was not fully evaluated. Thus, some problems with new equipment is inevitable during the start-up period of new plants.

The specific equipment and systems where performance in the pilot plants reflects a need to improve reliability, availability and maintainability for future plants are discussed below.

Valves Handling Multiphase Streams

Based on pilot plant maintenance records and anticipated future needs, developing methods to extend the working life of various valves is a high-priority problem for three reasons. First, almost all of the coal gasification processes now being considered for commercialization operate at high pressure and require valves for process control; second, the useful life of many key valves in all of the pilot plants was short; and, third, some valve improvements may require new concepts followed by time-consuming development and demonstration testing.

The most severe application was pressure letdown of slurries. The solids in these slurries were hard and abrasive and the mixture generally contained gases which expanded rapidly during letdown and thereby tended to intensify the erosive effect of the solids. Although improvements were made by the operating contractor that increased life by a factor of 10, the demonstrated maximum life of these severe service letdown valves remained on the order of 100 hours.

In addition to letdown valves, there were five other valve applications which require improvements to extend their useful life:

- Pressure differential control involving a hot gas containing abrasive particles
- Throttling of an abrasive slurry for flow control
- Bubble-tight block valves, check valves and pressure relief valves operating in slurry and particle-laden streams
- Lock hopper venting and cross pressurizing valves
- Lock hopper feeding and discharge valves.

As indicated at the 2nd Symposium on Valves for Coal Conversion and Utilization (7), DOE recognizes these valve problems and is conducting an ongoing lock hopper valve improvement program at the Morgantown Energy Technology Center. In addition, valve exhibits at a recent conference (8) showed the industry to be actively engaged in improving their products for coal conversion applications. One conclusion from the Valve Symposium was that actual environmental testing was considered mandatory to ensure an accurate assessment of valve performance. However, none of the valve manufacturers could provide a completely appropriate test facility and some applications were considered so complex that valves are likely to remain on the problem list for a long time.

Slurry Pumps and Slurry Systems

Slurry pumps and slurry systems required frequent parts replacement and cleanout in the pilot plants. This type of equipment is highly developed for ore and coal slurry transport; it is readily available and will probably be used in a variety of coal conversion plants after specific design changes have been made to adopt this equipment for the temperature and chemistry of coal service.

Primarily, two types of pumps were used in the pilot plants: reciprocating plunger pumps and single-stage centrifugal pumps. The reciprocating pumps were used for high-pressure coal feeding. The centrifugal pumps were used to circulate coal slurry in a high-pressure feed system and as transfer pumps for ash/water and char/water slurries.

The principal failure modes in the reciprocating pumps were valve erosion, packing failure, plunger scoring, and plugging due to solids accumulation in the pumps and piping. The principal failure modes in the centrifugal pumps were casing and impeller erosion, seal failure, and bearing failure.

At the beginning of the pilot plant program, the time-to-failure for both types of pumps was 10 to 100 hours. After a series of modifications, the time-to-failure was increased to between 100 and 1000 hours. In addition, plugging was reduced as operating know-how was accumulated. However, there remains considerable room for improvement and, to consistently go beyond 2000 operating hours, new pumping concepts and/or much more erosion-resistant materials will probably be required.

Multistage centrifugal slurry feed pumps, pipe feeders, screw pumps, and long, slow-stroke pumps are currently being considered as potential options for high-pressure coal feeding. However, operation in a realistic test environment is required to confirm to plant designers that these new concepts are actually improvements over existing designs.

Coal Grinding and Sizing Systems

Initially, 30 to 50% of the ground coal was below the minimum optimum size for the gasifiers of the three plants studied in detail. Consequently, an excessive quantity of solids entered the components downstream of the gasifier and caused mechanical problems.

Unwanted coal fines were sometimes eliminated by a double screening arrangement which extracted the middle cut for process use. However, the lower screen was delicate and often failed. The experience at all three pilot plants indicates future large plants will require a practical, durable, and cost-effective means for separating and utilizing coal fines. At least three options can be pursued:

- Improve coal grinding systems so that the quantity and quality of rejected fines equal the boiler requirements

- Include two types of gasifier in a process to handle an appropriately broad range of coal size
- Develop pelletizing equipment to convert fines into a size range that is optimum for the process.

Compressor Inlet Filtration Systems

Although most of the compressors were reciprocating machines because the pilot plant flow requirements were low, scaled-up plants will use rotating equipment to handle the larger flows. However, the basic problems caused by contaminated inlet gas will remain.

Many compressors used in the pilot plants ingested harmful amounts of liquid droplets and dust. This caused frequent failure of internal parts due to breakage and severe wear, particularly in recycle compressors.

Dry dust in reasonable quantities can usually be removed efficiently from the process stream by a pair of suction inlet filters; but, when the gas contains an excessive amount of liquid droplets, the filters tend to plug rapidly, even if the gas first passes through a vessel designed to knock out the droplets.

Accumulated particulate matter was found inside many compressors and this provided the base for attracting and holding corrosive liquids which eventually caused severe metallurgical problems. A basic cause for the limited service life of some compressor parts was the inability of the knockout vessels to remove the micron-sized droplets often found in many coal conversion process gas streams. New knockout concepts will be required to effectively separate these liquid droplets and particulate matter from compressor inlet streams.

Gasifier Effluent Cyclone

Cyclones were used in, or downstream of, the gasifiers in all pilot plants. These cyclones were employed to capture and return unreacted carbon to the gasifier and to prevent excessive solids from plugging downstream equipment. In performing these duties, the cyclones experienced severe erosion/corrosion damage which was often accompanied by plugging, especially in the case where the cyclone was located inside the gasifier. A measure of the wear rate in a gasifier effluent cyclone was provided at the HYGAS plant*. After 1350 tons of solids had been removed by the cyclone, the average wear was approximately 8 mils per 100 tons of removed fines. The implication is a potential high wear rate for a commercial plant cyclone - on the order of one inch per month.

Cyclones have two additional drawbacks. Separation efficiency decreases with the decreased flow that occurs during turndown, and returning solids to the gasifier require a long dip leg to overcome the pressure differential. On the other hand, cyclones can enhance the carbon conversion efficiency of the process by capturing and returning unreacted carbon to the gasifier. Cyclones also can limit the amount of unwanted carbon compounds from entering gas scrubbing systems. Since cyclones can provide effective performance at low cost, they merit the required additional development to extend cyclone life and enhance performance in coal gasification service. Improved service could be realized through a cyclone development program concentrating in five areas: 1) the development of cost-effective, wear-resistant liners and surface coatings, 2) the development of durable hot char recycle blowers to eliminate the dip leg and increase

*Cyclone Operating Conditions: 600°F, 100 psia gas containing 400 to 930 grains/acf at an inlet velocity of 45 to 60 fps.

carbon conversion efficiency, 3) the development of inlet configurations that minimize punch-through, and 4) the development of an improved design analysis for cyclone gas/solids flow.

Dynamic Seals

Proper design, installation and operation of dynamic seals could have eliminated a large number of equipment failures in the pilot plants. The mechanical seals, packing and piston rings were often the most vulnerable parts in troublesome pumps, compressors, rotary feeders and other dynamic equipment.

The root cause of seal failure was usually related to debilitating thermal effects. These effects cause surface distortion and vaporization of the fluid between the sealing surfaces that eventually lead to material failure through excessive rubbing contact.

Seals usually have full or partial clearance in the 0.5 to 5.0-micron range. If abrasive solids in this size range enter the clearance, they can quickly destroy the seal. Dynamic seals posed a problem in the pilot plants because all coal conversion processes contain an abundance of particles in this size range in a variety of different process streams. Therefore, if special precautions are not taken, the rate of seal failures in a commercial plant could become a major concern. To avoid this situation, a seal specialist should be involved in the project from the time the equipment is specified through plant start-up. His objective would be to achieve improved design of seal lubricant filtration and cooling systems, to increase standardization of replacement parts, to reduce number of failures due to improper installation and to oversee maintenance.

Other Critical Equipment

The items listed in this section were not incorporated in any of the pilot plants because this equipment was not essential for demonstrating the viability of the processes. However, this equipment may be needed to minimize the cost-of-product in a commercial plant. Equipment in this catagory is identified in References 4 and 5 and includes hot gas expanders, various types of heat exchangers, and hydraulic turbines for power recovery.

Hot gas expanders can be used to recover power from the gasifier effluent stream or regenerator off-gas as in the CO_2 acceptor pilot plant. The principal problems with this equipment are expected to come from erosion, corrosion and deposits. Heat exchangers offer an option for recovering power from a gasifier effulent stream. The problems with this equipment are expected to be associated with fouling, plugging, erosion and corrosion. Hydraulic expanders have been suggested as a means of recovering power from high-pressure slurry streams. Thus, this component can also be considered as an option for achieving pressure letdown. However, the erosion problems that plague the valves now used for pressure letdown will also present an equally difficult problem for an hydraulic turbine.

The three types of equipment listed in this section must, to a substantial degree, be demonstrated in large plants. Since some aspects of this equipment are sensitive to scale-up, provisions should be made to acquire necessary design varification data in the first full-size train of a commercial plant.

Maximizing Future Performance

The performance of demonstration and commercial plants can be improved compared to pilot plant experiences by planning actions indirectly related to component development. Four such planning actions are described below.

Assessing Commercial Readiness

Without performance constraints, components for every process requirement in the pilot plants were obtained without substantial difficulty from suppliers of off-the-shelf commercial products. For instance, the char slurry letdown valves for the SYNTHANE plant were purchased to meet the pressure, flow, solids loading, and temperature requirements; however, the initially installed valves failed within 24 operating hours. The label "commercially available," therefore, was misleading for this application.

Future plants may include components with little or no meaningful operating experience at the required service conditions. However, experience in the pilot plants suggests that many potential problems can be recognized at an early stage and early recognition of these problems basically determines the state of commercial readiness of the components. This can be achieved through the type of equipment design and performance audits that are often performed on industrial equipment to assess a prototype design or to aid in troubleshooting. These audits should contrast a realistic range of process requirements with the capabilities of a proposed component and should serve to highlight process areas where trade-offs between process requirements and machinery capabilities should be considered.

Preventing Repetitive Failures

Repetitive failures occurred relatively often in the pilot plants, because some of the equipment was not matched with the operating conditions and because pilot plant equipment replacement budgets were limited. The magnitude of this problem over a one-year test period is apparent from Reference 3, which shows that 70% of the equipment failures at the SYNTHANE pilot plant were repetitive and that wear-out increased substantially compared to the previous year.

The causes of repetitive failures in all pilot plants were basically similar. Prior to equipment selection, information on process stream composition during transient and equilibrium operation was incomplete. After the failure of a component, development of a complete understanding of the underlying causes of the failure in relation to equipment capabilities and process needs was not given a high manpower or budgetary priority. Since the causes of failure were often not remedied, the failures tended to be repetitive.

Complete operating requirements for the components to be used in demonstration or commercial plants are still not available for all equipment, and wear-out will continue at rates which are yet to be determined. To minimize the impact of this problem, a specialist dedicated to the supervision of equipment record keeping and failure analysis procedures should be assigned to each plant. His job scope should include responsibility for the implementation of a systematic method for collecting, analyzing and disseminating equipment data.

Organizing to Minimize Downtime

Plant organization had a strong influence on the time and cost to identify and resolve mechanical problems in the pilot plants. Plants staffed with qualified personnel organized in a simplified structure were more successful in quickly eliminating the causes of equipment problems. Plants with layered management and trainees in sensitive positions were slow to eliminate problem causes.

While high quality operation and maintenance should be expected in new, large plants, organizational simplicity will be difficult to achieve if high plant costs result in multiple ownership. The ineffectiveness that develops within a maintenance group when reporting channels are complex is a potentially costly and serious problem for multiple ownership plants. The ultimate result of this problem can be an accumulation of downtime and can cause a good process to be needlessly labeled "unpromising."

Minimizing Start-Up Time

The history of three pilot plants indicates that the start-up period for U.S. demonstration and commercial plants based on second-generation technology may be as long as five years. The calendar time between completion of construction and plant shutdown was six years for the CO Acceptor pilot plant and about four years for the SYNTHANE pilot plant. HYGAS completed the initial shakedown in two years but made substantial changes to the process and equipment over the next three years. As an additional point of reference, the existing 10,000 TPD SASOL 1 commercial plant in South Africa took about five years to become fully operational, including the time to complete the majority of changes and refinements made to the initial plant design.

Start-up time in the pilot plants was also influenced by weather and coal characteristics. The operations logs indicated a slowdown during freezing weather and a substantial learning-time curve for each new type of coal that was processed.

Machinery vibrations monitoring on a periodic basis has proven useful in assessing the condition of dynamic equipment; the use of this procedure could help shorten start-up time by earlier equipment problem identification. The basis for using this approach is that an analysis of vibration patterns provides a signature unique to each machine and that trends associated with signature changes can often be used to forecast avoidable problems. Vibrations monitoring of major mechanical components was demonstrated in the SYNTHANE plant (2). This pilot project showed that cost-effective data could be produced by the analysis of vibration measurements from sensors mounted on the housings of dynamic machinery. This procedure, which can be implemented by the operating contractor, appears to be one of the more promising ways of assessing the condition of a large number of components on a quantitative basis, thereby helping to minimize start-up time.

GENERAL PROCEDURE RECOMMENDATIONS FOR DEMONSTRATION/COMMERCIAL PLANTS

It is recommended that performance and reliability objectives for equipment be established during initial design studies. When components are identified, an engineering assessment relating the equipment capabilities to process requirements is also recommended. With this data, equipment objectives, component selections and process parameters can be iterated.

It is recommended that a data collection system be established for monitoring equipment performance. It is also recommended that an on-going system analysis be conducted to determine the reliability, availability and maintainability of the plant and its major subsystems. Finally, it is recommended that a failure analysis group be established in the plant to plan corrective actions and provide feedback data for the systems analysis.

Recommendations For Components Requiring New Design Concepts

The development of new design concepts is recommended for the following items because the current design concepts have proved inadequate or are nonexistent.

- Valves with longer life for slurry and hot, dirty gas service
- Slurry pumps with longer meantime-to-failure for high-pressure coal feeding
- Knockout vessels to improve the service life of process gas compressors by removing small liquid droplets
- Hot gas expanders for power recovery
- Heat exchangers with adequate erosion resistance and insensitivity of heat transfer characteristics to potential fouling in gasifier effluent streams
- Hydraulic turbine for power recovery and pressure letdown in high-pressure slurry streams
- Blowers for controlling hot char recycle between cyclone separation and gasifier
- Pelletizers to economically compress coal fines into a size range that can be consumed by the process without degenerating or agglomerating.

Recommendations for Components Requiring Improvement

The following components are recommended for development testing to improve their performance:

- Primary cyclones for hot gasifier effluent with substantial increase in wear life without sacrificing efficiency
- Valves and packing with increased life in positive displacement, high-pressure slurry pumps
- Casings and impellers with increased erosion resistance in low-head centrifugal slurry pumps
- Measurement requirements, instrumentation and techniques for expedient detection of process stream constituents entering key pumps and recycle compressors particularly during plant start-up.

Recommendations for Systems Requiring Improvement

The following systems are recommended for development testing to improve their performance:

- Coal grinding and sizing systems to meet the needs of the process with a minimum rejection of fines
- Seal and packing cooling and filtering systems to improve equipment availability and promote standardization
- Compressor inlet gas conditioning systems, including the appropriate integration of knockout vessels for submicron droplets, and solids filters coupled with an automatic cleaning system.

Recommendations for Component Development Facilities

The development of new components often involves resolving a number of problems that may be critical, but are peripheral to the principal objective. For instance, the separation efficiency of a gas/solids separation concept should be confirmed by cold testing in a controlled laboratory environment before committing the considerable time and cost required for hot testing.

Once the concept of a new component has been verified and secondary hardware problems have been resolved, testing under more realistic operating conditions, such as in a pilot plant, is recommended. Pilot plant testing is recommended if it provides critical data on the interaction between the process stream and the component. However, pilot plant testing is not recommended as a universally required step in development testing because tests of certain scaled-down components may produce misleading data due to the effects of misrepresented temperature gradients, tolerances, or similar potentially nonlinear characteristics. These special cases, together with components that have been commercialized and replicated for many different applications, may only benefit from advanced testing if the principal goal is to eliminate failure modes related specifically to coal conversion service. Improving the service life of the suction valve in a high-pressure reciprocating coal slurry feed pump is a typical example. For this and similar highly developed components, nothing short of full-scale testing in a plant environment appears adequate.

Due to lack of data for large-scale components, the first full-size process plant will undoubtedly become a test bed for some components. Therefore, special arrangements should be made in these plants to acquire design varification data on these components. To control over-all costs and refine each new type of multitrain process it is recommended that, when the second train is put into operation, the production of the first train be sacrificed in favor of using this line as a test bed for evaluating improvements to the components and to the process.

REFERENCES

1. Coal R&D, Casey Publishing Company, Vol 3, No. 23, Dec. 15, 1980.

2. McGee, J.P. and McCabe, J.T.; Review and Evaluation of Failures, Maintenance and Operability Data for CO_2 Acceptor Plant Mechanical Equipment; U.S. ERDA Contract EF-77-01-2615, Task 1; Mechanical Technology Incorporated, Latham, New York, MTI No. 78TR3, October 1977.

3. McCabe, J.T., Kramberger, F.E., Hao, B.R., Dubis, D., and Carson, S.E.; An Assessment of SYNTHANE Mechanical Equipment; DOE Contract DE-AT21-77-MC02615, Task 6; Mechanical Technology Incorporated, Latham, New York, MTI No. 79TR5, May 1980.

4. Albrecht, P.R., Kramberger, F.E., Recupero, R.M., Verden, M.L., and Rees, K.; An Assessment of HYGAS Mechanical Equipment; DOE Contract DE-AT21-77-MC02615, Task 9; Mechanical Technology Incorporated, Latham, New York, MTI No. 80TR10, October 1980.

5. Williams, N.R., et al; Survey of Industrial Coal Conversion Equipment Capabilities: Executive Summary; Oak Ridge National Laboratory, Oak Ridge, Tennessee, ORNL/TM-6810.

6. Workshop on Critical Coal Conversion Equipment, October 1-3, 1980, Huntington West Virginia, Sponsored by ESCOE, ORNL and U.S. DOE, METC; Contract No. EF-77-C-01-2468, FE-2468-88; January 1981.

7. The 2nd Symposium on Values for Coal Conversion and Utilization. October 1980, Morgantown W.V., Sponsored by U.S. DOE, METC in cooperation with the Valve Manufacturers Association.

8. Coal Technology '80; 3rd International Coal Utilization Exhibition and Conference, November 1980, Houston, Texas.

COAL PREPARATION FOR COAL GASIFICATION AND LIQUEFACTION

Dale R. Simbeck
Synthetic Fuels Associates, Inc.
Two Palo Alto Square, Suite 528
Palo Alto, California 94304

INTRODUCTION

The preparation or "cleaning" of coal has not grown with the increase of United States coal production. For example in 1967 the total U.S. coal production was 555 million tons, of which 350 tons were cleaned. (1) Ten years later in 1977, U.S. coal production had increased to 690 million tons of which only 265 million tons were cleaned. (2) The developing synthetic fuels industry may reverse this trend.

This paper explains how coal gasification and liquefaction processes can utilize coal cleaning to reduce the overall cost of producing synthetic fuels and improve plant operability. This paper addresses key aspects and issues for coal cleaning, particularly as applied with coal gasification and liquefaction:

- o overview
- o cost and general benefits
- o benefits for coal conversion.

COAL CLEANING OVERVIEW

Coal cleaning technology is often not well understood by engineers working in the other energy technologies. A popular misconception about coal cleaning is that it is the separation of clay and rock from coal by wetting and floating the coal in dense medium or jig processes. Coal cleaning is actually the separation of high ash coal particles from low ash coal particles by the differences in their specific gravity. High ash coal particles have specific gravities of 1.8 to 2.5 whereas low ash coal particles have specific gravities of about 1.3. The rejected solids produced in coal cleaning are high ash coal particles plus all the fine coal particles which are too small to be effectively recovered. It should be noted that minus 100 mesh coal is usually lost in steam coal cleaning whereas only the minus

300 mesh coal is lost in metallurgical coal cleaning. The much higher value of metallurgical relative to steam coal justifies the higher expense of recovering fine metallurgical coal down to the smaller particle size.

As stated in the introduction, the amount of coal being cleaned in the United States has significantly decreased over the last 15 years. There is a common misconception that this is due to the increased use of western coal. In reality, coal production east of the Mississippi has stayed relatively constant between 500 and 550 million tons per year. (3)

The primary reasons for the decrease in coal cleaning is the depressed demand for metallurgical coal and the increased cost of coal cleaning. The U.S. metallurgical coal production, of which essentially all is cleaned, is about 100 to 150 million tons per year, including exports. The demand for metallurgical coal depends on the world demand for basic steel which has been very weak during the last five years. The cost of coal cleaning has risen due to the higher value of run of mine (ROM) coal, general inflation, and more stringent pollution laws. The average f.o.b. mine price of U.S. coal increased from $5.00 per ton (about $0.20/MM Btu) in 1969 to $23.50 per ton (about $1.10/MM Btu) in 1979. This dramatic rise in coal price reflects the interfuel competition of coal with crude oil, which has risen in price by a factor of about 10 over the same period. The jump in the value of ROM coal means that the 10 to 15% Btu rejection during coal cleaning now represents a much greater cost. Table 1 shows representative cost of coal cleaning for new plants in 1969 and 1979. About 25% of the incremental cost of coal cleaning is due to Btu rejection in the 1969 plant, whereas over 40% of the coal cleaning cost is due to Btu rejection in the 1979 plant.

The drastic increase in coal cleaning plant capital, operating, and maintenance cost in Table 1 is due to more than general inflation. Environmental laws have also had a significant effect on coal cleaning practices. Some of the process changes due to environmental laws include:

- closed water circuits
- enclosed storage (silos)
- fine refuse dewatering (thickeners and filters)
- improved refuse disposal
- improved noise control.

The higher capital cost of coal cleaning plants is beginning to change the plant operation. Traditionally, coal cleaning plants were operated only two shifts per day, five days per week, and 44 weeks per year. This resulted in annual plant utilization factors of only about 40%. Many of the newer and more capital intensive coal cleaning plants are attempting to operate at higher plant utilization factors. There have been some problems caused by worker/union resistance and required equipment maintenance. Coal cleaning equipment is usually designed to operate only two shifts per day with the third shift for maintenance.

COAL CLEANING COST AND GENERAL BENEFITS

The important potential benefits of cleaning coal for any use are:

- reduction in variability of coal characteristics
- reduced ash
- reduction in pyritic sulfur
- increased heating value
- increased market value.

Run of mine coal characteristics constantly fluctuate. The magnitude of the fluctuation can be considerable. Figure 1 shows the day-to-day fluctuation of sulfur in the coal from a single underground mine in Western Pennsylvania. Other coal characteristics can also vary by the same magnitude. Core samples provide similar information for mines yet to be developed. Coal cleaning can be utilized to effectively reduce these fluctuations. It should be noted, however, that coal

Figure 1

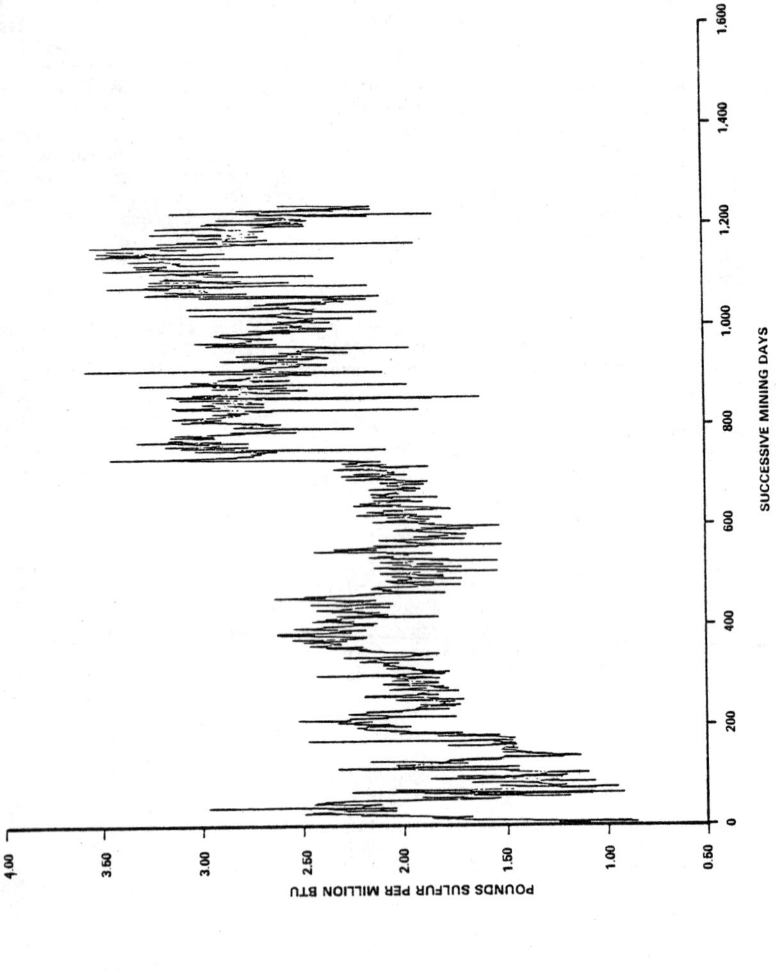

POUNDS OF SULFUR PER MILLION BTU FOR SUCCESSIVE MINING DAYS FOR HELEN MINE, JANUARY 1970 THROUGH DECEMBER 1975. (THOMAS, 1978)

blending can also be used to reduce fluctuations in coal characteristics at lower costs than coal cleaning.

Ash reduction is an important benefit of coal cleaning which coal blending cannot provide. Most coal cleaning now is done only for ash reduction because metallurgical coals are low in sulfur as mined. If a coal is high in pyritic sulfur (FeS_2), coal cleaning will reduce pyritic sulfur in proportion to the reduction in ash because pyritic sulfur only appears with the ash. Ash reduction results in increased heating value of the cleaned coal which leads to reduced transportation costs and higher efficiency in the final coal utilization process.

However, there is a potential pitfall in cleaning coal to increase heating value. Coal cleaning increases the surface moisture of the coal, especially of the minus ¼ inch coal. Deep cleaning of fine coal can easily increase the moisture content of cleaned coal to the point where there is no heating value increase due to the ash reduction.

Traditionally, more coal is cleaned during times when there is an oversupply of coal or where a "coal buyer's market" exists. Since 1979 the marketability of specific coal has become more important than the economics of coal cleaning. This situation presently exists for high sulfur coals in the Midwest and Eastern U.S. coal regions where there is an idle mine capacity estimated at over 100 million tons per year.

The cost of coal cleaning versus the benefits are very specific to the coal. The coal cleaning costs in Table 1 are merely representative costs to show the general increase in the cost of coal cleaning over the last ten years. To even grossly estimate actual coal cleaning costs and benefits requires potential market analysis plus the following information for a specific coal:

- washability (sink-float analysis)
- core samples
- coal and ash analysis
- ROM coal size distribution
- ROM coal value.

Washability analysis is the most important since this shows what the theoretical benefits and Btu rejection will be for various particle distribution and separation of specific gravities. It must be noted, however, that actual coal cleaning yield will always be poorer than the washability analysis. The Electric Power Research Institute, the National Coal Association, and several utility companies are funding the construction and operation of a 500 ton per day coal cleaning demonstration plant at Homer City, Pennsylvania. This unique plant will be equipped with many different types of coal cleaning equipment and will enable actual coal testing for various equipment configurations and separating to various specific gravities. The results of a specific coal test will enable accurate economic analysis of coal cleaning cost, yields, and benefits. (5, 6)

COAL CLEANING BENEFITS FOR COAL CONVERSION

The important potential benefits of cleaned coal to coal conversion system are:

- lower capital investment
- lower operating cost
- improved plant performance and reliability.

Consider the coal described in Figure 1. A coal conversion plant designed for the highest variation in ROM coal sulfur level would require a larger capital investment for overdesign. Coal cleaning would reduce the sulfur and ash variation and therefore reduce the plant overdesign. The reduction in ash and sulfur due to coal cleaning would further reduce the synthetic fuel plant capital investment by reducing the requirements for coal, ash and sulfur processing.

Cleaned coal can benefit the operating and maintenance cost of coal conversion plants as well as capital cost. Variation of feed coal heating value, ash and sulfur levels would adversely affect the operating and maintenance cost of coal conversion plants due to operating instability caused by feedstock variability.

The most important benefit of coal cleaning for coal conversion processes is the improved plant performance and reliability. The annual plant utilization factors have a significant effect on the cost of the coal conversion plant's products due to the large capital investments. The lower ash and sulfur levels of cleaned coal would improve plant operability. The reduction in fluctuations of feed coal characteristics by coal cleaning would have a major effect on plant performance. The benefits of coal cleaning on a coal-fired power plant's performance and reliability is well-documented. (7, 8) The benefits of coal cleaning on coal conversion plants could be even greater due to the larger capital investments and more complex coal processing associated with coal conversion relative to coal-fired boilers.

The total benefit of coal cleaning to coal conversion will be quite specific to the coal and the coal conversion technology. Generally, coal conversion processes which slag the coal ash, such as slagging gasifiers and some applications of direct coal liquefaction, would benefit the most from coal cleaning. A large amount of energy loss and capital cost is associated with slagging coal ash. Slag management and recovery is also a major operating and maintenance problem. Several direct coal liquefaction process developers believe the coal ash has catalytic effects. These benefits must be quantified and compared to the additional cost associated with processing high ash coals. High ash coals require larger solvent to MAF coal slurry ratios, solids/liquids separators, and residual unconverted coal gasifiers. Coal cleaning can significantly reduce the ash content of most coals. Non-slagging coal conversion processes as well as slagging processes could benefit from coal cleaning by the reduction of feed coal variability and reduction in coal components such as pyritic sulfur, clays, chloride, and other water solubles in ROM coal.

CONCLUSIONS

The use of coal cleaning for coal conversion processes must be carefully and objectively analyzed. Specific coal data such as washability (sink-float) analysis are required to estimate the cost of cleaning. The benefits of coal cleaning to coal conversion will also be technology specific; however, the most important benefit will be improved operability and reliability of the capital intensive coal conversion plant. Coal cleaning will also reduce the capital and operating cost of many coal conversion processes. In a literal sense, coke ovens and blast furnaces are coal conversion processes. It is interesting to note that essentially all the coal consumed in these existing coal conversion processes is cleaned. As new coal conversion systems such as coal gasification and coal liquefaction become a reality, they will also realize the benefits of coal cleaning.

Table 1

Representative Coal Cleaning Cost
Basis: 1,000 ton/hour ROM coal
4,000 hour/year operation
90% Btu recovery

	1969	1979
Capital Cost	$10 million	$35 million
ROM Coal Cost	$0.20/MM Btu	$1.10/MM Btu
Clean Coal Cost		
Coal cost*	$0.22/MM Btu	$1.22/MM Btu
Operating and Maintenance	0.02	0.07
Capital related	0.03	0.10
Total	0.27	1.39
Incremental cost of cleaning	$0.07/MM Btu	$0.29/MM Btu

*ROM coal cost divided by Btu recovery $(\frac{0.20}{.9})$ + 0.22

References

1. United States Bureau of Mines, Minerals Yearbook, 1967.

2. United States Bureau of Mines, Minerals Yearbook, 1977.

3. Lin, K., Coal Data 1977, National Coal Association, Washington, D.C.

4. Richardson, C. V. and G. F. Nielson, 1980 Keystone Coal Industry Manual, McGraw-Hill, New York.

5. Clifford, K., et al., EPRI Journal, June 1979, pp. 6-13.

6. Trerice, D., EPRI Journal, November 1980, pp. 40-41.

7. Vivenzio, T. A., Impact of Cleaned Coal on Power Plant Performance and Reliability, Electric Power Research Institute (EPRI), Report CS-1400, April 1980.

8. Buder, M. K., et al., "The Effects of Coal Cleaning on Power Generation Economics," presented at the 41st annual American Power Conference, April 1979.

9. Thomas, R. E., "Interpreting Statistical Variability," presented at the EPA Symposium on Coal Cleaning to Achieve Energy and Environmental Coals, Hollywood, Florida, September 1978.

SURVEY OF COAL GASIFICATION - COMBINED-CYCLE PROJECTS

Joseph J. Yancik
National Coal Association

John R. M. Alger
General Electric Company

INTRODUCTION

In August 1980, the National Coal Association (NCA) published a report titled, "Coal Synfuel Facility Survey", which listed all the existing and proposed research, development, demonstration and commercial facilities in the United States which are capable of converting coal to alternative energy fuels. (1) Table 1 presents a numerical summary of the coal gasification facilities identified by the survey. As shown in this summary table, 23 coal gasification facilities were in operation, four were under construction and 26 had been proposed with some of them being in the active planning stage.

None of the 23 operating plants involved the integrated operations of a gasification plant with a combined-cycle power generation system. Nor did any of the four plants under construction involve combined-cycle power generation. Only four of the 26 announced projects were integrated combined-cycle power generation plants. Of these four, "Cool Water", Southern California Edison's demonstration project, was well into the planning stage and proceeding on schedule toward a construction start in July 1981. The other three projects were commercial scale plants. The previously announced New England Energy Park project was subsequently submitted to the Department of Energy (DOE) and awarded a feasibility study grant under the "round one" solicitation for the Alternative Fuels Production Program. The two other projects, Central Maine Power Company and Florida Power Corporation, surfaced when they were announced winners of feasibility grants under the "round one" solicitation.

One project listed in the NCA survey but not initially categorized as an integrated combined-cycle power generation project is the KILnGAS project sponsored by Allis Chalmers and a consortia of

TABLE 1

COAL GASIFICATION FACILITIES [1/]

GENERAL STATUS AS OF JULY 9, 1980

	IN OPERATION			UNDER CONSTRUCTION			PROPOSED AND IN PLANNING STAGE			STAND-BY, MOTHBALLED OR DISMANTLED			STATUS UNKNOWN		
	HBG*	MBG	LBG	HBG*	MBG	LBG	HBG*	MBG	LBG	HBG*	MBG	LBG	HBG*	MBG	LBG
COMMERCIAL PLANTS	0	0	5	0	0	2	10	5	2	2	1	7	0	0	0
DEMONSTRATION PLANTS:	0	0	0	0	1	0	3	2	2	1	0	1	0	0	0
PILOT PLANTS (Capacity 25 TPD or more):	2	2	2	0	0	0	1	1	0	3	0	2	0	0	1
PROCESS DEVELOPMENT UNITS AND SMALL PILOT PLANTS:	3	5	4	0	0	1	0	0	0	1	1	0	0	0	0
TOTAL NUMBER OF FACILITIES BY RANK OF GAS:	5	7	11	0	1	3	14	8	4	7	2	10	0	0	1
TOTAL NUMBER OF GASIFICATION FACILITIES:	23			4			26			19			1		

* HBG = High-Btu Gas; MBG = Medium-Btu Gas; LBG = Low-Btu Gas.

1/ Coal Synfuel Facility Survey, National Coal Association, Washington, D. C., August 1980.

TABLE 2

INTEGRATED COAL GASIFICATION COMBINED-CYCLE POWER GENERATION PROJECTS

STATUS AS OF FEBRUARY 1981

	In Operation	Under Construction	With Firm Construction Schedule	In Planning or Feasibility Study Phase	Proposed
COMMERCIAL PLANTS:	0	0	0	13	8
DEMONSTRATION PLANTS:	0	1*	1	0	0
PILOT PLANTS:	0	0	0	0	0
TOTALS:	0	1*	1	13	8

* Technically not a combined-cycle project. Initially integrated with a boiler with plans for later integration with a combined-cycle system.

utility companies. This project was identified in the NCA survey as a low-Btu coal gasification-steam boiler power generation project. In the survey reported on in this paper, the project is included as a combined-cycle power generation project because future plans call for the installation of a gas turbine to be fueled by cleaned coal fuel gas.

In summary, as of July 1980, there were five projects identified which indicated a growing interest by the electric power utilities and others in coal gasification for combined-cycle power generation. Since this past July, the interest in this technology appears to have increased substantially. Between July 1980 and February 1981, an additional 18 projects involving coal gasification integrated with combined-cycle power generation were announced. The appearance of widespread geographic interest by electric power utilities and industrial companies was undoubtedly accelerated by the opportunity for financial assistance through feasibility study grants under the Alternative Fuels Production Program of the DOE. The purpose of this paper is to present a summary status report on all 23 USA projects, as well as an update on selected foreign projects.

SURVEY PROCEDURE

A survey questionnaire very similar to the one used in the first NCA "Coal Synfuel Facility Survey" was mailed to all known project sponsors. The mailing list was compiled from public news sources, press announcements and a list of proposals submitted to DOE under the Alternative Fuels Production Program solicitations. Data taken from the questionnaires are summarized in tables that classify the projects according to status: Table 3 lists plants with firm construction schedules, Table 4 shows plants in the planning or feasibility study phase, and Table 5 lists proposed plants. The latest information from public sources is used for the few projects for which questionnaires were not available. Every effort was made to include all the projects which have been publicly announced by their sponsors. The basic data reported in the tables include the project sponsor or sponsors, gasification process along with any other major synfuel product processes involved, the types and quantities of fuel gases produced plus other synfuel products and by-products, the capacity of the combined-cycle power generation system, the type and quantity of coal planned for use in the facility and the present status of the project.

For comparison purposes, similar information was compiled by the authors on Western European projects (Table 6) and on Japanese projects (Table 7) involving coal gasification plants integrated with combined-cycle power generation systems. The data on the Japanese project was provided by Marubeni, Inc.

REVIEW OF COAL GASIFICATION COMBINED-CYCLE TECHNOLOGY

Before discussing the results of the survey data, some background on the combined-cycle power generation technology and terminology used in this paper is appropriate. Figure 1 is a generic flow diagram of the processes involved in the conversion of coal to raw synthesis gas which, after cleaning and purification, can be used to generate electric power via a combustion turbine, steam (the combination is termed cogeneration), and fuel gas for process use (all three outputs are defined as trigeneration). Additional products through the conversion of clean synthesis gas to methanol and/or other chemical products is defined as polygeneration.

The terminology and basic process steps are illustrated in Figure 1. Coal is pretreated as required (dewatering, drying, crushing, or slurrying) and then fed into the gasifier along with steam,

TABLE 3: COAL GASIFICATION COMBINED-CYCLE PROJECTS

PLANTS WITH FIRM CONSTRUCTION SCHEDULES

PROJECT TITLE	SPONSOR/ OPERATOR	SITE OR AREA	PROCESS	SYNFUEL PRODUCTS QUANTITIES PER DAY	SYNFUEL MARKET USES	COAL TYPE USED	COAL USE	STATUS
Cool Water Coal Gasification Demonstration Project	Southern California Edison, Texaco Inc., EPRI, Bechtel and General Electric	Cool Water Power Station, Daggett, CA	Oxygen-Blown Texaco Coal Gasification Process with General Electric Combined-Cycle System	.MBG: 67 MMSCFD .Sulfur: 4.3 TPD	.Power Generation Combined Cycle, 100 MW	Utah Bituminous and other test coals	1,000 TPD	Final engineering design initiated in February 1980. Construction scheduled for start in July 1981. Systems to be integrated with combined-cycle unit for start-up in late 1983.
KILnGAS Demonstration Project*	Allis Chalmers, State of Illinois, Illinois Power Company and 11 other utilities	Wood River Power Station, East Alton, IL	KILnGAS Process	.LBG: 409 MM Btu/hr .Sulfur: 16.7 TPD .Ash: 95 TPD	.Power Generation: Fuel Gas for a 40 MW boiler; future plans for gas turbine operation	Illinois No. 6	600 TPD	Demonstration plant construction begun in January 1981. Plant scheduled to begin operations in mid-1983.

* Technically not a combined-cycle project, but included here as a project that demonstrates the integration of a coal gasifier with a power generating system.

TABLE 4: COAL GASIFICATION COMBINED-CYCLE PROJECTS

PLANTS IN PLANNING OR FEASIBILITY STUDY PHASE

PROJECT TITLE	SPONSOR/ OPERATOR	SITE OR AREA	PROCESS	SYNFUEL PRODUCTS QUANTITIES PER DAY	SYNFUEL MARKET USES	COAL TYPE USED	COAL USE	STATUS
Central Arkansas Energy Project	Arkansas Power and Light Company	Redfield, AR	Texaco Coal Gasification Process	.MBG: 400 MMSCFD .Sulfur 1,900 TPD	.Cogeneration Combined-Cycle Power Generation	Bituminous Illinois No. 6	7,100 TPD	Selected for negotiations under first round solicitation of the Department of Energy alternative fuels production program. Contract still under negotiation.
Central Maine Power Integrated Gasification Combined-Cycle Plant	Central Maine Power Company, Texaco Inc., General Electric and Stone and Webster	Sears Island, ME	Texaco Coal Gasification Process with GE STAG[R] Combined-Cycle System	.MBG .Sulfur	.Power Generation, 480 MW	Appalachian Bituminous	5,000 TPD	Selected for negotiations under first round solicitation of the Department of Energy alternative fuels production program. Feasibility study underway.
Missouri Combined-Power Cycle Coal Gasification Energy Centers	Consumer Energy Corporation	Reger, MO and Yates, MO	Texaco Coal Gasification Process with ICI Methanol Synthesis	.MBG: 245 MMSCFD .Methanol Fuel: 211,225 ST/Y .Sulfur: 71,963 ST/Y Products per project	.Power Production Combined Cycle, 1.85×10^6 MWH/ year .Chemical Feedstock	High Sulfur Missouri Bituminous	3,800 TPD per project	Preliminary feasibility studies have been completed and limited engineering studies begun. Financial commitments have been secured.
Dow Chemical Commercial Project	Dow Chemical USA	Plaquemine, LA	Texaco Coal Gasification Process	.LBG	.Power Generation Combined Cycle, 70 MW	Lignite or Bituminous	1,200 TPD	100 TPD pilot plant began operations in 1980. Based on successful operation, commercial-sized plant could begin construction in 1985.
Higgins Coal Gasification Repowering Project	Florida Power Corporation	Oldsmar, FL	BGC Slagging Gasifier	.MBG .Ammonia: 15.7 TPD .Sulfur: 115 TPD (100% capacity factor)	.Power Generation Combined Cycle, 8.98×10^6 KWH/ day	Illinois Bituminous	4,000 TPD	Selected for negotiations under first round solicitation of the Department of Energy alternative fuels production program. Feasibility study and engineering design begun in October 1980.

TABLE 4 (CONT'D): COAL GASIFICATION COMBINED-CYCLE PROJECTS

PLANTS IN PLANNING OR FEASIBILITY STUDY PHASE

PROJECT TITLE	SPONSOR/ OPERATOR	SITE OR AREA	PROCESS	SYNFUEL PRODUCTS QUANTITIES PER DAY	SYNFUEL MARKET USES	COAL TYPE USED	COAL USE	STATUS
Gulf States Utilities Combined-Cycle Gasification Project	Gulf States Utilities and Westinghouse Electric Corporation	Lake Charles, LA		.MBG	.Power Generation Combined Cycle			Gulf States is seeking an exemption under PIFUA for construction of Nelson #7 unit as a peakload powerplant. If exemption is issued, plans include converting plant to a combined-cycle operation. In the final phase of the project, Westinghouse would design and construct a coal gasification system. Operation of combined-cycle unit depends upon ongoing feasibility studies and Department of Energy approval and funding of project.
Massachusetts Municipal Wholesale Electric Gasification Project	Massachusetts Municipal Wholesale Electric Company	Ludlow, MA	Texaco Coal Gasification Process	.MBG	.Power Generation Combined Cycle		5,000 TPD	Selected for negotiations under second round solicitation of the Department of Energy alternative fuels production program. Negotiations underway.
NASA Lewis Coal Gasifier Cogeneration Power Plant	NASA Lewis Research Center	Cleveland, OH	Air-Blown Pressurized Fluidized-Bed Gasification		.Cogeneration Combined Cycle	Eastern Bituminous	240 TPD	Feasibility studies completed July 1980. Environmental assessment completed February 1981.
New England Energy Park (NEEP)	EG&G Synfuels, Inc.	Fall River, MA	Texaco Gasification Process	.MBG: 55.1 x 10^{12} Btu/year .Methanol: 6.2 MM BBL/ year	.Power Generation Combined Cycle, 470 MW .Methanol: Chemical Feedstock/Gasohol	High Sulfur Eastern Bituminous	10,000 TPD	Selected for negotiations under first round solicitation of the Department of Energy alternative fuels production program. Feasibility study underway.

TABLE 4 (CONT'D): COAL GASIFICATION COMBINED-CYCLE PROJECTS

PLANTS IN PLANNING OR FEASIBILITY STUDY PHASE

PROJECT TITLE	SPONSOR/ OPERATOR	SITE OR AREA	PROCESS	SYNFUEL PRODUCTS QUANTITIES PER DAY	SYNFUEL MARKET USES	COAL TYPE USED	COAL USE	STATUS
New York State Power Authority Combined-Cycle Project	New York State Power Authority	Buffalo, NY Area		.MBG .Methanol .Coke	.Electric Power Generation, 600 MW			Feasibility study proposal submitted to Department of Energy in response to second round solicitation under the alternative fuels production program, but project not selected for negotiations. Proposal resubmitted to Department for consideration under solicitation for innovative direct combustion systems. Awaiting final selection under solicitation.
PG&E/Texaco San Ardo Cogeneration Project	Pacific Gas & Electric Company and Texaco Inc.	San Ardo, CA	Oxygen-Blown Texaco Gasification Process	.MBG: 175 MMSCFD	.Cogeneration Power Generation Combined Cycle		17,000 TPD	Selected for negotiations under second round solicitation of the Department of Energy alternative fuels production program. Negotiations underway.
U.S. Department of Navy Combined-Cycle Project	U.S. Department of Navy and Pope, Evans and Robbins	Norfolk, VA		.MBG	.Cogeneration Power Generation Combined Cycle, 60 MW			Feasibility study underway and scheduled for completion in July 1981.
Fairmont Trigeneration Project	Westinghouse Electric Corporation	Fairmont, WV	Westinghouse Fluidized-Bed Coal Gasification Technology	.MBG: 15 MMSCFD	.Trigeneration Combined Cycle, 10 MW Gross	All Ranks	300 TPD	Selected for negotiations under second round solicitation of the Department of Energy alternative fuels production program. Negotiations underway.

TABLE 5: COAL GASIFICATION COMBINED-CYCLE PROJECTS

PROPOSED PLANTS

PROJECT TITLE	SPONSOR/ OPERATOR	SITE OR AREA	PROCESS	SYNFUEL PRODUCTS QUANTITIES PER DAY	SYNFUEL MARKET USES	COAL TYPE USED	COAL USE	STATUS
Bouse Coal Gasification Project	Arizona Public Service Company	Bouse, AZ	Slagging Lurgi Gasifier with Lurgi Methanol Synthesis	.MBG: 180 MMSCFD or .Methanol: 13,000 BBL/D	.Power Generation Combined Cycle or .Combustion Turbine Power Plant	Bituminous	3,800 TPD	Feasibility study proposal submitted to Department of Energy in response to first and second round solicitation under the alternative fuels production program, but project not selected for negotiations. Project is in an inactive status.
Coal Gas for Edgar Station Combined-Cycle Plant Project	Boston Edison Company	Weymouth, MA	BGC/Lurgi Slagging Gasifier	.MBG: 750 MMSCFD .Phenol: 45 TPD .Ammonia: 33 TPD .Naptha: 120 TPD .Sulfur: 440 TPD	.Boiler and Combined-Cycle Plant Power Generation	Bituminous Eastern High Sulfur	12,000 TPD	Feasibility study proposal submitted to Department of Energy in response to second round solicitation under the alternative fuels production program, but project not selected for negotiations.
Consumer Power Combined-Cycle Gasification Project	Consumer Power Company, Gilbert/Commonwealth, Westinghouse Electric Corp.	Bay City, MI	Westinghouse Single-Stage Fluidized Bed Gasifier	.MBG: 1.3 MMSCFD	.Power Generation Combined Cycle		2,160 TPD	Feasibility study proposal submitted to Department of Energy in response to second round solicitation under the alternative fuels production program, but project not selected for negotiations.
Foster-Wheeler Low-Btu Gasification Demonstration Project	Foster-Wheeler Corporation, Allegheny Power Company	Springdale, PA	Esso Fluidized Bed Combustor	.LBG	.Power Generation	Bituminous		Site selection for demonstration made in July 1980. Project still pending DOE decision.

TABLE 5 (CONT'D): COAL GASIFICATION COMBINED-CYCLE PROJECTS

PROPOSED PLANTS

PROJECT TITLE	SPONSOR/ OPERATOR	SITE OR AREA	PROCESS	SYNFUEL PRODUCTS QUANTITIES PER DAY	SYNFUEL MARKET USES	COAL TYPE USED	COAL USE	STATUS
Beaver Synfuels Project	Portland General Electric Company	Clatskanie, OR	Westinghouse Gasification Process	.MBG: 216 MMSCFD .SNG: 10 MMSCFD .Methanol: 1,750 TPD .Ammonia: 90 TPD .Sulfur: 50 TPD	.Power Generation .Chemical Feedstock .Local Fuel Gas	Subbituminous	10,000 TPD	Feasibility study proposal submitted to Department of Energy in response to second round solicitation under the alternative fuels production program, but project not selected for negotiations. Company reassessing alternative funding for study.
Puget Power Cogas Project	Puget Sound Power and Light Company and Boeing Engineers and Constructors	Washington State		.MBG: 100 MMSCFD .Methanol	.Power Generation Combined Cycle and Methanol for Turbine Power Generation			Feasibility study proposal submitted to Department of Energy in response to first round solicitation under the alternative fuels production program, but project not selected for negotiations.
SDG&E Blythe Gasification Project	San Diego Gas and Electric Company	Blythe, CA		.MBG .Methanol	.Power Generation Combined Cycle			Feasibility study proposal submitted to Department of Energy in response to second round solicitation under the alternative fuels production program, but project not selected for negotiations.
Integrated Industrial Synfuel Combined Cycle Cogeneration Plant (ISCP)	Texaco Inc., and General Electric with Brown & Root (Bardi)	Lynn, MA and Louisville, KY	Texaco Coal Gasification and General Electric STAG[R] operating in cogeneration or trigeneration modes	13.5 billion Btu per day in fuel gas equivalent	.Industrial power, steam and fuel gas	Eastern High Sulfur	Approximately 200,000 tons per year	Feasibility study proposal submitted to Department of Energy in response to second round solicitation under the alternative fuels production program, but project not selected for negotiations. Schedule and funding plans now under evaluation.

TABLE 6: WESTERN EUROPEAN COAL GASIFICATION COMBINED-CYCLE PROJECTS

PROJECT TITLE	SPONSOR/ OPERATOR	SITE OR AREA	PROCESS	SYNFUEL PRODUCTS QUANTITIES PER DAY	SYNFUEL MARKET USES	COAL TYPE USED	COAL USE	STATUS
Shell Nederland Coal Gasification Project	Shell Nederland	Moerdijk, The Netherlands	Shell-Koppers Process Coal Gasification	.MBG: 21 billion SCF/year	.Combined-Cycle Power Generation, 80 MW	.Hard coal	0.33 million short tons/ year	Plant presently under construction with completion scheduled for 1983-1984. Operations to begin in 1984-1985.
Texaco Gasification Project	Texaco	Power Station Site, Rheinpreuben, Moers-Meerbeck, Federal Republic of Germany	Texaco Process Coal Gasification	.MBG: 23 billion SCF/year		.Hard coal	0.55 million short tons/ year	Project planning to continue until 1983 with construction scheduled for 1983-1985. Plant operations scheduled to begin in 1985.
Saarberg-Otto Coal Gasification Project	Saarberg-Otto	Saarland, Federal Republic of Germany	Saarberg-Otto Gasification	.MBG: 28 billion SCF/year	.Combined-Cycle Power Generation, 60 MW	.German hard coal	0.44 million short tons/ year	Project planning to continue through 1983 with construction scheduled for 1983-1984. Plant operations scheduled to begin in 1985.
VEW Gasification Project	Vereinigte Elektrizitats-werke Boun	Undecided	Entrained Bed Gasification to produce low-Btu gas and char	.Low-Btu Gas .Coke	.Combined-Cycle Power Generation, 600 MW	.Hard coal	2.0 million short tons/ year	Engineering study for project begun. No construction schedule has been announced. A one ton/hour pilot plant has been successfully operated for two and one half years.
Rotterdam Energy Council Pilot Combined-Cycle Power Station	Rotterdam Energy Council	Power Station Site on the Maasvlakte near mouth of Rhine, The Netherlands	Texaco Entrained Bed Gasification Process					Design study scheduled for completion end of 1980. Depending upon planning permission and financing, plant operations could start-up in 1985.
Swedish State Power Board Gasification Project	Swedish State Power Board and Svanska Petroleum Technology	Gothenberg, Sweden	Texaco Partial Oxidation Process	.MBG	.Cogeneration Combined-Cycle Power Generation, 130 MW			Preliminary engineering studies underway. Facility to include a 150-ton-per-day pilot coal gasification plant.

TABLE 7: COAL GASIFICATION FACILITIES IN JAPAN

PROJECT TITLE	SPONSOR/ OPERATOR	SITE OR AREA	PROCESS	SYNFUEL PRODUCTS QUANTITIES PER DAY	SYNFUEL MARKET USES	COAL TYPE USED	COAL USE	STATUS
High Btu Coal Gasification Pilot Plant	M.I.T.I. (Japanese Government) with Electric Power Development Company (operator and project manager)	IWAKI City, Fukushima Prefecture, Japan	Fluidized-Bed Gasification	.Methane: 70 MSCFD	.Town Gas .Industrial Fuel Gas .Chemical Feedstock .Gas Turbine Fuel for Electric Power Generation	.Taiheiyo Low-Sulfur Subbituminous	12 TPD in a coal/ residual oil slurry (approximately 50-50 ratio)	Pilot plant under construction with operations to begin in February 1982. A 100 ton-per-day demonstration plant is planned following successful pilot plant operation.

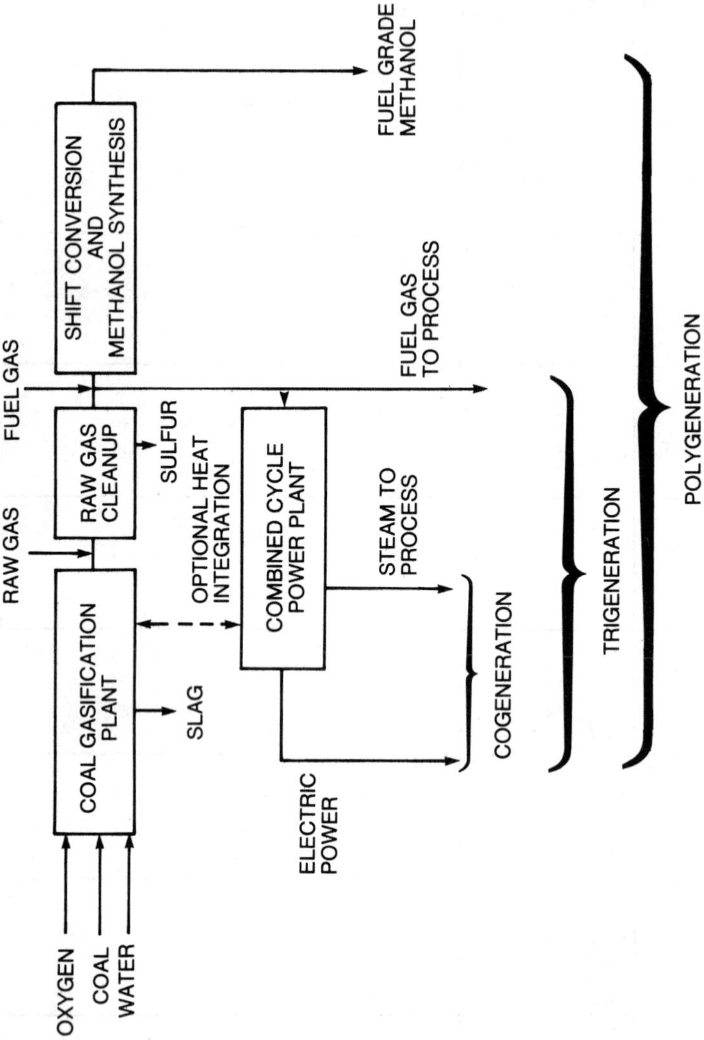

FIGURE 1. BASIC CONFIGURATION OF COAL GASIFICATION COMBINED CYCLE POWER PLANTS

air or oxygen. Raw fuel gas is produced along with by-product slag. Heat recovery, depending upon the gasifier, offers an opportunity to produce steam integrated with the combined-cycle power plant. Raw gas clean-up of particulates and sulfur containing compounds is accomplished next with particulate scrubbers and acid gas removal systems. Combustible by-products such as tar and coal char may be returned to the gasifier feed system for energy recovery in the gasification process, although this loop is not shown. Clean fuel gas is fed to a combined-cycle power generation system where the gas is combusted in the gas turbine, producing electric power and the gas turbine exhaust products pass through a waste heat boiler for heat recovery to produce steam. The steam is sent through a steam turbine-generator producing more electric power.

As described so far, the plant produces electric power as its singular output, with saleable by-product sulfur and slag suitable for land fills or construction purposes. In a cogeneration configuration, steam is withdrawn for industrial process use or for district heating, along with electric power production.

The trigeneration technology is illustrated for those cases where fuel gas may also be withdrawn for process use (such as in industrial heat-treating, glass production or oven baking operations). Trigeneration, therefore, includes cogenerated electric power, steam, plus fuel gas for industrial process uses. Polygeneration is a term coined to cover additional outputs. Methanol in this illustration is produced in a separate shift conversion and methanol synthesis loop. Advanced options for methanol synthesis in such plants include recycle of gas products to the gas turbine after partial conversion to methanol. These options are discussed in the paper in this session by Louks of EPRI. (2)

As described in the preceding paragraphs, coal gasification combined-cycle electric power generation technology involves two basic systems: a coal gasifier with a gas clean-up system and the combined-cycle power generator system which is a combination of combustion turbine with a heat recovery system and bottoming steam turbine. Coal gasification systems have been demonstrated on a large commercial scale in South Africa and on a lesser scale in many plants in other countries around the world outside the USA. The gasifiers used in these plants, however, have a number of environmental, coal feed and product gas specifications which limit their widespread use in the USA with its variety of coal types, emission standards and synfuel gas needs. In recent years, advanced gasifier designs have been pilot plant tested and these designs appear to overcome many of the serious limitations of the first generation gasifier. These advanced or improved gasifiers are now ready to be demonstrated in commercial plants.

Combined-cycle power generation, on the other hand, has been successfully used for about 20 years with natural gas or distillates as fuels for the combustion turbines. In recent years successful combined-cycle base load operation at high thermal efficiency (over 40%) and high availability (over 90%) has been demonstrated. What remains to be demonstrated is the gasifier scale-up and the integration of the coal gas plant feeding clean low- or medium-Btu fuel gas to a combined-cycle power generation system.

To date, the largest coal gasification plant integrated with an electric power generating unit was the pioneer project carried out by the West German utility STEAG AG, a subsidiary of Ruhrkohle AG. The plant demonstrated simultaneous operation of a first generation 60 ton per hour coal gasification system supplying a 170 MW unit consisting of a pressurized boiler which fed an expander gas turbine and steam turbine for power generation. System operation began in 1973, and over 8,600 hours were accumulated prior to the system's shutdown in 1978. As shown in Figure 2, the power generation system was different than the typical gasification combined-cycle systems proposed for U.S. projects. However, it did involve the integrated operation of a coal gasification system and a form of combined-cycle system similar to that for a pressurized fluid bed boiler combined-cycle. The STEAG plant consisted of coal gasification, gas cooling and clean-up essentially as previously described, except that the clean fuel gas and oil produced by cleaning the gas from the Lurgi dry ash gasifiers were fully combusted in a pressurized boiler from which steam was produced and fed to steam turbine generators. The clean combusted flue gas products were then fed to expander gas turbines to produce additional power with turbine exhaust for feed water preheating.

RESULTS OF THE SURVEY

A numerical summary of the results of the coal gasification combined-cycle NCA survey is reported in Table 2. A total of 23 projects were identified. These numbered: one demonstration plant under construction and one scheduled for a construction start in July 1981, 13 commercial plants in the planning or feasibility study phase and eight proposed commercial plants. The details available on each of these projects can be found in Tables 3, 4 and 5.

For comparative purposes, the authors identified six European projects of which four were demonstration plants and two were commercial plants. The status on the four demonstration projects were: one with a firm construction schedule, two in the planning stage, and one in the proposed category. The two commercial plants fall into the proposed category. Details on these projects are provided in Table 6. One Japanese pilot plant project was identified. The plant is under construction and due to begin operating early in 1982. Details are given in Table 7.

The best known integrated coal gasification combined-cycle project is the Cool Water demonstration plant which will combine a Texaco coal gasification system with General Electric's Stag[R] combined-cycle power generating system to produce 100 MW of power. This project is scheduled to begin construction in July 1981 and to start up in late 1983. The latest developments on this project are discussed in the paper presented by Ahner in this session (3). The KILnGAS demonstration plant project initially will demonstrate the integration of a coal gasification plant with a steam turbine power generation system. Future plans call for the installation of a gas turbine as the low-Btu gasification system includes all the process steps required to fully demonstrate the production of gas turbine quality gas suitable for combined-cycle power systems.

Of the 13 commercial projects in planning or in the feasibility study phase, five will only produce power, four plan cogeneration, one involves trigeneration and three plan to integrate methanol production with electric power generation. Of the eight proposed commercial projects, three involve only power generation, two involve trigeneration, and three involve production of methanol as well as power. Summing these two categories together gives eight plants solely producing power, four with cogeneration, three with trigeneration and six with methanol production in addition to power. It is too early to tell whether or not these statistics represent any sort of preference trends by utilities and industrial companies. However, these statistics seem to clearly reflect the local and regional market needs along with the advantages of the flexibility offered by integrating coal gasification systems with combined-cycle power generation systems.

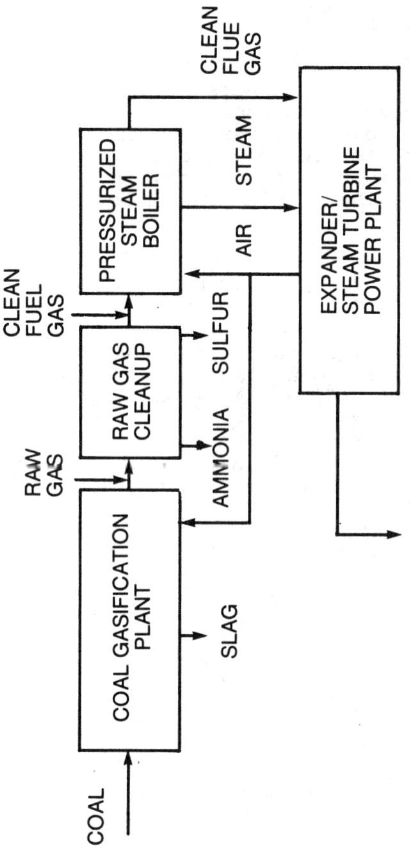

FIGURE 2. STEAG COAL GASIFICATION COMBINED CYCLE POWER PLANT

CONCLUDING REMARKS

A very rough approximation of the total generating capacity involved in the 21 commercial projects can be made by summing the projected coal use and converting this total to power generated. Using a ratio of 1000 tons of coal per day to produce 100 MW of power, the calculated generation capacity is approximately 9500 MW. This sum of 9500 MW compares to a total planned addition of coal-fired generating capacity by 1990 of 128 GW. (4) As a rough approximation then if all proposed projects were built, about 7.4% of the new capacity in 1990 would be from integrated coal gasification combined-cycle power generation units.

Events may overtake some of the projects identified in this survey and preclude their proceeding to commercialization. Nevertheless, it seems likely that in the next several years, as the design and economic data are developed for those projects which move forward to construction, there will be increasing interest in the commercialization of coal gasification combined-cycle power plants.

REFERENCES

1. "Coal Synfuel Facility Survey", National Coal Association, Washington, DC, August 1980.

2. Louks, B. M., "Methanol: An Opportunity for the Electric Utility Industry to Produce Its Own Clean Liquid Fuel", Electric Power Research Institute, Palo Alto, CA, ET-8 Session Paper, March 1981.

3. Ahner, D. J., "Cool Water Integrated Coal Gasification Combined-Cycle Plant", General Electric Company, Schenectady, NY, ET-8 Session Paper, March 1981.

4. Unpublished National Coal Association Data.

COOL WATER INTEGRATED COAL GASIFICATION COMBINED CYCLE POWER PLANT

D.J. AHNER, A.S. PATEL, R.C. SHELDON
GENERAL ELECTRIC COMPANY
COOL WATER PROJECT

INTRODUCTION

The Cool Water, Coal Gasification Combined Cycle (CGCC) Project will demonstrate an advanced power generation concept integrating the Texaco Coal Gasification Process (TCGP) with a General Electric combined cycle power plant. The project is currently sponsored by the following participants:

Southern California Edison

Texaco

Electric Power Research Institute

General Electric Company

Bechtel Corporation

The plant will be built at the Cool Water site of the Southern California Edison system near Daggett, California. Initial operation is planned for late 1983.

The state-of-the-art technology, low environmental impact and high efficiency potential of this new power generation approach, make the demonstration of the CGCC timely and essential for the advancement of electric power from fossil energy. This integrated coal gasification/combined cycle concept has potential for commercial application by the late 1980's with current state-of-the-art technology. Subsequent advancements in gas turbine operating conditions and gasification technology offer further economic, performance and environmental advantages to power generation.

Achieving specific technical and economic objectives is essential for the successful commercial application of this power system. These include:

a. The full scale design and reliable operation of those major equipment components and subsystems which will be incorporated into the commercial plant design.

b. Achieving a performance level which can be reliably and convincingly extrapolated to competitive levels through proven and/or low technical risk system modifications in the commercial plant design.

c. Demonstrating flexible plant operation and control throughout all operating modes consistent with power grid maneuvering requirements and plant operator capabilities.

d. Demonstrating low environmental impact.

e. Developing economic and operating experience data bases which can be applied to commercial plant capital and O&M cost projections and plant operations training and procedures.

This paper will describe the design features being incorporated into the Cool Water demonstration plant configuration. Design requirements, anticipated performance, plant integration characteristics, operational requirements and anticipated environmental performance will be discussed.

PLANT CONFIGURATION

Figure 1 indicates the arrangement of the major plant components. A ground coal/water slurry and oxygen are fed to the gasifier reactor where gasification occurs at a pressure of 600 psia and at temperatures in the range of 2000 to 2800°F. An intermediate BTU gas is produced having a heating value of approximately 260 to 280 BTU/SCF (HHV dry). After cooling, the gas is scrubbed to remove particulate and sulfur compounds to make it suitable for direct use in a gas turbine combustor.

All of the fuel plant output is consumed in the gas turbine combustors. The hot turbine exhaust gases are used to superheat all high pressure steam and economize the major portion of the steam system feedwater. The superheated steam at 925°F, 1350 psia generates power in a conventional steam turbine nominally rated at 56 MW.

Oxygen is supplied at pressure to the gasifier from a nominal 1000 T/day air separation plant incorporating motor driven air and oxygen compressors.

Sulfur compounds are removed from the gas by a low temperature full scale Selexol® unit with Claus® sulfur conversion and tail gas treatment subsystems.

A further description of the plant subsystems and program status and organization is discussed in Reference 1.

DESIGN REQUIREMENTS

Because of the significant sensible heat in the made gas (25 to 30% of the coal input energy) it is essential that this energy be recovered and efficiently converted to power. In addition, due to the significant amounts of unreacted steam in the raw gas, efficient low temperature heat utilization is an important requirement.

The high temperature reducing atmosphere of the raw gas with sulfur gas constituents introduces material considerations in the raw fuel gas steam heat recovery boilers (synthesis gas coolers, SGC) requiring that maximum metal temperature limits be observed. Maintain-

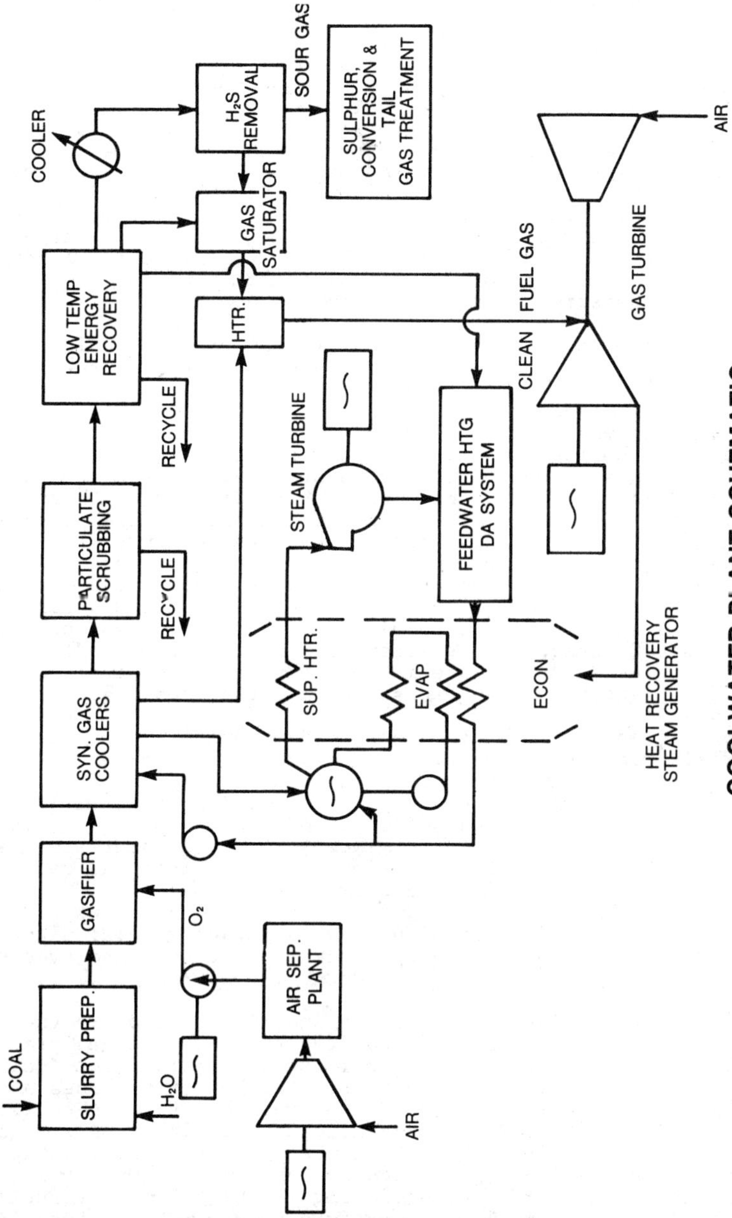

COOLWATER PLANT SCHEMATIC
FIG. 1

ing saturated steam conditions in these components provides an effective design which can be practically integrated with the gas turbine heat recovery steam generator (HRSG) producing superheated steam at efficient steam conditions.

The relative availability of high and low temperature heat in the gasifier and gas turbine exhaust systems requires a proper selection of the steam equipment configuration and steam conditions in order to efficiently match the individual heating duties with the capabilities of the respective heat sources. Studies indicate a performance benefit for operation at higher steam pressure levels, to obtain greater steam production, within the constraints of the HRSG superheating capability. Such considerations specify the HRSG & SGC superheater and evaporator surfaces. The division of economizing loads between the two heat sources is designed to take thermal advantage of the relatively large quantities of lower temperature heat in the HRSG exhaust, while the SGC surface satisfies minimum fuel gas temperature requirements of downstream gas cleanup equipment.

Variations in gasifier operation (e.g. temperature, cold gas efficiency, etc.) resulting from utilizing different coal feedstocks or changes in slurry conditions can have a significant influence in the design in order to assure reliable operation of the steam system under these conditions. Variations in the proportion of chemical and sensible heat in the fuel gas, changes the relative steam loads in the SGC and HRSG resulting in changes in steam turbine temperatures and flow. Reliable steam turbine operation requires that throttle steam conditions (e.g. temperature, pressure and flow) are properly coordinated such that last stage moisture and temperature limits are satisfied at all sustained operating points. Such considerations, for example, result in the requirement for variable steam pressure operation at reduced plant loads. Significant imbalance of SGC steam production and HRSG superheat or economizing capability at any operating point can result in unacceptable steam turbine throttle conditions, economizer steaming, throttle steam over-temperature or inefficient quenching of SGC steam through excess feedwater subcooling.

Such conditions must be accommodated by providing sufficient equipment design margin and proper control coordination for the anticipated range of operating conditions to insure acceptable and reliable operation. This results in equipment selection which will trend in the direction of reliable operation at off-design point conditions, at some compromise in optimum performance. The Cool Water steam system design has incorporated features into the equipment and control to demonstrate this flexibility and coordination.

There are similar considerations for larger commercial plant steam system designs although the operating conditions, and even configuration (e.g. reheat), may be considerably different.

A significant portion of the latent and low temperature energy in the raw gas can be efficiently recovered through fuel gas preconditioning prior to combustion. Moisturizing and heating the fuel gas through a low temperature direct contact water heater provides a means of transferring this energy at high availability in the basic cycle. This moisture contributes significantly to lowering the formation of NOx in the turbine exhaust. Current gas turbine combustion designs incorporate moisture injection to achieve low NOx levels, as required at the Cool Water site (~ 32 ppmv in the exhaust). Fuel gas moisturizing through use of low temperature energy versus steam injection from turbine extraction, results in significant cycle performance improvement. The intermediate BTU value of the gas shown in Table 1 and the added fuel gas moisture at the gas turbine combustors results in a significant increase in gas turbine output and additional fuel flow requirements to bring the moisture to turbine inlet conditions. Proper coordination and control of this moisturizing and heating is important in meeting acceptable gas turbine combustion system fuel moisture superheat and flow control requirements and in preventing interactions in the plant load control loops.

TABLE 1

COOL WATER COAL GASIFICATION PLANT

TYPICAL GAS COMPOSITIONS

Component	Vol. Percent (Dry) Raw Gas	Clean Gas
H_2	34.48	35.67
CO	43.31	44.79
CO_2	19.83	17.26
CH_4	0.05	0.05
N_2+Ar	2.16	2.23
H_2S	0.16	0. ppmv
COS	0.01	60 ppmv
TOTAL	100.00	100.00

The use of intermediate BTU fuel gas in the gas turbine has several design implications. This fuel will require modification of the turbine fuel nozzles, fuel gas piping arrangement and fuel gas control valve design. However, due to the combustion characteristics of the gas and the relatively small variation of composition over the expected operating conditions and fuel types, only minor modification to the on-base combustion liners are anticipated. A large diameter fuel gas piping harness arrangement with an off-base fuel valve skid is being designed for the Cool Water unit.

The relatively high fuel hydrogen content results in adiabatic flame zone temperatures somewhat higher (e.g. $\sim 50^{\circ}F$) than those resulting from the use of natural gas or distillate oil. This necessitates somewhat more moisture at a given load to achieve a given NOx level over conventional fuels. Combustion liner air cooling pattern modifications are being made to maintain long combustor life and proper hot gas temperature profiles.

The gas turbine combustor fuel and moisture flow is five times greater as compared to the use of natural gas or oil. Due to this increased turbine mass flow, gas turbine pressure ratios and turbine/compressor thrust mismatches will be somewhat higher, resulting in maximum gas turbine load limit considerations. Proper matching of the fuel plant capability within these constraints, which occur at lower ambient conditions, is an additional system design parameter. Maximum load limit conditions are not anticipated to be a significant factor over the load range at Cool Water.

The gas turbine will have dual gas/oil fuel capability, however; operation on oil fuel is only planned for unit startup and shutdown. However, for commercial plant application this dual fuel capability is of value in maintaining plant load carrying capability, particularly through short term outage of fuel plant equipment.

Sufficient margin is being incorporated in the Cool Water gas turbine HRSG and steam turbine designs to accommodate increased loading anticipated from potential process improvements and increased fuel processing capability with the Cool Water plant design.

Plant operation at high capacity factors is anticipated at Cool Water, with a target average of 77% over a six and one half year Project operating period. To achieve this goal, equipment modifications, redundancies and operating and maintenance procedures are being implemented at Cool Water. To demonstrate combined cycle running reliability consistent with project availability goals and commercial power generation base load requirements, a combined cycle 1500 Mean Time Between Failures (MTBF) goal with high turbine starting reliabilities and a preventive maintenance program is being implemented to aid in achieving project availability goals.

PLANT PERFORMANCE

Planned plant operation on a variety of coal feedstocks in addition to extended runs under various process conditions (e.g. slurry ratios, preheat, etc.) will result in a range of fuel gas output up to the gas turbine maximum of approximately 840 MM BTU/HR at the site conditions of 80°F and 2000' altitude. The gasifier temperature and cold gas efficiency corresponding to an individual operating condition changes the relationship of sensible to chemical energy, causing some variation in gas and steam turbine outputs, coal and oxygen inputs and, thus, overall plant performance. Table 2 indicates the range of performance anticipated under different fuel plant operating conditions at Cool Water. In addition, the performance level estimated for future commercial plant designs at higher gas turbine operating temperature and advanced gasification technology is also shown.

TABLE 2

GASIFICATION COMBINED CYCLE

PERFORMANCE ESTIMATES*

		Net Output (MW)	Net Heat Rate (BTU/KwHr)
Cool Water Plant			
Initial Operation		98.0 - 95.2	10,600 - 10,910
Projected During Plant Life		104.8 - 106.9	10,320 - 10,530
Large Scale Reference Plant Designs**			
Gas Turbine Inlet Temps.	1985°F	~500 - 1,000	9,200 - 9,500
	2100°F	~500 - 1,000	8,650 - 8,850
	2600°F	~500 - 1,000	8,250 - 8,450

* Coal pile to bus bar, including all auxiliary support requirements (O_2, fuel plant, etc.)

** Reference 3

The Cool Water plant configuration has not been optimized to achieve a maximum performance level. The relatively small size of the system, in addition to a finite program operating period, confines the selection of equipment and operating conditions, resulting in compromises in performance. The steam system configuration has been specified at lower level, non-reheat conditions. The modifications for more advanced reheat steam systems mainly effect the gas turbine heat recovery steam generator and steam turbine designs. Operation requirements with reheat steam turbines are known and have relatively little impact on fuel plant equipment, configuration or operating conditions and are, thus, not considered essential for demonstration of the concept.

In order to achieve operating flexibility, some fuel plant equipment is being included with additional design margins for demonstration purposes. Subsequent refinement of this system equipment will result in improved cost/performance benefit in follow-on designs.

The oxygen supply at Cool Water will be provided by an on-site air separation plant producing gaseous products for commercial sale. This requirement results in gaseous purity requirements beyond that necessary for gasification needs. Studies have indicated performance incentives and TCGP process capabilities to utilize oxygen of lower purity. There are additional performance benefits in thermal integration with the air separation plant. These features will not exist at Cool Water but can be readily implemented in subsequent commercial applications further improving performance and costs.

Large scale reference plant configurations incorporating the Texaco oxygen blown coal gas process have been studied and reported. The performance of these plants is based on projections of fuel plant performance as given in Reference 2 on Ill #6 coal which will be a test coal in the Cool Water Program. The results of the study conducted by General Electric (Ref. 3) is indicated as the Large Reference Plant Design in Table 2. This reference plant incorporates current state-of-the-art gas turbines with a 1450 psig 935/935°F reheat system and the basic Texaco coal gasification fuel system features to be demonstrated at Cool Water.

PLANT OPERATION AND CONTROL

An important aspect of the project is to develop procedures for, and demonstrate plant operation including startup, shutdown, power maneuvering and reliable functioning under contingency conditions. Plant operation in several configuration modes will be possible including:

a. Gasification system with the combined cycle;

b. Gasification system with the gas turbine in simple cycle; and

c. Stand-alone fuel plant supplying an on-site modified natural gas fired boiler.

An integrated plant control, including an equipment protection system, is an integral part of the station control equipment being designed for the project.

The Cool Water station control and operations concept is indicated in Figure 2. The two basic functions in this approach are:

1. Integrated Plant Control

2. Data Acquisition System

INTEGRATED PLANT CONTROL SYSTEM

The basic concept of the integrated plant control is to coordinate the hierarchy of the individual subloop controls accomplishing the necessary communication between equipment components to insure responsive and reliable overall plant operation.

The control and the plant data acquisition system are being designed based on the following considerations:

a. Minimize development risks;

b. Provide high control system running reliability and availability; and

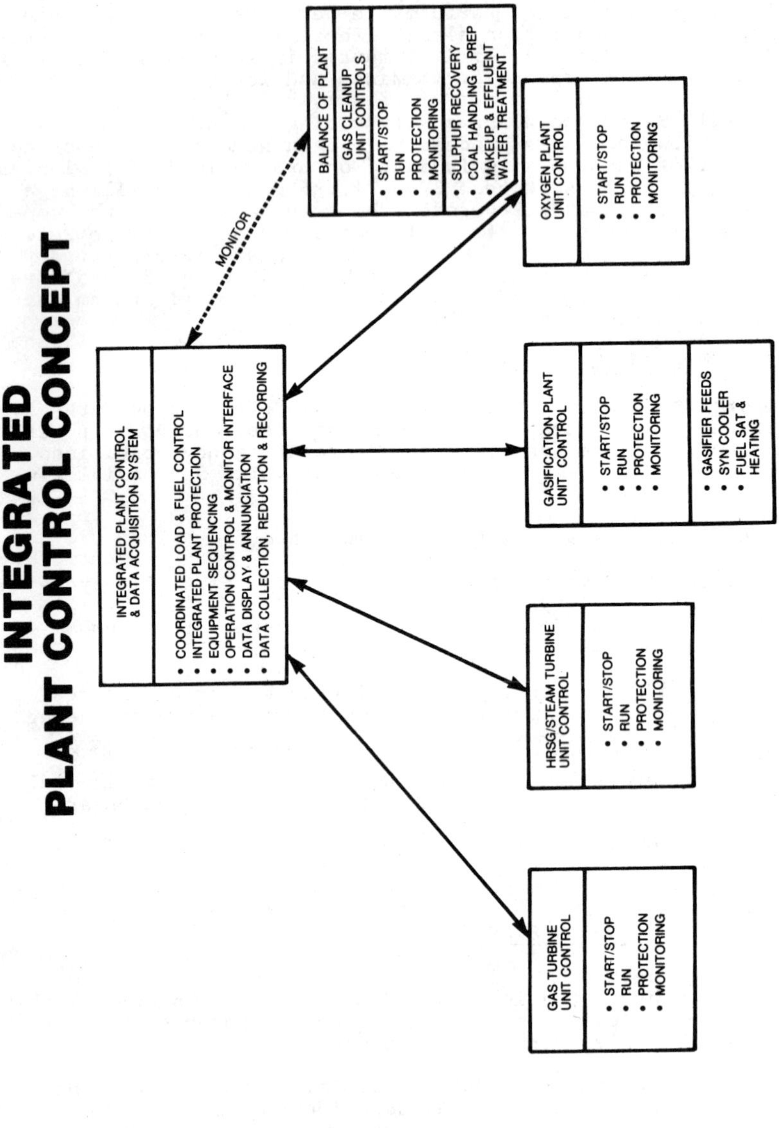
INTEGRATED
PLANT CONTROL CONCEPT
INTEGRATED PLANT CONTROL & DATA ACQUISITION SYSTEM
• COORDINATED LOAD & FUEL CONTROL
• INTEGRATED PLANT PROTECTION
• EQUIPMENT SEQUENCING
• OPERATION CONTROL & MONITOR INTERFACE
• DATA DISPLAY & ANNUNCIATION
• DATA COLLECTION, REDUCTION & RECORDING
MONITOR
BALANCE OF PLANT
GAS CLEANUP UNIT CONTROLS
• START/STOP
• RUN
• PROTECTION
• MONITORING
• SULPHUR RECOVERY
• COAL HANDLING & PREP
• MAKEUP & EFFLUENT WATER TREATMENT
OXYGEN PLANT UNIT CONTROL
• START/STOP
• RUN
• PROTECTION
• MONITORING
GASIFICATION PLANT UNIT CONTROL
• START/STOP
• RUN
• PROTECTION
• MONITORING
• GASIFIER FEEDS
• SYN COOLER
• FUEL SAT & HEATING
HRSG/STEAM TURBINE UNIT CONTROL
• START/STOP
• RUN
• PROTECTION
• MONITORING
GAS TURBINE UNIT CONTROL
• START/STOP
• RUN
• PROTECTION
• MONITORING

FIGURE 2

c. Provide a simple and flexible system with adequate operator visual displays.

To eliminate potential problems with control system redesigns, maximum use will be made of existing system equipment controls. This includes the combined cycle power plant equipment and other existing plant subsystem controls. The power plant controls will be a combination of electronic analog and microprocessors with a control room arranged tentatively as in Figure 3.

A commercially available distributed system will be utilized in the control of many components in the fuel plant. A distributed control system utilizing microprocessors affords greater flexibility for control loop configuration changes and modifications than is available with a conventional analog system. This is a highly desirable feature for a new process system design which has limited operating experience.

The system control is accomplished by several microprocessor units located throughout the plant close to the equipment location and linked to a central operator's console, having CRT and keyboard, by a communications line (data highway) consisting of a multiplexed digital information transmission system. Dual, redundant data highways and several CRT keyboards will be used for system security and reliability.

The integrated station control system will perform the functions of the overall plant operation including loading/unloading and operating mode changes. It will interface with, and coordinate, the operation and the control of the fuel gas plant and the combined cycle plant through their respective subsystem controllers. The integrated station control system will perform the function of closed loop, automatic control of the plant megawatt output and the fuel gas system pressure. A flexible automatic coordinated load/pressure control system will be implemented to enable plant operation with varying degrees of gas turbine lead and turbine follow. This logic will be used during plant testing to establish the plant load following capabilities and characteristics. Adequate operator control functions and information with video display via a keyboard will be provided to aid the operator in effectively performing his duties. After startup, the operator will be able to keep the entire plant under surveillance and make on-line control adjustments as necessary from a single point location at the integrated station control console.

An integrated protection system will be designed to recognize the tripping of major equipment and to take appropriate and timely action. It will issue coordinated signals, directed at specific equipment to keep the plant equipment within operating limits to allow rapid restoration of operations once the fault condition has been cleared. These signals will, as appropriate, initiate startup of standby equipment, effect transfer of operating modes, change controller setpoints and initiate equipment shutdown.

A goal of the control and protection system will be to attain a high running reliability. This will require isolation of control and hardwired protection circuits with sufficient instrumentation redundancy to insure high reliability and equipment protection.

DATA ACQUISITION SYSTEM

A computerized plant data acquisition system will be implemented for data collection, display and reporting. The computer is not utilized in the control of the plant, but is utilized for such functions as operator startup/shutdown assist and display of major plant process loop conditions. The computer will have capacity for storage and

COOL WATER STATION CONTROL ROOM ARRANGEMENT

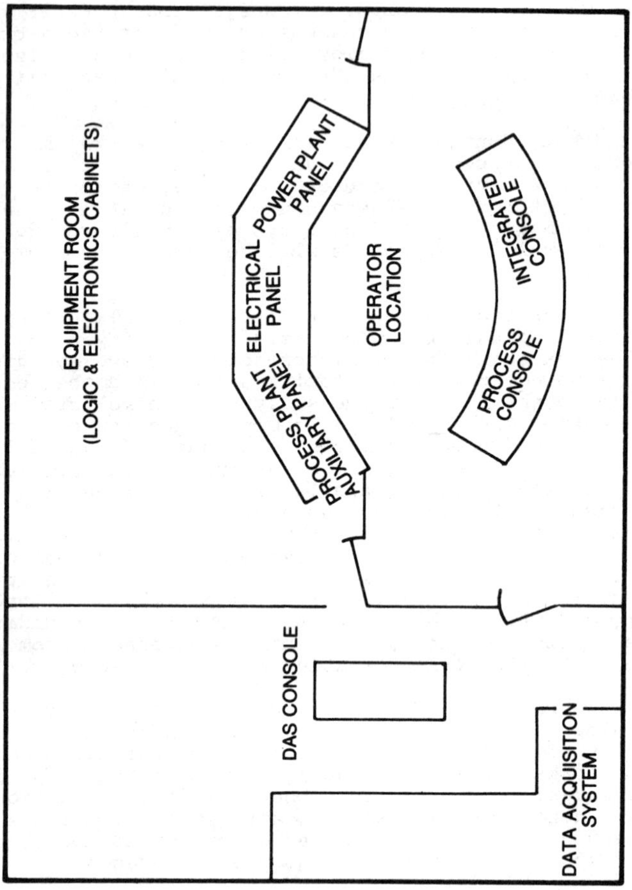

FIGURE 3

retrieval of large volumes of data. The data acquisition system will extend and improve the operator/plant interface. However, the plant operation and control will be independent of the computer.

The data acquisition system will provide support for:

a. Engineering analysis of plant operation and performance by way of several logs (periodic, trip, demand, etc.);

b. Plant test implementation and performance (steady state and dynamic) evaluation; and

c. Operator guidance through color video display.

Plant maneuvering and load following capability will be tested as part of a comprehensive overall Project Test Program. Each major component's response characteristics will be checked by subjecting it to controlled changes. All pertinent variables will be recorded in the plant data acquisition system to permit subsequent analysis and evaluation.

CONTROL ANALYSIS

To aid in the control system development and to more fully understand the plant process variable interactions, a dynamic simulation study of the Cool Water configuration is being conducted as a joint technical effort among Program participants. This simulation incorporates dynamic digital mathematical models of all major system components programmed in a flexible software system that provides expedient modification and modular model development, while selected subsystem studies proceed in parallel.

Initial results have been encouraging, indicating no major control problems or instability characteristics which cannot be solved by state-of-the-art control logic and hardware. The relatively large volumes of fuel gas in the fuel system heat exchangers, scrubbers and piping at intermediate BTU heating values, and at high pressure relative to the gas turbine combustion, give an inherent system storage capability which results in small fuel system pressure variations over a wide range of plant maneuvering. Proper implementation of coordinated fuel pressure/gas turbine control appears to result in a responsive power plant with flow and pressure excursions well within design limits.

ENVIRONMENTAL REQUIREMENTS AND PERFORMANCE

Extensive testing of the TCGP process has indicated that the various effluent gas, liquid and solid waste streams have a low environmental impact, requiring a minimum of subsequent treatment before disposal. The raw gas contains no detectable heavy hydrocarbons and there are very low ammonia levels in the cleaned gas.

The Cool Water plant site emission requirements, design capability emissions and current Federal New Source Performance Standards for the program coal are shown in Table 3. The Cool Water site NOx requirement of 140 #/HR is the lowest level in the U.S.

The GCC plant offers unique opportunities for NOx attenuation since large quantities of inert combustion moderator are available either as fuel water vapor, where water availability is of concern, or nitrogen from the air separation plant. Use of low temperature process heat for introducing fuel gas moisture at Cool Water greatly aids in satisfying the NOx requirement while improving overall plant efficiency and increasing specific power output. Further optimization of this system offers potential for reduction of NOx below the

Cool Water site requirement. Turbine extraction steam for NOx control will be available at Cool Water as supplement or backup in the event of any malfunctions in the fuel moisturizing or heating system components.

TABLE 3

COOL WATER ENVIRONMENTAL IMPACT

(#/10^6 BTU of Coal Feed)

Power Plant Emissions	Cool Water Design Capability	Site Limits	Conventional Steam - NSPS
SO_2	0.035	0.2	0.35
NOx	<0.14	0.14	0.6
Particulate	<0.01	0.01	0.03

Based on 0.7% S Coal

The plant sulfur system is designed for 97% sulfur removal on a continuous basis. The sulfur produced is of commercial grade with prospective use in various industrial applications. The order of magnitude reduction of SO_2 level below current standards is expected to be of particular advantage for application in very low level or non-attainment areas.

The particulate emission levels at Cool Water are projected to be very low. The fuel gas particulate is anticipated to be on the order of 3 to 5#/HR. Additional contribution to stack particulate arises from the combustion of sulfur products and depends on the measurement techniques to be employed in establishing SO_2 levels. However, the plant is expected to operate below the 10#/HR site limit.

The vitrified ash contains virtually no leachable constituents and will be temporarily stored at the site.

Water from the particulate scrubbing sections is clarified and reused for process and in the slurry preparation section. Water vapor condensing from the cooled raw gas is also recycled in the process. Waste water from bleed streams is degassed in a closed system (gases are routed to the sulfur recovery plant) and sent to a lined evaporation pond at the site. The use of low quality recycle water for slurry preparation and the subsequent decomposition of a major fraction of this water in the gasifier, affords an opportunity for the reduction of effluent waste water streams on larger plants where a higher degree of water treatment is justified. Chlorides, alkali and fine particulates in this waste water bleed stream can be further concentrated and low quality water recycled for use in the slurry preparation.

SUMMARY

The Cool Water Project is a 100 MW demonstration of an advanced power generation concept incorporating the integration of the Texaco Coal Gasification Process with a modified General Electric combined cycle power plant. Many of the design configurations and equipment features which will be incorporated in subsequent larger scale plants are being implemented in the Cool Water design in both the fuel system and combined cycle and will be demonstrated there.

The demonstration of reliable operation on a variety of feedstocks and in various modes of integrated combined cycle and standalone fuel plant operation represents a significant challenge in the design of a versatile and flexible control system. Extensive use of a microprocessor based distributed system control is being made at Cool Water to provide the required control, system monitoring and operations assist capabilities. The load following capability of this plant will be demonstrated in addition to the degree of coordinated control required to achieve acceptable response.

The Cool Water configuration has not been optimized for performance; however, average heat rates on the order of 10,300 BTU/KwHr are anticipated over the major life cycle of the plant with subsequent large scale unit designs indicating levels of 9,200 to 9,500 BTU/KwHr with current state-of-the-art gas turbines and 8,250 to 8,450 with advanced turbines.

The potential of this coal to power concept in reducing the environmental power generation impact with coal fuel is a major feature to be demonstrated at Cool Water with emissions reduced significantly from that currently existing with conventional coal fired power plants.

The Cool Water demonstration represents a significant step forward in bringing a new high potential clean electric power from coal concept to commercial reality.

REFERENCES

1. "The Cool Water Coal Gasification Program, A Demonstration of Gasification-Combined Cycle Technology", F. Brian Walter, H.C. Kaufman, T.L. Reed, Presented at Conference on Synthetic Fuels, 10/13-16/80.

2. "Economic Studies of Coal Gasification Combined Cycle Systems for Electric Power Generation", Fluor Engr. Constructors, EPRI AF-642, January 1978.

3. "Texaco-Based Gasification Combined Cycle System Performance Studies", General Electric Co., EPRI AP-1429, June 1980.

IGCC EXPERIMENTAL SIMULATION

James C. Corman
General Electric Company
Corporate Research and Development
Schenectady, NY

ABSTRACT

The Integrated Coal Gasification Gas Turbine Combined Cycle (IGCC) system has been identified as a viable coal conversion concept for utility application. Several projects that will demonstrate this technology on a commercial scale are moving toward the construction stages. An IGCC simulation facility has been designed and constructed that handles 24 tons of coal per day. This facility serves a dual function: (1) obtaining technical information on the performance of the individual components – coal gasifier, gas cleanup, and turbine simulator – that comprise the IGCC concept and (2) simulating the total IGCC system in an operational control mode that permits the evaluation of system response to imposed load variations characteristic of utility operation. The technology data base developed in this experimental facility demonstrates that an IGCC can be designed so that the fuel plant and power generation system operate in a compatible manner and the overall system meets the constraints imposed by environmental considerations.

INTRODUCTION

The utility industry is under increasing pressure to convert a significant fraction of new generation additions to coal, and to do so while meeting environmental requirements. The gasification/gas cleanup fuel plant coupled with a gas turbine/steam turbine combined cycle power generation subsystem has been identified as an economically and environmentally acceptable coal conversion system (References 1 and 2).

In order to increase power plant efficiency and reduce the projected cost of electricity, it is necessary to interactively couple the fuel plant and generation subsystems. In this configuration the integrated coal conversion system must operate under utility power generation system constraints. In order to match varying load conditions and contribute to grid network stability, the conversion system must demonstrate excellent dynamic control and response.

A second consideration for any advanced coal utilization system is that it meet current environmental regulations and have the potential for meeting future, more stringent constraints. The IGCC system is unique in that the cleanup system can be designed to control both gaseous and liquid effluents.

The Integrated Gasification Combined Cycle (IGCC) has an excellent possibility of meeting the needs of the utility industry and gaining wide acceptance. Because of this likelihood, decisions have been made to proceed to commercial-scale plant designs (Reference 3) and to construct a demonstration plant (Reference 4).

The ability of an IGCC to match the environmental, power generation, and control requirements for use in utility service will be conclusively shown with the operation of the first demonstration plant. A Process Evaluation Facility (PEF) simulating the critical features of an IGCC plant has already been constructed and experimental results developed. The data are verifying the ability of IGCC systems to satisfy the identified requirements.

IGCC SYSTEM CHARACTERISTICS

The IGCC concept includes two major subsystems: fuel processing, comprising the gasifier and cleanup components, and power generation, comprising the gas turbine/steam turbine combined cycle. The fuel processing subsystem gasifies coal in order to transform it into an energy form in which potential contaminants to the environment and to the power generation equipment can be removed effectively. The gaseous product of the coal-oxidant-steam reaction can be "scrubbed" to remove those contaminants prior to combustion.

All coal conversion processes include some steps that cause a loss of thermodynamic availability. As a result, the cleaned fuel has a lower total energy content than the coal supplied. The improved efficiency of the gas turbine/steam turbine combined cycle used for power generation compensates for the inefficiencies of fuel conversion to provide very attractive overall system performance.

A simplified schematic diagram of an open cycle gas turbine/steam turbine combined cycle is shown in Figure 1. In this concept a number of open cycle gas turbines are operated in parallel. The thermal energy in the exhaust from the turbines is utilized in a heat recovery steam generator (HRSG), and the steam is manifolded into a central steam turbine. With currently available equipment, each gas turbine would

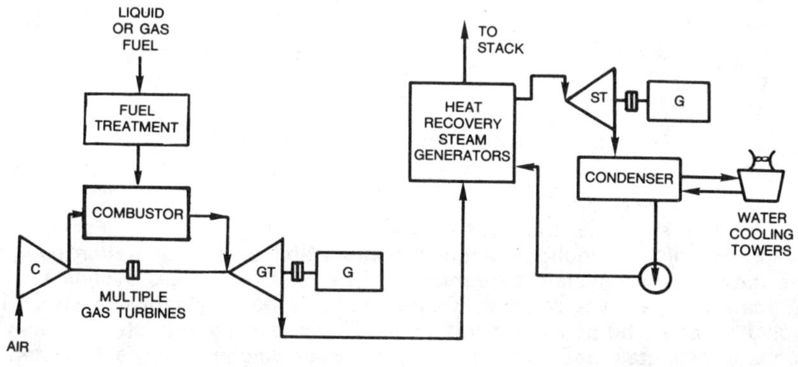

Figure 1. Gas Turbine/Steam Turbine Combined Cycle

produce approximately 70 megawatts, with the steam bottoming plant producing approximately one-third of the total plant output. Several commercial installations of this type are in operation.

One advantage of this cycle is that it has a very attractive conversion efficiency from fuel to electricity. Commercial combined cycle installations operate with power plant efficiencies as high as 42%. The current requirement of the gas turbine for a clean petroleum liquid or natural gas fuel represents an operational drawback since these fuels are not likely to be available for large-scale electric generation in the future. A second drawback is that the efficiency of the overall conversion process is directly dependent on the turbine inlet temperature, which is, in turn, limited by the materials and cooling schemes currently used for the hot gas path of the expansion turbine.

Gasification and gas cleanup fuel plants have the potential to overcome the first drawback by providing a coal-based fuel supply. The second drawback can be overcome by a new cooling concept for the gas turbines (i.e., water cooling) permitting significantly higher inlet temperatures which, in turn, will result in improved performance (References 5 and 6). The increase in gas turbine firing temperature through improved hot gas path cooling techniques will allow a continuing evolution of improved IGCC performance.

A schematic diagram of an integrated gasification combined cycle system is shown in Figure 2. The basic power generation components — open cycle gas turbine, heat recovery steam generator, and bottoming steam turbine — are similar to present-day commercially available equipment. However, a fuel processing subsystem is introduced to take coal as delivered to the plant site and produce a fuel which is acceptable to the gas turbine and also produce combustion by-products which are acceptable to the environment.

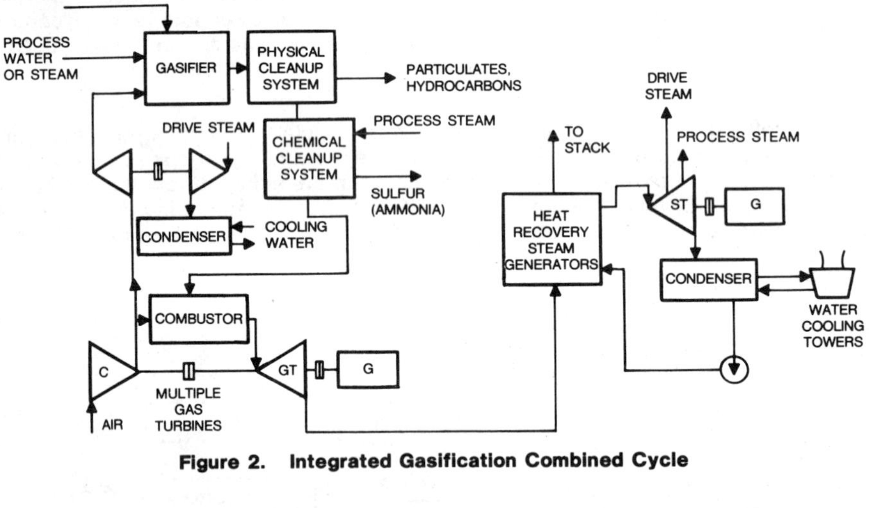

Figure 2. Integrated Gasification Combined Cycle

The oxidant supply to the gasifier can be either air or oxygen. For either oxidant, integration of the functions of the fuel plant and power generation equipment improves the projected system performance. The air-blown IGCC system shown in Figure 2 requires the more detailed integration since both oxidant and steam flows are coupled between the power generation subsystem and the fuel plant. In this configuration, approximately one-sixth of the compressor discharge air is extracted from the gas turbine and delivered to the gasifier through a steam-driven boost compressor. Coal, process steam, and compressed air are reacted in the gasifier to form a gaseous fuel. Since air is used as the gasification oxidant, the gaseous fuel has a heating value of approximately 150 Btu/scf (about one-sixth that of natural gas). When the fuel gas exits from the gasifier, it still contains the same contaminants as the parent coal, although most of the coal ash is removed in the gasification process. The gas then goes through two cleanup stages: the first, a physical cleanup stage in which the contaminants potentially harmful to the gas turbine are removed by liquid scrubbing; the second, a chemical cleanup stage in which the potential contaminants to the environment — sulfur and nitrogen compounds — are removed from the gas stream. At that point, a clean gaseous fuel is available for gas turbine combustion. The processes are integrated in that compressed air is extracted from the gas turbine compressor for use in the gasifier, and steam is extracted from the steam turbine to drive the boost compressor and for use as process steam in the chemical cleanup system.

INFLUENCE OF GASIFIER TYPE

A number of studies have been performed to compare the performance of alternative IGCC systems utilizing various gasifiers (References 7-9). These studies indicate that several of these systems are potentially competitive with conventional steam plants.

There are three basic types of coal gasifiers, although there can be many permutations and combinations of these. Each type is characterized by the way coal is brought into contact with the reactant gases (Figure 3). In a fixed bed the flows of coal and reactant gases are countercurrent. In an entrained bed they are cocurrent. The fluidized bed acts as a well-stirred reactor with the coal being added either from the top or with the reactant gases.

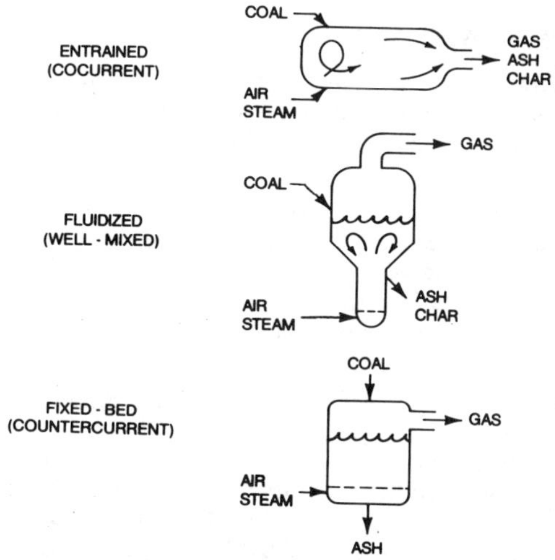

Figure 3. Basic Gasification Reactor Types

Each gasifier type has advantages and disadvantages in the fuel supply subsystem for steady-state operation of power generation systems. The fixed bed is most efficient in converting coal heating value into gas fuel value (i.e., making a rich gas), but it also produces liquid hydrocarbon as a by-product, which makes downstream gas cleanup more difficult. The entrained bed does not produce liquids as a by-product, but a larger fraction of the coal heating value is converted to thermal energy resulting in extremely high gas exit temperatures. The fluidized bed has uniform reactor temperature characteristics, but the coupling of reactant gas flow and bed dynamics places limits on the ability to reduce load.

These gasifiers have a few common characteristics when used in power generation system configurations. Each can be operated using air or oxygen and steam as reactants. In most arrangements, the gasifier pressure level is tied to that of the gas cleanup system so that process gas volume flow rates are low and contaminant concentrations are high to improve cleaning effectiveness and reduce vessel volumes.

The gasifier types also have unique characteristics that can have major effects on the way the system is controlled. The fixed bed and fluidized bed systems operate with high carbon inventories while the entrained bed has almost none. Therefore, very rapid load swings can be effected for short times in the fixed and fluidized bed systems by modulating gas flows only and effecting coal feedrate changes more slowly. In the entrained system, the reactant stoichiometry between coal, steam, and oxidant must be maintained precisely. This requirement puts a burden on control of the most difficult operation for any pressurized coal gasifier — coal feeding.

The gasifier type also influences the manner in which the plant is to be integrated. In a fixed bed, little or no steam is generated in the fuel plant because of the low gas outlet temperatures (about 1000°F). At the other extreme, the single stage entrained flow reactor has a high gas outlet temperature (about 2400°F) and requires substantial steam generation. The fluidized bed system is between these two, with an exit temperature of about 1800°F. As a result, in most configurations the fixed bed gas plants are steam importers while fluidized bed and entrained bed plants export steam to the power conversion equipment. Because the steam systems exhibit longer response times to load changes than the gas turbines, control logic for the alternative types of plants will be very different.

CRITICAL FACTORS FOR IGCC OPERATION

The IGCC system must be able to satisfy the dynamic load requirements imposed by the utility grid and be compatible with environmental regulations. These considerations place critical constraints on the design of the components and the overall system.

The IGCC control system must permit the power plant to match transient load requirements and to maintain stable conditions in the fuel plants during accidents such as loss of load. The development of an acceptable control logic requires detailed knowledge of the steady state and dynamic performance of the individual components and the interactions of the fuel plant and the power generation subsystems.

Each component of an IGCC system must exhibit desirable operating features. The gasifier must be able to handle a variety of coals and respond to variations in operating conditions. The gas quality must remain acceptably steady during these variations. It is also essential that gas composition be characterized as a function of gasifier events (e.g., coal loading, bed movement, and ash rejection).

The gas cleanup component is critical since it is the interface between the gasifier and the gas turbine combustion system. The main requirement of the gas cleanup component is that it remove the sulfur and nitrogen compounds from the fuel gas prior to combustion. It must also remove particulates and alkali metals from the gas stream in order to prevent erosion and corrosion of the gas turbine hot gas path parts.

The fuel gas is utilized in a pressurized combustion system. This combustion process must remain efficient and stable through a wide range of loads and fuel gas compositions. The combustion system must also be designed to limit NO_x generation from both thermal and fuel-bound nitrogen sources.

IGCC SIMULATION FACILITY

A need for technical understanding of the component and system performance of an IGCC system has led to the establishment of a process evaluation facility for IGCC simulation. This PEF-scale facility has the capability to simulate the operational characteristics of a total fuel plant/power generation system.

In order to ensure that the PEF matched the characteristics of a commercial-scale system, its configuration was based on a full-scale IGCC system that was developed to a detailed design stage (References 10 and 11). This facility simulates all the critical component features of the proposed commercial system (Figure 4).

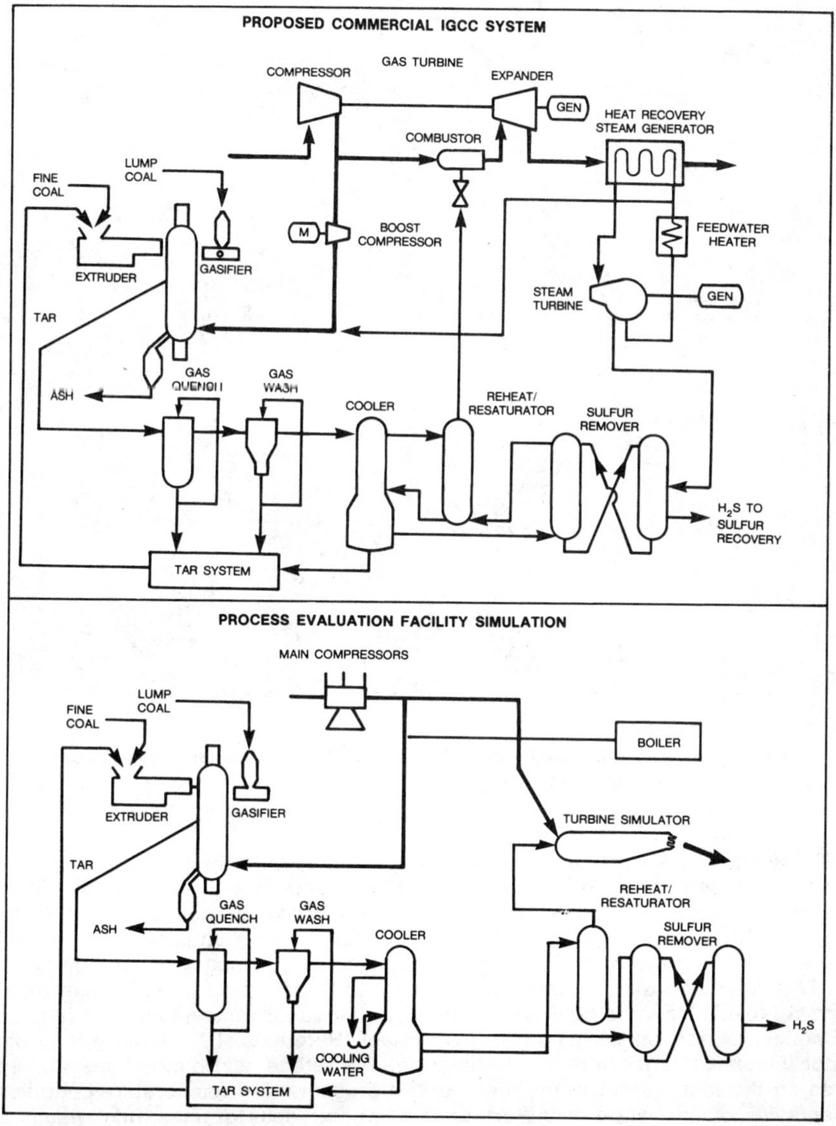

Figure 4. Schematic of Integrated Gasification Combined Cycle Processes

The PEF includes a fixed bed gasifier, a low-temperature gas cleanup system, and gas turbine simulators. Auxiliary facilities include a high-pressure air supply system and steam boilers as well as a computer data acquisition and control system that provides sophisticated data monitoring, analysis, and control functions. This total system is currently operational (Figure 5).

Figure 5. Integrated Gasification Combined Cycle – Process Evaluation Facility Simulation

The gasifier is a pressurized, advanced fixed bed, air-blown reactor (Figure 6) that is able to process a wide variety of coals at a coal feedrate of approximately one ton per hour (Reference 12). The low-temperature cleanup facility (Figure 7) can remove solid particulates, hydrocarbon mist, and sulfur compounds from the raw gas that exits from the gasifier, and can resaturate the cleaned gas with the condensed water. This final step also reintroduces light hydrocarbons and provides gas reheat by thermal recycle from the cool-down process. The sulfur removal process is based on a Benfield potassium carbonate scrub system (Reference 13). Each gas turbine simulator (Figure 8) consists of a combustor and first-stage nozzle airfoil cascade and operates on the total output of the fuel plant at pressure and temperature conditions representative of the state-of-the-art and advanced gas turbines that would be employed in IGCC application. In an actual IGCC system, the air and steam supplies

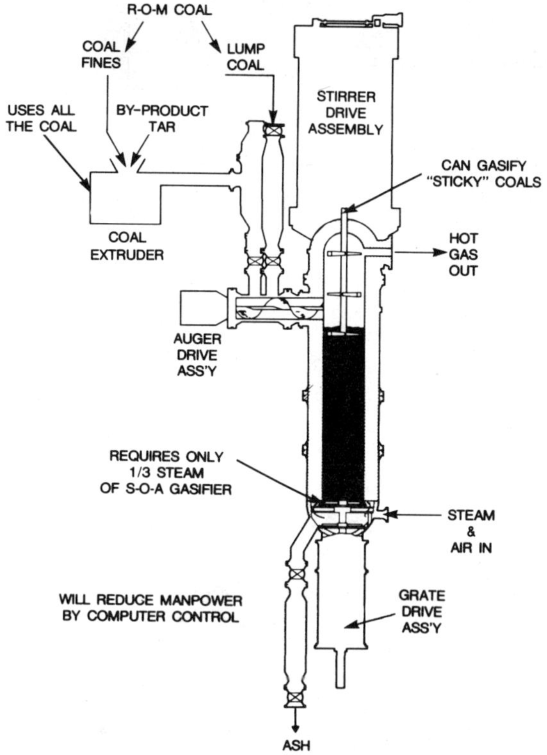

Figure 6. Advanced Fixed Bed Gasifier

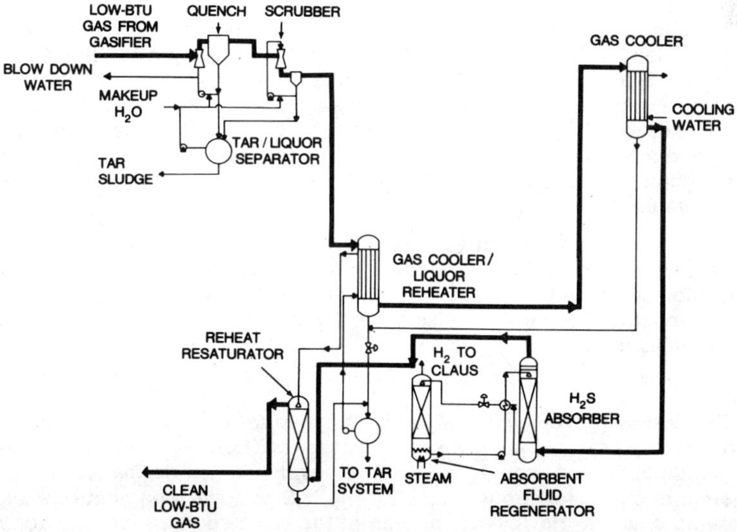

Figure 7. Coal Gas Cleanup System

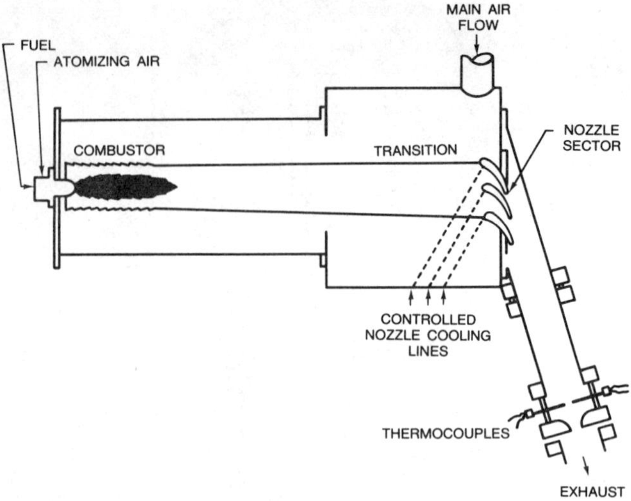

Figure 8. Turbine Simulator

to the fuel plant are extracted from the power generator equipment. In the experimental facility, the independent steam and air supplies are computer-controlled to simulate these extractions.

RESULTS FROM IGCC EXPERIMENTAL SIMULATION

The operation of the PEF IGCC simulation facility has resulted in the development of experimental information both on the individual components and on the system response to imposed transients.

Gasifier

The advanced fixed bed gasifier was designed to meet the operational requirements of IGCC operation. The stirrer, which traverses the entire reactor shaft, has permitted successful operation on two highly swelling coals, Pittsburgh No. 8 and Illinois No. 6. The combination of the top stirrer and the rotating grate has allowed the steam requirement for the dry ash fixed bed reactor to be significantly reduced (Figure 9). A major portion of the operation has been achieved at a steam/air ratio of 0.2. As a result, steam/water carryover to the cleanup system is reduced, simplifying the condensate handling system and allowing for the possibility of a zero water discharge IGCC plant design. Thus the reduced steam demand also increases the fuel gas higher heating value (Figure 10).

The use of bed agitation does however introduce variables into the fuel gas system which could affect IGCC operations. The heating value variation as a function of stirrer position is shown in Figure 11. Disruption in the stirrer action also influences gas composition (Figure 12). This off-design operation of the stirrer causes shifts in the CO/CO_2 ratio.

Gas Cleanup

The Benfield sulfur removal system is designed to remove H_2S through a chemical reaction with the potassium carbonate absorbent. The operating point is designed to be at a condition which would limit CO_2 removal. The performance of the sulfur removal system is characterized in Table 1. The H_2S removal was about 99% efficient in the absorbing column. However, a portion of the H_2S absorbed into the condensate steam was stripped back into the fuel gas during resaturation.

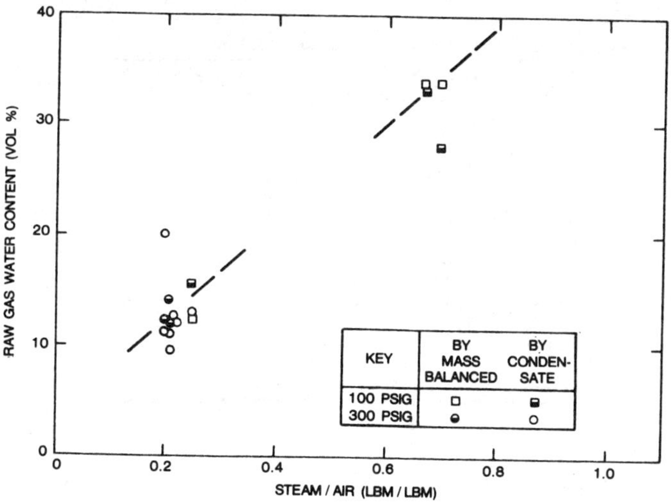

Figure 9. Water Content in Fuel Gas

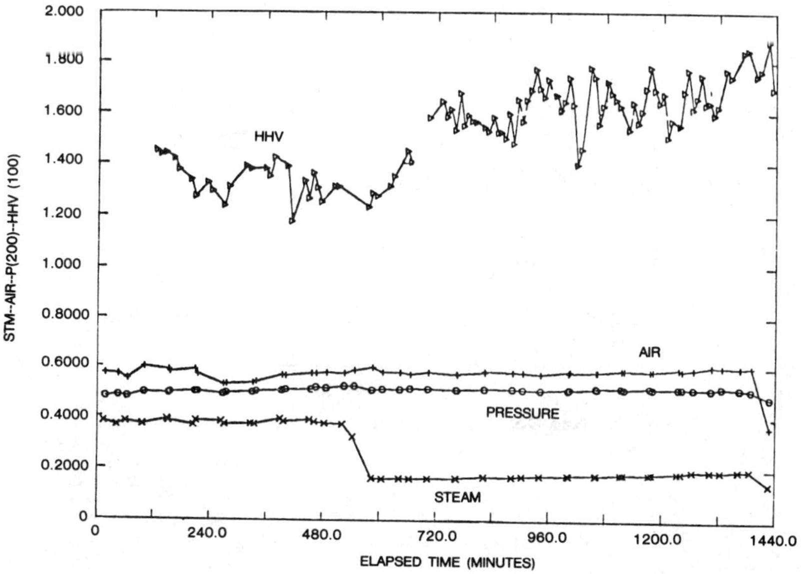

Figure 10. Gasifier Fuel Gas Heating Value Variation with Steam Supply

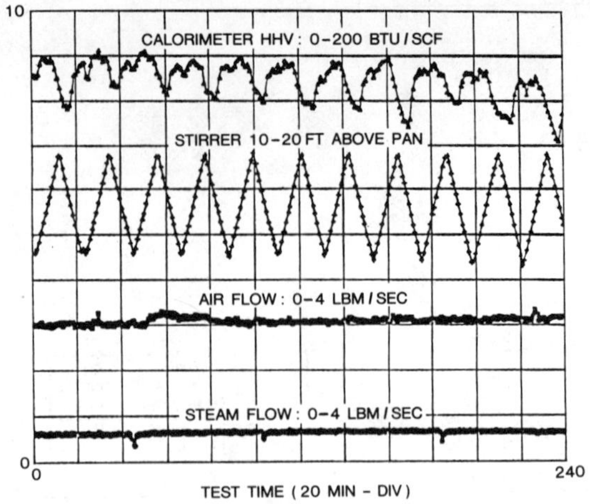

Figure 11. Gas Heating Value Variation with Stirrer Position

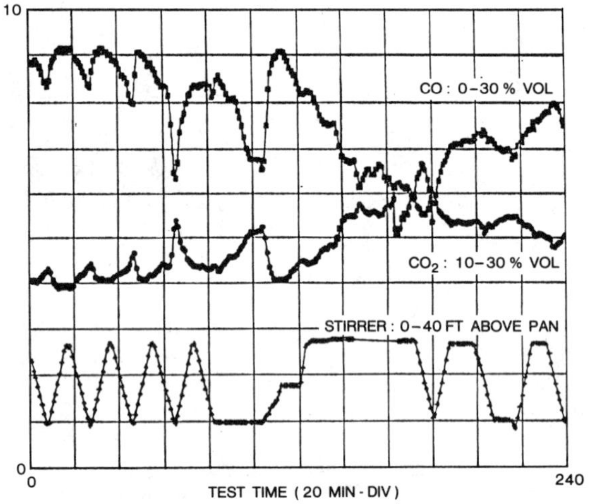

Figure 12. Gas Composition Variation with Stirrer Disruption

Table 1

GAS REMOVAL CHARACTERISTICS

	H_2S (ppm)	COS (ppm)	NH_3 (ppm)
Benfield Inlet	5500	475	1380
Washer Exit	63	360	20
Resaturator Exit	340	380	1285

COS was a major contributor to total sulfur content in the final gas stream. This result emphasizes the need to operate at a condition where COS hydrolysis could be achieved.

No specialized NH_3 removal component was incorporated into the cleanup system. However, approximately one half of the NH_3 in the gas stream exiting from the gasifier was removed with the condensate during the gas cooling process. The remaining NH_3 was removed in the Benfield column and was discharged with the acid gases. The NH_3 in the condensate stream was stripped back into the gas stream during resaturation.

The extent to which particles and alkali metals are removed by the cleanup component is shown in Figure 13. The loading was reduced from 570 ppm to 3.5 ppm; the alkali content from 320 ppb to 21 ppb.

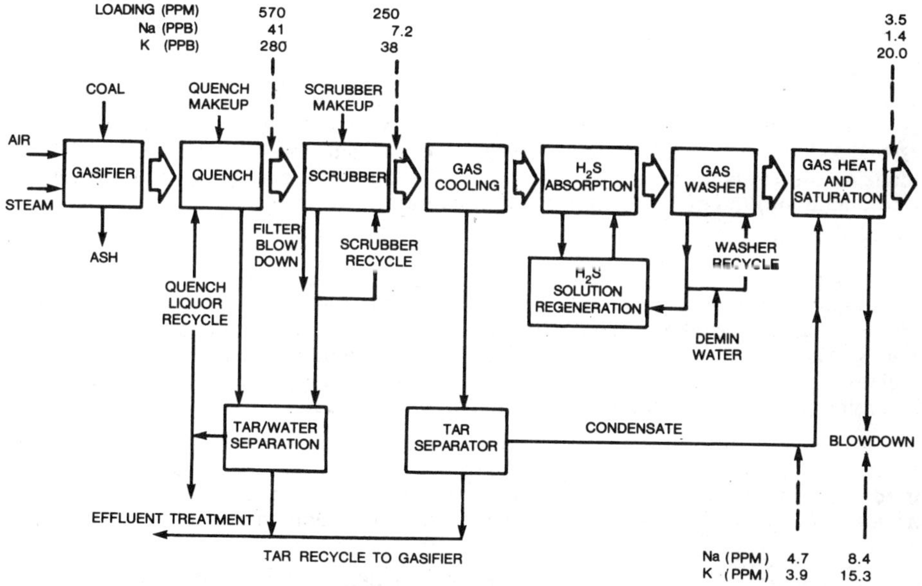

Figure 13. Average Particulate and Alkali Concentration vs Location in the Cleanup System

The reheat/resaturator reintroduces condensate into the gas stream after sulfur removal. This step permits closure of the system water balance and also adds light hydrocarbons to the gas, increasing the heating value by 5-10 Btu/ft^3. The resaturation step was accomplished with no operational problems.

A summary of the cleanup component design vs performance characteristics is shown in Table 2. The particulate removal is well within the design specifications. The particulates remaining in the gas stream are in a size range ($<1\mu$) that eliminates concern for airfoil erosion. The alkali level (3.5 ppb) remaining in the gas as delivered to the turbine simulator was low enough to meet a gas turbine limit of 20 ppb in the combustion product which relates back to an allowable level of 100 ppb in the delivered gas. There was no potassium carryover noted from the absorbing column.

Table 2

COAL GAS CLEANUP SYSTEM PERFORMANCE

	Design	Experimental
Particulate Scrubbing	30 ppm 1% (max) > 8μ	3-5 ppm 99+% < 1μ
Alkali Metal Scrubbing	~0.1 ppm	~0.02 ppm
Total Sulfur Removal	90%	92%
-H_2S	88%	99%
-COS	50%	20-30%
CO_2 Loss	50%	20-55%
Hydrocarbon Utilization	99% (including CH_4)	>90% (condensible hydrocarbon)

The total sulfur removal was within design specifications. The H_2S capture was excellent at levels greater than 99% in the Benfield column. If tighter control of sulfur was required, more emphasis would be placed on COS hydrolysis. The results on H_2S/CO_2 selectivity indicated that the CO_2 removal could be reduced below 50% by optimizing the system operating parameters. The light hydrocarbons were effectively reintroduced into the gas stream through resaturation.

Turbine Simulator

The gas from the fuel plant was introduced into a turbine simulator to evaluate the acceptability of coal-derived gas as a gas turbine fuel. Two simulators were utilized: one at 10-12 atmospheres, 2000-2100°F turbine inlet temperature and the other at 12-16 atmospheres, 2600-2800°F turbine inlet temperature. Both turbine simulators operate on the total fuel gas flow and achieve 8 pounds per second of combusted product flow through the sonic nozzle cascade.

There were no combustion problems noted from the normal ± 10% variation in gas heating value. During these investigations, the fuel gas water content was also varied from approximately 10 to 30% with no resulting combustion problems. There was no evidence of induced dynamic oscillations between the fuel system and the turbine simulator.

As would be expected from the gas composition shown in Figure 13, no evidence of erosion, corrosion, or deposition of the hot gas path parts was noted.

IGCC System Simulation

The fuel supply subsystem was coupled to the turbine simulator and the total IGCC PEF simulation placed under closed loop computer control. In this mode of operation, the total gas flow from the fuel subsystem was used by the turbine simulator with no gas bypass to the flare. The variations in fuel gas flow required to maintain constant firing temperature in the turbine simulator (i.e., constant simulated output) could introduce variations in fuel system pressure. This potential system instability was countered by changing the gasifier steam/air blast to maintain fuel plant pressure within tolerable limits.

In this closed loop operational mode, the turbine simulator was stabilized at a set point firing temperature. A controlled transient was introduced through a signal to simulate a loss of load demand. The firing temperature of the turbine simulator was reduced by 200°F over a 2-minute time period. Under computer control, the fuel system tracked this transient and restabilized at the new set point with no adverse effects. The transient simulation is shown in Figure 14.

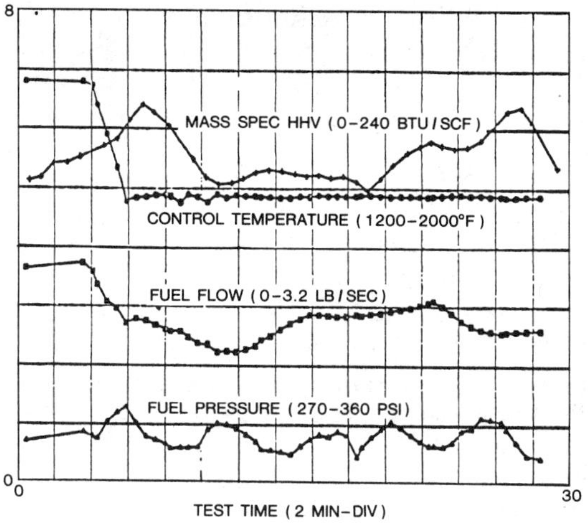

Figure 14. Simulated IGCC Operation

This brief transient experiment does not prove conclusively that IGCC systems can be dynamically controlled, but it does indicate stability of this fuel plant/turbine simulator system.

CONCLUSION

A simulation of an IGCC power generation concept has been developed at a PEF scale. This experimental facility is totally functional, with the gasifier, gas cleanup, and turbine simulator components operating in a closed loop computer-control mode. This facility has the capability to simulate the type of operation expected when an IGCC system is exposed to utility load cycles.

In a transient operation, the PEF IGCC simulator has been shown to be stable. The individual subsystems have operated in a manner compatible with requirements imposed by this application. The gasifier has demonstrated an ability to handle a variety of coals and produce an acceptable quality of fuel gas at both design and off-design points. The gas cleanup component has demonstrated an ability to remove 90% of the sulfur from the fuel gas and to remove alkali metals and particulate from the gas stream prior to combustion. The turbine simulator has shown that this coal-derived fuel would be acceptable to current and future gas turbines.

The PEF IGCC simulation will continue to be utilized to generate information for analytical simulation model verification and for establishing component and system conditions for demonstration plant operation.

REFERENCES

1. J.C. Corman and G.R. Fox, "Performance and Economics of Advanced Energy Conversion Systems for Coal and Coal-Derived Fuels," ASME Paper, J. Eng. Power *10,* 252 (April 1978).
2. B.D. Pomeroy et al., "Comparative Study and Evaluation of Advanced Cycle Systems," General Electric Corporate Research and Development, EPRI AF-664 (February 1978).
3. J.R.M. Alger, "Summary of IGCC Projects Worldwide," presented at the Eighth Energy Technology Conference, Washington, DC, March 1981.
4. D.J. Ahner, "Cool Water Integrated Coal Gasification Combined Cycle Plant," presented at the Eighth Energy Technology Conference, Washington, DC, March 1981.
5. A. Caruvana, "Development of High-Temperature Turbine Subsystem Technology to a Technology Readiness Status," Phase I, Final Report, Contract EX-76-C-01-1806, General Electric Gas Turbine Division, Schenectady, NY (July 1977).
6. A Caruvana, G.B. Manning, W.H. Day, and R.C. Sheldon, "Evaluation of a Water-Cooled Gas Turbine Combined Cycle Plant," ASME 78-GT-77.
7. "Economic Studies of Coal Gasification Combined Cycle Systems for Electric Power Generation," Fluor, EPRI AF-642 (January 1978).
8. B. McElmurry and S. Smelser, "Economics of Texaco Gasification — Combined Cycle Systems," Fluor, EPRI AF-753 (April 1978).
9. R.P. Shah et al., "Performance and Cost Characteristics of Combined Cycles Integrated with Second Generation Gasification Systems," ASME 80-GT-106.
10. A. Caruvana, "Development of High-Temperature Turbine Subsystem Technology to a Technology Readiness Status," Phase I, Final Report, Contract EX-76-C-01-1806, General Electric Gas Turbine Division, Schenectady, NY (July 1977).
11. J.C. Corman et al., "Energy Conversion Alternatives Study," NASA-CR-134949 (3 vols.), General Electric Corporate Research and Development, Schenectady, NY (December 1976).
12. D.E. Woodmansee, J.K. Floess, and A.H. Furman, "Gasification of Illinois No. 6 Coal in an Advanced Fixed Bed Reactor," presented at AIChE meeting, November 1979.
13. S.G. Kimura et al., "Low-Btu Gas Purification for Combined Cycle Power Generation," presented at AIChE meeting, April 1979.

PRESSURIZED FLUIDIZED BED COMBINED CYCLE FOR UTILITY APPLICATION

by
Robert D. Brooks and Richard Roberts
Energy Systems Programs Department
General Electric Company
Schenectady, New York

INTRODUCTION

The oil embargo of 1973, which threatened U.S. national security and led to dramatically increased oil prices, has significantly changed the electric power generating options available to the nation's electric utility industry. For existing oil-fired plants, operating costs continue to rise at a rate largely determined by factors outside the utilities' control. The prospect of externally-caused supply interruption further reduces the appeal of oil as a boiler fuel. America's coal re-emerged in national energy policymaking in the 1970's as the preferred fuel for fossil-fired electric power generating applications. With the Powerplant and Industrial Fuel Use Act of 1978, the Government has prohibited large new boilers from using oil or natural gas, thus adding a regulatory mandate for coal use to the already-existing economic and security-of-supply incentives. Coincident with this pressure to convert to coal, concern about the environmental consequences of increased coal burning was reflected in the issuance of more stringent New Source Performance Standards under the Clean Air Act.

The General Electric Company recognized the growing need for cleaner and more efficient coal-fired power plants. Viewing pressurized fluidized bed combustion technology as an attractive means to respond to this need, GE undertook, beginning in 1974, a series of activities to advance this promising technology toward commercial application. These activities, which built upon early work conducted in the United Kingdom, (1) included development efforts to resolve key technical issues and the preparation of a preliminary commercial plant design (626 MWe) incorporating the Pressurized Fluidized Bed combustor/boiler into a Combined Cycle (2) (PFB/CC). The PFB/CC power plant, representing the extension of conventional utility steam plant practice, is now seen to offer four important advances:

- Superior enviornmental performance, even when compared to oil firing, due to the inherently low emission of SO_2 and NO_x
- Higher net plant efficiency because of the combined cycle configuration
- Simpler and potentially more reliable in operation as compared with add-on chemical cleanup equipment
- Lower investment cost and lower cost of electricity than conventional coal plants equipped with stack gas scrubbers (3)(4)

Considerable success has been achieved in PFB/CC development work in the past few years. A sound design data base now exists for the PFB combustor/boiler, and a pilot-scale facility is now in operation at Grimethorpe, U.K., under the auspices of the International Energy Agency. The problems of gas turbine durability and hot gas dust removal in the PFB exhaust stream are now well understood. The next step in development of the PFB/CC power plant is commercial demonstration at an electric utility to establish operational suitability and economics under actual use conditions.

PFB/CC MODULE DESCRIPTION

The PFB/CC development work by GE and others has established an engineering basis for the design of a full-sized power plant module (200-250 MWe) suitable for utility operation. Figure 1 shows the essential features of the PFB/CC module, the simplicity of the plant arrangement and the absence of add-on chemical cleanup processes. The PFB/CC module may be used singly or combined with others to provide greater capacity. For a 650 MWe plant, three PFB combustor/boilers, with individual gas turbines and particulate removal equipment, would supply steam to a single large steam turbine- generator. The operating parameters of a single module are summarized in Table I.

The PFB combustor/boiler replaces the boiler in a conventional steam plant. Coal and sulfur-absorbing dolomite are crushed to the range of 1/8-1/4" top size, dried of surface moisture, and pneumatically fed to the PFB combustor. Compressed air, entering the bottom of the combustor, suspends and mixes the bed of coal and dolomite, resulting in complete carbon burnout and reduced emissions. The combustion temperature in the PFB is regulated between 1400 F and 1750 F by generating steam in tubes submerged in the combustion zone. A modern reheat steam turbine generator produces about 85% of the net electrical output. Additional steam tubes located in the gas stream above the bed cool the PFB combustion gases to the range of 1000 F to 1450 F. After removal of particulates, the combustion gases are expanded through a gas turbine to drive the compressor and generate the remaining 15% of the electric power output. The gas turbine exhaust passes through a heat recovery feed water heater prior to release. In some locations a baghouse filter will be required on this effluent to meet local suspended particulate standards.

Removal of particulates from the combustion gases is accomplished by conventional centrifugal separation. Two approaches currently appear attractive:

- A number of high efficiency cyclones operating in a series/parallel array
- A smaller number of large cyclones followed by a large number of centrifugal swirl tubes packaged in a single pressure vessel

Projected performance of the two approaches is similar. The eventual choice will be made on the basis of packaging simplicity, operating reliability, and cost. Effective cleanup of particulates in the combustion gas, and limiting the gas turbine inlet temperature to 1450 F rather than the current oil-fired rating of 1985 F, is projected to provide 20,000 to 30,000 hours of durable operation in the PFB environment.

TECHNOLOGY SUMMARY

Development of the PFB technology at General Electric has been sponsored by the U.S. Department of Energy (DOE), the Electric Power Research Institute (EPRI), the New York State Energy Research and Development Authority (NYS-ERDA), and GE itself. Important advances have been achieved in PFB combustor performance, hot gas cleanup, gas turbine materials durability, and engineering issues affecting power plant integration. The work in hot gas cleanup and materials development was conducted in GE's Schenectady laboratories. A series of key PFB tests were performed at the U.K. National Coal Board/Coal Utilization Research Laboratory at Leatherhead, England. These results were corroborated by work performed at the Exxon Miniplant in Linden, New Jersey under the U.S. Environmental Protection Agency (EPA) and DOE sponsorship. Even more significantly, an 80-MWth PFB pilot plant has begun operation at Grimethorpe, England, under the joint sponsorship of the U.S., the United Kingdom, and the Federal Republic of Germany in a major International Energy Agency (IEA) pro-

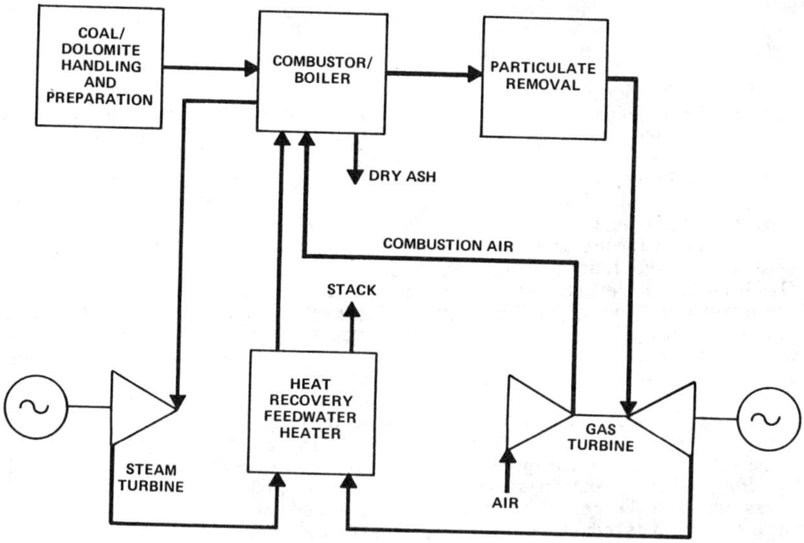

FIGURE 1. PRESSURIZED FLUIDIZED BED COMBINED CYCLE DIAGRAM.

TABLE 1. PFB/CC SINGLE MODULE DESCRIPTION.

NET MODULE OUTPUT (MWe)	248
STEAM TURBINE (MWe)	222
GAS TURBINE (MWe)	39
PARASITIC LOAD (MWe)	13
STEAM THROTTLE CONDITIONS (psia/°F/°F)	3500/1000/1000
CONDENSER BACK PRESSURE (inches Hg)	2.0
PFB BED TEMPERATURE (°F)	1750
PFB BED PRESSURE (psia)	165
GAS TURBINE INLET TEMPERATURE (°F)	1450
COAL: ILLINOIS #6 (HHV = 12235 Btu/lb)	

gram. Additionally, the petrochemical industry has accumulated 18 years of relevant operating experience with PFB combustion, hot gas cleanup, and power recovery turbines.

COMBUSTOR

The work at Leatherhead and at Exxon has established the attractive level of SO_2 and NO_x emissions inherent in the PFB. More recent work at MIT (5) has provided some insight into the reducing reactions which account for the very low NO_x formation at high pressure. Figure 2 compares the low emission levels obtainable with the PFB to the 1979 EPA New Source Performance Standards for utility boilers.

Combustion tests have shown the PFB process to be suitable for use with a wide variety of coals, without the need to impose operating restrictions such as those associated with pulverized coal furnaces (e.g., iron oxide content). This provides added flexibility to the utility user with respect to source of coal supply, as well as allowing minimization of fuel cost as coal market conditions change over the life of the power plant. The PFB has achieved combustion efficiency in excess of 99%, without fines recycle. The dry ash from this process has potential for structural materials, road building, and agricultural applications.

The large PFB pilot plant at Grimethorpe as the key coal utilization project of the IEA went into operation in September 1980. The 80 MWth combustor design will provide the very necessary design data (6) for the first utility PFB/CC power plants. The combustor, built by Vereinigte Kesselwerke, A.G. (VKW) and illustrated in Figure 3, features water wall containment of a single bed within a refractory-insulated pressure vessel. This construction assures safe containment of the combustion process and exposes the pressure vessel to only clean compressed air at 500-600 F. The combustor/boiler currently being tested has a bed cross section of 2 meters x 2 meters (6.6 ft x 6.6 ft) and can accommodate a bed depth up to 4.5 m (14.8 ft); 4 m (13.2 ft) of free-board is provided above the bed. A full size utility module (200-250 MWe) would require two such PFB combustor/boilers, each with approximately 4 m X 4 m (13.2 ft x 13.2 ft) cross section and the same bed depth and freeboard. The coal injection pattern at Grimethorpe can be varied by using from 4 to 9 nozzles to evaluate the optimum arrangement with respect to process uniformity. Two waste solids removal ports are provided in the grid plate with parallel lockhopper arrangements so that either may be used. Not shown in Figure 3 is the above-bed tube bundle which may be used to reduce gas outlet temperature. The operating parameters for the Grimethorpe PFB are summarized in Table 2.

The Grimethorpe PFB experiment is expected to provide accurate data on emissions, temperature distribution, combustion efficiency, and operating characteristics at a size scale sufficient to establish a direct design baseline for the first prototype power plant. Combustion testing with a wide variety of U.S., U.K., and F.R.G. coals is scheduled to include 2850 hours of operation by mid-1983. The modest scaleup for utility application should allow direct use of the pilot plant results with high confidence of success.

GAS TURBINE/COMBUSTOR INTERFACE

The major technical uncertainty in the PFB/CC power plant concept is the ability to achieve the necessary gas turbine durability when operating with the presence of dust in the combustion gases. A typical PFB efflux dust loading and particle size distribution is shown in Figure 4. The erosion resistance of the GE heavy duty MS7001 gas turbine in the PFB gas stream has been estimated (7) based on coal-fired locomotive turbine tests, GE erosion rig tests, and PFB cascade experiments conducted by GE at NCB/CURL and Exxon. The turbine erosion tolerance depends strongly upon gas velocities which are directly related to firing temperature, as shown in Figure 4. Boericke (7) has concluded that 98% of the dust must be under 10 microns with dust loadings in the range of 100 ppm to insure adequate erosion resistance at a turbine firing temperature of 1670 F.

The hot gas cleanup approach currently proposed for PFB/CC utilizes three stages of conventional cyclones arranged in series as depicted in Figure 5. The estimated

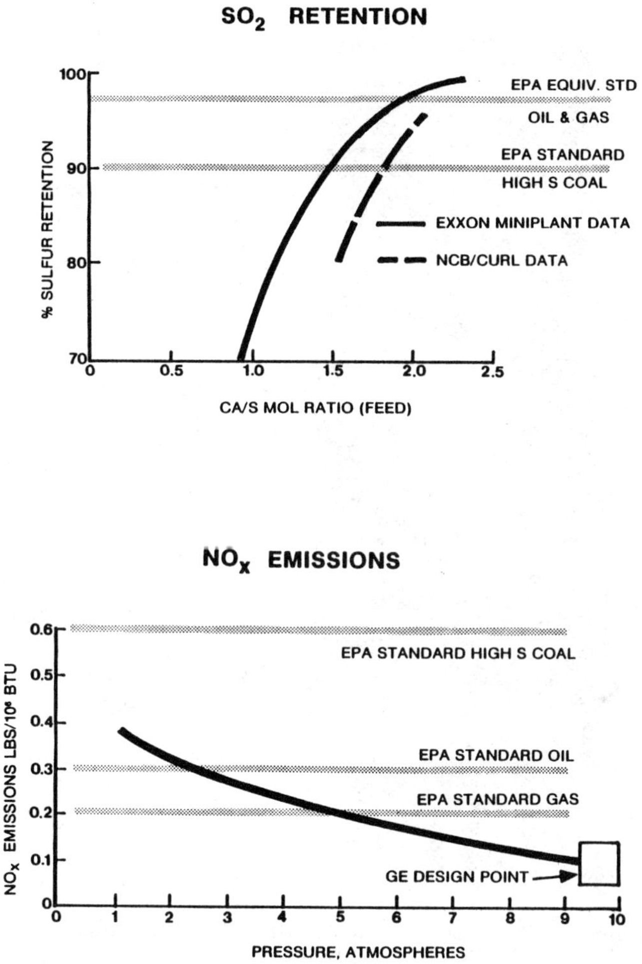

FIGURE 2. PFB/CC EMISSION PERFORMANCE BETTER THAN EPA STANDARDS.

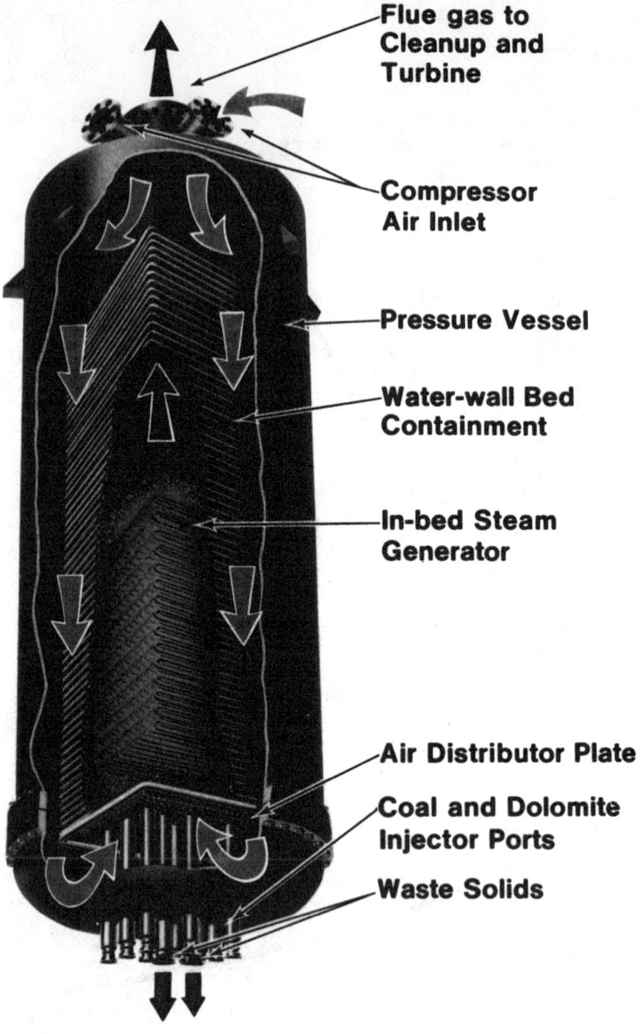

Figure 3: PFB COMBUSTOR DESIGNED BY VEREINIGTE KESSELWERKE, A.G.

FEATURE		RANGE
AIR MASS FLOW, lbs/sec	—	19.8 — 68
COMBUSTOR PRESSURE, bar	—	6 — 12
BED TEMPERATURE, °F	—	UP TO 1750
EXCESS AIR, %	—	10 — 100
FLUIDIZING VELOCITY, fps	—	3.3 — 12.1
COAL-TOP SIZE, inches	—	.062 — .25
COAL FEED RATE, tons/hr	—	UP TO 12.4
COAL HEAT INPUT, MWth	—	UP TO 80
DOLOMITE FEED RATE, tons/hr	—	UP TO 5.2

Table 2: OPERATING PARAMETERS FOR THE GRIMETHORPE PFB EXPERIMENTAL FACILITY.

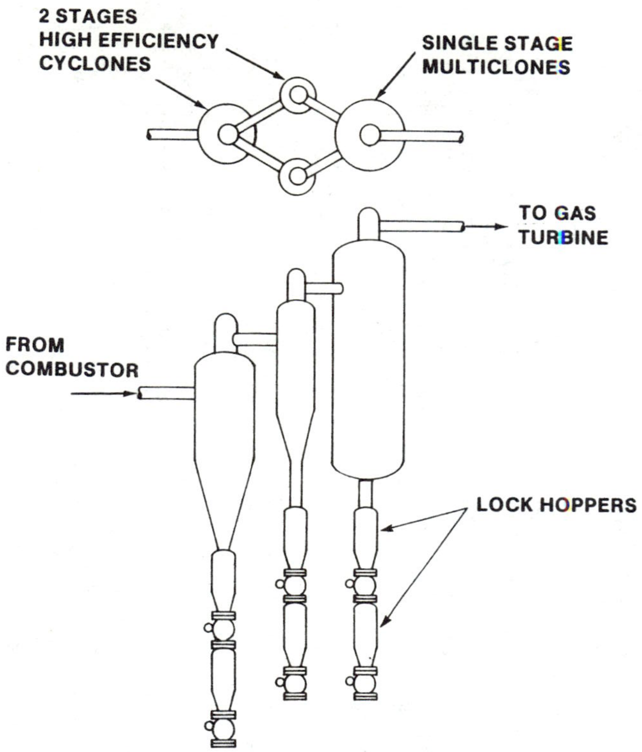

Figure 5: PFB/CC HOT GAS CLEANUP USING CYCLONE

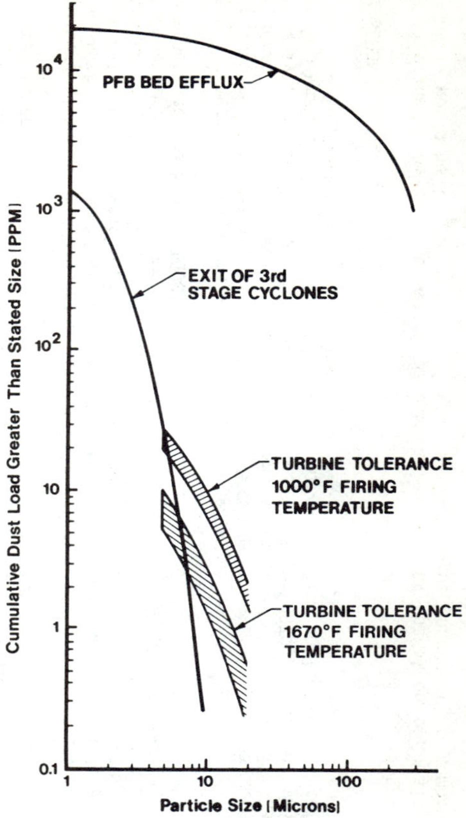

Figure 4: PROJECTED CYCLONE PERFORMANCE BASED ON PFB CLEANUP RESULTS.

performance of this cyclone arrangement, as shown in Figure 4, meets GE's criteria for avoiding erosion. Multi-cyclone final-stage separators have been employed in the petrochemical industry to protect power recovery turbines in fluid catalytic cracking installations for approximately 18 years. The typical design shown in Figure 6 has evolved over this time involving much experimentation and full-scale commercial testing. (8) There are now reported to be 29 units in operation at 1200-1400 F and 2-4 bar, with the largest connected to a 24,000 horsepower turbine. (9) Turbine blade life has approached five years, but the more normal practice is to reblade the turbine during scheduled shutdowns about every three years. The power recovery unit operates nearly continuously between scheduled shutdowns. In this application the multiclones remove catalyst fines (selected clay materials) from the exhaust gases of a PFB combustor used to regenerate the catalyst by burning off coke deposits. The dust loading and particle sizes are comparable to those found in the efflux of a PFB coal combustor. This commercially proven separator approach appears to be readily adaptable to the coal-fired PFB/CC application.

Extensive investigation of the hot corrosion phenomena in the PFB gas stream has indicated that 20,000-30,000 hours hot parts life should be obtainable with the proposed gas turbine inlet temperature limit of 1450 F and existing levels of air cooling. Metal temperature was found (10) to have a significant influence on the performance of materials in the GE 1000-hour PFB exposure tests. Some of these results are summarized in Figure 7. At 1250-1350 F metal temperature, the current production alloys, IN 738 and FSX 414, exhibited only mild surface oxidation with isolated instances of sulfide penetration. Buildup of ash deposits on the air foil surfaces reached a stable value in the NCB/CURL and Exxon cascade tests, with neither monotonic increase nor plugging observed in 1000 hours. This suggests that fouling of the gas turbine will not be of major consequence in the PFB/CC.

GE is building a larger scale (12 lb/s) gas turbine materials cascade under NCB (IEA Grimethorpe) Ltd. and EPRI sponsorship for the Grimethorpe PFB facility. This will allow evaluation of larger stator blades in a representative PFB environment. GE has also built a coal-fired PFB test rig in Schenectady, N.Y., dedicated to materials exposure testing in order to observe long-term effects. This rig went into operation in February 1981 and is expected to provide the first 10,000-hour materials exposure data by the end of 1983.

GAS TURBINE RELIABILITY

General Electric's heavy duty gas turbines are recognized throughout the world for sound design, rugged construction, and high reliability. Over 370 operators in 48 states and 70 foreign countries have selected GE gas turbines. These operators have more than 2800 GE units installed, totaling over 45 million kilowatts. Over 64 million hours of operation have been logged, and more than 120 of these GE gas turbines have completed over 100,000 hours of operation. Forty-one have logged over 150,000 hours each. More than 250 Model MS7001 gas turbines have been sold since the first installation in 1970.

The General Electric PFB/CC plant will be based on the MS7001 heavy duty industrial gas turbine shown in Figure 8. This machine is available with a regenerative cycle combustor housing which permits the compressor discharge air to be ducted to a heat recovery regenerator in the turbine exhaust and then returned to the on-board combustors at 1000 F (max). For the PFB external combustor, the wrapper and ducting will be redesigned to handle 1500 F combustion gases.

General Electric's high reliability combined cycle program has as its goal to make gas turbine based combined cycle plants (STAG ®) more reliable than any other form of power generation. Recent engineering and manufacturing improvements have substantially increased the dispatch availability (11) of the MS7001 STAG (®) cycle, as shown in Figure 9. This availability now equals fossil-fired boiler/steam turbine plants in base load use.

UTILITY DEMONSTRATION

An expedient, low-cost method of demonstrating utility operation of a single PFB/CC module would be to convert an existing steam turbine generator to combined

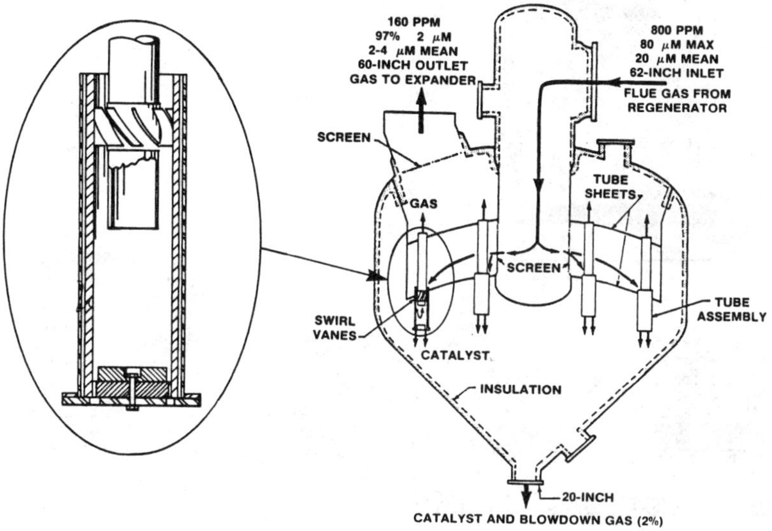

FIGURE 6. SINGLE STAGE MULTICLONE ASSEMBLY

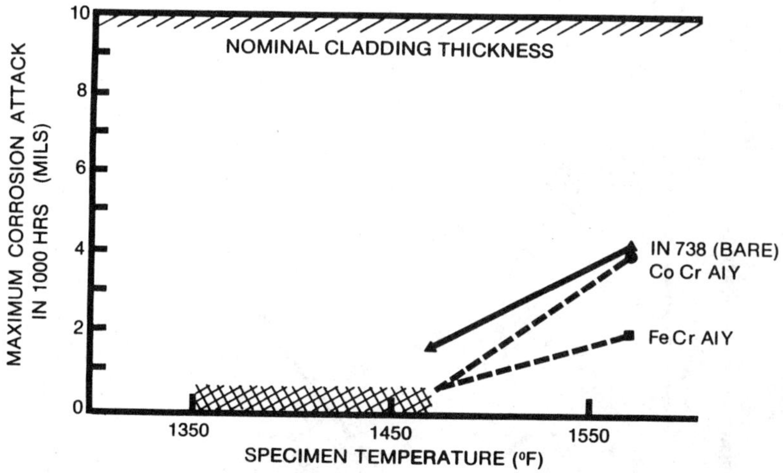

FIGURE 7. IMPROVED GAS TURBINE MATERIALS IN PFB ENVIRONMENT.

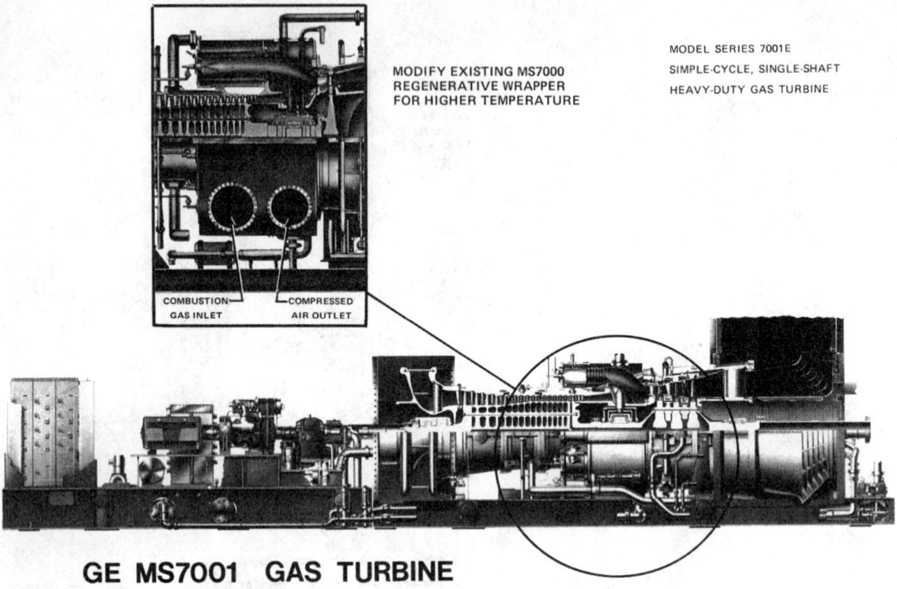

Figure 8: GE MS7001E GAS TURBINE

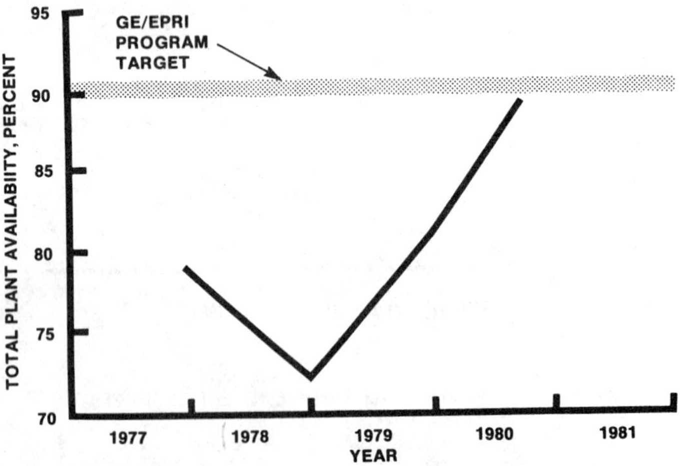

Figure 9: COMBINED CYCLE AVAILABILITY HISTORY OF GE MS7000 PLANTS

cycle operation. This could be accomplished by adding the PFB combustor with its coal and dolomite feed equipment, the particulate cleanup train, the gas turbine generator, and unit controls to interface with an existing utility steam turbine installation. An artist's sketch of such an installation is shown in Figure 10 for retrofit of an existing 150-200 MWe steam turbine generator unit. The components are compact, due to the lower volumetric flow at pressurized operation, and less than an acre of land is required for equipment siting.

The project time table to demonstration operation appears particularly attractive, as shown in Figure 11. Three years are estimated to obtain the project financing. A rather long period of four years is shown for obtaining the necessary permits, and engineering based on recent utility experience. In some localities, this may be reduced by a year, but a decision to proceed with hardware procurement is dependent upon funding availability. Three years for on-site construction would be consistent with current combined cycle experience. Checkout and startup would be accomplished by early 1988 when commercial operation would begin.

FUTURE PFB/CC TECHNOLOGY IMPROVEMENTS

GAS TURBINE INLET TEMPERATURE GROWTH

Historically, the power generation industry has reduced electricity costs to the consumer by increasing plant thermal efficiency. In analagous fashion, the potential exists for improving the thermal efficiency of the PFB/CC by increasing gas turbine inlet temperature up to the limiting bed temperature, currently estimated to be 1750 F. In fact, the upper limit for bed temperature has not been extensively investigated, and 1750 F has been accepted based upon successful operation at that temperature. At some future time, as more capable gas turbine materials and particulate removal equipment becomes available, it would be appropriate to correlate the effect of increasing bed temperature as a function of coal properties (particularly ash fusion temperature) and dolomite fraction. The relationship between gas turbine inlet temperature and plant net efficiency is illustrated in Figure 12. As indicated earlier, gas turbine inlet temperature is now limited to about 1450 F by currently available gas turbine materials and hot gas particulate removal technology. The PFB/CC is operable down to about 1000 F gas turbine inlet temperature. Cycle efficiency as high as 40% for new plants in the 500-600 MWe range is potentially achievable with 1750 F bed temperature.

Technology improvements in two areas will be needed to permit gas turbine firing at the full bed temperature limit. The first of these is improved hot gas cleanup to reduce particulate loading below that obtainable with high efficiency cyclones. General Electric has elected and is currently working on two advanced concepts - the electrostatically enhanced cyclone (12) and the electrostatically enhanced granular bed filter.(13) Both devices are now being tested at small scale with support from DOE.

A principal advantage of the electro cyclone, in addition to enhanced particulate removal, is the possibility of substituting a small number of quite large units for a larger number of conventional cyclones. It is currently estimated that a 12-foot diameter unit would provide performance comparable to an array of 10-inch diameter multiclones. A substantial improvement in plant capital and operating costs is expected.

The electrostatic granular bed filter, shown schematically in Figure 13, employs an annular bed of gravel moving downward between louvered container walls. The dirty gas flows into the center of the bed container and then back out through the upper part of the bed to the outlet. The gravel, with particulates attached, falls into a hopper and is returned by a screw feeder and rotary blower to the disengagement chamber for cleanup and recycle to the bed. This is an adaptation to PFB/CC operating conditions of a device now offered commercially by EFB, Inc. for lower temperature and pressure applications. An especially attractive feature of this device is the ability to remove very fine dust particles, thus eliminating the need for a low temperature baghouse filter to meet the EPA particulate standard. Because operability and geometric scale-up are such great issues in the development of hot gas particle removal equipment, it will be necessary to test at pilot scale (1/10 commercial size) or larger before considering these devices for utility application.

FIGURE 10. PFB/CC200-250 MWe ENGINEERING PROTOTYPE MODULE.

	1981	1982	1983	1984	1985	1986	1987	1988
OBTAIN FINANCING	■	■	■					
ENGINEERING & PERMITS	■	■	■	■				
EQUIPMENT MANUFACTURE				■	■			
CONSTRUCTION				■	■	■	■	
CHECKOUT/ START UP							■	■
COMMERCIAL OPERATION								■ →

FIGURE 11. PROJECT TIMETABLE TO UTILIZE COMMERCIAL OPERATION.

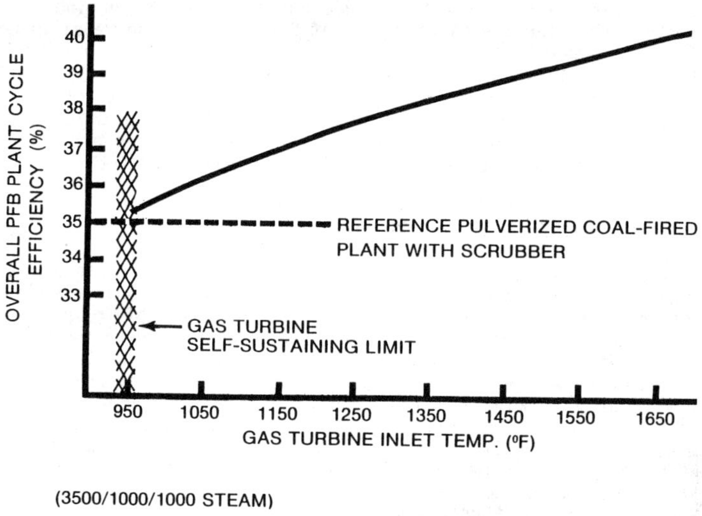

FIGURE 12. PFB/CC PLANT NET EFFICIENCY.

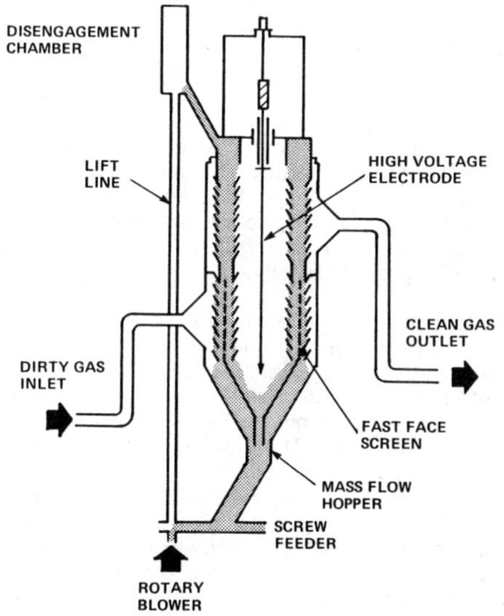

FIGURE 13. ELECTROSTATIC GRANULAR BED FILTER.

The second area of needed technology advancement is the qualification of improved corrosion resistant coatings or claddings for the gas turbine. GE findings (14) from the turbine materials cascade test at Exxon, where the metal temperature of 1500-1590 F corresponded to an equivalent gas inlet temperature in excess of 1700 F, indicate that iron-based (GE 2541) and cobalt-based (PFB-6) cladding materials could provide the needed protection. The coal-fired PFB gas turbine materials test project underway at GE is expected to confirm the long-term durability of those alloys for a variety of application processes. These materials are also to be included in the gas turbine cascade experiment to be operated at the Grimethorpe PFB Experimental Facility.

ADVANCES IN STEAM TURBINE THROTTLE CONDITIONS

There has been a continuing increase in steam turbine inlet (throttle) temperature and pressure when viewed from an historical perspective. This trend, along with the introduction of multiple stages of reheat, has been directly responsible for the past trend towards more efficient and lower cost electric power generation by the nation's utility industry. It is expected that this trend will continue in the future as better materials and fabrication procedures reach commercial use. General Electric is currently working in this area under the sponsorship of the Electric Power Research Institute. (15) An advantage of PFB/CC is that advances in steam throttle conditions or number of reheats can be readily accommodated within the cycle. Currently obtainable bed temperature is well above forseeable increases in steam throttle temperature, so the cycle arrangement itself does not impose a limit. Development of the PFB/CC is thus fully in accord with technology advancements in the steam cycle and can apply such advances to improve net power plant efficiency.

ADVANCED PFB COMBINED CYCLE

PFB/CC has often been characterized as an attractive near-term coal technology, but lacking the potential for very high cycle efficiency, and by implication low operating cost, offered by the integrated gasification combined cycle, carbonate fuel cell or magnetohydrodynamics cycles. A hybid cycle suggested by Raymond Hoy (16) could make use of a PFB combustor operated in conjunction with a PFB devolatilizer/ gasifier. This cycle is schematically represented in Figure 14. This hybrid approach to gasification-combustion would, by the addition of a PFB devolatilizer operating in parallel with the PFB combustor, allow afterburning of the gaseous fuel to raise the gas turbine firing temperature to the 1800-2100 F range and could have an efficiency of 43%. By this approach, gas turbine inlet temperature is not limited by the upper limit of PFB coal combustion. The waste dolomite stream from the PFB combustor would supply heat to the devolatizer. Three-fourths of the coal would be fed to the combustor and approximately one-fourth to the devolatilizer, which might operate at a temperature of 1800-1900 F. Considerable pilot plant data exist for a devolatalizer/gasifier component of this type from the COGAS program at Leatherhead. (17) A power plant design study would be required in order to accurately verify the attractiveness and practicality of this advanced concept.

The PFB/CC power plant concept, in both the direct-coal-fired and hybrid variants, offers a fundamental improvement in net plant efficiency without requiring a corresponding increase in conversion cycle (gas or steam turbine inlet) temperature. Figure 15 compares fuel efficiency with conversion cycle temperature needs for four advanced coal-fired options. Clearly on a thermodynamic basis there is sufficient growth potential for the PFB/CC that it should move aggressively towards utility application. It compares favorably with the Integrated Gasification Combined Cycle for near-term technology level and has the potential for significant cycle efficiency improvement if sufficient effort can be applied for advancing the technology. Carbonate Fuel Cells and MHD offer dramatic improvements in performance but still require pilot plant results on which to base large power plant designs.

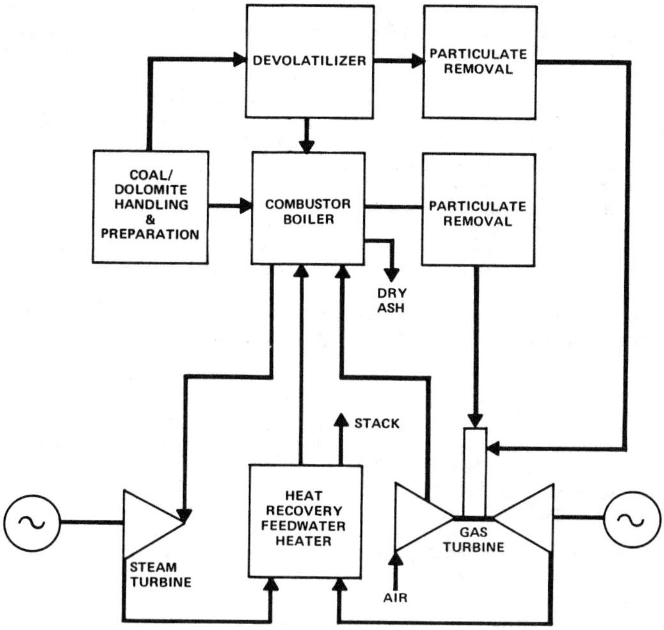

FIGURE 14. ADVANCED PRESSURIZED FLUIDIZED BED COMBINED CYCLE DIAGRAM.

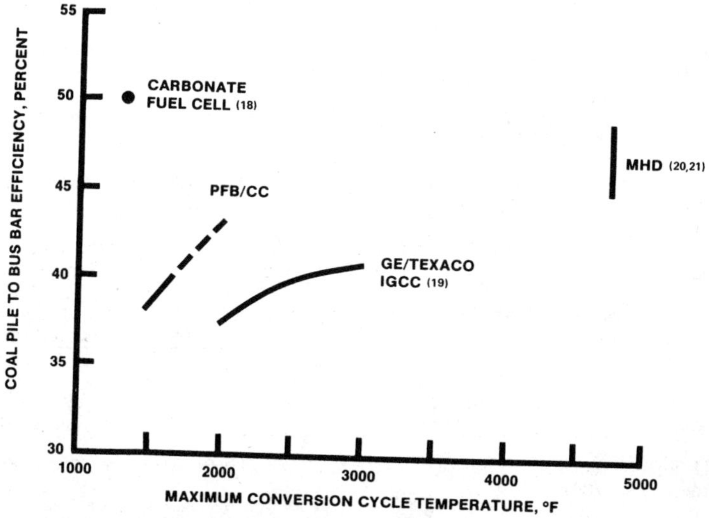

FIGURE 15. NET PLANT EFFICIENCY FOR COAL FIRED COMBINED CYCLE PLANTS.

SUMMARY

The burning of coal in a pressurized fluidized bed combustor is the first new method of burning coal since the introduction of pulverized coal firing in the 1930's. Following extensive testing in the U.S. and England, sufficient definition and experimental resolution of critical PFB/CC issues has been accomplished to justify application of a coal-fired PFB/combined cycle in an utility setting. Current gas turbine materials and currently available hot gas particulate removal equipment together act to limit gas turbine inlet temperature to about 1450 F. When combined with a modern steam cycle, the corresponding PFB/CC power plant is projected to provide a net plant efficiency of about 38.5%. Even with gas turbine firing temperature limited to 1450 F, this plant should have sufficient performance and cost advantages over pulverized coal plants with scrubbers to be an attractive option for utilities. In addition, the plant performance is expected to be fully competitive with pulverized coal plants using advanced scrubbers, AFBs, and early IGCC plants. Only by utility operation can the relative suitability of these different approaches be resolved with respect to capital and operating costs, availability, and environmental impact.

Improvements in hot gas cleanup and gas turbine materials, currently under development, are expected to permit the gas turbine inlet temperature to be increased to the 1600-1700 F range. This improved PFB/CC cycle would offer even lower power costs and improved environmental impact. A gas turbine pilot plant test program is however required to qualify higher temperature operation before this technology will be ready for utility application.

Further improvement in PFB/CC plant efficiency is possible, involving a parallel PFB devolatilizer, heated by the spent dolomite from the PFB combustor. This development allows the possibility of a plant which might be competitive with very advanced IGCC, fuel cell, and MHD plants. Preliminary power plant design studies are required to assess the concept's practicality and identify critical issues before proceeding with necessary development work.

REFERENCES

1. Gibson, J., "Present Status of Fluidized-Bed Combustion," Appl. Energy 3, 87-99 (1977).

2. Brooks, R.D., and Roberts, R., "Pressurized Fluidized Bed Combined Cycle Power Generation," IGT Symposium, Louisville, KY, May, 1979.

3. Wysocki, J., and Rogali, R., "Preliminary Assessment of Alternative PFBC Power Plant Systems," C5-1451, Final Report, EPRI Contract RP 1645-2.

4. Cain, W.C. et al. "Environmental R&D Needs for Advanced Power Plants," CEP 76, 62-69, October 1980.

5. Beer, J.M. et al. "No Formation and Reduction in Fluidized Bed Combustion of Coal," Proc. 6th Intl. Conf. on Fluid. Bed Combustion, Atlanta, GA.

6. Carls, E.L. et al. "The IEA Grimethorpe Pressurized Fluidized Bed Combustion Experimental Facility," Proc. 6th Intl. Conf. on Fluid. Bed Combustion, Atlanta, GA.

7. Boericke, R.R. et al. "Assessment of Gas Turbine Erosion by PFB Combustion Products," Proc. 6th Intl. Conf. on Fluid. Bed Combustion, Atlanta, GA.

8. "Shell Power Recovery System," Feb. 1980, Shell Development Co., PO Box 2463, Houston, TX 77001.

9. Mathers, W.G. and Schonewald, R., "Fluid Catalytic Cracker Power Recovery Expander Applied to Pressurized Fluidized Bed Combustion," Proc. 6th Intl. Conf. Fluid. Bed Combustion, Atlanta, GA.

10. McCarron, R.L., "Corrosion/Erosion in PFB Gas Turbine Application," 5th Ann. DOE Conf. on Matls. for Coal Conversion and Utilization, Oct. 1980, Gaithersburg, MD.

11. Kunkel, R.G., "High Reliability Gas Turbine Combined Cycle Development Program - Phase 1 Final Report," EPRI Contract RP1187-3.

12. Boericke, R.R. et al. "Electrocyclone for High Temperature, High Pressure Dust Removal," AIAA, 81-0393.

13. Presser, A.M. and Alexander, J.C., "Nonplugging Retaining Structure For Granular Bed Filters For High Temperature and High Pressure," 3rd Symp. On Transfer and Util. of Particulate Control Technology, Orlando, FL, March 1981.

14. Grey, D.A. et al. "High Temperature Corrosion/Erosion Combustors," Proc. 6th Intl. Conf. on Fluid. Bed Combustion, Atlanta, GA.

15. "Engineering Assessment of Low Heat Rate Pulverized Coal Power Plants," EPRI Contract RP1403-2.

16. Hoy, H.R. and Roberts, A.G., "Pressurized Fluidized Bed Combustion of Coal for Combined Cycle Power Generating Systems," Fluidized Combustion Conference, January 1981, Energy Research Institute, University of Cape Town, So. Africa.

17. Hoy, H.R., personal communication.

18. Peterson, J.R., and Blurton, K.F., "Commercial Applications of Molten Carbonate Fuel Cell Systems," Energy Technology VI, Washington, D.C., Mar. 1979.

19. Oliva, J.J. and Shemo, S.D., "Texaco-Based Gasification Combined Cycle System Performance Studies, "EPRI AP-1429, June 1980.

20. "Evaluation of Phase 2 Conceptual Designs and Implementation Assessment Resulting from the Energy Conversion Alternatives Study (ECAS)," NASA TMX-73515, April 1977.

21. Staiger, P.J., and Abott, J.M. "Summary and Evaluation of the Parametric Study of Potential Early Commercial MHD Power Plants," Proc. of 7th Intl. Conf. on MHD Electrical Power Generation, MIT, June 16-20, 1980.

ATMOSPHERIC FLUIDIZED BED COMBUSTION FOR UTILITY APPLICATIONS

TENNESSEE VALLEY AUTHORITY'S RESEARCH AND DEVELOPMENT IN THE ATMOSPHERIC FLUIDIZED BED COMBUSTION PROCESS

Michael D. High
Division of Energy Demonstration and Technology
Tennessee Valley Authority

INTRODUCTION

A major responsibility of the Tennessee Valley Authority (TVA) is to help stimulate economic development of the 7-state Tennessee Valley region. An integral part of this mandate involves providing enough energy to meet the needs of industrial and residential customers today and in the future. Over the 48-year history of the agency, this goal has been continuously met through the use of innovative technologies. Today, the quest continues as the agency is actively developing several new energy technologies for possible future use. These activities will benefit both the TVA power system and other utilities throughout the Nation.

One of these key technologies now under development by TVA is atmospheric fluidized bed combustion (AFBC).

Like other coal technologies (fuel cells, synthetic fuels from coal, AFBC), the concept of AFBC is not new. In the early 1920's a patent was sought for a fluidized bed device that could process gas from coal. Today, there are hundreds of fluidized bed processes in use at chemical plants and oil refineries throughout the world. However, the commercial use of coal in AFBC boilers to produce steam has been limited for the most part to small-scale industrial boilers with steam capacities of 50,000 to 60,000 pounds per hour. Despite significant research, the use of utility-scale AFBC plants has yet to occur.

STATUS OF AFBC TECHNOLOGY

In the 1970's, the thrust of utility AFBC research accelerated with a combination of bench-scale research, hot/cold modeling, and some pioneering successes. The Department of Energy (DOE) built the Rivesville 30-MW AFBC unit to demonstrate that AFBC could generate electricity.

The Rivesville AFBC design was initiated in late 1972, applying technology learned from Pope, Evans and Robbins' (PER's) 0.5 MW bench-scale combustor located at Alexandria, Virginia. Rivesville represented an initial design of a multicell boiler to be used as a basis for larger demonstration and utility scale plants, and was the first major step toward DOE's commercial size central station power program. From operating experiences at Rivesville, we have learned that:

- An excessive in-bed heat transfer surface will limit turndown and can cause excessive bed cooling which quenches the fire. Scale-up of this data from very small units is inappropriate.
- High freeboards are needed to improve performance in sulfur capture, combustion efficiency and sorbent utilization.
- Instrumentation and control flexibilities are necessary to control the process.
- Corrosion and erosion of boiler materials were not as severe as expected.
- Progressive startups of slumped beds can be successfully accomplished by transferring hot material from adjacent active beds.
- Pilot plant designs must include flexibility to test optional design features and changes to auxiliary systems. For example, several failures of the coal feed system occurred before a workable solution was found.
- Pilot plant and demonstration facilities must be independent of other operating units. For example, failures of the host utility's boiler, upon which the Rivesville unit was dependent, also forced the AFBC test facility to shut down.

Even though funding restraints prohibited the inclusion of many desired features, the Rivesville AFBC boiler achieved more than 800 hours of commercial steam supply at a maximum rate of 187,000 pounds/hour to the host utility's turbogenerator.

A significant advancement over the Rivesville experience was the Electric Power Research Institute's (EPRI) Babcock & Wilcox (B&W) 6' x 6' AFBC unit. This unit included design changes and provided much needed operating and performance data.

Results from the 6' x 6' led to installing higher freeboards for longer combustion and reaction times and to the recycling of unburned carbon and unreacted or partially reacted limestone elutriants; hence, separate carbon burnup beds could be eliminated. Even with the success of this unit, it became readily apparent that a larger plant having the capability of retrofitting design changes was needed to test new hardware and to resolve uncertainties.

Performance of the AFBC process over the last few years has shown marked improvements. For 90 percent sulfur removal, the technology has progressed from a predicted calcium-to-sulfur ratio of about 5.0 (4.5 for 85 percent removal) in early 1977 to a ratio of 2.7 today. The predicted ratio for TVA's pilot is 2.0.

Similar improvements in nitrogen oxide emissions have also been observed. The Rivesville design basis showed 0.6 pounds of NO_x per million Btu's in early 1977; current data from the EPRI/B&W 6' x 6' indicate 0.3 pounds/million Btu's. The predicted level in the 20-MW pilot plant is less than 0.3 pounds/million Btu.

ADVANTAGES OF AFBC

Studies of the AFBC process have confirmed the advantages of AFBC over conventional burning of pulverized coal coupled with flue gas desulfurization. These major advantages are:

- Capture of sulfur dioxide in the boiler simplifies plant operations.
- Dry waste product is easier to handle and is potentially useful.
- Overall plant efficiency is improved.
- Many types and grades of coal may be burned.

TECHNICAL UNCERTAINTIES IN AFBC

Several problem areas of technical uncertainties are evident from previous AFBC operating experiences of DOE, EPRI, and others. Some of the principal concerns are:

- Feeding and distributing coal and limestone in the boiler.
- Excessive wear and erosion of feed lines.
- Selection of boiler materials to withstand corrosive and erosive environments.
- Handling of hot, used bed material.
- Control of the plant during startup, turndown, and load following.
- Performance of cyclone dust collectors during load variations and in a recycle process mode.

TVA'S HISTORY IN FLUIDIZED BED COMBUSTION

In mid-1975, TVA participated with the Environmental Protection Agency in the Energy Conversion and Alternatives Study (ECAS)--a study conducted to compare the economic and performance characteristics of conventional pulverized coal-fired steam plants with flue gas desulfurization (scrubbers), AFBC power plants, and pressurized fluidized bed combustion (PFBC) power plants. A final report was published in November 1977.

Development activities began in earnest in September 1976 when the TVA Board of Directors authorized $4 million to prepare a preliminary conceptual design of a demonstration plant and to perform technical support research projects. TVA purchased preliminary designs from Combustion Engineering (CE), Foster-Wheeler, and B&W, and a conceptual design and study of the pilot plant from B&W. The demonstration plant preliminary designs were funded by both TVA and DOE and were completed in November 1978.

In April 1979, initial funding for the design and fabrication of a 20-MW pilot plant was authorized. A turnkey contract was subsequently placed with B&W.

In August 1979, final conceptual designs and cost estimates for a 200-MW demonstration plant were initiated by contracts with three major suppliers of utility steam generation equipment. The work of the three boiler manufacturers was completed in January 1981. TVA's evaluations of these designs and plant cost estimates are currently in progress.

In September 1979, TVA approved the construction and operation of a 20-MW AFBC pilot plant to be funded by TVA, congressional appropriations, and EPRI. Construction is scheduled for completion in March 1982, at which time a 5-year test program will begin.

TVA'S 20-MW AFBC PILOT PLANT

Now under construction, TVA's $65 million 20-MW AFBC pilot plant facility is located on the Shawnee Steam Plant reservation near Paducah, Kentucky. Site preparation and initial construction began in April of 1980. Testing and operation is slated for mid-1982.

TVA's pilot plant will be a key element in resolving many of the uncertainties concerning the development of utility-size AFBC power plants. It will be used to test and evaluate operating control modes, procedures, and equipment; to investigate key systems for performance and reliability; to demonstrate adequate emission controls, combustion efficiency, and limestone utilization for a variety of process and equipment configurations; and to train operating personnel in proper operation and maintenance procedures. Pilot plant testing will be a cooperative effort between TVA and EPRI.

Since AFBC is a completely different type of coal combustion process, it will obviously require different modes of operation than a conventional steam plant. Like any generating facility, a commercial-scale AFBC plant must be able to respond to changes in electricity demand. Startup, turndown, load control, shutdown procedures, and safety-related systems will be tested and developed.

The pilot plant will be a flexible test facility capable of testing various mechanical subsystems: feed systems (overbed and underbed), fluidizing velocities, and a variety of NO_x control techniques; and will be extremely well instrumented to obtain vital data for future AFBC designs.

The main fluidized bed area will have 216 square feet of active area with both evaporation and superheat duties. The 72-square-soot startup section will not contain inbed heat transfer surface. The steam conditions are 2,400 psig and 1,000^{o}F. The nominal design conditions include 8-foot fluidizing velocity, 4-foot bed depth, and 1,550^{o}F bed temperature.

The pilot plant design incorporates recycle of char to the main bed; a feedpoint spacing of one per 18 square feet; a water-cooled, bubble cap distribution plant; and a combustor design for a two-second gas residence time.

Areas of concern that will be resolved in the TVA pilot plant include: coal/limestone feed systems, materials corrosion, convection pass erosion, carbon and limestone recycle, cyclone performance, and control and load following techniques.

COAL/LIMESTONE DISTRIBUTION

In the area of coal/limestone distribution, several alternate feed systems will be tested. These include: overbed stoker feed system, separate overbed limestone feed system, and underbed pneumatic feed system.

The distribution of coal and limestone to the fluid bed has been a significant roadblock to full-scale AFBC development. Supplying these materials to a small test unit is no problem; however, distributing coal and limestone to a full-scale unit with thousands of square feet of bed area is something of a mechanical nightmare. To date, no feed system has been proven to be reliable and durable enough to withstand the tortures of prolonged operation. Conveying line configuration, allowable velocities with respect to moisture and erosion, splitter effectiveness, and feed nozzle design are among several items which must be better understood prior to design and operation of a full-scale plant. These factors are being addressed in a coal feed test facility.

MATERIALS CORROSION

Previous research has indicated that standard fossil plant materials will perform satisfactorily in an AFBC boiler. However, it must be noted that coal/limestone feeding and distribution, system control and operations, and bed performance and thermal efficiency could greatly impact the performance of boiler materials. Testing during pilot plant operation will address these potential problems.

Another problem in materials involves the fabrication of the weld between the 2-1/4 Cr-1 Mo ferritic superheater tubes and the stainless steel superheater tubes. Conventional practice is to use high nickel filler materials, but experience in AFBC indicates that high nickel materials should not be used due to rapid sulfidation. In the pilot plant this issue was avoided by using a two-pass superheater with both banks fabricated from all stainless steel. Dissimilar weld samples will be tested in a flue gas slip-stream in the pilot plant to provide data for full-scale commercial units.

EROSION

To achieve acceptable bed performance (SO_2 capture and carbon burnup, and acceptable combustion efficiency), recycle will be needed. However, recycle will cause higher dust loadings and higher velocities which could result in potentially more erosive conditions in the convective heat transfer region. AFBC recycle will require dust loadings and gas velocities greater than those found in conventional steam plants. In the pilot plant, erosion test racks will be placed in the convection pass to determine actual erosion rates.

RECYCLE

The original design for the TVA pilot plant incorporated a carbon burnup bed; but because of recent research, the TVA pilot plant will recycle unburned coal and limestone to the main combustor without the use of a carbon burnup bed. Combustion efficiencies of about 99 percent are expected with the use of recycle. The TVA pilot plant will initially be constructed with 750°F recycle and recycle rates of up to five times the coal feed rate. Provisions are made for 1,550°F recycle to be added later. It is predicted that with recycle, a Ca/S ratio of two or less can be obtained.

As might be expected, flexibility will be the key to TVA's 20-MW AFBC pilot plant. The plant will be operated in a wide variety of operating conditions for optimum value. To monitor these conditions, a direct digital and analog control system and data acquisition computer are featured in the design of the facility.

Because of the limited duration of the 20-MW AFBC pilot plant as a test unit, it is not economical to install a turbogenerator for the production of electricity. The operation and testing of this facility will provide us with valuable experience which will later be used to design commercial-scale units.

PILOT PLANT TEST PROGRAM

The major objective of the pilot plant project is to make the best use of the facility as a link between small-scale technology development units, larger demonstration plants, and utility-scale commercial plants. A test program, which will evaluate the advantages and uncertainties of AFBC technology, has been formulated. Six major tasks provide for the activities required to accomplish the test program objectives. A test plan will serve as a control document for this work.

The six major tasks are:

TASK I - TEST ENGINEER TRAINING

Activities are being conducted at various facilities to indoctrinate and train the test engineers in AFBC theory and practice. Computer systems, controls, sampling, and data reduction techniques are also stressed.

TASK II - PROCEDURE DEVELOPMENT

A set of standard procedures is being developed by the personnel who will be conducting the tests. These include test procedures and equipment operating procedures.

TASK III - COMPUTER SOFTWARE DEVELOPMENT

Various data reduction routines are being developed to efficiently and effectively handle the large amounts of data produced during the test program.

TASK IV - COLD MODE TESTING AND CALIBRATION

A preoperational shakedown phase will condition the pilot plant for light-off and steam production. This will accomplish the contractor operability tests and generate base-line performance data.

TASK V - HOT MODE TESTING CAMPAIGNS

This is the major thrust of the test program. The basic design and design modifications will be tested and evaluated to develop and demonstrate AFBC technology and enhance larger scale demonstration and commercialization.

TASK VI - DATA REDUCTION, ANALYSIS, AND RESULTS PRESENTATION

Data packages for each test and initial data analyses will be prepared at the test site. The major data analyses and reporting will be accomplished by the project management staff.

Work on the test program began in 1979 with the development of a test plan which is now nearing completion. The test plan is conceived to be an adaptive control document which allows for design modifications as test results are obtained and analyzed; therefore, test plan changes are planned and scheduled into mid-1985. The first three program tasks are also underway. Cold mode/shakedown testing is scheduled to begin in January 1982. Hot mode testing should begin in mid-1982; all tests should be completed in early 1986; and final results and reporting should be completed in mid-1986.

TECHNOLOGY TRANSFER AND COMMERCIALIZATION

Before the utilities will accept AFBC technology as a coal-fired alternative for electricity generation, their management staff must be convinced that the investment risks are more favorable than, or at least competitive with, existing coal-fired alternatives that are commercially available.

Specific AFBC operating experience and valid experimental data for engineering evaluation are essential prerequisites to the development of a comparative analysis of options for utility management decisions. Such an analysis must adequately address the more significant utility issues that influence the total cost of producing electricity, namely:

- Fuel flexibility and plant thermal efficiency
- Operating reliability and maintenance
- Environmental impact and waste product control
- Operating flexibility versus system needs
- Time required for total plant construction
- Capital investment and warranties
- Personnel and training levels required
- Operational health and safety aspects

The AFBC pilot plant will be operated at utility steam conditions. It will simulate the various transient maneuvers that a coal-fired plant is typically subjected to in the utility operating environment where load demands fluctuate extensively and frequently. While carefully operating the pilot plant using a preplanned test program, an extensive amount of experimental data will be obtained from an elaborate array of instruments, sensors, probes, etc. These data, covering all facets of the AFBC technology, will provide the real knowledge essential to adequately address the issues set forth above that are of paramount concern to the utility decision makers.

TVA will definitely transfer this technology to the utility and manufacturing industries as the 5-year test plan is carried out. The methods by which this will be accomplished are now being studied.

STATUS OF TVA'S 200-MW AFBC DEMONSTRATION PLANT

TVA began design activities for a 200-MW AFBC demonstration plant in September of 1976. We are working toward the final design, construction, and operation of a central station AFBC plant to establish this option for future coal-fired electric generation plants. TVA is investigating three potential applications for this technology: additional capacity, replacement power, or retrofit of an existing plant. Each of these will economically meet EPA standards for emissions control while utilizing our vast resources of medium and high sulfur coals.

Final conceptual designs of the AFBC boiler and auxiliary equipment were completed by the boiler manufacturers in January 1981. TVA's efforts in the Phase II studies and cost projections leading to the preparation of a bid specification are scheduled to be completed by October 1981. We are presently assessing the three designs, designing the balance of plant (BOP), and preparing a draft AFBC demonstration plant boiler specification. This draft will be reviewed by the boiler manufacturers prior to its finalization. Cost estimates and design specifications for major systems and equipment of the BOP will be completed by October 1981. An environmental impact statement (EIS) is also nearing completion.

Present planning is based on a demonstration plant schedule which will allow a reasonable amount of time to evaluate the pilot plant test results and will reduce the risk of engineering decisions based on extrapolations of earlier data or theoretical predictions alone. This extra time should be sufficient to optimize the demonstration plant design and provide a logical path for acceptance by other utilities.

On this timetable, completion of the demonstration plant is projected for early 1988 with utility acceptance a few years later. TVA could accelerate the schedule so that operations could begin as early as 1986. Results from the pilot plant would serve to confirm, but not substantially alter, the demonstation plant design if this accelerated schedule were followed.

Because of the inherent environmental benefits of AFBC, this plant should easily meet the New Source Performance Standards as set by the EPA. The EIS will contain the latest project information and be submitted for review and comment to interested individuals, groups, and agencies in mid-1982, prior to seeking TVA Board authorization to proceed with detailed design and construction.

OTHER FLUIDIZED BED COMBUSTION ACTIVITIES

TVA has also contracted with CE and Lurgi Chemie to prepare a preliminary design and cost estimate of a 200-MW demonstration plant and an 800-MW commercial-size circulating fluidized bed combustor (CFBC). This design activity was completed in December 1980. The CFBC has several advantages over the bubbling bed AFBC, two of which are: it utilizes fewer coal and limestone feed points; and it shows potential for higher carbon utilization with resulting lower Ca:sulfur ratios to meet the SO_2 capture requirements. The largest potential disadvantage is the higher auxiliary power requirement and its associated costs.

TVA's fossil plants, like many throughout the country, are getting old and will be needed to provide "intermediate" and "peaking" power. Many of the units are not designed to efficiently perform in this manner. Faced with these conditions, some of the more promising options are to replace the present boilers with an AFBC boiler by retrofit or add-on, or build an AFBC stand-alone plant to replace an old, inefficient, and environmentally unacceptable plant.

TVA has initiated a study to determine the feasibility to retrofit an aging pulverized coal (PC) boiler with an AFBC boiler. A short duration study conducted by Stone & Webster Engineering Corporation and Riley Stoker Corporation has concluded that there are no major problems which would prevent removing an existing PC boiler and replacing it with an AFBC boiler. Based on this study, a project authorization request is planned to obtain approval to proceed with a conceptual design feasibility AFBC retrofit study for the TVA site determined to be the best candidate for a retrofit demonstration. This conceptual design study would be the foundation for a detailed boiler design and construction contract.

While our projects continue to reach maturity, we are addressing some of the technical problems related to AFBC. One of the problems with AFBC is in the development of coal/limestone feed systems. Hoping to get answers to these problems, we have recently initiated a program that involves the testing of a Fuller-Kinyon feed pump at an existing TVA steam plant. TVA has built a coal feed test facility at the Watts Bar Steam Plant. Construction was completed in December 1980 and a six-month testing program is underway.

The Watts Bar coal feed test facility will be the only large-scale demonstration of a feed system that is applicable to the 200-MW AFBC demonstration plant. Also, since the 20-MW pilot plant will use a Fuller-Kinyon pump, the test facility will provide the added advantage of proving the effectiveness of the pump and splitter prior to construction.

Other technical supports include work with the Oak Ridge National Laboratory (ORNL). In a program with the Department of Energy (DOE), TVA and ORNL are working together on a number of task programs: an AFBC bench-scale combustor to test coal and limestone; a cold flow model for slumping tests; modeling and simulation; materials research; and other miscellaneous technical support activities.

TVA is also investigating the recycling of elutriated particulates (fly ash and unburned carbon) from the AFBC process. General Atomic is performing a series of experiments on its 16-inch combustor to investigate the various recycle parameters.

Fundamental to all of these specific needs is mathematical modeling. Steady state and transient models are being evolved from DOE work at the Massachusetts Institute of Technology and other sources to be used as the basic tool to predict the effects of operation and design parameters on the performance characteristics of the AFBC pilot plant under steady state and transient conditions. These models will highlight gaps in available knowledge and guide in development of the pilot plant test plan.

TVA has worked closely with DOE. We have been quite involved with reviews and planning for the Rivesville AFBC facility and have been a part of the monthly management reviews for the Component Test Integration Unit (CTIU) at Morgantown, West Virginia.

TVA has also followed the technological developments of pressurized fluidized bed combustion. A contract with ORNL produced an assessment of the state-of-the-art of PFBC systems, and we were a part of a DOE task force to evaluate cost estimates for their proposed PFBC facilities.

SUMMARY

Atmospheric fluidized bed combustion appears to be a promising technology for development in the years to come. Before AFBC can be used for full-scale commercial service, however, many problems need to be resolved. With its engineering capabilities and a large power system, TVA is an excellent candidate and proving ground for developing this technology.

METHANOL--AN OPPORTUNITY FOR THE ELECTRIC UTILITY INDUSTRY TO PRODUCE ITS OWN CLEAN LIQUID FUEL

M. J. Gluckman and B. M. Louks
Electric Power Research Institute

INTRODUCTION

The U.S. electric utility industry is facing a new and critical dilemma as it prepares to enter the twenty-first century. Severe capital shortages coupled with inflationary interest rates and long permitting and construction lead times have dramatically increased the incentives to keep the existing 200,000 MW of oil and natural gas fired capacity on-line. Second, recent conservation measures together with rapidly increasing utility rates have resulted in a sharp decrease in load growth projections for most sections of the country. A decreased requirement for new large base loaded units coupled with the capital problems mentioned above, have increased the need to consider smaller, more flexible equipment such as combined cycles and simple cycle gas turbines for system expansion. Such units require ultra-clean liquid or gaseous fuels that are available on an as-needed basis.

All of the above factors point in the direction of a greater need for a secure supply of clean liquid fuels for electric power generation in the near and intermediate term future.

The industry, however, senses an escalating difficulty in being able to secure the critically needed supply of liquid fuel. Legislative and political moves such as the Industrial and Power Plant Fuel Use Act of 1978, pending "oil backout" legislation and political uncertainties in the Middle East have made the vulnerability of the electric utility industry clear. Similarly, recent major price increases for imported crude oil together with deregulation of domestic oil and gas supplies have increased the difficulty associated with securing a long term supply of reasonably priced petroleum derived liquid fuels for electric power generation. Competition for petroleum derived liquids to supply the profitable transportation fuel market has further compromised the supply of liquid fuels for power generation.

The difficulties associated with the acquisition of petroleum derived liquids for power generation have stimulated interest on the part of the electric utility industry in synthetic liquid fuels produced from coal and oil shale. It seems clear, however, from actions taken to date, that both shale oil and coal derived liquid fuels production will be dominated by the existing petroleum and chemical industries. In particular, it appears highly unlikely that the electric utility industry will participate to any major extent in the capitalization or the operation of these large and highly capital intensive liquid fuel production facilities.

It is likely that some fraction of the output (probably the heavier products) of these synthetic liquid fuels plants will be available to the electric utility industry. However, they will be sold at market prices that are competitive with the equivalent petroleum products and their availability will be subject to the world demand for transportation fuels.

All of this leaves the electric utility industry in a particularly vulnerable and uncomfortable position with respect to future supplies of liquid fuels. The question that must be addressed is how can the utility industry control its own destiny by ensuring a long term supply of clean liquid and gaseous fuels at a reasonable cost without having to make major capital investments in facilities producing large fractions of their output as non-utility fuels?

The purpose of this paper is to present the authors' opinions concerning potential answers to the above question by indicating a variety of future opportunities for the utility industry to coproduce relatively small quantities of methanol either in new base loaded gasification-combined cycle power plants, or in conjunction with intermediate Btu fuel gas facilities, constructed to retrofit or repower existing oil and gas fired stations. It is important to point out the fact that coproduction of methanol and either electricity or fuel gas is not being presented as the only solution to the liquid fuel supply problem. It should rather be considered as one option that potentially appears to offer attractive economics, as well as supply security.

METHANOL AS A UTILITY FUEL

Environmental constraints and requirements within the last decade have contributed significantly to increased capital requirements for new and existing generating facilities and to major difficulties in siting new power plants. In the light of such constraints, the premium to be placed on ultra-clean fuels can be high.

Although methanol has not been traditionally used as a power generation fuel, it is extremely clean and could, therefore, contribute significantly to decreasing emission problems currently facing the industry. Methanol contains no sulfur, no particulate matter and no nitrogen. It also burns at lower temperatures than petroleum derived liquid fuels or natural gas. It would therefore be anticipated to produce significantly lower nitrogen oxides emissions than petroleum based liquids and other synthetic fuels.

In 1979, EPRI in conjunction with Southern California Edison Company, burned 2,500,000 gallons of methanol in one of two TPM FT4C-1DF gas turbines at SCE's Ellwood station.(1) By the end of the test program, 523 hours of methanol-fired operations had been logged. Results from these tests were compared with operating results from the same facility firing natural gas and a distillate JET A type fuel. Table 1 presents some of the emission results from these combustion trials. The results of Table 1 show that methanol without water injection significantly decreases NO_x emissions over the NO_x emissions from either distillate fuel or natural gas with no water injection. If water is injected with the methanol at the rate of 0.22 lbs. water per lb. of fuel, the NO_x emissions would be further reduced to 18 ppm.

Hot section inspections conducted after the test series indicated cleaner components in the methanol fired machine than were found in the distillate fired turbine. This suggests that firing methanol could improve liquid fired turbine reliability.

In summary, the results of these tests suggest that methanol would be an excellent turbine fuel both from an operational and an emissions viewpoint.

Methanol is not currently used as a turbine fuel due primarily to its high cost when produced from natural gas ($10.00/$10^6$ Btu - $13.00/$10^6$ Btu) and low availability. Studies by EPRI and others indicate that the cost can be reduced if methanol is produced from coal.

The purpose of this paper is to surface potential opportunities that the electric utility industry could consider to produce methanol together with electricity and/or fuel gas from coal at substantial additional reductions in cost and provide security of supply.

A word of caution must be injected at this point. Some of the technologies described in this paper have not yet been proven at full commercial scale. All of the cost estimates, however, have been prepared on the basis of mature technology and are therefore representative of what the fifth or sixth large plant will cost. The first plant will probably cost significantly more than is indicated by the results shown in this paper. However, it must be kept in mind that the primary objective of this study is to show relative differences between various plant and product costs, rather than absolute costs for any single system. The relative differences presented are considerably more reliable than any of the absolute costs.

DESCRIPTION OF SYSTEMS EVALUATED

Two separate circumstances have been evaluated to demonstrate potential opportunities for utilities to coproduce methanol in conjunction with the construction and operation of some other energy producing facility. The first situation involves the need to add a large, base-load power plant to the system. If this need is met by constructing an oxygen-blown gasification-combined cycle plant (see Figure 1), the cleaned, intermediate Btu fuel gas can be passed through a methanol synthesis reactor prior to entering the combined cycle power plant as shown in Figure 2. If the fuel gas is not shifted prior to entering the synthesis reactor, and if the unconverted gas leaving the methanol reactor is not recycled, approximately 25% of the volumetric flow of feed gas will be converted to methanol in a

---

(1) "Test and Evaluation of Methanol in a Gas Turbine System," EPRI AP-1712, February 1981.

TABLE 1

EMISSIONS FROM VARIOUS FUELS FIRED IN A TPM FT4C GAS TURBINE

	Methanol	Distillate Fuel		Natural Gas	
	No Water Injection	No Water Injection	0.55 Lbs. Water Per Lb. Fuel	No Water Injection	∿0.35 Lbs. Water Per Lb. Fuel
NO_x @ 15% O_2, ppm	45	207	56	124	65
NO_2 @ 15% O_2, ppm	10	10	10	50	40
CO @ 15% O_2, ppm	70	50	60	175	220
Hydrocarbons @ 15% O_2, ppm	10	5	5	216	280
Aldehydes @ 15% O_2, ppm	1.8	--	0.05	10.6	12.1
SO_2 @ 15% O_2, ppm	0	13	13	0	0
Solid Particulates, lb/10^6 Btu	.003	--	.008	--	--
Heat Rate, Btu/kWh	11,722	11,863	12,014	11,863	--

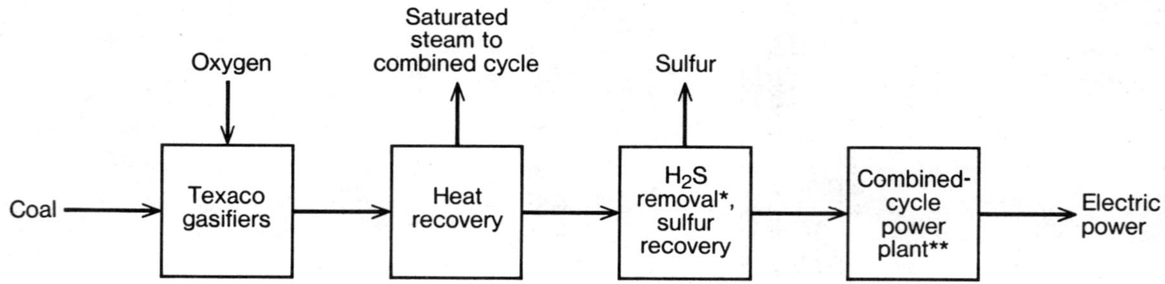

* COS hydrolysis and Selexol™
**With 2000°F combustion turbine

Figure 1. "Electricity only" gasification–combined cycle power plant.

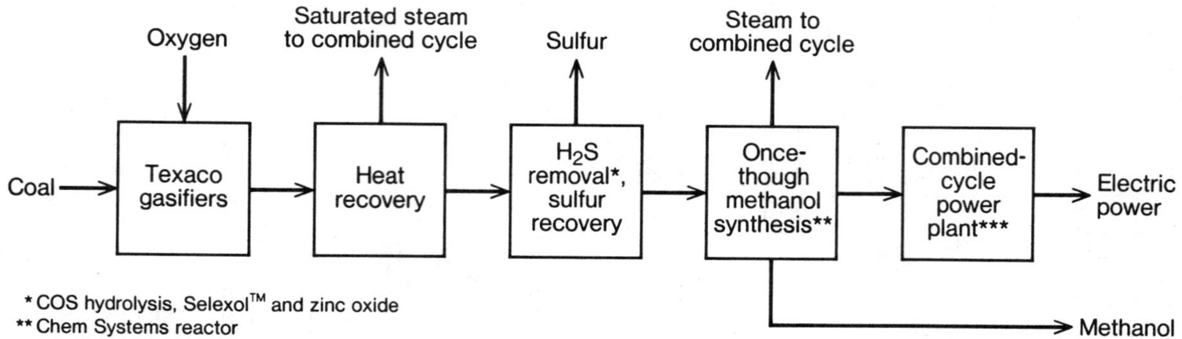

* COS hydrolysis, Selexol™ and zinc oxide
** Chem Systems reactor
*** With 2000°F combustion turbine

Figure 2. "Electricity plus once-through methanol" plant.

liquid phase ebullated bed system under development by Chem Systems, Inc. This mode of operation has been classified as "once-through" synthesis. Major advantages to be associated with this "once-through" mode of operation are:

- A reduction in capital cost due to the elimination of the requirement to shift the feed gas, to scrub CO_2 from the gas, as well as elimination of the unconverted gas recycle loop.

- Higher efficiency of conversion due to the ability to utilize the steam generated by the exothermic heat of reaction in the combined cycle power plant, as well as reduction in energy required to scrub CO_2 and to recycle unconverted gas.

- The ability to take advantage of the economics of large scale systems in a plant producing a relatively small quantity of methanol.

The second situation considered involves the requirement to produce a clean gaseous fuel to be supplied "over-the-fence" to retrofit and/or repower existing oil and gas fired units. Under this condition, a coal gasification plant producing a sulfur-free intermediate Btu fuel gas could be considered as shown in Figure 3. A potential problem associated with this concept is that if the existing generating units operate at a capacity factor of 50%-70%, and if no other customers for fuel gas can be found, the gas plant will have to operate at a similar low capacity factor, thereby sharply increasing the cost of the fuel gas produced and utilizing the invested capital in an inefficient manner.

A potentially attractive solution to this problem would be to coproduce methanol with the fuel gas. The methanol would provide liquid storage capability such that the gasification plant could operate at a high capacity factor (∿90%), thereby substantially reducing the capital investment per unit of fuel produced. This would also lead to a reduction in the total cost of fuel to the utility. Two coproduction cases have been considered. Figure 4 depicts a plant in which all of the clean fuel gas is passed "once-through" the Chem Systems liquid phase synthesis reactor. In this case, approximately 25% of the feed gas is converted to methanol, the rest being supplied "over-the-fence" as fuel gas product. The other alternative shown in Figure 5, represents a plant in which approximately 30% of the fuel gas from the particulate scrubber is shifted, has acid gas removed and is totally converted to methanol in a conventional ICI recycle synthesis loop. The remaining 70% of the fuel gas is supplied "over-the-fence" to the generating units after sulfur has been removed.

Finally, Figure 6 depicts a dedicated coal based methanol plant in which all of the fuel gas is shifted and converted to methanol in a conventional ICI recycle synthesis loop. This system has been evaluated in order to determine the cost of producing methanol in a large dedicated facility--probably owned and operated by a non-regulated private company.

TECHNICAL CONSIDERATIONS

For all cases, an Illinois No. 6 high sulfur coal is gasified at pressure in oxygen-blown Texaco gasifiers. All plants are mine-mouth and are self-sufficient with respect to process energy requirements. Detailed plant designs and itemized cost estimates will not be presented in this paper. Sources of all of the detailed design and cost information used will be indicated in the summary tables of results.

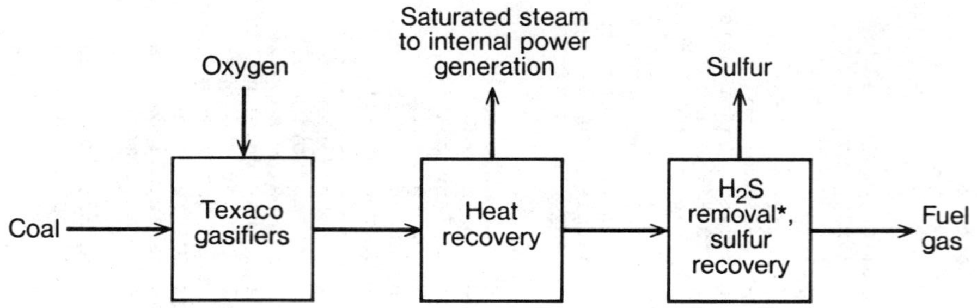

*COS hydrolysis and Selexol™

Figure 3. Fuel gas plant.

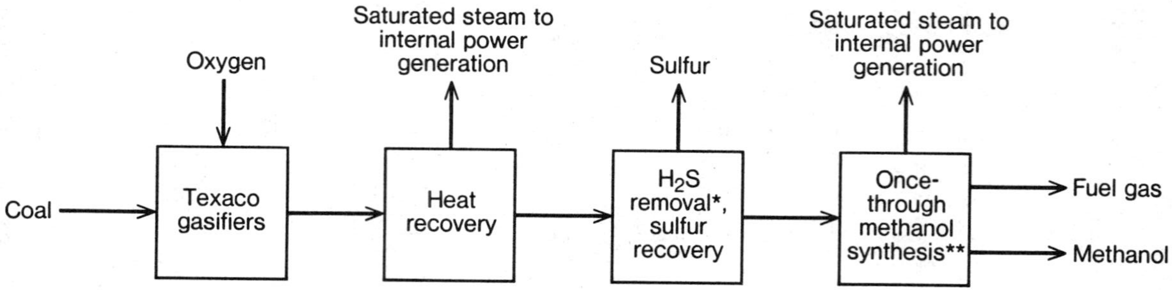

* COS hydrolysis, Selexol™ and zinc oxide
** Chem Systems reactor

Figure 4. "Fuel gas plus once-through methanol" plant.

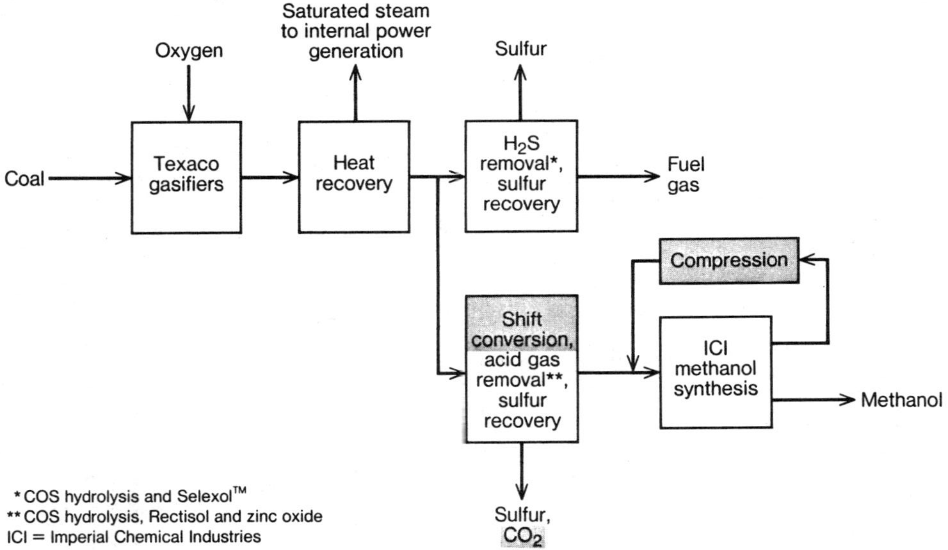

* COS hydrolysis and Selexol™
** COS hydrolysis, Rectisol and zinc oxide
ICI = Imperial Chemical Industries

Figure 5. "Fuel gas plus conventional methanol" plant.

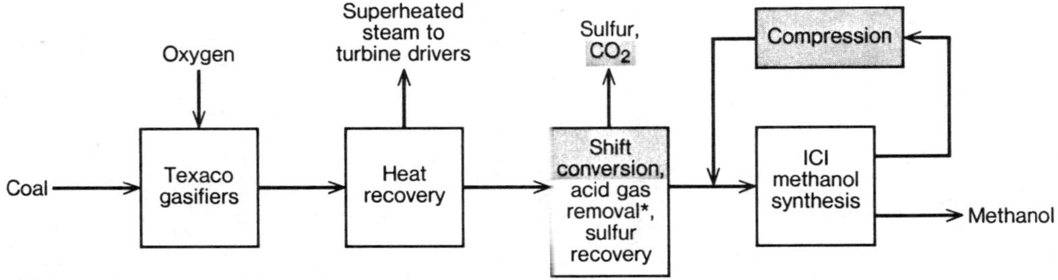

* COS hydrolysis, Rectisol and zinc oxide
ICI = Imperial Chemical Industries

Figure 6. Dedicated methanol plant.

Table 2 presents design and performance characteristics for the six systems being evaluated. Coal feed rates to all of the coproduction plants have been adjusted from those used in the original references such that the quantity of electricity or fuel gas being produced is the same as that in the non-coproduction case.

The last quantity shown in Table 2, thermal efficiency of methanol synthesis, is of particular interest. For the dedicated methanol plant, the conversion efficiency of coal to methanol is 57.9%. For the system coproducing electricity and "once-through" methanol, the conversion efficiency of coal to methanol is 68.9%. This increased efficiency strongly suggests a synergistic interaction between the GCC plant and the "once-through" methanol synthesis subsystem. Therefore, one would anticipate a reduction in the cost of methanol produced in this manner to methanol produced in a dedicated plant. For the two methanol/fuel gas coproduction cases, the efficiency of methanol synthesis appears to be the same as that in a dedicated plant, indicating the absence of process synergism. This suggests that methanol coproduced with fuel gas would cost more than methanol coproduced with electricity.

TECHNOLOGY STATUS

All of the plant sections employed in the designs of the six systems being evaluated have been operated for many years at commercial scale with two exceptions, i.e., the Texaco gasification section and the Chem Systems "once-through" methanol synthesis scheme.

The Texaco coal gasification process has been under development for approximately 10 years. Much pilot plant work has been conducted at Texaco's Montebello laboratory in two 15 ton/day gasifiers. Ruhrehemie has operated a 165 ton/day gasifier in Oberhausen, West Germany for 3 years and a 200 ton/day unit is currently under start-up at TVA's Muscle Shoals, Alabama facility. A number of Texaco coal gasifiers of unknown size are being designed for Tenessee Eastman and Dow Chemical Company has been operating a 150 ton/day air-blown Texaco gasifier for some time. Finally, detailed design for a 1,000 ton/day Texaco gasification system is underway for the Cool Water project sponsored by Southern California Edison Co., Texaco, EPRI, General Electric Co. and Bechtel. Start-up of this facility is planned for early 1984.

Therefore, although it is evident that the Texaco coal gasification technology is well on the way to commercial availability, it cannot yet be considered to be a technology that can be applied with normal commercial risk. Successful operation of the Cool Water plant is an essential next step for the large scale commercial development of this technology. It must also be pointed out that all of the performance and cost estimates presented in this paper have been based on "mature" Texaco technology. Larger gasifiers and gas coolers operating on more concentrated coal/water slurries than will be employed for Cool Water have been assumed.

Development work on the Chem Systems Liquid Phase Methanol Process has been underway since 1975.(2) Initial evaluations were conducted in a one inch diameter tube capable of handling approximately 40 SCFH of feed gas. This bench scale unit was scaled-up to a 3.62 inch diameter reactor process development unit capable of synthesis gas feed rates of 1500 SCFH (equivalent to between 0.25 tons/day and 0.5 tons/day methanol). Plans to demonstrate a scaled-up version (a two

(2) "Liquid Phase Methanol," EPRI AF-1291, December, 1979.

TABLE 2

PLANT DESIGN AND PERFORMANCE CHARACTERISTICS

	Gasification-Combined Cycle Power Plant	Gasification-Combined Cycle Power Plant With "Once-Through" Methanol	Gasification Based Fuel Gas Plant	Gasification Based Fuel Gas Plant With "Once-Through" Methanol	Gasification Based Fuel Gas Plant With Conventional Methanol	Gasification Based Dedicated Methanol Plant
Source of Design and Cost Data	(1)	(1)	(2)	(3)	(3)	(4)
Coal Feed Rate, Tons/Day A.R.	10,000	13,650	7,560	11,000	11,440	16,420
Methanol Produced, Tons/Day	--	3,260	--	2,390	2,485	10,930
10^9 Btu/Day (HHV)	--	61.5	--	46.3	48.1	211.8
FOEB/D[5]	--	10,513	--	7,915	8,230	36,205
Electric Power Produced, MW	1,106	1,106	--	--	--	--
Fuel Gas Produced, 10^9 Btu/Day (HHV)	--	--	122.2	122.2	122.2	--
Net Thermal Efficiency, % (HHV)	37.0	45.5	71.9[7]	68.1[7]	66.2[7]	57.9
Thermal Efficiency of Methanol Synthesis[6], % (HHV)	--	68.9	--	59.8	55.2	57.9

(1) Results of a detailed design and cost estimate prepared by Fluor Engineers & Constructors, Inc., under EPRI RP-239, Task 15. Publication anticipated by June, 1981.

(2) Results of a detailed design and cost estimate prepared by Fluor Engineers & Constructors, Inc., under EPRI RP-239, Task 18. Publication anticipated by July, 1981.

(3) Designs and cost estimates for these configurations prepared by the EPRI staff on the bases of references (1), (2) and (4).

(4) Results of a detailed design and cost estimate prepared by Fluor Engineers & Constructors, Inc., under EPRI RP-832-4. Publication anticipated for July, 1981.

(5) FOEB/D = Distillate fuel oil equivalent barrels per day. A fuel oil equivalent barrel is assumed to have an energy content of 5.85×10^6 Btu.

(6) Higher heating value of methanol expressed as a percentage of the higher heating value of the additional coal required to produce methanol.

(7) Based on producing fuel gas at the plant boundary at 750 psia.

foot diameter reactor feeding 62,500 SCFH of gas) of this process development unit are currently underway by Chem Systems and Air Products. For the designs presented in this study, "once-through" liquid phase methanol synthesis reactors having a capacity of approximately 5 x 10^6 SCFH of synthesis gas have been assumed. Therefore, much development work is necessary before the Chem Systems Liquid Phase Methanol Process can be considered to be commercial. It is for this reason that the coproduction of fuel gas and methanol employing the conventional, commercially proven ICI recycle methanol synthesis system was evaluated. For this design, the only non-mature large scale subsystem is the Texaco gasification section.

Finally, it must be pointed out that the two coproduction designs employing the Chem Systems "once-through" methanol synthesis concept do not depend completely on the development of the Chem Systems reactor. The authors have been assured that commercially proven methanol reactors (such as the ICI and Lurgi systems) can also be designed to operate in the "once-through" mode. This statement must be treated with caution at this time as yield data and costs have not yet been specified for this mode of operation. The Advanced Power Systems Division of EPRI plans to study the use of commercial methanol synthesis reactors operating in the "once-through" mode in 1981.

COSTS OF METHANOL FROM THE VARIOUS SYSTEMS

Two different costs of methanol to electric utilities will be presented: (1) A "price" for methanol produced by a non-regulated company in a dedicated coal to methanol plant, and (2) The "cost" of producing methanol by regulated utility companies. In the latter case, the cost (also referred to as "revenue requirement") includes required returns on common equity, preferred stock and debt. In the former case, the price includes a minimum required return on common equity with 100 percent common equity funding. This price is free to be adjusted up or down with the competitive market price for alternative fuels. On the other hand, the cost of methanol produced in a regulated utility plant is "fixed," since the utility must base its rates on actual costs. In addition, the cost (in constant dollars) in a regulated utility plant will usually decline with time, since unrecovered capital investment decreases.

Tables 3 and 4 present, respectively, the major bases used in calculating the revenue required for methanol and other bases that are usually of interest in defining the calculation method.

All of the plant costs were taken directly or derived from studies performed for EPRI by Fluor Engineers and Constructors, Inc. Costs for the GCC electric power plant, the GCC electric power plus "once-through" methanol system, and the fuel gas plant were taken directly from these studies and escalated from mid-1978 to mid-1980 at a rate of 13%/year. Costs for the dedicated methanol plant were based on the Fluor study, but were slightly adjusted to reflect consistent contingency allowances with the other cases being evaluated. Costs for the plants producing fuel gas plus methanol were generated by the authors using the data from the Fluor studies and including methanol synthesis sections in the fuel gas plant designs. These plant costs for systems having the capacities shown in Table 2, are presented in Table 5.

The following important observations can be made from the data in Table 5:

1. The cost per daily barrel of distillate fuel oil equivalent methanol in the electricity plus "once-through" methanol plant is 20% less than that for the dedicated methanol plant. This

TABLE 3

MAJOR ECONOMIC BASES

- Cost of coal - $1.14/million Btu plus 0.2% per year real cost increase.

- General inflation rate - 6% per year.

- Plant financing and required rates of return:

    - Dedicated methanol - 100% common equity, 15.01%/year rate of return*.

    - All other plants - 35% common equity, 13.5%/year rate of return.
      - 15% preferred stock, 8.5%/year dividend.
      - 50% debt, 8.0%/year interest rate.

    - Plant start-up date - 1990

    - Construction Period - 4 years for all systems.

*Derived from constant dollar rate of return of 8.5%/yr., e.g., (1.085)(1.06) - 1 = 0.1501

TABLE 4

OTHER ECONOMIC BASES

- Plant book lives:

  Dedicated methanol - 25 years

  All other plants - 30 years

- Plant tax lives:

  Dedicated methanol - 13 years

  All other plants - 22 years

- Investment tax credits:

  Dedicated methanol - 10% of plant investment, taken in year of expenditure.

  All other plants - 10% of 3/7 of plant investment, normalized over tax life.

- Depreciation for tax purposes - Sum of years digits for all plants

- Appropriate amounts were added to investment in plant for land, royalties, start-up costs and required working capital

TABLE 5

ESTIMATED PLANT COSTS[1]

Mid-1980 Dollars

	Million Dollars	$ Per FOEB/D[4] Methanol	$ Per Kilowatt
GCC Electric Power Only	1,117		1,010
GCC Electric Power Plus "Once-Through" Methanol	1,438	30,500[2]	
Fuel Gas Only	423		
Fuel Gas Plus "Once-Through" Methanol	696	34,500[3]	
Fuel Gas Plus Conventional ICI Methanol	761	41,100[3]	
Dedicated Methanol	1,377	38,000	

(1) These costs do not include prepaid royalties, working capital, land costs, start-up costs and allowance for funds used during construction, which are all essentially directly proportional to plant costs.

(2) Cost of electricity plus methanol plant minus cost of electricity plant: This quantity divided by the distillate fuel oil equivalent methanol production.

(3) Cost of fuel gas plus methanol plant minus cost of fuel gas plant: This quantity divided by the distillate fuel oil equivalent methanol production.

(4) FOE = Distillate fuel oil of heating value equivalent to that of methanol produced.

indicates that anticipated capital savings due to elimination of shift conversion, CO_2 scrubbing, and recycle synthesis are potentially available.

2. As would be expected, capital savings for methanol synthesis in the fuel gas plus conventional ICI methanol plant are not available.

3. Capital savings for methanol synthesis in the fuel gas plus "once-through" methanol plant are approximately half of what they are in the electricity plus "once-through" methanol case.

4. A most interesting observation, however, is that for all of the methanol coproduction plants, the capital required per unit for methanol synthesis is lower than or the same as that required for dedicated methanol production at approximately four times the production quantity. This implies that in methanol coproduction plants producing relatively small quantities of methanol, the economics of large scale plants are achieved.

METHANOL COSTS

Dedicated Methanol Plant - The prices required for methanol are plotted on Figure 7. They were first calculated using a general inflation rate of 6% per year, then de-escalated to constant 1980 dollars. The significance of the price in 1980 dollars ($6.63 per million Btu) can be stated as follows. If a contract were written in mid-1980 for product delivery beginning in 1990, and an initial mid-1980 selling price of $6.63 per million Btu, indexed to a general rate of inflation were agreed to, actual selling prices would be those represented by the upper curve if inflation was at 6% per year, and the seller would realize a 15.01 per cent per year discounted cash flow rate of return on common equity. On the other hand, if inflation was zero, the 1980 price would not change, and the seller would realize an 8.5 per cent per year return.

There is an advantage to thinking of prices and costs in constant 1980$. That is, the values can be compared with known costs of other energy products at the same time, e.g., mid-1980. For example, the marginal cost of producing finished oil products from the highest priced foreign crude oil in mid-1980 was in excess of $7/million Btu.(3) Therefore, it is easy to see that the methanol could be competitive.

Electricity Plus "Once-Through" Methanol Plant - In this case, it was first necessary to calculate revenue required for electricity produced in the GCC plant generating electricity only. These revenues were then credited as co-product revenue to the GCC plant producing both electricity and methanol, and the methanol cost determined accordingly.

Costs of electricity for the plant producing electric power only are presented in Figure 8 for annual plant capacity factors of 70 and 90 per cent. Although base loaded power plants generally do not operate at average capacity factors much greater than 70%, the 90% capacity factor case was also evaluated due to the novel nature of a plant coproducing electricity and methanol. The premium to be placed

---

(3) Reflects a landed Algerian or Libyan crude oil price of about $38 per barrel, and an approximate refining cost of $5 per barrel.

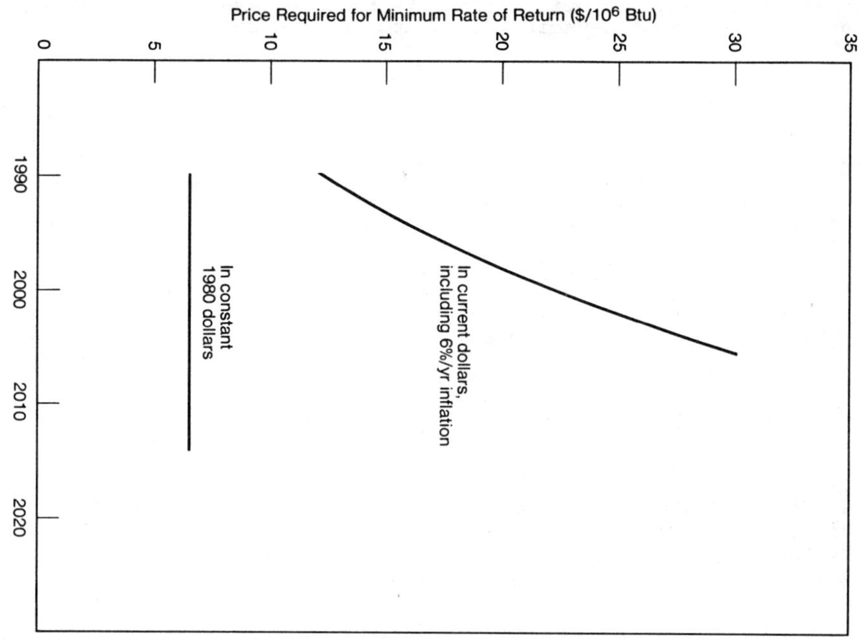

Figure 7. Price required for methanol produced in dedicated methanol plant by nonregulated company.

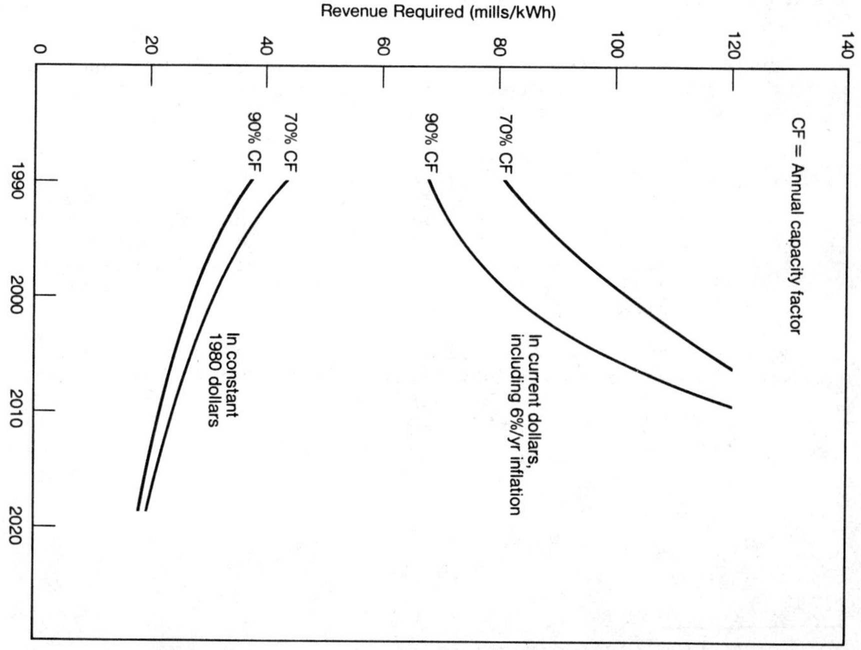

Figure 8. Year-by-year revenue required for electricity—electricity only plant.

on the ability to produce a clean liquid fuel for intermediate and peak load generation might indicate minimum system operating costs by maximizing the capacity factor for the coproduction plant.

The resulting costs of methanol produced in the plant coproducing electricity and "once-through" methanol are presented in Figure 9 at the two capacity factors. As observed, at either capacity factor, the costs (in 1980$) decline from a range of 5 to 6 $/million Btu to less than $3 per million Btu. The levelized costs(4) are $4.51 and $3.95/million Btu for 70% and 90% capacity factors respectively, considerably below the $6.63 value for purchase of dedicated methanol production.

Costs of Methanol Produced in Fuel Gas and Methanol/Co-Production Plants - The method for calculating the revenue required for methanol produced in these plants involved first calculating the required revenue for a unit of fuel gas. These results are presented in Figure 10. Fuel gas revenues were then credited to the co-production plants as co-product revenues, and revenue required for methanol determined from the known total revenue requirement. Year by year costs for methanol produced by "once-through" or by conventional recycle synthesis in conjunction with fuel gas production are presented in Figures 11 and 12 respectively.

The levelized costs of methanol in 1980 dollars are $4.71 and $5.32 per million Btu for the "once-through" and conventional synthesis schemes respectively. Both costs are appreciably less than the $6.63 per million Btu for purchased methanol.

Summary of Costs - All of the methanol costs calculated are summarized graphically in Figure 13. Also included in this figure is the purchase price of dedicated methanol if the competitive market price of clean fuels increases at the rate of 1 per cent per year. Levelized costs for methanol are shown in Table 6.

The methanol prices and costs plotted in Figure 13 and presented in Table 6 indicate a number of characteristics that should be of interest to the electric utility industry:

- The selling price of methanol in constant 1980 dollars, produced in a dedicated methanol plant owned and operated by a non-regulated company will stay level, at best, or will increase sharply if petroleum derived fuels continue to experience real price increases. In contrast to this, the cost of methanol in constant 1980 dollars, produced by an investor owned utility will always decline with time.

- Methanol coproduced with fuel gas or electricity by a utility by any of the methods discussed in this paper, has the potential to be between 20% and 40% lower in cost than coal derived methanol purchased on the open market.

(4) A levelized revenue requirement (cost) is one which if held constant will yield the same return on common equity as the varying values from which it is derived.

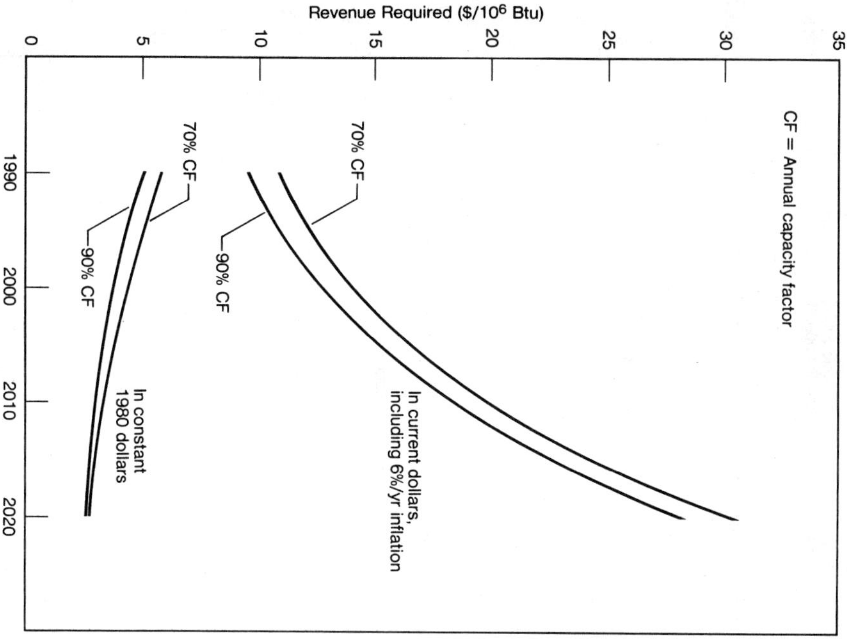

Figure 9. Year-by-year revenue required for methanol produced by "once-through" synthesis in electricity plus methanol plant.

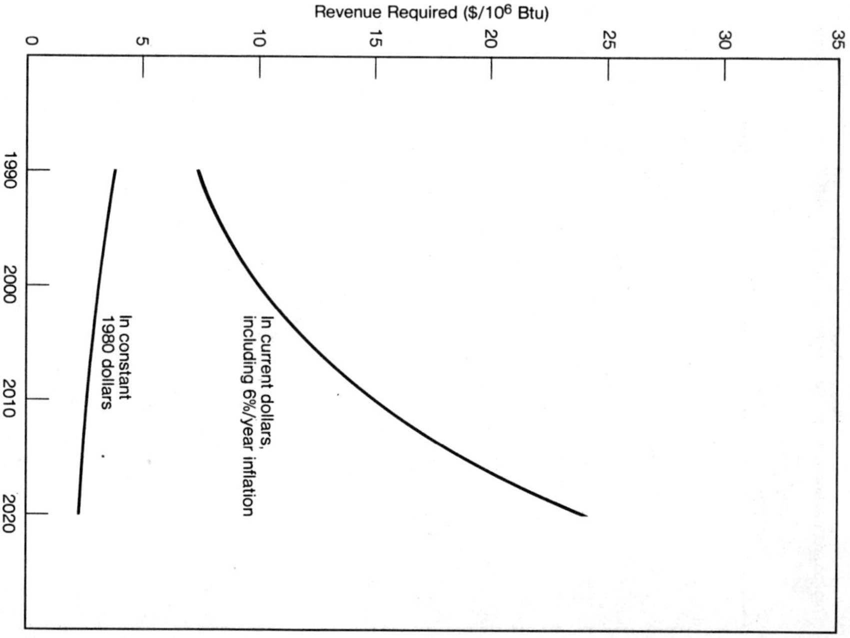

Figure 10. Year-by-year revenue required for fuel gas from coal (90% capacity factor).

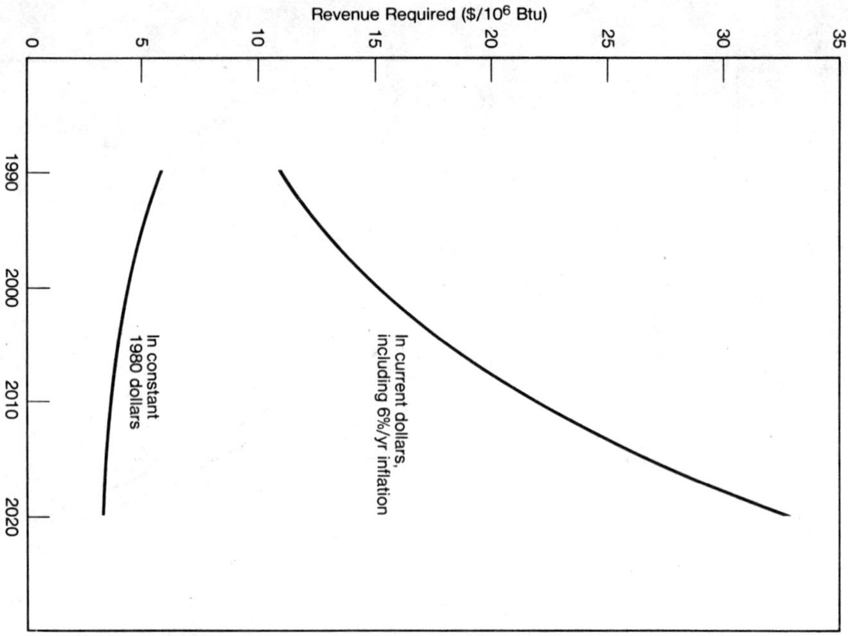

Figure 11. Year-by-year revenue required for methanol produced by "once-through" synthesis in fuel gas plus methanol plant.

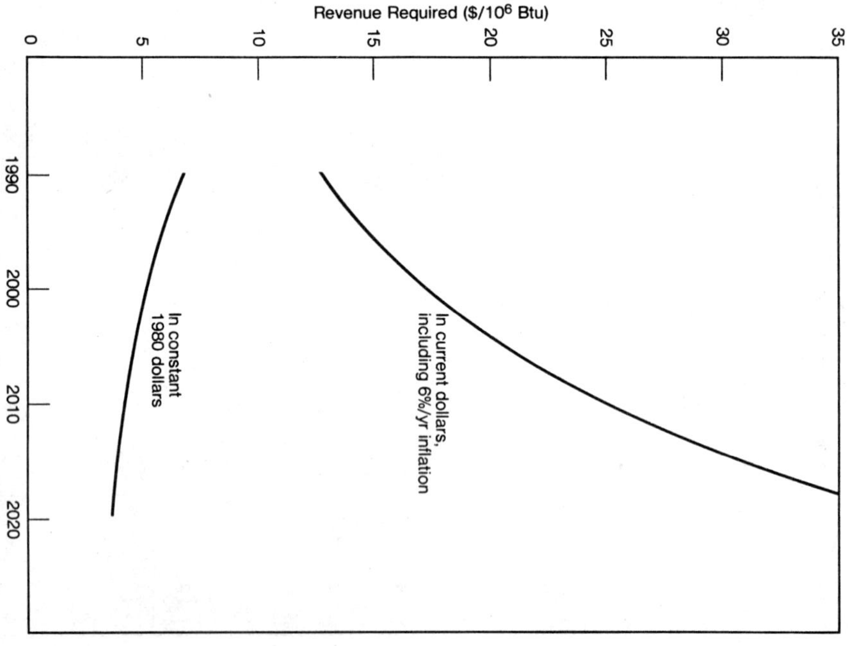

Figure 12. Year-by-year revenue required for methanol produced by recycle synthesis in fuel gas plus methanol plant.

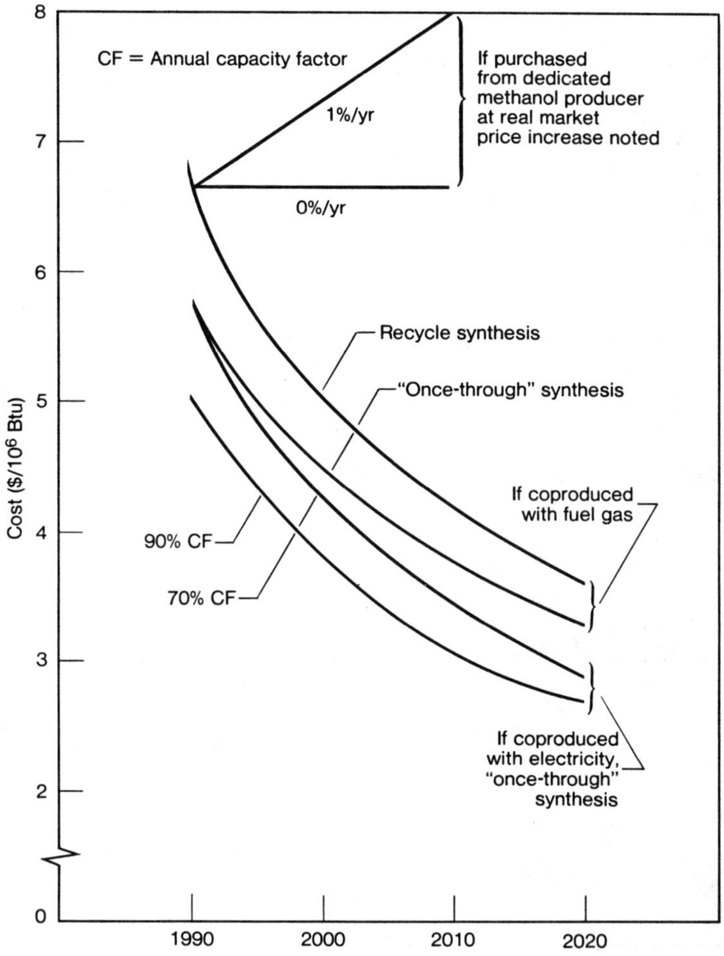

Figure 13. Costs of methanol in constant 1980 dollars.

TABLE 6

LEVELIZED COSTS OF METHANOL

IN CONSTANT 1980 DOLLARS*

	Dollars Per Million Btu	Percentage Reduction In Methanol Cost Over Dedicated Methanol Price With No Real Increase in Clean Fuel Market Price
Purchase of Dedicated Methanol From Non-Regulated Producer:		
No Real Increase in Clean Fuel Market Price	6.63	--
1%/Yr. Real Increase in Clean Fuel Market Price	7.27	(10)
2%/Yr. Real Increase in Clean Fuel Market Price	8.00	(21)
Methanol Co-Produced By Investor Owned Utility:		
With Electricity		
"Once-Through" Synthesis	3.95(1), 4.51(2)	40(1), 32(2)
With Fuel Gas		
"Once-Through" Synthesis	4.71	29
Conventional Recycle Synthesis	5.32	20

*A levelized price (or cost) is one which if constant over the life of a plant results in the same rate of return on common equity as does the varying prices (or costs) from which it is calculated.

(1) At 90% capacity factor.

(2) At 70% capacity factor.

- The lowest methanol costs in any of the coproduction schemes evaluated always occur when the methanol is synthesized via the "once-through" route. This implies strongly that there is a large incentive for "once-through" methanol synthesis processes to be developed.

CONCLUSIONS

In general, this work has indicated the potential benefits to be derived by the electric utility industry from coproduction of methanol with either fuel gas or electricity. Based on the capital shortages facing this industry, it is unlikely that many dedicated coal to methanol plants will be owned by power companies. This means that if the industry is contemplating purchasing methanol from non-regulated producers, when and if it is available, they must anticipate prices that will inflate at petroleum rates. By coproducing small quantities of this liquid fuel in large baseload GCC power plants or in gasification based fuel gas plants, the industry can actually achieve declining constant dollar costs for liquid fuels at a considerable saving in cost, while at the same time ensuring a secure supply.

For a company contemplating construction of a new coal gasification-combined cycle system, it would be beneficial to consider coproduction of methanol fuel. The systems studied have indicated that such a strategy could reduce the cost of liquid fuel to the utility by up to 40%. If a coal gasification plant producing fuel gas for retrofitting existing oil and gas fired capacity is contemplated, coproduction of methanol could reduce liquid fuel costs by as much as 30%.

The systems described in this paper appear to offer attractive opportunities for the industry to reduce both capital requirements and operating costs while meeting stringent environmental control requirements. However, it must be kept in mind that the gasification technology employed in these studies has only been demonstrated at one-tenth the scale required and the Chem Systems "once-through" methanol technology has only operated in a 3.62 inch diameter reactor. Therefore, these types of systems cannot be considered to be commercially proven at this time. The results of this work strongly suggest that the Cool Water Demonstration Plant and, to a lesser extent, rapid demonstration of the "once-through" methanol concept (by Chem Systems, Lurgi or ICI) are key requirements to enable these types of plants to be available to the industry by the early to mid-1990's.

ACKNOWLEDGEMENT

The authors greatfully acknowledge the assistance of Dr. S. Kohan in assembling the capital cost estimates and organizing the financial analyses presented in this paper.

CASE STUDY OF CONVERSION TO COAL - OIL - MIXTURE

Lawrence E. Page
McDonnell Douglas Corp.
Plant Engineering
St. Louis, MO

ABSTRACT

McDonnell Douglas Corporation, St. Louis, MO and CoaLiquid, Inc., a Kentucky corporation, began a joint effort to utilize a coal-oil-water mixture as the primary fuel in an industrial boiler. The Plant Engineering Department was directed to survey the feasibility and aspects of such a test burn. McDonnell Douglas would like to share with all interested parties the following points of interest of our test burn which commenced on September 3, 1980.

1.0 INTRODUCTION - COALIQUID, INC.

In June 1979, CoaLiquid, Inc., Louisville, KY, completed a commercial scale, preparation plant in nearby Shelbyville for a coal-oil-water emulsion. The dedication and official opening was on August 31, 1979. The facility is capable of producing 2,500 GPH and expandable to up to 5,000 GPH, with little additional expense. It marked the end of over three years of research and developmental work, and the beginning of a new era of fuel-oil from coal as a commodity in the marketplace. The technology employed to produce a stable mixture of 50% coal, 40% oil and 10% water by weight had been conceived by Eric C. Cottell of Long Island and is covered by U.S. Letters Patent No. 3,941,552.

2.0 PROCESS DESCRIPTION

CoaLiquid fuel is produced by the mixing of pulverized coal, oil, and water into a slurry, and then passing the slurry through an ultrasonic reactor system.

2.1 Pulverization

The most economical means of producing the pulverized coal was through the use of a cage mill. The cage mill will pulverize coal to the desired fineness, with coal remaining at stockpile moisture levels. Cage mills use gravity for discharge. In order to achieve the desired fineness, i.e., 80 to 90 percent minus 200 mesh, it was necessary to install three cage mills in series. The output of the cage mills is then processed over vibrating screens, with a screening size of 165 mesh. The particles passing through the screens are conveyed into processing and the oversized material recycled to the pulverization system. From the test burn study this system proved to be unsatisfactory. Further elaboration will follow later in this report.

The pulverized coal is not handled in an inert environment but rather conveyed in a fuel rich/air lean atmosphere through all stainless steel conveyors. Bucket elevators, vibratory conveyors, and motors are all explosion proof and spark resistant. There have been no difficulties or accidents associated with the pulverized coal handling. To assure good flow of bulk materials, CoaLiquid found it desirable to line all storage bins and chutes with either stainless steel or polyurethane. This has provided more than ample capacity for moving the balky material.

2.2 Weigh System

In determining the weight of coal to be used in a given batch of the CoaLiquid material, a set of three load cells were installed on a storage bin immediately above the premixing tanks. The signals generated at each of the load cells is totalized and transmitted to a batch scale controller in the main control room. The batch scale system should yield coal concentration control to within 1/10 of 1%.

2.3 Mixing

As far as premixing the material is concerned, the "continuous batch" concept is utilized. Under this concept, one tank is being filled with coal, oil, and water and a twin impeller mixer is used to begin blending the materials. While ingredient addition and blending are occurring in one tank, in the other tank the premixed material is being discharged through an in-line agitator into the ultrasonic device. Material flows through the ultrasonic device on a continuous basis straight to the storage tanks being the stable thixotropic fuel.

2.4 Pumps

For moving the C-O-M, CoaLiquid has found two types of pumps to be most satisfactory. They are first, positive displacement, rotary gear type and secondly, the progressive cavity, positive displacement type. Both types of pumps are used in the processing facility. A progressive cavity, positive displacement type, Moyno, Model CDO is used to transfer C-O-M from the tank transporter to our storage tank. All indications are that this pump is very reliable. To transfer the C-O-M through the test burn pumping system we use

Viking Model No. LL124 centrifugal rotary gear type pumps. They were not our first choice, but their availability was the dictating factor. These pumps begin showing signs of excessive clearances due to abrasiveness of C-O-M after 350 hours of operation. Although this time interval is relatively short, we have continued with these pumps due to the fact the time and cost to rebuild is minimal. In designing process piping, consideration should be given to the avoidance of gate valves and short radius, 90° elbows. Wherever possible, long radius elbows should be used and any form of trap or sharp bend eliminated. Contrary to the above, such is not the case in our piping; however, precautions are taken to blow out lines with steam or shop air after shutting down the system. The piping selected is of no special alloy, simply standard, Schedule 40 piping. Pipe diameters are somewhat larger than would be employed in a conventional fuel oil pipe system, because of the relatively higher viscosity of the C-O-M. As has been proven, ball valves and butterfly valves are most desired in C-O-M handling.

2.5 Metering

For metering, CoaLiquid found it necessary to insure no meter fuel contact. By installation of a simple diaphragm between the meter and fuel flow, it proved relatively simple to insure proper metering without equipment loss due to entrapped coal.

2.6 Stability

An extensive study was conducted by the University of Kentucky, Institute for Mining and Minerals Research for the Tennessee Valley Authority. The test was to determine stability for long term storage. The C-O-M, CoaLiquid, was stored in a four-inch diameter vertical pipe which was 21 feet tall. Samples were taken from the pipe at various heights over a span of six months and checked for viscosity and percent solids. Conclusions of the test performed were the following:

- C-O-M does not separate for at least a period of six months.
- No separation occurred in thixotropic emulsion.
- Mild extremes of heat and cold (7-140°F) do not permanently affect the emulsion.

2.7 Fuel Analysis

In conjunction with our 150,000 gallon test burn, an independent concern, Commercial Testing & Engineering Co., Chicago, IL, conducted an ultimate and mineral analysis of C-O-M with results as follows:

ULTIMATE ANALYSIS

	% Weight As Received
Carbon, C	80.17
*Hydrogen, H	7.99
Nitrogen, N	0.87
Chlorine, Cl	0.08
Sulfur, S	0.95
Ash	2.90
Oxygen(diff)	7.04
	100.00
Btu/lb.	15,289
Water, % by weight	5.99

*Includes hydrogen of water

MINERAL ANALYSIS

	% Weight Ignited Basis
Silica, SiO_2	49.71
Alumina, Al_2O_3	21.06
Titania, TiO_2	1.10
Ferric oxide, Fe_2O_3	16.91
Lime, CaO	2.12
Magnesia, MgO	1.70
Potassium oxide, K_2O	2.85
Sodium oxide, Na_2O	1.12
Sulfur trioxide, SO_3	2.91
Phos. pentoxide, P_2O_5	0.06
Undetermined	0.46
	100.00

The comparison of viscosities of the C-O-M and No. 6 fuel oil, a product which is more familiar to all industry, should be of interest and is shown by the following graph.

TYPICAL COMPARISON OF VISCOSITY CURVES OF NO. 6 FUEL OIL & COALIQUID

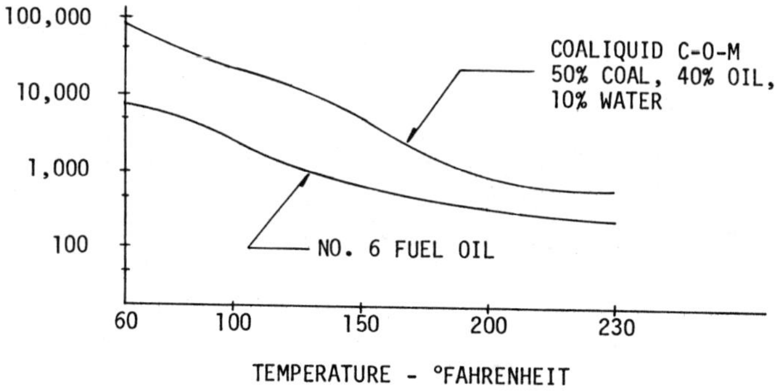

3.0 TEST BURN

McDonnell Douglas Corp. and CoaLiquid, Inc. jointly endeavored the aspects of conducting a 500 gallon test burn at the St. Louis main power plant facility, Building 5. A Riley-Stoker, Class P11 #19 * NW, industrial boiler was selected to conduct this test burn. Upon completion of the initial 500 gallon burn, McDonnell Douglas chose to continue this effort and conduct a 150,000 gallon test burn. The target date to begin this burn was set for 1 October 1980, and was met. In preparation of this burn the storage system, transfer pumps, heat exchangers, and piping to Riley Boiler was purchased, fabricated and installed for the purpose of transferring C-O-M. The boiler proper was not modified other than removing piping sections from flyash unloading system to provide capability of manually unloading 3 sections, firebox, air preheater, and mechanical precipitators; thereby being able to weigh all collections at periodic intervals. Control panels were installed with appropriate valving and monitoring devices for burner control.

The services of an outside independent testing company, KVB, Inc., Minneapolis, MN, were acquired to sample furnace and stack emission gases and perform particulate count by weight during test burn at periodic predetermined intervals.

3.1 Transporting

At the commercial preparation plant, Shelbyville, KY, the insulated truck tanker is loaded to capacity, 4800 gallons. The C-O-M is heated by means of steam coils in the plant storage tanks to 110 - 130°F before loading. The 6 hour trip to MAC - St. Louis, MO during winter condition arrives at approximately 80°F. In the past month the tanker was modified to accept 15 lbs. per sq. in. of steam which while unloading will increase the C-O-M temperature 10°F. The time required to unload has averaged 1-1/2 hours. The transfer pump utilized is a progressive cavity, positive displacement Moyno, Type CDO. It is noteworthy to emphasize from CoaLiquid, Inc.'s past experience that this type pump or the positive rotary gear, positive displacement type be used for satisfactory performance over the conceivable wide temperature ranges.

3.2 Storage

The storage tank capacity is 9,086 gallons. Our tank is uninsulated which is not recommended in a permanent facility. Two steam coils were fabricated and installed in the tank. One coil, running lengthwise in the tank, is located approximately one foot off the bottom; the other is positioned vertically in front of the pump suction port. The C-O-M temperature is maintained at a minimum of 90°F continually. We do experience difficulties with temperature stability when the C-O-M level is low and we take on a load from the transport tanker. Under these conditions during test burn periods, there is a recovery period with which we have experienced difficulties in maintaining optimum conditions.

3.3 Stability during storage

As previously mentioned, CoaLiquid has presented information and data in this regard. The stability has proven to be as predicted. In one instance, it was noticed that there was sediment in the bottom of the storage tank. Reviewing the situation with CoaLiquid, the question was asked if we had elevated the temperature of the C-O-M at any time. Such was the case in our situation. It has been noted, that separation occurs when the temperature of the C-O-M in the storage tank exceeds 175°F. It is suggested that an air lance system be incorporated in bottom of storage tanks to agitate the material in the event of similar circumstances.

3.4 Pumping System

We utilized two Viking Model No. LL124 Centrifugal pumps in parallel, to transfer C-O-M from storage tank to burner control panels and subsequently to the burner guns. After discharging from the pump, the C-O-M pressure is regulated by means of a by-pass loop back to the storage tank. The C-O-M temperature is elevated by a set of single pass heat exchangers. Temperature regulation is attained by use of a Jordon steam temperature regulation valve. The C-O-M temperature to burner gun is maintained at 230° - 240°F.

3.5 Control Panel

The firing operation in conjunction with the test burn is totally manual. This is not as we conceive it being in the future; this situation is due to fuel oil system of our boiler being a "standby" only from original design. Incorporated in the control system is valving and piping for a recirculating line to enable bringing C-O-M temperature to desired range on initial start-up. Pressure and temperature gauges monitor operating parameters of the C-O-M and steam which used is for atomization. It is noteworthy to bring out the aspect that upon completion of utilizing C-O-M or when maintenance is required, piping should be blown out with shop air or steam at 100 lbs. per sq. in. or greater.

4.0 BOILER CHARACTERISTICS AND COMPONENT SYSTEMS

Boiler #2, manufactured by Riley Stoker Corp. was selected because of its specifications which are:

<u>Class</u> P11 #19 * NW
<u>Build</u> 1941
<u>Tensile Strength</u> 70,000
<u>Working Pressure</u> 200 LB/In^2
<u>Heating Surface</u> Boiler 6,650 ft^2
Waterwalls 2,360 ft^2

Capacity
-Max. Continuous 63 $\overline{M}$ BTU/HR
or 65,000 LB/HR

<u>Heating Surface</u> - Airheater
5,500 ft^2
<u>Operating Pressure</u> 165 LBS/IN^2
<u>Furnace Volume</u> 3,510 ft^3
<u>History</u> 1941-1966 Pulverized Coal-Primary Fuel #6 Fuel Oil Standby

1966-Present Natural Gas - Primary Fuel #2 Fuel Oil - Standby

The coal burning capabilities of this boiler was the primary reason for its being selected for the 150,000 gallon test burn. The retrofitting of this boiler was minimal. CoaLiquid control panels after being installed were tied into existing "standby" fuel oil systems. After a few test runs, steam line size for atomizing was increased to eliminate line loss. The ash vacuum unloading system was isolated and essentially eliminated. This was to be able to control flyash removal. We wanted to remove all collected flyash manually and weigh. This would provide the most accurate method of comparing input versus output. The coal nozzle, or more commonly known as coal tube, associated with burning of pulverized coal was removed and replaced with a diffuser cone. More will be discussed concerning this further into the report.

4.1 Burner Atomizing Guns

Reviewing the test burn and looking back, the atomizing gun was and still remains the aspect of vital interest with regard to utilization of our C-O-M, CoaLiquid, for the purpose set down, that being burning as a primary fuel in industrial boiler applications or equivalent. I will categorize each burner gun and refer to it as a phase.

4.1A Phase I

The burner gun referenced in this phase was designed and provided by CoaLiquid, Inc. The burner gun with external steam atomizing was installed and lit. The flame was uncontrollable in regard to #2 Boiler firebox. We experienced flame impingement on the back tubewall of the boiler. Modifications were made to the burner gun by reducing the C-O-M passage diameter from 1/2 inch to 3/8 inch. This modified gun produced 6,000 lbs/hr with C-O-M supply pressure equal to 30 psig and temperature of 210°F. This burner gun was utilized in burning 450 gallons of C-O-M. The fact that we were able to only produce the limited steam load and the flame was nearly impinging the back tubewall, design changes were a must. Refer to this burner gun as No. 1 in Table I.

TABLE NO. I

BURNER GUN TYPICAL PERFORMANCE

BURNER GUN	MAX. CAPACITY (LBS/HR)	FLYASH L.O.I (&)	ASH IN/OUT (%)
Type 1	6,000	-	-
Type 2	44,000	F - 11.77 C - 68.26	1.1
Type 3	37,500	F - 5.22 C - 67.22	1.1
Type 4	56,500	F - 9.92 C - 22.94	0.2

TABLE NO. III

% COMBUSTIBLES

12-30-80 TEST BURN

Sample Origin	KVB	E. A. INC.
Firebox	13.82	9.92
North Leg	21.42	21.58
South Leg	25.34	24.38

4.1B Phase II

The No. 2 type burner guns were designed and fabricated. This burner gun similar in design to No. 1 burner gun, atomized the C-O-M with steam externally and featured a cone positioned directly in the atomized C-O-M discharge flow. The cone positioning with reference to the burner nozzle was variable and could be moved in or out. This allowed for fine control of flame pattern and thus better combustion.

Using two No. 2 type burners, a load was produced varying between 25,000 and 44,000 lb/hr. The C-O-M pressure to the burner gun was 40 psig with corresponding temperature of 210°F. The burner gun nozzle angles were varied, but did not improve combustion or increase output. During these burn periods, a Riley Stoker Service Engineer reviewed burn conditions, and accordingly made adjustments for optimum performance. To mention a few of these items, primary air, furnace draft, register settings, C-O-M passage areas and the use of high pressure air instead of steam for atomizing purposes. At this point, I would like to present some facts and figures. We burned 4,898 gallons of C-O-M and collected 660 lbs of flyash, 100 lbs in the firebox and 560 lbs from the mechanical collectors.

CoaLiquid (LBS)	=	9.25 lbs/gal x 4,898
	=	45,306 lbs
Ash Content (C-O-M)	=	lbs x % ash
	=	45,306 x .033
	=	1,495 lbs

By lab analysis a Loss-On-Ignition Test was run with the following results. Firebox and Collectors were 11.77% and 68.26% unburned carbon respectively.

Therefore, lbs of ash collected = Total - Unburned Carbon
= 660 -(100 x .117) + (560 x .6826)
= 660 -(11.7 + 382.2)
= 660 - 393.9
= 266.1 lbs

And therefore comparing ash input vs ash collected = %
266.1/1,495 = 17.8%

Calculating

Loss-Due-To-Unburned Carbon
Carbon Content (C-O-M) = (C-O-M) lbs x 77.92%
= 45,306 x .7792
= 35,302 LBS

Therefore % Loss-Due-To-Unburned Carbon = 393.9/35,302
= 1.1%

At this segment of the test burn we had used 51,679 gallons of CoaLiquid. All parameters and conditions evaluated, the overall operation was impressive. However, being unable to attain higher steam output along with extremely high percentages of unburned carbon being collected, the consensus was the externally atomizing type burner gun be abandoned. This aspect was enumerated by the initial test data results compiled by KVB, Inc. The first set of test runs for purposes of measuring gas samples and emission particulate count was conducted on October 8th and 9th, 1980. After receipt and review of the compiled test data, our performance was by no means satisfactory.

4.1C Phase III

For the #2 Boiler Standby Fuel Oil System, ENCO #10 atomizing guns are used. An attempt was made previously to utilize this gun burning the C-O-M, but the oil passages in the venturi clogged. It is to be noted that the ENCO #10 Gun venturi and nozzle was sized for boiler capacity with either No. 2 or 6 fuel oil. The steam and oil supply lines were interchanged which produced a steam load of 37,000 lb/hr. Although this was an unconventional maneuver, it led to an extensive testing program associated with ENCO atomizing gun venturis and nozzles. The data presented regarding the No. 3 burner gun does not reflect improved performance, however, such is not the case. A typical test burn associated with this gun reflected the following: A burn consumed 1,359 gallons of C-O-M and collected 210 lbs of flyash, 50 lbs from the firebox and 160 lbs from the collectors.

1,359 Gal CoaLiquid = 12,570 lbs thus ash content equals 414 lbs.

By lab analysis Loss-On-Ignition Test results being Firebox, 5.22% and Collectors 67.22%

Ash Collected - Unburned Carbon = 100 lbs.

Comparing Ash $_{(In)}$ vs Ash (collected) = 24.1%

And Loss-Due-To Unburned Carbon = 1.1%

During test burns with the No. 3 type gun, characteristics such as flame intensity and pattern, firebox appearance, stack smoke intensity and fineness of flyash collected indicated definite improvement.

4.1D Phase IV

The capacity of ENCO #10 type nozzle and venturi configuration was maximum. The subsequent step was to use the ENCO Type 20 atomizing gun. Initial concern in this regard was the modifications necessary to make this feasible. The facts were that no changes were required. This being the case, the Type 20 nozzles and venturis were obtained and test burn continued. I would like to emphasize although the thermal content relationship between C-O-M, CoaLiquid and No. 6 fuel oil is 90-92%, the sizing of ENCO's atomizing guns is not directly proportional. This information is being gathered at this time.

Enumerating test burn findings regarding sizing of Type 20 atomizing gun, when utilizing a venturi and nozzle configuration sized for No. 6 fuel oil and rated 70,000 lbs/hr per unit, a load of 48,000 lb/hr was produced with two burner guns.

The burn consumed 1,395 gallons of C-O-M at a rate of 279 GPH with boiler capacity producing 40,000 lb/hr. Total flyash collected was 104 lbs, 4 lbs in the firebox and 100 lbs in the collectors.

Loss-On-Ignition results were Firebox 3.67% and Collectors 32.57% respectively.

Ash Collected - Unburned Carbon = 71 lbs.
Comparing Ash (in) vs Ash (Collected) = Ratio %
= 16.7%

And Loss-Due-To-Unburned Carbon = 0.3%

These results from boiler test burn comparatively represented considerable improvement. Efforts were then concentrated to eliminate conditions which had been tolerated previously. One of these were the sparklers in the firebox. The coal fineness of the C-O-M was analyzed and found not to be within fineness test standards, i.e., 80 to 90 percent minus 200 mesh. CoaLiquid, Inc. recognized this to be fact. To alleviate this processing deficiency, an air classifier was purchased and incorporated in the Shelbyville commercial production plant system. The improved pulverization due to air classifier showed immediate positive results when burning the first load produced. Sparklers were eliminated in the firebox. Another aspect which we were experiencing and was of major concern, was the deterioration of burner gun nozzles due to abrasiveness of the C-O-M. There was a "washing" action taking place at the nozzle orifices which affected the included angle. As the abrasive "wash" continued, the angle of the atomized C-O-M discharge decreased which proportionally affected combustion. Refer to Table II for data of service life of different nozzle materials.

MATERIAL	SERVICE, HOURS
316 S.S.	20 - 30
440C, 60R	60 - 80
Tungsten, Carbide Filler	40 - See Note(1)
Tungsten Carbide	Presently Being Tested, See Note (2)
Cobalt Filler	Unavailable
Note(1): Configuration failed but verified tungsten carbide to be advantageous. Note(2): Has Been Utilized For 150 Hours - No indication of Wear.	

Table II Nozzle Service Life

Table II reflects materials used to date, there are numerous other considerations in our future plans. The fact of the matter is we cannot state which nozzle material and configuration will be recommended.

Many test burns were conducted using different nozzle configurations based on information we gained from each performance. The primary criterion was combustion with maximum capacity being secondary. I would like to emphasize that during the test burns we were unable to repeat the performance and attain a Loss-On-Ignition of 32.57% or lower as specified previously. In an attempt to improve combustion the coal tube was removed. There are more details in this regard under combustion. The test burns led to the findings which nozzle drill configuration and angle provided best performance in our industrial type boiler. For an example, we burned 646 gallons of C-O-M at a rate of 218 GPH while producing a load of 26,000 LB/HR. Total flyash collected was 45 lbs., 4 lbs in the firebox and 41 lbs in the collectors. Loss-On-Ignition results were Firebox 9.92% and Collectors 22.9%.

Ash Collected - Unburned Carbon = 35 LBS
Comparing Ash (in) vs Ash Collected = 17.7%
And Loss-Due-To Unburned Combustibles = 0.2%

As of this date, this is typical of test burn characteristics. The maximum capacity produced is 56,500 LB/HR. In the near future we will be testing with a larger capacity nozzle and venturi configuration to meet M.C.R., maximum capacity rating.

5.0 COMBUSTION

The combustibles determination was based on Loss-On-Ignition analysis of flyash by independent laboratory, Environmental Analysis, Inc.; St. Louis, MO.

Upon completion of test burn, ash samples were pulled and sent for analysis. The L-O-I was used as the instrument to dictate improved performance. To present the overall review of the L-O-I's, initially the results of the analysis were in the low seventy percentile. Presently the ash samples L-O-I are in the low twenty percentile.

For unbiased data, as mentioned previously, KVB, Inc. was contracted to perform and evaluate furnace and stack emission gas samples. All KVB, Inc. data referred to here is that which was acquired on the second set of test runs conducted on 17th and 18th of December 1980. The multiclone ash was combusted by fractions obtained from the sieve analysis. While combustibles in fractions 325 mesh and larger showed slight improvement, the combustibles in the fines (that through 325 mesh) showed the most significant decrease to 30%. This decrease coupled with an increase in ash fineness primarily due to air classifier resulted in an overall combustibles of 43%.

The flyash from the stack particulate test yielded combustibles at 27.4%.

Combustion showed decisive improvement at this time due to a number of reasons. Specifically the pulverized coal nozzle was removed. To introduce swirl of primary air at burner gun nozzle, a diffuser cone was installed. The nozzle included angle was varied, induced furnace draft was decreased, increasing retention time, and comparing excess O_2 with L-O-I analysis. To verify above, a set of ash samples from test burn of December 30, 1980 was sent to KVB, Inc. and Environmental Analysis, Inc. for L-O-I determination. Even though there was a difference in firing temperatures between the two employed methods, the results are comparable. Refer to Table III.

6.0 BOILER EFFICIENCY

A boiler efficiency test had not been performed recently. To have comparative figures, the Riley Stoker Co.'s original performance data was used which is as follows:

	Pulverized Coal Unit With Airheater		
Load (LBS/HR)	45,000	60,000	65,000
Efficiency	85.6	85.1	84.5

The calculated boiler efficiency of tests monitored by KVB, Inc. and performed on December 17 & 18 was 85.13%. Boiler efficiency was calculated using the ASME Test Form for Abbreviated Efficiency Test, Revised, September, 1965. This reflects a 2 percent gain over the October results of 83.17%. This increase in efficiency was the result of lower combustibles in each ash category and lower amounts of ash quantities. Losses due to unburned carbon amounted to 0.55% of the total fuel energy input.

7.0 EMISSIONS

The stack particulate loading for Boiler #2 firing C-O-M was 1.152 lb per 10^6 BTU. We realize if CoaLiquid became the primary fuel, the addition of a more efficient collector would be necessary to maintain loading levels below the 0.47 lb/10^6 BTU regulation limit. In conjunction with the American Air Filter Co. the present schedule calls for the installation of a test frame bag house. This testing will verify particulate emission data and also evaluate bag material which is compatible with C-O-M firing.

The nitric oxide emissions were reported in terms of NO_2 as required by Federal Regulations. The trend was seen towards lower emissions at higher firing rates, 0.67 lb/10^6 BTU at 28K lbs/hr to 0.49 lb/10^6 BTU at 53K lb/hr. Nitric oxide concentration often increases with boiler load. However, in cases similar to the operation of our boiler, the increased oxygen (excess air) at lower loads results in increased nitric oxide.

CONCLUSION

The test burn and/or affiliation with the C-O-M, CoaLiquid, to date has been regarded as very successful. Efforts by McDonnell Douglas will continue researching and developing methods for overall improved boiler performance specifically in regard to aspects mentioned in this case study report. Copies of KVB, Inc. reports referenced are available upon your request.

FLORIDA POWER & LIGHT COMPANY
SANFORD COM PROJECT*

M. C. Cook-Vice President
L. D. Slepow, PE-Project General Manager
A. S. Mendelssohn, PE, CCE-Project Control Supervisor
Florida Power & Light Company

Florida Power & Light Company is conducting a unique and comprehensive test using a coal and oil mixture, or COM, as an alternate boiler fuel in our Sanford Unit No. 4, to determine the problems associated with burning coal in a boiler which originally was designed to burn oil. Florida Power & Light wants to decrease its dependence on oil for several reasons. First, the price of oil has been rapidly increasing when compared to the price of coal, and now has become very expensive for our customers. Also the availability of oil is declining and is subject to the whims of foreign governments. Finally, the Federal Government is strongly encouraging and may soon require some form of conversion to coal. Any one of these is sufficient to motivate FPL to reduce its oil consumption. Within our system, there are nine 400 MW units and four 800 units, all designed to burn oil, and all are candidates for an alternate fuel. Our management has felt it necessary to find alternative fuels for these units. We are presently investigating a variety of coal options as future means of firing these boilers. COM is one of the potential alternatives. However, all the problems associated with COM are not known.

Because no electrostatic precipitator existed on the unit, it was necessary for FPL to request a variance from the Florida Department of Environmental Regulation and the Federal EPA to allow us to test burn this coal-oil mixture utilizing the pollution control equipment already at the facility. We received a variance to burn a coal-oil mixture for 120 days at an equivalent power of 400 MW or a total of 1,152,000 megawatt hours of operation. This variance is for one year, starting April 20, 1980 when the first COM was burned.

*Adapted from a paper published in the 1981 Transactions of the American Association of Cost Engineers and reprinted with permission.

The schedule for this project has been relatively short. FPL management decided in mid-September of 1979 to conduct a demonstration of coal-oil mixture. Engineering was started in mid-October with the selection of Bechtel Power Corporation as the Engineer/Constructor. Construction began in November of 1979 using the fast track approach and was completed in mid-April 1980. Sanford Unit No. 4 was shutdown for four weeks in March 1980 in order to make appropriate modifications to the unit. The first COM, a 10% by weight mix, was burned on April 20, 1980. The concentration of coal was increased in steps to a maximum of 50% which was reached on September 10, 1980 and marked the end of Phase I of the test program. The entire test program will be discussed later.

The Sanford Site has three operating units at the present time. Unit No. 3 is 165 MW and is designed to burn either oil or natural gas. Units No. 4 and 5 are 400 MW each and were designed to burn only oil. Unit No. 4, the one on which we are doing the demonstration, has been altered to burn either oil or COM.

A simplified flow diagram that describes the preparation of COM is shown in Figure 1. The coal is brought in by rail cars from mines in the state of Virginia, using a newly constructed railroad siding. The coal, which is of very high quality, has a heat value of over 13,000 BTU's per pound, and has a low sulfur and ash content, less than 1% and 10% respectively. The coal is unloaded using portable equipment such as undercar unloaders, conveyors and car shakers. This coal is then moved to either the main coal pile, where it is leveled, or to the active storage pile, or placed directly into the hopper and up a conveyor for feeding into the facility.

The COM preparation facility is illustrated in Figure 2.

The coal that enters the preparation facility is stored in one of four coal silos. As needed, it flows to the pulverizers through volumetric feeders, which regulate the quantity of coal. The coal is pulverized to a consistency of 80% through a 200 mesh screen. The air that is used to transport the coal enters each pulverizer through a duct on one side of the facility. At the end of the duct there is a gas heater which heats the air to the appropriate temperature.

The coal is then transported by air from the pulverizers to the top of the facility and through cyclone separators which separate out the pulverized coal from the air. The pulverized coal goes into the storage bin located directly below. The air and the remaining coal particles go into the baghouse filters located at the top of the facility. The coal-free air is then exhausted from the baghouse filters through blowers at a lower level of the facility. The pulverized coal that is collected on the filters is knocked off and stored in hoppers at the bottom of the baghouse. It is then piped to the pulverized coal storage bin. The pulverized coal storage bin is a large cylindrical stainless steel lined cone designed for smooth flow of the powdered coal. From the bin, the powdered coal initially flowed through a large knife gate valve and through a gravimetric feeder, which regulated the amount of coal going into the mix tank. Problems developed, however, when an uneven flow of powdered coal, resulting from pulsations within the pulverized coal storage bin, caused unstable operations within the gravimetric feeder. These pulsations were eliminated by removal of the knife gate valve at the base of the bin. The gravimetric feeder was designed to handle 42 tons of powdered coal per hour, the quantity required to produce 10,000 barrels of 50% COM per day. At loadings greater than 22 tons per hour, it was found that the coal would flow like a liquid, filling the inside of the feeder, and causing damage to the system and shutdown of the unit. The pulverized coal system was modified to remove this feeder and eliminate the storage of coal in the pulverized coal storage bin. Coal feed is now measured at the volumetric feeders for the

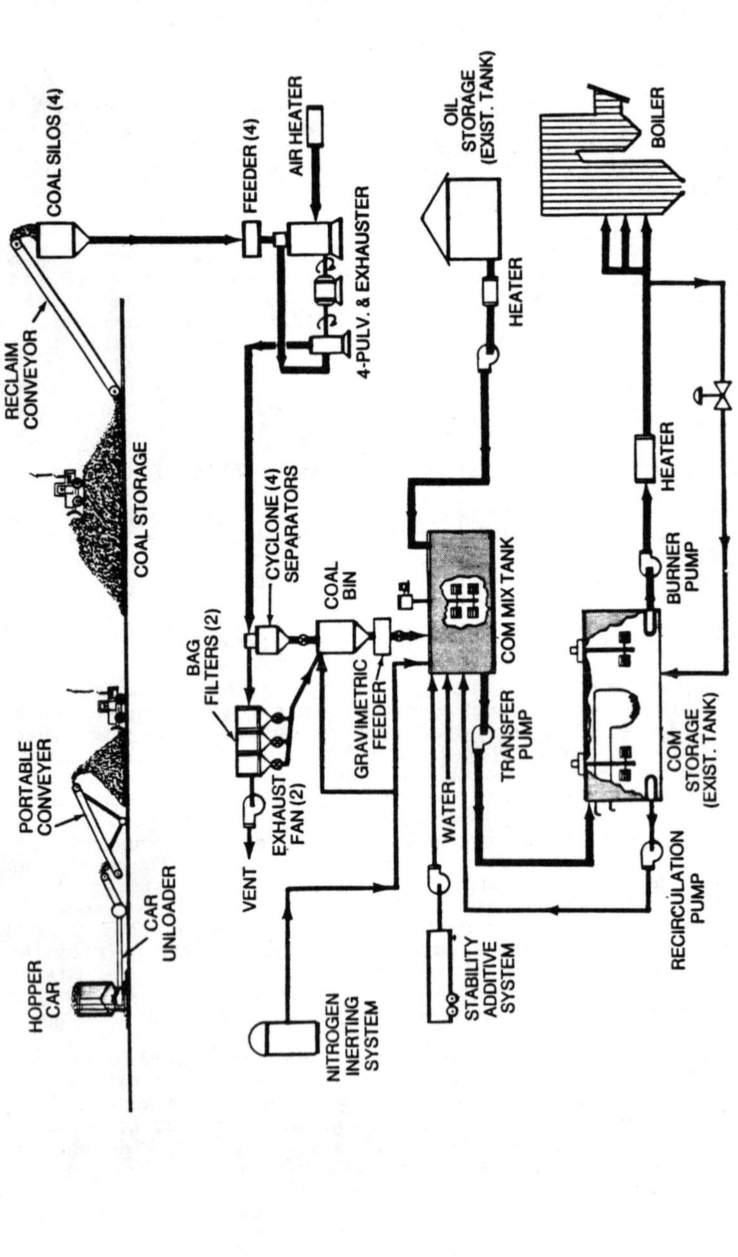

COM = COAL/OIL MIX

FLORIDA POWER & LIGHT CO.
SANFORD STATION
COAL/OIL MIX FACILITY

FIGURE 1

HOW THE COAL-OIL MIX FACILITY WORKS:

1. **Conveyor** - Transports coal to top of Coal-Oil Mix (COM) preparation facility.
2. **Silos** - Receive coal for storage and feeding into pulverizers.
3. **Feeders** - Regulate amount of coal by volume going into pulverizers.
4. **Pulverizers** - Crush coal into fine powder. Hot air is mixed with pulverized coal to dry and transport it through coal piping.
5. **Piping** - Carries pulverized coal to top of the COM facility.
6. **Cyclone Separators** - Separate pulverized coal and air.
7. **Baghouse Filter** - Cleans the air (separated in Step No. 6) by removing any remaining coal dust prior to releasing the air from the process.
8. **Storage Bin** - Stores pulverized coal prior to mixing.
9. **Mix Tank** - Mixes pulverized coal and oil into COM. Also added here: an additive to help keep the coal suspended in the oil. After mixing, the COM is piped to a storage tank and then to the power plant boiler.
10. **Nitrogen Storage Tank** - Part of the COM safety system. Nitrogen is used in the pulverized coal storage bin and the COM mix tank to provide an inert atmosphere.

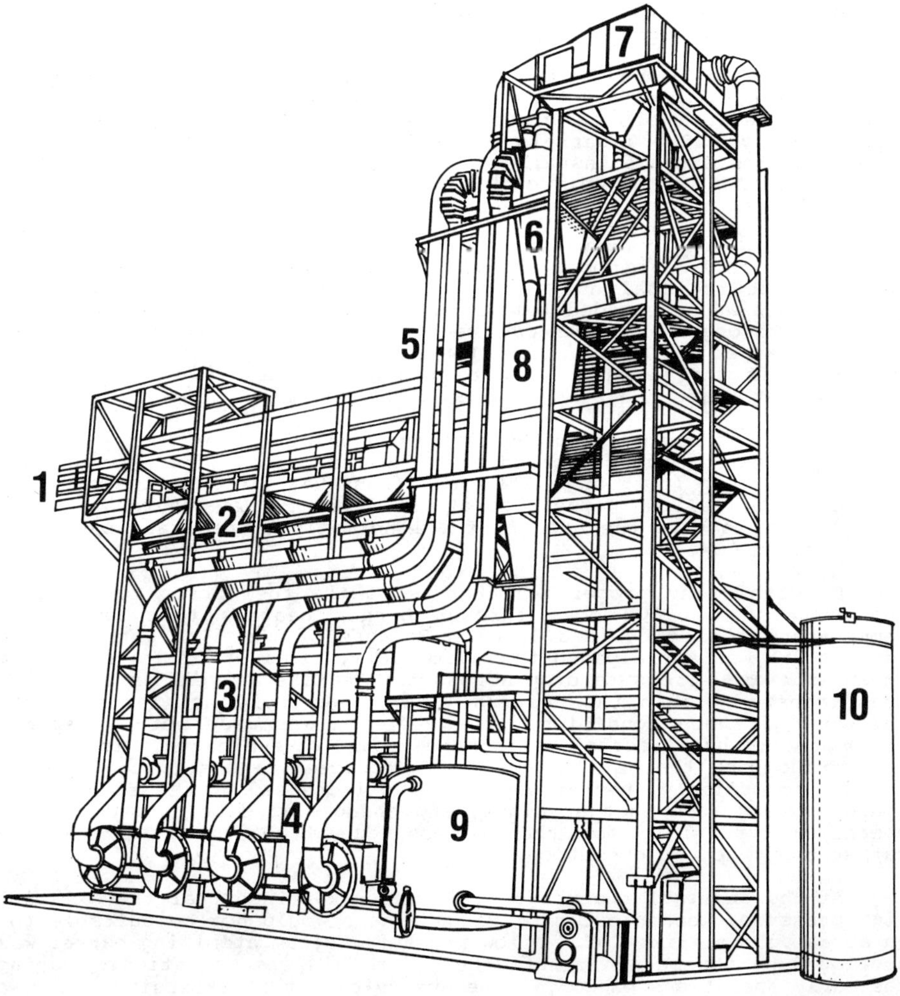

FIGURE 2

pulverizers and, after grinding, is conducted through the empty storage bin into the mix tank.

The 15,000 gallon mix tank, located at ground level, receives the pulverized coal from the storage bin, as well as oil from a storage tank within the power plant and mixes it together to form the coal-oil mixture. A problem with inadequate mixing caused excessive strainer plugging at higher percentages of coal loadings. Adjustments in the mix tank included increasing the speed of the mixer blades and changing the level of fluid in the mix tank. These changes provided adequate mixing, even at the higher coal loadings that were achieved with the elimination of the gravimetric feeder.

To produce a 50% by weight barrel of COM requires approximately 200 pounds of pulverized coal and 0.6 of a barrel of oil.

In addition to the coal and the oil, an additive is also provided to insure stability of the mixture. Coal has a tendency to settle when mixed with oil. The additive helps to prevent this.

Both the pulverized coal storage bin and the mix tank are injected with nitrogen and inerted to prevent explosion.

The COM is pumped from the preparation facility, at a maximum rate of 300 gpm, to a 55,000 barrel tank that has been modified for COM storage. Because of the higher density of COM, we can only store approximately 50,000 barrels of the mixture in the tank. We have insulated the tank and installed four paddle mixers that slowly rotate to keep the COM in motion, which helps maintain the suspension of the coal particles in the mixture.

The COM must be operated within a certain temperature range. If it gets too cool it becomes too viscous and difficult to pump. If it gets too hot the coal has a tendency to settle out of the oil. To control this we have added temperature control coils in the bottom of the tank through which we can heat the COM with steam should the temperature get too low, or cool the COM by running cold water through the coils if the temperature gets too high. Also, we have suction heaters that raise the temperature of the COM as it leaves the tank.

The COM is routed from the tank through a 12" line to the suction of the new burner pumps located near the tank area. These pumps take the COM and pump it through COM fuel lines that tie into the existing fuel oil supply line in the area of the daytank. The daytanks, which are used to measure the amount of oil normally burned, are by-passed when burning COM.

To obtain the proper temperature of the fuel prior to entering the burners, the COM goes through burner heaters that raise its temperature to the desired level. Because of the fins within these heaters, it was anticipated that some plugging might occur, resulting in an increased pressure drop across the heaters as well as a loss in heat transfer within the heaters. Since we could not forecast the severity of these conditions, the decision was made not to replace these heaters until actual performance could be evaluated. It has now been determined that at the higher COM concentrations, plugging does occur. Accordingly, heater capacity was increased by the addition of spare heaters from another plant, and changes were being made in the operating conditions for the heaters; these changes have restored satisfactory heater operation.

At the onset of the project, it was determined that the existing high pressure, mechanical atomizing burners would not be suitable for an abrasive fuel like COM. A low pressure, steam atomizing burner was developed by the burner manufacturer for this demonstration. During the early operations on COM, it became evident that excessive wear was occurring on the front nozzle of the burner tip. This caused poor atomization and combustion of the coal-oil mixture, and required frequent tip replacement. Through a series of unique field design modifications, this erosion was reduced very substantially. Further modificiations in tip design, as well as the use of more exotic, wear-resistant materials for tip constructon, have been pursued to essentially eliminate the wear and improve combustion. Adjustments in the temperature and pressure of the COM and steam entering the burner have also been evaluated.

The whole operation of burning COM is under the direction of the operators in the control room. The only modifications in the control room itself were the installation of a few switches and gauges.

When not burning COM, we can revert back to oil which comes from two large storage tanks located at the facility.

Even though we have not installed full size electrostatic precipitators for this demonstration, we have incorporated into our test program two pilot units which are used to gain design information on a precipitator while burning COM. Each one handles less than 1% of the exhaust gases from the boiler. One represents a conventional rigid frame type. The other represents a new ionizing type of precipitator.

Figure 3 illustrates the process functions after the COM is burned.

We had to modify the lower portion of the boiler to allow the installation of a bottom ash system. This system is basically a dumpster type bin connected to the bottom of the boiler. When the bin is full, the operator empties it for eventual disposal in an ash storage area.

Below the economizer we have installed hoppers to collect ash in this area. The ash is removed when necessary, using pressurized air, to take it to a fly ash silo.

The dust collectors, which were original equipment on the unit, are still being used for this project. When burning oil, the carbon reinjection system takes the oil ash from the dust collector hoppers back to the boiler. During COM operations, however, the coal ash predominates. Thus we have installed a system to remove the fly ash that collects there and, using pressurized air, transport it to the fly ash silo.

The fly ash is stored in the silo until it is removed by a commercial cement truck that has been modified for this operation. The ash is taken to an ash disposal area located near the Sanford Reservoir, which provides cooling water for Units No. 4 and 5.

Although the coal used for this demonstration has a relatively low ash content, it was anticipated that ash would build up in the boiler during operation. It was not known, however, where this ash would collect in the furnace and the quantities that would be involved. As the ash built up on the walls of the furnace, it insulated the walls, causing less heat to be transferred. This resulted in the gases exiting the furnace and entering the superheater section of the boiler having a higher temperature than normal. When the design metal temperature of the superheater tubes was reached, it was necessary to reduce the load on the unit in order to keep superheater temperatures within allowable limits. After reaching the 30% COM concentration, the problem became so severe that the unit was shut down and wall blowers were added to remove the ash from the furnace waterwalls.

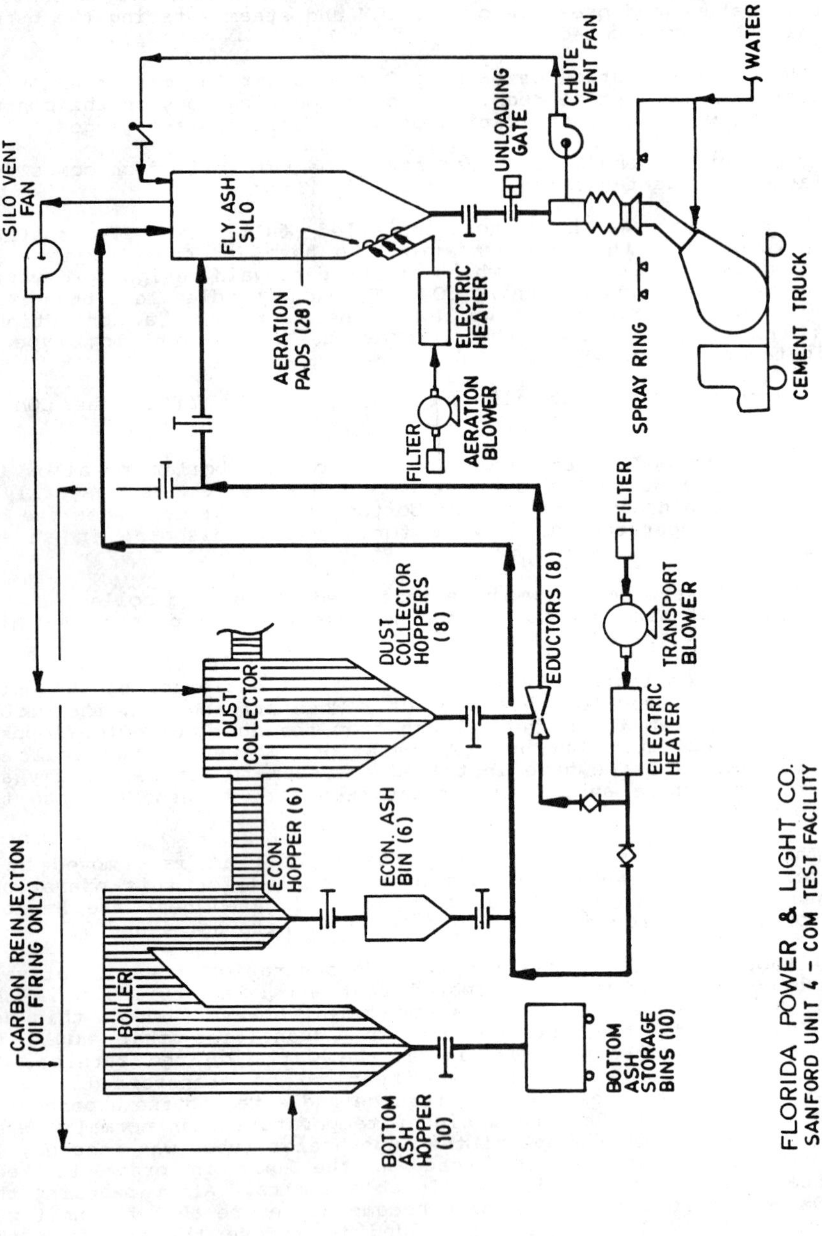

FLORIDA POWER & LIGHT CO.
SANFORD UNIT 4 - COM TEST FACILITY
ASH REMOVAL SYSTEM

FIGURE 3

After the installation of the wall blowers a 40% by weight COM was fired in the boiler. The boiler was tested at various loads, as at previous concentrations, and was capable of obtaining full power without overheating the superheater. The performance of the boiler at this percentage of COM was very similar to that of normal oil operations. A greater amount of superheater spray was required, however, due to higher temperatures entering the superheater.

During the test of 45% COM, slagging of the furnace walls was observed at the higher loads. It is assumed that this was caused by not having sufficient air for complete combustion. This condition occurred because the capacity of the forced draft fans was not sufficient to meet the required quantities of air. Complete burnout of the coal was not obtained and what appeared to be a sticky type of slag was formed on the waterwalls. While the wall blowers can remove ash deposits from the walls, they were unable to remove this sticky slag. With the 45% COM we were unable to operate at full load.

Following the 45% test, the unit continued to burn COM as the percentage of coal was gradually increased to 50%. The lack of sufficient combustion air, which caused severe slagging, continued to prevent the unit from achieving full load.

Some preliminary insights into boiler behavior and performance has been obtained through analysis of the data collected to date. We have found that the superheater temperatures are slightly higher on COM than on oil. Boiler efficiency appears to remain nearly the same on COM as on oil. The increase in excess air for COM firing has slightly reduced efficiency. Burner modifications reducing air pressure drops and providing better combustion could possibly bring COM efficiency even closer to that of oil. Chemical analysis of the coal being burned indicates that slagging should not occur in the furnace. Slagging, however, has taken place and further investigation as to its cause is underway.

In order to have sufficient air for combustion, the concentration of coal has been reduced to approximately 40%. At this concentration, full load can be achieved without exceeding any design limits. The unit has been turned over to the System Dispatcher for loading. Optimization of conditions at this level are underway to see if the plant can be restored to full load at higher COM concentrations.

Another major concern that developed during the preparation of COM was the fallout of the coal particles in the COM storage tank. This problem occurred at low COM concentrations and was caused by several different conditions. First, we initially had some coarse coal from the pulverizers. Adjustments to the classifiers beyond what had been recommended by the manufacturer eliminated the coarse coal. Second, the temperature of the COM within the tank increased, causing the stability additive to become ineffective. This increase was a result of the high temperatures needed for mixing and the return of higher temperature COM from the combustion control regulators. The cooling of the tank, as previously described, improved operations. Finally, it was found that the measurement of the stability additive into the mixing process was not accurate due to the properties of the additive itself. An alternate means of regulating the quantity has been developed. At higher COM concentrations another cause of instability appeared. The introduction of contaminants into the system counteracted the additive and produced additional fallout of coal. Improvements in the strainer operation decreased the probability of a reoccurrence.

As indicated earlier, FPL received a variance to burn coal-oil mixture using the existing pollution control equipment on the unit. The variance allowed for a particulate emission level on Unit No. 4 of 5,150 lbs/hour and an opacity of 100% while burning COM.

Preliminary results from environmental testing indicated that particulate emissions for a 50% COM at full load will be about 2,000 lbs/hour, or less than 40% of the allowable level. Opacity for these same conditions would be about 60%. Any permanent conversion to COM, or coal in any form, would require the addition of an electrostatic precipitator or other particulate removal system which would bring emissions in line with those obtained on oil.

To obtain the maximum amount of information from this demonstration, a four phase test program was developed. The first phase of the test program was the increase in coal concentration to the 50% level and the collection of data at the various percentages. Fine tuning of controls and flow parameters were not conducted due to time limitations of the testing at each concentration.

The second phase of the test program is an optimization phase and is being conducted with a COM concentration at approximately 40% so that reliable operation can take place at all loads. This extended operation has permitted adjustment of several parameters pertaining to COM supply and burning. These include atomizing steam flow, COM burner temperature, COM pumping temperature, and excess air. While these parameters are adjusted, observations are being made as to their effect on the boiler's heat transfer distribution, slagging and performance.

During Phase II, the effects of erosion on the various pumps, valves, etc., through which the COM flows, are also being monitored. Other areas being tested include improved stability techniques for COM and a detailed evaluation of the ash that is generated, its distribution and its potential value as a by-product of burning COM.

Phase III of the test program calls for the sustained burning of coal-oil mixture at the maximum concentration that will permit full, reliable operation. Additional material wear analysis will take place during this phase. More detailed analysis of operating performance will also be conducted.

The final phase of the test program includes the detailed evaluation of erosion and corrosion of all components involved in the production, transport and burning of COM. In particular, an extensive evaluation will be made of the inside of the boiler. Final analysis of all data and preparation of a complete report will also take place.

The cost for doing a demonstration of coal-oil mixture is not small. The COM Preparation Plant and modifications to Sanford Unit 4 required $12 million in capital expenditures. An additional $7 million is needed to operate both facilities during the test. This total of $19 million, which does not include the costs for coal, oil or additive, is partially offset by the savings in fuel costs from the substitution of COM for oil.

Coal-oil mixture appears to work as a substitute for oil in Florida Power & Light Company's 400 MW units. The remainder of the demonstration will better define the problems associated with it and provide insight into the solutions that need to be developed. But a 50% COM only displaces about 40% of the oil used. The high capital and operating cost of COM preparation facilities, plant modifications and emission control equipment offsets much of the economic advantage of coal vs. oil. Since the coal concentration in COM is limited to only 50% because of viscosity problems, it would be preferable to burn coal without mixing it with oil. Based upon the Sanford Plant's performance on COM, it would appear that, at some units, it may be possible to burn some form of coal directly or in a water slurry, thus displacing a greater quantity of oil. FPL is now doing a detailed feasibility study looking at the potential conversion of one of its 400 MW oil-burning units at Sanford to pulverized coal. If the regulatory, technical and economic viability can be demonstrated in these studies, a conversion would be likely.

"CASE STUDY OF CONVERSION OF COAL-CAPABLE BOILER PLUS INVESTIGATION OF CONVERSION OF OIL-DESIGNED BOILER"

Bruce F. Paul

EMHART Machinery Group / USM Machinery Division

Coal, wood, municipal solid waste, coal/oil mixes - Emhart's USM Machinery Group has investigated a broad range of alternate fuels for use in their Beverly, Mass. cogeneration plant. The Beverly manufacturing facility encompasses 95 acres of land (*), and has approximately 1,500,000 square feet of floor space. The cogeneration plant (*), one of the first to employ extraction turbines in Massachusetts, supplies all the energy needs of the facility including that of seven other companies located at the complex, and has exchanged power daily with Massachusetts Electric under a unique Energy Management Agreement; we now ship them "coal"power around the clock.

Alternate fuel studies which were completed in January, 1980, indicated that for the Beverly facility the conversion back to using pulverized coal was the soundest approach. That this conversion could be completed in the shortest time frame and at the least cost, and resulted in the largest cost decrease. Coal was not new to the Beverly complex as it had been burned in two of the existing boilers as late as 1958. Further, we are fortunate in that we are burning coal using the Wormser Coal Fired Fluidized Bed Unit on an oil fired boiler in our Drop Forge building (a separate operation).....(more about this Wormser unit later).

The cogeneration plant has three extraction/condensing turbines with a total output capacity of approximately 9 megawatts and three field set B&W boilers, with a total output of 210,000 lbs/hr, and the normal peak steam load is 100,000 lb/hr. Fortunately, the coal mills and other basic coal-handling equipment for the number 1 - 100,000 pound boiler were in place, and were only in need of updating and overhaul. This one boiler normally carries the full facility load for approximately 90% of the year, and the other two boiler units are used as back-up capacity.

With the alternate fuel studies completed, a goal was set to convert from dependence on #6 - 2.2% sulfur oil to coal, as rapidly as possible.

Of prime concern was the possible time involved in obtaining the necessary environmental approvals. A basic decision had been made from the beginning that any alternate fuel conversion would be made in an environmentally sound manner, and further, that we would avoid if possible, receiving a prohibition order as outlined in the F.R. in Part III of the Powerplant & Industrial Fuel Use Act Rules.

As the program proceeded every indication suggested that the initial decision to convert to coal in an environmentally sound manner, significantly contributed to the prompt approvals from all parties involved. Also, the "climate" in Massachusetts was extremely supportive of efforts to reduce the state's dependency on foreign residual imports. Early in February of 1980, a working meeting was held with the Secretary of The Massachusetts Department of Energy Resources acting as host, and participants from D.O.E., E.P.A., Massachusetts Environmental Affairs and USM/Beverly. The purpose of the meeting was to establish a positive working relationship between the various parties, in order for the project to proceed rapidly. At this meeting, which lasted just over an hour, each agency's concern and desires relative to the proposed coal conversion project were identified, needs established and a working relationship was developed. In principal, understandings were reached on the applicable regulatory requirements that had to be addressed on the acceptable approaches to their solution, and on communication routes.

The meeting was invaluable in that it set the stage for the project proceeding rapidly, engineering being performed, approvals being sought, and funding avenues having been arranged.

The technical approach to conversion centered on the environmental regulations that were to be applied to this project. Conversion back to coal was agreed to be a change back to an alternate fuel, and that it did not involve major boiler revision and reconstruction.

The applicable regulations for our facility were (*):

	OIL	COAL
Sulfur	1.21#/10^6 BTU	1.21/# 10^6BTU
NO_x	N/A	N/A
Particulate Emissions	.15#/10^6 BTU	.10#/10^6BTU
Fuel Ash Content	---	---

The sulfur level was based on our approved variance to operate our facility on 2.2% sulfur #6 fuel oil or 1.21# sulfur / 10^6 BTU. Since Pennsylvania Bituminous coal was available with sulfur content equal to or below this level (on the average), this sulfur regulation is met by fuel source control.

Therefore, the major modification to the plant was to add particulate removal equipment to meet the 0.10# particulate/10^6 BTU.

Electrostatic precipitators, baghouses and electrified filter beds, for particulate control, were all investigated. Based on the state of the art, the size of our facility, our fuel source and our typical industrial load curves, as well as the favorable experience with baghouse particulate control on the Wormser Fluidized Bed previously mentioned, we selected the baghouse approach

Further investigation led us to purchase off-line pulse jet baghouse modules with a total 5/1 cloth to air ratio. We chose this approach as a balance between on-line bag pulsing and reverse air cleaning. Although we clearly recognized the apparent improved bag life and operating characteristics of the reverse air approach, we felt these advantages were outweighed by the increased bag area (2-1/2 times) we felt required by reverse air, and the greater volume of cleaning equipment required to maintain above the sulfur dew point temperature.

The bags selected were 14/oz. teflon coated woven glass, based on our Wormser Fluidized Bed experience; however, we are planning on testing a cell of tufted teflon coated glass bags. We plan to perform conformance particulate testing to verify compliance, as required within the next 30 days, but based on start-up observations we are not concerned. Based on the anticipated bag loading and the design, we expect the actual emission rate to be under .05# particulate/10^6BTU input.

Since our three boilers are on a common breach, the ducting was designed so that we could bypass the baghouse (*) while operating on oil or alternately use the baghouse for flue gas cleaning while on coal. Further, ID fan capacity was designed into the baghouse ducting to overcome the baghouse and related ducting pressure drops and, therefore, minimize any affect of the system on the boiler. The main bypass damper is a dual-faced pressurized unit to ensure that there is zero leakage of particulate directly to the stack while operating on coal. All of the damper controls and dampers on the ID fans are tied back to the boiler operator and are interlocked to operate in safe modes.

Internal to the power plant we were fortunate to have retained all of the coal handling and pulverizing equipment for #1 boiler, and the only modifications undertaken were:

1. Add air cannons to solve historical bunker coal bridging problems (*);

2. Change to screw feeders and add dust collection on internal conveying systems to control fugitive dust.

   All other coal related equipment was overhauled or rebuilt.

The coal receiving and storage areas were expanded to permit more rapid car unloading and cleaner handling. This was accomplished by enlarging the apron conveyor area and adding additional yard conveying equipment (*), as well as the purchase of a front-end loader for yard work.

During the design of the particulate control system, we became concerned with the bag blinding due to oil droplet contamination. Normally the baghouse would be bypassed during oil operation, but during the transfer to or from oil or if an oil fired burner is added for coal flame stability, as during boiler turn-down, then oil droplet bag contamination became a concern. Therefore, in order to overcome potential problem a ground limestone injection system is being added; this unit will permit injection of limestone directly into the breach prior to the baghouse. Based on surveys, when in operation this system should provide bag protection during transitional periods with oil operation. It is our understanding that clean coal ash can also be used for this purpose. The equipment for injection is now being received (*), and this will be installed in the very near future. Meanwhile, ground limestone was initially injected manually into the breach to precoat the bags prior to use.

The regulatory agencies have been extremely cooperative, and have worked closely with us toward a common goal of converting our facility to coal as rapidly as possible, in an environmentally sound manner.

On March 10, 1979 we formally applied to the Massachusetts Department of Environmental Quality Engineering for conditional approval to convert our number one boiler to coal, and that conditional approval was received on May 15, 1980.

On January 19, 1981 we applied for final operational approval subject to test verification for the burning of coal in number one boiler. We received final approval verbally on February 24, 1981.

Since we already had a significant portion of the equipment necessary to burn pulverized coal, our converstion costs are low compared to what I would expect the norm to be.

The approximate conversion cost was as follows: (*)

Pollution Control Equipment Dampers, et al	$250,000
Foundations, duct work and breach modifications, insulation and baghouse erection	350,000
Fugitive dust controls, screw conveyors, vertical conveyors, coal handling equipment, etc.	-150,000
Misc. ash handling equipment bunker air cannons	50,000
Recorders, controls, opacity meters, etc.	30,000

Conversion Costs (cont'd)

Automatic limestone injection system & storage	\$ 30,000
Equipment, overhauls, R.R. siding rebuilding, demolition and piping changes, etc.	100,000
Total	\$960,000

Without the retained coal pulverizing and storage equipment we have concluded that the total cost of the facility reconversion still could have been kept under two million dollars ......a good investment.

Funding was obtained by utilizing Massachusetts Industrial Revenue Bonds with the cooperative effort of the Beverly City Council, and the Beverly Industrial Development Finance Agency, as well as the Massachusetts Industrial Finance Authority.

Coal quality and delivery reliability has been continuously posed to us as a lingering problem. As part of our program to convert back to coal, the Chief of the cogeneration plant and myself visited approximately 15 mines that had bid to supply coal to our specifications. We have had quality analysis made and sample car lots shipped from approximately five mines. The competition to sell us coal has been good, and we are presently purchasing coal for use and inventory from two different mines, and are sampling a third.

A summary coal specification follows: (*)

Ash	13% Max. (formerly 9%)
Ash Fusion	2450 F minimum
Volatile	23% minimum
BTU/#	13,000 Dry
Hardgrove Grindability	80 minimum
Size	2 X 0 - 10% Fines
Sulfur	1.21#/10^6 BTU
Moisture	6% nominal

Coal shipments are via Conrail or D & H to the B & M R.R.; shipments are presently split between the two source railroads in order to better observe delivery performance. To date the railroads involved have made what appears to be extraordinary efforts to provide good service and have always had cars available. Likewise, USM is a good customer since the cars are unloaded and returned promptly, unlike shipments through some points. Obviously, the railroads view coal as an important expanding revenue source in the northeast, and wish to develop the market. Of some interest is the fact that more than 40% of our delivered cost of coal is for rail transportation.

In a subjective manner our analysis of coal and mine visits

convinced us that the 9% ash limit was imposing an undesirable restriction on our coal supply. It not only restricted our supply, it also made a significant number of coal mines marginal sources even when operating with washing equipment. Further, additional delivered costs/BTU were being imposed by the state ash limit regulation without appreciable benefit, since we were also responsible for particulate emissions.

Therefore, on November 3, 1980 we applied for an exemption to this ash limiting regulation, and received approval for same on January 6, 1981.

Burning coal instead of oil is harder work, causes more equipment maintenance (in our case), takes more operating personnel, and can be very dirty. We have worked very hard to develop an attitude in our power plant crew that is positive and supportive. Coal does have its benefits and perhaps the most important one is its domestic availability and its cost per BTU. The curves shown (*) was obtained from the D.O.E., and is a summary of projected costs of oil vs. coal in constant 1980 dollars. If the curves hold true then the future real cost of coal will remain significantly lower and more constant than oil.

The coal use should displace 75,000 barrels of #6 fuel oil or 90% of use at a current cost of $36/barrel, this represents approximately 18,000 tons of coal at a cost of $50/ton delivered. The added costs associated with coal operations have been conservatively estimated to be $17.00/ton or 65¢ per million BTU, these costs include ash handling and maintenance, etc. Based on these numbers, the projected cost savings are at least $1,500,000 during the next one year period. Further, several used pulverizers have recently been obtained , and the other two boilers will be converted to coal by year-end; these boilers on the common breach will utilize the same pollution control equipment. This added conversion cost will be under $200,000, and will convert the facility to 100% coal firing.

It has been just over one year since a decision was made to convert to coal, it is perhaps important to review the significant dates occurring during this conversion period in order to put the time schedule in perspective.

(1)	Alternate Energy Studies Completed	Jan. 15, 1980
(2)	Meetings w/involved Regulatory Agencies - design initiated	Feb. 5, 1980
(3)	Application for Conditional Environmental Approval	March 20, 1980
(4)	Conditional Environmental Approval	May 15, 1980
(5)	Design approach & conversion Cost estimate completed & Funding Program initiated	June 5, 1980
(6)	Beverly I.D.F.A. Bond Approval	July 31, 1980
(7)	City of Beverly I.D.F.A. Bond Resolution - began ordering equip.	Aug. 12, 1980

(8)	General contractor started work	Aug. 15, 1980
(9)	Final drawings submitted to D.E.Q.E.	Nov. 12, 1980
(10)	Loaded first R.R. car w/coal	Jan. 8, 1981
(11)	All necessary equipment rec'd.	Jan. 16, 1981
(12)	Submitted final data for approval to start-up	Jan. 29, 1981
(13)	Final approval to start	Feb. 24, 1981
(14)	Operating on coal	March 1, 1981

Earlier I mentioned that we had an operating fluidized bed coal unit at our facility. In 1976, USM was approached by the New England Council - in particular its then President, and now Massachusetts Governor, Ed King, to determine if we would work cooperatively with Wormser Engineering to permit them to install a preproduction fluidized bed unit on one of our oil fired boilers.

A prototype development unit was installed at our Drop Forge facility and its oil fired boiler was converted to coal.

The Wormser Fluidized Bed unit (*) is unique in that it is a modular package that can be used to convert an oil designed boiler to coal(even high pressure), in an environmentally sound manner without derating the boiler. (It is the reverse of an oil burner conversion unit to a coal boiler). The unit has two fluidized beds (*) with the coal being burned in the lower bed of sand on top of a distributor plate. The lower bed temperature is maintained at an optimum burn temperature to minimize NO_x by the use of circulating water tubes located in the fluidized bed itself.

The hot gases from the lower bed pass through a water-cooled upper distributor plate into a secondary fluidized bed. This fluidized bed of limestone performs a primary function of absorbing the sulfur, which converts the limestone to gypsum. A secondary function offurs in this bed in that carbon carry-over tends to burn in the bed, thereby increasing the overall efficiency.

The particulate carried over from the upper bed is carried through the boiler and removed, in our case, in a baghouse provided by American Air Filter.

The Wormser Fluidized Bed is capable of 90% sulfur removal from 3.1% sulfur coal by using 23 # of limestone per million BTU input, and operates with 97-1/2% carbon utilization.

The coal and limestone for the unit is loaded and stored in separate bins. The coal is crushed, dried and pneumatically conveyed into the lower bed; the limestone is pneumatically conveyed to the upper bed.

The gypsum from the upper bed is bled off and pneumatically conveyed to the baghouse. The combined gypsum and ash from the collector is screw conveyed to a dumpster for disposal.

The entire system operates unattended with all systems being automatically controlled, and the unit has a 100% turn-down capability; this is accomplished by slumping the fluidized bed exposing the water tubes. While slumped the lower bed has been shown to maintain its temperature sufficiently to permit reliable restart after a three hour shutoff by reinitiating fluidizing air and injecting more coal.

We have an automatic telephone dialing system on the unit which alerts our main facility, if the unit malfunctions or is not operating properly. Our biggest problem to date has been that of foreign substances in the coal.

The unit in our Drop Forge underwent a continuous 1000 hours commissioning test run last spring, and is now being used for a series of evaluation tests on various coals, as well as that for sulfur removal efficiency. Meanwhile, it is the prime source of heat for our Drop Forge Shop.

Wormser Engineering in Middleton, Massachusetts, has recently begun to market these Fluidized Bed Conversion Units. USM is currently evaluating using a Wormser unit as a coal conversion unit at a second facility.

The boiler under consideration for conversion uses 16,000 barrels per year of 1% sulfur #6 fuel. This fuel, which currently costs $39.00/barrel in our area, would be displaced by approximately 4,000 tons of coal at a cost saving of approximately $420,000/year.

Part of the evaluation in the conversion of this particular unit, involves the installation of proper coal, limestone, and ash storage and handling equipment, as well as particulate collection equipment. Preliminary figures for entire modification for conversion and start-up appear to be about $1,200,000, equally divided between the 20,000 lb/hr Wormser Fluidized Bed Conversion Unit and the related receiving, handling, storage and particulate equipment.

(*) denotes slides

PHYSICAL CONVERSION OF OIL-DESIGNED POWER PLANTS

DAVID H. KREGG, VICE PRESIDENT, AND
JOHN PHILIPP, DIRECTOR, POWER TECHNOLOGY DIVISION
BURNS AND ROE, INC.

INTRODUCTION

The utility companies in the United States are the largest consumers of primary energy resources, which they transform into electricity, the most versatile form of energy available to consumers. A 1980 survey of oil-fired utility boilers (50 MW and larger) performed by the Mitre Corporation for the DOE identified 220 boilers with a total capacity of 30,165 MW as coal-designed and burning oil, and 245 boilers with a total capacity of 63,279 MW as oil-designed and firing oil. The survey indicated that 80 percent of the oil-designed capacity is in units larger than 200 MW, and over 50 percent is in units 400 MW or larger (Figure 1). Also, 83 percent of oil-designed capacity is less than 20 years old (Figure 2). Of oil-designed capacity, 65 percent is in California (21,601 MW), Florida (12,656 MW), and New York (6,560 MW).

In general, both regulatory requirements and economics favor the reduction of petroleum use for the generation of electricity. Severe limitations on the use of oil and natural gas in new installations, including electric power generating facilities, were imposed by the Power Plant and Industrial Fuel Use Act which is part of the National Energy Act of 1978. Pending legislation on oil "back-out" includes a requirement for a minimum 50 percent reduction in oil use by utilities by 1990. These policies to encourage the use of coal or other alternate fuels are consistent with the overall objective of conserving oil and natural gas and restricting their use to situations where substitutions are impractical.

CONVERSION TO DUAL COAL-OIL FIRING

Increased use of coal by electric utilities may be achieved by adding new coal fired steam generating capacity or by converting existing oil fired units to fire coal in some form. Several technologies

SIZE DISTRIBUTION OF OIL-DESIGNED UNITS

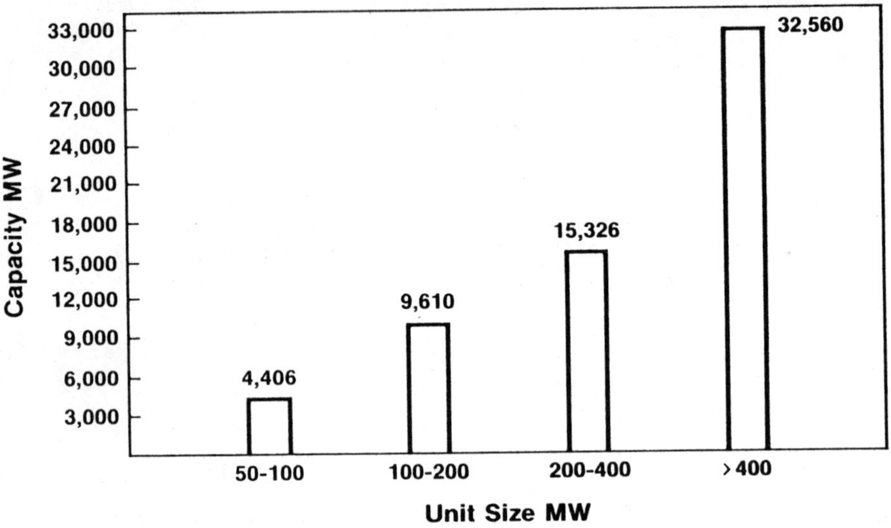

FIGURE 1

AGE DISTRIBUTION OF OIL-DESIGNED UNITS

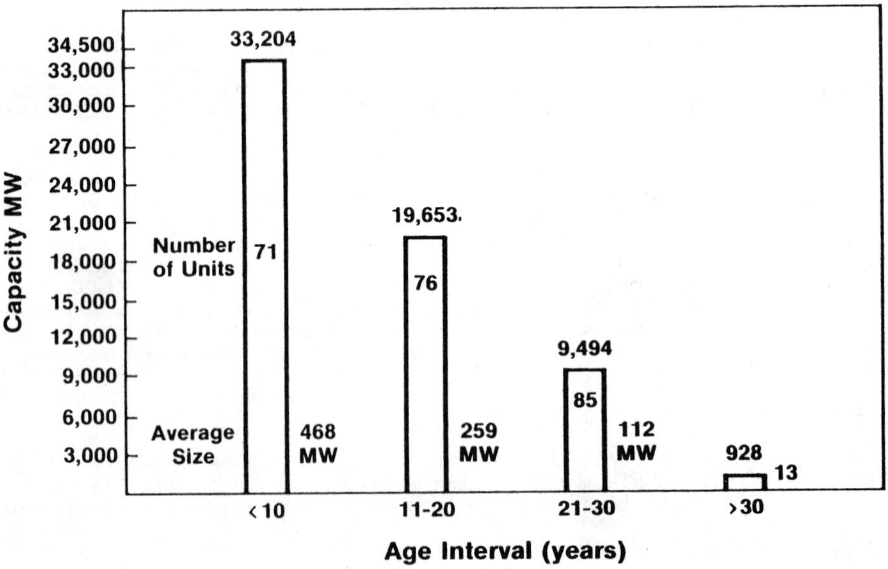

FIGURE 2

may be considered for such a conversion. Of these, an option that can be economically attractive is dual coal-oil firing in which pulverized coal firing capability is added to the boiler, and the existing means of firing oil are retained. Provisions are included for coal supply, soot blowing, particulate control, and ash removal. The converted boiler is operated on coal up to its maximum coal capability and switched to oil for a higher load. The annual consumptions of coal and oil depend on the unit load regimen and coal capability as shown on Figure 3. Boiler design parameters and coal characteristics determine the coal capability.

Such a conversion was implemented by the State Energy Commission of Western Australia. Kwinana Units 5 and 6, each originally designed to fire oil and rated at 200 MW, have been in successful operation firing coal and oil alternately since April 1978 and March 1979, respectively. Their operating records are shown in Figure 4.

The Kwinana boilers were determined to have capability of 120 MW when firing coal, a limitation established on the basis of gas velocities to minimize erosion from ash laden gas in the convection passes. The coal in Western Australia is relatively clean, with about 8000 Btu/lb, up to 7.5% ash, 0.76% sulfur, and 30% moisture.

FURNACE HOPPER AND CIRCULATION

Figure 5 shows the boiler before and after conversion. The original boiler had a flat bottom with a single lower drum that received water from three circulating pumps taking water from the upper drum. The furnace bottom was changed to incorporate a bottom ash hopper and two drums to provide water circulation to the furnace walls. These drums were located 90 degrees from the original drum location to allow sufficient room for the ash hopper and crusher. A new header was added on the discharge side of the circulating pumps, and three pipes from the header were installed to distribute water to each bottom drum. Figure 6 shows the back wall of the furnace and the cut line for adding the furnace walls for the hopper bottom. It also shows the wall area used for the inlet of the gas recirculation that is necessary for the oil design.

BURNER MODIFICATIONS

The furnace has three oil guns in each corner; two levels of coal burners were added between the oil guns. Since it was not possible to raise the elevation of the top burner, the length of the burner panel determined where the furnace hopper would start. The modification to the burner panel required a similar change to the windbox and also some platform changes.

EXCAVATIONS

The boiler is supported by six columns, each on a pile cap. The piles under each cap were driven to refusal about 40 feet below grade. because of limited headroom for pile driving equipment on the pit periphery, the pit for Unit 6 was constructed with two stages of sheet piling. The first level used the maximum length, without welding, that could be driven on the centerline of the columns. After excavation of the sand in the sheet piled area, a second stage of sheet piling was installed by placing the pile driver inside the furnace area where there was adequate headroom. Figure 7 shows the sheeting arrangement. The area under the Unit 5 boiler was excavated with the unit in service. A trench was dug on the same lines as the first stage sheeting on Unit 6, and reinforced concrete was placed in the trench. The area inside the concrete was then excavated and sheeting was driven at the lower level so that the excavation could proceed.

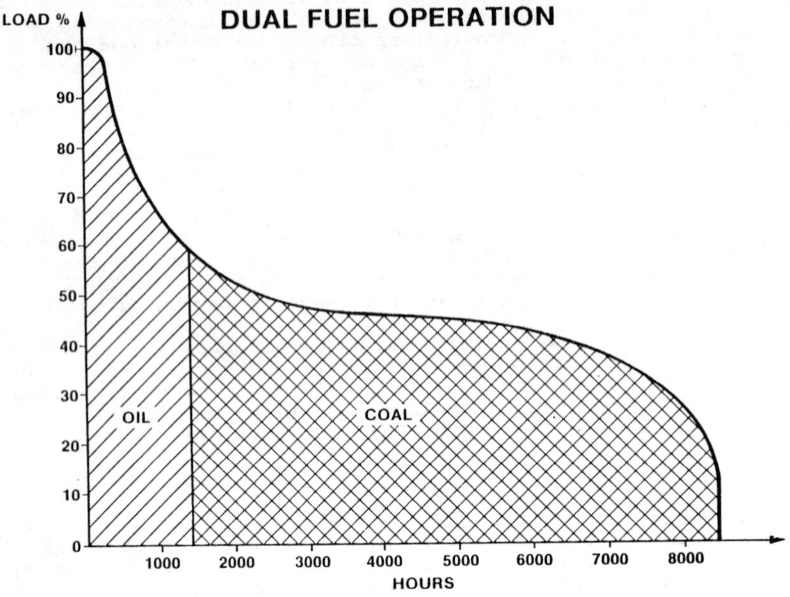

FIGURE 3

STATE ENERGY COMMISSION
WESTERN AUSTRALIA
KWINANA STATION

		OPERATING PERIODS	
		JULY 1, 1979 - DEC. 31, 1980	
		UNIT #5	UNIT #6
IN-SERVICE HOURS	HRS	11,972	11,380
SHUTDOWN RESERVE HOURS	HRS	275	139
TOTAL SHUTDOWN HOURS	HRS	1,192	1,784
UNIT AVAILABILITY	%	90.95	86.45
GENERATION	MWH	1,106,658	1,047,250
AVERAGE UNIT LOAD	MW	92.43	92.03
FUEL CONSUMED:			
COAL	TONS	643,755	599,116
OIL	TONS	14,495	17,393
COAL	10^9BTU	10,899	10,151
OIL	10^9BTU	535	641
TOTAL FUEL	10^9BTU	11,434	10,792
PERCENT COAL (BTU BASIS)	%	95.20	94.06
AVERAGE HEAT RATE	BTU/KWH	10,332	10,305

FIGURE 4

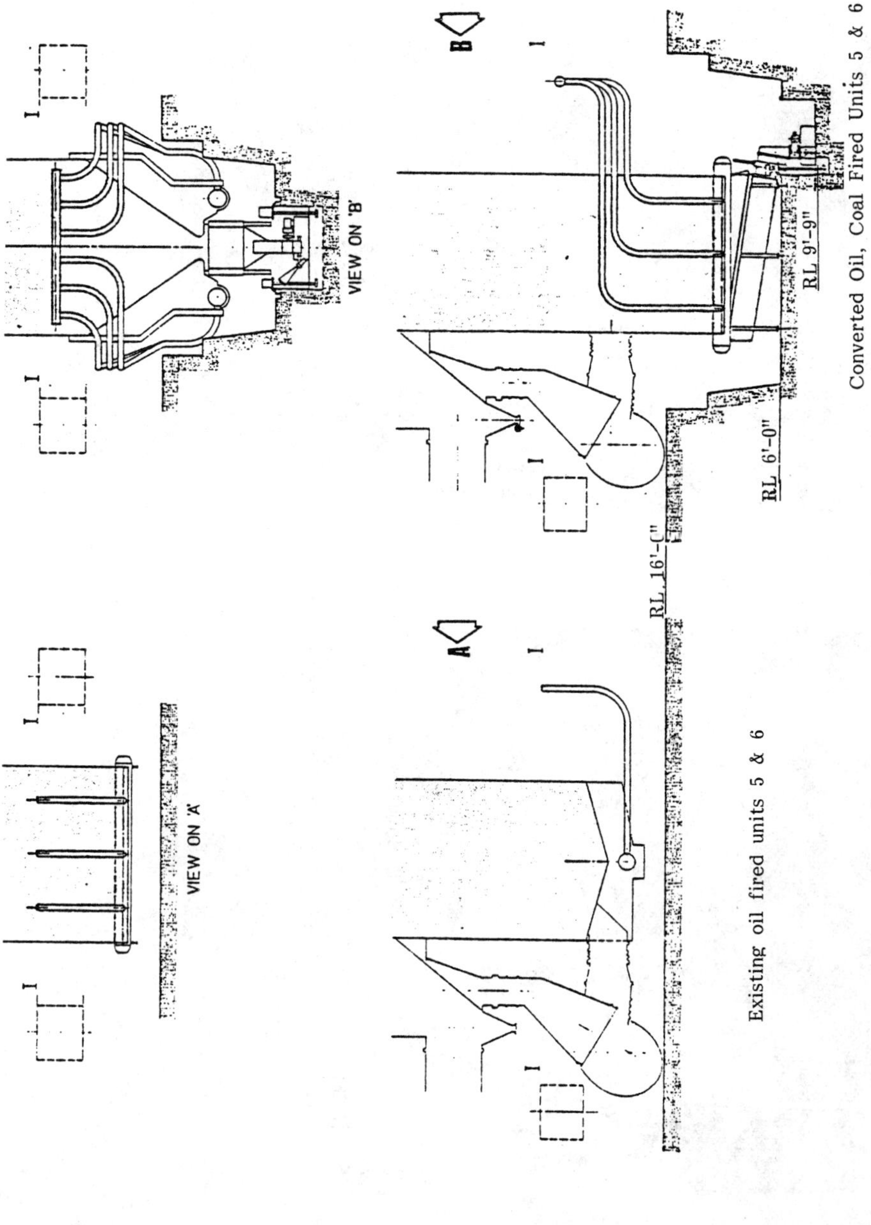

BOILER MODIFICATIONS FOR DUAL COAL-OIL FIRING

Figure 5

FIGURE 6. FURNACE WALL

FIGURE 7. EXCAVATION SHEETING

SOOT BLOWERS

One row of soot blowers was installed directly above the burners, and wall openings were provided for a second row. Tubes installed through the convection pass for installation of the extended IR blowers are cooled with forced draft air. Additional ID blowers were installed in the superheater, reheater, and convection passes.

PULVERIZERS

Each unit was furnished with two Raymond bowl mills with a combined capacity of 140 MW with the design coal. No spare mills were provided. Although this affects the availability of the units on coal, they can use oil during periods of mill maintenance. The actual operating mode is to run at 120 MW between 8 a.m. and 8 p.m., and at 60 MW overnight. Whenever the unit operates on one pulverizer, oil is fired for flame stabilization. The pulverizers are behind the stack, and the piping to the burners is about 200 feet long. Investigation indicated that at velocities above 70 feet per second coal settling would not occur in these long lines.

PRECIPITATORS

Each unit was equipped with a two-pass, three-field precipitator. Another field can be added if required. Gas flow is perpendicular to the centerline of the boiler. The design allows for a 2-inch water gauge pressure drop from the air heater outlet through the precipitator and ID fans to the stack. The units are equipped with electrically heated fly ash hoppers with provision for mounting vibrators. The precipitator control room is mounted above the gas inlet duct; remote controls are located in the main control room.

INDUCED DRAFT FANS

Although replacement centrifugal fans were recommended by the boiler manufacturer, the existing axial fans were used successfully.

COAL AND ASH PROVISIONS

Figure 8, an aerial view of the site, shows the new coal handling facilities. A rail unloading facility was constructed to bottom-dump two coal cars at a time. The coal is conveyed to a crusher and then to a lowering well where an active coal pile is maintained. A hopper below the active pile is connected by conveyor to the plant bunkers. Coal is also moved from the active pile by bulldozer to a 7-week storage pile. The two coal silos per unit have a combined capacity of about 9 hours at the 120 MW rating. The conveying system is set to refill the silos automatically when they reach a given level. A single conveyor belt system supplies the two units.

Bottom ash is conveyed from the furnace by a hydraulic system to an on-site disposal area . Fly ash from the economizer and precipitator hoppers is extracted hydraulically to a sump and then pumped to an abandoned quarry.

FIGURE 8: AERIAL VIEW, KWINANA STATION

CONVERSION OF THE BRAYTON POINT STATION

JOHN F. KASLOW
SR. V. P. - NEW ENGLAND ELECTRIC SYSTEM

Coal conversion is a high priority objective at New England Electric System as we attempt to take every reasonable step toward reducing our customer's energy costs and to reduce our Company's and the New England region's dependence on expensive, imported fuel oil. We have a relatively simple criteria in looking at converting our generating units from oil to coal - the conversion must be economic for our customers and must be accomplished in an environmentally sound manner. Despite being simple and straightforward, these criteria produce some interesting challenges!

Our Brayton Point coal conversion project highlights some of the challenges a utility faces in a major coal conversion and I hope to share our experiences with you today.

But first, let me briefly describe our Company and the Brayton Point station. New England Power Company (NEP) - which owns and operates Brayton Point - is the generation and transmission subsidiary of New England Electric System (NEES). Three retail subsidiaries provide electric service to over one million customers in Massachusetts, Rhode Island and New Hampshire. New England Power Company has 4017 MW installed capacity - with 2824 MW fossil fired and burning oil in recent years. Of the fossil fired capacity, 1728 MW was designed for, and did in fact, burn coal at one time.

The Brayton Point station is a four-unit station - our largest - and is located on the northern reaches of Mt. Hope Bay in southeastern Massachusetts. Table 1 describes the unit characteristics.

TABLE 1

UNIT	CAPACITY MW	COMMERCIAL DATE	TYPE SERVICE	FUEL	ANNUAL OIL CONSUMPTION MILLION BBLS
1	250	1963	Base Load	Coal/Oil	3
2	250	1964	Base Load	Coal/Oil	3
3	650	1969	Base Load	Coal/Oil	6
4	450	1974	Int. Load	Oil	3

Units 1, 2 and 3 are very efficient base load units which burned coal until 1969 - as did seven other units on our System. The Brayton Point units were equipped with 1960's vintage electrostatic precipitators and were designed to burn a high quality coal as noted in Table 2.

TABLE 2

DESIGN COAL CHARACTERISTICS

	UNIT 1 & 2	UNIT 3
Moisture %	6.50	3.20
Volatile Matter %	23.90	38.20
Fixed Carbon %	63.10	51.40
Ash %	6.50	7.20
Sulphur	0.80	2.44
BTU/lb.	13,600	13,580
Grindability	90	58
Ash Softening Temp. ^0F	2,375	2,120

With this backdrop let's look at our Brayton Point coal conversion project. The reason for wanting to convert to coal is very straightforward. Table 3 shows what has happened to oil prices at Brayton Point in recent years.

TABLE 3

1971	$ 1.81/bbl.
1972	$ 2.54/bbl.
1973	$ 3.53/bbl.
1974	$10.33/bbl.
1975	$12.29/bbl.
1976	$11.37/bbl.
1977	$13.27/bbl.
1978	$11.40/bbl.
1979	$15.00/bbl.
1980	$20.41-32/bbl.

While the chronology of oil price increases is significant, the chronology of environmental regulation is perhaps more significant. In 1969, when the units were converted to oil, there were no stack gas emission limits in effect - save for smoke color limits. The NPDES water discharge permit program was not in place - nor were there any restrictions on disposal of fly ash.

Following the Clean Air Act of 1970 and its Clean Water counterpart - the regulatory rules changed dramatically. Since we were burning oil, they didn't pose very serious problems - but how they complicated going back to coal!

The fuel sulphur limit under the Massachusetts State Implementation Plan (SIP) was established at 0.55#/MMBTU and the particulate emission limit at 0.12#/MMBTU. On oil we could meet these limits easily but on coal we knew they would require very costly modifications. A new generating unit installed in 1974 - and its closed cycle cooling system - occupied the real estate that had formerly been devoted to our wet fly ash sluice basins. This too posed a problem.

In 1977 we received a coal conversion order from the Department of Energy for Unit 1-2-3 under its Energy Supply and Environmental Coordination Act (ESECA) authority. Under the SIP conditions that prevailed at that time, we concluded that the fuel sulphur requirements would make coal conversion impractical and we informed DOE that unless we could burn a higher sulphur coal, we would contest the conversion order.

In a parallel effort we had been seeking authorization to increase our allowable oil sulphur content at Brayton Point - and at our Salem Harbor station. Air quality modeling coupled with extensive monitoring had indicated that a fuel sulphur content of 1.21#/MMBTU could be supported. This pursuit became intertwined with our coal conversion effort as did a freak primary standard violation of the particulate standard a few miles from the station.

I should also point out that our coal conversion effort - the largest in the country at that time - became very visible. This visibility led to the involvement of a number of parties and finally resulted in a task force being formed to explore the issues. Included on the task force were a Company representative and representatives from EPA, our state environmental agency, DOE and an independent party whose role evolved into that of a mediator.

After some eighteen meetings of the task force, agreement on the conversion groundrules was reached. The agreement was facilitated by an approval of our long standing request to burn a higher sulphur oil and a substantiation that the primary standard particulate violation was not the result of emissions from the Brayton Point station.

The principal elements of our final conversion agreement were:

(1) An agreement by the Company to meet a more stringent particulate emission limit (0.08#/MMBTU versus the SIP limit of 0.12#/MMBTU).

(2) An agreement by the state to base fuel sulphur compliance on a thirty-day average versus an instantaneous maximum value and to permit twenty-four hour fuel sulphur levels considerably higher than the thirty-day average.

(3) A best efforts commitment by our state environmental agency to grandfather Brayton Point from statewide SIP revisions for a ten-year period.

I personally believe that the mediated task force approach worked well in this case and the mutual respect and understanding that developed among the members prevails today. I don't claim that this approach can solve all environmental issues - however, it worked in this case.

While my description of the background of the regulatory process is lengthy, it is, nonetheless, important to an understanding of the project.

Now to the physical part of the project -

Once the basic agreements had been reached and a SIP change was completed in May of 1979 we set out on a crash program of engineering and construction with a view and a goal of having the three units permanently converted in March, June and October of 1981. Stone & Webster was retained as Engineer/Constructor in June of 1979 and the first equipment order was placed in July 1979.

The physical plant modifications can be separated into three categories. Those related to environmental backfitting represent the largest segment. Secondly, some modifications are related to the changed coal quality versus the original design coal; and, finally, some modifications are incorporated for operational and reliability purposes.

Looking at the physical changes in more detail, we find that those dictated by environmental backfitting requirements to be most consequential in terms of cost and schedule impact.

Of the $180 million project cost, roughly 50% is devoted to environmental control equipment. The principal modifications in this area are new precipitators added in series to the existing percipitators. Whereas the original Unit 1, 2 and 3 precipitators had design efficiencies of 98%, 98% and 99% respectively - using the old, less stringent ASME measurement method, the upgraded precipitators will have an efficiency of 99.3% using the current - and more stringent - EPA measurement method. Physically, the collecting area of the Unit 1 and 2 precipitators is being increased from 133,600 square feet to 345,000 square feet. The Unit 3 precipitator collecting surface is being increased from 335,000 to 944,000 square feet. Because of space constraints, the series percipitator additions are being constructed on the opposite side of the stacks.

When these units burned coal in the 1960s, the coal bottom ash and fly ash were sluiced via sea water to a large settling basin. Because the old settling basin became the site of the Unit 4 closed cycle cooling system - and because we hope to promote the resource use of fly ash, a new dry ash handling system is being installed as part of the conversion. Bottom ash will be sluiced to decanting bins - and all ash will be used or disposed of off site. Other environmentally related modifications include enclosing the inclined coal conveyor to reduce fugitive coal dust and collecting runoff from the coal pile and treating the runoff in the station waste water treatment system.

As noted earlier, the original design coal for the station was a premium coal - not available today at reasonable cost for steam purposes. Table 4 illustrates the pertinent characteristics of the original design coal versus the coal we are buying today.

TABLE 4

Characteristic	Original Design Coal Unit 1 & 2	Original Design Coal Unit 3	Modified Design Coal
Moisture %	6.50	3.20	8 maximum
Volatile Matter %	23.90	38.20	26 - 36
Fixed Carbon %	63.10	51.40	-----
Ash %	6.50	7.20	10 maximum
Sulphur	0.80	2.44	1.21 #/million BTU avg.
BTU/lb.	13,600	13,580	13,000
Grindability	90	58.6	60 minimum
Ash Softening Temp. 0_F	2,375	2,120	2450 minimum

While the new design coal influences our redesigned coal handling and ash systems, it was a factor in our decision to replace the pulverizers on our 650 MW Unit 3. New B&W type MPS-89 mills are being installed to permit full unit capacity with one mill out of service.

Beyond these major modifications there are a host of station electrical supply, distribution and instrumentation changes - and numerous equipment and piping relocations necessitated by the total scope of the modifications. These are typical of the "hidden cost" items which are not obvious until detailed engineering is undertaken.

Another interesting facet of our coal conversion at Brayton Point is the delivery and unloading of coal. The station is equipped with a coal unloading tower - which originally had an unloading capacity of 600 tons/hour. When Unit 1-2-3 last burned coal routinely in 1969, the unloading tower was supplemented with deliveries of coal by means of a self-discharging collier. As we viewed the logistics of coal delivery and unloading in the 1980s - with no self-discharging colliers in the coal hauling trade - we found that the single unloading tower could barely handle the projected 3,000,000 tons/year burn. Rail and truck delivery were not available to us as alternatives - or for backup purposes. Additionally, since Unit 4 burns oil, the dock logistics - with both oil tanker and conventional coal collier traffic to contend with - appeared to pose a significant problem. Another consideration was the reliability exposure if the single coal unloading tower was out of service. Finally, as we viewed the available fleet of colliers in the domestic trade, it was clear that the fleet was small in number and certainly not of modern vintage.

The combination of these factors led to our decision to refurbish our existing coal unloading tower and to build a new self-discharging coal collier. The collier - which we hope to build as a coal-burning, steam-propelled ship - will be capable of handling up to 75% of our annual coal needs at Brayton Point. The balance will be unloaded by the coal tower from conventional colliers and/or barges.

The self-discharging collier will have an unloading rate of 3500 tons/hour versus the rebuilt tower unloading rate of 800 tons/hour. This will allow us to reduce congestion at our station dock and facilitate the handling of oil deliveries for our No. 4 unit.

While a project of this magnitude has all the complications one would expect in any major undertaking, attempting to complete the project with a peak construction force of over 1,000 people working on all three units in parallel - with the units in operation - makes for an interesting life for the Project Manager and the Station Superintendent. Thus far, the interferences between construction and operation have been less onerous than one might have expected. This is due, in large measure, to the common and central goal in the organization to complete the conversion on schedule and within budget.

We eagerly await the completion this year of the Brayton Point conversion and the substitution of close to 3,000,000 tons of domestic coal/year for 12,000,000 barrels/year of increasingly expensive foreign oil. We know our electric customers are awaiting the impact of the lower fuel cost.

UTILITY PERSPECTIVE OF MHD POWER PLANTS

J. W. MOYER, J. W. GRISWOLD, M. C. WEHREY
SOUTHERN CALIFORNIA EDISON COMPANY

Presented by Heinz Pfeiffer

ABSTRACT

Among the advanced energy conversion concepts being pursued by industry and the government, Magnetohydrodynamics (MHD) may have the potential for significantly improving the efficiency of thermal electrical power generation. MHD may also provide means of using coal in an environmentally acceptable manner, either in new MHD plants or through a retrofit approach.

With approximately 70% of its total generating capacity in the form of oil- and gas-fired generating stations Southern California Edison views MHD as a potential coal conversion option. If the concept of retrofitting MHD to existing oil- and gas-fired stations proves feasible it may be possible to convert selected stations to coal in an environmentally acceptable manner, while increasing their capacity.

The Department of Energy's goals are currently focused on the engineering and construction of the Engineering Test Facility: the first commercial scale MHD demonstration plant. As the level of MHD awareness is increasing among electric utilities, efforts are being made to increase their participation in the national program.

Edison's early involvement with MHD is the basis for the views expressed here on this promising, yet challenging, technology.

INTRODUCTION

In a recent projection (1), the Electric Power Research Institute has shown that to maintain an intermediate economic growth of 3% per year, in terms of Gross National Product, the present installed capacity of the utilities across the nation would have to more than double by the year 2000. This capacity would be needed to produce 6 trillion kilowatt-hours. The coal generated fraction of this energy would be in the order of 2.6 trillion kilowatt-hours in the year 2000, not taking into account the energy produced from coal-derived fuels. These numbers point out a strong need for coal fueled technologies, existing and new.

Table 1 shows the potential shortfall looming on the energy scene of the country in the year 2000. A shortfall very dependent on the amount of nuclear generation installed by that date.

TABLE 1 - U.S. ELECTRICAL CAPACITY IN YEAR 2000

	LOW NUCLEAR		HIGH NUCLEAR	
	CAPACITY (GW)	GENERATION (10^9 KWH)	CAPACITY (GW)	GENERATION (10^9 KWH)
NUCLEAR	150	920	300	1840
COAL-FIRED (CONVENTIONAL AND ADVANCED)	470	2610	470	2610
OIL, GAS AND COAL LIQUIDS	224	313	224	313
HYDRO	100	420	100	420
GEOTHERMAL	16	105	16	105
SOLAR, WIND, AND BIOMASS	10	44	10	44
STORAGE	70	-25	70	-25
TOTAL SUPPLY CAPABILITY	1040	4387	1190	5307
SHORTFALL	239	1463	89	543
TOTAL REQUIREMENT	1279	5850	1279	5850

Looking at a closer horizon, and taking as an example the Southern California Edison Company, one finds that approximately 6000 MW of new resources are needed by the year 1990. Table 2 shows how Edison hopes to meet these requirements and points out that, even with the most aggressive program in the nation, renewable or alternate resources may only satisfy approximately one-third of these needs. Table 3 shows the existing resources of Edison and illustrates its current dependence on gas- and oil-fired resources.

It is also important to address an often minimized or overlooked need: to maintain and preserve the existing capacity of electric utilities. Those with a large fraction of gas- and oil-fired resources are faced with escalating fuel costs and availability concerns, particularly when their oil fuels have to be of the low sulfur variety currently used by Edison.

MHD should be part of the future generation resource of the country. MHD could also be the key to operating existing gas- and oil-fired resources with a reliable source of fuel, at reasonable fuel costs and while meeting environmental requirements. The retrofit or repowering approach has been the focus of our efforts in the area of MHD and is the underlying reason for our current interest in MHD.

TABLE 2 - SCE NEW GENERATING CAPACITY - YEAR 1990

		MW
NUCLEAR		2,340
PURCHASES		1,400
RENEWABLE/ALTERNATE		1,900
WIND	120	
GEOTHERMAL	420	
SOLAR	310	
FUEL CELL	130	
COGENERATION	300	
HYDRO	620	
	1,900	
COAL		1,000
TOTAL		6,640

TABLE 3 - SCE EXISTING GENERATING CAPACITY - YEAR 1980

	MW - SUMMER
THERMAL - GAS AND OIL	10,366
THERMAL - COAL	1,653
THERMAL - NUCLEAR	349
HYDRO GENERATION	875
FIRM PURCHASES	2,009
TOTAL RESOURCES	15,252

WHY MHD?

As summarized on Table 4, we view MHD with having long-term benefits and also near-term potential. Direct coal combustion and high plant efficiency have been the traditional drivers of MHD R&D. The conservation of resources resulting from high efficiencies, particularly operating resources (fuel and cooling water for inland plants) is also a positive aspect of MHD.

Environmental benefits can be derived from high efficiencies in terms of reduced gaseous and solid discharge, and also from the "built-in" sulfur control afforded by the seeding process used with the open-cycle option.

TABLE 4 - SCE VIEWS ON MHD

LONG-TERM BENEFITS:

- DIRECT COAL COMBUSTION
- HIGH PLANT EFFICIENCY
- MEETS ENVIRONMENTAL STANDARDS
- TECHNOLOGY SPIN-OFFS

NEAR-TERM POTENTIAL:

- COAL COMBUSTORS
- COAL CONVERSION OF OIL FIRED PLANTS WITH INCREASE IN CAPACITY

Technology spin-offs are expected from the MHD program and could have a positive impact on electric utility operation. The spin-offs could include slagging combustors, high temperature heat exchangers, magnets for energy storage, gasification technology and progress in combustion kinetics and slag and materials chemistry.

Slagging coal combustors are also part of MHD's near-term potential. The direct use of slagging coal combustors with oil-fired boilers could considerably simplify the conversion of oil-fired stations to coal.

When an increase of capacity is needed at an existing station site the retrofit of an MHD train would be preferred. This use of the MHD technology is part of MHD's near-term potential.

MHD OPTIONS AND DESIGN CHOICES

Open-cycle and closed-cycle are the two MHD cycle configurations currently under development. Although open-cycle MHD (OCMHD) is considerably more developed than closed-cycle MHD (CCMHD), there are some features of CCMHD that should justify utility interest. Table 5 summarizes the main differences between the two options. Studies sponsored by Southern California Edison (2) (3), have surfaced the most critical aspects of the operations of CCMHD generators and have also assessed the feasibility of the CCMHD retrofit concept.

Southern California Edison was also party to a major study effort which included Chas T. Main, Inc., as Architect Engineer. The study was focused on the feasibility of retrofitting an OCMHD train to Etiwanda Unit 1, an existing oil and gas 132 MW generating station of the Edison system (4). The study led to the identification of the configuration shown on Figure 1. During the conceptual engineering efforts, several design choices had to be made, as shown in Table 6.

TABLE 5 - THE MAIN DIFFERENCES BETWEEN OPEN AND CLOSED CYCLE MHD

OPEN CYCLE	CLOSED CYCLE
• USES A SEEDED COMBUSTION GAS	• USES A SEEDED INERT GAS
• NEEDS 4500°F TEMPERATURE	• NEEDS 3100°F TEMPERATURE
• NEEDS HIGH TEMP. AIR PREHEAT OR OXYGEN ENRICHMENT	• NEEDS MORE COMPRESSOR POWER
• CHANNEL OPERATION IS BETTER UNDERSTOOD	• CHANNEL ENERGY EXTRACTION MAY BE HIGHER
• NEEDS PRESSURIZED SLAGGING COAL COMBUSTOR	• SLAGGING COMBUSTOR MAY OPERATE AT LOWER PRESSURE
• NEEDS A SEED RECOVERY AND REGENERATION PLANT	• NEEDS A CESIUM RECOVERY AND PURIFICATION SYSTEM
• MAY PROVIDE "BUILT-IN" SULFUR EMISSION CONTROL	• NEEDS A "CLEAN" ARGON HEATER
• MORE R&D AND TESTING DONE TO DATE	• MAY LEAD TO LOWER O&M COSTS

FIGURE 1 - OCMHD RETROFIT CONFIGURATION

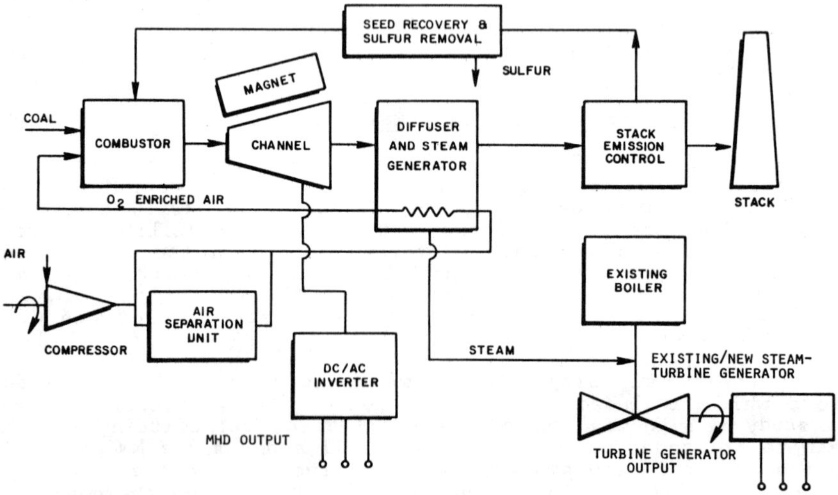

TABLE 6 - CONCEPTUAL DESIGN OF AN OCMHD RETROFIT INSTALLATION

DESIGN CHOICES	DECISIONS
• EXISTING OR NEW STEAM GENERATOR	• NEW STEAM GENERATOR
• SEPARATELY FIRED AIR PREHEATERS OR OXYGEN ENRICHMENT	• OXYGEN ENRICHMENT
• CHANNEL PERFORMANCE	• MODERATE PERFORMANCE
• CHANNEL TERMINALS	• TWO TERMINAL CHANNEL
• FEEDWATER HEATING	• RETAIN EXISTING TRAIN

Perhaps of most significance were the moderate performance requirements placed on the MHD train and summarized on Table 7. These moderate requirements along with the selection of a new steam generator and oxygen enrichment maximized the chances of success of the installation. Although no modifications of the feedwater train were allowed, as shown on Figure 2, the projected performance of the retrofit installation was attractive as shown on Table 8.

TABLE 7 - MODERATE REQUIREMENTS FOR OCMHD RETROFIT

- RELAX MHD GENERATOR PERFORMANCE REQUIREMENTS
- UTILIZE "STATE OF THE ART" MHD GENERATOR
    - MODEST ENERGY EXTRACTION
    - LOW PRESSURE RATIO
    - LOW CURRENT AND POWER DENSITIES
    - TWO-TERMINAL OUTPUT
    - MODEST MAGNETIC FIELD
    - LOW STATIC PRESSURE AND HEAT TRANSFER RATES
    - LOW TEMPERATURE FEEDWATER MHD CHANNEL COOLING
- BUILD RETROFIT AROUND STATE OF ART MHD GENERATOR
- CONCENTRATE ON COAL FIRING AND ENVIRONMENTAL ADVANTAGES

FIGURE 2 - STEAM/CONDENSATE FLOW FOR OCMHD RETROFIT

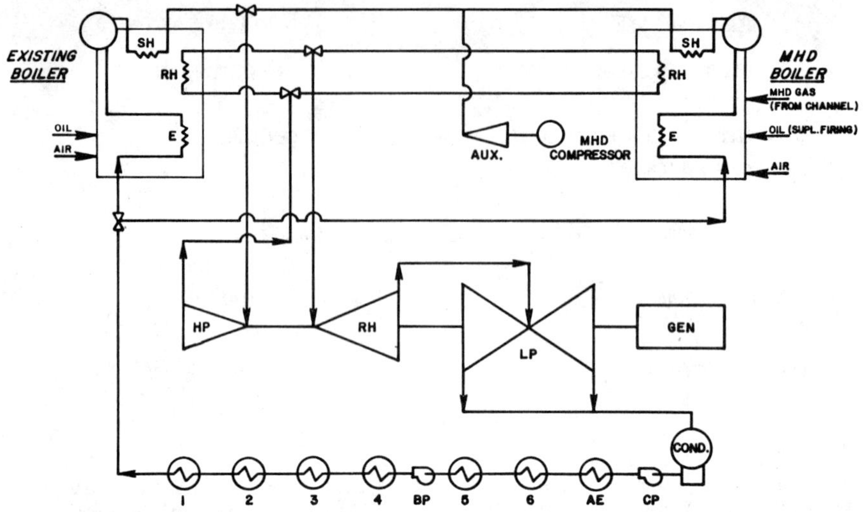

TABLE 8 - PERFORMANCE SUMMARY - ETIWANDA OCMHD RETROFIT

		WITH MHD		
	EXISTING	OPTION 1	OPTION 2	OPTION 3
NET STATION OUTPUT, MW_E	132	194.5	227.4	303.2
MHD OUTPUT, MW_E		61.5	69.1	145.8
STEAM TURB. OUTPUT, MW_E		133.0	158.3	157.4
BOILER STEAM FLOW, LBS/HR	945,000	1,161,000	1,190,000	1,324,000
FUEL INPUT, OIL, BBL/HR	211	NA	NA	NA
FUEL INPUT, COAL, T/HR	NA	75.3	85.8	107.5
HEAT RATE, BTU/KWHR	9,605	9,059	8,834	8,295
STATION EFFICIENCY, %	35.5	37.7	38.6	41.1
STATION HEAT RATE IMPROVEMENT	BASE	5.7%	8.2%	13.6%

UTILITY REQUIREMENTS

The ground rules set for the Etiwanda OCMHD retrofit study illustrate a typical set of utility requirements for a demonstration project on a commercial scale. Not shown on Table 9 are the requirements for adequate space at the generating station site and access to a coal source.

TABLE 9 - OCMHD RETROFIT STUDY GROUND RULES

- NO DEGRADATION OF EXISTING STATION EFFICIENCY
- NO DEGRADATION OF EXISTING STATION AVAILABILITY
- STATION WILL OPERATE DURING MHD SYSTEM SHUTDOWN
- MINIMUM DOWNTIME DURING RETROFIT INSTALLATION
- RETROFIT INSTALLATION WILL COMPLY WITH EMISSION REGULATIONS
- NO ALTERATION TO EXISTING FEEDWATER SYSTEM
- 100% COAL FIRING
- UTILIZE FULL EXISTING TURBINE GENERATOR CAPACITY

The introduction of commercial MHD installations in the utility market will take place in an increasingly restrictive operating environment. Trends resulting from environmental, siting, economic and operating issues have been identified and are shown on Table 10.

TABLE 10 - FUTURE UTILITY ENVIRONMENT

A. ENVIRONMENTAL AND SITING

- AIR QUALITY: SEVERE LIMITS FOR NO_x, SO_2 AND PARTICULATES
- WATER: SEVERE REGULATIONS ON USE AND CONSUMPTION
- LAND USE: COMPETITION FOR AVAILABLE LAND
- WASTE DISPOSAL: STRICT REGULATIONS
- SOCIO-ECONOMIC: WILL DOMINATE TRADITIONAL ECONOMIC AND ENGINEERING CONSIDERATIONS
- LICENSABILITY: STRINGENT AND LENGTHY REGULATORY PROCESS

B. ENGINEERING AND DESIGN

- PERFORMANCE: INCREASED RELIABILITY AND EFFICIENCY
- ECONOMICS: LOWER CAPITAL COSTS AND SHORTER CONSTRUCTION SCHEDULES
- FUEL SUPPLY: FUEL USE ACT - SHIFT TO COAL AND SYNTHETIC FUELS
- OPERABILITY/CONTROL: LESS HUMAN DEPENDENCE - INCREASED OSHA REGULATIONS

These trends illustrate the constraints within which the Engineering Test Facility designeers will have to work and the need for a strong interaction with the utilities and regulatory agencies during the design and evaluation process.

An often expressed concern relative to MHD is the durability of critical components of the MHD train, particularly of the channel electrodes. Operational analysis data have shown that, in order to provide an MHD plant availability of 75% the MHD channel has to have an availability of about 90% (5). In our opinion tradeoffs will have to be made between performance and availability. We believe that performance and efficiency will take second place when availability and reliability of the MHD installations are in question.

UTILITY ROLE

MHD R&D was confined for a long period of time to laboratories and universities and was, except for a few exceptions, unknown to utilities. The "MHD community" is expanding and the dialogue has now been open with electric utilities and other potential users. Utilities are asked to participate in an advisory capacity to the Advanced Power Train effort. Utilities will participate in design reviews of the Engineering Test Facility. More recently a National Users Group on MHD has been formed to assure that the requirements and viewpoints of electric utilities are incorporated in the MHD program. The group will also coordinate, cooperate and exchange information with the utility industry and with other public or private organizations concerned with the development of MHD.

CONCLUSION

MHD is still a fledging technology and in need of demonstration. Encouraging results have been obtained and point the way to operable facilities. Key answers are expected from tests at the Component Development and Integration Facility and at the Coal Fired Flow Facility. However, the ultimate test of MHD as a practical and competitive technology will take place at the Engineering Test Facility. Also critical, will be the timing of this test. Slippages of the ETF schedule will adversely affect the commercialization of MHD. CCMHD should also receive more attention in the national program because of its potential advantages in the areas of operation and reliability. As previously discussed shortfall of generating capacity may appear as early as the year 2000 or before. This period may well be the "commercialization window" for MHD. A steady RD&D program with concrete objectives and continuous dialogue with utilities will be needed to make MHD a successful commercial technology.

REFERENCES

(1) EPRI, "Overview and Strategy, 1981-1985 Research and Development Program," October 1980, P. 1700 SR.

(2) STD Engineering Corp., "Evaluation of Integrated Closed-Cycle MHD as an Option for Retrofitting an Existing Oil-Fired Steam Generating Power Plant (Etiwanda No. 1) and Converting to Coal-Firing," November 24, 1980.

(3) General Electric Co., Advanced Energy Systems, "Study of Closed-Cycle Magnetohydrodynamic Retrofit to Etiwanda Plant," December 8, 1980.

(4) C. T. Main, Inc., "Study of Magnetohydrodynamic (MHD) Retrofit at Southern California Edison's Etiwanda Station Unit No. 1," October 31, 1979.

(5) EPRI, "Operational Analysis of Open-Cycle MHD," July 1980, AP-1463.

AN INDUSTRY PERSPECTIVE OF MHD POWER PLANTS

F. D. Retallick, Manager MHD Programs
Westinghouse Electric Corporation
Advanced Energy Systems Division
Pittsburgh, PA 15236

The industry perspective of open cycle coal fired MHD power plants is one of cautious optimism. After initial exploratory research of MHD by industry, the technology is now in the stage of government supported development, and several more years of effort will be required before industrial investment for facilities and operations necessary to commercialization can be expected. Present Department of Energy planning includes a prototype demonstration unit for operation in the early 1990s. Successful operation of the demonstration unit could develop a viable commercial industry before the turn of the century.

The subjects to be addressed in this paper are the present understanding of the promise for MHD power plants, the major requirements for success, and the recommended route.

The Promise of MHD

The open cycle, coal fired, combined MHD-steam power plant offers the potential to reduce the amount of coal burned to produce electricity, and as a consequence can reduce the total pollutant output from coal fired power plants. In addition an alternate means of removing sulfur from the combustion gas stream is possible. These overall advantages, however, must be interpreted in terms of advantages addressing the individual needs of a Utility operator. The MHD-steam power plant must, in the most simple case, be related to the conventional coal steam plant. These conventional plants with continuing steady improvements will be with us for an extended period of time. The MHD-steam plant in the eventual test will have to show an improvement in the cost of generating electricity over this proven technology in order to be accepted.

An improvement in the cost of electricity can derive from several characteristics such as efficiency, capital cost, operating cost, construction time, reliability maintenance, operating ease, fuel cost, and waste disposal. It is important that each of these characteristics be evaluated and understood as a basis for the decision on commercialization.

Major Requirements for Success

The prime motivation for MHD has been its potential for a high efficiency power plant when combined with a steam turbine generator unit. Several studies have been performed over the years which result in the present assessment that early commercial configurations using oxygen have the potential for an overall power plant efficiency of 42-43.5% (3) (4) and more advanced configurations with integral preheat have the potential for 45-48% (1) (2) (5). These numbers can be compared to a maximum of 36% (6) with conventional supercritical steam plants incorporating flue gas desulfurization. Thus, on the basis of conceptual performance analysis, which is the starting point for all new developments, a significant benefit has been identified. It has also been established that the incremental gain in efficiency for an MHD power plant is a function of power output, thus making the larger base load plants the most beneficial applications.

This later fact leads to the conclusion that it will not be possible to demonstrate the absolute level of efficiency predicted for 1000 MWe units through demonstration units of 250-500 MWt, rather it will be necessary to prove the predictability at this level and to exhibit an understanding which provides confidence in extrapolation to larger designs. Table 1 lists some of the more significant characteristics that will need verification to gain the required confidence in efficiency. Probably, the most important is the power extraction capability of the MHD channel. Good progress has been made in showing that the power extraction can occur as predicted and it is this characteristic that has had the major emphasis in the MHD program to date.

It appears possible at this point in time to assemble a technical program aimed at verifying performance characteristics required for success. This is a significant accomplishment but it is not sufficient to answer the question of commercial viability. Commercial viability requires a basic evaluation of the advantage of MHD in its eventual power plant application.

The development of a new power plant concept can be for several reasons related to Utility needs. In the past the simple gas turbine was introduced because of the low capital cost, quick start characteristics which made it a front runner for peaking applications; The combined gas turbine steam cycle with good efficiency and operational flexibility fulfilled a need in intermediate load units; Coal gasification combined cycles are under investigation because they could fulfill a need in the absence of oil and gas as power plant fuels; Nuclear power plants opened a new fuel source for use in power generation; In terms of advanced concepts, the fuel cell offers potential for diversified local installation of power generation capability and modular power plant construction capability; Solar power again offers the potential of a new energy source.

In addition to benefits such as indicated above, all power plant concepts must show competitive economics in their area of application.

The major MHD advantage is primarily economics, derived from efficiency. Therefore the effort in development must be aimed at assuring this advantage. Coal is the United States most abundant energy source. A mature technology for burning coal and generating electricity at a lower efficiency, than by MHD exists, for which costs can be confidently predicted. With MHD, it must be shown that superior economics to this established base will prevail. Understanding this is important, for it allows an evaluation of what must be accomplished, starting from the basic efficiency increase of the MHD power plant. Figure 1 indicates the basic efficiency advantage of MHD. The levelized fuel costs of a conventional coal steam plant are compared to the early MHD plants using oxygen and the advanced units with integral preheat. Respectively, they represent an approximate potential fuel savings of 17% and 28%. As coal costs increase, the absolute value of fuel costs will increase and will represent a larger fraction of the total cost for generation, thus enhancing the advantage of MHD. The importance of the 17% and 28% fuel cost advantage is best understood by relating it to the total cost of electricity (COE) which has three major ingredients - fuel costs, capital costs and operating and maintenance costs. By utilizing consistent COE calculations for reference conventional coal fired plants (6) (12), and assuming that the capital costs and

TABLE 1: EFFICIENCY VERIFICATION CHARACTERISTICS

- MHD Extraction Capability
- MHD Component Cooling Requirements
- Efficiency of Oxygen Generation (Early Commercial)
- Preheat Temperature Capability (Advanced Configuration)
- Energy Requirements for Coal Processing
- Energy Requirements for Seed Processing

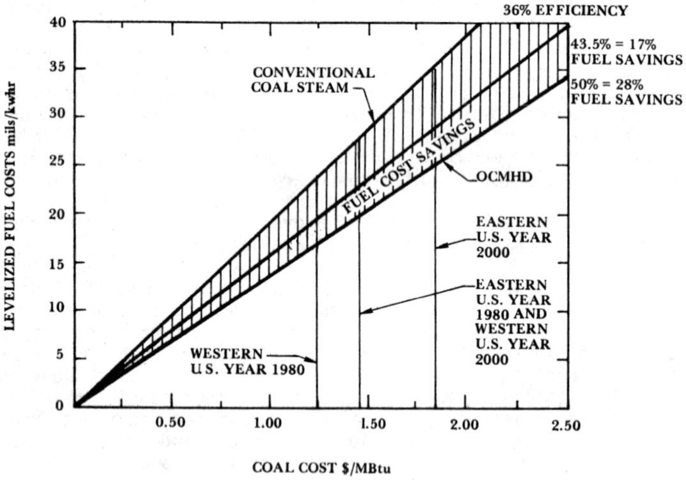

Figure 1. Comparison of Fuel Costs for MHD and Conventional Coal Steam Plants

maintenance costs for MHD will be equal to conventional plants, the fuel advantage translates into a maximum COE advantage as shown on Table 2.

A further refinement in the understanding of development requirements can be introduced by assessing the impact of capital cost, power plant capacity factor and coal cost on this benefit. Figures 2 and 3 presents an analysis of these factors for both the early and advanced MHD plant. The basis for these charts is the selection of a necessary 5% and 10% improvement in the cost of electricity for the early and advanced MHD power plants. The tradeoff between capital cost and decreased capacity factor relative to the two reference conventional plants are presented to maintain the specified 5% and 10% cost of electricity advantage when the cost of coal is 1.05$/MBTU and 1.75$/MBTU. As an example the curves indicate that in the case of coal costing 1.05$/MBTU, a capital cost increase of MHD over the ECAS reference coal plant of 10% and 8.5% is the maximum that can be allowed for the early and advanced MHD plants so long as the plant capacity factor is maintained at the same level as the conventional plant. On the other hand if the MHD plant has a 2% lower capacity factor than the ECAS reference coal plant, the maximum allowed capital cost increment is 5.5% and 4%. The curves shown for the case where coal cost is 1.75$/MBTU indicate the major importance of coal cost in the solution and the enhanced benefits of MHD when coal cost is high.

If we look at this analysis of requirements against the results of MHD power plant studies to date we begin to understand the importance of early consideration of these power plant characteristics. Figure 4 indicates plant capital costs from several MHD studies performed for Midtown USA compared to the range of conventional plant capital costs encountered in the U.S.A. It is evident that the more recent studies all show higher capital costs for MHD to the extent that they will seriously degrade or wipe out the MHD cost benefit. This fact is borne out by Figure 5 which presents the cost of electricity from these same studies.

In most of the MHD power plant studies to date the assumption has been that the plant availability will be the same as a conventional plant. This is a desirable objective but it must be interpreted into requirements in order to assure success of MHD. Several aspects of this requirement were investigated in an Operational Analysis Study of Open Cycle MHD (7). Probably the most significant unknown element in accomplishing a high availability system in the early MHD plants with oxygen, is the MHD channel lifetime and failure rates. In the advanced integral preheat MHD, the preheater system is also a major technical challenge. Table 3 indicates an effort to allocate lost operational time in an MHD plant to its various hardware elements. The attempt here was to determine, based on experience, what lost time needed to be allocated to various systems and how much could be allocated to the MHD channel (duct). This preliminary attempt indicated that 9.2% of the plant non-operational time could be allocated to the channel in the early configuration and 5.1% in the advanced configuration.

These two numbers are then utilized to derive requirements in terms of life-time and change-out time for the channel. Table 4 summarizes the findings. It indicates that early commercial MHD plants will require MHD channel lifetime requirements of 1300-3900 hours depending on the partial failure characteristics of the channel and the number of hours needed to replace the channel. Channel replacement times in the 100-200 hour range are required. On the same basis, the advanced MHD plant will require channel lifetimes of 2400-7400 hours. The large range of lifetime requirements indicates the importance of understanding and developing design knowledge. Several other considerations will eventually have to be accounted for, such as life-time probability characteristics, and yearly opportunities for shutdown as dictated by Utility power demand curves.

The Recommended Route

The above discussion serves to outline an approach to assessing the requirements for MHD power plants. It determines the potential benefit to the customer and derives the requirements necessary for realizing this benefit. In the case of an MHD power plant it can be seen that several requirements must be met in order to approach the theoretical benefit. The development of MHD is a total systems task which depends on an ever expanding series of accomplishments, until it finally

TABLE 2: MAXIMUM COST OF ELECTRICITY ADVANTAGE OF MHD RELATIVE TO CONVENTIONAL COAL PLANTS

	EPRI REFERENCE COAL PLANT (6)		ECAS UPDATED REFERENCE PLANT (12)	
	NEAR TERM MHD	ADVANCED MHD	NEAR TERM MHD	ADVANCED MHD
COAL COST OF 1.05$/MBTU	6.4%	10%	9.9%	15.7%
COAL COST OF 1.75$/MBTU	8.4%	13.8%	12.2%	17.9%

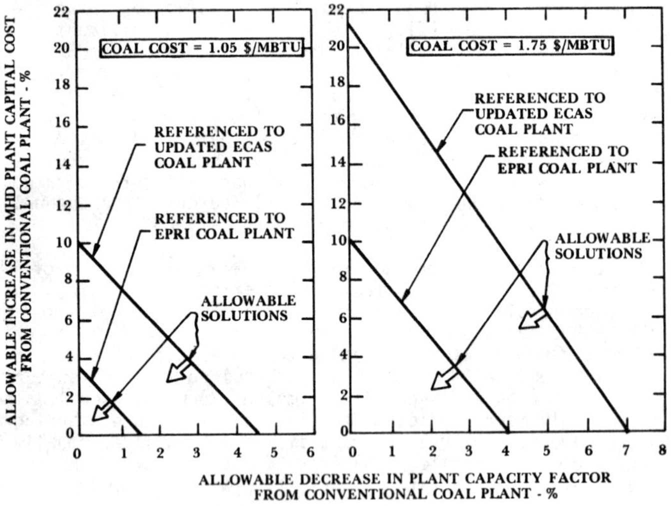

Figure 2. Early Commercial MHD Requirements to Maintain a 5% COE Advantage

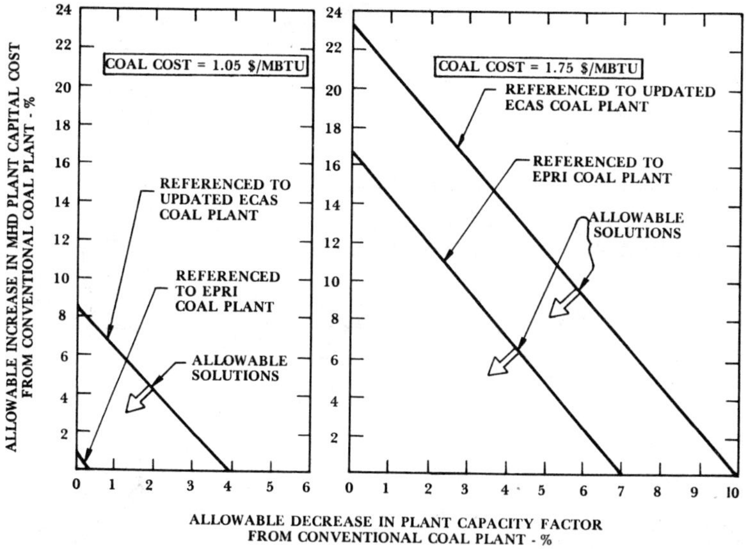

Figure 3. Advanced MHD Requirements to Maintain a 10% COE Advantage

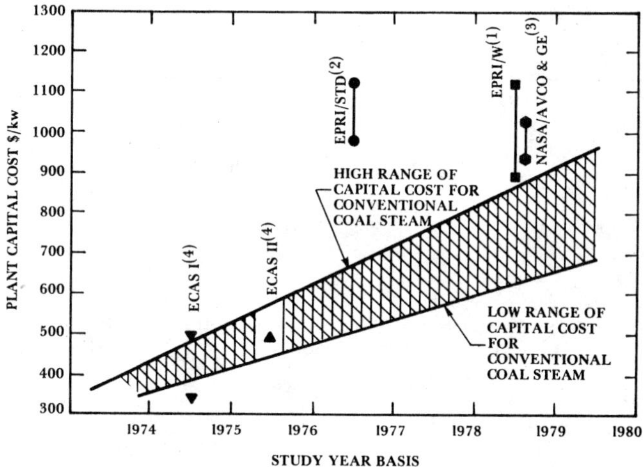

Figure 4. Comparison of OCMHD and Conventional Coal Steam Plant Capital Cost Estimate Trends Over Different Study Periods

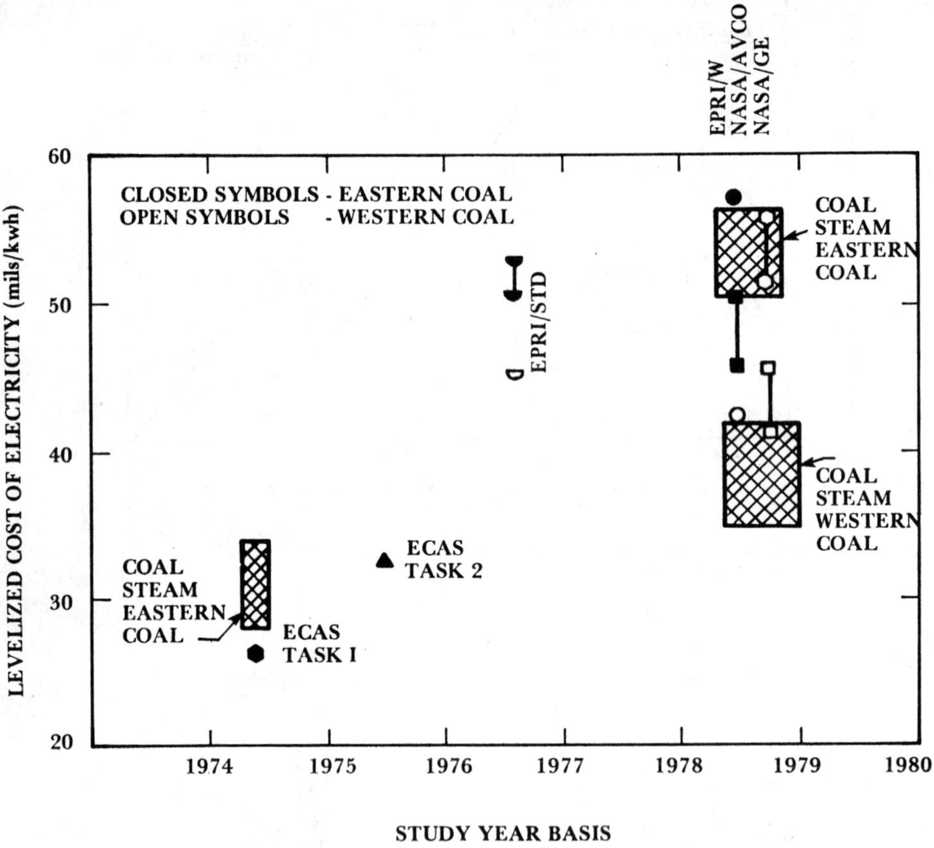

Figure 5: Comparison of Cost of Electricity for MHD and Coal Steam Plants

TABLE 3: INFERRED MAXIMUM DUCT UNAVAILABILITY OR "EQUIVALENT OUTAGE RATE" (EFOR) TO PROVIDE MHD PLANT AVAILABILITY OF 75%

Calculation Steps[1]	Advanced MHD	Early Commercial MHD
• Maximum Allowed Plant EFOR	24.9%	24.9%
• Deduct EFOR's of "Conventional" Components of MHD Plant		
- Coal Processing, Combustor	3.5	3.5
- Air Preheat, Slag, Ash Particulate Removal	2.2	2.2
- Boiler, SPHT, Reheat	5.8	5.8
- TG, Condenser, Transformer	4.4	4.4
• Deduct EFOR's of Postulated High-Availability MHD System		
- Magnet, Inverter	1.3	1.3
- Turbocompressor (3-1/2 Size)	0.1	0.1
- Seed Inj/Recover/Regen.	0.7	0.7
• Deduct EFOR's of Oxygen Supply	---	0.7
• Sum of EFOR's Allotted to Duct and HTAH	9.8%	9.2%
Estimated EFOR forHTAH	5.0	0
• Net Allowed Duct EFOR	5.1%	9.2%

[1]All implied sums (or subtractions) actually derived by multiplying (or dividing) $\overline{A}$'s ($\overline{A}$ = 1 - EFOR).

TABLE 4: REQUIRED DUCT MEAN LIFE FOR AN OCMHD PLANT WITH PLANT FOR = 0.16 AND EFOR = 0.25

PLANT CONCEPT	DUCT EFOR %	EFOH/FOH	MINIMUM MTTFFO 1000 HR: 200 Hr. Change Time	MINIMUM MTTFFO 1000 HR: 100 Hr. Change Time
Early Commercial MHD	9.2	1.00	3900	2000
		0.30	2600	2300
Advanced MHD	5.1	1.00	7400	3700
		0.30	4800	2400

FOR = Forced Outage Rate

EFOR = Equivalent Forced Outage Rate

MTTFFO = Mean-time to Failure Forced Outage

encompasses the entire power plant system. It is the type of development program which must have a very well thought out plan and long term dedication aimed at meeting requirements necessary to eventually accomplishing the overall objectives related to the cost of electricity. Such a plan will provide a means for constant assessment to determine if accomplishments are progressing as necessary or whether modifications to the program are needed.

This detailed plan for MHD is presently being developed in the form of a component and system development program; in the form of component development facilities; and in the form of a recognized need for a prototype demonstration facility. Detail programs are being generated on the Heat Recovery/Seed Recovery System and the Advanced Power Train System. Integration and proper sequencing of all elements under a unified program of objectives and requirements will be managed by DOE. This integration function is required to keep all elements of the MHD systems development in their proper perspective and to focus attention on areas of greatest uncertainty.

Summary

MHD power plants compete directly with conventional coal fired power plants. An improvement in the cost of electricity is the major advantage offered but can be eroded rapidly by capital costs in excess of the conventional plants or by losses of availability due to system reliability. Future coal cost projections are of major importance since rising costs will greatly enchance the benefits of MHD. No single accomplishment is going to dictate success for an MHD power plant, but rather a continuing evaluation of system applicability resulting in successful performance and operating experience of a prototype demonstration unit will be required. Reaching this goal will require an integrated approach involving the coordinated efforts of Industry, Utility and Government.

References

(1) Open Cycle MHD Systems Analysis, EPRI AP-1316, final report on RF640-1, Westinghouse Electric Corporation, EPRI, February 1980.

(2) Open Cycle MHD Systems Analysis, EPRI AF-1230, final report on RF640-2 STD Research Corporation, EPRI, November 1979.

(3) Parametric Study of Potential Early Commercial MHD Power Plants, DOE/NASA/0051/79/1, NASA CR-159633, prepared by AVCO Everett Research Laboratory, Inc., December 1979.

(4) Parametric Study of Potential Early Commercial MHD Power Plants, DOE/NASA/0052-79/1, NASA CR-159634, prepared by General Electric Space Division, February 1980.

(5) Energy Conversion Alternatives Study (ECAS), NASA Lewis Research Center, final reports for contracts NAS3-19406 and NAS3-19407, February 1980.

(6) Technical Assessment Guide, EPRI PS-1201-SR, EPRI, July 1979.

(7) Operational Analysis of Open-Cycle MHD, EPRI AP-1463, final report on RF639-1 Westinghouse Electric Corporation, July 1980.

(8) Environmental Development Plan Magnetohydrodynamics, prepared by U. S. Department of Energy, DOE/EDP 0045, May 1979.

(9) Magnetohydrodynamics: An Engineering Perspective, Zygielbaum, P. S., A. C. Dolbec, IEEE Paper No. 80 JPGC 816-9, IEEE/ASME/ASCE Joint Power Generation Conference, September 1980.

(10) Open Cycle MHD Technology Status and Development Perspective, Draft Report Prepared for EPRI by Westinghouse Electric Corporation.

(11) Magnetohydrodynamics: A Promising Technology for Efficiently Generating Electricity from Coal, Report to the Congress prepared by the Comptroller General of the United States, EMD-80-114, February 11, 1980.

(12) Modification of the ECAS Reference Steam Power Generating Plant to Comply with the EPA June 1979 New Source Performance Standards, DOE/NASA/0107-2, NASA-CR-159853, prepared by Burns and Roe, Inc., August 1980.

THE OUTLOOK FOR CONVENTIONAL LOWER 48 STATE GAS PRODUCTION AND TOTAL GAS SUPPLIES TO THE YEAR 2000

William T. McCormick, Jr.
Executive Vice President
Michigan Wisconsin Pipeline Company

INTRODUCTION

The Gas Supply Committee of the American Gas Association recently completed a two-year study on the long-term outlook for natural gas. Based, in part, on in-depth data from pipeline and exploration companies, the study clearly demonstrates the potential for natural gas as a major contributor to future energy supplies.

Although the U.S. currently consumes only 78 quadrillion Btu (quads) of energy, consumption is expected to increase to at least 100 quads by the year 2000. For comparison, estimates from a number of recent energy studies are presented in Figure 1. Estimates range from a low of around 75 quads in the year 2000 to a high of approximately 150 quads, with most of the forecasts in the 95-125 quad range. Using the forecast of 100 quads, and the current market share of roughly 25%, gas requirements would be 25 quads by the end of the century. With the proper regulatory environment and price incentives, however, gas consumption could rise to 30 quads of energy, or 50% today's level.

PROVEN RESERVES

Following the post World War II growth in the economy, proven reserves of natural gas grew from roughly 165 trillion cubic feet (Tcf) in 1947 to a peak of 292 Tcf in 1967. Since that time reserves have steadily declined to the current level of 195 Tcf. In fact, only once (1972) since 1967 have reserves

ALTERNATIVE PROJECTIONS OF PRIMARY ENERGY CONSUMPTION IN THE U.S.: 1980-2010

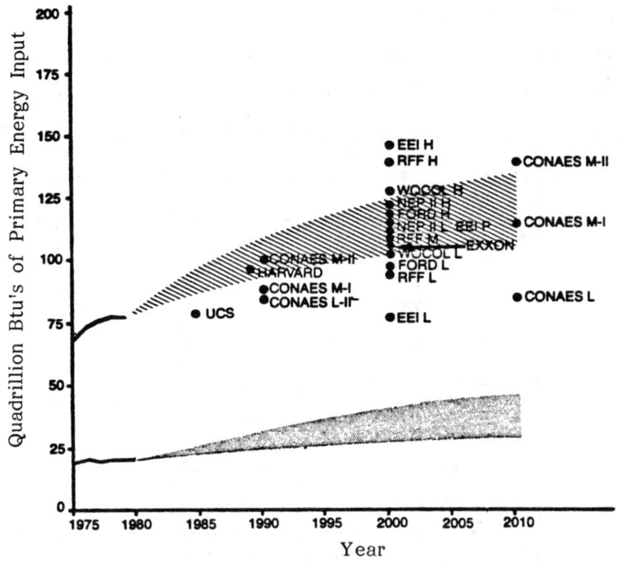

LEGEND: General range of forecasts of U.S. energy consumption
Total gas demand assuming 25-33% market share

UCS - Union of Concerned Scientists
HARVARD - Harvard Business School
EEI - Edison Electric Institute
EEI H - High Scenario (3.8% GNP growth)
EEI P - Preferred Scenario (3.1% GNP growth)
EEI L - Low Scenario (1.7% GNP growth)
CONAES - Committee on Nuclear and Alternative Energy Systems National Academy of Sciences
CONAES M-II - Scenario III_3
CONAES M-I - Scenario II_3
CONAES L - Scenario I_3
RFF - Resources for the Future
RFF H - Upper bound of consumption projection
RFF M - Consumption projection
RFF L - Lower bound of consumption projection
WOCOL - World Coal Study
WOCOL H - Case B (high coal use)
WOCOL L - Case A (low coal use)
NEP II - National Energy Plan II
NEP II H - Low World Oil Price Case
NEP II L - High World Oil Price Case
FORD - Ford Foundation
FORD H - Upper estimate (3% GNP growth)
FORD L - Lower estimate (2% GNP growth)
EXXON - Exxon Corporation

Figure 1

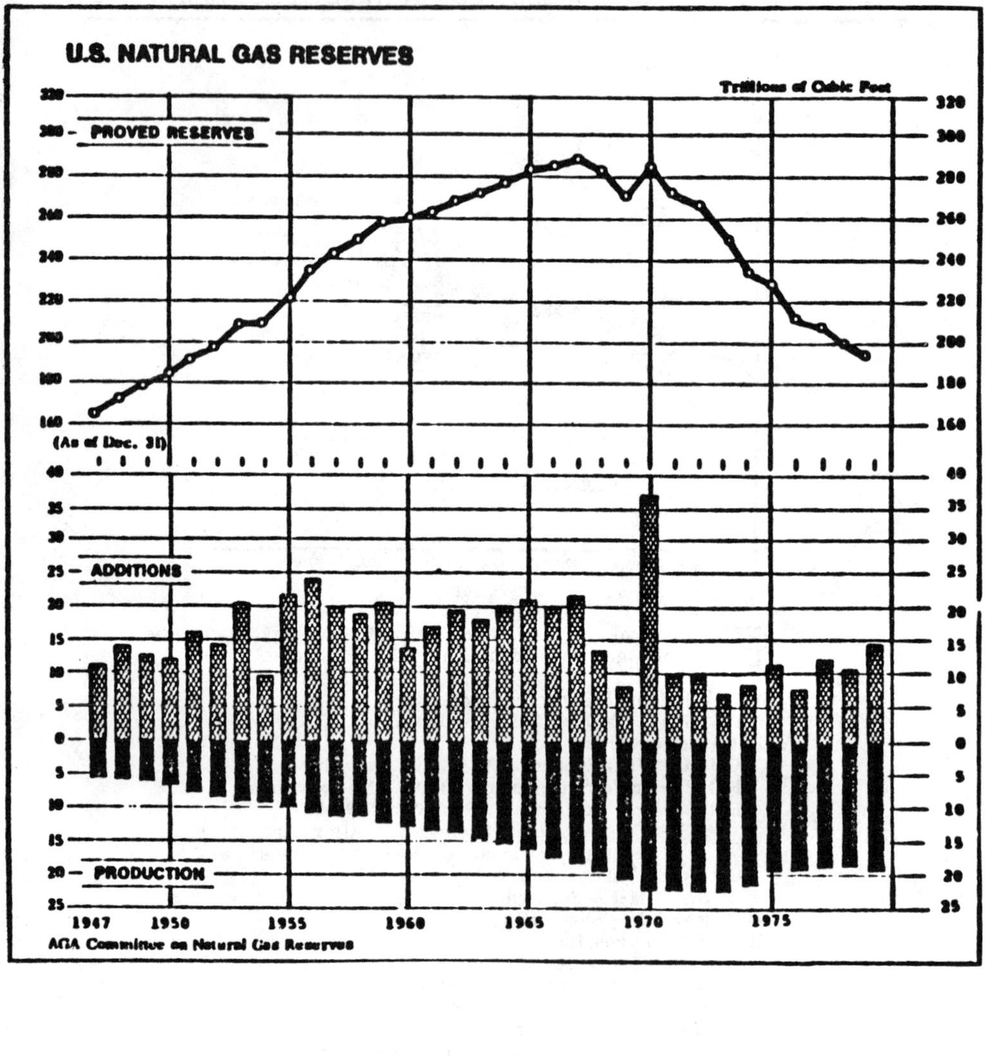
U.S. NATURAL GAS RESERVES
Trillions of Cubic Feet
PROVED RESERVES
(As of Dec. 31)
ADDITIONS
PRODUCTION
320
300
280
260
240
220
200
180
160
40
35
30
25
20
15
10
5
0
5
10
15
20
25
1947
1950
1955
1960
1965
1970
1975
AGA Committee on Natural Gas Reserves

FIGURE 2

FIGURE 3

RECENT DRILLING STATISTICS

	1976	1977	1978	1979	1980[2]
Seismic crews	262	308	352	400	521
Rotary rigs operating	1,656	2,001	2,259	2,177	2,843
Footage drilled (millions of ft.)	182	211	227	239	286
Gas well completions[1]	9,085	11,378	13,064	14,681	15,620
Oil well completions[1]	17,059	18,912	17,775	19,338	27,194
Success Ratio	.66	.67	.66	.68	.70
U.S. Reserve additions	7.6	11.9	10.6	14.3	N/A

[1]Exploratory and Developmental

[2]1980 Estimates based on ten months of data

References: API, DOE Oil and Gas Journal

FIGURE 4

POTENTIAL CONVENTIONAL U.S. GAS RESOURCES
(Trillion Cubic Feet-Including Alaska)

	Potential Resources					
	Year Of Est.	New Fields	Old Fields	Total Potential	1979 Proved Reserves	Total Remaining Resources*
USGS	1974	322-655	162	484-817	195	623-956
National Academy of Sciences	1974	530	118	648	195	787
Exxon Base	1974	342-942	56-321	423-1143	195	562-1282
Potential Gas Comm.	1979	820	199	1019	195	1199

*As of December 31, 1979. Estimates are corrected for resources consumed since the date of estimate.

increased, and this was due primarily to the addition of Prodhoe Bay, Alaskan reserves. Figure 2 depicts proven reserves along with reserve additions and production from 1947 to 1979.

The steady decline in proven reserves results from not finding nearly as much gas as has been produced. Excepting the Alaskan finds, each year since 1967, production has exceeded reserve additions. Because of this situation, the gas industry, during part of the 1970s, experienced serious delivery problems. Despite record demand for gas, the decline in reserves was so great that the industry was unable to maintain its previously high level of production.

DRILLING ACTIVITY

Beginning in 1973, rising gas prices did stimulate a renewed interest in drilling activity. The average wellhead price for newly discovered natural gas rose from roughly 20 cents in 1972 to over $2.60/MMBtu by the end of 1980. It is important to note, however, that despite the substantial rise in new gas prices, these increases have not kept pace with crude oil. Moreover, while gas prices can be expected to track more closely with oil prices in the future, gas is exptected to continue to be less expensive on a Btu basis in end-use markets.

As a result of rising prices under the NGPA, record levels were achieved for wells drilled, footage drilled and well completions. Between 1976 and 1980, footage drilled and gas well completions have increased 57% and 72%, respectively. At the same time, there has been a steady rise in the success ratio from 66% in 1976 to 70% in 1980 (see Figure 3).

RESERVE ADDITIONS

Unfortunately, although somewhat higher, reserve additions have not kept pace with the increased level of drilling activity. Despite the increase in well completions, only in 1979 have reserve additions begun to rise substantially. Between 1978 and 1979, for example, reserve additions rose 35%. Furthermore, although reserve additions data for 1980 is not yet available, preliminary indications are that continued growth in reserve additions is expected. Total proven reserves in the U.S. were estimated to be 195 Tcf as of December 31, 1979. This is in addition to potential resource base of approximately 800 Tcf (see Figure 4).

GAS SUPPLY FORECAST SURVEY RESULTS

As previously indicated, much of the study input for the conventional lower 48 state gas supply forecasts was obtained through the cooperation of the participating oil and gas companies. Individual company estimates of lower 48 state reserve additions and gas production over the coming decade were obtained in 1979 from 13 companies in all, as well as from the U.S. Department of Energy. Although most companies were reluctant to provide specific detail on their methodology for calculating reserve additions and production, a number of general observations can be made:

1. Most oil and gas producers (with the exception of Mobil and Exxon) relied on the U.S. Geological Survey reserve base estimates modified according to their own subjective analysis;

2. Gas transmission companies (with the exception of Michigan Wisconsin) relied primarily on probabilistic assessments of drilling activity and productivity per foot drilled for reserve and production estimates; and

3. Consistent with the NGPA, deregulation of new wellhead gas prices is assumed to occur in 1985.

Summaries of the individual company forecasts of reserve additions and annual lower 48 state gas production through 1990 are shown in the scatter diagrams, Figures 5 and 6, which plot forecasted reserve additions versus production for 1990 and 1985, respectively. There are several conclusions which can be drawn from the data supplied by the individual companies.

1. Lower 48 State Reserve Additions - The range of average annual reserve additions (lower 48 state) over the 1980-1990 period is between 7.0 and 19.0 Tcf per year, with the median estimate approximately 12.5 Tcf/year. With the exception of two estimates (Tennessee Gas Transmission's of 7.0 Tcf/year and Gulf's of 19.0 Tcf/year), all of the company estimates are between 8.0 and 15.0 Tcf/year.

2. Lower 48 State Gas Production - Production estimates from lower 48 reserves ranges between 10.9 Tcf and 19.2 Tcf for 1990. Again, with the exception of Gulf Oil and Tennessee Gas, all of the 1990 estimates are between 13.0 Tcf/year and 16.0 Tcf/year. Although many of the company forecasts suggest significantly reduced reserve additions by 1990, production is not expected to decline as much. Consequently the reserve to production ratio (R/P) is expected to decline to an average of 8.0 or less from the current level of 9.3.

3. Correlation Between Production and Reserve Additions - Among the various forecasts, there is a greater correlation between annual average reserve additions and production in 1990 (see Figure 5) than in 1985 (see Figure 6). This results primarily from greater variations in underlying estimates of the R/P ratios, then to differences in the assumed rate of production of newly discovered reserves.

FIGURE 5

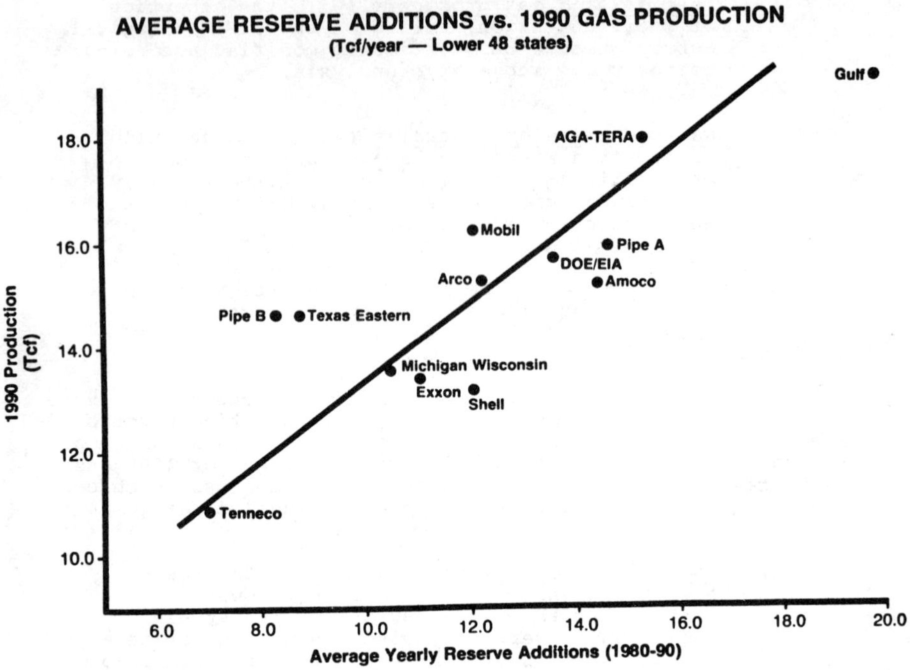

FIGURE 6

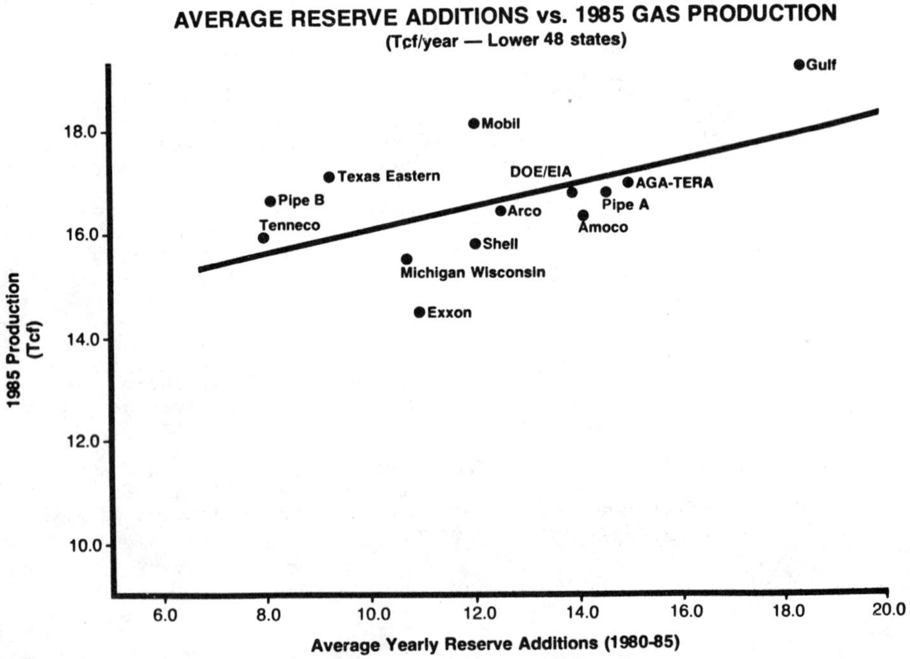

GAS SUPPLY COMMITTEE FORECAST OF LOWER 48 STATE CONVENTIONAL GAS PRODUCTION

As a result of phased deregulation of new gas prices, new reserve additions will undoubtedly slow the precipitous decline in U.S. gas production which characterized the 1970s. However, even with rising gas prices, including deregulation in 1985, production from lower 48 sources is still expected to slowly decline over time. Overall, the A.G.A. Gas Supply Committee sees annual reserve additions of between 10 and 15 Tcf/year through 1990. Based on these estimates, production from lower 48 conventional sources will likely range from 16-18 Tcf in 1985 and roughly 15-17 Tcf in 1990 (see Figure 7).

Although these analyses focused only on the 1980-1990 period, this trend will continue through the end of the century. In view of the unlikely discovery of major new gas deposits, such as the Alaskan North Slope, within the Continental U.S., lower 48 reserve additions will decline to approximately 8-12 Tcf by 2000. At the same time, conventional production will also decline to the 12-14 Tcf/year level. Consequently, in order for the gas industry to maintain even its current market share of 25%, additional supplies of natural gas will be necessary. This includes both supplemental sources such as Canadian, Mexican, and Alaskan as well as non-conventional supplies.

SUPPLEMENTALS

Although most of the emphasis has been on conventional supplies from lower 48 sources, supplemental supplies will be needed to achieve even the lower-end of supply estimates of 18.5 Tcf by 1990 and 19.3 Tcf by 2000 (see Figure 8). One of the major supplemental sources will be Alaskan gas. With completion of the Alaskan gas pipeline in 1985, 1.0-1.7 Tcf/year of gas will become available.

To facilitate combining the various source estimates into a meaningful overall outlook, four national policy scenarios were developed by the A.G.A. Gas Supply Committee. These policy scenarios are not meant to characterize our current policy nor are they policy recommendations. They are simply policy options which illustrate the availability of different gas supplies under alternative federal policies. The scenarios are:

1. Self Sufficiency - This scenario assumes Federal policies aimed at encouraging all domestic gas sources, but at the same time discouraging imports. Naturally, Alaskan Gas and Coal Gasification are key sources, as well as additional R&D to develop new domestic gas sources.

2. North American Focus - This scenario is also based on emphasizing secure sources of gas supply except that Canadian and Mexican sources are encouraged. Imports of LNG are, however, restricted.

FIGURE 7

Sensitivity of Lower 48 State Gas Production* to Annual Reserve Additions

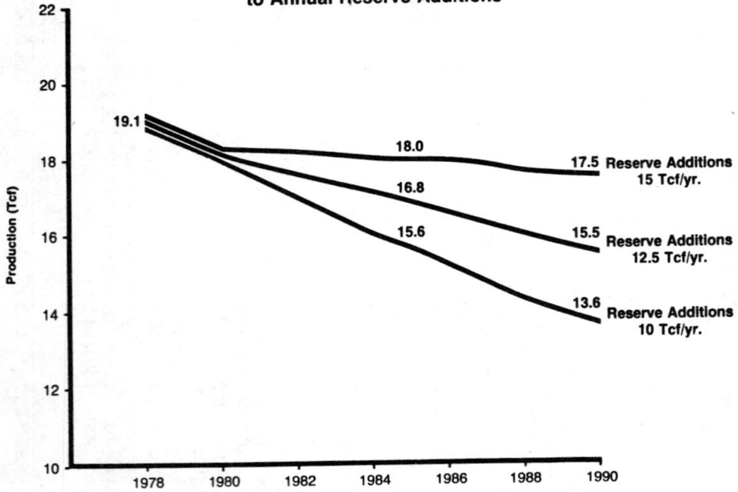

*R/P ratio assumed to decline from 9.3 in 1978 to 7.5 in 1990

FIGURE 8

PRODUCTION POTENTIAL OF NATURAL GAS SUPPLY SOURCES (in Tcf)

	1979	1990		2000	
		Low[10]	High[11]	Low[10]	High[11]
Lower 48[1]	19.9	15	17	12	14
SNG[2]	0.2	0.1	0.5	0.1	0.5
Alaskan[3]	--	0.8	1.4	1.5	3.0
Canadian[4]	1.0	1.0	1.7	1.0	2.0
Mexican[5]	--	0.1	1.0	0.1	2.0
LNG[6]	0.2	0.7	2.0	0.7	4.0
Coal Gasification[7]	--	0.2	1.0	1.5	3.5
Tight Formations[8]	--	0.4	1.4	1.5	5.0
Nonconventional[9]	--	0.2	0.8	1.0	2.5

[1]National Energy Plan II, Executive Office of the President, 1979.

[2]Synthetic natural gas made from naphtha or natural gas liquids in existing plants.

[3]Includes pipeline deliveries from Prudhoe Bay (range of one to two pipeline systems) and LNG deliveries from Southern Alaska.

[4]Volumes are dependent on Canadian National Energy Board (NEB) export policies.

[5]Volumes are dependent on PEMEX export policies.

[6]World gas option.

[7]U.S. encouragement/support of proposed programs.

[8]Includes Western light sands and Devonian shale. Volumes depend on price incentives and productivity.

[9]Includes gas from coal seams, in situ coal gasification, peat and oil shale gasification, geopressured gas, biomass, gas from urban wastes and gas from animal residues. Volumes depend on the level of technology development.

[10]Estimate is based on existing technology with only minor improvements and on an indifferent and unsupportive regulatory climate.

[11]Estimate is based on a vigorous technology growth with broad based government and industry support coupled with a positive regulatory environment which encourages new gas supply projects.

3. Moderate World Imports - This scenario modifies the previous scenario to allow only a limited increase in the use of LNG. In addition to developing domestic sources, however, restrictions would be placed on such imports so that they would not unduly affect our domestic gas supply.

4. World Conventional Gas Emphasis - This scenario emphasizes the vigorous development of all natural gas resources. As one might expect, imports of relatively less expensive gas would undercut domestic R&D programs and limit the full development of some domestic gas sources.

The relative impact of these alternative scenarios on gas sources and availability of gas supply, for the year 2000, is presented in Figure 9.

CONCLUSION

In summary, the A.G.A. Gas Supply Committee's best estimate is that between 23.0 and 30.0 Tcf of gas will be available in the year 2000. Thus, there is little doubt supplies will be adequate for natural gas to maintain its current market share. Although natural gas use conceivably could grow to satisfy as much as one-third of total U.S. energy requirements, positive efforts need to be undertaken immediately. Conventional lower 48 state production which currently accounts for almost 95% of all U.S. gas requirements, is expected to steadily decline over time, and by the year 2000, satisfy less than 60%. Supplementals from Canada, Mexico, and Alaska along with coal gasification and some non-conventional supplies will, therefore, be needed to fill the gap. Unprecedented high levels of capital investment will be required to support all of these supply projections.

The major uncertainties in these estimates are those associated with future Federal and State regulatory policy decisions. Technological uncertainties are also a factor but timely investment in research, development, and demonstration programs can reduce this uncertainty to acceptable levels.

FIGURE 9

GAS SUPPLY SCENARIOS IN YEAR 2000 (TCF)

	SELF SUFFICIENCY	NORTH AMERICAN FOCUS	MODERATE WORLD IMPORTS	WORLD CONVENTIONAL GAS EMPHASIS
LOWER-48	12-14	12-14	12-14	12-14
SNG FROM LIQUID HYDROCARBONS	0.3	0.3	0.3	0.1
ALASKAN	3.0	3.0	1.5	3.0
CANADIAN	1.0	2.0	2.0	2.0
MEXICAN	0.1	2.0	2.0	2.0
LNG IMPORTS	0.7	0.7	2.5	4.0
COAL GAS	3.5	3.5	1.5-2.5	1.5-2.5
TIGHT FORMATIONS	1.5-5.0	1.5-4.0	1.5-3.0	1.5-3.0
MISC. NEW TECHNOLOGIES	1.0-2.5	1.0-2.5	1.0-2.5	1.0-2.5
TOTAL	23.1-30.1	26.0-32.0	24.3-30.3	27.1-33.1

THE OUTLOOK FOR REMAINING U.S. GAS RESOURCES BASED ON RECENT EXPLORATION AND DEVELOPMENT ACTIVITY

Harry C. Kent

Director, Potential Gas Agency
Mineral Resources Institute
Colorado School of Mines

The subject of the potential resources of natural gas in the United States continues to be of concern and interest to both the natural gas industry and to the general public. There have been recent evaluations of the potential from nonconventional sources prepared by task groups working for the National Petroleum Council. A number of newspaper and magazine articles and television news programs have addressed the subject of gas resources within the past year. Some of the reports in the public press have given credibility to very large estimates for both conventional and nonconventional resources while the estimates from industry and the Federal Government tend to be much more conservative. I propose to examine briefly the status of knowledge of conventional resources of natural gas in the United States as represented by the work of the Potential Gas Committee and recent estimates released by the Resource Appraisal Group of the United States Geological Survey.

The Potential Gas Committee has been involved in the estimation of the potential supply of conventional natural gas in the United States since the early 1960s. The Committee has published a series of reports giving the results of its estimates for each of 12 areas of the country. At the present time, the Committee is working on the re-evaluation of the potential based on information available through the end of the year 1980. Completion of this work and publication of the revised estimates is slated for this coming May. However, some trends are beginning to emerge from the deliberations of the Committee and I will comment briefly on those.

The U.S. Geological Survey published the first report of the Resource Appraisal Group dealing with undiscovered recoverable resources of both oil and gas in the United States, again from conventional sources, as Circular 725, which was published in 1975 and represented the information available through the end of 1974. The

Resource Appraisal Group has recently been working on a revision of these estimates and the preliminary results from this revision have just been released as USGS Open-File Report 81-192. This preliminary open-file report was prepared during 1980 and reflects information available through the end of the year 1979. A complete report giving the final results of the USGS appraisal and further details is scheduled for publication in the near future.

NATIONAL TRENDS

Based on information available at this time, the following national trends in estimates of potential gas resources appear to be emerging from both the work of the Potential Gas Committee and of the U.S. Geological Survey Resources Appraisal Group. The Potential Gas Committee summary is based on preliminary reports submitted by the chairmen of the various Work Areas of the Potential Gas Committee and must be considered as preliminary as no final estimates have, as yet, been released. For purposes of preliminary comparison, some estimates based on the 1978 report of the Potential Gas Committee are used. Comments concerning the U.S. Geological Survey's resource estimates are based on the information contained in Open-File Report 81-192 and comparisons are made to Circular 725. Table 1 gives the comparative estimates for onshore and offshore regions of the lower 48 states and Alaska from the U.S. Geological Survey's two reports and from the 1978 report of the Potential Gas Committee.

Table 1--Comparison of Estimates
(Conventional natural gas - Tcf)

	USGS		PGC*
Year-end	1974	1979 (mean)	1978
Lower 48 Onshore	345	390	473
Lower 48 Offshore	63	102	171
Alaska Onshore	32	37	48
Alaska Offshore	44	65	128
	484	594	820

* Potential Gas Committee estimates are the total of the Possible and Speculative estimates and are comparable to the USGS estimates of undiscovered recoverable resources.

Potential Gas Committee

The work of the Potential Gas Committee indicates, based on 1980 information, that more resources exist in the deeper parts (depths greater than 15,000 feet) of sedimentary basins than was previously estimated. This reflects the generally encouraging and favorable results of greater drilling to deeper depths in onshore basins together with the availability of more geophysical information, especially seismic surveys, of better quality. There has been a very definite increase in the number of wells drilled below 15,000 feet. In 1977, there were 472 wells drilled below 15,000 feet in the entire United States, including the offshore regions. In 1980, Petroleum Information estimates that 743 wells were drilled to depths below 15,000 feet. In each of the years 1978 and 1979, between 600 and 700 deep wells were drilled.

The second trend which appears to be emerging from the work of the Potential Gas Committee is that less resources are estimated to lie in offshore regions than was previously thought. This trend seems to be partly the result of the maturity of the Gulf Coast region, in terms of both exploration and production, resulting in transfers of potential resources to proved reserves and production. In addition, there have been disappointing results from offshore exploration and drilling in the Baltimore Canyon area of the Atlantic continental margin and in the Gulf of Alaska, both areas where it was previously thought there was considerable potential for natural gas.

Finally, the Potential Gas Committee work would indicate that there are probably smaller resources of natural gas in Alaska than the Committee had previously believed. This is based on the previously-mentioned disappointment in the Gulf of Alaska and the re-evaluation of onshore basins based on more geologic, geophysical and drilling information.

U.S. Geological Survey

Overall, the U.S. Geological Survey estimates prepared in 1980 are 23 percent higher than those prepared in 1975. However, some of this increase is the result of extension of the offshore area evaluated for resource potential from a water depth of 200 meters (1975) to a depth of 2,500 meters (1980). (For comparison, the Potential Gas Committee estimates include the area only to a water depth of 1,000 meters.) If we separate the USGS estimates for onshore and offshore areas, there is a net increase of 13 percent in the estimates for onshore areas and a net increase of 56 percent in the estimates for offshore areas.

AREA TRENDS

The following sections will review the developments in each region of the United States which have a bearing on the estimation of potential gas resources. This regional analysis will be based on the 12 Work Areas established by the Potential Gas Committee and references to comparable USGS regions will be made where appropriate. Figure 1 shows the geographic distribution of potential supply of natural gas as estimated by the Potential Gas Committee in 1978. It is divided into the three categories recognized by the PGC, Probable, Possible and Speculative. Briefly, Probable resources are associated with the further development of fields which have already been discovered. They include both extensions and new pool discoveries. Possible resources are estimated to occur in new fields to be discovered in basins or provinces where natural gas production has been established from comparable formations. Speculative resources are those resources estimated to exist in provinces or basins where no gas production has been established or from formations which have not previously been established as productive. Table 2 gives the total estimated potential supply of natural gas by depth increments as prepared by the Potential Gas Committee for 1978.

Area A

The prospects for natural gas resources in onshore areas of the Appalachian region appear bright to both the Potential Gas Committee and the U.S. Geological Survey. U.S. Geological Survey estimates of undiscovered recoverable natural gas have been increased sharply in 1980 compared to 1975. Much of the interest in this area is in the development of the concept of the Eastern Overthrust Belt. Part of this postulated belt lies beneath metamorphic rocks, and if the sedimentary rocks beneath the basin are not metamorphosed, a virtually untested basin exists with an enormous volume of prospective rocks. Offshore, there have been 29 tests drilled since 1978. Twenty-three (23) of these have been drilled in the Mid-Atlantic (Baltimore Canyon) area and 6 in the South Atlantic (Southeast Georgia Embayment) area.

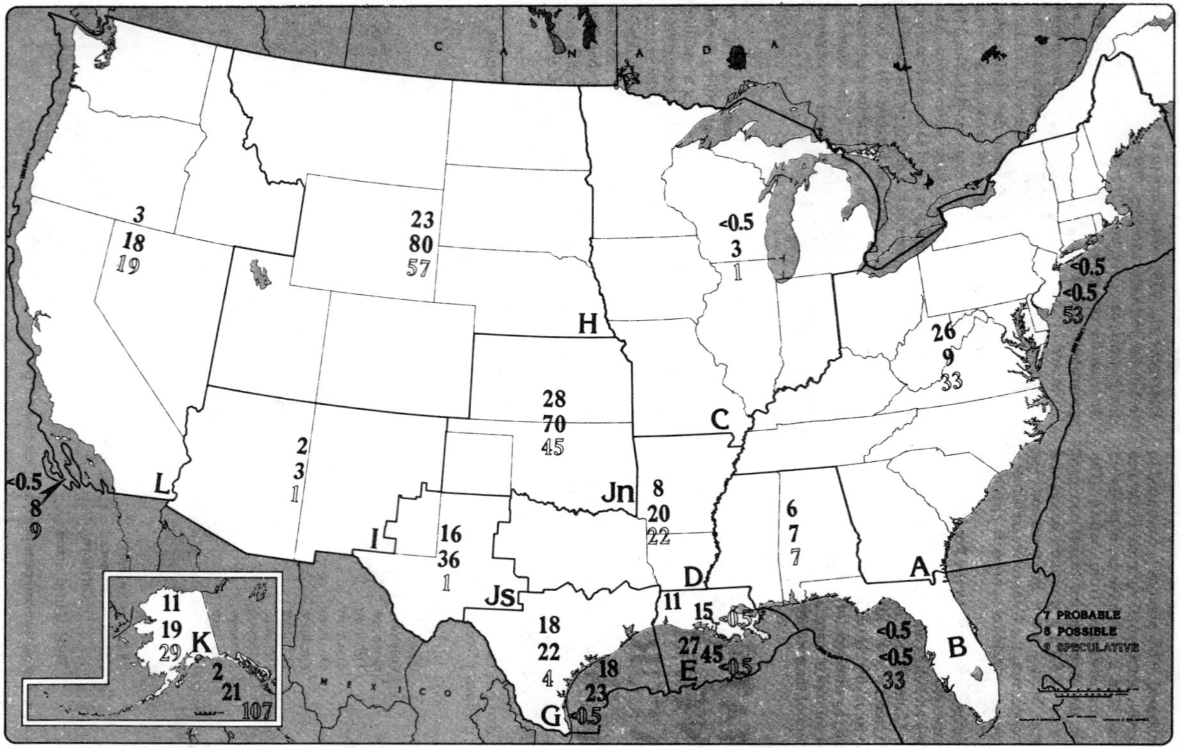

Figure 1--Geographic distribution of potential supply as estimated by the Potential Gas Committee, 1978

Table 2--Total estimated potential supply of natural gas by depth increments, 1978 (trillion cubic feet)

	Probable	Possible	Speculative	Total
Onshore (Drilling Depth)				
Lower 48 States				
less than 15,000 feet	112	188	115	415
15,000 to 30,000 feet	29	95	75	199
	141	283	190	614
Alaska				
less than 15,000 feet	11	19	29	59
15,000 to 30,000 feet	--	--	--	--
	11	19	29	59
Total Onshore	152	302	219	673
Offshore (Water Depth)				
Lower 48 States				
less than 200 meters	44	52	46	142
200 to 1,000 meters	1	24	49	74
	45	76	95	216
Alaska				
less than 200 meters	2	11	84	97
200 to 1,000 meters	--	10	23	33
	2	21	107	130
Total Offshore	47	97	202	346
Total Lower 48 States	186	359	285	830
Total Alaska	13	40	136	189
Total United States	199	399	421	1,019

Significant hydrocarbon indications were present in five of the Mid-Atlantic wildcats, all of which were located within a few miles of each other. All other tests were unsuccessful. The results of this drilling have, in the view of the Potential Gas Committee, diminished the natural gas potential of the Atlantic continental margin. The USGS evaluation extends to deeper water, where potential resources may exist in a postulated reef trend, and the USGS has more than doubled its estimate of undiscovered recoverable resources in the offshore area.

Area B

There is little change in the picture for the southeastern states compared to previous years, and the Potential Gas Committee's estimates of potential resources are not expected to change greatly. The deep trend of the Jurassic Cotton Valley Formation is of interest as companies attempt to extend the area of production from Louisiana into Mississippi.

Area C

Exploration for Middle Silurian Niagaran pinnacle reefs continues to be the main exploration emphasis in the north-central states and this exploration occurs specifically in the Michigan basin. The Potential Gas Committee does not see any developments which might cause any great change in the estimates of potential resource. However, the U.S. Geological Survey has increased its mean estimate of undiscovered recoverable resources for this area from 1 Tcf to 5 Tcf.

Area D

There have been no recent developments in this area which would significantly change the estimates of potential resources.

Area E

This is an actively developing area which has produced about 25 percent of the cumulative U.S. gas production and is presently producing about 35 percent of the U.S. annual production. Maturity of the area means that the evaluation of the resource potential is fairly firm and that a significant amount of gas is added to proved reserves each year, thus reducing the potential resource. Recently the deep Tuscaloosa trend (Lower Cretaceous) has been the focus of extensive exploratory drilling activity. This trend extends in a belt approximately 20 miles wide and 300 miles long from Texas to Mississippi, across the northern portion of south Louisiana. Even after five years of furious drilling activity, the trend is still supporting active exploration and drilling; however, the success ratio has dropped from nearly 100 percent initially to 20 percent in 1980. The ultimate potential of this trend is not yet completely evaluated but it appears significant.

Area G

This area is also a mature area from the standpoint of exploration and drilling activity. Re-evaluation of this area, plus transfers from potential resources to proved reserves, will probably result in decreases in the estimates of potential resources.

The USGS includes much of the prospective area of Potential Gas Committee Areas B, E and G in its estimate of undiscovered recoverable resources for the Gulf Coast region. The mean estimate for the onshore part of this region has been decreased to 124 Tcf compared to 133 Tcf in 1975. The offshore estimate has been increased from 50 Tcf to 72 Tcf, but much of this increase is probably the result of the extension of the area included in the evaluation.

Area H

The Rocky Mountain states and adjacent High Plains area are currently the site of intensive exploratory activity. There are significant indications of increased natural gas potential in the Green River and Wind River basins of southwestern and west-central Wyoming and in the Uinta and Piceance basins of eastern Utah and western Colorado. The potential for additional resources in Montana appears bright. There have also been significant gas discoveries in the deeper part of the Williston basin in western North Dakota. The Western Thrust Belt continues to be the most publicized area of new gas discoveries in the United States. Major finds have been announced within the past year in the Central portion of the Thrust Belt in southwestern Wyoming and northern Utah. These discoveries are tending to confirm optimistic estimates of gas potential. There have been some promising discoveries in the Northern portion of the Thrust Belt in Montana but this area is still in the very early stages of exploration. The Southern portion of the Thrust Belt in Utah is virtually unexplored.

Area I

Other than the drilling of a deep well in Pinal County, south-central Arizona, and a deep well in the Rio Grande rift south of

Albuquerque, there have been no extensive exploratory efforts in western New Mexico or Arizona for the past two years. The Arizona well was intended to test the concept that the Western Thrust Belt extends across southern Arizona. It was hoped that the well would enter a favorable section of sedimentary rocks at a depth of about 12,000 feet beneath thrusted granitic rocks. Unfortunately, the well was drilled to a depth of about 18,000 feet, still in granitic rocks, and there were no indications of hydrocarbons. Another deep exploratory well is planned in the extreme southeastern part of Arizona. The deep well in New Mexico was intended to test the possibility of hydrocarbon accumulation within the Rio Grande rift zone and was also unsuccessful. Further testing of the rift zone is planned.

Areas H and I of the Potential Gas Committee combined correspond essentially to regions 3 (Colorado Plateau and Basin and Range) and 4 (Rocky Mountains and northern Great Plains) of the USGS. USGS estimates for these regions have been increased by 92 Tcf (mean) from 1975 to 1980, an increase of 214 percent. Most of this increase can be attributed to new evaluations of potential of the Western Thrust Belt and the Wyoming-Utah-Colorado basins.

Area J-North

Exploratory drilling activity is high in Area J-North, especially in the deep part of the Anadarko basin. At year-end 1980 there were about 350 locations which had either been drilled to below 15,000 feet or had been projected to drill below 15,000 feet. The members of the Potential Gas Committee have been optimistic about the potential of this area for a number of years and they continue to be optimistic. The USGS, however, has reduced the estimate of the undiscovered potential of their Mid-continent region, which includes this area, by almost 40 percent in their 1980 report compared to their 1975 report.

Area J-South

The Permian basin continues to be the location of development of existing gas resources, both discovered and undiscovered, but there have been no major developments which would greatly affect the estimates of gas resource potential. The USGS has reduced its estimate of the undiscovered potential for this region by almost 40 percent.

Area L

In the Pacific Coast area, exploratory developments have not significantly affected the Potential Gas Committee's outlook for potential gas supply. The discovery of gas in northwestern Oregon in 1979 has been encouraging and has spurred new exploratory drilling and further evaluation of sedimentary basins in both Oregon and Washington. The USGS mean estimate of undiscovered recoverable gas resources for the Pacific Coast region has been increased very slightly.

Area K

The Potential Gas Committee is considerably less optimistic now about the potential supply of gas from Alaska than it was a few years ago. Negative results of wildcat drilling and geophysical surveys in the National Petroleum Reserve-Alaska, the Kodiak area, the Lower Cook Inlet, and the northern Gulf of Alaska have reduced what was once considered potentially productive acreage. The USGS, on the other hand, appears now to be more optimistic about Alaska, both onshore and offshore, than they were previously. It appears likely, however, that the 1980 PGC estimates of Alaskan potential, even with the possibility of some reductions from their 1978 estimates, will still be larger than USGS estimates.

SUMMARY

If we look at the entire United States, the outlook for additional supplies of conventional natural gas still appears to be promising. The Potential Gas Committee estimates would indicate that more gas remains to be found than has been discovered in the past. The mean estimate of undiscovered recoverable resources by the U.S. Geological Survey Resource Appraisal Group would also indicate that at least as much gas could be found and recovered in the future as has been discovered in the past. If we look at the total of present proved reserves and potential supply, it would appear that possibly twice as much gas could be produced in the future from conventional sources as has been produced in the entire history of the industry.

EOR BY CHEMICAL FLOODING

Robert P. Murray
Cities Service Company

My topic for this panel discussion is Chemical Flooding for improved oil recovery. I hope to convey to you what we in the oil industry perceive in the use of chemicals to improve our country's oil supplies, some of the mechanisms of displacement and problems associated with these floods, an update of industry activities and the potential target.

Since I am the first speaker, let me lay some groundwork that is common to all displacement processes, whether using chemical, miscible or thermal methods. These restraints are also true with secondary and primary recovery methods.

I know some of these terms may be foreign to you, so I would like to take a minute and talk about a few basic parameters governing the production of oil from underground reservoir rock. You have been exposed to a great amount of rhetoric that has done more to confuse you than to explain Mother Nature's limitations on oil production, plus the compounding effect of economics.

First - Oil is not found in large underground lakes, even though the industry has spoken of oil pools since Colonel Drake's discovery in 1859. The vast majority of the world's oil is locked up in the pore spaces of reservoir rock that may have no more porosity (i.e., space between the rock grains) than you have in your concrete driveways or the brick on your house. Depending upon the size, distribution, and packing of the rock grains, the porosity can vary from about 1 percent to about 40 percent of the total rock volume. This is the space available for the accumulation of water, oil and gas. Modern theories of hydrocarbon sources (oil and gas) point toward generation within the carbon rich shales and then migration toward good reservoir rocks of sandstone or carbonates overlain by a cap rock of very tight shale, chert or chalk, and some structural or depositional anomaly.

So, in order to have an oil or gas accumulation, three things are needed: 1) source rock; 2) reservoir rock; and 3) a trap, either stratigraphic or structural, to prevent further migration of the hydrocarbons, and hopefully enough reservoir volume to contain a commercial deposit.

Second - Permeability of the reservoir rock is probably one of the most important factors in determining the commercial viability of recovering oil from any deposit. Although it is of prime importance, it is probably the most misunderstood factor, even within the oil industry itself! Permeability is a measure of how well the pore spaces are interconnected. If the pore throats between pore spaces are large, the permeability is large and vice versa. You might visualize it this way - consider this building or any building with a large number of rooms and interconnecting hallways. The rooms represent pore space - large rooms hold a lot of people, etc. The hallways represent permeability (i.e., the interconnection between pore spaces). Now, if we had a fire drill we could empty the rooms or pore space in a hurry if we have nice, wide halls. But we could be in real trouble if the halls are crooked and narrow and filled with obstructions. This access between pores determines the rate at which oil and gas can be produced. To further complicate the problem, Mother Nature says that the number of fluids present must share this same flow capacity. In other words, if only oil were present it could utilize all of this flow space or capacity. But this is not the usual case; in fact, it would be a very rare phenomenon in nature. There is usually gas associated with any oil deposit so gas interferes with oil flow; and invariably in nearly all accumulations, water is also present which must also share in the flowing capacity and further reduces the flow capacity of the commercial product.

Third - The availability of any energy source is governed by economics. Nature has her constraints which are mainly physical factors. But economics (supply and demand and government decrees) further compound the problem as to the commercial viability of hydrocarbon production. Exploitation must be economically viable or it will never come to fruition.

CHEMICAL FLOODING - What is it and what are we talking about in terms of improved oil recovery? At our present state of technology, we are talking about three processes or the combination of processes. Those three are: 1) the use of polymers (long chain molecules that enhance the viscosity of the fluids driving oil toward the producing wells); 2) the use of surfactants (soap-like materials) to reduce the interfacial tension between water and oil to mobilize and recover more oil (i.e., reduce the residual oil saturation); 3) the use of caustic (basic or alkaline-like) solutions to create surfactant-type materials by reaction with the in-situ or in place hydrocarbons.

POLYMER FLOODING involves the addition of a polymer material to water to create a thicker or more viscous fluid. The technical significance is that a thick or viscous fluid can push a thin or less viscous fluid readily. For example, if we push water with molasses, the water moves easily and the molasses slug stays intact. This would be a very efficient process, one that would displace almost all of the water ahead of the molasses bank. If the situation were reversed, the process would be very inefficient. The thin water would have a strong tendency to finger or channel through the molasses bank and leave much of the molasses behind. Water is a mobile fluid and molasses is a less mobile fluid.

We can measure the mobility of various fluids in the laboratory by displacing them through reservoir rock samples. If we divide the mobility of the pushing fluid by the mobility of the fluid being pushed, we can arrive at a number called mobility ratio. When the mobility ratio is less than or equal to 1, we have good displacement and a favorable mobility ratio. Unfavorable mobility ratios will be greater than 1 and will result in the premature breakthrough of the driving fluid through the fluid being displaced and into the producing well. This effect is demonstrated in Figure 1.

Although additional oil can be recovered after the breakthrough of the driving or injected fluid, it eventually becomes uneconomical to continue the process. The production of injected fluid continues to increase in volume and the oil production continues to diminish.

Polymers are called good mobility control agents. When used with a conventional waterflood, the premature breakthrough of water is reduced and more "water-free" oil is produced since the pattern is being displaced in a more uniform manner. Polymers are also used to reduce the flowing capacity of high permeability sections of reservoirs that have already been depleted of oil and are now channeling water.

Polymers do not recover more oil from individual pores - some residual oil is left as after a waterflood. The advantages of polymer flooding are simplicity of operations, low additional investment and moderate chemical cost. Polymer flooding is a fairly simple addition to our known waterflood technology. Certain precautions must be taken however to prevent or minimize the mechanical, chemical or bacterial degradation of the polymers.

SURFACTANT OR MICELLAR FLOODING has the potential of becoming the most extensively used chemical flooding process. It involves the addition of a very complex surfactant or micellar fluid to the reservoir. These soap-like materials are capable of recovering nearly all of the oil from any given pore that is contacted. Your wives and mothers have been doing this for years by adding detergents or soaps to water for washing dishes and clothes.

Surfactants are capable of reducing the interfacial tension between water and oil to very low values and increasing oil recovery. Figure 2 shows a globule of oil trapped in a pore space in the rock and water flowing past it from left to right. By adding surfactants to the water, the globule of oil is broken into many small droplets and displaced by the driving fluids. Figure 3 shows the results of several laboratory experiments. Interfacial tension is shown on the horizontal scale decreasing to the right. Oil recovery is shown on the vertical scale and increases upward. At very low interfacial tensions, oil recovery is very good.

Early industry attempts at surfactant flooding were largely unsuccessful. Continuing research revealed many problems that had been overlooked or not known during the early tests. Figure 4 represents a cross section or vertical slice so that we can see the various slugs that are injected from the injection well to the producing well. Some of the problem areas can be identified

FIGURE 1

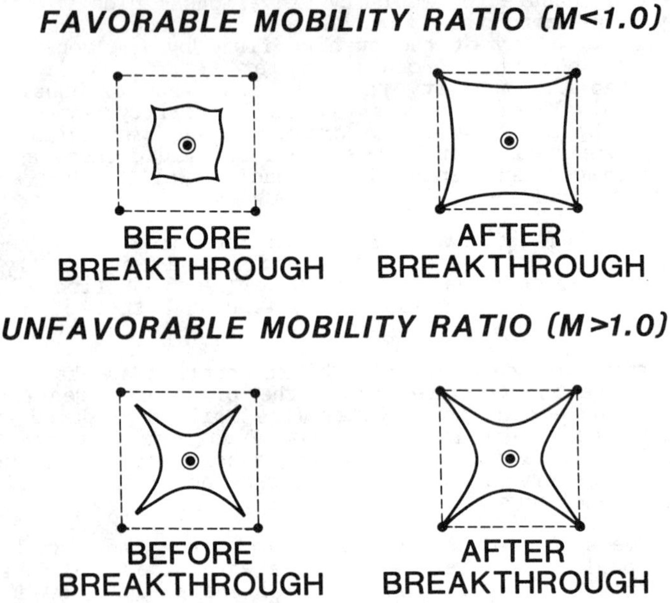

FIGURE 2

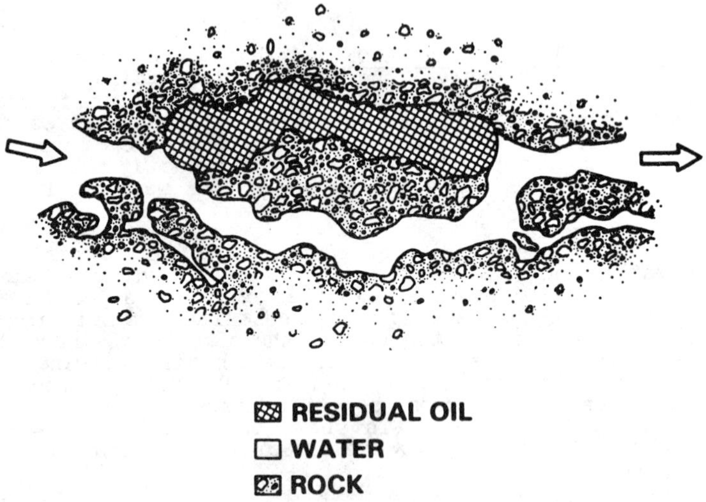

A POSSIBLE CONFIGURATION OF A TRAPPED OIL DROPLET

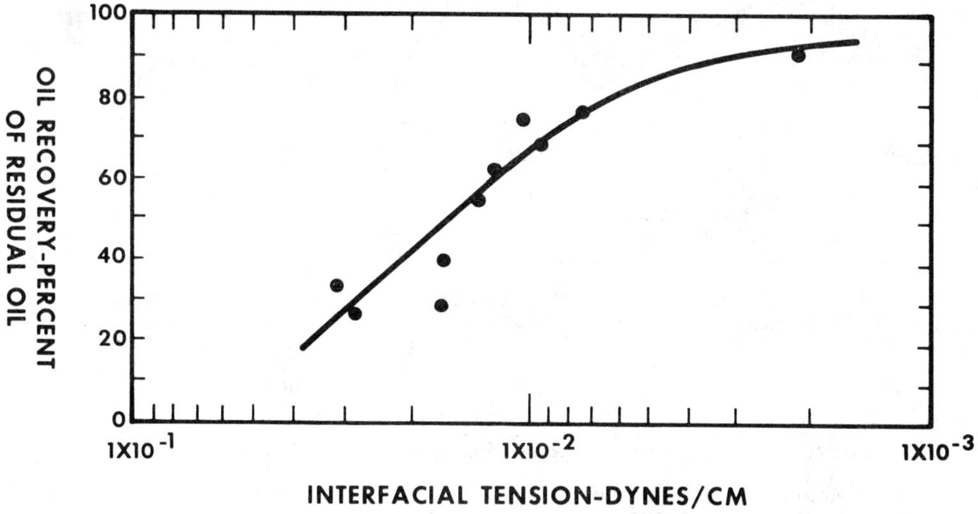

FIGURE 3

MICELLAR FLOODING

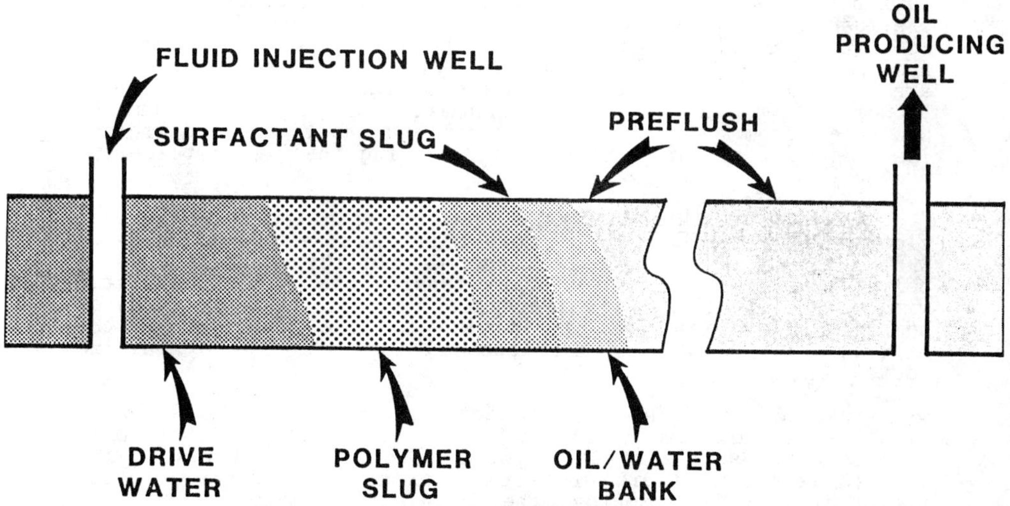

FIGURE 4

with this illustration.

Many oil fields or reservoirs contain some very salty waters that may have large amounts of hard ions such as calcium, magnesium, iron and barium. Since many of the ingredients in the surfactant slug are of the same type used in detergents and soaps, this hard water hinders the effectiveness of the slug as well as the following polymer. A preflush material of low salinity brine water is the first fluid injected to push the undesirable high salinity formation water toward the producing well and get it out of the way to avoid contact with the surfactant and polymer slugs. This is pretty much today's technology - research efforts continue toward developing surfactants and polymers that are more tolerant to saline environments.

If the surfactant slug is well designed, the residual oil saturation behind the surfactant slug is being reduced to a minimum and oil is being mobilized and pushed ahead toward the producing well. The surfactant slug is designed with a favorable mobility ratio to the mobilized oil/water bank. In a similar manner, the mobility ratio of the polymer slug is designed to be favorable with the surfactant slug and the drive water mobility ratio with the polymer is also designed to be favorable.

Another problem area with these systems is in the adsorption of the various chemicals by components in the reservoir rock. These materials and clays can denude or strip the surfactants and polymers from the various slugs. It is only through laboratory tests using actual reservoir rock that we can design the slugs to remedy this problem. At best, this design is a scientific guess.

The idea is to protect the surfactant slug from salinity effects, poor mobility control and adsorption - to transport it as an integral bank to effectively displace oil. Modern researchers continue to address these problems daily. If surfactant slugs can be designed to effectively displace oil without a preflush, the economics of a project are greatly improved. Mobility control is of prime importance, and the problem of adsorption effects the economic design of any given project. We are no longer injecting inexpensive fluids as we did in waterflooding. We are injecting expensive chemicals that must be protected from a hostile environment and should exhibit minimum loss, whether physical or chemical, during the oil displacement process.

ALKALINE OR CAUSTIC FLOODING is very similar to surfactant flooding. I'm not going to dwell on this very long since I wanted to emphasize the importance of mobility control by using polymers and the low interfacial tension phenomenon associated with oil displacement by surfactants. All of these processes or combination of processes are trying to take advantage of the natural forces that help displace oil while trying to minimize cost.

Instead of injecting surfactants from the surface, in alkaline or caustic flooding we are injecting lower cost caustic-like materials such as sodium hydroxide to react with certain selected acidic crude oils to develop very low interfacial tensions and improve the displacement process. Many of the same principles of displacement and similar problems exist with surfactant and alkaline flooding. The primary difference is in trying to create surfactant-like materials in the reservoir by injecting lower cost materials rather than injecting higher cost surfactants from the surface. Since this process requires a special type of crude oil in order to be effective in bringing about low interfacial tensions, most of the projects will probably be confined to some of the acidic crude oils of California, West Texas and Wyoming.

CURRENT INDUSTRY ACTIVITY - Figure 5 shows the distribution of chemical projects completed by industry through February 1981. According to our company survey, 37 projects were completed in Canada and the United States in chemical flooding. Figure 6 shows the results in terms of process, technical and economic success. Active projects total 69 as shown on Figures 7 and 8. Results are not yet available as to technical and economic success. You will note that polymer projects predominate because of their modest extension of technology beyond waterflooding and their lower cost. Figures 8 and 9 show the 26 planned projects for 1981. The planned projects are pretty well equally divided between the three chemical processes. You will also note that pilot tests far exceed the larger field scale tests.

This is to be expected since the technology is still emerging. Before any company is willing to commit large sums of money to any field project, they want as much assurance as possible that the process will work under field conditions. That is why, I think, during the next few years any organization implementing these processes to field development will have to have a strong research involvement.

Figure 10 shows the potential areas in the United States and Canada for the various types of processes - chemical, thermal and miscible. In our company we believe that chemical methods will be used primarily in the shallow sands of Kansas, Oklahoma, Texas and the Illinois Basin. Thermal processes will predominate in California, Wyoming, Alberta, Saskatchewan, Louisiana, Arkansas and Mississippi. Miscible processes will predominate the carbonate reservoirs of West Texas, New Mexico and Southern Oklahoma. In terms of the process undetermined areas of the Gulf Coast, many of these reservoirs have strong natural water drives and are fairly high in reservoir temperature. Not enough investigation has been completed to determine the best process to use under these conditions.

Figure 11 shows the potential enhanced oil recovery distribution by recovery process. Numerous surveys and studies have been conducted to determine the enhanced oil recovery potential for the United States. Because of differences in assumptions concerning recovery, price and costs, the investigators have published estimates ranging from 25 to 60 billion barrels of oil. However, they are in much closer agreement concerning the distribution of reserves by process - approximately 40 percent by chemical, 29 percent by thermal and 31 percent by miscible means.

I hope my talk has given you a little better insight into what your oil industry is doing to extend our Nation's oil reserves through chemical flooding efforts. We are attempting to bring about a smooth transition between our oil and gas energy resources of today and the more exotic energy resources of the future.

All processes have limitations and problems - physical, environmental and financial. If these limitations and problems are honestly addressed by both the critic and the investigator, I see nothing but a bright future for the solving of the United States energy problem. We are a nation founded on the pioneering spirit, a love of freedom, and the opportunity to function individually and collectively for the common good. As long as we aspire to these noble efforts, we can prevail.

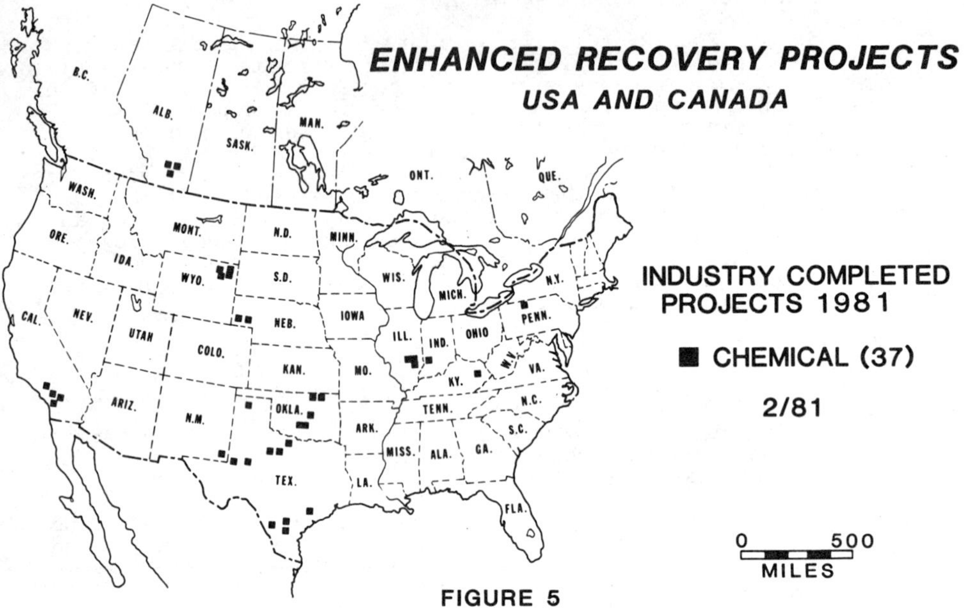

FIGURE 5

1981 CHEMICAL PROJECTS
CANADA AND U.S.A.

COMPLETED PROJECTS

	TECHNICALLY SUCCESSFUL	TECHNICALLY UNSUCCESSFUL	ECONOMICALLY SUCCESSFUL	TOTAL
SURFACTANT / POLYMER	3	6	1	10
POLYMER	2	11	7	20
ALKALINE	2	5	0	7
	7	22	8	37

2/81

FIGURE 6

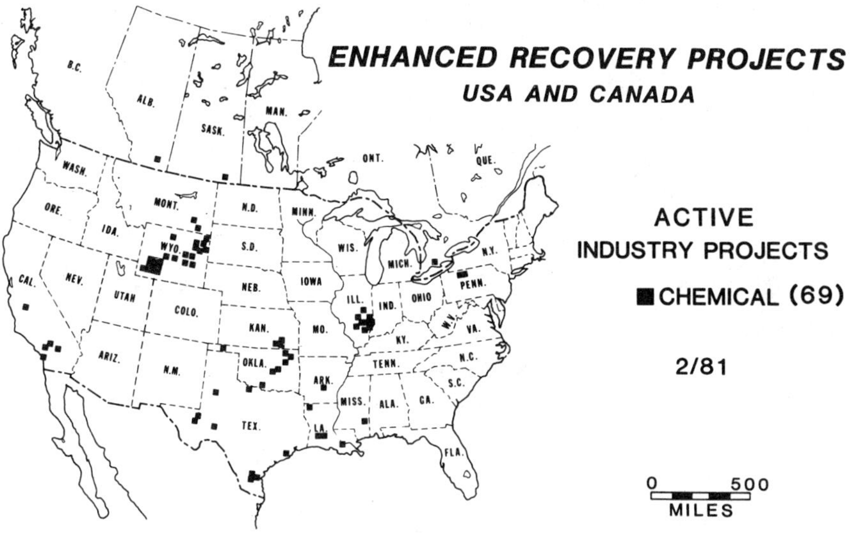

FIGURE 7

1981 CHEMICAL PROJECTS
CANADA AND U.S.A.

ACTIVE PROJECTS

	PILOTS	FIELD SCALE	TOTAL
SURFACTANT / POLYMER	26	1	27
POLYMER	30	5	35
ALKALINE	6	1	7
			69

PLANNED PROJECTS

	PILOTS	FIELD SCALE	TOTAL
SURFACTANT / POLYMER	7	2	9
POLYMER	7	1	8
ALKALINE	8	1	9
			26

2/81

FIGURE 8

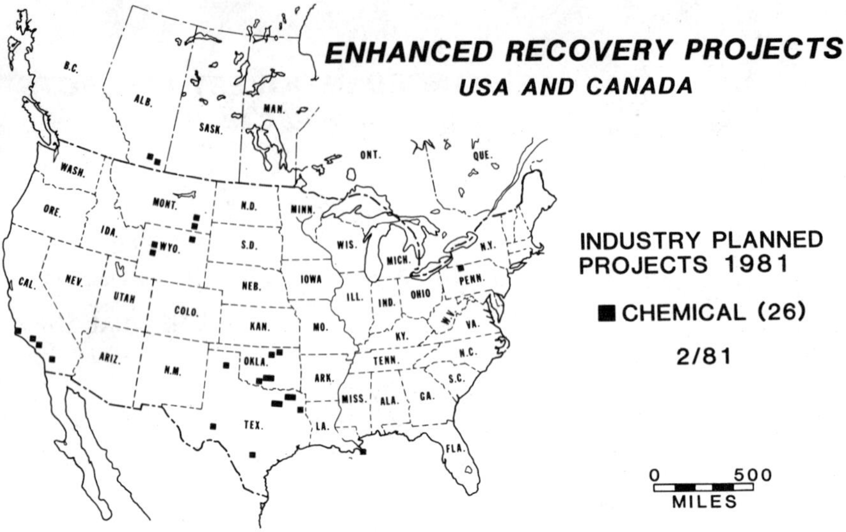

FIGURE 9

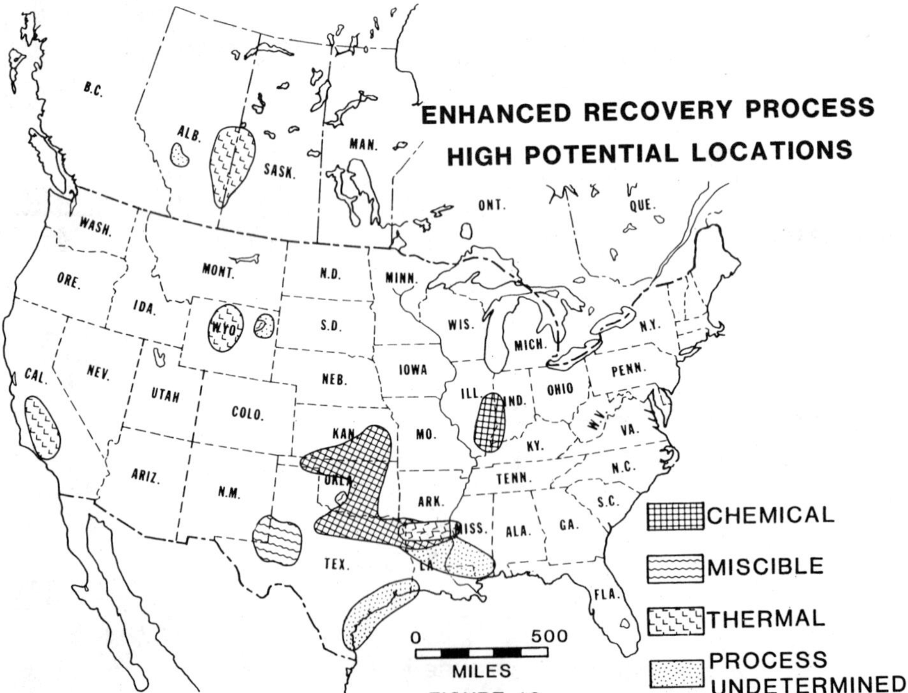

FIGURE 10

POTENTIAL EOR DISTRIBUTION
BY RECOVERY PROCESSES

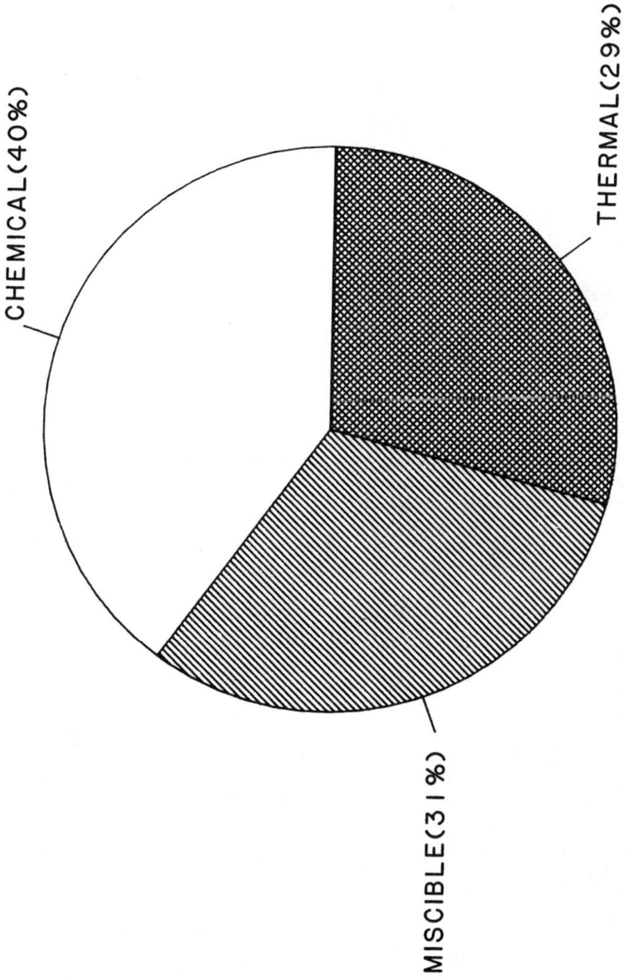

FIGURE 11

ENHANCED OIL RECOVERY BY CARBON DIOXIDE FLOODING

JOHN H. GOODRICH
GRUY FEDERAL, INC.

I. DESCRIPTION OF THE CARBON DIOXIDE FLOODING PROCESS

Enhanced oil recovery is a term coined in recent years to encompass techniques for recovering oil that cannot be recovered by application of more conventional practices and techniques. The three basic natural driving forces by which oil is expelled and recovered from underground hydrocarbon accumulations contained in porous reservoir rocks are:

Drive Mechanism	Average Recovery Efficiency (percent of original oil-in-place)
Depletion	5 - 20
Gas Cap Expansion	10 - 40
Water Influx	30 - 80

The wide variations in "average" oil recovery occur because of the influence of variations in reservoir rock quality, in-situ oil composition and properties, geologic configuration of the accumulations, and operating practices. Supplemental recovery techniques by which natural reservoir energies and drives are supplemented through injection of hydrocarbon or other gas and/or water may increase recoveries substantially but still leave residual (immobile) oil saturations ranging from 20 to 60 percent of reservoir pore space because of interfacial and capillary forces in the rock/fluid system. Enhanced recovery techniques are designed to mobilize and displace this residual oil. Injection of a fluid miscible with the residual oil is one such technique.

A miscible displacement of fluid from porous material occurs when no interface exists between the driving fluid and the driven fluid because of the miscibility, or mutual solubility, of the two fluids. This is in contrast to immiscible displacements such as water or low-pressure gas driving oil, where identifiable interfaces exist. In the miscible displacement of oil by another fluid, interfacial tension and related capillary forces between the two fluids and the rock are absent, so the driving fluid can totally displace the residual oil in place in the rock entered by the driving fluid. Because of this, miscible displacement has obvious technical potential for enhanced oil recovery.

The first efforts at achieving miscible displacement began in the 1950's using hydrocarbon driving fluids. For practical oil field applications, three technically different miscible displacement processes using hydrocarbon driving fluids have been conceived and applied:

- LPG or LPG-slug process (propane)
- High-pressure dry gas process (methane)
- Enriched-gas process (methane mixed with LPG)

The first of these processes, in which propane was generally used as the solvent or miscible agent, is a "first contact" miscible process. "First contact" miscibility means that the fluids are mutually soluble without any compositional alterations. The other two, in general, require multiple contacts between the driving fluid and the reservoir oil before miscibility is achieved. Carbon dioxide flooding began as a variation of solvent-type miscible flooding (1). However, recent work has shown that "miscibility" between carbon dioxide and reservoir oils is an elusive and complex subject, most relatable to the multiple-contact processes.

The use of carbon dioxide as an agent for enhanced oil recovery has been the subject of research by many investigators for more than 20 years. Presently, the spectrum of published research (2) (3) appears broad enough to provide the basis for a general understanding of the phenomenon termed "miscible displacement" when carbon dioxide is used to displace oil from porous materials, either in the laboratory or in the field. The main points of this general understanding can be summarized as follows:

(1) When high concentrations of carbon dioxide are mixed with oil, mass transfer of components between the carbon dioxide and the oil results in complex phase behavior in which as many as four separate fluid and solid phases may coexist.

(2) The two predominant phases in the system can generally be described as a more-volatile carbon dioxide-rich phase that also contains some hydrocarbons and a less-volatile hydrocarbon-rich phase that also contains some carbon dioxide.

(3) At relatively low temperatures (less than about 120°F), both the more-volatile carbon dioxide-rich phase and the less-volatile hydrocarbon-rich phase are liquids at high pressure. As pressure is reduced, vapors evolve predominantly from the carbon dioxide-rich phase.

(4) At higher temperatures (above 120°±F), the entire system will be in the vapor phase at high pressure (and high carbon dioxide concentration). As pressure is reduced, a

hydrocarbon-rich liquid phase condenses, while the carbon dioxide-rich phase remains a vapor. This seemingly paradoxical behavior is probably related to the critical temperature of the carbon dioxide-rich phase, which has been measured to be in the vicinity of 120°F.

(5) The mechanisms by which multiple-contact miscibility develops between carbon dioxide and oil are controlled to some degree by the temperature-dependent phase behavior. At higher temperatures, the mechanism is clearly a vaporizing gas mechanism in which hydrocarbons are extracted into the carbon dioxide-rich phase with countercurrent carbon dioxide condensation into the hydrocarbon-rich phase. Miscibility apparently also develops by a vaporization or extraction mechanism at low temperature. However, the carbon dioxide-rich phase is a liquid at "miscible" conditions, as is the hydrocarbon-rich phase.

Conditions necessary for attainment of miscibility between injected carbon dioxide and in-situ oil is a relatively strong function of reservoir (rock) temperature and a somewhat weaker function of reservoir oil composition. Higher pressure is required to attain miscibility as temperature increases. Increased pressure is necessary to attain miscibility at any given temperature as the in-situ oils become more dense (i.e., contain a higher molecular weight C_5 + fraction). When "miscibility" is attained, the injected carbon dioxide is capable of displacing and/or extracting most of the oil in place, leaving a residual oil saturation of only 3 to 5 percent of pore space. Some lesser benefits may be obtained by injection of carbon dioxide under non-miscible conditions because solution of carbon dioxide in the oil results in swelling (expansion) of the oil volume and a reduction in oil viscosity. Both effects contribute to increased recovery.

In principle, the additional oil recovery to be expected from carbon dioxide injection and the volume of carbon dioxide required can be calculated using simple algebraic equations.

In practice, many of the variables contained in the equations cannot be evaluated a priori except in extremely simple cases that only roughly approximate field conditions. Hence, mathematical computer models called reservoir simulators are generally used to design and evaluate carbon dioxide flooding projects. Ideally, these models account for the various interactions of carbon dioxide and oil and in addition provide for evaluating the effects of such factors as:

- well configurations or patterns
- geologic and petrophysical data
- reservoir fluid flow properties
- viscous and gravitational instabilities
- operating strategies

on the performance to be expected from a project. Pilot-scale field projects are frequently necessary to provide the information needed to calibrate the less-than-ideal models now in use.

II. CURRENT STATUS OF CARBON DIOXIDE FLOODING

In addition to the extensive research summarized above, a number of field projects involving carbon dioxide injection have been undertaken. The National Petroleum Council has published guidelines for applicability of carbon dioxide flooding as follows:

Property	Value
Minimum oil gravity	27° API
Maximum oil viscosity	10 centipoise
Minimum depth	2300 feet
Maximum temperature	250°F

These guidelines were based both on research and field results. However, the properties of the various field projects for which information is available provide extensions to these guidelines and suggest broad applicability of the process.

A recent report published by the U. S. Department of Energy (4) presented information compiled from public sources for 19 carbon dioxide injection projects in 16 different oil fields located in 5 states. The projects for which data are available are listed in Table 1, which shows that full-scale, commercial-intent projects have been undertaken in seven fields: three in west Texas, two on the Gulf Coast, and two in Arkansas. In addition, 12 tertiary pilot scale projects have been or are being conducted in 9 fields. Of these, 8 are in 6 fields in west Texas, 3 are in West Virginia, and 1 is in Mississippi. It can be seen that three of the full-scale projects and eight of the pilot projects are in various fields in west Texas (Figure 1).

The preponderance of projects, both full-scale and pilot, in west Texas should be no surprise. Source carbon dioxide in large quantities has long been available in the region as a by-product of natural gas processing, a fact which encouraged early application. In addition, the region contains many large fields in which waterflooding, while efficient and commercial, is expected to leave hundreds of millions of barrels of residual oil as a target for tertiary recovery. Summaries of the three full-scale projects in this province are given in the material that follows. Also presented are similarly abbreviated discussions of the other projects in Table 1.

FULL-SCALE PROJECTS

The three most mature full-scale carbon dioxide injection projects listed in Table 1 are in the Crossett (North Cross), Twofreds, and Kelly-Snyder (SACROC) fields of west Texas. The North Cross project is a clear-cut secondary recovery application in a carbonate reservoir whose permeability is too low to allow economical waterflooding. Carbon dioxide injection has been continuous, including recycling of produced carbon dioxide-contaminated gas. Production response has been excellent. The Twofreds project is similar, because the preponderance of carbon dioxide injection and response has been in the east reservoir of the Bell Canyon sandstone, in which waterflooding was unsuccessful and was terminated early. Carbon dioxide is being injected continuously in the project. The SACROC project is also a secondary recovery application; however, it was implemented by choice in a limestone reef reservoir in which waterflooding was being successfully applied. In view of the early implementation of alternating water and carbon dioxide injection, the initial water injection into later phases of the project, and the relatively small slug of carbon dioxide injected into each pattern before continuous water injection was applied, it could be argued that this project is as much a waterflood as it is a carbon dioxide misci-

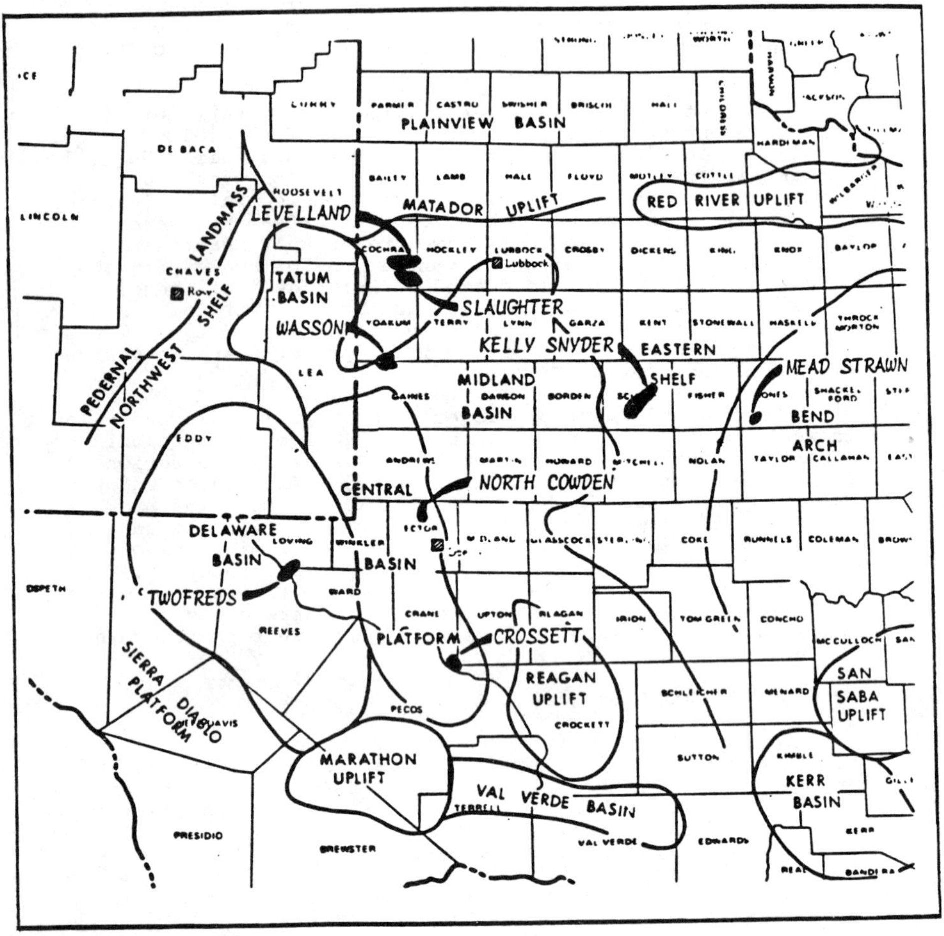

Figure 1--Carbon dioxide projects in west Texas

ble flood. All of these projects fit within the NPC guidelines of applicability and may, indeed, have been influential in defining them.

Each of the two Gulf Coast projects, at South Gillock and Weeks Island, fits within the NPC guidelines of applicability. However, they are conceptually different from the three west Texas projects discussed above. Both were designed to inject carbon dioxide at the gas/oil contact of a dipping reservoir to achieve gravity-stabilized downward miscible displacement of residual oil in a watered-out oil column below a gas cap. The South Gillock project was aborted before meaningful results could be observed. The Weeks Island project continues, with what appears to be favorable initial response.

The two Arkansas projects, at Lick Creek and Ritchie, obviously fall far below the minimum API gravity guideline and far above the maximum viscosity guideline for carbon dioxide miscible flooding. On the basis of available information, it is doubtful that miscible conditions are being achieved in either project. Nonetheless, the favorable response obtained in each project suggests that carbon dioxide injection may be a viable method for enhanced oil recovery in thin, shallow, intermediate-viscosity oil reservoirs in which waterflooding (conventional or enhanced) is inefficient and in which thermal methods cannot be applied economically.

PILOT PROJECTS

Ten of the 12 pilot tests listed in Table 1 are on field-scale spacing, with several hundred feet separating injection and production wells. The other two, the ARCO and Shell projects in Wasson, are small scale "mini-tests." Five of the field-scale projects are in sandstone reservoirs, three in West Virginia and one each in Mississippi and west Texas. The other five field-scale and two mini-test pilot projects are in carbonate reservoirs in west Texas. All of these projects fall within the NPC guidelines.

All three West Virginia projects are in sandstone reservoirs having reservoir temperatures below the critical temperature of carbon dioxide (88°F). Laboratory tests established that liquid carbon dioxide (1000± psi, 74-78°F) would achieve miscible displacement efficiencies. Results from the field project at Granny's Creek, now concluded, verified that injection of carbon dioxide could successfully mobilize and drive tertiary (waterflood residual) oil under actual reservoir conditions. Response was minimal, however, because carbon dioxide escaped from the project pattern area. Causes of this are apparently geological and not easily remedied. Carbon dioxide injection into the project has recently started again using a revision of the original pattern. Oil production response attributable to carbon dioxide flooding is again being observed. The carbon dioxide project at Griffithsville has apparently been suspended after considerable water injection but before any carbon dioxide was injected. While reasons for the suspension are not related to the merits of the carbon dioxide process, there are no carbon dioxide injection and response data for this project. Carbon dioxide injection is underway in the third project, at Rock Creek, after extensive water injection to increase reservoir pressure. Tertiary oil response from two pattern production wells is being observed.

TABLE I
SUMMARY OF DATA FOR PAST AND ONGOING CO_2 INJECTION FIELD PROJECTS

FULL-SCALE PROJECTS

FIELD / STATE OR REGION		CROSSETT (NORTH CROSS) WEST TEX.	LICK CREEK ARK.	KELLY SNYDER(SACROC) WEST TEX.	RITCHEY ARK.	SOUTH GILLOCK TEX. GULF COAST	THOFREDS WEST TEX.	WEEKS ISLAND LA. GULF COAST
RESERVOIR NAME		DEVONIAN	MEAKIN	CANYON REEF	BAKER	FRIO	BELL CANYON	"S" SAND
ROCK TYPE		CHERTY DOLO.	SANDSTONE	LIMESTONE	SANDSTONE	SANDSTONE	SANDSTONE	SANDSTONE
POROSITY %		22	33	3.9	30.7	29.5	19.5	26
PERMEABILITY md		3-5	1200	19.4	2800	1156	32	3500
INITIAL WATER SATUR. %		35	32	21.9	20	22.5	43.5	7
AVG. THICKNESS (NET PAY) FT.		96	12	139	9	39	25	250±
AVG. DEPTH FT.		5300	2500	6700	2600	9000	4820	12,700
RESERVOIR TEMP. °F		106	118	130	126	214	104	225
STOCK TANK OIL GRAV. °API		44	17	41.8	15-16	38	36	32
INITIAL RES. PRESS. PSIA		2328	1200	3137	1250	4354	2385	5100
BUBBLE POINT PRESS. PSIA		2328	85	1820	100sst.	4354	2285	5100
BUBBLE POINT OIL VISCOS. C'POISE		—	160	.38	195	.64	1.467	.34
FORMATION VOLUME FACTORS	INITIAL VOL/VOL	1.986	1.05	—	1.05	1.585	1.179	1.7
	BUBBLE POINT VOL/VOL	1.986	1.05	1.5	1.05	1.585	—	1.7
CO_2 MISC. PRESS. PSIA		1650a	?	1600a	?	?	2285b	5100b
PROJECT SIZE	AREA ACRES	1120	900	51,000	220	5940	4392	5±
	NO. WELLS PROD	19	38	305	15	30±	40	1
	NO. WELLS INJ	9	16	659	1	6	22	1
TYPE PROJECT		SECONDARY	SECONDARY*	SECONDARY	TERTIARY	SECONDARY*	TERTIARY?	SECONDARY*
OPERATOR		SHELL	PHILLIPS	CHEVRON	U.S.O.R./ PHILLIPS	AMOCO	HNG	SHELL
START DATE		1972	1976	1972	1969	1971	1974	1978
PRESENT STATUS		ONGOING	ONGOING	ONGOING	RESUMED1978	TERMINATED	ONGOING	ONGOING

TABLE 1 (CONT.)

PILOT PROJECTS

FIELD / STATE OR REGION	GRANNY'S CREEK WEST VA.	GRIFFITHSVILLE WEST VA.	LEVELLAND WEST TEX.	LITTLE CREEK MISS.	MEAD WEST TEX.	NORTH COWDEN WEST TEX.	ROCK CREEK WEST VA.	SLAUGHTER WEST TEX.	WASSON WEST TEX.			KELLY-SNYDER(SACROC) WEST TEX.
RESERVOIR NAME	BIG INJUN	BEREA	SAN ANDRES	TUSCALOOSA	STRAWN	GRAYBURG	BIG INJUN	SAN ANDRES	SAN ANDRES			CANYON REEF
ROCK TYPE	SANDSTONE	SANDSTONE	DOLOMITE	SANDSTONE	SANDSTONE	SDY.DOLO.	SANDSTONE	DOLOMITE	DOLOMITE			LIMESTONE
POROSITY %	18	11	11	23	9.4	9	21.7	10	8.5			3.9
PERMEABILITY md	2-4	14	3	90	12	5.3	19.3	8.1	3			19.4
INITIAL WATER SATUR. %	?	?	23.4	56	40	15	50±	13	15			21.9
AVG. THICKNESS (NET PAY) FT.	20	23	80	29	9	92	32	89	66			139
AVG. DEPTH FT.	2100	2350	4750	10,600	4500-4900	4400	2050	4950	5100			6700
RESERVOIR TEMP. °F	77	78	102	248	135	94	74	105	107			130
STOCK TANK OIL GRAV. °API	45	43	30	39	41	36	43.5	32	33			41.8
INITIAL RES. PRESS. PSIA	?	?	1620	4840	1807	1800	?	1710	1805			3137
BUBBLE POINT PRESS. PSIA	?	?	1620	2150	1500	1800	?	1710	1805			1820
BUBBLE POINT OIL VISCOS. C POISE	2.5	2.8	2.3	.30	1.3	1.46	3.1	1.38	1.18			.38
FORMATION VOLUME FACTORS: INITIAL VOL/VOL	?	?	1.209	1.32	1.29	1.25	?	1.228	1.312			—
FORMATION VOLUME FACTORS: BUBBLE POINT VOL/VOL	?	?	—	?	—	1.25	?	1.228	1.312			1.5
CO_2 MISC. PRESS. PSIA	1000a	1100a	1050±a	4500a	850a	1050±a	1000a	1050±a	1250a			1600a
PROJECT SIZE: AREA ACRES	6.4		13.2	31	44.5	12	20	13.2	(1)425	(2)0.7	(3)0.7	80±
PROJECT SIZE: NO. WELLS PROD.	1	9	2	3	2	2	2	2	15	1	1	2
PROJECT SIZE: NO. WELLS INJ.	4	16	6	1	4	6	6	6	8	1	1	6
TYPE PROJECT	TERTIARY	TERTIARY	TERTIARY	TERTIARY	TERTIARY	TERTIARY	TERTIARY	TERTIARY	SEC.?	TERT.	TERT.	TERTIARY
OPERATOR	COLUMBIA	GUYAN	AMOCO	SHELL	UNION	AMOCO	PENNZOIL	AMOCO	ARCO	ARCO	SHELL	CHEVRON
START DATE	1975	1977	1973	1973	1964	1973	1976	1972	1972	1973	1977	1974
PRESENT STATUS	CONCLUDED	SUSPENDED?	ONGOING	CONCLUDED	CONCLUDED	ONGOING	ONGOING	ONGOING	TERM.	CONCL.	ONGO.	CONCLUDED

The carbon dioxide injection pilot project in the Strawn sandstone reservoir of the Mead field in west Texas was concluded in 1968. In this project, a small slug of carbon dioxide was followed by a slug of carbonated water, which was followed by ordinary brine. Results indicated that miscible recoveries were attained, and oil recovery from the carbon dioxide pilot area was approximately 100 percent greater than that realized from a comparable waterflood area.

The Little Creek project in Mississippi, concluded in 1977, was conducted at a higher reservoir temperature (248°F) than any other project. Continuous carbon dioxide injection was highly successful in mobilizing, banking, and displacing a low (21 percent) immobile waterflood-residual oil saturation.

Of the seven west Texas carbon dioxide injection pilot tests in carbonate reservoirs, two are concluded: the field-scale tertiary pilot test conducted by Chevron in the SACROC unit of the Kelly-Snyder field and the mini-test conducted by ARCO in the Willard unit of the Wasson field. A third project, ARCO's "Phase I" project in the Willard unit, was terminated because of a field accident. The four remaining projects, three field-scale and one mini-test, are ongoing in various states of maturity.

In Chevron's pilot test in the SACROC unit, carbon dioxide injection successfully mobilized and displaced waterflood residual oil in a two-pattern pilot. Response of the patterns was unbalanced; one responded well and the other poorly. Oil production was still increasing when carbon dioxide injection was stopped.

The ARCO mini-test in the Willard unit at Wasson demonstrated, through wire-line log, pressure-core, and production-sampling tests, that carbon dioxide mobilized and displaced waterflood residual oil from the San Andres reservoir rock. Data from this test were used to calibrate a simulation model study, which indicated that full-scale tertiary flooding could be economically feasible.

Published data from ARCO's completed Phase I test at Wasson and the four ongoing projects--three field-scale by Amoco in Levelland, North Cowden, and Slaughter and a mini-test by Shell in the Denver unit of Wasson--are insufficient to permit any meaningful summary of results in terms of oil production response. The Amoco projects are expected to require 2 to 3 more years to evaluate. The Shell mini-test is essentially completed; evaluation of full-scale project installation will require simulator studies to scale up mini-test results.

CONCLUSIONS

We can conclude from the data given for the field tests that carbon dioxide injection is technically capable of displacing oil from reservoir rocks over a wide range of geological and physical conditions. Projects have been conducted at depths ranging from 2100 feet to 12,700 feet over a reservoir temperature range from 74°F to 248°F in rocks whose lithologies included unconsolidated Gulf Coast sands, cemented sandstones, limestones, and cherty dolomites. These projects included reservoirs whose stock tank oil gravities ranged from 15° to 45° API. This spectrum of technical applicability is much less restrictive than any of the other enhanced oil recovery methods being developed or field-tested.

The capacity to move and recover oil on a field-scale (as distinguished from laboratory cores or packs of porous material) is amply demonstrated for "secondary" conditions by the North Cross, Twofreds, and SACROC projects. In addition, the potential of carbon dioxide injection under "tertiary" (i.e., waterflood-residual oil) conditions has been demonstrated for a wide variety of reservoir conditions by the projects in West Virginia and Arkansas and by the Little Creek, Mead-Strawn, Weeks Island, and west Texas carbonate reservoir projects. It thus seems that the limitations to widespread application of carbon dioxide flooding are mainly economic rather than technical.

III. FUTURE POTENTIAL

The broad spectrum of experience in field applications and the research efforts that supported these projects, as summarized earlier, has advanced the technology of carbon dioxide flooding to what can be considered a takeoff point for widespread, large-scale application limited principally by the local availability of carbon dioxide. This observation is supported by recent announcements of 15 or more projects by various companies and by reports of another 15 or more projects in various stages of planning (see Table 2). As might be expected, 5 of the announced projects and 11 of the planned projects are in the Permian Basin of west Texas and New Mexico.

As further support for the "takeoff stage" observation, at least three of the west Texas projects, ARCO's Willard unit project and Shell's Denver unit project in the Wasson field and Amoco's North Cowden field project are full-scale expansions of pilot projects listed in Table 1. In addition, Conoco's Ford-Geraldine field project is based on the Twofreds project results and experience as an analog. It is reasonable to infer that field experiences also provide the technical basis for west Texas projects planned at Slaughter, Levelland, McElroy, and Anton-Irish and for the Mississippi projects listed.

As mentioned above, carbon dioxide supply is a major factor limiting widespread application of carbon dioxide injection. In the Permian Basin area, three major pipeline projects to supply carbon dioxide from natural source fields (5) have recently been announced. These are:

Companies	Source	Terminus
Amoco	Bravo Dome, New Mexico	Slaughter field
ARCO	Sheep Mountain, New Mexico	Wasson field
Shell-Mobil	McElmo Dome, Colorado	Wasson field

The original intent of these source field development and pipeline projects was to supply carbon dioxide to projects planned by the companies themselves. However, recent news releases indicate that each source field/pipeline project will provide carbon dioxide reserves in excess of the operator's needs. The excess is indicated to be available for sale to other operators. A similar situation prevails around the Jackson Dome source field (5) in Mississippi. Hence, it seems likely that the projects announced to date may be the tip of the iceberg in terms of future projects.

TABLE 2

NEAR FUTURE CARBON DIOXIDE FLOODING PROJECTS

ANNOUNCED PROJECTS:

Operator	Field	State/Region
Santa Fe Energy	Raymond	Montana
Santa Fe Energy	Reeves	West Texas
Conoco	Maljamar	New Mexico
Conoco	Ford-Geraldine	West Texas
Texaco	Paradis	Louisiana Gulf Coast
Texaco	Bay St. Elaine	Louisiana Gulf Coast
Pennzoil	Tinsley	Mississippi
Citronelle Operators	Citronelle	Alabama
Tenneco	Dillinger Ranch	Wyoming
ARCO	Wasson	West Texas
Shell	Wasson	West Texas
Amoco	North Cowden	West Texas
Gold King	Rankin	Texas Gulf Coast
Sabine	Oil Springs	Kentucky
Mitchell Energy	Two Fields	North Texas
Halliburton	Two Fields	Texas Panhandle

PROJECTS REPORTED BEING PLANNED:

Chevron	Brookhaven	Mississippi
Inexco	Central Hilight	Wyoming
Superior	MacKelmo Creek	Utah
Southland Royalty	McElroy	West Texas
Sun Texas	Wasson	West Texas
Sun Texas	Levelland	West Texas
Sun Texas	Slaughter	West Texas
Shell	Jordon	West Texas
Shell	Flanagan	West Texas
Shell	Wasson 72	West Texas
Shell	Monahans	West Texas
Shell	Mallalieu	Mississippi
Amoco	Slaughter	West Texas
Amoco	Levelland	West Texas
Amoco	Anton-Irish	West Texas

REFERENCES

1. Holm, L. W.: "Carbon Dioxide Solvent Flooding for Increased Oil Recovery," Trans., AIME (1959) Vol. 216, p. 225.

2. Holm. L. W.: "Status of CO_2 and Hydrocarbon Miscible Oil Recovery Methods," J. Pet. Tech. (Jan. 1976).

3. Stalkup, F. I.: "Carbon Dioxide Miscible Flooding: Past, Present, and Outlook for the Future," J. Pet. Tech. (Aug. 1978).

4. Goodrich, John H. (Principal Investigator): "Target Reservoirs for CO_2 Miscible Flooding," Vol. I: Discussion of CO_2 Injection Process, Vol. II: Full Scale and Pilot Field Tests; U.S. Dept. of Energy Report DOE/MC/0834]-17, October 1980.

5. Hare, M., Perlich, H., Robinson, R., Meghji, S., and Zimmerman, F.: "Sources and Delivery of Carbon Dioxide for Enhanced Oil Recovery," U.S. Dept. of Energy Report FE-2515-24, December 1978.

ENHANCED OIL RECOVERY BY THERMAL METHODS*

H. M. Stoller**

Sandia National Laboratories

INTRODUCTION

While a variety of enhanced oil recovery methods hold promise for commercial applications, thermal methods are established and commercially-viable for the recovery of medium to high viscosity crude oils and have the potential to be employed to recover a broader range of oil resources. The 1980 EOR Survey conducted by the Oil and Gas Journal (1) (2) reported that of the EOR production of 332,821 barrels of oil per day [BOPD], thermal methods accounted for 255,610 BOPD.

Three thermal methods are employed. These are cyclic steam, steam drive, and in situ combustion. The first two accounted for 243,000 BOPD of the thermal production, almost all of this coming from the heavy oil fields of California.

The primary purpose of introducing heat into the reservoir is to reduce the viscosity of the in-place oil. At ambient temperature, the viscosity of these heavy oils can be a million times more viscous than water, making the oil immobile in the reservoir rock. Viscosity is temperature dependent; a temperature increase of a few hundred degrees can result in a four to five order of magnitude reduction in viscosity. Heated to these temperatures and with adequate reservoir pressure, the oil can then flow through the reservoir at a rate sufficient to justify commercial development.

There are other forces at work, other than viscosity reduction, which aids oil production when thermal recovery methods are employed. The injection of steam, hot water and air serves to maintain or reestablish a driving pressure within the reservoir. Flowing steam can remove heated oil via a viscous drag effect. A comprehensive experimental study (3) also identified as recovery mechanisms

*Work performed for US/DOE, Contract No. DE-AC04-76DP00789.

**Presently Vice President, Vedette Energy Research, Inc.

the thermal expansion of the heated oil, the steam distillation of the lighter components of the crude oil and gas-drive and solvent extraction effects.

The perspective of this paper is derived from a laboratory engaged in the development of advanced technology to improve thermal recovery processes, not from a producer involved in the production-application of thermal methods. Hence, other overview perspectives recently presented are of interest (4-7). Other pertinent publications include an assessment of the worldwide potential for EOR including thermal methods (8) and a survey of critical research needed to improve the state-of-the-art (9).

EOR THERMAL METHODS

Thermal methods are by steam injection processes or in situ combustion. Steam injection processes, primarily in the State of California, are now economically viable. The success of in situ combustion projects has been limited.

The first step of the steam process is steam stimulation [also known as steam soak and cyclic steam] of the production well. Steam is injected into the well for a period of time. The well is then shut-in to allow the heat to distribute. The well is then placed on production. This cycle is repeated until the production response declines, as the recoverable oil in the vicinity of the wellbore is produced, to the point where the process becomes uneconomic. The process is then usually converted into a steam drive although the steam soak technique can be the only enhanced production method employed.

The steam soak process has proven attractive, partly because the immediate response allows an early evaluation of the reservoir and the high oil production rates from the initial soak cycles. While the recent EOR survey (1) indicates about as many steam soak projects as steam drive projects, the most significant recent development in steam injection is the trend to steam drive, with cyclic steam stimulation becoming an important adjunct, rather than a separate oil recovery process (4).

Conceptually, steam drive is a recovery process analogous to waterfloding [Figure 1]. Ideally, injected steam forms a steam saturated zone around the injection well. As the steam moves away from the well, its temperature decreases as it expands in response to the pressure drop. At some distance from the well, the steam condenses and forms a hot water bank. In the steam zone, oil is displaced by steam distillation and steam drive. In the hot water zone, thermal expansion of the oil and reduction in oil viscosity are the major production mechanisms.

Actual performance of the steam drive can be considerably different from the ideal situation. When steam is injected, it may form a fingerlike channel through the easiest conduit to reach a producing well. With time and continued injection, steam, being lighter than oil, travels upwards in the reservoir and blankets the oil - a situation known as gravity override. Fingering and gravity override imply a critical problem in steam drive usage, namely the inability to control downhole steam profiles.

Where steam encounters oil, recovery efficiencies approaching 100% are achieved. Gravity override and fingering reduces the amount of oil contacted by the steam and hot water, but overall recovery efficiencies exceeding 50% have been achieved in reservoirs having good characteristics for thermal recovery. The vast amount of energy left in the heated rock after the steam drive has been com-

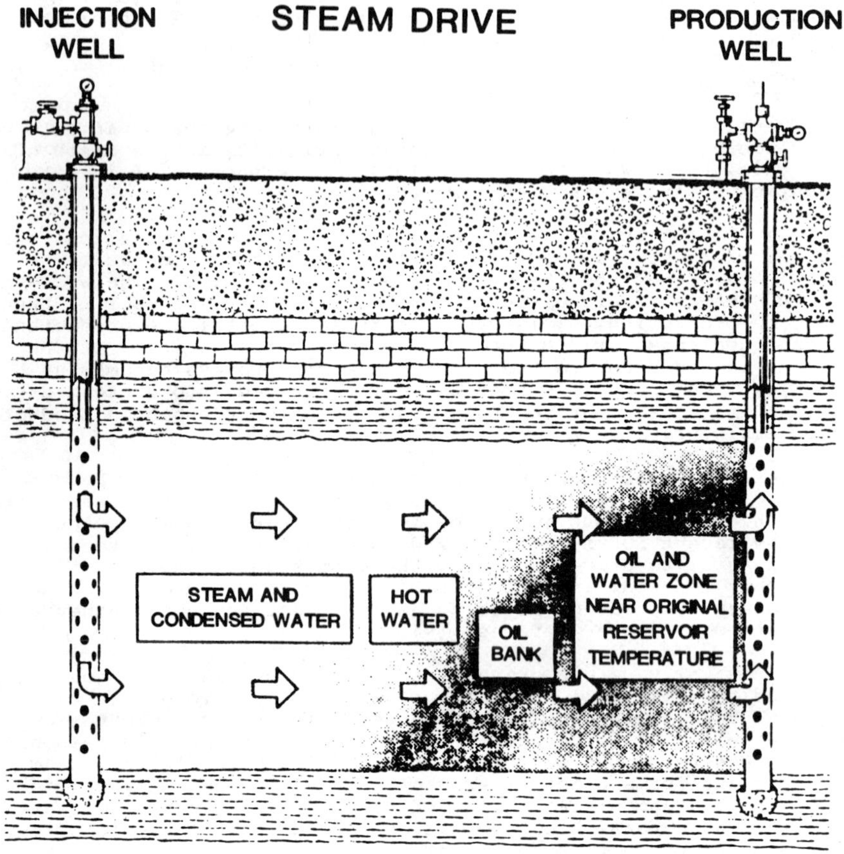
INJECTION WELL
STEAM DRIVE
PRODUCTION WELL
STEAM AND CONDENSED WATER
HOT WATER
OIL BANK
OIL AND WATER ZONE NEAR ORIGINAL RESERVOIR TEMPERATURE

FIGURE 1

pleted has led to heat scavenging concepts such as a post-steam drive waterflood.

Because of the viscosity reduction effect, steam drives have been primarily directed at heavy oils. However, the steam distillation recovery mechanism is most effective on lighter oils, since they contain a greater fraction of steam distillable components, leading to a greater recovery by steam in light oils. One argument against the use of steam drive in waterflood-depleted light oil reservoirs is the thermal inefficiency attributed to the heating of large volumes of reservoir rock relative to the residual oil saturation. Yet interest in steam distillation drives appears to be building.

Oil/steam ratios, be they for steam soak or steam drive operations tend to converge to a value determined only by the reservoir and steam properties and time. The need to pretreat the production wells to enable a steam drive process to progress, the early economic return from cyclic steam, and the greater overall payout from steam drive have resulted in a marriage of the steam soak and steam drive processes.

In the in situ combustion process, ignition is started at an injection well and the combustion zone is propagated to the producing wells. The in situ combustion method is based on the combustion of the coke desposit laid down by the crude oil. Figure 2 shows the mechanisms and the temperature distribution of the in situ combustion process.

Many factors affect the application and limitations of the in situ combustion recovery process. The fuel concentration that is deposited on the reservoir rock, along with the air flux, determines if sufficient heat can be generated in the reservoir rock to support the combustion process. Oils that leave very high deposits [> 3w/o for 10° API] make the process uneconomic because of a high air-oil ratio; conversely an insufficient fuel concentration [<1w/o for > 40° API] may not allow the combustion process to propagate. The air required to sustain combustion, a function of fuel concentration, injection pressure and air flux, is a key economic parameter because of compressor costs. Formation thickness has a significant influence on heat loss; ten feet being deemed the minimum acceptable.

A tabulation of twenty-four in situ combustion field projects indicated total recovery [primary plus combustion] of oil in place averaged around 45-50% (10). A good indicator of economic fireflood operation is the ratio of injected air to produced oil. Successful floods have air-oil ratios of less than 20,000 SCF/BBL.

Variations of the dry forward combustion process have been developed to improve economics. Wet combustion, also known as COFCAW [combination of forward combustion and waterflooding] is a cocurrent process in which water is injected with air. In dry combustion more than half the heat is contained between the combustion front and the air injection well. Water can be used to transfer this heat forward increasing the size of the steam zone downstream of the advancing combustion zone. This water injection can have a significant impact on the air/oil ratio. Heat scavenging, via waterflood after in situ combustion is another approach to more effectively use the heat contained in the reservoir rock.

STATUS OF THERMAL PROCESSES

A. CURRENT U.S. ACTIVITIES

The biennial survey of EOR activities by the Oil and Gas Journal presents a comprehensive report on all EOR processes, including

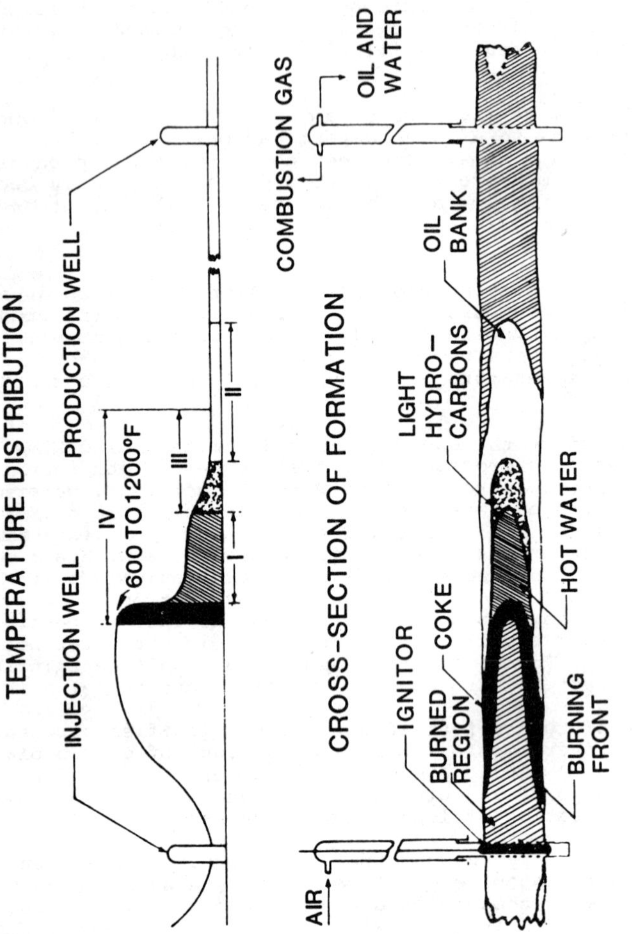

MECHANISMS OF THE IN SITU COMBUSTION PROCESS

FIGURE 2

thermal. Figure 3 clearly shows the dominant influence of steam recovery methods over in situ combustion; almost 90% of the reported thermal projects are steam, either drive or soak. The growth in steam projects is equally impressive; over the past decade, the number of steam drive projects has increased at a rate exceeding 10% a year while the number of in situ combustion projects has stayed essentially constant since declining from its 1970 peak. More impressive has been the surge in steam drive projects, the largest contributor to the growth in steam-enhanced recovery. From a number of 22 in 1973 to almost 70 in 1979, the number of steam drive projects has increased an average rate of 20% per year.

Production figures are equally impressive. The 1980 survey attributes 255,610 BPD to thermal EOR methods, with over 96% of this production resulting from steam methods. The steam production is split about equally between the two steam methods.

The Getty/Kern River field project continues to be the largest thermally-aided recovery project, a combination of steam drive and steam soak, with a daily production of 65,000 barrels of oil of which 52,250 are attributed to the thermal processes. The Texaco/San Ardo project as well as the Shell/Mount Poso project are large steam drives credited with 22,500 BOPD and 20,000 BOPD, respectively. The largest in situ combustion projects all have production about an order of magnitude less: these include Getty's project at Bellevue, Louisiana [2775 BOPD], and Gulf's project in Cotton Valley, Mississippi [1800 BOPD].

Besides the number of steam projects and the magnitude of steam-aided enhanced recovery relative to in situ combustion, the 1980 survey offers additional insight into the application and differences of these two thermal methods. Over 90% of the steam projects are being conducted in the State of California; about half of the combustion projects are being conducted in other states. Of those projects for which a profitability determination could be made, 90% of the steam projects were deemed profitable; this assessment applied to two-third's of the combustion projects. With respect to reservoir characteristics, no steam drive project is being conducted in a reservoir deeper than 2500 feet, [only three steam soaks were] and with the exception of Texaco's Shiells Canyon project, all steam projects are in heavy gravity oils, $<25°$ API. Conversely, almost half of the combustion projects are in reservoirs deeper than 2500 feet and about one-fourth are in lighter gravity oils, $>25°$ API.

It is clear that in the State of California the operators have developed the expertise to make steam-aided enhanced recovery a viable commercial process. Outside of this state, neither steam nor combustion have established themselves as major enhanced recovery processes with respect to being major contributors to domestic U.S. production.

B. <u>IMPACT OF THE TERTIARY ENHANCED RECOVERY PROGRAM</u>

An important development in EOR was the initiation in 1979 by the Department of Energy of the Tertiary Enhanced Recovery Program. This program is essentially in effect until October 1981 when U.S. oil prices are scheduled to be decontrolled to world oil prices. This allows producers who invest in qualified EOR projects to receive the world market price for <u>any</u> crude oil sold on or after January 2, 1980, to the extent of 75% of certain qualified EOR expenses incurred on or after August 22, 1979, limited to $20 million per project. This permits producers to receive market prices for crude oil produced from properties other than the property on which the EOR project is being conducted. The Tertiary Enhanced Recovery Program, while domestic crude prices are controlled, is a mechanism which

FIGURE 3

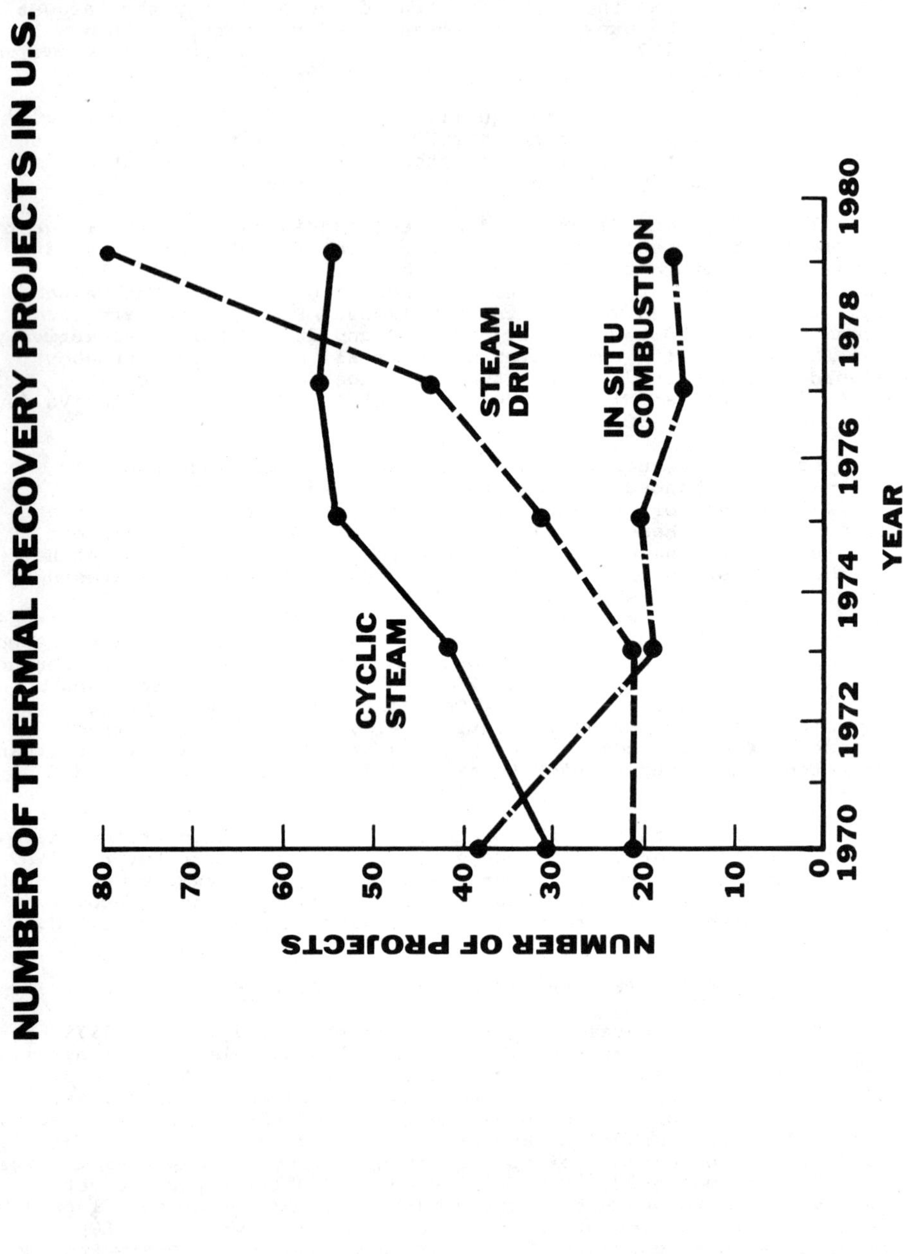
NUMBER OF THERMAL RECOVERY PROJECTS IN U.S.
NUMBER OF PROJECTS
0
10
20
30
40
50
60
70
80
1970
1972
1974
1976
1978
1980
YEAR
CYCLIC STEAM
STEAM DRIVE
IN SITU COMBUSTION

has provided financial incentives for industry to initiate a significant number of new EOR projects.

In its 1980 EOR Survey, the Oil and Gas Journal listed 72 new EOR projects planned for the U.S. Of these, 26 were listed as steam projects and six were indicated to be combustion projects. More recently, as of December 1980, 250 EOR projects had been submitted to the Economic Regulatory Administration [ERA] under the Tertiary Enhanced Recovery Program (11). Of these, 97 or almost 40% were thermal projects. Under the categories defined by the ERA, 53 were conventional steam drive, 17 were unconventional steam drive and 27 were in situ combustion [Table 1].

TERTIARY ENHANCED RECOVERY PPROGRAM

Thermal Projects by State

Type	Total	CA	TX	LA	Others
Conventional Steam	53	49	1	2	1
Unconventional Steam	17	15	0	0	2
In Situ Combustion	27	10	9	3	5

TABLE 1

Of the 70 steam drive projects, 64 are to be conducted in California with only four other states involved. Conversely, combustion projects are proposed for seven states with California and Texas sharing two-third's of the proposed projects.

The number of projects proposed under the Tertiary Enhanced Recovery Program would indicate an accelerating interest in EOR. Whether this interest will be maintained when the special financial incentives of the program disappear remains to be seen. Equally uncertain is the impact of the Windfall Profits Tax [WPT].

Under the WPT, oil subject to the tier 3 tax rate includes "newly discovered oil," "heavy oil" and "incremental tertiary oil." Tier 3 oil is subject to a 30% tax on the difference between the removal price and the inflation-adjusted base price of $16.55. Under the WPT, "heavy oil" means all crude oil produced from a property that has a weighted gravity of 16° API or less. "Incremental tertiary oil" is defined as the excess crude oil that is produced during the period for which a qualified tertiary recovery project is in effect over the base level of production for the property. These provisions should provide some incentive for thermal recovery processes relative to tier 1 crude oil categories; their impact of EOR is difficult to forecast.

C. THERMAL PROJECTS - WORLDWIDE

In addition to the U.S., Venezuela and Canada are very active in thermal EOR projects. The USSR may also have a major program in thermally aided oil production (12). In a variety of other oil producing countries, thermal recovery methods are being tested, but they do not constitute major contributors to oil production totals.

The 1980 Oil and Gas Journal EOR survey reported 39 EOR projects currently in Venezuela, all thermal, the majority being cyclic steam, with a production of 154,400 BOPD. Cyclic steam is considered a standard production technique while steam drive is in the pilot stage only. Combustion projects have been presented as having diversified results and generally unattractive (13). Cyclic steam has proven to be a long producing recovery technique since subsidence effects, upon production, have been quite effective in maintaining reservoir pressure.

The 1980 survey lists 30 thermal projects for Canada equally divided between steam and combustion methods. Resulting production is indicated to be 16,000 BOPD. The Canadians are making extensive use of wet combustion, as 10 out of 16 combustion projects employ this method.

It has been reported that steam injection is adding up to 200,000 BOPD to the oil production of USSR (11). While this magnitude of production is unconfirmed, Dafter (8) presents an impressive list of Soviet activities in thermal EOR.

The 1980 EOR survey also lists two thermal projects in France and one in the Congo as representing activity in the Eastern Hemisphere. Dafter's survey indicated a slightly broader range of activities with projects also being conducted in Germany, Holland, and Indonesia.

Table 2 summarizes the active thermal projects worldwide for which reasonably firm data exists. The thermally aided production of about 500,000 BOPD is about 50% greater than that attributed to miscible recovery methods; both recovery approaches dwarfing that attributable to chemical methods (8).

ACTIVE THERMAL RECOVERY PROJECTS

TYPE	LOCATION	NO. OF PROJECTS	ENHANCED OIL PRODUCTION (MBOD)
Combustion	U.S.A.	29	10.3 - 19.1
(Wet & Dry)	Canada	14	4.7 - 5.8
	Venezuela	3	.3
Total		46	15.3 25.2
Steam Soak	U.S.A.	56	97.9 - 117.2
	Canada	13	7.6 - 7.8
	Venezuela	33	138.7
	Others	3	22.0 - 22.3
Total		105	266.2 286.0
Steam Drive	U.S.A.	58	143.4 - 160.5
	Canada	5	3.5 - 9.4
	Venezuela	3	21.9
	Others	6	7.4 - 15.8
Total		72	176.2 - 207.6
Hot Water	Ven. & Trin.	3	2.9 - 3.4
Other Thermal	U.S.A.	2	-- --
Thermal Total		228	460.6- 522.2

TABLE 2

POTENTIAL IMPACT OF THERMAL METHODS ON OIL PRODUCTION

A. GENERAL FACTORS

It is difficult to make a quantitative assessment of the impact that enhanced oil recovery methods in general and thermal recovery methods in particular will have on domestic as well as worldwide oil production. A myriad of factors will influence this outcome. Included among these are: demand - how effective will conservation efforts be in reducing consumption?; supply - at what rate will OPEC and other large oil producing countries continue to develop their resources?; price - at what rate will worldwide crude oil prices increase and how effective will these increases act to spur developments.....in conservation methods, in bringing new hydrocarbon resources on line, and in new technologies?; unconventional oil sources - will Canada, the U.S.A. and Venezuela act in a concerted fashion to develop their resources of heavy oil, tar [oil] sands and oil shale?; alternative energy sources - at what speed can these technically be developed and economically brought on line?; and, technology - can the improvement and application of new technology significantly expand the production of known hydrocarbon resources.

It is reported that worldwide EOR is almost 2% of total oil production, with thermal production in the majority (8). Contrary to the U.S., where a number of studies have attempted to predict the impact of EOR, it does not appear that any study has been conducted on how new production techniques could affect the world's recoverable reserves. Yet current world reserves are estimated to be 650 billion barrels and with average oil recovery efficiencies on the order of 35% [through primary and pressure maintenance modes], it is clear that an increase in recovery efficiency through the application of enhanced recovery techniques could greatly increase crude oil reserves.

The potential for unconventional oil resources may be even greater. World estimates of the heavy oil and tar sands resources total 3,000 to 5,000 billion barrels, to which only modest recovery efficiencies [2-10%] are being achieved. Equally impressive is oil shale. Estimates of this resource worldwide are 3,000 billion to 4,000 billion barrels, the majority being in the U.S. For both resources, the application of heat has been shown to be an effective recovery method.

The potential impact of EOR, particularly thermal methods on domestic as well as worldwide production could be truly significant. Whether it will become so remains to be seen.

B. U.S. PRODUCTION POTENTIAL

Of the 300 billion barrels of crude oil resources known to exist in the U.S., about 100 billion barrels are heavy oil, < 25° API, the most likely target for thermal recovery methods. The National Petroleum Council, the Office of Technology Assessment and the Department of Energy have all made estimates on the impact of thermal recovery methods on domestic oil production. It appears that an addition of 10 billion barrels to proven reserves through price incentives and technology improvements is a reasonable consensus.

Geographically, heavy oil reservoirs are found in many states. The current emphasis on the use of thermal recovery methods in California is one challenge facing application of thermal methods today; how to broaden the usage of thermal methods into heavy oil reservoirs in other states.

As previously stated, current steam drive methods have an economic depth limitation; this presently appears to be 2500 feet. However a significant amount of the heavy oil resources, about half, lie deeper than 3000 feet. A subsequent challenge in thermal methods is to extend the economic depth of steam drive technology or broaden the applicability of in situ combustion.

Even where suitable reservoir conditions may exist, environmental regulations are playing an increasing role in limiting the growth in application of thermal methods. A continued growth for thermal recovery must concurrently address the pollution issues which accompany the process.

While the detailed plans of individual companies are not known with respect to forecasting trends, specific actions do provide insight for the future. One recent development is of significance. In late 1979, the Shell Oil Co. purchased the Belridge Oil Co. for $3.6 billion. With this acquisition, Shell increased its U.S. reserves by 44% and its production. The reason - the approximately 3 billion barrels of oil believed to be in place in the San Joaquin Valley. Shell has now upgraded its estimates of proved oil reserves to 598 million barrels of hydrocarbon liquids. Of these reserves, 474 million barrels are in the Tulare formation, which Shell plans on exploiting via steam injection. Shell plans on increasing production from the current 36,000 BOPD to 77,000 BOPD by 1987.

While this project points out the massive capital expenditures involved to obtain a relatively small production increase, it certainly indicates the confidence that certain producers have in steam injection technology. If followed by others, the U.S. will see a significant increase in thermally aided oil production.

TECHNOLOGY TRENDS IN THERMAL METHODS

There are a variety of developments underway which could advance the usage of thermal recovery methods. Most are related to steam drive technology. These developments are in the areas of: improving the efficiency in application of surface steam generation; diagnostics for reducing process uncertainties; and new approaches to thermal recovery methods.

A. SURFACE STEAM GENERATION

1. Alternate Fuels

Currently, steam drive processes require about one barrel of crude consumed for every three barrels produced. Thermal losses in steam delivery put an economic limit on the reservoir depth which can be exploited. Gravity override leads to early steam breakthrough and lower-than-desired sweep efficiency in the reservoir. These areas, the major limitations in steam drive efficiency, are, not too surprisingly, the ones of greatest interest in seeking technological improvements.

Alternate fuels or improved efficiency in fuel combustion, is receiving increased attention. Concepts under investigation range all the way from cogeneration to solar-generated steam. Cogeneration is a natural with the steam drive process; where the waste heat from electricity generation is used to create steam. The State of California is actively encouraging this approach and collaborative studies between electric utilities and oil producers are underway.

Coal as an alternate fuel is attractive as a cheaper fuel, freeing up more crude oil, and providing some environmental control advantages. The lack of coal in California and hence the need to import are countering disadvantagees. At any rate, fluidized bed combustion [FBC] as a new approach to steam generation may be the most viable approach. FBC, with the addition of a limestone sorbent, can control sulfur oxide emissions, thus eliminating the need for scrubbing equipment. FBC may also have some advantages in size - the units could be portable; it should be possible to use untreated water; and this approach may achieve a reduction in the emission of oxides of nitrogen. Prototype fluidized bed combustors for oil field applications are currently under development.

Coal gasification, agricultural wastes and solar energy are also receiving study and some development as alternate fuel sources. Coal gasifiers could be coupled with existing steam generators and may have environmental emission advantages; this approach however is feasible only when applied to large scale operations. Waste products would probably employ FBC technology and are attractive because of cost and low sulfur content; availability of adequate quantities and in the right location are major disadvantages. Solar-aided EOR has received limited study, but with Exxon's recent abandonment of a trough-type collector system, the future appears limited.

2. Thermally Efficient Well Completions

Thermally efficient well completion is an active area as a means of increasing the efficiency of steam delivery systems. An uninsulated injection string, in addition to subjecting the well casing to high thermal stresses, can result in a degradation of 80% quality steam at the surface to on the order of 50% quality steam at 2500 foot depths. Hence for reservoirs in the depth range of 1500 to 3000 feet, it appears economical to consider insulated injection strings. Calcium and sodium silicate have been employed extensively in Venezuela, which domestic use appears on the upsurge. New products with improved thermal efficiencies are now entering the market place.

3. Steam with Additives

Gravity override and channeling decreases the sweep efficiency of steam drives. The industry has sought a blocking agent that could be emplaced in steam-swept zones to force the steam down into unaffected parts of the reservoir. Some of the most recent efforts have been sponsored by the DOE and one field project, that conducted by CORCO/Petro-Lewis, is showing preliminary signs of success. A steam-foam-polymer combination is showing relative long-term stability at steam temperatures apparently restricting steam flow in steam channels leading to increased oil production. Successful development of this technology could significantly increase the recovery efficiency of steam drive.

B. DIAGNOSTIC METHODS FOR PROCESS CONTROL

One problem in the application of thermal recovery methods is the inability to discern physical effects occurring in situ thus restricting the ability to control the process. Some application is normally made of monitoring/diagnostic wells primarily for temperature instrumentation but cost is a limiting factor. Recently some remote sensing techniques have been applied to thermal recovery processes with some indication they can at least discern the thermal front and hence the extent of the thermal process.

Tiltmeters which in effect measure the displacement of the earth's surface as a function of an internal volumetric change, have been employed in several feasibility experiments.

One experiment, done in the Canadian oil sands, measured the response of the reservoir to a steam drive. The measurements indicated a pulsation effect, with the formation inflating, apparently as a result of steam buildup, followed by a rapid formation deflation with the cycle then repeating. This indicates steam drive may not be a smooth continuous process.

Accoustic profiling techniques have also been applied to in situ combustion processes to locate the combustion front.

Electric and electromagnetic techniques are also under investigation. One of the most promising is the controlled source audio-frequency magnetotelluric [CSAMT] technique which is an electromagnetic induction prospecting method which measures electrical and magnetic fields at audio frequencies. These quantities are then used to determine apparent resistivities of subsurface features. The technique can distinguish resistivity contrasts that exist in the affected-process area and surrounding strata. Sandia National Laboratories has used this technique to map a steam-drive in a tar sands experiment and to map an in situ combustion [fireflood] in a shallow, heavy reservoir. Figure 4, six months after ignition of this latter test, presents isoresistivity contours for 128 Hz; the low resistivity areas are representative of heated ground waters, absence of oil and hot carbonaceous material.

Obviously these techniques are in an exploratory phase, but their potential to aid in process control makes them worthy of additional investigation.

C. <u>NEW APPROACHES IN THERMAL RECOVERY PROCESSES</u>

While revolutionary changes in production processes seldom occur, there are innovative approaches in thermal recovery methods that have outstanding potential. The first of these is the combination of mining with thermal recovery and the second is the development of the downhole steam generator.

The use of mining techniques as an oil recovery method, primarily the use of gravity drainage, has been explored in a few locations [e.g., USSR, Getty in California and Conoco in Wyoming]. An interesting twist to this approach is the combination of mining with application of heat. Underway in the Kern River field in California is an experiment to evaluate the feasibility of this approach. The extent of the production potential of this approach is difficult to define, but it is an interesting approach to put the heat where it can be most effectively used, at the bottom of the reservoir where gravity, with respect to steam rise and oil drainage, works in the right direction.

The downhole steam generator has the potential to significantly expand the use of steam drive by increasing the economic viability of deep reservoirs, reservoirs reached via long, directionally drilled wells, and reservoirs in environmentally restricted areas. While industry has experimented with a variety of methods for generating heat and, in some cases, steam at the bottom of a wellbore, the recent resurgence of interest in downhole steam generators derives from Project DEEP STEAM, a project funded by the Department of Energy and conducted by Sandia National Laboratories. This project has successfully developed low pressure, indirect contact generators where only steam is injected into the reservoir and high pressure, direct contact generators where both steam and combustion gases are injected into the reservoirs. Oxygen-fired and air-fired units have been developed and successfully tested on a variety of fuels, LPG, natural gas, and #2 fuel oil, with current efforts focusing on lower volatile fuels. In 1980, Sandia with the cooperation of Chevron successfully tested a high pressure device, in a surface configuration, for four months at a shallow, Kern River Field reservoir. Currently, tests are about to be initiated with the City of Long Beach in the Tar Zone of the Wilmington Field, a reservoir at a depth of 2300 feet. Depending on the response of the reservoir to the combination of steam and combustion the economics appear very favorable.

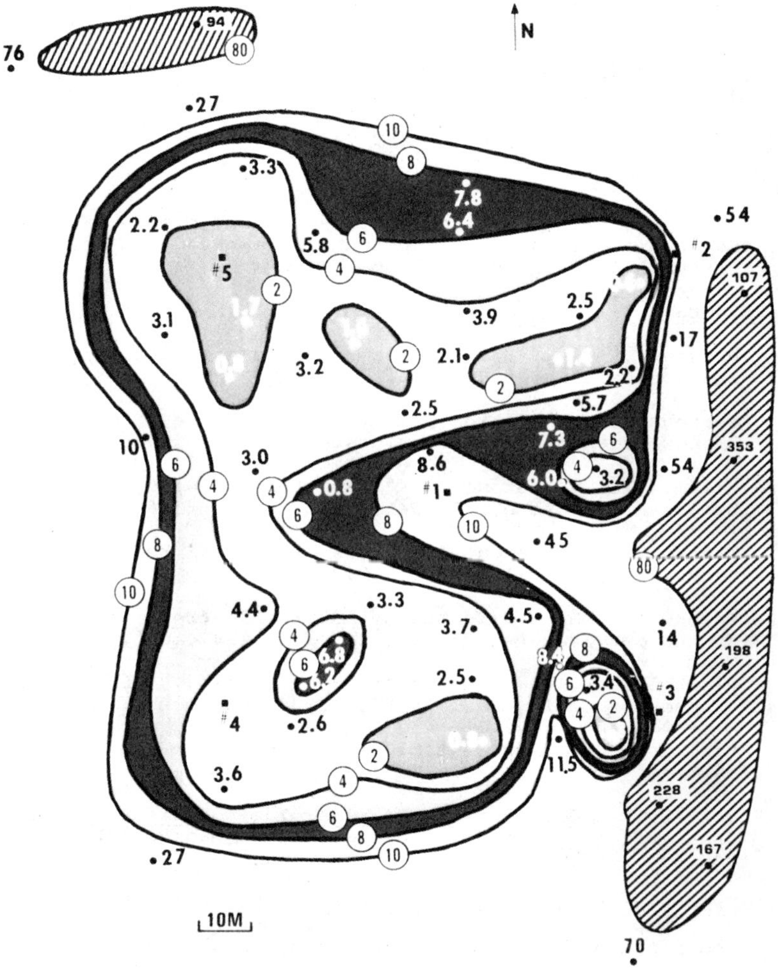

CSAMT RESISTIVITY MEASUREMENTS
AT BETC "IN-SITU" COMBUSTION EXPERIMENT, 128Hz

FIGURE 4

SUMMARY

A variety of opinions exist on the future of thermal oil recovery processes ranging from a pessimistic view that they are a diminishing factor because the best reservoirs for thermal floods have already been exploited to the optimistic based on the availability of resources, the gaining of experience, the increasing in prices and the development of new technologies. One fact does exist - the US and the world have tremendous resources of heavy oil for which thermal methods may be the only recovery process. The need exists to develop this capability; and if successful, the impact on availability of hydrocarbons will be great.

ACKNOWLEDGEMENTS

Because of space limitations, many sources were employed in the preparation of this paper but not referenced. A complete list of references will be supplied upon request.

REFERENCES

1. Matheny, S. L., "EOR Methods Help Ultimate Recovery," Oil and Gas Journal, Mar. 31, 1980.

2. Personal communication, S. L. Matheny, Production Editor, Acknowledgement of corrections to Mar. 31, 1980,

3. Willman, B. T., et al. , "Laboratory Studies of Oil Recovery by Steam Injection," J. Pet. Tech., 23, 681-690, July 1961.

4. Prats, M., "A Current Appraisal of Thermal Recovery," J. Pet. Tech., 30, 1129-1136, Aug. 1978.

5. Farouq Ali, S. M. and Meldau, R. F., "Current Steamflood Technology," J. Pet. Tech., 32, 1332-1342, Oct. 1979.

6. Blackwell, R. J., "Status of EOR Technology," Seventh Energy Technology Conf., Mar. 24-26, 1980, Washington, DC.

7. Stalkup, F. I., "Potential of Enhanced Oil Recovery Applications, Ibid.

8. Dafter, R., Scraping the Barrel - The Worldwide Potential for Enhanced Oil Recovery, Financial Times Management Reports, London, 1980.

9. Symposium on Technology of Enhanced Oil Recovery in the Year 2000, sponsored by Div. of Fossil Fuel Extraction/DOE and Lewin and Associates, Williamsburg, Va., June 1979.

10. Faroug Ali, S. M., "A Current Appraisal of In Situ Combustion Field Tests," J. Pet. Tech., 24, 477-486, Apr. 1972.

11. List of EOR Projects supplied by Bartlesville Energy Technology Center/DOE.

12. Oil and Gas Journal, p. 64, Mar. 26, 1979.

13. Presentation by Evanan Romero, INTEVEP, to U.S./Venezuela EOR Working Group, Bartlesville Energy Technology Center, March 1980.

PHOSPHORIC ACID FUEL CELL POWER PLANT

J.J. Buggy, B.R. Krasicki and B.L. Pierce
Westinghouse Electric Corporation
Advanced Energy Systems Division
Pittsburgh, Pennsylvania

INTRODUCTION

For the past twenty years, Westinghouse Electric Corporation has been working on various fuel cell technologies. These include first generation phosphoric acid, second generation molten carbonate and third generation solid oxide fuel cells. Studies have shown that the phosphoric acid fuel cell is near-term and can become a viable power source by the mid-eighties for both utility on-site power generation and commercial co-generation. In particular, the gas-cooled phosphoric acid fuel cell has a high efficiency and a potential for high reliability while remaining cost effective. It is simple to operate and maintain and is suitable for a range of fuels. Further, since fuel cells can be designed to be modular and are virtually pollution free, they can be sited adjacent to residential, commercial or industrial areas. All of these characteristics provide a firm basis for the market acceptance of phosphoric acid fuel cells.

The fuel cell selected for development by Westinghouse is the Distributed Gas Cooled (DIGAS) concept originated by Energy Research Corporation (ERC). Westinghouse and ERC have formal working agreements whereby Westinghouse has full access to ERC technology. Westinghouse is engaged in the design and development of this basic fuel cell to be used in residential and commercial on-site integrated energy systems (OS/IES), utility all-electric power plants and industrial/utility cogeneration applications.

Cogeneration is a logical application of the phosphoric acid fuel cell. In such applications thermal and electrical requirements are met by utilizing the waste heat produced by the fuel cell. Studies have shown that the overall system efficiency of the gas-cooled fuel cell plant can approach 90 percent, depending upon the application. Westinghouse is convinced that phosphoric acid fuel cell power plants can contribute to the solution of the U.S. energy problems. The hydrogen-rich fuel that feeds the cell can be processed from almost any organic-based chemical, including biomass fuels such as digester gas from anaerobic digestion of raw sewage. Other sources of hydrogen gas, such as electrochemical hydrogen separators, etc., can significantly reduce the fuel conditioning costs for the overall plant.

TECHNOLOGY

A fuel cell, like a battery, converts chemical energy directly and efficiently into direct current electricity. Batteries store a fixed amount of energy; in contrast, fuel cells do not store energy but are continuously supplied with fuel and oxidant from outside the cell. Like a battery, a fuel cell is composed of an anode, a cathode and an electrolyte. In the phosphoric acid fuel cell, hydrogen is fed to the anode and oxygen is fed to the cathode. When the fuel and oxidant come into contact with the phosphoric acid electrolyte, a chemical reaction occurs:

$$2H_2 + O_2 \longrightarrow \text{DC power + Heat} + 2H_2O$$

This conversion is efficient (40-46 percent) because it involves an electrochemical reaction rather than a thermal cycle.

The phosphoric acid fuel cell (PAFC) power plant is comprised of three major systems: The Fuel Processing System (FPS), the Fuel Cell DC System (FCS) and the Power Conditioning System (PCS). The function of the FPS is to produce hydrogen for the phosphoric acid fuel cells from the intended source fuel. Basically, the FPS steam reforms the source fuel to produce H_2, CO_2, and CO; then shift converts the CO to produce more H_2 and CO_2. The purity of the input H_2 stream to the fuel cells is not critical with regard to water and CO_2 content. But, phosphoric acid fuel cells are somewhat intolerant of CO and it must therefore be limited to a low CO percentage. Impurities in the source fuel, particularly sulfur, can have a deleterious impact on the catalysts used in the steam reformer and they must be considered in the FPS design. The FPS will typically consist of the following components:

- Cleanup
- Catalytic Steam Reformer
- Combustor
- Steam Generator
- Shift Converters
- Air and Exhaust Gas Circulators
- Heat Exchanger for source fuel heating, shift cooling, fuel cell anode gas cooling, and exhaust gas energy recovery
- Piping, Valves, and Instrumentation and Controls

Candidate source fuels include hydrogen, methanol, ethanol, methane (natural gas), light coal liquids, synthesis gas (i.e., coal gas, SNG), naphtha and biomass fuels. The nature of the source fuel or fuels used for the PAFC Power Plant will influence the size of the above components and the FPS configuration. The integration of the FPS with the fuel cell system will also impact the FPS function for alternate or multiple fuel capability at a later time. Higher efficiencies can be obtained with integration but at the expense of complexity and higher costs.

The overall function of the FCS is to process the hydrogen-rich fuel into direct electrical current for the PCS. The FCS consists of fuel cell assemblies, a low pressure boiler subsystem and recirculation ducts.

The commercial design of the fuel cell assembly features the DIGAS fuel cell concept for power generation. Figure 1 illustrates the DIGAS stack which uses the same air for cooling and for feeding oxygen to the cathode side of the cell. The fuel cell reject heat from the DIGAS stacks can be used to drive auxiliary rotating equipment within the fuel cell power plant. For a cogeneration application, the reject heat can also be used in an industrial facility.

In Figure 1 a key component of a fuel cell stack is the bipolar plate, an imperious matrix. The plate serves four primary functions: it separates the chemical reactants in adjacent cells, it conducts electrical current, it directs fuel to one cell and oxidant to the next, and it removes waste heat generated within the cell.

Other components of the fuel cell stack include the cooling plates which allow for additional heat removal from the cell. The fuel cell which produces dc power generates an appreciable amount of heat. The same air stream which supplies oxygen to the cathode can be used for the cooling channels. The process is referred to as DIGAS (distributed gas) cooling and represents a major step in simplified fuel cell design. The DIGAS approach significantly reduces demands for water treatment and manifolding connections, and permits coolant pressure to be independent of temperature.

Four stacks are combined to form a complete assembly (see Figure 2). The DIGAS concept permits the cathode gas to bathe the entire stack and be exhausted downward in the center section of the assembly. In this configuration, no separate cooling manifolds are necessary. Depending on the actual design, a fully assembled module would produce approximately 350-400kW_e (dc). This approach to module design makes fuel cell systems extremely flexible. Adding capacity is accomplished simply by adding more assemblies to the total array. Similarly, individual assemblies can be replaced easily without disturbing the entire system. Given the rapdily changing environment in which utilities find themselves, this feature should be attractive.

The next major system for a PAFC power plant is the Power Conditioning System (PCS). In an all-electric utility application, the PCS provides the means for conditioning the dc power from the fuel cells to ac power of a quality acceptable to the utility distribution system. The electrical conversion system or PCS will require adaptation to the needs of the user. Every system, whether dc or ac, would require some power conditioning to control such factors as ripple, phase and harmonics.

ELECTRIC UTILITY POWER PLANT

Conceptual systems studies conducted on both all-electric and cogeneration applications and the status of the fuel cell technology indicate that the fuel cell systems can be on-line within this decade. One system which has been examined in some detail is a 7.5MW dc all-electric system (Figure 3). Guidelines and requirements which governed the design were selected to assure the most appropriate system for a commercial utility prototype. A modular design was sought to prove the flexible capacity of fuel cell systems. A highly automated system was desired to reduce manpower requirements. Because fuel cell power plants appear to be excellent candidates for siting adjacent to rapidly increasing load areas, a relatively small (5 to 10MW_e) system capable of being sited in a utility substation was required.

The conceptual design described is for an all-electric application using methanol as fuel. Relatively minor modifications are necessary to accommodate a cogeneration application should that selection be made by the utility or industrial customer.

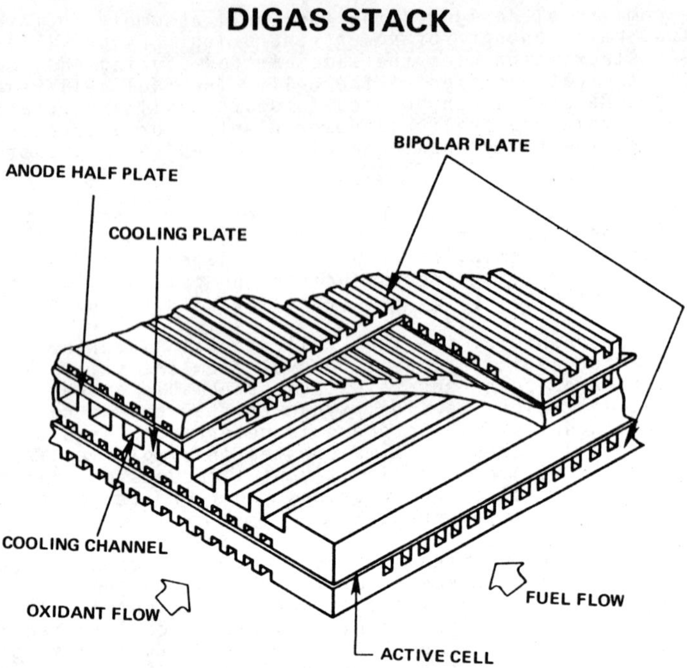

Figure 1. DIGAS Stack

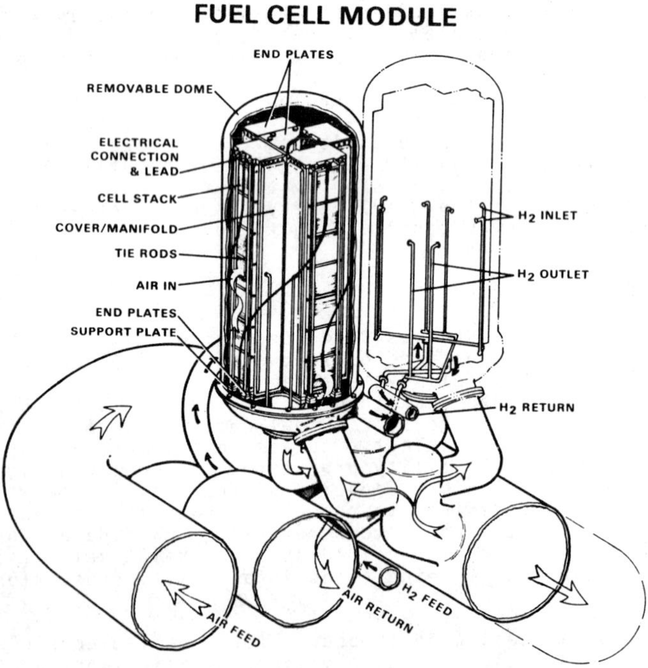

Figure 2. Fuel Cell Module

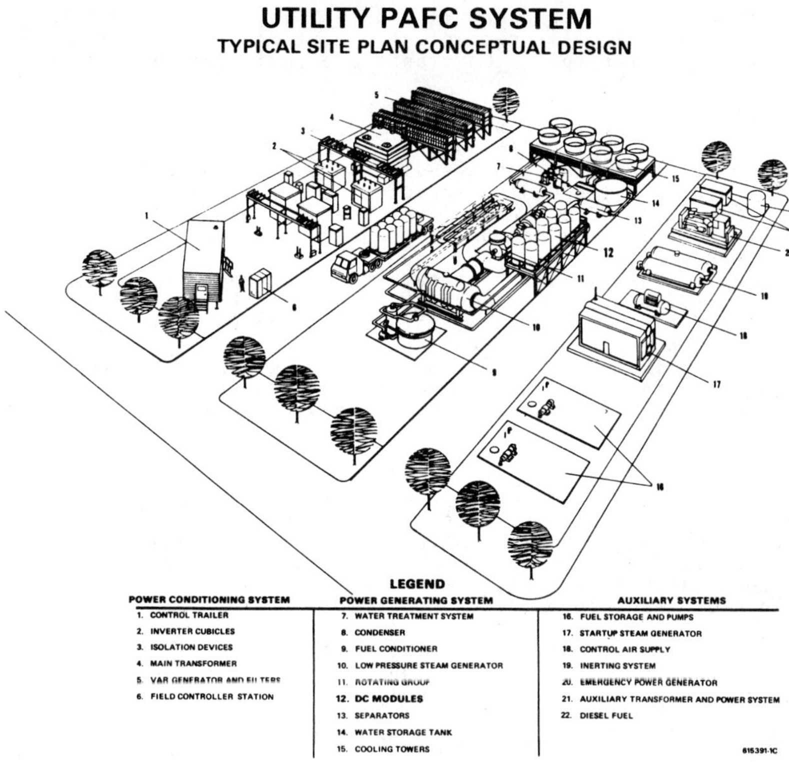

Figure 3

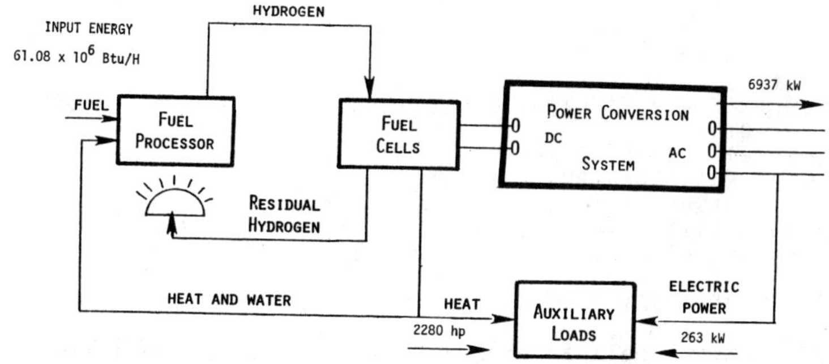

$$\frac{61.08 \times 10^6 \text{ Btu/H}}{6937 \text{ kW}} = 8800 \text{ Btu/kWH}$$

Figure 4. Fuel Cell Power Plant Block Diagram

The control trailer shown in Figure 3 is specific to the commercial prototype fuel cell power plant. Features for use of a central control facility for follow-up plants are envisioned in the future.

For the 7.5MW dc power plant, approximately two megawatts of auxiliary electric power can be saved by using the fuel cell reject energy in the power plant (Figure 4). The amount of useful process heat to be recovered (and the resulting reduction in electrical output) depends on the desired process heat temperature. For example, more than 85 percent of the energy required to power the auxiliary loads is provided by waste heat from the fuel cell. Unreacted hydrogen is used to provide energy for the Fuel Processing System (FPS). The overall heat rates for the PAFC plant range between 7300-9300 depending upon the degree of FPS integration and the type of fuel. The low heat rate makes the PAFC plant a candidate for base load operation. Partial power operation is slightly more efficient than full power; this makes fuel cell plants excellent candidates for intermediate or load follow operations as well. Water recovery systems and dry cooling towers would eliminate the need for a supply of water.

The commercial prototype conceptual design was based on consideration of available components, simplicity in design, methanol fuel and electrical requirements. Available components were considered whenever possible which include off-the-shelf equipment such as steam turbines and other equipment that required only minor development. Some major system equipment such as the fuel cell assemblies will, however, require development effort.

Table 1 presents the commercial prototype plant characteristics. The data presented here are the results for the 7.5MW dc all-electric fuel cell power plant using methanol as the fuel.

TABLE 1: PLANT CHARACTERISTICS

Gross Electrical DC Output (kW)	7500
Gross Electrical AC Output (kW)	7200
Parasitic Losses (kW)	
Pumps	23
Fans (Cooling Tower)	210
Blowers (Fuel Processing)	11
Vacuum Pump	3
Air Compressor & Dryer	1
Controls	15
Net Electrical AC Output (kW)	6937
Input Energy (HHV=9758 Btu/lb) Btu/hr	$61.08 \cdot 10^6$
Overall Plant Efficiency	0.39
Heat Rate (Btu/kW-hr)	8800
Design Current Density (amp/ft^2)	300
Fuel Utilization Factor	0.8
Design Cell Voltage (volts/cell)	0.66

COGENERATION APPLICATION

Cogeneration applications of the PAFC system are attractive. In such applications, thermal demands can be met by utilizing the waste heat produced in the fuel cell. Because fuel cell systems are modular and virtually pollution-free, they can be sited adjacent to a thermal load without compromising the environment. Studies have shown that overall system efficiency of a gas-cooled fuel cell can approach 90 percent, depending upon the application.

The generic design for a cogeneration facility builds on the basic fuel cell system described earlier. The primary variation is that the waste heat exhausted from the fuel cell assembly is captured and used to meet thermal load demands. Another factor to be considered is the type of power, ac or dc, required by the load.

In the earlier discussion of the all-electric fuel cell system, it was stated that the plant layout design lends itself to cogeneration. The compact power generating system plus a minimal amount of power conditioning equipment would suffice in cases where dc power is needed. The steam used to drive auxiliaries in the all-electric application can simply be re-routed to a process heat load as required by a cogeneration application. Even given this commonality, every cogeneration application would require a site-specific evaluation because requirements, including process temperature and the ratio of electric-to-thermal power, vary.

Recently, several cogeneration studies have been conducted by Westinghouse. These include:

- Water pollution control facility - using the waste digester gas, the PAFC provides electricity for pumping equipment and heat for digester tanks, grease melting and space heating.

- Chlorine plant - using the relatively pure hydrogen by-product gas, the PAFC provides electricity to a chlor-alkali electrolytic reduction cell and heat for the caustic evaporator.

- Aluminum industry - the system provides electricity for aluminum oxide reduction and heat for bauxite refining.

- On-site integrated energy system (OS/IES) - this PAFC can provide electrical and thermal energy for commercial applications such as apartments and hospitals.

For the sewage treatment facility, the objective was to utilize the methane gas generated in the treatment process to provide as much usable power as possible. Thermal loads were required in the digester tanks, in the grease melting operation, and for space heating. A system was designed which provided 100 percent of the heat needed throughout the year. In addition, the system generated 6.9MW_e ac which was used to power process equipment including pumps. Varying PAFC operating conditions were examined to maximize the electrical vs. thermal trade-off.

Another cogeneration application, for a chlorine plant, was governed not by process needs, but by available fuel. In that case, a relatively pure hydrogen by-product gas is used as the fuel cell feedstock. Because of the purity of the hydrogen, the electrical efficiency of the system was projected to be 46 percent, and the total output, 7.9MW_e dc. The required dc electricity was to be used in the process while 31 million Btu/hr of usable heat was to be dispatched to the thermal load. In this application, the overall efficiency was projected to be 89 percent.

Finally, studies were made using the fuel cell to provide power for commercial systems. In this case, heat is provided to power space heating and air conditioning systems and electricity is generated for on-site electrical demand. Such systems would be applicable for apartment complexes, shopping marts, or hospitals where high-temperature process heat is not necessary. A system using gas-cooled fuel cells to generate 120 KW ac and one-half million Btu/hr has undergone extensive examination. Overall efficiencies of 85 percent have been projected. As in the industrial cogeneration studies, this on-site integrated energy system appears to be extremely attractive.

Cogeneration always ensures a more efficient use of the input fuel. In the case of gas-cooled fuel cells, efficiency appears to be extremely high because the waste heat from the fuel cell requires very little processing. Because the reject heat would be in the form of uncontaminated air and steam, it could be piped safely to thermal loads. Thus, one simple design for an efficient all-electric fuel cell power plant could be transformed quite easily to an extremely efficient cogeneration application.

For the aluminum processing application, natural gas was available as a fuel. The key concerns with integration of a PAFC plant and an aluminum process included matching the fuel cell voltage/amperage with the aluminum processing circuit voltage/amperage. The application of the PAFC plant in a cogeneration mode gave a good heat rate of 8430 Btu/KW-hr LHV. The overall fuel efficiency was projected to be 85 percent.

CONCLUSIONS

The study revealed a number of new advantages for a DIGAS cooled power plant. These are as follows:

- Low heat rate at rated conditions - potential for base load plant.
- Low heat rate at part power - potential for intermediate and peaking operation.
- No significant environmental impact.
  - No vapor plume
  - No significant contaminants, such as NO_x, SO_2 and CO
  - No outside water source required.
  - Adaptable to co-generation applications with minimal design changes.
- Fuel flexibility.

STATUS OF MOLTEN CARBONATE FUEL CELL TECHNOLOGY

Leonard G. Marianowski and John B. O'Sullivan
Institute of Gas Technology
IIT Center
Chicago, Illinois, U.S.A.

ABSTRACT

A molten carbonate fuel cell integrated with a coal gasification power plant is one of the most promising coal-using technologies because of its high efficiency, acceptable cost, and environmental acceptability. The high-temperature heat available from the fuel cell may be used in a bottoming cycle and/or industrial cogeneration applications. However, for the molten carbonate system to achieve these goals, continued developmental work is required which must take account of the operating conditions of the application. This paper discusses the progress made in improving cell performance and life and in producing inexpensive cell components. Performance is enhanced by alternate electrolytes and electrode compositions and increased pressures; life by control of contaminants, electrode-electrolyte behavior, and electrolyte loss; and cost by multicell stack designs and new means of fabricating components, the electrolyte tile, and substrate materials. The status, direction, and priority of future research and engineering efforts will also be discussed.

INTRODUCTION

Molten carbonate fuel cell research has been under way in the U.S. and at IGT for over 20 years. Because the system has been aimed primarily at the industrial and utility markets, its development has lagged behind the phosphoric acid system (first generation) which benefited from significant technical advances made during the TARGET program(1) of the '60's and the DOE-EPRI-UTC program(2) of the '70's. Interest in the molten carbonate system, sometimes called a "second generation system," began to increase during the energy crisis of the mid 1970's because of several inherent and unique characteristics which make it attractive for larger system applications where cost, fuel flexibility, and efficiency are of prime concern:

- Cost — No noble metal electrode catalysts are required. The remaining cell components utilize readily available and cost-effective materials.

- Fuel Flexibility — Because of its elevated operating temperature any solid, liquid, or gaseous hydrocarbon can be used to form a primary feed containing H_2, CO, CO_2, and H_2O.

- Efficiency — ECAS(3) and many subsequent studies have shown total system efficiencies of 50% using coal as the hydrocarbon fuel and utilizing the available high quality fuel cell heat in a steam bottoming cycle.

A total fuel cell system contains three major subsystems — a fuel processing section, the fuel cell section, and a DC to AC inverter section. Such a system configuration for a multihundred-megawatt base-load plant is shown in Figure 1.

Recently, the range of potential application has been expanded to include much smaller gas- or oil-fired cogenerating systems wherein the high-temperature reject heat supplies all or part of the process heat requirements.(4) These systems obtain benefits not only from heat utilization but also from the fuel cell efficiency in power generation. This yields net energy savings of 35% to 50% for some major energy consuming industries (Table 1).

Table 1. EFFECT OF COGENERATION ON ENERGY SAVINGS WITH MOLTEN CARBONATE FUEL CELL

Process	Total Energy Savings	Energy Savings Due to High Fuel Cell Efficiency	Energy Savings Due to Cogeneration
Chlorine/Caustic	49.0	19.0	30.0
Newsprint	50.0	18.5	31.5
Magnesium	46.0	18.0	28.0
LDPE	36.0	16.5	19.5
Carbon-Steel Castings	34.0	14.0	20.0
Copper	45.5	12.0	33.5
Cement	8.0	3.0	5.0
Aluminum Fabrication	37.7	8.5	29.2

This paper will review the status and direction of present research at IGT and, where pertinent, efforts of General Electric Corp. (GE) with whom we are participating in a major, DOE-funded commercialization effort. The performance goals which are believed necessary to assure a substantial penetration in the central station market are given in Table 2. The goals for cogeneration systems, especially as regards thermal cycles and cost, have not been adequately defined. A summary of the present technology status is shown in Tables 3 and 4. Comparing the present status to the goals shown in Table 2 indicates that, although a great deal of progress has been made, additional developmental efforts are required to translate promising laboratory results to prototype units from which realistic engineering and cost data can be generated. In general, the performance (power output) of the molten carbonate fuel cell is adequate to meet system requirements. As a result, the development should center on improving the endurance and providing manufacturable cell components for the engineering evaluation of prototype multicell fuel cell stacks. As stated earlier, the molten carbonate materials costs are reasonably low. However, the manufacturing costs of the electrolyte tile and the cell separator plate are high for present

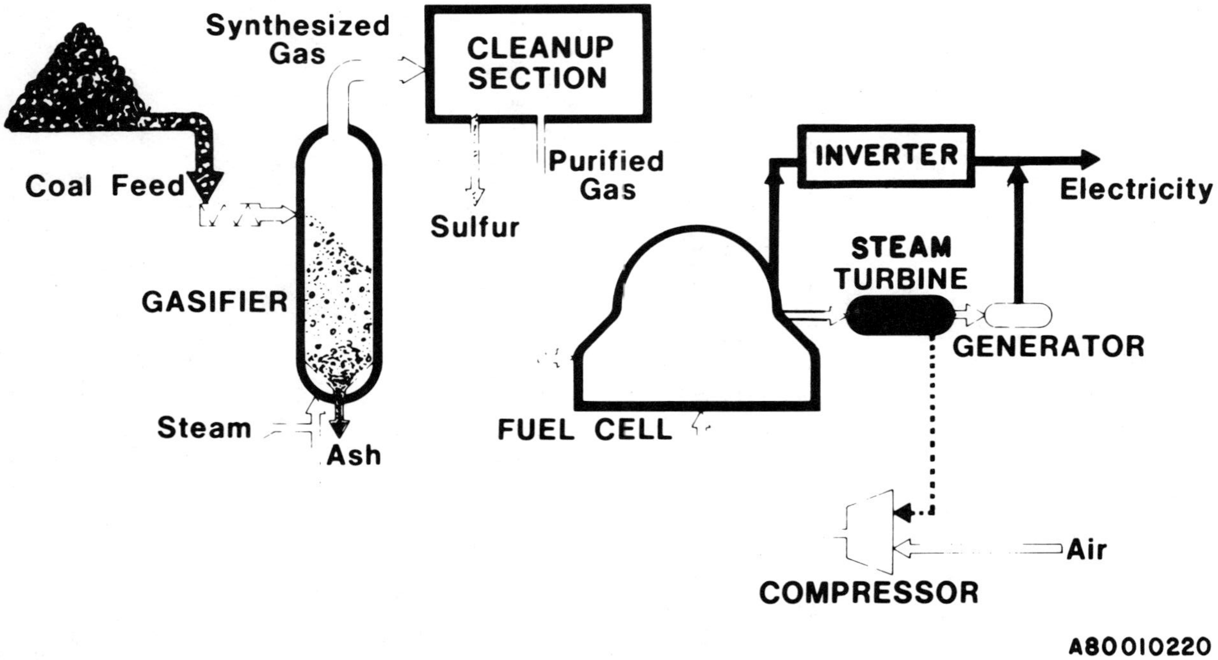

Figure 1. SCHEMATIC OF COAL-FUELED MOLTEN CARBONATE FUEL CELL SYSTEM

Table 2. TECHNOLOGY CATEGORIES AND GOALS

Performance	650 mV (End of Life) 160 mA/cm^2 650°C 10 atm 75% Fuel Utilization 50% Oxid. Utilization
Cost	$100/Stack kW (78$)
Life	40,000 hours (< 4mV/1000 hr) 20 plus thermal cycles
Scale-up	700 Cell Stack 2 ft by 2 ft or larger

Table 3. ENDURANCE TESTING

Configuration	Hours	Terminations	Decay, mV/1000 hrs
Boiler Plate	15,000	Wet Seal Failure	~3*
Boiler Plate	10,000	Voluntary	9
Boiler Plate	10,000	Voluntary	8
Sheet Metal	3,000	Gas Crossover	<1
Boiler Plate	3,700	Voluntary	12

* Periodic additions of carbonate.

Table 4. TECHNOLOGY STATUS SUMMARY 1959 TO 1980, LIFE LIMITATIONS

Item	Fix	10,000 to 15,000 hr Demonstration	40,000 hr Potential
Anode Sintering	Cr Stab.	Yes	Good
Carbonate Loss	Component Mod. Pressure	No	TBD
Wet Seal Corrosion	Aluminization	Yes	Excellent
Separator Plate Corrosion	Nickel or Nickel/SS	Yes	Excellent
Performance Decay	Carb Mgt. Stable Comp.	10 mV/1000	2.5 mV/1000 Desired
$LiAlO_2$ Stability	Doping & Pretreat	Yes	Good
Electrolyte Tile Cracking	Crack Arrestors Microstructural Improvements Bubble Barriers	No	TBD
Cathode Strength	Reinforcement Tape Casting Modified Thickness	No	TBD

cell configurations. In addition, the electrolyte tile, probably the most critical cell component, needs further material and configuration development to achieve the endurance, cost, and system operational goals.

EXPERIMENTAL HARDWARE

The initial experimental screening of materials and concepts has been done primarily in circular cells of 3 cm^2 area. These are termed "laboratory cells" and are extremely reproducible and provide the unambiguous data needed to establish the fundamental relationships. After components have been qualified in laboratory cells, they are tested in larger cells, 94 cm^2 area, to determine performance and endurance, effect of process variables such as gas composition, temperature, and pressure, and stability of metallic cell components. These are termed "boiler plate cells" and can test critical components in a more realistic manner. Our baseline performance, which closely approaches the goals noted in Table 2, has been established in this configuration. Presently, we are in the process of developing the data base and experience which will allow the baseline performance to be obtained in cost-effective cell configurations. This configuration, termed "sheet metal," uses components that are more realistic in terms of cost and mass producibility for gas distribution, sealing, and current collection. Two sizes of "sheet metal" cells have been designed. One is the same as the boiler plate cell, 94 cm^2, and another has been scaled to 929 cm^2 (1 ft^2) preparatory to construction of cell stacks. No experimental testing of these 929 cm^2 sheet metal cells has yet been performed.

ENDURANCE

The laboratory-scale cells have surpassed 40,000 hours of life, thus establishing that critical components can be designed to perform adequately. Three boiler plate cells, 94 cm^2, have attained in excess of 10,000 hours. Performance of two cells is shown in Figure 2. The average degradation rate was 10 mV/1000 hr. These rates are roughly double the goal established for the system. A large part of this degradation is believed to result from a gradual electrolyte loss through vaporization. The approaches being taken to reduce this loss will be described in a later section dealing with the electrolyte tile. Most of the endurance testing has been performed at 1 atmosphere pressure; however, boiler plate cells operated at 5 and 10 atmospheres have exceeded 1000 hours of operation. Reproducibility of baseline (i.e., boiler plate) performance in sheet metal hardware has been mostly successful as demonstrated by that shown in Figure 3.

Operation at elevated pressure has a number of positive attributes — reduced electrolyte loss by evaporation, reduction in system volumes and thus reduction in cost of materials, and enhanced performance. This performance increase, typically in excess of 80 mV, exceeds that of 45 mV predicted by thermodynamics. Most of this additional increase is attributed to reduced polarization and enhanced reactant solubility. However, there are also drawbacks; one is the necessity for accurate pressure control to prevent an excessive differential across the tile; another is the loss of reactants by the formation of inert methane. That reaction ($2CO + 2H_2 \rightarrow CO_2 + CH_4$) is strongly pressure-dependent. Fortunately, the kinetics are not too rapid and equilibrium is seldom attained. Of greater concern is the possible deposition of carbon leading to blocking of the fuel distribution which leads to blocking of operation of the cells. That reaction ($2CO \rightarrow C + CO_2$) is also favored by a pressure increase. Prevention requires close monitoring and control of the fuel stream composition.

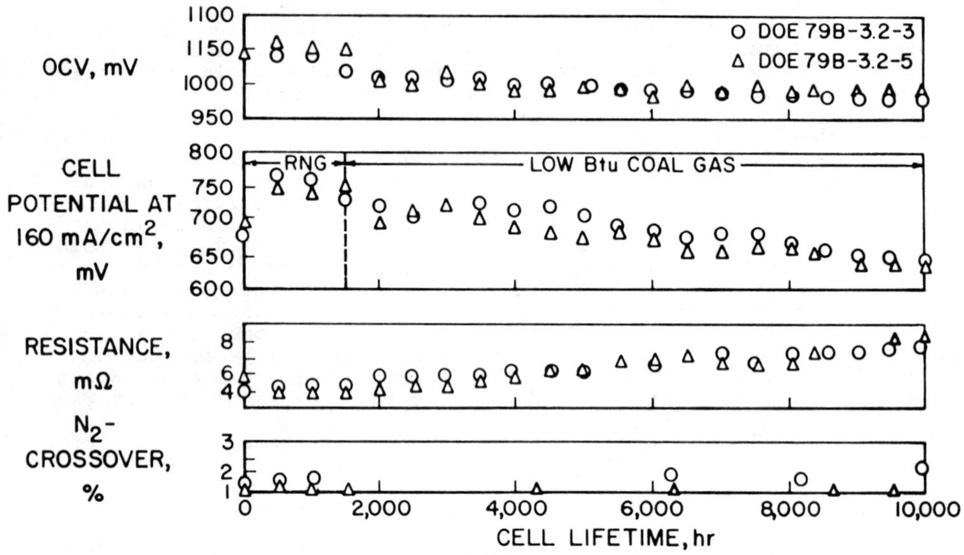

Figure 2. 10,000 HOUR ENDURANCE TESTING

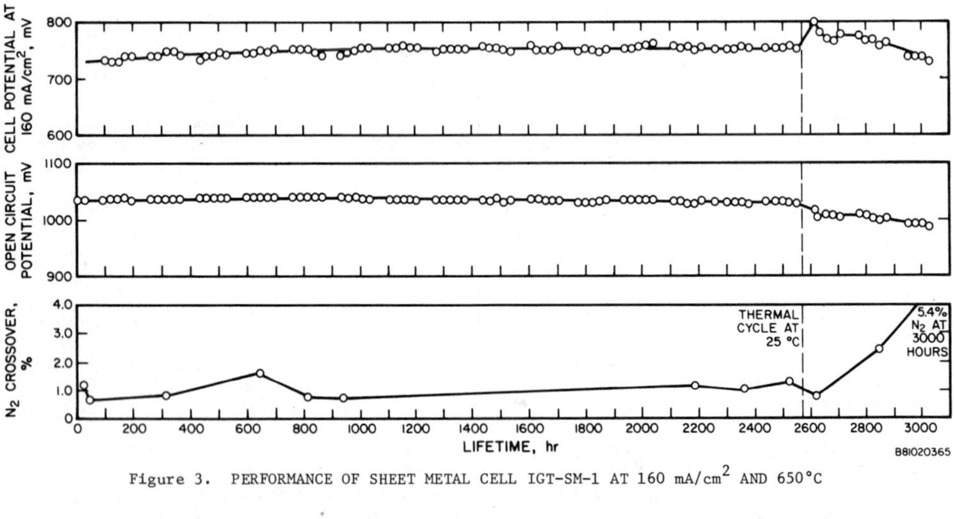

Figure 3. PERFORMANCE OF SHEET METAL CELL IGT-SM-1 AT 160 mA/cm^2 AND 650°C

FUEL COMPOSITION

The composition of the fuel stream entering the fuel cell depends upon the primary hydrocarbon fuel and the process selected for production of the hydrogen-rich stream. Steam reforming of methane yields a hydrogen-rich stream while an air-blown coal gasifier yields an inherently lower hydrogen stream further diluted by nitrogen. Fortunately, molten carbonate systems are not poisoned by carbon monoxide as are phosphoric acid cells. In fact, via the water gas shift production of hydrogen, carbon monoxide is consumed as fuel. However, the materials currently used in construction are extremely sensitive to the presence of sulfur-containing contaminants of the fuel, primarily H_2S and COS. Present emphasis is on reducing these contaminants to a tolerable level via a cleanup system, rather than using materials of construction that are tolerant of the untreated levels. Based on experimental results generated in an EPRI-funded program, it is believed the increase in materials cost would exceed that of gas cleaning. However, there is still a need to establish the long-term impurity level limits and the impact of short-term sulfur increases resulting from process upsets. Sulfur-containing contaminants primarily affect the anode performance, but if levels are not high and do not last long, the degradation observed is reversible once the level is reduced. This recovery is shown in Figure 5. Sulfur species not only cause harm when in the anode stream; if they are in the cathode stream, as SO_2, for instance, they react with the electrolyte, are transported through the electrolyte to the anode, where they are reduced, and behave as if originally present in that location. (See Figure 6.) Operation at an elevated temperature, say 700°C instead of 650°C, increases tolerance to H_2S and COS. This improvement in performance is shown in Figure 7. For COS, this improvement in anode tolerance was less than for H_2S at a comparable concentration. This could be used as a short-term control response to an upset, but the long-term disadvantages of increased corrosion, vapor loss, electrode sintering, and other effects are felt to outweigh continual operation above 650°C.

ELECTRODES

Based on experimental results, the endurance and performance of the chromium-stabilized nickel anode and the lithiated nickel oxide cathode are satisfactory for the present 40,000-hour goal. However, it is desirable to reduce the amount of nickel (one of the highest cost-per-pound materials used in the system) used for the anode and to improve the strength and structural integrity of the nickel oxide cathode. Efforts to reduce the amount of nickel are proceeding in two parallel paths — using less expensive alternative metals and reducing the amount of nickel by plating the nickel on inorganic structures. Initial efforts in both areas are showing promise. The cathode strength is being improved both by using alternative fabrication processes such as tape casting and by using metallic reinforcing.

ELECTROLYTE TILE

As stated earlier, the electrolyte tile is probably the most critical molten carbonate fuel cell component. It is a composite structure wherein a mixed alkali carbonate is contained by capillary forces within a finely divided inorganic matrix. This matrix material is presently $LiAlO_2$, though some effort is being expended by others on alternatives such as $SrTiO_3$. The chemical stability of $LiAlO_2$ is not in question; the physical stability is. Such a high surface material, in the range of 15-30 m^2/g, has the inherent tendency to sinter and reduce surface area. This increase in particle size, often accompanied by a change in phase from β to $\gamma+\alpha$, permits the electrolyte to have a greater and undesirable mobility. Methods of preparation yielding higher surface areas can result in an effec-

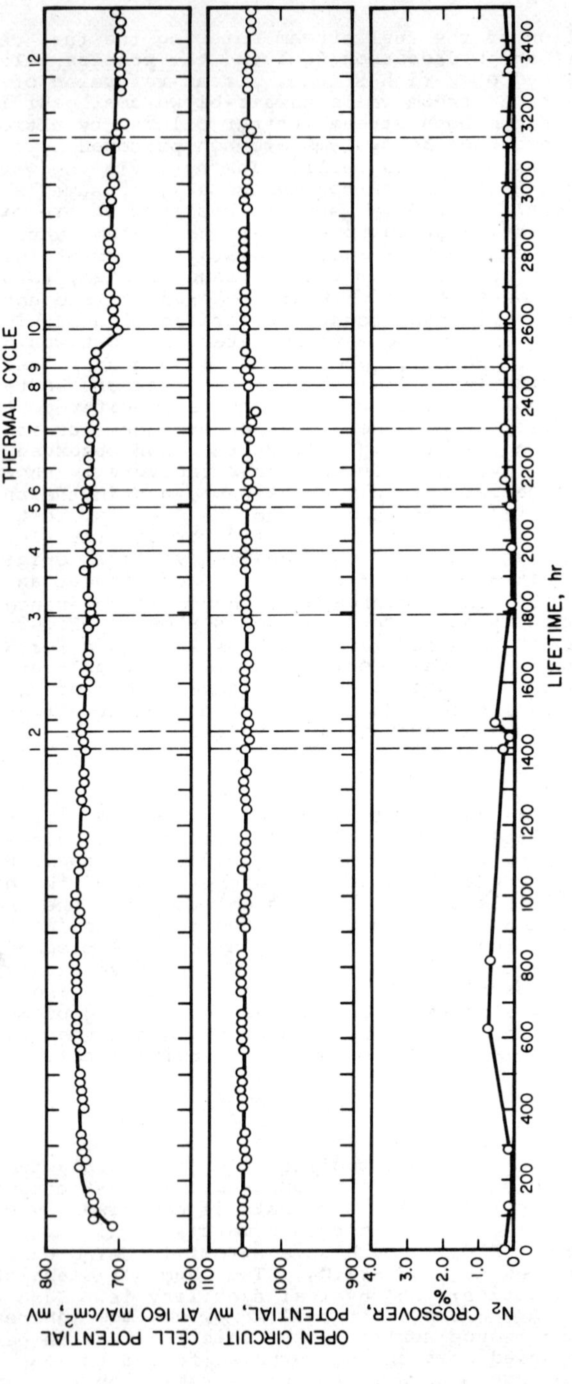

Figure 4. THERMAL CYCLING OF CELL IGT-BS-2 WITH TILE FROM SPRAY-DRIED POWDER (AD60-4A)

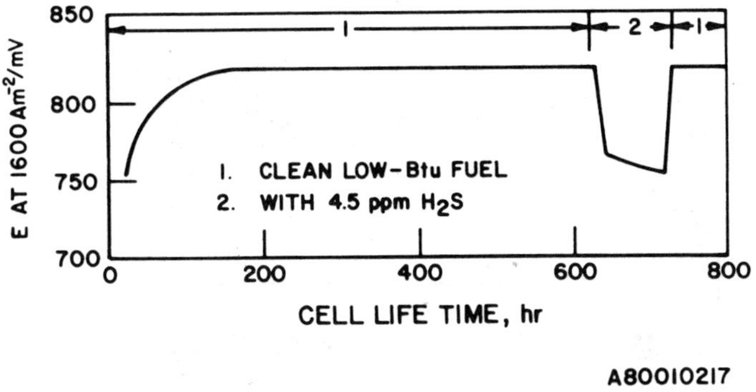

Figure 5. RECOVERY OF CELL VOLTAGE ON REMOVAL OF H_2S FROM FUEL

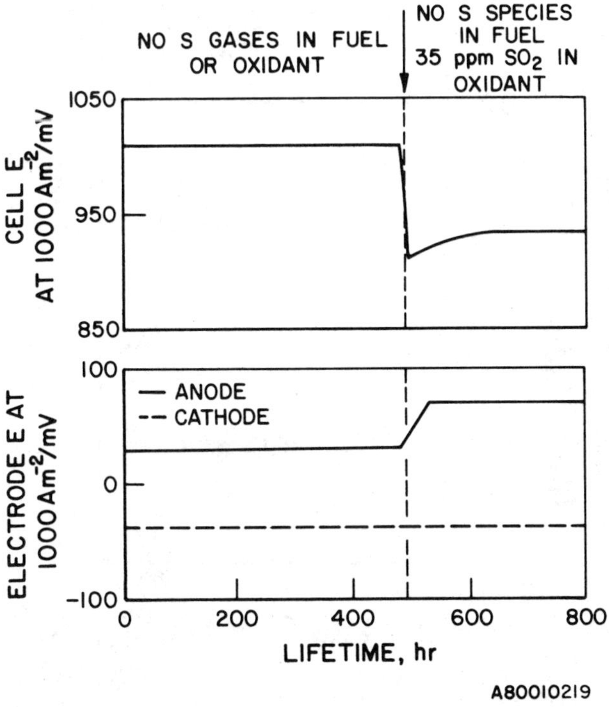

Figure 6. EFFECT OF SO_2 IN OXIDANT ON ANODE AND CATHODE POLARIZATION

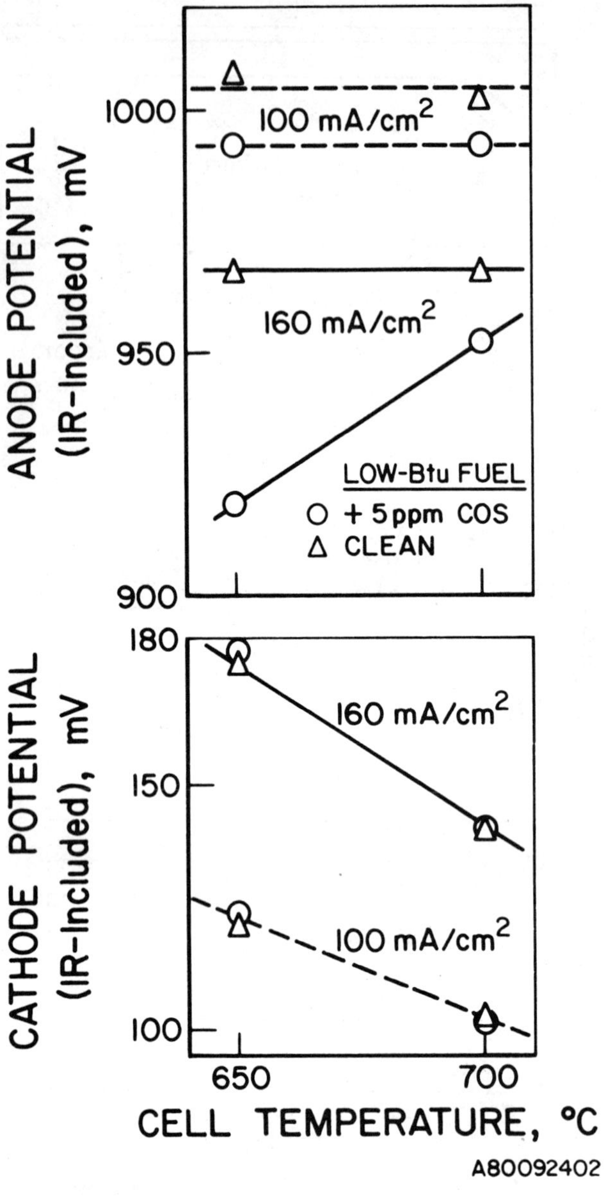

Figure 7. EFFECT OF TEMPERATURE ON ANODE TOLERANCE OF CELL EPRI-24

tive immobilization for longer periods of time. If the log-log relationship between surface area and time, shown in Figure 8, is followed, then a minor increase in the initial area may yield a satisfactory life above some final value of surface area. Attempts to find dopants to reduce the sintering rate and/or affect the phase composition shift have not been completely successful.

The other major life variable of the electrolyte tile is the loss of the electrolyte, both through reaction with other components and by evaporation. This eventually leads to gas crossover resulting in inefficiency, an increase in local cell temperature, and a waste heat load. The relative losses via reaction and evaporation have been determined (Figure 9). It is projected that by preloading the electrodes with electrolyte before cell assembly, operating at pressure, and using a sodium-lithium carbonate mixture, sufficient electrolyte will remain for a 40,000-hour life. The sodium-lithium carbonate has a measured vapor pressure at 650°C that is approximately one-fifth that of the potassium-lithium carbonate in current use.

Alternative electrolyte compositions are being examined for performance improvements over the 62% Li_2CO_3/38% K_2CO_3 normally used. Highest performance has been obtained for high Li/K compositions and a high-lithium ternary mixture. However, the 52% Li_2CO_3/48% Na_2CO_3 mixture with the lower vapor pressure also has a slight reduction in performance. Since lithium is lost preferentially, all compositions will eventually drift toward sodium rich, if sodium is present, since it has the lowest vapor pressure. A more complete mapping of performance and component loss is required to establish the performance path the cell will take with time.

THERMAL CYCLING

The ability to sustain thermal cycles wherein the constrained electrolyte tile freezes and contracts without degradation is a basic requirement of a potential system. Fortunately, those synthesis efforts which have yielded higher surface area $LiAlO_2$ via a more reproducible and less expensive process have also resulted in a higher strength tile. This has enabled one cell to be cycled through the melting point 12 times without significant performance losses (Figure 4). The performance reduction at the unscheduled cycle 10 (about 2600 hours) resulted from a known change in operating conditions, rather than a tile failure. Further testing of broader temperature excursions and greater frequency is in process.

Comprehensive analytical studies have been completed to identify the source and magnitude of the thermally generated stresses causing tile cracking. Both the cell design and the materials used and the means of tile fabrication have a major influence on tile integrity. Neither approach by itself is sufficient to result in a functional tile which will prevent gas crossover. Therefore, in addition to using cell designs that minimize thermally generated stresses, we are improving the tile microstructure, using reinforcement materials, bubble barriers, and crack arrestors.

The technique of acoustic emission has been adopted to assist in evaluating many conceptual solutions in out-of-cell testing. A large increase in the magnitude and frequency of acoustic events at temperatures below the melting point, about 400°C, is clearly demonstrated by this means. Though there is some question in the interpretation of these events, this technique is being used to evaluate the effect of the inclusion of particles of various materials as crack arresters. The size and number of the particles and their chemical-physical bonding to the electrolyte are all variables needing quantification. Some materials have produced a clear decrease in both frequency and magnitude of acoustic events. It is

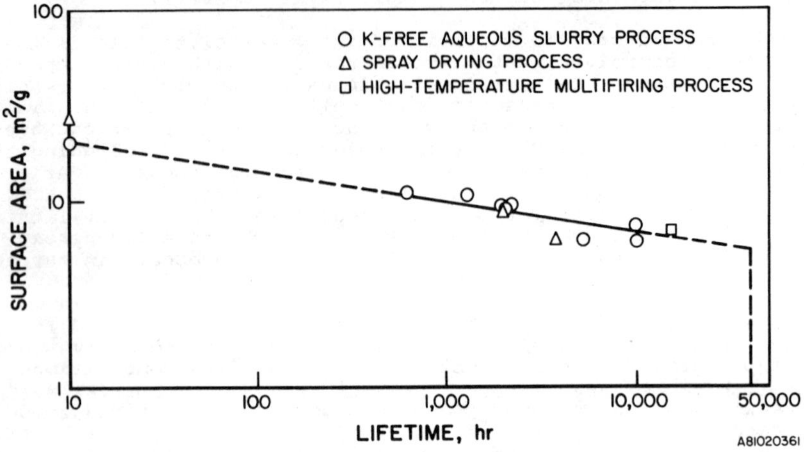

Figure 8. $LiAlO_2$ SURFACE AREA LOSS IN 94 cm^2 CELLS

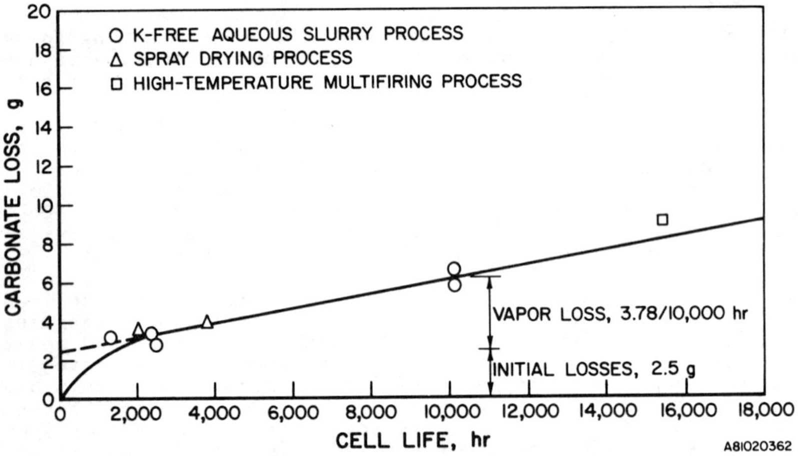

Figure 9. CARBONATE LOSS CURVE FROM STANDARD 94 cm^2 BOILER PLATE FUEL CELLS

hoped that relating all these will also improve the physical explanation of the events. This screening will be followed by in-cell evaluation of the better candidates.

CELL SEPARATOR

In a bipolar cell stack, each cell is separated from the next by a separator which serves as an electrical connection, a gas separator, and a gas distribution plate. The environment within which the cell separator must survive is extremely adverse. The ability to withstand the presence of reductants on one side and oxidants on the other, under a potential and exposure to an electrolyte, and yet maintain conductivity and be inexpensive and easily manufactured or formed — this is a severe set of requirements. For the short term, say 2000 hours, 316 stainless steel seems to adequately meet the above requirements. This material, in 10 mil sheets, is being formed into the corrugated structure that provides for gas distribution without an excessive pressure drop. The electrode strength dictates the spacing of the ribs and this, in conjunction with the material properties and thickness, controls the height. These rib dimensions are 120 mil pitch and 58 mil height. Figure 10 is a cross-section showing the tolerances maintained during pressing. For lifetimes exceeding 20,000 hours a stable material is required on the anode side, and the logical though expensive choice is a composite with nickel. This may not require a large and costly amount and alternatives may also be found, such as those being sought for the anode. It is likely that the process for forming and subsequent treatment of such a composite will have a pronounced effect on its performance and stability in the cell environment.

A further critical environment is the wet-seal area where the separator contacts the electrolyte tile. In stack designs, exposure of the separator in this area to the reactants will create an electrochemical cell that results in rapid corrosion unless insulated. Alumina or aluminum (later oxidized) or NiAluminide are satisfactory; however, methods of fabrication strongly influence adherence and density and thus stability.

SCALE-UP

Prior to the eventual manufacture of large-scale cells and stacks, some emphasis on small stacks of 1 ft^2 active area can provide an insight into the fundamental problems inherent in cell stacking. The selection of external gas manifolding provides for a design alternative to the internal manifolds being designed by GE. The cell package and stack design to be constructed in early FY82 are shown in Figures 11 and 12. That 20-cell stack will produce in excess of 2 kW and will be thermally self-sustaining.

General Electric will be producing cells larger than 4 ft^2 to more closely approximate, if not represent, preproduction prototypes. They have successfully pressed and handled electrolyte tiles 50 by 50 inches. Based on uniformity and reproducibility, a final cell size somewhere between 4 and 16 ft^2 will be selected for an eventual 70 cell stack which provides one-tenth of the output of an ultimate production stack.

CONCLUSIONS

The progress of the DOE and EPRI funded programs has been substantial, and a number of very productive concepts and major accomplishments have been realized. Though it would be presumptous to state that all the problems are recognized and being addressed, those that are being addressed seem to have identifiable and promising technical solutions. Admittedly, the economics of these solutions are less certain.

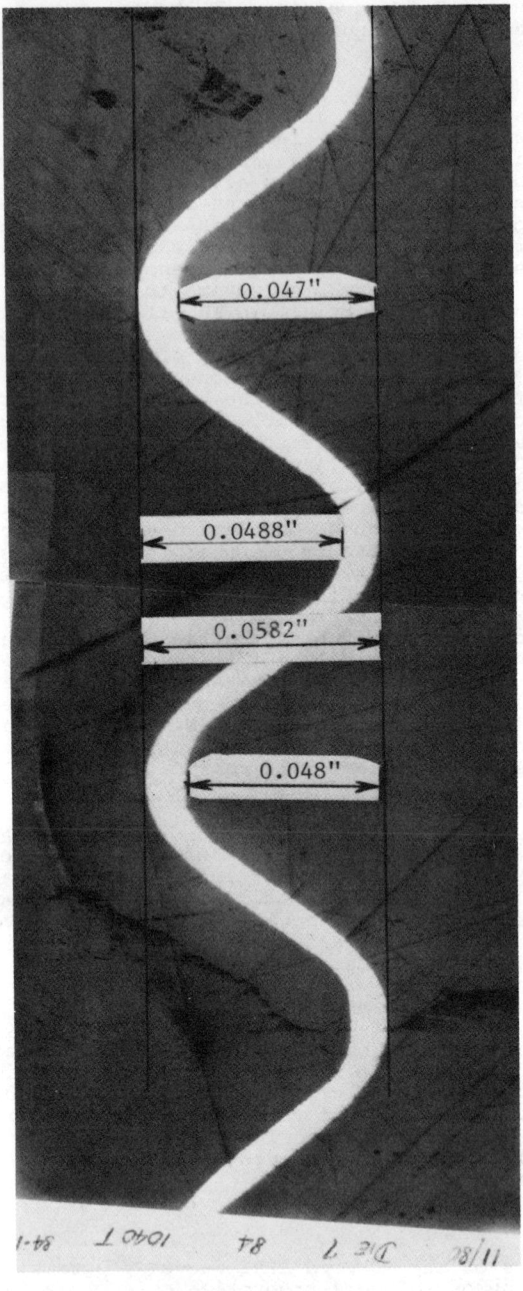

Figure 10. CROSS SECTION OF BIPOLAR PLATE FOR MCFC STACK
(Material: Type 316 S.S.)

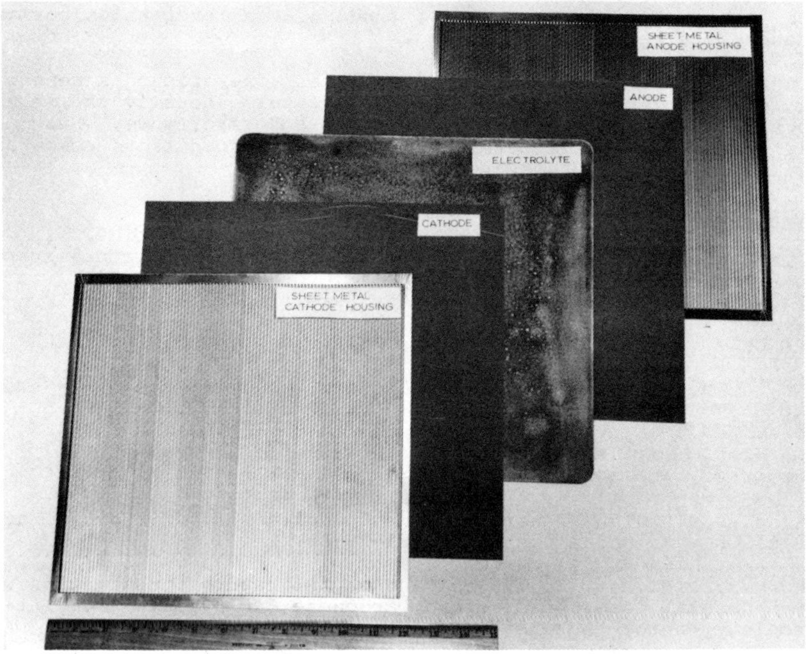

Figure 11. DESIGN FOR 2-kW CELL STACK USING 1-ft^2 SHEET METAL

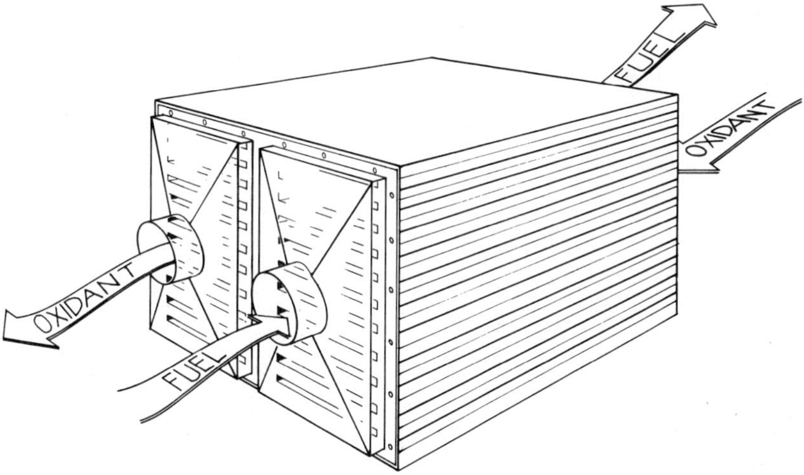

Figure 12. X-CONFIGURATION EXTERNAL GAS MANIFOLDING

There are a number of creative and technically competent firms involved in the establishment of an efficient and economic means of power generation via molten carbonate fuel cells. What is critically needed is a continued assurance of funding at a level which permits pursuance of alternative solutions.

This will be especially true once stack operation has been of sufficient magnitude and duration to adequately identify the remaining problems and their relative magnitude and importance. Such a stable and aggressive program can provide for expeditious commercialization in the early 1990's.

ACKNOWLEDGMENT

The information reported here resulted from contractual support by General Electric Co., the U.S. Department of Energy, Electric Power Research Institute, and the Gas Research Institute.

REFERENCES

1. Noyes, R., Ed. 1977. "Fuel Cells for Public Utility and Industrial Power," Noyes Data Corp., Park Ridge, N.J.

2. Handley, L.M. 1978. "Status of 40-kW and 4.8-MW Fuel Cell Programs," National Fuel Cell Symposium, San Francisoco, CA.

3. King, J.M. 1975. "Energy Conversion Alternative Study," Phase II Final Report under contract NAS3-19586.

4. Bowman, R.M., et al. 1980. "Industrial Energy Conservation with the Natural Gas-Fueled Molten Carbonate Fuel Cell," 15th Intersociety Energy Conversion Engineering Conference, Seattle, WA.

43(3000)/MCFC-Paper

FUEL CELL POWER PLANTS: UTC PERSPECTIVE

JAMES W. BABCOCK AND WILLIAM J. LUECKEL
POWER SYSTEMS DIVISION
UNITED TECHNOLOGIES CORPORATION
SOUTH WINDSOR, CONNECTICUT

1981 will be a significant year in the evolution of fuel cell power plants at United Technologies. The 4.8-MW electric utility demonstration power plant is in the final stages of construction and installation at the Consolidated Edison, New York City test site, and will begin testing later this year. The delivery of the components and subsystems for a similar, but improved, system is underway to the Tokyo Electric Power System and the construction of the system in Tokyo should be completed by the end of the year so testing can begin early in 1982. Deliveries of 40-kilowatt fuel cell power plants to be used in a comprehensive field test conducted by the natural gas industry will begin later this year.

These events will be the most visible evidence of progress in the evolution of UTC fuel cell power plants. The equipment delivered for these test programs will be capable of verifying nearly all of the long-publicized characteristics of fuel cell power plants with respect to performance, environmental characteristics, and other operational features. However, this equipment will not represent the manufacturing cost necessary for a self-sustaining fuel cell business. Consequently, there is a less visible, but very significant aspect of the activity directed to improvement of the technology so cost objectives can be achieved. The program support received from DOE, EPRI, GRI, and individual utilities is critical to pursuing these objectives.

The 4.8-megawatt and 40-kilowatt test systems are the current vehicles used by UTC to demonstrate the status of the technology and to measure the value of fuel cell power plants to gas and electric utility customers. Serving these utility markets has been the goal of UTC throughout the development history of fuel cell power plants. In the late 50's, the fuel cell concept was identified as potentially having the characteristics of a superior commercial electric power generator: superior in terms of performance and reliability. The more difficult question then, as now, is could the concept be developed in practical configurations and be produced at a cost less than the value as perceived by utility customers? We concluded in 1960 the answer was yes and we believe the answer is still yes.

The research and development cycle for fuel cell power plants, as for any new technology, requires the design and construction of demonstration test vehicles at key milestone points throughout the cycle. These are needed for two purposes: to assess the development status against operational and physical specifications, and to permit system evaluation by the ultimate user in the environment where use is intended. Only in this way can the user make a judgment as to the value of the development against which estimated manufacturing and operating costs can be measured.

The evolution of the systems being tested today started in 1959 when we acquired the development rights to the Bacon cell technology from England. The first development task was to verify the feasibility of a controlled reaction between hydrogen and oxygen in a series stack of cells to produce a useful voltage and to demonstrate that water and heat produced by this reaction could be removed from the reaction sites. This test was successfully conducted in 1960. (Figure 1.)

At this time, there was a fortuitous match between fuel cell capability, though very limited by today's standards, and a real market as measured by a customer's perceived value of the concept being greater than the cost to produce it. The customer was NASA and UTC developed the Bacon cell power plant to meet the high performance and high reliability requirements of the Apollo mission. (Figure 2.) Continuous improvment, i.e., lower temperature, longer life, and lower cost cells have provided hydrogen-oxygen fuel cell power plants for undersea use (Figure 3) and for the Space Shuttle Orbiter Vehicle (Figure 4). The value of the space program to longer range commercial objectives was that it established that fuel cells could be developed and reproduced, albeit in small quantities, to meet a customer need.

The space programs, though extremely valuable in providing program momentum, did not contribute directly to commercial technology. The utility requirement is for a system fueled by conventional hydrocarbon fuels producing useful utility voltages at 60 Hertz.

The first reduction to hardware of a fuel cell operating on hydrocarbon fuels was the adaptation of the space cell technology to a 500 watt, 28 volt d.c. system developed under Army auspices. (Figure 5.) This demonstrated the fuel cell to be capable of meeting a requirement for portable, quiet power using petroleum distillate fuels. It was also used to test a very specialized commercial market (Figure 6), but was found lacking because the cost projections exceeded the perceived value even in this specialized, premium market.

The gas industry requirement for an efficient generator compatible with an on-site environment was the focus of development effort in the late 1960's. In 1966 the same basic cell technology of the 500-watt PC-9 was adapted to a 4-kW system that was the first natural gas fueled power plant. It demonstrated initial compatibility with the load variations required by a utility generator by providing power to a Columbus, Ohio home. (Figure 7.)

At this point, two major technology deficiencies existed. The cell was not tolerant of carbon bearing gases without prohibitively expensive cleanup components and there was not available a power conditioning system to convert low voltage d.c. to utility quality a.c. consistent with the performance and cost goals that had been established. Both of these deficiencies were accommodated in the 12-kW power plant (Figure 8) that was extensively tested by the gas utilities in 1971-73. The cell electrolyte was changed from potassium hydroxide to phosphoric acid to provide tolerance to carbon bearing gases and the self-commutated solid state inverter was developed specifically for the fuel cell.

The field test established the technical feasibility of using a natural gas fuel cell to generate electric power at the site where power is required. Did the perceived value of this configuration exceed the cost for which it could be manufactured? The conclusion was "no" primarily because of the endurance limitations of the power plant and the realization that single-family residences, for which the 12.5-kW power plant was sized, did not represent a good initial fuel cell market. The peak-to-average load ratio in a home requires a capital cost for the generator that distorts the economics of on-site power generation.

Figure 1. 250-Watt Hydrogen-Oxygen Fuel Cell Powerplant Tested in 1960

Figure 2. Three powerplants each rated at 1.5 kilowatts installed in the service module of the Apollo Spacecraft.

Figure 3: kW powerplant developed for undersea applications.

Figure 4: 12kW powerplant has been developed for the Space Shuttle program. Three of these will be installed in each orbiter vehicle.

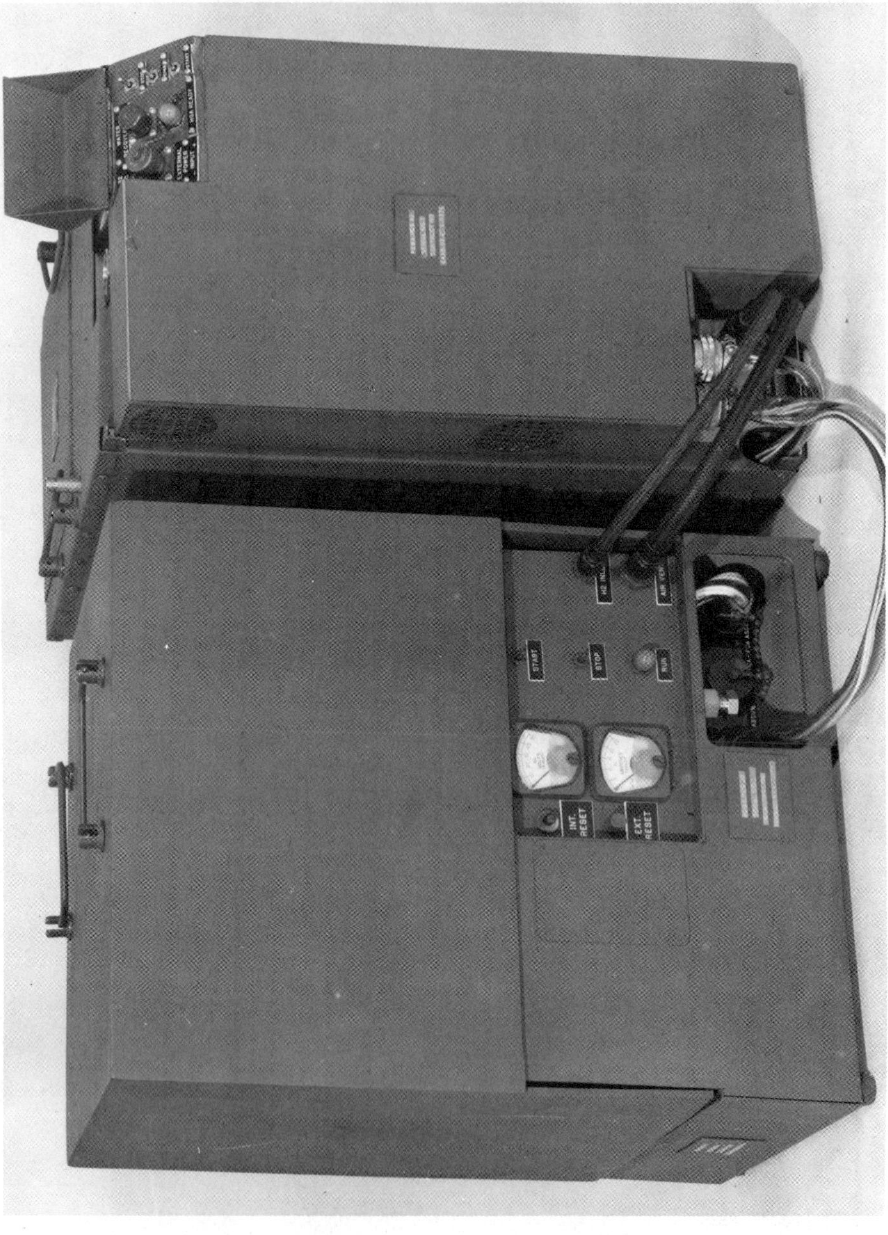

Figure 5. Initial UTC experience using distillate fuels was gained in 1965 with 500 watt, 28 volt powerplant.

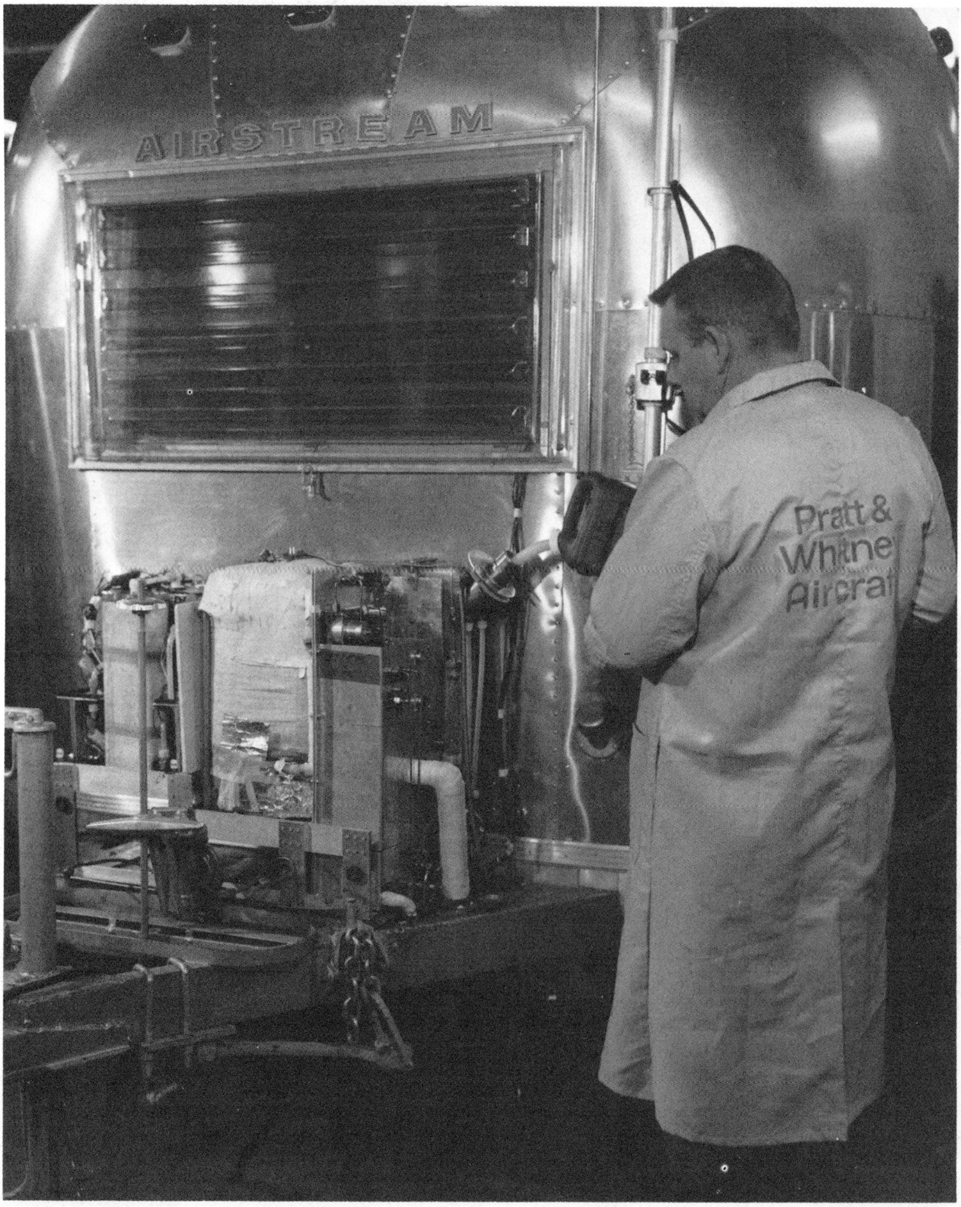

Figure 6. Tests were conducted with the 500 watt powerplant installed on an Airstream Trailer to provide housekeeping power.

Figure 7: Initial field test of natural gas powerplant was conducted in Columbus, Ohio. Four kilowatt powerplant (left) required gas compressor and breadboard inverter (between powerplant and instrumentation panel).

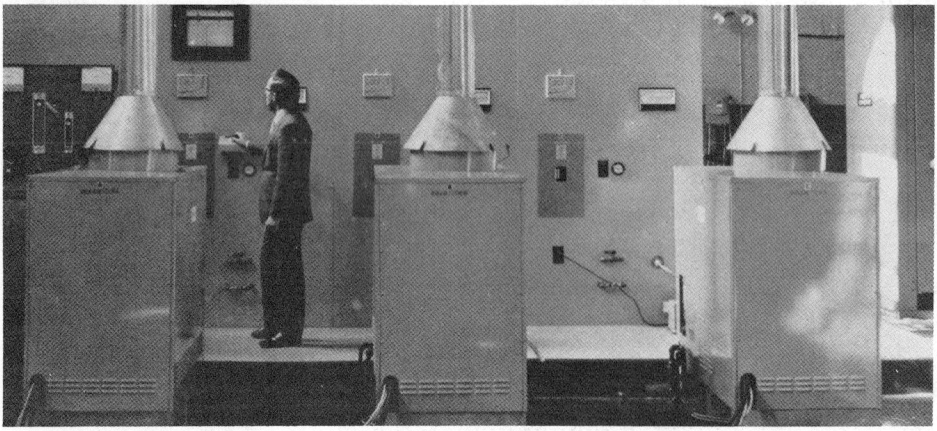

Figure 8: Three 12.5 kilowatt powerplants installed in Public Service Electric and Gas system in 1972. This was initial use of phosphoric acid technology and specially developed inverter.

There were many positive results from this field test, especially the high degree of acceptance by utilities and utility customers of the concept of on-site power generation. The program, therefore, moved forward on both technology and power plant fronts. Market studies indicated a need for a 40-kW power plant and an experimental test vehicle rated at 40-kW was built. This was used to demonstrate compatibility with pipeline gas, 40% electrical efficiency, water recovery so dependence on an unreliable water supply was eliminated, and during the latter portion of the 18,000 hour test period, fuel utilization in excess of 80% when heat was recovered. This system was also the first integration of the inverter and the d.c. module subsystems in a single physical package.

All of the improvements evolved during this period, including a two-phase water cooled, extended endurance cell were incorporated into the present revision of the 40-kW power plant (Figure 9). This unit has been undergoing tests at UTC since April 1980. This model will be used in a forthcoming field test sponsored by GRI, utilities, and DOE. The first units are now being constructed for this test.

Will this model meet the market test of perceived value exceeding the cost to manufacture it? We believe that there could be limited sales of this power plant so the answer is a qualified "yes." However, there will be a premium price associated with such sales which will undoubtedly limit the market to a few for evaluation testing under special circumstances. Development of the technology base for a truly commercial product forming the basis for a sustained business is a necessary activity to be conducted concurrently with the field test program.

There has been a similar evolution of fuel cell systems to meet electric utility industry requirements that either cannot be satisfied at all or as well by more conventional generators. The initial market interest in this program came from combination gas and electric utilities who supported the gas industry program. They identified a need for environmentally benign, efficient generators that could be sited anywhere in the utility system, including urban sites that could not otherwise be used for generation. Thus, with considerable support from the utility industry, scale-up of the technology was undertaken and testing began in 1975 of a breadboard 1-MW power plant (Figure 10). During 2000 hours of testing, this system provided utility quality power to the 13,800 volt distribution grid of Connecticut Light & Power at our South Windsor facility. In addition to scale-up of all the subsystems and the superior performance characteristics inherent in fuel cells, this test demonstrated operation on both gas and liquid fuels and the reactive power control capability of the inverter within the 1000 kVA rating of the equipment.

The next step in the electric utility program was to design and build a 4.8-MW system for installation and test by a utility. This is being accomplished in a program sponsored by UTC, DOE, and EPRI: Con Edison as the host utility is a major program participant (Figure 11). The manufacture of the power plant is complete and all of the systems have been delivered by UTC. Initial testing will checkout all subsystems except the cell stacks using a simulated power section source of d.c. It is scheduled to start in March 1981. Later this year the cell stacks will be installed and the system test program will begin. The principal objectives of the program are to verify the suitability of factory assembled pallets, truck transport to the site, installation, and operation by the utility, and the capability to site and operate fuel cells in an urban area.

The Consolidated Edison installation does not address the question of the value of the fuel cell relative to the cost of producing it. That has never been a goal of the program. The costs associated with the power plant at Con Edison are those of a development power plant and a development test program. However, the manufacturing experience gained in this program has enabled UTC to contract with Tokyo Electric Power Company on a fixed price, commercial basis for the delivery of a second 4.8-megawatt demonstrator. The subsystems will be delivered on preassembled pallets to Tokyo throughout 1981 and testing will begin early in 1982. For the first time, a commercial customer has perceived the value of the power plant to be more than the cost of producing it, although the circumstances are unique and it is doubtful if any U. S. utility would or could place such a high value on fuel cell power plants at this time.

FIGURE 9. 40 KILOWATT NATURAL GAS POWER PLANT ON TEST AT UTC IN 1980. SEVERAL OF THESE WILL BE FIELD TESTED IN THE GAS INDUSTRY PROGRAM.

FIGURE 10. INITIAL MAJOR SCALEUP OF FUEL CELL TECHNOLOGY WAS ACCOMPLISHED IN 1975 WITH 1 MEGAWATT BREADBOARD SYSTEM TESTED AT UTC WHILE CONNECTED TO CONNECTICUT LIGHT AND POWER GRID.

Figure 11. 4.8 megawatt electric utility demonstrator under construction at Consolidated Edison site in New York City.

The perspective at UTC is still cautiously optimistic. We believe an ever-increasing segment of the utility markets will perceive the value of the fuel cell to be greater than the costs to manufacture it. The inherent operational characteristics of fuel cells, especially efficiency and the lack of an impact on the environment, are viewed as important parameters in both dispersed applications and integrated on-site energy systems. Perhaps as, or more, important are the economic advantages of a generation option that can be acquired and installed in about two years, as will be the case at the TEPCO site. Thus, we are working with DOE and the utility industry through EPRI, GRI, and the respective utility users groups to develop a technology of sufficient value to users so a self-sustaining business will result.

NUCLEAR PLANT RISK ASSESSMENT VARIOUS APPLICATIONS

SAUL LEVINE
NUS CORPORATION

1. INTRODUCTION

I will talk today about the history of the development and various applications of Probablistic Risk Assessment (PRA) techniques in the field of nuclear power. The history starts essentially with the Reator Safety Study, WASH-1400, "An Assessment of Accident Risks in U.S. Commercial Nuclear Power Plants." This study was initially sponsored by the U.S. Atomic Energy Commission and completed under the sponsorship of the U.S. Nuclear Regulatory Commission. However, it was performed by an independent, ad hoc group under the direction of Professor Norman C. Rasmussen of MIT. Because it has been done many times before, I will not describe the study today.

BACKGROUND

Because WASH-1400 predicted the risks from potential accidents in a 100 nuclear power plant industry of light water cooled reactors to be quite small compared to other risks in society, it immediately became a target of extraordinary attack by critics of nuclear power. Not only were the techniques and results of the study criticized, but allegations were also made questioning the honesty of the study and the integrity of its authors.

This study became so politicized that the application of PRA techniques did not spread widely or rapidly in the nuclear community. Nonetheless, there were significant applications made over a period of several years and I will discuss some of these later. Before doing that, it would probably be useful to spend a few minutes on the political history of WASH-1400.

2. HISTORY

Figure 1 presents a series of key events related to WASH-1400 and the accident at the Three Mile Island (TMI) reactor. As I will show later, the TMI accident had a key impact on the growth of the application of PRA techniques. Work on WASH-1400 started in September 1972 and a draft report was published two years later in August of 1974. The draft was widely circulated for comment to many individuals and organizations representing a wide variety of viewpoints, from ardent supporters to ardent critics of nuclear power. Some 1800 pages of comments from about 90 sources were received and analyzed. A final report was published in October of 1975.

The publication of the draft and final reports were causes for both immediate and continuing controversy, principally because the findings did not match the preconceptions of many people, including critics of nuclear power as well as members of the nuclear community. Critics could not accept the findings that nuclear risks appeared to be smaller than the accidents risks from many other technologies in society. The U.S. Nuclear Regulatory Commission found that WASH-1400 predicted accidents whose probabilities were too high and whose consequences were too small to match its preconceived ideas; yet it presented no analyses to support that viewpoint.

The controversy continued unabated for some time and it became necessary for the NRC to form an independent advisory committee under Professor Harold Lewis of the University of California, Santa Barbara to perform an independent assessment of key aspects of WASH-1400. The Lewis report[1], published in September of 1978, although critical of some portions of WASH-1400 was in general a ringing endorsement of the use of this methodology and recommended its application to a fundamental re-examination of the technical framework of NRC's regulatory process. Unfortunately, the report was grossly misinterpreted by the NRC which issued a policy statement in early 1979 regarding WASH-1400 and its techniques which completely misled both lay readers and many technical people.

The Commission policy statement essentially destroyed the viability of the broad scale application of these techniques within the NRC.

This remained true until the Three Mile Island accident happened in March of 1979. In the hectic days that followed the Three Mile Island accident, it was noted that WASH-1400 contains an accident sequence in which the relief valve on the primary coolant system opened on high system pressure, but

---

1) "Risk Assessment Review Group Report to the U.S. NRC", NUREG/CR-400, September 1978.

HISTORY OF WASH 1400

- WORK STARTED IN 1972
- DRAFT REPORT PUBLISHED IN 1974
- FINAL REPORT PUBLISHED IN 1975
- CENTER OF CONTROVERSY
- LEWIS REPORT
- NRC POLICY STATEMENT
- THREE MILE ISLAND
- KEMENY REPORT
- ROGOVIN REPORT
- LEWIS SCIENTIFIC AMERICAN ARTICLE

FIGURE 1

failed to close when pressure was reduced. This is exactly what happened during the Three Mile Island accident and was the component failure that led to the accident. At that time the NRC staff began to understand that there were accident sequences considered in WASH-1400 which could realistically happen even though they departed from and went beyond the conventional design basis accidents analyzed in the normal licensing process. This understanding began the rebirth of WASH-1400. Then along came the TMI investigation by the President's Commission on Three Mile Island headed by Professor Kemeny and the Rogovin investigation performed for the NRC, both of which noted the applicability of WASH-1400 to what had occured at Three Mile Island and both strongly encouraged the use of PRA techniques. Finally in March of 1980, Professor Lewis published an article in the Scientific American which essentially demanded the use of PRA techniques as a rationale in safety decision making.

3. APPLICATIONS

Going on to Figure 2, it describes an accident sequence called TMLQ, which was analyzed in WASH-1400. This accident sequence had a relatively low probability of occurrence and was conservatively deemed to be a core melt accident because analytical techniques did not permit its finer analysis. The reactor analyzed in WASH-1400 was a Westinghouse designed reactor.

Figure 3 shows the same accident sequence for the Babcock and Wilcox TMI reactor in which the TMLQ sequence is not applicable and is replaced by a TMQ sequence. The design of the TMI reactor required the lifting of the relief valve whenever main feedwater was lost in order to prevent the reactor from having to shut down on loss of main feedwater. Thus, when a loss of main feedwater occurred in that reactor, the additional failure of auxiliary feedwater was not required to lift the relief valve. This situation resulted in a small size loss of coolant accident with a frequency of 6×10^{-2} and an estimated emergency core cooling system failure probability of 10^{-2}, this brings the core melt probablity to about 6×10^{-4} which is about 10 times higher than that estimated in WASH-1400.

The same figure shows the post TMI situation in which the L has been put back into the TMQ sequence by a simple design change in the reactor to require the use of anticipatory scram signals to make the reactor scram on loss of main feedwater. Thus, the use of auxiliary feedwater can be used to prevent the relief valve from opening and auxiliary feedwater would have to fail for the relief valve to be opened. This has the effect of reducing the probability of the accident sequence significantly.

Figure 4 shows the result of a study made by the NRC of the reliability of auxiliary feedwater systems upon loss of main feedwater for 25 different auxiliary feedwater systems for pressurized water reactors. As you can see there is a wide variation in the reliability of these systems; and it is known from the WASH-1400 accident sequences that the reliability of these systems should be high, such as those shown in the lower

WASH 1400

TMLQ* - 10^{-8}

T - FREQUENCY OF TRANSIENT INVOLVING LOSS OF MAIN FEEDWATER (MFW)

M - PROBABILITY OF STARTING MFW IN 30 MINS.

L - PROBABILITY OF LOSS OF AFW ON DEMAND

Q - PROBABILITY OF RELIEF VALVE STICKING OPEN

T - 3

M - 10^{-2}

L - 4×10^{-5}

Q - 10^{-2}

*TMLQ - SMALL LOCA OR CORE MELT, DEPENDING ON SIZE OF RELIEF VALVE AND CAPACITY OF HPSI SYSTEM

Figure 2

B & W REACTORS

PRE TMI - 2	PST TMI - 2
T - 3	T - 3
M - 1	M - 5×10^{-1} - 10^{-1}
Q - 2×10^{-2}	L - 10^{-2} - 10^{-5}
TMQ - 6×10^{-2}	Q - 2×10^{-2}
- SMALL LOCA	TMLQ* - 3×10^{-4} - 6×10^{-8}

*TMLQ - SMALL LOCA OR CORE MELT, DEPENDING ON SIZE OF RELIEF VALVE AND CAPACITY OF HPSI SYSTEM.

Figure 3

RELIABILITY OF PWR AUXILIARY FEEDWATER SYSTEMS

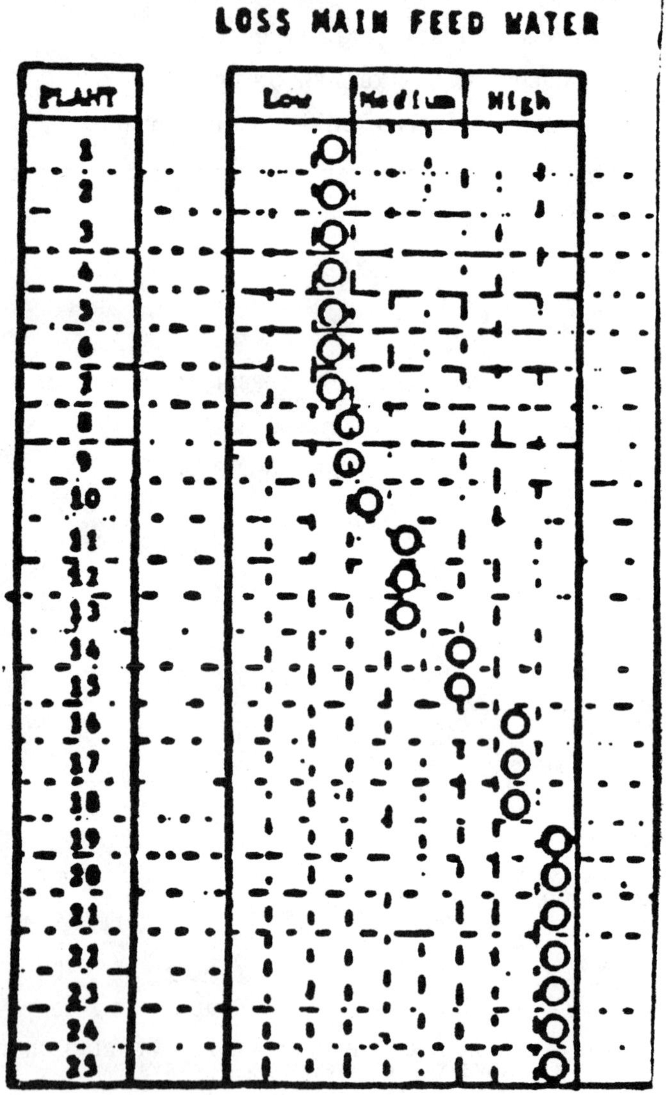

FIGURE 4

left part of the diagram. Unfortunately, a large number of systems were much lower in reliability and required improvement; the NRC has issued bulletins and orders to effect these improvements.

One of the most interesting topics concerned the divergent opinions expressed by 4 NRC Staff members several years ago. They raised 15 issues supposedly related to safety and not being handled adequately by the Regulatory Staff. It took a few days to understand the issues because they were continually changing as discussions with those having the divergent opinions continued. The analysis of these issues, once understood, took only a few minutes each.

They fitted into one of several categories. About five of the issues related to procedural matters which had no safety impact. The remaining ten fit into one of two categories. One category involved was accident sequences that had very small releases of radioactivity to the environment which would result in negligible public health impacts and, therefore, were not of significant concern to safety. The remaining sequences fit into a category of accidents having significant potential releases of radioactivity; but the items of concern in these accident sequences were so far down the chain of probabilities that they could not effect the outcome of the accident in any significant way.

Another important example relates to NRC's list of generic safety issues. There had existed for some time 133 unresolved or generic safety issues that had arisen in the licensing process. These were examined by the use of probabilistic techniques and it was determined that about 20 of these were of any direct safety significance and the others were of very small concern. Thus, the 133 items were reduced to about 20 in a one month analysis.

Figure 5 describes the preparation, at the request of Congress, of a plan for safety research directed to improving the safety of reactors. A typical regimen was followed in examining various sources of suggestions on how to improve safety. Some 200 suggestions were grouped into 16 topics and then analyzed according to a set of criteria which, for the first time, involved some estimate of their quantitative risk reduction potential. Figure 6 shows the matrix which includes the breath of support, that is the breath of opinion that supported these items, as well as quantitative estimate of the risk reduction potentials achievable. The applicability to the number of reactors in which the change could be made and the cost were also included to help make the selection because these also help to determine the value of the changes. Each of these areas was estimated by assigning value of high, medium, or low. The estimates of risks reduction potential were made judgementally based on our knowledge of WASH-1400. It should be noted that the 4th item from the top, Accident Response, pertains exactly to the item of greatest significance in the Three Mile Island accident; that is the response of operators to accidents, the question of the design of the man/machine interface, the information displayed to the operator and the use of computer assisted techniques to help

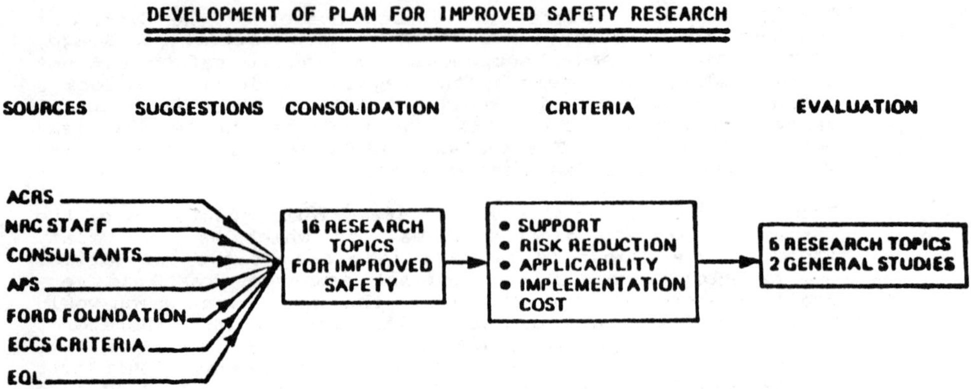

FIGURE 5

IMPROVED SAFETY RESEARCH - CRITERIA EVALUATION RESULTS

	BREATH OF SUPPORT	RISK REDUCTION	APPLICABILITY	COST
VENTED CONTAINMENT	H	H	H	M
DECAY HEAT REMOVAL	H	H	M-H	M
ECCS SYSTEMS	H	M	M	L-M
ACCIDENT RESPONSE	H	H-M	H	L
SEISMIC DECOUPLING	M	H-M	L-M	M-H

CRITERION VALUE:
H = HIGH
M = MEDIUM
L = LOW

FIGURE 6

the operator understand abnormal events that could occur in the reactor. This item was identified as being of major importance to safety long before the accident; this knowledge came from, among other sources, WASH-1400 which identified operator errors as a principal contributor to accident risks.

I will now discuss some examples of how to think about reducing potential reactor accident risks by the use of PRA techniques. Figure 7 shows a typical curve of probability versus consequence for potential reactor accidents which we know is determined by a small number of dominant accident sequences. If one were to try to lower the curve, one would then run into a greater number of accidents and the reduction of the curve by a large amount would become more and more difficult. Therefore, one has to be careful in estimating what level of risk reduction is really achievable below current levels of risk. It is probable that more than a factor of ten would be exceedingly hard to achieve.

Figure 8 indicates some risk reductions that could be achieved for the PWR reactor analyzed in WASH-1400. There are three sequences listed, and the first two need not be discussed here in great detail except to say that they are significant contributors to the risk in WASH-1400 and simple design or testing changes at essentially no cost can reduce their contribution to risk by about one to two orders of magnitude so their contributions would not be important. This would leave the accident sequence TMLB' as the principal contributor to risk. TMLB' is a sequence in which main feedwater is lost and not recovered, followed by the failure of auxiliary feedwater to start as required, and the loss of all AC power for more than three hours. If electrical power were to be recovered in less than 3 hours, the reactor could recover from the accident. The next column shows this accident sequence modified by the addition of one of three systems arbitrarily labelled X, Y, or Z and costing \$10M, \$25M, \$50M respectively. System X, for instance, could be a self powered auxiliary feedwater system that would perform the same function as the normal auxiliary feedwater system, L, but be self powered with its own electric power supply independent of normal AC power systems. System Y would be a self powered containment spray system such that its operation would present containment failure due to overpressure, thus, significantly reducing the likelihood of a large release of radioactivity. System Z is a filtered vented containment system that would prevent containment rupture by venting the pressure to the atmosphere. This last system would result in releasing all the noble gases to the environment but would filter out the bulk of the most dangerous radioactivity, resulting in essentially no health effects; however, there are basic questions about the use of a filtered vented containment in a PWR because the vent size would have to be of the order of 18 feet in diameter and this raises significant questions of engineering feasibility. Comparing the three systems, X is a system which prevents the core from melting and costs the least. Y is a system which, if a core melt accident were to occur, would prevent any significant release of radioactivity to the environment. Z is a system which, if the core were to melt would prevent an uncontrolled release of radioactivity but would release noble

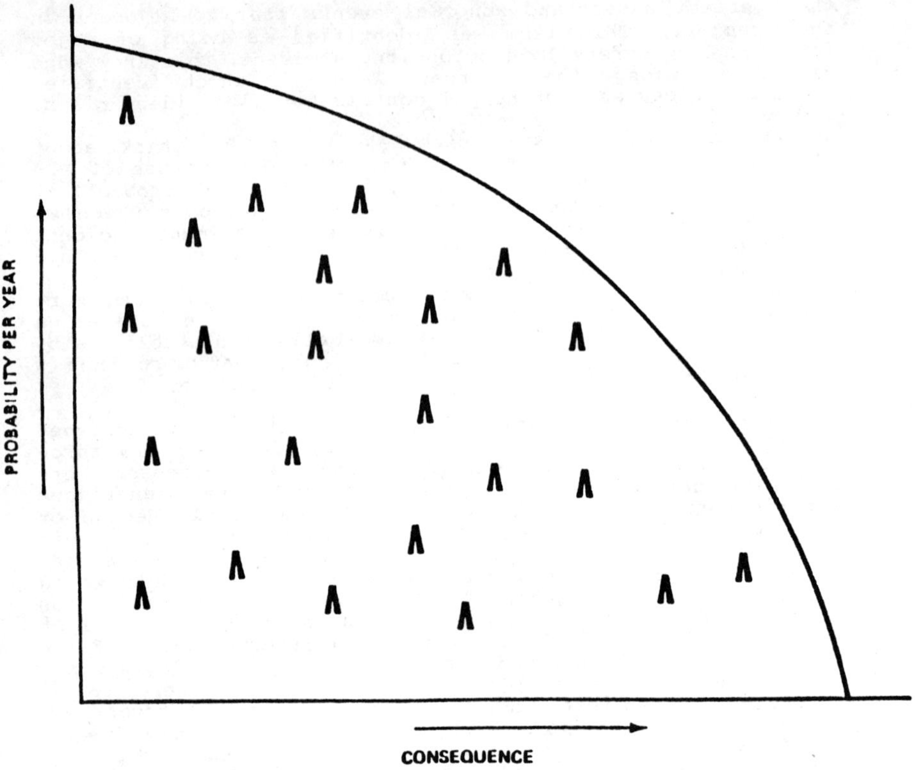

FIGURE 7

RISK REDUCTION POTENTIAL

PWR WASH 1400	RISK REDUCTION MEASURES	COST
S_2C	DESIGN CHANGE	~0
V	TESTING CHANGE	~0
TMLB′	TMLB′X	~10M
	TMLB′Y	~25M
	TMLB′Z	~50M*

*BASIC QUESTIONS ABOUT FEASIBILITY OF
A FILTERED VENTED CONTAINMENT

FIGURE 8

gases which have almost no health effects. With all three systems having about the same risk reduction potential and marked differences in cost, the choice of which direction to follow is very clear.

4. SAFETY GOALS

As I said before, WASH-1400 died and was reborn as a result of the Three Mile Island accident. As shown in Figure 9, before Three Mile Island only a few large risk assessment studies had been done. Aside from WASH-1400 there were risk assessments done of the Clinch River Breeder Reactor and of a high temperature gas cooled reactor, and in Germany, Professor Birkhofer[1] conducted a study of the Biblis water cooled reactor. In the United Kingdom, a study was conducted of Canvey Island, an island in the Thames River which has some dozen large chemical plants which were shown to have quite high risks. Since TMI, risk assessment has become a buzz word in Washington and there are a large number of risk assessments ongoing now, only some of which are listed here. One should be concerned about the proliferation of risk assessments and the ability to find the people who can do such studies in a competent way.

To make full use of the results of such risk assessments it is necessary to establish quantitative criteria for acceptable levels of risk or "safety goals", because the risk assessments now being done will present their results in form of typical WASH-1400 curves. But it is not clear how one will decide what levels of risk will be acceptable unless a safety goal has been established. Many organizations and individuals,[2] including myself, have recently begun proposing suggested safety goals that can be used to help judge the adequacy of reactor safety. An examination of these various goals indicates that most reactor designs can meet most of the elements of these goals. It is my belief that safety goals that will be established will be at such levels that reducing the risks of accidents in nuclear power plants are not likely to be required as a general matter.

5. CONCLUSION

I believe that the application of quantitative risk assessment techniques can and should be broadly applied to help in safety decisionmaking. This is now beginning to happen in the field of nuclear power. These same techniques can also be usefully applied to other technologies. In such applications, they can offer better insights than conventional reliability analyses in safety decisions. They are able to do this because they tie plant failures to public health consequences and thus the

1) "The German Risk Study", August 15, 1979.

2) NUREG-0739, C. Starr, E. L. Zebroski, AIF, V. Joksimovic, G. H. Kinchin, P. A. Corkerton, et. al., S. Levine.

RISK ASSESSMENT STUDIES

- PAST
  - LMFBR (CLINCH RIVER)
  - HTGR
  - FRG (BIRKHOFER)
  - CANVEY ISLAND
- CURRENT
  - RSSMAP (OCONEE,CALVERT CLIFFS, GRAND GULF)
  - OYSTER CREEK
  - CRYSTAL RIVER
  - BIG ROCK POINT
  - INDIAN POINT
  - ZION
  - LIMERICK
  - OCONEE (NSAC)
  - LASALLE
  - OTHERS
  - IREP

FIGURE 9

major contributors to risks can be identified and corrected if it is deemed necessary. The same application of these techniques can, of course, also produce plant reliability analyses to help achieve the desired degree of plant availability. I forsee a bright future for the growing application of PRA techniques to many industries, not only to estimate public risks, but to improve plant reliability and give good cost benefit insights to proposed plant modifications.

"NUCLEAR WASTE MANAGEMENT OVERVIEW"

Robert. W. Kupp, Vice President
The S. M. Stoller Corporation

1. Introduction

Perspective is from a given point of view; in the eye of the beholder. It is shaped by a person's background, feelings and, most importantly exposure to information and knowledge. In past years, as I have attempted to survey - that is to understand the perspective of - my friends, neighbors, non-nuclear associates and, more generally, the lay public, I find these groups almost universally have genuine concern, in some cases to the point of paranoia, regarding the hazards of the current and future status of "radioactive nuclear waste". I hope, during the course of this paper and the subsequent questions period, to at least develop some understanding of the disposal question which may illustrate the wide variation in a perception of the problem of "permanent" radioactive waste disposal.

It is not surprising that nearly everyone, but relatively few in the nuclear business, views nuclear radiation with such fear. It was born dangerous, starting with the discovery of X-Rays. Many early doctors died of radiation-induced cancer, as did Mme. Curie, the discoverer of radium. An important fraction of the New Jersey Radium dial painters died of bone cancer and, of course, in most modern times, the most violent of nuclear radiation - the atom bomb -leaves everyone terrified. In addition to history, regularly the media bombard the public with items which are typified by the TV presentation by NBC a few years ago: "Danger -Radioactive Waste", in which the simplistic questions are asked, "Is radiation bad? Can it last for tens of thousands of years?" The answer of course, is "Yes", and the thought is left right there. With this background, and no arguments to the contrary, it is not surprising that nuclear waste disposal is a hazard that is viewed by the public, the press, and the politicians as, perhaps an unsolvable technological and sociological problem - "it's just too hazardous."

Even among the scientific community, and those working in the nuclear radiation or waste disposal field there are questions on the subject - generally as to "How safe?" or "What if?" Concurrently with this Conference, there is a technical meeting at Gatlinburg, Tennessee, entitled, "The Uncertainties Associated with the Regulation of the Geologic Disposal of High Level Radioactive Waste". On this narrowly focused question there are aproximately sixty papers. The resolution of this many questions would logically indicate that we have a long way to go; for as you attempt to answer any involved, technological and problematical, question, many more are raised. This approach of first asking and then answering "all of the questions" results in a very extensive R&D program as currently funded at about $650 million per year. The program if continued as currently planned will result in the first "high level" nuclear disposal repository becoming available about the turn of the century.

An alternative scientific view of the high level waste program is that we have already investigated the most probable of the geologic disposal routes and although proceeding with it on the basis of our current knowledge might have certain risks, those risks are commensurate with the risks normally present in a technological society and are perhaps even an order of magnitude lower and, therefore, "why don't we get on with it?"

The perspective then can be seen from three different views - (1) the public: "An incredible risk we should not take,";(2) the researcher and developer: "We can certainly find a solution - all we have to do is study it further"; (3) the pragmatist: "We already have sufficient information at hand and the risk is not significant." If we could fully understand the basis, the arguments, the fallacies in each of these perspectives, we might be in a better position to consider how best to expend the limited funds available to protect our world from the varied technological risks present in our society.

2. Quantification of the Nuclear Waste Problem

There are basically two major types of nuclear waste which will require disposal. The first of these is "low-level radioactive waste" which has historically been disposed of by shallow-land burial at isolated nuclear disposal sites. About half of these wastes are from the medical/industrial sector of our economy, and the other half from our nuclear power industry. Although there is a current and projected shortage of such burial sites, this is principally a public acceptance problem for the actual area required and the site characteristics are not particularly severe. The scientific perspective on these wastes is that they are relatively easily handled as compared to our other waste disposal problems in our society; and, hence, in the remaining time we will consider the more controversial high-level radioactive waste question.

When nuclear fuel is "burned" in a reactor about 6% of the uranium originally present is split into fission products, or captures neutrons and is converted into higher isotopes, principally plutonium. Both the fission products and the plutonium are radioactive as was the original uranium to a lesser degree. It is characteristic of a radioactive species that it has its maximum radioactivity at the time of its birth and thereafter it decays; that is, reduces its radioactivity until it eventually becomes a non-radioactive isotope. This process may take from a fraction of a second to hundreds of thousands of years, depending upon the particular isotope in question.

In order to assess the difficulty of disposal of these wastes, two primary questions should be asked: (1) "What is the quantity of these wastes?" and (2) "What is their toxicity?". Mankind, more specifically the technological man, has dealt with toxic materials for hundreds of years and radioactivity is merely another albeit somewhat newer form. Its toxicity to man is related not only to the type of radioactive decay and its rate, but also the chemical element in question, e.g., does the element stay in the body?, Where? Fortunately, we probably know more about the hazards to biological life from radiation than we know about the hazards of almost any other trace chemical or material. We can quantify reasonably precisely an upper bound of effects, based on information gained from high exposure levels. Although for very low radiation levels there is some question whether a linear extrapolation of these radiation effects accurately reflect potential impacts, it is generally considered to be a conservative approach. There may be a threshold below which the effects are minimal, but this hypothesis has neither been proven or disproven.

When spent nuclear fuel is discharged from a reactor, it looks pretty much the same as when it went in. That is, it is still an assemblage of metallic tubes containing principally uranium, 94% of which has not undergone a change with most of the remaining 6% being fission products and some transmuted plutonium. Even the volume of these assemblies is relatively modest, consisting o the order of about 400 cubic feet of "hardware" per plant year of operation, which for the 70 nuclear plants in operation over their lifetime would amount to one football field stacked 15 feet high. In this regard, it has been the Federal Government's policy, as defined by the Administration approximately four years ago, that futher treatment of this fuel should not take place because such treatment would represent a nuclear proliferation hazard (nuclear weaponry) and, hence, these fuel assemblies should be disposed of as waste.

Prior to that ruling it was planned that, because of the energy potential these assemblies represent they would be chemically reprocessed; that is, the uranium and plutonium components of the spent fuel would be recovered and the remaining fission products be disposed of as waste. The actual volume of these fission products is about one cubic meter per year for each plant after it is in a "suitable" form for long-term "perpetual" disposal. If we multiply this for the 30 year plant lifetime, and the number of existing plants (70), we find that the same football-field analogy used above would be covered to a depth of about one-and-a-half feet. Because these voumes are so small, essentially negligible in a waste oriented society, it is easy to contemplate the handling of these wastes in a very deluxe way so as to minimize the probability of any future effect on mankind.

It has been stated that although the volumes are small, the toxicity is so incredibly high that a solution to the disposal problem is just not reasonable to contemplate. Such irrational thoughts have been put forward that ten year old nuclear waste would have sufficient lethal doses in them to kill over 10^{10} people (drops to 10^{7} after the wastes have aged a hundred years). If this analogy were applied to the annually used quantities of other chemicals, such as barium and arsenic, they would have similar lethal characteristics, 10^{10} to 10^{11} people. Chlorine is 10^{14} [(1)]. It should be obvious that the number of lethal doses in a given quantity of material is not a very realistic way to compare hazards, since optimally packaged lethal capsules are not generally circulated to everybody in the world.

A classic presentation approach, to represent the hazards, is that shown in Figure 1,[(2)] which indicates the volume of water necessary to dilute a particular material to "drinking water tolerance". Very toxic materials, of course, require more water for dilution. Three major components of radioactivity in discharged fuel are fission products, plutonium, and the structural portions of the assembly. Fission products are the controlling hazard for 600 years, after which the higher isotopes are more important. It is significant that at about this same point in time that the nuclear wastes represent an equivalent hazard to that of the natural uranium that was originally mined and went into the reactor to generate those fission products. In other words, beyond the 600-year point, the world has less nuclear radiation hazards present in it, as a result of that uranium having been placed in a reactor, that it would have had if the uranium remained as ore. This concept of relative toxicity to that of uranium ore has been presentedin many technical publications.[(3) (4)] The time of this equivalent hazard to that of uranium ore is useful, in considering the integrity requirements of a disposal scheme. It is obvious that if the fission products had excellent environmental isolation for about 500 to 1000 years and isolation equivalent to that of uranium beyond that time, the wastes would pose no more of a hazard than uranium ore did in the first place. This kind of realistic hazards analysis puts into better perspective the risk to future generations and destroys the notion that wastes are "incredibly hazardous for hundreds of thousands of years."

3. Radioactive Disposal Alternatives

Over the years there have been a great many alternatives considered for the disposal of nuclear waste. Conceptually, there are two fundamental ideas: (1). Dispersal in our environment, if the concentrations of radioactivity would be so low as to be unimportant; or (2). The isolation or separation of the radioactivity from man or all other biological species. The first of these can be dismissed with a fairly superficial analysis and is not being considered by any country in the world. The second - isolation from man - has many alternatives and subheadings, most of which have been discussed, analyzed and opinionated upon by international committees, environmental agencies and a broad spectrum of scientists and the public. The front runner of all these alternatives is that of geologic disposal; that is, permanent disposal in a suitably isolated geologic formation that has been permanent on a geologic time scale so that the implaced radioactivity will decay in isolation to a insignificant level. The principal geologic disposal approach involves a mined cavity, which will be reviewed in some detail after some of the other alternatives[(5)] are touched upon.

Deep Drill Hole - A drill hole below any water table. The waste would cause local melting which would eventually solidify, forming a rock fission product mixture of indeterminate stability..

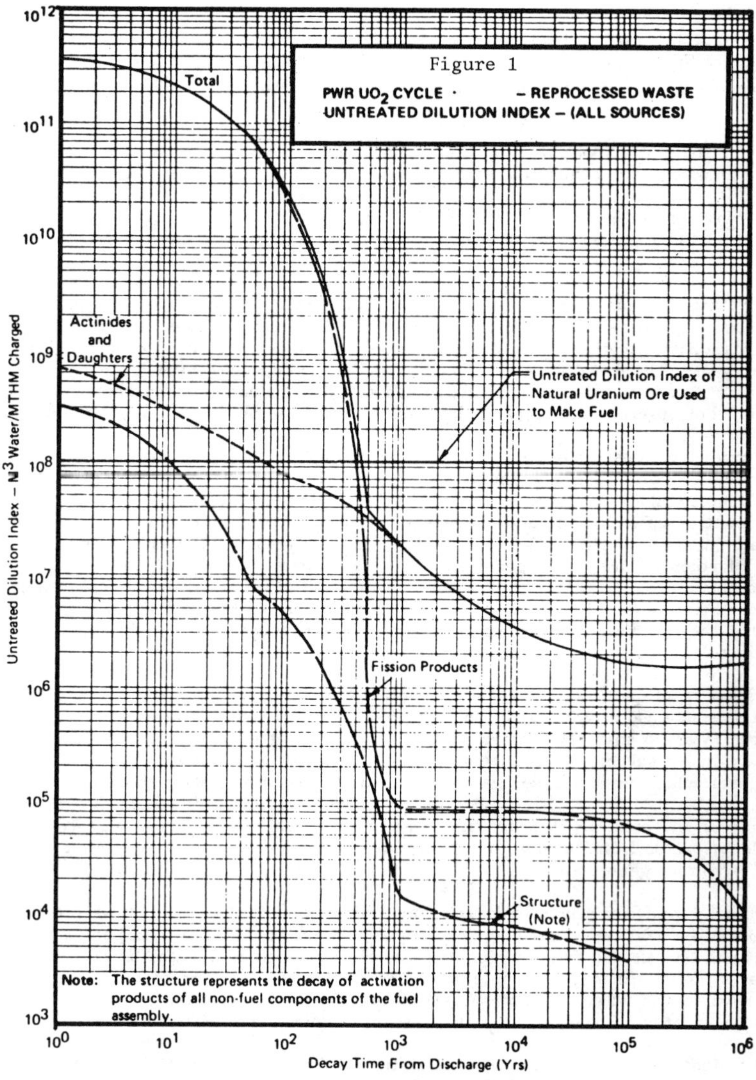
Figure 1
PWR UO_2 CYCLE · – REPROCESSED WASTE
UNTREATED DILUTION INDEX – (ALL SOURCES)
Total
Actinides and Daughters
Untreated Dilution Index of Natural Uranium Ore Used to Make Fuel
Fission Products
Structure (Note)
Note: The structure represents the decay of activation products of all non-fuel components of the fuel assembly.
Untreated Dilution Index – M^3 Water/MTHM Charged
Decay Time From Discharge (Yrs)
10^{12}
10^{11}
10^{10}
10^9
10^8
10^7
10^6
10^5
10^4
10^3
10^0
10^1
10^2
10^3
10^4
10^5
10^6

River Delta - Relatively shallow burial in a river delta area on the theory that the ground is building rather than eroding. The problem is that riverbeds change and ocean levels drop.

Antarctica - Disposal in the Antarctica icecap has a self disposal appeal in that the wastes melt their way down through the icecap to the underlying rock. The icecap's longevity, a function of the earth's climatology, has gone through many cycles.

Ocean Floor - The ocean floor is building but how to get down to assure satisfactory burial is the question.

Ocean Trench - This idea has great scientific appeal, for if the wastes could be imbedded in a deep trench, they would then be transferred under the earth's crust as the underlying earth platelet goes "down the trench hole". The eddies and turbulence at those locations are not known.

Extra Terrestrial - Shoot it into space, that is into an earth or solar orbit or into the sun. The aborted mission is the problem.

Transmutation - Use of neutron or other atomic particle radiation to tranmutes the waste into short lived or non-radioactive isotopes. How much radiation can you use to change radiation?

Mined Cavity - This approach has over the years received by far the major attention as the most promising radioactive disposal alternative. The principle questions regarding such an underground facility is, where should it be located?, and what kind of host rock or geology should be present? The world wide front runner and a candidate for more than twenty years is that of salt.[6] In addition to being easily mined salt has a number of inherent characteristics which make it both useful and demonstrably desirable as a waste-disposal host environment.

The conditions desirable for a waste depository are those of geologic stability, an absence of seismic conditions, volcanos, etc. and an absence of ground water. Water transport is the major mechanism by which buried radioactive material might be moved from the deep repository location up to the surface of the earth, where it could contribute to the radiological dose to man. Most salt fields or domes have been in their present location for tens to hundreds of millions of years. During that period there is de facto evidence that no substantive water penetration occurred - the salt is still there. As an ancillary, but not unrelated aspect of this no water condition is the hypothesis that had there been significant seismic activity, it is probable that the fracturing of the earth would have introduced water paths which would have dissolved the salt.

Another advantage of salt is that it is a plastic material; it has no cracks under pressure, and once the waste is in place and the salt area backfilled, it would tend to creep into a homogeneous mass so that the wastes would be "permanently sealed" within the salt matrix. There are extensive salt fields and domes within the United States - more than 50,000 square miles; and even when the high level wastes are distributed appropriately to allow for their heat release, only about three square miles of repository would be required for the lifetime requirements of the existing 70 reactors. Also, because salt is so extensively available from shallower locations or from the sea, it is considered improbable that these deeper areas would be "mined" by future generations.

A typical repository design looks like most mines except it is much more precise in that it is developed in a grid array in a homogeneous region. A couple of shafts penetrate say 1500 feet to latteral tunnels and drifts which are sufficiently sized to handle "trucks" with shielded transfer casks. In the final tunnels holes would be drilled on a pattern of 6 to 8 feet center to center to receive the waste canisters. The total area of the tunneling might cover about 3 square miles - an area large enough for the lifetime requirements of 70 reactors.

This typical design would be characteristic not only of salt, but of other kinds of rock repositories with relatively minor differences in the geometry to account for the differences in heat conductivity and structural requirements. The individual cyclinders of high level waste would be spaced at different grid patterns, depending upon the salt or rock temperature characteristics. The heat released by the fission products causes a maximum temperature effect over the first few hundred years, and this is one of the design concerns. It would normally be thought that a rock repository would be more stable, except that with this temperature change, failure or cracking might occur. Rock is also more impervious to water; it does not dissolve, but on the other hand, cracks might cause water to penetrate into the region of the repository. Some of the pro and con arguments of rock versus salt, as well as other technological and technical features being investigated have been summarized in a recent Department of Energy report. This document, "Status of Technology for Isolating High Level Radioactive Wastes in Geologic Repository" [3] is about as concise a report as is available to describe the current U.S. program. Even so, the document is a hundred pages long, and has 300 technical references.

As the major hypothesized route from a waste repository to the biologic environment of the earth is water transport, it is obvious that the best form of the waste would be as a water insoluble material. Historically it has generally been thought that the product would be glass, in which the fission isotopes would be incorporated in the melting process. Research and development of a suitable glass has been underway for over two decades. Glass, by any normal standard, is totally insoluble in water; but in a many thousands-of-years time frame somewhat is soluble. It may also devitrify (become crystaline) which would make it somewhat more soluble. These deficiencies have led some researchers to the concept of a synthetic rock, which is substantially more insoluble and does not have the potential problems of devitrification. The difficulty with synthetic rock is that it must be carefully tailored to a particular mixture of elements in a fairly precise ratio; hence getting a high quality on a consistent basis from a somewhat non-homogeneous mixture of nuclear wastes might present a problem.

In any case, the argument of what material has not been resolved to date; although one must ask the question, "Is glass good enough?". It would seem to be, since the Department of Energy policy, formally implemented during the previous Administration, is proceeding with plans for disposal of spent fuel assemblies, a material form which is substantially more soluble than glass.

The combined effects of the waste form, auxiliary "over pack" containment, the geographic location, the geology of the area, the hydrology and limnology, the seismic characteristics, and the design parameters that enter into the repository all have some input into the final bottom line - what is the risk assessment of this kind of operation?

4. Nuclear Disposal - Risk Analysis and Assessment

There are several conceptual problems which make the nuclear waste disposal program difficult for the government regulatory authorities, society and even the technical community to deal with. The first of these has to do with the "what if" syndrome. When we talk about tens or hundreds of thousands of years, a time frame difficult to contemplate, it may seem that anything can happen, and it is impossible to conceive of a test to prove that something will not happen. Secondly, there is no standard, or deminimis value, or satisfactory level of risk in the nuclear power industry. Therefore, the administration, researcher, regulator each seek perfection, since something can always be made better, e.g., synthetic rock is better than glass; double containment is better than single; triple or quadruple is even better. As you combine these "what if" and "no level of radiation is low enough," the result is a program and a technology which gets more and more complicated and, perhaps, even undo-able. Let me just illustrate a few of the "what if" -syndrome situations as they have occurred in the past.

In talking about some reasonably improbable events, which might lead to the disruption or release from a geologically stable high-level waste repository, an early NRC document [7] states:

> "A less likely event (a major meteor strike on the repository, which has an average occurrence rate of once every ten trillion years) could lead to a major release."

To make a statement as "A less likely event could lead to a major release," for general public consumption, for an event which would happen every ten trillion years - in round numbers: two thousand times the estimated age of the earth; a thousand times the age of our solar system and possibly our galaxy, and at least hundreds of times greater than the lifetime of most stars is a gross distortion of any logic that one should apply to this kind of question. This is, perhaps, an ultimate example of a "what if" principle, but certainly illustrates the point in an official document which then generates a public concern.

If I perhaps were a little more clever, I would ask the Johnny Carson-kind of question: "How hazardous are these wastes?" and recite four or five one-liners which might get some attention on the national news media. Even without being clever, however, if you examine the biological toxicity of these wastes, it can be calculated that after 600 years of aging an individual would have to ingest (eat) approximately half a pound of the buried waste to incur a 50% probability of suffering a lethal cancer,[1] not too funny but also not too hazardous.

Because of this relatively low, specific toxicity level, most risk analyses result in a very low man dose or risk to man - it is just very difficult even to hypothesize a mechanism for a significant level of radioactivity to be evolved from a deep geologic formation. There have not been many analyses to date, because the mechanisms are extremely hypothetical and the step wise calculations are not that simple. The results are generally a negligible dosage to man, even if a major disruption of the repository is hypothesized.

To illustrate that with a specific example, I will briefly review certain of the results developed in a detailed migration pathways analysis reported by EPA in 1979.[8] A repository was assumed in which an aquifer was below the repository and a potable aquifer above it. Extremely low-velocity water flowed in the repository strata. The potable aquifer went to wells or surface streams and then through the aquicultural chain to man. In order to develop a "measurable" scenario an "unacceptable" geologic location must be assumed and immediately after the repository is full, water from the lower aquifer penetrates the repository and flows into the upper potable aquifer. In one case a relatively few people directly drink this water but in, a more reasonable population-dose scenario water is carried to a much larger population via surface streams and the agricultural food chain. For this latter case the report concludes "hence, the fifty-year dose commitment calculated for "reasonable scenarios"; i.e., by contaiinated surface water and irrigated food pathways, is negligible in comparison with radiation doses from natural sources." This is for the "instantaneous failure (the time span because of dissolution and transsport delays is over a period of several hundred to more than 100,000 years depending on which isotope is considered.) for a repository that was sited in an "unacceptable" location.

In this same example it is further assumed that the total flow from the upper aquifer flows into the Colorado River one mile from the repository. It was then calculated that the current level of radioactivity in the Colorado River is ten thousand times greater than the radiation from this high level waste repository failure. Similar studies in Sweden and by Battelle Northwest Laboratory have shown that the most exposed man would receive only a small fraction of natural background.[9]

This example illustrates one of the major difficulties of coming to a conclusion as to what is satisfactory in the nuclear regulatory field. That is, although the probable results and, in fact, even the highly improbable results, are very good or quite satisfactory, in the maximum "what if" case, a few people have some risk. Although improving the risk for a few may sound rational, the seeking of perfection in one societal-risk area, results in other societal risks being inappropriately higher. There is no standard for dollars expended per life saved. I might recommend the October IAEA Journal for an extensive dicussion of these relative risk/cost benefit concepts.[10] There are just not enough resources and money to go around. There is no administrative or regulatory procedure to deal with the societal cost-benefit effects, even one narrowly focused on radiation risks per se.

It seems clear that this inability of society to focus on and then understand radition risks from the nuclear power industry relative to other radiation risks or to other "forced" risks of a technological world is a major reason for the emotionalism of the nuclear waste disposal issue. We don't have a perspective because we haven't a suitable observing post. In order to start on that perspective let me pass on just a couple of related data points — First on radiation and then on risks.

EPA published in the Federal Register[11] a summary of radionuclides being released as air pollutants. From the tables summarizing the data a simple multiplication indicates that the most hazardous of the radionuclide releases comes from drinking water, on the order of 40 deaths per year; coal power plants are next, with 20 deaths per year (coal contains up to a few parts per million of uranium); the nuclear industry - primarily that of uranium mining - is 8 deaths per year, while the phosphate industry is very close, with 6. Although not included in this release, because it is not a "man-made environmental" pollutant, natural background radiation results in an estimated 3,000 deaths per year and medical diagnostics (X-ray), 2,600.[12] I might comment that the nuclear waste disposal is only a small fraction of the nuclear industry component above.

One other radiation source which is not very well understood or perceived is radiation risk associated with energy-efficient homes and buildings, the accumulation and buildup of Radon within the structure. This risk situation does relate for if our paramount concern is that of radiation to the population, the question should at least be asked as to whether conservation or nuclear generation is more hazardous. This question is put in detailed perspective by a recent paper entitled "Are Energy-Efficient Homes More Radioactive than Nuclear Meltdowns?".[13] The author concludes the answer to that question is "Yes". As part of the backup data radiation levels in private dwellings were reported. Seven tight homes in Schenectady, New York, e.g., those having on the order of a tenth of an air change hour (as compared to older homes which have at least one to two air changes per hour), had radioactivity levels between three and forty times natural radiation background. Utilizing as the radiations effects basis the EPA tabulation above or the NRC tabulation below, an imputed risk can be calculated of from 9,000 to 90,000 deaths per year, if all homes in the United States were built to those very tight specifications.

This risk evolved from this example of conservation is 100,000 to a million times more hazardous than the waste disposal hazard from 100 nuclear plants based on an summary by the Nuclear Regulatory Commission. The following table summarizes those NRC findings.

Table 1[14]

Comparison of Potential Excess Mortality
Of Nuclear Versus Coal Power Generation per 0.8 GWY (e)

Fuel Cycle Component	Nuclear	Coal
Resource recovery (mining, drilling, etc.)	0.32	0.3 - 8.0
Processing	0.073-1.1	10
Power generation	0.13-0.3	3-100
Fuel Storage	~0	~0
Transportation	0.01	1.2
Reprocessing	0.057-0.065	—
Waste management	0.001	~0
TOTALS	0.59-1.7	15 - 120

The above nuclear values also include asmall component for deaths as a result of a maximum accident as part of the power generation number. Note that the waste management is only a small fraction of the relatively small nuclear total. Even though the averages are low you might say that the risk isn't worth taking because of the catastrophic possibilities of a nuclear incident, either in the generation or in the waste disposal operation, although for the latter it is very difficult to even hypothesize a catastrophic risk. There are, however, other catastrophic power generation risks which can be easily hypothesized. A dam rupture with a probability of failure of once every one to ten thousand years per dam could have fatalities of from 20,000 to 750,000 for certain of the more hazardous locations in the United States.[15] This is to be compared to a nuclear reactor accident, with a probability of occurance of from 1,000 to 10,000 less likely and a maximum number of fatalities of 10,000 to 20,000.

A true perspective on a nuclear program, including waste disposal can only be achieved by examining comparative risks of the alternatives, and relating those to alternatives which society is taking and must take in a technological age. If we are going to look at radiation, consider some of the above examples of radiation risks. If we are going to look at waste disposal, consider that, instead of being concerned that radiation lasts for thousands, or hundreds of thousands of years, compare actual toxicity to other wastes and consider that they have an infinite half-life, that is, they last forever. Materials which immediately come to mind are copper, mercury, lead, arsenic (the latter is spread over the growing areas of our country as a herbicide) and chemicals such as PCBs which eventually end up in the local city dump to leach out to the environment over the next ten, hundred, thousand years?

We do have major technical waste disposal problems in our country, but clearly nuclear waste is not one of them. First nuclear wastes have always been recognized as a hazard which something should be done about; second, the quantities are so small that we can afford to do something about it; and third, by any rational standard, even the most improbable failures of disposal approaches used and being considered would be a nominal societal risk.

(1) Cohen, B. L. "The Disposal of Radioactive Wastes from Fission Reactors." Scientific American, 236(6), 21-31. June 1977.

(2) Arthur D. Little, Inc. (eds.). Technical Support of Standards For High-Level Radioactive Waste Management. Task A Report. Source Term Characterization/Definition. Cambridge, Massachussetts: March-July 1977. EPA 520/4-79-007A.

(3) Klingsberg, C. and J. Duguid. Status of Technology for Isolating High-Level Radioactive Wastes in Geologic Repositories. Washington, D. C.: October 1980. DOE/TIC 11207 (Draft).

(4) U.S. Department of Energy (eds.). Final Environmental Impact Statement. Management of Commercially Generated Radioactive Waste. Volume 1. Washington, D. C.: October 1980. DOE/EIS-0046F, UC-70.

(5) Schneider, K. J. and A. M. Platt (eds.). High-Level Radioactive Waste Management Alternatives. Richland, Washington: May 1974. BNWL-1900, Volume 1-4.

(6) Kupp, R. W. and E. W. Copeland, Jr. Preliminary Site and Hazards Evaluation Nuclear Waste Storage and Reprocessing Facility. New York, New York: June 1960 (Vitro Corporation of America).

(7) U.S. Nuclear Regulatory Commission (eds.). Environmental Survey of a Reprocessing and Waste Management Portion of the LWR Fuel Cycle. Washington, D.C.: October 1976. NUREG-0116.

(8) Arthur D. Little, Inc. (eds). Technical Support of Standards for High-Level Radioactive Waste Management. Task C Report. Assessment of Migration Pathways. Cambridge, Massachusetts: March-July 1977. EPA 520/4-79-007C.

(9) Lennartson, J. R. (ed.). Nuclear Energy Digest. Westinghouse Corporation, Pittsburg, Pennsylvania: 1979.

(10) International Atomic Energy Agency (eds.). International Atomic Energy Agency Bulletin. 22(5/6). Vienna, Austria: October 1980.

(11) "Environmental Protection Agency." Federal Register. 44(249), 76738. Washington, D.C.: December 27, 1979.

(12) Sagan, L. "Radiation and Human Health." EPRI Journal. 4(7). Electric Power Research Institute, Palo Alto, California: September 1979.

(13) Hurwitz, H. Jr. "Are Energy Efficient Homes More Radioactive Than Nuclear Meltdowns?" Paper delivered at the Forum on Nuclear Radiation Risks, International Institute of Safety and Health, Washington, D.C., September 22-23, 1980. Updated Version, November 21, 1980.

(14) U.S. Nuclear Regulatory Commission, Office of Nuclear Material Safety and Safeguards (eds.). Final Generic Environmental Impact Statement on Handling and Storage of Spent Light Water Power Reactor Fuel. Washington, D.C.: August 1979. NUREG-0575, Vol. 1, Executive Summary.

(15) Solomon, K. A., C. Whipple and D. Okrent. More on Insurance and Catastrophic Events: Can We Expect De Facto Limits on Liability Recoveries? Santa Monica, California: March 1978. P-5940 (Rand Corporation).

FUSION - BEGINNING THE ENGINEERING PHASE

MICHAEL ROBERTS, DIRECTOR
DIVISION OF PLANNING AND PROJECTS
OFFICE OF FUSION ENERGY
OFFICE OF ENERGY RESEARCH

An Overview of Fusion

Fusion is the ultimate physical process by which energy may be produced; it is the primal energy source of the universe. On earth, all forms of energy are one form or another of fusion (solar) initiated energy or fission energy. The motivations for pursuing fusion energy are that fusion offers an unlimited fuel supply, has environmental effects projected to be acceptable and potential flexibility in application. The questions associated with fusion have been the achievability of scientific, engineering and economic feasibility.

For more than twenty years, the United States has conducted a magnetic fusion energy research and development program. For most of those years, the program was conducted at a low level in today's terms (i.e., ∿ 10% of current resources). With Project Independence, however, the program's resources grew rapidly and remarkable progress resulted, so much so that in the last two years there has been a fundamental change in the program. Until 1979, the question was "can we achieve scientific feasibility?". The question now as generally perceived by the U.S. and world fusion community is not whether we can achieve it, but how best do we exploit scientific feasibility into a practical system?

We believe the answer to this question will be found in the planning and execution of the engineering development phase just beginning now.

The objective of the engineering phase is the demonstration of fusion's engineering feasibility. This demonstration will come as a result of success in two complementary ventures - full systems integration test in the Fusion Engineering Device (FED) of all those technologies required to generate energy from fusion and supporting component development tests and analyses of those specific issues requiring more extensive exploration. While these achievements will be difficult, we do not see insurmountable barriers.

This paper presents the basis for understanding the decision to enter the engineering phase of fusion energy development.

The Elements of Magnetic Fusion Energy

There are two processes for generating energy from atomic nuclei - fission and fusion. At one end of the periodic table of elements, there are the heavy elements which, when broken apart, transfer mass into energy through the fission process. At the other end of the periodic table, the lightest elements can be held together so that they combine or fuse to produce a heavier particle with a mass loss transfered into energy by fusion. The "holding together" of the light atoms can be accomplished by any of three methods - gravitational, inertial or magnetic confinement. The first method describes our sun as a fusion furnace, the second method describes the extremely rapid heating of a small fusion fuel pellet by a very intense light or particle beam and the third describes the means of confining the electrically charged (ionized) atoms by the use of especially configured magnet field line geometrics. The development of fusion as an energy source is principally pursued via the magnetic confinement approach.

The practical implications of this elementary fusion process can be better appreciated by considering the process in its three steps.

1. Fuel Preparation - The fuel cycle most likely to be used first will be the Deuterium-Tritium one in which these two isotopes of hydrogen are extracted from water and produced from lithium respectively. The former is relatively straightforward while the latter is most easily done as a breeding function in fusion reactors themselves.

2. Fusion Itself - The fuel constituents must be held together with sufficient mass density, interaction lifetime, internal heat and gross volume that fusion takes place. Excluded from this fusing environment is any impurity; this exclusion requires some form of wall between the fusion chamber and outside world. Magnetic fields are used to provide the confinement, one or more heating schemes for the particle temperature, pellet fuelers for the maintenance of steady or increased density and some magnetic or mechanical means of isolating the plasma from the walls. Magnetic field configurations can be either ones with closed field surfaces, toruses typified by the tokamak configuration, or non-closed or open field lines typified by the mirror geometry.

3. Reactions from the Fusion Process - After D-T fusion occurs, the resultant products are two particles - a 14 Mev neutron and a 3.5 Mev alpha (He) particle. The neutron carries the principal usable energy and must be slowed down in a neutron absorber to transfer the kinetic energy into thermal heat energy. In so doing, the neutron activates the structures through which it passes leading to the remote maintenance and materials requirements of fusion. The alpha or ash particle provides the self heating for fusion and remains in or on the chamber.

The greater part of the fusion program has been fully occupied with the second step for more than two decades. In recent years, an increasing fraction of the effect has been directed toward developing materials for fusion service and preparing for experimentation with burning plasmas with alpha particles.

The engineering initiative is directed at accomplishing the second and especially the third steps in a practical and usable fashion resulting in an engineering data base. Subsequent work would focus on the first step as well.

The policy basis for fusion R&D readiness

The Department of Energy's basic policy for fusion was developed in 1978; the policy is to develop the highest potential of fusion energy in an environmentally acceptable and economically competitive manner. Fundamental to this policy, which calls for a maintenance of the program's strength and breadth, is the concept of proceeding toward an energy producing device without foreclosing the options for significant improvements in the current technical approaches.

We are proceeding with two concurrent, complementary and coordinated efforts toward an energy producing system that will form the basis for subsequent commercialization. One effort is the engineering approach aimed at developing individual component and integrated system experience with the practical aspects of creating and managing a burning plasma. The other effort is a scientific investigation with two principal goals: 1) resolution of those remaining technical uncertainties crucial to the achievement of the burning plasma engineering objective; and 2) development or discovery of reactor concepts that promise more optimum characteristics than presently projected for current concepts.

Given this program strategy, there remain three critical questions whose answers will shape the outcome of the fusion R&D effort for some years: 1) is the program technically ready to shift into the engineering phase? 2) what is the appropriate pace of the program? and 3) what are the more particular technical, organizational, programmatic and institutional characteristics of the next major step in the program?

In order to address these questions most properly, the Department chartered in 1980 a high level fusion review by the Energy Research Advisory Board (ERAB), led by Dr. S. J. Buchsbaum. The August, 1980 ERAB report begins with the review group's first conclusion that "recent progress in plasma confinement justifies confidence that demonstration of scientific feasibility of magnetic fusion...is near" and "that a device containing burning...plasma can be built and operated successfully" and presents the following principal recommendations that "the magnetic fusion program can, and should, embark on the next logical phase toward its goal of achieving economic feasibility of magnetic fusion. To this end, a broad program of engineering, experimentation and analysis should be undertaken under the aegis of a Center for Fusion Engineering (CFE)." The CFE should undertake the design and construction of a Fusion Engineering Device to provide a focus and test capability for developing fusion reactor technologies and a broad based program in plasma confinement should be conducted, based on five new major elements.

In 1980, there were three other intensive reviews of magnetic fusion readiness - by the Fusion Committee of the Atomic Industrial Forum, the International Fusion Research Council of the International

Atomic Energy Agency and the Fusion Advisory Panel of the Energy Research and Production Subcommittee of the House Science and Technology Committee. All four concluded that fusion was ready to move into the engineering phase. This judgment and many of the specific supporting ideas have been incorporated both into Public Law 96-386 - the Magnetic Fusion Energy Engineering Act of 1980 - signed into law by the President on October 7, 1980 and, by addendum, into the Department's fusion policy in October as well. With the change of Administrations the decision to enter the engineering phase at such a pace as to double the budget in the next 5-7 years is being reviewed. Fusion is technically ready for engineering development; the pace of that development, however, must be determined on the basis of an analysis of the whole range of alternatives available to the country.

Recent progress as a basis for judgment

Across the program during these past two years, there have been scientific successes whose achievements serve as the basis from which to move into engineering. In the mirror confinement program, the new tandem mirror principle was proven in the TMX device at Lawrence Livermore Laboratory leading to support of a major scale-up in a follow-on device, the Mirror Fusion Test Facility-B (MFTF-B) project. A new idea for further enhancement of the reactor potential called the thermal barrier has extended the tandem mirror principle. In the tokamak confinement program, a host of experimental successes were accompanied by new and advanced tokamak ideas promising to lead to more attractive tokamak reactor concepts.

The attainment of reactor level ion temperatures in PLT at Princeton Plasma Physics Laboratory was followed by operation at near reactor level betas (critical engineering efficiency of plasma pressure compared to magnetic field pressure) in the ISX-B device at Oak Ridge National Laboratory (ORNL) where pellet fueling was also demonstrated. In Princeton, the Poloidal Divertor Experiment (PDX) was commissioned to investigate an active density and purity control technique; first, continued operation indicates positive successes with the divertor scheme, an observation recently corroborated and extended in the ASDEX device in Garching, Germany.

In the Wendlestein VII stellarator, also at Garching, recent successful operation with the combined features of both stellarator and tokamak have led to renewed promise of successful steady-state toroidal reactor operation. In the Doublet III device at the General Atomic Company, reactor level plasma currents in the mega-ampere range were achieved. In the 'bumpy torus' confinement program, there was a sufficient basis of technical achievement and confidence developed from the EBT-S device at ORNL to initiate a proof-of-principle experiment called EBT-P.

These individual achievements are dramatic, crucial to the actual advances of the program and are truly satisfying but there is also another way of considering progress that is rather productive and illuminating. Even though there appear to be many different confinement configurations, they all can be characterized by a small set of common parameters - namely ion and electron confinement measures (particle density times confinement time - $n\tau$ -), ion and electron temperatures, electron density (including fueling), beta, purity control and alpha particle effects. With regard to the first seven parameters, those with which we have had experience, we can now say that conditions adequate for reactor level operation can be achieved; more specifically, all seven are adequate in tokamaks and all but the first two are adequate more generally in both open and closed magnetic geometry concepts. With regard to the eighth parameter, alpha particle effects, it appears likely that this parameter will be adequate for a reactor. Alpha particle effects will be studied directly in the TFTR and then fully in the FED.

With regard to technological gains, there have been a number of important achievements of which a few examples must suffice. An innovative new accelerator concept called the radio frequency quadrupole, conceived in the Soviet Union, was developed, tested and proven by Los Alamos National Scientific Laboratory (LANSL) for the Fusion Materials Irradiation Test (FMIT), recently renamed the Mike McCormack Fusion Materials Test Facility, project being built at Hanford Engineering Development Laboratory (HEDL). This achievement is a most exciting technological success with regard to the inherent science, the furtherance of the FMIT project and perhaps most interestingly, to the accelerator field and a host of improved applications totally apart from fusion, including medical technology. For TFTR, the high energy neutral beam plasma heating system upon which the entire project depends was successfully tested as a full prototype recently. Using these advances and others in the areas of practical superconducting magnets, microwave tubes and materials, an improved tokamak reactor concept called Starfire evolved within work done at the Argonne National Laboratory.

One way of demonstrating graphically the dramatic change in the fusion program over the last decade is shown in the figure in which a comparison is made between the size of a tokamak reactor projected in the various studies and the actual size of experiments being used at the time these projections were being made. This vast disparity so evident to workers in the field in the early '70's has been reduced and effectively eliminated by the combination of increasing the sizes of plasmas actually being worked on and reducing the projections through increased understanding. We are now at the point where the size of the next experimental device (FED) is effectively equal to that of the energy producing reactor (FED)!

Major judgments

Traditionally and currently, the fusion community has been fully involved in the planning and replanning of the fusion program. The plans, consistent with P.L. 96-386, are formulated on the basis of seven fundamental judgments.

- The first is that the next major step in the fusion program should be toward engineering and not power production per se or more basic R&D. Simply stated, this judgment reflects the fact that the questions left to be answered are predominantly engineering rather than physics in nature and are of sufficient concern that it would be imprudent to set out today to build a power plant.

This judgment is the unanimous view of each of the four groups mentioned above. The technical basis of readiness is described in each group's report and in a comprehensive and current statement by Clarke.(1)

(1) An Interpretive Overview of the United States Magnetic Fusion Program, J. F. Clarke; a program overview article in the Special Fusion Issue of the Proceedings of the IEEE, 1981.

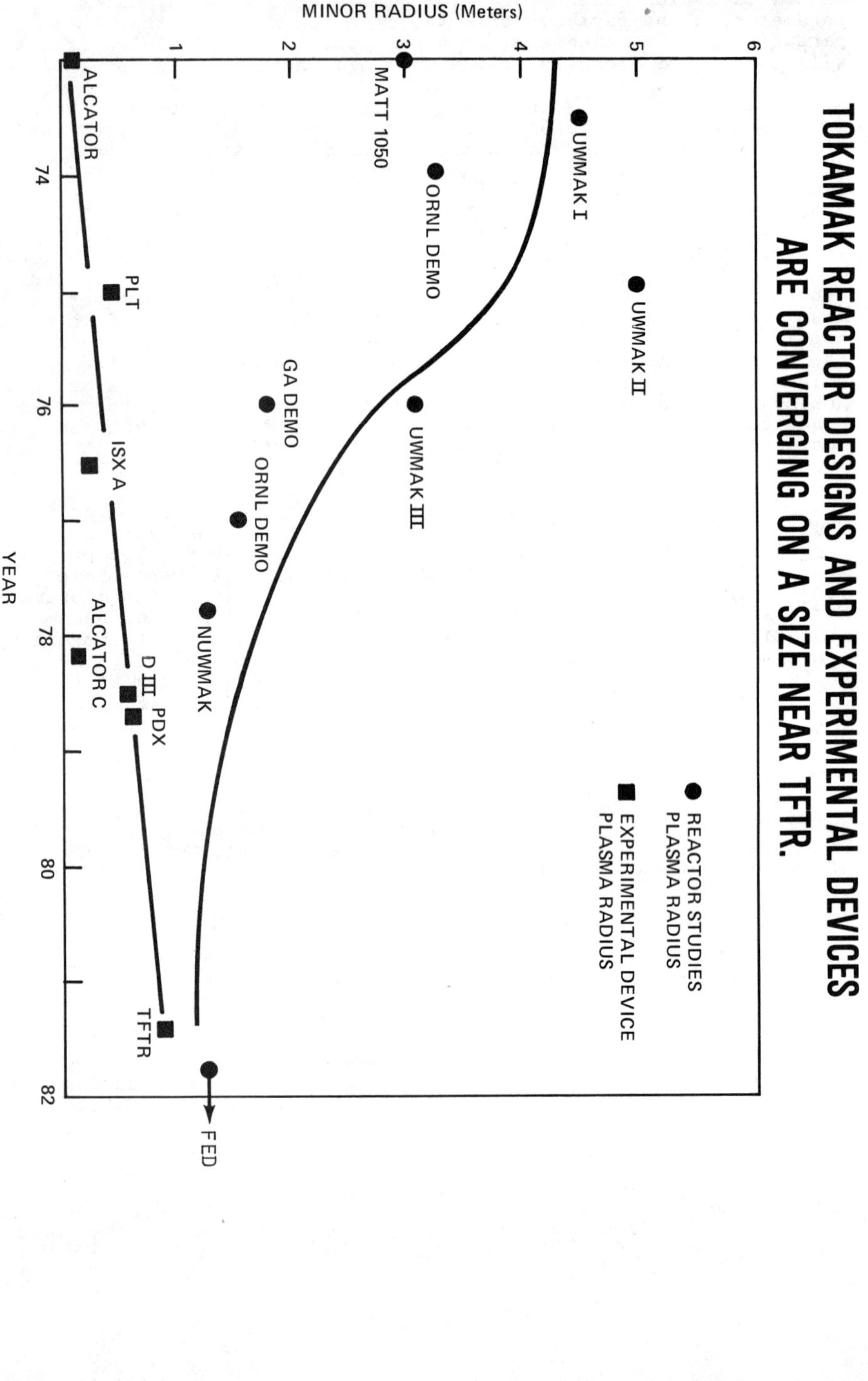
TOKAMAK REACTOR DESIGNS AND EXPERIMENTAL DEVICES
ARE CONVERGING ON A SIZE NEAR TFTR.
REACTOR STUDIES PLASMA RADIUS
EXPERIMENTAL DEVICE PLASMA RADIUS
MINOR RADIUS (Meters)
6
5
4
3
2
1
UWMAK I
UWMAK II
UWMAK III
MATT 1050
ORNL DEMO
GA DEMO
ORNL DEMO
NUWMAK
FED
ALCATOR
PLT
ISX A
ALCATOR C
D III
PDX
TFTR
74
76
78
80
82
YEAR

FIGURE 1

- The second is the basic fusion strategy developed to implement this initiative. Over the past two years, we have built a strategy around the fundamental characteristics of fusion that have emerged from the successful efforts in the mid-to-late 1970's. This strategy is built upon two basic ideas - the generic nature of fusion engineering and the convergence of the various fusion confinement concept physics bases.

With regard to the engineering, the tokamak based Fusion Engineering Device will serve as the source of a great deal of generic engineering information focused on energy production in an integrated systems environment. When examining the various systems in a fusion energy producing device, it is clear that the majority of those systems is in common for any of the currently likely reactor concepts. Those systems in the Balance-of-Plant are highly similar, those systems in the auxiliary parts of the central fusion plant, such as cryogenics, tritium handling and some electrical systems have most engineering essentials in common and those systems specific to the confinemnt concept have common points but differ in some important elements and design consideration. As a result, we believe the tokamak based FED will provide operating experience as well as design and construction experience in systems common to nearly any reactor concept.

The overall engineering development program consists of two pieces - the FED centered system integration part and the complementary, off-FED ancillary test-facilities-oriented broad research program in engineering issues. These two parts of the engineering program could be led and coordinated by the proposed Center for Fusion Engineering.

With regard to the physics, there is now a significant and growing convergence of scientific understanding which underlies the various confinement concepts and leads to a commonness of physics understanding and a clarifying to a single set of technological requirements for a generic reactor. Only a few years ago, the program was organized around each of the dozen or so reactor concepts. Two principal factors have caused the fundamental change to the current program. The cost of pursuing each approach has increased significantly as the line of investigation proceeds to more advanced tests, and the basic understanding about each has increased significantly with these investigations. As a result of both forces, there has been both a natural focusing on fewer, promising concepts and in the last two years the evolution of a rapidly growing body of common understanding to fusion.

- The third judgment has to do with the size of the next step in the program. The FED is a large and costly step, made so by the simultaneity of two threshhold requirements. The first is the creation and maintenance of a burning plasma, and the second, necessary threshhold is the removal of heat energy at a high enough temperature to convert to useful power. Putting both of these together requires a significant and costly step in the sense of a "quantum jump".

  A corollary to these judgments is that the fusion program, while a nuclear based technology, is not a replication of the fission program; indeed, the two development paths are rather different and specific to their intrinsic technological characteristics.

. The fourth judgment has to do with the setting of the next major objective in the program - a major fusion assessment. We believe that the combination of the FED experience in design, fabrication, assembly and operation complemented by the broad engineering program and the output of the science and technology base program will be the basis of the major national assessment of fusion's engineering practicality as a means of determining its "highest potential" for subsequent demonstration and deployment. Our judgment today is that the outcome of the assessment will be to proceed directly with the commitment to the operation of a utility-size energy producing plant.

. The fifth judgment is that the science and technology facilities are in or being put in place now for the 1980's so that nearly all the information required for the FED will be available without other new facilities. To be sure, there will be modifications and evolutions of present and currently approved new facilities as R&D information is obtained and digested, but the number of additional facilities needed is believed very few.

. The sixth judgment deals with the scope of the engineering device or more generally the balance between system integration and component development in the engineering program.

There are three alternative strategies for implementation of the engineering development program goals:

1. Focus on one, all encompassing Engineering Test Facility (ETF) for component testing, systems integration, and materials testing.

2. Pursue a program of an FED for component testing and systems integration coupled with a complementary engineering development program accomplished using test facilities other than FED.

3. Place emphasis on existing and modest scale devices and test stands for component development while deferring systems integration and fusion power handling to a later and presumably, larger step.

With regard to strategy 1, ERAB concluded that the previously planned approach to develop and test components and build an integrated or high performance reactor (ETF) was inappropriate at this stage. They said:

> "Some of the goals of the ETF are overly ambitious, resulting in part from the desire to use the ETF as the ultimate material testing facility, as well as a test bed to pin down engineering design parameters. Such goals result in stressful requirements: the required fluence levels are very high; component downtime is very low; burn times are very long. As a result of these multiple goals, the complexity, cost and risk of failure are high."

Similarly, strategy 3, using existing experiments as modified for component development and simulation for systems integration, has been discarded as too risky for demonstrating engineering feasibility. Upgrades of the largest fusion devices, MFTF-B and TFTR, pursued alone in the absence of a systems integration facility, would not be adequate to meet the requirements of achieving confirmation of engineering feasibility.

The ERAB recommendation of strategy 2, focussing on FED as the technically based strategy of choice, was chosen as Department Policy in October, 1980.

. The seventh and final judgment is that of the pace of the fusion program, a factor closely linked to resources and relative priorities. The implementation of the strategy supporting the policy described above is highly dependent upon funding provided to the fusion program. Our planning therefore has considered a range of possible courses of action all supporting the basic strategy but differing in pace, breadth, risk, yearly cost, and total cost. An extensive discussion of the strategic planning issues, the plans and the costs can be found in the Record (No. 67) of a Hearing on Strategic Planning in the Fusion Program held by the House Committee on Science and Technology, Subcommittee on Energy Research and Production on December 10-11, 1979. The three plans include a range of dates for serious initiation of a Fusion Engineering Device between 1982-1986, with operation 5-8 years later respectively, the major program assessment between 1990 and 2001 respectively and operation of a Demonstration Plant between 1995 and 2010 respectively.

These planning ideas formed a significant input to the process leading to the adoption of P.L. 96-386 in which the technically based pace having as goals the assessment by the early 1990's and operation of a fusion demonstration plant by the year 2000 was directed by the Congress and President subject to the yearly availability of funds. As stated above, with the change of Administrations, the decision with regard to pace is being reviewed; following this review, the plans will be reformulated as appropriate.

In addition to these broad statements, there are other crucial considerations in the implementation plans. One of the most important is the mix of participants which continues to evolve. Currently, the national laboratories play the predominant role with support by industry and university personnel. As we move into the engineering phase, industrial participation will increase while the laboratories focus on the resolution of the remaining technical issues and then proceed to advanced ideas leading to optimized fusion systems. Another important consideration is the extent and scope of our already active international cooperation program encompassing each of the major fusion blocs and international organizations.

In summary, it is clear from the judgments of each of the four intensive reviews that the fusion program is technically ready to proceed into the engineering phase and the principal practical question remaining is that of affordability in the near term.

FUSION - TECHNOLOGY AND ENGINEERING CHALLENGES

Dr. J. R. Gilleland
General Atomic Company

INTRODUCTION

Successful experiments conducted throughout the world's scientific community are moving fusion power research toward an important new phase of development. The scientific feasibility of fusion is being demonstrated in a number of plasma confinement devices, and fusion's theoretical potential as a nearly ideal and limitless source of energy is being confirmed. It is therefore logical that fusion move into the era of engineering and technology development. Only thus can the advantages of fusion be realized on an acceptable timescale. As discussed in the previous paper, the demonstration of engineering feasibility will require a balanced program of technology development, with construction and operation of a reactor (the Fusion Engineering Device) as the centerpiece of the effort. In the following, a brief sketch of the technologies and engineering challenges inherent in fusion will be given.

THE ENGINEERING DISCIPLINES AND TECHNOLOGIES OF FUSION

All of the basic technologies required for fusion have been developed and used, to some degree, in other high-technology ventures, although considerable extrapolations are often required. From the engineer's viewpoint, the uniqueness of fusion lies in the diversity and special mix of its technologies. The high degree of interaction of the systems has led to a need for "fusion engineers" whose expertise spans a broad range of technologies and disciplines. Some of the more interesting are discussed below.

PLASMA ENGINEERING

Use of a magnetic bottle as the pressure vessel for containment of a hot, reacting plasma is the foundation of "magnetic fusion." While the concept is extremely elegant, its employment requires a high level of technical sophistication. Plasma engineering encompasses a knowledge of all those disciplines required to generate and control the reactor plasma in a magnetic bottle. Examples are control of the shape of the magnetic field and of the density and temperature profiles within the plasma. This is accomplished through the use of carefully programmed magnetic field coils and of finely tuned fuel heating techniques. Our capabilities in this discipline are the result of twenty-five years of research.

MAGNET TECHNOLOGY

For most of the leading approaches to fusion power, the use of superconducting magnets is highly desirable as a means of minimizing the recirculating power. The shape of the coils and variation of their currents is dictated by the type of magnetic bottle (e.g., tokamak, mirror, Elmo Bumpy Torus, etc.). Most concepts require magnetic fields at the conductor in the range from 8 to 12T (8,000 to 12,000 Gauss). For coils with a conductor field below 10T, the choice of NbTi superconductor is feasible and probably advisable. For fields above 10T, Nb_3Sn superconductors is being contemplated. Several significant development or demonstration efforts are under way. Of fundamental importance is the high-field magnet program of the Department of Energy, which is aimed toward exploring the limits of stable high-field operation for titanium- and tin-based superconductors. The Large Coil Project at Oak Ridge National Laboratory (ORNL) will demonstrate reliable operation of a toroidal set of large superconducting magnets, each of which features a unique engineering approach. All of the coils are being fabricated by industrial suppliers in the United States, Germany, and Japan. Figure 1 shows the vacuum vessel and coil-holding "spider" under assembly. First tests are scheduled for 1982. Figure 2 shows a large superconducting NbTi magnet under construction at Lawrence Livermore Laboratory (LLL) for the MFTF-B mirror experiment.

PLASMA HEATING TECHNOLOGIES

Fusion reactors may be categorized as operating in either a "driven" or in an "ignited" mode. In a driven system, a continuous source of heat must be supplied to the fuel in order to maintain the fuel burn. In the ignited mode, the fuel continues to burn on its own after an initial heat-up. In either case, large plasma heating systems (tens of megawatts) are required. Two main classes of systems, neutral beams and radiofrequency heating, are being developed.

Neutral beams have served a workhorse role in demonstrating scientific feasibility. A high-energy positive ion beam of hydrogen is neutralized upon passage through a chamber of gas. Because it is neutral, the beam can then penetrate the strong magnetic fields of the reactor bottle and deposit its energy in the plasma.

High-power radiofrequency heating systems are also being successfully tested over a wide range of frequencies. These heat the plasma in a manner which is crudely analogous to the operation of a microwave oven. Further development of both techniques is considered essential, as each has its own set of ancillary applications or advantages.

NUCLEAR TECHNOLOGY

The high-energy neutrons produced by the fusion reactions are stopped in a "blanket" that surrounds the plasma and produces high-temperature coolants for electricity production or process heat applications. Some portion of the blanket is devoted to producing tritium, a necessary constituent of the fuel. Examples of these systems are shown in the following paper by Dr. Charles Baker. In this area of technology, fusion benefits greatly from previous experience in the nuclear industry; a vast store of engineering and testing tools is directly transferable. Nevertheless, neutron spectrum differences and electromagnetic effects present challenging new development problems. Argonne National Laboratory (ANL) has recently undertaken a broad program of blanket module development and testing. They plan to work closely with industry in this effort.

The ability to process and handle tritium efficiently is a key to practical and safe plant operation. The Tritium Systems Test Assembly now under construction at Los Alamos Scientific Laboratory (LASL) will demonstrate the key tritium systems (see Fig. 3) required for operation of a reactor.

Fig. 1 View of Large Coil Test Facility, Oak Ridge National Laboratory. Structure base for 6-coil torus being lowered into 11 meter diameter vacuum tank, August 1980.

Fig. 2 Large superconducting magnets for MFTB-B, under construction at Lawrence Livermore Laboratory.

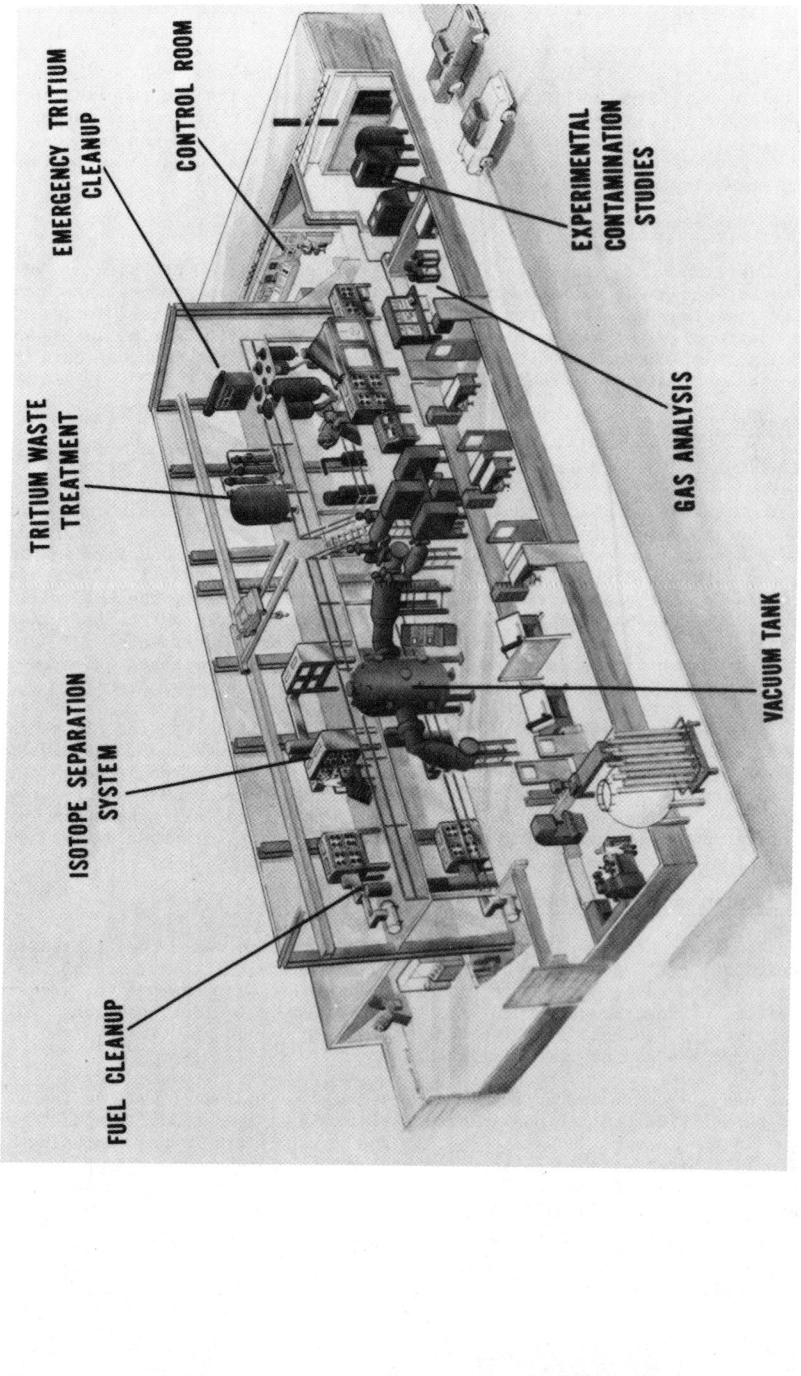

Fig. 3 Tritium test assembly, under construction at Los Alamos Scientific Laboratory

A small tritium test module is being designed by General Atomic Company for use on Princeton's Tokamak Fusion Test Reactor. This experiment will serve as a verification test for a tritium production in an integrated tokamak environment. It is a cooperative program in which the Electric Power Research Institute (EPRI) is providing support for a test to be conducted on a Department of Energy-sponsored machine.

The geometry of all fusion reactor blankets is characterized by the necessity for a relatively large number of penetrations. These holes are required to allow for vacuum pumping, plasma heating, and plasma diagnostic or sensing purposes. The shielding for neutrons that stream down these penetrations can add cost and operational awkwardness to a reactor. Programs enhancing our ability to calculate neutron streaming are under way.

DIAGNOSTICS ENGINEERING

The development of fundamentally new plasma diagnostics has been one of the keys to the successful advancement of fusion research. Experimentalists have invented ingenious new tools for measuring temperatures, densities, ion species mix, radiation levels, and many other effects as a function of time and space. The challenge of the coming decade will be to "harden" these diagnostics so they can operate reliably in a true reactor environment.

MATERIALS TECHNOLOGY

Most fusion reactor designs call for neutron fluences of the order of 10 $MW\text{-}yr/m^2$ into the blanket. Since the fusion neutron spectrum has an energy component higher than that encountered in fission, we cannot rely solely on our previous nuclear experiences for prediction of long-term materials performance. Materials behavior under relevant fluence and spectral conditions is being investigated with the Rotating Target Neutron Source at LLL and will be extended with the Fusion Materials Irradiation Test (FMIT) at Hanford. The FMIT will be able to subject small specimens to damage rates that are four to five times those anticipated for reactors. It is hoped that these experiments will serve not only to search for fundamental effects, but also to permit better correlations of future fusion reactor performance with fission simulations.

In addition to radiation damage, physical and chemical erosion of the chamber wall next to the plasma is of concern. Ideally this surface should be made from a low atomic number material to minimize the effect of sputtered atoms on the plasma performance, yet capable of withstanding high temperatures and relatively high stresses. Although workable designs have been found, most require maintenance operations sometime during the life of the reactor. Further development is required to allow maintenance-free operation.

REMOTE MAINTENANCE TECHNOLOGY

The fusion reactor will generate far less radiation waste than a fission plant. The basic fuel cycle generates no radioactive products. Tritium, one of the constituents of the fuel, emits soft radiation and is of short half-life. Nevertheless, pieces of reactor structure will be activated by incident neutrons. Although these activation products are also short-lived, it is not practical to run a fusion plant without the use of remote maintenance.

The use of robots and of custom-designed remote equipment will definetly be a part of normal plant operations. Fortunately, remote operations in space, remote handling in the fission industry, and now the rapid growth in manufacturing robotics will provide the necessary technology base for fusion. The Fusion Engineering Device will be the first to require a true integration of reactor-relevant remote maintenance techniques into its basic design criteria.

THE CHALLENGE OF INTEGRATION

Although the above list of engineering disciplines and technologies is far from complete, it does illustrate effectively what a diversity of technologies is involved. The key challenge lies in their proper integration. In contemplating the systems integration task, one is immediately struck by the multiplicity and degree of interactive effects. These interactions are the result of several simultaneous mechanisms. For example, mechanical structures must be designed to withstand stresses which result from gravity, thermal gradients resulting from plasma particle and radiation deposition, from differential nuclear heating, from differential induction heating, and from electromagnetic body forces. Further, because the properties of the structural materials are a function of temperature, chemical environment, and radiation history, the design solution is time dependent. Unique to fusion are the electromagnetic effects, which can dominate and couple the design of several major systems. Fortunately, the room for possible engineering solutions is large and credible designs do exist. Examples of this type of integrations are shown in Figs. 4-6. Figure 4 shows the largest integrated fusion system in operation. This tokamak, called Doublet III, was constructed by General Atomic Company and has been successful in confirming many aspects of plasma engineering. Figure 5 is the main vacuum chamber of the Mirror Fusion Test Facility now under construction at Lawrence Livermore Laboratory. Note the scale of hardware involved. Princeton's Tokamak Fusion Test Reactor (Fig. 6) is scheduled to go into operation in 1982. It is the first U. S. device designed with remote maintenance considerations in mind. It will be the first machine to burn tritium, rather than stimulate reactor performance with another type of fuel. The amounts of tritium involved are much smaller than needed in a reactor, and thus TFTR will not attempt to demonstrate reactor-scale nuclear technology.

All of these projects, however, have as their primary focus the study of key plasma engineering issues. Incorporation of reactor relevant technologies and engineering practices has been of secondary importance. Such an engineering integration is now overdue. On the basis of our recent experimental successes, national and international assessment groups have concluded that a device should be constructed that will have as its prime mission the integration of reactor technologies in a forceful demonstration of fusion's engineering feasibility. The Fusion Engineering Device will provide precisely that type of integrative focus.

(1) Engineering Test Facility Interim Design Description Document, July 1980, by Engineering Test Facility Design Center (Oak Ridge National Laboratory).

(2) INTOR Group, International Tokamak Reactor: Zero Phase, STI/PUB/556, IAEA, Vienna, 1980.

(3) Report on the Department of Energy Magnetic Fusion Program, Buchsbaum, et. al., August 1980.

Fig. 4 Doublet III Tokamak Facility in operation at General Atomic Company, San Diego.

Fig. 5 Main vessel for MFTF-B

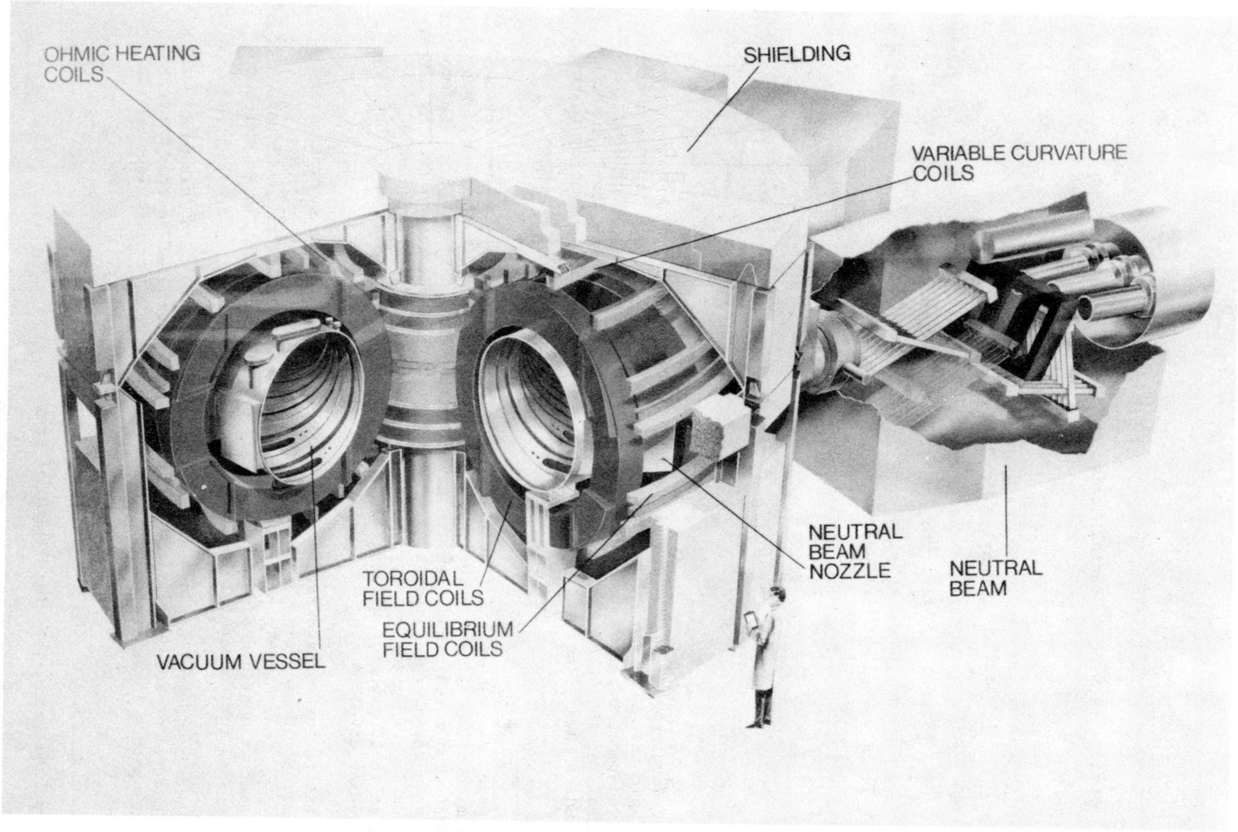

Fig. 6 Cutaway of the Tokamak Fusion Test Reactor, under construction at Princeton Plasma Physics Laboratory.

Fusion - Advanced Reactor Design and Potential Application*

Charles C. Baker

Argonne National Laboratory
Argonne, Illinois 60439

Abstract

Recent developments in commercial reactor concepts based on magnetic confinement fusion are reviewed. New ideas in tokamaks, mirrors and other concepts have improved reactor performance and have resulted in increased confidence in the potential for the successful development of commercial fusion energy. Recent reactor studies have placed increased emphasis on maintenance and safety features.

1.0 Introduction

This paper presents a brief overview of recent developments in reactor concepts for magnetic confinement fusion. Much of the material in this paper has been taken from a recent comprehensive review of the same subject [1]. The emphasis here is on recent developments with respect to reactor concepts for the commercial application of fusion energy. Companion papers by M. Roberts and J. Gilleland in these proceedings describe current and near-term fusion technology and engineering developments.

The last few years have seen a substantial increase in the understanding of the engineering and technology issues regarding the development of commercial fusion energy. Considerable progress has been made in developing new ideas which has increased our confidence that practically safe and economic fusion reactors can be developed for commercial application.

There are a wide variety of magnetic confinement concepts. This paper will highlight the recent developments in tokamak and tandem mirror reactor concepts. Recent developments in several additional alternate concepts will also be briefly described.

---

* Work supported by the U.S. Department of Energy.

A variety of elements can undergo fusion reactions. The deuterium/tritium (DT) fuel cycle has the highest reaction rate together with a high energy release per reaction (17.6 MeV), and, therefore, the largest power density for a given plasma density. In addition, the DT reaction has the lowest plasma temperature requirement of any of the fusion fuels. It is generally thought that the first commercial fusion reactors will operate on the DT fuel cycle, thus DT reactors will be the emphasis in this paper.

The non-DT fusion fuels, often called "advanced" or "alternate" fuels, are of interest because they offer the potential features of no tritium breeding, reduced tritium handling requirements, reduced neutron activation of structural materials, and a larger fraction of the energy released by charged particles which may permit the use of direct energy conversion into electricity. On the other hand, the smaller cross sections and higher temperature requirements imply more difficult plasma physics confinement problems and the need for larger magnetic fields. The physics and technology issues of alternate fuels are receiving increasing attention [2,3].

It is important to also note that there are several potential end uses of fusion energy. The focus of this paper is on electricity production, usually in the form of a base-loaded central station power plant. However, there is continuing interest in the application of fusion energy to synthetic fuel production [4-6] (e.g., hydrogen) and the production of fissile fuel and electricity in fusion-fission hybrid reactors [7-9].

2.0 Tokamak Reactor Designs

2.1 Overview of Tokamak Reactor Designs

The tokamak concept represents today the mainline experimental device in the field of magnetic confinement fusion research throughout the world. It is the approach which has achieved the most promising results to date, and is expected to be the concept by which energy breakeven will first be demonstrated for fusion energy in the next few years.

Studies of tokamak reactor concepts have been underway for more than ten years. More than 30 studies have been carried out within the United States as well as several studies in the Soviet Union, United Kingdom, Japan and Italy (see Ref. 1 for a more detailed discussion and list of references of past reactor studies). The progression of tokamak reactor concepts is schematically illustrated in Fig. 1, which illustrates the various stages of reactor design studies using representative reactor concepts (UWMAK-I [10], UWMAK-III [11], NUWMAK [12] and STARFIRE [13]). Clearly, there has been substantial improvements in the tokamak concept in terms of reducing its size and power output. This improvement [14] has resulted because of better estimates of the plasma size required for ignition (minor radii of $1 \rightarrow 2$ m), expected stable values of β in the range of $5 \rightarrow 10\%$, and first wall/blanket lifetimes of 10-20 MW-yr/m^2 which permit higher wall loadings (see Fig. 1). These developments in physics and technology have resulted in tokamak reactor plant outputs in the range of $600 \rightarrow 1200$ MWe which are compatible with current utility requirements.

The major reactor parameters for several selected tokamak reactor designs are listed in Table I. The most recent study is STARFIRE [13] which has incorporated several ideas developed in a variety of recent design efforts. These ideas have resulted in improved reactor simplicity, maintainability and safety. The major features of the STARFIRE concept are shown in the isometric view in Fig. 2 and the key features are listed in Table II.

A major effort has been devoted in STARFIRE to enhancing reactor maintainability and improving plant availability. The approach was to select design features and develop a design configuration that reduced the frequency of failure and shortened the replacement time. Relevant examples are: steady state operation with lower hybrid current drive; limiter/vacuum system for impurity control and exhaust; vacuum boundary located at the shield with all mechanical seals (no welds); all service connections (e.g., for high pressure coolant) are located

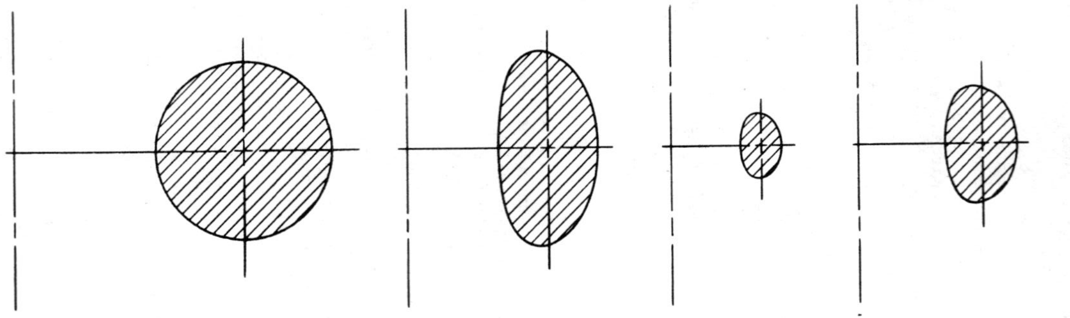

	UWMAK-I 1973	UWMAK-III 1976	NUWMAK 1978	STARFIRE 1980
R (m)	13	8.1	5.1	7.0
a (m)	5.0	2.7	1.1	2.0
P_{NW} (MW/m^2)	1.25	2.5	4.0	3.6
THERMAL POWER (MW)	5000	5000	2100	4000

Fig. 1. Evolution of commercial tokamak reactors.

Table I. Summary of Major Parameters for Some Selected Tokamak Reactor Studies

Parameter	UWMAK-I	UWMAK-III	ORNL[15] DEMO	HFCTR[16]	NUWMAK	STARFIRE
R (m)	13	8	6.2	6	5.2	7.0
a (m)	5	2.5	1.5	1.2	1.1	1.94
P_{th} (MW)	5000	5000	2150	2470	2300	4000
P_e (MW)	1450	2000	825	775	620	1200
Material	316 SS	TZM	Mod. 316 SS	TZM	Ti	Mod. 316 SS
Coolant	Li	He/Li	He	Flibe	Boiling H_2O	Press. H_2O
Breeder	Li	Li	Li	Li	$Pb_{38}Li_{62}$	$LiAlO_2$
P_{MW} (MW/m^2)	1.25	2.0	2.7	3.4	4.0	3.6
$B_{T,max}$ (T)	390	8.7	7.1	13.1	11.9	11.1
Burn Time	5400	1800	1260	500	224	Continuous
Off Time	390	100	60	90	21	--
Duty Cycle	93%	95%	95%	85%	91%	100%
$\langle\beta\rangle$ (%)	5.2	5.8	10.0	4.0	6.0	6.7
I_p (MA)	21	16	3.9	6.7	7.2	10.1
b/a	1.0	2.0	1.6	1.5	1.6	1.6
T Burnup	7.2%	0.83%	∿ 3%	Not Avail.	1.5% (∿ 20%)	42%
Power Cycle	Li/Na/ Steam	Direct Cycle/ Li/Na/He	He/ Steam	Flibe/ Steam	Steam	H_2O/ Steam
Impurity	Double Null Divertor	Double Null Divertor	--	Limiter/ Vac. Halo	Gas Puffing	Limiter/ Vacuum
Heating	NB	RF	NB	NB	RF	RF

outside the vacuum boundary (shield); optimized modular design; all superconducting poloidal coils are outside the toroidal field (TF) coils; conservative TF coil design; fully remote maintenance permitting some repairs during reactor operation; "remove and replace" maintenance approach (failed parts are replaced with spare parts and the reactor is operated while repairs are made in the hot cell) that minimizes downtime; combining components for simplicity (e.g., TF coil room-temperature dewar provides support for the poloidal coils and shield); and providing redundancy where it is justified. These features as well as potential future improvements in component reliability provide optimism that the plant availability goal of 75% can be achieved.

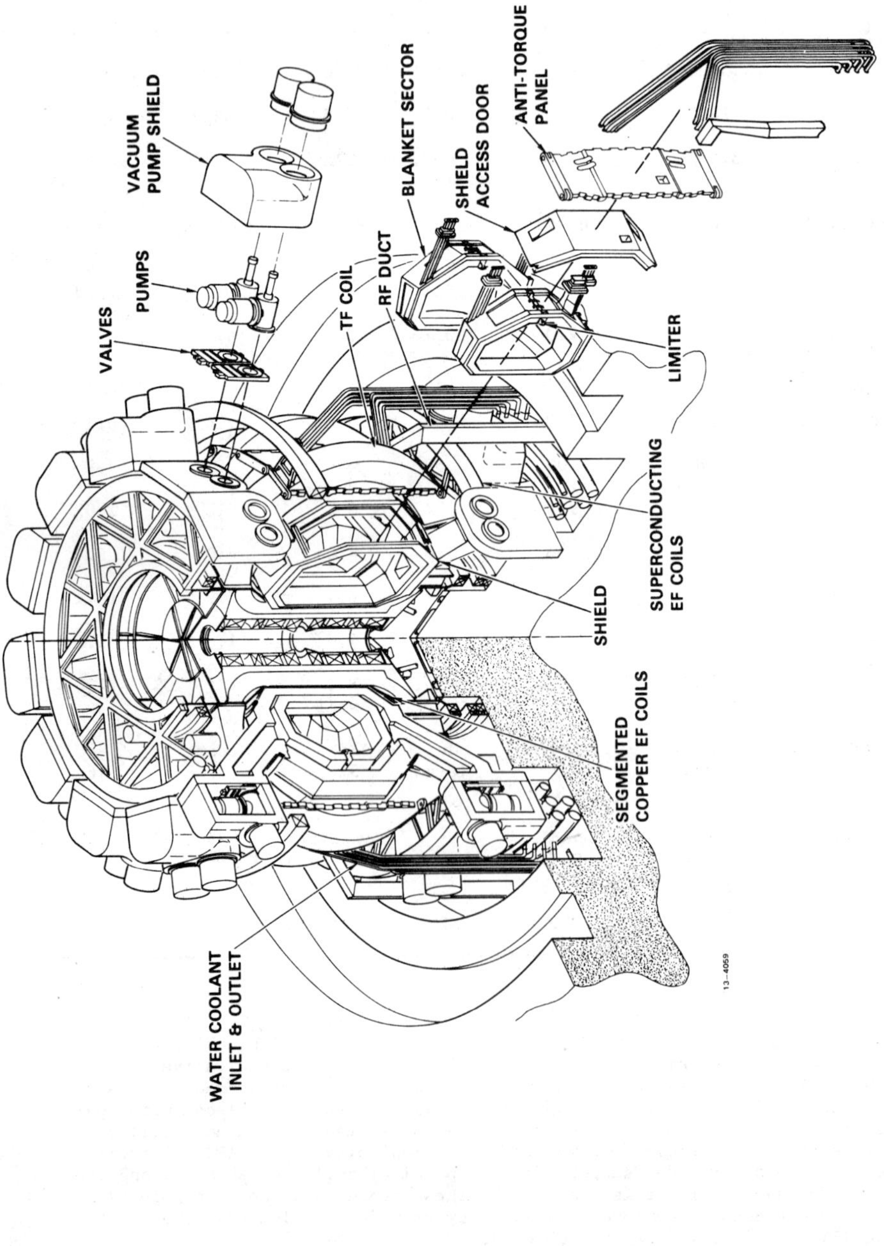

Fig. 2 STARFIRE reference design - isometric view.

Table II. Key Features of STARFIRE

- Steady-state plasma operation
- Lower hybrid radio frequency power for plasma heating and current drive
- Electron cyclotron radio frequency assisted startup
- Limiter/vacuum system for plasma impurity control and exhaust
- All superconducting poloidal coils outside toroidal field coils
- Vacuum boundary at the shield, mechanical seals
- Total remote maintenance with modular design
- Water-cooled, solid tritium breeder blanket with stainless steel structure
- All materials outside the blanket are recyclable within 30 yr
- Less than 0.5 kg of vulnerable tritium inventory
- Minimum radiation exposure to personnel
- Conventional water/steam power cycle with no intermediate coolant loop and no thermal energy storage

Safety and environmental considerations have played a major role in the STARFIRE design effort. A solid tritium breeder was selected in preference to liquid lithium in order to minimize the stored chemical energy. The impurity control and exhaust system was selected and designed so that the tritium fractional burnup is maximized and the vulnerable tritium inventory in the fueling and vacuum pumping systems is minimized. Furthermore, the reactor design was developed to contain the tritium inside multiple barriers and to minimize the size of potential tritium releases. The shield was designed and all reactor materials selected so that recycling of all materials outside the blanket is possible in less than 30 yr. Radiation exposure of personnel has been minimized by the use of extensive remote maintenance operations and by providing adequate shielding. The use of resource-limited materials was minimized. Mechanisms for fast reactor shutdown and auxiliary cooling systems have been incorporated into the design. The beryllium coating on the first wall and limiter provide an inherent safety feature that terminates the plasma burn if the metal temperature reaches $\sim$ 900°C. Calculations show that the reactor will be automatically shut down in less than one second, if a hot spot forms on a small area of the first wall, without the need for any active control system. No major damage, other than some first wall coating ablation, will occur. No runaway accident that could pose a major risk to the public was identified for STARFIRE. Furthermore, no plausible scenario could be formulated for the release of radioactive materials from the blanket (excluding corrosion products in the primary coolant loop) to the outside of the reactor building.

NUWMAK [12] is a reactor concept which shows that tokamaks can have high power density, a high degree of modularity, and moderate size. A cross-section of NUWMAK is shown in Fig. 3. The power density (10 MW/m^3) and electrical power output (660 MW) are chosen as typical of a full scale reactor operating in a base-loaded mode. The TF coil set is unique in that just eight superconducting coils are used. A set of 16 small water cooled copper trim coils that do not encircle the vacuum chamber correct the field ripple to below 2%. The blanket is constructed of the titanium alloy, Ti-6Al-4V, and is designed to minimize thermal cycling, to provide internal energy storage, and to eliminate the need for an intermediate heat exchanger. A lithium-lead eutectic, $Li_{62}Pb_{38}$, with a melting point of 464°C is used as the tritium breeding and thermal energy storage material. The latent heat of fusion for $Li_{62}Pb_{38}$ provides the required energy between plasma burns. Boiling water at 300°C, 1250 psi is the coolant and this further reduces thermal fatigue problems.

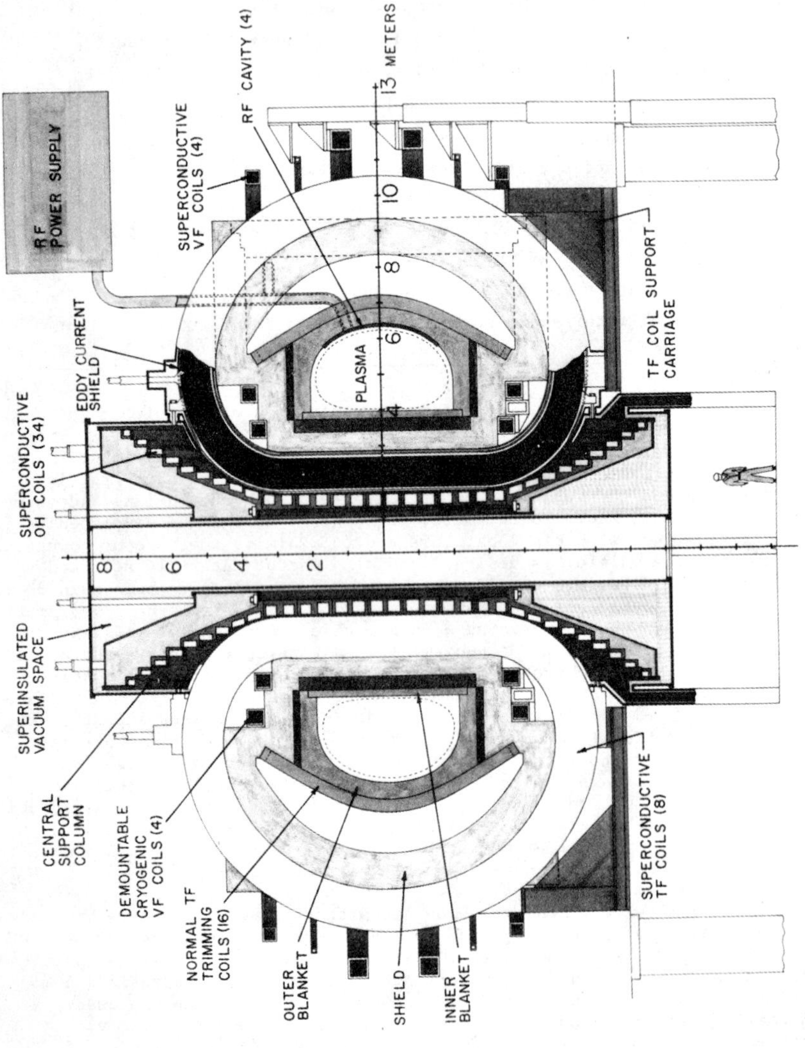

Fig. 3. Cross-sectional view of NUWMAK.

2.2 Steady-State Operation

Most tokamak reactor designs in the past were pulsed, albeit with very long pulses (typically 200-1000 seconds) and high duty cycles (> 95%). Nevertheless, pulsed operation implies certain penalties, such as the need for thermal and electrical energy storage, large ohmic heating coils, and first wall thermal fatigue concerns which dictate lower first wall loadings and result in reduced reliability. Fortunately, theory and experiments indicate the possibility that toroidal plasma currents may be maintained in tokamaks with noninductive external momentum sources to the electrons [17]. This suggests that steady state may be an achievable mode of operation for tokamaks.

Steady-state operation offers many technological and engineering benefits in commercial reactors. Among these are the following: component and system reliability is increased; material fatigue is eliminated as a serious concern; higher neutron wall loads are acceptable; thermal energy storage is not required; the need for an intermediate coolant loop is reduced; a significantly reduced probability of plasma disruptions; electrical energy storage is significantly reduced or eliminated; and a full-size ohmic heating solenoid is not needed, and external placement of the EF coils is simplified. The penalty for steady-state operation comes primarily from potential problems associated with a noninductive current driver; in particular, the electrical power requirements, the capital cost, and reliability and engineering complexity of the current driver.

In STARFIRE, a lower-hybrid rf system is utilized for the dual purpose of plasma heating and current drive. It was found that 66 MW of power at 1.7 GHz is dissipated in maintaining the plasma current which would require 150 MW of electrical power, as compared to the gross electric plant output of 1440 MW. It has been estimated that steady-state operation can result in a cost savings of at least 30%. This assumes the same availability for steady-state and pulsed reactors. It is likely that a steady-state system will result in a more reliable reactor with a higher availability. In addition, much larger savings are potentially realizable if the performance of the lower-hybrid current driver can be further improved or substantially better alternatives for the current driver are developed.

2.3 Impurity Control and Ash Removal

One of the most difficult problems confronting fusion research is to develop workable, attractive concepts for controlling the influx of impurities into the plasma and for the removal of the fusion reactor products (helium in the case of the DT fuel cycle) from the reactor. This is essential in order to achieve long burn times or steady-state operation.

Most earlier reactor studies examined the use of magnetic divertors which are systems to divert a portion of the magnetic flux, either poloidal flux in the case of a poloidal divertor or toroidal flux in the case of a bundle divertor. The idea of using non-divertor options were explored in NUWMAK and STARFIRE.

NUWMAK considered the concept of periodic gas puffing and a trapped ring of impurities acting as a halo around the plasma which would radiate the plasma alpha energy (transported to the halo from the plasma interior) to the first wall. This would result in depositing the alpha power (390 MW) uniformly on the first wall with a surface loading of ~ 1 MW/m^2. NUWMAK also explored the concept of fueling the center of the plasma with injected tritium pellets while puffing in deuterium gas at the plasma edge. This concept would significantly reduce tritium pumping at the plasma edge and, hence, increase the fractional burnup of tritium.

The STARFIRE study has developed in some detail the concept of using mechanical limiters and vacuum pumping ports for impurity control and ash removal. The limiter/vacuum system is a concept with many inherent features that can simplify a commercial power reactor:

- It is a mechanical system that does not require magnets.
- It has minimal requirements on space; the limiter fits naturally into the scrape-off region.

- Because of its location inside the first wall, the surface area available for the limiter is relatively large, thus permitting operation at reasonable heat fluxes.
- The limiter can be replaced simultaneously with the first wall with no special maintenance requirements.
- The limiter/vacuum system can be designed to dramatically reduce radiation streaming.
- The system is simple and inexpensive.

The limiter/vacuum system can be employed in any toroidal system (e.g., EBT, RFP, stellarators) as well as possibly the tandem mirror concept.

Figure 4 shows the location of the STARFIRE limiter concept. The limiter consists of 96 segments that form one toroidal ring centered at the midplane and positioned at the outer side of the plasma chamber. Each of the limiter segments is 1 m high and ∿ 0.6 m wide. The limiter slot, which is the region between the limiter and first wall, leads to a 0.4-m high limiter duct that penetrates the 0.7-m thick blanket. The limiter duct opens into a plenum region that is located between the blanket and shield and extends all the way around the torus. This plenum region is large enough so that it spreads the radiation leakage from the limiter duct into a larger surface area of the bulk shield. The conductance of the plenum region is large enough to permit locating the vacuum ducts in the bulk shield sufficiently removed from the midplane so that radiation streaming from the limiter duct in the blanket to the vacuum pumps is acceptable.

One of the major improvements that resulted from the limiter/vacuum system is a substantially higher tritium burnup in STARFIRE (42%) compared to previous tokamak reactor designs (see Table I). This is a direct result of the higher reflection coefficient for tritons compared to α-particles. This has resulted in a significant decrease in the tritium inventory in the vacuum pumping and fueling system. This illustrates the important point that the tritium burnup fraction is more a function of the type of impurity/ash removal system than the type of magnetic confinement concept.

2.4 Plasma Heating Technology for Tokamaks

The two principal methods considered for heating tokamak plasmas to ignition temperatures are neutral beam and radio frequency (rf) heating.

If neutral beams are used to heat ignition size plasmas at full densities ($\sim 10^{14}$ particles/cm^3), then neutral beam energies of about 300 keV will be required to penetrate the plasma [14]. This will require the development of negative ion beams in order to achieve acceptable beam efficiencies. Fortunately, there are some recent concepts which may reduce the beam energy requirements. Plasma modeling studies indicate that alpha particle heating in the center of the plasma will reduce the need for full beam penetration. If low density startup scenarios are feasible, then beam energies of ∿ 150 keV may be adequate [18], thus permitting the use of the more developed positive ion beam technology. It has been proposed to create a top-bottom asymmetry in the toroidal field ripple of a tokamak using suitable auxiliary coils [19] which will assist the penetration of the beam.

RF heating offers an attractive alternative to neutral beam heating with important reactor advantages. These include reduced neutron streaming effects, easier penetration of the blanket and shield, much freer access to the reactor to facilitate maintenance, reduced penetration shielding costs, removal of much of the system's components from the reactor building, and (except for electron cyclotron resonance heating) a largely existing technology at the power levels and efficiencies required for reactors.

UWMAK-III employed an rf heating system delivering 100 MW at 60 MHz for 15 s. NUWMAK's plasma heating system is based on fast magnetosonic waves (92 MHz). Approximately 80 MW of power will ignite NUWMAK in about one second of heating. The overall system efficiency is about 50%. A wedge shaped coaxial cavity (with

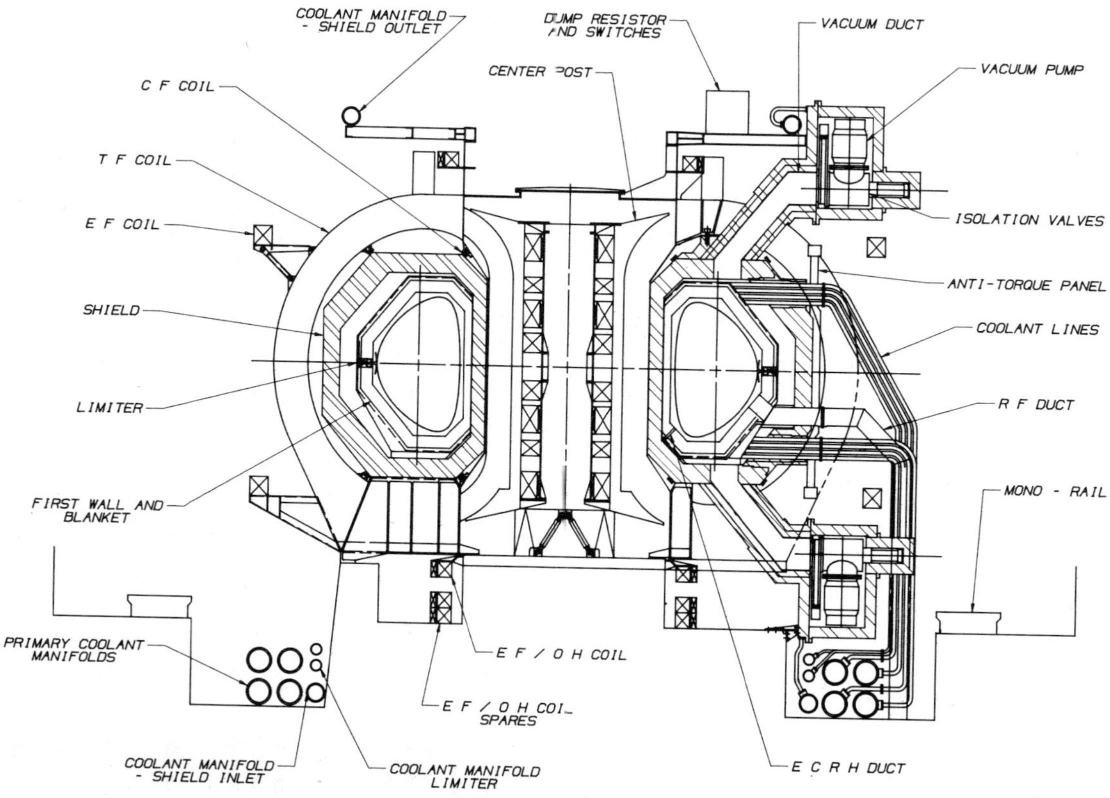

Fig. 4. STARFIRE reference design - cross section.

a total of four in the torus) is mounted flush to the first wall and is fed by coaxial transmission lines.

STARFIRE's plasma heating system employs lower hybrid wave power of 90 MW at 1.7 GHz which is also used for current drive to achieve steady-state operation. No launching structure internal to the first wall is required. There are 12 waveguide modules between each TF coil, each module has a cross section of 0.66 m x 0.78 m. Averaged over the total antenna area, the wave intensity at the plasma is $\sim$ 1.5 kW/cm^2. Higher intensity could possibly result in nonlinear plasma responses. The power supplies and rf system elements are located outside the reactor building.

2.5 Startup Requirements

The power supply, electrical energy storage, and poloidal magnet design considerations related to the startup and shutdown of large tokamak reactors has been recognized as an area of key concern [20]. One of the major motivating factors in examining steady-state operation is the desire to significantly reduce, if not eliminate, the need for large OH coils and associated energy storage systems. If the reactor is capable of truly steady-state operation, then one can consider a much broader range of startup and shutdown scenarios than with pulsed reactors.

A good example of a scenario for startup for steady-state operation is provided by STARFIRE. The total startup period takes 24 minutes. The maximum reactive power of the poloidal coil power supply is set during the startup and is 290 MVA. This is only about 20% of the estimated value that would be needed if STARFIRE were operated in a pulsed mode. The maximum draw of power from the grid is determined by the sum of rf and poloidal field coils' instantaneous powers. For a value of 150 MW of rf input electrical power and assuming a 95% conversion efficiency for the poloidal coil power supply, the maximum power from the grid is about 250 MW. The utility grid would supply this power during a few times each year.

There has also been a significant reduction in the magnet, power supply and energy storage technology requirements for pulsed tokamak power reactors. This has resulted because of two developments. The first is the use of rf-induced breakdown during the startup phase. Such techniques would reduce the startup voltage requirements from typically $\sim$ 300 V to $\sim$ 50 V as well as reduce the rate of change of the magnetic field of the OH coil to 2 $\rightarrow$ 3 T/s, which is well within the state-of-the-art for pulsed superconducting coils.

The other key development is the recognition that startup periods for pulsed reactors of $\sim$ 10 s, instead of 1 $\rightarrow$ 2 seconds, will significantly reduce technology requirements. This will make it possible to use conventional technology, such as motor-generator-flywheel energy storage and SCR-type power supplies, in place of more advanced technologies such as homopolar generators and inductive energy storage.

2.6 Energy Conversion and Tritium Breeding

Most previous reactor studies have emphasized liquid lithium blanket concepts for tritium breeding with either liquid lithium or helium as the coolant (see Table I). In two of the most recent studies, STARFIRE and NUWMAK, attention has been focused on water as a coolant and on non-liquid lithium blankets.

A major effort of STARFIRE has focused on the use of solid compounds for breeding tritium. One of the major considerations with solid breeders is the development of an efficient tritium recovery scheme to keep the tritium inventory in the blanket to a low level. Periodic removal of the breeder appears to be an unacceptable option because it entails an intolerably high tritium inventory that could reach $\sim$ 40 kg/GW of fusion power for annual replacement. Another approach for tritium recovery is continuous circulation of the solid breeder. This approach presents very difficult engineering problems in tokamak geometries. A nonmobile solid tritium breeder blanket with in-situ tritium recovery appears to be the preferred approach. Low-pressure (∿ 0.1 MPa) helium is circulated through formed channels in the highly porous solid breeding material.

The ∿ 40-cm tritium-breeding zone consists of a packed bed of α-$LiAlO_2$ with 1-cm diameter stainless steel coolant tubes spaced appropriately throughout the zone (see Fig. 5) to maintain a maximum breeder temperature of 850°C. The tube spacing increases from ∿ 2 cm at the front of the breeder zone to 5 → 10 cm at back. The coolant inlet temperature is 280°C with an outlet temperature of 320°C. The relatively low temperature of the austenitic stainless steel tubes (< 400°C) and the oxide film on the water side of the tubes provide an adequate tritium barrier for inleakage into the coolant. The $LiAlO_2$ is perforated with ∿ 2 mm diameter holes through which low-pressure helium passes to recover the tritium from the breeder. The $LiAlO_2$ is ∿ 60% dense to facilitate percolation of tritium (as T_2O) to the helium purge channels.

The power conversion system is utilized to convert the reactor thermal energy to electrical power. Two separate heat removal circuits are utilized, one for the first wall/blanket and the other for the limiter. The power deposited in the limiter (200 MW) is used for feedwater heating while the recoverable power (3800 MW) from the first wall and blanket is used to produce steam at 299°C and 6.3 MPa. The steam is then used in a turbine-generator unit for producing 1440 MW of electric power. The net electrical power is 1200 MW with 240 MW recirculating power for the rf system, coolant pumps, and other reactor subsystems. This system corresponds to a rather conventional pressurized water reactor (PWR) balance of plant system.

The NUWMAK concept is based on the idea of using a breeding material that operates at its melting point. The energy stored in the latent heat of fusion of the material eliminates the need for an external thermal energy storage system which would otherwise be required due to NUWMAK's pulsed operating mode. The blanket is cooled by boiling water which also helps to minimize temperature cycling. The breeding material is the eutectic $Li_{62}Pb_{38}$ which undergoes a solid-liquid phase change at 464°C. The use of the boiling water coolant and internal blanket energy storage permits the use of a simple direct boiling water reactor (BWR) cycle.

In both of the above examples, the use of relatively conventional power conversion systems results in net plant efficiencies of typically ∿ 30%. This is lower than some previous studies, e.g., UWMAK-III which was 40%. However, this disadvantage is expected to be more than offset by significant reductions in the cost of the balance of plant.

3.0 Mirror Reactor Concepts

3.1 Overview of Mirror Reactor Development

Magnetic fusion devices of the mirror type are characterized by open magnetic field line geometries, that is, magnetic flux passes through a mirror-type device and intersects material walls outside the reaction chamber. In order for such a device to be an adequate container of fusion plasma, it is essential that end leakage of the plasma be strongly inhibited. The earliest magnetic mirror configuration was a simple solenoid with increased magnetic field strength near its ends (see Fig. 6).

Unfortunately, plasma in a simple mirror is grossly unstable to sideways motion because the magnetic field weakens in directions perpendicular to the coil axis. To solve the gross, or MHD, stability problem, the simple mirror configuration was replaced by the minimum-B mirror. From the center of a minimum-B magnetic field -- produced by a pair of solenoids and Ioffe bars, a baseball coil (shown in Fig. 6), or a yin-yang coil -- the field strength increases in all directions. Conceptual design studies of mirror reactors based on the minimum-B geometry [21, 22] revealed that end losses from such a reactor would severely limit its plasma Q (fusion power divided by trapped injected power). The subsequent search for enhanced-Q mirror machines has led to work on two new concepts: the field-reversed mirror and the tandem mirror.

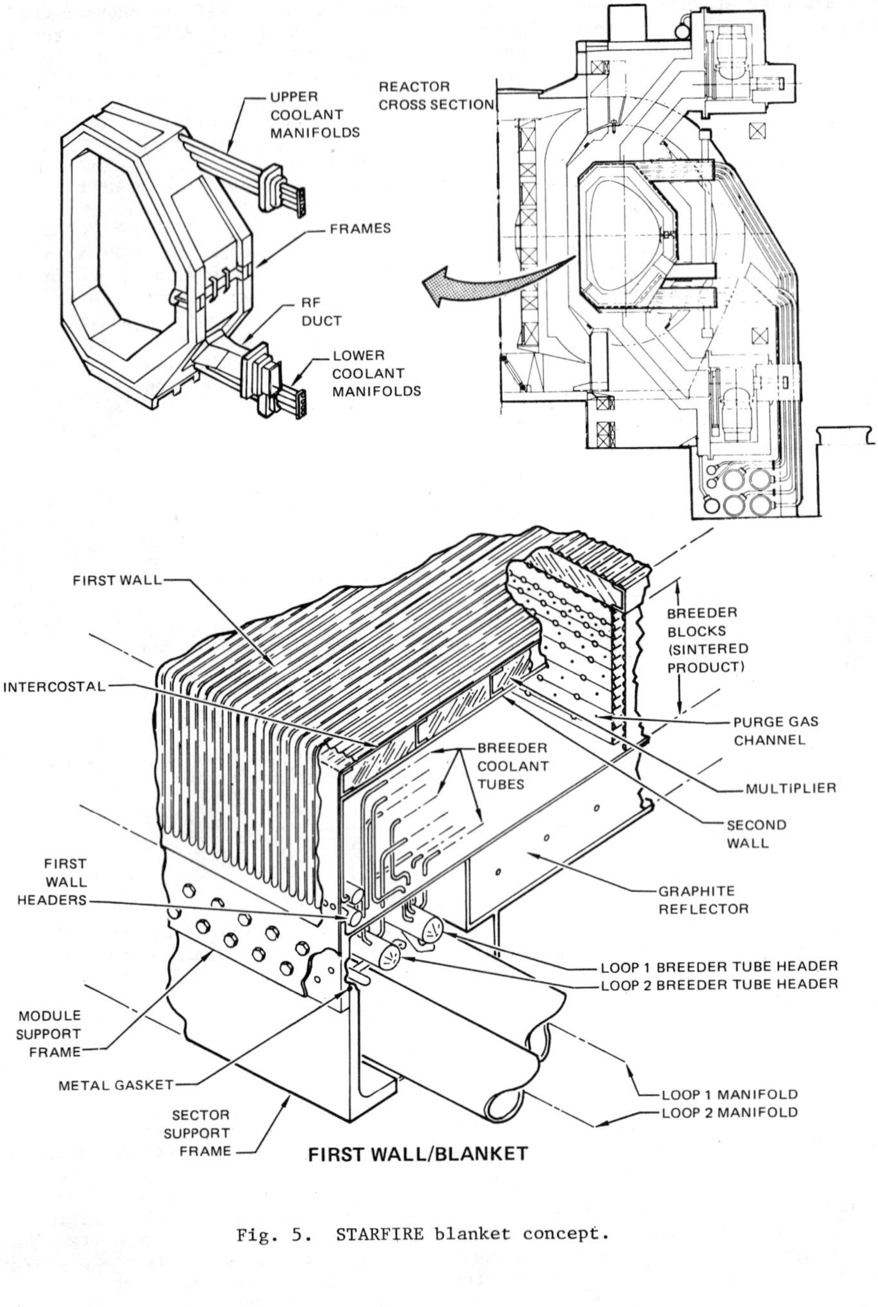

Fig. 5. STARFIRE blanket concept.

By field-reversed mirror confinement, we mean the confinement of plasma in a toroidal region of closed magnetic field lines generated by diamagnetic plasma currents in a nearly uniform background field (Fig. 6). The field-reversed mirror confinement concept offers the exciting possibility of fusion electric power reactors in small sizes. Conceptual design studies of mirror reactors based on field-reversed confinement have been carried out, including a multicell reactor producing 75 MW (electric) of net electric power [23] and a single cell pilot reactor producing ∿ 11 MW (electric) [24]. In the pilot reactor a single field-reversed D-T plasma of toroidal geometry is created by a coaxial plasma gun at one end of the reactor. The plasma is subsequently trapped in a shallow minimum-B magnetic well (produced by niobium-titanium superconducting coils) and sustained in steady state by the injection of neutral beams of deuterium and tritium at 200 keV. A cylindrical blanket and shield surround the plasma. The blanket is helium cooled and contains lithium oxide for the breeding of tritium fuel. Large tanks at the ends of the reactor house direct energy converters and vacuum pumping systems for the plasma end leakage. The reactor is designed to accommodate a range of plasma sizes (12- to 22-cm major radius) and, accordingly, a range of fusion power (22 to 42 MW). The pilot reactor with the smaller plasma is just barely a net power producer: 90% of the gross electric power must be recirculated to operate the reactor, and the net electric power is only 1.2 MW. With the larger plasma, the net electric power is 10.7 MW.

Most recently, a conceptual design study sponsored by the Electric Power Research Institute has considered a moving ring version of the field-reversed mirror reactor, wherein field-reversed plasma rings are created, transported through a cylindrical burn chamber, and exhausted [25]. The plasma rings, formed by a coaxial plasma gun, are magnetically compressed to ignition temperature while they are being injected in to the reactor's burner section. The D-T ice pellets refuel the rings during the burn at a rate that maintains constant fusion power. A steady train of plasma rings moves at constant speed through the reactor under the influence of a slightly diverging magnetic field. The aluminum first wall and breeding zone structure minimize induced radioactivity: "hands-on" maintenance is possible on reactor components outside the breeding blanket. Helium removes the heat from the Li_2O tritium breeding blanket and is used to generate steam. The reactor produces a constant, net power of 376 MW (electric).

The tandem mirror confinement concept (TMR) was invented in 1976 by Fowler and Logan [26] and independently by Dimov, et al. [27] in the Soviet Union. The tandem mirror is now the mainline effort of the mirror fusion program. The basic concept entails the improved axial confinement of a long cylindrical fusion plasma within a solenoid by means of strong electrostatic potentials at the ends, produced by mirror-confined, end-plug plasmas (Fig. 6).

The first conceptual fusion reactor design based on the concept was published [28] in July 1977. The reactor had a central cell 100 m in length and produced 1000 MW of net electrical power. Drawbacks of the first design included a somewhat modest plasma performance (Q = fusion power/total injected power trapped by the plasma) $\simeq 5$, and a requirement for high technology components for the end plugs (17-T magnetic coils and 1.2-MeV neutral-beam injectors).

A major new invention for tandem mirrors -- the thermal barrier concept -- was reported [29] in April 1979. This invention followed from the realization that the optimal use of electron heating in the tandem mirror involved the establishment of a hotter electron population in the plugs than in the central cell because the confining electrostatic potential increases with the electron temperature in the plugs, while cooler central cell electrons allow a higher fusion power density for a given magnetic field strength. However, in the normal tandem mirror, electron flow between the plugs and central cell is so high that only small temperature differences can be established, even with electron heating localized in the end plugs. The new concept introduces a barrier between the plug and central cell that effectively reduces the passing of central cell electrons into the plug. Basically, the thermal barrier consists of a region of much reduced magnetic field strength, plasma density, and plasma potential. The depressed positive plasma potential is an electrostatic barrier to electrons.

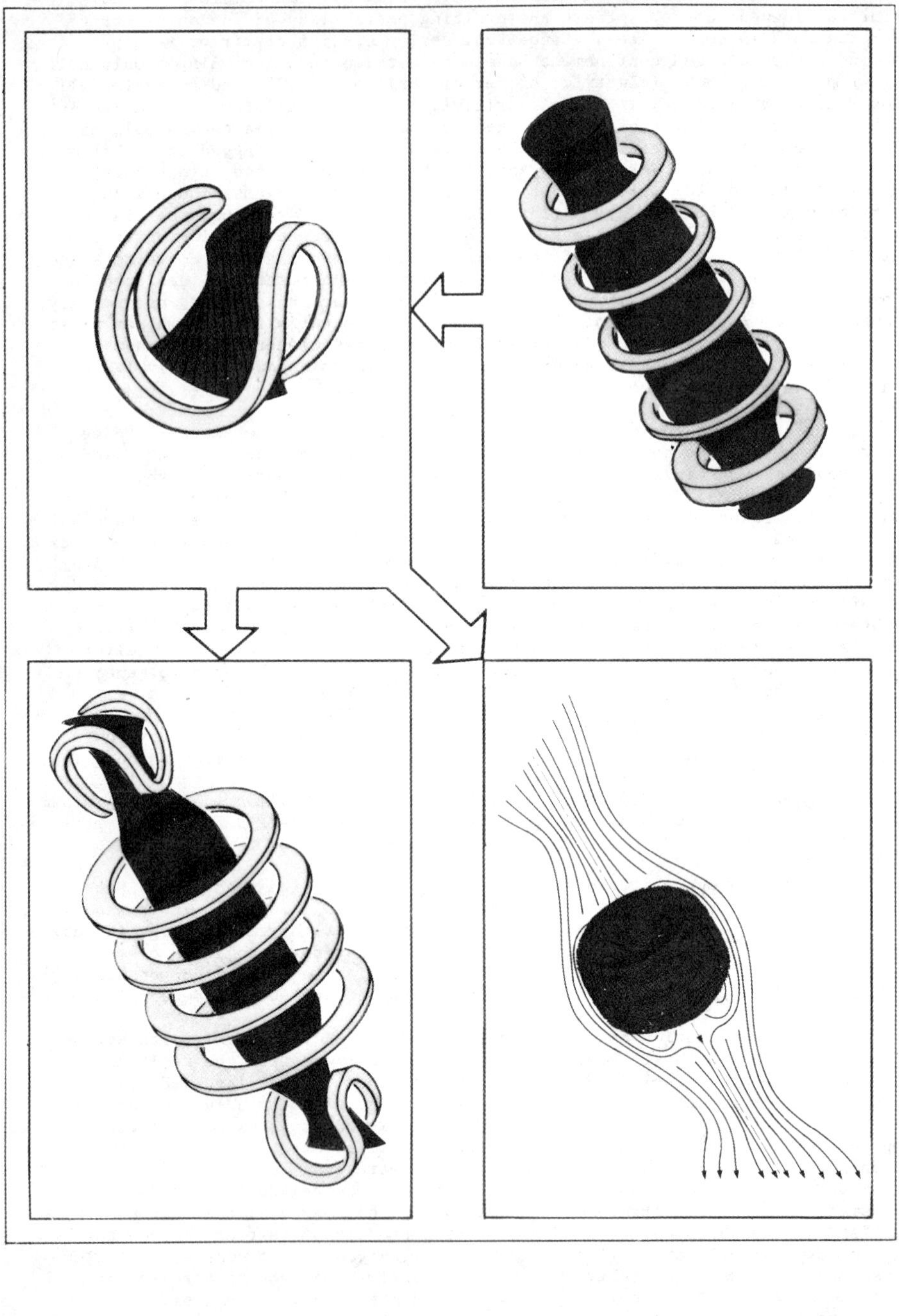

Fig. 6 Evolution of mirror confinement concepts.

3.2 Tandem Mirror Reactors and Thermal Barriers

A preliminary conceptual design of a power reactor based on the tandem mirror with thermal barriers was reported [30] in September 1979. An overall view of the reactor is shown in Fig. 7. The D-T fusion plasma is contained in the 56-m-long central cell and produces 1770 MW of fusion power. With $Q \simeq 10$, the reactor produces $\sim$ 500 MW of net electricity. Because the central cell plasma is near or at ignition, the power output of the reactor can be increased by increasing the central cell length and retaining the same end plug systems. The central cell consists of twenty-eight 2-m-long modules, each containing an annular blanket region, a magnet shield region, and two niobium titanium solenoidal magnets. The entire central cell resides in a vacuum trench, which allows the module-to-module seal to be made by an annular metal inflatable cushion. When the cushion is deflated, these seals allow the removal of individual modules by an overhead crane. Blanket coolant galleries are shown below the vacuum trench. The end plug magnets are housed in large cylindrical vacuum vessels at each end of the reactor. The plug plasmas, contained in the plug yin-yang coils, are each sustained by a low-current, 400-keV neutral beam (shown only on the far end in Fig. 7). Also shown is the gyrotron tube system for microwave heating of the electrons on the plug side of the thermal barrier. The small neutral beams indicated onthe end wall of the plug vacuum vessel are the barrier region beams for charge-exchange pumping of the barrier and fueling of the central cell.

In the TMR, the fusion power is produced in the cylindrical geometry central cell. The design goal for the central cell is that it be compact, simple to fabricate using mass production techniques, and easily maintained. Fortunately, the central cell is largely decoupled from the plug regions, and its design can be optimized somewhat independently of the complexities of the end plugs. A basic design philosophy adopted for the TMR central cell is that of axial modularity.

Figure 8 shows the tube-type blanket as designed at the University of Wisconsin [31]. The blanket consists of banks of seamless tubes running circumferentially around the plasma in the central cell. The blanket will be divided into modules, the length of which will depend on the separation between central cell magnets. The tubes in each module will be welded to tube sheets at the top and the bottom of the central cell. These tube sheets are in turn welded to semicylindrical manifolds, which distribute the coolant uniformly among the tubes. Each module will have a single supply and single return header attached to the manifolds at the top and bottom. This design provides the flexibility of running the coolant from the top to the bottom or vice versa. Since radiation damage, in particular swelling, is a strong function of the temperature of the structure, the capability of periodic reversal of the coolant flow might mitigate the problems of radiation damage. The unique feature of this design is the elimination of welds in the severe radiation zone. All welds are confined to regions shielded by at least 50 cm of lithium-lead alloy.

4.0 Elmo Bumpy Torus Reactor

The EBT concept is a toroidal array of simple magnetic mirrors. The promise of a steady-state reactor that operates at or near D-T ignition emerges from this combination of simple mirrors and toroidal geometry. The creation of an rf-generated, low-density, and energetic electron ring at each position between mirror coils (i.e., midplane location) is needed to stabilize the bulk, toroidal plasma against well-known instabilities associated with simple mirror confinement.

The EBT was first examined as a reactor over four years ago [32]. An EBT configuration that is stabilized by energetic electron rings combines a number of unique features that describe a fusion reactor with the following attractions: steady-state operation in an ignited or a high-Z mode; large aspect ratio to give an open and accessible geometry; an engineering assembly that is comprised of relatively simple and compact modules; ease of maintenance, modular construction, and a relatively simple magnet system. Although the earliest EBT designs predicted relatively large power plants, the attainment of high magnetic aspect ratios in systems with lower physical aspect ratios through the use of aspect-ratio-enhancement (ARE) coils indicates [33] that smaller reactors may be possible while

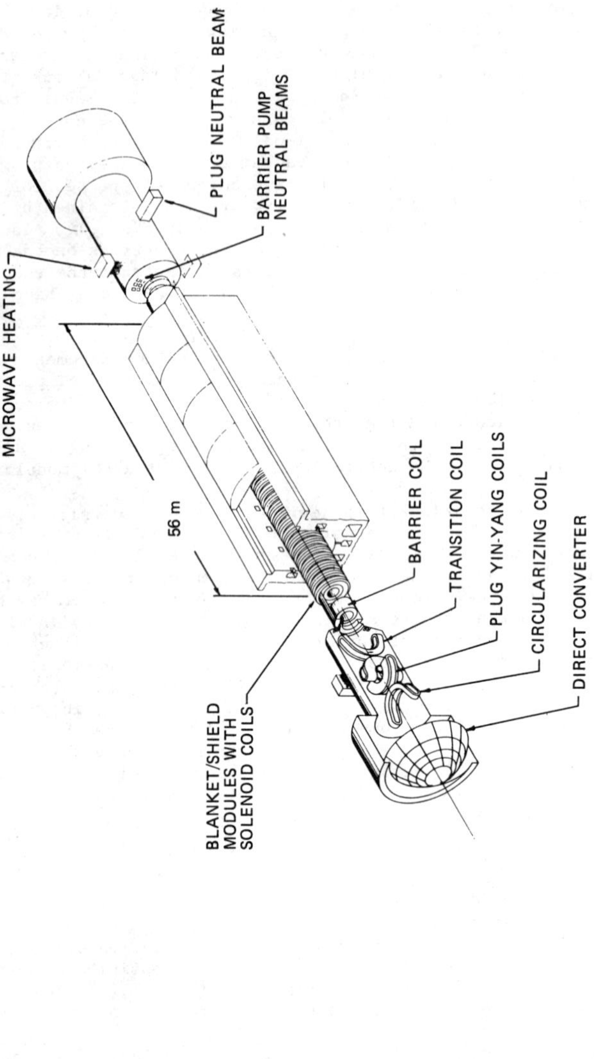

Fig. 7. The TMR with thermal barriers.

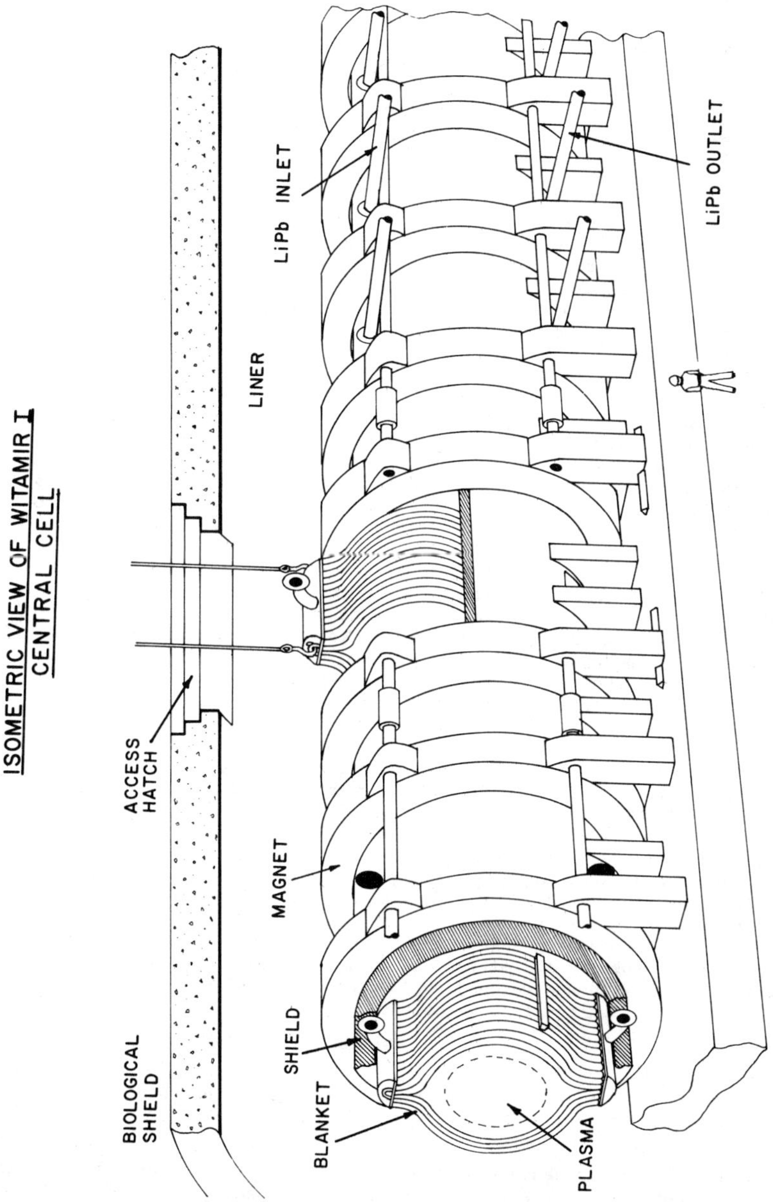

Fig. 8. University of Wisconsin TMR design.

simultaneously maintaining the above-mentioned reactor features. The use of ARE coils is illustrated in Fig. 9.

5.0 Alternate Reactor Concepts

There are a number of other magnetic confinement concepts [1,34] which have been considered as reactor concepts in addition to those discussed in the previous sections. A summary list is shown in Table III, [1] where they are grouped as toroidal, compact toroids, linear, and very dense plasma concepts. While these concepts have been investigated in much less detail than the tokamak and tandem mirror, they do illustrate the potential variety of magnetic confinement reactor concepts. In particular, concepts such as the field-reversed mirror and spheromak offer the possibility of relatively small fusion power plant sizes.

6.0 Summary

A summary of the design developments for tokamaks discussed in Sec. 2 is given in Table IV. In general, these design developments have emphasized less complex, more reliable components and systems. Several of these developments are applicable to other reactor concepts; e.g., nondivertor impurity control, rf heating, maintenance concepts and simplified energy conversion systems.

Some of these ideas are reasonably well supported by current experiments and theory (e.g., projected size of tokamaks for ignition) while some of the other concepts (e.g., non-divertor impurity control, steady-state operation, solid tritium breeders) require much more research and development. As these concepts are developed, as well as what is sure to be further improvements in the tokamak which are not envisioned today, then the tokamak will continue to improve as an attractive commercial reactor concept. The improvements in the more recent tokamak reactor studies compared to the earlier designs has been substantial.

The development of the tandem mirror reactor concept with thermal barriers has resulted in a significant improvement in the attractiveness of mirror systems as commercial reactors. Reactor performance in terms of the plasma Q has been improved while maintaining reasonable demands on key technology systems such as neutral beams and superconducting magnets. Additional features include steady-state operation, linear geometry, reduced neutron streaming in regions of high technology components, and the use of direct energy recovery systems.

Several additional concepts have been surveyed which offer a wide variety of reactor features and sizes. Some of these concepts (e.g., spheromak and the field-reversed mirror) offer the possibility of relatively small reactor sizes, typically 10's of megawatts of thermal energy.

In general, considerable progress has been achieved in the last few years in developing a better understanding of the technology requirements of magnetic confinement fusion reactors. Studies have emphasized less complex designs with much improved maintenance features as well as improved safety features. Several promising new concepts have been suggested for further improving the commercial attractiveness of the tokamak and mirror concepts as well as several alternate concepts. A firm basis has been developed for moving with confidence into the reactor engineering phase of fusion energy development.

References

1. C. C. Baker, G. A. Carlson and R. A. Krakowski, "Trends and Developments in Magnetic Confinement Fusion Reactor Concepts," Nucl. Technol. - Fusion, 1 (1981).

2. R. W. Conn, et al., Proceedings of the 8th International Conference on Plasma Physics and Controlled Nuclear Fusion Research, Brussels, July, 1980, IAEA-CN-38/V-5.

3. C. C. Baker, et al., "The Impact of Alternate Fusion Fuels on Fusion Reactor Technology," Argonne National Laboratory, ANL/FPP/TM-128, (November, 1979), also, "D-D. Tokamak Reactor Studies," Argonne National Laboratory, ANL/FPP/TM-138 (1981).

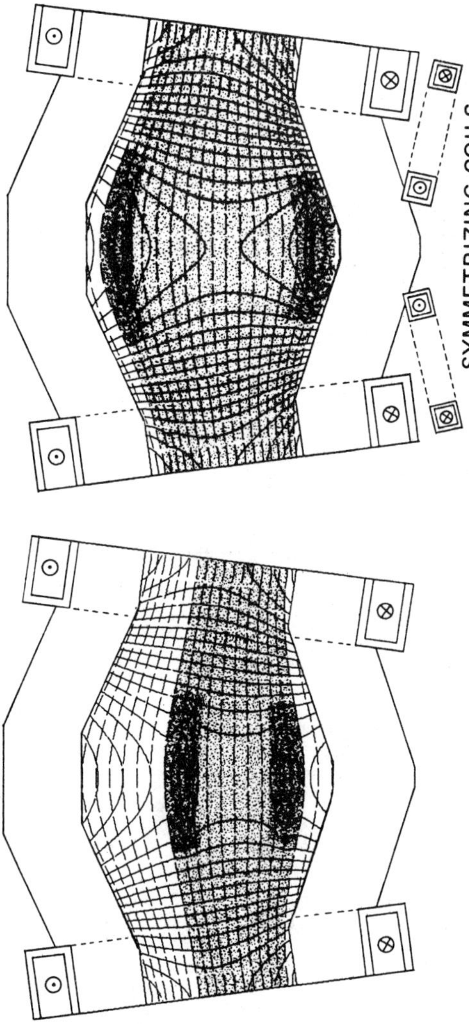

Fig. 9. EBT concept with Aspect Ratio Enhancement (ARE) coils.

Table III. Summary of Alternative Concepts for Magnetic Fusion

I. Toroidal

A. Steady state

- Stellarator
- Torsatron
- Bumpy torus (EBTR)
- Toroidal bicusp (Tormac)
- Surface magnetic confinement (Surmac)

B. Long pulsed

- Reversed-field pinch (RFPR)
- Ohmically heated torus (OHTE)
- Ohmically heated tokamak (Riggatron)

C. Pulsed

- Theta-pinch (RTPR)
- High-beta stellarator (HBS)
- Belt-shaped screw-pinch (BSPR)

II. Compact toroid

A. Stationary

- Spheromak
- Field-reversed mirror (FRM)
- Triggered-reconnected adiabatically compressed torus (TRACT)
- Electron-layer field-reversed mirror (Astron)
- Slowly imploding liner (LINUS)

B. Translating

- Spheromak
- Field-reversed theta pinch (CTOR)
- Moving-ring field reversed mirror (MRFRM)
- Ion-ring compressor

III. Linear

A. Steady

- Tandem mirror (TMR)
- Multiple-mirror solenoid

B. Pulsed

- Linear theta pinch (LTPR)
- Laser-heated solenoid (LHS)
- Electron-beam heated solenoid (EBHS)

IV. Very dense (fast-pulsed linear) systems

- Fast-imploding liner (FLR)
- Dense plasma focus (DPF)
- Wall-confined shock-heated reactor (SHR)
- Dense Z-pinch (DZPR)
- Passive liners

Table IV. Summary of Tokamak Design Advances

Design Features	Impacts
Reduced physical size and reactor output	More compatible with current plant sizes and lower unit costs
Steady-state operation	More reliable operation, lower energy costs
	Higher wall loadings - smaller reactors, lower capital costs
	No energy storage - lower capital costs
	Simplified energy conversion system - lower capital costs
	Reduce size of, or eliminate, OH coil
Non-divertor concepts for impurity control and ash removal	Lower capital costs
	Less complexity - more reliability
	Improved access for maintenance
	Higher tritium burnup, lower tritium inventories
	Reduced neutron streaming
Plasma heating technology	
Reduce neutral beam energy	Use existing positive ion beam technology
rf heating	Reduce neutron streaming
	Easier interface with blanket/shield
	Better access for maintenance
	Reduced shielding costs
Startup and shutdown	
rf assisted startup	Reduced OH voltage requirements
	Reduce dB/dt for OH coil
	Reduce volt-second requirement
Steady-state operation	Permits long startup times with very small or no electrical energy storage
Longer startup times (∿ 10 s) for pulsed operation	Can use conventional power supply and energy storage technology
Maintenance	
Modularized first-wall/blanket concept	Simplified maintenance operations with minimum in-reactor operations
EF coils outside TF coils	Better access without moving or disconnecting coils
Reduce number of TF coils	Better access
Location of vacuum boundary at shield	Reduced requirements on first wall
	Simpler mechanical vacuum seals
	Easier access
Energy conversion	
Solid tritium lead/lithium breeders	Improved safety
H_2O coolants and PWR and BWR power conversion	Simplified BOP systems
	Reduce BOP costs

4. J. Fillo, Proceedings of the Third Topical Meeting on the Technology of Controlled Nuclear Fusion, Santa Fe, New Mexico, May, 1978, p. 193.

5. E. E.Kintner and C. M. Stickley, Proceedings of the Third Topical Meeting on the Technology of Controlled Nuclear Fusion, Santa Fe, New Mexico, May, 1978, p. 143.

6. J. F. de Paz, Y. A. Gohar and H. L. Schreyer, "Development of a Falling-Bed Fusion Blanket System for Synthetic Fuel Production," 15th Intersociety Energy Conversion Engineering Conference, Seattle, Washington (August, 1980).

7. B. R. Leonard, Jr., Nucl. Technol. 20, 161 (1973).

8. L. M. Lidsky, Nuclear Fusion, 15, 151 (1975).

9. W. C. Wolkenhauer, Proceedings of the Third Topical Meeting on the Technology of Controlled Nuclear Fusion, Santa Fe, New Mexico, May, 1978, p. 132.

10. B. Badger, et al., "UWMAK-I," University of Wisconsin, UWFDM-68, Vols. I and II (1973).

11. B. Badger, et al., "UWMAK-III," University of Wisconsin, UWFDM-150 (1976).

12. R. W. Conn, et al., Proceedings of the Third Topical Meeting on the Technology of Controlled Nuclear Fusion, Santa Fe, New Mexico, May, 1978, p. 1049; B. Badger, et al., "NUWMAK," University of Wisconsin, UWFDM-330 (1979).

13. C. C. Baker, et al., "STARFIRE - Commercial Tokamak Fusion Power Plant Study," Argonne National Laboratory, ANL/FPP-80-1 (1980).

14. D. Steiner, Proceedings of the Third Topical Meeting on the Technology of Controlled Nuclear Fusion, Santa Fe, New Mexico, May, 1978. p. 440.

15. D. Steiner, et al., "ORNL Fusion Power Demonstration Study: Interim Report," Oak Ridge National Laboratory, ORNL/TM-5813 (1977).

16. D. R. Cohn, et al., "High Field Compact Tokamak Reactor (HFCTR)," MIT Plasma Center, RR-78-2 (March, 1978).

17. "Steady-State Current Drive in Tokamaks Workshop Summary," U.S. Department of Energy, ET-0077 (February, 1979).

18. Y.-K. M. Peng, et al., Oak Ridge National Laboratory, ORNL/TM-6150 (1977).

19. D. L. Jassby and R. J. Goldstone, Nuclear Fusion, 16, 613 (1976).

20. J. N. Brooks and R. L. Kustom, Nucl. Technol., 46, 61 (1979).

21. R. Moir, ed., "Standard Mirror Fusion Reactor Design Study," Lawrence Livermore National Laboratory, UCID-17644 (January 1978).

22. D. J. Bender, ed., "Reference Design for the Standard Mirror Hybrid Reactor," Lawrence Livermore National Laboratory, UCRL-52478 (May 1978).

23. G. A. Carlson, W. C. Condit, R. S. Devoto, J. H. Fink, J. D. Hanson, W. S. Neef, and A. C. Smith, Jr., "Conceptual Design of the Field-Reversed Mirror Reactor," Lawrence Livermore National Laboratory, UCRL-52467 (May 1978).

24. G. A. Carlson, K. R. Schultz and A. C. Smith, Jr., "Field-Reversed Mirror Pilot Reactor," Electric Power Research Institute, EPRI AP-1544 (September 1980).

25. A. C. Smith, Jr., et al., "The Moving Ring Field-Reversed Mirror Reactor," Proceedings of the 4th Topical Meeting on the Technology of Controlled Nuclear Fusion, King of Prussia, Pennsylvania, October 14-17, 1980 (to be published).

26. T. K. Fowler and B. G. Logan, "The Tandem Mirror Reactor," Comments on Plasma Physics, 2, 167 (1977).

27. G. I. Dimov, V. V. Zakaidakov, and M. E. Kishinevsky, "Open Trap with Ambipolar Mirrors," Fiz. Plasmy. 2, 597 (1976).

28. R. W. Moir, W. L. Barr, G. A. Carlson, W. L. Dexter, J. N. Doggett, J. H. Fink, G. W. Hamilton, J. D. Lee, B. G. Logan, W. S. Neef, Jr., M. A. Peterson, and M. E. Rensink, "Preliminary Design Study of the Tandem Mirror Reactor," Lawrence Livermore National Laboratory, UCRL-52302 (1977).

29. D. E. Baldwin, B. G. Logan and T. K. Fowler, "An Improved Tandem Mirror Fusion Reactor," Lawrence Livermore National Laboratory, UCID-18156 (1979).

30. G. A. Carlson, B. Arfin, W. L. Barr, B. M. Boghosian, J. L. Erickson, J. H. Fink, G. W. Hamilton, B. G. Logan, J. O. Myall, and W. S. Neef, Jr., "Tandem Mirror Reactor with Thermal Barriers," Lawrence Livermore National Laboratory, UCRL-52836 (September 1979).

31. G. L. Kulcinski, et al., "A Commercial Tandem Mirror Reactor Design with Thermal Barriers - WITAMIR-I," University of Wisconsin, UWFDM-375 (October, 1980).

32. D. G. McAlees, N. A. Uckan, E. S. Bettis, C. L. Hedrick, E. F. Jaeger, and D. B. Nelson, "The Elmo Bumpy Torus Reactor (EBTR) Reference Design," Oak Ridge National Laboratory, ORNL/TM-5669 (November, 1976).

33. N. A. Uckan, D. G. Batchflor, E. S. Bettis, R. A. Dandl, C. L. Hedrick, and E. F. Jaeger, "The Elmo Bumpy Torus (EBT) Reactor," Proceedings of the 7th In International Conference on Plasma Physics and Controlled Fusion Research, August 23-30, 1978, International Atomic Energy Agency, IAEA-CN-37/1-3, III, 343.

34. R. A. Krakowski, et al., "Reactor Systems Studies of Alternative Fusion Concepts," Proceedings of the 8th Conference on Plasma Physics and Controlled Nuclear Fusion Research, Brussels, July 1-10, 1980, International Atomic Energy Agency, IAEA-CN-38/V-4 (1980).

Acknowledgements

This paper represents a review of the work of many fusion scientists and engineers who are too numerous to list here. The reader should consult the references for an indication of the many contributors. The author expresses his appreciation to many of his colleagues for supplying graphic art materials for this paper. Particular appreciation is extended to G. Carlson (LLNL) for providing the material on tandem mirror reactors, to R. Krakowski (LASL) for the material on alternate concepts, and to N. Uckan (ORNL) for the figure on EBT ARE coils. Appreciation is also extended to M. A. Abdou, J. N. Brooks and D. A. Ehst at ANL.

HYDROTHERMAL ENERGY DEVELOPMENT PROJECTS

Edward G. DiBello

EG&G Idaho, Inc.

ABSTRACT

The development of hydrothermal energy for direct heat applications is being accelerated by twenty-two demonstration projects that are funded on a cost-sharing basis by the U.S. Department of Energy, Division of Geothermal Energy. These projects are designed to demonstrate the technical and economic feasibility of the direct use of hydrothermal resources in the United States. Engineering and economic data for the projects are summarized in this paper. The data and experience being generated by these projects will serve as an important basis for future direct heat development.

IINTRODUCTION

The application of hydrothermal energy for direct use by the private sector in the United States has been quite limited to date. The reasons most commonly given for the limited development of this alternative energy source are the lack of: (a) inexpensive and reliable exploration and drilling techniques, (b) knowledge of the resource base, (c) established geothermal laws and regulations, and (d) technical and economic data. Therefore, baseline information regarding direct applications is needed to assist prospective users in defining the engineering and economic requirements of direct use projects.

To stimulate development and build the necessary data base, the U.S. Department of Energy, Division of Geothermal Energy, in 1977 and 1978 issued two Program Opportunity Notices for cost-shared direct use projects, and 22 projects (Figure 1) were initiated as a result. The type and complexity of these projects vary from space heating and grain drying to industrial processing. As development proceeds, the projects will demonstrate the feasibility of design concepts and transfer technology to the private sector. In addition, they will assist in the building of an infrastructure of engineering and other services that will promote future hydrothermal development.

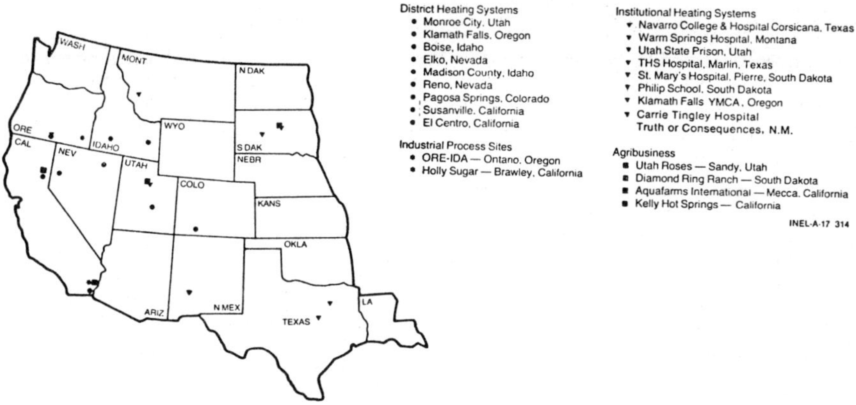

Figure 1. Direct Use Projects

MANAGEMENT

Each project is managed by a principal investigator who is assisted by a private-sector technical team. A flow diagram for a typical project is shown in Figure 2.

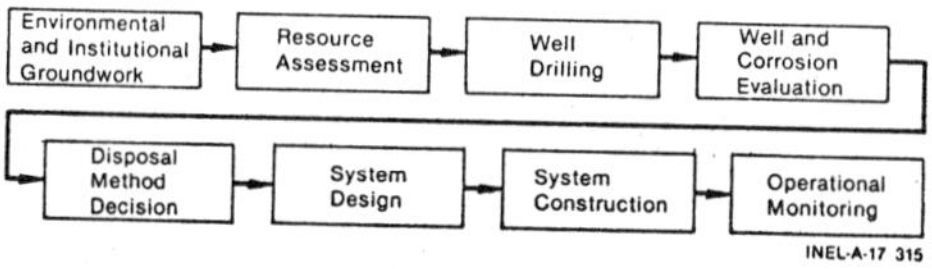

Figure 2. Project Flowchart

ENVIRONMENTAL/INSTITUTIONAL

The projects are complying with all local, state and federal environmental and institutional requirements. Some of these requirements have resulted in project delays, changes in scope, and increased costs. However, to date, no project has been terminated due to unreasonable environmental or institutional constraints. Typical issues that are applicable to future development projects include: a) well designs, well locations, and disposal methods must be selected that prevent the mixture of geothermal fluids with irrigation or potable water supplies, b) disposal of geothermal fluids may require treatment and additional cooling prior to the discharge of the fluids to surface water, c) replacement flow or other compensation may be required to compensate existing well or spring owners if new geothermal production causes interference with the established use, d) state laws and regulations guiding resource development and ownership should be closely reviewed in the initial stages of the project.

A major environmental consideration for all projects involves the disposal of the geothermal fluid. The methods of disposal usually cannot be selected until the specific water chemistry is known. Therefore, this decision is normally made after the production well has been completed and tested. Of the fifteen projects that have reached this decision point, nine projects have been granted permits for surface disposal. Six projects will inject fluids into underground aquifers (3).

WELL DRILLING

Thirty-three wells will be completed by 1981 under the auspices of this program, representing a total of 85,788 ft (26,148 m) drilled. The average production well depth for 20 projects is 2378 ft (725 m). Six projects, as presently planned, will use injection wells for disposal with an average depth of 2683 ft (818 m).

Project experience has highlighted the need for drilling techniques specifically designed for geothermal applications. In general, oil field techniques were adapted to the specific geothermal requirement. Improvement in drilling fluids, logging and testing methodology will aid future drilling operations. The majority of drilling sites have been in areas remote from drilling centers. This has complicated supply problems and increased costs. Table 1 summarizes representative costs of several wells drilled under this program.

Table 1. Well Cost Summary(1)

Project Name	Date Well Completed	Well Depth Meters (ft)	Drilling Cost/Well (x 10^3)	Cost/Meter (Foot)
Philip School	2/23/79	1300 (4266)	$317.0	$243 ($74)
Madison County	6/28/80	1524 (5000)	$483.0	$317 ($97)
Monroe City	4/10/79	457 (1500)	$245.0	$536 ($163)
St. Mary's Hospital	4/21/79	663 (2176)	$320.0	$482 ($147)
Utah Roses	12/8/79	1527 (5009)	$305.0	$199 ($60)
Warm Springs Hospital	12/5/79	457 (1498)	$186.0	$407 ($124)

Of the nineteen wells drilled to date under this program, twelve wells have encountered a resource suitable for direct applications. This equates to a success rate of approximately 63%. Four wells (21%) require additional testing and three wells (16%) did not encounter a geothermal resource (3). These figures compare favorably with the 47.4% success ratio reported in 1980 by the Petroleum Information National Geothermal Service for geothermal exploratory wells (2).

UTILIZATION SYSTEMS

Each direct use design involves a geothermal energy extraction system which consists of four basic processes: production of a geothermal fluid, transmission to an application site, heat extraction for direct use, and disposal of the energy expended fluid. The production and transmission portions of the system consist of a well or well fluid and associated pumping equipment and piping. Conventional fan coil space heating units, plate or shell and tube heat exchangers, or absorption refrigeration equipment are examples of typical heat extraction systems. As previously mentioned, disposal options include injection into suitable aquifers or surface disposal to rivers, streams or wetlands.

Figure 3 shows a simplified design for a typical direct heat application, whether district heating or industrial processing. All projects use existing technology and commercially available hardware. However, some specific geothermal design considerations include:

a) the characteristics of the resource should be well established prior to expending any significant design effort.

b) If the fluid corrosion and scaling qualities are not well known, approximately three months of materials testing should be completed prior to selecting piping and specifying other equipment.

c) Of the twelve projects that have reached a design stage, ten have incorporated isolation heat exchangers to minimize the exposure of system components to the geothermal fluids. Flat plate heat exchangers appear to be particularly adaptable for geothermal service.

c) System retrofit has sometimes been more complicated than originally expected. Reasons have included lower than expected resource temperatures, lack of as-built drawings, and poor material condition of existing systems.

Detailed design and construction information will be available in the final reports to be issued on each project. If all twenty of the projects that have encountered resources come on-line, the annual energy replaced will be 27.2 MW-Yr (813 x 10^9 Btu/yr). Assuming 100% conversion efficiency, this is equivalent to 1.46 x 10^5 barrels of crude oil per year.

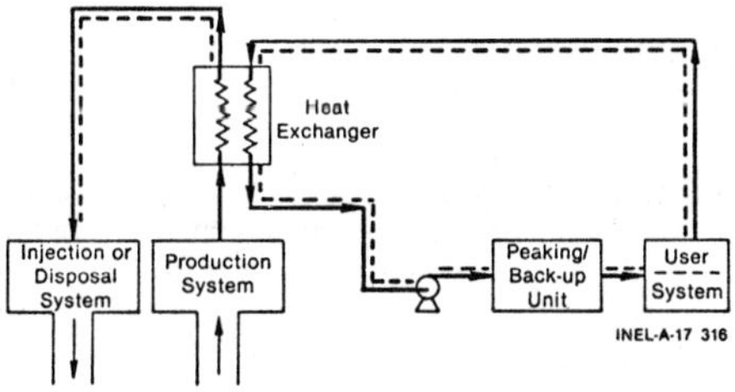

Figure 3. Direct Use System

PROJECT COSTS

Due to the site-specific nature of hydrothermal development, the project costs and resultant energy delivered costs vary from project to project. Major factors which determine the economics include resource development costs, distribution/disposal costs, system load factor, pumping costs, and financing costs. As would be expected, the definition and development of the resource and disposal system costs comprise a significant portion of the overall capital cost of a typical geothermal project. Based on a review of ten project costs summaries, the resource and disposal costs (resource assessment, well drilling, testing and disposal) averaged 40% of the of the total project costs.

A summary of economic data for several projects is presented in Table 2.

Table 2. Project Economic Summary

	$/MBtu	$/Installed Kw	Simple Payback (yrs.)
Boise City	$5.59	$284	6.1
Pagosa Springs	$5.66	$167	9
Elko Heat Company	$7.03	$446	9
Diamond Ring Ranch	$8.65	$411	8
St. Mary's Hospital	$9.07	$440	9

SUMMARY

Several projects, Klamath Falls YMCA. Diamond Ring Ranch, Philip School, and St. Mary's Hospital, are on-line. In two cases, Philip School and St. Mary's Hospital, the initial Department of Energy investment has spurred additional commercially-funded use of the hydrothermal resource.

As a result of these and similar projects, institutional barriers are being tested, private firms and organizations are gaining experience, and public awareness of hydrothermal energy is being increased. In addition, valuable environmental, technical, operational, and economic data are being generated. This information will be available in final reports to be issued upon project completion. The Department of Energy is also preparing a generic analysis on the overall program that will be available to those interested in hydrothermal direct use development.

ACKNOWLEDGEMENTS

This effort was supported by the Department of Energy, Contract No. DE-AC07-76ID01570.

REFERENCES

1) Childs, F. W., et. al., Progress in Direct Heat Applications Projects; Geothermal Resources Council, Transactions Vol. IV; September 1980.

2) Special Report, 1980 Annual Review; Petroleum Information - National Geothermal Service, January 30, 1981.

3) Geothermal Direct Heat Applications Program Summary, Semi-Annual Review Meeting, Las Vegas, Nevada; November 1980.

Preliminary Evaluation of an Advanced Binary Power Plant for Big Creek Hot Springs

T. W. Lawford and O. J. Demuth
EG&G Idaho, Inc.

ABSTRACT

A technical assistance project was performed to scope the feasibility and costs of constructing an 11 MW(e) net geothermal power plant at Big Creek Hot Springs in north-central Idaho. This plant would support reopening of the Blackbird Cobalt Mine, which is located approximately 13 miles south of Big Creek Hot Springs. This paper addresses that portion of the total technical assistance package which developed a new advanced power cycle, sized and costed power plant and field system components and evaluated operation and maintenance costs, leading to estimates of overall power costs. Other portions of the total technical assistance package which are not addressed in this paper included an evaluation of the geothermal potential of Big Creek Hot Springs, a suggested exploration program and an appraisal of the institutional factors influencing geothermal development at Big Creek Hot Springs.

BACKGROUND

Idaho Power Company, the utility supplying the geographic area of the Blackbird Mine has indicated that they can supply only a small portion of the projected power needs of the mine, and that power is available only on an interruptable basis. This, combined with the proximity of the mine to the Big Creek Hot Springs area prompted the request for geothermal technical assistance.

PLANT CONCEPTUAL DESIGN

The Earth Science Laboratory of the University of Utah Research Institute (ESL) cooperated on this project with an evalution of the geothermal potential of Big Creek Hot Springs (BCHS) and a suggested exploration strategy. The ESL estimate of the producible temperature from the BCHS resource was 300°F, based upon quartz conductive geothermometry. This is an extremely marginal geothermal temperature for economic electric power generation, as the low plant performance expected at this temperature requires large geothermal fluid flowrates to be delivered to the plant. Figure 1 illustrates the relationship between conventional geothermal plant performance, measured as brine effectiveness in watt hours per pound of geothermal fluid, and geothermal resource temperature. Figure 1 also illustrates that at 300°F, the performance of dual flash steam systems is so low that only binary systems need be considered. Furthermore, previous work had indicated that conventional binary plants would be uneconomic at 300°F unless exceptionally large flowrates could be obtained from the geothermal production wells.

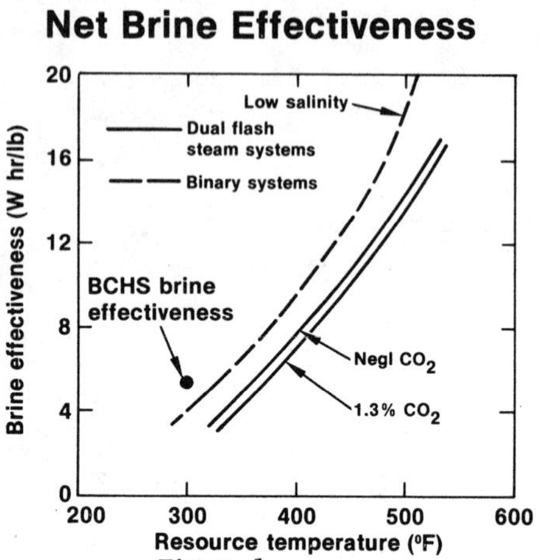

Figure 1

Previous work at EG&G on improved binary cycle performance through the use of unusual mixtures of hydrocarbon working fluids had been especially encouraging, so this approach toward improving the overall economics was pursued. Hydrocarbon mixtures have the desirable properties of boiling and condensing at increasing and decreasing temperatures, respectively, which significantly reduce the thermodynamic losses of cycles using them. Figure 2 shows the cycle selection process for working fluid and heater outlet temperature to maximize system

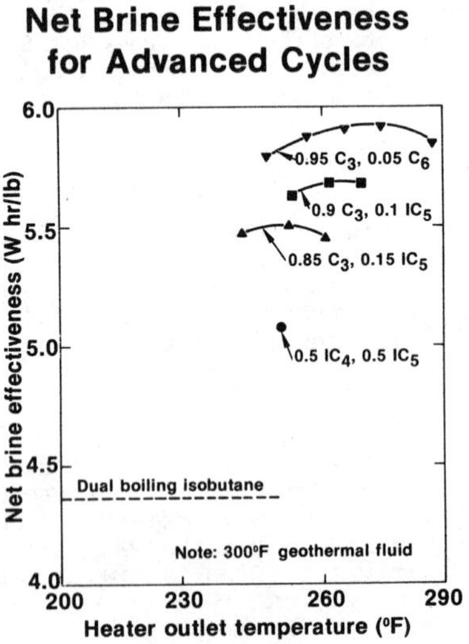

Figure 2

performance. Also shown in Figure 2 is the performance level of a dual boiling isobutane cycle (similar to the 5 MW(e) Raft River Pilot Plant). The most efficient advanced cycle developed here has a performance, as measured by net brine effectiveness, almost 40% above that base level. This performance point is also noted on Figure 1. A cycle using a working fluid mixture of 95% propane, 5% hexane with a boiler outlet temperature of approximately 270°F was selected for this application.

The cycle heat balance developed for this process, as shown in Figure 3, was then used to size major system components, and to determine the total geothermal fluid production requirements and parasitic power loads.

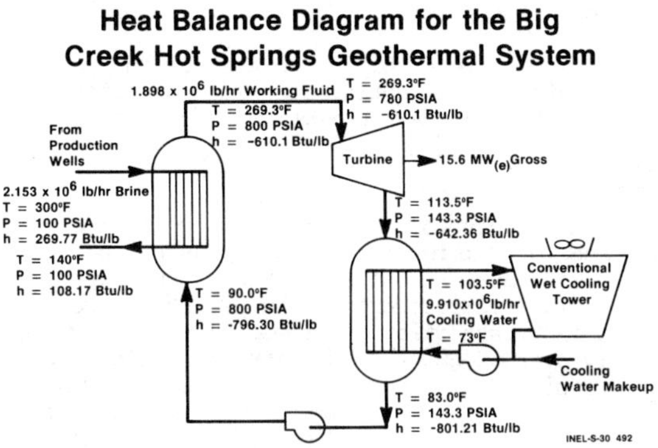

Figure 3

A conceptual layout of the process area was developed, as shown in Figure 4. The whole area is relatively compact, occupying a space only 80 by 100 feet.

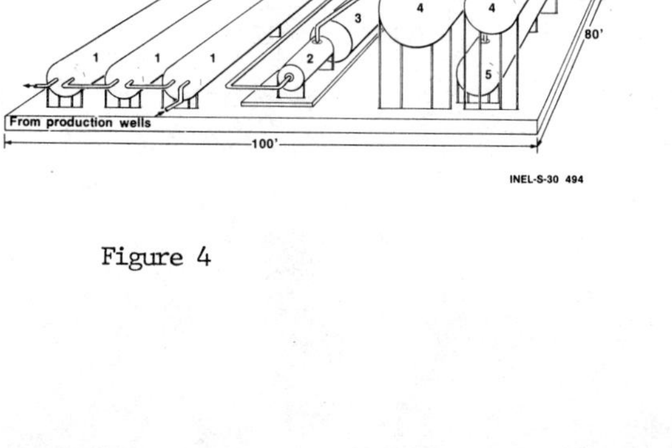

Figure 4

GEOTHERMAL FIELD SYSTEM

ESL judged that the most likely location for production well drilling is near the hot springs themselves. This is in extremely rugged terrain approximately one mile from and 800 feet higher than the likely plant site adjacent to Panther Creek and an access road. Based upon the geological faulting of this area, ESL also estimated that the production wells would need to be 6000 feet deep, and would have pumped production flowrates of from 200,000 to 400,000 lb/hr. Because of the terrain in the production area, a well field concept was used similar to that used in the Geysers area of northern California, whereby multiple wells (up to six) are directionally drilled from a single drill pad. Therefore, depending upon the well flowrates actually achieved, either one or two well pads containing a total of 6 to 11 wells, would be adequate to supply the 2.1×10^6 lb/hr required by the plant.

The injection well field would preferably be located in the valley adjacent to the plant, where the elevation differential between production and injection wells should be sufficient to provide reinjection without pumps.

POWER GENERATION COSTS

Although the advanced concept developed for this plant provide a significant improvement in performance over more conventional plants, the overall power generation costs are working against two significant handicaps: the small size of the plant and the marginal temperature of the geothermal resource. Using EG&G's geothermal plant cost data file, the power plant cost was estimated at $25,500,000, or slightly over $2,300/net kW. The production and injection wells, piping, and associated equipment would cost $15-26,000,000, depending upon well flowrates actually achieved. In order to develop power generation costs, factors for "Total Fixed Cost of Capital" were used of 17% for the plant and 25% for the wells and well field equipment. These are typical of investor owned utilities and independent geothermal field developers, respectively. Table 1 shows total power costs for the two projected average well flowrates, ranging from 131 to 176 mills/Kw-h. While these power costs are significantly above those presently being charged by Idaho Power, there may be little or no alternative to this geothermal power plant.

Price of Power (mill/kW-h)

	Average Well Flow Rate (lbm/hr)	
	200,000	400,000
Field System Capital Costs	83.6	49.0
Field O&M Costs	11.1	7.7
Plant Capital Costs	56.7	56.7
Plant O&M Costs	7.0	7.0
Transmission Line Costs	1.8	1.8
Well Replacement/Redrilling	15.5	8.6
Total	175.7	130.8

INEL-S-30 493

Table 1

PART V

TECHNOLOGY FOR RENEWABLE RESOURCES

Renewable energy resources are those that are derived directly or indirectly from solar radiation. Since all fossil and nuclear fuels are finite, much of our energy supplies must come from renewable resources in future years. Although the exact sizes of the fossil energy resources are uncertain, clearly a significant transition to renewable supplies must come about during the next century.

This section examines renewable energy sources of passive solar, active solar, central station solar, biomass, grain alcohol, hydropower, ocean thermal (OTEC), wind, and photovoltaics.

PASSIVE COMPARATIVE REPORT

John W. Spears
Stanley J. Sersen
Paul C. Miller
Automation Industries, Inc.
Vitro Laboratories Division
14000 Georgia Avenue
Silver Spring, Maryland 20910

1. INTRODUCTION

The National Solar Data Network (NSDN) has been continuously monitoring the performance of solar energy systems in buildings throughout the country on a 24-hour basis for the past three years. The data collected from this program is analyzed and performance evaluation reports are published monthly and seasonally for each site. The passive systems in the NSDN are instrumented with an average of 90 sensors to measure weather; heat loss and gain; auxiliary energy and electrical energy consumption; operation of windows, doors, vents, and movable insulation; temperatures and relative humidity of the rooms; and surface and gradient temperatures of all thermal storage masses.

This paper will report on the performance of three passive solar energy systems in the NSDN during the 1979-1980 heating season. In addition, a three day cycle of both storage charging and discharging is evaluated for each site. All three passive structures analyzed are of the direct gain type. Energy storage is in water walls and slab, slab and interior mass, and lastly by a high mass wall with a fan forced rock bed under the house slab.

2. SYSTEM DESCRIPTIONS AND PERFORMANCE

Living Systems Passive Solar System Description

The Living Systems solar energy provides space heating to an approximately 1,700 square foot single family dwelling located in Davis, California.

The passive solar space heating system is a direct gain type system, illustrated schematically in Figure 1. Incident solar energy is admitted to the building through both the large south-facing window wall (approximately 200 square feet, 61 square meters) and the overhead clerestory (approximately 80 square feet, 24 square meters). Solar energy storage is provided by blue painted water filled tubes containing nearly 2,500 gallons (9,463 liters) placed near the south window wall and under the clerestory. Additional storage is provided by the six-inch (152 MM) thick concrete slab floor of the building. Distribution of collected solar energy is primarily through natural convection. The building envelope is well insulated with R-19 insulation in the walls and R-30 insulation in the roof. The effective R-value of the window areas with curtains and shutters in place is in the range of R-2 to R-10. All glass surfaces are double-glazed with minimum window area in non-south facing walls. Manually operated insulating curtains which cover the south window wall, along with manually operated insulating shutters for the skylight glazing, are used to reduce night time heat losses. Therefore, no operating energy (other than human) needs to be applied to the system for proper operation. Auxiliary space heating energy is provided by a gas-fired furnace.

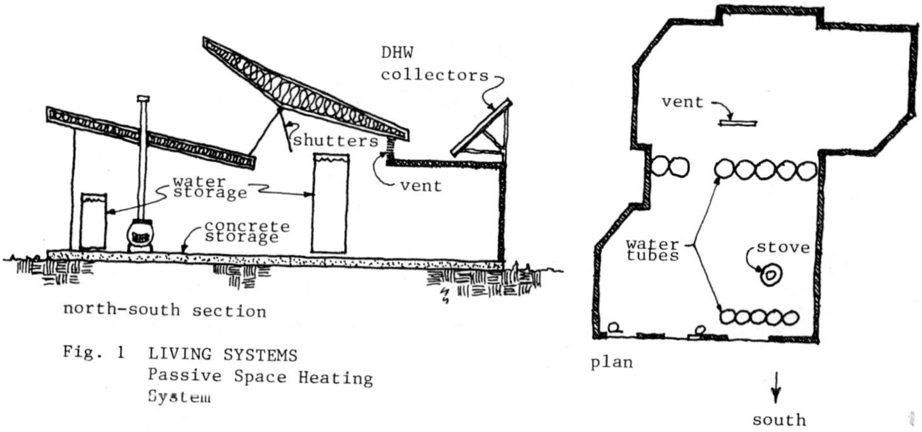

Fig. 1 LIVING SYSTEMS
Passive Space Heating
System

Living Systems Performance

Seasonal Summary

Solar energy supplied 19.40 MMBTU of the 24.28 MMBTU equipment heating load for a solar fraction of 80%. The total energy savings of 19.40 MMBTU is equivalent to 5,684 kwh of electricity, 24,872 cubic feet of natural gas, or 199 gallons of No. 2 oil. The total heating degree-days from October through February was 2,005 as compared to the long-term average of 2,099. As can be seen in Table 1, the total incident solar energy on the south-facing glazing was 49.23 MMBTU, which was about 6% above the long-term average of 46.24 MMBTU.

Performance Summary

The Living Systems site was occupied during the heating season and the passive space heating system operated very well. Although more solar energy was available than normal, the overall outdoor ambient temperature was about 5% lower than normal, corresponding to more cold, clear days.

During every month, except January, the amount of solar collected was less than the total solar energy used. The total solar energy used is equal to that collected minus the change in stored energy. For the first three months of the season the storage gave off more energy than it absorbed. A high storage temperature was observed during October with a three degree temperature degradation each month until the end of January (see Table 1). As the weather turned colder than normal in December and the storage mass had cooled down substantially, more electric consumption (portable space heaters) and wood-burning occurred to maintain the occupant's comfort (see Table 2).

During all months except October, the measured total available incident radiation was higher than the long-term average. Even with the 39% higher than average insolation during December the storage temperatures still declined. As a

TABLE 1. PASSIVE THERMAL PERFORMANCE - COLLECTION & STORAGE LIVING SYSTEMS

MONTH	INCIDENT SOLAR RADIATION		HEATING DEGREE-DAYS		SOLAR ENERGY		AVERAGE STORAGE TEMP.	AVERAGE BUILDING TEMP.	MAXIMUM BUILDING TEMP.	MINIMUM BUILDING TEMP.
	MEASURED	L.T. AVERAGE	MEASURED	L.T. AVERAGE	COLLECTED	USED				
UNITS	MMBTU's (GJ)		NONE		MMBTU's (GJ)		°F (°C)	°F (°C)	°F (°C)	°F (°C)
OCT	13.39 (14.14)	14.45 (15.26)	86	101	1.95 (2.06)	2.34 (2.47)	75 (24)	74 (23)	79 (26)	68 (20)
NOV	11.08 (11.70)	10.47 (11.05)	405	360	5.88 (6.21)	5.92 (6.25)	72 (22)	70 (21)	73 (23)	67 (19)
DEC	9.43 (9.96)	5.77 (6.09)	586	595	5.21 (9.90)	5.72 (6.04)	69 (20)	67 (19)	72 (22)	62 (17)
JAN	7.22 (7.62)	5.95 (6.28)	531	617	2.99 (3.16)	2.85 (3.01)	66 (19)	65 (18)	67 (19)	61 (16)
FEB	8.11 (8.56)	9.60 (10.01)	397	426	2.51 (2.65)	2.58 (2.72)	68 (20)	66 (19)	69 (20)	65 (18)
MAR	*	*	*	*	*	*	*	*	*	*
APR	*	*	*	*	*	*	*	*	*	*
SUM	49.23 (51.98)	46.24 (48.82)	2005	2009	18.54 (19.57)	19.41 (20.49)	--	--	--	--
AVG	9.85 (1.04)	9.25 9.76)	--	--	3.71 (3.92)	3.88 (4.10)	70 (21)	68 (20)	72 (22)	64 (18)

*Data not available due to Site Data Acquisition System (S.D.A.S.) down.

TABLE 2. PASSIVE THERMAL PERFORMANCE LOSSES & AUXILIARY LIVING SYSTEMS

MONTH	BUILDING HEAT LOAD	U * A * T	INFILTRATION LOSSES	AUXILIARY ENERGY			EQUIPMENT HEAT LOAD	SOLAR FRACTION OF DEMAND
				FIREPLACE	INTERNAL GAINS	ELECTRIC OR FOSSIL USED		
UNITS	MMBTU (GJ)	MMBTU (GJ)	MMBTU (GJ)	MMBTU (GJ)	MMBTU (GJ)	MMBTU (GJ)	MMBTU (GJ)	PERCENT
OCT	2.34 (2.47)	1.34 (1.41)	0.99 (1.05)	0.0	0.7 (0.7)	0	2.34 (2.47)	100
NOV.	6.90 (7.28)	4.92 (5.19)	1.99 (2.10)	0.24 (0.25)	0.74 (0.78)	0	6.16 (6.50)	96
DEC	7.97 (7.89)	5.73 (6.05)	2.25 (2.37)	1.33 (1.40)	0.93 (0.98)	0	7.05 (7.44)	81
JAN	6.63 (7.00	4.57 (4.83)	2.05 (2.16)	1.86 (1.96)	1.92 (2.03)	0	4.71 (4.97)	60
FEB	5.70 (6.02)	3.76 (3.97)	1.94 (2.05)	1.45 (1.53)	1.66 (1.75)	0	4.03 (4.25)	64
MAR	*	*	*	*	*	*	*	*
APR	*	*	*	*	*	*	*	*
SUM	29.54 (31.19)	20.32 (21.45)	9.22 (9.73)	4.88 (5.15)	5.95 (6.28)	0	24.29 (25.65)	--
AVG	5.91 (6.24)	4.06 (4.29)	1.84 (1.94)	3.72 (3.93)	1.19 (1.26)	0	4.86 (5.13)	80

*Data not available.

result of the increased use of auxiliary energy in December, January and February, the solar fraction for each month decreased (see Table 2).

Kalwall Corporation Passive Solar System Description

The Kalwall Corporation solar-energy system is a direct gain passive space-heating system (Figure 2). The heated space (approximately 10,000 square feet) is a portion of the commercial warehouse in Manchester, New Hampshire. Solar energy is admitted through vertical double-glazed Kalwall panels covering the entire south and east walls. Storage is provided by the dark colored six-inch thick concrete floor of the building and the contents of the warehouse. Summer overheat protection is provided by both an overhang on the south roof edge and by natural ventilation. Auxiliary energy is provided by two liquid-to-air heat exchangers heated by the building boiler. Control thermostats for the auxiliary system are set at approximately 60°F.

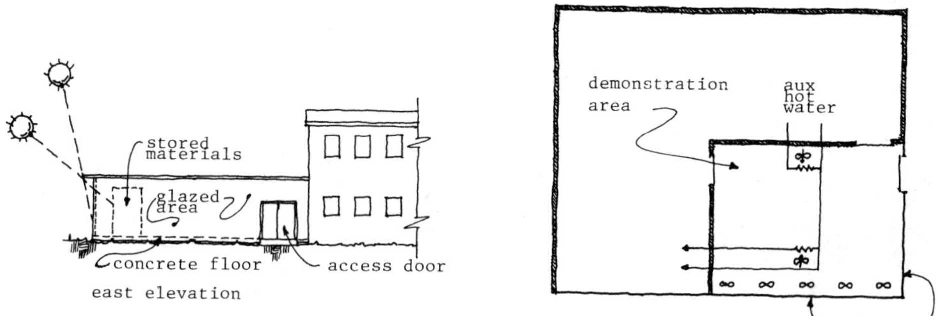

Fig. 2 KALWALL CORPORATION Passive Space Heating System

The system was designed to satisfy approximately 50 percent of the annual space heating demand. However, the design performance projections were made before additional insulation was added to the building roof. Consequently, revised design performance projections would be expected to predict a solar contribution of greater than 50 percent. Results of previous performance analysis for the near-average winter of 1977-78 have shown that the system does perform above these design projections.

Kalwall Corporation Performance

Seasonal Summary

Solar energy supplied 179.36 MMBTU of the 226.52 MMBTU equipment heating load for a solar fraction of 79%. The total energy savings of 179.36 MMBTU is equivalent to 52,552 kw of electricity, 229,948 cubic feet of natural gas, or 1,843 gallons of No. 2 oil. The total heating degree-days from October through April was 5,162. This was considerably lower than the long-term average of 6,744 for the respective months. As can be seen in Table 3, the total measured solar radiation falling on both the south and east facing glazing was 359.34 MMBTU. This compares closely with the long-term average of 370.04 MMBTU.

Performance Summary

The Kalwall Corporation site was in full use during the heating season and the passive space heating system operated very well. Since this site is a warehouse, comfort levels for the workers is much lower than what would be required for a residence. As a result, building temperatures and storage temperatures are allowed to go lower than would normally be experienced (see Table 3). This was allowed to happen from October through January, but from February through April higher building

TABLE 3. PASSIVE THERMAL PERFORMANCE - COLLECTION & STORAGE KALWALL CORPORATION

MONTH	INCIDENT SOLAR RADIATION MEASURED	INCIDENT SOLAR RADIATION L.T. AVERAGE	HEATING DEGREE-DAYS MEASURED	HEATING DEGREE-DAYS L.T. AVERAGE	SOLAR ENERGY COLLECTED	SOLAR ENERGY USED	AVERAGE STORAGE TEMP.	AVERAGE BUILDING TEMP.	MAXIMUM BUILDING TEMP.	MINIMUM BUILDING TEMP.
	MMBTU's (GJ)		NONE		MMBTU's (GJ)		°F (°C)	°F (°C)	°F (°C)	°F (°C)
OCT	46.71 (49.32)	42.41 (44.78)	456	487	24.18 (25.53)	23.36 (24.66)	71 (22)	68 (20)	76 (24)	62 (17)
NOV	41.91 (44.25)	51.51 (54.39)	585	810	21.91 (23.13)	21.98 (23.21)	67 (19)	67 (19)	71 (22)	64 (18)
DEC	46.28 (48.87)	40.70 (42.97)	1005	1246	33.74 (35.63)	34.17 (36.08)	59 (15)	61 (16)	67 (19)	53 (12)
JAN	56.23 (59.37)	51.58 (54.46)	1209	1376	33.65 (35.53)	34.28 (36.19)	57 (14)	60 (15)	63 (17)	56 (13)
FEB	72.10 (76.13)	55.36 (58.45)	1130	1187	31.80 (33.57)	31.37 (33.13)	61 (16)	63 (17)	67 (19)	58 (14)
MAR	54.04 (57.06)	73.16 (77.25)	844	1014	21.02 (22.19)	21.69 (22.90)	62 (17)	64 (18)	69 (20)	60 (16)
ARP	42.07 (44.42)	55.36 (58.45)	512	624	12.78 (13.49)	12.51 (13.21)	65 (18)	67 (19)	71 (22)	62 (17)
SUM	359.34 (379.42)	370.04 (390.71)	5285	6744	170.08 (179.58)	179.36 (189.38)	--	--	--	--
AVG	51.33 (54.20)	52.86 (55.81)	--	--	25.58 (27.01)	25.62 (27.05)	63 (17)	64 (18)	69 (20)	59 (15)

TABLE 4. PASSIVE THERMAL PERFORMANCE LOSSES & AUXILIARY KALWALL CORPORATION

MONTH	BUILDING HEAT LOAD	U * A * T	INFILTRATION LOSSES	AUXILIARY ENERGY: FIREPLACE	AUXILIARY ENERGY: INTERNAL GAINS	AUXILIARY ENERGY: ELECTRIC OR FOSSIL USED	EQUIPMENT HEAT LOAD	SOLAR FRACTION OF DEMAND
UNITS	MMBTU (GJ)	MMBTU (GJ)	MMBTU (GJ)	MMBTU (GJ)	MMBTU (GJ)	MMBTU (GJ)	MMBTU (GJ)	PERCENT
OCT	24.50 (25.86)	22.47E (23.73)	2.03E (2.14)	NA	NA	1.15 (1.21)	24.50 (35.87)	95
NOV	30.76 (32.47)	28.30E (29.88)	2.46E (2.59)			8.77 (9.26)	30.76 (32.48)	71
DEC	39.96 (42.19)	37.15 (39.22)	2.81 (2.87)			5.79 (6.11)	39.96 (42.19)	86
JAN	42.45 (44.82)	39.30 (41.49)	3.15 (3.33)			8.62 (9.10)	42.45 (44.82)	81
FEB	42.72 (45.11)	39.43 (42.63)	3.24 (3.42)			11.35 (11.98)	42.72 (45.11)	73
MAR	29.29 (30.93)	26.38 (27.85)	2.91 (3.07)			7.60 (8.02)	29.29 (30.93)	74
APR	16.84 (17.78)	15.55 (16.42)	1.29 (1.36)			4.33 (4.57)	16.84 (17.78)	74
SUM	226.52 (239.17)	208.58 (220.02)	17.94 (18.94)	NA	NA	47.61 (50.27)	226.52 (239.17)	--
AVG	32.36 (34.17)	29.79 (31.45)	2.56 (2.70)			6.8 (7.18)	32.36 (34.17)	79

temperatures were maintained by the increased use of auxiliary energy. During this time the energy stored from the summer was slowly used, as evident in the decreasing storage temperature (see Table 3). In February the combination of the 11.35 MMBTU of auxiliary energy, coupled with the higher than normal 72.10 MMBTU of solar energy, brought the average storage temperature up from a low of 57°F (14°C) to 61°F (16°) (see Tables 3 & 4). As a result of the increase usage of auxiliary energy, the solar fraction decreased. Even with the extra auxiliary used, savings were still very high and the measured contribution from solar energy was higher than expected.

Hullco Construction Passive System Description[3]

The Hullco Construction solar energy system is a combination sunspace and mass wall system used for heating a single-family residence located in Prescott, Arizona. The 1,056 square foot, south-facing building, illustrated in the drawings of Figure 3 is a combination greenhouse and direct gain system. Incident solar energy enters the building through approximately 400 square feet of double-glazed prefabricated Kalwall panels. The sliding glass doors between the greenhouse and the living area admit incident solar energy directly into the master bedroom and the living room areas. Solar energy collected but not used to satisfy the immediate building space heating demand is stored directly in the massive walls and floors of the building or indirectly in the 670 cubic feet of 3 to 5 inch diameter rock located under the floor of the north half of the building. Stored energy is released to satisfy the building space heating demand during periods of time when incident solar energy is not available.

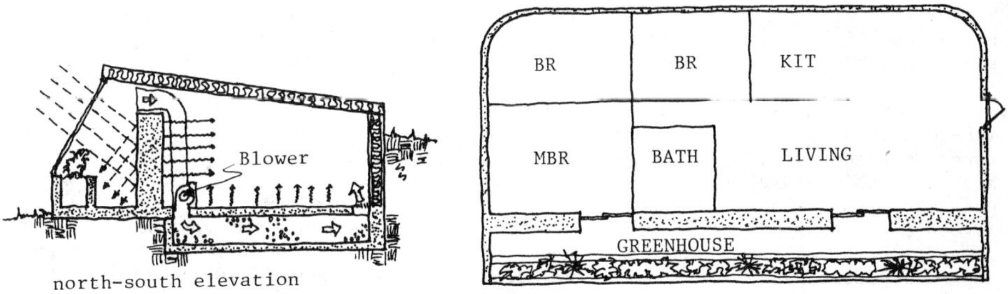

Fig. 3 HULLCO CONSTRUCTION Passive Space Heating System

plan

south

Direct storage of collected solar energy is provided by the walls and floor of the building. The brick floor and the north wall of the greenhouse provide solar energy storage for the greenhouse. The 12-inch thick, sand-filled concrete block greenhouse north wall acts as a Trombe wall, storing collected energy from the greenhouse during the day and releasing the collected energy by radiation of the building at night. The 4-inch thick building concrete slab floor acts as direct solar storage, particularly in the living room and master bedroom areas where the floor is covered with Red Mexican tile. Additional storage is provided by the 8-inch thick solid grouted concrete block exterior building walls on the north, east, and west perimeter of the building.

Indirect storage of collected solar energy is provided by the 670 cubic feet of rock storage located under the north side of the building and by the 4-inch thick concrete slab floor poured on top of the rock. Solar energy collected in the greenhouse is transferred to the rock bed from vents at the top of the greenhouse through under-floor ducts by two one-third horsepower blowers located at the east and west ends of the greenhouse. After transferring energy to the rocks, the air returns to the greenhouse directly from the rock bed. The greenhouse fans operate when temperatures near the top of the greenhouse reach approximately 90°F.

Energy stored in the rock is released through the carpeted concrete floor to the house.

Auxiliary energy for space heating can be supplied by either electric radiant heat panels or by the wood-burning stove located in the living room. The ceiling-mounted electric radiant heat panels located in each room are controlled by individual room thermostats.

Hullco Construction Performance

Seasonal Summary

Solar energy supplied 54.26 MMBTU of the 59.34 MMBTU equipment heating load for a solar fraction of 91%. The total energy savings of 54.26 MMBTU is equivalent to 15,898 kw of electricity, 69,564 cubic feet of natural gas, or 553 gallons of No. 2 oil. The total heating-degree days from October through May (excluding November, due to SDAS being down) was 4,152 as opposed to the long-term average of 3,824. As can be seen in Table 5, the measured incident solar radiation of 137.66 MMBTU was below the long-term average of 160.16 MMBTU.

Performance Summary

The Hullco Construction site was operational throughout the heating season and performed very well. The house was unoccupied from November 30 through March 5. During this period the auxiliary heating thermostats were set back. This allowed the solar system to supply all of the space heating requirements of the building by allowing the building temperatures to "float" and the house to equalize. By the end of December the house had equalized, energy in storage left from the summer had been used, and the building temperature stablized at $64^{\circ}F$ ($18^{\circ}C$).

After the house was reoccupied in March, storage and building temperatures were continually increased by use of auxiliary energy. These much higher temperatures were needed to meet the new occupant's comfort level. In April auxiliary heat was used only during the periods of low insolation. However, in May, auxiliary energy was used everyday even though there were many days with high levels of solar radiation. The main source of this heat was the radiant heat panel located in the bathroom which was in the heating mode much of the time. With the increased use of auxiliary energy the solar fraction dropped from almost 100% to 85%.

Comparison of Performance

When designing a passive solar system there are a few problems common to every structure regardless of its purpose or design. The design must take into account the amount of heated space involved, the mass of the storage system, the area of the glazing and the type of weather conditions common to the location where the system is to be built. All of these factors affect the performance of the systems in this report. Using design data and some of the seasonal data for each site it is possible to produce some normalized ratios. These ratios make it possible to compare the performance of these sites even though they are greatly different in their types of storage, their location and their purpose. These normalized figures are put on a per square foot of floor space per day basis for comparison purposes.

Figure 4 is a bar graph which illustrates these normalized figures for the systems involved. Looking at the solar energy incident/square foot floor space-day bar, Hullco is seen as overwhelming the other sites. The first reason for this is that Hullco has a high design solar fraction, thus a larger amount of glazing area was used. Second, Hullco experiences a cold winter (4,152 heating degree-days as compared to 2,000 at Living Systems) so it needs a large collection system to maintain the heated space at a comfortable temperature. Finally, more solar energy was available to the system this season than usual. Living Systems had approximately one-third as much solar energy available per square foot of floor space per day as Hullco. Even though this system also has a high design solar fraction it experiences a much more mild climate than Hullco and doesn't need as much energy

TABLE 5. PASSIVE THERMAL PERFORMANCE - COLLECTION & STORAGE
HULLCO CONSTRUCTION

MONTH	INCIDENT SOLAR RADIATION MEASURED	INCIDENT SOLAR RADIATION L.T. AVERAGE	HEATING DEGREE-DAYS MEASURED	HEATING DEGREE-DAYS L.T. AVERAGE	SOLAR ENERGY COLLECTED	SOLAR ENERGY USED	AVERAGE STORAGE TEMP.	AVERAGE BUILDING TEMP.	MAXIMUM BUILDING TEMP.	MINIMUM BUILDING TEMP.
UNITS	MMBTU's (GJ)		NONE		MMBTU's (GJ)		°F (°C)	°F (°C)	°F (°C)	°F (°C)
OCT	23.46 (24.77)	25.80 (27.24)	93	254	3.82 (4.03)	3.42 (3.61)	76 (24)	76 (24)	78 (25)	76 (24)
NOV	*	*	*	*	*	*	*	*	*	*
DEC	16.83 (17.77)	21.40 (22.59)	899	818	10.74 (11.34)	11.71 (12.36)	74 (23)	69 (20)	73 (23)	65 (18)
JAN	15.21 (16.06)	22.25 (23.50)	837	865	9.16 (9.67)	9.55 (10.08)	68 (20)	64 (18)	67 (19)	61 (16)
FEB	18.74 (19.79)	21.52 (22.72)	696	686	9.32 (9.84)	9.01 (9.51)	67 (19)	64 (18)	67 (19)	58 (14)
MAR	22.52 (23.78)	24.47 (25.84)	775	642	7.73 (8.16)	7.67 (8.09)	70 (21)	69 (20)	72 (22)	65 (18)
APR	21.17 (22.35)	23.02 (24.31)	780	394	6.74 (7.12)	6.66 (7.03)	73 (23)	73 (23)	77 (25)	70 (21)
MAY	19.73 (20.83)	21.70 (22.91)	372	165	6.66 (7.03)	6.24 (6.58)	74 (23)	75 (24)	77 (25)	73 (23)
SUM	137.66 (145.35)	160.16 (169.11)	4152	3824	51.84 (54.74)	54.26 (57.29)	--	--	--	--
AVG	19.66 (20.75)	22.88 (24.16)	593	546	7.41 (7.82)	7.75 (8.18)	72 (22)	70 (21)	73 (23)	67 (19)

*No data available for November.

TABLE 6. PASSIVE THERMAL PERFORMANCE LOSSES & AUXILIARY
HULLCO CONSTRUCTION

MONTH	BUILDING HEAT LOAD	U * A * T	INFILTRATION LOSSES	AUXILIARY ENERGY FIREPLACE	AUXILIARY ENERGY INTERNAL GAINS	AUXILIARY ENERGY ELECTRIC OF FOSSIL USED	SYSTEM OPERATING ENERGY	EQUIPMENT HEAT LOAD	SOLAR FRACTION OF DEMAND
UNITS	MMBTU (GJ)	MMBTU (GJ)	MMBTU (GJ)	MMBTU (GJ)	MMBTU (GJ)	MMBTU (GJ)	MMBTU (GJ)	MMBTU (GJ)	PERCENT
OCT	5.04 (5.32)	3.19 (3.37)	1.85 (1.95)	0.0	1.49 (1.57)	0.0	0.20 (0.21)	3.62 (3.82)	94
NOV	*	*	*	*	*	*	*	*	*
DEC	12.30 (12.98)	7.32 (7.73)	4.98 (5.26)	0.0	0.63 (0.66)	0.01 (0.01)	0.11 (0.16)	11.82 (12.48)	99
JAN	10.13 (10.70	6.12 (6.46)	4.01 (4.23)	0.0	0.45 (0.47)	0.0	0.08 (0.08)	9.63 (10.16)	99
FEB	9.49 (10.02)	6.23 (6.58)	3.26 (3.44)	0.0	0.43 (0.45)	0.0	0.12 (0.13)	9.13 (9.64)	96
MAR	11.39 (12.06)	6.81 (7.19)	4.58 (4.83)	2.03 (2.14)	1.25 (1.32)	0.25 (0.16)	0.14 (0.15)	10.09 (10.65)	96
APR	9.21 (9.72)	5.61 (5.92)	3.60 (3.80)	0.53 (0.56)	1.44 (1.52)	0.35 (0.37)	0.19 (0.20)	7.73 (8.61)	86
MAY	8.94 (9.49)	5.45 (5.75)	3.49 (3.68)	0.0	1.49 (1.57)	0.91 (0.96)	0.17 (0.18)	7.32 (7.72)	85
SUM	66.50 (70.21)	40.73 (43.00)	25.77 (27.21)	2.56 (2.70)	7.18 (7.58)	1.52 (1.60)	1.01 (1.06)	59.34 (62.66)	--
AVG	9.50 (10.03)	5.82 (6.14)	3.68 (3.88)	0.37 (0.39)	1.03 (1.09)	0.22 (0.23)	0.14 (0.15)	8.48 (8.95)	91

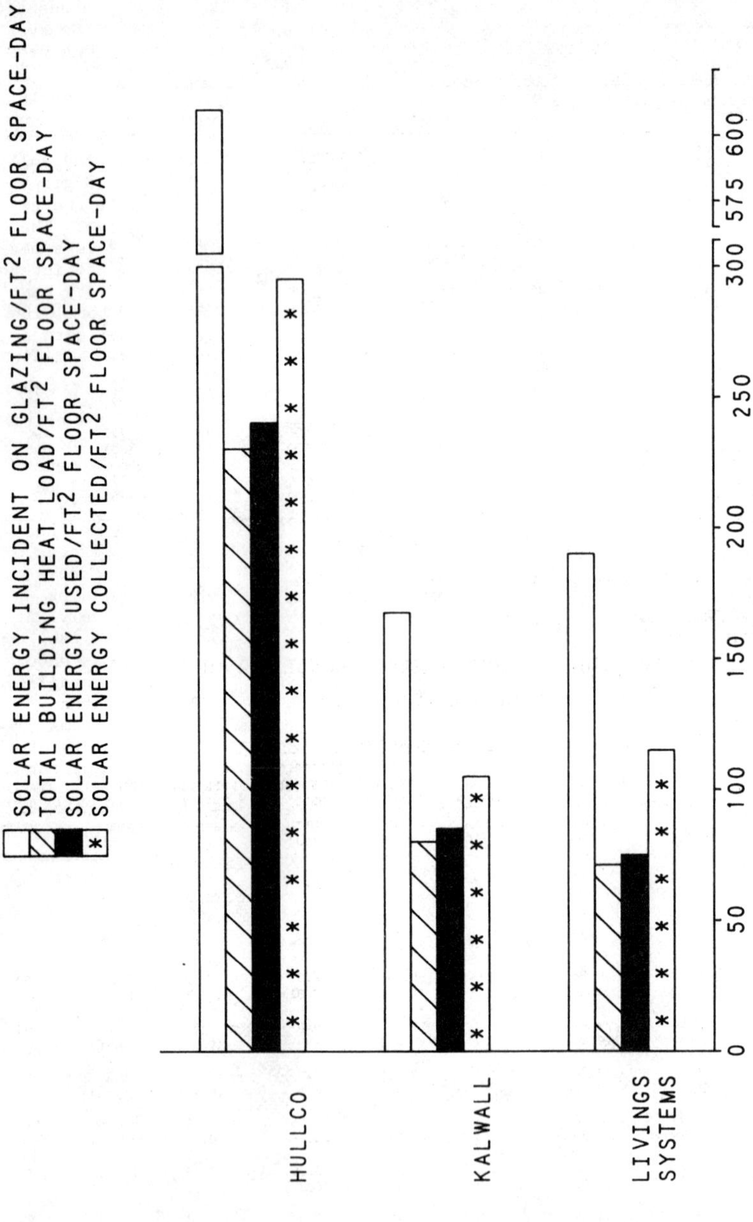
SOLAR ENERGY INCIDENT ON GLAZING/FT2 FLOOR SPACE-DAY
TOTAL BUILDING HEAT LOAD/FT2 FLOOR SPACE-DAY
SOLAR ENERGY USED/FT2 FLOOR SPACE-DAY
SOLAR ENERGY COLLECTED/FT2 FLOOR SPACE-DAY
HULLCO
KALWALL
LIVINGS SYSTEMS
0
50
100
150
200
250
300
575
600

FIGURE 4

to achieve a high solar fraction. Finally Kalwall has the lowest ratio. This might seem like an unusual design since the warehouse experiences a cold winter (5,285 heating degree-days) until the conditions are examined. This system is designed to meet only 50% of the load. Since the heated space is a warehouse, the heated space temperature is kept lower than in a residential building. So even though Kalwall has a large floor space, the reduced heating load resulting from the lower space heating temperatures combine to produce a reasonable solar fraction which can be met my the system despite a relatively small solar aperature to floor area ratio.

The other three ratios in Fig. 4 deal with the collected energy, used solar energy, and total load energy per square foot floor space per day. This graph indicates that more solar energy is used than is collected. This extra energy is energy that was left from the summer and reflects a well designed system. The portion of the total load bar which is above the solar energy used bar is a measure of the other energies used to meet the load. These energies include internal gains from the occupants and household appliances as well as the auxiliary heat energy. Hullco's load is seen as much larger than either of the other building's load. This is explained by the occupants desire to keep the house very warm and is reflected in the building's warm average temperature (Hullco's average building temperature, 70°F; Living Systems, 68°F; Kalwall 64°F).

Sizing the amount of storage material needed in a passive house is a critical element in the design of the building. If too little material is used, the living space will experience large temperature savings from the periods of solar energy to the periods of darkness. If too much material is used, the system will be sluggish in responding to solar energy. If storage temperatures drop due to low solar insolation it will require an excessive amount of time to re-charge the storage. Figures 5 & 6 are graphs of typical discharging and charging cycles of the reported sites. Looking at Fig. 5, it can be seen that during a few days of low solar insolation the building and storage temperatures for each building slowly decline until the building and storage temperatures are about equal. At this point the system is said to be in equilibrium. The slow decline of the systems to their equilibrium states suggests that each of the systems has enough storage mass to prevent large temperature swings. When solar energy is available to the systems, the building temperatures increase fairly rapidly, with the storage temperatures rising more slowly. This rapid recovery of the systems suggests that the storage masses are not so large as to hinder the recovery of the systems. Combination of both these graphs proves that the storage systems for all these sites were properly sized.

The periods graphed for Hullco in Fig. 5 & 6 provide a unique examination of a well-sized storage system which uses several types of storages. During the period presented, the residence of Hullco was unoccupied. All the thermostats in the house were set back to their minimum and the solar system was allowed to heat the building alone. Building temperatures during the discharging period fell off at a slow steady rate of about 1°F per day. During the charging cycle though, the building temperatures rise about 5-6°F in only two days. From these graphs it is apparent that the Trombe wall controls the room temperature while the slab and rock bed temperatures serve to dampen any large temperature swings. In the graphs of the other sites, the storage reaction to low solar energy is affected by the use of auxiliary energy. Even with use of auxiliary energy both systems react well to the cloudy weather.

Summary

The passive sites in this report have been operating for two or three years and have performed very well. Monthly and Seasonal reports are available for each site since it started. (See References.)

After monitoring these systems for two or three years several trends became evident. First the simplist systems work the best. Kalwall has had no system problems and is the simplist system. Living Systems has had problems with the movable insulation latches and seals.

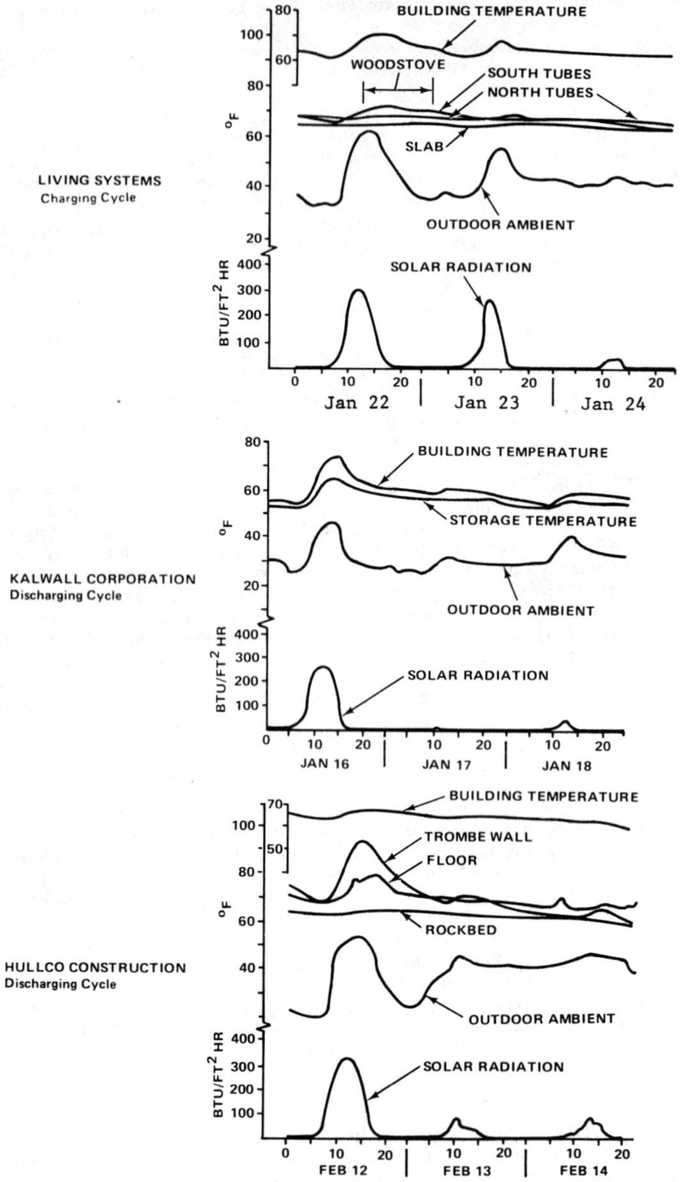
LIVING SYSTEMS
Charging Cycle
BUILDING TEMPERATURE
WOODSTOVE
SOUTH TUBES
NORTH TUBES
SLAB
OUTDOOR AMBIENT
SOLAR RADIATION
°F
BTU/FT² HR
Jan 22
Jan 23
Jan 24
KALWALL CORPORATION
Discharging Cycle
BUILDING TEMPERATURE
STORAGE TEMPERATURE
OUTDOOR AMBIENT
SOLAR RADIATION
JAN 16
JAN 17
JAN 18
HULLCO CONSTRUCTION
Discharging Cycle
BUILDING TEMPERATURE
TROMBE WALL
FLOOR
ROCKBED
OUTDOOR AMBIENT
SOLAR RADIATION
FEB 12
FEB 13
FEB 14

FIGURE 5.

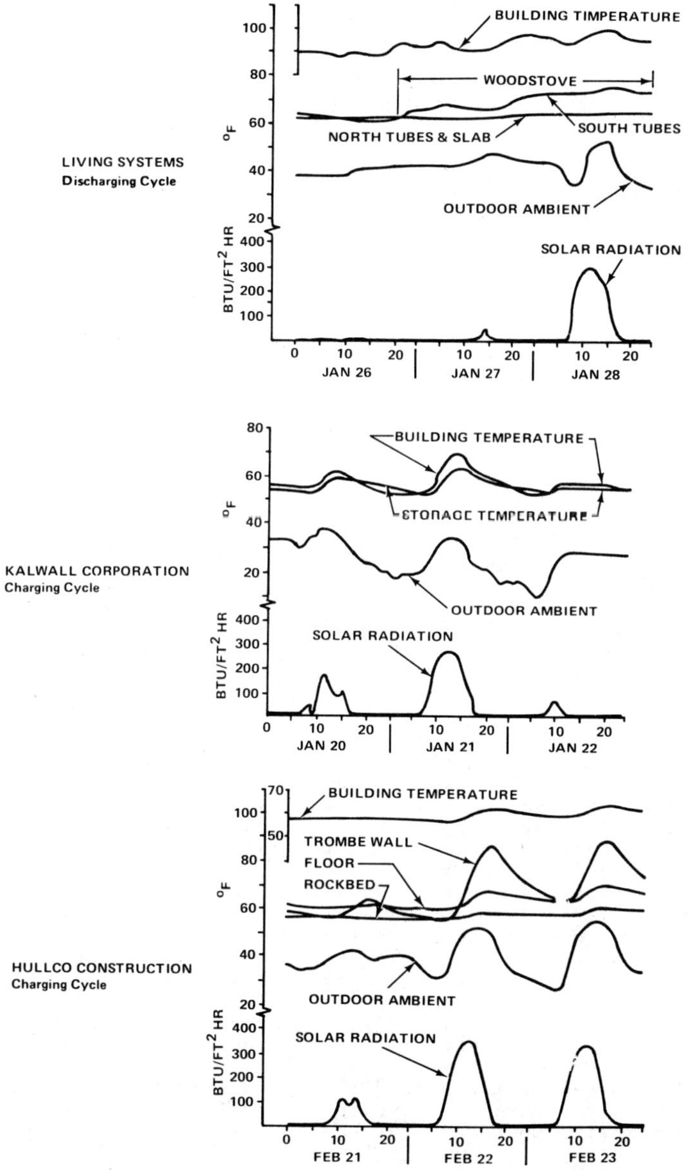
LIVING SYSTEMS
Discharging Cycle
BUILDING TIMPERATURE
WOODSTOVE
SOUTH TUBES
NORTH TUBES & SLAB
OUTDOOR AMBIENT
SOLAR RADIATION
°F
BTU/FT² HR
JAN 26
JAN 27
JAN 28
KALWALL CORPORATION
Charging Cycle
BUILDING TEMPERATURE
STORAGE TEMPERATURE
OUTDOOR AMBIENT
SOLAR RADIATION
JAN 20
JAN 21
JAN 22
HULLCO CONSTRUCTION
Charging Cycle
BUILDING TEMPERATURE
TROMBE WALL
FLOOR
ROCKBED
OUTDOOR AMBIENT
SOLAR RADIATION
FEB 21
FEB 22
FEB 23

FIGURE 6.

DETAILED PERFORMANCE PREDICTION OF A LARGE PASSIVE SOLAR BUILDING BY TRANSIENT SIMULATION

SYED F. AHMED
DAVID F. HILL
BURT HILL KOSAR RITTELMANN ASSOCIATES
BUTLER, PA 16001

FRANK W. GRAUMAN
BOHLIN POWELL LARKIN CYWINSKI
ARCHITECTS
WILKES-BARRE, PA 18701

KENNETH E. ARGALL
BURT HILL KOSAR RITTELMANN
ASSOCIATES
BUTLER, PA 16001

ABSTRACT

The performance of a large passive solar building has been predicted by Transient Simulation using TRNSYS 10.1 computer program. Two buildings were simulated -- one passive solar, and the other a reference building. The net contribution of solar energy relative to the non-solar building, as evidenced by the reduction in auxiliary heating energy for solar building, has been predicted to be 62.4 x 10^6 KJ per year. Also predicted are hourly heat fluxes and temperatures during certain winter and summer months.

INTRODUCTION

The passive solar building whose performance has been simulated is located at Shelly Ridge development in Philadelphia, PA. The Philadelphia Girl Scouts Council, who own the building and the Shelly Ridge development, called for a design which would make strong and demonstrative use of solar energy. The concern for energy included economic interests but were largely directed at the facility's educational purpose and the Girl Scouts desire to symbolize their commitment to native and environmental issues. Consequently it was decided to emphasize passive solar design in the building, the passive approach was selected both because it appeared practical and because its use encouraged more occupant involvement.

Since the 526 m^2 building is dominated by a single main space, it was evident that its energy requirements would be affected chiefly by external energy influences, indicating that solar strategies would indeed be most helpful in offsetting the heating load. Passive cooling was rejected since Philadelphia's humid summers and lack of diurnal temperature swing offer little opportunity for application. Because the summertime activities in the area are largely outdoors, air conditioning was not considered. Instead cooling is provided by increased ventilation when the outside weather conditions are favorable.

BUILDING DESCRIPTION

The plan view of the building is shown in Figure 1. The building's total area is 526 m^2. Because of its compact size, it was determined that it could be well served by vertical collection areas. A Trombe Wall approach was selected for the main hall because it could function as a heat collector without losing the ability to darken the hall for audio visual presentations. A direct gain lobby space was designed to give access to the southerly views and to make the solar features a significant part of the entrance sequence. This lobby, semi-circular in plan, is designed to function as a sundial, which calls attention to the building's solar nature, while also illustrating the sun's apparent daily movement.

In plan, the building is sliced on a diagonal creating the southerly surface. This triangular arrangement maximized the available surface while minimizing the northeast and northwest exposures. A tree line and the low roof eaves on the northwest side provide protection from the winter winds.

The direct gain space in the lobby area, i.e., the sundial, is triple glazed, and has a net glazing area of 51.8 m^2; no night insulation was provided. The direct gain spaces in the program area, that is the openings in the mass wall, are double glazed with a night insulation of "U" value 4.088 $KJ.m^{-2}.hr^{-1}.°C^{-1}$. These spaces also must be covered in the day for audio visual presentations, so the night insulation has been retained on the openings.

The mass wall (Trombe Wall) in the program space (the main) hall is the unvented type. Some TRNSYS studies had earlier indicated only a marginal improvement in the performance of a vented over an unvented mass wall, and for this reason, as well as because of the difficulties in the use of backdraft dampers in the vented wall due to leakage, an unvented mass wall was more practical.

The mass wall consists of a structural timber grid with 10 cm thick brick infill panels. The 10 cm thickness will deliver the heat to the space sometime in early evening when the building may still be occupied, rather than at night when it will usually be unoccupied. By using brick as an infill material, the 10 cm thickness is structurally possible. This approach also permits the substitution of clear glazing for brick, thus providing a mix of direct gain and mass wall areas in an integrated grid pattern.

The exterior glazing is a translucent material such as Kalwall or Sunwall in the grid quadrants of mass wall which are brickfilled. Glass is used in the lobby area and in the mass wall only where a "window" to the exterior exists.

The shading is provided by a canvas awning deployed in spring and removed at the beginning of winter.

In addition to passive solar gain, the south glazing provides significant daylighting to almost all the spaces. High windows in the ceiling in the main hall face the light colored sloping ceiling. Daylighting is supplemented by dormers in both the deeper portion of the main hall and in the loft.

CLIMATIC DATA

The climate data used for Philadelphia, PA, was 1969 which has been termed as a Test Reference Year (TRY), by the National Oceanic and Atmospheric Administration (NOAA). Hourly weather data for dry bulb temperature, solar radiation, dew point temperature and wind speed was used.

BRIEF DESCRIPTION OF TRNSYS PROGRAM [1]

The TRNSYS program uses a modular simulation technique. Because every system can be defined as a set of components connected in a certain manner to accomplish a specific task, it is possible to simulate the performance of the system by collectively simulating the performance of the interconnected components. The modular simulation technique greatly reduces the complexity of system simulation because it essentially reduces a large problem into a number of smaller problems each of which can be more easily solved independently. In addition, many components are common to different systems and provided that the performance of the components is described in a general form they can then be used in many different systems with little or no modification. This feature makes modular simulation most attractive.

With a program such as TRNSYS which has the capability of interconnecting components in any desired manner, solving differential equations, and facilitating information output, the entire problem of system simulation reduces to a problem of identifying all of the components and formulating a general mathematical description of each.

The simulation for the present study was done by calculating various quantities every 15 minutes for the whole year, i.e., the year was divided into 35038 time steps, each of which was evaluated separately.

ASSUMPTIONS MADE FOR THE SIMULATION

The following assumptions were made consistent with the actual use pattern of the building.

1. Heating temperature set point for first stage heating was 18.1°C. For second stage heating the set point was 17.9°C. For cooling by high ventilation, the set point was 27.5°C. Therefore, between 18.1°C and 27.5°C no heating or cooling was desired.

2. Since the actual occupancy schedule of the building was very complex and extremely variable, an average occupancy schedule was used which shows variable occupancy of from 2 to 40 people, and the variation thereof every 48 hours. The ventilation schedule was also made consistent with the number of occupants at any given time.

3. The lighting control function was such that it varied both with the occupancy and with the summer and winter season. In summer since most of the activities are outdoors, the lighting needs were reduced.

4. Shading devices employed were canvas awnings deployed in the spring and removed at the start of winter.

5. The walls and roof of the building were of lightweight construction. The mass present in the building was the floor slab, the fireplace, and the mass in the sundial space.

6. Since the building is a complicated structure, as shown in Figure 2 to Figure 6, solar radiation was calculated at seven different surfaces. The effects of sun on the heat gain from all surfaces of the building were considered.

7. A reference building, called non-solar building, was also evaluated. The non-solar building was similar to the solar building except that the south facing glass wall was replaced by a wall of typical construction with a few windows of a type which normally are present in any building.

8. The effects of sun on all surfaces and windows in the non-solar building were considered, because any building, solar or not would have those solar effects anyway.

RESULTS OF SIMULATION

The results of transient simulation for the non-solar building are shown in Table I. The total auxiliary energy required per year is 228.800 x 10^6 KJ. This amounts to 434,980 KJ.m^{-2} per year. The ventilation fan dumps 52.52 x 10^6 KJ per year. The number of hours of heating required are 1913 and the hours of fan operation are 721. The solar heat gain from all surfaces of the building is 167.700 x 10^6 KJ per year.

Table II shows the monthly summaries for various heat fluxes for the solar building. It should be noted that the mass wall area is 90 m^2, the direct gain space with triple glazing and no night insulation is 51.8 m^2 and the direct gain space with double glazing and night insulation is 38 m^2. From Table II it is seen that the total auxiliary heating energy required by the building is 166.400 x 10^6KJ. Compared to the non-solar building (Table I) the decrease in auxiliary energy requirements are (228.800 x 10^6 - 166.400 x 10^6 = 62.4 x 10^6 KJ). This saving is the net usuable solar energy during the heating months.

Also shown in Table II is the net solar contribution by the mass wall, the two kinds of direct gain spaces and solar heat gain from other building surfaces. The number of hours of heating required is 1384 per year. For obvious reasons, the ventilation fan operates more in a solar building and is on for 1078 hours compared to 721 hours for the non-solar building. The total heat dumped by the fan is 83.890 x 10^6 KJ per year.

Figure 7 shows a series of six days in January. These days consist of a sequence of days which are sunny, cloudy, average, etc. Shown plotted is the solar radiation, the room temperature, the ambient temperature, the temperature of the inside surface of the mass wall, net energy contributed by the mass wall and the net energy contributed by the direct gain spaces. The same information is shown for six days in the month of March. This is given in Figure 8.

Shown in Figure 9 are two days, one cloudy, and one sunny, occurring in sequence, for the solar radiation, ambient temperature, room temperature, temperature of the inside surface of mass wall, the net solar gain from mass wall, and the net solar gain from direct gain spaces. It will be seen that solar gains from the solar aperture during these months are very small compared to January and March, as shown in Figure 7 and Figure 8. This is because a large shade is used on the solar aperture during the summer months.

CONCLUSIONS

It has been predicted that the use of the passive solar techniques described, will save 62.4×10^6 KJ of heating energy per year, a saving amounting to 17,333 KW hrs. of electricity. The potential for cooling by the high ventilation economizer has been shown to be 83.890×10^6 KJ per year, and using 804 KW hrs. of energy for fan operation. The use of the economizer represents (on an equivalent basis) 6630 ton-hours of cooling.

In addition to the energy savings listed above, there are further energy savings due to the use of daylighting. This aspect of the project will be reported later.

The passive solar building that has been simulated is due for the start of construction in March 1981. After completion the building will be instrumented in a limited manner to check the actual performance with the predicted performance.

ACKNOWLEDGEMENTS

The study presented was performed under Cooperative Agreement No. DE-FC02-80CS30338 between the U. S. Department of Energy and Girl Scouts of Greater Philadelphia in the Passive Commercial Buildings Program of U.S. DOE. The authors wish to thank Mr. P. Richard Rittelmann, and Mr. David Linamen of Burt Hill Kosar Rittelmann Associates; Mr. Peter Bohlin and Mr. James Bell of Bohlin Powell Larkin Cywinski Architects, for their valuable assistance throughout the study. Special thanks are due to Ms. Lawther for typing this manuscript.

REFERENCES

1. "TRNSYS Version 10.1, A Transient Simulation Program" Engineering Experiment Station Report 38-10, June 1979, University of Wisconsin, Madison, WI.

2. Bugler, J. W., "The Determination of Hourly Insolation on an Inclined Plane Using a Diffuse Irradiance Model Based on Hourly Measured Global Horizontal Radiation," Solar Energy, Volume 19, No. 5, 1977.

3. Balcomb, J. D. and McFarland, R. D., "The Effect of Design Parameter Changes on the Performance of Thermal Storage Wall Passive Systems", proceedings of the 3rd National Passive Solar Conference, San Jose, CA, January 1979.

4. Balcomb, J. D. et al, "Passive Solar Design Handbook", Volume II, U. S. Department of Energy, DOE/CS-0127/2, January 1980.

5. Mazria, R., "The Passive Solar Energy Book", Rodale Press, Organic Park, Emmaus, PA 18049.

6. Wray, W. O., and Balcomb, J. D., "Sensitivity of Direct Gain Space Heating Performance to Fundamental Parameter Variations", Los Alamos Scientific Laboratories LA-UR-78-2570, August 1978.

7. Wray, W. O., and Balcom, J. D., "Trombe Wall vs. Direct Gain: A Comparative Analysis of Passive Solar Heating Systems", presented at 3rd National Passive Solar Conference, San Jose, CA, January 11-13, 1979.

8. Balcomb, J. D., McFarland R. D., "The Simulation Analysis of Passive Solar Heating Buildings - The Influence of Climate and Geometry on Performance", presented at the ISES American Section Annual Meeting, Orlando, FL, June 1977.

9. Trombe, F., "Maison Solaires", Techniques de l'Ingenieur (3), 1974, C777.

10. ASHRAE Handbook of Fundamentals, 1977.

11. Utzinger, D. M., "A Method of Estimating Monthly Average Solar Radiation on Shaded Surfaces", proceedings of 3rd National Passive Solar Conference, San Jose, CA, January 1979.

NOMENCLATURE FOR FIGURES

AUX: Hours when auxiliary energy was required
QN-DG: Net solar gain from all direct gain solar apertures (KJ)
QR: Net solar gain from the mass wall (KJ)
RAD: Horizontal solar radiation (KJ . m^{-2})
TA: Ambient temperature (°C)
TR: Room temperature (°C)
TW: Room side surface temperature of mass wall (°C)

TABLE I SUMMARY OF RESULTS
NON-SOLAR BUILDING

(SOLAR APERTURE REPLACED WITH WALL OF REFERENCE MATERIAL WITH SOME WINDOWS)

	AUXILIARY ENERGY REQUIRED FOR HEATING KJ x 10^{-6}	SOLAR CONTRIBUTION KJ x 10^{-6} TOTAL SOLAR CONTRIBUTION FROM OTHER BUILDING SURFACES KJ x 10^{-6}	HEAT DUMPED BY FAN KJ x 10^{-6}	HOURS OF OPERATION FAN HOURS	HOURS OF OPERATION HEATING SYSTEM HOURS
J	58.470	9.341	0	0	502
F	47.130	9.751	0	0	399
M	31.530	16.100	0	0	260
A	5.265	15.890	0.920	10	39
M	0.145	19.680	5.503	66	1
J	0	18.160	12.390	161	0
J	0	15.260	12.730	193	0
A	0	18.080	15.460	218	0
S	0	15.310	5.115	67	0
O	7.509	13.490	0.410	4	58
N	26.210	8.982	0	0	206
D	52.500	7.644	0	0	446
TOTAL	228.800	167.700	52.520	721	1913

TABLE II SUMMARY OF RESULTS
PASSIVE SOLAR BUILDING

TOTAL SOLAR APERTURE = 170 m^2
MASS WALL = 80 m^2, DOUBLE GLAZED DIRECT GAIN = 38 m^2, TRIPLE GLAZED DIRECT GAIN = 51.8 m^2

	AUXILIARY ENERGY REQUIRED FOR HEATING KJ x 10^{-6}	SOLAR CONTRIBUTION KJ x 10^{-6}							HOURS OF OPERATION	
			DIRECT GAIN SPACE							
		NET SOLAR FROM MASS WALL	NET SOLAR FROM DOUBLE GLAZED SPACE	NET SOLAR FROM TRIPLE GLAZED SPACE	TOTAL NET SOLAR FROM DIRECT GAIN SPACE	TOTAL SOLAR CONTRIBUTION FROM THE SOLAR APERTURE	SOLAR GAIN FROM OTHER BUILDING SURFACES KJ x 10^{-6}	HEAT DUMPED BY FAN KJ x 10^{-6}	FAN HOURS	HEATING SYSTEM HOURS
J	46.680	5.089	4.324	3.637	7.691	12.780	4.958	0	0	384
F	36.150	4.919	4.134	3.589	7.723	12.642	5.741	0	0	305
M	17.380	9.424	7.632	7.482	15.110	24.534	10.340	0.525	2	146
A	3.730	1.130	3.411	3.052	6.464	7.594	11.440	1.392	16	28
M	0	1.497	3.140	2.859	6.000	7.497	15.270	8.247	99	0
J	0	2.302	3.577	3.588	7.165	9.467	14.330	17.53	221	0
J	0	2.117	3.330	3.397	6.727	8.844	11.840	17.470	265	0
A	0	2.619	4.240	4.371	8.610	11.229	13.240	21.010	292	0
S	0	1.507	4.904	4.987	9.890	11.397	10.180	9.914	124	0
O	2.180	10.410	8.104	8.365	16.470	26.880	8.040	7.797	58	18
N	15.780	6.389	5.073	4.901	9.974	16.363	4.923	0	0	126
D	44.470	3.434	3.092	2.370	5.462	8.896	4.079	0	0	376
TOTAL*	166.400	50.840	54.960	52.600	107.600	158.440	144.400	83.890	1078	1384

*May not add up exactly due to rounding

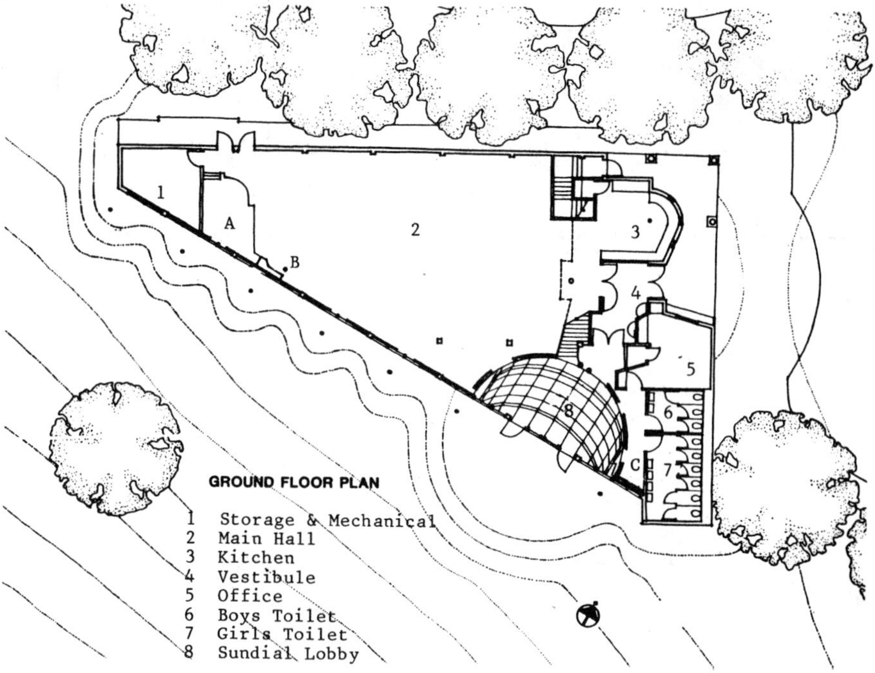

1 Storage & Mechanical
2 Main Hall
3 Kitchen
4 Vestibule
5 Office
6 Boys Toilet
7 Girls Toilet
8 Sundial Lobby

A Stage
B Fireplace
C Curtain

Figure 1 Plan View of the Building

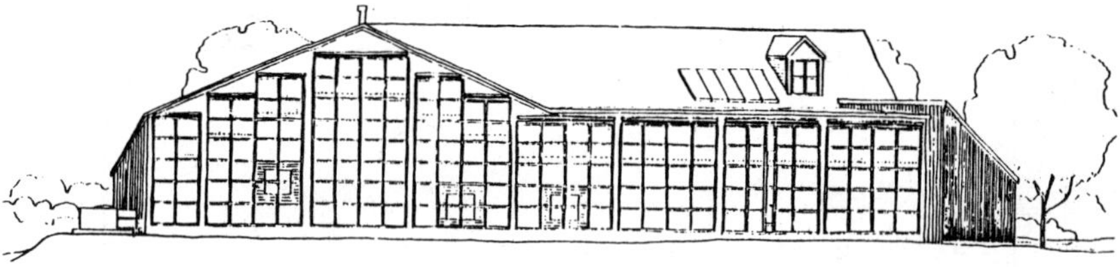

Figure 2 South Elevation of the Building

Figure 3 Northeast Elevation

Figure 4 Transverse Section Through Lobby

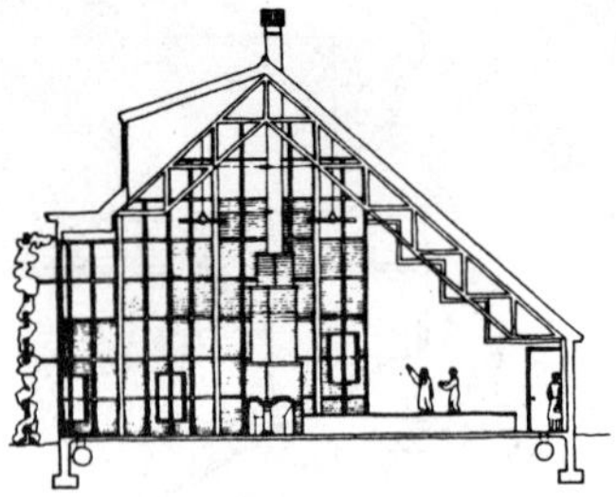

Figure 5 Transverse Section Through Main Hall

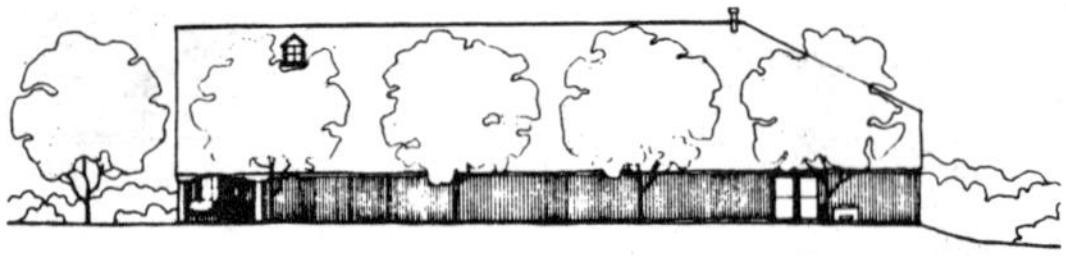

Figure 6 Northwest Elevation

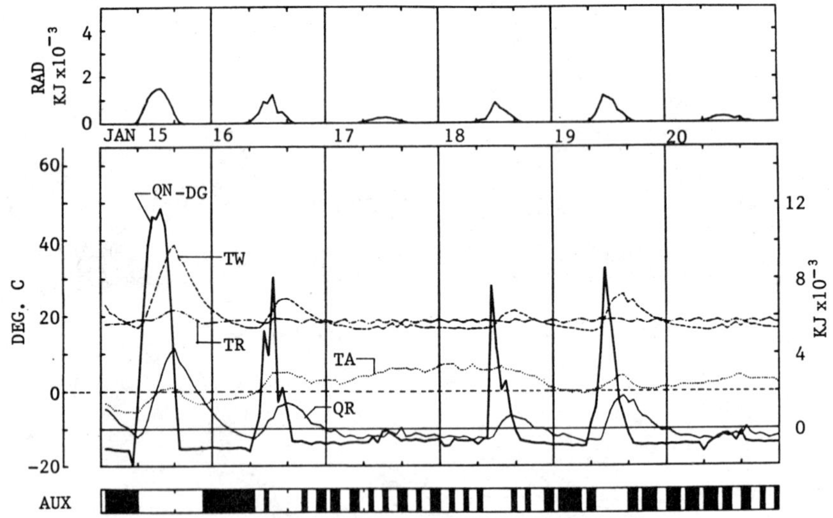

Figure 7 Predicted Performance of the Passive Solar Building for Six Days in January

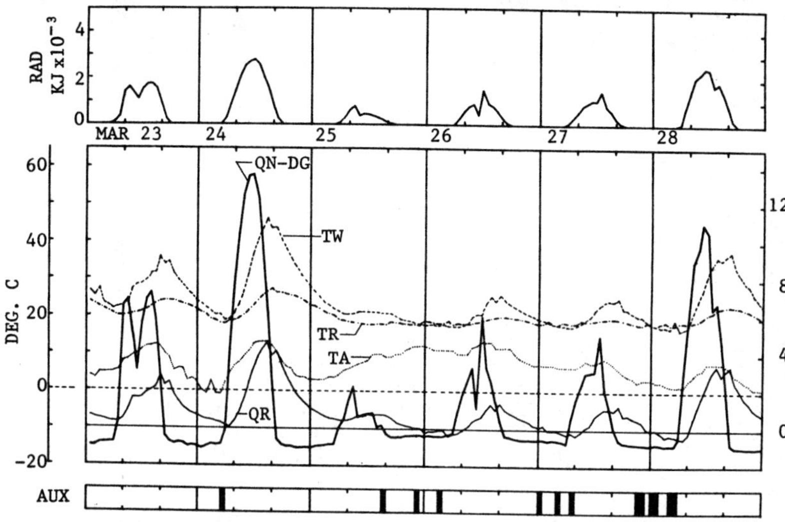

Figure 8 Predicted Performance of the Passive Solar Building for Six Days in March

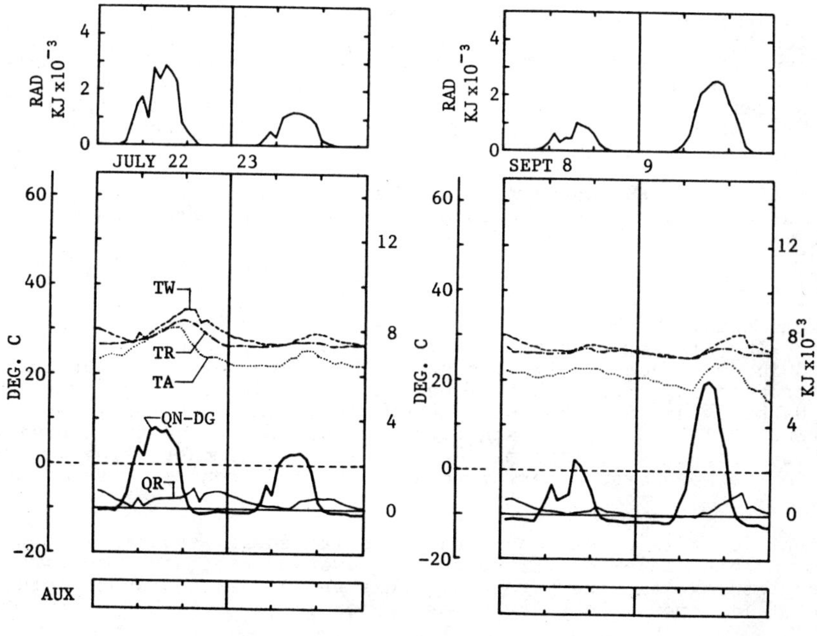

Figure 9 Predicted Performance of the Passive Solar Building for Two Days in July and September

SOLAR ENERGY CONTROL SYSTEMS INSTALLATION AND MODIFICATION: EFFECTS ON SYSTEM PERFORMANCE

Dr. R. E. Waterman
Energy Systems Department
Automation Industries, Inc.
Vitro Laboratories Division
14000 Georgia Avenue
Silver Spring, Maryland 20910

ABSTRACT

The performance of solar energy space heating systems, particularly those systems that employ heat pumps for auxiliary energy, is a sensitive function of control system installation. This paper examines the cause and effect relationship between overall system performance and differing control system sensor replacementsin three solar homes. Control system hysteresis induced by sensor location is examined in detail. the effect of modifying the space heating delivery system in one of the homes by means of a timer is also discussed.

Information from the National Solar Data Network (reference (1) and (2)) was used extensively to support the analysis presented in this paper.

1. INTRODUCTION

Problems arising from improperly installed control systems are widely known. The list of controller related problems, some of which are described in references (3)-(7), tend to fall into two major categories: catastrophic failures such as freeze-ups, and degraded performance problems by studying the behavior of control systems in three solar homes. References (1) and (2) contain detailed information on the energy usage in these homes. This information is useful in demonstrating the correlation between system performance and various control systems problems.

Section 2 of this paper describes the solar energy systems in the three homes and the placement of the control system temperature sensors that measure storage tank temperature. This section also discusses the efrects of improperly locating these sensors.

Section 2 deals with control systems problems arising from misplacements of temperature sensors for the space heating delivery system and the solar energy collection system. Section 4 addresses the effect on performance due to the addition of a timer in the space heating delivery system.

2. SYSTEM DESCRIPTION/SENSOR PLACEMENT

The solar energy system for each of the three homes is shown in Figure 1. The solar energy subsystem operates as follows: Temperature sensor (CC) measures collector plate temperature. Sensor (C) measures storage tank temperature. The difference of the temperatures is used to control a pump that circulates the water in the storage tank through the collectors. If (CC) is 17°F greater than (C), then the circulation pump is turned on. If (CC) drops to less than 3°F greater than (C), then the pump is turned off. The hysteresis logic (different values for turning the pump on and off) is designed to prevent inadvertent pump cycling. This is shown in Figure 2. A more complete description of the operation of differential controllers can be found in reference (8). The space heating control system operates as follows: A two stage thermostat, set nominally at 70°F, is located in the dining room of the home. When the temperature in the room drops below 69°F, a signal is sent to either the heat pump or the solar heating system to supply heat. If the room temperature continues to drop to below 66°F, then the strip heaters (i.e., resistance heaters) are activated in addition to any other subsystem that is operating. The decision on whether to use the heat pump or the solar system is made as follows: If the storage tank temperature, as measured by sensor (H), is rising and is greater than 90°F then the solar space heating system is activated. If the storage temperature falls to below 90°F, then the heat pump is turned on. Note that if the tank temperature is rising and is between 90°F and 93°F, then both the heat pump and solar system will be operating. Figure 3a shows typical design operating points for the heat pump and solar system as a function of storage tank temperature.

Sensors (C) and (H) are supposed to measure storage tank temperature but are strapped to pipes that carry water from the solar storage tank. The tank is surrounded by approximately 15 inches of fiberglass insulation. The location of sensors (H) and (C) relative to the insulation is shown in Figure 4 for each of the three homes. The major effect on mounting a sensor outside of the insulation is to induce additional hysteresis in the control logic. For example, when water is flowing in the heating loop in home #1 (see Figure 4), then sensors (H) and (C) both indicate a temperature that is close (within about 5°F) to the actual tank temperature. On the other hand, when there has been no flow for about 30 minutes, these sensors, separated from the tank by the insulation, indicate a temperature that is closer to the air temperature in the basement (typically about 60-63°F during the winter). The hysteresis behavior due to sensor placement is manifested as follows: The controller that selects solar heating is set to turn on when sensor (H) indicates a temperature of about 83°F. When the storage tank temperature rises to about 90°F, sensor (H) is reading about 83°F since there is no flow in the heating loop. Solar heating is now activated and the resultant flow causes sensor (H) to read close to 90°F. Later as the energy in the tank is depleted, the temperature in the storage tank drops. Since there is flow in the heating pipe now, sensor (H) continues to indicate actual tank temperature. Finally, when sensor (H) drops below about 80°F, the solar heating system is turned off. Thus the fact that (H) indicates markedly different storage tank temperatures, depending on whether or not there is flow, causes

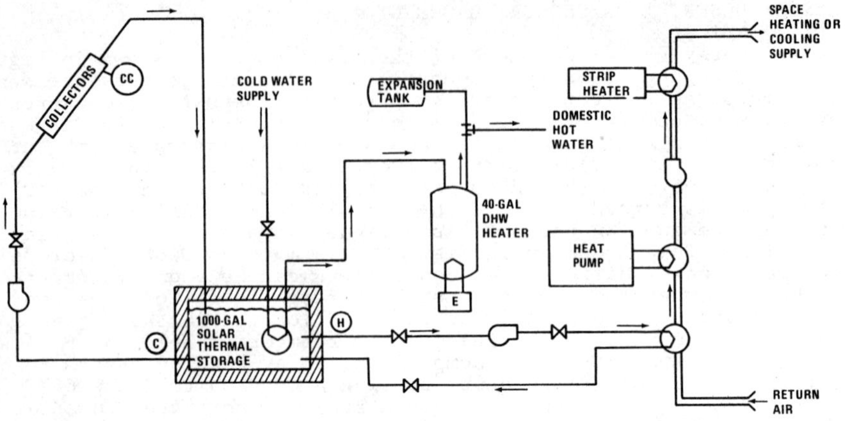

Figure 1. Solar Energy System Schematic

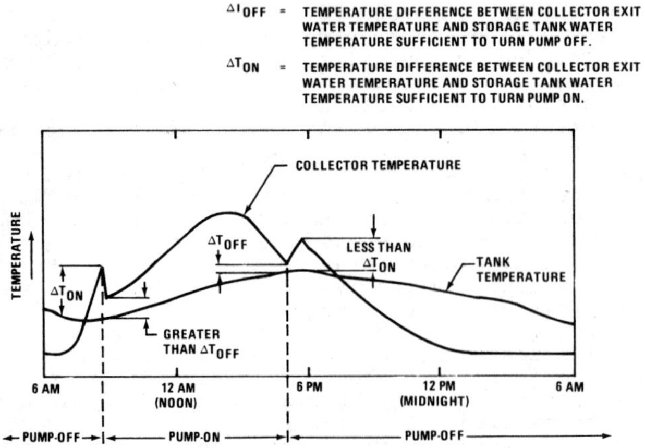

Figure 2. Typical Operating Cycle for a Solar Heating System

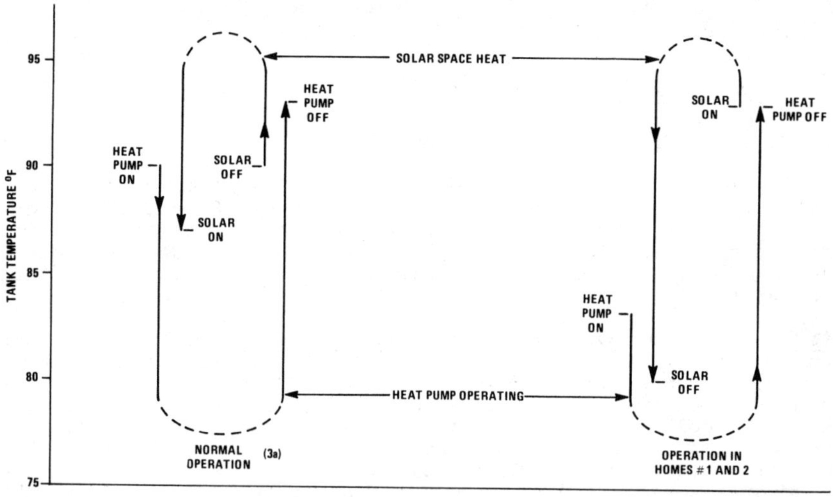

Figure 3. Space Heating Control Cycle

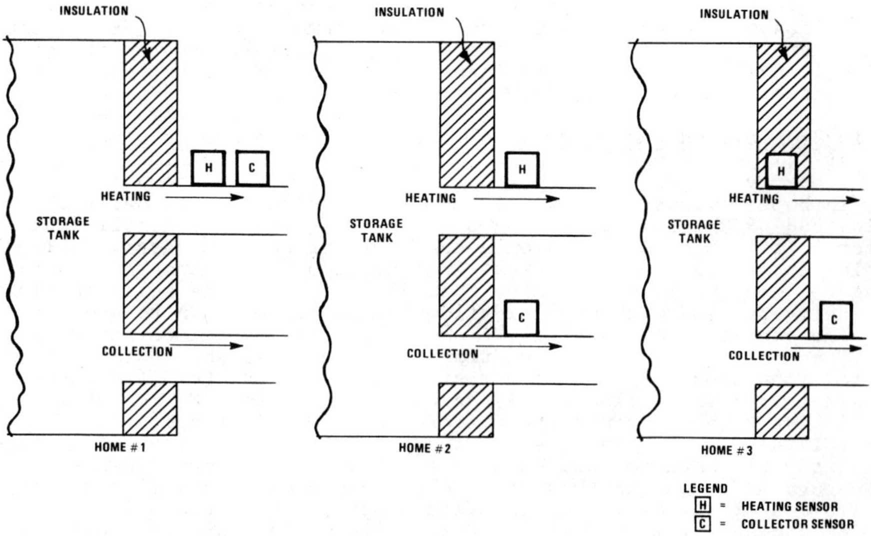

Figure 4. Sensor Placements

the solar space heating system to have on-off points that are quite different than those design points shown in Figure 3a. The situation just described is shown in Figure 3b.

Similar hysteresis effects occur for each of the homes and each of the two sensors (H) and (C). For the most part, it is this installation induced hysteresis that causes the performance anomalies described in the next section.

3. EFFECTS OF TEMPERATURE SENSOR PLACEMENT

This section is devoted to variations in overall system performance that arise from different placements of the sensor (C) and (H) described in the previous section. Each sensor placement configuration is presented in the form of a scenario that describes the sensor placement, operating conditions, consequent problems, and which homes are affected by the problem.

Scenario 1: Start up cycling of solar energy collection system due to placement of temperature sensor (C). This problem affects homes #1, 2, and 3. As shown in Figure 4, homes #1, 2, and 3 have the solar storage tank insulation between sensor (C) and the tank surface. When there is no flow in the collector pipe, sensor (C) reads considerably lower than the actual tank temperature. For example, with the tank at 85°F and the collectors at 91°F, sensor (C) may indicate about 73°F. Since the collectors (sensor (CC) in Figure 1) appear to be 17°F hotter than the tank, the circulation pump will turn on. Shortly thereafter the collectors will be cooled to about 87°F. Since fluid is now flowing in the collector loop, sensor (C) will indicate the actual tank temperature of 85°F. At this point the temperature difference between the collectors and sensor (C) is less than 3°F causing the pump to turn back off. This initial attempt at energy collection has actually resulted in a loss of energy from the tank. Furthermore, on the next attempt at pump start up, solar energy collection may initially represent an additional small loss due to the inability of sensor (C) to correctly measure actual tank temperature.

Scenario 2: Delayed turn off of collector loop pump due to placement of sensor (C). This problem affects only home #1 and results in significant energy losses under certain circumstances. As shown in Figure 4, in home #1 sensor (C) is placed on the space heating delivery pipe instead of the collector loop pipe. This leads to the following behavior. It is mid-afternoon on a sunny winter day and passive solar gain plus internally generated heat from household appliances are satisfying the heating load (for more than 30 minutes). The solar heating system delivery pipe with no fluid flowing has cooled down to about 95°F. The storage tank is at 110°F, and the collector temperature is at 112°F and is dropping as insolation decreases. At this point, the collector loop pump should turn off since the temperature difference between the tank and collectors is actually less than 3°F. But the pump will not turn off since the difference between the collectors and sensor (C) is 17°F. In fact, even if sensor (C) remained at 95°F, the collectors would have to cool down to about 98°F in order to cause the pump to turn off. During this time, which can amount to well over an hour, 110°F storage tank water is being pumped through the collectors and being cooled. This results in a significant loss of previously collected solar energy. The loss can be interrupted if the solar space heating system is activated during this time since in this case the temperature indicated by sensor (C) will immediately rise to the actual tank temperature, causing the pump to turn off.

Scenario 3: Delayed turn off of solar space heating system due to placement of sensor (H). This problem affects homes #1 and 2. As shown in Figure 4, homes #1 and 2 have the solar storage tank insulation between sensor (H) and the surface of the storage tank. When there is no flow in the space heating pipe, sensor (H) reads lower than the actual tank temperature. Assume that it is desirable to have the solar heating system operate when the tank temperature is greater than 90°F. Since sensor (H) reads lower than actual tank temperature, the controller will have to be set to activate when (H) reads approximately 83°F. Suppose that the solar heating system has been operating for some time and consequently the storage tank temperature is decreasing. The same controller that turned on the pump at 83°F will turn off the pump when sensor (H) is less than 80°F. But, since there is flow now in the heating pipe, sensor (H) indicates actual tank temperature. Thus, the solar heating system will attempt to operate all the way down to a tank temperature of 80°F. The controller will allow the heat pump to activate at about 3°F above this point, or at 83°F. This is shown in Figure 3b. The ability of the solar heating system to deliver heat to the house when the tank temperature is 80°F is shown in Figure 5. As can be seen, the rate is about 8500 BTU per hour. This may be far less than the heating demand of the house. With the heat pump "locked-out" by the control system, the house temperature drops until the strip (resistance) heaters are activated. This causes a large and obviously unnecessary amount of auxiliary energy to be used to meet the space heating demands of the house.

The three problems discussed above could be eliminated by properly locating the sensors and modifying the control system set points. The questions of proper set point selection will not be addressed further here.

4. SPACE HEATING CONTROL SYSTEM MODIFICATION

This section discusses the effect of modifying the solar energy space heating system by means of a timer. The timer is used in addition to the space heating control logic described earlier in section 2. Why add a timer to the system? To adequately answer this question a brief explanation of the auxiliary heating system is necessary.

The auxiliary heating system consists of two components, the heat pump and the strip heaters. The strip heaters are simply resistance heaters that supply one unit of heat for each unit of electricity used. The heat pump on the other hand is a more complex device whose efficiency, as measured by its coefficient of performance (COP), varies inversely with outdoor temperature. Thus, the heat pump runs least efficiently at night when it is colder outside.

This brings us back to the question of the timer on the control system. Suppose that our solar space heating system cannot satisfy the heating demand of the house for a full 24 hour period (due to limitations on the size of the solar system and other factors). Suppose further that we have collected a certain quantity of solar energy and stored it in the 1000 gallon storage tank shown in Figure 1. If we now have a choice as to when to deliver this energy to the home to satisfy at most a part of the daily heating load, then the question arises: during what part of the day will solar provide the greatest benefit as measured by electrical energy savings?

This question seems to have an easy answer. When we are not using solar, we will presumably rely on the heat pump for heating. The heat pump

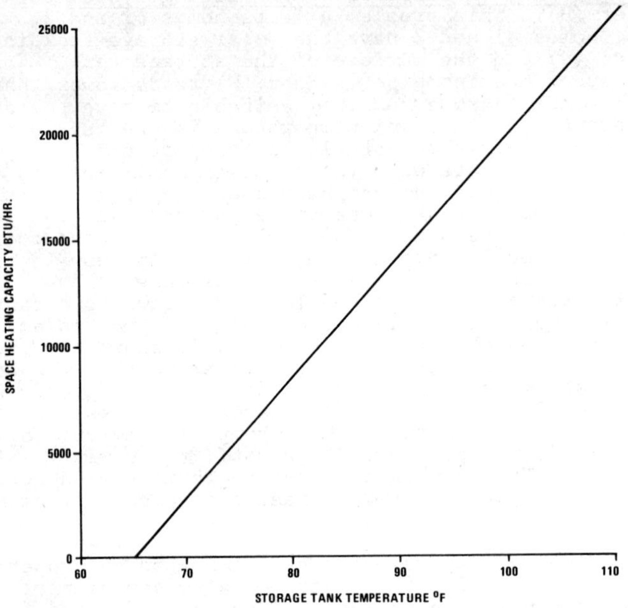

Figure 5. Solar Heating Capacity at Different Storage Tank Temperatures (Home #3)

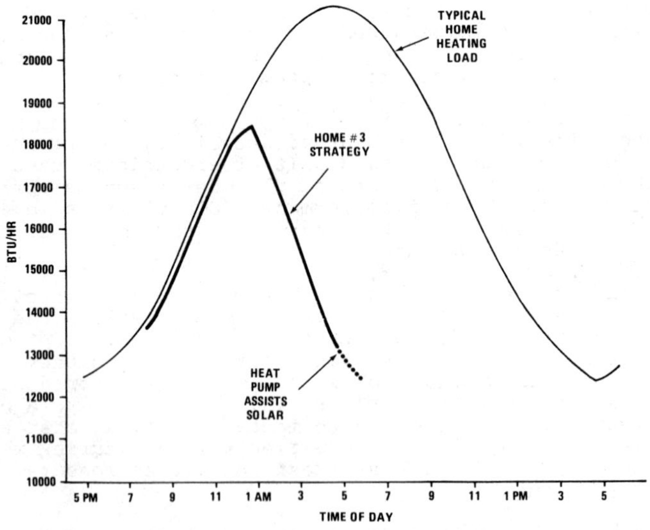

Figure 6. Space Heating Strategies

is most efficient during the day and least efficient at night. Therefore, why not install a timer that causes solar energy to be used for space heating only at night? In home #3 the space heating control system was modified to do just that. The solar space heating system was allowed to begin operating only after 8 p.m. (provided the storage tank was above 90°F) and was forced to stop operating no later than 10 a.m. the following morning.

The installation of the timer was done prior to the 1979-80 heating season. Energy usage in home #3 was then compared to energy usage for the two preceding winters. Energy usage increased sharply following the installation of the timer.

What happened? The answer is displayed in Figures 5 and 6. In Figure 5 one sees that as storage tank temperature decreases, the ability of the solar space heating system to deliver heat also decreases.

Examination of Figure 6 shows that at 8 p.m. when the solar space heating system begins to operate, the heating load of the home is increasing. As heat is delivered from the storage tank the tank temperature drops and so the rate at which heat can be delivered also drops (Figure 5). Thus, just as the heating load is rising, the ability to deliver heat is dropping.

The sad part of this story occurs at about 1 a.m. when the tank temperature has dropped to about 98°F. The tank is still too warm to effect a changeover from solar to the heat pump (see Figure 3a). The heating load is now about 19,200 BTU per hour. However, with the storage tank at 98°F, Figure 5 shows that the solar system can only deliver 18,500 BTU per hour. This causes the strip heaters to activate to account for the deficit. The situation is even worse during the next three hours. Finally, at about 5 a.m., when the tank temperature drops below 90°F the heat pump is activated along with the solar system, and so the strip heaters are no longer needed. The operation of the solar system and the heat pump between 90°F and 87°F is shown in Figure 3a.

Although this particular attempt to optimize the use of the collected solar energy was not very successful, the question remains as to the best way to combine a solar energy space heating system and an auxiliary system consisting of a heat pump and strip heaters. The author has investigated this questions for a number of different control strategies and heating load profiles. The results will be reported on in a subsequent paper.

5. CONCLUSIONS

The three problems discussed in section 3 were all attributable to sensor locations. Without a careful examination of the system installation, one might incorrectly conclude that the system was poorly designed or that the control system had somehow failed. In the present study, the fact that there were three homes to compare with each other made the problems fairly easy to diagnose.

The problem of the timer might easily have been attributed to difficulties with the heat pump or poor design of the solar energy system. Again, the author was fortunate to have three winters of data, the first two of which demonstrated that the system was reasonably designed and had the potential for significant savings.

The present study demonstrates that common symptoms of degraded performance can sometimes be attributed to minor installation errors or control system modifications rather than the outright failure of a control system component. It is hoped that these lessons will be of value to those working in the field maintaining existing solar energy systems.

ACKNOWLEDGEMENT

This work was sponsored by the U.S. Department of Energy, Office of Conservation and Solar Applicaitons, under contract DE-AC01-79CS30027.

REFERENCES

1. "Solar Energy System Performance Evaluation for J.D. Evans, Inc. House A," U.S. Department of Energy, SOLAR/1012-79/14.

2. "Solar Energy System Performance Evaluation for J.D. Evans, Inc. House B," U.S. Department of Energy, SOLAR/1013-79/14.

Reference 3 through 7 are from the Proceedings of the Solar Heating and Cooling Operational Results Conference, Colorado Springs, Colorado, U.S. Department of Energy CONF-781102, November 1978:

3. J.C. Bartlett, "Comparative Evaluation of the Thermal Performance of Solar Energy Systems", pp. 379-383.

4. J.C. Bartlett, "Evaluation of Solar Energy Control Systems", pp. 419-423.

5. D.W. Abrams, "Installation and Operation Problems Encountered in Residential Solar Systems", pp. 259-261.

6. H.R. Sparkes, K. Raman, "Lessons Learned on Solar System Design Problems from the HUD Solar Residential Demonstration Program", pp. 251-256.

7. M. Cash, "Hardware Problems Affect the Performance of Solar Heating and Cooling Systems, ;;. 263-367.

8. R.J. Schlesinger, "Operating Cycle For a Typical Solar Heating System", Solar Energy, pp. 26-28, February/March 1976.

PREFERENCES AND CONCERNS OF POTENTIAL USERS IN THE SELECTION OF SOLAR THERMAL SYSTEMS FOR INDUSTRIAL AND SMALL UTILITY APPLICATIONS

James B. Gresham, Science Applications, Inc.
Thomas A. Kriz, Solar Energy Research Institute

ABSTRACT

To achieve widespread application in the industrial and utility sectors, solar systems must be economically competitive. Economic viability is, in turn, determined by a number of supporting criteria, ranging from system reliability to dispatch characteristics to how the system supports the main product line. In addition, solar systems possess some inherent attributes that may render some of the traditional supporting criteria inappropriate or require their redefinition. This paper discusses those criteria and their relation to the solar investment in three steps. First, the main concerns and preferences of the potential users, as identified in recent SERI studies, are identified. Second, the equitability of the resulting decision criteria for solar investments are examined. Finally, the implications of these criteria for solar energy's penetration into these markets are discussed.

INTRODUCTION

Since early 1978, SERI (the Solar Energy Research Institute) has been extensively involved in evaluation of the technical and economic feasibility of solar thermal power and process heat systems. Evaluations conducted in two recent studies have the common purpose of comparing different possible collector types and system configurations to help insure that the allocation of government funding supports the most promising concepts. Our primary purpose has been to judge the potential of these systems from the point of view of the end user. Two application areas and the appropriate user groups have been addressed. The potential use of solar thermal systems in the generation of electricity for utility applications was examined in the first study (1). The second study, to be completed this year, investigates the use of solar thermal systems to provide heat for industrial processes (2). During the course of these studies we have consulted with nearly fifty industrial and utility decision makers in order to obtain a comprehensive understanding of their concerns and preferences. The purpose of this paper is not to report on the results of the system evaluations, but rather to convey the results of our interactions with industry and utility decision makers. We feel that an understanding of their concerns regarding use of solar energy is an important step on the long path toward eventual commercialization. In addition to the background provided by our own work, we have drawn from several recent studies concerned with the problems of commercial acceptance of solar energy systems. The Gas Research Institute (3) addressed the integration of solar energy with existing gas fired systems, and the potential impact on the gas market. Insights West under subcontract from SERI broadly surveyed the attitudes of industry toward the use of solar energy (4). A concurrent study at SERI (5) has examined the dynamics of the decision process in industry, concentrating on the decision criteria surrounding the primary economic factors. Sandia Laboratory has explored the utility industry's acceptance of a particular solar thermal concept in a series of interviews (6). The major contributions of the authors' analyses has been the coupling of a comparison and evaluation of the most feasible systems from the technical, i.e., system design and performance viewpoint, with an evaluation as it is made in the market place.

The presentation of this paper follows in four parts: (1) general background material concerning the solar systems examined and the decision analysis approach exercised to model user preferences; (2) a brief schematic of the study's overall methodology, concentrating on the process used to elicit end user preferences; (3) results in the form of a summary of user concerns and preferences; and (4) an evaluation of these results with regard to their impact on the implementation of solar thermal systems. This last point at once raises questions pertaining to the appropriateness of often used decision criteria, and at the same time illustrates the impact of these criteria on the development of solar energy systems for commercial application.

BACKGROUND

Solar thermal systems directly use the heat in solar radiation, converting or concentrating it to provide useful energy. Several examples of the types of systems examined in our studies are currently being demonstrated throughout the country. Southern California Edison, in conjunction with the Department of Energy and the California Energy Commission, is funding a 10 MW_e pilot plant for the generation of electricity using the so-called "power tower" or central receiver system at Barstow, California. Georgia Power Company is constructing a field of parabolic dishes at Shenandoah, Georgia, to test the capability of that technology to generate electricity. Numerous examples of cost-shared and wholly owned solar thermal systems are currently providing process heat to a broad spectrum of industrial users. One of the more well-known applications is at the Campbell Soup plant in Sacramento which combines a parabolic trough and flat plate system to provide hot water for can washing. Many variations of the flat plate collectors, widely used for domestic hot water heating, are capable of providing low temperature process heat. Two systems using unconventional forms of collectors, the shallow solar pond and the salt gradient pond, have been examined for their potential to provide heat and generate electricity.

A simple schematic of how a solar thermal system can be used to provide heat to an industrial process is shown in Figure 1.

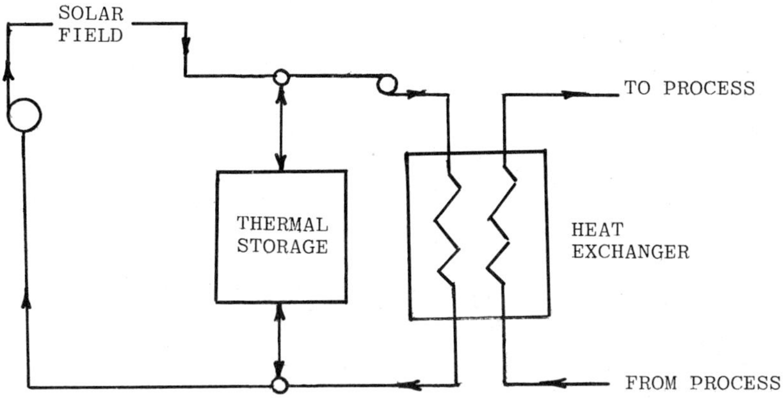

FIGURE 1. PROCESS HEAT SCHEMATIC

Sunlight concentrated via the solar collectors provides the heat source in place of or in conjunction with a conventional gas or oil fired boiler. A working fluid in the solar system is heated, then passed through a heat exchanger where the thermal energy is transferred to the process heat medium (steam, hot air or hot water). One unique aspect of the solar system is the optional addition of storage to capture excess or unused (e.g., weekend) energy. A simple example of storage is a hot water tank or a tank of high temperature molten salt. The schematic for generation of electricity is analogous, although somewhat more complex, since a variety of engine types were examined (the Stirling, Brayton, and Rankine cycles).

The solar thermal systems were examined in the fuel saver mode only; i.e., replacing the burning of conventional fuels. Alternate sources such as photovoltaics, wind energy conversion, biomass and synthetic fuels were neither examined nor used as a basis for comparison. An hourly simulation model was used to evaluate the performance of the various concepts. These simulations provided volumes of performance data. However, to effectively use this data in evaluating system acceptance, we needed to define quantitative and measurable attributes germane to the eventual end user's purchase decision.

The approach selected to methodically and quantitatively assess how a decision maker would evaluate a solar system was multi-attribute decision analysis (MADA). A straightforward and simple explanation of MADA is given in Appendix C of (1). The essence of the procedure as applied here is described from this Appendix:

> The problem was first reduced from an objective to a set of independent criteria that are measured by quantitative or subjective attributes. Because the complexity

of the formal analysis does not allow the consideration of more than ten attributes, it is necessary to select attributes that are: (1) important to the decision, (2) independent, (3) measurable, (4) differentiable characteristics of the options being considered, and (5) familiar to the decision maker.

During the second step, interviews were conducted with decision makers or their surrogates to obtain the proper data that could be used in a simulation of a decision maker's thought process. A simple method of questioning was developed for use in this study, permitting assessment of utility preferences over the scale of each attribute and the relative weighting of each attribute in a short series of lottery-type questions.

In the third and last step, results of the interviews were used to calculate the coefficients of a multiplicative form of the utility function. Given the actual attribute values of a system, the value of this function is an absolute and quantitative measure of the utility or preference of the system to the decision maker. An ordering of these values then provided a ranking.

A more comprehensive explanation of MADA is contained in (7). Details of how utility functions quantify user preferences and how the relative importance of each attribute was assessed is also given. One point worth stressing here is the weights assigned to the attributes depend not only on the importance of the criteria being measured, but also on the range of values to be judged. For example, cost is significant to the purchaser of a car, but if all models under consideration are within $100 of each other in price, then cost becomes a second order consideration. This point is crucial in avoiding misunderstanding of the relative importance of environmental concerns as mentioned later.

Preliminary to discussing the results of the interviews conducted during our studies, we will next describe how the interviews were structured and with whom they were conducted.

INTERVIEW METHODOLOGY

The first step towards defining the decision process to be used in evaluating the system is, of course, to define the users whose viewpoints are to be modeled. Although representatives of large utilities were among the interviewees, the electric power study concentrated on sizes (.1 - 10 MW_e) more appropriate to smaller utilities. Thus, the majority of interviewees were from small to mid-size utilities, including small investor owned, public and municipal utilities and rural cooperatives. To gain overviews and inputs from special interest groups we also interviewed utility consultants, trade journals, public utility regulatory bodies and R&D personnel. The smaller size systems (less than 1 MW_e) were more relevant to a different end user group in need of an independent source of electricity because of unreliable or high cost sources. Such users include small communities, the minerals industry, military bases, small or dispersed government institutions (e.g., the National Park Services) and the agriculture industry. For thermal applications in industry, the relevant end user group includes any industry with significant thermal needs occurring below 1100°F. Three major categories of industrial users, differentiated by varying degrees of knowledge concerning the decision to purchase solar energy were sampled. The first and by far the largest group includes those who have little familiarity with the possibilities of solar energy related to their operations. The second group consists of five companies currently in the throes of assessing alternative solar thermal systems. These companies are vying for participation in a DOE-sponsored demonstration program. The final category includes companies who have installed solar thermal systems, usually cost-shared with the government. This last group has actual hands on experience and indeed may become the prime source of information for other companies considering solar energy.

Initial contacts with these end user groups were used to evolve a working decision model appropriate to each group. This model was fine tuned to final sets of criteria through preliminary interviews. Three models were developed: one for the utilities (1-10MW$_e$ plants); one for the small electric power users (.1-1 MW$_e$); and the final for process heat users. Table 1 summarizes, in descending order of importance, the criteria selected for each group. Also included is the "bottom line" or overall criterion most often used as the best single gauge for acceptability.

Utilities	Small Electric Users	Industry
Capital Cost	Payback	Payback
Capacity Factor	Capital Cost	Rate of Return
O&M Costs	Capacity Factor	Reliability
Safety		Capital Cost
Environmental Effects		Availability
R&D Costs		
Application Variety		
	Bottom Line Criterion	
Levelized Bus Bar Energy Costs (BBEC)	Payback	Payback/ROR

TABLE 1. DECISION CRITERIA

The variation of decision criteria reflect not only a shift in emphasis for the user group, but also our experience in weeding out unimportant criteria and those difficult to deal with in a quantitative manner. The MADA models built to include these criteria provided the basic structure for the interviews. While such rigid models were not appropriate to all members of each group, the models did provide a focus for the interviews and were necessary to establish a consistent basis for comparison from one decision maker to another. Most interviews started with a discussion of the decision we were trying to analyze and an untethered description of how this decision was currently being made. Interspersed throughout the interviews and summarized at the end were qualitative or subjective concerns which did not fit into the decision model, but certainly were significant, and often gave us information on attributes which for one reason or another we could not assess. A prime example of this last category is the general area of risk. However, enough has been said of a general nature; now to the actual results uncovered.

RESULTS

Although these results were illuminating, they were not surprising. Overall, decision makers were receptive to the possibility of solar energy. There was broad agreement that solar energy might ultimately play a large role in providing energy to all sectors of our economy. However, it is not surprising that few agreed that long-run potential provides a defensible basis for making near-term investment decisions. As is evident from the ordering of criteria in Table 1, economic factors dominate the ultimate decision. The utilities stressed that capital cost and capacity factor were most significant, with O&M costs close behind. These three criteria determine the $\overline{BBEC}$, utilities' bottom line criterion. Reluctance to pay more than \$2000/KW_e, 1990 installed cost (1980 dollars) often removed many of the solar systems, ranging from \$1500 to more than \$3400/KW_e, from consideration for utility applications. The more subjectively quantified attributes, safety, environmental effects, R&D costs, and range of application were relatively unimportant. The first two attributes (as in the case of the \$100 difference in car prices mentioned earlier) did not vary significantly from system to system, hence they did not attract a great deal of attention. R&D costs and variety of applications were simply not of significance to utility decision makers. Small power users and industry both used payback (simple, after taxes) as the primary criteria. Capital cost was more significant to the small power users because it reflects how they traditionally assess their investments and also because it is not well reflected in their primary attribute (payback period). Industrialists place a greater reliance on a combination of rate of return and payback that internalizes the effects of initial costs. Further, the small power users often either do not directly control their own funds (as in the case of government institutions) or have severely limited capital (as in the case of farmers). If we generalize from the three sets of decision criteria, user concerns seemed to cluster about three issues: risk, system performance, and economics.

No criterion addressing risk appears in our first two models because we were unable to effectively differentiate between the advanced systems (not yet built), the performance of which we were predicting into the 1990s. Reliability was included in the last model in an attempt to quantify its role in the purchase decision, even though data is still not available to defensibly differentiate between the systems based on that criterion. As applied here, risk includes not only the issues of whether or not a solar system will perform as expected, but such problems as permit requirements or pricing policies by gas or oil suppliers to customers no longer using large and predictable volumes of fuel as well. First, we treat the risks of this last category.

Risk is a factor to be minimized for almost any investment decision. A high-level of risk implies a high degree of uncertainty which, in turn, implies a greater potential for disaster. If only because they are new, solar thermal systems are associated with a large amount of risk. Every aspect in which the solar thermal system differs from the conventional system increases the level of perceived risk. The reduction of that risk can, in turn, only come about through experimentation and experience (preferably hands-on).

For decision makers, risk can encompass a wide range of possibilities, ranging from institutional arrangements to the economic effects of environmental problems to technical problems of system integration. On the institutional level, a key consideration of risk for industry stems from the fact that it often purchases natural gas at a reduced price as an interruptible customer. Were a firm to make a large commitment to solar energy, the gas utility would become what might be termed an interruptable supplier. That in turn could result in the firm losing access to gas at the cheaper rate.

Electric utilities are justifiably concerned about environmental problems (witness nuclear power plant construction delays) from the standpoint of risk management. Although solar energy is generally considered an environmentally benign technology, the case is far from proven, and the reaction of agencies like the EPA to a large solar thermal power plant is yet to be observed. Specifically, the impacts of a large solar collector field on runoff or local plant ecology may affect the ability to speedily obtain permits.

At the technological level, risk takes on the guise of system reliability. The distinction between risk and performance blurs somewhat and in reality system performance is part and parcel of the risk concern. It is such a large concern that we treat it separately here. Questions of system reliability usually must be answered at the engineering department level before any serious consideration will be given a new system. Feelings about this issue were very consistent across both industry and utility decision makers. A typical comment was, "First, our engineering ensures that the design will meet our reliability standards and only then do we examine the cost effectiveness of it." Questions about whether a cheaper but less reliable design would be considered seriously often produced a negative response. While utilities plan for a givon amount of unscheduled down time for each generating unit, industrial plants often do not. The question of what constituted an acceptable level of unscheduled down time generally brought the response,"None!" Instead, preventative maintenance might be scheduled during plant down times. An oft voiced comment from manufacturers was that an unnecessary energy investment was generally peripheral to their main production, hence of lower priority. The possibility that such an off-line system could interrupt production was viewed as a distinct negative point. "We produce widgets, not BTUs" capsulized this sentiment.

Utilities, on the other hand, produce power as their business, so alternative energy investments are given considerable attention. Reliability is a problem of utmost importance because the solar system becomes part of their main line product. Further, utilities typically must work with state PUCs, who tend to take a dim view of speculating rate payer's money on unproven technologies. We found a consistent concern about several of the unique and as yet unproven concepts associated with some of the solar thermal systems (for example small distributed Brayton and Stirling cycle engines, each located at a separate collector). The utilities felt far more at ease with central generating concepts using conventional steam cycle turbines.

While questions of risk and technical performance generally must be answered first, there is no doubt that economic considerations are premier. In each of our three decision models, economic measures carried the most weight by far. Generally, one of two evaluation approaches is used. Small power users and the majority of industry use simple payback (see also (3) and (4)). The remainder of industry and the utilities use some form of rate of return analysis. Levelized bus bar energy cost, rate of return, and life cycle costing techniques are similar in that they involve the time value of money and the entire lifetime of the system. These techniques contrast with payback, and herein lies an important distinction. A simple example of this distinction is illustrated in Table 2.

	Scenario 1	Scenario 2
Cost	$1,000,000	$1,000,000
Economic Life	20 Years	20 Years
Escalation in Income (or Energy Savings)	0%	20%
First Year Savings	$100,000	$39,000
Payback Period	10 Years	10 Years
Rate of Return	7.75%	15.75%

SIMPLE (NON-DISCOUNTED) PAYBACK VS. RATE OF RETURN

TABLE 2.

This table could well illustrate the difference in results between an alternative energy investment in the decades of the 1950s and 1960s (0% fuel escalation) and one in the 1970s (20% fuel escalation.) Payback is the same for both investments, rate of return has doubled in the 20% escalation case. The comparison is an extreme example of current comparisons of energy saving investments to other investments where income may track inflation (6-8% over the 1970 decade). The point is still valid: payback does not take into account the full economic implications of an investment. However, payback does give an idea whether the money invested is returned to the company before it has gone bankrupt or in time to invest in other near term opportunities. Keeping in mind the dichotomy of these thoughts we move to the next section.

INTERPRETATION

One important result of the industry/utility consultation has been the gradual emergence of two related but distinct conclusions. First, solar thermal systems embody, as energy delivering technologies, some new and unique aspects that must be considered in any complete evaluation. Decision makers should be made aware of these aspects and how some traditional outlooks should be changed to accommodate them. Second is that industrial and utility decision makers have developed evaluation procedures that are generally reflective of and well suited to their particular needs and that any ultimate commercial acceptance of solar technologies is dependent upon satisfying those procedures in the near term. In other words, it is necessary that, through education and increased familiarity, decision makers come to understand some of the unique factors of solar technologies. On the other hand, it is vital that the solar equipment manufacturers and government agencies realize solar systems must conform to industrial standards of achievement before commercial acceptance can occur.

How are solar systems unique? There is no doubt that solar thermal systems mark a departure from traditional energy conversion systems. The first, and most obvious difference, is that these systems do not "burn" fuel. Instead, they redirect and concentrate thermal radiation from the sun. This leads us to the second difference: because solar systems are fueled by sunlight, they do not (without some form of storage) operate at night or during overcast periods and, hence, are not naturally suitable to be a base load type of system. Third, because sunlight is the fuel, variable costs consist only of operation and maintenance. Fourth, because a solar collector array must be distributed over a large, outdoor area, the systems integrate somewhat differently with existing equipment. Fifth, based on the above differences, ideal uses for solar thermal systems may be somewhat specialized (for example, preheat water for a boiler). A sixth, and most basic difference, is that solar thermal systems are new to industrial/utility decision makers and, as such, must gradually earn their confidence.

This section examines how these unique solar aspects relate to the rationale behind the three primary concerns of risk, performance and economics as discussed in the previous section. We also speculate on how these concerns can be accommodated by changes in R&D and commercialization strategies.

To this point we have noted how risk works to the detriment of solar energy. There are some risk reducing aspects to be noted briefly. A 10 MW_e solar power plant could be constructed in as little as 1-2 years (of course, the permit problems alluded to earlier must be solved) as compared to 6-12 years for nuclear or coal-fired plants. Admittedly, 10 MW_e is a small plant, but the capability to construct small and modular plants may be a distinct advantage. The modular idea was particularly appealing to industry. The ability to start with a small plant and, if the results proved encouraging, move to larger systems was seen as a good way to minimize economic exposure. Another risk reducer is the perhaps overworked reminder that fuel supplies may be interruptible. But a more tangible point for industry may be that fuel prices are so uncertain that long range plans are impacted by this variable. Once performance records are established, solar systems will have a more stable impact on cash flows, though the initial impact might be large.

In order to make the product attractive, sponsors of solar technologies are forced to come to grips with the individual issues such as risk. For the utility market with its lengthy planning horizons and need for large scale demonstrations, resolution will be a lengthy process, and there are few prospects for near-term commercial inroads. The industrial market offers better near-term prospects. The key is to concentrate on systems and applications that do not entail a big transition from current ones. Systems that embody a large number of established concepts (low technical risk) will be far more readily accepted. Advertising solar technology as being a "whole new way of doing things" may be useful for a residential market. However, in an industrial market, a better approach is to make the new idea appear as much like the old one as possible. One advantage of a system which uses simple, recognizable technologies (e.g., steam as a transport fluid instead of molten salt) is that a lower risk premium may be assigned. This is important because often a company may assign a premium in the form of higher required rates of return to projects perceived as risky. Therefore a costlier, but more straightforward system, may be a preferred alternative. This point was established in (6) where utilities preferred water and steam cycles in the near-term even if other fluids were more economical in the long run.

Regarding the performance issue, the prime concern stems directly from the intermittent nature of solar energy. This is an issue which is not as well understood by industry as it is by the utilities. While they are quick to ascertain that a 100% backup is necessary, industrialists still require stringent reliability standards. To wit, a system that allowed individual collector lines to be shut down while the rest of the system remains functional has obvious advantages. However, even if the solar system does not provide energy continuously, as long as the economic requirements are met and control problems are manageable, interruption may not be a show stopper. The place of storage in mitigating this problem should be mentioned here. Although energy cycled through storage is generally more expensive than that used directly from the collectors, storage can be used to even out transients in energy supply. The use of waste heat recovery tanks in common with solar storage will lower the costs of storage and at the same time effect a useful conservation measure. Cogeneration is an analogous situation for the utilities.

But the bottom line is still economic feasibility. As mentioned previously, rate of return calculations are economically more correct (see (8)) than payback. Sponsors of solar technologies are well aware of this fact and of the advantage of using ROR to show the economic possibilities of solar energy. However, a communication barrier often arises when they attempt to convince a firm that relies upon payback that solar energy can be a useful option. The problem is to show that an after-tax return on investment calculation is neither especially difficult to perform nor understand and is not merely a numerical trick designed to enhance the promise of a solar system. Rather, it is a reliable method that can easily account for all measurable influences on an investment and evaluate their impacts represented as a single number: percentage rate of return.

It is not our attempt to dismiss payback as short-sighted, hence incorrect. The necessarily short planning horizon that confronts capital poor industrialists and small power users provides ample rationale for using payback as an important economic indicator. Perhaps it is more a burden on the solar manufacturers and R&D community to recognize and adapt, than to expect this decision criteria to change. We can only make limited suggestions here. Is it possible to make a system which may not last as long (say 10 years), but is cheaper and yet reliable and has a quick payback? The question has not yet received sufficient attention. In the initial stages of commercialization, is it important to provide systems with lifetimes exceeding the expected lifetime of many of the potential applications? Will we fall into the trap into which pocket calculator manufacturers have run? Potential buyers hesitate to purchase because a new model, cheaper and more powerful, will come along well before the current model is no longer useful.

CONCLUSION

What can be learned from reading this paper? Three major points can be culled from our work. First, we sketch a procedure for evaluating solar technologies stressing the importance of incorporating user concerns. By referring to the brief list of references, the reader may gain valuable insights into how to perform a comparative evaluation of innovative technologies. Second, the paper provides a summary of important concerns and the insights we gained into the decision process used by utilities and industry in evaluating alternative energy concepts. This information should be valuable to manufacturers, proponents and government R&D centers because it will help define the atmosphere they will encounter as they venture into the marketplace. Finally, and most importantly, the decision criteria are examined critically. We hope decision makers will reassess their decision process in light of what has been said here. We also hope that some of the suggestions made will be useful to those firms trying to design and market solar systems. Herein lies a common ground of unbiased (at least not intentionally biased) information for use in the interaction between the suppliers and the users of solar energy.

ACKNOWLEDGEMENT

The authors would like to acknowledge Ken Brown of Science Applications, Inc. who conducted the initial interviews with the utility industry and provided an excellent critique of the first draft of this paper. We would also like to thank Neil Woodley of Westinghouse for his review and suggestions which added measurably to the quality of the paper. Finally, we would like to acknowledge the many, many people we interviewed, who gave us several hours of their valuable time.

REFERENCES

(1) Thornton, J.P., et al, Comparative Ranking of 0.1-10 MWe Solar Thermal Electric Power Systems, Vol. 1, Summary of Results, 1980, SERI/TR-351-461, Solar Energy Research Institute, Golden, CO.

(2) Thornton, J.P., Kriz, T., Gresham, J., Herlevich, F.A., Hooker, D.; Comparative Ranking of Solar Thermal Systems for Thermal Applications in Industry, to be released Sept. 1981.

(3) GAS Research Institute, Solar-Augmented Applications in Industry, 1979, Chicago, Ill.

(4) Wilson, V., Insights West; Solar Industrial Process Heat Survey of Industrial Applications and Attitudes, to be released. Solar Energy Research Institute, Golden, CO.

(5) Perwin, E., et al, Decision Criteria of Potential Solar IPH Adopters. To be published, Solar Energy Research Institute, Golden, CO.

(6) Fish, M., Utility Views on Solar Thermal Central Receivers, SAND 80-8203, 1980, Sandia Laboratories, Albuquerque, NM.

(7) Kriz, T., Decision Analysis: A Tool to Guide the R&D Selection of Alternative Energy Sources. SERI/TP-731-706, May 1980. Solar Energy Research Institute, Golden, CO.

(8) Stermole, F., Economic Evaluation and Investment Decision Methods, 1974. Investment Evaluations Corporation, Golden, CO.

SOLAR PONDS AS A SOURCE OF LOW TEMPERATURE HEAT*

D. A. Neeper and K. A. Meyer
Los Alamos National Laboratory
Solar Energy Group
Los Alamos, NM 87545

ABSTRACT

Salt-gradient solar ponds are a promising source of low-temperature thermal energy, and they are also potentially attractive for electric power generation. The Solar Energy Group of the Los Alamos National Laboratory has provided the technical monitoring for a number of DOE-supported solar pond research and development efforts. This paper reviews the progress of salt-gradient solar pond development, both in the United States and abroad. Technical problem areas and their implications for commercialization are discussed. Potential market applications for solar ponds are examined.

INTRODUCTION

The general term, solar pond, refers to large areas of water constructed so as to collect useful thermal energy by exposure to sunlight. So-called shallow solar ponds are large plastic bags that can be filled to several centimeters depth with water in the morning and drained at night. Deep ponds may be several meters in depth and serve to store the heated water as well as to collect the incident radiation. Deep ponds usually maintain a layer of heated water at the bottom, and employ a gel, membranes, or a salt gradient to prevent upwelling of the heated water. Salt-gradient ponds have received the most attention, and are the subject of this paper. Hereafter, we use the term "solar pond" or "pond" as an abbreviation for salt-gradient solar pond.

To date, US researchers have concentrated on development of small (less than 400 m^2) experimental salt-gradient ponds in order to establish the feasibility of the technology. More extensive pond development work has been done in Israel, where the emphasis is on the development of large ponds for the generation of electrical power. The Israelis have operated ponds up to 7500 m^2 in area. Detailed information regarding the Israeli developments is not available, since some of the information is proprietary.

*This work was performed under the auspices of the US Department of Energy, Office of Solar Applications for Buildings.

Thermal modeling of large solar ponds indicates that solar collection efficiencies of at least 20% are potentially available. Published figures on US pond performance indicate that achieved efficiencies have been in the range of 9% to 12%. Some of this lack in efficiency is probably due to the higher fractional side losses in small ponds, but much of the decrease can be attributed to factors such as nonoptimal layer thicknesses and loss of water clarity. The objectives of continued pond research and development include understanding pond hydrodynamics and developing methods of water clarity control in order to improve pond performance, thereby reducing the cost of energy produced. Other areas of pond technology needing study include liner materials, salts, heat removal, covers, instrumentation, and maintenance procedures.

There are many potential market applications for solar ponds. Because of the large thermal storage, the energy may be used days or weeks after it is collected; thus, the sun need not be shining when energy is wanted. The yield of the pond depends upon the seasonal amount of sunshine, but, unlike other solar customers, the pond user need not be concerned with the weather for any particular day or week. The energy can be collected during a season and used during a few days, as for grain drying. The pond can deliver heat continuously, 24 hours per day, as required by a base-load electric generating plant. Ponds should also be suitable for large-scale space heating and cooling, low-temperature process heat (canneries, laundries), desalination, gasohol production, and remote-site electric generation. A single pond might serve for crop drying, warming a livestock shelter, and providing domestic heat for the farmhouse.

A DESCRIPTION OF THE SALT-GRADIENT POND

Figure 1 shows a schematic diagram of a solar pond. Solar radiation penetrating to the floor of the pond (about 30% of the total incident energy)

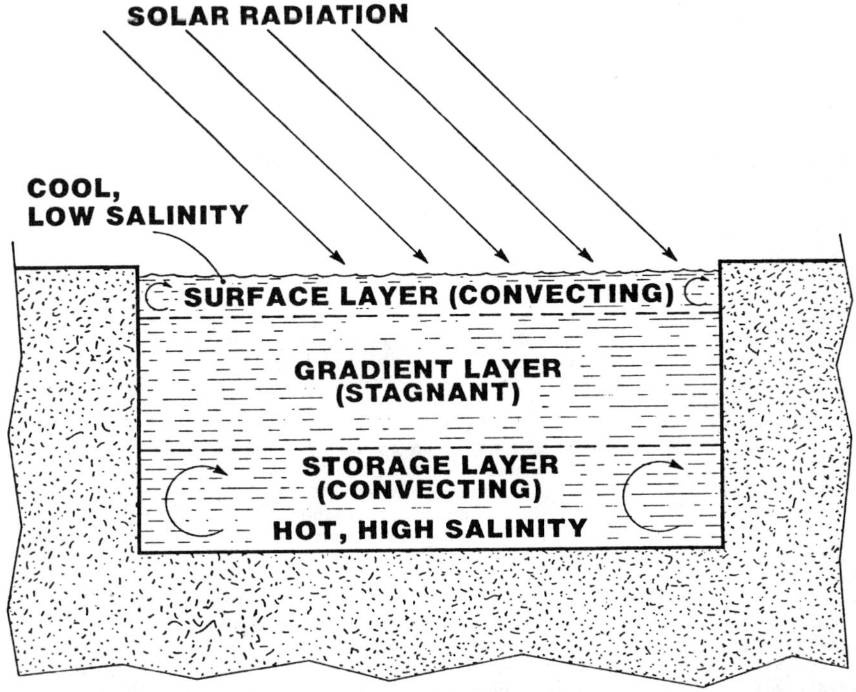

Fig. 1. Salt Gradient Solar Pond.

heats the bottom water layer. If convection is suppressed, thermal losses will be small, and a layer of hot water becomes available for useful heat extraction. If fluid circulation should develop in the gradient layer, the hot water would be brought to the surface and would rapidly cool, destroying the pond's effectiveness. Convection is suppressed in solar ponds by progressively dissolving more salt as one goes deeper in the pond. This makes the lower layers of fluid more dense than those above, thereby preventing upward flow.

The pond can be described as a system of three layers of water. The surface layer is a circulating region of constant temperature and salinity. This layer is believed to be formed by upward salt transport, surface heating and cooling, and wave action. Its temperature is near that of the ambient air. Surface layer thicknesses in the range of 0.1 to 0.4 m are presently common. Means for controlling this thickness have not been developed. This surface convective zone is undesirable because it is effective neither as insulation nor thermal storage and decreases the insolation reaching deeper layers.

The second layer is the nonconvecting gradient layer. This is the insulating layer of the pond and is maintained stagnant by having the density increase with increasing depth. Because the temperature of the gradient layer increases with depth, from the ambient temperature at the top to the temperature of the hot storage layer at the bottom, the salinity must also increase with depth in order to maintain the required density profile. Optimum gradient layer thickness depends on the desired storage temperature, solar transmission properties of the water, and thermal conduction of the water. Typical gradient layer thicknesses range from 0.6 to 1.0 m.

The third layer is the bottom, high temperature, thermal storage layer. This is a constant temperature, constant salinity, circulating region. The desired thickness of this layer depends on the temperature and the amount of thermal energy to be stored. Useful heat output is extracted from this layer, either by pumping some of the hot brine through an external heat exchanger, or by installing a heat exchanger in the storage layer itself.

A more complete description of solar ponds can be found in an article by C. E. Nielsen (Ref. 1).

THE US SALT-GRADIENT POND PROGRAM

Prior to 1981, federal support for salt-gradient pond research has been distributed among a few small (less than \$100k/yr) projects. Some important research, using experimental ponds, has been conducted at the Ohio State University and the University of New Mexico pond projects.

The OSU program has included the construction of three ponds, starting with a pond of 24 m^2 in area and 1 m depth, subsequently progressing to a pond 200 m^2 in area, and 2.5 m depth, and finally to a pond of 400 m^2 in area with 3 m depth. Observations were made on the behavior of the ponds over yearly cycles. Particular attention was given to the variation, with time, of the temperature and salinity profiles. Of particular interest was the movement of the boundaries between layers. A key result of the OSU studies was that, although the region boundaries moved, the ponds maintained some amount of insulating gradient layer. This made possible continuous thermal storage in the lower layer, albeit at less than optimal level because of periodic narrowing of the gradient layer. Operation of these ponds also provided experience in pond maintenance. Problems of water clarity associated with algae, bacteria, and dirt were encountered and, to some extent, overcome.

The UNM program included a salt-gradient pond with a 105 m^2 effective collecting area and depth of 2.5 m. A combined analytic and experimental study of the stability of the gradient layer was undertaken. Analytic results agreed reasonably well with pond data in predicting local instabilities in the gradient. The pond was operated for a year with heat extraction corresponding to a load generated by a 185-m^2 residence in Albuquerque, NM. An improved filling procedure for establishing the salinity gradient for a new pond was developed.

A key contribution of the UNM pond was the demonstration of reasonably stable operation under load. Furthermore, when heat extraction was stopped, the storage layer temperature exceeded 100°C, proving that a sodium chloride pond could reach boiling conditions. Annual efficiency under load was about 9%.

Other US ponds include the Living History Farm pond in Iowa. This demonstration pond has an area of 75 m^2 and a depth of 3.6 m. Water transparency contributed to limiting its collection efficiency to 8-9%. A 156-m^2-area, 3.6-m-deep pond was built by the Ohio Agricultural Research and Development Center at Wooster, Ohio. The purpose of this pond was to supply heat to a greenhouse. Maximum efficiency achieved was 12%. The largest US pond (2000 m^2 in area and 3.0 m deep) was built by the town of Miamisburg, Ohio, for the purpose of swimming pool heating and hot water supply. (See Fig. 2). Its performance has been monitored by Mound Laboratory.

Fig. 2. Miamisburg, Ohio, Solar Pond.

In addition to the research efforts and small ponds described above, there is a combined US-Israeli study under way to determine the feasibility of converting a portion of the Salton Sea into a large solar pond to be used in the generation of base load electricity. Currently DOE, the California Energy Commission, Southern California Edison, Jet Propulsion Laboratory, and Ormat Turbine Company of Israel are involved in systems studies related to this program. Other laboratories, including Los Alamos, will soon be conducting research in support of this program.

The Los Alamos National Laboratory's program has included acting as technical monitor for the Department of Energy on a number of the previously mentioned projects and doing basic pond research. Our research emphasis has been in the area of computer modeling and simulation of the pond heat flow, salt flow, and layer interface motion.

THE ISRAELI POND PROGRAM

In the late 1950s, Israel began development of salt-gradient ponds using salt mixtures consisting mostly of magnesium chloride. Technical feasibility was established when the temperature of a 600-m^2 pond reached 96°C. Work continued until the mid-1960's, when a change of government combined with the availability of cheap oil caused a cutoff of pond funding. Work resumed after the 1973 oil embargo, and since that time, four solar ponds have been tested at various locations in Israel.

In 1975, an 1100-m^2 pond was constructed at the Dead Sea. This pond was operational for one year, during which time it reached a maximum temperature of 103°C. Heat extraction experiments were carried out, and the pond achieved a collection efficiency of 15% during operation. A pond of 1100 m^2 was constructed at Eilat on the Red Sea in 1977. It reached a maximum temperature of 87°C. A 1500-m^2 pond was built in 1977 at the Ormat Turbine Plant in Yavne. During the summer of its first year in operation, the pond supplied hot brine continuously (24 hours per day) at 90°C to the organic fluid boiler of a 6-kW_e turbine, while water at 29°C from the surface of the solar pond was used to cool the turbine condenser. Electric power was generated both day and night from the pond.

The latest Israeli facility is a pond having an area of 7500 m^2. It was constructed in 1978 at Ein Bokek, near the Dead Sea. Temperatures in this pond have reached 93°C. This pond is currently driving a 150-kW_e turbine on an intermittent basis.

At the present time, Israel is the world leader in the development of salt-gradient ponds. Emphasis has centered on the production of electricity. The Israelis currently feel sufficiently pressed by energy needs that they are taking acknowledged risks in developing large ponds without a full understanding of the relevant fluid dynamics.

POTENTIAL APPLICATIONS AND TYPICAL COSTS

Ponds have a number of potential markets with a variety of applications. In the agricultural sector, low temperature applications include service hot water, farm shelter and greenhouse heating, and crop drying. Higher temperatures (90°C) would make ponds applicable to ethanol production. In the buildings sector of the economy, ponds could supply service hot water, swimming pool heating, and energy for space conditioning. Industrial applications of ponds include supplying hot water or hot air for washing, drying, or process heat requirements. Higher temperature ponds could be used to provide both base load electricity and peaking power. Current estimates indicate that when all potential applications are included, ponds can probably displace more than one quad, but less than ten quads, of conventional fuel annually.

An extensive discussion of pond costs and the resulting energy costs are given by Jayadev and Edesess (Ref. 2). Capital cost estimates range from \$5 to \$60/m^2, depending on the site, application, and various assumptions regarding construction costs. The low end of the cost range applies where land, salt, and water are free and no liner is required. For a pond costing about \$60/m^2, the cost of salt accounts for about \$30/m^2, while the liner costs about \$15/m^2.

The cost of energy from a pond depends not only on the capital cost, but also on the average insolation, the conversion efficiency of the pond, and the cost of maintenance. Estimated energy costs vary from \$0.60/MBtu for a salt works in Texas (free salt and no liner) to \$12/MBtu for winter space heating in the midwest. For comparison, we note that the price of fuel oil is around \$7/MBtu, and it is often burned at efficiencies below 70%.

TECHNICAL PROBLEM AREAS

The technical feasibility of salt gradient ponds as a source of thermal energy has been demonstrated. The economic viability of ponds depends, to a large extent, on their achieving thermal efficiencies of 15% to 25% at average operating temperatures ranging from 40°C to 85°C. In order to achieve reliable, consistent operation at the desired conditions, a number of technical problem areas must be addressed.

Adequate water clarity is essential for efficient operation. The radiant solar energy that reaches the pond's storage layer depends on the water clarity. Surface dirt, dust, chemical precipitates, algae, and bacteria all act to reduce water transparency. Methods for cleaning the surface and the body of the pond need further development, as do techniques for control of biological growth.

Heat can be extracted either by means of in-pond heat exchangers or by the circulation of pond brine to external heat exchangers. In-pond heat exchangers appear to be suitable only for small ponds. Criteria for heat exchanger size and location in the pond are needed. The circulation of hot brine to external heat exchangers appears to be the most appropriate technique for use with large ponds, but the effect of inlet and outlet flows on pond behavior is not well known.

There are two areas of concern regarding salt control in ponds, namely, gradient maintenance and environmental effects. There is a continuous diffusion of salt upward that acts to weaken the gradient. In a typical NaCl pond, this diffusion is about 10 Kg/m^2 yr. To maintain the gradient, the pond operator must flush the surface with low salinity water. The discharged salt brine may be concentrated and returned to the storage layer or new salt must be supplied. If the salt is not recycled, an environmentally acceptable means of disposal must be provided. Salt disposal presents no problem at locations near the ocean or saline lakes.

Many proposed pond applications are located at sites that will require a pond liner to prevent loss of brine into the soil. Liners must withstand high temperatures (110°C) in the presence of corrosive brines, be stable under ultraviolet exposure, and have sufficient mechanical strength to support installation stresses and working loads. Cost is an important factor because liners currently account for about one-quarter of the initial pond cost.

Pond performance depends strongly on maintaining an insulating gradient layer of adequate thickness. The upper and lower convective layers continuously interact with the gradient layer, frequently encroaching on the gradient and reducing its thickness. Loss of gradient layer thickness results in a larger heat loss from the storage layer, a lower operating temperature, and a reduction in the pond's efficiency. This layer motion is just one aspect of the general area of pond hydrodynamics. An understanding of pond hydrodynamics is essential for reliable modeling of pond performance. Effective performance modeling is necessary for accurate prediction and successful design.

The Los Alamos National Laboratory is developing an understanding of those aspects of pond hydrodynamics that govern layer motions. We feel that our extensive background in various areas of fluid mechanics and our expertise in computer simulations enable us to contribute in this area.

We now believe that the interface between nonconvecting and convecting regions has the form of an intermittent boundary layer in which bursts or plumes of fluid transport salt and heat between the adjacent regions. This view is supported by related oceanographic data, as well as some solar pond data. Our computer simulation, based on a numerical model of this boundary layer, has given good agreement with experiment. An implication of this model is that periodic external intervention may be required in order to maintain a sufficiently thick gradient layer.

We feel the technical areas noted above must be addressed before we can forecast with confidence the role of ponds in supplying a significant fraction of our energy needs.

CONCLUSION

Initial research projects have shown that solar ponds are feasible as a source of thermal energy at temperatures less than 100°C. System studies indicate that ponds may be economical energy sources for space heating, low temperature industrial process heat, agricultural process heat, and, in some instances, electrical power generation. Achieving the necessary performance goals, at low cost, will require technical advances in a number of problem areas, including the understanding and control of layer motion.

REFERENCES

1. Nielsen, C. E., "Nonconvecting Salt Gradient Solar Ponds," Solar Energy Handbook, edited by W. D. Dickinson and P. N. Cheremisinoff, Marcel Dekker, Inc., 1979.

2. Jayadev, T. S. and Edesess, M., "Solar Ponds and Their Applications," 7th Energy Technology Conference, 1980.

SOLAR ENERGY WATER DESALINATION IN THE UNITED STATES AND SAUDI ARABIA

Werner Luft
Solar Energy Research Institute

ABSTRACT

Five solar energy water desalination systems are described. The systems will each deliver 6000 m^3/day of desalted water from either seawater or brackish water. After the system definition study is completed in July 1981, two systems will be selected for pilot plant construction. The pilot plants will have capacities in the range of 100 to 400 m^3/day.

1.0 BACKGROUND

In October 1977, Saudi Arabia and the United States signed a Project Agreement for Cooperation in the Field of Solar Energy (SOLERAS) under the auspices of the United States-Saudi Arabian Joint Commission on Economic Cooperation. The objectives of the agreement are to:

- cooperate in the field of solar energy technology for the mutual benefit of the two countries, including the development and stimulation of solar industries within the two countries;
- advance the development of solar energy technology in the two countries; and
- facilitate the transfer between the two countries of technology developed under this agreement.

The Solar Energy Research Institute (SERI), as the Operating Agent, is responsible for implementing SOLERAS in accordance with directives of the SOLERAS Executive Board who has approved a five-year technical program plan.

As part of this technical program plan, an area of Industrial Solar Applications for solar technology has been identified. The objectives of the Industrial Solar Applications program are to introduce solar energy technologies into industrial applications and foster the establishment of domestic industries using renewable energy sources, thereby lessening industrial dependence on fossil fuels and minimizing deleterious effects on the environment. A specific objective is to demonstrate the use of solar energy in desalinating water.

Water desalination is needed in both Saudi Arabia and the United States. In Saudi Arabia, water is needed principally for municipal and agricultural applications. In the United States, desalination is mainly required to control river salinity and provide potable water to selected communities that have critical water quality problems or water shortages.

Conventionally-powered desalting plants have been in operation for several years. At the beginning of 1977, about 1500 land-based, fossil-fueled or electric-powered, desalting plants with a minimum capacity of 100 m^3/day were in operation or under construction throughout the world. These plants are capable of producing nearly 4 million cubic metres of fresh water daily for municipal or industrial uses. Distillation processes account for 77% of the total plant capacity; the balance is almost entirely membrane processes (1).

In 1977, Saudi Arabia had a conventional desalination plant capacity of 87,000 m^3/day. Projects are underway for plants to be constructed during the next five years with a total capacity of 1.36 million m^3/day. By the year 2000, it is planned that fossil fuel powered plants will provide 8-12 million m^3/day of desalted water. Most of these plants are located along the Arabian Gulf and the Red Sea coast.

In the United States at the beginning of 1977, 510 conventional desalination plants provided about 0.4 million m^3/day of fresh water.

Despite this widespread desalination activity, no significant efforts have been made to replace the conventional power plants associated with desalination of water with solar energy systems. Most solar desalination work in the world has been restricted to simple solar stills of limited output. To remedy this situation, the SOLERAS Executive Board authorized the initiation of a solar energy water desalination project. The objective is to advance the technical and economic feasibility of large-scale solar-powered desalination of brackish water and seawater.

Because of the differences between desalting seawater and brackish water, two distinct systems will be developed. One system will desalt seawater to provide potable water. The second system will desalt brackish water for either (a) clean-up and reuse or (b) potable applications. The two systems could find applications in both the United States and Saudi Arabia.

Estimates by the Saline Water Conversion Corporation (SWCC) in Saudi Arabia indicate that steam from conventionally-powered dual-purpose power-desalting units now being designed, constructed, and operated in the coastal areas of Saudi Arabia for seawater desalination would cost in the range of \$0.50 to \$1.50 per gigajoule. Solar energy costs presently are considerably higher and are likely to remain higher. Thus, it is unlikely that solar-powered desalination plants could be economically competitive in these locations.

For inland locations the situation is quite different, especially for communities not connected to the national electric power grid. The absence of electric grid power makes the solar-powered alternative much more attractive. Inland solar-powered desalting plants would desalt brackish water rather than seawater as the coastal plants would do.

Huge fossil fuel powered plants with individual process trains of up to 200,000 m^3/day and total capacity of 8-12 million m^3/day of desalted water are being planned for the year 1990 for coastal areas in Saudi Arabia. The water needs for small inland communities would be considerably less than those for large coastal communities. Six thousand m^3/day of desalted water could supply the needs for a community of 15,000 to 30,000 people in Saudi Arabia, could irrigate 1.2 km^2 of greenhouse area, or could be used for medium-sized industrial application.

Seventeen states in the United States and the Virgin Islands have critical water quality problems or water shortages. In a study for the Office of Water Research and Technology, U.S. Department of the Interior, 37 communities were identified with problems that can be solved by using desalting technology. The population in these communities ranges from a few hundred to over a million. Twelve of the communities depend considerably on inland brackish groundwater with total dissolved solids ranging from 1,000 to 35,000 mg/litre (2).

In addition, four U.S. islands must desalinate either brackish groundwater or seawater to provide potable water. Also, U.S. river basins such as the Colorado River Basin, the Brazos River Basin of Texas, and the Arkansas and Red River Basins of Oklahoma have salinity problems.

2.0 PROJECT PLANS

To accomplish the objective of the SOLERAS solar energy water desalination project, a 3-phase activity is planned. The phases are as follows:

Phase 1: Preliminary System Design and Cost Analysis

Phase 2: Detailed Pilot Plant Design and Construction

Phase 3: Pilot Plant Operation and Training of Personnel

Phase 1: System analyses and economic analyses will be performed by several companies on a solar energy desalination system of their choice for either seawater or brackish water desalination. The systems will each be for an average daily product water capacity of 6000 m^3. The main criterion for the analysis will be the product water cost. Each system will be designed for a specific site and application. The site, application, and technology will have broad applicability to general water desalination needs in either the United States or Saudi Arabia. It is the intent of this project to encourage innovation without unduly affecting performance and reliability. Subsystems and their interfaces will be defined during Phase 1 and product-water cost projections will be made for commercial plants of a range of capacities.

Finally, a development plan for Phase 2 will be generated including detailed cost estimates for the design and construction of a pilot plant with a capacity of 100 to 400 m^3/day using the technology of the baseline system.

Phase 2: Of the several systems designed in Phase 1, one system in each category (brackish and seawater desalination) will be chosen for pilot plant construction. The criteria for selection will include levelized cost per unit of product water for the commercial-sized plant, design and construction cost for the pilot plant, consistency in cost between the commercial-sized plant and the pilot plant, maturity of system design and projected plant reliability. Each pilot plant will have a product-water output capacity of 100 to 400 m^3/day. The pilot plants will be designed in detail and constructed on specific sites.

The size of the pilot plant was selected to be within the budget limitations of the SOLERAS program and is of a capacity that provides useful technical and economic data for the planning, design, and construction of a commercially-sized plant. A pilot plant delivering 400 m^3/day of desalted water would provide water to 2,000 people or could provide irrigation water for about 8,000 m^2 of greenhouse agriculture. If the ratio of the ultimate plant capacity to the pilot plant capacity becomes too great, less useful technical and economic information for application to the full scale plant can be extracted from the pilot plant construction and operation.

Phase 3: The pilot plants will be operated and performance measurements made to provide the information essential for designing commercial-sized desalting plants. Local personnel will be trained in the operation and maintenance of the plant so they can make performance measurements.

The schedule for Phase 1 is from October 1980 to July 1981. Phase 2 is expected to start in October 1981 with the pilot plant construction completed by July 1983. Phase 3 will start at the completion of Phase 2 and will continue until the end of 1983.

3.0 TECHNOLOGY CONSIDERATIONS

Water desalination processes can be divided into four categories, namely: membrane, distillation, crystallization, and chemical processes. Presently, plants using distillation provide most product water. It is projected that in 20 years, the majority of desalination plants will use either membrane or crystalization technology because these processes are more energy efficient (3).

Of the membrane processes, reverse osmosis and electrodialysis represent the technologies most advanced at this time. The crystallization processes currently being developed include: vacuum-freezing vapor-compression, secondary refrigerant freezing, and eutectic freezing.

Some studies regarding the technologies, system designs, and cost for large-scale solar energy water desalination have been ongoing since 1977 (4,5). One conclusion from these studies was that depending on the escalation rate of fuel costs, solar energy water desalination would become economic somewhere in the time span from 1980 to 1990 assuming solar thermal collector costs of $\$180/m^2$. Another conclusion was that the lowest product water cost would be obtained from plants that depend partially on solar energy and partially on conventional fossil energy.

The selection of an optimum solar energy water desalination system is affected by many factors. The feed water characteristics, the product water requirements, the solar collector type and size, required water recovery ratio, the plant utilization factor, the site, the performance factor, the brine disposal method all affect the selection of the desalination process in a complex manner.

As examples, consider the effect of feed-water salinity upon the product water costs for membrane and distillation processes. Figure 1 shows a hypothetical example. For water recovery ratio A, the membrane process provides less expensive product water than the distillation process below a given feed-water salinity. Increasing the water recovery ratio decreases the feed water salinity for which the cross-over in product water cost occurs.

The product water recovery ratio may be dictated by economic reasons, such as minimum product water cost or it may be dictated by the scarcity of the feed water. The product water recovery ratio in its turn is often the determining factor in selecting methods for control of scale formations.

It is because of the complex interaction among many factors that the Phase 1 definition study is being implemented.

4.0 PHASE 1 IMPLEMENTATION

The combinations of desalination and solar technologies that were covered in the 36 proposals received for the Phase 1 definition study are shown in Table 1. This table shows eight solar energy technologies and eight desalination technologies that were represented by the proposals. The numbers in Table 1 indicate the number of proposals incorporating each of the solar energy/desalination technology combinations. Since some proposals offered a coupling of several solar technologies or of several desalination technologies, the total number (55) shown in this table is higher than the number of proposals received.

The most frequent solar energy technology was line-focus collectors, particularly in combination with reverse osmosis. Photovoltaic solar energy conversion, particularly in combination with other solar energy technologies, was the second most frequent option.

The five companies that have been awarded contracts for Phase 1 and their team members are shown in Table 2. The technologies involved in the five systems, the water type, and projected plant locations are given in Table 3. Table 3 shows that these five contracts represent six different desalination technologies (seawater and brackish water reverse osmosis are regarded as two different processes), and five different solar energy technologies.

The Boeing system uses a 25,000 m^2 heliostat field with a central receiver operating at a heat transfer medium (air) temperature of 815°C. The energy storage capacity of 65 MWh is obtained using 40,000 kg alumina bricks operating over a temperature range of 408°-815°C. Energy conversion is achieved using a 14.6 MW Brayton gas turbine connected to a 1.1 MW electric generator and to a compressor for the incoming air going to the central receiver. Backup power is provided by a boiler.

The desalination subsystem consists of two stages of reverse osmosis units in series operating at 2.4 MPa and 5.9 MPa. Alternative systems are being considered. The feedwater is pretreated and stored in 24,000 m^3 tanks. The brine is disposed in evaporation ponds having a surface area of 162,000 m^2. The water recovery ratio is 0.90.

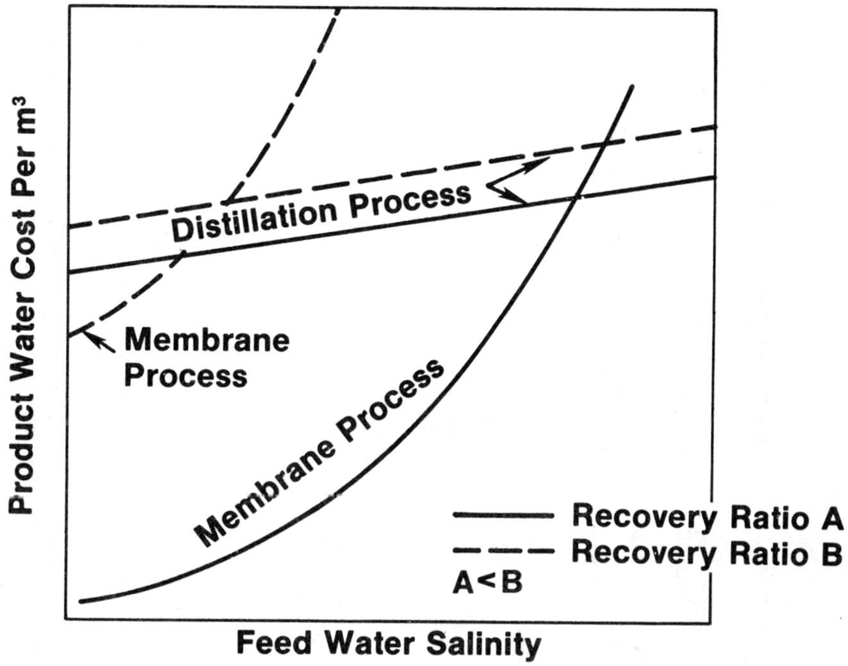

Figure 1. Effect of Feed Water Salinity on Product Water Cost

Table 1. MATRIX SHOWING COMBINATION OF DESALINATION AND SOLAR ENERGY TECHNOLOGIES

	Solar Energy Technology								
Desalination Technology	Central Receiver	Point Focus	Line Focus	Evacuated Tube	Photo-voltaic	Wind	Solar Pond	OTEC	Total
Electro-dialysis					4	1			5
Reverse Osmosis-Seawater	3	4	6		1	2	1		17
Reverse Osmosis-Brackish	1	2	5		2	1			11
Multieffect Distillation	1		2						3
Multistage Flash Distillation	1		2		2	2	1	2	10
Vapor Compression		1	2	2					5
Freezing		1			1		1		3
Solar Still					1				1
Total	6	8	17	2	11	6	3	2	55

Note: Numbers in chart are the number of proposals in each category. Several proposals offered a combination of technologies; therefore, a higher number than the number of proposals is shown.

Table 2. CONTRACTORS FOR PHASE 1

Prime Contractor	Team Members
Boeing Engineering & Construction Co.	Resources Conservation Co. International
Catalytic, Inc.	Science Applications, Inc.
Chicago Bridge & Iron Co.	Foster-Miller Associates Inc. Arabian Chicago Bridge & Iron Co.
DHR, Inc.	Science Applications, Inc. Ionics, Inc. Al-Radwan
Exxon Research & Engineering Co.	Permutit Co., Inc. Ecodyne-Unitec Div. Martin-Marietta Badger Energy, Inc. Saudi Investment Development Center

The Catalytic solar energy collection subsystem consists of three types of solar thermal collectors, having a total area of 64,000 m^2. The collectors include high-temperature point-focus Omnium-G thermal collectors, medium-temperature line-focus Fresnel thermal collectors, and low-temperature Winston thermal collectors. In addition, 12 wind generators provide a total of 2.4 MW of electric power.

Energy storage is provided using a high-temperature air thermal storage system over the range of 290°-430°C, and medium-temperature and low-temperature thermal storage with a range of 45°-120°C and 180°-290°C, respectively. The medium and low temperature thermal storage systems use a liquid medium. The total capacity for the thermal storage system is 60 MWh. The electric storage capacity is 725 kWh.

Energy conversion is obtained through a steam turbine with a 560 kW electric generator and through the use of a power recovery turbine. Backup power is obtained through a motor with a 207 kW electric generator.

The brackish water is pretreated and uses 18,000 m^3 storage tanks. The desalination subsystem consists of two stages of reverse osmosis units in series, operating at 2.9 MPa and 5.6 MPa and operating in series with a multiple-effect vertical tube evaporator. The brine is disposed in 93,000 m^2 surface area evaporation ponds. The water recovery ratio is 0.98.

The Chicago Bridge and Iron system uses 37,000 m^2 distributed point-focus thermal collectors with two axes tracking. Energy storage is obtained through two tanks containing HITEC molten salt operating over a temperature range from 286°-565°C and having a capacity of 148 MWh. The desalination is obtained through freezing. The primary freezing unit is driven by a compressor and is augmented by an absorption freezing unit.

The energy conversion subsystem uses a steam turbine with a 560 kW electric generator and a turbine driving the 1,216 kW primary refrigeration compressor. Backup power is obtained from a 7.5 MW boiler. There is no waste disposal subsystem as the brine is rejected directly into the sea.

The DHR system achieves the solar energy collection by using 87,000 m^2 of 60 power line-focus thermal collectors and 39,000 m^2 of flat-plate photovoltaics. Thermal energy storage is provided by a single tank of 28 MWh capacity and using Caloria HT-43 as a heat transfer medium. The storage temperature range is 272°-300°C. Energy conversion is obtained from a Toluene turbine with a 4.14 MW electric generator and through the use of a power recovery turbine with a 411 kW electric generator. Backup power is obtained from a motor with a 5.5 MW generator.

The desalination subsystem provides for feedwater storage and pretreatment and uses a one-stage reverse osmosis system in series with an electrodialysis unit. There is no waste disposal subsystem as the brine is discharged directly into the sea.

Table 3. **WATER TYPES, PLANT LOCATIONS, AND TECHNOLOGIES FOR FIVE SYSTEMS**

Prime Contractor	Water Type	Plant Location	Desalination Technology	Solar Energy Technology
Boeing	Brackish water	Upton County, Texas, United States	Reverse osmosis, 2 stages in series, 2.4 and 5.9 MPa.	Heliostats and central receiver
Catalytic, Inc.	Brackish water	Brownsville, Texas, United States	Reverse osmosis, 2 stages in series, 2.9 and 5.6 MPa, and vertical tube distillation	Wind generators and line- and point-focus thermal collectors
Chicago Bridge & Iron Co.	Seawater	Al Jubayl, Arabian Gulf, Saudi Arabia	Indirect Freezing	Point-focus thermal collectors
DHR, Inc.	Seawater	Uquair, Arabian Gulf, Saudi Arabia	One stage reverse osmosis in series with electrodialysis	Line-focus thermal collectors and photovoltaics
Exxon	Seawater	Rabigh, Red Sea, Saudi Arabia	Two stages of reverse osmosis in parallel with 24-effect distillation	Heliostats and central receiver

The Exxon system uses a 27,000 m^2 heliostat field with a central receiver operating at a heat transfer medium temperature of 566° C. Energy storage is obtained through two thermal storage tanks having a capacity of 99 MWh and operating over the temperature range from 260°-566° C.

The energy conversion is obtained through a noncondensing steam turbine with an 800 kW electric generator and through the use of a power recovery turbine. Backup power is provided by a motor with a generator. The desalination is achieved through a two-stage reverse osmosis unit producing 4100 m^3/day product water with 800 mg/L total dissolved solids and having a water recovery ratio of 0.50 in parallel with a 24-effect distillation unit providing 2400 m^3/day water with essentially zero total dissolved solids and having a water recovery ratio of 0.35. There is no waste disposal subsystem as the brine is directly discharged into the sea.

6.0 REFERENCES

1. U.S. Department of the Interior. Desalting Plants Inventory Report No. 6. 1977 (Oct.).

2. Boyle Engineering Corporation. 1978 (Dec.). Evaluation of Technical Material and Information for Potential Desalting Demonstration Plants. Document PB-290-338. Final report for the Office of Water Research and Technology.

3. Fluor Engineers and Constructors, Inc. 1978 (Sep.). Desalting Plans and Progress, An Evaluation of the State-of-the-Art and Future Research & Development Requirements. 2nd Ed. PB-290-785. Final report for the Office of Water Research and Technology.

4. Bechtel Corporation. 1977 (Dec.). Technical & Economic Assessment of Solar Distillation for Large-Scale Production of Fresh Water. Final report prepared for Sandia Laboratories, Livermore, Research and Engineering Operation. Contract No. 87-9814.

5. A Feasibility Study on the Use of Solar Energy for Desalination of Water. Prepared for the Texas Water Development Board, State of Texas Contract No. 14-70028, 31 May 1977.

ECONOMIC VIABILITY OF SOLAR THERMAL CENTRAL RECEIVER POWER PLANTS

John E. Montague
Martin Marietta Denver Aerospace

ABSTRACT

Solar thermal central receiver (STCR) power plant technology, as soon to be demonstrated by Solar One in Barstow, CA, can be a significant contributor to the nation's energy needs in the "New Fuels Era - 80's." In addition to further technology demonstrations, the only perceived barrier to the widespread commercialization of STCR technology in the utility sector is its economic viability. This paper compares the levelized busbar cost of electricity from a 100 MWe STCR plant, with 18 hours of storage, and various heliostat costs, to electricity costs from new conventional fossil-fired and nuclear power plants. In order to be as realistic as possible, published plant costs and utility fuel cost/escalation data, as well as "typical" utility economic parameters were used. Sensitivity analyses are also included to illustrate the effects of varying plant costs, fuel costs, escalation rates and alternative utility economic parameters.

Significant conclusions from this study include:

o For the base case assumptions (such as real fuel escalation between 1% and 2%), the STCR plant is competitive with coal-fired and nuclear power plants with second-generation heliostat costs as high as $\$132/m^2$ ($\$12.25/ft^2$).

o Using "first generation" heliostat costs (i.e., 10 MWe Pilot Plant), STCR power plants exhibit a significant cost advantage over oil-fired power plants.

o As would be expected, the viability of STCR plants is quite sensitive to the assumed escalation rates for conventional fuels.

o As would be expected, the viability of STCR plants is quite sensitive to the assumed escalation rates for conventional fuels.

o The inclusion of an additional 15% investment tax credit applicable to solar significantly enhances the economic advantages of STCR power plants in utility applications.

INTRODUCTION

Solar thermal central receiver (STCR) technology encompasses a variety of specific concepts and applications. Concepts include the use of water/steam, liquid sodium, molten salt and air as heat transport fluids generating electricity through Rankine and Brayton power conversion cycles and/or procss steam. Applications include large electric power plants in a utility environment and a variety of industrial process heat (IPH) uses, such as enhanced oil recovery, metal smelting, oil refining and gypsum drying.

The concept to be evaluated in this discussion is a solar central receiver plant using molten salt (60% $NaNO_3$, 40% KNO_3) as a heat transport and thermal storage medium providing source energy for a steam-Rankine power cycle. This concept of using molten salt as both the heat transport fluid and the thermal storage medium provides a means for cost-effective energy storage, enabling high capacity factors (~75%) and thus baseload capabilities. The STCR plant design, sized at 100 MWe with 18 hours of thermal storage capacity was optimized and evaluated for a Barstow, CA, location, which has an annual average insolation value of 7.3 kWh/m^2-day. However, this design can be equally attractive for a wide area of the southwestern United States, as well as other locations in the world, where annual average insolation is relatively high. Figure 1 depicts the area where annual average insolation is greater than 5.5 kWh/m^2-day, and thus the projected near-term U.S. market area of STCR plants. Later discussion will cover the impact of the variation in annual insolation on the projected cost-competitiveness of a STCR plant.

The following discussion will provide an overview of the STCR plant concept with cost and performance estimates before evaluating its economic feasibility in comparison to conventional (i.e., oil, coal, and nuclear) power plants in an investor-owned utility financial environment. The solar central receiver system based on molten salt technology is being supported by the Department of Energy (DOE) and Sandia National Laboratories-Livermore for a number of applications, and is documented in literature (Ref 1, 2, 3, 4).

SYSTEM DESCRIPTION

The general concept of a molten salt STCR power system can be best explained using a simplified system schematic, shown in Figure 2. Starting at the cold salt storage tank, cold salt at 288°C (550°F) is pumped to the receiver unit, located on top of a central tower. Incident solar insolation is reflected by the surrounding field of heliostats onto the absorber surfaces of the receiver, heating the salt to 566°C (1050°F) which then flows to the hot storage tank. From the hot storage tank, the salt is pumped through the steam generator equipment, which is a series of heat exchangers that heat feedwater from the steam cycle with the hot salt to generate superheated steam and also reheat steam from the high-pressure stage of the turbine. The turbine is a commercial design, with main steam conditions of 538°C (1000°F), 12.41 MPa (1800 psig) and 538°C (1000°F) reheat.

The collector subsystem has been optimized to provide sufficient energy to charge 1800 MWHe of storage capacity on summer solstice while the plant is producing 100 MWe net during daylight hours. This set of conditions resulted in a design point receiver output power of

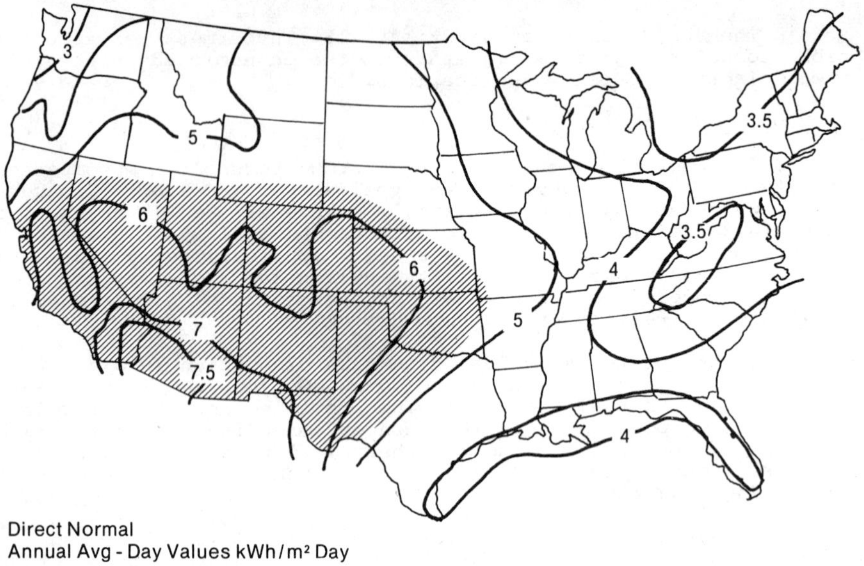

Figure 1 Annual Average Insolation

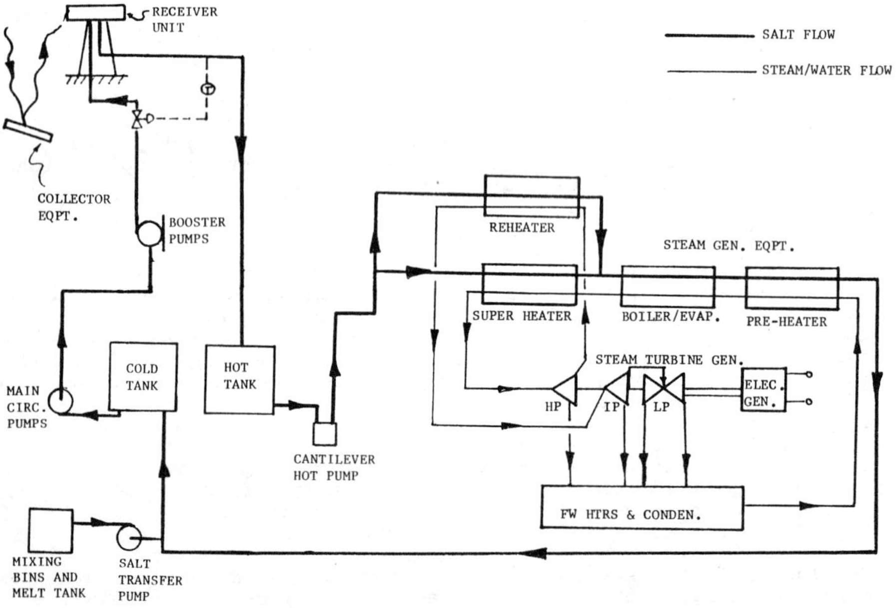

Figure 2 Molten Salt STCR System Schematic

842 MWt, a solar multiple near 3.0, and a two-module collector field configuration, as shown conceptually in Figure 3. A total of 20,644 second generation heliostats, each 57.4 m^2 (618 ft^2) in area are required, or 10,322 heliostats per module. The two-module configuration was selected on the basis of minimizing the cost of providing thermal energy to the plant, and included the cost of the piping from the two receivers to the plant located centrally between the two modules. This configuration also provides additional reliability and flexibility in plant design

The collector fields are divided into 4 quadrants to represent the portion of heliostats redirecting energy into each cavity of the quad-cavity receivers shown in Figure 4. The side tube walls in the receiver are common between adjacent cavities so that solar energy heats both sides of the tubes. The design point (noon, day 172 with insolation of 950 W/m^2) power output of each receiver is 421 MWt.

The final solar-peculiar subsystem is the thermal storage subsystem, which consists of 4 large tanks; 2 cold salt storage tanks and 2 hot salt storage tanks. The hot salt storage tanks are sized approximately 27.4 m (90 ft) in diameter and 18.3 m (60 ft) high, with the cold salt storage tanks being slightly smaller to account for the increased density of the cold salt. The tank shells are made of carbon steel, with internal insulation being used in the hot salt tank to maintain the shell temperature below 316°C (600°F). The thermal storage configuration was optimized considering soil-bearing capacities, design stresses, heat losses and insulation thicknesses and costs.

A complete description of the plant design is contained in Reference 1.

SYSTEM PERFORMANCE

The annual net electric power generation of the 100 MWe-18 hr STCR plant was determined by computer simulation of an entire year using actual solar insolation and weather data on a 15-minute interval basis. The insolation and weather data used were actual measurements taken at Barstow, CA during 1976. Figure 5 is the annual energy stairstep based on the simulation. The total net annual energy is 660,215 MWHe, resulting in a plant capacity factor of 0.75 and an annual average efficiency of converting insolation to net electrical energy of 20.1%. It should be noted, that in the absence of actual operating data in the utility environment, forced outages due to mechanical failure have not been considered in determining the capacity factor.

In order to extend the economic analyses to areas other than Barstow, CA, the computer simulation of plant operation was performed for 4 other locations: Fresno, CA, Phoenix, AZ, El Paso, TX, and Ely, NV, using the SOLMET Typical Meteorological Year (TMY) data base. These sites have annual average insolation values ranging from 6.21 $kWh/m^2\cdot day$ to 7.32 $kWh/m^2\cdot day$. A summary of the 100 MWe-18 hr STCR plant output at these locations, as well as the baseline Barstow, CA site, is provided in Table 1. As can be seen in the table, the annual electric output is not directly proportional to the annual average insolation. This occurs because of the size of storage (18 hours), which means that at certain locations with several consecutive clear days during the summer months some available energy must be discarded because of fully charged storage from the previous day. For the Barstow insolation data, approximately 3.5% of the total available energy is lost during the summer months. Thus, a site may have lower annual insolation than Barstow, such as Phoenix, but the distribution of insolation during the year allows more efficient use of the available energy, and a higher capacity factor. The capacity factors would be directly proportional to the insolation for STCR plants with smaller storage capacities (0 - 8 hrs).

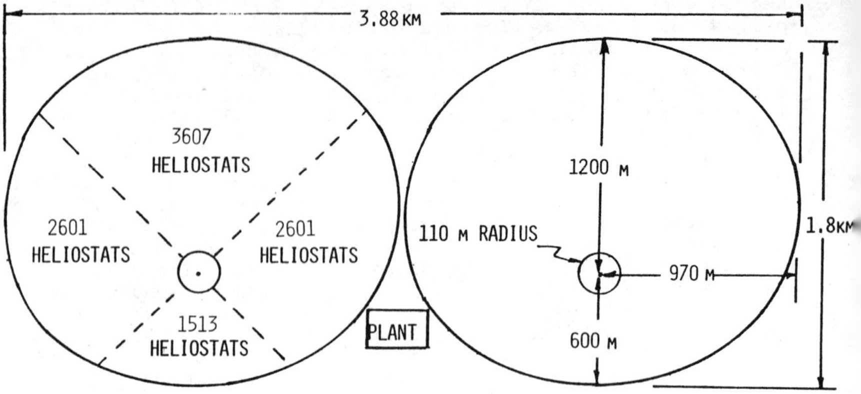

NUMBER OF HELIOSTATS PER MODULE = 10,322
HELIOSTAT AREA PER MODULE = 592,586 M^2
LAND AREA PER MODULE = 2.75 KM^2
TOTAL PLANT LAND AREA = 7.0 KM^2

Figure 3 100 MWe-18 hr STCR Plant Collector Configuration

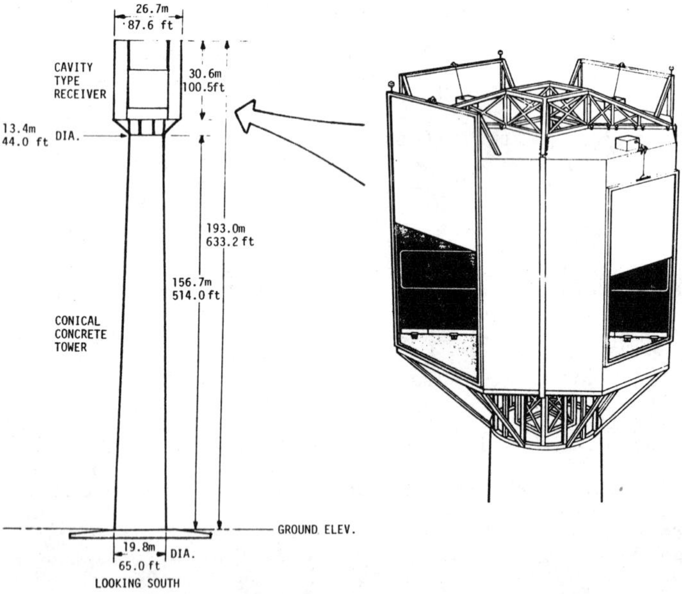

Figure 4 Quad-Cavity Receiver Configuration

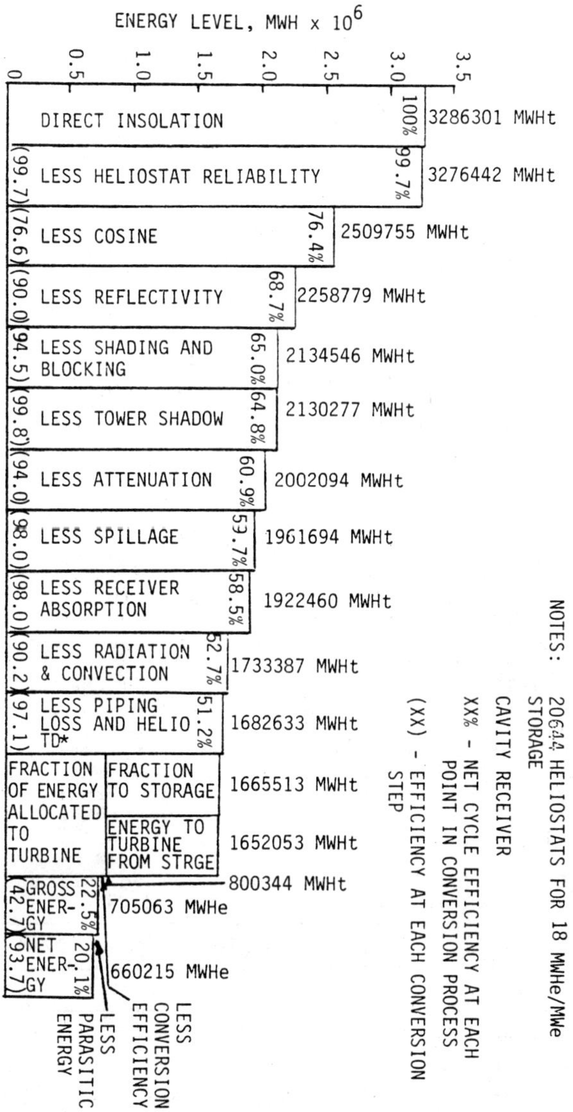
ENERGY LEVEL, MWH x 10^6
0
0.5
1.0
1.5
2.0
2.5
3.0
3.5
DIRECT INSOLATION 100% 3286301 MWHt
(99.7) LESS HELIOSTAT RELIABILITY 99.7% 3276442 MWHt
(76.6) LESS COSINE 76.4% 2509755 MWHt
(90.0) LESS REFLECTIVITY 68.7% 2258779 MWHt
(94.5) LESS SHADING AND BLOCKING 65.0% 2134546 MWHt
(99.8) LESS TOWER SHADOW 64.8% 2130277 MWHt
(94.0) LESS ATTENUATION 60.9% 2002094 MWHt
(98.0) LESS SPILLAGE 59.7% 1961694 MWHt
(98.0) LESS RECEIVER ABSORPTION 58.5% 1922460 MWHt
(90.2) LESS RADIATION & CONVECTION 52.7% 1733387 MWHt
(97.1) LESS PIPING LOSS AND HELIO TD* 51.2% 1682633 MWHt
FRACTION OF ENERGY ALLOCATED TO TURBINE
FRACTION TO STORAGE 1665513 MWHt
ENERGY TO TURBINE FROM STRGE 1652053 MWHt
800344 MWHt
(42.7) GROSS ENERGY 22.5% 705063 MWHe
(93.7) NET ENERGY 20.1% 660215 MWHe
LESS CONVERSION EFFICIENCY
LESS PARASITIC ENERGY
NOTES: 20644 HELIOSTATS FOR 18 MWHe/MWe STORAGE
CAVITY RECEIVER
XX% - NET CYCLE EFFICIENCY AT EACH POINT IN CONVERSION PROCESS
(XX) - EFFICIENCY AT EACH CONVERSION STEP

Figure 5 100 MWe-18 hr STCR Plant Annual Energy

Table 1 Variation in Plant Performance with Location

	Fresno	Ely	Phoenix	El Paso	Barstow
Annual Average Insolation (kWh/m^2 ·day)	6.21	6.34	6.90	7.32	7.33
Net Electric Output (MWHe)	581,726	604,413	661,149	685,675	660,215
Annual Capacity Factor	0.664	0.690	0.755	0.783	0.754

STCR SYSTEM COST

The construction cost estimate for a 100 MWe-18 hr storage standalone STCR plant is shown by subsystem in Figure 6. The total construction cost of \$220M (1980 \$) equates to \$2200/kWe, using a mass produced collector cost of \$97/m^2 (\$9.00/ft^2). As can be seen in the figure, the major portion of the plant cost, and thus the primary driver in the resultant cost of useful energy, is the collector subsystem, which consists of heliostats, foundations, wiring, and associated controls. Therefore, considerable emphasis has been placed on using realistic heliostat cost estimates. An important consideration in developing these estimates is the annual production rate, since most heliostat designs are easily mass producible. The curve of installed heliostat cost (normalized to a reflectivity of 1.0) versus cumulative heliostat production shown in Figure 7 has been developed from a number of DOE-sponsored studies performed by Battelle Laboratories, General Motors, McDonnell Douglas and Martin Marietta Denver Aerospace. The starting point on the curve is the actual cost of heliostats installed at the Central Receiver Test Facility in Albuquerque, NM by Martin Marietta Aerospace in 1976-78. As shown in the figure, the forces of technological innovation and mass production combine to significantly reduce heliostat costs, achieving the \$97/m^2 figure with production rates as low as 50,000 units per year, expected in the 1987-1990 time frame. To provide a realistic economic evaluation of STCR plants, three heliostat costs have been used in this analysis: \$217/m^2, representing near-term, low production (5,000 units/yr); \$132/m^2, representing medium-production (15,000-25,000 units/yr); and the baseline mass-produced (50,000 units/yr) cost of \$97/m^2. The effect on the total construction cost of a 100 MWe-18 hr STCR standalone plant on these various heliostat cost assumptions is shown in Table 2.

Table 2 100 MWe-18 hr STCR Construction Cost Estimates (1980 \$)

	Heliostat Cost, \$/m^2		
	\$217	\$132	\$97
Collector Subsystem	\$265.7M	\$161.5M	\$118.8M
Balance of Plant	101.4M	101.4M	101.4M
Total Cost	\$362.1M	\$262.9M	\$220.2M
Collector Percentage	72.4%	61.4%	53.9%

The cost of a STCR power plant is also significantly dependent on the amount of storage that has been chosen. Although the actual expense of the storage subsystem (i.e., tanks and storage media) is a relatively small percentage of the total plant cost, the amount of

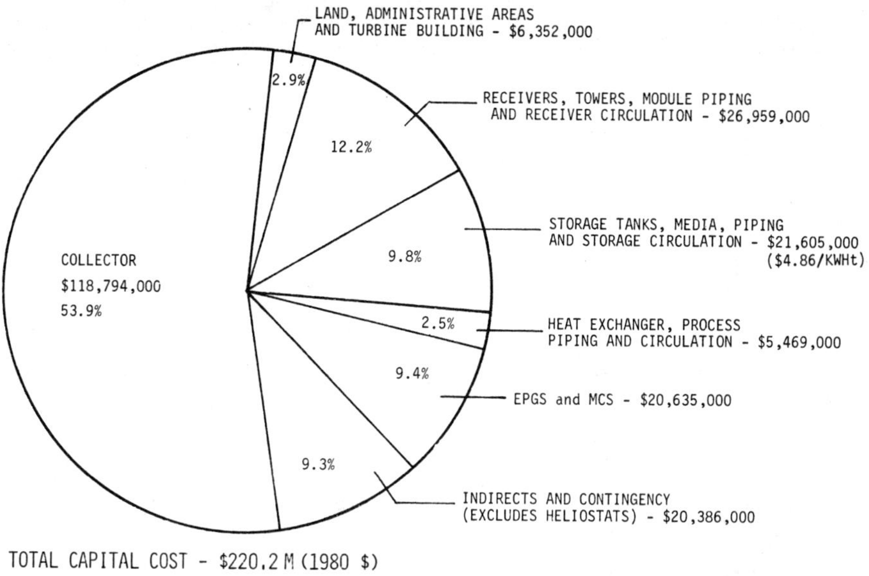

Figure 6 100 MWe-18 hr STCR Plant Construction Costs (1980 \$, \$97/m^2 Heliostat Cost)

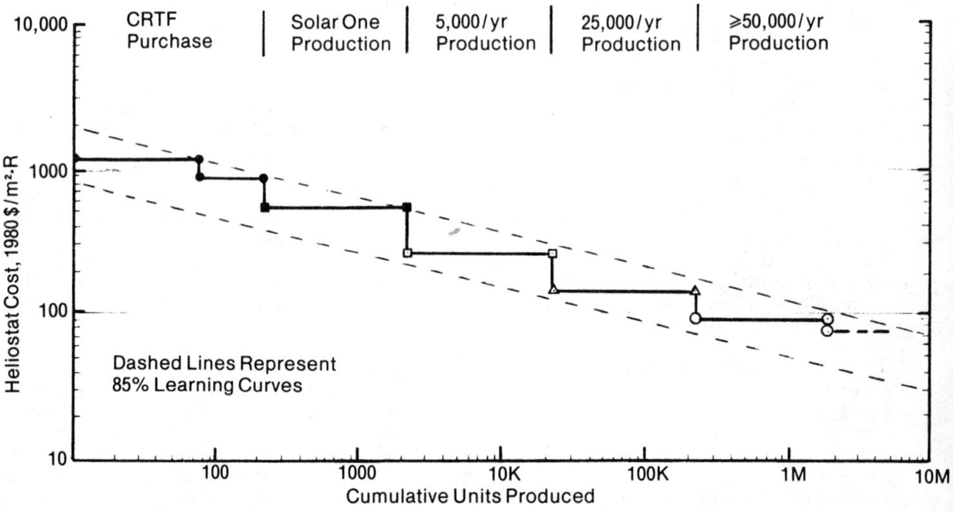

Figure 7 Heliostat Cost Trends

storage desired drives the number of heliostats required to charge the storage capacity. The storage size of 18 hours baselined in this analysis was chosen on a series of optimizations which indicated a minimum levelized busbar energy cost of electricity between 12 and 18 hours of storage. However, for given applications, smaller amounts of storage may be desirable. As shown in Table 3, decreasing the amount of storage (and thus the total number heliostats required), significantly impacts the cost of the plant, at the penalty of reducing the annual capacity factor. The plant costs in the table are based on 57.41 m^2 heliostat size, $97/$m^2$ heliostat cost, and the capacity factors shown are based on Barstow, CA insolation data.

Table 3 100 MWe STCR Plant Costs with Varying Storage (1980 $)

	Storage Size			
	0 hrs	3 hrs	18 hrs	33 hrs
Number of Heliostats	7208	10,322	20,644	30,966
Heliostat Cost	$41.9M	$59.4M	$118.8M	$178.2M
Balance of Plant	52.2	62.1	101.4	145.6
Construction Cost	$94.1M	$121.5M	$220.2M	$323.8M
$/kWe	$941.	$1215.	$2202.	$3238.
Capacity Factor	0.27	0.37	0.75	0.95

A significant portion of the capital cost of a large power plant to a utility is the Allowance for Funds During Construction (AFDC). These expenses are normally included as part of the capital cost of the plant and are capitalized at the beginning of operations. The amount of these funds varies depending on anticipated construction time span and allowed AFDC rate set by the utility's regulatory agencies. A significant advantage of STCR power plants over conventional plants is foreseen in their shorter construction periods (4-5 years) as compared with conventional plants (7 years for coal, over 10 years for nuclear). However, the AFDC charge for a STCR plant is still significant, typically being from 22 to 30 percent of the construction cost estimate. For the baseline 100 MWe-18 hour STCR plant with a construction cost estimate of $220 million, AFDC charges would be $48-$66 million, yielding a total capital cost of $268 to $286 million ($2680-$2860/kWe).

As of the present time, there is no operating and maintenance (O&M) experience for a STCR standalone plant, and so O&M cost estimates have been developed based on reliability analyses for the various solar components and operating histories of conventional power plants. The annual fixed O&M cost estimate for a 100 MWe-18 hour STCR plant has been estimated at $2.5M, of which approximately $0.9M is for operating personnel.

The cost of electricity from any type of power plant, whether the plant is fossil-fueled or solar, depends on the economic characteristics of the utility that builds and operates the plant. These economic characteristics, usually defined as allowable rates of return and capitalization fractions, vary from utility to utility and, for a given utility, change from year to year. To provide a consistent and realistic basis for the economic evaluation of solar and conventional power systems, the economic factors shown in Table 4 will be used as those of a typical southwestern utility. In addition to the parameters shown in the table, several other assumptions were used. Construction cash flows were assumed to follow a typical "S-shaped" spend plan over the construction period, which were baselined at 4 years for

Table 4 Economic Parameters

System Life		30 Years
Year of Initial Operation		1989
Capitalization Structure		
	Ratio	Rate
Bonds	48%	11.0%
Preferred Stock	12%	11.0%
Common Stock	40%	16.0%
Composite Rate		13.0%
Tax Structure		
Federal Tax Rate		46.0%
State Tax Rate		9.0%
Composite Income Tax Rate		50.86%
Other Factors		
Ad Valorem Tax Rate		1.04%
Investment Tax Credit		10%
AFDC Rate		10.31%
Escalation Factors		
Capital Escalation Rate		8%
Operating Cost Escalation Rate		8%
GNP Deflator (inflation)		8%

solar, 5 years for coal, and 7 years for nuclear plants. Capital escalation was calculated up to the year of construction outlay; allowance for funds during construction was then computed at 10.31% per year. Straight line depreciation for 30 years was used for book depreciation. A period of 28 years was used for tax depreciation purposes, with the double-declining balance depreciation method used for 14 years, then switching to straight-line for the remaining period. Similarly, normalized accounting practices were applied to the investment tax credit. The net effect of many of these parameters is summarized as the levelized fixed charge rate (FCR); for these assumptions a fixed charge rate of 0.184 has been calculated.

The cost of electricity has been calculated using the above parameters and required revenue methodology (Ref 5, 6) to arrive at both the year-by-year busbar energy cost and the levelized busbar cost of energy, expressed in cents per kilowatt-hour. The levelized busbar energy cost (BBEC) accounts for all costs incurred during the lifetime of the system and is a constant value over the operating period which has the same present value as the actual year-by-year revenue requirements. In all cases, the busbar energy costs have been evaluated using a Martin Marietta Denver Aerospace developed computer model that allows for a wide range of input selection.

The levelized busbar energy cost for the 100 MWe-18 hr STCR plant using the above economic parameters and performance estimates are shown in Table 5 for the three heliostat cost scenarios and the various locations previously discussed. For clarity, the performance and cost at the Barstow location will be baselined for the remaining analyses.

COMPARISON OF STCR AND CONVENTIONAL ENERGY COSTS

For utility applications, the primary competitive energy sources are conventional power generating units: gas, oil, coal, and nuclear. With the implementation of the Fuel Use Act of 1978, the gas and oil options are effectively eliminated for utility use, leaving only coal and nuclear as choices. In these analyses, the energy costs resulting

Table 5 100 MWe-18 hr STCR Levelized Busbar Energy Cost (1980 cents/kW-hr)

Heliostat Cost	Fresno	Ely	Phoenix	El Paso	Barstow
$217/m^2$	12.75	12.28	11.22	10.82	11.24
$132/m^2$	9.37	9.02	8.24	7.95	8.26
$97/m^2$	7.99	7.69	7.03	6.77	7.04

from a hypothetical oil plant are included simply for comparison purposes. Table 6 summarizes the cost and performance assumptions used in the cost of energy comparisons. The data, for 500 MWe units, has been collected from a number of sources, including DOE and EPRI studies, as well as utility projections.

Table 6 Conventional Power Plant Cost Assumptions (1980 $)

	Oil	Coal	Nuclear
Capitalized Cost, $/kWe	$1015	$1150	$1600
Range	830-1210	930-1330	1400-1800
Fixed O&M, $/kWe	$5.00	$7.78	$12.75
Variable O&M, ¢/kWh	0.6	1.2	0.35
Net Heat Rate, Btu/kWh	9000	10,400	10,400
Fuel Cost, $/MBtu	$5.00 ($29.00/bbl)	$1.70 ($34.00/ton)	$0.56
Fuel Escalation	10%	9%	9%
Range	8 - 11%	7 - 10%	7 - 10%

Levelized busbar energy costs for both the 100 MWe-18 hr STCR plant and the conventional plants described in the above table are shown in Figure 8 for varying hours of annual operation. Due to the capital intensity of the STCR plants, the slopes are relatively steep until they reach the predicted hours of operation, 6600 in this case, representing the 0.75 capacity factor not considering scheduled outages. It is a recognized fact that the conventional plants will not achieve 8760 hours of annual operation; the curves are intended to show the relative availability of conventional plants as compared with STCR plants. These considerations do not alter the basic result shown in the figure, however--that at heliostat costs of under $150/m^2$ the STCR plant is definitely competitive with coal plants, and at the goal of $97/m^2$, STCR provides electricity at costs comparable to nuclear plants.

The curves shown in Figure 8 are based on levelized busbar energy costs, where present-value methodology reduces the effect the costs incurred in the later years of the plants. The comparison between solar and conventional technologies on a year-by-year basis is shown in Figure 9, where the curves are the actual yearly revenue requirements (busbar energy cost) and the horizontal lines represent the value of the levelized BBEC in 1989 dollars. For the solar plants, the high revenue requirements in the first few years for debt service and return on equity significantly weight the levelized BBEC to a higher value. Although these initial yearly revenue requirements are higher than those of the conventional plants, the reverse is soon the case. It should be noted that the discontinuity in the yearly revenue requirements occurring in the solar curves is due to the difference in

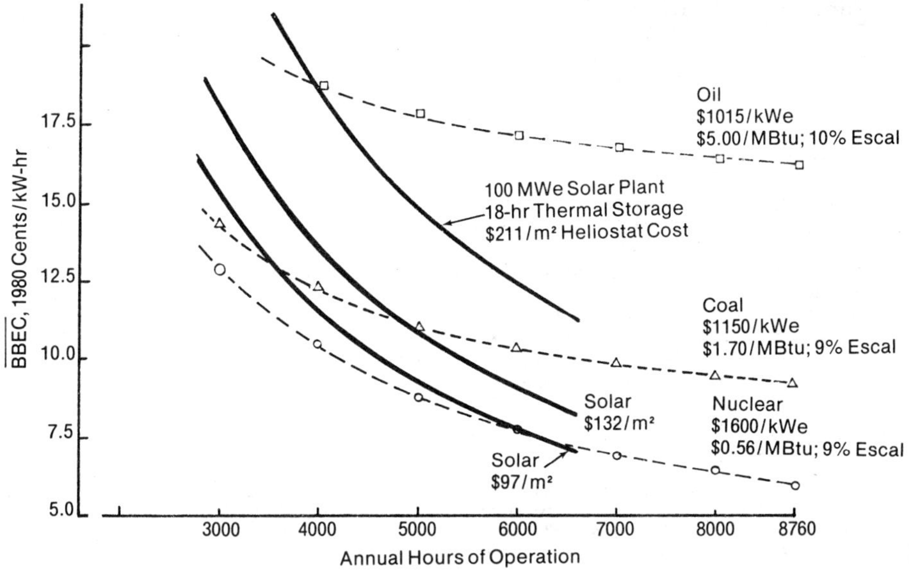

Figure 8 Levelized Busbar Energy Costs, 1989 Start-up

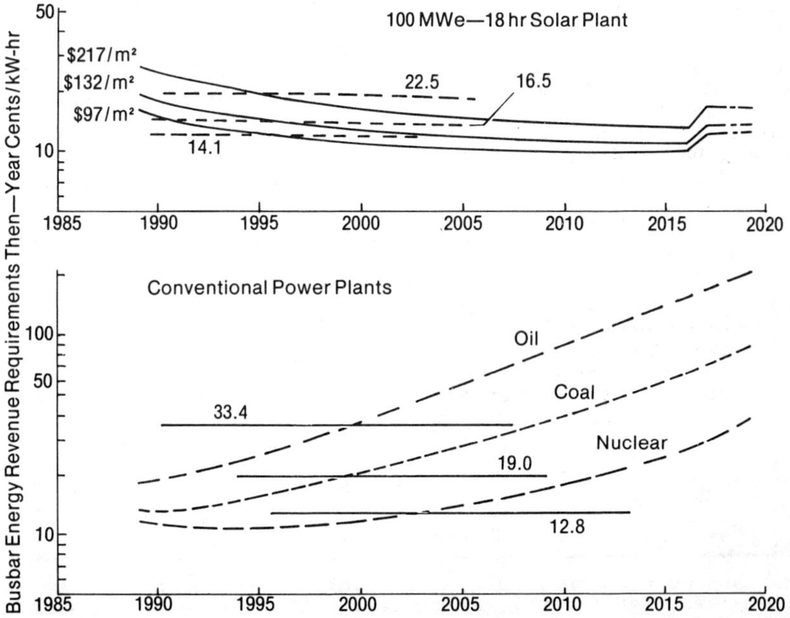

Figure 9 Yearly Busbar Energy Costs, 1989 Start-up

tax life (28 years) as opposed to book life (30 years). The disparity between the levelized BBEC and the later-year actual energy costs for the conventional plant is readily apparent in the lower figure. A definite advantage of STCR plants that can be recognized is that future revenue requirements (cost to the ratepayer) can be forecasted with a large degree of certainty as opposed to the high uncertainties associated with future fuel costs, particularly over a 20-30 year time frame.

SENSITIVITY ANALYSES

The previous figures show that, for a given set of assumptions, a 100 MWe-18 hr STCR plant can be competitive with conventional generating facilities. Various sensitivity analyses have been performed to identify those assumptions that may significantly impact the analyses. The effects of varying amounts of storage, discussed earlier, is shown in Figure 10 for $97/m^2 heliostat costs. Solar plants that may be intended to be used as daytime fuel-savers, i.e., 0-3 hrs of storage, show slightly higher levelized BBEC than the baseline 18-hr storage case, but still very competitive economics. Likewise, the 33-hr storage case has a higher energy cost due to inefficient use of heliostats during summer months (discarding energy due to already charged storage).

Figure 11 examines the change in the levelized BBEC for coal and nuclear plants with variations in plant costs and fuel escalation rates. In the figure, the range of plant capital costs given in Table 6 are used, and the percentage value is the assumed fuel escalation rate. For the coal case, the total energy cost variation over the range of parameters considered is ±15%, and less than ±5% for the nuclear case.

As the allowable rates of return and resultant cost of capital and fixed charge rates vary widely from utility to utility (and year to year), the effects of these variations on the comparision between STCR and conventional plants were evaluated. The results for this analysis are shown in Table 7 for the range of financial parameters that have been encountered in discussions with various utilities, again for plants coming on-line in 1989. No attempt was made in this analysis to "match" escalation rates with the variation in costs of capital, although there is historical evidence of linkage between these parameters.

Table 7 Effect of Cost of Capital on Levelized Busbar Energy Cost (1980 cents/kW-hr)

		Cost of Capital			
		11.5%	12.0%	13.0%	15.0%
Fixed Charge Rate		.165	.170	.184	.217
Solar	$97/m^2	6.39	6.55	7.04	8.34
	$132/m^2	7.47	7.66	8.26	9.83
	$217/m^2	10.11	10.39	11.24	13.44
Coal		9.66	9.58	9.51	9.61
Nuclear		6.16	6.21	6.44	7.14

As shown in the table, the levelized energy costs for the solar plants are significantly affected by an increase or decrease in the cost of capital (and resultant fixed charge rates). Assuming a baseline of 13% cost of capital, the levelized BBEC for the solar cases vary as much as -10% to +20% over the range of costs of capital.

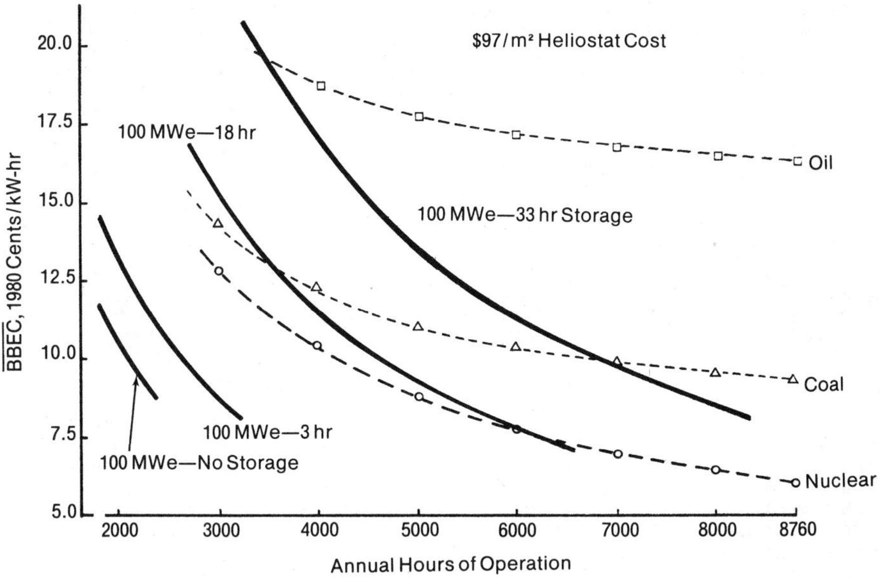

Figure 10 Levelized Busbar Energy Costs vs. Hours of Storage

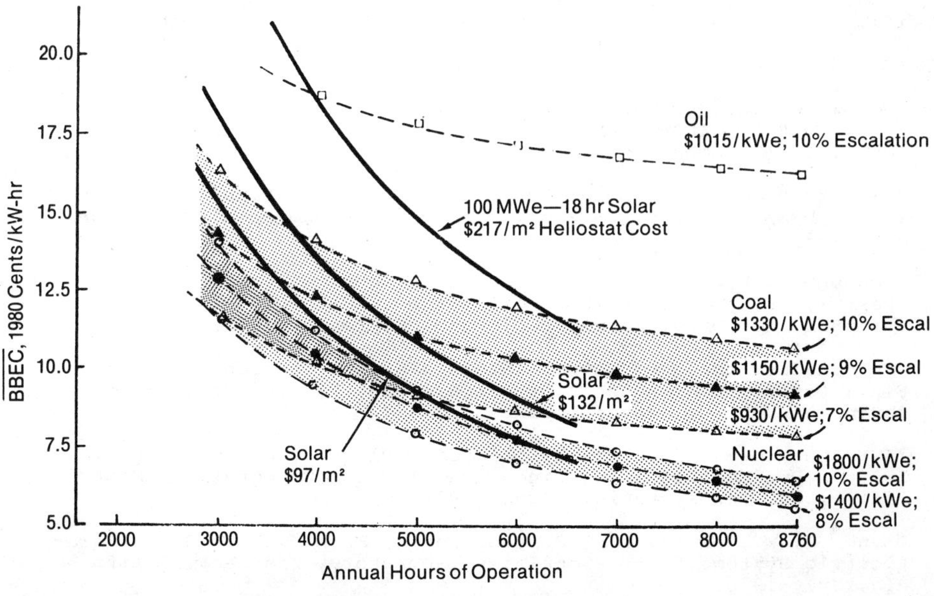

Figure 11 Variations in Conventional Plant Energy Costs

However, for the conventional plants, the variation is much less--as little as ±1% for the coal plant. This occurs because the increase in capital recovery revenue requirements due to higher costs of capital is offset by the decrease in the present value of fuel expenses due to use of higher present value discount factors, which are assumed to be the after-tax cost of capital. Of course, the converse is also true for lower costs of capital. Because there is no offsetting influence of fuel expenses, the levelized BBEC for STCR plants are much more sensitive to the assumed financial parameters. Thus, if the cost of capital or fixed charge rate for a solar plant could be reduced through some method of innovative financing, significant reductions in the costs of energy could be realized. For example, an additional 15% investment tax credit, such as that allowed industry under the 1978 National Energy Act, would reduce the fixed charge rate from 0.184 to 0.166, thereby reducing the levelized BBEC by over 8%. Other means of reducing the cost of capital for STCR plants that have been advanced are such items as tax-free solar bonds and reduced depreciation periods.

SUMMARY

In summary, a 100 MWe solar thermal central receiver plant with 18 hours of storage has been shown to be a cost-effective alternative for baseload applications in a utility environment. Large amounts of storage are cost effective due to the use of molten salt, with its high specific heat and low cost, as both the heat transport fluid and storage medium. With 18 hours of storage, capacity factors as high as 75% can be predicted before scheduled outages, yielding levelized busbar energy costs just over 7 cents/kW-hr in 1980 dollars using heliostat costs of $\$97/m^2$. It is projected that these heliostat costs will be achieved with mass-production techniques in the 1987 - 1990 time frame.

Also, due to the capital intensive nature of STCR plants, inclusion of additional investment tax credits, such as the 15% Energy Tax Credit available to industry under the 1978 National Energy Act, can significantly enhance the economic viability of STCR technology.

REFERENCES

1. "Solar Central Receiver Hybrid Power System," Final Report, Vol I-III, DOE-ET-21038-1, Martin Marietta Aerospace, Denver, Colorado, September 1979.

2. "Saguaro Power Plant Solar Repowering Project," Final Report, Vol I-III, DOE/SF-10739-2, Arizona Public Service, Phoenix, Arizona, July 1980.

3. "Conceptual Design of Advanced Central Receiver Power System, Phase I," Final Report, DOE Contract No. EG-77-C-03-1724, Martin Marietta Aerospace, Denver, Colorado, September 1978.

4. "Internally Insulated Thermal Storage System Development Program," Final Report, Sandia Contract No. 83-3638, Martin Marietta Aerospace, Denver, Colorado, December 1979.

5. "A Methodology for Solar-Thermal Power Plant Evaluation," EPRI ER-869, Project 648-1, Westinghouse Electric Corporation, East Pittsburgh, Pennsylvania, August 1978.

6. Doane, J. W., et al; "The Cost of Energy From Utility-Owned Solar Electric Systems," Jet Propulsion Laboratory, Pasadena, California, July 1976.

SOLAR REPOWERING AN 82 MW REHEAT STEAM TURBINE WITH WATER/STEAM CENTRAL RECEIVER TECHNOLOGY

JAMES E. BROWN
El Paso Electric Company
El Paso, Texas

REINER W. KUHR
Stone & Webster Engineering Corporation
Boston, Massachusetts

W. GERALD PARKER
Westinghouse Electric Corporation
Pittsburgh, Pennsylvania

ABSTRACT

The development of solar thermal power system technology for utility applications is an important outgrowth of the desire to reduce usage of conventional oil and natural gas fuels in the generation of electrical energy. Solar repowering utilizing the central receiver technology has been identified as the most promising near-term application of this technology. The concept consists of locating solar hardware adjacent and connected to existing units to displace a portion or all of the fossil fuel normally used during daylight hours. The ability to operate on fossil fuel is retained, thus providing full backup capability and maximum operational flexibility during periods of inclement weather as well as for economic dispatch if required to meet load demand.

A recent assessment performed for the Department of Energy has established the technical feasibility, utility system impact, and economic attractiveness of demonstrating the solar repowering concept. El Paso Electric Company's Newman Unit 1 represents the first generation of utility repowering applications, characterized by an excellent solar resource, availability of unencumbered, flat land allowing location of the receivers adjacent to the turbine building, a substantial remaining unit lifetime, a reheat steam turbine and gas-fired boiler capable of withstanding solar transients without thermal storage, and the absence of major environmental or institutional constraints. These conditions allow for a simplicity of design that minimizes technical risk and maximizes the likelihood of a successful demonstration project that will develop utility industry confidence in this emerging technology.

The repowering concept developed by El Paso Electric Company for solar repowering Newman Unit 1 utilizes water/steam central receiver technology to provide main steam to the high pressure stage and reheat steam to the intermediate stage of the turbine-generator. Water/steam technology was selected for this "first-of-a-kind" demonstration on the basis of minimizing technical risk by using a technology that builds on conventional boiler experience, conventional control philosophy and familiarity to utility operators--all prerequisites to initial utility acceptance of solar repowering as a viable energy option.

The solar subsystem, sized to provide 41 MWe, employs an integrated 160° north heliostat field redirecting sunlight onto the main and reheat receivers located atop a 500-foot tower. A solar repowering fraction of 50 percent was selected for this demonstration unit as the minimum size considered acceptable to adequately demonstrate the engineering, operating, and maintenance aspects of solar repowering. The solar-repowered unit will displace the equivalent of 133,000 barrels of oil per year, with an evaluated cost/value ratio of 1.5 to 2.3 for fuel oil escalation rates of 12 and 8 percent, respectively, for this "first-of-a-kind" demonstration unit. Based on mass-produced heliostats, a commercial unit is expected to have a cost/value ratio of less than 1.0.

The development plan for this project calls for operation of the repowered unit in late 1985, assuming that a typical utility-oriented design and construction program is initiated by mid-1981. The capital cost of this demonstration unit is estimated at $165 million (1985 dollars).

SOLAR REPOWERING CONCEPT

The Power Plant and Industrial Fuel Use Act of 1978 restricts the use of oil or natural gas in new electric utility units or in new industrial boilers with a fuel heat input rate of 100 million BTU's per hour or greater, unless exemption is granted by the Department of Energy. The Act similarly encourages, and in some cases requires, that existing oil and natural gas facilities be converted to coal or to an alternate fuel such as solar energy. In complying with the intent of this Act, utilities are evaluating alternative repowering options for their existing gas and oil-fired units. One option being considered is the use of solar energy to displace a portion or all of the gas and/or oil that would normally be used for electric power generation; this option is called "solar repowering."

Solar repowering, as illustrated in Figure 1, consists of modifying existing units to operate on solar energy when available. The ability to operate on fossil fuel is retained, thus providing full backup capability and maximum operational flexibility during periods of inclement weather. The potential for conventional electric power generation at night is retained, thus eliminating the need for costly, multi-hour energy storage systems.

The energy from the sun may be used in any one of several technically feasible concepts for repowering either steam, combustion turbine, or combined cycle electric power generating units. For the steam electric generating units, the solar energy can be used to produce either steam to drive the turbine-generator or to provide feedwater heating. The solar repowering concept in the context of this paper will primarily be limited to providing steam to the turbine-generator of a steam Rankine unit.

This solar repowering concept utilizes the central receiver technology and consists of the addition of a solar collector field, a central receiver (boiler), and potentially a thermal energy buffer storage subsystem to existing generation facilities; the integration of the solar hardware with the existing systems; and the appropriate refurbishments/modifications to the existing generating unit which primarily involves the turbine-generator and the instrumentation and control system. The principal solar/fossil interfaces between the existing unit and the solar subsystem consist of steam and feedwater supply interfaces from the solar and fossil steam generators; control interface between the fossil and solar subsystems; and power supply interface to the heliostat field, receivers, valves and pumps.

Steam generated by the solar subsystem will be mixed with any steam provided by the existing fossil steam generator prior to admission to the turbine. Attemperation of the solar-generated steam will ensure that the temperatures are maintained within turbine design limits. Solar-generated steam will be used for most of the flow, with fossil steam generation to replace any steam flow reduction due to intermittent cloud cover and for economic dispatch. A thermal energy buffer storage subsystem may be included in the solar generated steam flow path if required to protect the turbine-generator during intermittent cloudy day operation or if shown to be economically more attractive than using the existing fossil boiler to offset insolation transients.

BACKGROUND

The development of solar thermal power system technology for utility applications in the United States is primarily being undertaken by the Department of Energy (DOE) and the Electric Power Research Institute (EPRI). The primary objective of programs sponsored by DOE is to provide a sound technological and industrial base which will result in rapid commercialization of solar thermal technologies and thus contribute significantly to conserving our dwindling fossil fuel supplies.

The earliest practical demonstration of solar thermal power in a utility application will be Solar One, the 10 MWe Pilot Plant currently under construction near Barstow, California. Solar One will be operated as part of the Southern California Edison system and is scheduled for operation in 1981. This plant will provide a technical demonstration of the conversion of solar energy to electric power using central receiver hardware and is a necessary step toward commercialization.

The next logical step toward commercialization is to consider solar repowering of existing gas and oil-fired electric power generating units to develop the necessary design, installation, operation and maintenance experience and economic confidence, necessary to justify utility investment in solar hardware. A number of system design studies and market evaluations of solar repowering utility-sized generating units have been and are currently being performed. In the mid-1970's Foster-Wheeler Corporation (1), under contract to the National Science Foundation, studied a large matrix of potential solar "hybrid" systems in which solar energy could be employed in various portions of a steam plant such as for feedwater heating and for combined evaporation/superheating; their conclusion was that using solar energy for evaporation/superheating was the most desirable approach. Public Service Company of New Mexico (PNM) utilized the results of this study to initiate, in 1974, an inhouse evaluation of repowering their existing gas and oil-fired units with solar energy. Encouraged by the results of their preliminary systems analysis and subsequent market assessments, PNM, in 1977-1978,

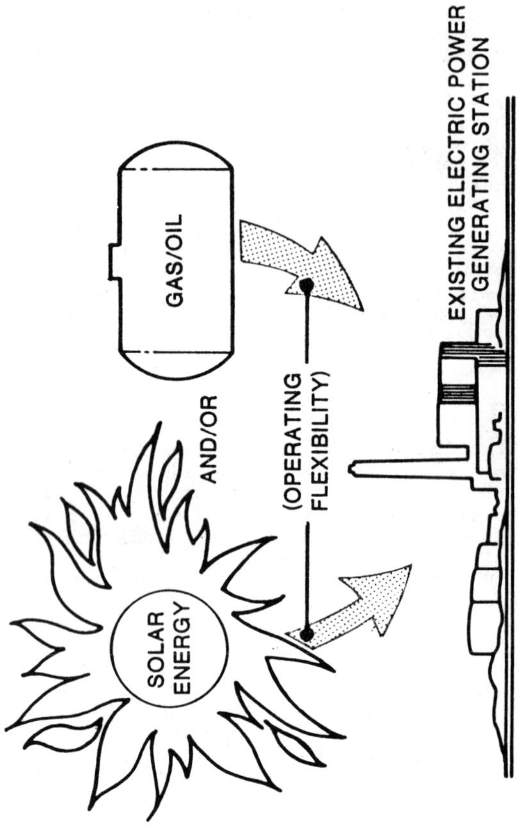

Figure 1. Solar Repowering Concept

performed a technical and economic assessment (2) of solar repowering under sponsorship from DOE, EPRI, Western Energy Supply and Transmission Associates, and several utilities (Arizona Power Authority, Public Service Company of Oklahoma, San Diego Gas & Electric Company, and Tucumcari Power Plant). PNM concluded that a substantial market exists in the Southwest for solar repowering; the concept may be economically competitive with other equivalent generation alternatives, assuming DOE central receiver capital cost goals are met and substantial fossil fuel escalation continues over the next 30 years; and the technical aspects of solar repowering appear to be relatively straightforward since reasonable engineering approaches were identified as possible solutions for major design problems.

In 1978, the MITRE Corporation initiated (3) studies under contract to DOE of the solar repowering market for both utility and industrial process applications and concluded that solar repowering can potentially save the equivalent of one-half million barrels of oil per day. Likewise, in 1978 and 1979, the Solar Energy Research Institute performed a Repowering Strategy Analysis (4) for DOE which concluded that technical uncertainties are the dominant impediments to commercialization of solar central receiver systems and that solar repowering should be undertaken by DOE to develop a market for central receiver technology.

The studies gave sufficient promise for the solar repowering concept marketability that in 1979 DOE awarded multiple system design contracts (5) for utility and industrial process applications. The Newman Unit 1 Solar Repowering Program was funded by DOE for the period of September 30, 1979, to July 15, 1980. The principal objective was to develop a conceptual design and cost estimate for solar repowering Newman Unit 1 that had the potential for construction and operation by 1985, made use of available solar thermal technology, and provided the best economics for this application. El Paso Electric Company (EPE) as the prime contractor was supported by Stone & Webster Engineering Corporation, Westinghouse Electric Corporation, the Texas Energy and Natural Resources Advisory Council, the Regional Development Division of the Office of the Governor of Texas, and a Utility Advisory Council consisting of thirty-two members representing investor-owned, municipal, state, federal, district, and rural electric cooperative utility systems. EPE, through its study, demonstrated the technical feasibility and economic attractiveness of solar repowering reheat steam turbines using conventional water/steam technologies familiar to the utility industry in general and to plant operators of existing units specifically. The EPE design approach for a water/steam central receiver system is described in this paper.

By the end of 1981, DOE has proposed to select an approach to demonstrate solar repowering. This approach may involve two utilities and two industrial process applications. These demonstration units would be in operation by 1985 and form the foundation for the future commercialization of solar central receiver systems.

DEMONSTRATION UNIT SELECTION

The El Paso region is in the zone of highest solar insolation in the nation facilitating year-round research, development, and demonstration of solar energy applications. The annual variation of solar insolation in the El Paso region is also the lowest in the nation. The EPE system, which has a total generating capacity of 1033 MWe, has three local electric generating stations in this region. There is sufficient land available to apply solar repowering to all local EPE gas and oil-fired units, representing 922 MWe or 89 percent of the total system.

EPE selected Newman Unit 1 for the solar repowering demonstration program for the following reasons: (1) widespread market potential exists for solar repowering of reheat steam turbines similar to Newman Unit 1; (2) more than 14 km^2 (3,500 acres) of unencumbered, flat land is available adjacent to the Newman Station; (3) the remaining economic life of Newman Unit 1 favors dispatch of the solar-repowered unit relative to the balance of the EPE system; (4) no apparent major institutional or environmental constraints exist; and (5) the operating history of the Newman Unit 1 turbine-generator has demonstrated the capability to sustain cyclic operating conditions that would result from solar application.

Newman Unit 1 has an 82 MWe (net) tandem-compound, double-flow, reheat steam turbine. It was built in 1960 for baseload duty using natural gas as the primary fuel (oil as the alternative fuel source). The unit is currently operated as an intermediate load unit; the 1979 capacity factor was 46 percent. The Allis-Chalmers turbine-generator utilizes 10.1 MPa/538°C (1,450 psi/1,000°F) main steam and 3.0 MPa/538°C (425 psi/1000°F) reheat steam to the intermediate stage. The Babcock & Wilcox natural convection boiler is rated at 254,240 kg/hr (560,000 lb/hr) and has a pressurized water-cooled radiant furnace, a two-stage drainable type superheater, and a drainable reheater.

NEWMAN SITE DESCRIPTION

Newman Station is located in a rural area at the north end of the city of El Paso, 24 km (15 miles) northeast of the downtown area and 19 km (12 miles) from the El Paso Solmet weather station. There are no commercial buildings and only one residence within a 3 km radius. Annual mean weather data show an average temperature of 17.4°C (64.4°F), average precipitation of 19.8 cm (7.8 inches), average sunshine of 3,583 hours (83 percent of possible sunshine), and direct normal insolation for the typical meteorological year of 7.26 kW-hr/m^2 day. Average wind speed is 4.24 m/sec (9.5 mph) from the north and mean sky cover (tenths) is 3.8, sunrise to sunset.

Solar repowering will have a beneficial impact on air quality since it will displace the use of fossil fuels and reduce the resultant pollutant emissions. The air quality monitoring unit nearest Newman is in downtown El Paso. Although El Paso air quality is in violation of ambient air quality standards for several pollutants, air quality at Newman Station is in compliance. There is no surface water at the site; however, water is plentiful from nearby wells.

The site is accessible by road from all directions, and a freeway is being completed with a major interchange planned 6.4 km (4 miles) from the generating plant. A railway siding is located 9.6 km (6 miles) to the southeast. Newman Station is near, but not directly beneath, two Federal airways.

PREFERRED SYSTEM DESCRIPTION

In general, the primary design considerations for solar repowering existing utility units are:

- Unit Selection: Balance-of-System Economics/Operation
- Land: Cost/Availability/Constraints

- Solar Repowering Fraction: Size of Collector Field
- Solar/Fossil Interface: Intermittent Cloudy Day Operation
- Thermal Energy Buffer Storage Capability: Cost versus Benefit
- Solar Reheat Capability: Cost versus Benefit
- Solar Technology Selection: Heliostats/Receivers/Thermal Energy Storage
- Existing Unit Considerations: Facilities, Boiler, Turbine-Generator and Control

The first three design considerations (Unit Selection, Land and Solar Repowering Fraction) primarily influence the resulting economics of operating and dispatching the repowered unit while the remainder of the considerations influence the repowered unit performance throughout the year under varying weather conditions. For example, if the unit selected for repowering has a low fossil fuel incremental operating cost relative to the balance of the system, then the repowered unit will not only operate when solar energy is available but will also be economically dispatched on fossil energy. In addition, the variation in the repowered unit operating economics with respect to the solar repowering fraction (size of collector field) is less than for a unit with a high fossil fuel incremental operating cost. Likewise, if land availability dictates a solar repowering fraction of less than fifty percent, then for most units the resulting operating economics are not favorable to solar repowering.

El Paso Electric Company addressed these design considerations in selecting Newman Unit 1 and in establishing the Preferred Configuration for demonstrating the solar repowering concept.

The Preferred Configuration for solar repowering Newman Unit 1 is illustrated in Figure 2. This design utilizes water/steam central receiver technology to provide main steam to the high pressure stage, 10.1 MPa/537°C (1450 psi/1000°F), and reheat steam to the intermediate stage 2.97 MPa/538°C (425 psi/1000°F) of the turbine-generator. Fossil energy is used to supplement solar-generated steam for intermittent cloudy day operation and for economic dispatch when solar energy is not available.

Table 1 summarizes the characteristics of the Preferred Configuration. EPE selected a solar repowering fraction of 50 percent for this demonstration unit as the minimum size considered acceptable to adequately demonstrate the engineering, operating, and maintenance aspects of solar repowering. There is little economic incentive for considering higher repowering fractions for a demonstration unit.

Figure 3 is a simplified flow schematic of the concept. Steam generated by the solar subsystem is mixed with the steam provided by the existing fossil steam generator prior to admission to the high pressure and intermediate stages of the turbine. Attemperation of the solar-generated steam ensures that the temperatures are maintained within turbine design limits. Solar-generated steam is used for most of the flow, with fossil steam generation to replace any steam flow reduction due to intermittent cloud cover and for economic dispatch when solar energy is nonavailable.

The feedwater supplied to each steam generator matches the steam flow and pressure requirements of each unit by means of a coordinated

TABLE 1. SOLAR-REPOWERED NEWMAN UNIT 1

•Unit Type	Reheat Steam Turbine
•Unit Rating	82.3 MWe
•Solar Repowering Percentage*	50 Percent
•Plant Operating Scenario	Maximize Solar Benefit Fossil Operation Full-Time and only on Cloudy Days Economic Dispatch Fossil Energy
•Collector Subsystem	
-Field Configuration	North Field (160° ARC)
-Field Area	370 Acres
-Heliostat Area	226,938m^2
-Number of Heliostats	3981 (57m^2/heliostat)
•Primary Receiver	
-Type	External (Pumped, Recirculation Boiler/Screened Tube Concept)
-Size	12.6m Dia x 15.7m Long (240° ARC)
-Outlet Temperature	1020°F
•Reheat Receiver	
-Type	External
-Size	12.6m Dia x 15.7m Long (210° ARC)
-Outlet Temperature	1020°F
•Tower Height	
-Number of Towers	1
-Primary Receiver C/L	155m
-Reheat Receiver C/L	139m
•Electric Power Generation Subsystem	
-Cycle	Steam Ranking (Reheat)
-Net Unit Efficiency	40 Percent
-Turbine Inlet	1450 psi/1000°F/1000°F
-Heat Rejection	Wet Cooling Tower
•Fossil Boiler	
-Type	Gas/Oil
-Rate Load Efficiency	84.4 Percent
-Hot Standby (%)	28
-Startup Energy	100 MBTU
-Warm Standby	15 MBTU/Startup

*Based on an Insolation Level of 950 Watts/m^2

Figure 2. Solar Repowered Newman Unit 1

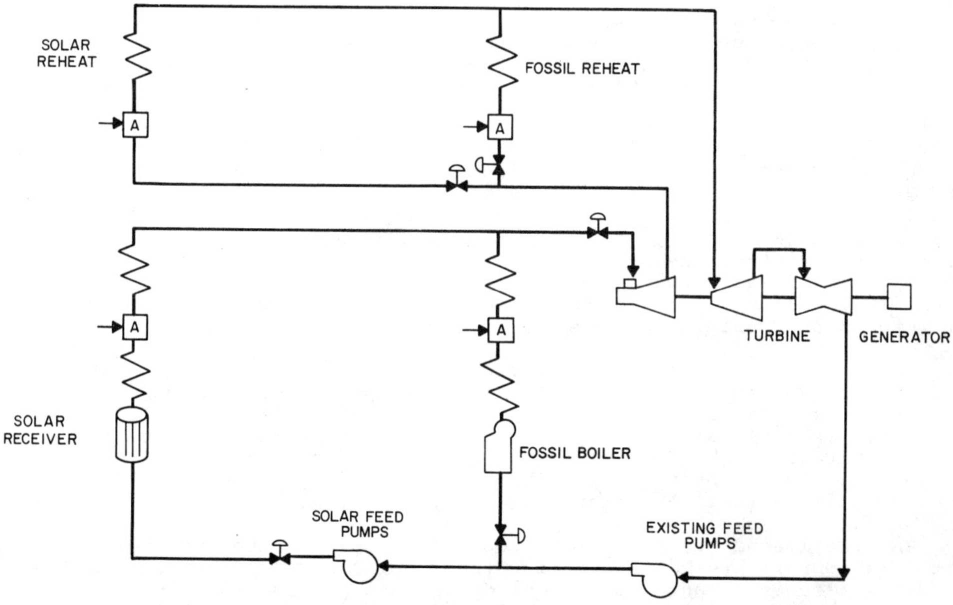

Figure 3. Simplified Flow Schematic

control system. The control system of the existing unit is modified and interfaced with the solar system by means of a master-control system.

The solar subsystem is sized to provide 41 MWe (50 percent repowering) at noon summer solstice based on a direct insolation level of 950 watts/m^2. The heliostat field is located north of the unit. The receiver tower is as close as possible to the turbine building to minimize feedwater and steam piping distances, thus reducing thermal losses, pressure drops, and costs. Existing transmission and natural gas pipeline rights-of-way transect this field location. Transmission lines will be relocated, and pipeline rights-of-way will be maintained as exclusion areas.

The collector subsystem consists of a 160-degree array of heliostats. The heliostats employed in the collector field are the Second Generation Heliostats being developed by the DOE which were selected as representative of the class of configurations that will be available for 1985 solar repowering applications.

The receiver subsystem provides a means of transferring the incident radiant flux energy from the collector subsystem into superheated steam. The receiver subsystem consists of primary and reheat receivers to intercept the radiant flux reflected from the collector subsystem, a single tower structure to support the two receivers, and associated feedwater and steam piping. The external central receiver concepts (primary and reheat) are based on the water/steam pumped recirculation central receiver technology being developed by DOE. The receiver subsystem also incorporates the pumps, valves, and control system within the tower structure necessary to regulate flow, temperature, and pressure, and includes the required control system components necessary for safe and efficient operation, startup, shutdown, and standby.

The control subsystem is used to sense, detect, monitor, and control all system and subsystem parameters necessary to ensure safe and proper operation of the entire integrated repowered plant. The control subsystem consists of computers, peripheral equipment, control and display consoles, control interfaces, and software.

The fossil boiler subsystem provides a fossil energy source that is used to enhance performance and/or maintain normal plant operation during periods of reduced or no insolation. The fossil boiler subsystem consists of the existing Newman Unit 1 fuel storage, fuel handling, boiler, and related equipment. It also consists of any additional fuel supply, fuel storage and transfer facilities; and the required controls necessary for safe and efficient operation, startup, shutdown, and standby of the fossil boiler subsystem. Essentially all the existing Newman Unit 1 remains intact after being repowered with a solar steam supply system.

The electrical power generating subsystem (EPGS) provides the means for converting to electrical power the thermal output from the receiver and fossil energy subsystems. The output from the EPGS is regulated for integration into the El Paso Electric Company transmission network. The EPGS consists of the existing balance-of-plant equipment at Newman Unit 1, and the piping and related equipment required to interface the solar steam supply system.

The estimated construction cost for solar repowered Newman Unit 1 is approximately $165 million (1985 dollars). This capital estimate assumes plant operation by the end of 1985, and includes direct costs, indirects, distributables, escalation, contingency, allowance for funds used during construction, and owner costs.

Solar-only operating and maintenance costs for the solar-repowered Newman Unit 1 are estimated to be approximately $1.7 million per year in 1985 dollars, or about 1 percent of the total capital cost.

ENVIRONMENTAL CONSIDERATIONS

The major environmental impacts associated with solar repowering Newman Unit 1 are the potential impacts on air and water quality and on terrestrial resources, human resources, and the ecology. Preliminary assessments have been made of these considerations using available information for the Preferred Configuration discussed in the preceeding section. It appears, at present, that there will be no major environmental impacts resulting from construction or operation of the solar-repowered Newman Unit 1.

The major environmental considerations can be summarized as follows:

Air Quality - Operation of the solar-powered unit will result in a net reduction in air emissions associated with not burning 450-550 x 10^{12} J/year (450-500 x 10^9 BTU/year) of natural gas or oil at the Newman Station and will thus have a positive effect on local air quality.

Hydrology - Additional consumptive water use will consist only of domestic use for station personnel and for heliostat cleaning. Surface water flows through the heliostat field area will be rerouted. This will not adversely affect local hydrology or other local water users.

Water Quality - No new liquid discharges are anticipated from the solar-repowered facility.

Vegetation - Vegetation will be cleared from approximately 1.5 km^2 (370 acres) at the heliostat field site; however, the species present are not unique to the region and do not represent critical habitat.

Endangered Species - Based on available information, no endangered or threatened species of plants or animals are known to occur on the site; some endangered birds may pass through the region during seasonal migration.

Land Use - Land is available for construction of the heliostat field; future land use plans do not conflict with the proposed project.

Socioeconomics - It is anticipated that the necessary craftsmen will be available locally and will not strain existing services. Positive benefits will include added wages and salaries, tax revenues, and decreased unemployment (within a Surplus Labor Area). Local traffic congestion may occur during construction and would be the subject of further study.

Archaeology - Numerous small archaeological sites are indicated in the area proposed for the collector field. Although some survey work has been completed, the significance of the sites is not known (though expected to be minor) at this time and will require a subsequent field study.

Aesthetics - The collector field will be visible for several miles in this undeveloped industrial area but should not represent a major visual impact. Concerns related to possible ground glare have been reviewed and are considered minimal. The primary receiver centerline height is 155 m (509 feet) and will be visible over the flat terrain for about 8 km (5 miles).

SAFETY CONSIDERATIONS

The primary safety consideration which must be addressed in the design and operation of solar-repowered Newman Unit 1 is the potential solar reflectance hazard imposed by a large field of heliostats. Several hazardous conditions potentially result from the effects of solar reflectance from individual or multiple heliostats in the collector subsystem. These potential hazards include, for example, an eye hazard for those personnel looking at, and who happen to be near the focal point of, several heliostats, a skin hazard (burns) from concentrated solar reflectance near the focal point, a glare hazard to airline pilots and passengers, and a distractive glare, nuisance glare, misdirected light, or spurious reflections hazard affecting nearby residents and highway travelers. A preliminary review of these safety considerations did not identify any hazards that would preclude the safe construction and operation of solar-repowered Newman Unit 1.

Most of these reflectance hazards are of concern primarily to the construction, testing, operating and maintenance personnel; and to visitors, authorized or unauthorized, of the solar-repowered facility. Techniques which can be used to eliminate, reduce the frequency of or mitigate the severity of, some of these potential hazards include: the use of fencing to enclose the collector field; requiring eye protection, protective clothing and/or gloves when working near the heliostat collector field or the reciever at the top of the tower; proper instruction of personnel on the methods to avoid these hazards; proper design of the controls for the collector subsystem (particularly for quick and safe emergency shutdown conditions); proper storage of all combustible materials in places inaccessible to misdirected radiation; and the use of safety and warning devices or signs.

ECONOMIC CONSIDERATIONS

The integration of solar-repowered units into electric utility systems raises a number of questions as the value of the repowered units, problems they might introduce, and requirements that should be placed upon them. In addition to technical feasibility, economic and reliability impacts are a major concern when evaluating the attractiveness of solar repowering Newman Unit 1. These include the cost of repowering, the quantity of fossil fuels displaced, a capacity credit for unit life extension, and the reliability of the solar-repowered unit.

A cost/value analysis was performed to evaluate the solar repowering of Newman Unit 1 on the EPE system. The analysis was performed utilizing the methodology developed by Westinghouse as part of EPRI Contract RP 648-1 entitled "Requirements Definition and Impact Analysis of Solar Thermal Power Plants."

The reference unit used for performing the economic analysis is based on the Preferred Configuration. The capital cost for this "first-of-a-kind" demonstration unit is estimated at $165 million (1985 dollars) with total anticipated operating and maintenance costs (solar and fossil) for the first year of 3.3 million dollars.

The repowered unit was modeled for a 1985 to 2014 dispatch on the EPE system. The fossil boiler at Newman Unit 1 will operate using either natural gas or fuel oil. EPE currently has gas supply contracts extending into the 1990's. Between 1985 and 1990, the Newman Unit 1 boiler is projected to burn natural gas. It is assumed that after 1989 the unit will burn fuel oil.

The operating scenario for the fossil boiler is important in assessing the economic benefit of solar repowering. Since the solar-repowered Newman Unit 1 will be a "first-of-a-kind" demonstration unit, a conservative operating strategy for the fossil boiler has been selected to permit the development of operator confidence and experience with the solar subsystem without jeopardizing the integrity of the existing equipment or the ability of the unit to produce power. The operating strategy consists of:

Solar operation initiated August, 1985

8/85 to 12/86, the fossil boiler produces 41 MWe minimum when the unit is operating on solar; the unit is also economically dispatched on fossil fuel.

1/87 to 12/87, the fossil boiler produces 23 MWe minimum when the unit is operating on solar; the unit is also economically dispatched on fossil fuel.

Beyond 1987, the fossil boiler operates only when required to offset solar insolation transients on cloudy days or when economical to dispatch on fossil fuel.

The solar-repowered unit will result in a total lifetime energy displaced of approximately 1.69×10^4 MJ (16×10^{12} BTU's) of gas/oil and 0.84×10^4 MJ (8×10^{12} BTU's) of coal over its 30-year life.

The results of the cost/value analysis indicate a ratio of 2.27 for the demonstration unit. Similar analyses performed for commercial units indicate cost/value ratios of 0.8 or less. The higher cost/value ratio for the demonstration unit is due to higher costs for solar components (such as heliostats) and restricted plant operation in the early years (due to testing and establishment of operating confidence/experience).

CONCLUSIONS

The following conclusions result from this assessment of solar repowering Newman Unit 1:

1. Solar repowering of reheat steam turbines utilizing water/steam central receiver technology is technically feasible, economically attractive and does not require development thermal energy storage subsytems.

2. Solar-repowered demonstration unit can be constructed on the EPE system in 55 months using water/steam technologies familiar to the utility industry.

3. No major environmental problem or safety hazard was identified that would preclude solar repowering Newman Unit 1.

4. Solar repowering concept is potentially economically competitive with conventional means of electric power generation if projected capital costs for mass-produced solar hardware can be realized.

REFERENCES

1. "Studies of the Direct Input of Solar Energy to a Fossil-Fueled Central Station Steam Power Plant," R. J. Zoschak and S. F. Wu, Research Division, Foster Wheeler Corporation, Livingston, New Jersey, 1975.

2. SAN/1068-4-1, "Technical and Economic Assessment of Solar Hybrid Repowering," prepared by Public Service Company of New Mexico as part of Contract No. EG-77-C-03-1068, September 1978.

3. Solar Repowering/Industrial Retrofit Program Element Plan-Review Draft, prepared by San Francisco Operations Office for Solar Thermal Branch, Department of Energy Headquarters, January 1980.

4. A Government Role in Solar Thermal Repowering: Draft Report of the Repowering Strategy Analysis, prepared by Solar Energy Research Institute, December 1979.

5. Newman Unit 1 Solar Repowering Program, Contract No. DE-AC03-79SF10740, El Paso Electric Company, September 30, 1979.

HOW TO GET INTO SOLAR... AND MAKE MONEY

Warren J. Mackensen, P.E.
Sippican Solar Systems, Inc.
Marion, Massachusetts

ABSTRACT

The startup and day-to-day operation of a solar business is discussed from an experiential point of view. The paper discusses the problems and solutions surrounding the startup and operation of Sippican Solar Systems, Inc., a corporation founded in 1976 devoted exclusively to solar contracting. The firm is located in Marion, Massachusetts, and performs contracting services throughout New England.

WHY START A SOLAR BUSINESS?

Sippican Solar Systems was started primarily for personal reasons, for the same reason that many small corporations are started. The owner, Warren J. Mackensen, started his business because of a desire to be employed independently. Previous employers (the Navy, and a consulting engineering firm) kept Mr. Mackensen away from home more than he liked.

The sequence of events started when his oil fired water heater was found leaking on a Saturday morning in October 1975. Having read about solar energy, a replacement water tank was ordered with an internal heat exchanger. Having the solar tank in his basement now, Mr. Mackensen went about surveying the industry for collectors to complement the water tank. The industry was quite young in 1975, and two collectors were chosen in kit form to be assembled in his basement. After many hours, the collectors were put on the roof and by gosh, they worked! The idea of a solar business was born.

PHYSICAL PLANT

For two and a half years, the author's house, basement, and garage were used for business purposes. Today, Sippican owns a 5,200-square-foot warehouse/showroom/office building on U.S. Route 6 in Marion, Massachusetts. The building is 130 feet long by 40 feet wide and is divided into thirds lengthwise.

The one-third with frontage on Route 6 is the office and showroom area (2,000 sq. ft.). The offices are heated and cooled with a Wescorp Model EC-44 water-source heat pump utilizing a 215-foot-deep artesian well as a source of 50° geothermal energy. The water is discharged to an adjacent brook. At 9.5¢ per kilowatt-hour, last season's annual heating and cooling bill was $800. No night setback is used. The heat pump COP, being a water-source unit, has been measured at 4:1 utilizing permanently installed instrumentation.

The middle third of the plant is a truck lane, extending through the full length of the building, with garage doors at either end. All four Sippican trucks are parked indoors at night.

The last third of the plant adjacent to the truck lane, is the warehouse for collectors, tanks, and stock.

Most of the inventory is assigned part numbers for internal control and auditing purposes. Items over $30 have control tags on them, such as collectors, tanks, pumps, controls, heat exchangers, air handlers, and so forth. These control tags are turned in to the office for cost accounting purposes when the material is loaded on a truck for a customer.

An adequate physical plant has been instrumental in the organization of the business, even though it has increased the overhead.

BRAND AFFILIATION

Solar manufacturers are very interested in helping new dealers get started. A new solar contractor should immediately ally himself with a particular brand of solar collector. The author feels that help from a manufacturer was a key element in the success of his business, as the solar learning curve was greatly accelerated with the additional experience of a manufacturer.

PRODUCT MIX

A new entrepreneur should determine the distinctive competence of his or her firm. What can the firm do best in comparison with other competitors?

For one thing, a new firm should stay out of tangential and diversionary product lines. The author deliberately stayed out of wood stoves, thermal curtains, and other energy saving devices which were tangential to the solar business. A large percentage of an entrepreneur's time can be spent on product lines that do not bring in the profits commensurate with the time expended.

If the decision is made to start an energy store, then certainly such products as wood stoves and thermal curtains are most appropriate. One solar contractor in northern Vermont has found that his energy store complements the slack business in the winter period. The author has found in his own contracting business, however, that there is no slack period through the winter months.

No attempt has been made to concentrate on one market over another. Sippican accepts every domestic hot water contract that comes along. Price cutting or "specials" are not used to close DHW sales. In 1980, 14% of Sippican's work was DHW packages. Also in 1980, 7% of Sippican's work was residential space heating systems. Sippican has pursued commercial installations and government grant work (DOE Program Opportunity Notices, HUD Demonstration Grants, Schools and Hospitals Program, and the Federal Buildings Program). A major portion (69% in 1980) of Sippican's work was government related. Sippican actively bids every project in the six New England states.

Sippican accepts service and repair work on solar systems installed by others. One system repaired by Sippican was located on the top of a ten-story high-rise housing for elderly in Battle Creek, Michigan. Sippican has learned valuable lessons on how not to install a solar system by working on systems poorly installed by others. Sippican's philosophy is, "When no one else can, Sippican."

SALES STRATEGY

Well-established sales procedures call for the immediate response to a sales lead. Although some initial screening may be achieved over the phone with a solar system inquiry, the salesman should arrange an appointment with the inquiring customer as soon as possible.

Home and energy shows are becoming better and better sources of serious leads. Although shows can be costly and demanding, a contractor's presence in the market place is important for future sales. An inexpensive company brochure should be readily passed out at these shows. The author has experienced that future buyers of solar systems seem to be organized within their homes and maintain a file entitled "solar." They place the pamphlet in the "solar" file and call as much as two years later. This has happened often to Sippican.

ADVERTISING AND PROMOTION

Sippican has been careful in the expenditure of advertising dollars. Many solar contractors, Sippican included, have found that mass media advertising (radio and newspaper) does not produce the expected results. Marketing textbooks tell one that mass advertising should be used for mass merchandising. Solar is not yet a mass consumed product. It is a "planned purchase."

Sippican Solar has found yellow pages advertising to be the most important. When a consumer decides to purchase a solar system, most of the time the consumer is not brand conscious or company conscious. He is looking for a product. A reasonable listing in the yellow pages will generate the call for a sale. No newspaper or radio advertising is contemplated for 1981.

The successful solar contractor must listen to the customer's needs. A solar contractor should be prepared to complete the plumbing and electrical work and take out the required permits, including a building permit, under the solar contract. The homeowner does not want to go to city hall to get permits. The solar contract should be all encompassing, even if the solar contractor must subcontract the electrical and plumbing work. The need of the consumer is for a complete solar system.

With regard to new construction and solar space heating systems, the need of the customer is for a complete heating system. Sippican Solar has found that one of the keys to a successful solar space heating sale is the willingness to install the complete heating system. Generally, Sippican subcontracts the conventional heating work, but exercises complete control and responsibility for the conventional heating system. It must be very clear to the owner (the customer) that the solar contractor has full responsibility. The solar contractor in turn must demonstrate this responsibility very clearly. For example, when the customer calls with a "no heat" complaint, the solar contractor must never tell his customer to call the conventional heating contractor.

WARRANTIES

Sippican provides a full one-year warranty with a limited five-year warranty carried by the manufacturers of the collectors and the tanks. System failures have been virtually nil beyond the one-year warranty period. Where failures have occurred outside the warranty period that were related to an installation defect during original installation, Sippican has covered the repairs under a "goodwill" warranty (unwritten) that generates much praise for the integrity of the company.

PRICING

Sippican has found that some competitors in its geographical area (mostly new entrants into the solar marketplace) underprice their services, causing eventual demise. Sippican has always priced its services based on the hard numbers of a profit and loss statement, not on gut feeling. Prices have been raised when necessary in spite of competition. The company stands solidly today because tough pricing decisions were made in a timely fashion.

RESEARCH AND DEVELOPMENT

Although Sippican has conducted some installation-related research work and some systems development, by and large R & D is tangential to a solar contracting business.

In 1977, Sippican developed an "optimum" way to install the mechanical components on a water tank. In 1981, the same "optimum" method is still used. Pre-built "appliance modules" from manufacturers are not used.

Uniformity in installation work is important to lessen the possibility of developing a new learning curve. In general, Sippican avoids the new and innovative products recently introduced to the marketplace. Sippican lets other contractors do the field testing that the manufacturer is sometimes unable to finance. Sippican has owned some "better" solar mousetraps that failed. In short, there is no requirement for a solar contractor to buy the newest product in the marketplace; thus, Sippican now avoids doing so to stay profitable.

BUSINESS ORGANIZATION

A key to successful management of Sippican has been the considerable time expended in developing a good organization. Any task or job within the company that is repetitive in nature has been reduced to a "procedure." The following description will provide a brief overview of a few of the procedures in effect at Sippican.

PROJECT ADMINISTRATION

Two folders are generated in the office for each project. One is an "office folder", and one is the "field folder." The office folder contains the original copy of the contract, the solar energy survey form used to sell the job, the building permit information, and other paperwork which becomes a permanent record of the job.

The field folder contains a copy of the sales contract so that the installers in the field have a complete record of the agreements made between the home owner and the contractor. The field folder additionally receives a blank solar system installation startup checkoff list to be used after the system is started up. The field folder also contains a standard parts list if the system is a domestic hot water package. The parts list has been invaluable in pulling stock from the shelves.

TRUCKS AND TOOLS

With four trucks, Sippican found it necessary to color code each tool box and the tools therein. Thus, at the end of the day, tools are readily returned to their proper tool box. In this fashion, tool boxes remain fairly complete with the proper tools.

Tools have been further arranged in each truck according to their use on the job. For example, a "roof kit" contains the sockets and speed wrenches that are commonly used on the roof. Additionally, the roof kit contains chalk lines and other paraphernalia that is used on each roof.

The mechanical and electrical tool boxes always go immediately to the basement.

An "insulating kit" goes to the roof and another one to the attic. These kits contain the tools necessary for insulating pipes.

There are several "sweat kits" on each truck along with their respective six pound propane torches. Each solar installation has three general areas where simultaneous sweating of pipes is required: the roof, the attic, and the basement. By having triplicate sweat kits, work can proceed simultaneously at all three locations.

Checklists on the door of each truck insure that everything is on the truck that is necessary to complete a job.

PERSONNEL

Sippican Solar has derived most of its people, other than the initial startup core, by writing the New England Fuel Institute (NEFI) and appraising the school that a need for a solar technician exists. NEFI runs a four week solar technician hands-on course in Watertown, Massachusetts. Sippican interviews these solar technicians, including asking them technical questions, and then hires accordingly. Of the fifteen installers working at Sippican Solar Systems, eleven are New England Fuel Institute graduates. One solar technician is a female.

The office staff at Sippican comprises one secretary, one bookeeper, one engineer, one marketing director, one salesman, one production scheduler/purchasing agent, and the chief executive officer, for a total of seven staff people (all full-time).

Turnover at Sippican has been virtually non-existent. People come to Sippican because they want to work in the solar industry. They have been "sunstruck" before they arrive.

COMPANY NEWSLETTER

SUNFLARES is published internally as a house organ on a weekly basis. The newsletter provides newsy items to Sippican's 21 employees and their spouses. The newsletter is inserted in everyone's paycheck envelope (so it receives 100% distribution), and it includes the following week's work schedule. People are the backbone of any company, and they deserve to know what is going on.

COST ACCOUNTING

In January 1981, Sippican started purchasing computerized cost accounting services. The information derived from this service will be critical to the future profitable operation of the business.

COMPANY GROWTH

Sippican growth has exceeded the solar industry's growth rate. Sippican's annual sales volume is listed below:

Year	Annual Sales	Increase
1976 (six months)	$ 4,790.	-
1977	66,338.	-
1978	205,608.	310%
1979	354,670.	172%
1980	1,061,953.	299%

The stated growth rate goal in Sippican's company policy manual is 33% per year. The dedicated solar people at Sippican have far exceeded management's goal and without their dedication, this success story could not have been written. The author is thankful to all who have made this paper possible.

BIO-ENERGY: A TAXONOMY OF LAND USE IMPACTS

Virginia Kay Barber Parsons
Los Alamos National Laboratory
Los Alamos New Mexico 87545

ABSTRACT

The energy crisis has prompted research, development and implementation of alternatives to foreign petroleum, among which are the bioconversion technologies. The sun's energy, captured and stored by green plants or passed on through animals, can be converted to gaseous, liquid, or solid fuel through a variety of processes.

With the exception of aquatic biomass, the feedstocks for the processes originate on the farm or in the forest and therefore are land extensive. Although current land use practices in rural areas are amenable to biomass implementation, the realization of an intensive rural energy program based on bioconversion would impact existing uses. Land quality and availability, competition for land with other uses, and the uses or abuses of the land that affect water or air quality will be important issues.

This paper identifies potential land use impacts from each of the four major bio-energy technology classes: anaerobic digestion, fermentation, gasification, and direct combustion. Each technology class is discussed with respect to each of the five major land use categories in the US: production of food and fiber; energy, industry, and commerce; housing and community; transportation and recreation, and open space.

INTRODUCTION

Many people have a misconception of the term "land use." I would like to define it at this time. In the legal sense, land is not only the surface of the earth but everything over and under it(1). It is the natural environment and all its attributes within which all production takes place. Minerals, soil, air, water, and biotic resources are tied together in the general context of land. An effective means of grasping this concept is to consider land as a bundle of sticks tied together with twine. Each stick is a property right which can be owned or managed separately. Land use in the US can be divided into five broad categories. These categories are production of food and fiber; energy industry and commerce; housing and community; transportation; and recreation and open space(2). If deployed to their maximum potential, the technologically available bio-energy conversion processes will have an impact on these existing land-use practices at the local level. In some instances, these bio-energy activities will be compatible with or enhance each category of existing use. In others, there are direct or indirect adverse environmental consequences. This paper identifies impacts to land use at the microlevel for each technology class and with respect to each land use category. Those generic impacts that affect all categories will be discussed first.

GENERIC IMPACTS

Management of a bioconversion facility is key to the impact it may have on the environment. Each of the biomass technologies in any given locale could present an environmental hazard. The feedstocks could be collected and the by-product disposed of in an irresponsible manner. Facilities can be mismanaged to allow for discharge of pollutants. This section addresses the overall threats to air, water, and land quality and availability in such worse-case situations.

Degradation of the environment severely affects a wide range of uses. Agricultural production, industry, urbanization, and recreation are dependent upon available supplies of clean air, water, and land.

During the past decade, there has been a growing concern about excess CO_2 in the atmosphere. Increased CO_2 could create a greenhouse effect, thereby warming the earth's surface and melting polar ice caps. Although research programs have been launched, scientists are still a long way from a consensus on how the earth, atmosphere, ocean, and biological systems disperse CO_2.

The primary concern has been with CO_2 from the burning fossil fuels or a rapid release of CO_2 that has been trapped for millenia under the ground. There is also controversy surrounding the relationship of biomass to CO_2 release. Plants absorb CO_2 for photosynthesis and release it when they decay. Each of the bioconversion technologies emits this CO_2 during the conversion process. Many people believe there will be a net balance in CO_2 with bio-energy, perhaps even a decrease. There is little difference between a tree dying in the forest and releasing CO_2 in a relatively short time period or being burned and releasing it immediately(3). In addition, if much more land is brought into production, there is an increase in demand for CO_2. On the other hand, deforestation can have the opposite effect. CO_2 issues will remain in controversy until further research answers this global question.

An anaerobic digestion system involves collection of manure; storage and preparation, if necessary, processing, conversion or transportation, and energy end use. Residuals from these processes can include leachates of salts, heavy metals from storage, and improper recycling of sludge, sulfur oxides, and carbon dioxide from conversion to electricity. Water from the digester can increase biological oxygen demand (BOD) if disposed of improperly. The added nutrients from the waste water enhance growth of microorganisms that place a demand on the oxygen supply of the existing system. Mismanagement of the effluents from anaerobic digestion could threaten local air quality and water quality and availability.

There is a trade-off, however, between these potential threats to the environment from bioconversion of animal waste and existing practices. Odor, inherent in any large confined-animal operation, will be decreased by digestion, rather than the current practice of open lagoons. Containment and treatment will have positive impact on existing water-quality. Total nitrogen availability will be enhanced by the adoption of this waste-recycling energy resource. The fertilizer, which would otherwise leach into the environment, can be used as a soil amendment.

Fermentation of grain to produce alcohol fuel requires water and a heat source. Coal-fired stills emit sulfur particulates, and CO_2. Irresponsible disposal of large quantities of stillage would increase BOD in surface water. However, solar drying of stillage on concrete slabs or wet feeding to livestock would mitigate this potential pollution problem.

Home wood burners are a potential fire hazard. They also emit carcinogenic and photochemically reactive residues. Because home wood fires are not regulated and are becoming more popular, many areas are currently experiencing local air quality problems(4).

Direct combustion of wood or residues produces particulates and hydrocarbon emissions. Even though these emissions are far less than for fossil fuels, and control technology exists to mitigate impacts, mismanagement could create visibility and air standards attainment problems(5). Residue removal and wood-cutting activities disturb the soil. The initial adverse impact is fugitive dust, which affects visibility. Indirect consequences are increased turbidity and pollution in surface water from soil transport by runoff. Cleaning of grates, boiler water disposal, and leaching of organic acids in ash can also adversely impact water quality.

Gasification of residue or wood creates many of the same collection and processing problems as does direct combustion. In addition, oxides of sulfur, nitrogen, hydrogen sulfide, hydrogen cyanide, and ammonia can be emitted if solid waste is gasified(6). Phenols, trace metals, and leachates of char and ash in the water from clearing and storage piles could be problems if mismanaged.

IMPACTS TO FOOD AND FIBER PRODUCTION

Near-term feedstocks for bio-energy will originate on the farm or in the forest, and energy conversion will most likely occur in close proximity to the feedstock. Production of biomass, collection and transportation, conversion, and energy end-use will have an effect on the existing agricultural system.

Anaerobic digestion provides an excellent opportunity to the farmer for resource recovery. Very little land is necessary for the facility, which fits in easily with existing operations. Protein, on-farm energy, and enriched fertilizer are potential benefits. In addition, there is a chance for increased productivity by the incorporation of an integrated appropriate technology system. A greenhouse could be fed CO_2 from the conversion system, and water from the digestor could be pumped into an existing agricultural operation to augment the existing irrigation system.

Demand for land for grain to produce alcohol fuel will compete with existing uses. A direct impact could be displacement of land currently devoted to food and fiber. For example, on the one hand, if corn is the existing product, the benefit in dried distiller grain would actually improve the local protein market(7). In effect, dried distiller's grain is a more concentrated protein than was the corn; the starch has been removed. Protein is not, however, the only nutrient necessary to animals. Carbohydrates are important, too. On the other hand, soy beans may be the present crop, but this crop is not a good candidate for alcohol fuel production. If the farmer decided to shift this existing use to corn production, there would be a regional loss in protein, because soy beans are much higher in protein than is corn.

An alternative to crop shifting is increase in land brought into production. The land is, in some cases, available, but quality is an important variable. Any expansion of cropped acreage will expose more land to soil erosion and make it a greater source of non-point pollution. As the quality decreases, the danger from pollution and soil degradation increases(8). Land quality in the US varies widely, sometimes within any given county. Table I shows land categorized by county and differentiated into eight land capability classes. Figure 1 illustrates the amount of land, by capability class, for each region.

Direct combustion technologies provide the farmer with an opportunity to co-fire residues in existing facilities. For example, propane or natural-gas grain dryers can be modified with minimum up-front costs to accept residues. Fuel savings can be substantial(9). Removal of residues, however, can easily be mismanaged to produce a chain of adverse environmental consequences. An explanation follows.

In natural soil formation, native vegetation extracts chemicals from the deep soil and returns them to the surface as the vegetation decays. Modern agriculture has altered this process. Crops are removed from the land, and therefore most of the nutrients are removed. Necessary minerals are returned to the soil artificially and residues left to <u>bind</u> the soil.

Removal of residues, whether in the field or forest, directly affects both soil loss and soil fertility. Soil fertility is maintained by nutrient input, fragmentation and transpiration by soil fauna, chemical decomposition, and output via root uptake and remineralization. Crop residues are a reservoir of plant nutrients. Leaving them in the soil results in better water infiltration and less evaporation; therefore, soil productivity is enhanced(10).

Any land practice that exposes soil to erosive forces of wind or rain represents an environmental hazard. The degree of impact is greatly dependent on soil texture, but surface geology and climate are also important factors. Finely textured soils are generally more prone to erosion than are sandy soils. Sandy soils tend to absorb surface runoff, whereas clay soils are easily suspended and carried away(11). Slope/length and slope/gradient, cropping management, and

TABLE 1

LAND CAPABILITY CLASSES

Class	Characteristics
Land suited for cultivation:	
I.	Suited to a wide range of crops; nearly level; low erosion hazard; productive soils; can be intensively cropped; favorable climate.
II.	Some limitation on suitable crops; requires conservation practices to prevent deterioration or improve air and water relationship within soil.
III.	Limitations restrict: (a) amount of clean cultivation; (b) timing of planting, tillage, and harvesting, and (c) choice of crops; requires conservation practices more difficult to apply and maintain than those on class II land.
IV.	May be suited to only two or three common crops; yields may be low in relation to inputs over a long period; management and conservation measures more difficult to apply than for those on class III land.
Land generally not suited for cultivation:	
V.	Nearly level; limitations which are impractical to remove may include wetness, frequent overflow, stoniness, climatic limitation.
VI.	Continuing limitations which cannot be corrected may include steep slope, stoniness, severe climate; unusually intensive management necessary if used for common crops.
VII.	Unsuited for cultivation; impractical to supply pasture improvements or water controls.
VIII.	Cannot be expected to return significant benefits from management for crops, grasses, or trees.

SOURCE: Zeimetz, Kathryn. "Growing Energy: Land for Biomass Farms." Washington: US Government Printing Office (June 1979) p. 20.

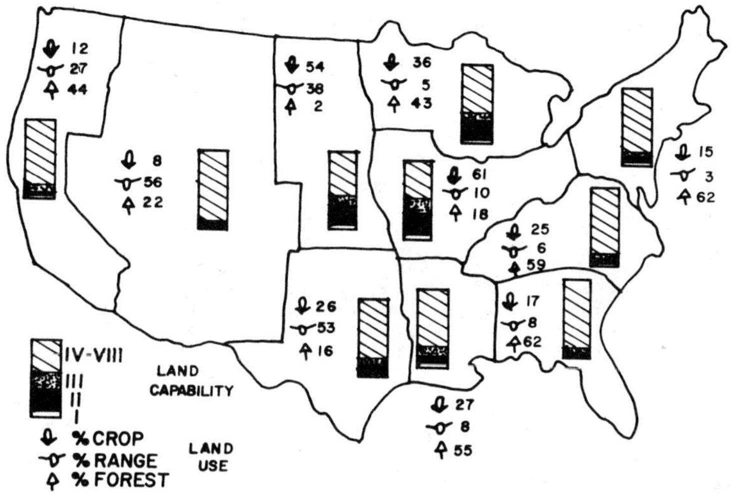

Fig. 1. Land Capability and Agricultural Land Use in the United States

---

source: Zeimetz, Kathryn. " Growing Energy: Land for Biomass Farms." Washington: US Government Printing Office, June, 1979, p. 22,26.

frequency and length of wind or rain storms are important considerations. Therefore, the necessity for leaving some or all of the residues is very site-specific, because it depends on the condition of the soil.

IMPACTS TO ENERGY, INDUSTRY AND COMMERCE

Industry tends to cluster around an energy supply, and commercial services follow shortly thereafter. An example is the recent development in the Great Plains because of the coal supply. Effects on land for the energy facility itself are often secondary to the effects of change in settlement patterns throughout the entire area(12). Bio-conversion fits very well with the existing industry and commerce in agricultural areas. Services and industry tend to support food, fiber, and timber production. Agricultural processing operations generate residues that can and are being used for energy production. An example would be a cannery for peaches. The pits, which are a waste disposal problem, could be burned in a direct combustion facility or co-fired with fossil fuels. The forest products industry is another area where the opportunity exists to turn a liability into an asset. Wood waste can be burned in conventional boilers with very little retrofit or capital improvement costs.

Bio-energy production could affect existing energy production. For example, in an area where citizens see the present extractive use, such as coal mining, as an environmental hazard, bio-energy may replace it. Another possibility is that this small-scale, self-sufficient resource may eliminate dependence on large metropolitan-based utilities. This would reinforce trends in rural growth(13).

IMPACTS TO HOUSING AND COMMUNITY

As a viable energy supply followed by commerce and industry is brought into an area, new residents follow. This can be a boom and bust situation. Boomtown effects of bioconversion are not the same as those that accompany coal development. Small-scale, renewable energy systems do not have the same character as a large, centralized, concentrated, and temporary fossil fuel resource supply. Residents in coal boomtowns arrive and leave *en masse*. Renewable energy would make for a slower, more controlled growth--a more viable community. People would come in to support, maintain, and supply machinery and labor. Only the seasonal nature of the feedstock would suggest migrant workers, an existing labor condition in most farming towns.

Total land devoted to each rural use has remained relatively unchanged in the past 50 years. Even though one million acres of land are converted from cropland yearly, there is a counterbalance of land to or from forest pasture, and other agricultural uses(14). The issue is not so much quantity but quality of land and location.

In urbanizing areas, competition for land for perishable goods such as eggs, milk, and fresh vegetables will intensify. Food production, energy development, and urbanization cannot occur in the same place at the same time without creating conflicts. Land prices will escalate as the tug-of-war for property increases.

IMPACTS TO TRANSPORTATION

Alcohol fuels could help relieve existing liquid fuel markets by increasing the supply of liquid fuel at the local level. At present, gasohol is being sold in many locations. Eventually, automobile manufacturers may redesign engines to run on pure alcohol.

With bio-conversion, residue collection would be an added activity at harvest time, and transportation of residues could contribute to existing farm-labor health and safety problems. There would also be an increase in air pollution from this activity. The large bulky machinery and tractors and trailers loaded with large bales could interfere with local traffic patterns if centralized units are deployed and long distances must be traveled to deliver feedstock.

As rural communities become more self-sufficient, there could be a decrease in transportation links. With more agricultural diversity, integrated use of resources, and an increase in part-time subsistence farming, there would be less need to "go to town."

IMPACTS TO RECREATION AND OPEN SPACE

Concern for the environment has become a major public concern in the past decade. Land is being considered as a public resource, a threatened natural resource of concern to everyone. Bioconversion deployment will create a challenge at the local level to maintain renewable energy supply while recognizing dangers inherent to the future of the land.

Whole tree harvesting can disturb water tables, terrestial and aquatic systems, and degrade the landscape(15). Aesthetic, cultural, and recreation values of the land can be diminished. Bringing more land into production can also bring about visual degradation. An example is channelizing to improve drainage in poorly drained fields. On the other hand, an increase in trees or vegetation on barren land will be an aesthetic plus factor.

Soil disturbance can add to bacterial and viral concentrations in water. As the soil runs off into streams, so do the viruses and bacteria within it. High turbidity, eutrophication, and nutrient depletion accompany erosion. Residue removal could adversely affect fish populations by increasing pesticide levels and altering BOD(16).

CONCLUSION

In conclusion, these are the following major environmental concerns associated with bioconversion: (1) soil loss and soil fertility depletion from residue removal and whole-tree harvesting; (2) increased competition for land; (3) air quality impacts from facility emissions; and (4) water quality impacts from erosion that is associated with increased crop residue removal and whole-tree harvesting.

Bio-energy production at the local level will vary, depending upon resource availability. Management of the residuals from the conversion processes will determine the potential for local impacts. Whether the land use impact is beneficial or adverse depends on the attitude of the producer. In the case of on-farm energy production, the incentive for recycling water, CO_2, feed, and fertilizer may reduce the potential for "mismanagement of residuals." For the local energy entrepreneur, there may be an economic advantage to recycling rather than polluting with the residuals. Because of these site-specific differences, the degree and type of impact will vary from region to region. Be it adverse or beneficial, informing the bio-energy producer and regulating mismanaged facilities are local concerns. State and local planners and agricultural extension personnel are the people within existing agencies who have the know-how and public outreach programs to handle such responsibility.

ACKNOWLEDGEMENTS

The author wishes to acknowledge the continuing support of the Department of Energy, Assistant Secretary for Environment, Office of Environmental Assessments, Technology Assessments Division, who have funded the research reflected in this paper.

REFERENCES

1. Tiffany, H.T. Real Property, Chicago: Callaghan and Co. (1912) p. 515-16.

2. US Congress, House Sub-committee on Energy and the Environment Hearings. Land Use and Resource Conservation, (March-April 1975,) p. 89.

3. Schneider, Stephen H. "The CO_2 Problem: Are There Policy Implications, Yet?" Climatic Change (March 1980) p. 205.

4. US Department of Energy. "Environmental Readiness Document: Bio-mass Energy Systems." Washington: US Government Printing Office, p. 32.

5. Leonard, Ellen. "Wood Burning for Power Production." Los Alamos National Laboratory report LA-7924-MS (August 1979) p. 9.

6. US Department of Energy. "Environmental Readiness Document: Biomass Energy Systems." Washington: US Government Printing Office (September 1979) p. 24.

7. Hertzmark, Donald I. "A Preliminary Report on the Agricultural Sector Impacts of Obtaining Ethanol." Golden: Solar Energy Research Institute (July 1979) p. 13.

8. Zeimetz, Kathryn A. "Growing Energy: Land for Biomass", Washington: US Government Printing Office (June 1979) p. 19.

9. Flaim, Silvio and Urban, David. "The Cost of Using Crop Residues in Direct Combustion Applications." Golden: Solar Energy Research Institute report SERI/TR-353-513 (March 1980) p. 21.

10. Flaim, Silvio. "Soil Fertility and Soil Loss Constraints on Crop Residue Removal for Energy Production." Golden: Solar Energy Research Institute report SERI/RR-52-324 (July 1979) p. 6.

11. Zonzogni, William C., et al. "Pollution from Land Runoff." Environmental Science and Technology (February 1980) p. 149.

12. US Congress, House Sub-committee on Energy and the Environment hearings, Land Use and Resource Conservation (March-April 1975) p. 93

13. Trethgarthen, Timothy. Food, Fuel and Shelter, Boulder: Westview Press (1978) p. 67.

14. Zeimetz, K. "Growing Energy" p. 15.

15. Carlisle, A. and Methven, I.R. "Environmental Consequences of Intensive Forestry and the Removal of Whole Trees." Madison: Workshop on Biological and Sociological Basis for a Rational Use of Forest Residues for Energy and Organics (May 6, 1979) pg. 48.

16. Dance, E.W. and Hynes, H.B.N. "Some Effects of Agriculture Land Use on Stream Insect Communities." Environmental Pollution, 22 (1980) p. 19, 20.

ELEMENTS AFFECTING THE FEASIBILITY OF MANURE-TO-METHANE PLANTS

Richard P. Arber
and
Frederick K. Marotte
CH2M HILL, INC.

INTRODUCTION

Escalating costs and dwindling supplies of natural gas, coupled with overdependence on foreign oil, has directed national attention to alternative energy sources. One alternative under going careful research and development is energy production from biomass (i.e. grains, trees, livestock manure, etc.). Since biomass is a renewable resource, energy production through bioconversion holds promise of helping the United States achieve its goal of energy independence.

For any technology to be economically feasible, it must be technically achievable and cost competitive. This paper will define and discuss the important factors that affect the economic feasibility of producing methane gas, an alternative energy source, through the biological conversion (anaerobic digestion) of cattle manure.

The technology to produce methane gas through bioconversion of organic materials such as manure is not new. In the United States, there are over 3,000 wastewater treatment plants using anaerobic digestion for waste disposal and methane gas production. For nearly a century, many of these wastewater treatment plants have utilized this produced gas as an energy source. As a result of many years of research and experience, the technology of anaerobic bioconversion has been fine tuned. The important question to be answered now concerns the cost of producing methane gas at manure-to-methane plants.

Methane production at sewage treatment plants is considered a byproduct which is used to help offset the cost of treatment. In a commercial bioconversion facility, sale of the produced gas and other byproducts must generate enough revenue to support the entire facility.

PROCESS DESCRIPTION

The production of methane gas through bioconversion is an anaerobic biological process (anaerobic means "without oxygen"), that breaks down organic material thereby producing methane, carbon dioxide, and small amounts of hydrogen sulfide, ammonia, and water. The process described in this section and shown on Figure 1 was tailored specifically for bioconversion of cattle manure. The basic design concepts also apply to other animal wastes and other biologically degradable biomass.

The initial steps in the process are collection, transportation, storage, and preparation of the raw cattle manure. Due to the large amount of manure required to produce a significant amount of methane, the application of bioconversion will probably be restricted to handling wastes generated either at dairies or confinement feedlots.

Collection of manure from a concrete surface is quite easy. The important thing is to collect fresh manure. With more typical dirt feedlots, collection of high-quality manure is a greater problem. Care must be taken to collect only fresh manure. Manure with large quantities of dirt or which has become thoroughly decomposed will produce very little gas.

The bioconversion plants in Lamar, Colorado, and the Imperial Valley in California are quite similar in that both will be processing manure from dirt feedlots. Although the following description applies specifically to the plant in Lamar, Colorado, the same basic unit processes will probably be used in the Imperial Valley plant.

Raw manure arrives at the bioconversion plant in 20-yard semitrailer dump trucks. The manure is dumped into a hopper and transported through a shredder and into a manure storage silo. The silo is large enough to allow 4 days storage of manure. This will allow the plant to operate without interruption on holidays and long weekends.

An automatic removal system from the manure storage silos will automatically feed another manure shredder located directly over the manure slurry tanks. In the slurry tanks, heated water is added to the manure to obtain a 10-percent manure slurry. The slurry is then pumped at a continuous controlled rate to the anaerobic digesters. The anaerobic digesters are large tanks constructed of either steel or concrete. They provide approximately 20 days' detention time for the manure slurry. The temperature of these digesters is maintained at approximately 95 degrees F to ensure optimum biological activity. The contents of the digesters are mixed by withdrawing gas from the top, compressing it, and then reinjecting it into the bottom of the digesters.

Mixing helps the anaerobic digestion process by maintaining uniformity in the digester, and by reducing the accumulation of hay and other light material at the top of the digester, where it interferes with operation. The excess gas produced is then compressed for use. For the plant in Lamar, Colorado, this gas will be delivered directly to a gas-fired power plant for steam generation. In the case of the Imperial Valley bioconversion plant, it is anticipated that carbon dioxide and hydrogen sulfide

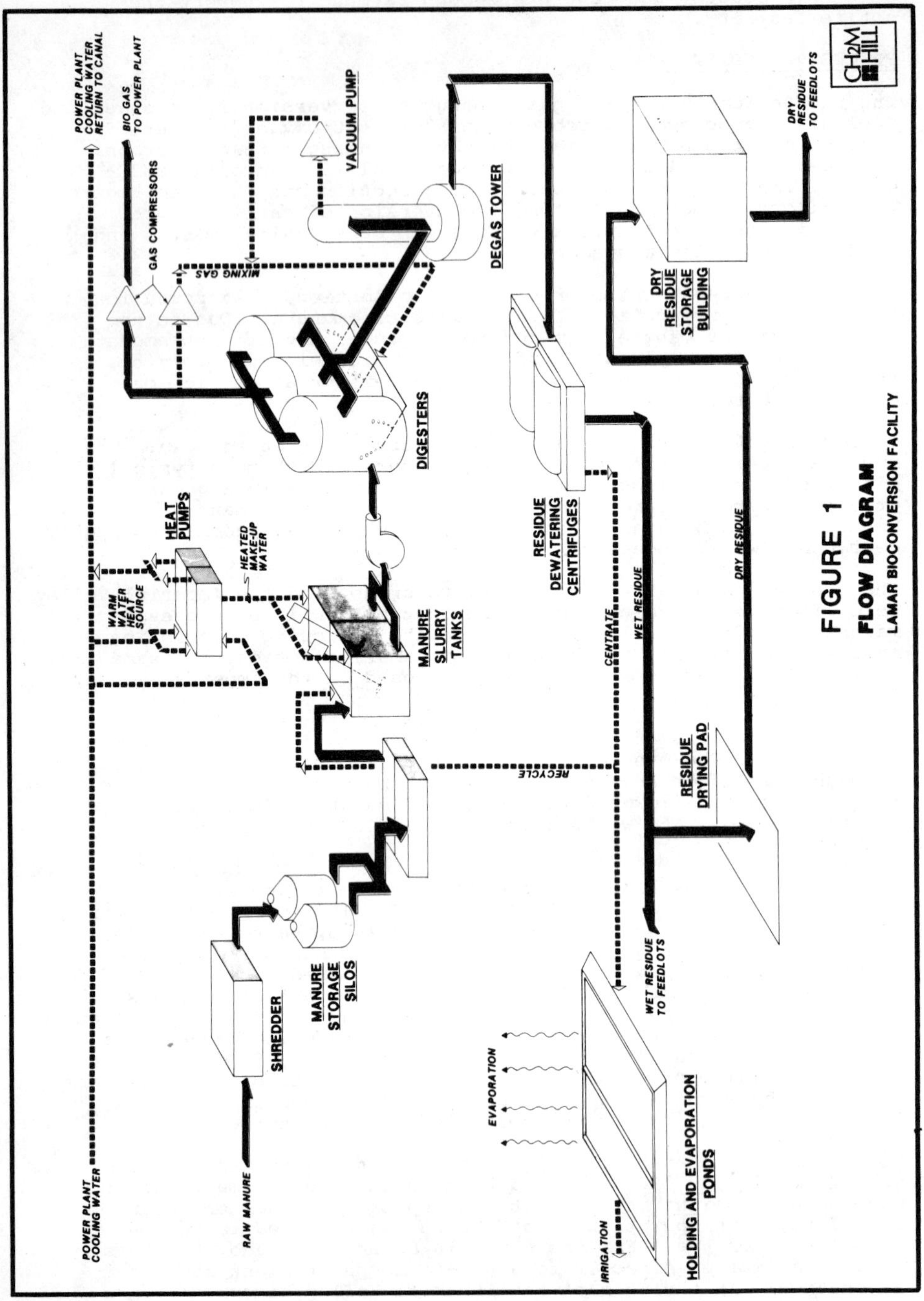

FIGURE 1
FLOW DIAGRAM
LAMAR BIOCONVERSION FACILITY

will be removed from the gas to provide a high-quality fuel supply suitable for residential use.

The effluent slurry from the digesters passes through a vacuum degasifier tower to remove excess hydrogen sulfide, carbon dioxide, and methane. The degasification system is primarily for odor control, in that it reduces the amount of hydrogen sulfide in the effluent. In remote areas where the odor of hydrogen sulfide will not be considered objectionable, it may be possible to eliminate the degas system.

Depending upon the end use of the residue solids, the effluent slurry can then be either dewatered or pumped to the final use end point. If the residue is to be used for refeeding to cattle, it can be dewatered in either a centrifuge or belt press to approximately 30-percent solids (70-percent moisture). The wet residue can then be conveyed either to the feedlots for refeed in a wet form, or to a drying pad or other drying device to remove the remaining water and produce dry cattle food.

A portion of the remaining liquid stream can be recycled back to the manure slurry tank to displace some of the clean makeup water. The remaining liquid must be disposed of either through land application as a fertilizer, evaporation in a nondischarging evaporation pond, or some other suitable method.

To provide some indication as to the size of a commercial bioconversion facility, the following design criteria apply to a plant with a capacity of 50,000 head:

- The manure processing rate would be approximately 350 tons of raw manure per day.
- A 20-cubic-yard dump truck full of manure would need to pass through the plant gates every 8 minutes to allow the necessary amount of manure to be delivered within 40 hours per week.
- The digester volume to provide approximately 20 days' detention time would be provided by three 135-foot-diameter digesters, with a 35-foot water depth.
- Storage of 3 months' dry residue would require a building 80 feet wide and 750 feet long.

FACTORS AFFECTING FEASIBILITY

For any energy production project to be considered feasible, it must be technically achievable, environmentally acceptable, and have favorable economics in terms of projected energy costs.

As mentioned earlier, the technology of methane production through the bioconversion of organic material is well developed. There is ongoing research examining different temperatures of digestion, different detention times, and other operating aspects. This research will provide information to better optimize the design of a bioconversion plant; however, it will probably not have a significant impact on the evaluation of overall feasibility.

The environmental impact of a bioconversion facility must be reviewed in terms of a specific plant location. The socio-economic impacts and environmental impacts are too site-specific to be generalized. It can be assumed, however, that the processing of cattle wastes will have no more environmental impacts than current handling practices at existing feedlots. It can be further assumed that the economic impact of a bioconversion facility will generally be positive, in that these facilities will provide employment opportunities in the agricultural areas of the country.

The key to the success of a bioconversion facility is therefore economic feasibility. The important cost parameters in a bioconversion plant are:

- Raw materials (manure)
- Capital
- Operations
- Maintenance
- Residue disposal, if any
- Liquid waste disposal, if any

Not including any subsidy-type programs, the plant income must be sufficient to pay the above-listed costs. The potential revenue-producing items in a plant are:

- Sale of produced gas;
- Sale of residue, as in cattle food or fertilizer; and
- Sale of the liquid effluent as a fertilizer.

Since there are no commercial manure-to-methane facilities in operation, accurate definition of the above-listed costs and revenue-producing items is difficult. Of course, many of the items can be defined in the context of similar facilities, such as anaerobic digestion at sewage treatment plants. Unfortunately, the undefined, or at least poorly defined cost items, have very large impacts on the overall economic feasibility.

One important consideration in evaluating the feasibility of a bioconversion plant is the raw material cost and availability. In many locations, manure disposal is normally a problem for feedlot operators; therefore, purchase of excess manure by local bioconversion plants should not be a problem. There may be problems, however, in obtaining only the freshest manure to ensure high methane yields. Feedlot operators have not been very receptive in changing their manure-removal procedures to improve the quality of the manure. Feedlot operators have also been reluctant to enter into long-term contracts to provide a specified quality of manure to bioconversion plants. These problems, coupled with normal fluctuations in feedlot inventories, are considerations affecting the feasibility of this technology.

The primary and most reliable source of revenue for manure-to-methane plants is the sale of produced gas. The actual income from this gas will depend upon the availability and cost of other energy sources in a particular project area. For example, produced biogas replacing expensive propane or diesel fuel will have more favorable economics than that of produced biogas replacing an inexpensive supply of natural gas.

Further, biogas purified to allow residential use will have a much higher value than biogas burned in a gas turbine or steam power plant. To be truly economically feasible, the cost of producing biogas should be competitive with the cost of the energy source it replaces. In terms of evaluating income from gas production, the revenue produced by the sale of gas should be equal to or less than the cost of replacement energy.

The liquid effluent from a bioconversion plant has a potential for either generating income for the facility or creating a disposal cost. This liquid contains relatively high concentrations of nitrogen, phosphorus, and potassium, and is therefore valuable as a fertilizer; unfortunately, the effluent also contains high concentrations of other dissolved salts. In parts of the country where salinity is a concern, land application of this liquid to obtain the benefit of its fertilizer value may not be feasible due to high salt concentrations. In areas where high concentrations of salts in irrigation water do not present a problem, it may be possible to sell the effluent as a fertilizer. In the arid western states, where salinity is a problem, the only proven disposal option available is the discharge of water into solar evaporation ponds. However, studies are in progress examining various possibilities for agricultural application of this liquid to reduce disposal costs.

The final, most ill-defined byproduct of the bioconversion process is the solid material produced when the slurry is dewatered (bioconversion plant residue). Currently, there is a considerable amount of controversy surrounding the value of this material. On the most positive side, this material is roughly equivalent to alfalfa hay, and could serve as a substitute for alfalfa in a cattle food ration. In this light, there is a very strong potential for generating revenue through the sale of this bioconversion residue.

Considering the idea of refeeding residue from a slightly different perspective, many feedlot operators are aware that cattle in confinement feedlots will typically eat small amounts of raw manure. In fact, many feedlot operators feel that raw manure has a nutritional value equal to or greater than the bioconversion residue. Therefore, one of the major questions posed by feedlot operators has been why they should pay more to buy the residue than they were paid for selling the manure. This attitude leads one to question the income potential of the sale of residue for cattle food.

Since raw manure has long been recognized as a valuable fertilizer, the processed bioconversion residue should also be considered valuable fertilizer. In areas where salt is a problem, the dewatered residue would have the advantage of having a lower salt content than raw manure. Conversely, this material would have somewhat less fertilizer value. It would, however, serve as a good soil conditioner with a moderate fertilizer value.

Presenting the most negative side, disposal of residue may result in the same problems as disposal of manure generated by feedlots. In this instance, disposal would represent a cost rather than income.

Therefore, the disposal of residue could represent a large source of revenue for a bioconversion plant or a large disposal cost to the bioconversion plant. The actual value or cost associated with this residue will not be known until the bioconversion industry has been established and sufficient residue produced to fully determine its value and marketability. The value and/or cost assigned to the handling of this residue is one of the major unknowns in the evaluation of feasibility for a commercial bioconversion plant.

CONCLUSIONS

The production of methane gas through bioconversion is not new; it has been applied throughout the world for nearly a century. The real question regarding the feasibility of this technology does not involve whether anaerobic digestion of organic material will produce methane gas, but rather whether this gas can be produced and priced competitively with other energy sources.

The four major factors affecting the feasibility of commercial manure-to-methane plants include:

- The availability of the raw material.
- The income generated through the sale of gas.
- The cost or revenue associated with disposal of the liquid effluent.
- The revenue or cost associated with the disposal of the bioconversion residue.

The factor posing the greatest impact is residue disposal. If refeed of this material to animals becomes viable, the income generated would have a very positive influence on the feasibility of this technology. Unfortunately, the cost or revenue produced through disposal of this residue will not be known until the industry becomes well established.

One of the areas needing further definition is the issue of the value of the bioconversion residue. Perhaps the only way to firmly establish this value is to construct a bioconversion plant to provide adequate quantities of residue to allow a market to develop, if the potential for this development exists.

MICROBIAL CONVERSION OF BIOMASS TO METHANE *

by
David P. Chynoweth
Institute of Gas Technology
Chicago, Illinois

ABSTRACT

Laboratory studies at the Institute of Gas Technology (IGT) have investigated the anaerobic digestion of a variety of feedstocks including sea kelp, water hyacinth, terrestrial herbaceous and woody plants, sewage sludge, municipal solid waste, and biomass-organic waste blends. The results of these and other studies will be used to illustrate key factors which influence methane production rates and yields, including feed organic composition, nutrients, inoculum, temperature, retention time, feed concentration, particle size, and mixing. A new process recently developed at IGT which combines biological and thermal operations for conversion of biomass to substitute natural gas is described. Current and future applications and research needs for improvement of bioconversion processes will be discussed.

INTRODUCTION

Microbial methanogenesis is a natural process occurring in anaerobic environments such as ocean and lake sediments and animal digestive tracts. Man is actively engaged in harnessing and optimizing this process for decomposition of wastes and conversion of wastes and biomass into useful fuels. Methanogenic decomposition, as illustrated in Figure 1, is a process occurring only under strict anaerobic conditions where mixed populations of bacteria decompose organic matter to carbon dioxide and methane. The first step, carried out by nonmethanogenic bacteria, is called hydrolysis, and involves cleavage of interpolymeric bonding (which contributes to plant structure) and subsequent hydrolysis of polymers such as carbohydrates, proteins, liquids and nucleic acids to small, soluble molecules which can be transported into bacterial cells for further metabolism. The hydrolytic products are further metabolized to acetate, hydrogen, and carbon dioxide by a relatively

* This work was sponsored in part by the Gas Research Institute and the U.S. Department of Energy (SERI).

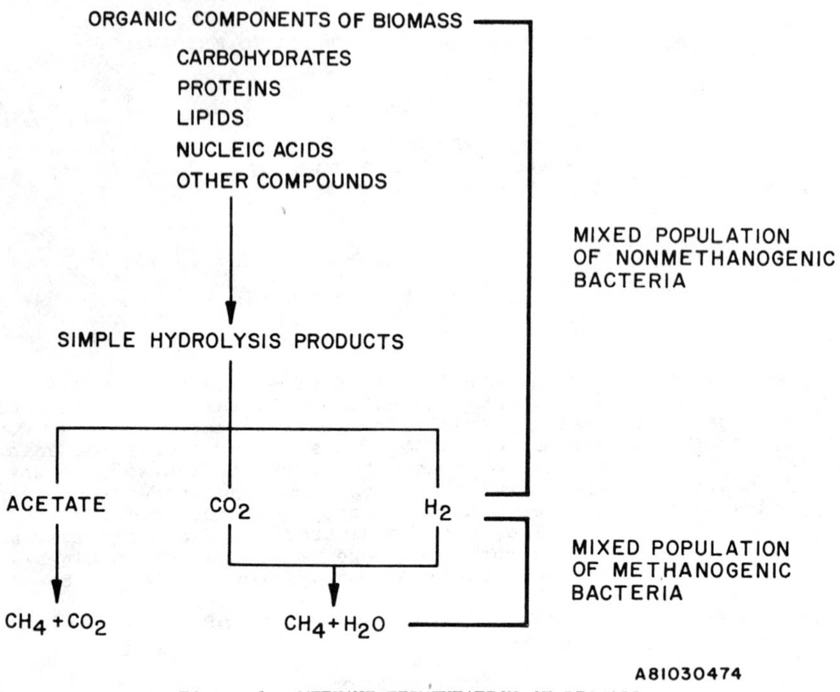

Figure 1. METHANE FERMENTATION OF BIOMASS

hearty and fast-growing group of nonmethanogenic bacteria. Acetate and hydrogen (along with carbon dioxide) are subsequently converted to methane by a fastidious, slow-growing population of methanogenic bacteria. A balanced intermicrobial association between these two groups of bacteria is required for a stable high-performance fermentation.

A generalized scheme based on this microbial process is known as anaerobic digestion and is illustrated in Figure 2. Upon harvest or collection, biomass or wastes are chopped and ground for size reduction, and possibly subjected to some kind of pretreatment to enhance biodegradation. The feed is added directly or as a 5% to 10% slurry to a primary digester having a typical volume of 300,000 ft^3 at a loading of 0.1 lb VS/ft^3 -day and hydraulic retention time of 15 to 20 days. In this continuously-mixed reactor, which is usually operated at a mesophilic temperature of 35°C, a mixed population of anaerobic bacteria effect conversion of about 50% of the organic matter (volatile solids) to methane and carbon dioxide, at a mole percent ratio of 60/40. This is equivalent to a methane yield of 4 SCF/lb organic matter added and a methane production rate of 0.4 vol/vol-day.

Solids in the effluent may be settled in an anaerobic secondary digester and recycled (directly or following posttreatment), and processed as fertilizer or animal feed. Supernatant can be recycled, used as fertilizer, or processed into a form suitable for disposal. Process gas can be utilized directly or treated to remove carbon dioxide and traces of hydrogen sulfide.

A number of variations in this scheme are possible. Methane yields from this process are limited by the fact that usually only about 50% of the feed organic matter can be hydrolyzed under anaerobic conditions. Thus, current research in this field is focusing on various pre- and posttreatment schemes to improve conversion efficiencies. Process rates are limited by the slow growth rates of the methanogenic bacteria. Research to overcome this limitation is focusing on cell recycle or attachment of cells to fixed particles to enhance their maintenance within the reactor. Several variations in reactor design and operation under development to promote kinetics and yields and to simplify design include thermophilic digestion, continuous-batch operation, plug-flow operation, use of fixed or expanded bed systems, and microbial phase separation.

Anaerobic digestion has been practiced at small and intermediate scales for several years as a method of waste reduction and conversion to a useful fuel. Its use is widespread in developing countries such as India and the Republic of China for production of biogas from animal wastes and crop residues. In developed countries the process is used for treatment of domestic sewage sludge and is under development for conversion of animal wastes and agricultural residues, industrial wastes, municipal solid wastes, and biomass grown on a large scale specifically as an energy resource.

The major criteria used for evaluation of the performance of anaerobic digestion on different feeds or a new process concept are methane yield, methane production rate, organic reduction, culture stability, thermal efficiency, and process economics. Methane yield is the quantity of methane produced relative to the quantity of organic matter added, and is usually reported as SCF/lb volatile solids (ash-free dry weight) added. Methane production rate refers to the quantity of methane generated reported as volumes of methane per volume of reactor per day. Organic reduction efficiency is significant because it directly influences methane yield and is inversely related to the quantity of process residues that must be

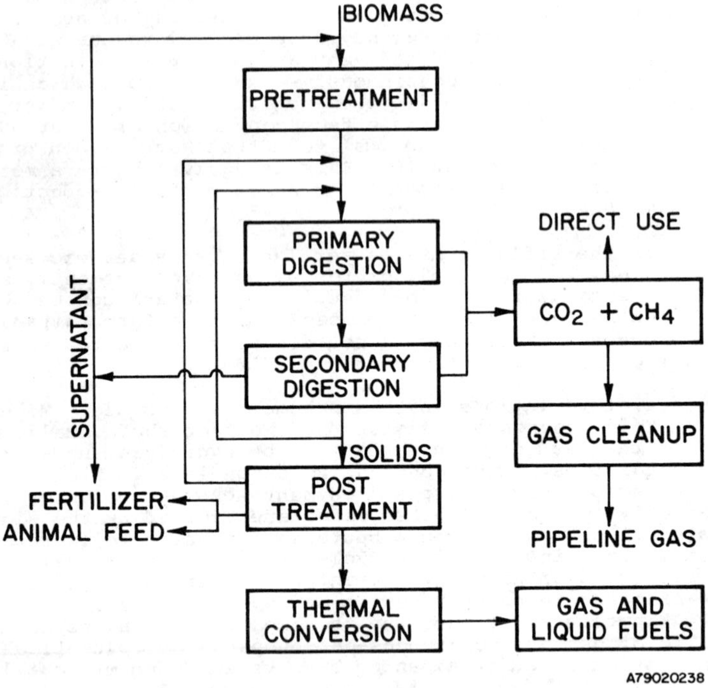

Figure 2. ANAEROBIC DIGESTION PROCESS SCHEME

processed or disposed of. This parameter is determined by a process materials balance in terms of organic (volatile solids) reduction. Certain combinations of feedstocks and operating conditions result in microbial population imbalance and related reduction or failure of digester performance. This condition is characterized by reduced gas production, increase in volatile acids concentration, and reduction in pH. Finally, process evaluation must consider the energy balance and economics relative to the overall biomass production or waste collection and treatment system. These criteria and factors influencing them are discussed in greater detail in the remainder of this paper.

FACTORS AFFECTING METHANE YIELDS

In any anaerobic digestion process that is not kinetically limited, the two major factors affecting methane yields are feedstock composition and inoculum.

COMPOSITION

Calculation of maximum theoretical methane yields is useful in evaluation of digester performance and provides a basis for establishing target yields for experimental work. If the organic fraction of biomass is assumed to be 100% biodegradable under anaerobic conditions, the upper theoretical yield can be calculated using heating value or compositional data. In the former method, the upper yield is equivalent to the heating value expressed as methane. In the latter method, compositional data are used to calculate the empirical formula and a stoichiometric equation is calculated for conversion of the organic fraction to methane and carbon dioxide. Because a fraction of the feed organic matter in all bacterial fermentations is converted to bacterial cells, theoretical yields should be corrected for cell maintenance. This factor may vary considerably for different anaerobic systems and feedstocks and has not been extensively investigated (particularly for mixed populations growing on particulate feeds). For anaerobic digestion of soluble substrates in a continuously stirred (CSTR) reactor, cell maintenance factors of 20% for carbohydrates and 7% for protein were reported.(1) This factor could be substantially different in processes designed to increased cell residence times, e.g., attached film or cell recycle systems.

Over the past several years these techniques have been utilized to determine upper theoretical methane yields from several different biomass and waste feedstocks, including marine algae, freshwater plants, herbaceous and terrestrial land biomass species, sewage sludge, municipal solid waste, and industrial wastes. These data are shown in Table 1 on the basis of volatile solids and total wet weight. Upper methane yields for most biomass species are similar with a value of about 8.5 SCF/lb VS. Using a cell maintenance factor of 0.15, a methane yield of 7.2 SCF/lb VS can be expected if all organic components are biodegradable. Domestic sewage is the only biomass feedstock with a substantially higher yield, probably because of its high protein and fat content. These data also illustrate that theoretical yields on the basis of wet weight may vary substantially from species to species because of differences in ash and water content. Theoretical yields for several different harvests of kelp are shown in Table 2 and illustrate the fact that significant variation in theoretical yields within a species can occur depending upon growth conditions.

Ultimate methane yields from feedstocks are also influenced by the biodegradability of the organic components. Certain natural organic compounds such as lignin are refractory to decomposition under anaerobic conditions even at long residence times. This explains why organic matter accumulates as lake sediment and fossil fuels such as peat, oil, and coal. Generally, this refractory

Table 1. HEATING VALUES AND THEORETICAL METHANE YIELDS OF DIFFERENT BIOMASS SPECIES

	Heating Value, Btu/lb VS	Theoretical Methane Yield*	
		SCF/lb VS	SCF/wet lb
Aquatic			
Kelp	8170	8.24	0.598
Water Hyacinth	8460	8.60	0.354
Herbaceous			
Bermuda Grass	8420	8.18	3.36
Kudzu	8620	8.61	2.40
Napier Grass	8470	8.39	2.20
Cattail	8480	8.28	1.81
Woody			
Eucalyptus	8260	8.05	4.49
Hybrid Poplar	8480	8.44	7.05
Cottonwood	8470	8.43	5.85
Loblolly Pine	8760	8.76	5.46
Wastes			
Primary Sludge	12,600	12.7	0.354
Municipal Solid Waste	8180	8.20	5.22

*Not corrected for cell maintenance and synthesis.

Table 2. THEORETICAL METHANE YIELDS FOR SEVERAL KELP LOTS

Kelp Lot	Stoichiometric Yield, SCF/lb VS*
37	8.24
42	7.11
44	7.59
46	6.45
47	7.29
49	8.58

*Assumes that the reactants are kelp + H_2O and the products are CH_4 and CO_2.

property is related to the lack of enzymes in the anaerobic bacteria or an oxygen requirement by many enzymes to carry out the initial hydrolytic reactions. It has been established that certain compounds can complex with other compounds in such a manner to render them refractory to anaerobic decomposition. For example, lignin affects decomposition of cellulose in this manner.(2)

Variability in the content of major biodegradable components of a biomass feedstock can have a significant effect on the ultimate methane yield. Recently, it has been shown that the mannitol content in kelp is variable (in our studies between 5% and 24% dry weight) and has a significant effect on methane yields.(3) This relationship is depicted in Figure 3. An equation to predict this relationship was derived from this graph and has been further verified experimentally.

The biodegradability of a broad spectrum of biomass and waste feedstocks expressed as methane yield and reduction in organic matter (volatile solids) is presented in Table 3. These data approach ultimate yields since they were obtained from digesters operated at long retention times and low loadings. The highest yields are obtained from soluble biodegradable feeds such as glucose and certain industrial wastes. Most feeds are about 30% to 60% biodegradable with corresponding methane yields in the range of 3 to 5 SCF/lb VS added. Lowest yields have been observed for certain herbaceous and most woody species with a high lignin content.

The large-scale implications of these data are that unreacted residues represent unrecovered energy and a potential waste disposal problem, and the significance of developing pre- and posttreatment techniques for increasing biodegradability is apparent. Most of the treatment processes for improvement of biodegradability of lignocellulosic materials have been developed for improvement of animal digestibility or conversion of biomass to ethanol. The techniques fall into the categories of physical, swelling, polysaccharide hydrolysis, and delignification and are listed in Table 4. Research on application of these methods for anaerobic digestion has been minimal. These methods are typically costly and energy-intensive and still do not result in total biodegradability. (See Table 5.)

An alternative to increasing biodegradability is to produce fuel gas from the digester residues by thermochemical gasification. IGT has recently developed a new process that employs this concept which is illustrated in Figure 4. Residues from anaerobic digestion are dewatered to about 50% and subjected to gasification. Thermochemical product gases are recycled to the digester where they are scrubbed free of pollutants and biomethanated. Sensible heat from the gasifier and hot gases, and chemical heat from biomethanation reactions supply heat requirements of anaerobic digestion. Ammonia in the gas and nutrients in the ash are recycled to the bioconversion processes as needed. The two major advantages of this concept over techniques to improve biodegradability are total conversion and higher overall thermal efficiency.

INOCULUM

The lack of decomposition of certain components of feed substances may be attributed to the absence of organisms capable of their degradation (rather than the inherent refractory nature of the compounds). It is generally thought that for most types of organic wastes and cellulosic biomass feedstocks, a viable inoculum can be developed from any environment in which anaerobic methanogenic decomposition of a mixture or organic compounds is occurring

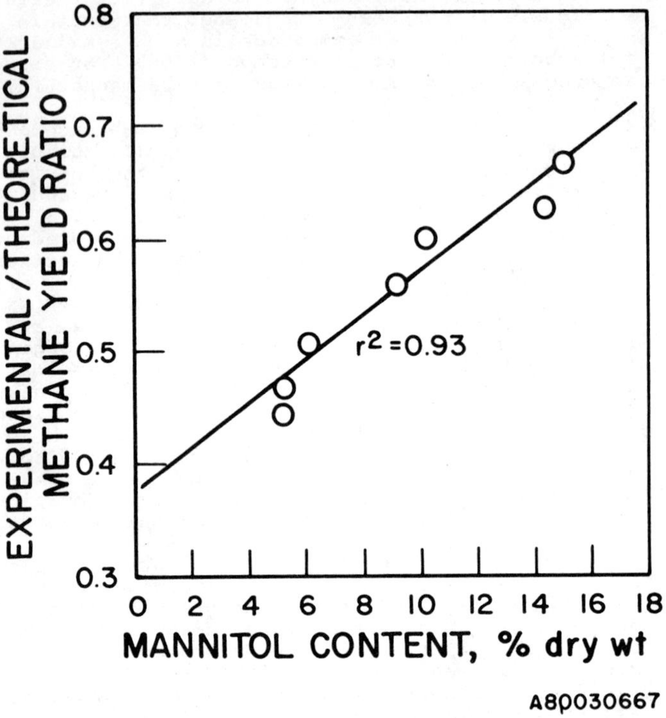

Figure 3. THE EFFECT OF MANNITOL CONTENT ON THE EXPERIMENTAL/THEORETICAL METHANE YIELD RATIO (3)

Table 3. BIODEGRADABILITY OF REPRESENTATIVE FEEDSTOCKS

Feedstock	Experimental Methane Yield, SCF/lb VS added	Volatile Solids Reduction, %
Aquatic		
Kelp	4.9	50
Water Hyacinth	3.0	35
Herbaceous		
Bermuda Grass	3.4	37
Napier Grass	2.6	26
Cattail	1.8	18
Woody		
Eucalyptus	0.31	2.0
Cottonwood	1.2	7.4
Loblolly Pine	0.27	1.6
Wastes		
Primary Sludge	8.3	72

Table 4. PRETREATMENT METHODS FOR IMPROVING BIOCONVERSION EFFICIENCY OF LIGNOCELLULOSE

Category	Method
Physical	Size Production
	Irradiation
	Thermal Stress
Swelling	Sodium Hydroxide
	Aqueous Ammonia
	Steaming at Moderate Temperatures
Polysaccharide Hydrolysis	Mineral Acids
	Autohydrolysis (oxidative and nonoxidative)
	Enzymatic Hydrolysis
Delignification	Alkaline (soda or Kraft processing)
	Acid (sulfite)
	Neutral (sulfite)
	Gaseous Sulfur Dioxide
	Microbiological (enzymatic or selective delignification with fungal system)
	Prehydrolysis (175°-220°C) and Extraction (caustic or aqueous dioxane)

Table 5. EFFECT OF PRETREATMENT ON ANAEROBIC BIODEGRADABILITY

Pretreatment Type	Feed	Digestibility, % Before	After	Method
Ball Milling	Aspen	38	76	a (4)
	Oak	8	50	a (4)
Irradiation (10^8 RADS)	Aspen	55	78	a (4)
	Spruce	3	14	a (4)
Temperature, 100°-175°C	Waste Activated Sludge	27 (35°C)	33 (100°C)	b (5)
			38 (135°C)	b (5)
			49 (175°C)	b (5)
200°-225°C	White Fir	<5	40	b (5)
Gaseous (SO_2)	Red Oak	1	60	a (4)
	Ponderosa Pine	0	46	a (4)
Alkali (NaOH)	Bagasse	29	73	c (6)
	Grass	38	71	d (7)
	Peat	3	27	e (8)
	Corn Stover	36	77	d (9)
	Basswood	5	55	a (4)
	Cottonwood	4	11	a (4)
	White Oak	4	20	a (4)

a. *in vitro* rumen fluid method.

b. Percent conversion to methane based on COD.

c. Percent carbohydrate utilization.

d. Methane yields.

e. Percent conversion to methane (ml CH_4/g COD).

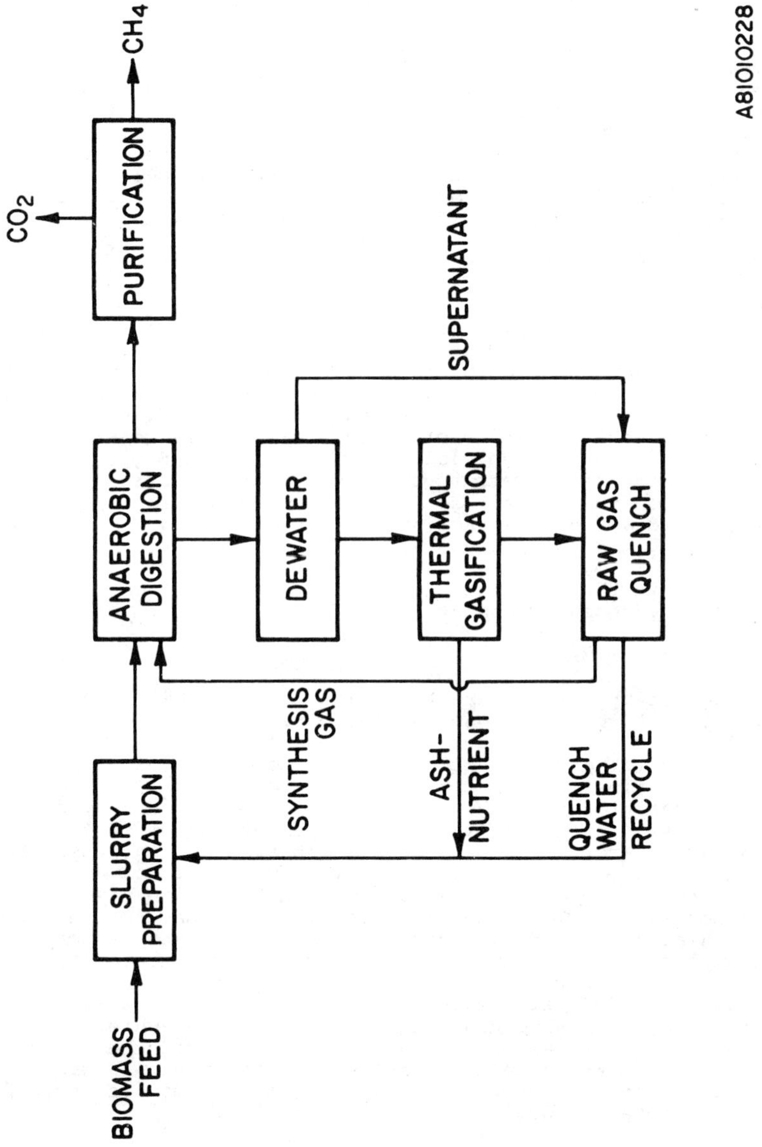

Figure 4. BIOTHERMGAS PROCESS SCHEME (10)

naturally, e.g., an anaerobic sewage digester, anaerobic lake sediments, or animal feces. Starting with such sources, an inoculum adapted to the feed under study can usually be developed within a reasonable period of time, i.e., two to four months. It would be logical to expect that the biodegradation potential of an inoculum would improve after an extended period of adaptation to a feed. In the case of kelp-adapted inoculum, this has not been the case. Improved yields observed over the past five years have not been related to the inoculum, but rather to variations in kelp composition. Because kelp contains substrates unique to marine algae (i.e., algin, fucoidin, and laminarin) and a high salt content, it was presumed that a marine inoculum derived from anaerobic sediments containing decomposing kelp might be capable of greater decomposition of kelp than a sewage sludge inoculum. We have evaluated several marine inocula and have not observed greater degradation than that observed with a sludge inoculum.

FACTORS AFFECTING METHANE PRODUCTION RATES

Process kinetics reflected as methane production rate are important to the design and operation of anaerobic digesters since rapid rates permit higher loadings or smaller reactor sizes. Representative methane production rates are reported for various feeds and operating conditions in Table 6 and range from 0.5 to 4 vol/vol of active reactor per day, for particulate feeds and as high as 10 vol/vol-day for conversion of soluble feeds. In general, higher rates result in reduced yields. Rates of most digesters are limited by the activity of methanogenic bacteria. Adjustment of design or operation to remove this limitation results in a hydrolysis-limited system. The major factors affecting the kinetics of the anaerobic digestion process are hydraulic retention time, feed concentration, temperature, nutrients, mixing, particle size, toxic substances in the feed, and feeding frequency. Each of these factors is briefly discussed below.

HYDRAULIC RETENTION TIME

The effect of hydraulic retention time on anaerobic digestion is illustrated in Figure 5. Long hydraulic retention times result in maximum volatile solids reduction and methane yields, but low decomposition and methane production rates. Compounds undecomposed at long retention times can be considered to be refractory to anaerobic decomposition. As the retention time is decreased, rates will increase due to the rapid throughput and metabolism of the feed. With further reductions, certain substrates and intermediates will wash out because of combinations of the following reasons: 1) organisms responsible for their degradation may have long generation times; 2) the solids retention time may not be long enough for degradation; and 3) easily metabolized substrates may prevent decomposition of more complex substrates via catabolite repression. The digester may become imbalanced and fail because of selective washout of slower growing methane-forming bacteria. Lower retention times and corresponding higher rates can be achieved by preventing microorganism washout by cell recycle or attached film techniques. At the retention time that substrates begin to wash out, rates may increase. Under these conditions yields are reduced because of substrate washout.

FEED CONCENTRATION

Increases in feed concentration, independent of HRT, should result in higher rates. At some upper concentration limit (~10%-12%) and at low residence times, additional feed will remain unreacted because of mass transport limitations or saturation of the organism's enzyme systems. It should be emphasized that, at reduced retention times, higher feed concentrations are required to support

Table 6. OPERATION AND PERFORMANCE DATA FOR DIFFERENT REACTOR CONFIGURATIONS

Reactor Configuration	Ref.	Temp., °C	Feed	Loading, lb VS/ft^3-day	Retention Time, days	Methane Yield, SCF/lb VS added	Methane Production Rate, vol/vol-day
CSTR	(3)	35	Kelp	0.10	15	4.9	0.49
CSTR	(11)	35	Dairy Manure	0.45	15	2.50	1.2
CSTR	(12)	55	Dairy Manure	0.94	4	3.36	3.14
Plug Flow	(11)	35	Dairy Manure	0.45	15	3.0	1.3
Upflow Sludge Blanket*	(14)	30	Potato Processing Wastewater	2.6	0.14	~3.7	~9.6
Packed Bed	(15)	35	Methane Phase for Dog Food Digester	0.42	2.7	3.2	0.42
Expanded Bed	(13)	35	Dairy Manure	0.86	1.2	2.9	2.8
Expanded Bed	(13)	35	Dairy Manure	3.5	0.3	1.3	4.6

*Advanced digester concepts.

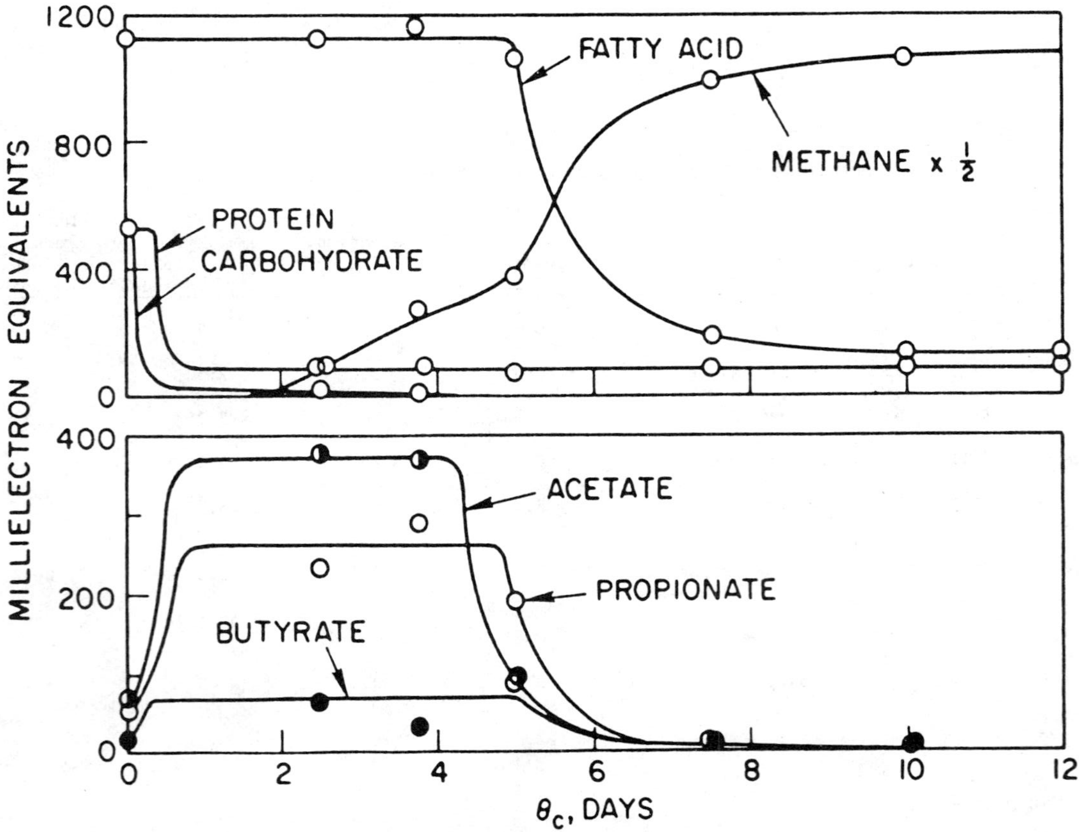

Figure 5. EFFECT OF SOLIDS RETENTION TIME ON THE RELATIVE BREAKDOWN OF DEGRADABLE WASTE COMPONENTS AND METHANE PRODUCTION AT 35°C (16)

the higher growth rates of organisms. A recent study by Wujcik and Jewell (17) has demonstrated effective batch fermentation of feeds with high solids concentrations (up to 30%). This "dry fermentation" concept reduces many problems associated with feed dilution such as heat requirements, volume of residue disposal, conservation of nutrients.

TEMPERATURE

Biological methanogenesis has been reported at temperatures ranging from 4°C (lake sediments) to 60°C (thermophilic dairy manure digesters). Most digesters are operated in the mesophilic (35°C) or thermophilic (55°C) range. Generally it has been observed that higher temperatures give higher reaction rates, thus permitting lower hydraulic retention times and higher loadings without reduction in conversion efficiency. However, in the range of 42° to 50°C, rates decrease until the thermophilic range is reached. (See Figure 6.) Above 60°C, rates drop off drastically.(18) Most digesters are operated at 35°C because of high sensitivity of bacteria in the thermophilic range and net energy considerations which are discussed below.

NUTRIENTS

Bacteria involved in methanogenic fermentations require nitrogen and phosphorus and other nutrients to achieve their maximum metabolic rates. We have found that in general C/N ratios exceeding 15, and C/P ratios exceeding 75 appear to be limiting. The relationship between C/N and methane production from kelp is illustrated in Figure 7. It should be emphasized that the biodegradable values of C, N, and P actually determine limiting conditions. However, total values for these elements such as those shown in Table 7 for various biomass feeds provide a rough estimate of potential nutrient limitations. These data suggest that many types of biomass feeds will require supplementary nutrients for bioconversion. The nutrient content of biomass is variable within a given species depending upon growth conditions. This variability and its effect on kelp digester performance are shown in Figure 7.

MIXING

The effects of mixing on anaerobic digestion have not been well documented. However, it seems reasonable to assume that, up to some point, mixing would enhance contact between organisms and substrates. Above that point, however, mixing could destroy organisms or obligate interspecies associations and thus decrease digester performance. Furthermore, effects of mixing could depend on digester size and configuration, feed composition and particle size, method of mixing, and a number of other factors. Studies in our laboratory have demonstrated that methane production rates are higher in two types of unmixed digesters (baffle flow and upflow solids blanket) than in continuously stirred reactors operated under the same conditions. Similar results were reported by Jewell _et al_.(13) for dairy manure. (See data in Table 6.)

PARTICLE SIZE

Particle size reduction should result in improved digester kinetics by increasing the surface area available for microbial attack. This effect is demonstrated for MSW digestion by Ghosh.(20) Particle size reduction below a certain point, however, becomes costly and energy intensive.

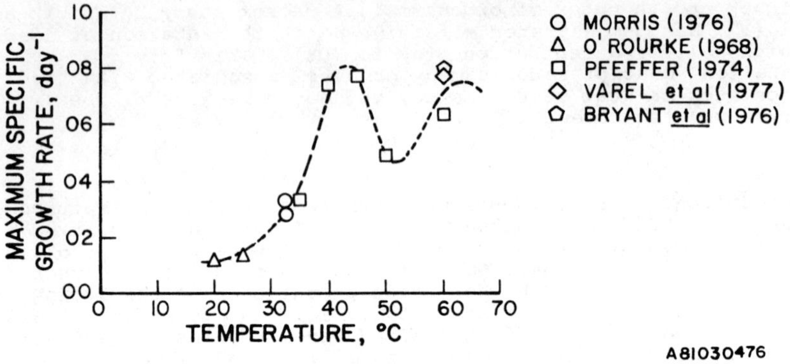

Figure 6. EFFECT OF TEMPERATURE ON MAXIMUM SPECIFIC GROWTH RATE (Curve Is Postulated) (18)

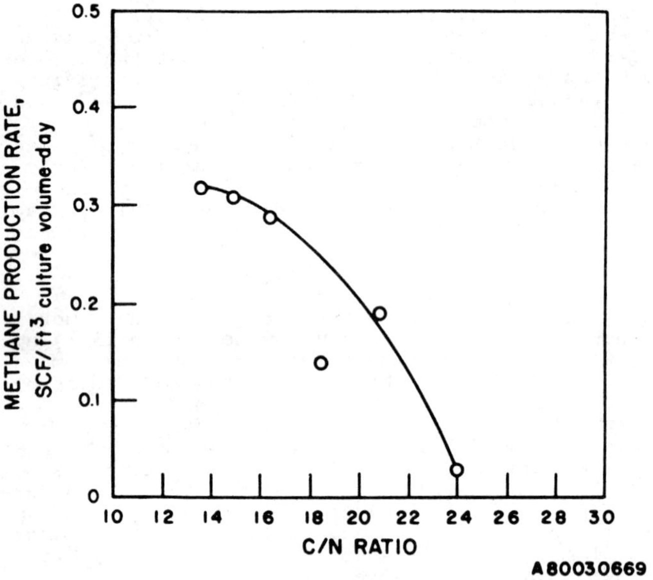

Figure 7. THE EFFECT OF C/N RATIO ON BIOMETHANATION OF KELP (19)

Table 7. CARBON-TO-NUTRIENT RATIOS OF DIFFERENT WASTES AND BIOMASS SPECIES

Feedstock	C/N	C/P
Aquatic		
Kelp	15	84
Water Hyacinth	10	94
Herbaceous		
Bermuda Grass	40	194
Napier Grass	41	527
Cattail	41	278
Woody		
Eucalyptus	490	446
Sycamore	178	2480
Loblolly Pine	432	2600
Wastes		
Primary Sludge	8	50
Municipal Solid Waste	76	204

TOXIC SUBSTANCES IN FEED

Biomass feeds may contain substances that are inhibitory to the biological methanogenesis. For example, sewage sludges may contain heavy metals or organic solvents. Chicken and pig manures sometimes contain inhibitory levels of ammonia. Pretreatment of biomass, especially woods, may result in formation of toxic compounds. In some cases, the bacteria may overcome the toxicity by adaptation. Other solutions to toxicity may include dilution with water or other feeds. Feed toxicity can be detected by appropriate bioassay techniques employing batch biodegradation tests at different concentrations.(21)

FEEDING FREQUENCY

The ideal feeding frequency for anaerobic digestion is continuous, because that mode of operation enhances culture stability. However, it has been demonstrated that the kinetics of digesters fed daily under low loadings and high retention times resemble that of a continuous culture. At some lower limit, culture instability could develop because of wide variations in growth rates of different populations. This phenomenon is currently under investigation in our laboratory.

DIGESTER DESIGN

Development and design of new and innovative digestion schemes should have the goals of maximizing methane yields and production rates, increasing process stability, decreasing processing energy requirements, and decreasing digester cost through simplification of design and operation. From the previous discussion it can be summarized that optimization of yields can be best achieved through selection of biodegradable feeds, pretreatment, and development of optimization of inoculum. Maximum process kinetics without sacrifice in yields may be achieved by schemes with long retention of organisms and unreacted solids, thermophilic temperature, minimum particle size, non-limiting nutrient levels, continuous feeding, and absence of toxic feed components.

Optimization of anaerobic digestion for high yields and rates may not be consistent with obtaining the highest net energy output. Ghosh (22) recently evaluated factors influencing net energy production. The two major heat requirements are feed heating and maintenance of digester temperature. Net energy production is maximized by operation at high feed concentrations and short hydraulic retention times. Figure 8 illustrates that thermophilic temperatures are justified at high feed concentrations on the basis of their marked effect on kinetics. Particle size reduction and other pre- or posttreatment schemes were not considered in this study but often are energy intensive.

Several innovative digester configurations (Figure 9) are under development with the objectives of increasing methane yields and production rates, increasing stability, increasing net energy output, and simplifying design and economics. Performance data for some of these schemes are illustrated in Table 6.

CSTR WITH CELL RECYCLE

This scheme, often referred to as the anaerobic contact process, results in improved kinetics through increase in retention of microorganisms and unreacted solids by settling and recycle of solids.

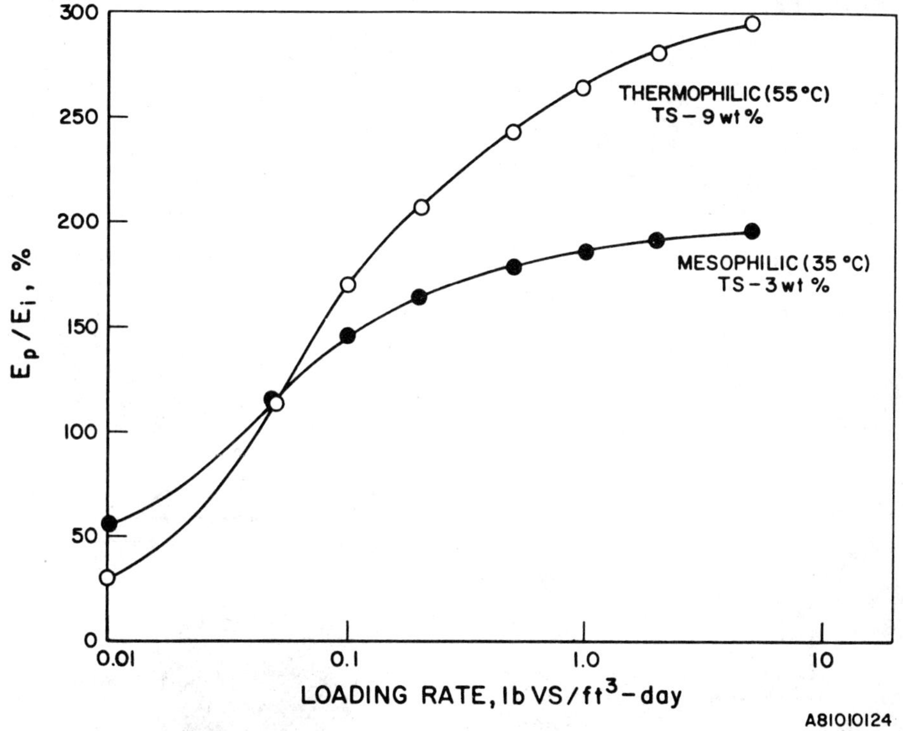

A81010124

Figure 8. COMPARISON OF NET ENERGY PRODUCTION RATIO FOR THERMOPHILIC (55°C) AND MESOPHILIC (35°C) DIGESTION OF A 60 wt % VS-CONTENT FEED (The feed slurries were at 15°C, and both digesters were assumed to exhibit a methane yield of 5 SCF/lb VS added. Total solids concentrations of the feeds for the thermophilic and mesophilic digestion were 9 wt % and 3 wt %, respectively.)

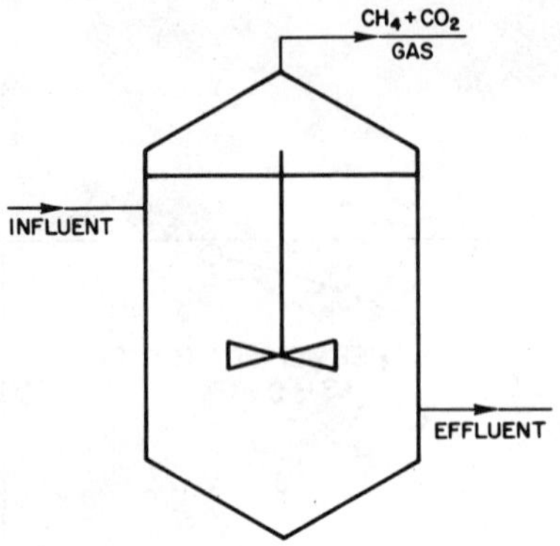

CONVENTIONAL DIGESTER (CSTR)

A81030467

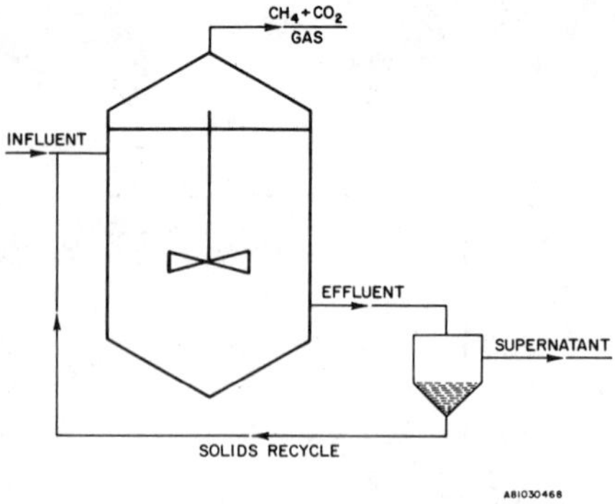

CONVENTIONAL DIGESTER (CSTR) WITH SOLIDS RECYCLE

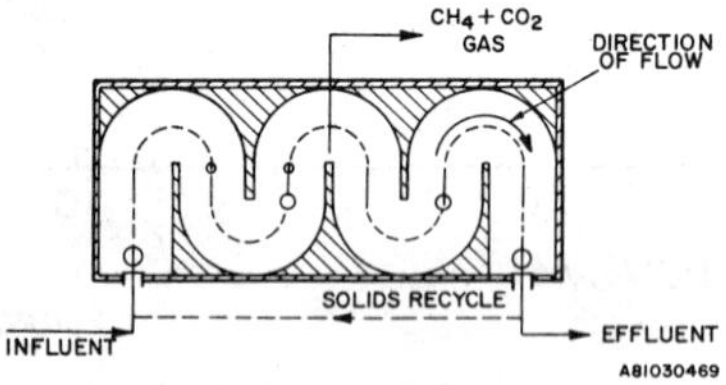

PLUG-FLOW DIGESTER
(Plan View)

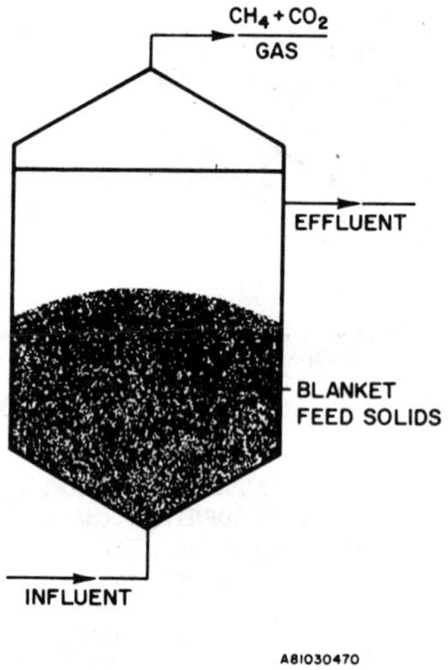

UPFLOW SOLIDS BLANKET DIGESTER

Figure 9, Part 1. BIOLOGICAL GASIFICATION PROCESS SCHEMES

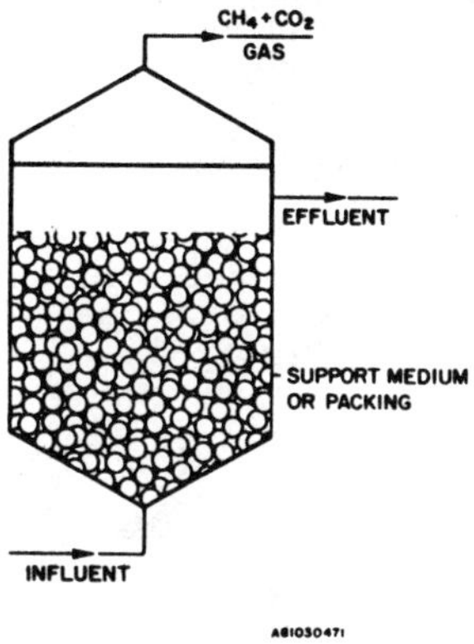

PACKED-BED DIGESTER

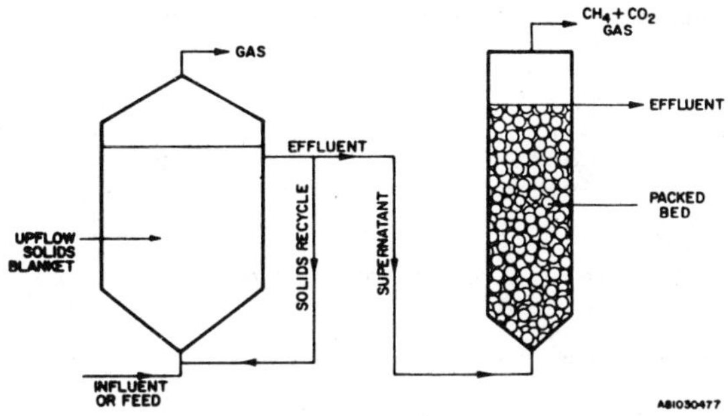

TWO-PHASE DIGESTION SYSTEM: UPFLOW SOLIDS BLANKET AS ACID PHASE AND PACKED-BED AS METHANE PHASE

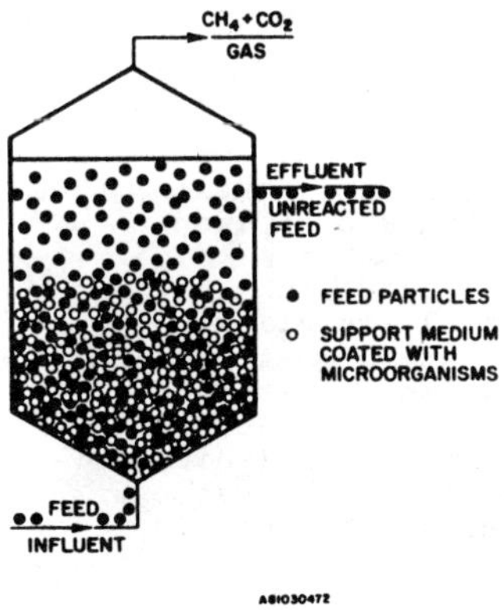

EXPANDED-BED DIGESTER

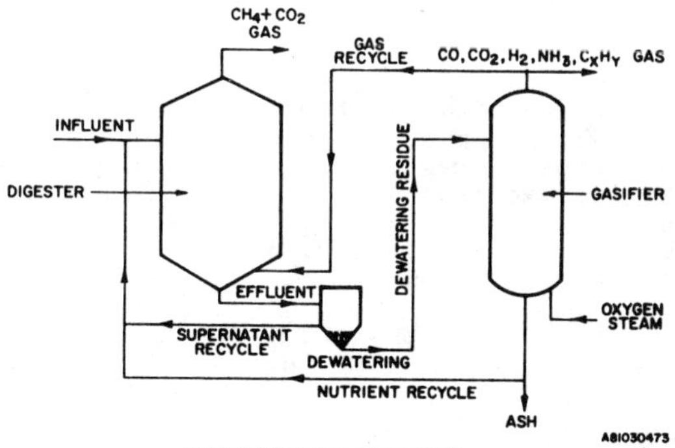

BIOTHERMGAS PROCESS

Figure 9, Part 2. BIOLOGICAL GASIFICATION PROCESS SCHEMES

PLUG-FLOW

A plug-flow reactor applied to anaerobic digestion is an oblong, unmixed reactor and receives feed at one end with effluent removal at the opposite end. While some vertical mixing of the digester contents occurs during gas production, longitudinal mixing is minimal, thus promoting phase separation of the digestion process. Feedstock hydrolysis and acid production occur initially in the digester while methane production occurs in the effluent end. Microorganism washout can occur in this system unless an inoculum is provided with the feed or solids recycle is incorporated into the operation. Data in Table 6 illustrate that the performance of plug-flow reactor was superior to that of a CSTR reactor. The major advantages are improved stability, higher conversion efficiencies, and simplicity and low cost.

UPFLOW SOLIDS BLANKET

The upflow solids blanket digester concept is based on the upward movement of soluble organic feed through a dense blanket of solids consisting of microorganisms and unreacted solids. As the liquid moves through the solids, increased solids residence times are achieved. This process achieves solids and cell recycle in a single digester. We believe that this system results in microbial phase separation with the acid-formers in the bottom and methanogens near the top. High rates and yields have been achieved with this system using soluble feeds (Table 6). We are currently evaluating the use of this process for anaerobic digestion of particulate feeds.

PACKED-BED

Attached film digesters such as packed- or expanded-beds are reactors which contain solid media upon which bacteria attach, thus preventing their washout with continuous feeding and effluent removal and promoting stable intermicrobial associations. The packed-bed digester consists of a filter bed filled with inert support material such as gravel, rocks, charcoal, or plastic media. This type of system results in rapid rate biomethanogenic conversion of soluble wastes. The ANFLOW process utilizes this concept for conversion of dilute raw sewage.(23) The packed-bed digester is also under consideration in this and other laboratories for the methane phase component of two-phase digestion. Feeds with high suspended solids levels are not suitable, as they rapidly clog the system.

EXPANDED-BED

A second type of attached film digester contains inert particles small enough to be hydraulically expanded. It has two major advantages over packed-bed systems: 1) a larger surface area is available for attachment of organisms; and 2) particulate feed can readily pass through the inert particles without removing them from the reactor. Data in Table 6 indicate that rapid rates are possible with this type of system.

TWO-PHASE

As discussed above, anaerobic digestion is mediated by two distinct microbial groups of bacteria which differ significantly with respect to physiology, nutritional requirements, growth, and metabolic characteristics, environmental optima, and sensitivity to environmental stress. In conventional digestion, it is impossible to optimize conditions for both groups of bacteria. Physical separation of the acidogenic and methanogenic phases permits: a) maintenance of appropriate densities of acid and methane formers, and b) maximization of rates through independent control of temperature, pH, oxidation reduction potential, biomass recycle, retention time, and other parameters for each phase. Phase separation should reduce instability of performance caused by fluctuation in feedstock loading, pH, and toxic feed components. As mentioned above, plug-flow and upflow solids blanket digesters result in phase separation in a single reactor. Other two-phase schemes isolate nonmethanogenic and methaogenic bacteria in separate reactors. The benefits of two-phase digestion of particulate feeds are not yet well documented by actual performance data.

BIOTHERMAL

Biothermal conversion combines biological and thermochemical gasification for total conversion of biomass and wastes, regardless of water or nutrient content. Residues from bioconversion are dewatered and thermally gasified to hydrogen, carbon monoxide, ammonia, and hydrocarbons. The product gases are quenched, scrubbed, and biomethanated in the anaerobic digester. Ammonia and ash are recycled to maintain required nutrient levels in the biological reactor. Reaction heat from biomethanation and sensible heat from the thermal process of are used for the anaerobic digestion energy requirements.

CONCLUSION

This paper has attempted to present an overview of the status of biological methanogenesis as a process for conversion of biomass and organic wastes to methane. The objective was to discuss the principal factors affecting methane yields and production rates and to describe conventional and nonconventional reactor designs and conversion schemes under development. This general conversion process shows promise for producing methane from organic matter at different scales ranging from the individual farm to large energy plantations consisting of several million acres. A number of applications were not discussed, including gasification of landfills and peat gasification. The scope of the paper has also not included detailed discussion of net energy and process economics. It is apparent that research efforts on this process should be directed toward isolation, characterization, pre- and posttreatment schemes to improve conversion efficiencies; optimization of the bacteria; application of the process to new feedstocks; and development of reactor and operating conditions to increase organism and solids retention, process stability, and simplicity of design.

REFERENCES

1. Speece, R. E. and McCarty, P. L., "Nutrient Requirements and Biological Solids Accumulation in Anaerobic Digestion," in Advances in Water Pollution Research, Proceedings of the First International Conference, Vol. 2, pp. 305-322 (1964).

2. Van Soest, P. J., "Comparison of Two Different Equations for the Prediction of Digestibility From Cell Contents, Cell Wall Constituents, and the Lignin Content of Acid-Detergent Fiber." Paper presented at the Annual Meeting of the American Dairy Science Association, Lexington, Kentucky, 1965.

3. Chynoweth, D. P. and Srivastava, V. J., "Methane Production From Marine Biomass." Paper presented at the International Symposium on Biogas, Microalgae, and Livestock Wastes — 1980, Taipei, Taiwan, September 1980.

4. Millet, M. A., Baker, A. J. Feist, W. C., Mellenberger, R. W. and Satter, L. D., "Modifying Wood to Increase its In Vitro Digestibility." J. Animal Sci. 31 (4), 781 (1970).

5. McCarty, P. L., Young, L., Owen, W., Stuckey, D. and Colberg, P. J., "Heat Treatment of Biomass for Increasing Biodegradability," at 3rd Annual Biomass Energy Systems Conference Proceedings, Golden, Colorado, sponsored by U.S. Department of Energy, October 1979.

6. Han, Y. W. and Callihan, C. D., "Carbohydrate Utilization Following Pretreatment by Cellulomonas sp." Appl. Microbiol. 27, 159 (1974).

7. Klass, D. L., Ghosh, S. and Conrad, J. R., "The Conversion of Grass to Fuel Gas for Captive Use." Proceedings of Symposium on Clean Fuels From Biomass, Sewage, Urban Refuse, Agricultural Wastes, Orlando, Florida, sponsored by the Institute of Gas Technology, Chicago, Illinois, 1976.

8. Buivid, M. G., Wise, D. L., Rader, A. M. and Kopstein, M. J., "Developments in Peat Biogasification." Presented at 179th National American Chemical Society Meeting, Division of Fuel Chemistry, Symposium on Utilization of Peat and Lignite, Houston, Texas, March 1980.

9. Pfeffer, J. T., "Methane From Agricultural Residues Process Conversion Efficiencies," in 3rd Annual Biomass Energy Systems Conference Proceedings, Golden, Colorado, sponsored by U.S. Department of Energy, October 1979.

10. Chynoweth, D. P., Srivastava, V. J., Henry, M. P. and Tarman, P. B., "Biothermal Gasification of Biomass." Symposium Papers: Energy From Biomass and Wastes — IV, pp. 527-554, sponsored by the Institute of Gas Technology, Chicago, Illinois, 1980.

11. Jewell, W. J., "Anaerobic Fermentation of & Agricultural Residues: Potential for Improvement and Implementation." Quarterly Progress Report, COO-EY-SO2-2981-12, for the Department of Energy, Contract No. EY-76-S-02-2981, January 1980.

12. Hashimoto, A. G., Chen, Y. R. and Prior, R. L., "Thermophilic Anaerobic Fermentation of Beef Cattle Residue." Symposium Papers: Energy From Biomass and Wastes, pp. 379-402, sponsored by the Institute of Gas Technology, Chicago, Illinois, 1978.

13. Jewell, E. W., Carpenter, H. R., Dell'Orto, S., Fanfori, K. J., Hayes, T. D. Leuschner, A. P., Miller, T. L., Sherman, D. F., Van Soest, P. J., Wolin, M. J. and Wujcik, W. M., "Anaerobic Fermentation of Agricultural Residues: Potential for Improvement and Implementation." Final Report for the Department of Energy. Report No. EY-76-S-02-2981-7. January 1978.

14. Lettinga, G., Van Velsen, A. F., Hobma, S. W., De Zeeun, W., and Lkapwijk, A., "Use of the Upflow Sludge Blanket (USB) Reactor Concept for Biological Wastewater Treatment, Especially for Anaerobic Treatment." Biotechnol. Bioeng. 22, 699-734 (1980).

15. Norman, J. and Frostell, B., "Anaerobic Wastewater in a Two-Stage Reactor of a New Design," Proc. 32nd Ind. Waste Conf., Purdue University, Lafayette, Indiana, 1978.

16. O'Rourke, J. T., "Kinetics of Anaerobic Treatment at Reduced Temperatures." Ph.D. Thesis. Stanford University (1968).

17. Wujcik, W. J. and Jewell, W. J., "Dry Anaerobic Fermentation." Biotechnol. and Bioeng. 10, 43-65 (1980).

18. Chen, Y. R. and Hashimoto, A. G., "Kinetics of Methane Formation," Biotechnol. Bioeng. Symp. 8, 289 (1978).

19. Chynoweth, D. P., "Anaerobic Digestion of Marine Biomass," Biomass and Alcohol Fuels Production, 185-192. Emmaus, PA: J. G. Press, 1980.

20. Ghosh, S., Klass, D. L., Conrad, J. R., Henry, M. P., Griswold, K. and Sedzielarz, "A Comprehensive Gasification Process for Energy Recovery From Cellulosic Wastes," in T. K. Ghose, ed., Bioconversion of Cellulosic Substances Into Energy , Chemicals, and Microbial Protein, Faridabad, Haryana, India: Thomas Press, 1978.

21. Owens, W. F., Stuckey, D. C., Healy, J. B., Young, Y.L. and McCarty, P. L., "Monitoring Biochemical Methane Potential and Anaerobic Toxicity," Water Research, 13, 485-92 (1979).

22. Ghosh, S., "Net Energy Production in Anaerobic Digestion." Paper presented at the Institute of Gas Technology Symposium on Energy From Biomass and Waste — Orlando, Florida, 1981.

23. Genung, R. K., Pitt, W. W., Jr., Davis, G. M. and Koon, J. H., "Energy Conservation and Scale-Up Studies for a Wastewater Treatment System Based on a Fixed-Film Anaerobic Bioreactor," Biotechnol. Bioeng. Symp. 10, 295-316 (1980).

OPERATION OF AN ANAEROBIC SLUDGE DIGESTION POWER GENERATION SYSTEM: CASE STUDY

Charles D. Malone
Plant Manager
Archie Elledge Waste Treatment Plant
Winston-Salem, North Carolina

Robert M. Stein
Vice President, Process and Operations Division
Associated Water and Air Resources Engineers, Inc.
Nashville, Tennessee

INTRODUCTION

The Archie Elledge Waste Treatment Plant, located in Winston-Salem, North Carolina, treats a combination of municipal and industrial wastewaters from the City of Winston-Salem. A two-stage biological process is utilized to provide secondary treatment, which consists of primary sedimentation, roughing trickling filters, and activated sludge. Excess sludge is stabilized by anaerobic digesters and dewatered on sand drying beds.

The average daily flow to the plant is 23 mgd. The organic loading to the system averages 70,000 lb BOD/day with a solids loading of 111,000 lb SS/day.

All electrical power used at the plant is produced on-site by diesel-driven generators. Table 1 presents the rated engine and generator capacity of the system. Under normal operating conditions, one 1,500 kW generator and one 600 kW generator supply the plant's electrical demand. The remaining three generators are maintained to provide backup capacity.

The diesel engines are multi-fuel units, and may be operated on digester gas, natural gas, fuel oil, or any blend of these three fuels. Ordinarily there is not a sufficient quantity of digester gas produced to satisfy the overall energy demand. Therefore, either fuel oil or natural gas is blended to supplement the digester gas. The choice of either fuel oil or natural gas is dependent upon the cost and availability of each. Additionally, it is necessary to inject a certain amount of "pilot oil" to ensure proper ignition. This normally amounts to 5 percent (as fuel oil) of the total fuel used.

The anaerobic digestion process consists of six primary and two secondary digesters. The volumetric capacity of the six primary digesters is 7.56 mil gallons while the capacity of the secondary

TABLE 1

RATED ENGINE AND GENERATION CAPACITIES

	Rated Engine Capacity, kW		Rated Generation Capacity, kW
1 Worthington engine at	533 kW	(715 bhp)	500 kW
2 DeLaval Enterprise engines at	645 kW	(865 bhp)	600 kW
2 DeLaval Enterprise engines at	1,579 kW	(2,118 bhp)	1,500 kW
Total	4,981 kW	(6,681 bhp)	4,700 kW

digesters is 2.99 mil gallons. Although the majority of organic conversion, solids reduction, and gas production occurs in the primary digesters, some limited solids reduction and associated gas production is observed in the secondary digesters. The principal function of the secondary digesters is to provide for solids separation and subsequent sludge thickening of the digested solids. A schematic illustration of the anaerobic digestion system is presented in Figure 1. Figure 2 depicts schematically the digester gas utilization system.

Typically, anaerobic digesters have been operated as a solids reduction and stabilization process. There has been little attention paid to gas generation because the alternative energy sources were relatively inexpensive. However, with the spiraling costs of hydrocarbon energy sources, the City of Winston-Salem embarked in 1977 on an extensive analysis of its overall energy utilization program.

The objective of the program was to evaluate the digester performance in terms of energy generation and to determine methods of optimizing gas generation without inhibiting the solids reduction process. The overall economics of the power generation system were evaluated in order to provide a realistic analysis of power generation costs and efficiencies to compare to alternative systems. The paper summarizes the results of the study, reviews programs which can be utilized to optimize digester performance, and reviews the economics of the power generation system.

DIGESTER PERFORMANCE

An intensive monitoring program was conducted on the operation of the digesters. The results of these analyses are presented in Table 2. These data indicate that the performance of the system is consistent with published data on anaerobic digester operation.

SYSTEM OPTIMIZATION

In order to optimize the performance of the digesters, several programs were implemented, as summarized in the following sections.

Digester Cleaning

It was found that there was a significant accumulation of scum and grit in the digesters. This resulted in a reduction of the effective volume of the digesters and a subsequent reduction in the efficiency of the digesters.

A program was implemented to periodically clean out the digesters. Prior to the implementation of the cleaning schedule, it was noted that there was a significant difference between the average solids concentration within the digester (≈8.0%) and the concentration of solids withdrawn from the digester (≈2.4%). The difference was the result of solids deposition, "coneing" during sludge withdrawal, and inadequate mixing. After cleaning, it was observed that the average concentration of sludge within the digester (2.3%) and that of the sludge withdrawn were more consistent (3.8%).

It is anticipated that the primary digesters will require cleaning every three to five years and the secondary digesters every two to four years.

Primary Screenings

The floatable matter on the primary clarifiers was designed to be skimmed off and pumped directly to the digesters. These skimmings consisted of grease, oil, hair, rubber, and plastic materials. These materials accumulated in the digesters and resulted in floating scum layers, which reduced the effective digester capacity.

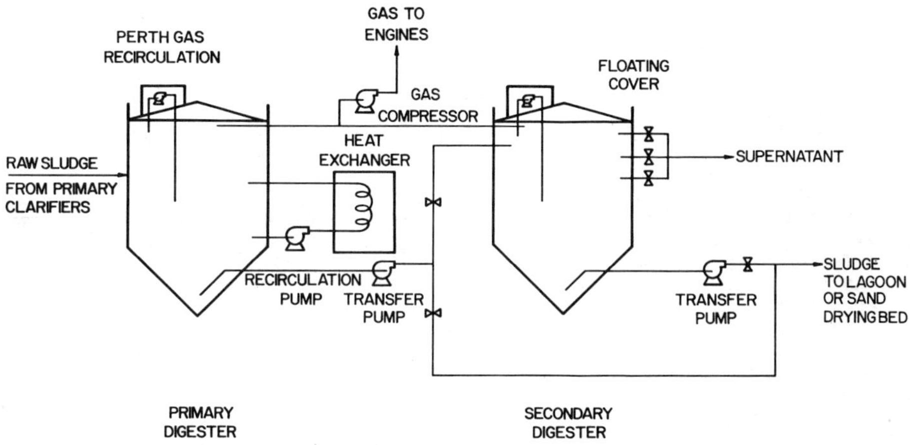

FIGURE 1 SCHEMATIC ILLUSTRATION OF ANAEROBIC DIGESTION SYSTEM

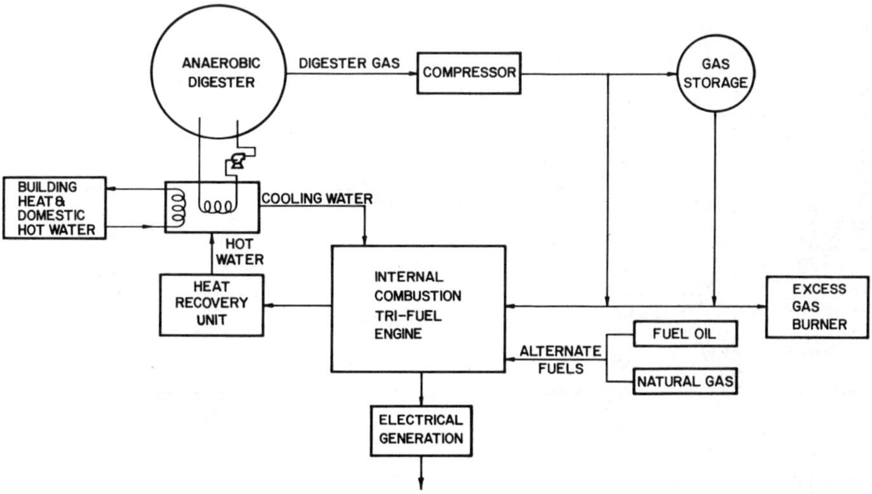

FIGURE 2 SCHEMATIC ILLUSTRATION OF DIGESTER GAS UTILIZATION SYSTEM

TABLE 2

SUMMARY OF ANAEROBIC DIGESTER PERFORMANCE AND CHARACTERISTICS

Digester Volumes		
Primary Digesters		
4 @ 90' DIA, Volume, MG	1.144	each
2 @ 100' DIA, Volume, MG	1.495	each
TOTAL	7.566	
Secondary Digesters		
2 @ 100' DIA Volume, MG	1.495	each
TOTAL	2.99	
Total Digester Volume, MG	10.556	
Raw Sludge Feed Rate, MGD	0.339	
Hydraulic detention Time		
Primary Digester, day	22.3	
Secondary Digester, day	8.8	
Raw Sludge pH	6.93	
Digester pH	6.96	
Digester Gas Composition		
Methane, %	63	
Raw Sludge Temperature, °F	71.6	
Digester Temperature, °F	96.8	
Influent Solids Loading, lb SS/day	67,020	
Influent Volatile Solids Loading, lb VSS/day	48,145	
Loading Rate, Primary Digester, lb VSS/ft^3/day	0.02	
Average Gas Production Rate, ft^3/day	372,000	
Ratio of Gas Produced/lb VSS Removed, ft^3/lb VSS/day	19.7	
Volatile Acids/Alkalinity Ratio	0.92	
Average Precent Volatile Solids Reduction, %	37.5	

Results for the Period of Jan. 1980 to Dec. 1980

A rotating screen was installed on the primary clarifier skimmings line to dewater the skimmings. This material is then hauled directly to a landfill. A two- to three-ft scum layer was measured on the digesters prior to installation of the rotary screen. However, since the installation of the screen, no detectable buildup of scum layers has been noted.

Improving Overall Operational Program

Generally, anaerobic digesters are operated solely to achieve solids reduction and stabilization. The Winston-Salem program was reviewed and revised to consider digester gas utilization and production. An intensive laboratory support program was developed and implemented to provide a basis of proper operation.

In anaerobic digestion, like other biological processes, best results are achieved when the external factors affecting operation (pH, temperature, feed rate, etc.) are relatively constant. Figure 3 presents the operational schedule that was developed to provide a more consistent schedule of feeding, digesting, and transferring of the sludges. Following this schedule has improved solids reduction since it allows for the maximum digestion time.

Each of the individual digesters is equipped with gas production meters and each of the engines has a meter which measures digester gas utilization. To obtain total gas production it is necessary to either sum the production from the eight individual digesters or sum the amount of gas used by the engines. However, the two methods of calculating gas production rarely agree. To correct the problem a gas flow meter with a totalizer was installed to measure the aggregate gas flow. The results from this meter indicated gas flows approximately 20 percent greater than previously measured.

Each of the digesters has a floating cover consisting of a 0.64 cm (0.25 in.) steel decking with a tar and gravel type roof above the decking. It has been found that there is significant heat loss through the digester covers, although the exact amount of heat loss is difficult to quantify. This becomes a major problem during winter operation, especially at night and on weekends, since if the digester temperature is allowed to fall during the winter, non-daylight hours, the gas production is reduced and there is a potential for digester upset.

Therefore, to maintain digester heat, both a 1,500 kW generator and a 600 kW generator are operated on weekends even though to provide power requirements a 1,500 kW generator is suffucient. This becomes a major operating cost since although there is adequate digester gas to operate the 1,500 kW unit, there is not enough gas to operate the 600 kW unit. This 600 kW unit is then operated primarily on purchased fuel.

To remedy this problem the six primary digester covers were insulated to an R-30 value with "blown insultion" to reduce the heat loss. The cost to insulate the 40,000 sq. ft area was $12,500.00.

Figure 4 presents a chronological summary of the specific gas production rate during 1979 and 1980. Two important factors can be noted from this data.

1. After implementation of the systematic operational schedule a more consistent digester operation is noted.

2. No reduction in the specific gas production rate is observed in the winter months after insulating the digester covers.

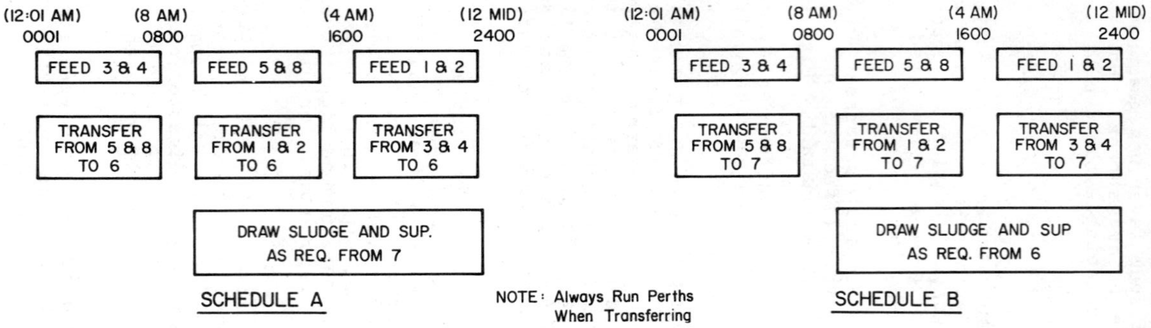

1980	1	2	3	4	5	6	7	8	9	10	11	12	13	14	15	16	17	18	19	20	21	22	23	24	25	26	27	28	29	30	31
JAN																															
FEB																															
MAR																															
APRIL																															
MAY	A	B	A	B	A	B	A	B	A	B	A	B	A	B	A	B	A	B	A	B	A	B	A	B	A	B	A	B	A	B	A
JUNE	B	A	B	A	B	A	B	A	B	A	B	A	B	A	B	A	B	A	B	A	B	A	B	A	B	A	B	A	B	A	
JULY	B	A	B	A	B	A	B	A	B	A	B	A	B	A	B	A	B	A	B	A	B	A	B	A	B	A	B	A	B	A	B
AUG	A	B	A	B	A	B	A	B	A	B	A	B	A	B	A	B	A	B	A	B	A	B	A	B	A	B	A	B	A	B	A
SEPT	B	A	B	A	B	A	B	A	B	A	B	A	B	A	B	A	B	A	B	A	B	A	B	A	B	A	B	A	B	A	
OCT	B	A	B	A	B	A	B	A	B	A	B	A	B	A	B	A	B	A	B	A	B	A	B	A	B	A	B	A	B	A	B
NOV	A	B	A	B	A	B	A	B	A	B	A	B	A	B	A	B	A	B	A	B	A	B	A	B	A	B	A	B	A	B	
DEC	A	B	A	B	A	B	A	B	A	B	A	B	A	B	A	B	A	B	A	B	A	B	A	B	A	B	A	B	A	B	A

FIGURE 3 DIGESTER OPERATIONAL SCHEDULE

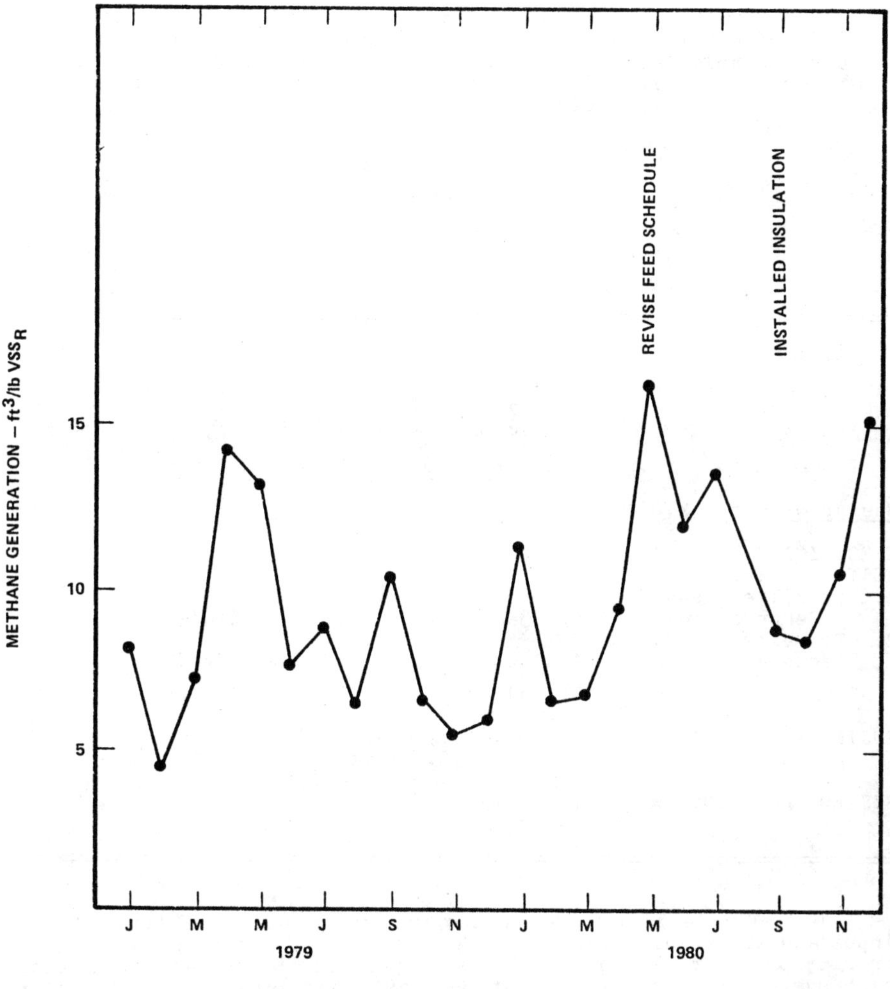

FIGURE 4 CHRONOLOGICAL SUMMARY OF SPECIFIC GAS PRODUCTION

In summary, it can be noted that the modifications to anaerobic digestor operation have improved both gas production and recovery and digester performance.

ENERGY BALANCE

To accurately evaluate the performance of the system, an energy audit was completed. This is summarized in Table 3. As noted, the plant has three sources of energy: digester gas, natural gas, and fuel oil. Fifty-five percent of the energy utilized at the plant is derived from digester gas.

TABLE 3

ENERGY BALANCE

Energy Input ($x 10^{10}$)					
NATURAL GAS =	3.8450	KJ/Yr	3.6446	BTU/Yr	23.5%
FUEL OIL =	3.4547	KJ/Yr	3.2746	BTU/Yr	21.2%
SEWAGE GAS =	9.0245	KJ/Yr	8.5541	BTU/Yr	55.3%
TOTAL	16.3242		15.4733		100.0%
Energy Output ($X10^{10}$)					
ELECTRICITY	4.9140	KJ/Yr	4.6578	BTU/Yr	
HEAT RECOVERED:					
1. To Raise Sludge Temperature	2.7435	KJ/Yr	2.6005	BTU/Yr	
2. To Maintain Digester Temperature	1.5186	KJ/Yr	1.4394	BTU/Yr	
TOTAL	9.1761		8.6977	BTU/Yr	

ENGINE - GENERATOR EFFICIENCY = (4.9140/16.3242)
30.1 Percent

SYSTEM EFFLCIENCY = (9.1761/16.3242)
56.2 Percent

The efficiency of the engine generator unit, in terms of fuel input and electricity output, is 30 percent. Ten percent of the electricity produced is required for the operation of the anaerobic digestion system (mixing, recirculation, gas compression, etc. Therefore, approximately 90 percent of the electricity produced is available for other plant uses. The performance engine-generation system is comparable to similar power generation units which indicate efficiencies of 21 to 36 percent.

In addition to the power generation, the system provides a significant amount of heat. This includes preheating of the raw sludge, maintaining the temperature of the digesters, providing building heat during the winter, and providing for the hot water plant. By including the heat captured for sludge preheating, and that required to maintain the digester temperatures, the plant obtains a 56 percent energy recovery; with the inclusion of all factors, the overall energy recovery appears to be approximately 65 percent.

ECONOMICS

In order to totally evaluate the feasibility of the system, an economics analyses is required. The factors included in evaluating the system costs were energy (fuel) costs, maintenance costs, manpower costs, operating costs, and amoritization. A summary of these costs is presented in Table 4. These indicate an average operating cost, in terms of electrical generation, of 0.98¢/1,000 BTU (power generation cost/electricity produced). This cost is 0.526¢/1,000 BTU in terms of total energy recovery (electricity plus heat).

TABLE 4

SUMMARY OF OPERATION AND MAINTENANCE COSTS FOR POWER GENERATION FACILITIES AND ANAEROBIC DIGESTERS[a]

ANAEROBIC DIGESTERS	
OPERATION:	
1/2 Operator Per Shift 3 shift/day, 365 day/yr = 4,380 MH/yr[b] =	$ 33,697.97
1/2 Lab Tech Per Day 5 day/wk, 52 wk/yr = 130 MH/yr[b] =	1,000.17
MAINTENANCE:	
1,099 MH/yr	8,455.27
Cleaning 2 Digesters	35,000.00
PARTS AND MATERIALS COST:	22,160.22
AMORITIZATION	74,625.00
TOTAL	174,938.63
POWER GENERATION FACILITIES	
OPERATION:	
1/2 Operator Per Shift 3 Shift/day, 365 day/yr = 4,380 MH/yr[b]	33,697.97
MAINTENANCE:	
2 men, 8 hr/day, 5 day/wk 52 wk/yr, = 4,160 MH/yr[b]	32,005.38
PARTS AND MATERIAL COSTS:	67,459.04
NATURAL GAS (36,446,000 cu ft)	130,946.67
FUEL OIL (233,900 gal)	140,799.91
LUBRICATING OIL (5,000 gal)	7,600.00
AMORITIZATION	44,635.00
TOTAL	$457,143.97

(a) All costs are for the period of Jan. 1980 to Dec. 1980
(b) Costs are calculated by applying avg cost/hr and fringe benefits ($6.52/hr x 1.18)

A summary of the energy costs for various sources for the period January 1980, to December 1980, is presented in Table 5. The results of this study indicate that the cost of digester gas and on-site power generation are comparable to alternate energy sources. The feasibility of on-site power generation is highly dependent upon two factors. These are:

1. The ability to maximize digester gas production and reduce utilization of purchased energy sources.

2. The ability to recover heat and utilize it for digester heating, heating of building and hot water production.

In reviewing the economics one must consider the additional costs for installing power generation facilities, backup power capability and alternate heating sources. It should be noted that many of the costs for operation of the anaerobic digesters are fixed and would appear even if there were no power generation. Based on these considerations, anaerobic digestion with accompanying power generation provides an economical method of wastewater plant operation since sludge stabilization would be required with or without on-site generation.

TABLE 5

SUMMARY OF ENERGY COSTS[(a)]

Natural Gas	0.359¢/1,000 BTU
Fuel Oil	0.430¢/1,000 BTU
Sewage Gas[(b)]	0.215¢/1,000 BTU
Commercial Power[(c)]	0.905¢/1,000 BTU

(a) Average cost for period of Jan. 1980 to Dec. 1980
(b) Based on allocating 100 percent of digester operational costs to gas generation
(c) Actual power cost with power correction factor, approved fuel charges, and billing demands included.

SUMMARY

Operation of anaerobic digesters provides a valuable by-product for use in wastewater treatment plant operation. The utilization of the digester gas is often neglected at many treatment plants. The results of the analysis at Winston-Salem indicates that digester gas utilization is an economical energy source for plant operations.

The anaerobic digester-power generation system is a cost-effective source of energy. The analysis indicated that more attention is required in the operation of the digesters as an energy generation system as compared to the stabilization of solids. In evaluating the economics of power generation, consideration should be given to the total costs of all alternatives. This would include the costs of demand charges and possible surcharges for electricity, along with costs for heating of the digesters and buildings.

In making an economic analysis of the system, one must be cognizant of the fact that sludge reduction and solid stabilization are an integral part of wastewater treatment with digester gas being a valuable by-product.

COMMERCIAL SMALL-SCALE GRAIN ALCOHOL PRODUCTION IN THE UNITED STATES

FRED S. LINDSEY
PRESIDENT, THE AGFERM CORPORATION

Thank you for the kind introduction Dr. Bentz. Good morning Ladies and Gentlemen. It is a pleasure to be here today to discuss one of my favorite topics - Small Scale Commercial Grain Alcohol or Ethanol Production. During this brief period we'll cover:

(1) My definition of "Small Scale" production facilities.

(2) Review the recent developments which have lead to the establishment of the new alcohol fuels industry.

(3) Analyze the probable scenario of the near future which will dictate energy economics.

(4) Review what is being built or is in the advanced planning stage in small plants.

(5) Examine critical factors of Considerations for production facilities.

(6) Look at the Economics of Scale on various sizes of plants.

(7) Evaluate the major factors which we believe will make the small plants economic in the future.

As Dr. Bentz metioned in his introduction, I have spent the last six years in engineering and economic studies for the development of ethanol plants. My firm has served as consultant to a large wet miller,(producting corn sweetner);a major food company;and numerous private investors or small firms evaluating the feasibility of converting existing food or alcohol beverage plants, and the building of new plants. Our major interest is not that of a consultant but as a developer and manager of ethanol plants. Since we are a small business firm, it is logical that we specialize in small plants because of capitalization requirements. My purpose here today is not to tout one size or scale over the other, but to share our thoughts on one segment of the fuel alcohol industry which we believe can play an important role in our efforts toward energy independence.

What do I mean by a Small-Scale commercial grain alcohol plant? I define this as a facility which uses grain or other biomass which can be converted to glucose or sugar which can then be fermented to produce between one and ten million gallons of anhydrous or 200 proof ethanol. The other categories of size are Middle-Scale, which ranges from ten to twenty million gallon capacity, and Large-Scale plants which produce twenty million gallons yearly or more.

You all know most of the events or developments which brought about the new fuel alcohol industry because they have occurred in just the past few years. Actually the root cause of our energy crisis has been a long term government intervention or regulation of the petroleum industry which made domestic production relatively unprofitable and forced our oil industry to move off shore. We exploited the resources of OPEC Nations and were wasteful with all energy products. In 1973-74 the OPEC embargo of oil signaled the dawn of a new energy era. Crude oil prices jumped from $3 to $12, we were forced to recognize the limits of fossil reserves. High energy costs caused inflation to spiral. Attention turned to the potential of renewable resources in biomass and solar. The escalating costs of imported oil has threatened to bankrupt the United States economy. Our $3 billion imported oil bill in 1970 has jumped to $80 billion in 1980. The dependence on foreign energy has jeopardized our national security.

Since 1973 a small band of fuel alcohol enthusiast in Nebraska has grown to involve hundreds of thousands across this land. Our initial concern was to find an increased market for agricultural grains. In 1954 a bushel of corn was equal in value to a barrel of crude oil. In 1974, the crude oil was three and a-half times more expensive. Today, oil costs eleven times more than corn. Our agricultural industry has made great strides in productivity gains per acre, so much so that the government was setting aside land in a soil bank and paying farmers not to plant crops on that land. In 1975 I was project manager for the first feasibility report on grain alcohol for the State of Indiana. We concluded at that time that grain alcohol was not economically feasible, but that it would become feasible by the time a major plant could be built because:

a. Improved engineering could make the plant energy efficient.

b. Synthetic alcohol prices were climbing because of the petroleum feedstock cost escalation.

c. Biochemical technology improvements were being developed which would allow the use of cheaper biomass material, rather than grains.

d. Energy costs would continue to escalate as domestic supplies dwindle.

These things have all come to pass, unfortunately. The small band of fuel alcohol enthusiast grew into the National Gasohol Commission, the Renewable Fuels Association and other groups. We had to fight the big oil companies and the big grain companies, as well as the bleeding hearts who argued that using corn to make alcohol would deny food to the starving millions. They didn't know, or refused to accept the fact, that about 72 percent of our corn is fed to animals. Those animals are inefficient converters of raw grain and most of the corn energy value ends up as manure on the floor. If we fermented the United States corn normally fed to animals we could produce over a billion gallons yearly of ethanol and still have a high protein feed supplement that is almost as nutritious for animals as the original corn. We export one third of our grain, about forty billion dollars worth, overseas to countries who use it primarily as animal feed. Russia alone imports over 43 metric tons yearly of grain, again, primarily for animal feed, United States embargo or not. Since the major value of grain for animals or people is the protein value, and since fermenting the starch from grain for alcohol does not affect the protein byproduct, doesn't it make sense to extract the energy as alcohol for domestic use and export the protein? In that way a $3 bushel of corn would produce a value of about $4.68 in alcohol plus $1.44 in animal feed. This more than doubles the value of the agricultural commodity. Pardon that detour in thought, but it is essential to understand the basics about the economics of fermentation alcohol.

In our search for alternative energy sources we quickly rediscovered ethanol or grain alcohol. The domestic market for grain ethanol had sharply declined over the years due to cheap gasoline and the process of making less expensive synthetic ethanol from cheap natural gas. Grain alcohol was relegated then primarily to the beverage alcohol market. Most of our existing distilleries were about 35-40 years old and were very energy inefficient. Many more distilleries had been shut down or demolished when the synthetic ethanol production started. As an example, there were 15 distilleries near Bardstown, Kentucky (Nelson County) twenty years ago. Now there are only seven. The most extensive beverage alcohol industry in the United States is in Kentucky but many of the Kentucky distilleries operate on a one-shift basis or shut down entirely for several months or more each year. As a result there is an excess capacity of 15 million gallons in these existing facilities. This capacity can be converted to fuel alcohol quickly and economically.

The Congress in 1979-80 passed extensive legislation to assist the development of fuel alcohol and other alternative energy sources. The National Alcohol Fuels Commission provided much of the leadership for this effort. President Carter declared our energy needs in terms of the "moral equivalent of War". If we must ever fight a war again I hope we exhibit a more "all-out" effort. Still, we made a great beginning. The consumers acceptance of Gasohol was very good. Excellent engineering and technology advances were made. Some major oil companies such as Texaco and Ashland Oil have begun to get involved in the production of ethanol. The Synthetic Fuel Corporation was established.

The moral commitment of government was finally evidenced by a variety of programs and incentives. It was this "government commitment" to encourage alternative energy that finally convinced many people to seriously pursue a variety of projects. Government commitment to domestic alternative energy is absolutely necessary to gain the support of the financial institutions because they realize that without that commitment the government can regulate the new energy industries out of business. Enormous sums of corporate and private investor dollars have been committed over the past year. Now much of this momentum has been jerked to a halt! Today, much of the progress in the fledgling industry and the credibility of government is threatened by the meat ax approach to bring our economy in line. We certainly must bring our budget into balance. I believe there has been a great waste in many governmental programs. But as Congressman Fithian of Indiana recently testified "Investment in Synthetic fuel production facilities is the government's most cost effective tool in adding to domestic energy production on which our National survival depend

We are now paying out $10 million every hour to foreign oil producers. Consider for a moment a possible future imaginary scenario as projected in the National Geographic's recent and very excellent special report on Energy (February 1981). This scenario projected that the Iran-Iraq war spilled over into Saudi Arabia, wrecking its oil production. That tragic event wll remove one-fifth of the oil available to the free world. Oil prices will skyrocket to $100 a barrel and the temporary oil glut of 1980 will quickly disappear. Although the United States currently gets a small amount of its imports from Saudi Arabia, the loss of the Saudi oil to the world will quickly cause a total loss to the United States of more than half it imports. Domestically the effects will be disastrous. Fuel shortages will curtail commerce and cause violence. Rationing will be slow and clumsy. Unemployment will double and inflation will pass 30 percent. Our defense forces will be severly curtailed because of the fuel shortage. "And all for the want of imported oil." I personally believe this scenario is very likely to occur. The Iran-Iraq war can boil over at any time. The Russian invasion of Afganistan may simply be a prelude to a quick march through Iran and Saudia Arabia to the Persian Gulf.

What role can the new alcohol fuels industry play in todays situation or that of the worsening scenario? The National Alcohol Fuels Commission's (NAFC) survey indicates planning for 340 ethanol plants, ranging from on-farm stills to 100 million gallons in size. If all were completed the total annual capacity would exceed 4.5 billion gallons. The NAFC estimates however, that only 850 million gallon capacity will be produced by early 1983, with likely increases of 1.4 billion gallons by 1985.

Lets look at the location and sizes of these announced plants. Please keep in mind that many of them are only in the planning stage and may never get off the ground. Figure 1 shows the number of plants of all sizes by state on the NFAC survey. Also shown is the map location of those small-scale commercial plants that I have been able to verify. Table 1 provides a lisiting of these Small-Scale plants, which might total about 104 million plus gallon capcaity. To my knowledge only six are now operational.

Now that we have looked at some of the projections, let's back up a minute to review the likelihood of success in getting these plants established and viably operating. There are many considerations that effect feasibility. You must be able to physically and financially build and operate the plant to produce products which can be sold for a profit to a market. There is not time here today to discuss all of the factors of feasibility, but I want to comment briefly on those that I believe to be the most critical. A famous Kentucky distiller once told his sons that what you needed to put a distillery into operation was "lots of water and lots of money". That is true as far as it goes. At this point in time I would say that the single most important factor is Financing. Regardless of the proforma for an individual plant, the economic climate in the

TABLE I

PARTIAL LISTING OF SMALL-SCALE COMMERCIAL ALCOHOL PLANTS

NAME	PLACE	CAP./GAL.	REMARKS
White Flame Fuel	Van Buren, AR	3mm	operating USDA - ACR
Organic Resources & Engineering	Maricopa, AZ	2mm	
Golden By Products	Turlock, CA	2mm	Almond Hulls
Colorado Agro-Energy, Inc.	Monte Vista, CO	2mm	Potato & Barley; constr.
Colorado Gasohol	Walsh, CO	3mm	Milo; ACR; constr.
Modern Fuels, Inc.	Loveland, CO	1mm	
United Energy Systems	Commerce City, CO	8.5mm	Sugar Beet
Ethylene Corp.	Jacksonville, FL	3mm	operating
Florida Ethanol Inc.	Mayo, FL	5mm	
Pryor Enterprises	Barnsville, GA	1.3mm	
Syncorp, Inc.	Roberta, GA	2.5mm	DOE-USDA Retrofit
United International, Inc.	Buena Vista, GA	4.9mm	
Idaho Fuel Alcohol Producers	Boise, ID	5mm	
Power Alcohol, Inc.	Blackfoot, ID	5mm	USDA II
Western Resources Recovery, Inc.	Twin Falls, ID	6mm	Potato
Magic Resources	Sun Valley, ID	2mm	Geo-DOE
Agrihol	Rupert, ID	5mm	Barley
Illini Energy Resources	Dekalb, IL	5.6mm	

PARTIAL LISTING CONTINUED

NAME	PLACE	CAP./GAL	REMARKS
Illini Energy Resources	Dekalb, IL	5.6mm	
Rochelle Corn Products	Rochelle, IL	5mm	
Rochelle Energy Development	Rochelle, IL	2.5mm	
Bryant & Toff	Coatsville, IN	1.5mm	operating
Portland Fuel Alchol Company	Portland, IN	3mm	USDA
Third Century Venture	Kokomo, IN	6mm	constr.
Agrifuels Corp.	Estherville, IA	5.5mm	USDA II
American Development, Corp.	Hamburg, IA	5mm	
Elgin Alcohol, Inc.	Elgin, IA	2.5mm	
Farm Fuel Products Corp.	Storm Lake, IA	2.3mm	USDA; ACR; Retrofit
Midwest Solvents Company	Atchinson, KS	3mm	operating
Bardstown Fuel Alcohol Company	Bardstown, KY	4mm	renovating USDA; ACR
National Distillers	Frankfort, KY	9.5mm	renovating
National Distillers	Lexington, KY	7.5mm	renovating
Xcel Corp.	Cynthiana, KY	3mm	renovating
Petrohol Associates	Fairfield, KY	5.4mm	
Goodwill Agri-Fuels Inc.	Goodwill, LA	3.3mm	USDA II
Michigan Agri-Fuels	MI	8mm	
Agri-Energy, Inc.	Crookston, MN	4mm	USDA II
Kraft	Melrose, MN	2mm	Retrofit; Vulcan; Cheese whey

PARTIAL LISTING CONTINUED

NAME	PLACE	CAP./GAL.	REMARKS
Alcohol Fuels of Mississippi	Vicksburg, MS	3mm	wood
Sun Belt Energy	Sardis, MS	4.4mm	DOE
Missouri Farmers Association	Mexico, MO	5mm	
AG-NRG, Inc.	Great Falls, MT	1mm	Barley
Energy Engineering Inc.	Hot Springs, MT	7.9mm	Geo.
Infinity Oil Co.	Mosby, MT	6mm	wheat
Montana Agri-Process	Glascow, MT	3mm	USDA II; Barley
Sun Praire Energy	Fairfield, MT	1.5mm	Barley
Agri-Hol, Inc.	NE		USDA
Alco, Inc.	Grafton, ND	3.5mm	Retrofit
Gasohol Corp.	Riverhead, NY	3.5mm	
Oklahoma Gasohol	Hydro, OK	3mm	constr.; ACR
Courtright	La Grande, OR	2mm	operating
Grande Ronde Commodities	Hot Lake, OR	2mm	
Technology International	Vale, OR	5mm	Geo.; DOE
Santee Production & Marketing Corp.	Greenleyville, SC	1mm	
SEPCO	Scotland, SD	1mm	
Smith Bowman	Reston, VA	2mm	Retrofit; ACR
Georgia Pacific Corporation	Bellingham, WA	2mm	operating wood liquor
Milbrew, Inc.	Juneau, WI	2mm	operating cheese whey
Mousinee Alcohol Company	Mousinee, WI	1mm	

Figure 1

PLANNED ALCOHOL PLANTS

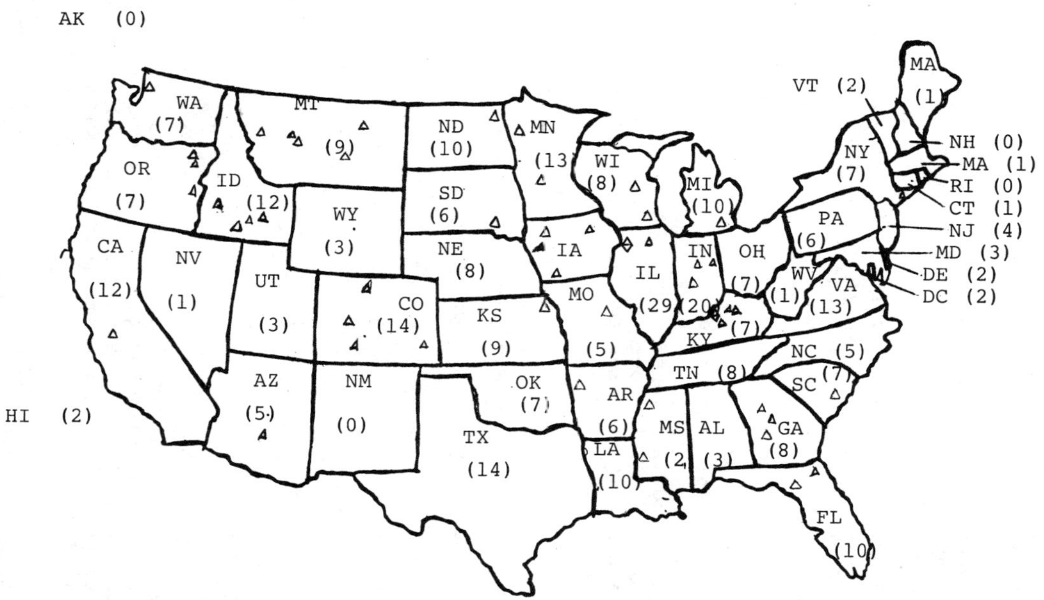

Legend: (5) Number of Planned Ethanol Plants of all sizes, per state

△ Location of Confirmed Small-Scale Commercial Plants

Source: US Naticnal Alcohol Fuels Commission (Rev)

United States today is very unstable. The prime lending rate has hit two record high marks with in the last year. Major industries and even major cities are on the edge of bankruptcy. All of the efforts that have been made over the last two years to help establish alternative or syn-fuels are now threatened as the Reagan administration tries to sort out its fiscal policy.

The loan guarantee program is now in chaos. The best example of this is the recent "freeze" by the Department of Agriculture on all Conditional Commitments for guarantees for alcohol plants. The awkwardness with which the transition teams'audit or program review has been conducted and the sensationalism of the press releases has severly damaged the credibility of all government commitments in the eyes of the banking industry. Some projects have been wiped out by the press releases which hinted of scandal just as stock issues were being sold. Our own project which is under construction in Kentucky has already suffered financial loss because of this uncertainty about the government reneging on it's commitment, even though we studied the project for two years and our application was reviewed by the government for over five months and our project was one of the first to be approved.

I wish I had never heard of a government loan guarantee program, but as long as it was there, the banks insisted that we qualify. Now that we are qualified, the lenders don't believe the government will honor its commitment. But enough of my own horror story. Lets assume that some stability returns to interest rates and the government bridges its "credibility-gap".

Look at the Economics of Scale reflected in Table II.

Table II

Plant Capacity	Investment	
MM Gal/YR	Fixed $	$/Gal
100	127	1.27
50	74	1.48
25	49	1.96
10	26	2.70
3	6	2.00

As you can see, the smaller the plant, the larger the investment on a per annual gallon capacity basis. The per gallon cost of the ten million gallon plant is more than double that of the 100 million gallon plant. But note that in the smallest size shown - 3 million gallons, the ratio is back nearer the 25 million gallon size.

It is our belief that the most economic alcohol plants will be those which are coupled to large wet-milling operations. The alcohol fermentation acts as both a profit center and a waste treatment for the weaker starch streams. This type plant has a flexibility of operation to respond to market conditions for the corn sweetner vs. alcohol fuels. The middle sized plants are <u>not</u> considered viable, in my judgement. They are too small for wet-milling economics and too large for cost reductions available to the small sized commercial plants located in agricultural communities. The exception to this would be those middle sized plants that may be structured from the renovation of existing breweries, sugar beet mills, food plants or distilleries. The Davy McKee Corporation study for the National Alcohol Fuels Commission showed a potential of 581 million gallons from these sources was possible by 1985 at a cost of about 755 million dollars - an investment ratio of about $1.30 per gallon annual production. I have been working on one project which would integrate and retrofit two distilleries to produce 10 million gallons for about $8 million, or 80¢/gallon, for the current owners.

The advantages of the small plants derive from the close tie to the agricultural community, reduction in transportation costs and the smaller financing challenge as noted in the table below:

Table III

ADVANTAGES OF THE COMMUNITY-SIZE DISTILLERY

- USES LOCALLY PRODUCED CROPS AND COAL
  - Direct purchase negotiation possible
  - Saves freight and broker costs of up to 25¢/bushel
- PERMITS ENERGY AND STORAGE SAVINGS ON THE FARM
  - Save cost of drying corn (10¢/Bu) by using plastic AG-BAGs for corn storage
  - Permits greater storage capacity on-farm
- ALCOHOL PRODUCTION CAN BE USED LOCALLY
  - 3 million gallon plant production can be used with about thirty mile radius (100 gallon per vehicle)
  - Saves 11¢/gallon freight costs
- RELATIVELY SMALL COST PERMITS LOCAL OWNERSHIP
  - $7 million cost financed by $2 million private investors and $5 million Industrial Revenue Bonds
- BY-PRODUCTS CAN BE USED LOCALLY
  - CO-2 for refrigeration, carbonation, water treatment Greenhouses and welding supplies, and teriary oilfield recovery.
  - Distillers Grains (DG) - ideal cattle & poultry feed If DG sold wet, can save one third of production energy
- SIZE IS MORE CONVENIENT FOR INDUSTRIAL INTEGRATION
  - Feedlots, Greenhouse, Solid Waste Recovery Centers
- SAVES TRANSPORTATION ENERGY AND COSTS
  - On raw materials and finished products

The close tie to agriculture can be a major factor for viability of the small plant. Farmers support Gasohol because they recognize it as the main way that they can participate in a decentralized form of domestic renewable energy to improve their own economics. You know, the farmer has always been thought of as being self-sufficient and independent. That is no longer true in this era of mechanized and specialized farms. The farmer is now as dependent upon foreign oil as any industry. Ninety three percent of the energy used for agriculture production is derived from petroleum. U.S. farmers used 7.6 billion gallons of petroleum in 1980. Also the majority of the fertilizer used in the corn belt is anhydrous ammonia which is made from natural gas. The farmer is at the end of the energy distribution system - be it pipeline or electric line, and therefore, he pays the highest cost for his BTU. Because he is at the end of the line, the farmer will be the first to feel the brunt of the next oil supply interruption.

In addition to energy integration schemes, the small plants can take advantage of skid mounted modular construction such as for the Johnston Fluidized Bed Boiler, or use the new innovative engineering such as that developed by Butler Research & Engineering of St. Paul, Minnesota. Butler uses a liquid/solids separation after fermentation and before distillation. This enables the use of smaller packed distillation columns. A mechanical vapor recompression evaporator recovers the solubles. The new biochemical technology improvements of biomass conversion holds the most promise for long term viability for the small-scale plant. High temperature yeasts and continuous fermentation techniques will also contribute to improved economies.

In conclusion, we need all of the domestic production of energy which we can develop and especially that which is renewable. We can not afford the gamble of depending upon foreign oil to sustain our national security. Domestic synfuels, solar, nuclear and biomass must all be developed with urgency. Alcohol fuel from grain or other biomass can play a small but vital role in our quest for energy independence. It is the only source which can achieve substantial increases in production in the near term.

THE GASOHOL INDUSTRY IN CANADA

R. C. QUITTENTON
EXECUTIVE DIRECTOR, THE INNOVATION PLACE
SASKATCHEWAN ECONOMIC DEVELOPMENT CORPORATION

We say in Canada that we run about 10 years behind the United States in many things. Certainly this is true of the gasohol industry. In 1971 Nebraska started the gasohol industry in the U.S. by reducing the gasoline tax 3 cents a gallon for fuel containing 10% alcohol from biomass. In 1981, just ten years later, we will have our first gasohol in Canada on the market in Manitoba. This is coming from the Mohawk Petroleum Company, a Canadian oil company that has made a corporate decision that liquid fuel from renewable feedstock is here and here to stay. Mohawk has started by buying a former whisky distillery in Minnedosa, Manitoba, and renovating it for fuel alcohol production. Either barley or corn will be used for feedstock, depending on relative price. The Minnedosa plant should be producing about 2,000,000 U.S. gallons a year, at a renovation cost of about $650,000 U.S. on top of the purchase price. Mohawk plans to double this capacity as operating experience is gained. Hereafter I will use American terminology unless I specify Canadian. An effective exchange rate of 20% is also used. Those of you in the gasohol business know this kind of an operation is not yet profitable without a subsidy. So we too have a subsidy in Manitoba, following your ten year leadership. The province has granted a tax credit of 4.6 cents Cdn. a liter on vehicle fuel containing 10% fuel alcohol produced anywhere in Canada from Canadian biomass. This is 46 cents Cdn. a liter for the alcohol or $1.39 per gallon of alcohol. This compares with the current average U.S. tax credit of about 90 cents a gallon, from the sum of federal and state credits, where these are given. The tax credit difference is caused by the difference in gasoline prices. Unleaded gasoline now goes for $1.11 a gallon wholesale in New York State (February 9/81) versus 55 cents in Ontario. This is because our gasoline is heavily subsidized. Not all of the U.S. States have given tax credits nor have all of the Canadian provinces. The only other province with a credit is Ontario. Ontario eliminates the gasoline tax of 4.6 cents a liter on any alcohol from biomass used for fuel. Note that this is for the alcohol only, and not for the blend. Thus this alcohol has a tax credit of only 14 cents a gallon, not enough to put anyone in business, and as a result there are no gasohol ventures planned yet for Ontario. This is really a classical Canadian response and Ontario is the most Canadian of all the provinces: don't get too far ahead or you might get lonely, but don't get too far behind or you might get left. So you can't knock Ontario for not trying and it doesn't cost them anything. Besides, they can get western Canada crude at $23 a barrel under the world price so why rock the boat. Most of the other provinces are too busy trying to get Britain to invade Canada to save them from Ottawa to get involved in biomass just yet. The federal government offers no tax incentives at all for fuel alcohol from biomass, unlike here in the U.S., where you have a 4 cents a gallon incentive for gasohol.

We do offer all kinds of tax subsidies however for fossil fuels. In 1980, the shareholders of Dome Petroleum, as one example only, paid 10 cents of their own money out of every dollar spent by Dome on frontier work for oil. In 1981, these shareholders will pay only 7 cents. The taxpayers will pay the other 93 cents as a direct subsidy. The whole synfuels industry in your country and mine is subsidized. You can't help but get a little fed up with those who point to subsidies for gasohol with scorn, but say nothing at all about the current subsidies for oil from coal, shale and tar sands, in both countries. The whole nuclear energy industry in Canada is a result of billions of tax dollars, subsidies, poured into the development of the CANDU reactor system, by Atomic Energy of Canada Ltd., a crown agency. The new coal contracts to ship coal from British Columbia to Japan are subsidized by about $9 Cdn. a ton for transportation infrastructure. The coal mines in Nova Scotia are about to get a direct subsidy of $100,000,000 Cdn. In the tar sands industry in Canada the governments, federal and provincial, give away much money in tax concessions as the private sector invests. For a tar sands venture costing $8 billion Cdn. for an output of 140,000 bbls a day of tar sands oil, this amounts to a subsidy of 7 1/2 cents Cdn. per liter for the transportation fuel produced, over the whole 25 year production period. How much are you paying here in subsidies for vehicle fuel from coal and oil shale? I get the impression from Canada that these operations are essentially all tax-supported. When you cost out these subsidies remember you are subsidizing them only for a portable fuel, for gasoline, diesel fuel or jet fuel. All our other fuel needs can be met very readily in both our countries from natural gas, coal or electricity. Thus the bunker fuel oil residue, the heavy distillates, the asphalts and the coke from natural or synthetic petroleum crudes have no real value to the taxpayer. We subsidize these fuels and these programs only, and I repeat only, for the portable fuel portion of the product. This comes to only one-half of the oil itself. Thus all the visible subsidies for synfuels have to be doubled to put them on the back of the portable liquid fuels produced, where they belong. Who needs to subsidize asphalt?

This subsidy issue for petroleum is particularly striking in Canada. We are now importing 50,000 bbls a day of Mexican crude at $38.50 a barrel (February 1981 price). Our federal government has fixed the well-head price of Canadian western crude at $17.75 Cdn. for January 1981 or about $15.20 U.S. delivered to eastern refineries. This is the price consumers pay at the pump, plus taxes and refining costs, etc. The difference between the Mexican crude of $38.50 and the fixed price of $15.20 or $23.30 is subsidized by all Canadians. Now Mexican crude is very heavy. Our refineries only get 4 barrels of vehicle fuel out of every 10 barrels of Mexican crude. This means the real subsidy, for gasoline, jet fuel and diesel fuel, the equivalent of fuel alcohol from biomass, is $58.25 per barrel of vehicle fuel. The point is, a barrel of fuel alcohol can go right into the pump. A barrel of Mexican crude cannot. The blunt fact is we would not import this Mexican crude if we did not need it to run cars, trucks, trains and planes. Now this $58.25 U.S. per barrel, or 43.8 cents Cdn. per liter direct subsidy leaves Canada. We get no economic spin-off from it at all. On the other hand, the 46 cents Cdn. per liter direct subsidy the government of Manitoba is now offering on fuel alcohol stays in Canada, chiefly in Manitoba. It is retaxed many times, for property taxes, income taxes, sales taxes. It generates jobs, productive jobs, and cuts welfare payments. In this very real and accountable fashion, governments get back about 40% of such productive subsidies. This brings the net fuel alcohol subsidy in Manitoba to 27.6 cents Cdn. per liter versus the 43.8 cents Cdn. per liter for vehicle fuel from Mexican crude. You don't have to be much of an economist to see where the bargain is. Gasohol is a real bargain for Canada. I suspect that if you sharpen your pencils and winkle out all of the hidden subsidies now going to vehicle fuel from fossil fuels, you will find gasohol is not a bad buy in the U.S. also.

But what about feedstocks you ask? Well, we now have some 34,000,000 hectares of developed land in western Canada assigned for production under the national quota system. There is, of course, additional land not registered under the quota system. However, of the registered land, about 23,000,000 hectares are in production and the rest is generally in summerfallow. Thus about one-third of western farm land is kept out of production each year. The argument is that this is to preserve moisture and to control weeds. The overriding reason however is that the western farmer has never been pushed for production because he cannot be sure of a market for his maximum output. Thus the 1980 barley quota, for off-board grains, the maximum allowable, was 827 kilograms per hectare, whereas the 10 year average yield was 2176 kilos per hectare, or nearly three times what the wheat board would accept. The same story applies for wheat. On the other hand, more and more evidence is accumulating that summerfallow is harmful, particularly since it leads to salination and wind erosion of soils. Unless this salination is stopped, thousands of hectares will be lost, perhaps forever. The situation is serious enough that last August it was proposed at the annual meeting of the Agricultural Institute of Canada that summerfallowing be legally banned in Canada. Well this caused a bit of an uproar as you might imagine, with visions of Mounties charging off over the Prairies on horseback looking for outlaw patches of summerfallow, doubtless to capture on shots of Polaroid, because the Mounties always get their quarry. The fact remains however, that the basic problem is an outlet for this stifled production capacity. These 11,000,000 hectares of summerfallow could produce 48,000,000 barrels of fuel alcohol a year, from barley, or some 130,000 bbls a day. This amount of fuel alcohol can replace at least 250,000 barrels of imported crude oil a day, since this crude only yields a half barrel of vehicle fuel, on the average. Thus quite conservatively, the imports of Mexican crude of 50,000 bbls a day could be replaced by Canadian fuel alcohol without straining the grain system. This would be a net gain to the Canadian economy of roughly $200,000,000 Cdn. a year, after allowing for the gasohol subsidy, and a reduction in our import bill of roughly $850,000,000 Cdn. a year.

This same fuel alcohol production, replacing 50,000 bbls a day of Mexican crude would also produce roughly 2,800,000 metric tonnes of distillers dried grains a year, holding at least 25% protein or some 700,000 tonnes of protein a year. This is a new addition to the supply of world protein, since this land is not now producing anything. Companies such as Archer Daniels Midland are upgrading this grain byproduct protein to direct human food. Most nutritionists claim the world food deficiency is not in starch or sugar, but in protein. Indeed, if we converted into gasohol all the feed grain in North America, corn and barley, that is now going into cattle rations we would increase the world protein supply by perhaps twice, since fuel alcohol produces about one tonne of protein from every 12 tonnes of feed grain, while cattle yield about one quarter of a ton of protein from 12 tonnes of feed grain. So if we were really serious in North America about addressing the world protein problem the first thing we would do would be to push the cow back onto grass, and to convert all the feed grain so released into gasohol and byproduct protein. Simultaneously, we would reduce the North American drain on the diminishing world pool of petroleum. The world would welcome both of these gains, for more protein and more oil. So I'm afraid I get turned off when I hear some learned type expounding eloquently on how gasohol will starve the world. Frankly that speaker doesn't have a clue. There are opportunities too for new directions in feedstock. The University of Canterbury in New Zealand is working with the fodder beet, a cross between a sugar beet and a mangold. They claim alcohol yields from this of 9000 liters per hectare, or twice that of sugar cane and four times that of corn. We have done some preliminary work in Canada with Danish fodder beets and obtained about 3600 liters of alcohol per hectare, well above corn or barley yields. The fodder beet is of interest because it grows almost anywhere, like the Jerusalem artichoke, and combines the high yields of the mangold with the high dry matter content of the sugar beet. It is however subject to nematodes, a problem that must yet be resolved. In all these feedstock considerations it is worth noting also that these crops consume CO_2 to grow. This CO_2 in turn comes from the combustion of the alcohol produced from the grain. Thus the CO_2 is recycled and does not build up in the atmosphere. I mention this since your President's Council on Environmental Quality has just reported in its "Global Energy Futures and the CO_2 Problem" that you should and I quote:

> "... assign a high priority to incorporating the CO_2 issue into U.S. energy policy planning".

This means CO_2 control has a cash value, just like SO_2 control. Would you like to speculate on the fuel cost of putting CO2 scrubbers on cars driven by fossil fuel?

Canada has by far the largest trade deficit in the developed world in terms of jobs per million dollars of trade. This is a major reason why we have about the highest unemployment rate in the developed world. Our exports are basically raw materials, resources, with low job inputs. Our imports are basically finished goods, with high job inputs. Wheat is a classic case. Very few jobs are generated in the production and export of a million dollars of wheat or barley, as grain. Upgrading feed barley to fuel alcohol generates more jobs than exporting it. It also reduces the load on elevator and transport systems. For these reasons I am disappointed that the Wheat Board is not yet promoting gasohol, as being of more benefit to the country and the farmer than export. There is a question of national interest here in my view. You should recognize also that a gasohol job can be generated for a fraction of the cost needed to generate a synfuels job from fossil fuels. It costs at least five times as much to generate a synfuels job as it does for a gasohol job. For example, it costs about $12 Cdn. in capital to produce a million annual kilojoules of energy from fuel alcohol versus $60 for a tar sands plant. Here it can be noted that the new and emerging technologies are very favorable for gasohol. The autocatalytic compression process to convert starch into sugar in three or four minutes, plus the continuous packed tower fermentation process to convert sugar into alcohol in two or three hours produce a plant with the same kind of rapid throughput as one gets from the huge synfuels from coal plants. The alcohol plant is, however, far more resistant to inflation as the wear on the alcohol plant is only a tiny fraction of the enormous replacements faced every day in the synfuels fossil fuel plant from the high temperatures, high pressures and erosive character of the fluid streams. And every day inflation drives up the cost of these replacements. The gasohol job can be generated also in under two years, versus the six to eight for the tar sands or oil shale job. This is why one can only be saddened when our Federal Minister for Energy, the Honourable Marc Lalonde says on fuel supply in the February 1981 issue of Canadian Business and I quote:

> "Don't forget that between now and 1987 there is very little you can do to increase Canadian supply, even if you throw billions and billions of dollars at it."

He is being poorly advised as you will recognize from your experience with gasohol in the U.S. If he were to launch a Canadian gasohol program analogous to yours we could replace between a third and a half of our petroleum imports within three years, or by 1984, not 1987. If this is not done, and there is any world crisis at all, we will be rationing gasoline in Canada by 1987 as Mr. Lalonde himself admits, when he says in the same article:

> "... that we would have to go to rationing with terrific speed".

Time and money are precious in both our countries. We must both press on with gasohol. And good luck to you! We need your leadership!

SMALL IS BEAUTIFUL:

ALTERNATIVE FUELS FROM POWER ENGINE WASTE HEAT

by

B. J. KIRKWOOD, P.E.
A. C. KIRKWOOD & ASSOCIATES
Consulting Engineers
Kansas City, Mo.

Presented at the eighth annual Energy Technology Conference and Exhibition, Washington, D.C., March 11th, 1981, as part of the session on "Integrated Diesels for Power Generation".

*   *   *

This paper might have been titled less prosaically as 'The Businessman's Lunch: Grog and Grub from Waste Heat', for we are considering the potential for producing tangible amounts of ethyl alcohol (ethanol), plus animal feed, by utilizing the normally-wasted heat from producing electricity. Specifically, we are relating this to the numerous small, diesel-engine powered generating plants in this country - most of which are located in rural, agricultural areas. And - hopefully - prudent businessmen (and agencies) will see attractive economic potential in such hybrid operations.

Most people today associate electric generation with super-large, steam power plants - usually two to four units at 300,000 kW to 800,000 kW each. That such units can be very efficient is a fact. That they are necessarily more efficient than units at, say, 1% of that size is not a fact. Similarly, in the burgeoning field of ethanol production most effort is being centered on plants of 20,000,000 to 50,000,000 gals/yr, as being the only possibility to achieve economy-of-scale in plant concepts and costs.

> THESIS: it is possible to realize attractive economies and other benefits from judicious mating of small ethanol plants with small, diesel-powered generating plants. Furthermore, in so doing it may be possible to meet a significant portion of the national goal for alcohol production, as a supplemental or alternative fuel vis-a-vis oil.

PERSPECTIVE ON DIESEL POWER GENERATION

This concept is not dependent upon building a new infrastructure of diesel plants so as to produce this 'waste heat' (although new units normally would also come about to meet growing electric loads and as replacements). Actually, a substantial number of such plants already exist, although as a proportion of total U.S. electric utility capacity they are almost insignificant, as per Table 1.

TABLE 1

U.S. GENERATING CAPACITY
(1979)

Type	GW	%	GWh	%
Steam - fossil	411.7	69	1,676,000	74.6
- nuclear	53.7	9	255,000	11.3
Hydroelectric/Geothermal	76.0	13	284,000	12.6
Combustion turbines	50.6	8	28,000	1.2
Diesel engines	5.5	1	4,000	0.2
Total	597.5		2,247,000	

GW = 1,000,000 kilowatts (kW)
GWh = 1,000,000 kilowatt-hours (kWh)

Diesel plants are only 0.9% of capacity, and produce an even lesser share (0.2%) of electric energy. Nevertheless - and of especial significance - this represents 3100 units in 920 plants. Of these plants some 35% are investor-owned and 10% are cooperatives, while 55% are municipally operated. And of these all, 80% are to be found in the East North Central, West North Central and West South Central FERC (ex-FPC) regions. There, too, and not so coincidentally, the bulk of grain is grown and cattle are raised. Most producing diesel units in these plants range between 2000 and 6000 kW each. That is about 3000 to 9000 horsepower - not exactly your typical automotive diesel, but still small by central station standards.

Although there are 3100 such units, they are only 0.9% of total capacity. Similarly, they use a very small portion of fossil energy consumed in U.S. electric production, illustrated by Table 2.

TABLE 2

FOSSIL FUEL CONSUMPTION - ELECTRIC UTILITIES
(Estimated 1979)

Type	Fuel Used	Btu	M BOE*
Steam - coal	529 M tons	$11,109.0 \times 10^{12}$	1763
- oil	480 M bbls	$3,024.0 \times 10^{12}$	480
- gas	3,315 M mcf	$2,983.5 \times 10^{12}$	474
Comb turb - oil	42 M bbls	242.6×10^{12}	42
- gas	140 M mcf	126.0×10^{12}	20
Diesels - oil	2 M bbls	11.6×10^{12}	2
- gas	35 M mcf	31.5×10^{12}	5

*MBOE = million bbls of oil equivalent.

Thus, these units have been consuming about 0.3% of total utility fossil fuel (about 0.7% of the oil and gas). In further perspective, however, this represents only about 0.11% of total gas and oil consumed in the U.S. Not only is this insignificant, but even a doubling or tripling (if justified by other societal benefits) would not noticeably impact availability or reserves of such fuels.

Why do such plants exist? Initially, most served electrically-isolated communities before transmission systems were widespread. But these plants were found to be economically competitive, too. The attributes realized are listed in Table 3.

TABLE 3

KEY ATTRIBUTES OF DIESEL-POWERED ELECTRIC PLANTS

- Low first cost - @ 1/2 to 2/3 of a coal plant
- High fuel efficiency - @ 33 to 38% vs steam @ 30 to 37%
- No transmission/transformation losses and costs
- Competitive operation and maintenance costs
- Reliability - high unit availability; no transmission outages
- Flexibility - "easy-on, easy-off"; as little as one minute to get on-line
- Dispersed, low-level pollution
- Local employment

Understandably, economics do not always favor such plants. Particularly troublesome today is their dependence on and the cost of gas and oil fuels. The 1978 Fuel Use Act does give such plants legal right to use these fuels, however, subject to their absolute availability and economic viability. Supplies of gas actually are improving in most cases. But its cost is now up 10-fold (in under ten years) whereas coal is up about four to eight times. Oil fuel also is up, by a factor of about nine. The future costs of all fuels is 'certainly uncertain' - except in direction, of course.

Faced with uncertainty like this, what are engine-users able to do to re-enhance the viability of such equipment? Table 4 lists some.

PERSPECTIVE ON ETHANOL

Little had been done to apply ethanol to fuel uses until the world's 1973 rude awakening to its oil vulnerability. Both methanol and ethanol have since then received growing attention, and in the U.S. in particular ethanol has been seen by many as having congruent energy and agricultural virtues (though others reach diametrically opposite conclusions for the same reasons). Ethanol can be used by itself as IC engine fuel (even in its 'impure' 160-190 proof state) or in a mixture with gasoline ("gasohol"), for which the alcohol must be virtually 200 proof. The attraction to many in this country (and a few others) is that it can be produced readily by fermenting grain carbohydrates, thus offering a new market for grain producers.

TABLE 4

POTENTIALS FOR ENHANCING DIESEL GENERATION

- Improve traditional efficiencies
  - Optimal operation, load matching
  - New equipment developments
- Expand unit thermo-mechanical cycles
  - 'Bottoming' units
- Add external waste-heat applications
  - District heating and cooling
  - Integrated energy systems
  - Industrial-process heat
  - Agricultural-product drying
  - Process/chemical applications (eg, ethanol)

Under President Carter's administration an objective was set to increase industrial ethanol production capacity from under 100 million gallons per year to 500 million gallons in 1981 and 8.5 billion gallons by 1990. (These are equivalent to about one percent and 10% of gasoline in those years.) Multi-pronged Federal efforts to encourage this rapid escalation have included gasoline tax abatement on gasohol, grants for studies and development, loans and guarantees for facilities and changes in regulations (such as by Treasury/AFT regulations). While substantial efforts have been channeled through Agriculture and Commerce, the heaviest thrusts have come through the Energy Department. Under Public Laws 96-126 and 96-304 substantial funds have been awarded for studies and 'cooperative agreements' (development) for ethanol concepts. These are summarized in Table 5.

TABLE 5

DOE FUNDS FOR ETHANOL STUDIES AND DEVELOPMENT
(1980 awards)

	No. Awards	$ Awarded	Potential Gals/Yr
Studies	53	24,717,000	1,420 million
Development*	8	99,952,000	232 million

*viz, through "cooperative agreements"

Additionally, loans and loan guarantees are offered under these and other laws. (It should be known, though, that the new Administration has proposed to not only cease such funding, but to cancel existing grants and loans; the fate of these programs is uncertain.)

Unfortunately, though, only one of these programs is known to propose use of power plant waste heat, although several such concepts were submitted for consideration. Also disturbing is that only twenty-three studies and three development programs receiving awards were in the sixteen 'grain belt' states (ie, essentially the FERC Regions previously cited).

Ethanol can be made from a variety of raw materials: coal; oil; natural gas; wood and other cellulose; dairy waste; potatoes; sugar cane and beets; and grains. Obviously, several of these resources are agricultural; hence the strong interest in the farm belt states where are located the preponderance of diesel-powered generating plants.

Some pundits, however, are concerned (not without reason) about the possible impact of diverting grains and other agricultural commodities to ethanol production. Will it result in higher prices for meat and grain products? Will it hurt our foreign grain trade? If all the 1990 ethanol objective of 8.5 billion gallons were to come from corn it would require 3.4 billion bushels, or about 60% of recent production. The National Alcohol Fuel Commission, a quasi-federal agency, in a recent study concluded the impact would be minimal. This is because much of this corn would only be 'side-tracked', for about 90% of all corn is grown for feed and much of the ethanol 'tailings' would be feed-usable stillage or distillers dried grains, with only the carbohydrate component lost 'in transit'; about one-third of the corn could become high-protein DDG. And beyond that is the fact that much of the ethanol could (and certainly would) come from other grains, such as rice, wheat, sorghum, etc., or from the other resources listed previously.

Besides the primary ethanol product, a grain-fermentation process yields a spent-grain by-product usable for animal feed, as just noted. This can be in the forms of wet stillage (merely dewatered after fermentation is completed), or distillers dried grains (DDG), dried by one or another processes so as to reduce its volume and weight and increase its usability. These components can be used for meat-animal feed (within certain limitations) or to feed fish. Additionally, the carbon dioxide and yeasts resulting from fermentation are marginally saleable.

ETHANOL COGENERATION CONCEPT

To produce ethanol via grain fermentation involves a familiar process of milling, cooking, fermentation and distillation, plus 'drying' if anhydrous alcohol (200 proof) is wanted. Traditional processes, chiefly for potable alcohol, have been rather energy-inefficient. Many efforts are underway to enhance productivity and efficiency, so it is difficult to project firm figures of energy requirements. Table 6 does provide a reasonable estimate.

TABLE 6

HEAT ENERGY REQUIREMENTS - ETHANOL PRODUCTION (2 MGY - 200 proof - DDGS)[1]

Stage	Btu/gal
Cooking and Fermentation	3,600
Distilling	28,000
Miscellaneous	9,400
Subtotal	41,000
Dehydration (for 200 proof)	20,000
Subtotal	61,000
Drying (for DDG)	21,000
Total	82,000

To meet these requirements via a diesel-cogeneration concept the approximate waste-heat quantities listed in Table 7 are available (relative to a one megawatt or 1 MW plant).

TABLE 7

WASTE HEAT REALIZABLE FROM ONE MW ENGINE LOAD

Source	Temperature	Btu/hr/MW
Exhaust*	750-900°F (or as 300°F, 50 psi stm)	1,700,000
Jacket Water	160-170°F	600,000
Lubricating Oil	150-160°F	300,000
Subtotal		2,600,000
Aftercooler**	100-120°F	200,000
Total		2,800,000

*Without excessive cooling of exhaust
**Cooling combustion air after turbocharging

Not all engines will provide the same amount nor division of the energy as listed, and it varies with engine load. And not all engines utilize aftercoolers, and the heat therefrom is highly dependent both on load and atmospheric temperatures. But these are representative figures.

As can be determined from the foregoing, it would appear that one megawatt (1 MW, or 1000 kW) of engine capacity could yield enough waste heat every hour to produce 46 gallons of dehydrated ethanol, on the basis of wet stillage, or 34 gallons with stillage drying. If only 190 proof alcohol is desired these quantities could

(1) "Small-Scale Fuel Alcohol Production"; pg A-21; Development, Planning and Research Assocs.; USDA; 1980

be increased about one-third. This would also depend greatly on the ethanol process itself, of course. The process diagram would be as shown roughly in Figure 1.

In a year's time the potential output from such a plant would depend upon a number of factors: the availability of the engine; the average load thereon; the concurrent availability and loading on the ethanol components; and, of course, the process design and heat exchange efficiencies; etc. Table 8 presents a reasonable estimate of what might be expectable (predicated on engine running capacity factor of 80% and availability of 85%, which are reasonable expectations).

TABLE 8

ETHANOL/DIESEL COGENERATION PLANT ANNUAL QUANTITIES
(One MW Plant)

Electricity	6,000,000 kWh/MW/yr
Ethanol - wet stillage	275,000 gals/MW/yr
- dry stillage	200,000 gals/MW/yr
Corn	80,000-110,000 bushels/MW/yr
Distillers dried grains	30,000 bushels/MW/yr

Most diesel-powered plants are in a size range of 10 to 30 MW. Thus, a rather fully-integrated plant of 10 MW of cogeneration capacity might produce 2,000,000 to 3,000,000 gallons per year.

And if the diesel plants in the United States were significantly developed for ethanol production, what might be achieved? Assuming that 25% of U.S. diesel unit capacity were so outfitted, and operated at an equivalent net 75% of the capacity cited for the one MW plant referenced above, the results might be as per Table 9.

TABLE 9

HYPOTHETICAL ANNUAL U.S. ETHANOL/DIESEL COGENERATION
PRODUCTION QUANTITIES

Capacity devoted to concept	1,400 MW
Electricity output	9,200,000 MWh
Ethanol output	300,000,000 to 425,000,000 Gals
Corn used	120,000,000 to 170,000,000 Bushels

If this were achieved by 1990 it would thus yield 5% of the national ethanol objective of 8.5 billion gals per year. While 170,000,000 bushels of corn (or its equivalent) would enter the process, it is not a total loss, as it would produce a significant amount of feed stillage, largely protein and roughage. And note especially that all this ethanol output is achieved without net consumption of heat energy; in fact, assuming the ethanol would have been produced anyway by conventional means, this will save 4.5 million BOE/yr in fuel input, plus approximately 7.8 million BOE/yr through substituting ethanol for gasoline.

FIGURE I

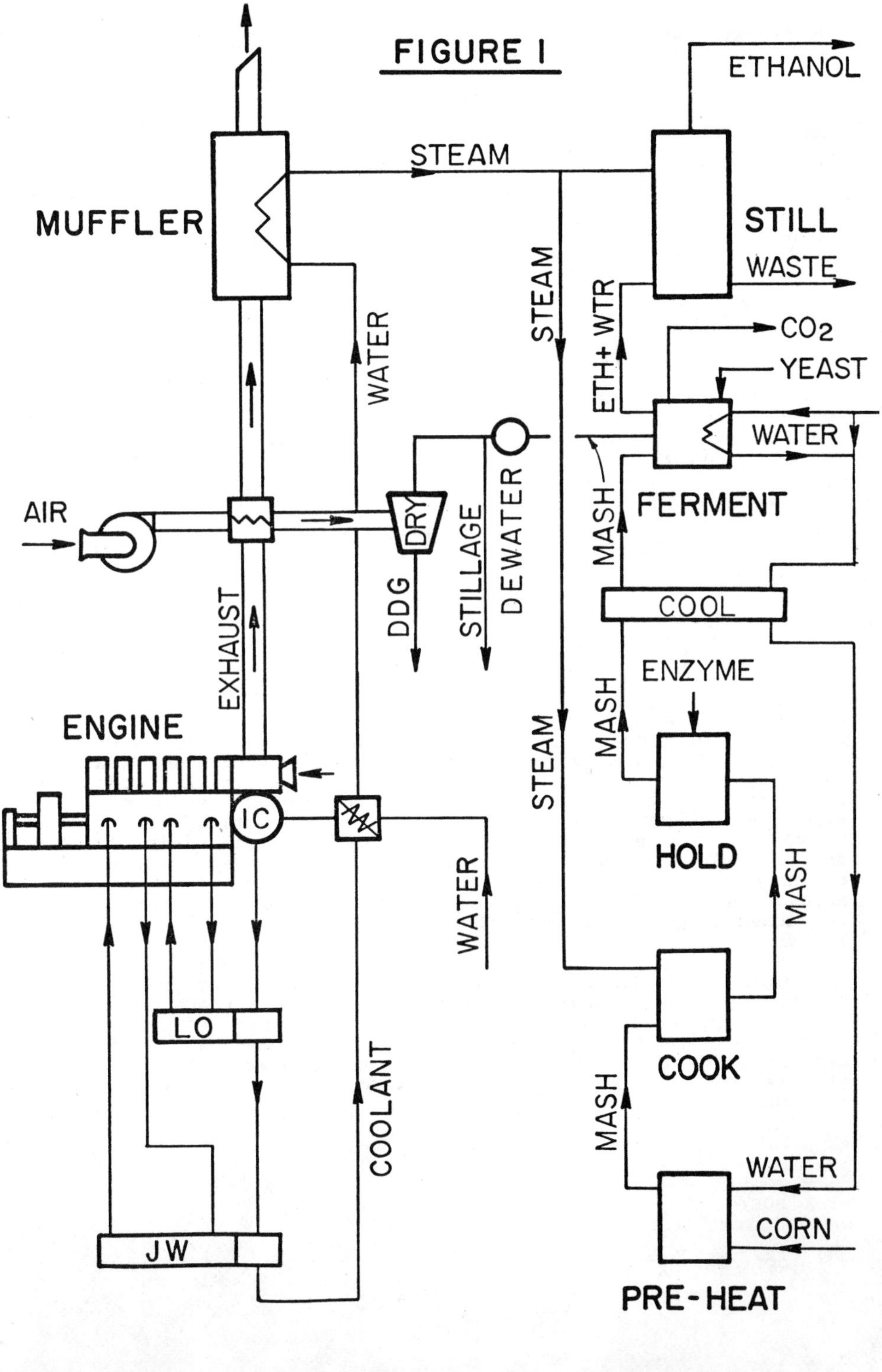

ETHANOL
STEAM
MUFFLER
STILL
WASTE
STEAM
ETH+ WTR
CO_2
YEAST
WATER
WATER
AIR
DRY
DEWATER
STILLAGE
MASH
FERMENT
DDG
COOL
EXHAUST
ENZYME
MASH
ENGINE
STEAM
IC
HOLD
WATER
MASH
LO
COOK
COOLANT
MASH
WATER
JW
CORN
PRE-HEAT

OTHER CONSIDERATIONS

Obviously a key factor in the equation has been omitted to this point: economics. Much conjecture is involved in attempting, in a non-specific analysis such as this, to establish meaningful estimates of costs, income and net profit. Depending upon the specific concepts and assumptions the computed economic conclusions can range widely - from handsome profits to devastating losses. In general, with costs and concepts typical of recent years a normal, (non-cogeneration), commercial plant could not produce alcohol at a cost competitive with gasoline. That is, there would be no incentive or justification to undertake ethanol production. By and large, then, to this point the key to profitable marketability has been the federal gasoline tax forgiveness (and that also available in some states).

As gasoline prices have risen dramatically, and processes, designs and production parameters have been improved, however, the potential has improved. One such step in this direction is the cogeneration concept, for it largely offsets a key cost component, heat energy. A generalized relationship of cost components is presented in Table 10.

TABLE 10

COST COMPONENTS IN ETHANOL PRODUCTION[1]

Grain	63%
Capital cost	10%
Operations, miscellaneous	17%
Energy	10%

It is apparent, then, that to reduce energy costs by one-half (as in a split-the-savings approach) would enhance concept viability by an attractive 5%. While this small-plant approach would likely cost more per gallon for capital costs, it could have some offsetting gains compared to a large plant. For one thing, it would not be so consumptive of grain grown in its immediate vicinity, which should yield slightly lower grain costs and lower transportation expenses. Likewise, mash or DDG should be more readily marketable. So, it is realistic to conclude that such a plant could at the least compete with conventional ethanol plants of 5 to 20 times its size.

Other considerations also include power plant fuel availability, reliability and cost, grain resources, markets for ethanol and stillage, U.S. and state policies and economic support, processes (and evolving changes therein), pollution, waste disposal, storage needs and many other. Each must be evaluated relative to the specific site and process to determine economic and functional feasibility.

(1) op. cit., pg VIII-12

CONCLUSIONS

Over 900 power plants in the United States utilize diesel engine generators in some degree. There are approximately 3100 units totaling 5500 MW in these plants. Because of the relatively high temperatures at which these units exhaust their cycle waste-heat there is real potential for usefully recapturing it to produce ethanol. About 100,000 gallons per year per megawatt of operating capacity is realizable. It is not inconceivable that these plants, which are largely in rural areas scattered across the nation, could provide the heat energy needed to produce as much as 400,000,000 gallons per year, about 5% of the nation's 1990 ethanol objective. In so doing these plants would save the equivalent of about 4.5 million bbls of oil per year required by ordinary ethanol process plants of the same output.

Assuming that such production would be profitable, the present operators of these plants and their associated communities would realize the economic benefits. Furthermore, the local agribusinessmen would have another viable market for grain output, possibly at better, and certainly more stable prices, and would in turn have an alternative source of high-protein animal feed. And there would be increased employment and business expenditures in these numerous smaller communities.

All of this is realizable because of the virtues of the diesel-powered electric plants operated in many such communities, which indeed are small by comparison to customary central station steam units a hundred times their size - whose prodigious quantities of waste heat are at too low a temperature and are too large in amount to be effectively recaptured in like manner. Coupled with other inherent virtues of these diesels, then, it can indeed be said that 'good things come in small packages.'

CORN COB GASIFICATION AND DIESEL ELECTRIC GENERATION

J.J. O'Toole, T.E. Wessels,* B.C. English,** and R.J. Blobaum***

*Energy & Mineral Resources Research Institute
**Center for Agricultural and Rural Development
Iowa State University, Ames, Iowa
***Roger Blobaum & Associates, Des Moines, Iowa

ABSTRACT

Throughout the Corn Belt, a rich potential source of biomass energy is available annually from crop residues. Each year, corn land in this ten-state area produces waste corn stover with a potential energy value greater than half a billion dollars; an equivalent of six barrels of oil per acre.

The proposal described here involves a program designed to evaluate the collection and transportation costs of corn cobs for the fueling of downdraft gasifiers in the production of low Btu gas.

A major use of this gas will be to fuel diesel/electric generating systems in the many municipalities which can be retrofitted for this purpose. At the present time, as reported by the American Public Power Association, the vast majority of these generating units are idle due to costly diesel fuel. A survey conducted by the Iowa Association of Municipal Utilities throughout the State of Iowa shows over 90% of these idle units to be in good operating condition and exceed a total of 300 megawatts.

This program will deal with the present state of corn crop practices and the economic factors influencing the supply of potential corn cobs and the costs incurred in the delivery and handling processes. The salient feature of this phase will be the installation of gasifier equipment and the retrofit of diesel engines in generating stations at two municipal sites in Iowa, where the technical and economic feasibility of the process will be demonstrated. These sites are selected on the basis of significantly different engine designs and major differences in available sources of their corn cob fuel.

A major effort will be involved in surveying the future economic and technological potential of biomass producer gas for several end-uses including diesel fuel, corn drying, low pressure steam and space heat in communities throughout the entire Corn Belt. The sensitivity to financial inducements, the motivations and processes of initiating change in agricultural community energy-saving options, including novel financing programs, will be assessed.

Economic models are expected to provide information on the readiness of farmers and municipalities to adopt this new energy program on the basis of technological and cost data obtained during the program.

INTRODUCTION

Biomass feedstocks are currently undergoing intensive examination for various energy end-uses. This biomass resource is produced in large quantities in agriculture and has the advantage of domestic production, wide geographic distribution, renewable supply, clean conversion, and immediate potential for near-term energy end-use applications.(1)

The largest biomass category, crop residues, includes huge volumes of straw and corn stover. It has been estimated that only 7 million tons of the 322 million tons of available crop residues are collected, a potential of over four Quads of energy, enough to fuel 100 one-gigawatt steam power plants. About half the captured energy in a corn crop is stored in the waste residue of leaves, stalks, and cobs. One acre yielding 100 bushels of corn contains waste residue energy equivalent to six barrels of oil.

Although the value of crop residues for conserving moisture, maintaining fertility, and controlling erosion is well documented; at least 60% can safely be removed from cropland without causing problems. The amount available for energy conversion is determined largely by farming practices, with less being available using conventional tillage than with reduced tillage where more residue is left on the surface.

Other constraints on the use of crop residues for fuel is the low Btu content per unit volume and the relatively expensive collection and sizing costs for many kinds of residue. Crop residue is also seasonal, becoming available at harvest time in the late months of the year.

SUPPLY OF THE CORN RESIDUE RESOURCE

Several recent studies have examined the technical and economic feasibility of using agricultural crop residues as an energy feedstock, mainly in combination with coal for fueling steam boilers in generating electricity. Nearly all these efforts deal with either wheat straw or corn stover, which produce the largest volume of biomass per acre.

A recent Iowa State University study examined the economic feasibility of using crop residues to supplement coal in generating electricity at 28 Iowa power plants and concluded that this would be feasible when coal and other energy prices doubled from 1975 levels. In 1979, a Solar Energy Ressearch Institute study of five systems for harvesting crop residues for energy conversion in Iowa and

Oklahoma concluded that large amounts of corn stover could be collected for less than $20 a ton.

Although none of the current studies deal specifically with corn cobs, this abundant residue is enhanced economically by its minimal handling and storage requirements. Cobs at 9.6 percent moisture have an estimated net energy value of 7,750 Btu per pound and up to 1,400 pounds are produced for every acre of corn. This represents only a fourth to a third of the total corn residue produced in a normal crop year.

Although cobs are too bulky to be shipped economically very far, they are a clean fuel, have a low ash content and contain very little sulphur. An important economic advantage as a fuel source is that they require no baling, cubing, or other special handling. They are easy to store and most of the cobs in a pile will stay dry and in good condition without cover in areas with normal amounts of rain and snow.

A reliable supply of cobs could be made available at energy costs competitive with coal to most small towns in the Corn Belt states (South Dakota, Nebraska, Kansas, Missouri, Iowa, Minnesota, Wisconsin, Illinois, Indiana, and Ohio). An average of 51 billion bushels of corn has been grown annually over the last six years in this ten-state region producing a residue of 38 million tons of cobs available for energy conversion.

Field collection of cobs in certain cases would require modification of corn havesting machinery. This would involve minor changes in harvesting procedures. Agricultural engineers have developed a cob-saver attachment designed to collect over 90 percent of the cobs during combining operations.

In farming practices where corn is harvested with mechanical pickers, stored and air dried in cribs and shelled at the farm or local elevator, cob collection is readily accomplished. An Iowa Department of Agriculture study shows this system also saves large amounts of fuel now being used to dry corn that is picked and shelled in the field.

Cob collection would impose no radical harvesting changes. The main incentive to induce cob collection during the harvest would be contract prices high enough to amortize the collection costs and provide delivery schedules that would minimize handling expense.

CONVERSION OF CORN RESIDUE

If this corn waste were burned under steam boilers, one acre could generate 2500 KWhrs, which approaches the residential per capita electric demand. the average small town in the Corn Belt with a population of 1000- 5000 people could be provided biomass fuels for electrical demand through the use of crop waste from 2,000-10,000 acres of corn, an area typically within two or three miles of the community.

Practical constraints, however, place limits on this model as follows:

1. Steam/electric generating plants are of large capacity and would require large volumes of corn waste hauled over long distances, making such fuel costs non-competitive.

2. Existing steam boilers cannot effectively or efficiently handle corn waste without costly modifications.

DIESEL/ELECTRIC GENERATION

The question of local energy conversion and end-use facilities for crop residues in midwestern communities is to a significant degree already answered by an existing number of functional diesel electric generating units now standing idle because of high diesel oil prices. A recent survey carried out by the Iowa Association of Municipal Utilities shows that more than 70 communities have a total in excess of 300 megawatts of local electric generating capacity. Much of this equipment, while operational, is used only for occasional peaking or emergency generation with the average capacity factor being less than two percent. The survey also indicates that this equipment for the most part is in a good serviceable state.

The technology to develop an alternative diesel fuel fortunately has been long established. Low power "producer gas" generated from coal, wood waste and crop waste, has a century-old history and continues to be used in Europe as fuel for a large range of engines both mobile and stationary. The special application of this process to the production of clean gas for diesel use is undergoing rebirth in this country . At present, however, only limited production facilities are domestically available.

TECHNICAL CONSIDERATIONS

Gasifiers can be either updraft or downdraft packed bed or fluid bed reactors. The downdraft gasifier is a moving, packed bed, vertical flow reactor, which employs concurrent gas flow. It has the potential to generate a gas containing substantially smaller amounts of tars and other higher hydrocarbons than can be obtained from an updraft gasifier. It is a more sophisticated reactor than the up-draft design and is marginally more expensive but is simpler than the fluid bed system. The generated low-Btu gas can be used as fuel for steam boilers, directly-fired dry kilns or internal combustion engine-generators.

The range of domestic diesel engine designs and models currently in use present a spectrum of considerations in converting or retrofitting existing types to low Btu gas. In this country, successful test runs of diesel engines on dual fuel, a combination of 80% producer gas and 20% diesel oil, were reported in 1947 and more recently in an extensive program of gasifier development at the University of California, Davis. At present, feasibility studies are in progress at several sites including Alaska, for generating producer gas from wood waste to be used as fuel in diesel/electric generators.

Field tests in California with downdraft gasifiers have successfully produced diesel quality producer gas from corn cobs, walnut shells and other crop residues.(3) Certain French and German firms, e.g., Duvant and Imbert, have continued since WW I to offer efficient producer gas/diesel equipment packages capable of continuous operation on 90% producer gas/10% diesel oil.

PROJECT ECONOMIC VIABILITY

The costs of corn cob conversion can be divided into three areas; on-farm transportation and storage, processing, and conversion. These three areas will have different costs depending on the location of the gasifier/conversion unit. If a central facility which processes corn accumulates cobs for disposal, then no on-farm costs and a minimal charge for transportation and storage will be incurred by this facility. On the other hand, if the cobs must be collected and stored, the costs in all three areas will have to be accounted for.

ON-FARM COSTS

The on-farm costs will vary by site. Parameters that need to be examined include (1) the area of land presently being picked, (2) the costs of converting from combines to pickers or retrofitting combines to collect cobs, (3) the participation rate of farmers under different incentive plans, (4) the percentage of farmers that presently use the cobs for other purposes, (5) the amount of cobs that presently remain on the farms and are not burned, (6) the present cropping pattern, (7) the costs of farmers' production, (8) the schedules and timing that the farmer presently practices, and (9) the size of farms in the collection area. It is evident in examining these parameters that the costs of cobs will be site specific and at the present time only a rough approximation can be made of the actual costs. A study conducted by the Center for Agricultural and Rural Development at Iowa State University (4) indicates that the farmer would have to receive between $9.00 and $11.00 per ton for crop residues, at 1980 prices.

TRANSPORTATION AND STORAGE

The transportation costs of moving residue from the field to the gasification site vary with distance. Distance also influences farmer participation rate and the density of potential cob supply within the area of interest. Assuming that the fuel would be received from within a ten-mile radius, the cost of transportation would be approximately $0.10 per ton-mile, or assuming an equal crop density within that ten mile radius, about $0.50 per ton delivered to the plant.

The farmer, if he shells his corn, must do so by April or May so that corn stored on the cob will not spoil. One can expect that no cobs will be available from May until October without storage facilities. Thus, it is assumed that a six month storage facility is needed. Storage can be obtained through two methods, on the farm or at the energy conversion site. Storage cost at the site are estimated in Table 1 .

Table 1. Corn Cob Storage Cost Estimates For a Six Month Capacity At a 1 Megawatt Gasifier Diesel Electric Site

Type of Cost	Total Cost	Life Span	Annual Cost	Quantity Cost
	Dollars	Years	Dollars	Dollars/ Ton
Bunker Size 180' x 150' x 24' Cost $38/running ft.	19,100	20	2,244	$0.37
Floor - Cement 6" thick at $46/cu. yd.	23,000	20	2,703	$0.45
TOTAL	42,100		4.947	$0.82

(Cost assumes a 10 percent interest rate. If a 15 percent interest rate is assumed, then cost per ton would be $1.08.)

PROCESSING AND CONVERSION COSTS AT GASIFICATION SITE

Each gasification site will accrue costs for acquiring a cob size- reducer (hog or hammer mill), with conveyor belts, a three-day storage shed, a conveyor from the storage to the gasifier unit, and, in addition, optional costs of converting the diesel engine to low BTU gas.

Table 2 estimates costs of processing cobs for the gasifier unit. The hog can be replaced by a hammer mill but in this analysis costs are assumed for a hog. In addition, the capacity of the hog is such that it will run an average of only two hours per day with increased usage projected on Friday and Monday. If cobs are below 3" in size then neither a hog nor hammer mill is necessary, but in this analysis it is assumed that a size reducer is required.

Costs in operating the gasifier unit designed for this project are described in Table 3. Fuel inputs and gas outputs of the gasifier/conversion unit are analyzed with and without diesel operation. Fixed costs and a portion of the variable costs reflect current experience and early 1980 prices. However, in Table 3 the cost of corn cob fuel is estimated from our previous study, which analyzed costs for collecting and transporting the entire corn residue including stalks, husks, leaves and cobs. The selective removal of cobs involves several equipment options and may be less costly and energy intensive per unit of biomass.

Table 3 indicates that the projected cost for corn cob fuel is a significant fraction of the variable cost center but until the actual market for corn cobs develops, the on-farm collection system, including equipment, techniques and labor cannot be accurately predicted. For this reason, we have used the estimates for complete corn residue collection in Table 4.

Table 2. Estimated Cost of Processing Cobs For the Gasifier Unit

Type of Cost	Total Cost	Life Span	Annual Cost	Quantity Cost
	Dollars	Years	Dollars	Dollars/ Ton
Equipment:				
Hog, 8-11 tons	16,000	20	1,880	0.31
Conveyors:				
To Hog 24 ft at $800/LF	19,200	20	2,256	0.37
From Hog to Storage	50,000	6	11,480	0.91
From Storage to Gasifier Unit	50,000	20	5,875	0.97
TOTAL	135,200		21,491	3.50
Building:				
Pole Building $6/sq.ft.	18,720	20	2,200	0.30
Electrical	5,000	20	588	0.10
TOTAL	23,720		2,788	0.40
Loader	20,000	6	4,592	0.77
Site Work	5,000		500	0.08
Operation & Maintenance:				
Building @ 3% of cost	NA	NA	712	0.12
Mechanical @ 5.5% of Mech. Costs	NA	NA	7,436	1.24
Front end loader	NA	NA	2,712	0.45
TOTAL	NA		10,860	1.81
Land (2 acres)	4,000		400	0.07
GRAND TOTAL	187,920		40,631	6.77

Table 3. Model Community Gasification/Diesel Electric Generation Energy Cost Estimate

Operating data from the Fuel Rate Calculations:

Energy Output: maximum, 1,000 KW; average, 750 KW
Gas Input for Electrical Power: 10.18 MM Btu/hr
Solid Fuel Input for Electrical Power: 1810 lb/hr
Fuel Oil Input for Electrical Power: 0.825 MM Btu/hr at 7.5% of Total Energy Input (5.9 gals/hr)

COSTS-GASIFIER PLANT		
Equipment		
Gasifier System	$259,610	$
Installation	64,800	
Capital Cost Total	324.410	
Annualized at 5% for 20 yrs		33,670
Diesel Retrofit @ $150/KW	150,000	
Annualized at 8.25% for 20 yrs		15,564
Insurance at 2.5% of Plant Cost		6,490
TOTAL FIXED COST/YR		$55,724
@ 6,570,000 kwh/yr, Fixed Cost = $.0085/kwh		
Operational Costs		
6570 hrs/yr, operating at 75% capacity		
Fuel handler, 1 man, 40 hrs/wk, w/fringes		31,200
Maintenance charge @ 5% of plant cost		16,200
Fuel Costs:		
Diesel oil @ $1.15/gal x 38,716 gal		44,523
Corn cobs at 18.80/T x 5979 T/yr		112,400
TOTAL OPERATIONAL COSTS		$204,323
6.57 MM kwhr/yr, operational costs/kwhr=	$.0311	
TOTAL COST	$.0396/kwhr	

COST OF HOT PRODUCER GAS FROM GASIFIER

Operating at 90% Capacity; with Gas Conversion Efficiency of 85%

1810 lbs cobs/hr x 8760 hr x .90 = 7135 T/yr @ $18.80/T =	(Fuel)	$134,138
Labor, Maintenance, Amortization, Insurance		87,560

TOTAL COST $221,698/(7135 T cobs x 15 MM Btu x .85 efficiency) = $2.44/MM Btu

Table 4. Total cost of delivery system

Type of Cost	Estimate Low	High
	Dollars per Ton	
On-farm	9.00	11.00
Transportation	0.70	0.70
Storage	0.82	0.82
Processing	6.77	6.77
TOTAL	17.29	19.29

PROJECT DEMONSTRATION

WORK STATEMENT

To demonstrate the range and degree of efficiency of corn cob biomass conversion to low BTU gas and its end use applicability, two rural community test sites have been identified and preliminary planning developed for gasifier installation. These towns have been selected in one case (a) where excess cobs are available on-site from a local source, and, (b) where cobs are only available in a diffuse pattern from individual farms, thus requiring that a collection and transportation system become established.

One test for low BTU gas end-use will involve the retrofit of two separate models of diesel engines in the municipal electric utilities at these two sites for operation with low Btu gas. These diesels have been selected on the basis of their uniquely different designs as well as their frequency of encountered usage.

A major consideration is to demonstrate the technical and economic feasibility of this operation from the standpoint of power generation and other potentially significant end-uses of low Btu gas and waste heat, and further, through appropriate surveys, to develop an index of technical capability for similar gasification and diesel retrofit installations at potential municipal sites throughout the Corn Belt.

Development of economic modeling for predicting resource availability, present and future fuel costs and the potential benefits of the gasification process to farmer producers, community and small business end-users and job creation potential of the over-all process is another major objective. Inputs into the models include resources used in agricultural production, resources required for cob use, socio-economic variables that have an impact on the analysis, and environmental constraints. Sensitivity analysis on some of the more significant variables such as energy prices, BTU value, moisture content, emission standards, farmer participation rate, different incentive plans for cob delivery, and inherent costs in cob processing will be run. Examples of output available from the model include agricultural production for the site, the quantity of cobs used by the gasifier unit, the cost of kwh from the system, the quantity of emissions, and the feasibility of cob use. An additional result from the model will be a statement on the energy balance situation. Several other producer gas end-uses besides electrical generation will be analyzed. Some of the probable end-uses include corn drying, steam for greenhouses and/or dairy operations, space heating, and additional drying of the fuel stock (cobs).

PARTICIPANTS IN THE DEMONSTRATION PROGRAM

The two Corn Belt communities volunteering for the program are Coon Rapids, Iowa, and Sanborn, Iowa, where the corn cob supply and cooperation in facilitating diesel engine retrofitting of appropriate capacity makes the project feasible.

The diesel manufacturers, Fairbanks-Morse Division of Colt Industries and Transamerica DeLaval, have provided protocols and cost estimates for retrofitting their selected models for use of the dual fuel mixture of low BTU gas and diesel oil. Iowa State University (the Energy & Minerals Resources Research Institute and the Center for Agricultural and Rural Development), and its subcontractors will manage and direct the over-all effort.

The fuel, gas production and economic parameters of the downdraft gasifier have been provided by the Biomass Corporation of Yuba City, California. The American Public Power Association is assisting in providing survey data and communication services.

BIBLIOGRAPHY

(1) "A Survey of Biomass Gasification," Vol. II, Solar Energy Research Institute, SERI/TR-33-239, July 1979.

(2) "Retrofit '79 Proceedings of a Workshop on Air Gasification," The Solar Energy Research Institute, Seattle, Wash., Feb. 2, 1979.

(3) "Fuel from Waste Products," R.O. Williams and R.C. Lang, Food Engineering, October 1979.

(4) "Economic Feasibility of Using Crop Residues to Generate Electricity in Iowa," Center for Agricultural and Rural Development, Iowa State University, CARD Report 88.

HYDROPOWER - AN ASSESSMENT OF THE PROSPECTS FOR DEVELOPMENT

James R. Hanchey
Acting Director, Institute for Water Resources

Historically, hydroelectric power has played a key role in the industrialization of society. From the early water wheels used in New England to mill grain, to the large electro-mechanical turbines and generators now in common use, hydropower has been employed as a means to accomplish the ends of civilization.

During the past decade, there has been considerable debate concerning the role that hydropower might play in a national energy system. Prior to the Arab oil embargo in 1973, little attention was focused on hydropower. Athough additional hydropower capacity was being slowly added to regional systems, the general consensus seemed to be that hydropower was a resource that had been substantially exploited and that there was little significant remaining potential.

A brief historical perspective might be useful in explaining the attitudes toward hydropower which prevailed at the beginning of the 1970's. The first hydroelectric power development in the United States was built in Niagara Falls in 1879. This development powered arc lamps for lighting in a nearby community. It was soon followed by the first central station for producing commercial power for incandescent lamps, built on the Fox River in Appleton, Wisconsin, in 1882. From that point on, many new sites were developed and older mechanical sites were transformed to produce electric power. This revolution continued until by 1920 about 5,000 MW of hydroelectric capacity had been developed in the United States. By 1930 hydropower supplied nearly 40% of the nation's electric energy.

Most early hydropower was privately built. The Federal government entered the development business by authorizing, through various statutes, the Corps of Engineers, the Water and Power Resources Service (formerly the Bureau of Reclamation) and the

Tennessee Valley Authority to develop public hydropower projects. Federal projects multiplied until by 1977 the Federal share accounted for nearly half of the total hydroelectric capacity. Most of the large hydroelectric power developments in the west were developed by the Corps and the Water and Power Resources Service, while the southeast region of the country was developed by the Corps and TVA. These projects were typified by large scale multi-purpose developments.

By the 1950's, however, hydropower's role in the U.S production of electric energy began to decline. By the 1970's hydro's share was less than 15% of the total electric generating capability. The major reasons for this were the lack of additional suitable sites, the lower cost of alternative sources of thermal generation, and a new regimen of environmental controls.

Hydropower depends on two major variables, streamflow and gradient. By the middle of the 20th century, most of the promising, easily accessible, economic sites located near load centers had already been developed. New sites were becoming harder to find, were located far from load sites and were too expensive to develop. This situation tended to depress new hydropower development.

In addition, the cost of electrical power generated by oil, gas and coal declined. For producers of electricity, economic sense dictated a choice of oil, gas or coal fired generation instead of hydroelectric power. Electric utilities began to build thermal plants to supply electricity, thereby reducing hydropower's relative share of the generating mix. In addition, overly optimistic projections of nuclear power generation further dampened the enthusiasm for hydropower development.

Finally, the environmental movement argued against development of large dams on free flowing steams. Though other sources of energy, it could be argued, would on a comparative basis be more detrimental to the environment than hydropower, hydro sites were large, visible and were the focal point for a number of environmental groups to register their protests against encroachment on the environment. Without the construction of new dams, the pace of hydroelectric power development was checked.

During the 1970's, several things happened that refocused national attention on hydropower. First, the OPEC oil embargo and subsequent increases in the price of oil and other fossil fuels has substantially altered the relative economics of hydropower as compared to fossil fuel fired thermal electric generation. Second, the "energy crisis," reflected in both shortages and prices has increased national concern for both conservation and renewable energy resources. Finally, many of the adverse environmental consequences of thermal electric generation are only now becoming clear. Thermal pollution of streams, acid rains, air pollution, and serious concern about the safety of nuclear power plants have caused many people to take a second look at the environmental consequences of hydropower development. The renewed interest in hydropower motivated Congress to initiate a comprehensive study of the hydroelectric power resources of the United States.

THE NATIONAL HYDROPOWER STUDY

Congress authorized the National Hydroelectric Power Resources Study in the Water Resources Development Act of 1976. The objective of the study is to examine the nation's undeveloped hydropower resources and to analyze the institutional and policy setting for hydropower planning, development and use. The Corps of Engineers was given the responsibility to conduct the study. The legislation authorizing the National Hydropower Study states that the study must develop an inventory of the physical potential and estimate the magnitude and regional distribution of demand for hydropower. In addition, the study must identify the social, economic, environmental, institutional and other policy issues affecting hydropower, assess their importance, and recommend policy modifications that will encourage the effective use of the nation's hydroelectric power resources.

The National Hydropower Study began in September 1977 and will be transmitted to Congress by September 1981. Although the study is not yet complete, enough has been accomplished to allow us to speculate with some degree of confidence on the future of hydropower in the United States. Before examining the future of hydroelectric power, it will be instructive to take a brief look at the existing situation.

THE HYDROPOWER RESOURCE - EXISTING AND POTENTIAL

Hydroelectric power provides about 14% of the nation's total electrical generating capability with an installed capacity of about 73,000 MW, including pumped storage. This amount of capacity produces an annual average energy generation of about 280 billion KWhr/year. On a regional basis hydropower accounts for over 40% of total electric capacity in the Pacific Northwest (including Alaska), 12% in the Upper Midwest, 10% in New England and less than 10% in the rest of the country.

Estimates of the potential hydropower resources in the United States vary according to the assumptions and methods used in the analysis. The Corps of Engineers, through the efforts undertaken as part of the National Hydropower Study, is developing estimates of the future potential of hydropower by aggregating individual sites across the country to form a national estimate. All sites along with physical, economic, environmental and institutional data will be included in a computerized data base for retrieval by interested parties.

The initial data base contained site information on approximately 50,000 existing dams from the Corps' National Dam Safety Program inventory. Information on potential dam sites that were undeveloped was added by the Corps, Water and Power Resources Service, United States Geological Survey, Soil Conservation Service, Federal Energy Regulatory Commission (FERC), Tennessee Valley Authority and other water development entities. Almost 20,000 undeveloped sites were added to the data base.

The information in the data base is being used to screen sites according to physical, economic, environmental and institutional criteria. This screening process is being carried out in stages using more strict criteria at each stage to sort and categorize the vast number of sites.

We have completed three separate screenings and are now in the process of formulating regional hydropower development plans. The active inventory now contains approximately 2,000 projects with an aggregate capacity of 55,000 MW and an annual energy potential of 140,000 GWH. These totals are for the Continential United States only. Estimates of the hydropower potential for Puerto Rico, Hawaii, and Alaska will be added later. Figure 1 shows the distribution of the potential by Electric Reliability Council Regions.

As these figures indicate, it would be possible to almost double the present installed capacity and to increase the average annual energy production by about 60%. Of particular note is the fraction of the sites in our current inventory that are existing Federal dams. Of the 1,983 potential projects, 469 are at existing

NATIONAL HYDROPOWER STUDY (REGIONAL SUMMARY)
ESTIMATES OF FEASIBLE DEVELOPMENT POTENTIAL AT
EXISTING FEDERAL PROJECTS

ERC REGION	PROJECTS (#)		CAPACITY (MW)		ENERGY (GW-H)	
	TOTAL	FEDERAL	TOTAL	FEDERAL	TOTAL	FEDERAL
NPCC	656	12	5,930	9	23,100	44
MAAC	48	13	960	83	1,830	290
ECAR	201	118	4,160	2,090	13,150	9,475
SERC	145	39	5,770	1,590	12,545	3,650
MAIN	101	33	710	420	4,020	2,760
MARCA	50	13	1,730	905	1,500	370
SWPP	111	76	3,505	2,010	8,430	4,930
ERCOT	52	25	605	310	1,240	680
WSCC	619	140	31,574	7,570	74,234	11,000
TOTALS	1,983	469	54,944	14,987	140,034	33,199

(Figure 1)

Federal dams. These Federal dams include projects with existing generation facilities as well as dams where no hydropower has been installed.

A second table, Figure 2, breaks the total inventory into undeveloped sites, where no dam exists, and existing dams. This breakout shows 1,405 existing dams and 578 undeveloped sites. As could be expected the largest part of the potential additional capacity would require the construction of new dams (33,639 MW of 54,944 MW total).

An interpretation of these figures is in order. The Western United States (WSCC) dominates the hydropower picture. This region has about 31% of the potential new projects, and about 67% of the potential capacity and 64% of the potential energy. Of particular note in the low energy to capacity ratio of the Federal projects is this region. This indicates that most of the Federal dams in the Western United States are already producing hydropower and that a large percentage of the available energy at these projects is already being realized. In these cases, the additional capacity is gained as the project is moved upward in the load curve and is operated as an intermediate or peaking plant at lower plant factors. This can be contrasted with the Federal dams in the ECAR region (primarily Ohio River Basin). In the ECAR region, the higher energy to capacity ratio indicates that many of the Federal dams do not now have hydropower installations. Many of these dams were built during a time that public power was not popular and there was not sufficient economic justification for hydropower development when compared to thermal generation alternatives. Today the economic picture has changed and most of these dams, many with penstocks and other appurtenant works installed, are very economical projects.

Another interesting statistic is reflected in the figures for the NPCC Region (New England - New York). Only 12 of the total 656 potential projects are at Federal dams. This is due to a large number of small dams which exist in New England and the relatively small number of Federal projects. Clearly this mix has implications for the role of the Federal government and the non-Federal sector in future hydropower development.

A brief look at Figure 2, which shows the break-down of existing dams and undeveloped sites in the inventory indicates a possible constraint to a large national hydropower development program. Although only 30% (578/1983) of the potential projects are at undeveloped sites, i.e., no existing dams, about 60% of the potential capacity and 64% of the potential energy is associated with these projects. This may limit the contribution that we can reasonably expect hydropower to make over the next 10-20 years. It should be noted, however, that if the WSCC is factored out of the chart the figures look a little better. Eliminating the WSCC figures, only 20% of the potential projects are at undeveloped sites and these undeveloped sites account for only 50% of the capacity and energy potential.

Having reviewed the existing resource, the next section discusses some of the factors that will influence the future of hydropower in the U.S.

FACTORS AFFECTING THE FUTURE OF HYDROPOWER

If all of the potential hydropower capacity that appears feasible were to be developed it would almost double the existing installed capacity from hydropower in the United States. Unfortunately, only a portion of this potential is likely to be developed. How much of the physical potential can be developed by the year 2000? The answer to this question depends on a myriad of

NATIONAL HYDROPOWER STUDY (REGIONAL SUMMARY)
ESTIMATES OF FEASIBLE DEVELOPMENT POTENTIAL AT
IDENTIFIED UNDEVELOPED SITES

ERC REGION	PROJECTS (#)		CAPACITY (MW)		ENERGY (GW-H)	
	TOTAL	UNDEVELOPED	TOTAL	UNDEVELOPED	TOTAL	UNDEVELOPED
NPCC	656	102	5,930	3,200	23,100	14,450
MAAC	48	1	960	420	1,830	520
ECAR	201	16	4,160	1,470	13,150	2,375
SERC	145	81	5,770	5,280	12,545	11,150
MAIN	101	0	710	0	4,020	0
MARCA	50	2	1,730	24	1,500	80
SWPP	111	47	3,505	1,185	8,430	3,060
ERCOT	52	33	605	455	1,240	950
WSCC	619	296	31,574	21,605	74,234	57,569
TOTALS	1,983	578	54,944	33,639	140,034	90,154

(Figure 2)

factors affecting hydropower development. These factors are being analyzed in depth as part of the NHS. The following discussion highlights a number of these factors.

Perhaps the primary factor affecting hydropower is cost--the cost of the hydropower plant itself and the cost of alternatives to hydropower. Cost analysis will differ depending on whether the developer is a Federal agency or a private concern, but the fundamentals remain the same. If a hydroelectric plant can supply electricity at the least cost it likely will be built.

Most hydroelectric projects are compared to a thermal alternative. Hydro projects are capital intensive with low operating cost while thermal projects have moderate capital costs and high operating costs, mostly attributed to fuel costs. The relative price shift of fuel versus the general inflation rate is a critical determinant of the feasibility of hydro projects. Consider the recent base price of $31 a barrel of oil from OPEC. This is equivalent to 43 mill/KWHr electricity. As oil costs and other thermal fuels continue to rise, hydroelectric projects will gain an economic advantage. Careful analysis of hydropower on a with and without basis is needed to make an optimal choice among alternatives. This type of systems analysis must include all costs that are easily measured in terms of dollars as well as those external costs (e.g. pollution costs, residual handling) that are not easily calculated. If all external costs are included in analyses, renewable resources, such as hydropower, will fare well in comparison to non-renewable, high residual thermal sources.

Many experts state that unless steps are taken immediately, we face an imminent water crisis whose magnitude will equal or exceed the energy crisis. Specific regions of the country are overdrafting their water resources now and prospects for reversing the situation are dim. Hydroelectric power plants do not consume water, but there is competition for water and water storage space behind reservoirs. For example, demand for irrigation could remove water from streams before it is run through hydro turbines. This situation may occur shortly in the Pacific Northwest. The flooding of natural rivers by hydro dams also removes stretches of free-flowing rivers from use. Likewise, the operation of a hydro facility can inhibit recreational use of a lake behind a dam by fluctuating the water level. Any estimates of hydro's future must consider the compromise that must be made to accommodate other water users. Water rights' battles dealing with hydropower will be a by-product of any future hydropower scenario.

Environmental considerations have both positive and negative aspects with regard to hydropower, although of all means of generating electricity hydropower is among the most benign, especially when generating facilities are added to existing structures. On the negative side, undeveloped hydropower sites require the flooding of a large section of natural streams, and the operation of some hydro facilities, may cause some environmental damage. On the other hand, thermal sources of energy exact a high cost on the environment. The extraction, transportation, use and disposal of fuels for thermal generating sources all highly impact the environment. Recent concern about acid rain and the disposal of nuclear fuel indicate a growing recognition of the environmental cost of energy. In general, and particularly for additions of hydropower at existing dams, the environmental balance favors hydropower.

There are certain characteristics of hydropower that affect it's future. Hydropower is a renewable resource constantly replenished by the hydrologic cycle powered by the sun. As such, hydropower is impervious to depleting fuel levels and soaring fuel costs. Hydropower is a highly efficient means of producing electricity. About 90% of the available energy in falling water is converted to electrical energy with modern turbines and generators. Coupled with this high efficiency is an equally impressive durability and reliability. Rarely do hydropower units fail to operate and many of the plants in existence today have been operating for over 70 years. Hydropower is extremely useful for electric utilities, because it can be started up very quickly and provides ready spinning reserve to meet load changes. Also, hydropower plants, particularly pumped storage, are the only large scale storage medium for electric energy available today.

Of course, the nature of hydro has its drawbacks. Since it depends on streamflow, hydropower is intermittent. The long-term availability of hydropower can not be determined except by probabilistic methods. This intermittent nature reduces the value of hydro, because backup systems may be needed to insure system reliability. During a drought, as in 1977 in the West, hydropower output can be reduced drastically.

Another potential factor that affects the future of hydropower is the ability of manufacturers to produce turbines and generators for the many small scale plants that could be developed by the year 2000. There is an interesting "Catch-22" in this situation. Domestic producers cannot increase production and lower prices without orders, yet the producers may not get orders until the prices go down.

The process by which private and federal hydropower projects are authorized and initiated can be a significant barrier to future development. Regardless of whether or not a project is large or small, good or bad, all private developers must go through the same arduous licensing process. The FERC has done a considerable amount of work in trying to remedy this situation and is attempting to insure that only those actions that are necessary to insure a sound, safe project be taken by potential developers.

FORECAST OF FUTURE HYDROPOWER DEVELOPMENT

In attempting to make a judgement concerning the prospects for future hydropower development, one is faced with the difficult task of integrating many diverse political, technological, and economic factors into a plausible scenario.

Certainly, the results of the comprehensive national inventory of hydropower potential sets a reasonable upper bounds on the amount of additional hydropower which could be developed within the next 20 years. However, even the most optimistic proponent of hydropower development could not expect that all of these projects will be pursued. Although these potential projects have survived a multi-stage screening process which took into account such factors as physical capability, economics, and environment, there remain legitimate differences among various interests as to the wisdom of development. The greatest degree of uncertainty concerns the number of the undeveloped sites that may be authorized for construction.

Ultimately, the future of hydropower is more likely to be determined by events that cannot be predicted very well at this time. The ultimate resolution of continuing tension in the Middle-East and the impact that this has on the continued availability of foreign oil is a factor that could strongly affect decisions concerning hydropower. The long gas lines following the oil embargo in 1973 have faded in our memory. To a certain extent we have returned to "business-as-usual" approach to the solution of our energy problems. Many scenarios concerning the Middle-East situation have been developed by our defense and energy strategists that would indicate a more extensive national effort to develop our indigenous resources. In such an environment, the priority given hydropower development would likely increase.

Another factor which will affect the amount and rate of hydropower development is the progress made in developing alternative energy technologies, such as synthetic fuels and solar installations. The promise of reasonably priced applications of these technologies has had a dampening effect on hydropower development. Although there is indeed reason for optimism concerning the eventual contribution that these technologies can make, significant problems remain to be solved before they can be expected to compete economically on a broad scale with conventional modes of energy production.

Figure 3 shows historical trends of the rate of hydropower development over the past 30 years. Superimposed on this graph is the estimate of feasible hydropower potential development resulting from the National Hydropower Study. As this graph shows a continuation of past trends would result in a total of about 100,000 MW of installed capacity by the year 2000, a period of 20 years. This increase could be realized by a combination of installations at existing dams and undeveloped sites. As the potential projects in our active inventory have all been screened against physical, economic and environmental criteria, it seems reasonable to assume that a large number of the existing dams will ultimately be developed. With respect to new dams, the picture is less clear. A prudent forecaster would not assume that the fundamental opposition to these type projects will disappear. However, there are a number of large scale multi-purpose projects which are in various stages of planning and it is reasonable to expect that many of them will be constructed. A conservative forecast is that a large percentage of existing dams will be retrofited with hydropower facilities and that the construction of new dams will continue, although perhaps at a slower rate than experienced over the past 30 years. The result of this would be an addition of about 30,000 to 35,000 MW of hydropower generating capacity by the year 2000. This would still leave a large undeveloped potential which is economically feasible. Whether these projects are ever seriously pursued cannot now be determined with any degree of confidence.

In summary, the future for additional hydropower development is promising. The results of our study show that hydropower can contribute to the solution of our energy problem by adding generating facilities to dams and other hydraulic structures which already exist. In addition, there are still a large number of undeveloped sites which are suitable for multi-purpose development that could add large amounts of capacity and energy to regional electrical systems, as well as contribute to solutions of other water problems such as flood control and water supply.

The job of developing these resources will not be easy. There is clearly a role for both the governmental and private sectors. FERC and DOE are leading the effort to mobilize the private sector with the cooperation of the other Federal water development agencies. The Corps and the Water and Power Resources Service are accelerating their planning, design, and construction activities to bring more power on line over the near-term. The combination of efforts of all interested parties should result in a significant amount of new hydropower construction over the next 10-20 years.

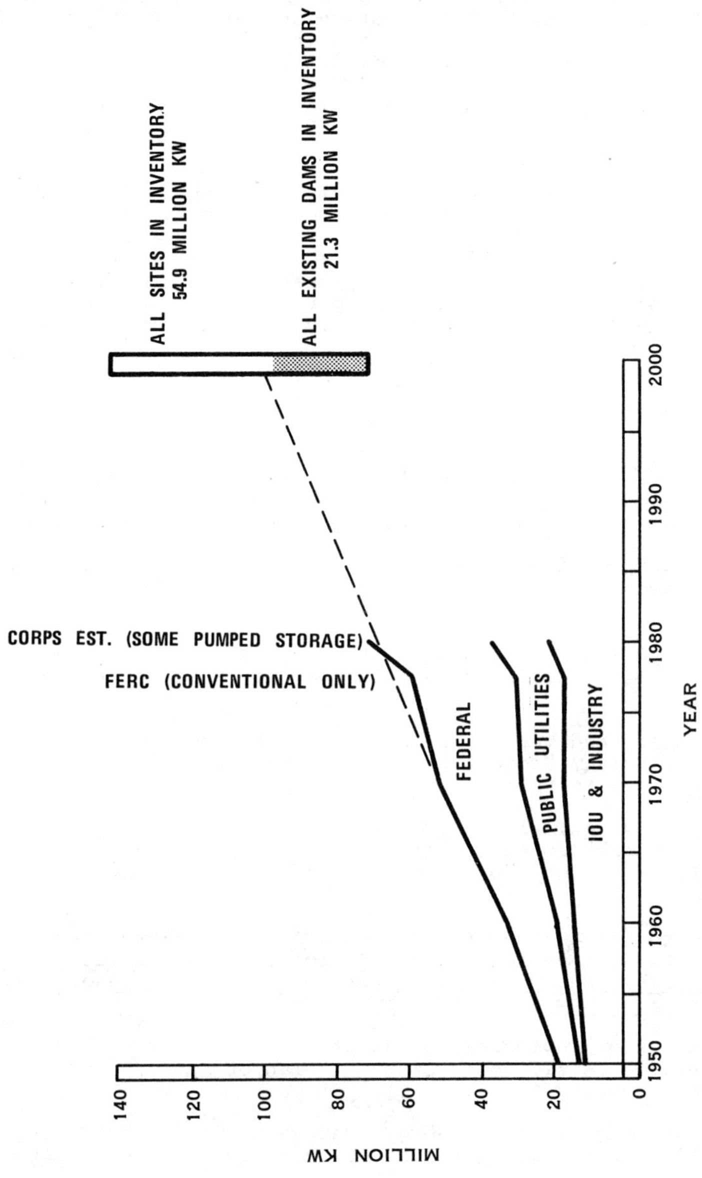

FIGURE 3
CONVENTIONAL HYDROELECTRIC CAPACITY
DEVELOPMENT TRENDS

APPENDIX - REFERENCES

1. Brown, Peter W., and Buxton, Anthony W., "Preliminary Analysis of Legal Obstacles and Incentives to the Development of Low Head Hydroelectric Power in the Northeastern United States," prepared for U.S. Department of Energy by the Energy Law Institute, Franklin-Pierce Law Center, Concord, N.H., March 12, 1979.

2. Federal Power Commission, Bureau of Power, "Origin and Development of the Hydroelectric Power Industry," Unpublished Paper, September, 1972.

3. Federal Power Commission, "Hydroelectric Power Resources of the United States," FPC-P43, Washington, D.C., November, 1976.

4. Harza Engineering Company, "The Magnitude and Regional Distribution of Needs for Hydropower, Phase I 1978 Electric Power Demand and Supply," Contract Report prepared for the Institute for Water Resources, Fort Belvoir, Virginia, April, 1979.

5. Hayes, Earl T., "Energy Resources Available to the United States, 1985-2000," Science, Vol 203, January 19, 1979, pp. 223-239.

6. Klotz, Louis H., "Waterpower--It's Promises and Problems," Center for Industrial and Institutional Development, University of New Hampshire, Durham, NH, October, 1979.

7. Northwest Energy Policy Project, "Energy Futures Northwest" Pacific Northwest Regional Commission, Vancouver, Washington, p. 24.

8. Rudman, R. L. et. al., "Strategic Planning for R&D," EPRI Journal, Palo Alto, California, Volume 4, No. 8, October, 1979, pp. 16-17.

9. Stobaugh, Robert and Yergin, Daniel, (ed.) "Energy Future--Report of the Energy Project at the Harvard Business School," Random House, New York, 1979, p. 214.

10. Trisko, Ralph L., et. al., "Hydroelectric Power Potential at Corps of Engineers Projects," IWR Research Report 75-R1, U.S. Army Engineer Institute for Water Resources, Ft. Belvoir, VA., July, 1975.

11. U.S. Army Corps of Engineers, Institute for Water Resources and the Hydrologic Engineering Center, National Hydroelectric Power Resources Study, Preliminary Inventory of Hydropower Resources, Volumes 1-6, Ft. Belvoir, VA., July, 1979.

12. U.S. Army Corps of Engineers, Institute for Water Resources, National Hydroelectric Power Study: "Plan of Study," Fort Belvoir, VA., January, 1979.

13. United States Department of Energy, Energy Information Administration, "Monthly Energy Review," October, 1979, p. 62.

14. Walsh, M. R., "The National Hydropower Study" in proceedings entitled: Conservation and Utilization of Water and Energy Resources, ASCE, New York, August, 1979, pp. 190-192.

SMALL-SCALE HYDRO EQUIPMENT

By

Warner W. Wayne, Jr.
Consulting Engineer

Stone & Webster Engineering Corporation

INTRODUCTION

This paper covers the principal types of small-scale turbo-generating units which are particularly suitable for low head hydro developments. Following are descriptions of proven and newly conceived turbine designs, as well as recent technological improvements which should help to reduce equipment costs. The latter include standardized package units that, in some cases, essentially represent a "water-to-wire" concept. References to specific manufacturers are solely for the purpose of equipment identification and are not intended as endorsements. In most cases, there are a number of different manufacturers that can provide the same basic type of hydroelectric machinery.

TRADITIONAL TURBINE DESIGNS

Francis-type reaction turbines and conventional vertical shaft propeller turbines with spiral cases have a long and proven record of efficient and reliable service and still represent potentially viable prospects for low head plants. Several examples of standardized designs covering small Francis units are cited below.

The James Leffel & Co. of Springfield, Ohio, a subsidiary of the Nohab Tampella Hydro Energy Group, offers standard vertical shaft Samson turbines of the Francis-type which are suitable for either an open flume setting or a pressure casing arrangement. The designs cover 13 runner sizes ranging from 17 to 74 inches (432 to 1,880 mm) for operating heads up to 26 ft (7.9 m) and rated outputs up to 1,100 kW.

The Norwegian Kvaerner Group has developed a series of Francis turbines that are predesigned and standardized to a certain degree. Runner diameters range from 31 to 79 inches (800 to 2,000 mm), heads from 23 to 164 ft (7 to 50 m), and unit outputs from 180 kW to 10,000 kW. These units are suitable for either horizontal or vertical shaft settings. A lower-cost cylindrical drum can be provided in place of the conventional steel spiral casing, if desired.

STANDARD TUBULAR UNITS

(1) Allis-Chalmers Corporation

The Hydro-Turbine Division, York, Pennsylvania, has developed a new series of axial flow tubular turbines and can now offer standard designs for three different equipment configurations. These comprise horizontal shaft TUBE units, vertical shaft axial flow turbines in the same size range as TUBE units, and small size inclined shaft axial flow turbines. All units are a propeller type and provide the highest feasible operating speed and maximum power output for a given runner size. This permits the use of a smaller powerhouse structure with attendant reductions in the civil costs. Since neither a spiral casing nor a concrete scroll case is required, the unit spacing is only about two-thirds that required for conventional vertical shaft propeller turbines of the same rated capacity. Further details about each of the three basic designs follow.

The standard horizontal shaft TUBE units are available in 10 different runner sizes ranging from 750 mm (30 inches) to 3,000 mm (118 inches) with unit outputs ranging from 50 kW at a head of 2 m (6.5 ft) to 5,000 kW at the maximum design head of 15 m (49 ft). The runners are equipped with either 3, 4, or 5 blades, depending on the rated head. A peak turbine efficiency of at least 90 percent can be expected. The packaged design normally includes an intake butterfly valve for flow control, an adjustable blade turbine runner, fixed guide vanes, hollow turbine shaft with guide and thrust bearings, a hydraulic operator for blade control, a geared type speed increaser, a synchronous generator with fan cooling, static excitation and voltage regulation equipment, electrical controls, and low voltage switchgear. Figure 1 is a photograph of the first standardized unit to be manufactured (late 1978) and shows the various components associated with this 1,500 mm (59 inches) diameter turbine and 1,000 kW generator. The package can be extended to include the main power transformer and high tension switchgear, thereby providing a complete "water-to-wire" concept.

The standard vertical shaft axial flow turbines also are available in the same 10 runner sizes as the TUBE units and cover the same range of operating heads and power outputs. However, the turbine configuration and general equipment arrangement are quite different, as shown in Figure 2.

The standard inclined shaft axial flow turbines are available in 12 runner diameters ranging from 12 inches (305 mm) to 72 inches (1,830 mm) in 6-inch increments. Based on design heads of from 5 to 20 ft (1.5 to 6 m), power outputs would range from about 10 kW to 1,000 kW. The equipment package includes a fixed blade turbine which is basically a modified low head propeller pump, a steel draft tube, a belt- or chain-driven induction generator, exciter, and other normal accessories. Since no flow control is provided, the turbine has a fixed point of operation on the efficiency curve.

KEY

1. INTAKE BUTTERFLY VALVE
2. TURBINE RUNNER (1500 mm DIAM.)
3. SHAFT
4. REMOVABLE TOP OF DRAFT TUBE
5. DRAFT TUBE LINER
6. SPEED INCREASER
7. ADJUSTABLE BLADE CONTROL MECHANISM
8. GENERATOR
9. EXCITER
10. GENERATOR TERMINAL CABINET
11. HYDRAULIC PUMPING UNIT
12. HYDRAULIC PRESSURE CYLINDERS
13. CONTROL CABINET

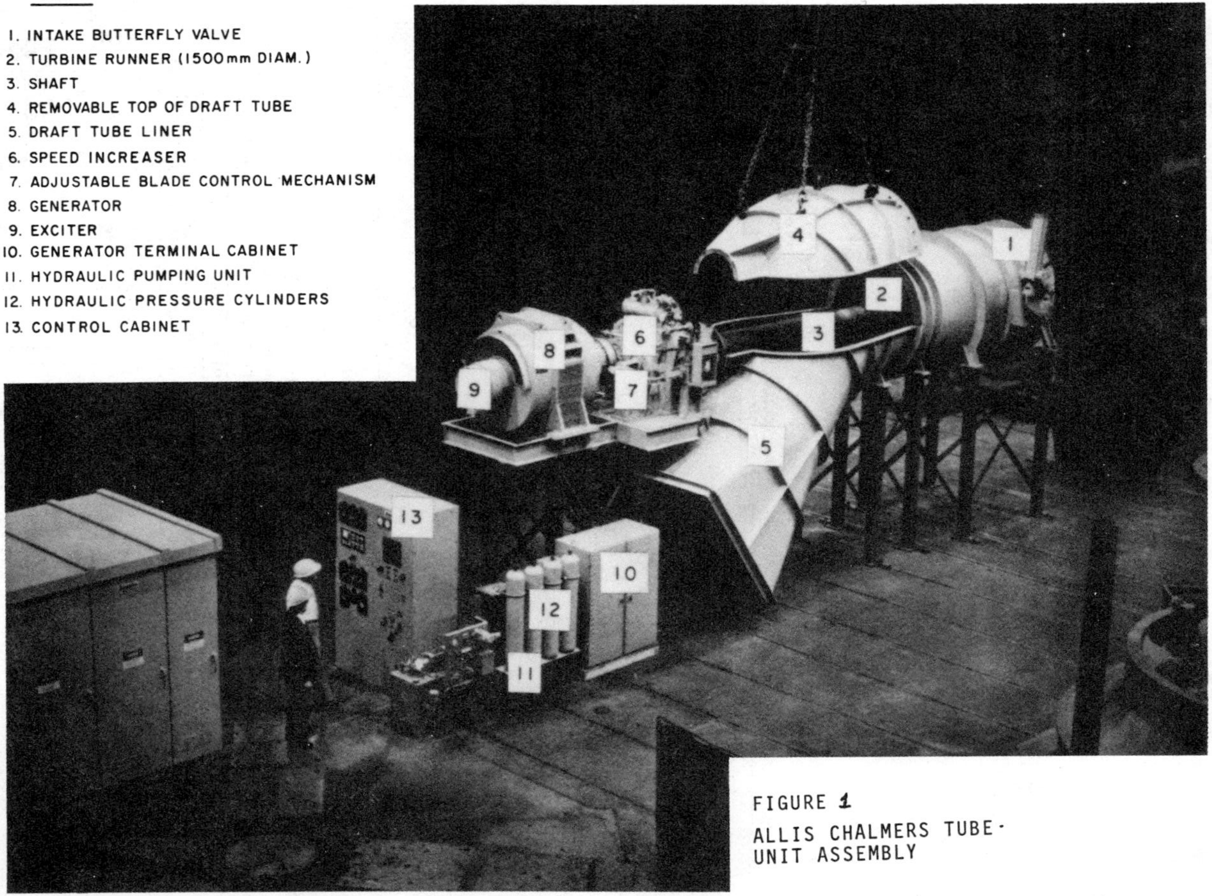

FIGURE 1
ALLIS CHALMERS TUBE-UNIT ASSEMBLY

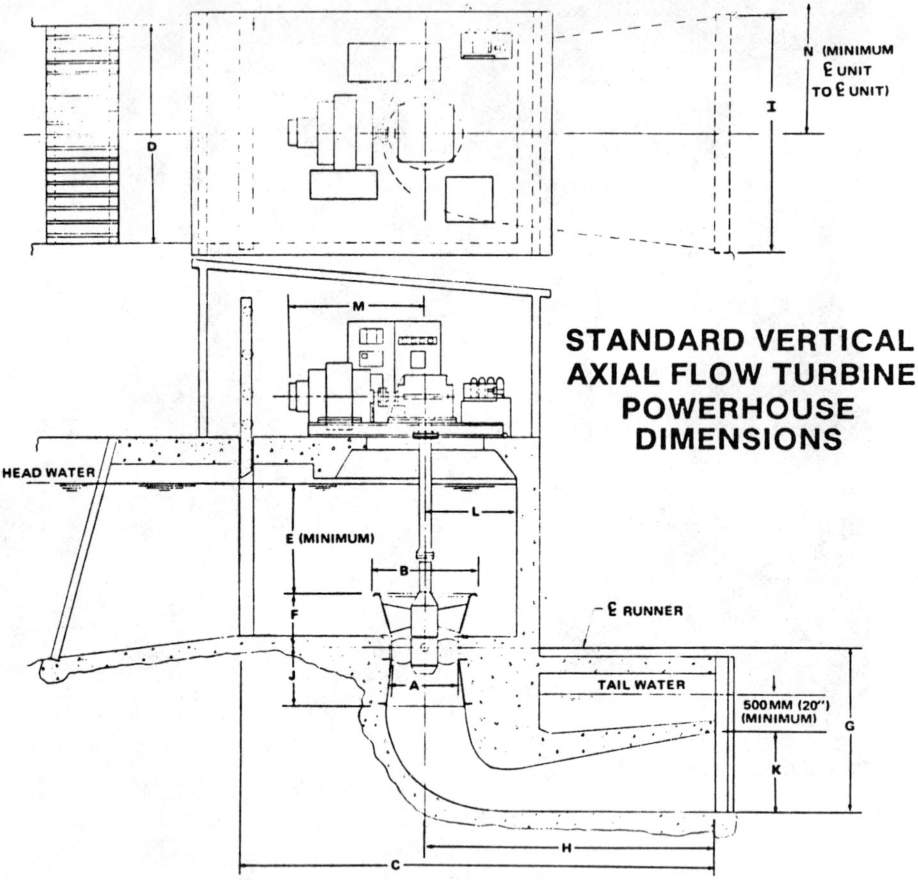

BASIC DIMENSIONS

A = Runner Diameter in millimeters (inches) = 1.00

All Other Dimensions Are In Proportion From Runner Diameter

DIMENSIONS	RUNNER SIZE									
A	750 (29.5)	1000 (39.4)	1250 (49.2)	1500 (59.6)	1750 (68.9)	2000 (78.7)	2250 (88.6)	2500 (98.4)	2750 (108.3)	3000 (118.1)
B	1.4	1.4	1.4	1.4	1.4	1.4	1.4	1.4	1.4	1.4
C	10.0	9.5	9 0	8.5	8.0	7.5	7.0	6.5	6.5	6.5
D	3.0	3.0	3.0	3.0	3.0	3.0	3.0	3.0	3.0	3.0
E	1.5	1.5	1.5	1.5	1.5	1.5	1.5	1.5	1.5	1.5
F	.78	.78	.78	.78	.78	.78	.78	.78	.78	.78
G	2.7	2.7	2.7	2.7	2.7	2.7	2.7	2.7	2.7	2.7
H	3.9	3.9	3.9	3.9	3.9	3.9	3.9	3.9	3.9	3.9
I	3.2	3.2	3.2	3.2	3.2	3.2	3.2	3.2	3.2	3.2
J	.9	.9	.9	.9	.9	.9	.9	.9	.9	.9
K	1.2	1.2	1.2	1.2	1.2	1.2	1.2	1.2	1.2	1.2
L	1.4	1.4	1.4	1.4	1.4	1.4	1.4	1.4	1.4	1.4
M	4.0	3.8	3.5	3.3	3.0	2.8	2.5	2.3	2.3	2.2
N	4.2	4.0	3.85	3.7	3.6	3.5	3.45	3.4	3.35	3.3

NOTE. Dimensions are approximate and may vary for specific applications.

FIGURE 2

(2) AB Bofors-Nohab

The Hydro Power Division located in Trollhatten, Sweden, has developed standard tubular turbine designs which can be used for either horizontal or vertical shaft arrangements similar to those shown in Figures 1 and 2, respectively. The pertinent characteristics of these standard units are shown in the following tabulation:

Runner Diameter mm (in)	Operating Head Range m (ft)	Capacity Range kW
700 (27.5)	3 - 30 (10-100)	100 - 750
900 (35.4)	3 - 30 (10-100)	100 - 1,200
1,150 (45.3)	3 - 20 (10-66)	150 - 1,500
1,400 (55.1)	3 - 18 (10-59)	200 - 2,000
1,700 (66.9)	3 - 17 (10-56)	300 - 2,000
2,000 (78.7)	3 - 14 (10-46)	400 - 2,000

Although the maximum output is presently limited to 2,000 kW, Bofors-Nohab expects to provide unit capacities up to 5,000 kW in the near future.

The standard turbine may be coupled direct, or through a speed increasing gear box, to an asynchronous generator. The runner blades are fixed and there are no wicket gates. Speed is held constant by the electrical system and the unit is turned on and off through head pond level controls.

(3) Hydro Energy Systems, Inc.

Hydro Energy Systems, Inc., of New York, New York, and Societe Neyrpic of Grenoble, France, have recently joined forces to manufacture, sell, and service standardized, packaged, low head hydroelectric units. These units consist of right-angle drive tubular turbines with the generators and auxiliaries located outside the water passage. They come in five standard sizes and are suitable for heads from 6 to 60 feet (1.8 to 18 m). Unit outputs range from 100 to 1,500 kW, depending upon the runner size and operating head. The generator may be mounted directly above the turbine in a vertical position, or to one side in a horizontal position. The right-angle gear is designed to provide a 3:1 speed ratio, thereby permitting the use of a higher speed and less costly generator.

The major advantage of this unit is its very compact design, which minimizes the size and cost of the civil works. In addition, the factory assembled package means faster and less costly field installation.

BULB UNITS

General Design Considerations

The term "bulb unit" refers to an axial flow unit which has the generator mounted inside a steel bulb located within the powerhouse water passage. Although some bulb units have been installed with moderately sloping shafts, the preferred setting is with a horizontal shaft, since this simplifies both construction and equipment maintenance activities. The generator should be placed upstream of the turbine so as to obtain better flow conditions and improved turbine performance.

There are numerous manufacturers of bulb turbines and generators, and they have the capability to provide both large and small size bulb units. Standardized small size bulb units are available from at least one supplier, as discussed next.

Fuji Package Type Bulb Units

Nissho-Iwai American Corp. of Los Angeles, California, can provide standardized Fuji package-type bulb units in 19 sizes covering a range of net heads from 17 to 59 ft (5 to 18 m) and outputs from 300 to 4,000 kW.

The bulb unit is fabricated from plate steel in two cylinders. The inner cylinder contains the generator, exciter, and bearings, and is supported by the struts from the outer cylinder. The outer cylinder forms an extension of the steel penstock, to which it is bolted. The downstream end is connected to the embedded draft tube by an expansion-type flanged coupling. The outer casing is horizontally split in two halves to facilitate transportation, assembly, and disassembly. The fixed blade runner has four stainless steel blades to minimize corrosion and cavitation effects. Water flow is controlled by movable wicket gates. Two oil pressure lubricated journal bearings support the turbine and generator assembly. A synchronous generator is standard, with an induction generator optional. General design details of the Fuji package unit are shown in Figure 3.

STRAFLOW TURBINES WITH RIM-TYPE GENERATORS

Introduction

Straflo is the proprietary name given by Escher Wyss, Ltd., of Zurich, Switzerland, to its axial flow turbine which has a rim type generator rotor mounted on the periphery of the turbine runner blades. The concept is not new, and 73 units of this type, all built by Escher Wyss, have been operating successfully for many years. These units are relatively small, design heads are low, and all have fixed blade propeller type runners. Escher Wyss has recently developed new bearings and seals which make it feasible to build units of this type in considerably larger sizes and for higher heads. Turbine runners with adjustable blades should also be possible. Escher Wyss anticipates that Straflo units will be suitable for the same head range for which tubular turbines and bulb turbines are normally considered.

Design Considerations

Since the outer rim of the turbine runner supports the generator rotor, a sealing system is required to prevent penstock water from entering the generator area. The complete generator, including stator and rotor, is entirely outside the water passage and thus easily accessible for maintenance. Cooling poses no unusual problems, since the generator is in the open. The greater inertia of the larger diameter rotor, as compared to a bulb unit, is a distinct advantage in terms of operating stability. With the generator outside the water passage, the "bulb" of the Straflo unit is only large enough to accommodate the runner journal and thrust bearings and the inner wicket gate bearings.

The compactness of the Straflo unit is probably its most important feature, when compared to bulb and tubular units, and makes it possible to achieve cost savings in the civil works, powerhouse crane, and accessory equipment.

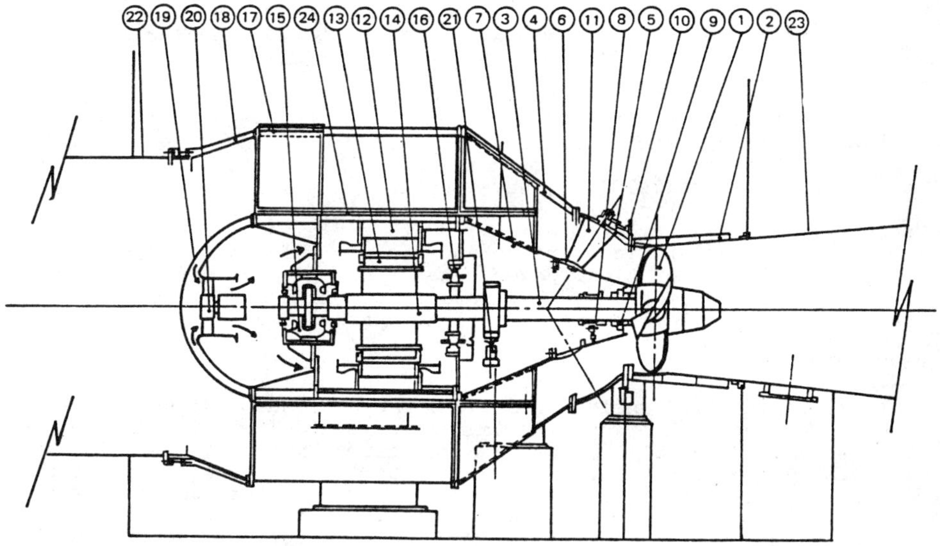

Sectional Drawing of Bulb Turbine & Generator

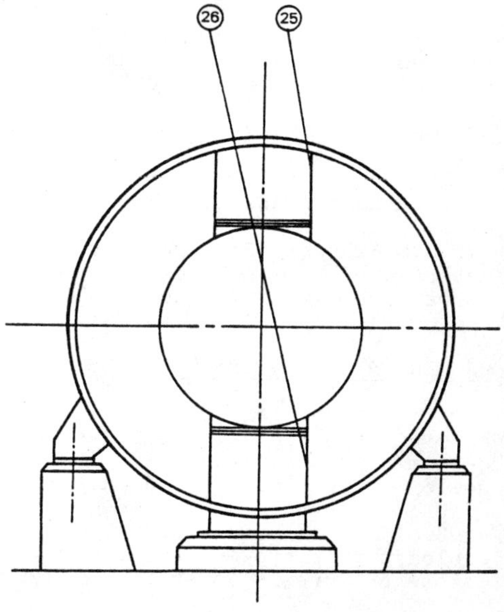

LEGEND

1 RUNNER
2 DISCHARGE RING
3 TURBINE SHAFT
4 CONIC RING
5 OUTER WICKET GATE RING
6 INNER WICKET GATE RING
7 BULB FOR TURBINE
8 GUIDE BEARING FOR TURBINE
9 SHAFT SEAL
10 REGULATING RING
11 WICKET GATE
12 STATOR
13 ROTOR
14 GENERATOR SHAFT
15 GUIDE (THRUST) BEARING FOR GENERATOR
16 EXCITER
17 OUTER CASING FOR GENERATOR
18 CONNECTING TUBE LINER
19 BULB NOSE
20 COOLING FAN
21 BRAKE
22 PENSTOCK
23 DRAFT TUBE
24 BULB FOR GENERATOR
25 UPPER STAY VANE
26 LOWER STAY VANE

FIGURE 3

Planned Installations with New Bearings and Seals

In cooperation with the City of Zurich, Switzerland, Escher Wyss is currently manufacturing a demonstration Straflo unit to be installed in the existing Hoengg plant. It will have adjustable runner blades, the new type hydrostatic seals, and the new hydrostatic rotor rim bearings. The runner diameter will be 3,000 mm (118 inches), head 4 m (13 feet), output 1,500 kW, and discharge 50 m^3/s (1,766 cfs). Commercial operation of this unit is planned for mid-1981.

The Tidal Power Corp. of Nova Scotia recently awarded a contract to Dominion Bridge and Sulzer Inc. (parent company of Escher Wyss) for a large Straflo unit to be installed in a tidal power plant at the mouth of Annapolis River, Nova Scotia. The unit will be rated 17.8 MW at a head of 5 m (16.4 ft). It will have a fixed blade runner with a diameter of 7.6 m (25.0 ft) and also will be equipped with the new hydrostatic bearings and seals. This is the first use of a Straflo unit for tidal power generation and also represents the first application of the new hydrostatic bearings and seals for a large size unit. Commercial operation is scheduled for 1983.

SCHNEIDER ENGINE

Dr. Daniel J. Schneider has developed a unique hydraulic turbine termed the "Schneider Lift Translator Engine." This turbine operates on a hydrofoil principle and presumably can be installed in canals, pipelines, rivers, and tidal estuaries to develop power from heads as low as 3 ft (1 m). Construction of the first demonstration project using this new turbine was completed in September 1980. This project is located on an irrigation canal near the Town of Richvale, California, and consists of a small unit rated 75 kW at a head of 10 ft (3 m). Figure 4 is a copy of a brochure which describes some of the main features of the Schneider engine.

Plans are being made to develop standard engine designs for five models which will cover a head range of 3 to 27 ft (1 to 8.2 m) and outputs from 2 to 4,000 kW.

PERFORMANCE COMPARISONS

Figure 5 shows a comparison of the expected performance curves for standard Francis, bulb, and tubular type turbines based on units with a maximum rated output of 1,130 kW at a head of 32.8 ft (10 m). The settings relative to tailwater, runner discharge diameters, and turbine speeds also are compared. These additional elements directly affect power plant costs and are an important part of any overall equipment evaluations.

OSSBERGER CROSS-FLOW TURBINES

The Ossberger cross-flow turbine is a radial impulse-type turbine with partial air admission and is suitable for both low and medium head installations. The intake water is distributed to a hollow cylindrical shaped turbine runner with peripheral blades and flows through the blades on the intake side, then across the central core and out through the blades on the discharge side. An exploded view showing the relationship of the various parts of an Ossberger turbine is shown in Figure 6.

Two Ossberger turbines rated 600 kW and 940 kW, respectively, at a head of 32 ft (9.8 m) are in commercial operation at the Goodyear Lake Dam on the Susquehanna River near Colliersville, New York. Two similar capacity units rated at a head of 115 ft (35 m)

Richvale, California installation under construction. Note the simple gate/throat housing and draft tube.

Simplified drawing of Schneider lifting hydrofoils.

Cutaway drawing of a twin gate installation showing main components and structure.

SCHNEIDER TECHNOLOGY MAKES SMALL WATERWAYS PROFITABLE PRODUCERS OF ELECTRICITY!

Schneider low-head power generating installations can be used in the most unlikely spots. Electrical power from creeks, streams and even irrigation canals is economically feasible with the unique Schneider Lift Translator. Lifting hydrofoils, patented in over thirty countries, wring an amazing amount of energy from low-head sites.

The Schneider system can operate where others cannot. Heads of five feet (or even less) can generate electricity with cost efficiency.

The Schneider Corporation goes beyond engineering and design. We are pioneers in low-head power generation and can provide expertise in site selection, financing, power sales and licensing to meet federal, state and local requirements. In short, we have state-of-the-art hardware PLUS experience in working with governmental and private agencies.

CHECK THESE BENEFITS:

- Simple in form
- Easily sized
- Advanced engineering
- Works where others won't
- Keeps landscape unspoiled
- Comparatively quick installation
- Financing guidance
- Experience in working with government & private agencies
- Assistance with power contracts

The Schneider Lift Translator Engine, designed and patented by Dr. Daniel J. Schneider, has been featured in POPULAR SCIENCE MAGAZINE, BUSINESS WEEK, QUEST, WALL STREET JOURNAL and other major publications.

FIGURE 4

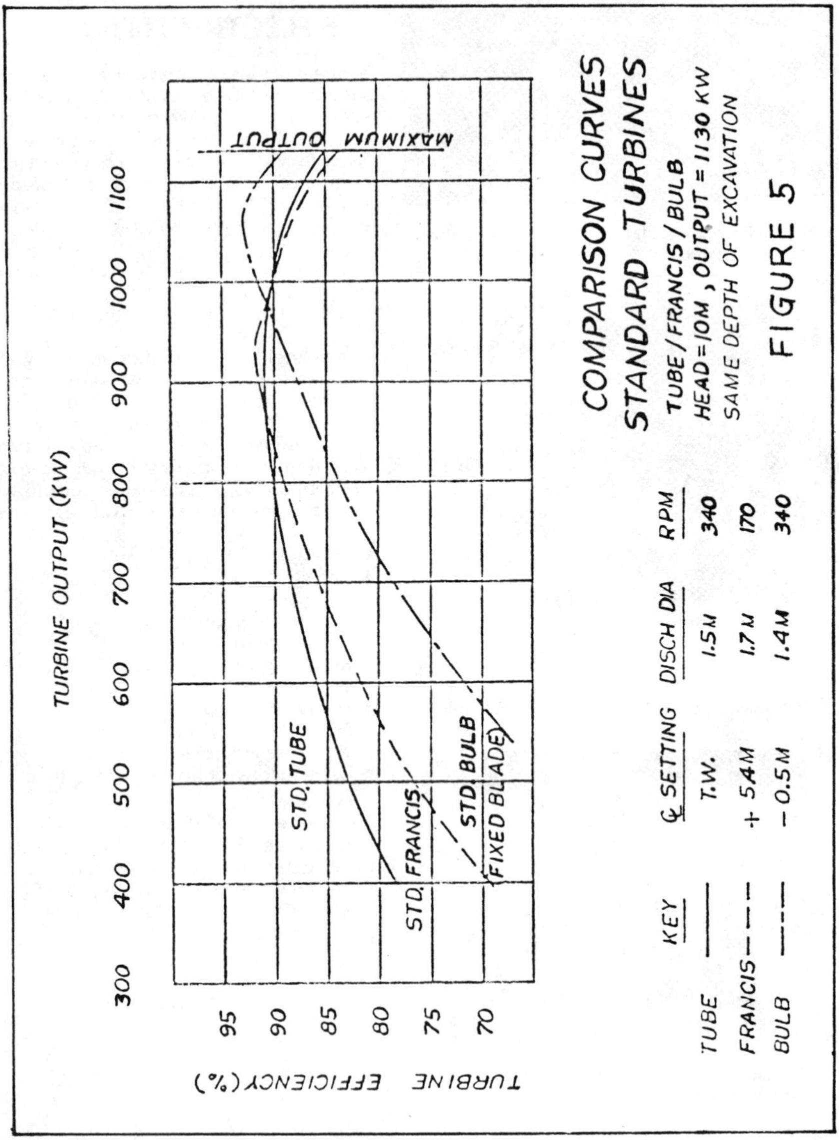

COMPARISON CURVES
STANDARD TURBINES
TUBE / FRANCIS / BULB
HEAD = 10M , OUTPUT = 1130 KW
SAME DEPTH OF EXCAVATION

FIGURE 5

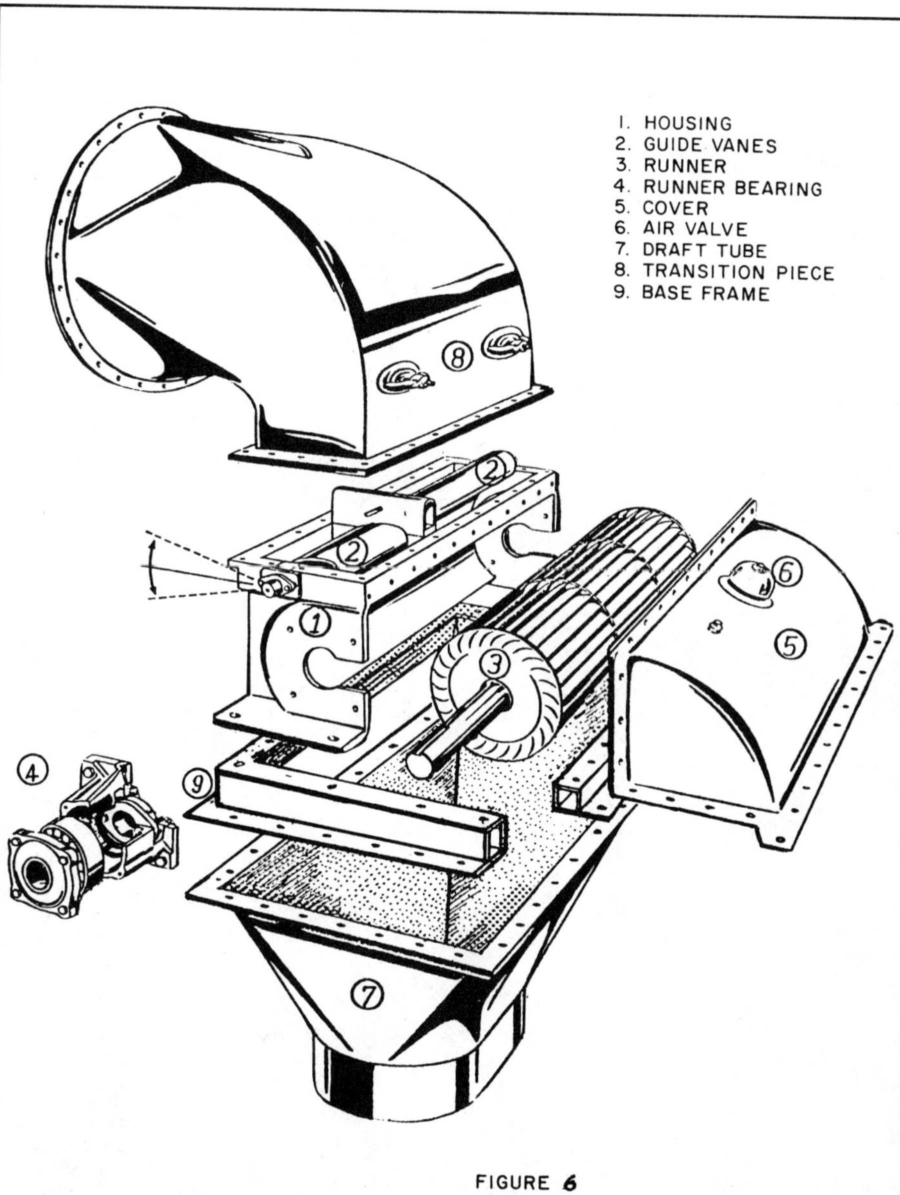

FIGURE 6
VIEW OF TYPICAL OSSBERGER TURBINE ASSEMBLY

are now being installed at the Cornell University hydro plant on Falls Creek in Ithaca, New York.

Ossberger Turbine Works of Weissenberg, Germany, is the only manufacturer of this type of cross-flow turbine. While hundreds of these units have been furnished to countries worldwide, they are still unique in North America. Design heads have varied from 6 to 562 ft (2 to 171 m) and outputs from 10 kW to 1,000 kW.

SUMMARY

Significant steps have been taken in recent years to improve the cost effectiveness of turbogenerating units for small scale low head hydro developments. Such steps include new turbine designs and configurations, development of standardized designs covering a broad range of flows, heads, and power outputs, and the offering of practically complete powerhouse equipment packages.

SMALL SCALE HYDRO (PUMPS AS TURBINES)

HOWARD A. MAYO, JR., P.E.
WARREN G. WHIPPEN, P.E.

ALLIS-CHALMERS CORPORATION, HYDRO-TURBINE DIVISION

INTRODUCTION

The two basic categories of small scale hydro are low head and high head applications. This facilitates dividing the hydraulic machinery into the two corresponding machinery categories of propeller pumps and mixed flow or centrifugal pumps. They each have somewhat different backgrounds and significantly different types of applications. This paper will address only propeller pumps and their low head turbine applications. This is where the major current interest and activity exists.

The use of axial flow low head pumps for power generation with the flow and rotation reversed has not been common practice. However, there is no question whether a pump can operate as a power generating unit. The concerns are more related to equipment application, efficiencies, operating range, cavitation, transients, etc.

There is also no doubt that modern technology is available. Specific new installations to demonstrate the technology and new application techniques are the current phase.

ECONOMIC CONSIDERATIONS

The increasing demand and cost of energy has emphasized the need and interest in renewable energy resources. This has been particularly evident in low density energy sources such as solar, wind and low head waterpower. The challenge facing us today is: TO DEVELOP WAYS OF MAKING LOW DENSITY AND SMALL INCREMENTS OF WATERPOWER ECONOMICAL.

Representative project cost trends were illustrated in the low head ranges by curves prepared by Tippetts-Abbett-McCarthy-Stratton(1) for "Typical Maximum Installed Capacity and Minimum Energy Cost for Small Hydro Plants at Existing Dams in Northeastern United States". The implication was that sites below 500 kW were not economical at that time.

Another study by Dr. R. A. Arndt, Director of St. Anthony Falls Hydraulic Laboratory, University of Minnesota(2) indicated costs were excessive for installations of less than 500 kW capacity. He has identified the cost per kilowatt hours produced annually for specific sites. Based on an assumed 20 year pay-back with a 10 percent inflation rate and energy at $0.03 per kW hour, Dr. Arndt advised that relative costs greater than $0.045 per kW hours produced annually were not feasible.

There are many sites where lower cost generating units are needed for small outputs and low heads. Also, lower cost generating units may be suitable for some of the higher output applications. The objective is to make more water power sites economically practical to develop.

The specific challenge is very simply to <u>DEVELOP LOWER COSTS</u>.

The cost breakdown of hydroelectric projects is represented for both a large and a small project based on specific installations using existing dams(3) (Figure 1). The Bonneville plant <u>addition</u> with a 575 million dollar cost for 626 megawatts of additional capacity has its largest cost in civil construction and equipment installation. Only 15% of the total cost is represented by equipment manufacture and delivery, and only 5% by Project Design and Management.

In contrast, at the 3.2 MW Sawmill Project, costing approximately $4 million, more than 50% is in equipment manufacture and delivery, and almost 10% in Project Design and Management. The civil construction and equipment installation costs have been reduced to almost 40% of the total. Extrapolating this trend emphasizes the need for lower cost equipment and that reductions in civil construction and equipment installation costs have a significant impact. From another viewpoint, it becomes evident that there is not very much money available for Project Design and Management as the projects become smaller.

The reason for considering pumps for power generation is the fact that they are lower cost. Pumps are produced by numerous manufacturers and are already standardized. The difference between pumps and conventional hydraulic turbines is simply that hydraulic turbines have conventionally had some means for efficient flow control and therefore, load control. Pumps usually provide only a single operating condition, one discharge for a given head and speed. It is primarily the flow control means which causes the higher cost of hydraulic turbines. Limited market and resulting custom equipment and

(1)E. O'BRIEN AND A. C. GEORGE and C. C. PURDY, "Evaluation of Small Hydroelectric Potential", TAMS, New York, NY, April 1979

(2)R. E. A. ARNDT, "Small Turbine Technology: A Selective Review", St. Anthony Falls Hydraulic Laboratory, University of Minnesota, August 1980

(3)G. E. PFAFFLIN, "The Importance of Standardization in the Development of Small Hydroelectric Sites", Allis-Chalmers Corporation, Hydro-Turbine Division, March 1979

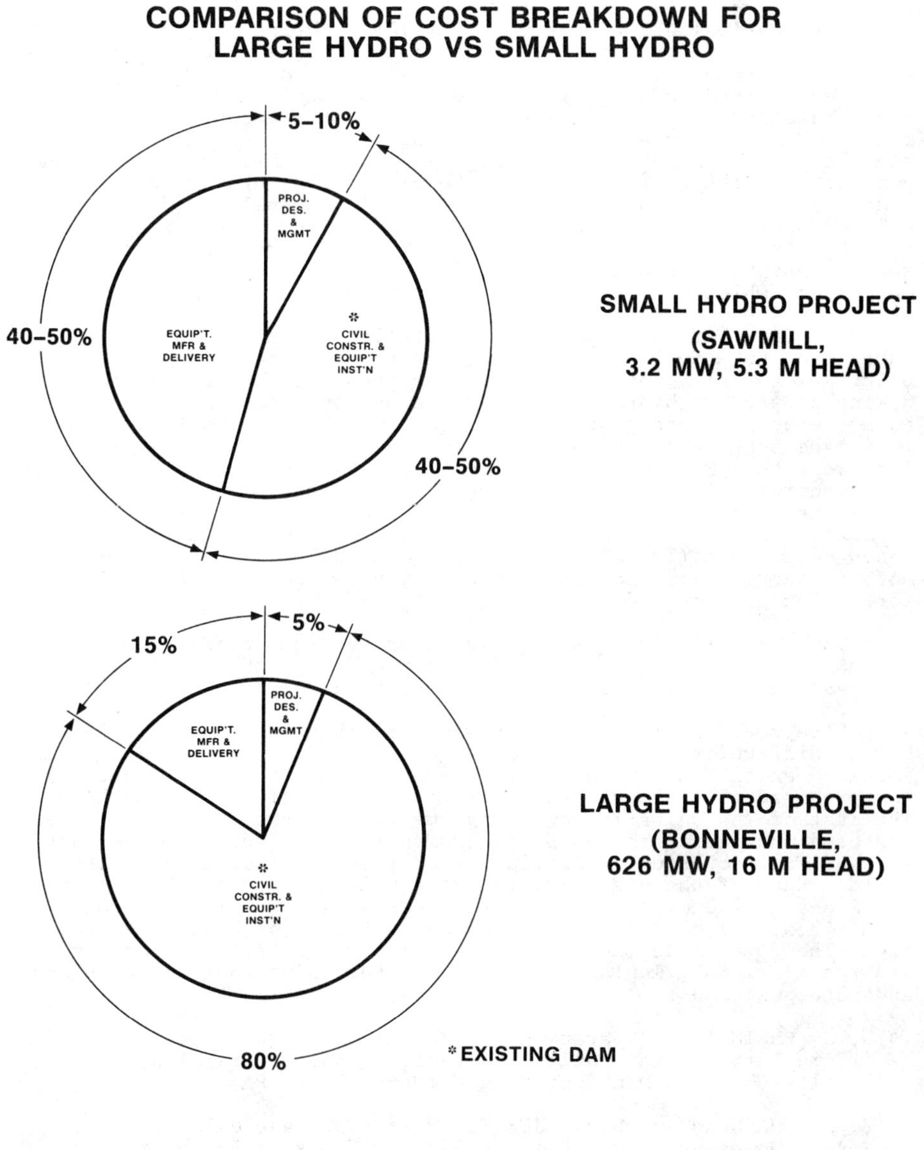
COMPARISON OF COST BREAKDOWN FOR
LARGE HYDRO VS SMALL HYDRO
5–10%
PROJ. DES. & MGMT
40–50%
EQUIP'T. MFR & DELIVERY
✲ CIVIL CONSTR. & EQUIP'T INST'N
40–50%
SMALL HYDRO PROJECT
(SAWMILL,
3.2 MW, 5.3 M HEAD)
5%
15%
PROJ. DES. & MGMT
EQUIP'T. MFR & DELIVERY
✲ CIVIL CONSTR. & EQUIP'T INST'N
80%
LARGE HYDRO PROJECT
(BONNEVILLE,
626 MW, 16 M HEAD)
✲ EXISTING DAM

FIGURE 1

construction design practices have obviously also contributed to higher cost machinery and projects, as well as longer deliveries and longer project schedules.

HISTORICAL INSTALLATIONS

In the low head category, one of the first significant installations was at the Traicao Plant in Brazil(4) (Figure 2).

The physical arrangement of this TUBE turbine type unit includes a siphon intake when the unit is viewed as a turbine and an elbow draft tube. The changes in area in both of these sections are designed to accommodate reverse flow at optimum efficiency. Both the intake and discharge are provided with stop log slots and trash racks. However, no intake gates are provided, so that each unit is started as a turbine by initiating flow over the siphon as a vacuum pump reduces the pressure within the siphon chamber. Once a significant amount of flow has been established, the balance of the air will be drawn out of the siphon by this flow and full operation of the unit is under the control of the adjustable runner blades. The runner design was based on conventional vertical Kaplan turbine technology which was only ten or fifteen years old at the time.

The first of three units was installed at this plant in 1938, subsequent to model test programs that started about 1933. These units are actually reversible pump/turbines which were originally designed based on hydraulic turbine technology, but then tested in both the pumping and generating modes (Figure 3). Modifications were made subsequent to the first model tests to provide performance improvements. New tests completed about 1945 demonstrated increased efficiencies. The original 75% best efficiency was increased to approximately 78%, and generating best efficiency was increased from approximately 75% to approximately 84%. These efficiencies were based on 25.4 cm (10 in.) and 30.5 cm (12 in.) models respectively. Note in particular, that at best efficiency the phi or speed is different for pumping and generating.

This installation was the basis for subsequent fixed and adjustable blade axial flow pumps built for flood control, irrigation, water supply, sewage treatment and industrial applications.

More recently in the 1960's when the Corps of Engineers was having difficulty making the Kaysinger Bluff Project economical, an inclined axis TUBE turbine type reversible pump/turbine (Figure 4) was proposed.(5) Almost ten years later, the units were purchased. The last of the units is now being put into operation and the name has been changed to the Harry S. Truman Plant. These are the largest and most powerful axial flow reversible pump/turbines in the world. Each of the six units is rated under a 15.2 m (50 ft.) head to discharge 138 m^3/sec. (4,860 $ft.^3$/sec.) when pumping and under 13 m (42.5 ft.) head, will have an output of 31.5 MW when generating. The five-blade runners are each 6.5 m (21 ft.) in diameter and these are full Kaplan units in that they have coordinated blade and gate positions.

(4)W. J. RHEINGANS, "Operating and Maintenance Experience with Pump-Turbines in Brazil", Hydraulic Products Division, Allis-Chalmers Manufacturing Company, York, PA

(5)F. J. CHAPURAN and D. J. JUETT, "Pumped Storage Using Inclined Axis Units at Truman Dam", U.S. Army Corps of Engineers, Fort Leavenworth, KS

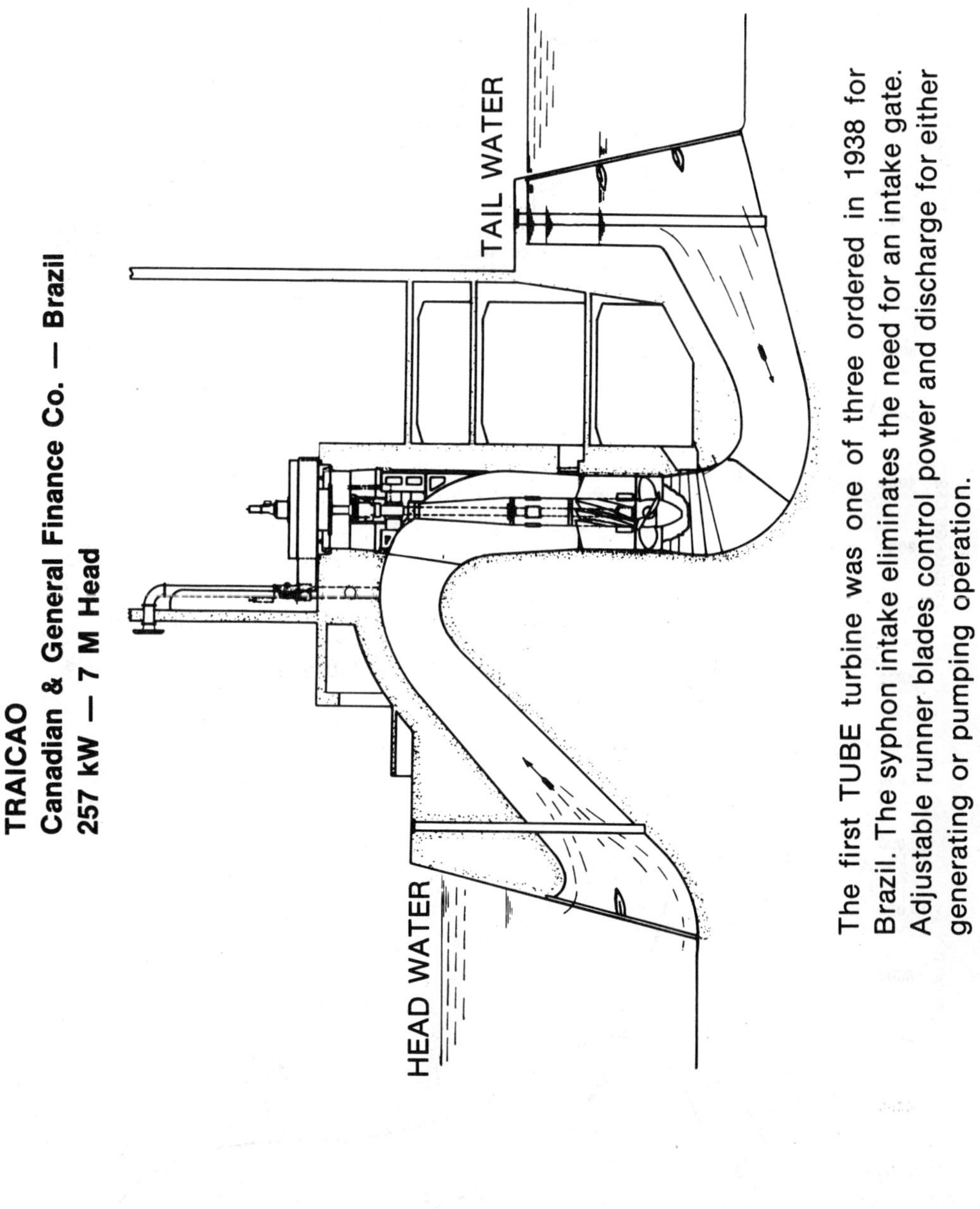
TRAICAO
Canadian & General Finance Co. — Brazil
257 kW — 7 M Head
TAIL WATER
HEAD WATER
The first TUBE turbine was one of three ordered in 1938 for Brazil. The syphon intake eliminates the need for an intake gate. Adjustable runner blades control power and discharge for either generating or pumping operation.

FIGURE 2

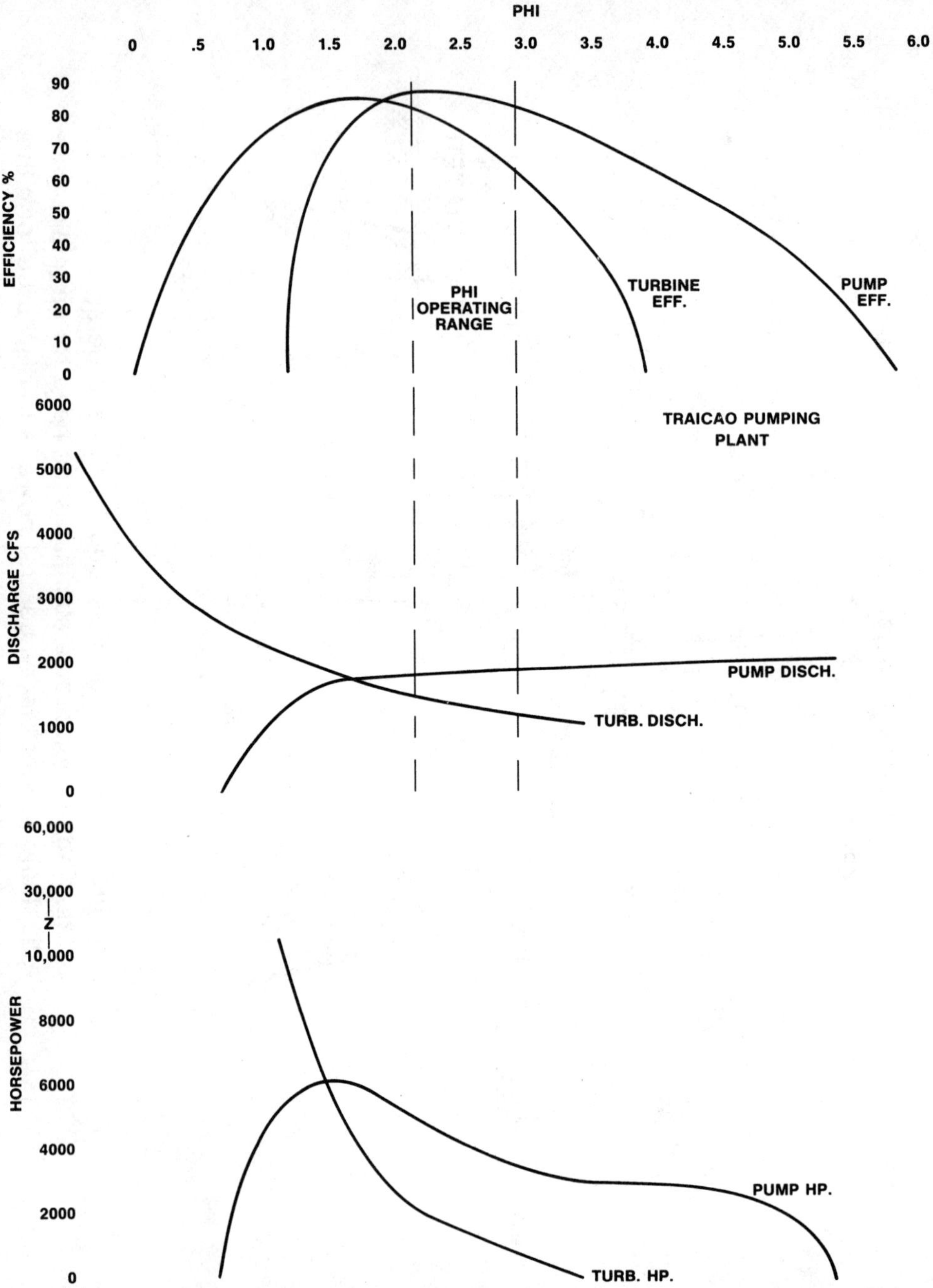
PHI
0
.5
1.0
1.5
2.0
2.5
3.0
3.5
4.0
4.5
5.0
5.5
6.0
EFFICIENCY %
90
80
70
60
50
40
30
20
10
0
PHI
OPERATING
RANGE
TURBINE
EFF.
PUMP
EFF.
TRAICAO PUMPING
PLANT
DISCHARGE CFS
6000
5000
4000
3000
2000
1000
0
PUMP DISCH.
TURB. DISCH.
HORSEPOWER
60,000
30,000
Z
10,000
8000
6000
4000
2000
0
PUMP HP.
TURB. HP.

FIGURE 3

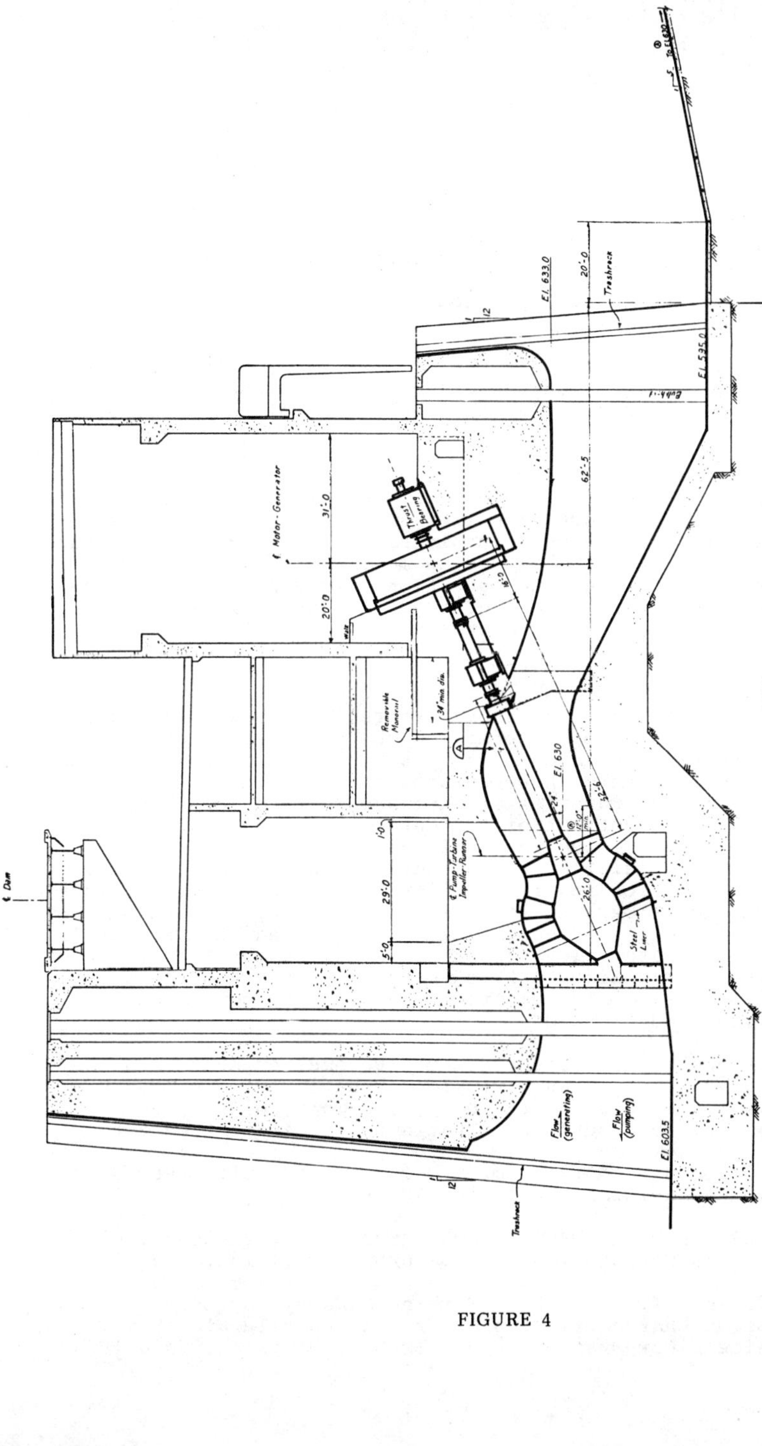

World's highest capacity tubular, reversible pump/turbines at Harry S. Truman Dam.

Installed by U.S. Army, Corps of Engineers in Missouri. Each unit is rated 31.5 Megawatts when generating under 13 meters (42.5') head and pumps 138 cms (4868 cfs) under 15.2 meter (50') head.

FIGURE 4

These two outstanding TUBE turbine examples are only a small part of the experience and modern technology on which a new line of low head hydraulic turbines has been based.

MODERN INSTALLATION

Kennebunk Light & Power Company has installed a unit rated 300 kilowatts under 5.5 m (18 ft.) net effective head at their Twine Mill site in Maine (Figure 5). This unit is designed to discharge 7.4 m^3/sec. (262 ft.^{3}/sec.) and the river is controlled so that the flow is essentially shut off unless it can equal or exceed this amount. The unit is started and stopped by the inlet butterfly valve. A parallel shaft speed increaser is provided to step up the turbine speed to the 925 RPM generator speed. The generator is actually an induction motor modified for the high turbine runaway speed. The necessary control and protective equipment is included in an adjacent cubicle.

Stationary guide vane and runner blade shapes of the turbine have been established based on the extensive experience and model test data developed over many years of building units for projects such as the two previously described. This technology is the heart of assuring the best possible performance without the added cost and delays which would be incurred should specific model tests be necessary for a specific application. Since model test data are available on both fixed blade propeller and adjustable blade propeller pumps and reversible pump/turbines, as well as hydraulic turbines, they may conveniently be used to modify conventional axial flow propeller pumps for turbine operation. This is a most effective means for reducing costs of equipment in the smaller capacity, lower head categories of power generation.

CAPABILITY NEEDED BY MANUFACTURERS

The Holyoke test flume in Massachusetts in 1879 was basically built to provide a means for determining the performance of full size hydraulic turbines on a comparable basis. Historically in the hydraulic turbine industry, those companies which did not build their own hydraulic laboratories or have access to a reliable hydraulic test facility went out of business. The importance of hydraulic tests became more and more critical as hydraulic turbines became larger and the value of efficiency became greater. Today, operating characteristics and reliability are major concerns of hydraulic turbine test programs, as well as efficiency.(6) Pump/turbines are tested in all four quadrants (Figure 6) and at various gate openings. Tailwater elevations are varied to obtain cavitation characteristics, hydraulic thrust is measured, air and water temperatures are controlled and measured. Results are continually recorded, analyzed and plotted.

The importance of accurate test data becomes evident when they are used to analyze the typical transient behavior of a pump, turbine or in this illustration (Figure 7), a pump/turbine. Computer programs(7)(8) have been developed that apply these test data to

(6) The Allis-Chalmers S. Morgan Smith Memorial Hydraulic Laboratory, Hydro-Turbine Division, Bulletin 54B5419-01

(7) S. A. CHACOUR and R. E. DEITZ, "CORA Hydraulic Transient PLUS", Allis-Chalmers Corporation, Hydro-Turbine Division, June 1978

(8) W. G. WHIPPEN and T.Y.W. CHEN, "Turbine Runaway, Hysteresis and Thoma's Sigma Characteristics of Francis Pump/Turbines", Allis-Chalmers Corporation, Hydro-Turbine Division, June 1978

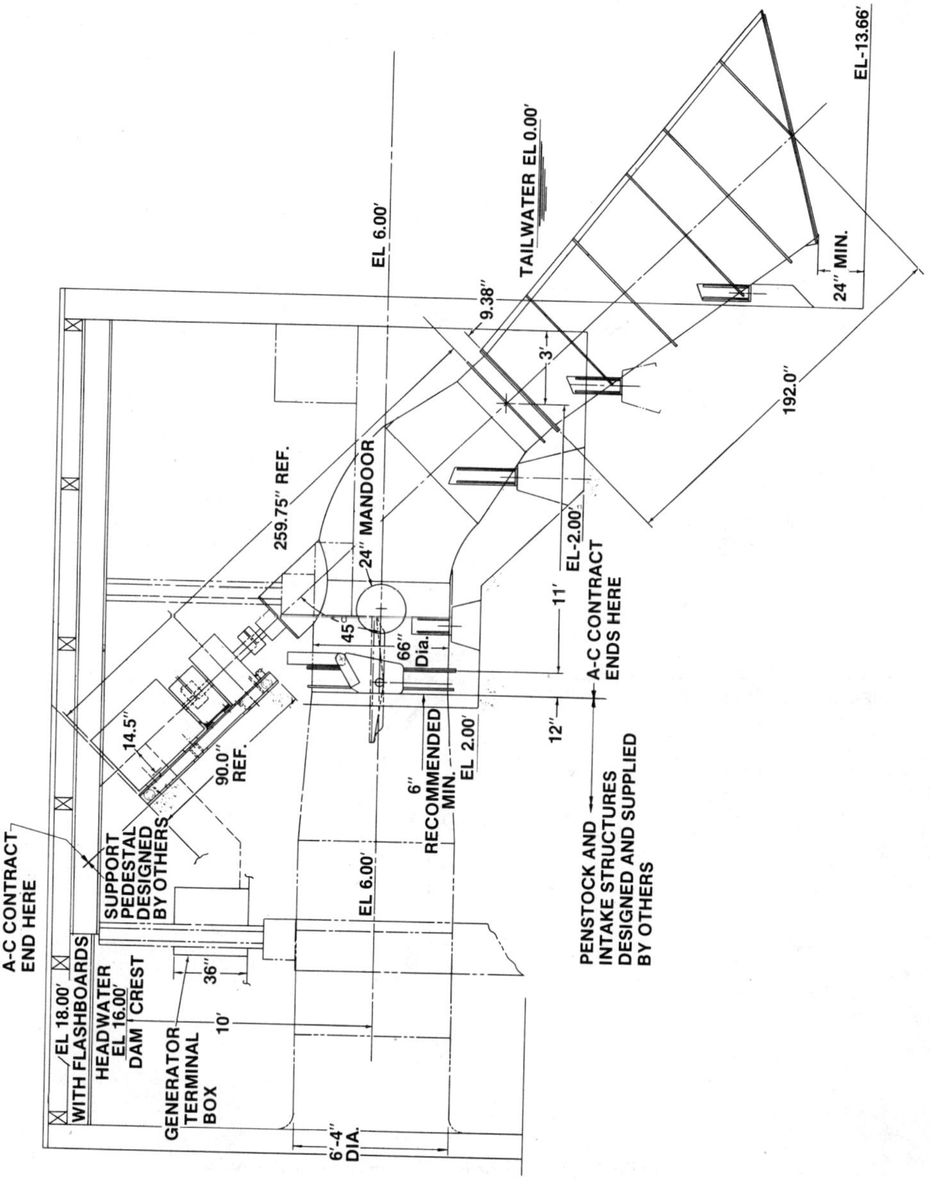
A-C CONTRACT END HERE
EL 18.00' WITH FLASHBOARDS
HEADWATER EL 16.00'
DAM CREST
GENERATOR TERMINAL BOX
36"
10'
6'-4" DIA.
SUPPORT PEDESTAL DESIGNED BY OTHERS
14.5"
90.0" REF.
259.75" REF.
EL 6.00'
24" MANDOOR
45°
66" Dia.
6" RECOMMENDED MIN.
EL 2.00'
12"
11'
EL-2.00'
A-C CONTRACT ENDS HERE
PENSTOCK AND INTAKE STRUCTURES DESIGNED AND SUPPLIED BY OTHERS
9.38"
3'
192.0"
24" MIN.
TAILWATER EL 0.00'
EL-13.66'

FIGURE 5

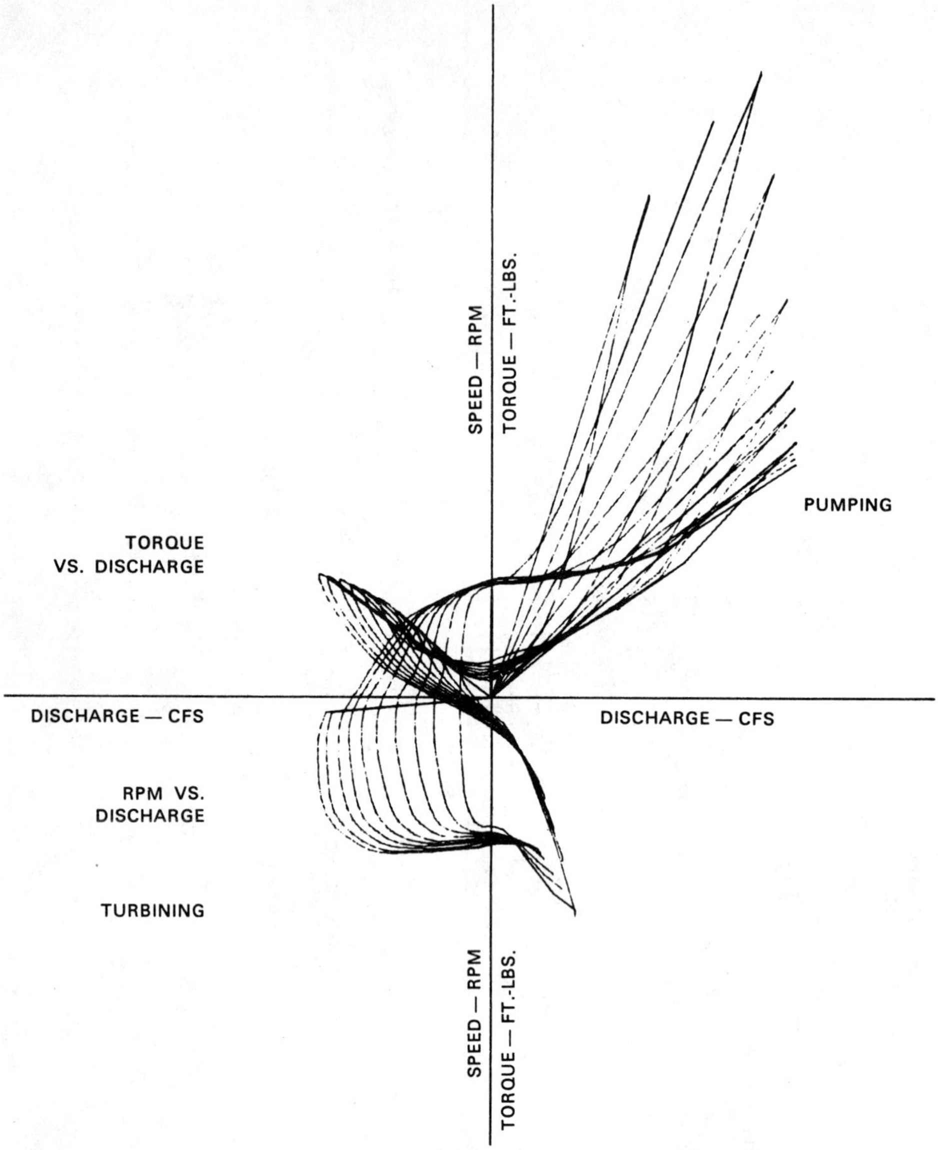

TYPICAL 4-QUADRANT SYNOPTIC CURVES FOR REVERSIBLE PUMP/TURBINE MODEL.

FIGURE 6

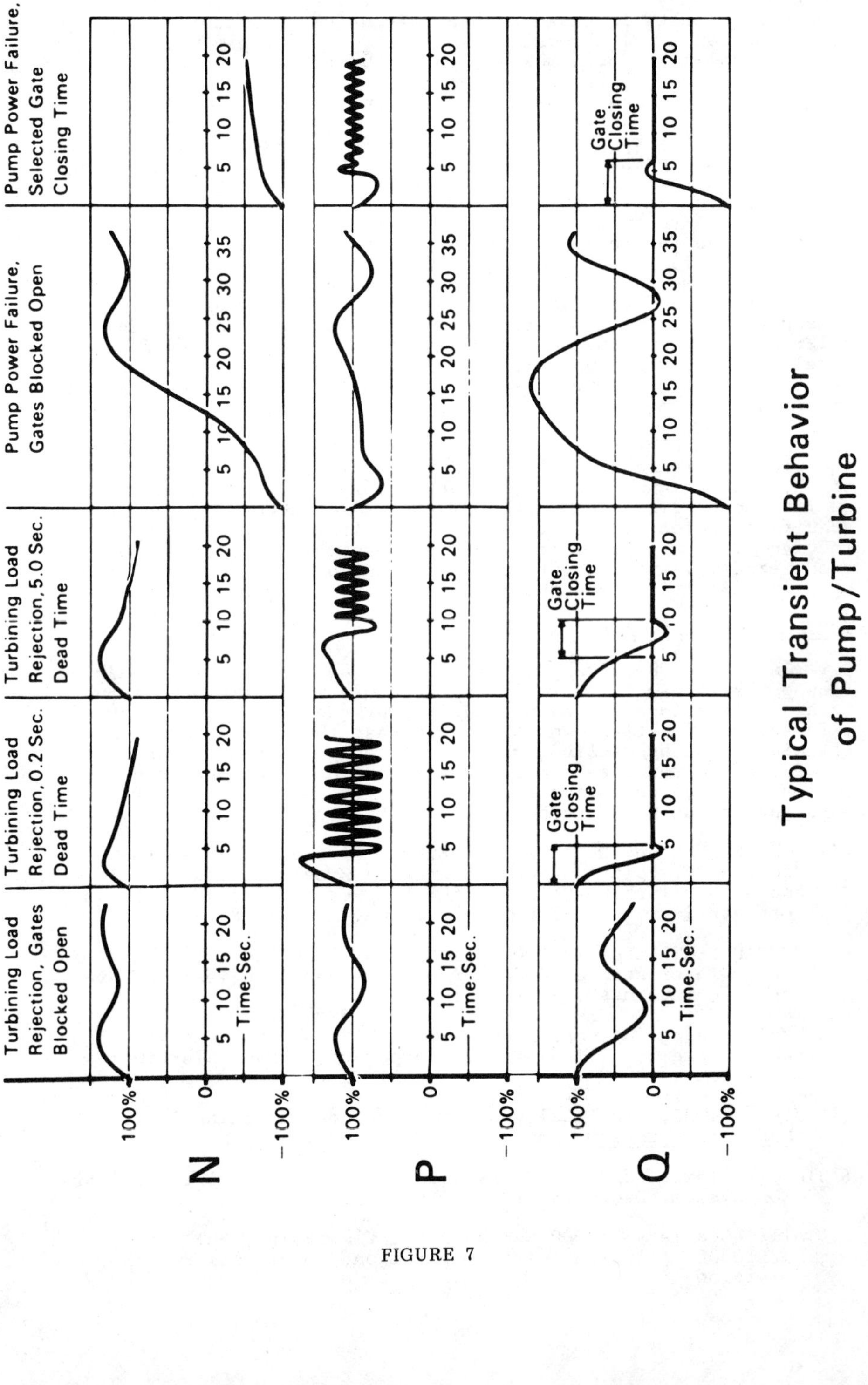

Turbining Load Rejection, Gates Blocked Open
Turbining Load Rejection, 0.2 Sec. Dead Time
Turbining Load Rejection, 5.0 Sec. Dead Time
Pump Power Failure, Gates Blocked Open
Pump Power Failure, Selected Gate Closing Time
N
P
Q
100%
0
-100%
Time-Sec.
Gate Closing Time
Typical Transient Behavior of Pump/Turbine

FIGURE 7

component and system designs. The ability to analyze and apply such test results is based on both extensive laboratory and field test experience. The results have been used for performance evaluations, valve and penstock design, valve and gate operating schedules, structural design of equipment and foundations, hydraulic design of most components and cavitation impact analysis(9).

The ability to predict and evaluate the results of cavitation under various operating conditions facilitates providing pitting protection in critical areas (Figure 8). Such information is not available from only a pump performance test. The International Standard for Hydraulic Turbine Model Tests is provided in "International Code for Model Acceptance Tests of Hydraulic Turbines"(10) Publications 193, 193A and Amendment No. 1. This code also provides measurement tolerances which are necessary to meet high standards of test accuracy and field performance predictability. A manufacturer should be able to relate the impact of manufacturing tolerances to performance results. Again, extensive experience in model testing, test analysis and application, field testing and relating the comparative results to actual model and prototype designs is a prerequisite to doing the job properly.

Field results based on an extensive technical foundation for equipment design and manufacture may be significantly jeopardized if the actual manufacture does not meet design requirements. Total Quality Control(11) is an effective tool for accomplishing the specified project objectives. Training is vital to assure an effective program(12). This particular program has been nationally recognized(13) and is continually monitored and adjusted to meet changing conditions and the required industrial quality level.

EQUIPMENT AND APPLICATION

In the low head ranges {below 12 m (40 feet)} standard axial flow pumps(14) as built for flood control, irrigation and industrial applications are being modified by the supply of suitable diffuser and impeller vane shapes so that they will operate more efficiently as hydraulic turbines. These pumps are currently fabricated of Corten steel to provide added corrosion resistance. Instead of the conventional bell intake, either an elbow or a straight conical draft tube is added (Figure 9).

(9) R. K. FISHER and D. R. WEBB, "Effect of Cavitation on the Discontinuity Point and on Alternating Pressures and Gate Torques on a Pump/Turbine Model in the Pump Cycle", Allis-Chalmers Corporation, Hydro-Turbine Division, June 1978

(10) BUREAU CENTRAL de la COMMISSION ELECTROTECHNIQUE INTERNATIONALE, "International Code for Model Acceptance Tests of Hydraulic Turbines", September 1977

(11) "Maintaining Customer Requirements With a Total Quality Assurance System", Allis-Chalmers Corporation, Hydro-Turbine Division

(12) J. R. FLOHR, "In Total Quality Assurance, Training is Vital", Management Review, June 1974, Pgs. 25-31

(13) Entire Issue of "Quality Assurance" devoted to Allis-Chalmers TQA System, December 1970

(14) Allis-Chalmers 15,000 Series Axial Flow Pumps, Bulletin 52B10036, Allis-Chalmers Corporation, Industrial Pump Division

PITTING PROTECTION

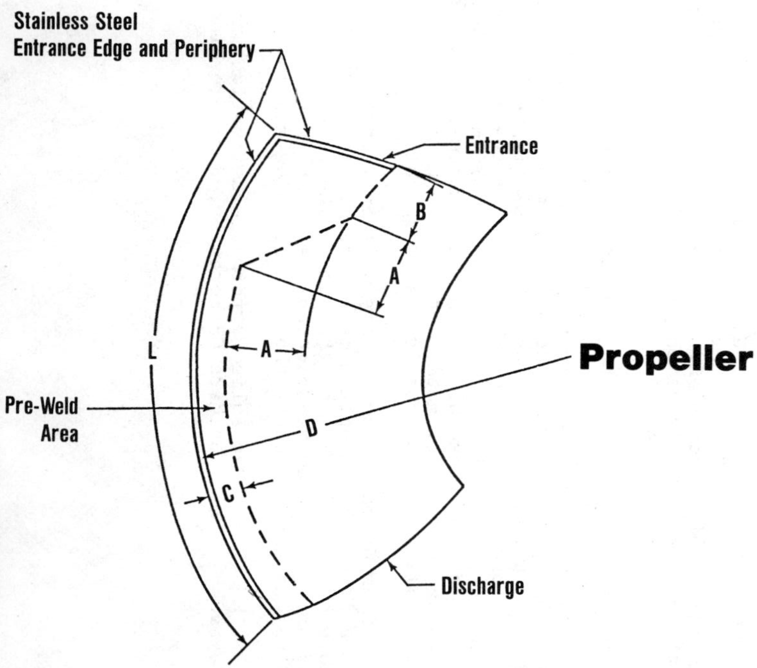

FIGURE 8

Dimensions/Materials/Parts

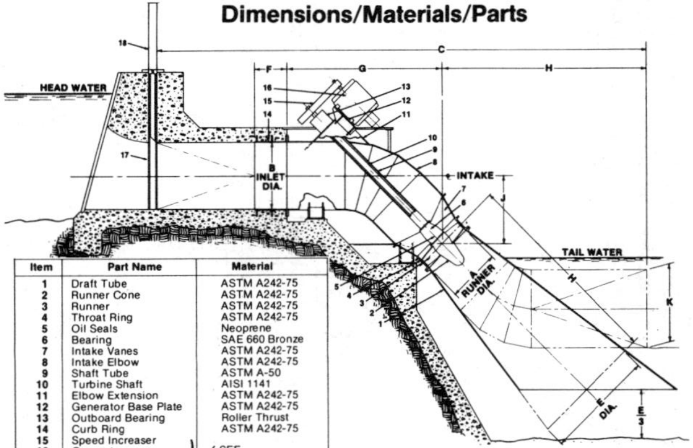

Item	Part Name	Material
1	Draft Tube	ASTM A242-75
2	Runner Cone	ASTM A242-75
3	Runner	ASTM A242-75
4	Throat Ring	ASTM A242-75
5	Oil Seals	Neoprene
6	Bearing	SAE 660 Bronze
7	Intake Vanes	ASTM A242-75
8	Intake Elbow	ASTM A242-75
9	Shaft Tube	ASTM A-50
10	Turbine Shaft	AISI 1141
11	Elbow Extension	ASTM A242-75
12	Generator Base Plate	ASTM A242-75
13	Outboard Bearing	Roller Thrust
14	Curb Ring	ASTM A242-75
15	Speed Increaser	SEE SEPARATE SPECIFICATION PAGES
16	Generator	
17	Intake Gate or Valve	
18	Hydraulic Power Unit	
19	Electrical Controls	
20	Power Transformer	

BASIC DIMENSIONS

A = Runner Diameter in Inches = 1.00

All Other Dimensions Are In Proportion From Runner Diameter

DIMENSIONS	RUNNER SIZE											
A	12	18	20	24	30	36	42	48	54	60	66	72
B	1.5	1.5	1.5	1.5	1.4	1.3	1.3	1.3	1.3	1.3	1.3	1.3
C	9.0	9.0	9.0	9.0	9.0	9.0	9.0	9.0	9.0	9.0	9.0	9.0
D	2.1	2.1	2.1	2.1	2.1	2.1	2.1	2.1	2.1	2.1	2.1	2.1
E	2.0	2.0	2.0	2.0	2.0	2.0	2.0	2.0	2.0	2.0	2.0	2.0
F	0.5	0.5	0.5	0.5	0.5	0.5	0.5	0.5	0.5	0.5	0.5	0.5
G	2.8	2.8	2.8	2.8	2.8	2.8	2.8	2.8	2.8	2.8	2.8	2.8
H	4.0	4.0	4.0	4.0	4.0	4.0	4.0	4.0	4.0	4.0	4.0	4.0
J	1.3	1.3	1.3	1.3	1.3	1.2	1.2	1.2	1.2	1.2	1.2	1.2
K	1.5	1.5	1.5	1.5	1.5	1.5	1.5	1.5	1.5	1.5	1.5	1.5
L	2.1	2.1	2.1	2.1	2.1	2.1	2.1	2.1	2.1	2.1	2.1	2.1

NOTE: Dimensions are approximate and may vary for specific applications.

FIGURE 9

The recommended arrangement is with the shaft at a 45° incline so that the generator is well above tailwater and so the intake as well as draft tube can be adequately submerged to prevent the entrance of air. The conventional pump discharge elbow is made slightly larger to provide a greater contraction for the intake flow as it approaches the runner through the stationary guide vanes. These vanes support an oil-lubricated bearing within the shaft tube which also provides a means for including intermediate guide bearings where necessary, to assure that the turbine shaft critical speed is higher than the turbine runaway speed.

At the intake, either a fixed wheel type gate or butterfly valve with hydraulic operator is used to provide tight shut-off and emergency closure. The intake gate or valve also controls the turbine speed as the unit starts up and until it reaches the normal operating speed. On the upstream end of the turbine shaft is a combination antifriction guide and thrust bearing which is designed to withstand the hydraulic thrust of the runner, weight of rotating parts and side thrust caused by a belt or chain drive. For smaller sizes, belts are used to step up the speed to a suitable induction generator speed. In most cases 900 or 1200 RPM induction generators are being used, since at these design speeds the induction generators can most economically be modified to accommodate the full runaway speed of the turbine.

Overall dimensions are provided on this standard dimension sheet. Some variation can be provided at additional cost, however, this is a complete, standard package generating unit. It includes the inlet valve or gate with hydraulic operator and oil pressure accumulator system, turbine with fixed blade propeller runner and conical draft tube, belt drive to 900 or 1200 RPM induction generator with terminal box, control and protection equipment for the generator and power transformer.

Sizing chart (Figure 10) provides a quick means for selecting minimum unit sizes based on the head and flow conditions available. The small sizes (12"-36" runner) can be used under higher heads and the number of runner blades can be increased for higher heads and settings. By oversizing the unit slightly, the efficiency and setting can be increased(15).

It is very convenient when a constant flow water source is available. For such applications a pump is a logical consideration. Most rivers, however, have a variable flow (Figure 11), therefore the use of pumps for power generation is not as usual. One approach which permits recovering 80% to 90% of the energy conventionally recovered but at lower cost, is to use at least two pumps of different sizes. This is not a completely new or unique approach. It has not been as common with custom-designed equipment due to the added cost of designing the second turbine and the fact that two small turbines usually cost more than one larger unit.

The flow duration curve shape will influence the size ratio, however a good starting place is to use 70% and 30% of the flow available or exceeded 25% of the time. The nearest standard generating units may then be selected. At least one other combination should be checked to select the one providing maximum flow utili-

(15)W. G. WHIPPEN, "Hydraulic Performance Evaluation and Potential Payback", Allis-Chalmers Corporation, Hydro-Turbine Division, October 1978

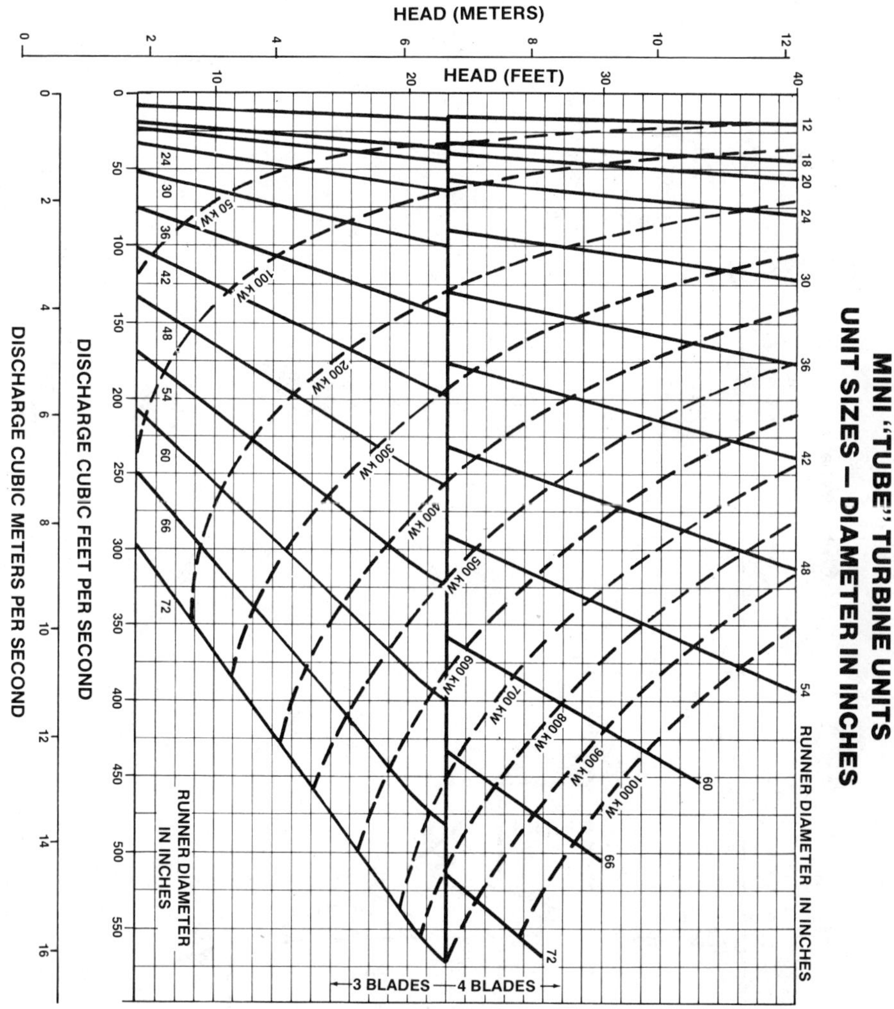
MINI "TUBE" TURBINE UNITS
UNIT SIZES — DIAMETER IN INCHES
HEAD (METERS)
HEAD (FEET)
DISCHARGE CUBIC FEET PER SECOND
DISCHARGE CUBIC METERS PER SECOND
RUNNER DIAMETER IN INCHES
3 BLADES — 4 BLADES
50 KW
100 KW
200 KW
300 KW
400 KW
500 KW
600 KW
700 KW
800 KW
900 KW
1000 KW

FIGURE 10

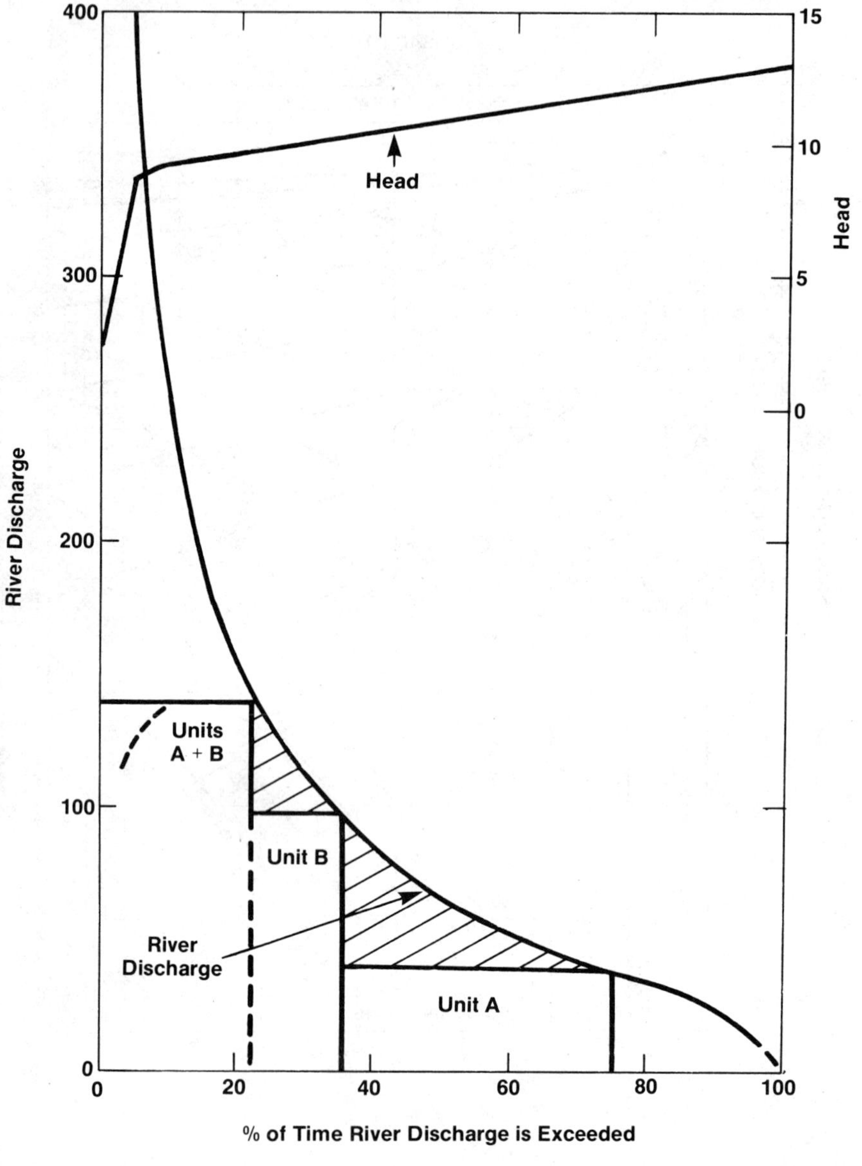
400
300
200
100
0
River Discharge
15
10
5
0
Head
Head
Units
A + B
Unit B
River
Discharge
Unit A
0
20
40
60
80
100
% of Time River Discharge is Exceeded

FIGURE 11

zation or lowest cost per KWH produced annually(16). Relatively simple and fast techniques can be adequate for initial and preliminary equipment selection and sizing. However, it is important that the basic data such as head and flow duration, as well as existing structures and site conditions, be accurate and complete. Low head units now commercially available are applicable to the area of head and output illustrated in Figure 12. The overlap with Standardized Horizontal TUBE Turbine Units(17) makes these lower head and small output sites more economical to develop. In the smaller sizes this is a two to one improvement and is a significant step toward our specific challenge to DEVELOP LOWER COSTS.

CONCLUSION

There is no question whether a pump can operate as a power generating unit. It has and is being done. Experience and modern technology needed for sound engineering application is readily available and is being used. As the market needs develop, modifications can and will be economically made to existing designs to improve performance, application ranges and operating flexibility. Research and development programs undertaken by the industry are helping to further reduce project costs so that low head and small capacity projects can be developed more economically. The challenge — TO DEVELOP WAYS OF MAKING LOW DENSITY AND SMALL INCREMENTS OF WATERPOWER ECONOMICAL — has been and is being addressed.

(16) R. K. FISHER, JR., "Selection and Optimization of Standardized Small Turbines Based on Considerations of Annual Energy, Allis-Chalmers Corporation, Hydro-Turbine Division

(17) Allis-Chalmers Standardized Hydroelectric Generating Units, Bulletin 54B10241-03

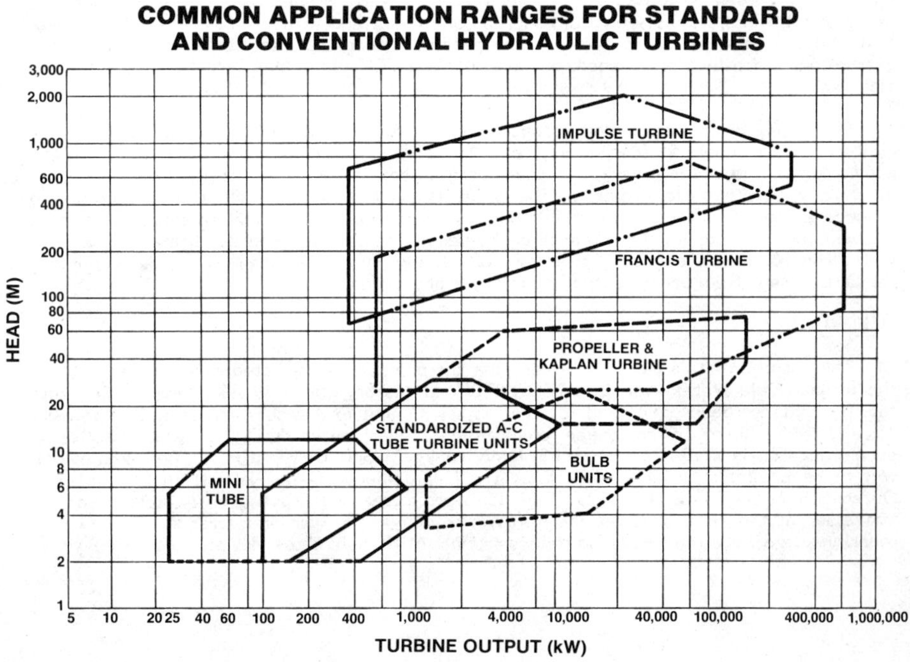

FIGURE 12

OTEC-1: DEPLOYMENT OF COLD WATER PIPE

NATHAN SONENSHEIN
GLOBAL MARINE DEVELOPMENT INC.

INTRODUCTION

The U.S. Department of Energy (DOE) has been developing since 1972 an alternative energy system known as Ocean Thermal Energy Conversion (OTEC). It extracts energy from the oceans by using the difference in temperature (35-45°F) between the solar warmed surfaces and cold waters at great depths (about 3,000 feet) in tropical ocean regions. OTEC is a many faceted program aimed at successful demonstration of commercial power plants of 400-500 MW_e net power output by 1985-1995. Such plants will produce electricity by transmission to shore by submarine cable and for manufacture at sea of energy-intensive products such as ammonia, hydrogen and aluminum.

There are two basic ways that OTEC plants can convert an ocean thermal gradient to electricity: through a closed-cycle process (the opposite process of a home refrigerator) using an intermediate working fluid such as ammonia; and through an open-cycle process that evaporates seawater to produce low pressure steam. Steam condensed, after work extraction, is freshwater. Hence, an open-cycle OTEC plant may also be a desalination plant.

OTEC plants are envisioned as operating on one of three platforms: a fixed facility built on land; a moored, floating ship; or a "grazing" plant ship that moves slowly on the ocean surface in areas where the thermal gradient is high. All three types generate electricity as the basic product. The plant ship, in addition, uses the electricity to produce an energy-intensive material that can be stored for shipment to shore. The most challenging aspect of all three modes is the design, fabrication and deployment of a Cold Water Pipe (CWP), the marine riser, or conduit by which cold ocean water is brought to the OTEC plant from great depths and in great quantities.

OTEC-1

The purpose of this paper is to describe the deployment in October-November 1980 of the largest Cold Water Pipe in OTEC's 100-year history for a floating engineering test facility that has been developed for testing OTEC heat exchangers and other components at a 1 MW_e level in an ocean environment. Designated 'OTEC-1', this facility is accommodated in a 1944 vintage U.S. Navy tanker, USNS CHEPACHET (TAO-78), which was made available in 1978 to DOE by the Department of the Navy from the National Defense Reserve Fleet at Suisun Bay, California and was subsequently activated and converted at the Northwest Marine Iron Works shipyard in Portland, Oregon. Transit to a mooring site off Keahole Point on the west coast of the Island off Hawaii for a three-year test period took place in July 1980. To reflect the new mission, DOE renamed the ship SS OCEAN ENERGY CONVERTER, but it should be noted that this facility does not contain a turbine-generator for converting thermal into electrical energy through the OTEC process because such conversion is not needed for testing heat exchangers and other components. Figures 1 and 2 illustrate the salient features of the ship system as converted and deployed off the Island of Hawaii.

PROGRAM STATUS

Global Marine Development Inc. (GMDI) of Newport Beach, California was awarded a contract in September 1978 by DOE to perform the function of System Integration Contractor with TRW, Inc. as its principal subcontractor. The OTEC-1 Program is being accomplished in three successive phases:

Phase I:	Design	November 1978 - May 1979
Phase II:	Conversion of Ship Deployment	May 1979 - November 1980
Phase III:	Operational Testing	December 1980 - September 1983

The GMDI contract covered Phases I and II. Rockwell International (Energy Technology Engineering Center) is now prime contractor for Phase III.

The OTEC-1 project has been highly challenging from both the marine engineering and naval architectural aspects. For example, this project included the design and installation in the test ship of a 1 MW_e ammonia power cycle of a size and complexity exceeding any prior marine experience; the design and development of a unique mooring system for long term mooring of a 1,7000 ton ship in 4,500 feet of water, a feat that establishes a world's record; and the design fabrication onshore, towing to a mooring site 18 miles offshore, and vertical attachment to the keel of a polyethylene CWP Assembly 2,150 feet long, representing another quantum jump in marine technology.

THE OTEC-1 COLD WATER PIPE

CWP DESIGN (Figure 3)

GMDI's design consists of a bundled configuration wherein three 48-inch diameter polyethylene pipes, each 2,100 feet long, are joined by polyethylene jacketed steel cables that run inside and down the length of each pipe, joining them at top and bottom. The pipe bundles are connected to a single weighted cable which stabilizes the lower end of the entire pipe system, whose total length is about 2,300 feet. The interface between the CWP and the ship is subject to severe stress due to the combined effects of ship motion, and wave and current forces on the CWP. A two-ring, steel-fabricated gimbal attachment was provided in the ship to allow a 30° transverse and longitudinal angular deflection at the interface. Combined with inherent flexibility of the polyethylene, the CWP can remain attached up to and including the maximum operating sea state limit for the site (40 foot waves, 60 knot winds and 2-3 knot currents).

CWP FABRICATION

Figure 4 shows clockwise:

(a) The assembly site at Kawaihae Harbor on the Island of Hawaii.
(b) The special gauge railroad track and carts on which the pipe was assembled from 100 foot long sections manufactured by Dupont of Canada and thermally butt-welded using that company's welding equipment and the upper buoyancy collar.
(c) The banding of the pipes with nylon straps, and;
(d) The lower transition section with a sonic reflector offsticker for use in locating bottom position after suspension from the OTEC-1 ship.

CWP DEPLOYMENT

The concept for deployment was bold yet simple: after fabrication on the railroad track at Kawaikae:

(a) Launch the assembly by pulling it with tugs into the harbor,
(b) Support the heavy transition end on a barge for the tow to site,

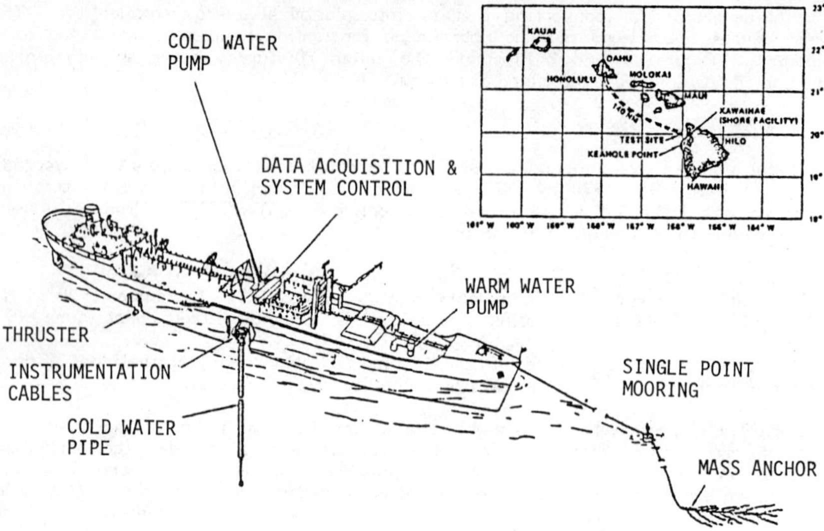

FIGURE 1. OTEC-1: SITE & AT-SEA TEST SYSTEM

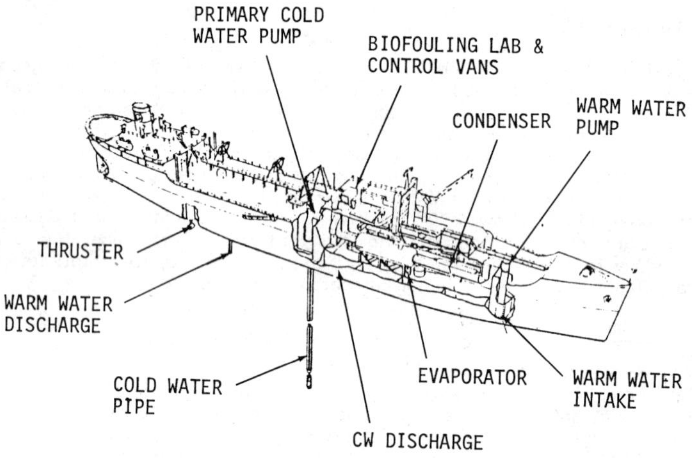

FIGURE 2. OTEC-1: POWER CYCLE

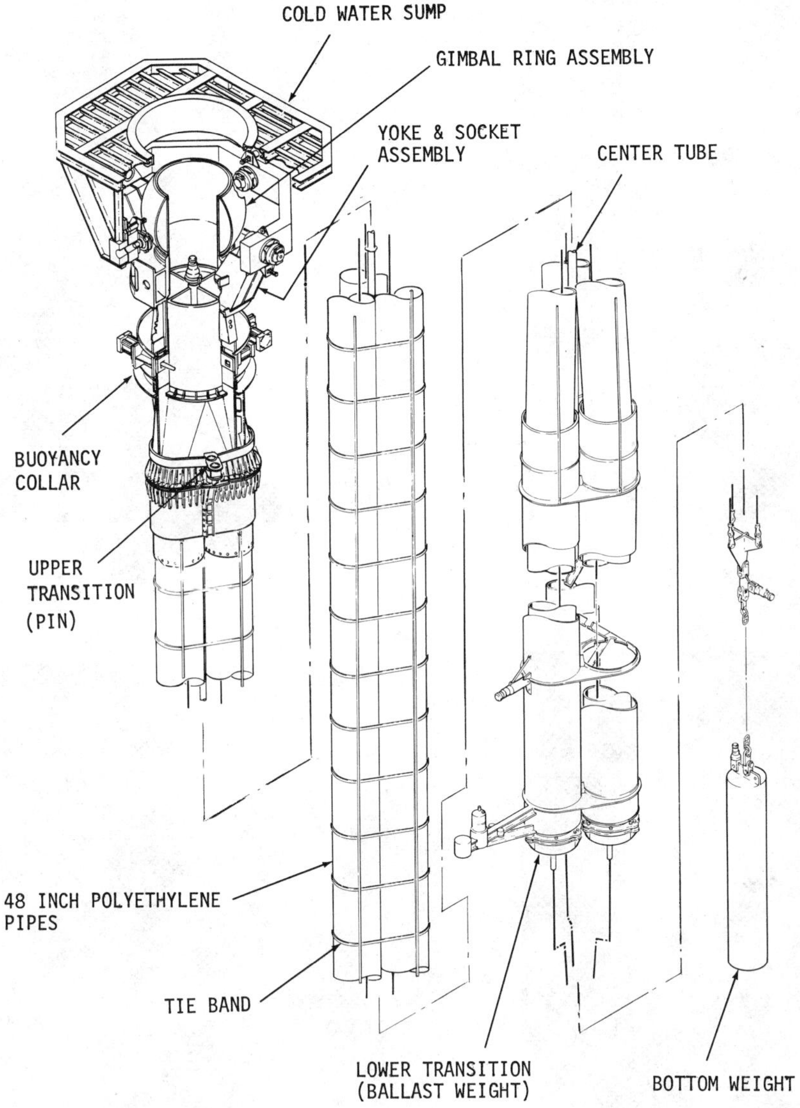

FIGURE 3. COLD WATER PIPE DESIGN

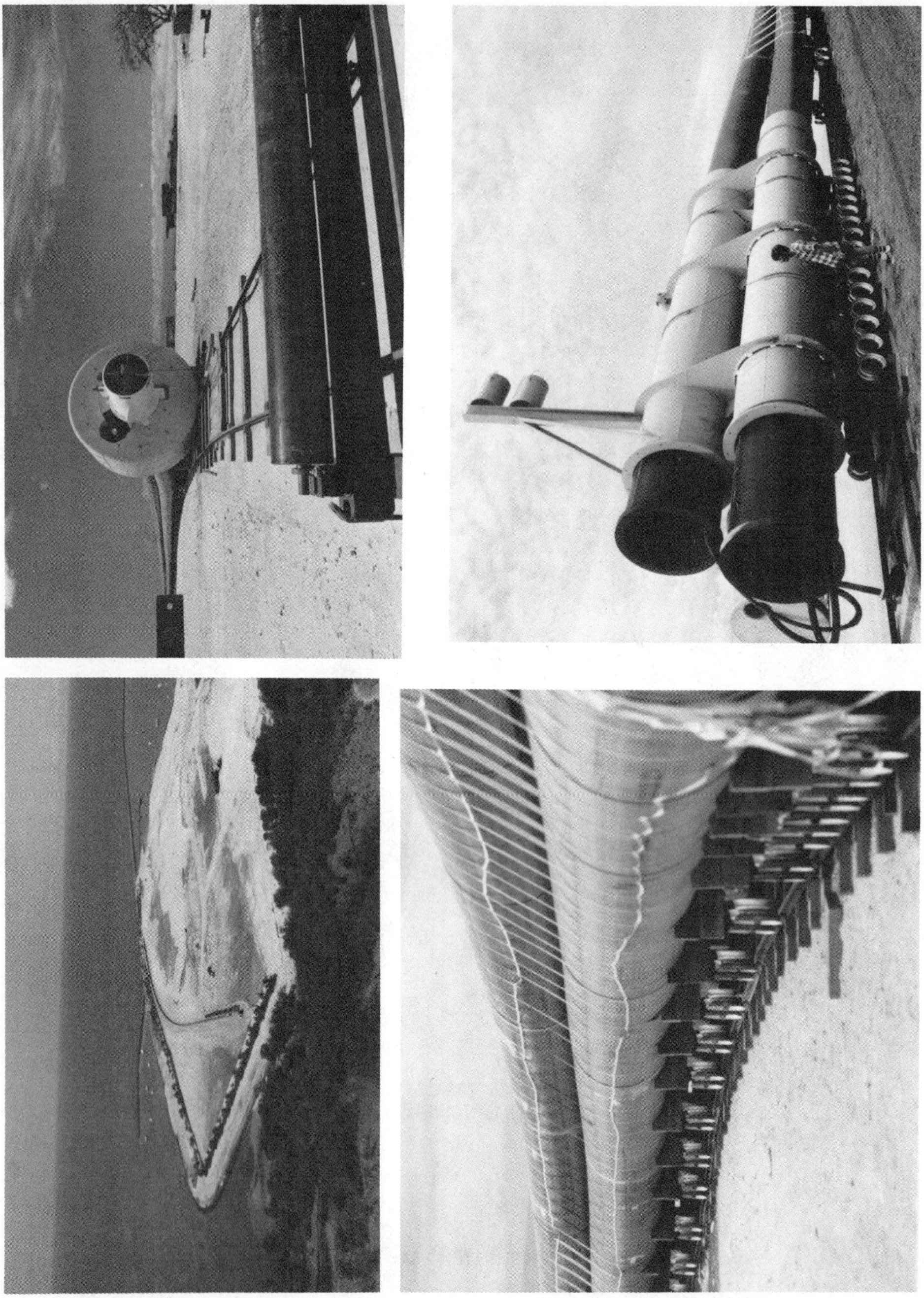

FIGURE 4 CWP FABRICATION

(c) Suspend the pipe at the site, using the bottom weight (previously positioned on a davit on the OTEC-1 ship) to pull the lower end down while a tug holds the buoyancy collar at the surface, and;

(d) Keelhaul the CWP and attach it to the ship's gimbals. Figure 5, 6, 7 and 8 illustrate the principal events in the deployment sequence.

Actual deployment of the CWP was carried out per the following schedule:

24 Oct 1980	Bottom weight loaded on ship at Kawaihae Harbor and CWP launch commenced.
29 Oct	CWP launch completed.
31 Oct	Ship moored at site, and CWP tow commenced.
1 Nov	CWP at site and upended.
4 Nov	CWP made up to gimbal in ship.

Excellent weather, described as "flat calm", facilitated these complex operations, which were carried out exactly as planned with one minor exception. The cable used for lowering the lower transition section of the CWP (Cable A in Figure 6) could not be released when its Regan Latch was actuated from the surface. Such latches are commonly used in offshore oil industry operations and were fitted on both Cables A and B. Some six days later, after several unsuccessful efforts to determine what was causing the hang-up of the simple latch and to effect release, the latch released itself for reasons not fully understood! Thus, deployment of the CWP was completed successfully and on 11 November 1980 "clear, sparkling, cracking cold" water was pumped through it at 80,000 gpm from the depths of Keahole Basin and into the OTEC-1 test facility to start DOE acceptance tests.

SUMMARY

The OTEC-1 program was a highly developmental effort in which major engineering challenges were met successfully in:

(a) The design and installation in the test ship of a 1 MW_e ammonia power cycle of a size and complexity exceeding prior marine experience, and

(b) The design and development of a unique mooring system for mooring a 17,000 ton ship in 4,500 feet of water, a world's record.

(c) The greatest achievement of the program, however, was the design, fabrication and deployment of the Cold Water Pipe Assembly. In the author's opinion, this represents the most significant ocean engineering advancement since the GLOMAR EXPLORER was delivered in 1974.

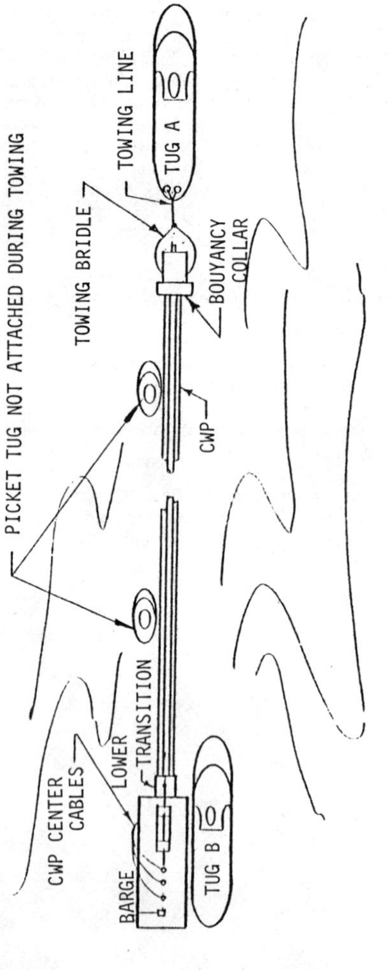

FIGURE 5 CWP TOWING ARRANGEMENT

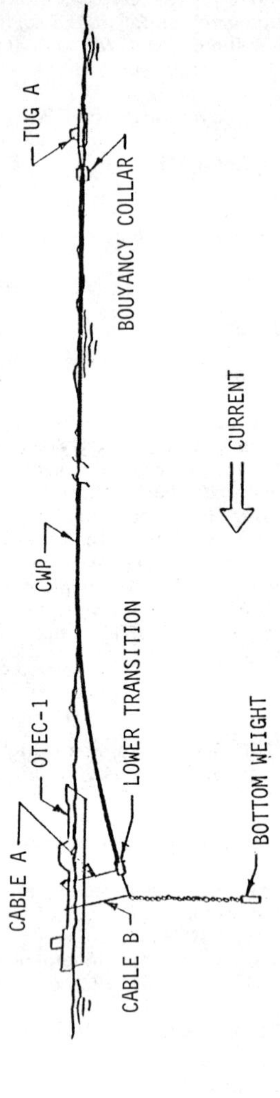

FIGURE 6 BEGINNING CWP DEPLOYMENT

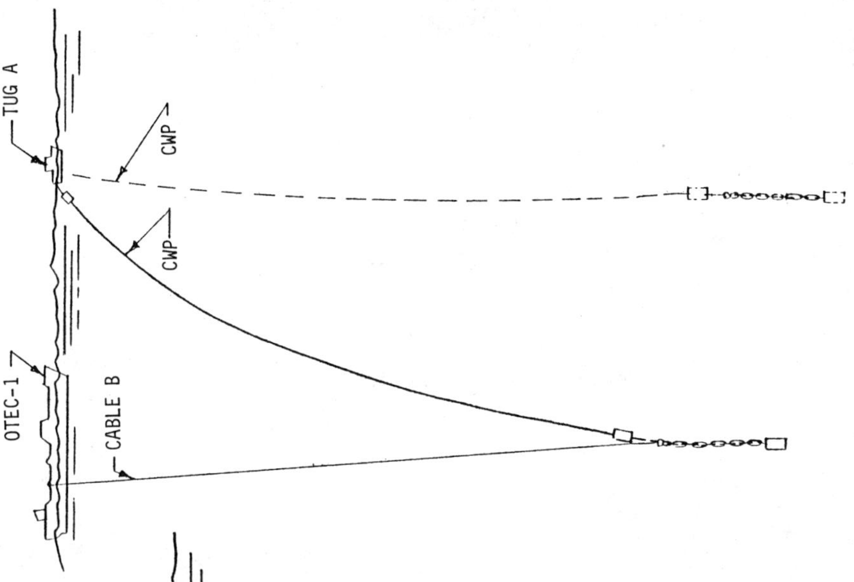

FIGURE 8 CWP - RELEASE OF CABLE B, READY FOR KEELHAUL

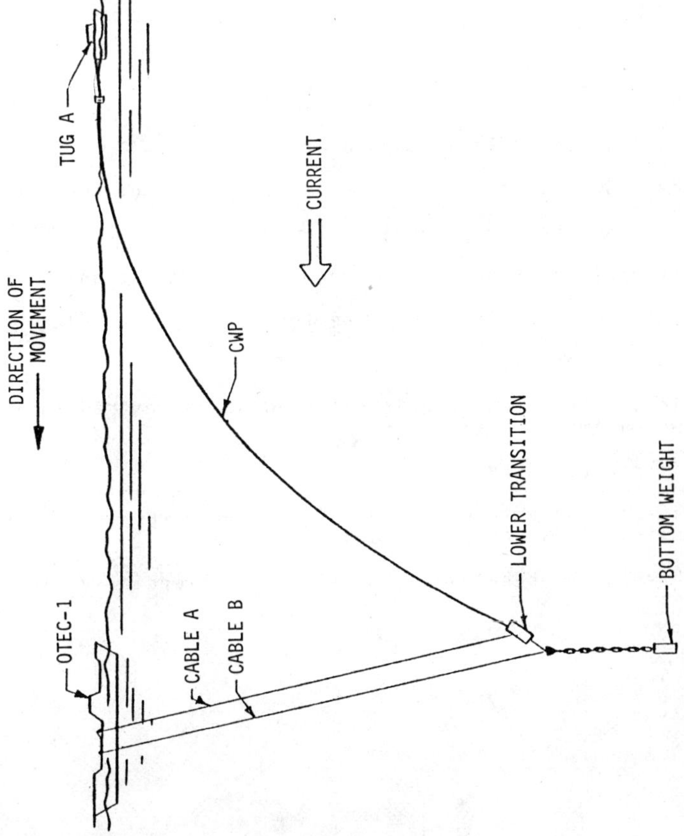

FIGURE 7 CWP - NEARING RELEASE OF CABLE A

ECONOMICS OF OTEC

B. Coffay

Westinghouse Electric Corporation

Introduction

OTEC is a good news/bad news story. The good news is that OTEC:

- is a solar based renewable energy source;
- has no known adverse environment impacts of consequence;
- is base-load by nature;
- is cheaper than fossil energy in many locations; and
- has been demonstrated to be technically feasible.

The bad news is that OTEC:

- in larger sizes, requires technological extrapolations, which imply risk;
- is capital intensive;
- is not thought to be the lowest cost power source for most mainland US locations; and
- hasn't yet been demonstrated in a commercially useful size.

This paper is aimed at trying to define the conditions which make building the first OTEC plant probable and attractive. Though many size plants will be discussed, the greatest emphasis will be on the smaller plants because:

- o small OTEC plants require less capital than larger OTEC plants to build and, therefore, stand a better chance of being built, all other things being equal;
- o small plants require less technological extrapolation to build;
- o small plants are most applicable to island communities, which have only fossil and OTEC as alternatives for base load power; and
- o many island communities have their economies crippled by oil costs and oil cost increases, resulting in very high energy production costs.

Two application areas for OTEC will be discussed in economic terms: electricity and water production. In addressing the two technology areas, my underlying economic assumptions are for US tax laws and regulations. The prime motivations for this selection are:

- o they are the most familiar;
- o they are probably most applicable to the audience; and
- o a plant built first in the US could cause other countries to be more interested in the technology. It could well be the basis for starting an OTEC industry in the US.

The economic analysis methods used in this paper could apply to non-domestic situations, but each international opportunity will have a different set of economic constraints.

Electricity Generation

In exploring who or what entity will likely build the first OTEC plant, we shall start by seeing what economic forces are at work in the utility industry itself. Let us begin by confirming that OTEC appears to be a lower-cost option for the generation of base-load electricity than is the oil-fired steam plant. This demonstration has its greatest applicability to islands that are far from a source of coal.

As input to the analysis, I have established two curves which give OTEC plant capital cost in 1980 dollars vs. OTEC plant size, and a similar curve for OTEC plant operating and maintenance (O&M) costs. These curves are shown in Figures 1 and 2 respectively. Note that both of these curves indicate an economic disadvantage in building smaller plants, as has been indicated by the work done to date for DOE. These curves are used to establish approximate numbers in the analysis that follows; they are not intended to be precise.

The method in comparing the oil cost with OTEC cost is essentially that given by the Electric Power Research Institute (EPRI). EPRI is an organization that is mostly made up of large, mainland utilities that have not yet developed guidelines for OTEC. Nevertheless, to the extent possible, I used their Technical Assessment Guide to establish what is termed the levelized busbar cost of electricity. The type of analysis selected was for the investor-owned utility, although a similar analysis can be performed for government-owned utilities.

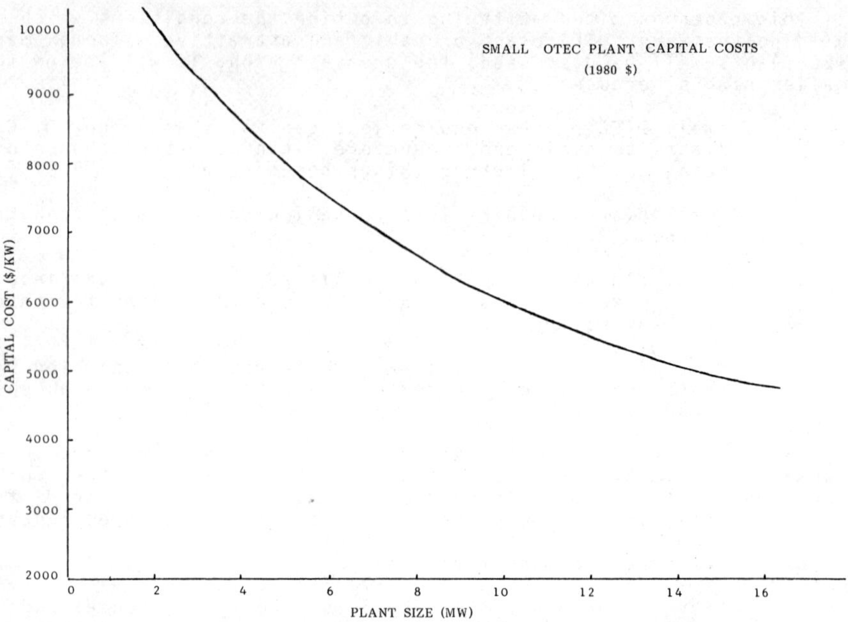

Figure 1

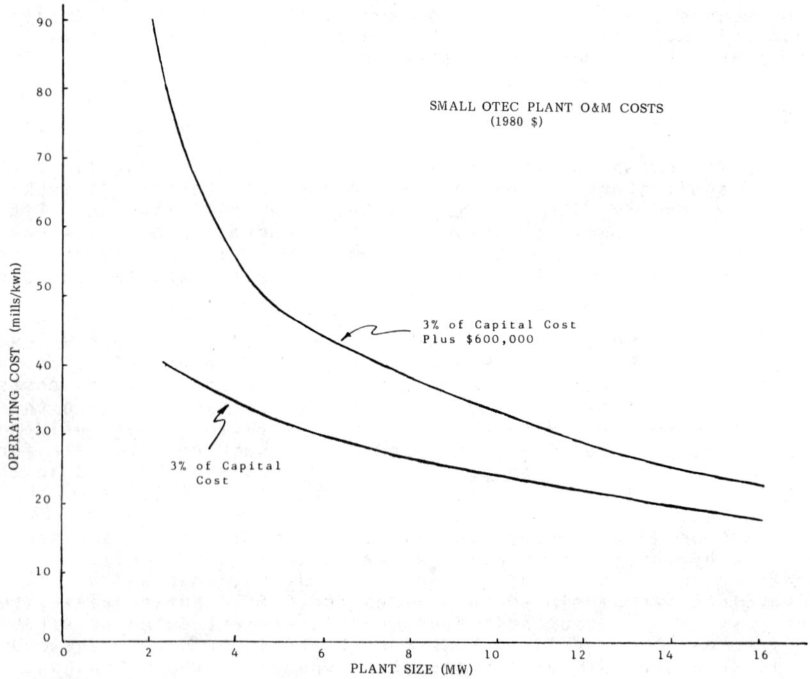

Figure 2

The conditions for this analysis are shown in Table 1: EPRI numbers are used except for the following:

- o the capacity factor is set at .8, and
- o the investment tax credit for OTEC is assumed to be 25% rather than 10% as for the oil plant.

Other key assumptions are:

- o plant start date = 1984
- o general inflation = 6%
- o cost of money = 10%
- o cost of fuel = $35/bbl (1980 dollars) in 1984.

The results of the analysis are shown in Figure 3. The levelized cost of electricity, which is EPRI's term for representing the effect of future cost of energy increases in today's dollars, is plotted vs the "as built" power plant capital cost. Three lines are plotted for fossil-fired plants with the only difference among the lines being the annual rate at which fossil fuels will increase. Three lines are also plotted for OTEC plants with the only difference among the lines being O&M costs.

Note that any annual rate of fuel price increase over 8.5% (includes general inflation at 6%) would seem to justify the selection of OTEC, as long as the plant size was about 10 MW or greater. This assumes that the "as built" cost of a fossil fired steam plant would be about $500/kW to $1000/kW. Rates of fuel price increase that are larger than the 8.5% maximum EPRI guideline would favor smaller OTEC plants. A 14% fuel price increase rate, for example, would reduce the minimum economic size for an OTEC plant to less than 5 MW. To place the 14% in context, it is less than half the annual increase experienced since 1973. It is also significant to note that the latest EPRI Technical Assessment Guide (7/79) projects the 1984 price for oil as $35/bbl. Since we are already there, this can be taken as a good indiction of how unexpectedly fast fuel prices are rising.

Some words of caution are called for here. As the current legislation is written, it is relatively certain that the full 25% tax credit applies to OTEC projects which produce electricity for one or more of the 50 states. Without further clarification, it is not clear that those credits would apply to the US island territories. Further, all tax credits will be allowed by the Department of the Treasury only after consultation with DOE. And lastly, the availability of the incremental 15% OTEC tax credit (over and above the ordinary 10% tax credit) has an expiration date of December 31, 1985. With all of these conditions, it is clear that the validity of the foregoing analysis is dependent on project location, timing and the concurrence of DOE. And, in turn, all of this is indicative of a certain limitation of the applicability for OTEC.

But to reaffirm the positive, OTEC will (from the utility analysis point of view) make electricity cheaper than oil-fired plants, even in sizes as small as 5 MW to 10 MW.

If it is true that OTEC plants make less costly electricity than oil-fired plants, how is it that there aren't any OTEC plants? Why won't a utility buy one?

Probably because the purpose of a utility is to provide reliable, low cost power. Since OTEC is unproven in commercial sizes, its reliability is not really known. Furthermore, the utility investor looks for high security. The utility investor does not expect or want a utility to take chances on unproven technology which could jeopardize

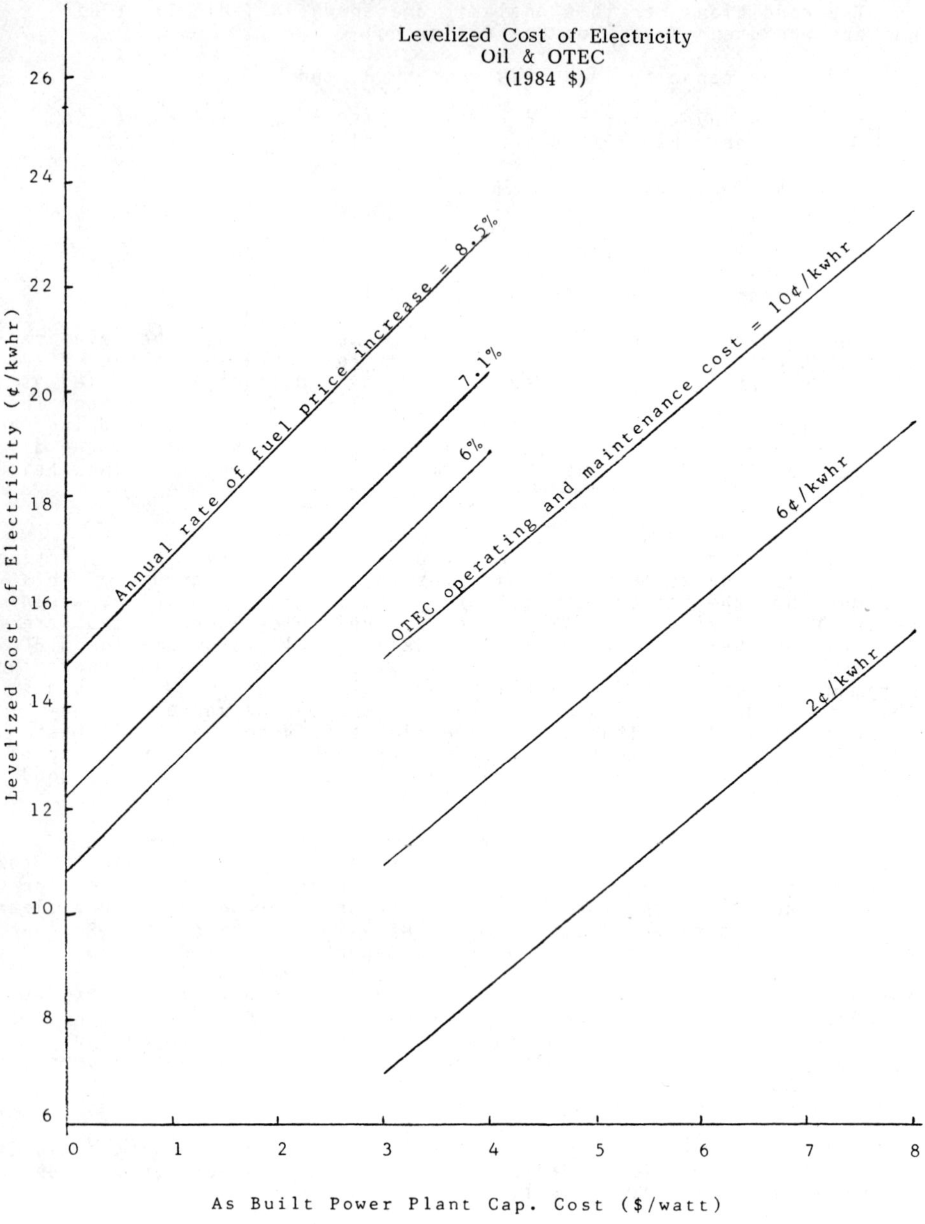
Levelized Cost of Electricity
Oil & OTEC
(1984 $)
Levelized Cost of Electricity (¢/kwhr)
26
24
22
20
18
16
14
12
10
8
6
Annual rate of fuel price increase = 8.5%
7.1%
6%
OTEC operating and maintenance cost = 10¢/kwhr
6¢/kwhr
2¢/kwhr
0
1
2
3
4
5
6
7
8
As Built Power Plant Cap. Cost ($/watt)

Figure 3

ASSUMPTIONS FOR EPRI REVENUE REQUIREMENT ANALYSIS
(1984 Dollar Values and Start Date)
Table 1

From EPRI Handbook:

1. Annual general inflation rate = 6%.
2. Average cost of money = 10%.
3. Debt Ratio (private utility) = 50%.
4. Discount rate (average cost of money) = 10%.
5. Income tax rate (local & federal) = 50%.
6. Accelerated depreciation used = sum of year's digits.
7. Tax credit for oil plant = 10%.
8. Cost of property taxes, insurance, etc. = 2% of capital cost.
9. Iowa State type S retirement dispersion allowance.
10. 30 year book life, 20 year tax life.
11. 1984 fuel cost = $5.50/10^6 Btu = $35/bbl.
12. Oil plant heat rate 10,000 Btu/kWh.
13. Oil plant O&M cost = 2.36 mills/kWh.

Other Values:

14. OTEC plant tax credit = 25%.
15. Capacity factor = 0.8.

the investment security. In addition, there are regulatory constraints; a high rate of return on investment is not normally permitted by the Public Utility Commissions, and therefore the allowble return is not commensurate with the risk of an investment in a first-of-a-kind facility. And, if all of the foregoing is not enough, the capital intensiveness of OTEC is a bit frightening to utility management whose backgrounds suggest capital costs that are 1/3 to 1/5 that of OTEC for a given capacity. A similar set of arguments applies to the public utility case.

If a utility isn't going to build the first one, who is? An equipment manufacturer?

Mini OTEC appears to be the level of commitment available from the OTEC equipment manufacturers. In other words, commercially significant OTEC plants will probably not be built entirely at the expense of the equipment manufacturers. The quantity of money involved is large, the risk is perceived as large, and the market to be captured is ill-defined and may be too far into the future to make confident projections. So the motivation to build an OTEC plant to get a leg up in the market-place is not strong enough to justify the risk.

Will the Government be the one to build the first plant?

The Program Opportunity Notice (PON) program has been started with every intention of being carried through to completion. This is a possible route to having an OTEC plant on line in a 5-year time frame. I, for one, hope that this program acquires the necessary support from industry and government to be completed because it is possible that this program will pave the way towards greater applicability of OTEC in larger sizes. It is probably the key to the mainland US market because a project of this size would not otherwise be explored (at least at this time) because of the large dollar requirements for a plant that demonstrates the commercial significance of OTEC to the US mainland market.

Will a risk-taker or speculatively inclined group of investors be the first ones to build an OTEC plant?

For this case, the perceived risk and reward must be compatible. There is not a regulated constraint to provide reliable, low cost power. If the project risk is high, so must be the potential reward. Let's look at this more closely. We shall see that the prospects for this approach are quite good under certain circumstances.

With a different set of assumptions than was used in the EPRI utility analysis, some screening curves were drawn to determine what the selling price of electricity would have to be from a privately owned electric plant to give a 50% internal (or discounted cash flow) rate of return. For these cases the plant is assumed to be financed by private sources with the assistance of the Mar Ad Title XI loan guarantee program (enabled by recent legislation).

Key assumptions in this analysis are:

- 1985 start-up date
- 18% interest on borrowings
- 10% annual increase in revenue
- 12% annual capital cost increase
- 2 year construction period
- 25% tax credit.

The balance of the assumptions are given in Table 2.

ASSUMPTIONS FOR PRIVATELY FINANCED DOMESTIC OTEC PLANT
Table 2

1. Rate of capital cost increases = 12%/yr.

2. Rate of O&M cost increases = 9%/yr.

3. Rate of increase of selling price = 10%/yr.

4. Cost of money, including insurance = 18%/yr.

5. Term of loan for capital for construction = 20 yr., mortgage type loan.

6. Initial capital requirement = 25% (75% financed)

7. Tax credits = 25%, 1 yr delay between expenditure and credit.

8. A tax-effective combination of double-declining balance and straight-line depreciation.

9. Asset life for depreciation purposes = 20 yr.

10. Plant start date = 1/1/85.

11. Project start date = 1/1/81; end date for analysis 12/31/2000.

12. Two years of design, two years of construction.

13. Design costs are expensed, construction costs are capitalized.

14. Capacity factors: 1985 = 0.3, 1986 = 0.6, 1987 = 0.75, 1988 and following = 0.8.

15. Capitalize interest during first year of operation.

16. Construction costs are 50% in each of two years of construction.

17. Design costs are 1% and 3% of capital cost for the two years of design.

18. Income tax rate = 46%.

The results shown in Figure 4 indicate that the selling price for electricity, required by the 50% return an investment ground rule, is rather high compared to current electricity production costs in most locations. It looks as though a selling price from an OTEC plant of, say, 10 MW would have to be 17¢/kWh or so. Of course, because of economies of scale, larger plants would require a lower selling price but a larger outlay of capital. Even the 10 MW facility in the scenario presented would require an equity of $15 million, and the total plant cost would be $60 million. These are large numbers for the first-of-a-kind facility. Thus, in my opinion, improvements in this scenario are to be sought before private capital can successfully be attracted to an OTEC project of this type.

In this paper, only the influence of the cost of borrowed funds and the influence of the equity requirement will be explored. The whole purpose of this sensitivity analysis is to try to find a set of conditions which make it likely that a "champion" of the OTEC cause could commit sufficient resources to build the first OTEC plant.

The cost of capital was varied in the base case with operating and maintenance costs set at 6¢/kWh. The results are shown in Figure 5. Note that, for a nominal 10 MW OTEC plant, the required selling prices to obtain a 50% return on investment are about 17¢/kWh for 18% cost of capital and 14¢/kWh for a 6% cost of capital. Of course, this represents a step in the right direction as far as attractiveness is concerned, but, for a privately financed case, 6% is unrealistically low. The conclusion is that, while the interest in borrowed funds is influential, lowering interest rates alone is not sufficient to make the base case attractive.

The next variable examined was the equity requirements at time of construction. Existing legislation allows for the possibility of an equity requirement of 12.5% rather than the 25% that was utilized in the base case. Figure 6 shows the influence of lowering the equity requirement as well as lowering the interest on borrowed funds. Lowering the equity requirement and keeping the interest on borrowings at 18% has the effect of reducing the required selling price to about 10¢/kWh for a 10 MW facility. This is a rather dramatic change, which also means that the amount of capital to be raised, exclusive of the borrowed funds, is $7.5 million rather than $15 million for the same 10 MW facility. If, in addition to reducing the equity requirement to 12.5% from 25%, the interest rate is lowered from 18% to a (possible?) 12%, the selling price to meet the 50% return on investment guideline is only 7¢/kWh. For some locations 7¢/kWh is very attractive in view of current power generation costs. Thus, there should be no difficulty in selling power at that price.

An OTEC electrical energy selling price of 7¢/kWh is still high for the states of the US in which OTEC has applicability, with the possible exception of Hawaii. The only US territories that normally have a higher cost of power generation than this are the US islands. As indicated earlier, the legislation which enables the tax credits for OTEC plants does so, unambiguously for only the 50 states. Tax credits may not be available for OTEC power projects for the US islands. If the possibility of using tax credits for island OTEC projects is eliminated, so also is the possibility that such plants will be deployed under the privately financed scenario that has been assumed. I do not believe that even the granting of a tax-free status to venturing firm interested in deploying an OTEC plant to produce electricity would be sufficient to overcome the non-availability of tax credits. Therefore, unless there is a future dramatic price increase in the cost of producing power (without an offsetting dramatic increase in the OTEC capital cost), I do not expect any commercial activity to build the first-of-a-kind OTEC plant to produce electricity with private capital under the no-tax-credit condition.

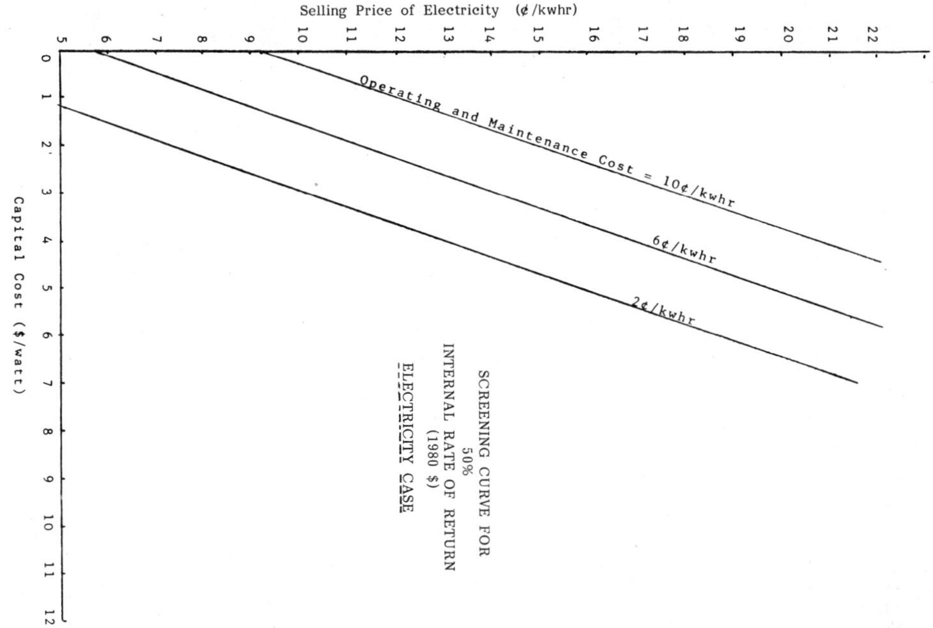

Figure 4

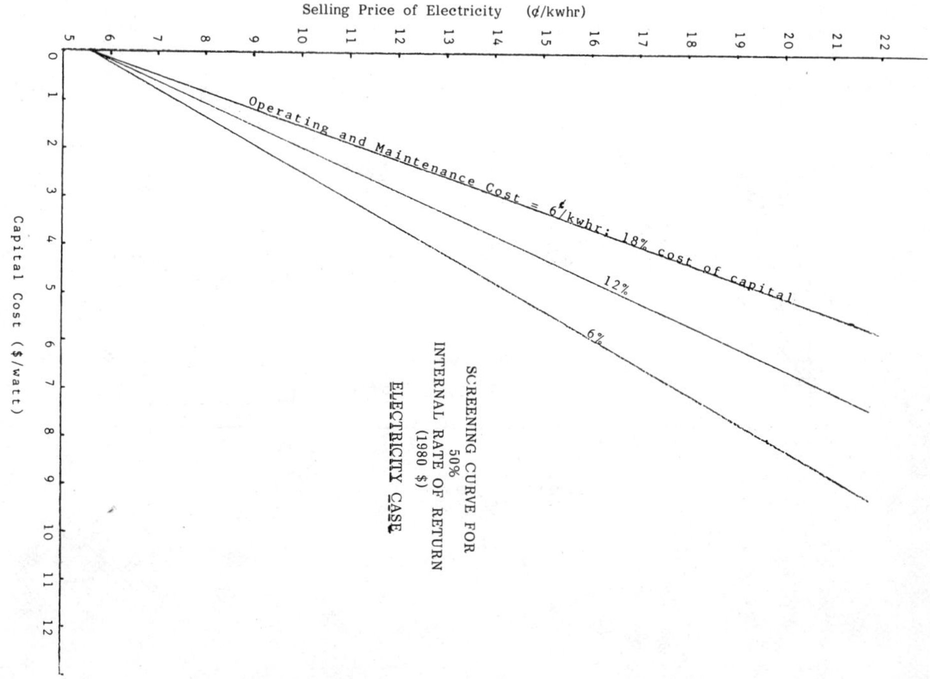

Figure 5

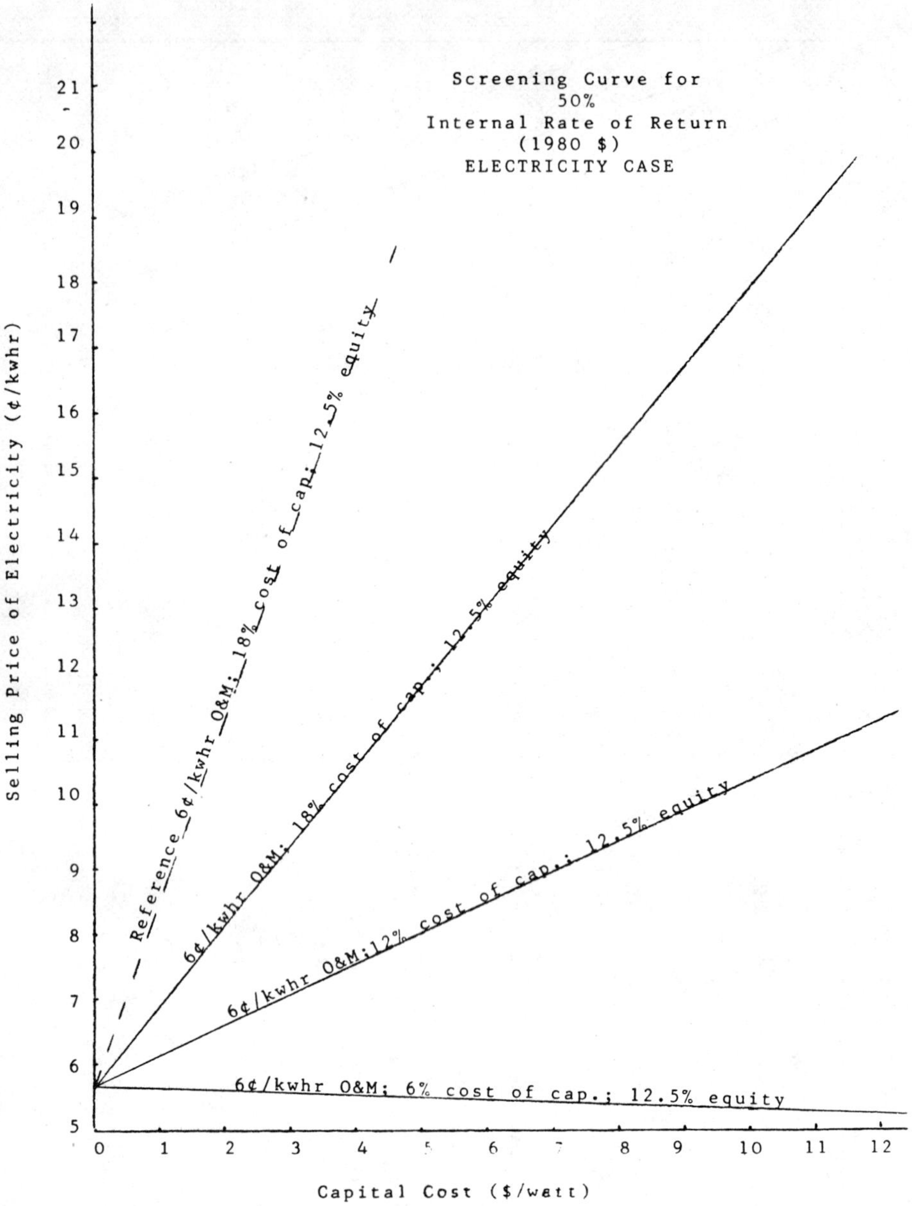
Screening Curve for
50%
Internal Rate of Return
(1980 $)
ELECTRICITY CASE
Selling Price of Electricity (¢/kwhr)
Reference 6¢/kwhr O&M; 18% cost of cap; 12.5% equity
6¢/kwhr O&M; 18% cost of cap.; 12.5% equity
6¢/kwhr O&M;12% cost of cap.; 12.5% equity
6¢/kwhr O&M; 6% cost of cap.; 12.5% equity
Capital Cost ($/watt)

Figure 6

On the positive side of this, it would behoove those interested in such a venture to watch most closely the developments and prospects for the state of Hawaii. Further, I would hasten to add that those projects which share capital commitments and risk with the appropriate government entities stand a much better chance of being attractive investment opportunities.

Water Production

What about making water as end product? Is that a more promising prospect for the private venture?

Now that electricity has been dealt with, let us look at the case for fresh water production from seawater. Again, the point of view will be that of a private investor seeking a balance between risk and return. The design that I have assumed is what Westinghouse has called a hybrid system in which a combination of OTEC closed cycle and flash evaporation is used. The OTEC system produces the power to run pumps used in the evaporation process; and the flash evaporator makes the fresh water. The technology of this concept has been presented in a report prepared for the US Department of Commerce which funded the development of the concept. My purpose here is to present economic screening curves for water production similar to those which were developed for electricity generation.

For a base case, all of the key assumptions for the analysis were exactly as they were in the privately financed electricity case (Table 2). The same rate of increase in capital cost, same rate of increase of selling price, same cost of money, and the same arbitrary 50% internal rate of return assumption are used. For the water case, I have made no estimates of capital cost vs. size or operating cost vs. size. But in the work done to date it appears that a capital cost of about \$10/(gal/day) and an O&M cost of \$1.25/(1000 gal) should be reasonable in the plant size range of 1 to 10 million gallons per day.

Given the foregoing assumptions and estimates of costs, the selling price of water to the distributor would have to be \$8 or \$9/(1000 gal), as shown in Figure 7, to obtain the 50% internal rate of return.

In the Department of Commerce report mentioned above it was shown that, within the United States, there is only one location where the cost of water meets or exceeds the \$8 to \$9 per thousand gallons: the Virgin Islands. Unfortunately, the foregoing analysis may be technically inaccurate for that location because the assumed 25% tax credit may not apply as stated earlier. However, the possibility of establishing a tax exempt status for a new corporation formed in the Virgin Islands to make fresh water was investigated from a purely economic point of view. The simplistic analysis that did not use the tax credits, assumed that a minimum of 12.5% of the equity is required at the time of construction (a critical assumption), and no income taxes are paid after 1988. Prior to 1988, any income or losses are assumed to be absorbed by a larger entity with a tax rate of 46%. Further, the transfer of ownership from the larger entity to the tax-free entity is assumed to take place without gain or loss; assets are transferred at book value. Table 3 summarizes the conditions.

The results of the analysis performed with the assumptions used in Table 3 are shown in Figure 8. A comparison of this case with that presented in Figure 7 shows that the economics are not as favorable because a higher selling price is required to meet the 50% internal rate of return criterion and a favorable ruling on a 12.5% equity project must be received. Still, the prospects for such a plant in the Virgin Islands are more promising than for any power generation OTEC plant that I am aware of (assuming, as has been done, that all

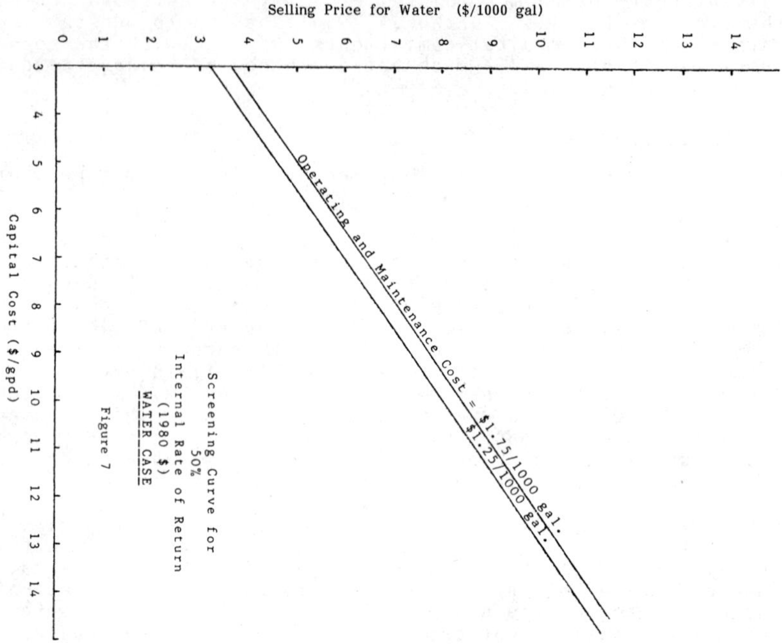

Figure 7

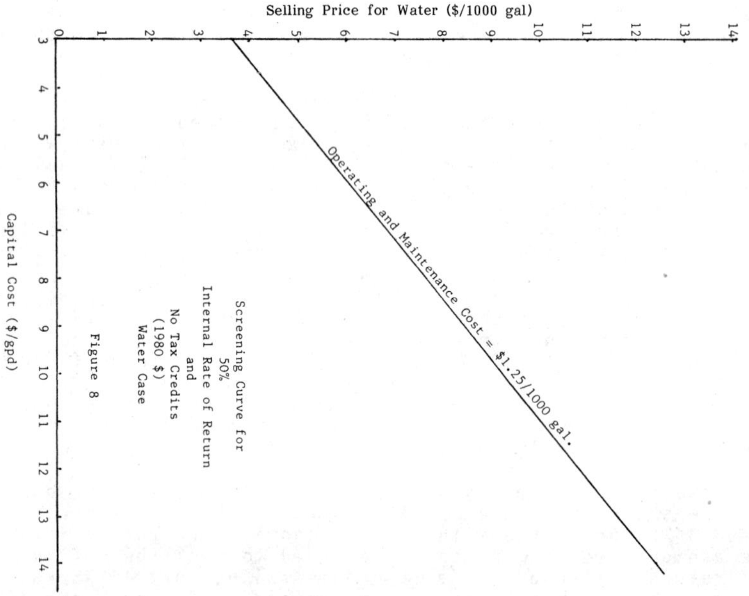

Figure 8

ASSUMPTIONS FOR PRIVATELY FINANCED DOMESTIC OTEC PLANT
Table 3

1. Rate of capital cost increases = 12%/yr.
2. Rate of O&M cost increases = 9%/yr.
3. Rate of increase of selling price = 10%/yr.
4. Cost of money, including insurance = 18%/yr.
5. Term of loan for capital for construction = 20 yr., mortgage type loan.
6. Initial capital requirement = 12.5% (87.5% financed).
7. Tax credits = 0%
8. A tax-effective combination of double declining balance and straight-line depreciation.
9. Asset life for depreciation purposes = 20 years.
10. Plant start date = 1/1/85.
11. Project start date = 1/1/81, end date for analysis 12/31/2000.
12. Two years of design; two years of construction.
13. Design costs are expensed, construction costs capitalized.
14. Capacity factors: 1985 = 0.3, 1986 = 0.6, 1987 = 0.75, 1988 and following = 0.8.
15. Capitalize interest during first year of operation.
16. Construction costs are 50% in each of two years of construction.
17. Design costs are 1% and 3% of capital cost for the two years of design.
18. Income tax rate = 46% for years 1981 through 1987. A tax free entity is assumed to be formed in 1988, tax rate then goes to 0%. No gain or loss of cash flow is allowed for in the 1987 to 1988 change of status transaction.

equity comes from private sources). The reason for this is primarily the fact that the Virgin Islands are desperate for fresh water and it appears that OTEC may be their best prospect for a reliable source of water at a relatively stable (not tied to fossil fuel) price. A key feature of the reliability of the water supply is to keep the ownership and operation of the facilities in private hands so that there is strong economic motivation to keep the production rate up.

There are two points worth emphasizing about the Virgin Islands case. First, it is not clear that the tax credits do or do not apply to an OTEC/fresh water plant for the benefit of the Virgin Islands. It may well be that the review of a potential project by DOE will yield an opinion which will provide the deciding factor in the matter of tax credits. It may also be that political factors will influence a ruling from the IRS on the tax credit issue. Second, the foregoing analysis provides only "screening curve" types of information for the purpose of determining whether or not a 50% internal rate of return is possible. It may well be that, under further examination, other measures of economic worth for the project are such that the project is undesirable no matter what the internal rate of return is. In short, the real conclusion in the matter of water production is only that the opportunity in the Virgin Islands is worthy of further examination.

Conclusions and Recommendations

In conclusion, let me emphasize the points I consider most important.

- OTEC plants owned by a utility are a cheaper source of electricity than an oil plant by a very conservative estimate of oil plant electricity costs. This indicates that a market potential in the US where oil and diesel are the main competitors for OTEC exists. Other nations will have similar findings, I am confident, even though the tax structure, cost of money, OTEC plant capital costs, and other costs will differ from those applicable to US plants.

- Given reasonable interest rates, privately borrowed funds and a favorable interaction with the US Federal Government, especially as pertains to tax credits, OTEC plants of the 5 to 10 MW range for the US have a possibility of becoming a reality in the next five years through venture-type projects. But the prospects for an OTEC plant to produce electricity in the next five years are best in a cooperative government/industry program.

- A private project in the Virgin Islands to make OTEC power, used by a flash evaporation process to produce fresh water, is apparently the best OTEC option for private industry in the near-term. Although such a program is more appealing than a power generation OTEC plant, it still may prove unattractive, under further study, due to the non-availability of tax credits or other factors which influence critical economic evaluators.

It appears that an international market for renewable energy projects may hold great promise because:

- very low interest loans may be available;

- fossil power options may be the only competition for baseload applications;

- current high costs of electricity make reasonable rates of return for an early OTEC plant possible; and

- favorably inclined governments can probably be found because of the devastating effect that the export of local currency for fuel has on relatively fragile economies.

It is recommended that all of these favorable aspects be analyzed in light of the particulars of the economic, social and political fabric of the candidate countries.

PRELIMINARY PERFORMANCE OF THE LOW-COST 17 METER DARRIEUS WIND TURBINE*

Mark H. Worstell
Sandia National Laboratories

The prime emphasis of the Federal Wind Energy Program is to establish the economic viability of wind turbines in a utility grid application and then to encourage private industry to initiate producing machines utilizing the knowledge gained from previous government sponsored work. This goal has begun to take shape in a number of research projects, wind turbine designs and construction, and subsequent testing.

The low-cost 17 meter (diameter) Darrieus wind turbine is representative of one part of this program. The main objectives of this project are to obtain realistic fabrication cost data based on current technology and to provide a low-cost system design suitable for continued production. Coupled with this, the low-cost 17 meter turbine may be used as a baseline for further cost reduction. The Aluminum Company of America (Alcoa) was selected to perform the design and fabrication of the 17 meter Darrieus wind turbine. Sandia National Laboratories, a government national laboratory, provided technical support in unique areas of structural and system design along with contract monitoring. Sandia National Laboratories has been involved in research, development, and testing work on the Darrieus wind turbine for the past eight years. The low-cost 17 meter program represents a merging of government and private industry toward a common goal.

The first low-cost 17 meter turbine (three more are to be erected in various locations in the country) was erected at the National Rocky Flats Test Center in Colorado. Figure 1 depicts this machine.

*This work supported by the U.S. Department of Energy under Contract No. DE-AC04-78AL04272.

FIGURE 1 Low-Cost 17-m Darrieus Wind Turbine

General specifications of this wind turbine are listed in Table 1. The drive train is illustrated in Fig. 2 which shows the transmission (located on the bottom), disc brake, and the generator (which also doubles as the starter motor).

Erected and first operated in August 1980, the low-cost 17 meter testing began with the direct primary emphasis of test data assimilation placed on structural response under operation in varying wind conditions. Structural integrity cannot be understated, though, while the Darrieus wind turbine gives the visual impression of a simple machine, it represents a substantial challenge in structural dynamics. This is primarily due to:

1. Varying windspeed altering the loading states.
2. A desired production machine lifetime of 30 years.
3. Economic viability.

Investigations into the performance of the low-cost 17 meter wind turbine have begun only recently. Performance in the context of this paper is to examine the output of the turbine rotor and the subsequent generator output, both as a function of ambient windspeed. The difference between these two will delineate the power loss of all turbine components that are between the power transducers, the first being a torque sensor located directly beneath the turbine rotor on the low speed shaft of the transmission, the second an electric power transducer connected across the output terminals of the generator. To measure the ambient windspeed, an anemometer was mounted at a 30 foot height 131 feet from the turbine centerline.

The low-cost 17 meter wind turbine was designed to operate in synchronization with the power grid.[1] Essentially this implies that the turbine rotor will turn at a fixed rotational speed with the generator producing electric power at the same voltage and frequency of its connection to the utility line. For a turbine operating in such conditions, the output power will level off for winds above a specified level due to aerodynamic stalling. The generator, when sized slightly larger than maximum turbine output, will constrain the turbine to operate at a constant speed. Figure 3 illustrates this for the DOE/Sandia 17 meter Darrieus wind turbine which has been in operation since March 1977.

The method of data collection used in the testing of the low-cost is known as the method of "BINS."[2] This proven technique basically samples windspeed and a certain number of power transducers simultaneously at a specified sampling rate. Measured power outputs are sorted out according to measured windspeed, and averaged to the previous values associated with that windspeed. The result then defines turbine power output as a function of windspeed.

The performance of the low-cost 17 meter wind turbine is illustrated in Fig. 4, the result of data taken to date. These data were taken with the anemometer upstream, a sampling rate of 10/sec on all transducers, and tare torque (friction torque) of the turbine rotor not included. Also, these data were corrected to depict turbine performance at sea level, corresponding to an air density of .0765 lb_m/ft^3. Although these data are essentially preliminary in nature, it should be representative of the actual performance. These curves can be expected to change slightly with further data accumulation which is necessary for a complete definitive description of turbine performance.

TABLE 1
Low-Cost 17 Meter General Specifications

Rotor Diameter	17.0 m (56.0 ft)
Rotor Height	25.1 m (82.5 ft)
Swept Area	284 m^2 (3060 ft^2)
Rotor Ground Clearance	2.7 m (9 ft)
Overall Height	29 m (95 ft)
Operating Speed	48.1 rpm
System Mass Less Foundation	12127 kg (26,741 lb)
Number of Blades	2
Blade Material	6.35 mm (0.25 in) Wall Aluminum Extrusion 6063-T6
Airfoil Section and Chord Length	0.616 m (24.24 in) Chord NACA 0015, Trailing Edge Radiused for 0.610 m (24 in) Chord Length
Blade Length	30.92 m (101.46 ft) per Blade
Blade Extrusion Mass	833.8 kg (1838 lb) per Blade
Blade to Blade Joints	2 Riveted Joints per Blade
Blade to Tower Connection	Cantilevered Stiffback Welded to Blade
Tower Material	Spiral Welded Steel Tube
Tower OD	0.762 m (30 in)
Tower Wall Thickness	4.77 mm (.188 in)
Speed Increaser	3 Stage Parallel Shaft
Speed Increaser Ratio	35.068:1
Motor/Generator	112 kW (150 hp) Squirrel Cage Induction
Brake	Single Disc Dual Independent Caliper
Brake Torque Capacity	60,000 Nm (44,000 ft-lb)
Brake Dissipation Capacity: Service	1.02 x 10^6 J (.75 x 10^6 ft-lb)
Brake Dissipation Capacity: Emergency	4.07 x 10^6 J (3 x 10^6 ft-lb)
Number of Guy Cables	3
Cable Angle (to Horizontal)	35°
Cable Diameter	22.2 mm (.875 in)
Cable Pretension	.135 x 10^6 N (30,300 lb)
Cable Length	48.8 m (160 ft)

FIGURE 2 Drive Train

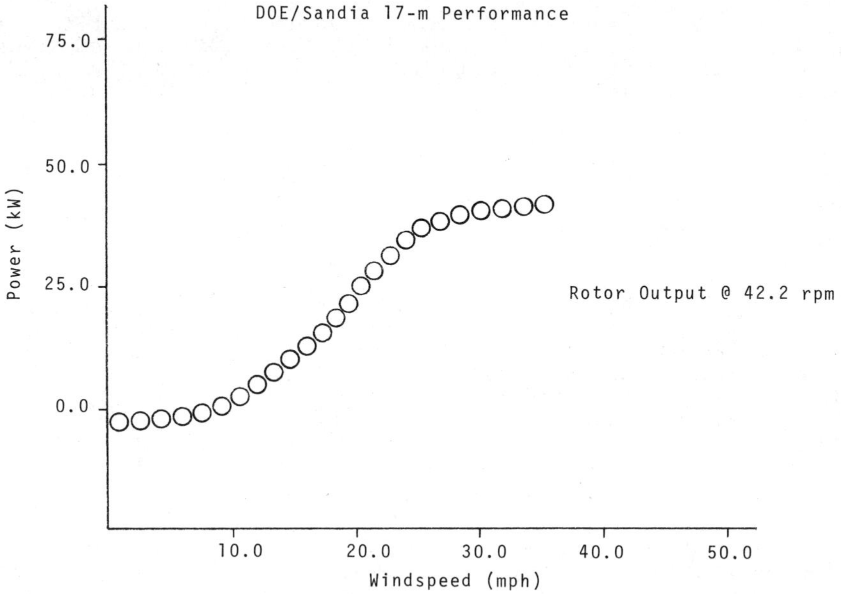

FIGURE 3

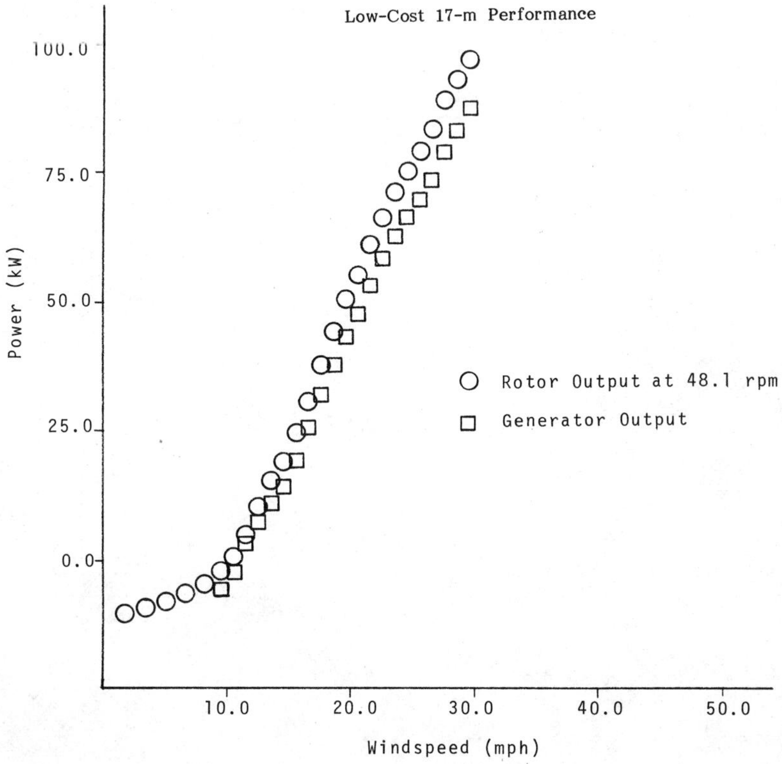

FIGURE 4

Figure 4 reveals that the highest recorded rotor output is approximately 100 kW, which is in line with theoretical predictions. The electrical output associated with this is 90 kW implying an overall drive train efficiency of ∿ 90% at this point. Drive train efficiencies this high can be expected when nearing the maximum predicted power output of the turbine, which is 100 kW_e. The crossover point for positive electrical output occurs at approximately 12 to 13 mph. The rotor power output curve exhibits a smooth, upward rise with the electrical output curve being less so, most likely due to the fact that the curve for generator output contains 30% less sample points. It is clear that data for winds above 30 mph are needed to further complete Fig. 4.

The data shown in Fig. 4 are the first obtained from the low-cost 17 meter wind turbine. The combined data that went into the rotor power output curve of Fig. 4 represent only 5.5 hours of operation. Typically, two to three times this is necessary to sufficiently depict complete turbine performance. The greatest area of data needed for the 17 meter is for winds above 30 mph. The current data base is expected to rapidly expand with continued testing of both this and the next low-cost 17 meter Darrieus wind turbines.

References

1. J. F. Banas, E. G. Kadlec, and W. N. Sullivan, "Application of the Darrieus Vertical Axis Wind Turbine to Synchronous Electrical Power Generation," SAND 75-0165, Sandia National Laboratories, Albuquerque, NM, March 1975.

2. R. E. Akins, "Performance Evaluation of Wind Energy Conversion Systems Using the Method of Bins - Current Status," Sandia National Laboratories, Albuquerque, NM, March 1978.

DEVELOPMENT OF THE UTRC BEARINGLESS ROTOR WIND TURBINE

M. C. Cheney
United Technologies Research Center

ABSTRACT

The Bearingless Rotor Wind Turbine was first studied at the United Technologies Research Center (UTRC) in 1975 under a Government (ERDA) contract to explore advanced wind turbine design concepts. A series of DOE development contracts followed, resulting in the fabrication of an 8 kW prototype in 1979. An improved version of this design was later developed by UTRC and operated in 1980. A DOE contract was also awarded to UTRC in 1979 to develop a 15 kW SWECS for agricultural and industrial applications. This paper presents an overview of the UTRC Bearingless Rotor Wind Turbine Concept, and summarizes the important performance and operating characteristics of the two models now under development.

INTRODUCTION & BACKGROUND

The UTRC small wind turbine was conceived with the principal goals of providing high reliability and low cost of energy production. This objective dictated the requirement for a simple design having high aerodynamic efficiency and a minimum number of parts. The major thrust of the UTRC design effort was to minimize the use of active system controls including wind speed sensors, direction sensors, blade pitch controls, rpm sensors, overspeed or shutdown controls, braking mechanisms, yaw motors, and start-up devices. The Bearingless Rotor conceived by UTRC and currently adapted to the tail rotors on the Sikorsky Blackhawk and S-76 helicopters was determined to be ideally suited to the fulfillment of these design requirements.

Several bearingless rotor control concepts were considered at the outset of the UTRC small wind turbine program in 1975 under an ERDA feasibility study, and these concepts were eventually evaluated in tests of a 4 ft model in the UTRC Pilot Wind Tunnel, Fig. 1. Three different concepts were tested, and eventually the hub mounted pendulum concept was selected to provide passive pitch control of a low stiffness bearingless rotor with free yaw. The electrical controls consist of a permanent magnet rpm pickup on the generator shaft and a solid-state speed switch to activate and deactivate the field excitation of the induction generator. Wind tunnel tests of a 10 ft model conducted under a subsequent DOE contract confirmed the high speed stall characteristics and passive speed control of the rotor. This installation is shown in Fig. 2. Results of the tests of the 4 ft and 10 ft model wind turbines are given in Ref. 1-3. Additional design information and a description of the system operating characteristics are given in Ref. 4-6. In 1977 a contract was received from DOE to design and fabricate a prototype 8 kW wind turbine which was to be delivered to the Rockwell Rocky Flats test site following a 2-year development program. This system was first tested in July 1979 at the UTRC test site in East Hartford and shipped to Rocky Flats in February 1980. In the fall of 1979 a DOE contract was awarded to UTRC to develop a larger 15 kW version of the UTRC wind turbine concept. The 15 kW design is based on the same concepts developed for the 8 kW unit but includes features to enhance reliability and reduce manufacturing costs for high volume production.

Fig. 1 4 ft. Model Wind Turbine

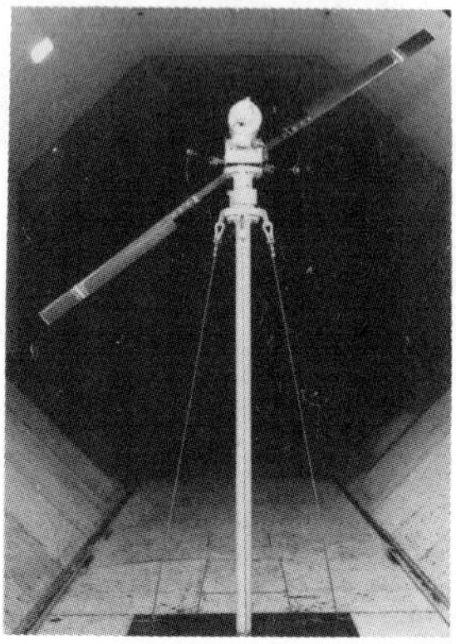

Fig. 2 10 ft. Model Wind Turbine

An improved version of the 8 kW prototype was fabricated in 1980 under a Corporate-sponsored program. This improved design was similar to the prototype but with changes to the main frame and flexbeam to reduce weight and cost. The design changes incorporated in this model will be described herein, and the performance results obtained to date will be presented.

Tests of the prototype 15 kW wind turbine will be initiated in April, 1981. This unit will be delivered to Rocky Flats following a three-month test period at East Hartford. Plans for marketing both the 15 kW and 8 kW systems are currently under study at United Technologies.

Concept Description

A schematic of the basic rotor components of the UTRC wind turbine are shown in Fig. 3. The blade is comprised of two distinct parts. The outer, or aerodynamic, portion of the blade is fabricated from fiberglass using a pultrusion process. The inner portion of the blade, called the flexbeam, is made up of unidirectional composite material. The dimensions of the flexbeam are dictated by the edgewise and flatwise frequency requirements, and the required torsional properties are determined from the pendulum control system. The free length of the flexbeam runs from the hub to approximately 20% of the radius where it is joined to the fiberglass blade. As shown in Fig. 3, the outer portion of the flexbeam is connected to the pendulum through a steel strap, called the flexstrap. The flexbeam extends through the hub and is connected to the opposite blade. The blade is attached to the flexbeam with the leading edge positioned 12 deg towards the tower to place the blade in a position for self-starting. In this static condition the pendulums are coned toward the tower approximately 45 degrees by virtue of their connection to the flexbeam through the flexstraps. For optimum performance it has been shown that the blade pitch angle should be approximately ± 2 degrees. This is achieved by virtue of the pendulums acting under the influence of centrifugal force as rpm increases. The pendulums begin to move towards the plane of rotation and, through the flexstrap connections to the flexbeam, provide a moment to the blades which twist the flexbeam thereby increasing the blade pitch. The pendulums, acting against the torsional spring of the flexbeam, are always in moment equilibrium with the flexbeam. When operating in conjunction with a utility, an induction generator is used which is brought on line by a tachometer switch which senses when the generator rpm is slightly higher than synchronous speed (1800 rpm). At this point the generator, under the torque of the rotor, slips slightly greater than synchronous speed thus generating electrical output. The rotor will remain at nearly constant rpm while on-line and will not break out unless excessive torque is sustained for a period of time. As the rotor speed drops due to inadequate winds, the tach switch will decouple the grid to avoid motoring.

In the event that grid excitation is lost or excess current is generated in high winds, the rotor can be brought to a stop with a brake that actuates automatically with the loss of grid power. However, redundancy in limiting this critical overspeed condition is provided by producing stall on the rotor blades at a specific combination of wind speed and rpm. At a tip speed of approximately 350 ft/sec the pendulums will be essentially in the plane of rotation and produce a pitch angle on the blades of approximately 5 deg. This angle, in combination with the aerodynamic angle produced by the wind speed, is sufficient to stall the blades. At this point the rotor tip speed is reduced and adjusts itself to a lower equilibrium which is approximately 275 f/s. The centrifugal force remains sufficiently high at this rpm to assist the blade in counteracting the root moment caused by blade thrust. The rotor will remain fully stalled until the wind speed drops below the stall point which is estimated to occur at approximately 35 mph.

CURRENT STATUS OF 8 KW PROGRAM

Prototype Design Features

Three significant design changes were made during the course of the field evaluation tests of the 8 kW wind turbine in East Hartford. These consisted of changes in the yaw axis location, modification of the upper guy wire attachment, and a reduction in the blade pre-cone angle. The changes were incorporated to correct a precessional problem experienced during the initial tests.

The prototype design consisted of a heavy weldment serving as the main supporting frame which contained the rotor shaft support bearings, the transmission platform, and the yaw bearings. The weldment was made up from ½ in. steel plate. A photo of the completed weldment is shown in Fig. 4. The rotor was 31 ft in diameter, consisting of two 15 in. chord blades with 8 deg precone. The flexbeam, which supports the blades 4 ft out from the hub, was hand fabricaed of unidirectional graphite/epoxy in a preconed pretwisted mold.

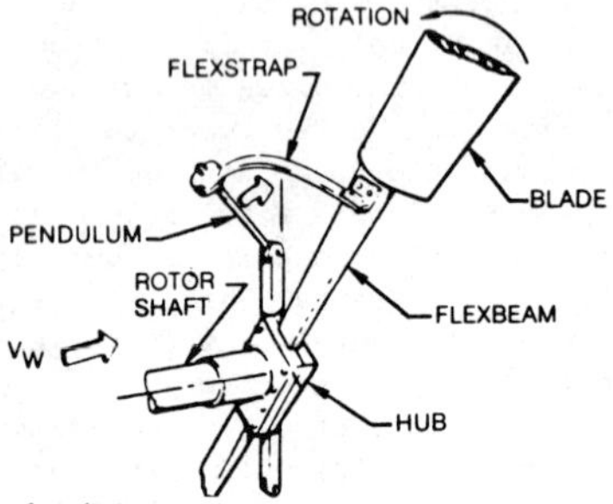

Fig. 3 Schematic of pitch control system

Fig. 4 Prototype Support Frame

A summary of the design parameters is given in Table I and a photograph of the completed installation is shown in Fig. 5. The initial shakedown tests were conducted using Sikorsky S-61 aluminum helicopter blades having an 18 in. chord which weighted 50% more than the fiber glass blades specified for eventual installation. The heavy aluminum blades were found to cause a low speed precessional phenomenon which was corrected by two hardware modifications. The cone angle was reduced from 8 deg to 3 deg and the guy wire upper attachment point was lowered 8 ft to increase the tip clearance resulting from the lower cone angle. The yaw axis was subsequently displaced 2 ft to increase the system static stability.

TABLE I

8 KW PROTOTYPE DESIGN PARAMETERS

Generator rating	15 kW
Diameter	9.45 m (31 ft)
No. of blades	2
Blade chord	38 cm (15 in)
Twist & taper	0
Airfoil	NACA 23112
Blade material	E-glass/epoxy
Flexbeam size	1.4 x 14 x 78.7cm (0.55x5.5x31 in)
Flexbeam material	Graphite/epoxy
Flexbeam pretwist	-16 deg
Hub offset	20.3 cm (8 in)
Design Tip Speed	53.3 m/s (175 f/s)
Pre-cone	8 deg
Tower height	55 ft
Wind turbine weight	2300 lb
Tower weight	2150 lb

Improved 8 kW Design

In 1980, several modifications were made to the prototype design to reduce weight and costs. The improved system is designated the Model 108. The two most significant changes were: 1) replacement of the hand fabricated preconed graphite/epoxy flexbeam with a fiber glass pultrusion; and 2) replacement of the heavy steel main frame weldment with an aluminum casting. The assembled components are shown in Fig. 6. The use of the aluminum casting reduced the system weight by 800 lb. The rotor diameter was increased to 32 ft to increase energy capture, and a single-phase induction generator was used in place of the three-phase generator used in the prototype. For the single-phase system, 100 microfarads of capacitance was added to achieve the rated power and rpm characteristics. Other changes included 1) reduction of tower height from 55 ft to 50 ft to increase the tower third mode bending frequency; 2) use of castings for the hub and pendulum components and the main yaw shaft; 3) replacement of the steel erection boom with an aluminum A-frame design which can be easily handled by two men; and 4) addition of a brake on the low torque generator shaft so that positive manual braking can be achieved at any speed. The brake is also activated automatically with loss of grid power. Although the rotor will operate safely in its high speed stall mode as described earlier, the ability to stop the rotor is considered a desirable feature during the current period of field testing.

Fig. 6 Model 108 Support Frame

Fig. 5 Prototype 8 kW Wind Turbine

A photograph of the complete installation at the UTRC test site is shown in Fig. 7, and a summary of its design parameters is given in Table II.

TABLE II

MODEL 108 DESIGN PARAMETERS

Generator rating	15 kW
Diameter	9.75 m (32 ft)
No. of blades	2
Blade chord	38 cm (15 in)
Blade material	E-glass/vinyl ester
Flexbeam size	1.65x16.5x78.7 cm (0.65x6.5x31 in)
Flexbeam material	S2-glass/vinyl ester
Pre-cone	0

TABLE II (Cont'd)

Blade static pitch	-12 deg
Design tip speed	54.7 m/s (180 f/s)
Tower height	50 ft
Wind turbine weight	1500 lb
Tower weight	2000 lb

Performance

A summary of performance measurements for the Model 108 is shown in Fig. 8. The design objective of 8 kW at 20 mph was achieved. Approximately 3 kW is generated at 12 mph, 13 kW at 25 mph. A maximum output of 15 kW would be generated at 30 mph. At wind speeds in excess of 30 mph, the rotor stalls with a corresponding drop in power output. However, power continues to be generated at all wind speeds up to 100 mph, at which point the maximum power of 15 kW would again be reached.

A curve of predicted energy capture is given on Fig. 9. 25,000 kWh would be produced annually with a mean wind speed of 12 mph. At a mean speed of 18 mph the annual output would increase to 54,000 kWh.

CURRENT STATUS OF 15 KW PROGRAM

A contract was awarded to UTRC in 1979 to develop a 15 kW version of the 8 kW design and to incorporate cost, reliability, and performance improvements identified during the course of the 8 kW program.

Design Features

A modified transmission has been designed which incorporates interfaces to directly attach the rotor, generator, and yaw bearing. This eliminates the need for a support frame and separate rotor shaft, which substantially reduces the cost of these components. The guyed steel tower has been replaced with a free-standing tapered aluminum tower. These improvements provide reduced weight, lower costs, fewer components, and a more attractive appearance.

A summary of the 15 kW wind turbine design parameters is given in Table III.

TABLE III

15 KW WIND TURBINE SPECIFICATIONS

Generator rating	23 kW
Diameter	14.6m (48 ft)
No. of blades	2
Blade chord	45.7 cm (18 in)
Flexbeam size	2.1 x 21 x 122cm (0.84x8.4x48 in)
Flexbeam material	S2-glass/vinyl ester
Pre-cone	0
Blade static pitch	-10 deg
Design tip speed	56.5 m/s (185 f/s)
Tower height	41 ft
Wind Turbine weight	2100 lb
Tower weight	1200 lb

Fig. 7 Model 108 Wind Turbine

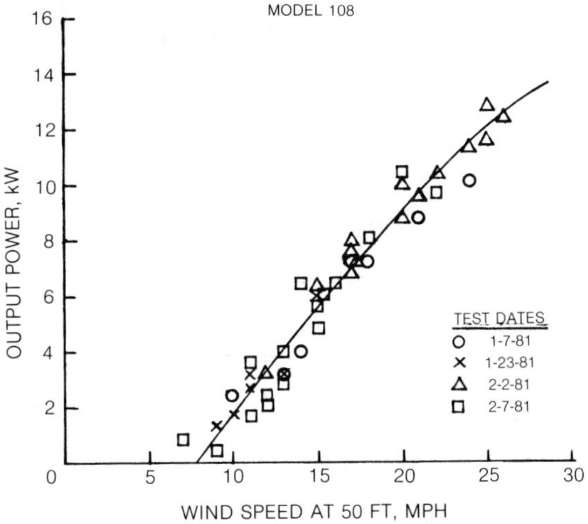

Fig. 8 Measured Power Output

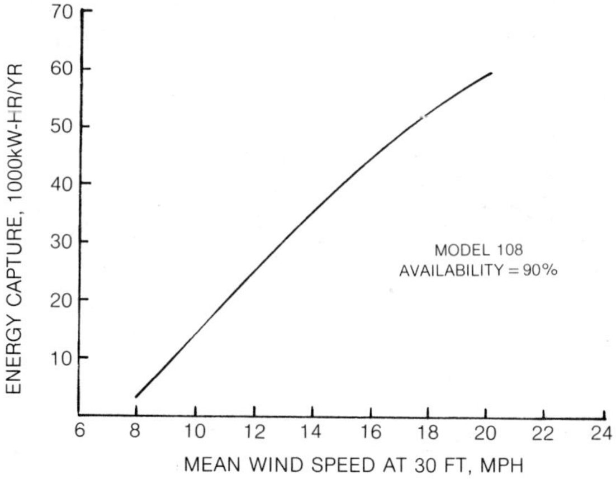

Fig. 9 Estimated Energy Capture

Performance

The power output and energy capture of the 15 kW wind turbine is approximately double that of the 8 kW system. The estimated power output versus wind speed is shown in Fig. 10. The output at 12 mph is approximately 4 kW with a maximum output of 23 kW at 30 mph. The projected energy captured is shown in Fig. 11 which shows a level of 52,000 kWh at a mean wind speed of 12 mph.

The prototype 15 kW wind turbine is scheduled for initial testing in April 1981 and is expected to be shipped to Rockwell's Rocky Flats test site during the summer of 1981. The specific market at which this size machine is directed is primarily large farms and small industrial complexes.

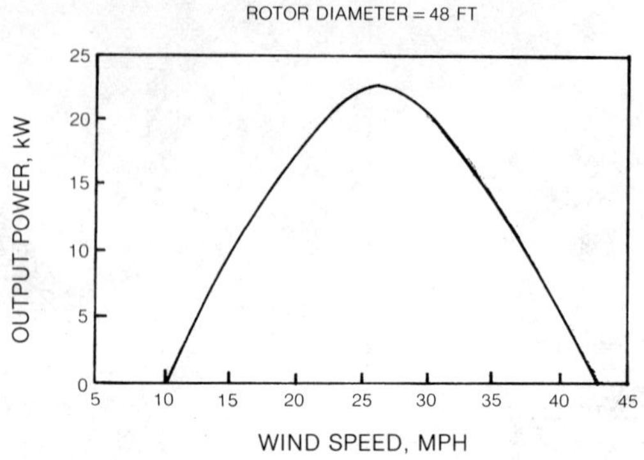

Fig. 10 15 kW Estimated Power Output

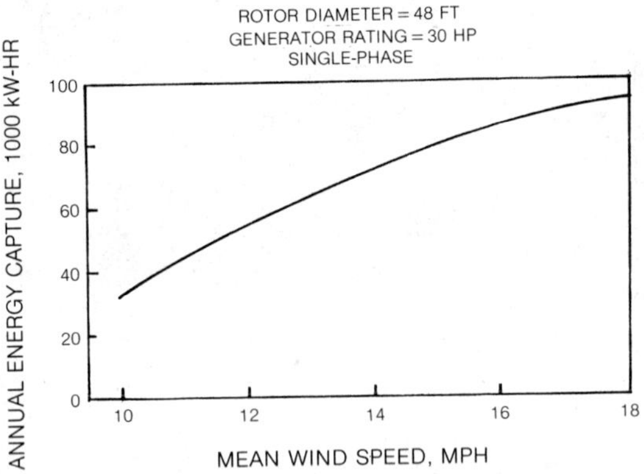

Fig. 11 15 kW Estimated Energy Capture

CONCLUDING REMARKS

The UTRC wind turbine has undergone an extensive development program to bring it to its current operational status. Wind tunnel experiments and full-scale field tests have demonstrated all important operating features. It is believed the design has advanced the state-of-the-art in wind energy conversion systems and through the simplicity of its passive control concept provides a potentially low cost, low maintenance wind turbine system.

During 1981, test data from field units will be evaluated, and maintenance requirements and operating characteristics will be monitored as additional running experience is gained.

REFERENCES

1. Cheney, M. C.: UTRC 8 kW Wind Turbine Tests. AIAA/SERI Wind Energy Conference, Boulder, Colorado, April 9-11, 1980.

2. Cheney, M. C.: UTRC 8 kW Wind Turbine. American Wind Energy Association Annual Conference. San Francisco, California, April 16-19, 1979.

3. Taylor, R. B. and M. C. Cheney: Development of an 8 kW Wind Turbine Generator for Residential Type Applications. Phase II Final Report. DOE/Rockwell International Contract PF68186F. February, 1981.

4. Cheney, M. C., et al: Development of an 8 kW Wind Turbine Generator for Residential Type Applications. Phase I Final Report. DOE/Rockwell International Contract PF68186F. June, 1979.

5. Cheney, M. C. and P. A. M. Spierings: Self-Regulating Composite Bearingless Wind Turbine. Solar Energy, Vol. 20, 1978.

6. Spierings, P. A. M. and M. C. Cheney: Design of a Self-Regulating Composite Bearingless Wind Turbine. Final Report, ERDA COO-4150 77/8 UC-60, January, 1978.

LOW COST, HIGH EFFICIENCY PHOTOVOLTAIC MODULES

D. R. ROBERTS
WESTINGHOUSE ELECTRIC CORPORATION
ADVANCED ENERGY SYSTEMS DIVISION

The Westinghouse Electric Corporation has been a major producer of electrical generation equipment for almost 90 years and we look forward to continuing in that role for many more years to come. Over the years, participation in this market has required the utilization of many energy resources such as gas, oil, coal, hydro, and nuclear along with a broad range of energy conversion technologies applicable to those resources. For the remainder of this century, and probably well into the next, we are convinced that the major contributors of electrical energy will continue to be coal and nuclear. However, we are equally convinced that the range of energy resources and conversion technologies will be increased to include alternate energy technologies such as synthetic fuels, fuel cells, and solar. We believe that these resources will make a meaningful contribution to the nation's energy supply by the end of this century, with ever increasing significance as we move to the twenty-first century.

Naturally, primary attention is being focused on those specific alternative energy technologies which we feel show promise for widespread application resulting in a significant contribution to the overall energy supply and a high volume business opportunity. In this respect, solar photovoltaic systems look particularly promising because of the wide range of applicability - from very small systems producing a few watts to very large systems producing many megawatts - from navigational aids to residential rooftops to utility central stations. The fact that photovoltaic systems convert sunlight directly to electricity with no moving parts, require very little maintenance and no trained technicians for operation, and have long life, results in these systems being particularly applicable to residences and small remote systems.

Similar to the other solar energy technologies, photovoltaic devices are not new; the photovoltaic effect was discovered in 1839 and solar cells have been used in light meters for years. Over the past twenty years the technology has been well demonstrated in power supplies for space craft, navigational aids, remote communications systems, and other special purpose applications. Truly widespread application, however, has been limited by the cost of the systems.

Over the past five years, extensive effort on the part of the Federal Government and private industry has been devoted to the price reduction of photovoltaic devices. This effort has been successful in reducing the price of photovoltaic modules from $50 to $100 per watt down to the range of $10 per watt today.

The DOE goals for future price reduction for photovoltaic modules and systems are shown in Figure 1. The Westinghouse effort is focused on achievement of the goals for 1986 and beyond. At the module and system prices shown for 1986, our calculations indicate an energy cost for residential application from four to eight cents per kilowatt hour. Given a widespread residential application and commensurate large production volumes, experience will further decrease price to the point where photovoltaic systems can be competitive in almost any application or location.

While most of the recent attention has been focused on reducing the photovoltaic module price, it is actually the installed system price which dictates the cost effectiveness and applicability of photovoltaic systems. To achieve low system costs requires a photovoltaic module which is not only low cost but also one which has a relatively high efficiency. In the overall photovoltaic program relatively little attention has been directed toward the achievement of high efficiency, even though it has a significant impact on system cost.

The achievement of higher efficiencies has a direct impact (Figure 2) on transportation, installation, interconnection, wiring, support structure, land, and maintenance. What we are talking about is the difference between 8 watts per square foot, 12 watts per square foot, or 16 watts per square foot, which correlates roughly to module efficiencies of 8, 12, or 16%. The costs for these aspects of overall system cost vary just about linearly with efficiency and doubling the efficiency therefore reduces these costs nearly one-half.

In the overall cost of the small remote applications which have made up most applications to date, the efficiency of the modules is not important. If you need a 50 watt system installed in a remote area, it doesn't make much difference if that requires five square feet or ten square feet because the cost of engineering, design, system hardware, travel to the site, and other logistics far outweigh the size of the modules required to produce the power. In widespread applications such as residential or central station, however, the actual area related costs can equal the cost of the hardware. In this kind of application the area required for a given power output becomes vitally important. In this regard, it should be noted that residential application is not quite as sensitive to efficiency as central station application because support structure and land costs are not involved - it goes on the roof.

At the present time Westinghouse has ongoing research and development programs in a wide variety of photovoltaic technologies which show promise for low cost modules but only the flat plate modules with single crystal silicon cells have shown sufficient promise for both low cost and high efficiency to warrant proceeding to the pilot production line stage of development. We are continuing our research on many fronts but are currently focusing on single crystal silicon technology.

When one examines a breakdown of the "cost-drivers" for single crystal silicon (Figure 3) it can be seen that three out of four (silicon, sheet, and encapsulation) are also reduced by increased efficiency. This implies that there is some measure of increased effort which can be devoted to processing, to increase efficiency, to minimize overall system price. For example, if one could increase overall module efficiency by improving cell efficiency by 25% through increased processing costs, this would reduce the cost of silicon, sheet, and encapsulation by an equivalent amount. This would indicate that processing cost could be increased by over 14 cents per watt and the 70 cents per watt would remain the same but there would be attendant savings in the overall system. This would not necessarily constitute a minimum system cost but it certainly is a step in the right direction.

The Westinghouse approach to a high efficiency, low-cost module is the use of single crystal silicon produced in the Westinghouse patented dendritic web process. The dendritic web process is a method by which we produce a long thin ribbon of single crystal silicon (Figure 4). This ribbon can be processed directly into solar cells without the need for slicing and attendant material losses and is about one-half the thickness of wafers from other technologies. This results in a factor of four reduction in the use of silicon.

The dendritic web is formed (Figure 5) by lowering a needle-like dendritic seed crystal into a slightly subcooled melt of silicon. This results in a button being formed at the surface of the silicon. After the button has grown to sufficient width it is withdrawn vertically from the surface. Under the proper conditions, a new dendrite will be formed at each end of the button, growing into the liquid. Between the two dendrites, a film of silicon is formed due to surface tension forces, similar to a soap film between the tines of a fork. When this film, or web, solidifies, it becomes a single crystal because it is bounded by single crystal.

After removal from the furnace and cleaning, the strips are cut into convenient lengths for processing (Figure 6). The length shown here is dictated by the hot zone in the diffusion furnace.

The cells are processed on the strips of web (Figure 7) thereby eliminating the need to handle individual cells. At the present time we are using a phosphorous diffusion for the front junction and a boron diffusion to form the back surface field; a photolithographic technique for grid definition; vacuum deposition of contacts followed by electroplated copper - which is a relatively expensive process sequence. While we are examining less expensive alternatives, our criterion is that the new processes not prohibit us from achieving our mid-80's goal of 18% solar cells and 16% modules.

Processing of the cells is followed by laser scribing to separate the cells and remove the dendrites (Figure 8). After testing, these cells are then connected in series strips (Figure 9) and finally encapsulated by vacuum lamination into a module with a glass face and a moisture proof back (Figure 10) along with an installation frame. It is at this point that a major benefit of the dendritic web process becomes obvious; the cells are rectangular and therefore have a very high packing fraction in the module, which leads to high module efficiency.

We therefore have defined a process (Figure 11) which greatly reduces the requirement for starting silicon, but at the same time is high quality single crystal material which permits high cell efficiency, and results in a rectangular cell which leads to high module efficiency and therefore decreased system costs.

DOE PHOTOVOLTAICS PROGRAM GOALS

YEAR	MODULE PRICE*	SYSTEM PRICE*	PRIMARY APPLICATION
1982	$2.80/peak Watt	$6–13/W	REMOTE/INTERNATIONAL
1986	$0.70/peak Watt	$1.60–2.20/W	RESIDENCES
1990-2000	$0.15-.50/peak Watt	$1.10–1.30/W	UTILITIES

*1980 CONSTANT DOLLARS

Figure 1. DOE Goals for Price Reduction of Photovoltaic Systems.

INCREASING EFFICIENCY DECREASES COST FOR:

- TRANSPORTATION
- INSTALLATION
- INTERCONNECTION
- WIRING
- SUPPORT STRUCTURE
- LAND
- MAINTENANCE

Figure 2. Systems Costs Reduced by High Efficiency.

PRICE ALLOCATIONS

SILICON	4.2¢/Watt
SHEET (WAFERS)	18.3
PROCESSING	22.0
ENCAPSULATION AND ASSEMBLY	25.5
	70.0 ¢/Watt

Figure 3. Allocations of Prices to Achieve the 1986 Goals.

Figure 4. Dendritic Web Produces a Long Thin Ribbon of Single Crystal Silicon.

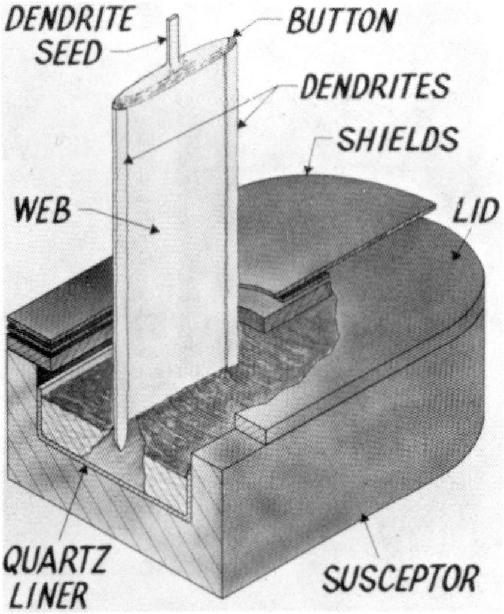

Figure 5. Dendritic Web Growth.

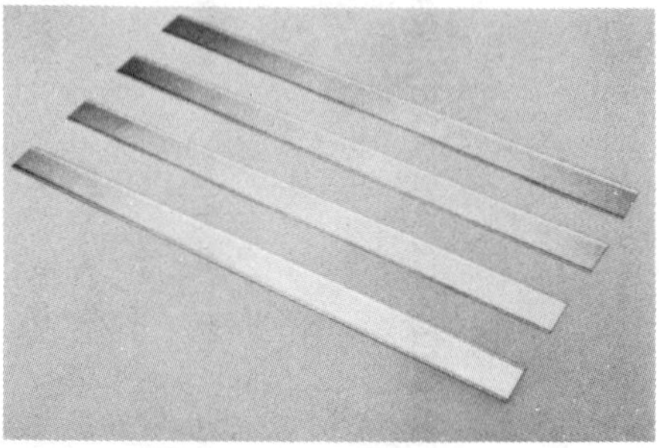

Figure 6. Strips of Dendritic Web Ready for Processing.

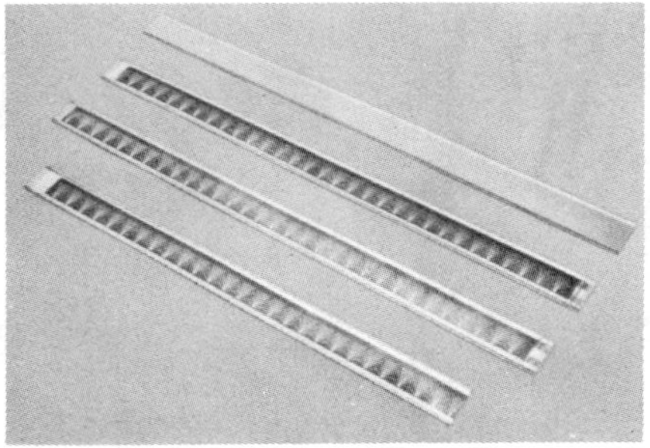

Figure 7. Strips of Dendritic Web Processed into Solar Cells.

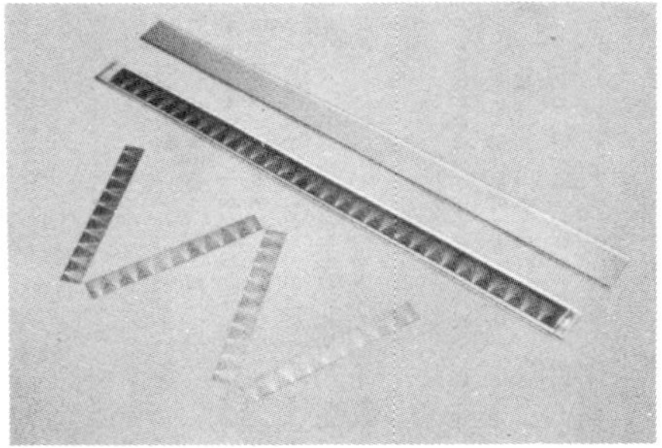

Figure 8. Dendritic Web Solar Cells after Separation.

Figure 9. Strip of Series Connected Solar Cells.

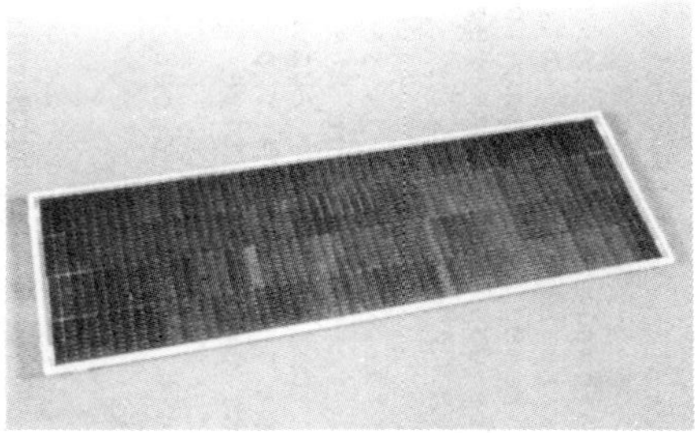

Figure 10. Silicon Dendritic Web Photovoltaic Modules.

At the present time we are forecasting module prices below 70¢ per peak watt by 1986 with overall module efficiencies in the range of 16 percent. Based on this promise, we have initiated efforts (Figure 12) which can eventually lead to large scale commercialization. With Westinghouse funds, we have installed a small pre-pilot production line for the purpose of transferring the technology from the research and development environment to a production environment and this will provide modules for limited demonstration. This is to be followed by a significantly larger, much more sophisticated semi-automated pilot line which is being jointly funded by DOE and Westinghouse which will permit large demonstration systems and further verification of forecasts of cost and performance.

Obviously, there are gates we must pass, but assuming that our results remain sufficiently encouraging during these phases, we would then foresee proceeding to a fully automated production line which would be duplicated as required for commercial production, dependent upon the market size.

At this time it is safe to say that photovoltaic systems will be capital intensive for the manufacturer as well as the purchaser. The sophisticated, fully automated production lines will be very expensive and the resultant photovoltaic systems are going to have a high purchase price relative to conventional energy technologies. We are convinced that system purchase price and later maintenance costs can be significantly reduced through the use of high efficiency, simple systems.

DENDRITIC WEB MODULE FEATURES

- REDUCED USE OF SILICON
- SINGLE CRYSTAL SILICON RIBBON
- HIGH EFFICIENCY CELLS
- RECTANGULAR CELLS
- HIGH MODULE EFFICIENCY
- LOW SYSTEM COST

Figure 11. Dendritic Web Photovoltaic Module Yields High Efficiency and Low System Cost.

AN APPROACH TO LARGE SCALE PRODUCTION

- PRE-PILOT FACILITY
- SEMI-AUTOMATED PILOT PRODUCTION LINE
- FULLY AUTOMATED PRODUCTION LINE

Figure 12. The Approach to Commercial Production.

SOLAR PHOTOVOLTAIC CONCENTRATOR SYSTEMS FOR SMALL AND LARGE ELECTRIC POWER SUPPLIES

Robert L. Sanders
Manager — Energy Systems Engineering
Motorola, Inc. — Government Electronics Division

INTRODUCTION

The ultimate goal of our national energy policy is to reduce dependence on both foreign and domestic nonrenewable energy resources. To achieve this goal, a long-term national commitment must be made to the development of domestic renewable energy resources.

Solar energy, while it cannot displace all other energy sources, is an important energy option. The sun is essentially a limitless source of zero cost fuel. Scientists, engineers, and architects are just beginning to devise new methods and components for the cost-effective use of this zero cost fuel.

Solar photovoltaic power, initially developed for the space program, is a technically viable alternative to conventional methods of terrestrial power generation. Work accomplished to date on development programs funded by the United States Department of Energy (DOE) has demonstrated technical viability. The high cost of the photovoltaic solar cell is the primary reason for the lack of economic viability.

One approach to enhance the economic viability of photovoltaics has been the development of solar collectors which concentrate sunlight onto the cells as a means of increasing their output power density. The intent of this approach is to reduce overall cost-per-unit-power by replacing solar cell area with lower cost areas of concentrating hardware.

Once industry has surmounted the hurdle of economic viability, it is expected that photovoltaics will experience widespread use in applications which require both small and large electric power supplies. This trend is already evident in small, remote stand-alone applications where flat plate systems are presently economically viable.

RENEWABLE ENERGY DEVELOPMENT

Renewable energy resources are those that are derived from solar energy sources either directly or indirectly. Primarily, renewable energy sources include direct solar heating and cooling, solar industrial process heat, electric power generation, ocean thermal energy conversion, hydropower, wind, waves, tides, wood, other biomass, and organic wastes.

The viability of each of these sources must be established in the 80's if we are to achieve the ultimate goal of our national energy policy. Both the government and industry must support enough research, development, production; and application experience to establish credibility. Once technical and economic feasibility are established, the government should provide stimulus to that particular energy source in the form of continued R&D support, tax rebates, and consumer protection policies.

Economic viability does not usually occur as early as technical viability. Typically, technical viability is proven, or disproven, in a 3 to 5 year time period during the development of the first generation design. Economic feasibility is established along with the first generation design, but economic viability is usually not established until the second or third generation design. Therefore, the time required to establish economic viability is typically 5 to 8 years.

Industry typically does not have the necessary resources to fund long term development in the absence of a viable market demand. The federal government has set a goal that 20% of the total United States energy consumption be supplied by solar energy by the year 2000. Probably only the government and the oil companies can afford to fund the long term development programs which will be required to achieve this goal.

The United States is dependent on foreign oil imports for 45% of its total oil consumption. This dependency came about over the past 20 years. The need to reduce this dependency to a more tolerable level by the year 2000 is a tremendous challenge. All avenues to solve this problem must be pursued, including the development of reliable, low-cost terrestrial photovoltaic systems.

SOLAR ENERGY APPLICATIONS

Solar energy from the sun can be used effectively today for domestic water heating, swimming pool heating, building space heating, and remote electric power generation. If the present government funding levels are maintained, this list is expected to expand over the next ten years to include solar cooling, distillation of sea water, irrigation pumping, food dehydration, industrial process heat, and widespread electric power generation.

The overall objective of the United States DOE National Solar Photovoltaic Conversion Program is to promote the development of low-cost, reliable, terrestrial photovoltaic systems, and to stimulate creation of viable industrial and commercial capability to manufacture and distribute these systems for widespread use in remote stand-alone, residential, commercial, industrial, and utility applications.

Flat plate photovoltaic modules are economically effective today in many remote stand-alone applications. These applications typically require small amounts of electric power, ranging from a few watts (W) up to 1 or 2 kilowatts (kW). The majority of these applications are in areas where there are no utility grids. The cost to bring in utility power lines for these applications is much higher than utilizing photovoltaic flat plate modules. Concentrator systems may eventually find use in larger remote stand-alone applications.

Residential applications experiments to evaluate the use of photovoltaic systems are just now being initiated. The residential system sizes range from 3 kW to 8 kW. Photovoltaics are not yet cost effective in residential applications. Flat plate systems are more likely to be used in residential applications than are concentrator systems.

Solar photovoltaic concentrator power generation systems are likely to be used primarily in commercial, industrial, and utility applications. These systems will generally range from 10 kW to 200 megawatts (MW). Some commercial and industrial applications are being evaluated under the DOE Program Research and Development Announcements (PRDA) projects.

Utility Central Power Station (CPS) conceptual designs have been formulated and it is anticipated that detailed applications experiment designs will be developed in the early 80's. A 1 MW photovoltaic CPS application project could be in place and operational by the mid-80's if it gets the go-ahead from the DOE in 1981.

PHOTOVOLTAIC POWER SYSTEMS

Photovoltaic solar cells were first used in 1958 to power the Vanguard I satellite. Other satellites and the manned space capsules also used large panels of solar cells to produce operating power. Vanguard I used 6 milliwatts of electricity, whereas Skylab consumed 11.5 kW. The technology used by scientists and engineers to accomplish the development of space solar cells has been applied to terrestrial solar cells.

Silicon solar cells for application in flat plate modules were originally circular in shape because of the economy associated with slicing the wafers directly from circular Czochralski - grown crystals. Now most manufacturers are using square shaped cells to take advantage of increased packing factor. A typical "one-sun" silicon solar cell is shown in Figure 1. The current carrying contact metalization pattern varies with manufacturer, but is usually characterized by widely spaced grid lines and redundant main collector bars.

A Jet Propulsion Laboratory Block IV flat plate module manufactured by Motorola is shown in Figure 2. Recent advances in the growth of silicon sheet-ribbon indicate that the cost of future solar flat plate modules will reduce from the present $11.50 per peak watt ($/Wp) to $0.85/Wp by 1990. A module simulating the use of ribbon cells is shown in Figure 3. Silicon sheet-ribbon is expected to be commercially available by the mid-80's.

A typical flat plate system is depicted in Figure 4. The system contains an array of modules, a blocking diode to prevent battery discharge, a shunt voltage regulator to prevent battery overcharging, a storage battery for storage of excess solar power and for supplying power to the load during the night and during bad weather, and the load. A DC to AC inverter is also required when supplying power to an AC load.

Widespread market demand for flat plate systems has been hampered primarily by the high cost of the silicon solar cells. Photovoltaic concentrator modules have been designed using a single concentrator solar cell to generate the same output power that 33 one-sun cells produce. A typical 75X silicon concentrator solar cell is shown in Figure 1. Concentrator cells are usually designed to optimize the performance requirements related to a specific concentrator module design. Factors such as concentration ratio, anticipated operating temperatures, type of optical system, and the flux distribution across the cell determine its ultimate design configuration. The high density, closely spaced grid lines characterize a concentrator cell. The gridding reduces ohmic losses and is optimized for specific concentration ratios. A concentrator module using two silicon cells to generate 80 Wp is shown in Figure 5.

The advantage of concentrator systems is the use of fewer of the high cost silicon solar cells. This is only an advantage if the additional cost of the required optics, tracking controls, and more complex structures

Figure 1. Silicon "One-Sun" And Concentrator Solar Cells

Figure 2. Motorola Flat Plate Modules — JPL Block IV

Figure 3. Motorola Flat Plate Module With Ribbon Cells (Simulated)

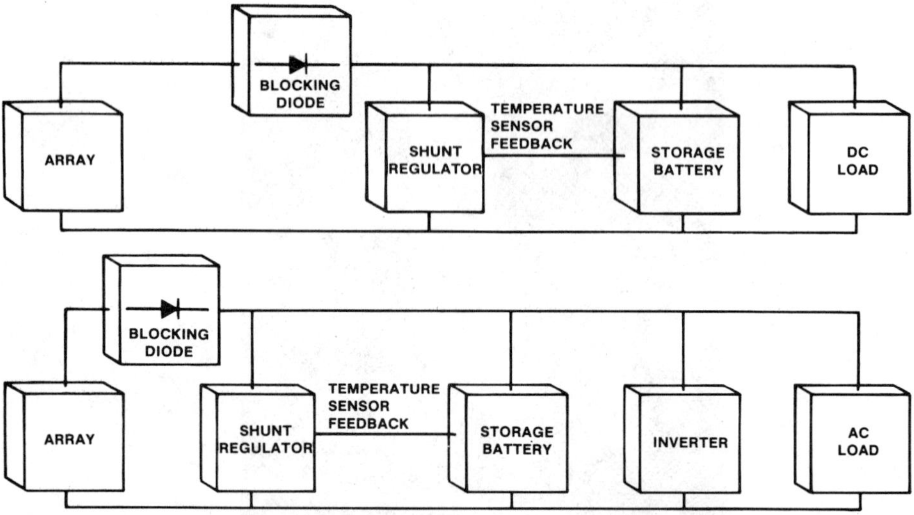

Figure 4. Typical Flat Plate PV System Block Diagram For DC and AC Loads

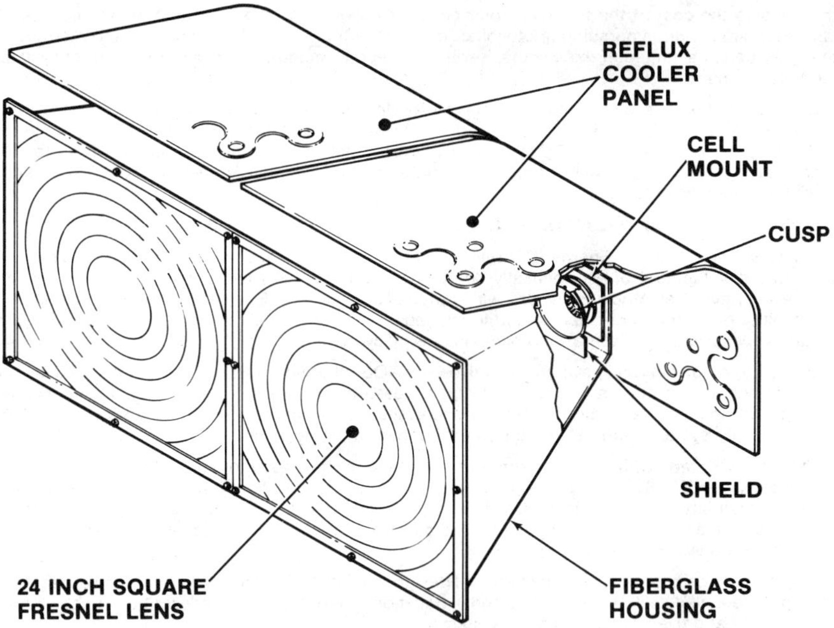

Figure 5. Motorola Fresnel Lens Concentrator Module

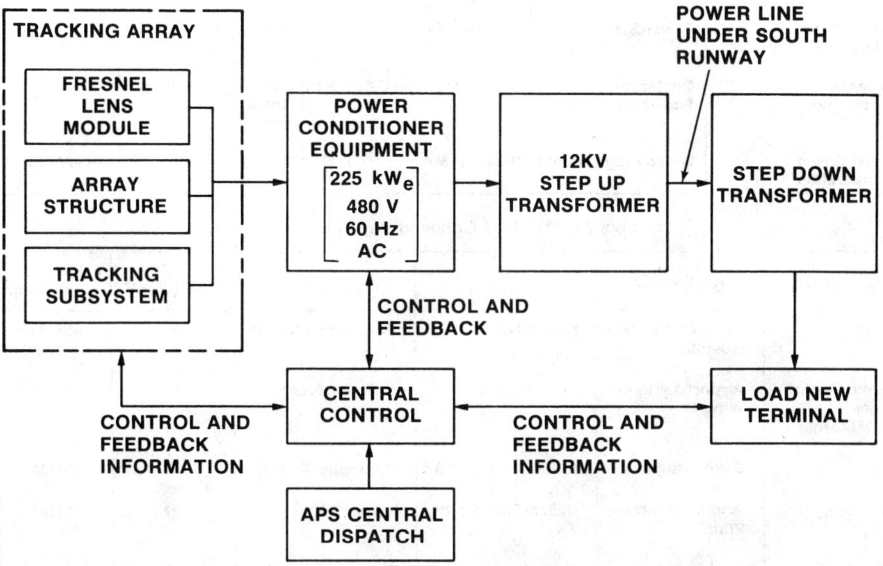

Figure 6. ASPCP System Block Diagram

more than offsets the cost of the replaced solar cells. However, regardless of comparative costs, another important reason to continue pursuing photovoltaic concentrator development is that eventually the supply of solar cells may become limited. If this occurs, it will then become mandatory to use concentrator systems to satisfy the market needs.

The concentrator system designed by Motorola for the Phoenix Airport Solar Photovoltaic Concentrator Project (ASPCP) is depicted in Figure 6. The array field design output is presently planned to be 225 kWDC when 18, two-axis, 12.5 kWDC circular arrays are installed. The array field output is fed into a DC to AC power inverter and the AC power output is transmitted to the American Airlines wing of the new terminal building at the Phoenix Airport.

PHOTOVOLTAIC APPLICATION PROJECTS

The U.S. Department of Energy (DOE) initiated a National Photovoltaics Program for the purpose of stimulating the development of economically competitive, commercially available photovoltaic systems to provide safe and reliable electric energy for a wide range of applications. Equally important, the program is to provide resolution of the technical, institutional, legal, environmental, and social issues involved in encouraging widespread adoption of photovoltaic energy systems.

In support of the national program objectives, the DOE Albuquerque Operations Office, with technical support from Sandia Laboratories, issued two Program Research and Development Announcements (PRDA) for the design, fabrication, installation, operation, and evaluation of on-site applications experiments. PRDA-38 was for flat plate systems and PRDA-35 was for concentrating systems.

The design criteria of these programs was to use the existing technology to develop systems which could be fielded by mid 1981. By designing, fabricating, installing, and operating these systems, the actual technical cost feasibility could be established. This is how every industry gets started. Once you have the first prototype system up and working, then you can begin to work on improving reliability and reducing cost by improved second generation designs.

Twenty-nine companies performed Phase I design studies. Nine systems were then awarded Phase II contracts for the detailed design, fabrication, and installation. These systems are described in Tables 1 and 2. Four are flat plate and five are concentrator systems.

Table 1. PRDA-38 Flat Plate Systems

Organization	Description	Location	Module Type	Size (kWp)	Phase II Completion
Lea County Electrical Cooperative	Shopping Center	Lovington, NM	Flat Panel	100	05/81
New Mexico State University	Power Station	El Paso, TX	Flat Panel	18	02/12/81
Science Applications Inc.	Center for Science and Arts	Oklahoma City, OK	Reflector Augmented Flat Panel	150	08/81
Solar Power Corporation	Beverly High School	Beverly, MA	Flat Panel	100	03/81

Table 2. PRDA-35 Concentrator Systems

Organization	Description	Location	Module Type	Size (kWp)	Phase II Completion
Acurex	Total Energy Hospital	Kauai, HI	Parabolic Trough	35	08/81
Arizona Public Service And Motorola	Airport Terminal Wing	Phoenix, AZ	Circular Fresnel	225	11/81
BDM	Office Building	Albuquerque, NM	Parabolic Trough	47	08/81
E-Systems	Airport Utility Plant	Dallas-Fort Worth, TX	Linear Fresnel	27	11/81
General Electric	Marine Exhibit Park	Orlando, FL	Parabolic Trough	110	03/82

CONCENTRATOR SYSTEM HARDWARE

The photovoltaic concentrator array developed at the Government Electronics Division of Motorola is depicted in Figure 7.

The major components of the array are:

- 150 concentrator modules.
- A space frame array structure with three subarrays.
- Elevation and azimuth drive motors.
- A Sun Sensor.
- A Hybrid Sun Tracker.
- Motor Drive Electronics.

The concentrator module is shown in Figure 8. It consists of a fiberglass housing, two 24 inch by 24 inch Fresnel lenses, a protective shield, and two cell mounts. A unique closed freon reflux cooler panel is used to keep the solar cell cool for maximum power output. Module characteristics are described in Table 3.

The cell mount shown in Figure 9 is the heart of the system. A three inch diameter 79X concentrator solar cell is soldered to a copper metalization pattern on an alumina substrate. Contact fingers to carry the DC current are soldered to the top of the solar cell and the copper surface in contact with the bottom of the cell. The current is then transferred from the contact fingers through conductor bus rings to the positive and negative output terminals. The nominal output current is 80 ADC and the nominal output voltage is 0.5 VDC, resulting in a nominal output power level of 40W per cell mount. This assumes an ambient temperature of 28° C, a cell temperature of 60° C, an insolation level of 1000 W/m^2, and a wind speed of 1.5 meter/second.

The space frame structure with the three subarrays is shown in Figure 10. The assembly of this structure is nearing completion at the Motorola GED plant. The structure is made of aluminum to minimize both weight and cost. Aluminum also avoids many environmental problems. The base structure contains four wheels which ride on a support rail supported by 25 concrete piers. The array rotates in azimuth and the subarrays rotate in elevation. The structure weighs approximately 8,000 pounds. The 150 modules that it must support weigh approximately 10,050 pounds.

Two 1/15 HP tracking motors, mounted on diametrically opposed wheels, drive the base structure in azimuth. One 1/15 HP tracking motor drives the subarrays in elevation. One 1 HP stow motor drives the subarrays in elevation when wind conditions dictate that the array be stowed to preclude damage. The stow position is with all three subarrays horizontal, with module lenses facing upward. The stow motor can also be controlled manually to speed up maintenance operations. The array tracks the sun throughout the sunlight hours and returns to the wake-up position after sunlight hours. The subarrays can be rotated into the stow position at anytime, independent of the azimuth position.

The Sun Sensor and the Hybrid Sun Tracker combine to provide precision tracking of the sun in both open loop and closed loop modes. The Hybrid Sun Tracker in the open loop mode positions the array when the sun is obscured, such as by clouds. The Sun Sensor determines the location of the sun in the open loop mode and control signals are processed through the Hybrid Sun Tracker to the motors.

The Motor Drive Electronics accepts control signals from the Hybrid Sun Tracker, translates these signals, and outputs drive signals to the array track and stow motors. Manual operation of the motors is available for maintenance purposes.

SYSTEM INTEGRATION

Motorola has been responsible for the system design and development. System details are described in Table 4. Arizona Public Service Company, a progressive Arizona Utility, is responsible for the site and construction design and will be the operators of the system.

SUMMARY AND CONCLUSIONS

Solar photovoltaic concentrator power systems are technically viable today. The economic viability of concentrator systems is yet to be determined as second and third generation designs attack the high cost components. Flat plate photovoltaic systems are already cost effective in remote stand-alone applications. Indications are that silicon ribbon-sheet solar cells, a technology well in-hand, will bring the cost of photovoltaics down into a range where it will be very cost competitive with other forms of energy by the year 1990. The means to harness the sun's energy is at hand, but only if sufficient development funding for both pure research and hardware application is maintained over the next decade.

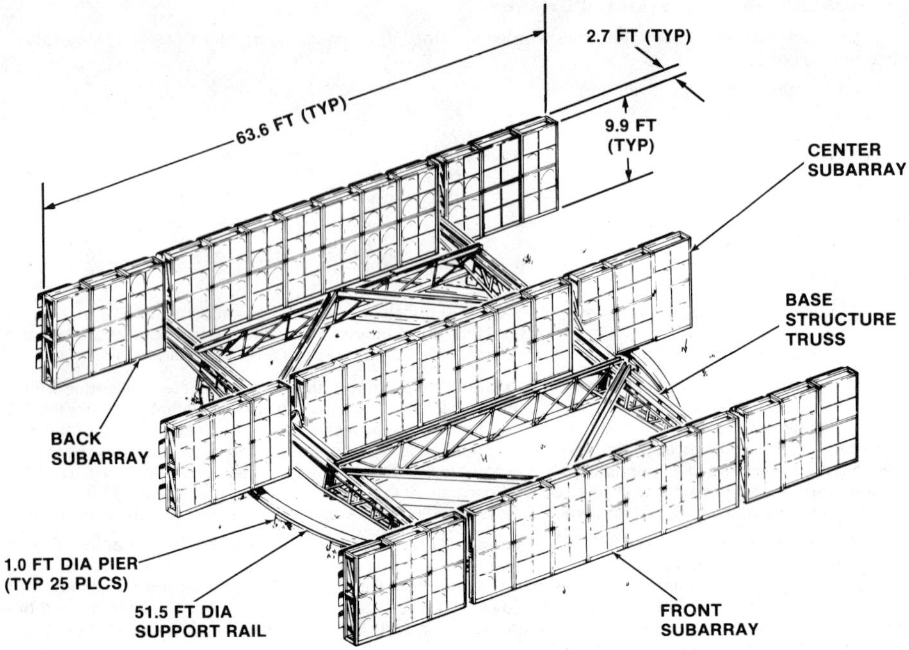

Figure 7. ASPCP 2-Axis Tracking Array Structure

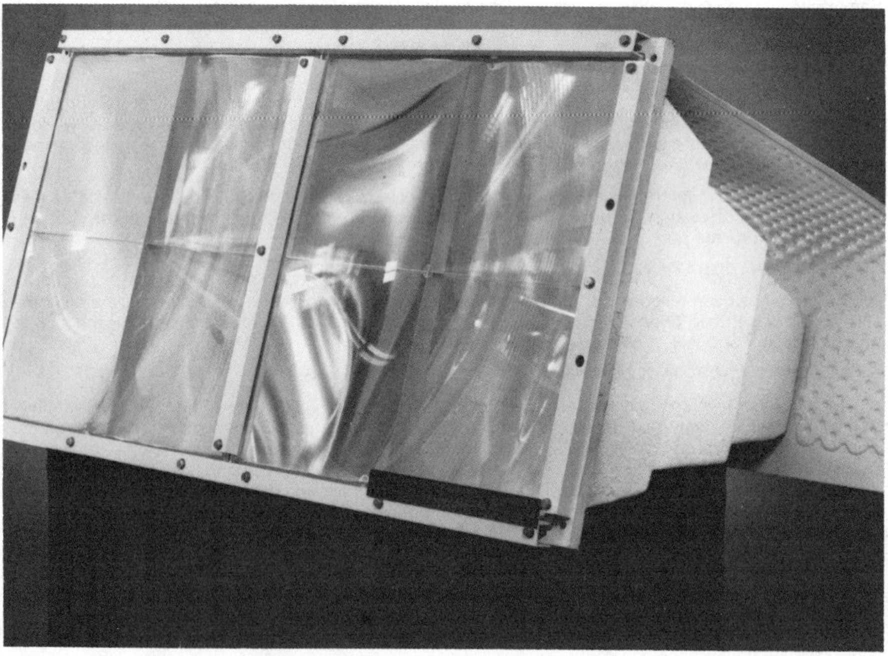

Figure 8. Concentrator Module

Figure 9. Concentrator Cell Mount

Figure 10. Space Frame Structure

Table 3. Motorola Module Characteristics

Weight	**67.0**	**Pounds**	**30.4**	**Kilograms**
Length	**53.3**	**Inches**	**135.4**	**Centimeters**
Width	**27.0**	**Inches**	**68.6**	**Centimeters**
Depth	**29.8**	**Inches**	**75.7**	**Centimeters**
Aperture Area	**4.03**	**Sq. Feet**	**0.754**	**Sq. Meters**
Geometric Concentration Ratio	**98X**			
Actual Concentration Ratio	**79X**			
Power Output	**83.3**	**Watts**		
Efficiency-Design Goal	**11.1**	**Percent**		

Table 4. System Characteristics

Land Area	**7.4.**	**Acres**
Number of Silicon Cells	**5,400**	
Number of Modules	**2,700**	
Number of Arrays	**18**	
Field Power Output	**225**	**kWDC**
Inverter Efficiency	**92**	**Percent**
System Losses	**4**	**Percent**
System Power Output	**198.3**	**kWAC**
System Power Output with Off-Axis Standard Deviation of 0.325 Degrees	**190.4**	**kWAC**

REFERENCES

1. "A National Plan for Energy Reserach, Development & Demonstration: Creating Energy Choices for the Future", ERDA 76-1, 1976.

2. Ronald C. Hodge, "Photovoltaic Concentrator Array Production Process Study", Final Report, SAND 79-7055/1, 1981.

3. Robert L. Sanders and Raymond Kendall, P.E., "Solar Photovoltaic Electric Power, Today's Realities – Tomorrow's Promises", 3rd World Energy Engineering Congress, Atlanta, 1980.

4. Robert L. Sanders, "Design of a Photovoltaic Solar Electric Power Plant", Mid-continent Area Power Pool Photovoltaic Workshop, Minneapolis, 1980.

5. Dr. David A. Freiwald, Presentation, 46th Symposium of the Mobility Operations Research Society, Newport, R.I., 1980.

SOME PHOTOVOLTAIC SYSTEM DESIGN CONSIDERATIONS

Gerald F. Hein
Solar Power Corporation
Woburn, Massachusetts 01801

A photovoltaic system is an electrical generator. As such, the photovoltaic system can be expected to provide a reliable, cost-effective source of electricity. To expect less would render the technology inappropriate; to expect more is to render one's perspective inappropriate. Although such a statement seems quite obvious, the simplicity becomes lost in the confrontation with a two-inch thick request for a proposal. The other extreme is represented by the request for solar electricity for a pump or a light or a refrigerator without provision of duty cycle, environment or solar radiation data.

The elements of a DC photovoltaic system are shown in Figure 1. An AC system would have an inverter placed between the load and the battery. An examination of each element of the system provides some of the characteristics of the system.

Solar Radiation (stachostic in nature) -- Data exists for about 1,000 stations throughout the world--large variances over relatively short distance from stations, large variances throughout the world.

Photovoltaic Array (typically silicon solar cells) -- Wired in series-parallel configurations to produce desired voltage and current--conversion efficiency on the order of 10% to 15%--encapsulation required to protect cells and contacts.

Regulator -- Voltage window required to protect the battery from overcharging or hydrogen gassing--monitoring and other processing may be added, for example, low and high voltage cutoffs, maximum power trackers, readout devices and so on.

Storage (typically lead-acid or nickel-cadmium battery cells) -- Lifetime is a function of charge-discharge cycles, temperature, and a number of other variables.

Load -- Typically, a water pump, refrigerator, light or any other device

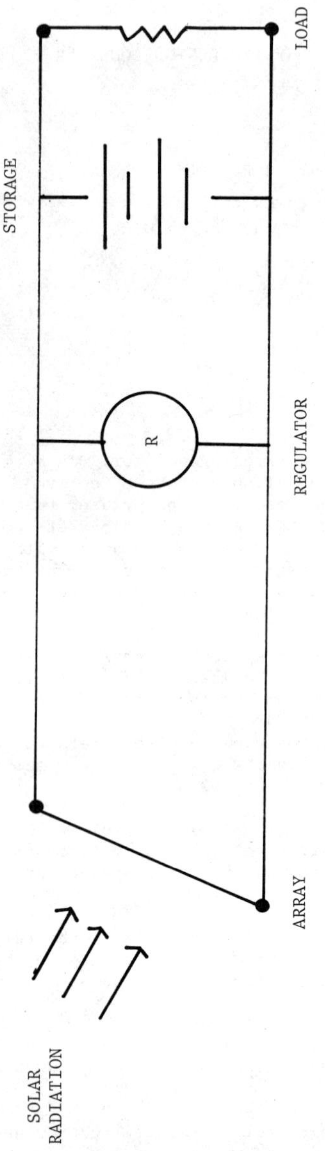

FIGURE 1

DC SYSTEM SCHEMATIC

powered by electricity.

Despite the simplicity of the examples, photovoltaic power systems are providing power reliably throughout the world at power levels in the hundreds of kilowatts. Some organizations are even planning multi-megawatt photovoltaic power plants. The sophistication and complexity rise rapidly, but the system schematics do not vary appreciably from that shown in Figure 1. The solar insolation is the same whether the array is one or one million watts. The array itself will vary because of efficiency, ground fault detection requirements, wiring losses, structural requirements, installation requirements and relative costs. Photovoltaic modules and arrays are provided by manufacturers, but installation by the manufacturer is neither desirable nor economical for smaller systems. For larger systems, clearing the array field, building a fence, pouring concrete foundations may all be expected in a large contract. The building of roads or their maintenance is not a reasonable demand upon a photovoltaic manufacturer. And yet, we are seeing these types of requirements more and more in requests for proposals. In one recent request, the customer expected an office building to be constructed even though the array was to be erected a large distance away from the building. As most who respond to proposals know, the response time is usually less than thirty days.

In the regulation, control and monitoring subsystem, specifications are often included for many elaborate and costly items that are not needed. If the manufacturer responds to the specification, there is a cost which must be passed on. Since each request is the creation of a different organization, these extra items tend to be one-of-a-kind designs, thus adding even more cost. If the manufacturer attempts to recover the full cost of these unique design requests, the sale will probably not be made. That is an unreasonable burden and yet it has become commonplace.

The battery storage subsystem can be of high quality or relatively low quality. Cells can be attached in series and parallel configurations. The photovoltaic manufacturers are generally not experts in electro-chemical storage, but sufficient experience and knowledge has been gained to differentiate battery products and optimum configurations. Generally, car batteries should not be connected in series-parallel configurations because of the cell mismatch that occurs in this type of battery. The best applications for the small batteries are small loads and small photovoltaic systems. Large systems that require storage should be designed with large two-volt cells connected in series to produce the desired voltage. Specifications for pure lead, lead-antimony or lead-calcium grids should be based solidly on the electro-chemistry of the environment rather than an arbitrary selection of a particular battery type.

In alternating current systems, the choice of an inverter is a cricital one. Communication systems will not tolerate a sloppy waveform. Filtering and RFI shielding will also be required. Efficiency is always important. The difference in total system cost between 70% and 85% efficiencies will result in more than a 20% cost difference. Tare (no load) losses are very important also, with such loads ranging up to 25%. These losses must be paid for in the system cost. Appliances such as lights, pumps, refrigerators and even some radio frequency receivers will tolerate sloppy waveforms which translate directly to cost savings without compromising reliability or performance. If inverter efficiency is critical, the product specifications should be tested and such a test is not a simple one. If the waveform produces numerous harmonics, the measurement of current and voltage output will include an induced error that may be attributed to the inverter noise. Thus, one would get a more accurate measurement in the energy domain rather than the power domain. However, the energy domain measurements may have to be measured in analog fashion.

Specifications for photovoltaic systems seem to include a widely varying set of warranty requirements. The photovoltaic module warranty specifications usually vary from one to five years. Such requirements are not especially difficult to deal with because the modules being manufactured presently have MTBF's that easily exceed five years. The most difficulty for the manufacturer is provided by the specifications that require that the module degrade no more than a certain minimal percentage over a five year period. Ambiguity is introduced because the specification usually fails to state whether the maximum deviation will be measured from initial module performance or the manufacturer's nominal rating which is usually associated with a variance larger than the percentage stated in the specification. Adding to the complexity is the fact that module degradation can only be measured by an expensive simulator that would not usually be available to the customer. If the module is placed in a harsh environment, degradation may be caused by deposits of various materials becoming embedded in the superstrate. Such an occurrence would not be included in the warranty. Manufacturers generally do provide warranties on their products. However, they attempt to avoid warranties that insure against degradation, e.g., no more than one percent per year. This type of warranty is similar to purchasing a light bulb and asking for a 1,000 hour warranty with no more than 0.1 lumen loss per hour. Although such warranties may be desirable, they are impossible to implement because of the inability of the customer to measure performance.

Another difficulty encountered more and more often is the expectation that the photovoltaic manufacturer will assume total responsibility for warranty claims including batteries, fences, portable instruments, tools and so on. This represents a cost that must be passed on especially when the user begins to demand 10 to 15 year guarantees on batteries. Manufacturers of batteries will normally provide a two year warranty.

The problem of how to reduce the cost of a photovoltaic system is one that must be solved by the manufacturers. However, government agencies and users of these systems can provide significant contributions to the effort by following some rather simple guidelines. Some of these are:

1. When writing specifications, use performance criteria rather than detail criteria.

2. Reliability should not be arbitrary, but should represent a realistic, achievable event that is related to cost.

3. Instrumentation and monitoring subsystems should measure only what is necessary to measure.

4. Warranties must be written so that the user and the manufacturer can implement them.

5. All specifications have a cost associated with them. Each should be questioned for need and function.

6. When specifying a photovoltaic system, limit the scope to the system rather than peripherals.

All of these guidelines can be summarized in a fundamental law of economics: All goods and services have a cost associated with them. Bearing this in mind, if one attempts to imagine how a specification can be met during the writing phase of planning and associates a cost with the specification, system costs will begin to decline. The best technique that one can use is check with three or four manufacturers during the specification phase. The cost is minimal and the sources are more independent from each other than any others available to the user.